WITHDRAWN
WRIGHT STATE UNIVERSITY LIBRARIES

Cell Death
and Diseases of the Nervous System

Cell Death and Diseases of the Nervous System

Edited by

Vassilis E. Koliatsos, MD
*The Johns Hopkins University
School of Medicine, Baltimore, MD*

Rajiv R. Ratan, MD, PhD
*Beth Israel Deaconess Medical Center
and Harvard Medical School, Boston, MA*

Foreword by

Dennis Choi, MD, PhD
*Washington University School of Medicine
St. Louis, MO*

Humana Press ✴ **Totowa, New Jersey**

WL
140
C393
1999

© 1999 Humana Press Inc.
999 Riverview Drive, Suite 208
Totowa, New Jersey 07512

All rights reserved.

No part of this book may be reproduced, stored in a retrieval system, or transmitted in any form or by any means, electronic, mechanical, photocopying, microfilming, recording, or otherwise without written permission from the Publisher.

All authored papers, comments, opinions, conclusions, or recommendations are those of the author(s), and do not necessarily reflect the views of the publisher.

For additional copies, pricing for bulk purchases, and/or information about other Humana titles, contact Humana at the above address or at any of the following numbers: Tel.: 973-256-1699; Fax: 973-256-8341; E-mail: humana@humanapr.com, or visit our Website: http://humanapress.com

This publication is printed on acid-free paper. ∞
ANSI Z39.48-1984 (American Standards Institute) Permanence of Paper for Printed Library Materials.

Cover illustrations: Background illustration is a portion of Fig. 256 from Ramon y Cajal's *Degeneration and Regeneration of the Nervous System,* first published in Spanish in 1913. The center illustration, courtesy of Vassilis E. Koliatsos, is a facial motor neuron undergoing apoptotic death after axotomy in the neonatal rat

Cover design by Patricia F. Cleary and Vassilis E. Koliatsos, MD.

Photocopy Authorization Policy:
Authorization to photocopy items for internal or personal use, or the internal or personal use of specific clients, is granted by Humana Press Inc., provided that the base fee of US $8.00 per copy, plus US $00.25 per page, is paid directly to the Copyright Clearance Center at 222 Rosewood Drive, Danvers, MA 01923. For those organizations that have been granted a photocopy license from the CCC, a separate system of payment has been arranged and is acceptable to Humana Press Inc. The fee code for users of the Transactional Reporting Service is: [0-89603-413-5/99 $8.00 + $00.25].

Printed in the United States of America. 10 9 8 7 6 5 4 3 2 1

Library of Congress Cataloging-in-Publication Data

Cell death and diseases of the nervous system / edited by Vassilis E. Koliatsos, Rajiv R. Ratan.
      p.   cm.
  Includes index.
  ISBN 0-89603-413-5 (alk. paper)
  1. Nervous system—Degeneration. 2. Cell death. I. Koliatsos, Vassilis E. II. Ratan, Raj R.
  [DNLM: 1. Nervous System Diseases—physiopathology. 2. Cell Death—physiology. 3. Disease Models, Animal. WL140 C393 1998]
  RC394.D35C44   1998
  616.8'047—dc21
  DNLM/DLC   98-22462
  for Library of Congress   CIP

Foreword

It is an honor to be asked to write a foreword for this timely book, which assembles contributions from authorities working in the area of pathological nervous system cell death, and has been expertly edited by Vassilis Koliatsos and Rajiv Ratan. That the inappropriate demise of brain or spinal cord cells is at the root of many neurological diseases has been appreciated since the early days of microscopic neuropathology, but it has only been in the last decade or so that pervasive therapeutic nihilism has lifted. In the journey of medical progress, we have reached the shores of a marvelous new land.

Three major scientific thrusts in particular have converged to produce the press of ideas covered here. First, burgeoning information about the fundamental nature of central nervous system cell–cell signaling, both the fast signaling mediated by conventional neurotransmitters and the usually slower signaling mediated by neuromodulators and growth factors. A central theme emerging in recent years has been the duality of these signaling mechanisms, which serve the nervous system in health, but which can become the very mediators of neuronal or glial cell degeneration in disease settings. Glutamate-mediated neurotransmission and excitotoxicity have been the defining and best-studied examples, but many other examples have also emerged. Second, delineation of the molecular underpinnings of programmed cell death, and an appreciation of their awesome power. Cells in the nervous system share with other cell types a disconcerting readiness to self-destruct, which can be unleashed by the interruption of critical inputs, or by damage to vital structures. It has become clear in recent years how easily diseases of the nervous system can disturb the precarious suspension of programmed cell death that we call life. Third, development of animal models of neurodegenerative diseases. The most dramatic advantages have occurred in devising transgenic models of genetically driven diseases, but major refinements have also taken place in a variety of rodent or large animal models of ischemia or trauma. These models have made it possible to peer into the pathogenesis of brain or spinal cord cell loss as it occurs, to develop specific hypotheses about underlying mechanisms, and to determine whether specific experimental maneuvers can ameliorate this loss. Such neuroprotective maneuvers often represent unique, necessary tests of hypotheses, and, furthermore, can point the way toward the exciting prospect of developing therapeutic interventions in humans.

As Koliatsos and Ratan point out in their Preface, the biology of cell death has attracted unparalleled interest in recent years. Although no single book can cover this biology comprehensively, Parts I and II of *Cell Death and Diseases of the Nervous System* provide a good overview of mechanisms relevant to neuronal cell death. Building on this conceptual background, the heart of the book, Part III, places the theme of abnormal nervous system cell death in the context of a selected set of neurological and psychiatric diseases. Finally, Part IV looks into the future, with several chapters discussing the approaches that might be used to attenuate such abnormal cell death. As intended by the authors and editors, *Cell Death and Diseases of the Nervous System* should indeed be a useful guide to researchers or clinicians seeking a broad perspective on the field of pathological neuronal cell death. However, the field is moving quickly and new topics are coming into view with exciting regularity—insights into how nonneuronal cell death may differ from neuronal cell death, novel death mechanisms, new disease links and therapeutic strategies—so a sequel seems inevitable. Vassilis, Raj, I hope you have a chance to rest...

Dennis Choi, MD, PhD

Preface

One of our professors used to say that a preface serves its purpose only if written with the expectation that it is the only part of the book the reader will have time to study. This requirement implies a text that can capture the essence of the book in a nutshell. Unfortunately for editors of the book like *Cell Death and Diseases of the Nervous System*, the amount of information about cell death is enormous. On the other hand, asking questions about cell death represents a "Columbus egg"-type of paradox—it seems more a bypass than a solution to the problem. An enthusiast might argue that, since many problems in neurology are related to the death of neurons, cell thanatology is the science *par excellence* of clinical neurology. A cynic might respond that it is just a neuroscience written from the opposite—the bottom line—end. Whatever the epistemology of the topic, we have tried to present it for what it is: a useful concept for thinking about brain disease based on a good deal of common ground, but one also marked by several dilemmas. To those limitations we should add our—and the contributors'—biases and we can expect that every reader has his or her own (death is, after all, the *ultimate* event, and as such it tends to mobilize idiosyncratic responses).

What is cell death? What is death in general? Ignoring for the moment the philosophical response, this question brings painful memories to clinicians—neurologists in particular—who in certain cases have gone through a multiple test protocol for brain stem reflexes and even used EEG recording to diagnose a patient's death based on the permanent functional silence of the nervous system. This, of course, is already too late for the clinician-therapist. It is of interest that the diagnosis of cell death in tissues is as equally involving as the diagnosis of a patient's death. In fact, we have just begun to develop techniques to determine neuronal death in tissues beyond any reasonable doubt. But again, by the time the diagnosis is established, it is too late to intervene. It is much more useful, at least from the viewpoint of therapeutic interventions, to think in terms of irreversible injury or decision point, i.e., the stage at which the neuron commits itself to a self-destructive process—in the case of programmed cell death or apoptosis—or the stage in which the accumulated damage inflicted upon the neuron by extrinsic agents is incompatible with organelle function or maintenance of the membranous compartments. This point of no return should be the operational definition of cell death and the target phenomenon of investigations aiming at prevention or treatment of disease.

There is very little doubt that, in recent years, the biology of cell death has attracted unparalleled interest. Although the current fad is an overdue response of the biomedical community to a relatively small number of seminal observations made by pioneer scientists, the recent massive recruitment into the field of talented, ambitious investigators has contributed significant advances to our understanding of cell death mechanisms. What is more important for the clinic, the previous progress in understanding has led to recommendations of ways to combat cell death—indeed, some of them employed by nature to regulate the physiological cell death that occurs during development. Based on the latter, one can argue that directly targeting cell death, instead of "upstream" causes, in treating neurological illness is not cracking the Columbus egg, but hitting the beast at the head with a smart—even convenient—weapon. On the other hand, precisely because nature uses cell death as a regulatory process, we must be cautious in assuming that cell death is always negative, or bad, for the organism; this may be naive, even in the case of disease. Nerve cells out of network or loaded with abnormal organelles and aberrant metabolites represent unproductive consumers of space and energy. Even worse, they are carriers of pathologic burdens potentially spreadable to neighboring healthy tissues in the form of transmissible agents, amyloidogenic foci recruiting nearby processes, or inflammatory triggers inviting phagocytic cells that may lyse innocent bystanders.

Capturing the excitement in the scientific community is definitely one of the causes of this book. Another cause that needs underlining is our commitment to the presentation of the great ideas in the field, rather than mere accounts of a series of high-tech experiments. The first of these commitments does not originate simply in the need to give appropriate credit to the innovators. It is primarily the result of our belief that this is the only secure way to promote the important, i.e., persisting, issues. It is pertinent here to state that we agree with Ron Oppenheim who has commented that the study of cell death in the nervous system is not a recent endeavor. Regardless of its popularization by rebaptizing the issues with molecular nomenclature—coinciding with the disappointment that the features of neurons that were the focus of attention in the last two decades have not been of great therapeutic utility—the field of cell death has been a busy and rewarding venture for a few faithful investigators for almost twenty years. The molecular era has helped popularize and rejuvenate the field, though the underlying ideas are quite old. In *Cell Death and Diseases of the Nervous System*, we have tried to underline the permanent and at the same time pass on the excitement about current and future discoveries.

Although it is usually difficult to trace precisely the developments that lead to major shifts in scientific directions, it can be argued that an underlying reason for refocusing on the bottom line—the death of neurons—is a continued frustration in the end-of-century Neurology after multiple waves of exacting science with very limited clinical gains. This tradition begins with the astute and highly replicable bedside observations in the Charcot era, then moves to the structural revolution with Nissl and Ramon y Cajal, continues on with the lesion-and-deficit experiments in the first half of the century and the connectional and single unit recording era in the 60s, then with the recent transmitter era in the 70s and 80s, which prepared what has been called the molecular movement, and is currently experiencing the cognitive revolution. Although most of the above approaches have yet to speak their last word, clinicians remain eager for treatment tools.

The organization of the material in the book follows a logical scheme designed to guide the study of the clinician-scientist. However, in keeping with our commitment to the presentation of ideas, the underlying thread is historical. The neuronal death chronicle starts with observations on the death of Rohon-Beard cells in developing skates by Beard at the end of the previous century and the experiments of Kallius, Ernst, and Glucksman during the third decade of this century. However, it was Hamburger and Levi-Montalcini who promoted the phenomenon of neuronal death to its present prominence as a major regulator event during development and who provided an adequate conceptual framework for its understanding. The historical importance of the studies of these two pioneer investigators becomes evident in Hamburger's confession that, to a neurobiologist of the Spemman school (Spemman was Hamburger's mentor), *regressive* (death) phenomena during the *progressive* process of development would hardly be conceivable. It is interesting, though, that few people in the neuroscience community paid attention to these historical developments, even after the issue was reintroduced in the scientific literature by Hamburger and a few other investigators in the 70s (*see* chapter by Burek and Oppenheim in Part I). At the same time, other pioneer investigations were being made in the invertebrate nervous system using the convenient tiny roundworm *C. elegans*, adding up to the discovery of a strict genetic program leading to a predictable death of 50% of neurons during development (the case *par excellence* of programmed cell death). The dynamics of this *instructive* process are very different from those working in the *selective* death of neurons during vertebrate development, but the underlying molecular details may be exactly the same. At about the same time (70s and 80s), pioneering observations were being made also in tumor cells and lymphocytes, with the discovery of the major morphological subtypes of dying cells (apoptosis and necrosis) (*see* chapters by Clarke and by Pittman and colleagues in Part I).

Despite the previous remarkable discoveries, neuroscientists were absorbed in their long honeymoon with the neurotransmitters in what can be characterized as the most influential single period in the history of neuroscience, because of its implicit promises for therapy through links with pharmacology. Our obsession with transmitters, and the shapes and sizes of different classes of neurons, in the 70s and 80s led to the simplistic notion that cells not responding to phenotypic probes (Nissl stains, immunocytochemistry, *in situ* hybridization), were practically dead. Only approximately 20 years after neurochemical transmitter research had become the predominant approach in neuroscience, we began to separate dedifferentiation from death. Another symptom of this neglect was that a crucial theory published by Stan Appel in the heyday of the transmitter era linking neurodegenerative disease with developmental neuronal death, although widely cited, was not useful in guiding research. In our defense, we did have an unambiguous way of labeling dead neurons *in situ* and we had no particular desire to occupy ourselves with hard-to-substantiate concepts.

Bob Horvitz's presentation to the broader neuroscience community of the experience with the *C. elegans* at the annual meeting of the Society for Neuroscience in Phoenix in 1989 had a major impact on an audience already excited by the prospect of treating neuronal degeneration with molecules naturally employed to combat cell death (*see* chapter by Hilt and colleagues in Part IV). These molecules (trophic factors) had made a remarkable appearance in 1989 with the discovery of the NGF family and the cloning of CNTF, the first cytokine with effects on the nervous system (*see* chapter by Koliatsos and Mocchetti in Part IV). It is perhaps the concept of cell death most linked in scientists' minds to the notion of substances with therapeutic potential, despite the fact that the prototypical trophic factor NGF was a direct product of developmental neurobiology and, indeed, the conceptual offspring of the same team of scientists who first studied embryonic neuronal death 35 years ago.

A remarkable development was the cloning of the genes *ced-3* and *ced-4,* which instruct cell death, as well as the gene *ced-9*, which prevents cell death in the *C. elegans* (*see* chapter by Royal and Driscoll in Part II). The mammalian proto-oncogene *bcl-2*, initially characterized by its ability to prevent death in myeloid precursors deprived of trophic factors, has been shown to have 23% nucleotide identity with *ced-9*, whereas the mammalian interleukin-1 beta-converting enzyme (ICE) gene is homologous to *ced-3*. These developments facilitate understanding the findings in mammalian systems in light of the rich experimental record on *C. elegans* and raise the possibility of discovering specific compounds to combat cell death, including neuronal cell death (*see* chapter by Rosen and Casciola-Rosen in Part I).

The recent discovery of 11 different missense mutations of the principal free radical scavenger, superoxide dismutase 1 (SOD1), in patients with familial ALS has brought to the fore the free radical theory of neuronal cell death (*see* chapter by Rabin and Borchelt in Part III). Exciting new mechanisms have been proposed for the interaction of free radicals, calcium, and the toxic messenger, nitric oxide, in keeping with the basic idea of multiple converging pathways for neuronal death, as suggested by Dennis Choi in the 1990 Dahlem workshop on neurodegenerative diseases (*see* chapters by Dykens and by Leist and Nicotera in Part I). In the meantime, tumor biology has continued to provide clues about coupled regulation of cell proliferation and death, a still evolving concept that may be especially pertinent to postmitotic, terminally differentiated cells like neurons (*see* chapter by Freeman in part I).

Very important observations have been made, in the last few years, by exposing neurons to various types of injury in vitro and in vivo, under conditions modeling human disease, in which investigators challenge ideas originating in general cell biology against the peculiarities of the nervous system. These models of neurological disease (especially animal models) are the best tools we have to test the clinical significance of more fundamental observations from cell lines or nonneuronal cells (*see* chapters by Elliot and Snider, Olney and Ishimaru, and O'Hearn and Molliver in Part II). The introduction of altered genes—especially genes known to cause familial forms of neurologic disease—into the genome of experimental mammals promises to revolutionize our ability to look into the mechanisms and to test drugs preclinically for the treatment of neurologic diseases. These transgenic animals—principally mice—have already provided excellent naturalistic models for such devastating diseases as ALS (*see* chapter by Rabin and Borchelt in Part III) and prion diseases (*see* chapter by Borchelt in Part III) and represent superb tools for the study of the principal mechanisms associated with other diseases, such as amyloidogenesis in Alzheimer's disease (*see* chapter by Koliatsos and Mocchetti in Part IV). Transgenic mice can also be used to examine the crucial issue of the interaction between genetic vulnerability and environmental triggers in the pathogenesis of neurologic disease (*see* chapter by Elliot and Snider in Part I).

The advent of *in situ* methods to label dying (apoptotic) neurons in tissues has allowed direct observations of neuronal death in the human brain. These observations will help clarify pathogenic mechanisms and, possibly, to redirect or redefine our therapeutic targets in various diseases of the nervous system (*see* chapters by Vornov, Shin and Lee, Dietrich, Back and colleagues, Wood, Burke, Ross and colleagues, and Gelbard in Part III). On the other hand, the progressive introduction of neurobiological tools to the neuropathological study of the human brain has reinvigorated a field that has been rather lim-

Preface

ited in its ability to explain mechanisms of human neurologic disease. This renaissance of neuropathology is founded not only on the availability of tools that can be used for labeling phenomena that occur *in situ* (*see* chapter by Ross and colleagues in Part III), but also on the application of systems neuroscience, principally the result of physiological and anatomical investigations on nonhuman primates or the combinations of molecular and systems approach (*see* chapter by Braak and Braak in Part III). The emergence of what has been called by some "smart neuropathology" may be especially helpful for a better understanding of the most elusive diseases of the nervous system, such as schizophrenia (*see* chapter by Arnold in Part III).

Continued work on animal models of neurological disease, combined with more specific observations on the brains of patients using imaging or tissue technique, will allow the development of specific ways to combat neuronal cell death, either with small molecules that can be given systematically (*see* chapters by Koliatsos and Mocchetti, Tymianski, Bar-Peled and Rothstein, and Ratan in Part IV) or larger peptides for intrathecal administrations (*see* chapter by Koliatsos and Mocchetti in Part IV). Targeting mechanisms of cell death for therapeutic interventions is becoming a realistic and powerful approach, supported by a constantly expanding and increasingly innovative biotechnology.

Some of our editorial principles: Our book primarily targets the clinician-scientist and the graduate student of biology and neuroscience, but also hopes to refresh the memory and reinforce the commitment of established basic scientists. We have placed a great emphasis on the teaching function of the book. This function is best served through the presentation of informed views (rather than dogmas or single findings) in a simple, clear, and diagrammatic fashion. The topics were also selected with the previous function in mind. Conscious of the fact that no book can compete with scientific journals, we have also decided not to reproduce the style and content of excellent topical volumes on the subject of cell death. Our hope is that *Cell Death and Diseases of the Nervous System* will remain the definitive reference on neuronal cell death for the clinician-scientist for several years to come.

In keeping with its teaching function, we have tried to make ours a user-friendly book. We have edited to ensure uniformity in the quality and mode of presentation, as well as a logical progression from one topic to another, and have provided extensive cross-referencing. Because of our strong emphasis on human disease, the overall schematic should facilitated the readers' understanding and treatment of neurological illness. In the chapters that cover the basic sciences (the first two parts of the book), there is a concluding passage on the implications for human disease. This is realized in detail in the clinical chapters, where the various authors comment on models pertinent to the diseases they discuss. For the sake of simplicity and uniformity, we have discouraged halftone illustrations, whereas a large number of line diagrams were included. The result, we sincerely believe, is one of the best organized and user-friendly books on clinical neuroscience a reader can find today.

Vassilis E. Koliatsos, MD
Rajiv R. Ratan, MD, PhD

Contents

Foreword .... *v*
Preface .... *vii*
List of Contributors .... *xvii*
List of Color Plates .... *xxi*

PART I. DEFINITIONS—CELLULAR AND MOLECULAR MECHANISMS

1. Apoptosis Versus Necrosis: *How Valid a Dichotomy for Neurons?*
   Peter G. H. Clarke 3

2. Asynchronous Death as a Characteristic Feature of Apoptosis
   *Randall N. Pittman, Conrad A. Messam,
   and Jason C. Mills* 29

3. Free Radicals and Mitochondria Dysfunction in Excitotoxicity
   and Neurodegenerative Disease
   James A. Dykens 45

4. Calcium and Cell Death
   Marcel Leist and Pierluigi Nicotera 69

5. Proteases: *Critical Mediators of Apoptosis*
   Anthony Rosen and Livia Casciola-Rosen 91

6. The Cell Cycle and Neuronal Cell Death
   Robert S. Freeman 103

PART II. ANIMAL MODELS

7. Neuronal Cell Death in *C. elegans*
   Dewey Royal and Monica Driscoll 123

8. Cellular Interactions that Regulate Programmed Cell Death
   in the Developing Vertebrate Nervous System
   Michael J. Burek and Ronald W. Oppenheim 145

9. Axotomy-Induced Motor Neuron Death
   Jeffrey L. Elliott and William D. Snider 181

10. Excitotoxic Cell Death
    John W. Olney and Masahiko J. Ishimaru 197

11. Neurotoxins and Neuronal Death: *An Animal Model of Excitotoxicity*
   Elizabeth O'Hearn and Mark E. Molliver 221

PART III. NERVE CELL DEATH IN HUMAN DISEASES

12. DNA Repair and Neurological Diseases
   Brian G. Fuller and Vilhelm Bohr 249

13. Cell Death and the Mitochondrial Encephalomyopathies
   Kevin M. Flanigan and Rajiv R. Ratan 275

14. Viral Encephalitis
   J. Marie Hardwick, David N. Irani, and Diane E. Griffin 295

15. Mechanisms of Cell Injury in Prion Diseases
   David R. Borchelt 325

16. Blood Flow, Energy Failure, and Vulnerability to Stroke
   James J. Vornov 343

17. Epilepsy and Cell Death
   Cheolsu Shin and Ki-Hyeong Lee 361

18. Trauma to the Nervous System
   W. Dalton Dietrich 379

19. Approaches to the Study of Diseases Involving Oligodendroglial Death
   Stephen A. Back and Joseph J. Volpe 401

20. Motor Neuron Disease
   Bruce A. Rabin and David R. Borchelt 429

21. Cerebellar Degenerations
   Katherine A. Wood 445

22. Parkinson's Disease
   Robert E. Burke 459

23. Huntington's Disease and DRPLA: *Two Glutamine Repeat Diseases*
   Christopher A. Ross, Mark W. Becher, and Vassilis E. Koliatsos 477

24. Cortical Destruction and Cell Death in Alzheimer's Disease
   Heiko Braak and Eva Braak 497

25. HIV-1 Infection of the CNS: *Evidence for Apoptosis of Nerve Cells*
   Harris A. Gelbard 511

26. Neurotoxicity of Drugs of Abuse
   Jean Lud Cadet 521

27. Schizophrenia
   Steven E. Arnold 527

PART IV. APPROACHES IN TREATING NERVE CELL DEATH

28. Trophic Factors as Therapeutic Agents for Diseases Characterized by Neuronal Death
   Vassilis E. Koliatsos and Italo Mocchetti 545

29. Early Experiences with Trophic Factors as Drugs for Neurological Disease: *Brain-Derived Neurotrophic Factor and Ciliary Neurotrophic Factor for ALS*
   Dana C. Hilt, James A. Miller, and Errol Malta 593

30. Approaches in Treating Nerve Cell Death with Calcium Chelators
   Michael Tymianski 609

31. Antiglutamate Therapies for Neurodegenerative Disease: *The Case of Amyotrophic Lateral Sclerosis*
   Osnat Bar-Peled and Jeffrey D. Rothstein 633

32. Antioxidants and the Treatment of Neurological Disease
   Rajiv R. Ratan 649

Index 667

Contributors

STEVEN E. ARNOLD • *Department of Psychiatry, University of Pennsylvania School of Medicine, Philadelphia, PA*

STEVEN A. BACK • *Department of Neurology, Children's Hospital, Harvard Medical School, Boston, MA*

OSNAT BAR-PELED • *Department of Neurology, The Johns Hopkins University School of Medicine, Baltimore, MD*

MARK W. BECHER • *Department of Pathology (Division of Neuropathology), The Johns Hopkins University School of Medicine, Baltimore, MD*

VILHELM A. BOHR • *Gerontology Research Center, National Institute on Aging, National Institutes of Health, Baltimore, MD*

DAVID R. BORCHELT • *Department of Pathology (Division of Neuropathology), The Johns Hopkins University School of Medicine, Baltimore, MD*

EVA BRAAK • *Department of Anatomy, Johann Wolfgang Goethe-Universität, Frankfurt/Main, Germany*

HEIKO BRAAK • *Department of Anatomy, Johann Wolfgang Goethe-Universität, Frankfurt/Main, Germany*

MICHAEL J. BUREK • *Department of Neurobiology and Anatomy, School of Medicine, Wake Forest University, Winston-Salem, NC*

ROBERT E. BURKE • *Department of Neurology, Columbia University College of Physicians and Surgeons, New York, NY*

LIVIA CASCIOLA-ROSEN • *Departments of Dermatology and Cell Biology and Anatomy, The Johns Hopkins University School of Medicine, Baltimore, MD*

JEAN LUD CADET • *Molecular Neuropsychiatry Section, NIH/NIDA, Addiction Research Center, Baltimore, MD*

PETER G. H. CLARKE • *Institut D'Anatomie, Lausanne, Switzerland*

W. DALTON DIETRICH • *Department of Neurology, University of Miami School of Medicine, Miami, FL*

MONICA DRISCOLL • *Department of Molecular Biology and Biochemistry, Rutgers University, Piscataway, NJ*

JAMES A. DYKENS • *Mitokor, San Diego, CA*

JEFFREY L. ELLIOTT • *Department of Neurology, Washington University School of Medicine, St. Louis, MO*

KEVIN M. FLANIGAN • *Eccles Institute of Human Genetics, University of Utah, Salt Lake City, UT*

ROBERT S. FREEMAN • *Departments of Pharmacology and Physiology, University of Rochester School of Medicine, Rochester, NY*

BRIAN G. FULLER • *Gerontology Research Center, National Institute on Aging, National Institutes of Health, Baltimore, MD*

HARRIS A. GELBARD • *Department of Neurology, Child Neurology Division, University of Rochester Medical Center, Rochester, NY*

DIANE E. GRIFFIN • *Department of Molecular Microbiology and Immunology, The Johns Hopkins University School of Public Health, Baltimore, MD*

J. MARIE HARDWICK • *Department of Molecular Microbiology and Immunology, School of Public Health, The Johns Hopkins University, Baltimore, MD*

DANA C. HILT • *Amgen Inc., Thousand Oaks, CA (present address: Guilford Pharmaceuticals, Baltimore, MD)*

DAVID N. IRANI • *Department of Molecular Microbiology and Immunology, The Johns Hopkins University School of Public Health, Baltimore, MD*

MASAHIKO J. ISHIMARU • *Department of Psychiatry, Washington University School of Medicine, St. Louis, MO*

VASSILIS E. KOLIATSOS • *Departments of Pathology (Neuropathology), Neurology, Neuroscience, and Psychiatry and Behavioral Science, The Johns Hopkins University School of Medicine, Baltimore, MD*

KI-HYEONG LEE • *Department of Neurology, Mayo Clinic, Rochester, MN*

MARCEL LEIST • *Universität Konstanz, Konstanz, Germany*

ERROL MALTA • *Amgen, Inc., Thousand Oaks, CA*

CONRAD A. MESSAM • *Department of Pharmacology, University of Pennsylvania, Philadelphia, PA*

JAMES A. MILLER • *Amgen, Inc., Thousand Oaks, CA*

JASON C. MILLS • *Department of Pharmacology, University of Pennsylvania School of Medicine, Philadelphia, PA*

MARK E. MOLLIVER • *Department of Neuroscience, The Johns Hopkins University School of Medicine, Baltimore, MD*

ITALO MOCCHETTI • *Department of Cell Biology, Georgetown University, Washington, DC*

PIERLUIGI NICOTERA • *Universität Konstanz, Konstanz, Germany*

ELIZABETH O'HEARN • *Department of Neuroscience, The Johns Hopkins University School of Medicine, Baltimore, MD*

JOHN W. OLNEY • *Department of Psychiatry, Washington University School of Medicine, St. Louis, MO*

RONALD W. OPPENHEIM • *Department of Neurobiology and Anatomy, Bowman Gray School of Medicine, Winston-Salem, NC*

RANDALL N. PITTMAN • *Department of Pharmacology, University of Pennsylvania School of Medicine, Philadelphia, PA*

BRUCE A. RABIN • *Department of Pathology (Division of Neuropathology), The Johns Hopkins University School of Medicine, Baltimore, MD*

RAJIV R. RATAN • *Department of Neurology, Beth Israel Deaconess Medical Center and Harvard Medical School, Boston, MA*

ANTHONY ROSEN • *Department of Medicine and Cell Biology and Anatomy, The Johns Hopkins University School of Medicine, Baltimore, MD*

CHRISTOPHER A. ROSS • *Neuropathology Laboratory, The Johns Hopkins University School of Medicine, Baltimore, MD*

JEFFREY D. ROTHSTEIN • *Department of Neurology, The Johns Hopkins University School of Medicine, Baltimore, MD*

DEWEY ROYAL • *Department of Molecular Biology and Biochemistry, Rutgers University, Piscataway, NJ*

CHEOLSU SHIN • *Department of Neurology, Mayo Clinic, Rochester, MN*

WILLIAM D. SNIDER • *Department of Neurology, Washington University School of Medicine, St. Louis, MO*

MICHAEL TYMIANSKI • *Neuroprotection Laboratory, Playfair Neuroscience Unit, Toronto Hospital, Toronto, Ontario, Canada*

JOSEPH J. VOLPE • *Department of Neurology, Children's Hospital, Harvard Medical School, Boston, MA*

JAMES J. VORNOV • *Guilford Pharmaceuticals, Baltimore, MD*

KATHERINE A. WOOD • *Trevigen, Gaithersburg, MD*

List of Color Plates

Color plates appear as an insert following p. 298.

Plate 1 *(Fig. 1A–D from Chapter 9).* Apoptosis in neonatal facial motor neurons after axotomy.

Plate 2 *(Fig. 4A–D from Chapter 11).* Cytologic changes in injured Purkinje cells 24 h after ibogaine administration.

Plate 3 *(Fig. 3A–G from Chapter 11).* Degeneration of Purkinje cells following administration of the indole alkaloid ibogaine.

Plate 4 *(Fig. 2A–F from Chapter 20).* Similarities in spinal cord and muscle pathology between G37R transgenic mice and patients with amyotrophic lateral sclerosis.

Plate 5 *(Fig. 1 from Chapter 22).* Apoptosis in dopaminergic neurons of the substantia nigra after striatal excitotoxic lesions.

Plate 6 *(Fig. 4 from Chapter 24).* Patterns of information flow along associative pathways relate to the distribution of neuropathological changes in brains with Alzheimer's disease.

Plate 7 *(Fig. 7 from Chapter 24).* Myelination patterns in human brain relate to the progression of neuropathology in Alzeimer's disease.

Plate 8 *(Fig. 3 from Chapter 23).* Intranuclear aggregates of huntingtin provide clues in pathogenesis of Huntington's disease.

Plate 9 *(Fig. 2 from Chapter 25).* Double staining of Bax cytoplasm and TUNEL (nucleus) in the brain of a child with encephalopathy.

Plate 10 *(Fig. 2A–D from Chapter 28).* Partial initiation in APP transgenic mice, of the neuropathological features of Alzheimer's disease.

I
Definitions—Cellular and Molecular Mechanisms

1
Apoptosis Versus Necrosis
How Valid a Dichotomy for Neurons?

Peter G. H. Clarke

INTRODUCTION

The word *apoptosis* was coined, from its Greek equivalent, in 1972 *(1)*, but its morphological identity as a distinct kind of cell death was recognized by Flemming as long ago as 1885 *(2)*. The word *necrosis* has been used in English, French ("nécrose"), and German for several centuries to mean the mortification of tissue, and 2000 years ago its Greek equivalent carried a similar meaning. Even today, the word is still frequently used in this general sense, but since 1980 it has been given a particular, cellular, sense, with the claim that virtually all cell deaths can be classified dichotomously as either apoptosis or necrosis *(3)*. Previously, this necrosis had been called "coagulative necrosis" by several authors including Kerr et al. *(1)*.

Modern research on apoptosis has been concentrated mainly in the areas of cancer and immunology. The very term was coined in a cancer journal *(1)* and much early research on its mechanisms focused on thymocytes *(4)*. The extraordinary boom in apoptosis research over the last ten years was triggered largely by the discoveries that apoptosis plays a central role in the development of a self-tolerant immune system by the removal of autoreactive cells *(5)*, and that an oncogene (*bcl*-2) involved in follicular lymphoma, the commonest haematological malignancy, acts by suppressing apoptosis rather than by regulating the cell cycle *(6)*. In the Medline database, of the almost 5000 papers mentioning apoptosis in the abstract in the last four years, almost 50% deal with cancer and more than 30% with some aspect of immunology (including leukemia and AIDS). But 7% deal with neurons, and most assume "apoptotic" neurons (and "necrotic" ones when mentioned) to have the main characteristics of apoptosis (or necrosis) as elucidated in the fields of cancer and immunology.

For these reasons, there has been a tendency to apply uncritically, in the context of neurons, a considerable series of assumptions that are based largely on research in other fields (*see* also chapter by Olney and Ishimaru). The purpose of the present paper is to evaluate these assumptions. For brevity, we shall limit discussion largely to the best-studied situations in which neuronal death occurs: normal development, axotomy, and experimental excitotoxicity. To facilitate this analysis, the author subdivides the current doctrine about apoptosis and necrosis into eight central tenets, to be evaluated individually in the following pages (Table 1).

From: Cell Death and Diseases of the Nervous System
Edited by: V. E. Koliatsos and R. R. Ratan © Humana Press Inc., Totowa, NJ

Table 1
Eight Widespread Tenets Related to the Apoptosis–Necrosis Dichotomy, and an Evaluation of Their Validity as Applied to Neurons

Tenet	Evaluation
1. There are just two morphological types of cell death.	False
2. In physiological situations and in mild pathology all cell death is apoptotic, but grossly pathological insults always lead to necrosis.	False
3. Apoptosis differs from necrosis in that it has a unique physiological role, the regulation of cell numbers.	Not true in the nervous system
4. Apoptotic cells occur in isolation, but necrotic ones in clusters.	Often the case, but not a universal rule
5. Necrosis involves spilling of cell contents, hence inflammation. Since apoptotic bodies are membrane-bound, apoptosis does not elicit inflammation, heterophagy being performed mainly by local cells.	Not proven in general Untrue in the special case of the nervous system
6. Apoptosis, unlike necrosis, is active, controlled and programmed, being regulated by specific death genes.	Not clearcut
7. Apoptosis, unlike necrosis, involves double-stranded DNA-breaks.	Not universally true
8. The cellular mechanisms of apoptosis and necrosis are fundamentally different.	Not clearcut

TENET 1: THERE ARE JUST TWO MORPHOLOGICAL TYPES OF CELL DEATH

The central claim of Wyllie et al. *(3)* was that there are two main types of cell death distinguishable morphologically. In their earlier review *(1)*, apoptosis had already been stated to involve nuclear and cytoplasmic condensation, accompanied by the clumping of chromatin along the inside of the nuclear envelope, while in the cytoplasm the organelles are preserved until the cell is broken into membrane-bound, ultrastructurally well-preserved fragments called apoptotic bodies that are generally phagocytozed. In the 1980 review, this was contrasted with necrosis, whose essential feature is said to be swelling: first, gross dilation of mitochondria, whose cristae break, and then of endoplasmic reticulum, which fragments into vesicles. At the same time, the nucleus also swells and its heterochromatin becomes coarser, forming small discrete masses on the nuclear membrane before its dissolution, which leaves a nuclear "ghost" *(1,7,8)*. Many authors state that the entire cell also swells, leading to loss of plasma membrane integrity *(7,9,10)*. How valid is this dichotomy in the nervous system?

Development

In normal development, it is simple. Here, several different morphologies occur which can mostly be grouped into three main types (Tables 2 and 3) *(11)*, although intermediate types are sometimes found *(11,12)*. The following discussion deals with neuronal death in vertebrate development that is either normal or disrupted in ways that

Table 2
Summary of the Three Main Types of Cell Death that Occur in Development (from ref. 11)

Various designations		Nucleus	Cell membrane	Cytoplasm	Heterophagic elimination
Type 1	Apoptosis; shrinkage necrosis; precocious pyknosis; nuclear type of cell death	Nuclear condensation, clumping of chromatin leading to *pronounced pyknosis*	Convoluted, forming blebs	Loss of ribosomes from RER and from polysomes; cytoplasm reduced in volume becoming electron-dense	Prominent and important
Type 2	Autophagic cell death	Pyknosis in some cases. Parts of nucleus may bleb or segregate	Endocytosis at least in some cases; blebbing can occur	*Abundant autophagic vacuoles*; ER and mitochondria sometimes dilated; Golgi often enlarged	Occasional and late
Type 3A	Nonlysosomal disintegration	Late vacuolization, then disintegration	Breaks	General disintegration; dilatation of organelles, forming "empty" spaces that fuse with each other and with the extracellular space	No
Type 3B	Cytoplasmic type	Late increase in granularity of chromatin	Rounding up of cell	*Dilatation of ER, nuclear envelope, Golgi and sometimes mitochondria, forming "empty" spaces*	Yes

Types 1, 2, and 3B, but not 3A, have been found in neurons. The most striking and reliable characteristic of each type is shown in italics.

Table 3
Situations in Which the Three Main Kinds of Neuronal Death Occur

Structure or cell type	Species	References
Type 1: Apoptosis-Like Neuron Death		
Cervical visceral motoneurons	Chick embryo	*(134)*
Spinal motoneurons	Chick embryo	*(135)*
Retina	Young mouse	*(136)*
Second order lateral line neuron	Metamorphic frog	*(137)*
Mesencephalon	Lizard (Gallotia) embryo	*(22)*
Type 2: Autophagic Neuron Death		
Axotomized spinal motoneurons	Larval frog	*(14)*
Axotomized spinal motoneurons	Chick embryo	*(138)*
Rohon-Beard neurons	Larval frog	*(139)*
Normal and target-deprived isthmo-optic neurons	Chick embryo	*(16,19)*
Cerebral cortex	Pre-/postnatal rat	*(17)*
Type 3B: Cytoplasmic Neuron Death		
Olfactory epithelium	Mouse and rat embryos	*(140,141)*
Ciliary ganglion cells	Chick embryo	*(13)*
Trochlear motoneurons	Duck embryo	*(142)*
Spinal motoneurons	Chick embryo	*(135)*
Superior colliculus	Neonatal rat	*(143)*
Mesencephalon	Lizard (Gallotia) embryo	*(22)*
Supraoptic nucleus	Ovariectomized young rats	*(144)*

are believed to exacerbate the problems encountered by neurons in normal development (e.g., axotomy, which deprives the neuron of retrograde trophic support for which neurons are believed to compete normally).

The first morphological type resembles apoptosis (Fig. 1A; Table 2). The hallmarks of this type include nuclear and cytoplasmic condensation, clumping of nuclear chromatin, and in some cases, fragmentation of the cell into apoptotic bodies. Moreover, like most apoptotic cells, type 1 dying neurons lose ribosomes from their rough endoplasmic reticulum, and polyribosomes disperse into free ribosomes *(13)*. The only striking morphological difference, as compared to standard apoptosis, is that the chromatin aggregates into several balls inside the nucleus and does not line the nuclear membrane *(11)*. The significance of this difference is unclear. It probably does not reflect a special property of neurons, but rather of developing cells in general. The author is not aware of chromatin margination occurring in the death of any kind of cell during normal development.

Type 2 is characterized primarily by the formation of numerous autophagic vacuoles (Fig. 2; Table 2). Autophagy occurs even in healthy cells, being associated with cytoplasmic turnover, but in type 2 dying neurons it is far more extensive and probably plays a major role in the destruction of the cell. The Golgi apparatus is often enlarged and has been found to exhibit greatly-enhanced nucleoside diphosphatase activity *(14)*; this makes sense, because the Golgi apparatus is the source of primary lysosomes, which are needed to provide hydrolytic enzymes for the autophagic vacuoles. In a few cases, autophagic dying neurons have been shown to be undergoing intense endocytosis, which may be

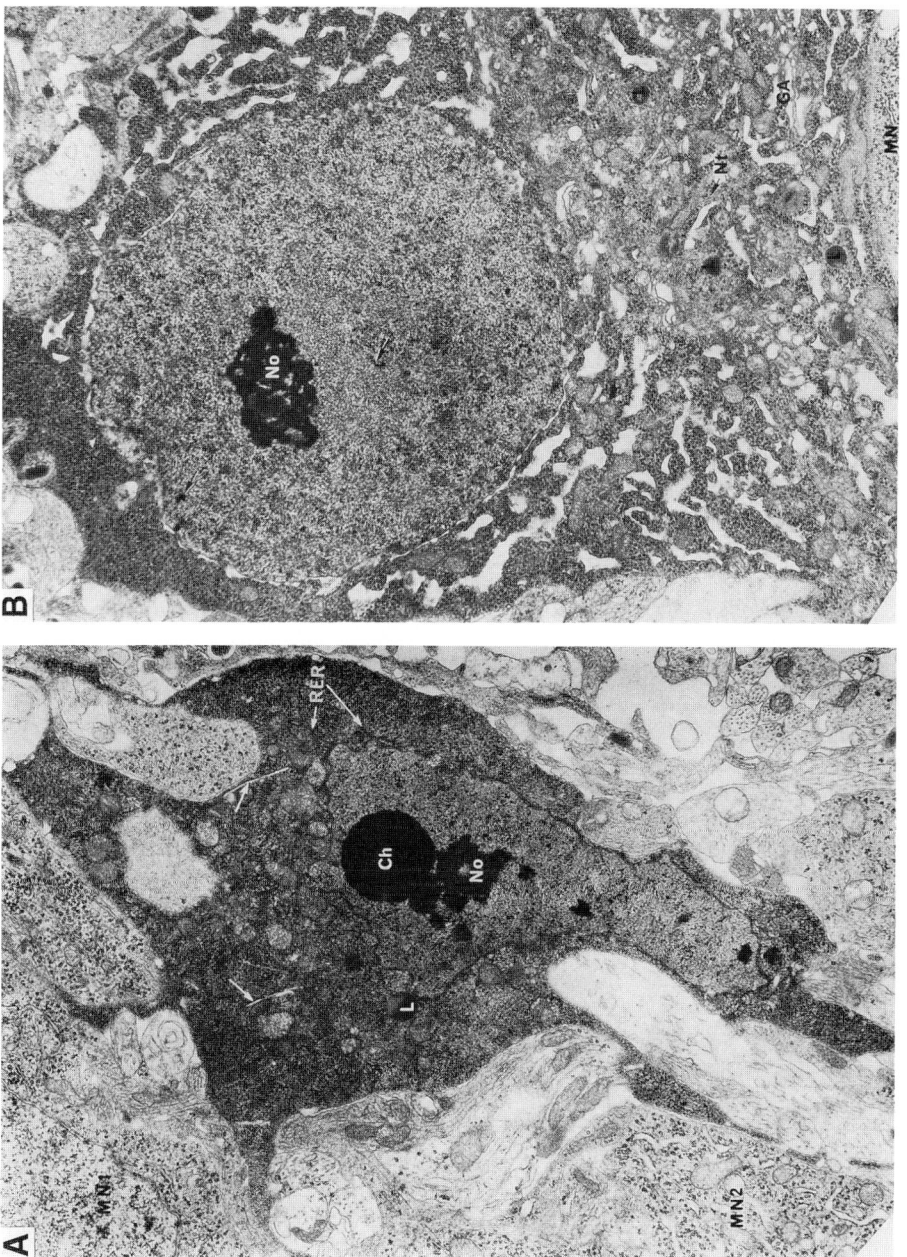

Fig. 1. Dying motoneurons from the lateral motor column of 6–8 d-old chick embryos. **(A)** Type 1 (apoptotic) degeneration. The nucleus is condensed and misshapen and contains chromatin masses (Ch). Polyribosomes have dissociated, but small cisternae of rough endoplasmic reticulum (RER) are still recognizable. **(B)** Type 3B degeneration. The rough endoplasmic reticulum, Golgi apparatus (GA), and nuclear envelope are markedly dilated. Polyribosomes still have a typical rosette-like arrangement. Original magnification: ×6000 (A) and ×4500 (B). (From Chu-Wang and Oppenheim (135), with permission)

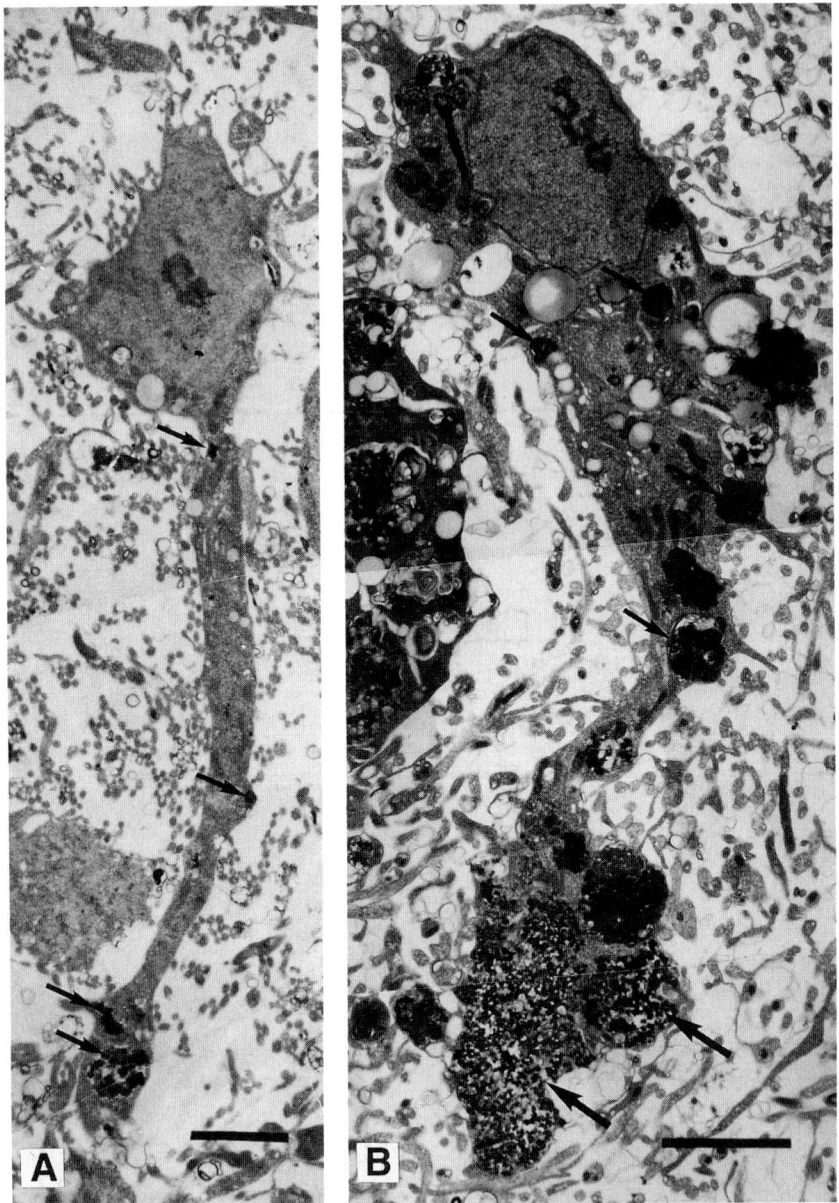

Fig. 2. Two type 2 (autophagic) dying neurons in the isthmo-optic nucleus of 14-d old chick embryos that had received an intravascular injection of horseradish peroxidase (to show endocytosis) about 3 h before fixation. The death of the neurons was provoked by blockade of axoplasmic transport in their axonal terminal region, the retina. **(A)** Early stage of degeneration. Several vacuoles, some peroxidase-labeled *(arrows)*, can be seen in the soma and main dendrite. **(B)** Advanced stage of degeneration. The vacuoles *(arrows)* are larger and more numerous, many containing myelin figures (membranous whorls), and peroxidase label. Bars = 2 μm. (From Hornung et al. *(16)* with permission.)

used to reduce the area of the plasma membrane *(15,16)*. The nuclei of autophagic dying neurons are sometimes pyknotic, but not so strikingly as in apoptotic cells. In two situations, there is evidence that nuclear chromatin ultimately leaves the pyknotic nuclei and is destroyed in autolysosomes of the same cell *(16,17)*. The autophagic type of neuron death has been well documented in only five situations (Table 3), but this may be because it is difficult to recognize, since these dying cells are hard to distinguish from phagocytes.

The third type of developmental cell death is characterized by swelling and has some of the features of necrosis. We previously divided it into two subcategories, types 3A and 3B (Table 2) *(11)*. Since type 3A is rare and has never been reported in neurons, it will not be discussed here. Type 3B (Fig. 1B) is what Pilar and Landmesser *(13)* called the "cytoplasmic" type of neuronal death. Initially, the cisternae of rough endoplasmic reticulum dilate, but, unlike apoptosis, the ribosomes remain attached to the cisternae and there is no dispersion of polysomes into free ribosomes. The Golgi apparatus may also be dilated and sometimes the mitochondria swell. This kind of cell death partly resembles necrosis with its most prominent characteristic of organelle dilatation. Two differences with respect to necrosis are that in the cytoplasmic type of neuron death mitochondrial swelling is less prominent and that chromatin margination has not been described.

Depending on the situation, a given developing neuron can die by more than one type of cell death. For example, ciliary ganglion cells normally die by the cytoplasmic type *(13)*; if they are deprived of afferents during development they still die by this type *(18)*, but if they are deprived of retrograde support from their target, they die by apoptosis. In normal development, neurons of the chick embryo's isthmo-optic nucleus can probably die by apoptosis or by autophagic cell death *(19)*, but following the removal of afferents they appear to die by apoptosis *(20)*, whereas after the interruption of retrograde support they all die by autophagic cell death *(16)*. The maturity of the neuron influences the type of cell death to some extent, because in several cases it has been reported that very immature neurons die by apoptosis whereas later in development the same class of neurons die by the cytoplasmic type of death *(21,22)*. This is not, however, a general rule since adult neurons can die by apoptosis (Fig. 3) *(23)*.

Although most dying neurons in development fall into one of the three main categories, some combine features of more than one of them or do not match any. Such exceptions can occur even in normal development *(11)*, but are more common in abnormal situations such as genetic mutants as, for example, in dopaminergic neurons of the substantia nigra of weaver mice. These die with a morphology resembling type 3B in their vacuolation of nuclear membrane and endoplasmic reticulum and in their abundance of ribosomes; but their condensed nuclei with irregular chromatin aggregates are quite different from those of type 3B (Fig. 4) *(24)*.

Axotomy of Adult Neurons

One of the more thoroughly studied cases of neuronal death is in central neurons of mammals after axotomy. In the peripheral nervous system and in the central nervous system of lower vertebrates, axotomy is generally followed by regeneration, albeit with some degree of neuronal loss. Even in the mammalian central nervous system, axotomized neurons sometimes survive in a shrunken state *(25)*, but in other cases they die in large numbers, and the question arises whether such dying neurons can be fitted into the categories of apoptosis and/or necrosis.

Classical descriptions of axotomy-induced neuronal death display features of delayed necrosis following a period of atrophy. The necrotic features include vacuolar dilatation

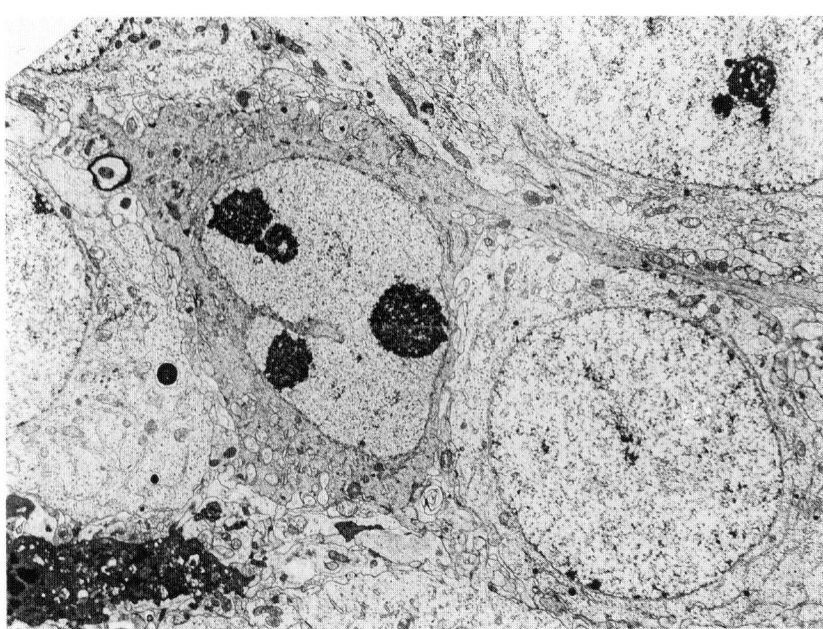

Fig. 3. Apoptotic neuron in the granule cell layer of rat hippocampus 3 h after the end of a 24 h period of perforant path stimulation. Note formation of multiple chromatin clumps and ribosomal disaggregation. ×4120. (From Sloviter et al. *(23)*, with permission.)

of endoplasmic reticulum and Golgi cisterns leading to the formation of large cytoplasmic vacuoles, and swelling of mitochondria. Studies on the thalamus and red nucleus show that these changes often occur in the presence of a normal-looking nucleus *(26,27)*, although the nuclear chromatin may become clumped at an advanced stage of degeneration *(28)* as occurs in necrosis. However, the protracted nature of such cell death, spanning many days or weeks, contrasts with the rapidity of most cases of necrosis and the same axotomized neurons with necrotic features can also show characteristics of other types of cell death. Notably, depletion of rough endoplasmic reticulum *(26–28)* (typical of apoptosis) and in some cases, an abundance of secondary lysosomes *(28)* (typical of autophagic cell death).

Recently, several groups have claimed that retinal ganglion cell death after optic nerve section in adult mammals is apoptotic. This diagnosis was based, to a great extent, on criteria now known to be nonspecific for apoptosis such as *in situ* staining for DNA fragmentation, and on very limited morphology, purely at the light microscopic level *(29–31)*. However, ultrastructural evidence of apoptosis in ganglion cells was reported in the case of experimental glaucoma *(31)*, and seems to have been present in an earlier study on the effects of optic nerve section *(32)*. Thus, adult retinal ganglion cells seem to respond to axotomy with apoptosis more often than do neurons in other parts of the brain, although they probably do not exhibit pure apoptosis.

We conclude that neuronal death after axotomy in adult mammals does not fit neatly into the categories of apoptosis and/or necrosis. It appears, rather, that a given dying neuron can display mixed features of apoptosis, necrosis, and autophagic cell death in varying proportions.

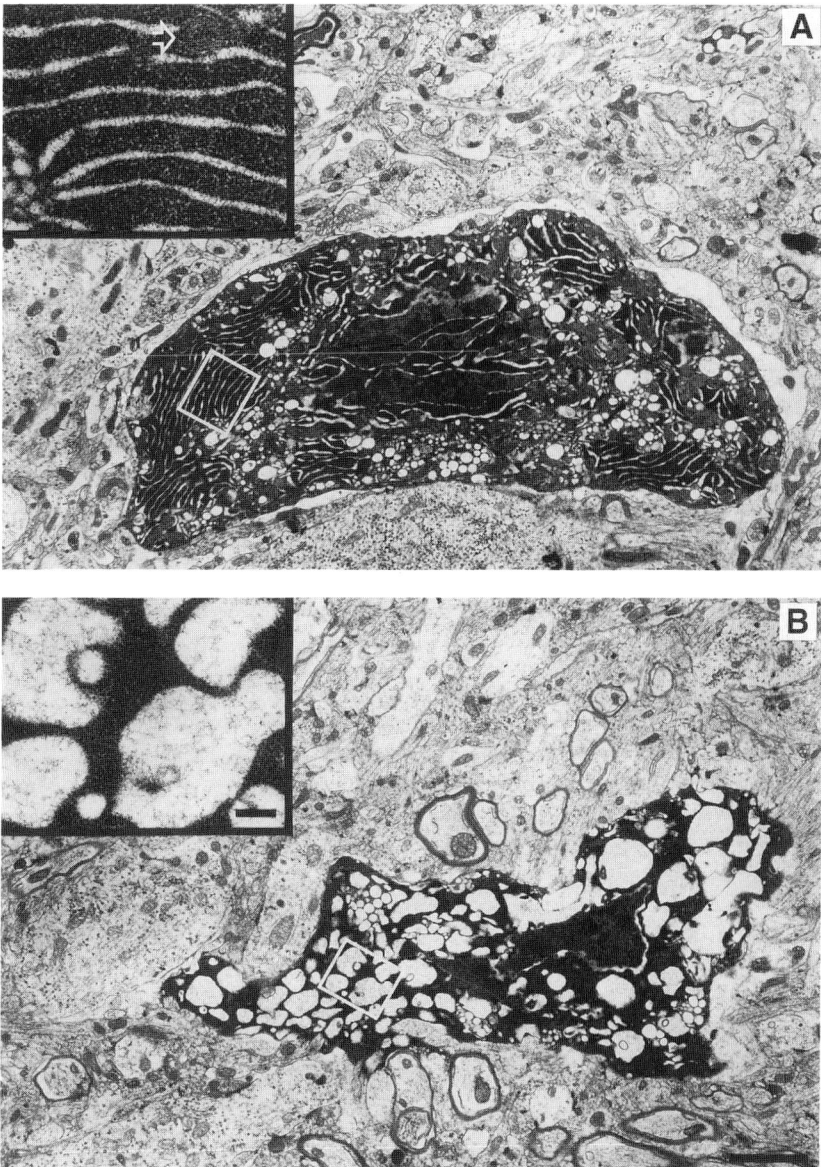

Fig. 4. Degenerating neurons in substantia nigra of 24 d postnatal weaver mouse. In each neuron, the irregularly shaped, condensed nucleus contains chromatin aggregates and is surrounded by a dilated nuclear envelope. The upper neuron **(A)** is characterized by dilated cisternae of endoplasmic reticulum (seen as layered structures) and contains round cytoplasmic vacuoles. In the lower neuron **(B)** the vacuoles predominate. Scale bars: 2 µm in A and B, 200 nm in the insets. (From Oo et al. *(24)*, with permission.)

Excitotoxic Neuron Death

Excitotoxicity means the killing of neurons by glutamate and by other substances that act at glutamate receptors *(33)* (also *see* chapter by Olney and Ishimaru). It has been intensively studied, partly because it causes neuronal death in several clinically important condi-

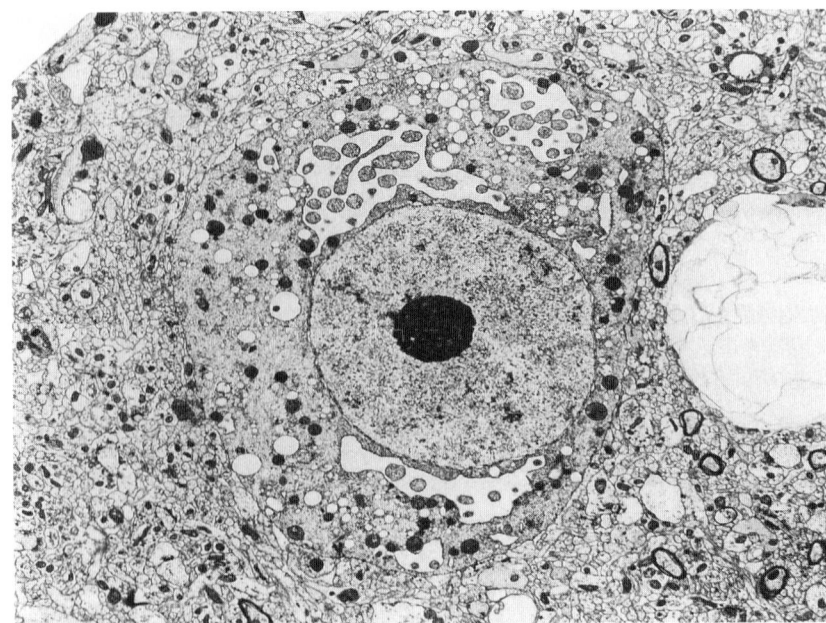

Fig. 5. Necrotic neuron in the hilar region of rat hippocampus after 12 h of perforant path electrical stimulation. Note abnormal cytoplasmic vacuolization and relatively normal nucleus. ×4120. (From Sloviter et al. *(23)*, with permission.)

tions including hypoxia/ischemia, stroke, and epilepsy *(34,35)*. Excitotoxicity is, in fact, one of the situations in which the apoptosis/necrosis distinction may hold fairly well *(36–38)*.

Necrosis-like death generally occurs rapidly after an excitotoxic insult to a neuron. Experiments in culture show that intense exposure to glutamate is followed within minutes by neuronal swelling, due to the influx of Na^+, which induces Cl^- entry followed by the osmotically induced entry of water *(9)*. If the stimulus is sufficiently strong, the excessive swelling apparently leads to cell death. This type of cell death occurs rapidly by what seems to be necrosis on morphological grounds and on the basis of the presence of intracellular debris scattered in the incubation medium *(36,39,40)*. It is not clear whether the loss of intracellular contents occurs in vivo after an excitotoxic stimulus, but there is considerable ultrastructural evidence that excitotoxic neuron death occurring in vivo can display a typical necrotic morphology (Fig. 5) *(23)*, including swollen mitochondria and rupture of the nuclear envelope and plasma membrane *(37)*; this effect can be mediated by the entry of Na^+, Cl^-, and water as it occurs in vitro *(41)*. Calcium entry is a major effector of necrosis in many situations *(42)*. Cultured neurons can die by necrosis in calcium-free medium in response to an acute excitotoxic stimulus *(9)*, but one cannot rule out internal calcium shifts from one membranous compartment to another.

Cells that survive the immediate effects of the excitotoxic insult may die 4–48 h later by what appears to be apoptosis, as characterized morphologically and in other ways discussed below. This has been described both in vitro (although only at the light microscopical level) *(36,39)* and in vivo, where convincing ultrastructural confirmation is available (Fig. 3) *(23,37)*. Unlike the acute, necrosis-like excitotoxic cell death, this delayed, apoptosis-like death depends on the influx of calcium *(34,36)*.

In ischemia-anoxia and epilepsy, neuronal death is caused largely by excitotoxicity, as a result of strong neuronal depolarization that leads to glutamate release. As might be expected, the situation is rather similar to that of experimental excitotoxicity. Both necrosis and apoptosis are reported to occur in these clinical conditions, the former tending to be prominent following more extreme ischemic or epileptic insults and at shorter postinsult times *(35,43)*.

Although it is fair to say that most evidence supports a simple dichotomy of apoptosis–necrosis in excitotoxicity-induced neuron death with little evidence for the autophagic type, the latter may have been missed. A possible, but debatable, example was illustrated by Van Lookeren Campagne et al. *(37)*. In addition, recent evidence indicates that intermediate forms of excitotoxic cell death occur on a continuum ranging from apoptosis to necrosis *(44)*. Another morphology of excitotoxic neuron death has been described that is neither apoptotic, nor necrotic, nor autophagic, but involves the early formation of numerous, mainly empty, vacuoles in the cytoplasm in the presence of well-preserved mitochondria, after which the nucleus and cytoplasm became translucent *(40)*.

Evaluation of Tenet 1

In view of the occurrence of at least three types of neuron death in development, the "mixed" nature of neuron death after axotomy in adults, and the occurrence of "mixed" and occasionally of nonapoptotic, non-necrotic neuron death as a result of excitotoxicity, Tenet 1 appears to be false. In this respect, neurons are not exceptional. Nonapoptotic, non-necrotic dying cells have been frequently reported among various non-neuronal cell types *(11,45–48)*.

TENET 2: IN PHYSIOLOGICAL SITUATIONS AND IN MILD PATHOLOGY ALL CELL DEATH IS APOPTOTIC, BUT GROSSLY PATHOLOGICAL INSULTS ALWAYS LEAD TO NECROSIS

The originators of the dichotomy have consistently maintained that necrosis is invariably the result of a gross insult to the cell, such as severe hypoxic, ischemic, or toxic damage, whereas apoptosis occurs in physiological situations or in pathology that is sufficiently mild to leave intact the active mechanisms involved in the apoptotic death programme *(3,8,49)*. This is widely accepted and does indeed seem to be true in many situations. Gross insults do often, perhaps always, lead to necrosis, and at least in excitotoxicity, apoptosis is more likely to occur if the stimulus is mild (*see* above and chapter by Leist and Nicotera). But this is not a rigorous rule, because the same stimulus can produce apoptosis in some parts of the brain and necrosis in others *(23,37)* and a stimulus that causes apoptosis in immature brains often causes necrosis in the adult *(44)*. Moreover, reducing stimulus concentration does not infallibly transform necrosis into apoptosis *(44)*. And, most strikingly, the occurrence, in normal development, of three kinds of cell death, including one having much in common with necrosis, is a major contradiction to Tenet 2. It must be concluded that Tenet 2 is false.

TENET 3: APOPTOSIS DIFFERS FROM NECROSIS IN THAT IT HAS A UNIQUE PHYSIOLOGICAL ROLE, THE REGULATION OF CELL NUMBERS

The summary of Kerr et al. (1972) review begins: "The term apoptosis is proposed for a…mechanism…which appears to play a complementary but opposite role to mitosis…". The consonance between the terms apoptosis and mitosis was deliberate and the former was coined from the Greek word for the falling of petals from flowers or of leaves from trees, so as to suggest the death of a replaceable component. Published in a cancer journal, the new term and the underlying idea were particularly relevant to the readership, but can they make any sense in the context of neurons? While it is true that cell death occurs in the germinal layers of the brain *(50,51)*, the cells thus affected have not differentiated to become neurons. Neuron death concerns, by definition, postmitotic cells.

The only way to impart any sense to Tenet 3 when applied to neurons is to interpret it as meaning that apoptosis during development serves to adjust neuronal numbers subsequent to excessive proliferation, so as to match a given population to the size of its axonal target territory and, in recent versions, to the number of its afferents as well (for a more topical discussion of developmental neuron death, *see* chapter by Burek and Oppenheim). This is, in fact, the oldest and most widely held interpretation of neuronal death, dating back to the classic experiments of Hamburger and Levi-Montalcini *(52)* on spinal ganglia and motoneurons. They reported that initially the spinal ganglia were all of similar size and the motor column uniform along its length, and that the subsequently larger numbers of spinal ganglion cells and motoneurons at brachial and lumbar levels were due to a retrograde influence from the periphery on the proliferation of neuronal precursors and on neuronal death *(52,53)*. It has long been clear that there is no such effect on proliferation *(54)*, but the notion that axial-level related differences in neuronal number were due to sculpting through regulation by the periphery of neuronal death (including apoptosis), and that this regulation is the very purpose of developmental neuronal death, became widely accepted. However, several of the early claims of Hamburger and Levi-Montalcini were false. Whereas they had thought neuronal death occurred almost exclusively at upper cervical and thoracic levels (i.e., those not involved in limb innervation), it is now known to occur at all levels in both spinal ganglia and motor columns *(55)*. Moreover, even before the period of neuron death, differences in neuron number related to axial level occur in both the spinal ganglia *(56)* and the motor columns *(57)*. Therefore, the sculpting of regional differences in neuronal number seems to be at most a minor purpose of neuronal death in the development of peripheral projections.

In central projections, a number-adjusting role for apoptosis is harder to test than in the periphery. A particular prediction of this hypothesis, i.e., that decreasing or increasing the number of afferent and/or target cells should, respectively, decrease or increase the amount of neuronal death, is generally fulfilled, and in a few cases the even stronger prediction of a proportional relationship has been validated *(58,59)*. But these results can be explained equally well on an alternative hypothesis.

The alternative hypothesis is that, far from being a means of eliminating normal but numerically superfluous neurons, neuron death in development (including apoptosis) is a means of eliminating abnormal neurons—ones that have made developmental errors. Such a notion goes back to Ramon y Cajal *(60)*, but until the last decade it seemed to explain only a tiny proportion of the death of 50% or more neurons that occurs in most structures. It is now clear in many vertebrate central pathways that initial neural connections are much more widespread than in the adult, and that the change involves the massive elimination of axons that are aberrant with respect to the adult pattern. In at least two situations, the change is known to be due to neuronal death *(61–63)*. In short, the role of neuron death in development is not simply to adjust numbers, as implied by Tenet 3. Whether this is even part of its role is uncertain.

TENET 4: APOPTOTIC CELLS OCCUR IN ISOLATION, BUT NECROTIC ONES IN CLUSTERS

As many authors have mentioned *(3,64)*, apoptotic cells tend to occur in isolation, and this is frequently true of neurons. For example, in development, even when 50% or more of the neurons in a given region die in a few days, only a few are visibly dying at a given moment. This is because a neuron's death and elimination is very rapid, taking only 1 or 2 h. The speed of the clearance is due partly to the fact that the apoptotic process is itself very rapid and partly to the efficiency of local phagocytic mechanisms. It is likewise true, and almost banal, that gross insults to the tissue that cause necrosis kill most or all the cells in a region at the same time, with the result that the dying cells occur in clusters.

But that is not to say that Tenet 4 is valid as a general rule. Apoptotic cells are not always isolated. Although clustered dying cells in normal development are not always apoptotic, in some cases they are so, as in the anterior and posterior marginal zones of mouse embryos *(12)*. Moreover, in several experimental situations, wholesale cell death has been provoked that includes many unequivocally apoptotic neurons. Examples include cell death due to an NMDA injection into the striatum and adjacent cortical tissues of 7-d-old rats *(37)* and hippocampal granule cell degeneration provoked by adrenalectomy in adult rats *(46)*. In an earlier study, it was found that the death occurring naturally in 50% of neurons in the ciliary ganglion was broadly necrosis-like (type 3B, or "cytoplasmic" cell death), but that increasing the cell death to 100% by target ablation transformed it into apoptosis *(13)*. This is the opposite of what would be expected on the basis of Tenet 4. We conclude that although Tenet 4 is frequently fulfilled, it is not valid as a universal rule.

TENET 5: NECROSIS INVOLVES SPILLING OF CELL CONTENTS, HENCE INFLAMMATION. SINCE APOPTOTIC BODIES ARE MEMBRANE-BOUND, APOPTOSIS DOES NOT ELICIT INFLAMMATION, PHAGOCYTOSIS BEING PERFORMED MAINLY BY LOCAL CELLS

It is frequently claimed that a special feature of apoptosis is that, in contrast to necrosis, it does not elicit inflammation *(1,3,8)* and neurobiologists often accept these claims as valid for the nervous system *(65)*. This is said to be important physiologically, and is attributed to the fact that in apoptosis the cell contents always remain membrane-bound, unlike in necrosis, which is said to involve disruption of the plasma membrane and hence the loss of diffusible cell contents that provoke an inflammatory response *(3,66)*. What is meant by an inflammatory response is not always clear, but most authors seem to have in mind "exudative inflammation" *(3)* including invasion by committed phagocytes from the blood, notably neutrophils. In contrast, apoptotic cells are held to provoke *local* cells such as resident tissue macrophages to phagocytose them owing to the expression at their surface of a number of lectins *(67)* and vitronectin-like molecules *(68)*. Let us first consider the evidence for these claims in a general context, after which special problems related to the central nervous system will be dealt with.

First, is it true that only necrosis elicits inflammation? In a sense, yes. In typical cases of apoptosis where scattered dying cells constitute less than 1% of the cell population, inflammation is indeed much less than in necrotic regions where most or all the cells are dying. But is this really due to the special features of apoptosis or is it simply because the disruption and cell death are far greater in the necrotic region? To the best of our knowledge, the answer is not known. At first sight, a possible test case might seem to be the massive cell death that occurs in certain regions during embryonic development, as for example, in the interdigital zones. Such regions certainly become packed with mononuclear phagocytes, but the conclusion that this represents "exudative inflammation" *(3)* induced by apoptosis would be premature for two reasons. First, although such cell death is usually referred to as apoptosis, in many (but not all) cases its morphological characteristics are closer to that of autophagic cell death *(11,12,69)*. Second, the phagocytes are probably not derived from the blood. They are more likely to be resident tissue macrophages or parenchymal cells *(70)*.

Second, is it true that necrotic cells spill their contents whereas apoptotic cells do not? In tissue culture, the claim that necrotic cells spill their contents is indeed true of neurons *(39)*, but ultimately *both* apoptotic and necrotic cells spill their contents. In the case of apoptosis, this is blamed on "secondary necrosis" that would supposedly not have occurred in vivo because phagocytosis would have occurred too rapidly *(71)*. This may be true, but the same excuse might be made for necrosis; moreover, cellular swelling might be restricted by the surrounding tissue framework.

In vivo evidence is therefore needed to show whether necrotic cells spill their contents more readily than apoptotic ones. Yet, the only in vivo evidence we know is the ultrastructural observation that the plasma membranes of necrotic cells (including neurons) sometimes appear ruptured *(37)*. Whether this accurately reflects the situation before fixation, and whether the resulting loss (if any) of cell contents is sufficient to provoke an inflammatory response, remains uncertain.

But in the central nervous system of adult mammals, there is an additional reason to doubt the validity of Tenet 7. Here, the acute inflammatory response to necrosis is radically different from that in other tissues, involving almost exclusively the activation of microglia, the resident macrophages of the nervous system. Even when the injury causes the blood–brain barrier to leak, the rapid recruitment of neutrophils that occurs in other tissues is in most cases minimal, and circulating monocytes are recruited only after a delay of several days *(72–74)*. In other words, necrosis in the (adult) central nervous system provokes a phagocytic response qualitatively similar to that which apoptosis is believed to provoke elsewhere. Minor differences can still not be excluded, e.g., increased activation of microglia in this or that condition, but the eliciting of exudative inflammation is clearly no basis for a dichotomy in this situation. We conclude that Tenet 5 is unproven even for peripheral tissues, and is untrue in the central nervous system of adult mammals.

TENET 6: APOPTOSIS, UNLIKE NECROSIS, IS ACTIVE, CONTROLLED AND PROGRAMMED, BEING REGULATED BY SPECIFIC DEATH GENES

Kerr et al. *(1)* stated that the morphological features of apoptosis suggested it was "an active, inherently programmed phenomenon," and postulated that it might depend "on expression of part of the genome, which is normally repressed in viable cells." To these authors, and to very many since, the term "programmed" referred to a hypothetical cell death program in the genome, and this notion became very popular when it became generally known that many cases of apoptosis could be blocked by inhibitors of macromolecular synthesis. Ironically, unknown to Kerr et al., there was already evidence well before 1972 showing that such inhibition could prevent cell death in metamorphosis *(75–77)*. Yet, one of the latter authors (Lockshin) had coined the phrase "programmed cell death" some years earlier *(78)* with no implication of a genetic program, but merely the idea that the cell death occurs predictably, without the need of a toxic stimulus *(79)*. The phrase "programmed cell death" is currently used in both senses leading sometimes to ambiguity.

Here we are concerned with the former sense—that of a genetic program. The notion that apoptosis depends on a genetic program became popular in the late 1980s because of three lines of evidence. First, it was shown that inhibitors of RNA or of protein synthesis prevented apoptosis in many situations. Following the initial in vivo and organ culture experiments mentioned above, experiments on cultures of dissociated cells made it clear that the prevention of cell death was not due to the impedance of phagocytes as had been suspected *(75)*, but must be due to inhibition of macromolecular synthesis in the class of cells whose death was prevented *(80,81)*, including neurons *(82,83)*. The second line of evidence came from genetic experiments on nematode worms where it was shown that certain genes (*"ced*-3*," "ced*-4*"*) were necessary for naturally occurring cell death (including neuronal death) whereas expression of another gene (*"ced*-9*"*) played a role in suppressing cell death *(84)* *(*for a topical discussion of the nematode death genes, *see* chapter by Royal and Driscoll). A possible objection was that this nematode cell death does not have the morphology of apoptosis and might be irrelevant to apoptosis. However, the third line of evidence came from experiments on apoptosis in vertebrates where molecular genetic manipulations permitted the identification of death-suppressing genes such as *bcl*-2 *(6)*, which is homologous to *ced*-9 of nematodes *(85)*, and of death-promoting genes such as *bax (86)*, *p53 (87)*, and the interleukin-1β converting enzyme (ICE) family of genes (or "caspases"), which are homologous to *ced*-3 *(85)* (*see* chapter by Rosen and Casciola-Rosen). The apoptosis-regulating role of these genes was first identified in non-neuronal cells and subsequently in neurons *(88,89)*.

Initially, the cherished hypothesis of many was that "cell death genes" would be unexpressed in normal, healthy cells, but would be switched on to cause apoptosis. This notion is now largely abandoned, partly because of accumulating evidence from immunology and cancer research that, while inhibition of protein or RNA synthesis does often delay apoptosis, the protected cells do still ultimately die apoptotically, even if the inhibition is maintained; in many cases, it fails to delay apoptosis and in several, it even triggers it *(90,91)*. Moreover, even when such inhibition does prevent apoptosis, this need not have been due to the blockade of death genes. In thymocytes, the degree of protection provided by different translational inhibitors is not correlated with the extent of inhibition of protein synthesis, suggesting that the death-preventing effects of the inhibitors were nonspecific *(92)*. But even when inhibition of protein synthesis is truly the cause of protection, this still does not prove a role for death genes (or death proteins). There is evidence that the prevention by cycloheximide of apoptosis in cortical neurons exposed to a form of oxidative stress is not due to inhibited synthesis of a "death protein," but to increased synthesis of the tripeptide glutathione owing to increased availability of cysteine *(93)*. Decreased utilization of cysteine for protein synthesis makes it available for glutathione, which protects against oxidative stress. In many other situations, the mechanism of protection by cycloheximide is probably different, but may still not be due to blocking *de novo* translation of death genes.

Furthermore, it turns out that the *ced*-3, *ced*-4, and *ced*-9 genes of nematodes, and most vertebrate cell death-related genes, are constitutively expressed even in cells that do not die *(94–96)*. The only "death gene" known to be switched on selectively from an essentially zero base level shortly before the onset of apoptosis *and* to trigger apoptosis when expressed ectopically is the *reaper* gene in Drosophila *(97,98)*. But even the *reaper* gene appears to be a regulator of apoptosis rather than an effector, because although apoptosis occurring naturally or due to X-radiation is mostly blocked in mutants lacking the gene, very high doses of X-radiation still kill some cells with typical apoptic morphology *(97,99)*. Yet another blow was delivered to the death gene switch-on hypothesis when it was shown that apoptosis could be triggered by extracellular signals in cells lacking a nucleus *(100)*.

Thus, the death gene hypothesis has had to be weakened considerably and now refers more to death proteins than death genes. But Tenet 6 concerns not only apoptosis, but necrosis. How good is the evidence that death proteins regulate apoptosis *more than necrosis*?

It is true that proteins such as bcl-2, bcl-x, bax, p53, and the ICE family have more often been shown to play a role in apoptosis than in necrosis. But they have been studied more often in apoptosis, and the greater number of positive results becomes less impressive if one bears in mind that virtually all cases in which death (or survival) proteins have been reported *not* to prevent neuronal death have likewise involved apoptosis *(101–103)*. The fact is that the role of death proteins in necrosis has rarely been addressed experimentally. However, it has recently been shown that bcl-2 inhibits necrosis in both nonneural (erythroleukemia) cells *(104)* and in neurons *(105)*. Moreover, inhibitors of protein synthesis can be very effective in preventing necrosis *(106)*. This has yet to be shown in neurons, but various nonapoptotic forms of neuron death have been shown to be thus prevented, including chromatolytic degeneration of axotomized mature motoneurons (shown 30 yr ago) *(107)*, and autophagic death of immature neurons deprived of retrograde support *(108)*.

In summary, in neurons as elsewhere, specific gene-products are undeniably important in the mechanisms of apoptosis. It is not clear that they are more important in apoptosis than necrosis and the notion that apoptosis is triggered and/or mediated by the switching on of dormant death genes is unwarranted. Tenet 6, therefore, seems unjustified.

TENET 7: APOPTOSIS, UNLIKE NECROSIS, INVOLVES DOUBLE-STRANDED DNA BREAKS

Experiments carried out in the late 1970s showed that DNA from apoptotic thymocytes was degraded into regular nucleosome-sized fragments (about 180 base pairs) and this was interpreted to be due to double-strand cleavage between nucleosomes owing to endonuclease activation *(4,109,110)*. Since this regular pattern was not found to occur in several cases of necrosis, where the DNA degradation was nonspecific, "DNA ladders" in agarose gels soon came to be widely taken as the biochemical hallmark of apoptosis, and endonuclease activation was held to be its distinctive lethal

Table 4
Cellular Events Common to Apoptosis and Necrosis

Cellular event	Role in inducing apoptosis (references)	Role in inducing necrosis (references)
Protection by protein- and RNA-synthesis inhibitors	Often *(82,83,145,146)*	Sometimes *(106)*
Protection by *bcl*-2	Often *(88,101,147)* but not always *(101)*	Sometimes *(105)*
Internucleosomal DNA cleavage	Sometimes but not always *(43,112–114,148)*	Sometimes but not always *(43,65,117,148)*
Rise in $[Ca^{2+}]i$	Often *(36,149,150)* but not always *(127)*	Often *(36,42,151)*
Oxygenated free radicals	Often *(38,129–131)*	Often *(38,131)*

mechanism *(111)*. Moreover, after several negative attempts, DNA ladders were eventually reported also in dying *neurons*, including NGF-deprived sympathetic neurons *(112)*, glutamate-killed cortical neurons *(113)*, and neurons dying naturally in the retinas of chick embryos *(114)*. In at least one report, it was indeed found that DNA ladders could be obtained after ischemia from brain regions containing many apoptotic cells, but not from regions containing mainly necrotic cells *(43)*.

But more recent evidence casts doubt on whether regular double-strand DNA cuts are a reliable hallmark of apoptosis. First, there is evidence that the DNA fragmentation in thymocytes is not due to double-stranded cuts, but to multiple single-strand cuts occurring preferentially in the internucleosomal regions *(115)*. Second, in many cases of neuronal and nonneuronal apoptosis, DNA ladders are not obtained; although this may be due to technical difficulties in some cases, there is convincing biochemical evidence in thymocytes and in liver cells that apoptosis can occur without DNA being fragmented at all *(100,116)*. Third, there are reports of DNA ladders being produced from necrotic cells *(117)* and there is evidence that this may be the case for neurons *(43,65)* although it is difficult to be certain that the DNA ladders were not due to undetected contamination by small numbers of apoptotic cells. Recent experiments outside the nervous system have emphasized the role of early DNA cleavage to large kilobase pair fragments in apoptosis, but this likewise fails to segregate clearly between apoptosis and necrosis, since it can occur in both types *(118)*. If, as is widely believed, DNA cleavage in dying cells reflects the action of a calcium-dependent endonuclease, it is logical that this should be produced by necrosis as well as apoptosis, since both kinds of cell death frequently involve raised levels of intracellular free calcium (*see* below and Table 4).

For technical reasons, however, gel electrophoresis is not often used in the nervous system for identifying apoptosis—partly because it is difficult in many situations to obtain cell samples with a sufficiently high proportion of dying cells. It is far more common to identify DNA breaks by labeling *in situ* in individual nuclei in histological sections, and it is somewhat alarming to see that this is frequently considered a criterion for apoptosis. While the specificity for apoptosis of DNA ladders may be debated, the nonspecificity of *in situ* labeling is an undeniable fact. Although it is true that the

DNA of apoptotic neurons (and other cells) is frequently labeled by these methods, so is the DNA of necrotic neurons *(35,37,119)*. This is not even surprising, given that necrosis has long been considered to cause DNA fragmentation. Worse still, *in situ* labeling does not always occur in apoptotic neurons *(120,121)*. Conceivably it may ultimately become possible in some cases to distinguish apoptosis from necrosis by comparing the results of different *in situ* labeling methods, and by using additional criteria including the morphology of the stained cells *(119)*. But the simple occurrence of *in situ* labeling for DNA fragmentation is neither a necessary nor a sufficient criterion for apoptosis. We conclude that whereas Tenet 7 may be valid for certain cell types in limited experimental conditions, it is not universally true.

TENET 8: THE CELLULAR MECHANISMS OF APOPTOSIS AND NECROSIS ARE FUNDAMENTALLY DIFFERENT

Much of the attractiveness of the apoptosis-necrosis dichotomy stems from the belief that it is fundamental, a dichotomy not just of appearance but of essential mechanism. This is the claim of the major reviews that launched it *(1,3)* and this is still the claim of its current defenders. There must be some truth in it, since the morphological characteristics of each kind of cell death are the results of cellular mechanisms, but it is important to realize that apoptosis and necrosis share, to some extent, common mechanisms (Table 4). And the real differences that exist between them need to be viewed in proportion to those among different kinds of apoptosis or among different kinds of necrosis.

As discussed in the context of Tenet 6, it is true that inhibition of macromolecular synthesis has been more often reported to prevent apoptosis than necrosis, but given the great variability of its effects on apoptosis and the fact that it does, in at least some cases, prevent necrosis, this does not seem to prove a mechanistic dichotomy between the two kinds of cell death.

As discussed also under Tenet 6, the fact that proteins such as bcl-2, bcl-x, bax, p53, and the ICE family have more often been shown to play a role in apoptosis than in necrosis may simply reflect their having been studied more often in apoptosis. It remains to be seen to what extent they play a role in necrosis, but the protective effect of bcl-2 against neuronal necrosis *(105)* serves as a warning that the involvement of cell death proteins and cell survival proteins may not be a sure basis for a mechanistic dichotomy.

As discussed in the context of Tenet 7, it is true that DNA ladders are more often obtained from apoptotic than from necrotic cells. But, given the numerous situations in which they are not obtained from apoptotic cells and the occasional ones in which they are obtained from necrotic ones, the occurrence of internucleosomal DNA-cleavage does not seem an adequate basis for a mechanistic dichotomy.

Of course, none of this excludes the possibility that some other cellular mechanism may turn out to be distinctive of apoptosis or necrosis (or of autophagic cell death, although this is rarely discussed) and crucial in the cell death process. But the fact is that none is currently established. Other well-studied death-mediating events, such as a rise in intracellular free calcium and the generation of free radicals are as equivocal in this respect as are death genes and DNA-fragmentation (Table 4). A rise in intracellular free calcium is believed to be causally involved in most cases of *both* apoptotic and necrotic neuron death *(36,38,42,122–126)*, although apoptotic neuronal death can in other cases be *prevented* by such a rise *(127)* and a modest sustained elevation in cytoplasmic free calcium protects immature neurons against cell death with certain features of apoptosis *(128)*. Calcium levels affect the generation of oxygenated free radicals and these are likewise involved in many cases of neuronal death with either apoptotic or necrotic features *(38,129,130)*; as is also true of other cell types *(131)* (for a more topical discussion of calcium and cell death, *see* chapter by Leist and Nicotera).

CONCLUSIONS

The thesis that all neuron death is either apoptotic or necrotic is not universally valid, although it seems to be a good approximation in some situations. Many of the assumptions that support this dichotomy are false, incomplete, or misleading. Although the present review deals mainly with neurons, many of the limitations of the apoptosis–necrosis dichotomy extend to numerous other kinds of cells *(7,10–12,45–48,132,133)*.

The criticisms of the dichotomy presented in this chapter are not just a matter of words, but of substance. Uncritical acceptance of a morphological dichotomy (Tenets 1 and 2) blinds us to structural diversity that may provide clues to disparate mechanisms of cell death in different situations. Uncritical acceptance of a mechanistic dichotomy (Tenets 5–8) limits the kind of experiments that will be done, as in the current dearth of investigation of genetic expression in necrosis, reflecting the presupposition that only apoptosis can be under the control of "death genes." In the therapeutic context, understanding the diversity of cell death, unhampered by dichotomous presuppositions, may be necessary for designing effective neuroprotection, tailored to the particular kind of neuronal death that needs to be prevented.

ACKNOWLEDGMENT

The author would like to thank Dr. Vincent Castagné and Menno van Lookeren Campagne for their helpful comments on the manuscript, and C. Vaclavik for typing it. This work was supported by the Swiss National Science Foundation under Grant 31-40709.94.

REFERENCES

1. Kerr JF, Wyllie AH, Currie AR. Apoptosis: a basic biological phenomenon with wide-ranging implications in tissue kinetics. *Brit J Cancer* 1972, **26**: 239–257.
2. Flemming W. Ueber die Bildung von Richtungsfiguren in Säugethiereiern beim Untergang Graaf'scher Follikel. *Arch Anat Physiol* 1885, (Jahrgang 1885): 221–224.
3. Wyllie AH, Kerr JF, Currie AR. Cell death: the significance of apoptosis. *Int Rev Cytol* 1980, **68**: 251–306.
4. Wyllie AH. Glucocorticoid-induced thymocyte apoptosis is associated with endogenous endonuclease activation. *Nature* 1980, **284**: 555–556.
5. MacDonald HR, Lees RK. Programmed death of autoreactive thymocytes. *Nature* 1990, **343**: 642–644.
6. Vaux DL, Cory S, Adams JM. Bcl-2 gene promotes haemopoietic cell survival and cooperates with c-myc to immortalize pre-B cells. *Nature* 1988, **335**: 440–442.
7. Majno G, Joris I. Apoptosis, oncosis, and necrosis: An overview of cell death. *Amer J Pathol* 1995, **146**: 3–15.
8. Wyllie AH. Cell death. *Int Rev Cytol* 1987, **Suppl. 17**: 755–785.

9. Rothman SM. The neurotoxicity of excitatory amino acids is produced by passive chloride influx. *J Neurosci* 1985, **5:** 1483–1489.
10. Levin S. A toxicologic pathologist's view of apoptosis or I used to call it necrobiosis, but now I'm singing the apoptosis blues. *Toxicol Pathol* 1995, **23:** 533–539.
11. Clarke PGH. Developmental cell death: morphological diversity and multiple mechanisms. *Anat Embryol (Berl)* 1990, **181:** 195–213.
12. Zakeri Z, Bursch W, Tenniswood M, Lockshin RA. Cell death: programmed, necrosis, or other? *Cell Death & Differentiation* 1995, **2:** 87–96.
13. Pilar G, Landmesser L. Ultrastructural differences during embryonic cell death in normal and peripherally deprived ciliary ganglia. *J Cell Biol* 1976, **68:** 339–356.
14. Decker RS. Retrograde responses of developing lateral motor column neurons. *J Comp Neurol* 1978, **180:** 635–660.
15. Clarke PGH. Labelling of dying neurones by peroxidase injected intravascularly in chick embryos. *Neurosci Lett* 1982, **30:** 223–228.
16. Hornung JP, Koppel H, Clarke PGH. Endocytosis and autophagy in dying neurons: an ultrastructural study in chick embryos. *J Comp Neurol* 1989, **283:** 425–437.
17. Gonzalez-Martin C, de Diego I, Crespo D, Fairen A. Transient c-fos expression accompanies naturally occurring cell death in the developing interhemispheric cortex of the rat. *Brain Res Dev Brain Res* 1992, **68:** 83–95.
18. Furber S, Oppenheim RW, Prevette D. Naturally-occurring neuron death in the ciliary ganglion of the chick. *J Neurosci* 1987, **7:** 1816–1832.
19. Clarke PGH. Identical populations of phagocytes and dying neurons revealed by intravascularly injected horseradish peroxidase, and by endogenous glutaraldehyde-resistant acid phosphatase, in the brains of chick embryos. *Histochem J* 1984, **16:** 955–969.
20. Clarke PGH. Neuronal death during development in the isthmo-optic nucleus of the chick: sustaining role of afferents from the tectum. *J Comp Neurol* 1985, **234:** 365–379.
21. Cunningham TJ. Naturally occurring neuron death and its regulation by developing neural pathways. *Int Rev Cytol* 1982, **74:** 163–186.
22. Monzon M, Yanes CM, Trujillo CM, Marrero A. Cell death in the normal development of Gallotia galloti mesencephalon (Reptilia Lacertidae). An ultrastructural study. *J Submicrosc Cytol* 1987, **19:** 71–76.
23. Sloviter RS, Dean E, Sollas AL, Goodman JH. Apoptosis and necrosis induced in different hippocampal neuron populations by repetitive perforant path stimulation in the rat. *J Comp Neurol* 1996, **366:** 516–533.
24. Oo TF, Blazeski R, Harrison SMW, Henchcliffe C, Mason CA, Roffler-Tarlov SK, Burke RE. Neuron death in the substantia nigra of weaver mouse occurs late in development and is not apoptotic. *J. Neurosci* 1996, **16:** 6134–6145.
25. Barron KD, Dentinger MP, Popp AJ, Mankes R. Neurons of layer Vb of rat sensorimotor cortex atrophy but do not die after thoracic cord transection. *J Neuropathol Exp Neurol* 1988, **47:** 62–74.
26. Barron KD, Means ED, Larsen E. Ultrastructure of retrograde degeneration in thalamus of rat. 1. Neuronal somata and dendrites. *J Neuropathol Exp Neurol* 1973, **32:** 218–244.
27. Barron KD, Dentinger MP, Nelson LR, Mincy JE. Ultrastructure of axonal reaction in red nucleus of cat. *J Neuropathol Exp Neurol* 1975, **34:** 222–248.
28. Matthews MA. Death of the central neuron: an electron microscopic study of thalamic retrograde degeneration following cortical ablation. *J Neurocytol* 1973, **2:** 265–288.
29. Garcia-Valenzuela E, Shareef S, Walsh J, Sharma SC. Programmed cell death of retinal ganglion cells during experimental glaucoma. *Exp Eye Res* 1995, **61:** 33–44.
30. Berkelaar M, Clarke DB, Wang Y-C, Bray GM, Aguayo AJ. Axotomy results in delayed death and apoptosis of retinal ganglion cells in adult rats. *J Neurosci* 1994, **14:** 4368–4374.

31. Quigley HA, Nickells RW, Kerrigan LA, Pease ME, Thibault DJ, Zack DJ. Retinal ganglion cell death in experimental glaucoma and after axotomy occurs by apoptosis. *Invest Ophthalmol Vis Sci* 1995, **36:** 774–786.
32. Barron KD, Dentinger MP, Krohel G, Easton SK, Mankes R. Qualitative and quantitative ultrastructural observations on retinal ganglion cell layer of rat after intraorbital optic nerve crush. *J Neurocytol* 1986, **15:** 345–362.
33. Olney JW. The toxic effects of glutamate and related compounds in the retina and the brain. *Retina* 1982, **2:** 341–359.
34. Rothman SM. Excitotoxic neuronal death. Mechanisms and clinical relevance. *Semin Neurosci* 1994, **6:** 315–322.
35. Charriaut-Marlangue C, Aggoun-Zouaoui D, Represa A, Ben-Ari Y. Apoptotic features of selective neuronal death in ischemia, epilepsy and gp120 toxicity. *Trends Neurosci* 1996, **19:** 109–114.
36. Choi DW. Excitotoxic cell death. *J Neurobiol* 1992, **23:** 1261–1276.
37. Van Lookeren Campagne M, Lucassen PJ, Vermeulen JP, Balázs R. NMDA and kainate induce internucleosomal DNA cleavage associated with both apoptotic and necrotic cell death in the neonatal rat brain. *European J Neurosci* 1995, **7:** 1627–1640.
38. Bonfoco E, Krainc D, Ankarcrona M, Nicotera P, Lipton SA. Apoptosis and necrosis: Two distinct events induced, respectively, by mild and intense insults with N-methyl-D-aspartate or nitric oxide/superoxide in cortical cell cultures. *Proc Natl Acad Sci USA* 1995, **92:** 7162–7166.
39. Ankarcrona M, Dypbukt JM, Bonfoco E, Zhivotovsky B, Orrenius S, Lipton SA, Nicotera P. Glutamate-induced neuronal death: A succession of necrosis or apoptosis depending on mitochondrial function. *Neuron* 1995, **15:** 961–973.
40. Regan RF, Panter SS, Witz A, Tilly JL, Giffard RG. Ultrastructure of excitotoxic neuronal death in murine cortical culture. *Brain Res* 1995, **705:** 188–198.
41. Olney JW, Price MT, Samson L, Labruyere J. The role of specific ions in glutamate neurotoxicity. *Neurosci Lett* 1986, **65:** 65–71.
42. Trump BF, Berezesky IK. Calcium-mediated cell injury and cell death. *FASEB J* 1995, **9:** 219–228.
43. Beilharz EJ, Williams CE, Dragunow M, Sirimanne ES, Gluckman PD. Mechanisms of delayed cell death following hypoxic-ischemic injury in the immature rat: evidence for apoptosis during selective neuronal loss. *Mol Brain Res* 1995, **29:** 1–14.
44. Portera-Cailliau C, Price DL, Martin LJ. Excitotoxic neuronal death is a morphological continuum from apoptosis to necrosis and involves programmed cell death. *Soc Neurosci Abstr* 1995, **21:** 301.
45. Schwartz LM, Smith SW, Jones ME, Osborne BA. Do all programmed cell deaths occur via apoptosis? *Proc Natl Acad Sci USA* 1993, **90:** 980–984.
46. Tounekti O, Pron G, Belehradek J Jr., Mir LM. Bleomycin, an apoptosis-mimetic drug that induces two types of cell death depending on the number of molecules internalized. *Cancer Res* 1993, **53:** 5462–5469.
47. Swanson PE, Carroll SB, Zhang XF, Mackey MA. Spontaneous premature chromosome condensation, micronucleus formation, and non-apoptotic cell death in heated HeLa S3 cells: Ultrastructural observations. *Am J Pathol* 1995, **146:** 963–971.
48. Takei H, Araki A, Watanabe H, Ichinose A, Sendo F. Rapid killing of human neutrophils by the potent activator phorbol 12-myristate 13-acetate (PMA) accompanied by changes different from typical apoptosis or necrosis. *J Leukocyte Biol* 1996, **59:** 229–240.
49. Kerr JF, Harmon BV. Definition and incidence of apoptosis. An historical perspective, in *Apoptosis: The Molecular Basis of Cell Death* (Tomei LD, Cope FO, eds.), Cold Spring Harbor Laboratory Press, New York, 1991: 5–29.
50. Lewis PD. Cell death in the germinal layers of the postnatal rat brain. *Neuropathol Appl Neurobiol* 1975, **1:** 21–29.

51. Blaschke AJ, Staley K, Chun J. Widespread programmed cell death in proliferative and postmitotic regions of the fetal cerebral cortex. *Development* 1996, **122**: 1165–1174.
52. Hamburger V, Levi-Montalcini R. Proliferation, differentiation and degeneration in the spinal ganglia of the chick embryo under normal and experimental conditions. *J Exp Zool* 1949, **111**: 457–501.
53. Levi-Montalcini R. The origin and development of the visceral system in the spinal cord of the chick embryo. *J Morphol* 1950, **86**: 253–283.
54. Jacobson M. *Developmental Neurobiology*, 3rd edn, Plenum, New York and London. 1991.
55. Oppenheim RW. Cell death during development of the nervous system. *Annu Rev Neurosci* 1991, **14**: 453–501.
56. Goldstein RS. Axial level-dependent differences in size of avian dorsal root ganglia are present from gangliogenesis. *J Neurobiol* 1993, **24**: 1121–1129.
57. Hollyday M, Hamburger V. An autoradiographic study of the formation of the lateral motor column in the chick embryo. *Brain Res* 1977, **132**: 197–208.
58. Herrup K, Sunter K. Numerical matching during cerebellar development: quantitative analysis of granule cell death in staggerer mouse chimeras. *J Neurosci* 1987, **7**: 829–836.
59. Linden R, Renteria AS. Afferent control of neuron numbers in the developing brain. *Dev Brain Res* 1988, **44**: 291–295.
60. Clarke PGH. Chance, repetition, and error in the development of normal nervous systems. *Perspect Biol Med* 1981, **25**: 2–17.
61. Clarke PGH, Cowan WM. Ectopic neurons and aberrant connections during neural development. *Proc Natl Acad Sci USA* 1975, **72**: 4455–4458.
62. O'Leary DDM, Fawcett JW, Cowan WM. Topographic targeting errors in the retinocollicular projection and their elimination by selective ganglion cell death. *J Neurosci* 1986, **6**: 3692–3705.
63. Catsicas S, Thanos S, Clarke PGH. Major role for neuronal death during brain development: refinement of topographical connections. *Proc Natl Acad Sci USA* 1987, **84**: 8165–8168.
64. Schweichel JU, Merker HJ. The morphology of various types of cell death in prenatal tissues. *Teratology* 1973, **7**: 253–266.
65. Rink A, Fung KM, Trojanowski JQ, Lee VMY, Neugebauer E, McIntosh TK. Evidence of apoptotic cell death after experimental traumatic brain injury in the rat. *Am J Pathol* 1995, **147**: 1575–1583.
66. Fesus L, Davies PJ, Piacentini M. Apoptosis: molecular mechanisms in programmed cell death. *Eur J Cell Biol* 1991, **56**: 170–177.
67. Duvall E, Wyllie AH, Morris RG. Macrophage recognition of cells undergoing programmed cell death (apoptosis). *Immunology* 1985, **56**: 351–358.
68. Savill J, Dransfield I, Hogg N, Haslett C. Vitronectin receptor-mediated phagocytosis of cells undergoing apoptosis. *Nature* 1990, **343**: 170–173.
69. Pautou M, Kieny M. Sur les mécanismes histologiques et cytologiques de la nécrose morphogène interdigitale chez l'embryon de poulet. *CR Acad Sci (Paris) Ser D* 1971, **272**: 2025–2028.
70. Coles HS, Burne JF, Raff MC. Large-scale normal cell death in the developing rat kidney and its reduction by epidermal growth factor. *Development* 1993, **118**: 777–784.
71. Wyllie AH. Cell death: a new classification separating apoptosis from necrosis, in *Cell Death in Biology and Pathology* (Bowen ID, Lockshin RA, eds.), Chapman and Hall, New York, 1981, 9–34.
72. Andersson PB, Perry VH, Gordon S. The kinetics and morphological characteristics of the macrophage-microglial response to kainic acid-induced neuronal degeneration. *Neuroscience* 1991, **42**: 201–214.
73. Perry VH, Bell MD, Brown HC, Matyszak MK. Inflammation in the nervous system. *Curr Opin Neurobiol* 1995, **5**: 636–641.

74. Hayward NJ, Elliott PJ, Sawyer SD, Bronson RT, Bartus RT. Lack of evidence for neutrophil participation during infarct formation following focal cerebral ischemia in the rat. *Experimental Neurology* 1996, **139**: 188–202.
75. Weber R. Inhibitory effect of actinomycin D on tail atrophy in Xenopus larvae at metamorphosis. *Experientia* 1965, **21**: 665–666.
76. Tata JR. Requirement for RNA and protein synthesis for induced regression of the tadpole tail in organ culture. *Dev Biol* 1966, **13**: 77–94.
77. Lockshin RA. Programmed cell death. Activation of lysis by a mechanism involving the synthesis of protein. *J Insect Physiol* 1969, **15**: 1505–1516.
78. Lockshin RA, Williams CM. Programmed cell death—II. Endocrine potentiation of the breakdown of the intersegmental muscles of silkmoths. *J Insect Physiol* 1964, **10**: 643–649.
79. Lockshin RA, Beaulaton J. Programmed cell death. *Life Sci* 1974, 15: 1549–1565.
80. Thomas N, Edwards JL, Bell PA. Studies of the mechanism of glucocorticoid-induced pyknosis in isolated rat thymocytes. *J Steroid Biochem* 1983, **18**: 519–524.
81. Cohen JJ, Duke RC. Glucocorticoid activation of a calcium-dependent endonuclease in thymocyte nuclei leads to cell death. *J Immunol* 1984, **132**: 38–42.
82. Martin DP, Schmidt RE, DiStefano PS, Lowry OH, Carter JG, Johnson EM Jr. Inhibitors of protein synthesis and RNA synthesis prevent neuronal death caused by nerve growth factor deprivation. *J Cell Biol* 1988, **106**: 829–844.
83. Oppenheim RW, Prevette D, Tytell M, Homma S. Naturally occurring and induced neuronal death in the chick embryo in vivo requires protein and RNA synthesis: evidence for the role of cell death genes. *Dev Biol* 1990, **138**: 104–113.
84. Ellis HM, Horvitz HR. Genetic control of programmed cell death in the nematode C. elegans. *Cell* 1986, **44**: 817–829.
85. Yuan JY. Evolutionary conservation of a genetic pathway of programmed cell death. *J Cell Biochem* 1996, **60**: 4–11.
86. Oltvai ZN, Milliman CL, Korsmeyer SJ. Bcl-2 heterodimerizes in vivo with a conserved homolog, Bax, that accelerates programmed cell death. *Cell* 1993, **74**: 609–619.
87. Yonish-Rouach E, Resnitzky D, Lotem J, Sachs L, Kimchi A, Oren M. Wild-type p53 induces apoptosis of myeloid leukaemic cells that is inhibited by interleukin-6. *Nature* 1991, **352**: 345–347.
88. Garcia I, Martinou I, Tsujimoto Y, Martinou JC. Prevention of programmed cell death of sympathetic neurons by the bcl-2 proto-oncogene. *Science* 1992, **258**: 302–304.
89. Freeman RS, Estus S, Horigome K, Johnson EM Jr. Cell death genes in invertebrates and (maybe) vertebrates. *Curr Opin Neurobiol* 1993, **3**: 25–31.
90. Martin SJ. Apoptosis: suicide, execution, or murder? *Trends Cell Biol* 1993, **3**: 141–144.
91. Rehen SK, Varella MH, Freitas FG, Moraes MO, Linden R. Contrasting effects of protein synthesis inhibition and of cyclic AMP on apoptosis in the developing retina. *Development* 1996, **122**: 1439–1448.
92. Chow SC, Peters I, Orrenius S. Reevaluation of the role of *de novo* protein synthesis in rat thymocyte apoptosis. *Exp Cell Res* 1995, **216**: 149–159.
93. Ratan RR, Murphy TH, Baraban JM. Macromolecular synthesis inhibitors prevent oxidative stress- induced apoptosis in embryonic cortical neurons by shunting cysteine from protein synthesis to glutathione. *J Neurosci* 1994, **14**: 4385–4392.
94. Merry DE, Veis DJ, Hickey WF, Korsmeyer SJ: bcl-2 protein expression is widespread in the developing nervous system and retained in the adult PNS. *Development* 1994, **120**: 301–311.
95. Shaham S, Horvitz HR. Developing Caenorhabditis elegans neurons may contain both cell-death protective and killer activities. *Genes Dev* 1996, **10**: 578–591.
96. Weil M, Jacobson MD, Coles HSR, Davies TJ, Gardner RL, Raff KD, Raff MC. Constitutive expression of the machinery for programmed cell death. *J Cell Biol* 1996, **133**: 1053–1059.

97. Steller H, Grether ME. Programmed cell death in Drosophila. *Neuron* 1994, **13:** 1269–1274.
98. White K, Tahaoglu E, Steller H. Cell killing by the drosophila gene reaper. *Science* 1996, **271:** 805–807.
99. White K, Grether ME, Abrams JM, Young L, Farrell K, Steller H. Genetic control of programmed cell death in *Drosophila*. *Science* 1994, **264:** 677–683.
100. Schulze-Osthoff K, Walczak H, Dröge W, Krammer PH. Cell nucleus and DNA fragmentation are not required for apoptosis. *J Cell Biol* 1994, **127:** 15–20.
101. Allsopp TE, Wyatt S, Paterson HF, Davies AM. The proto-oncogene bcl-2 can selectively rescue neurotrophic factor-dependent neurons from apoptosis. *Cell* 1993, **73:** 295–307.
102. Davies AM, Rosenthal A. Neurons from mouse embryos with a null mutation in the tumour suppressor gene p53 undergo normal cell death in the absence of neurotrophins. *Neurosci Lett* 1994, **182:** 112–114.
103. Sadoul R, Quiquerez AL, Martinou I, Fernandez PA, Martinou JC: p53 protein in sympathetic neurons: Cytoplasmic localization and no apparent function in apoptosis. *J Neurosci Res* 1996, **43:** 594–601.
104. Fukunaga-Johnson N, Ryan JJ, Wicha M, Nunez G, Clarke MF. Bcl-2 protects murine erythroleukemia cells from p53-dependent and -independent radiation-induced cell death. *Carcinogenesis* 1995, **16:** 1761–1767.
105. Kane DJ, Örd T, Anton R, Bredesen DE. Expression of *bcl-2* inhibits necrotic neural cell death. *J Neurosci Res* 1995, **40:** 269–275.
106. Popp JA, Shinozuka H, Farber E. The protective effects of diethyldithiocarbamate and cycloheximide on the multiple hepatic lesions induced by carbon tetrachloride in the rat. *Toxicol Appl Pharmacol* 1978, **45:** 549–564.
107. Torvik A, Heding A. Histological studies on the effect of actinomycin D on retrograde nerve cell reaction in the facial nucleus of mice. *Acta Neuropathol (Berl)* 1967, **9:** 146–157.
108. Catsicas M, Clarke PGH. Cycloheximide prevents neuronal death during embryogenesis. *Experientia* 1990, **46:** A78.
109. Skalka M, Matyasova J, Cejkova M. DNA in chromatin of irradiated lymphoid tissues degrades in vivo into regular fragments. *FEBS Lett* 1976, **72:** 271–274.
110. Matyasova J, Skalka M, Cejkova M. Regular character of chromatin degradation in lymphoid tissues after treatment with biological alkylating agents in vivo. *Folia Biol (Praha)* 1979, **25:** 380–388.
111. Buja LM, Eigenbrodt ML, Eigenbrodt EH. Apoptosis and necrosis: Basic types and mechanisms of cell death. *Arch Pathol Lab Med* 1993, **117:** 1208–1214.
112. Edwards SN, Buckmaster AE, Tolkovsky AM. The death programme in cultured sympathetic neurones can be suppressed at the posttranslational level by nerve growth factor, cyclic AMP, and depolarization. *J Neurochem* 1991, **57:** 2140–2143.
113. Kure S, Tominaga T, Yoshimoto T, Tada K, Narisawa K. Glutamate triggers internucleosomal DNA cleavage in neuronal cells. *Biochem Biophys Res Commun* 1991, **179:** 39–45.
114. Ilschner SU, Waring P. Fragmentation of DNA in the retina of chicken embryos coincides with retinal ganglion cell death. *Biochem Biophys Res Commun* 1992, **183:** 1056–1061.
115. Peitsch MC, Muller C, Tschopp J. DNA fragmentation during apoptosis is caused by frequent single-strand cuts. *Nucleic Acids Res* 1993, **21:** 4206–4209.
116. Cohen GM, Sun XM, Snowden RT, Dinsdale D, Skilleter DN. Key morphological features of apoptosis may occur in the absence of internucleosomal DNA fragmentation. *Biochem J* 1992, **286:** 331–334.
117. Fukuda K, Kojiro M, Chiu JF. Demonstration of extensive chromatin cleavage in transplanted Morris hepatoma 7777 tissue: apoptosis or necrosis? *Am J Pathol* 1993, **142:** 935–946.
118. Bicknell GR, Cohen GM. Cleavage of DNA to large kilobase pair fragments occurs in some forms of necrosis as well as apoptosis. *Biochem Biophys Res Commun* 1995, **207:** 40–47.

119. Gold R, Schmied M, Giegerich G, Breitschopf H, Hartung HP, Toyka KV, Lassmann H. Differentiation between cellular apoptosis and necrosis by the combined use of *in situ* tailing and nick translation techniques. *Lab Invest* 1994, **71:** 219–225.
120. Wood KA, Dipasquale B, Youle RJ. In situ labeling of granule cells for apoptosis-associated DNA fragmentation reveals different mechanisms of cell loss in developing cerebellum. *Neuron* 1993, **11:** 621–632.
121. Herrup K, Busser JC. The induction of multiple cell cycle events precedes target-related neuronal death. *Development* 1995, **121:** 2385–2395.
122. Randall RD, Thayer SA. Glutamate-induced calcium transient triggers delayed calcium overload and neurotoxicity in rat hippocampal neurons. *J Neurosci* 1992, **12:** 1882–1895.
123. Orrenius S, Nicotera P. The calcium ion and cell death. *J Neural Transm Suppl* 1994, **43:** 1–11.
124. Chiou GC, Hong SJ. Prevention and reduction of neural damage in ischemic strokes by w-(N,N'-diethylamino)-n-alkyl-3,4,5-trimethoxybenzoate compounds. *J Pharmacol Exp Ther* 1995, **275:** 474–478.
125. Ciutat D, Esquerda JE, Calderó J. Evidence for calcium regulation of spinal cord motoneuron death in the chick embryo in vivo. *Dev Brain Res* 1995, **86:** 167–179.
126. Lachica EA, Rubsamen R, Zirpel L, Rubel EW. Glutamatergic inhibition of voltage-operated calcium channels in the avian cochlear nucleus. *J Neurosci* 1995, **15:** 1724–1734.
127. Koh JY, Wie MB, Gwag BJ, Sensi SL, Canzoniero LMT, Demaro J, Csernansky C, Choi DW. Staurosporine-induced neuronal apoptosis. *Exp Neurol* 1995, **135:** 153–159.
128. Johnson EM Jr, Koike T, Franklin J. A "calcium set-point hypothesis" of neuronal dependence on neurotrophic factor. *Exp Neurol* 1992, **115:** 163–166.
129. Ratan RR, Murphy TH, Baraban JM. Oxidative stress induces apoptosis in embryonic cortical neurons. *J Neurochem* 1994, **62:** 376–379.
130. Greenlund LJS, Deckwerth TL, Johnson EM Jr. Superoxide dismutase delays neuronal apoptosis: A role for reactive oxygen species in programmed neuronal death. *Neuron* 1995, **14:** 303–315.
131. Donnini D, Zambito AM, Perrella G, Ambesi-Impiombato FS, Curcio F. Glucose may induce cell death through a free radical-mediated mechanism. *Biochem Biophys Res Commun* 1996, **219:** 412–417.
132. Farber E. Programmed cell death: necrosis versus apoptosis. *Mod Pathol* 1994, **7:** 605–609.
133. Zakeri Z, Lockshin RA. Physiological cell death during development and its relationship to aging. *Ann NY Acad Sci* 1994, **719:** 212–229.
134. O'Connor TM, Wyttenbach CR. Cell death in the embryonic chick spinal cord. *J Cell Biol* 1974, **60:** 448–459.
135. Chu-Wang IW, Oppenheim RW. Cell death of motoneurons in the chick embryo spinal cord. I. A light and electron microscopic study of naturally occurring and induced cell loss during development. *J Comp Neurol* 1978, **177:** 33–57.
136. Cunningham TJ, Mohler IM, Giordano DL. Naturally occurring neuron death in the ganglion cell layer of the neonatal rat: morphology and evidence for regional correspondence with neuron death in superior colliculus. *Brain Res* 1981, **254:** 203–215.
137. Wahnschaffe U, Bartsch U, Fritzsch B. Metamorphic changes within the lateral-line system of Anura. *Anat Embryol (Berl)* 1987, **175:** 431–442.
138. Clarke PGH, Martin AH. Effects of de-efferentation on chick spinal motoneurons: peroxidase-uptake, and activities of acid phosphatase and *N*-acetyl-B-glucosaminidase. *Cell Biol Int Rep* 1985, **9:** 676.
139. Lamborghini JE. Disappearance of Rohon-Beard neurons from the spinal cord of larval Xenopus laevis. *J Comp Neurol* 1987, **264:** 47–55.
140. Cuschieri A, Bannister LH. The development of the olfactory mucosa in the mouse: electron microscopy. *J Anat* 1975, **119:** 471–498.

141. Pellier V, Astic L. Cell death in the developing olfactory epithelium of rat embryos. *Dev Brain Res* 1994, **79**: 307–315.
142. Sohal GS, Weidman TA. Ultrastructural sequence of embryonic cell death in normal and peripherally deprived trochlear nucleus. *Exp Neurol* 1978, **61**: 53–64.
143. Giordano DL, Murray M, Cunningham TJ. Naturally occurring neuron death in the optic layers of superior colliculus of the postnatal rat. *J Neurocytol* 1980, **9**: 603–614.
144. Crespo D, Cos S, Fernandez-Viadero C, Gonzalez C. Ultrastructural changes in hypothalamic supraoptic nucleus neurons of ovariectomized estrogen-deprived young rats. *Neurosci Lett* 1991, **133**: 253–256.
145. Catsicas M, Péquignot Y, Clarke PGH. Rapid onset of neuronal death induced by blockade of either axoplasmic transport or action potentials in afferent fibers during brain development. *J Neurosci* 1992, **12**: 4642–4650.
146. Rabacchi SA, Bonfanti L, Liu X-H, Maffei L. Apoptotic cell death induced by optic nerve lesion in the neonatal rat. *J Neurosci* 1994, **14**: 5292–5301.
147. Farlie PG, Dringen R, Rees SM, Kannourakis G, Bernard O: *bcl-2* transgene expression can protect neurons against developmental and induced cell death. *Proc Natl Acad Sci USA* 1995, **92**: 4397–4401.
148. Portera-Cailliau C, Hedreen JC, Price DL, Koliatsos VE. Evidence for apoptotic cell death in Huntington disease and excitotoxic animal models. *J Neurosci* 1995, **15**: 3775–3787.
149. Franklin JL, Sanz-Rodriguez C, Juhasz A, Deckwerth TL, Johnson EM Jr. Chronic depolarization prevents programmed death of sympathetic neurons *in vitro* but does not support growth. Requirement for Ca^{2+} influx but not Trk activation. *J Neurosci* 1995, **15**: 643–664.
150. Trump BF, Berezesky IK. Calcium-mediated cell injury and cell death. *FASEB J* 1995, **9**: 219–228.
151. Tymianski M, Wallace MC, Spigelman I, Uno M, Carlen PL, Tator CH, Charlton MP. Cell-permeant Ca2+ chelators reduce early excitotoxic and ischemic neuronal injury in vitro and in vivo. *Neuron* 1993, **11**: 221–235.

2
Asynchronous Death as a Characteristic Feature of Apoptosis

Randall N. Pittman, Conrad A. Messam, and Jason C. Mills

INTRODUCTION

A basic feature of programmed cell death/apoptosis in vivo is that cells within a population die at very different times even following an apparently identical signal to die. Nowhere is this more obvious than in the developing nervous system where neurons in the same population often die over several days. In fact, "naturally occurring" cell death in the developing nervous system was overlooked for a long time because of its sporadic asynchronous nature and the relatively short period of time cells show morphological signs of death.

Asynchronous death of neurons in vivo may be attributed to differences between neurons such as birthdates, local environment (e.g., distance from vascular supply, contact with glia, or position within morphogenetic gradients), time of neurite outgrowth, size of arbors, access to growth/survival factors, number and types of targets contacted, and input received from other neurons. Any or all of these factors could contribute to the extended period of time during which cell death occurs in neural populations in vivo; however, in in vitro systems where most of these variables can be controlled, death still occurs asynchronously. For instance, sympathetic neurons in the developing superior cervical ganglion die over a 3–4 d period in vivo, whereas these same neurons in a more controlled in vitro environment induced to die at a very specific point in time by removing nerve growth factor (NGF) still die over a 2–3 d period *(1,2)*. By manipulating cells and culture conditions it is possible to rule out many of the cellular and environmental properties that could give rise to asynchrony. Therefore, one is left with the intriguing possibility that asynchrony associated with apoptosis/ programmed cell death may be inherent to the cell death program itself.

From an experimental standpoint, the principal problem associated with asynchronous death is that at any one point in time, individual cells in a population are at various stages along the cell death pathway; yet, standard biochemical measurements (e.g., energetic and metabolic parameters), and assays (e.g., protein phosphorylation) that are critical for understanding mechanisms underlying cell death measure parameters on the entire population of cells. Biochemical measurements taken on a population of cells with individual cells at various stages of cell death are difficult to interpret rela-

From: Cell Death and Diseases of the Nervous System
Edited by: V. E. Koliatsos and R. R. Ratan © Humana Press Inc., Totowa, NJ

tive to the death process itself and it is not clear how meaningful the data are unless changes occur very early in the process prior to cells becoming desynchronized along the death pathway.

Whereas asynchrony is a characteristic feature of apoptosis and programmed cell death both in vivo and in vitro, it is generally not associated with necrosis. Although necrosis can occur shortly after or many hours after an insult, in general, most cells die during a restricted time frame following a necrotic insult. There is a large amount of physiologically relevant neural death that is necrotic including much of the initial neural death following trauma, stroke, and excitoxicity. Asynchrony should not greatly affect biochemical measurements in these cases. Because necrosis exhibits very little asynchrony it will not be dealt with in this chapter. Rather, the focus will be on apoptosis and programmed cell death. Although not synonymous (for reservations, *see* chapter by Clarke), the terms apoptosis and programmed cell death will be used more or less interchangeably in this chapter to refer to cell death in which the cell actively participates in its own demise (that is, a self-planned execution).

Major issues to be covered in this chapter include the following:

1. The basic phenomenon of asynchrony at the population and cellular levels;
2. Biochemical/molecular stages of apoptosis;
3. Commitment to die and its importance;
4. Practical consequences of dealing with asynchrony at the experimental level;
5. Possible mechanisms underlying asynchrony; and
6. Potential ways of dealing with and/or overcoming the problem of asynchrony.

BASIC PHENOMENON OF ASYNCHRONY DURING NEURAL DEATH

Asynchrony was a characteristic feature associated with apoptosis when it was first defined *(see ref. 3)* and studies on "naturally occurring cell death" in the developing nervous system noted the extended period of time during which neuronal death occurred (*[4]*; for early literature, *see ref. 5*). Asynchrony appears to be a general feature of apoptosis in all systems *(6)*; therefore, it is somewhat surprising that there is very little specific information on asynchrony during apoptosis. For the purposes of this chapter, we will draw on neural models when possible with particular reference to model systems being used in our lab including undifferentiated and differentiated PC12 cells, and primary cultures of sympathetic neurons *(see refs. 1,2,7–9)*.

A time course of cell death typical of many cell types in vitro is shown in Fig. 1. Key points to take from this figure are that even with a very precise time of initiating cell death in a relatively homogenous population of cells, many cells die more than 48 h after the first cells begin dying, and at any one point in time only a very small percentage of cells are actually dying.

Figure 1 represents a static view of cell death in a population of cells. When viewed at the level of individual cells in vitro using time lapse video microscopy a more dynamic view of cellular events occurring during cell death is obtained. A fairly stereotyped set of morphological changes occurs in most cells during apoptosis in vitro. Both neuronal and non-neuronal cells undergoing apoptosis have at least two distinct morphological states including a quiescent state indistinguishable from control cells and a subsequent highly dynamic state of active blebbing in which large membrane blebs are extended and retracted (Fig. 2). This latter state consisting of dynamic whole cell blebbing is

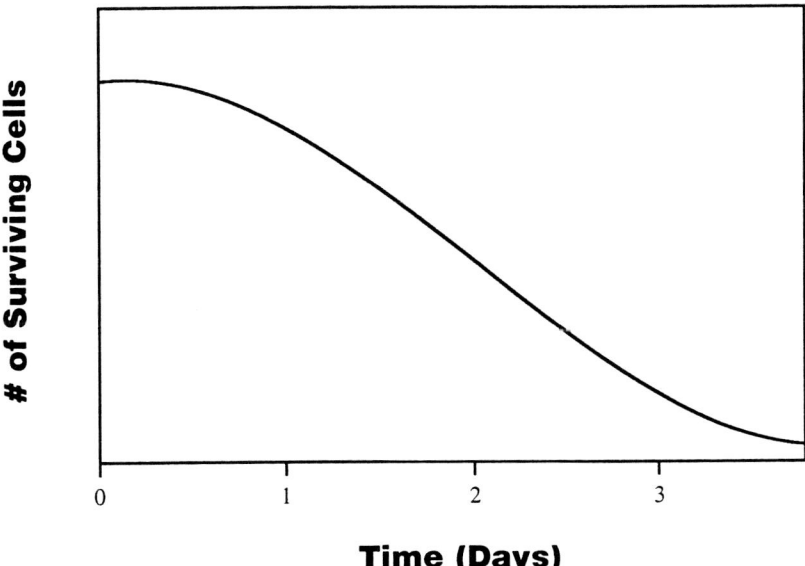

Fig. 1. Time course of neural cell death. This is an idealized curve representative of the cell death pattern shown by several in vitro model systems used in our studies. The general shape of the curve and timing typifies death of sympathetic neurons following removal of the NGF, undifferentiated and differentiated PC12 cells following removal of serum or NGF, and undifferentiated and differentiated PC12 cells following exposure to UV irradiation. Note that even in a relatively homogeneous population of cells exposed to the same apoptotic stimulus cells die over a protracted period of time.

often referred to as the "active phase" of apoptosis. The first signs of the onset of active apoptosis in neurons and differentiated PC12 cells are small irregularities in the shape of the cell body and neurites exhibiting areas of "thinning and beading" (Fig. 2B). Some cells such as undifferentiated PC12 cells exhibit a third morphological state that serves as a transition between normal morphology and active apoptosis. This state is characterized by membrane "bubbling" in which the surface of the cell appears to "boil" with very small blebs (Fig. 2A). For convenience, the morphologically normal state is referred to as phase 1, the morphological state characterized by membrane bubbling as phase 2, and dynamic whole cell blebbing as phase 3 or the "active phase" of apoptosis. Plotting these morphological changes for individual undifferentiated PC12 cells as a function of time following an apoptotic stimulus (Fig. 3) illustrates several initial observations made from analyzing morphological phases during apoptosis (*10,11*) as follows:

1. Phase 1 representing normal morphology and phase 2 representing membrane bubbling are highly variable in length from cell to cell within a population;
2. No correlation exists between the length of any one phase and the length of any other phase; and
3. The duration of phase 3 or the active phase of cell death is highly restricted (about 1 h). This highly restricted time span of active apoptosis occurs in vivo as well as in other in vitro systems and is very similar in length for most if not all cell types at different states of differentiation responding to a variety of apoptotic stimuli (also *see ref. 6*).

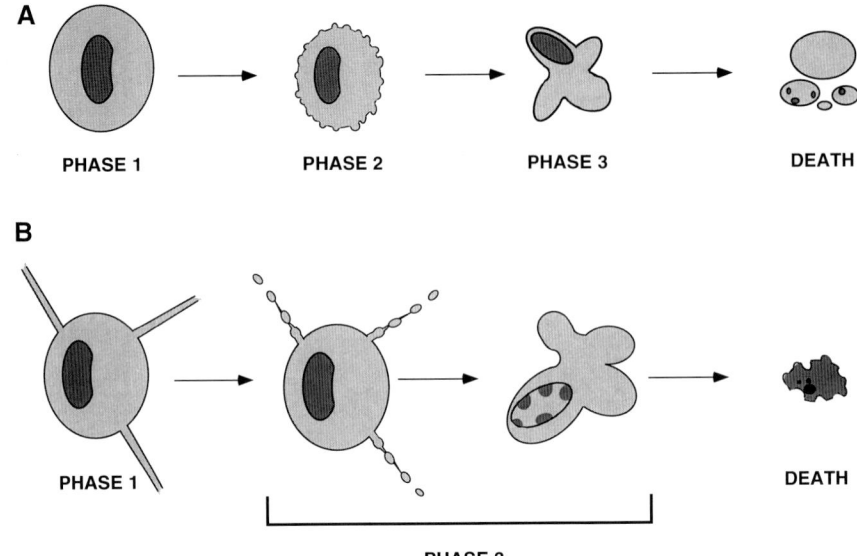

Fig. 2. Morphological phases of apoptosis. The figure is based on data from time lapse video microscopy studies performed on undifferentiated **(A)** and differentiated **(B)** PC12 cells. Morphological changes in these cells are probably representative of those occurring in many neural and non-neural cells following a variety of apoptotic stimuli. Cells remain in a morphologically normal state (Phase 1) for variable times following an apoptotic stimulus. The first morphological change observed in undifferentiated cells (A) following removal of serum is rapid formation and retraction of multiple small "bubbles" on the surface of cells (phase 2). This is followed after a variable time by cells exhibiting dynamic whole cell blebbing resulting in drastic changes in cell shape (phase 3). Phase 3 is often referred to as "active apoptosis" and lasts about an hour in most cells. The first morphological change observed in differentiated cells (B) following removal of NGF is a slight deformity in cell body shape and thinning and beading of neurites. A short time after these initial changes, neurites fragment and cells begin dynamic whole cell blebbing characteristic of active apoptosis (phase 3). Unlike undifferentiated cells, differentiated cells do not exhibit features characteristic of phase 2 (the slight distortion in the shape of the cell body and thinning and beading of neurites last for a restricted time, make a smooth transition into whole cell blebbing, and are considered to be the first changes associated with phase 3 rather than a separate phase 2). Note that in undifferentiated PC12 cells, chromatin condensation does not occur until very late in phase 3, or after cells die, while in differentiated cells condensation of chromatin occurs early in phase 3 well before cells die. Also, undifferentiated PC12 cells often form apoptotic bodies whereas differentiated PC12 cells and sympathetic neurons routinely form a single compacted mass of debris rather than multiple apoptotic bodies.

An unexpected observation made while characterizing morphological changes using time lapse video microscopy was that sister cells died at different times even though they were synchronized at mitosis (10). This occurred in proliferating PC12 cells for a subset of cells that passed the G1/S checkpoint prior to serum removal and continued on to undergo mitosis in the presence of an apoptotic stimulus. After cell division, sister cells entered phase 2 (membrane bubbling) at the same time, but remained in this phase for differing lengths of time and eventually died asynchronously. This observation supports the notion of the inherent asynchrony in apoptosis and suggests that, at

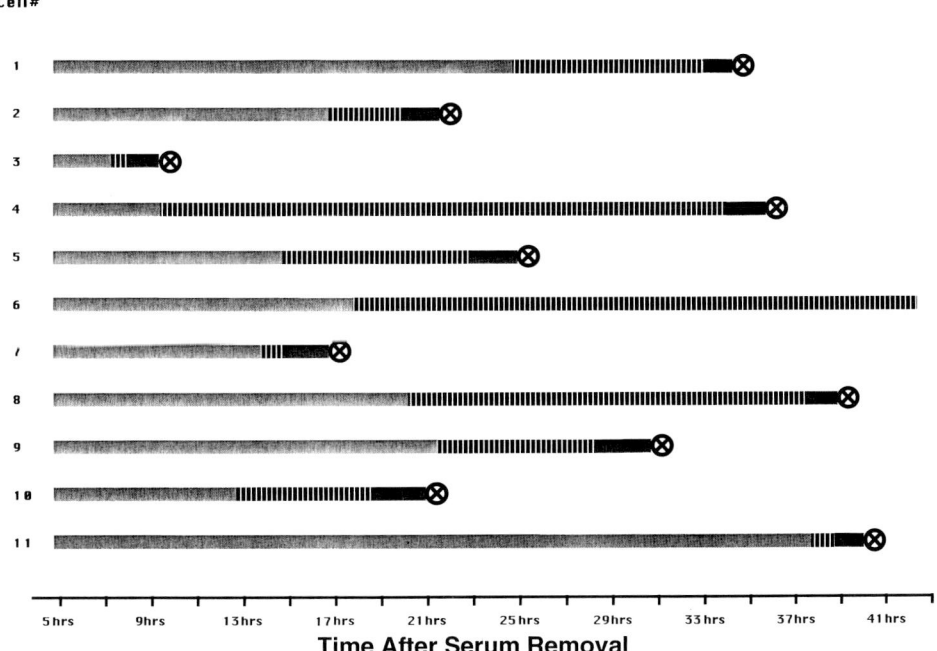

Fig. 3. Schematic representation of morphological changes undifferentiated PC12 cells undergo following removal of serum. Data are taken from time lapse video microscopy experiments with each line representing the fate of an individual cell as it goes through various morphological phases (*see* Fig. 2A). Phase 1 (normal morphology) is represented by the solid gray line; phase 2 (membrane bubbling), by the stippled line; phase 3 (whole cell blebbing/active apoptosis), by the solid black line; and cell death by the circle with an X. Note the asynchrony associated with cell death. Also note that phases 1 and 2 are highly variable in length, whereas phase 3 (active apoptosis) has a similar duration in all cells.

least in PC12 cells, although the stage of the cell cycle may be important for some aspects of apoptosis, it does not dictate when cells eventually die.

An obvious question is whether cells in vivo also progress through similar phases during apoptosis. The three-dimensional constraints present in vivo make it unlikely that identical morphological changes occur in vivo and in vitro. However, based on a large number of electron microscopy studies of cells undergoing apoptosis/programmed cell death in vivo *(3,12–14)* it is clear that many cells in vitro and in vivo undergo very similar nuclear and cytoplasmic changes including blebbing. Given the likelihood that the active phase of apoptosis is a highly conserved process (*[6]*; discussed below) it might be expected that similar biochemical and morphological changes occur in vivo and in vitro.

A MODEL OF BIOCHEMICAL AND MOLECULAR STAGES OF APOPTOSIS

Although the path a neuron takes toward death has not been determined at any level, it seems apparent that it is no longer realistic to think of apoptosis/programmed cell death as a simple linear path. One possible view of biochemical/molecular changes

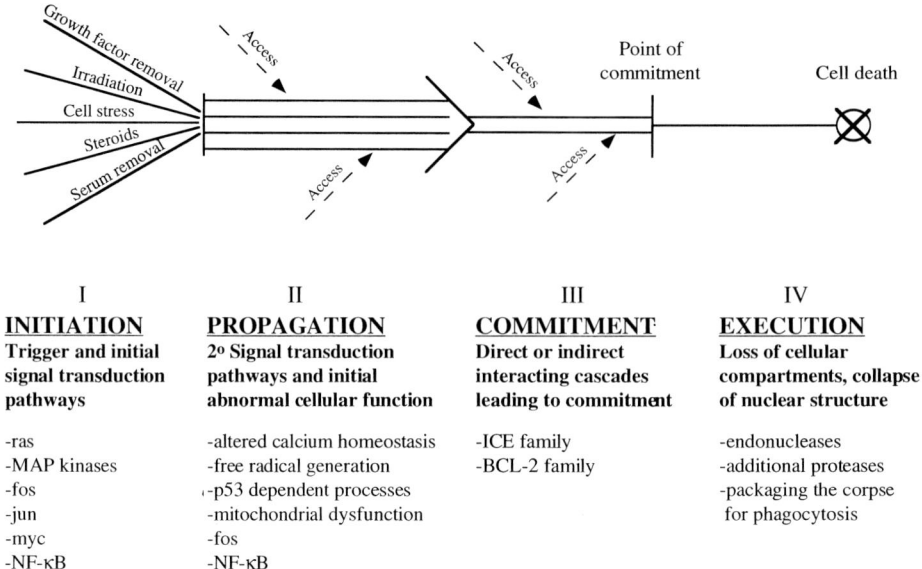

Fig. 4. Hypothetical model of biochemical and molecular stages of apoptosis. The cell death pathway is divided into four stages and although the stages are presented as a sequential series of events moving from initiation to propagation, commitment, and execution; in fact, the apoptotic pathway can be accessed ("triggered") at any point along this continuum (dashed "access" arrows along the pathway). The initiation stage consists of initial signal transduction pathways and activation of immediate early genes following an apoptotic stimulus ("trigger"). Propagation involves secondary signal transduction events, activation of genes requiring new protein synthesis and dysregulation of normal cellular processes. In some apoptotic cells, genes such as *c-fos* and *NF-κB* are immediate early genes whereas in other apoptotic cells induction occurs later in the process and new protein synthesis is required. Therefore, the same biochemical or molecular changes may occur at different places in the cell death pathway depending on the cell type and trigger used to induce apoptosis. The commitment stage represents the convergence of various apoptotic pathways onto the *bcl-2* and ICE/ced-3 families and ends at the point cells commit to die (hypothesized to reflect activation of a member of the ICE/ced-3 family). The commitment point represents an irreversible step for the cell—upstream of this point the cell death process can be reversed; however, downstream of this specific biochemical event the cell is committed to die. The execution stage probably corresponds to active apoptosis (*see* phase 3 in Figs. 2 and 3) and involves degradation of the cell. Note that the diagram is not intended to indicate the time spent in individual stages, and the execution stage consists of multiple pathways to degrade cellular components and prepare the corpse for phagocytosis.

associated with various stages a cell goes through during apoptosis is depicted in Fig. 4. If and how these stages relate to morphological phases depicted in Figs. 2 and 3 is unknown, but it is likely that the entire initiation stage and much of the propagation stage (Fig. 4) occur in morphologically normal cells (phase 1 in Figs. 2 and 3).

The initiation stage of apoptosis involves many of the commonly used "triggers" for initiating apoptosis and initial signal transduction pathways including induction of immediate early genes. It is beyond the scope of this chapter to examine the enormous number of triggers and pathways that have been identified that induce apoptosis but the example of removing NGF from PC12 cells or sympathetic neurons will be used to illustrate some of the steps and events in the initiation phase. It should be noted that early biochemical and molecular changes occurring after removing NGF (to induce apoptosis) represent normal cellular events. However, it is very likely that these normal cellular events are occurring in an abnormal context such as activating 2 or 3 biochemical pathways that are never simultaneously activated during normal cellular function. This represents a more refined version of the "conflict hypothesis" of apoptosis *(9,15–17)* in which major cellular conflicts such as signals to simultaneously proliferate and differentiate result in apoptosis. Removal of NGF from differentiated PC12 cells stops normal signaling through the trk tyrosine kinase pathways resulting in increased JNK (c-Jun N-terminal protein kinase) and p38 kinase activities and decreased ERK (extracellular signal regulated kinase) kinase activity *(18)*. Differential regulation of all three of these MAP (mitogen activated protein) kinases are required for apoptosis to occur. Even at this very early stage of apoptosis, interactions between parallel signaling pathways establish a complex mechanism of regulation. In differentiated PC12 cells and sympathetic neurons, the decreased binding of nuclear proteins to octamer sequences is also an early characteristic feature of apoptosis; following removal of NGF and experimentally decreasing this binding is sufficient to induce apoptosis *(19)*. In sympathetic neurons *c-jun* is induced in neurons at early times following removal of NGF and antibodies against c-Jun or a dominant negative c-Jun block cell death *(20,21)*.

In the proposed model, transition into the propagation stage is not clearly defined but is characterized by propagation of the initial apoptotic signal through secondary signal transduction mechanisms, gene expression dependent on new protein synthesis, and initial abnormal cellular functions. It is during this stage that cellular functions may begin to deviate from normal, and although changes in parameters such as free radical production, calcium homeostasis, or mitochondrial function may occur, they are not yet lethal and can be reversed (obviously, huge deviations of these functions from normal results in immediate necrotic death). It should be kept in mind that all, some, or none of these altered cellular functions may accompany apoptosis depending on the initiating trigger and initial biochemical pathways activated. In sympathetic neurons, following removal of the NGF, the propagation stage probably includes an increased production of reactive oxygen species *(22)* and in differentiated PC12 cells reversible changes in metabolic parameters such as lactate production and ATP levels occur in this stage *(23)*. In both sympathetic neurons and differentiated PC12 cells, induction of *c-fos* and *egr*-1/NGFI-A probably occur during this stage *(20,24)*. Typically *c-fos* is a primary response gene following cell stress; however, in sympathetic neurons following removal of the NGF induction of *c-fos* occurs much later in the apoptotic process and requires new protein synthesis *(20)*.

During early stages of cell death (initiation and propagation) different cell types often exhibit vastly different cellular, biochemical, and molecular changes depending on the cell death trigger and the internal environment within the cell. As a result, many "cell death genes" have been identified that are components of normal signal transduction pathways or immediate early genes that are typically induced under normal nonapoptotic conditions. Even initial cellular dysfunctions exhibited by various cells appear to be quite varied with some cells showing prominent changes in calcium homeostasis, while other cells exhibit changes in free radicals and still other cells modulate neither or both. The commitment stage, however, may represent a more restricted set of cellular changes in which secondary signal transduction pathways and altered cellular functions funnel into a small set of regulatory pathways that are similar but not identical in many different cell types. Whereas any given cell or trigger may exhibit very different early events in apoptosis, most cells and triggers eventually appear to modulate activities of members of the bcl-2 family and the ICE/ced-3 family. Therefore, at some point it appears that most or all apoptotic pathways converge onto these two families. The convergence of a number of different pathways onto the regulation (transcriptional, translational, or post-translational) of these two families defines the onset of the commitment stage. This stage of cell death is centered around activation, inactivation, and interactions within and possibly between members of the *bcl-2* family and ICE/ced-3 family. A large number of members of both families exist and individual cell types may have several different members of each family. The reason for the existence of such a large number of family members is unknown but may provide numerous points for regulating the process as well as enabling different cell types to monitor and couple to different early pathways involved in initiation and propagation of the cell death signal.

The commitment stage ends at the point when cells become irreversibly committed to die. Although the biochemical correlate to the point of commitment is unknown, it may represent activation and subsequent cleavage of key cellular proteins by a member of the ICE/ced-3 family. The final stage of apoptosis, execution, likely represents the "active" phase of apoptosis identified as morphological phase 3 in Figs. 2 and 3. As indicated above, the active phase of apoptosis is morphologically similar in different cell types responding to different triggers of cell death and has a very consistent duration for cells in vitro and in vivo. Besides killing the cell, a key function of the execution stage is to package the cell so that intracellular contents are not released (as occurs during necrosis) and thus avoiding an inflammatory response (a hallmark feature of apoptosis in vivo). Given the likely morphological and temporal conservation of the execution stage across species and tissue types, this process may represent a highly conserved biological process that is similar or identical in all cells. In this sense, apoptosis would be similar to cell proliferation where a variety of different molecules coupling through a large number of signal transduction pathways can alter cell proliferation (similar to initiation and propagation in apoptosis). However, once initiated, progression through the cell cycle (similar to execution in apoptosis) is essentially identical in all cell types.

Although this discussion of the stages of apoptosis has proceeded in an orderly manner from typical triggers initiating apoptosis through propagation, commitment and execution, it should be realized that the apoptotic machinery can be accessed ("triggered") all along the pathway. For instance, addition of a calcium ionophore may bypass the initiation stage and immediately access the propagation pathway, or overexpression of a member of the ICE family may access the apoptotic pathway during the commitment stage. Although activation of one or more ICE family members is hypothesized to occur late in apoptosis (near or at the commitment point) it is also possible that a self-regulating protease cascade of several ICE family members (both positive and negative regulators of apoptosis) occurs throughout the apoptotic process.

COMMITMENT TO DIE AND ITS IMPORTANCE

A key to understanding apoptosis will be to eventually define and biochemically characterize the point at which cells commit to die. It is at this point that cells make the transition from a reversible state to an irreversible state leading to death (Fig. 4). Events occurring upstream of the commitment point represent potential places to block the process through pharmacological or molecular intervention, and events downstream represent cellular processes responsible for degrading the cell and preparing the corpse for disposal.

A "pseudo" commitment point can be operationally defined by the point when a particular agent, such as cycloheximide or NGF, no longer reverses the process. However, the inability of a particular agent to reverse the process is based on a large number of variables and this should be considered when interpreting data using this type of commitment point. Although a commitment point can be defined for a population of cells, a more precise point of commitment can be made at the level of individual cells. A video "history" of cells undergoing apoptosis is generated and then protective agents are added to cultures at various times, and the fate of individual cells in the field is followed. When this type of analysis is performed on undifferentiated PC12 cells induced to die by serum removal (see Fig. 2A), adding serum back to cells rescues 100% of the cells in morphological phase 1, 87% in phase 2, and 0% in phase 3 (10). This is consistent with cells committing to die (based on serum rescue) very late in phase 2 near the transition to phase 3 (the active phase of apoptosis). The actual point cells commit to die (see Fig. 4) must occur somewhat later than this due to the time required for serum to affect cellular processes that can block cell death. Preliminary studies suggest that commitment of differentiated PC12 cells following removal of NGF occurs at or very near the time cells enter phase 3 (11). In both cases where commitment of individual cells is known, cells commit to die very near the time they enter active apoptosis which occurs about 1 h prior to death. This suggests that a cell that dies by apoptosis 36 h after the initial insult can be rescued up until about 35 h after the insult. From a clinical standpoint this clearly provides hope and promise for being able to reverse the apoptotic process well after it has initiated. It should also provide additional incentive for identifying the final series of biochemical events that commit cells to die.

PRACTICAL CONSEQUENCES OF DEALING WITH ASYNCHRONY AT THE EXPERIMENTAL LEVEL

Asynchronous death of cells in an organ or population of cells is probably advantageous to the organism for several reasons: it decreases the chances of aberrant physiological responses to massive death, it allows time for efficient phagocytosis of dead cells, it provides time to reverse the process before global damage is done to an organ or population of cells, and it should allow more control over the process (multiple steps and/or a longer time frame increase potential places for controlling cell death). However, from the standpoint of understanding apoptosis experimentally, asynchrony makes it difficult to use data obtained from a population of cells to get more than a gross idea of events occurring inside an individual dying cell. Biochemical parameters such as energetic and metabolic changes during apoptosis are especially hard to study given that most assays require large numbers of cells. We will briefly examine some of the difficulties encountered when trying to interpolate data obtained on populations of cells to events occurring in individual cells.

As indicated above, the active phase of apoptosis appears to have its onset very near the commitment point. In most cells, the active phase lasts about 1 h, and following many triggers of apoptosis (such as those depicted in Fig. 4), only 5–10% of cells are in this state at any one time. Early events in the active phase of apoptosis may be critical for committing cells to die and for generating initial signals and cellular changes that set the stage for degrading the cell. It is obvious that any biochemical measurement made on cells induced to undergo apoptosis reflects a composite of information from cells at various stages of apoptosis with only a small fraction in any particular stage. For instance, if an 80% decrease in ATP occurs at the time cells commit to die, then ATP levels for the population will still be approximately 95% of normal values (only 5–10% of the cells will be in this committed stage) and are likely to go undetected. With highly sensitive assays such as PCR or assays that remove contaminating material from healthy cells such as those that involve isolating "soluble" DNA from bulk chromatin (which effectively removes > 99% of the "normal" DNA) to monitor DNA laddering it is possible to measure changes occurring in a small number of cells. Nevertheless, the majority of biochemical measurements will reflect changes in the entire population of cells, only a fraction of which are actively dying. In general, the closer the temporal relationship between triggering apoptosis and the biochemical change being investigated, the more likely it is that a significant portion of the population will exhibit the change, and it will be detected as a change in the population (e.g., activation of an immediate early gene or a change in reactive oxygen species, or ATP soon after triggering apoptosis). Transient changes (particularly those occurring later in the overall process) are more difficult to detect.

POSSIBLE MECHANISMS UNDERLYING ASYNCHRONY

Asynchrony is a common feature of biological systems, however, in few, if any cases are underlying mechanisms understood. A well known example of asynchrony is the desynchronization of a population of cells that has been synchronized to a specific point in the cell cycle. Within one or two transits of the cell cycle, cells are again totally dispersed throughout all phases of the cycle. Asynchrony associated with cell cycle progression has been quantitatively modeled (25–28), but the mechanisms that govern asynchrony are not understood at a biochemical or molecular level. Three different possibilities will be briefly presented in this section to provide an idea of the types of mechanisms that might result in apoptosis being asynchronous. Potential mechanisms to be presented include a series of sequential biological events, a pathway containing a "bottleneck" or low-probability event, and a cellular or biochemical event having one or more stochastic components.

A series of biological events (occurring as sequential normal and/or skewed distributions) each depending on a preceding step should broaden the distribution of the overall effect and create asynchrony. The number of steps, their distribution and linkage, and cellular heterogeneity with respect to concentrations of components will determine the extent of asynchrony. Although unproved, an example of this type of biological event may be NGF-induced neurite outgrowth from PC12 cells in which a homogenous population shows asynchronous outgrowth of neurites over several days—

even though all cells are exposed to NGF at the same time. Different numbers of Trk receptors, efficiency of coupling to various signal transduction pathways, concentrations of second messenger systems, rate of translocation of components to the nucleus, etc., would be variables that affect the rate any one particular cell grows neurites.

Another means of generating asynchrony would be for a simple biological path to contain a bottleneck or "transition state" (low-probability state) that must be overcome. A general example could be a conformational change in a key molecule that once it occurs (either spontaneously or more likely through interactions with products of one or more apoptotic pathways) it now serves as a key substrate for a kinase, phosphatase, or protease. Once this reaction occurred the apoptotic process would move rapidly toward commitment. A subset of this model would be the build-up of a self-regulating cellular component (or particular state of a cellular component such as its phosphorylation state). An example of this would be the build-up of cyclins during progression through the cell cycle or phosphorylation of key regulators dependent on this build up such as retinoblastoma protein.

A final example of the type of mechanism that could underlie asynchrony during apoptosis would be for one or more of the cellular or biochemical changes to have a stochastic component. In the case of apoptosis this could represent a process such as weakening of compartmental boundaries through alterations in the cytoskeleton with subsequent "leakage" of molecules into an inappropriate compartment followed by a chance encounter and modification of a second molecule that is required for continued propagation of apoptosis. A similar stochastic scenario would be created if a protein (or RNA) required for cell death was produced at very low levels and had to diffuse and bind to another protein in a different part of the cell. The time required for the two components to reach each other by "chance" encounter and interact would be expected to vary greatly between cells and create asynchronous death. Clearly, asynchrony associated with cell death may be a composite of all three of these possibilities or have a totally different mechanism. The important point is that asynchrony is a fundamental characteristic of apoptosis and must be dealt with at both the intellectual and experimental levels.

POTENTIAL METHODS FOR DEALING WITH AND/OR OVERCOMING THE PROBLEM OF ASYNCHRONY

Obviously, highly sensitive techniques such as polymerase chain reaction (PCR) can be used to identify changes in apoptotic cells even in the face of asynchronous death. However, the power of cellular resolution and correlation with other cellular changes is lost (i.e., if two molecular changes are correlated based on PCR they could be occurring in different subsets of cells in different stages of apoptosis). Several in vitro approaches have been developed recently to overcome the problem of asynchrony when characterizing biochemical changes occurring during specific stages of apoptosis. Unfortunately, at present, none of these have been applied to in vivo populations of cells undergoing apoptosis. The two general in vitro approaches being used are to 1) "synchronize" the apoptotic population so that standard biochemical techniques can be used or 2) characterize changes at the level of individual cells so the phase of apoptosis can be identified. Both of these general approaches are directed primarily at characterizing cells in the active phase of apoptosis (phase 3, Figs. 2 and 3; execution, Fig. 4).

Several approaches have been used to try to increase synchrony in apoptotic cells including using cell-free systems and isolating the subset of cells in active apoptosis. Thus far, the most successful cell free systems have used cytoplasmic extracts from a population of cells undergoing apoptosis and then exposing normal nuclei to these extracts to characterize events involved in DNA fragmentation and nuclear breakdown *(29–31)*. In some cases, the percentage of apoptotic cells from which cytoplasmic extracts are made has been increased by sequentially synchronizing cells in S-phase and M-phase of the cell cycle *(29,32)*. These represent powerful systems for understanding mechanisms responsible for nuclear changes associated with apoptosis but are currently restricted to the nuclear breakdown phase of the active phase of apoptosis. In addition it is not clear how experimental destruction of compartmental boundaries in these experiments affects the actual events as they would happen inside an intact cell. A second method for determining biochemical parameters in cells undergoing active apoptosis involves using differential centrifugation to isolate an enriched population of cells in active apoptosis *(33)*. This technique is similar to older, "shake-off" methods for isolating mitotic cells. In short, this technique takes advantage of the fact that most cells round up as they begin to undergo the active phase of apoptosis, and these rounded cells can be isolated by applying shear to the cultures (most flattened cells remain attached to the substrate whereas most cells in active apoptosis are removed). Intact apoptotic cells are separated from cellular debris by differential centrifugation. These "enriched" cells in active apoptosis can then be used to measure various biochemical parameters or for immunoblots with the expectation that a large fraction of cells will be in the same stage of apoptosis. This technique holds considerable promise for many cell types but is not very effective for neurons in vitro, because most neurons routinely have round cell bodies and shearing forces required to remove apoptotic cells remove many "normal" neurons.

Another approach for analyzing biochemical changes during active apoptosis in a population of cells is to initiate cell death with an agent that accesses ("triggers") the cell death pathway near the commitment point. This should increase the fraction of the population showing changes, that is, there are fewer steps to desynchronize the population. This approach is probably not as "clean" as techniques described above; however, it can be used with neurons as opposed to some of the procedures. To illustrate this method an example will be given using three different initiators of apoptosis for NGF-dependent neurons—removing NGF, exposing cells to staurosporine, and increasing expression of ICE/ced-3 family members. By removing the NGF from the cells, it would access the apoptotic pathway during the "trigger" part of the initiation stage while treatment of cells with staurosporin (which exhibits a shorter time interval between exposure and active apoptosis and can induce cytoplasm from non-apoptotic to cleave DNA in cell free assays) should access the pathway in the propagation stage or early commitment stage. Microinjecting an ICE family member (or overexpressing the ICE family member transiently or under an inducible promoter) should access the pathway very near commitment. The percentage of cells in the active phase of apoptosis and therefore the possibility of detecting changes should be greater with staurosporin than following removal of NGF and greater with expression of an ICE family member than with staurosporin (assuming that cellular differences in transcription and translation of the ICE family member are not too great). Therefore, by accessing the apoptotic process closer to the point of commitment a larger fraction of cells should be in active apoptosis and biochemical changes occurring during this phase of cell death should be more easily measured.

Rather than dealing with a population of cells, a second way of overcoming problems associated with asynchrony is to approach cell death at the level of individual cells. Several approaches have been used to do this including the following:

1. Time lapse video microscopy to create a morphological history of the cell and to use this information as a frame of reference to correlate other changes;
2. *In situ* hybridization (rather than northern blots);
3. Immunohistochemistry (rather than immunoblots);
4. Imaging molecules such as calcium and free radicals in single cells (rather than measuring these parameters on populations of cells); and
5. Microinjecting and then following individual cells rather than transfecting a population of cells.

The power of this approach is particularly evident when biochemical changes are correlated with a specific phase of apoptosis. Two examples involving changes in calcium and activation of a phosphatase will be used to demonstrate the potential power of this approach.

Deregulation of calcium is thought to play a functional role in apoptosis including a potential role in activating a calcium-dependent endonuclease; however, the relationship between changes in calcium and the overall apoptotic process is not well understood. For instance, does calcium increase early in the process and affect a variety of cellular functions, or does it increase late in the process, or is it regulated differently in different cell types or following different initiators of apoptosis. By knowing that cells move through very specific morphological phases during apoptosis (Figs. 2 and 3) it should be possible to correlate changes in calcium in individual cells with the phase of apoptosis based on the morphology of cells showing altered calcium levels. Intracellular free calcium increases in undifferentiated apoptotic PC12 cells very near the time cells enter phase 3 *(10)*. This is also the time cells commit to die; therefore, based on analysis of single cells, increases in intracellular free calcium can be correlated with a specific phase of apoptosis and in close proximity to the point where cells commit to die.

A different approach using single cell analysis can be taken by following activation, inactivation, or post-translational changes in proteins using antibodies. Again, the power of this type of analysis is increased considerably by knowing the morphological phases a cell goes through during apoptosis and how these correlate with other biochemical and/or cellular events (such as commitment to die). An example of this type of analysis is our recent study following changes in the phosphorylation state of tau during apoptosis after removing NGF from terminally differentiated PC12 cells *(11)*. Using antibodies specific for both phosphorylated and non-phosphorylated epitopes, these experiments show quantitative decreases in phosphorylation at two different sites on tau that occur as cells enter the active phase of apoptosis (phase 3; which also appears to be the time these cells are committing to die). These studies can be viewed as providing information on tau during apoptosis, but they can also be viewed as tau acting as a "reporter" protein providing information on a much larger event. In this case, the process being "viewed" would be activation of a phosphatase or inactivation of a kinase which could be affecting a large number of cellular proteins at a very critical point in the apoptotic process. Changes in the phosphorylation state of tau appear to be due to activation of a PP2A-like phosphatase that occurs just downstream of the commitment to die. Activation of this phosphatase cannot be reliably detected in mass cultures of these cells because only 5% of the cells are in the active phase of apoptosis at any one time, but is easily detected using single cell analysis. It should be possible to use other endogenous or exogenous "reporters" to investigate a variety of biochemical changes occurring during different stages of apoptosis.

In summary, asynchronous death of cells is a characteristic feature of apoptosis/programmed cell death both in vivo and in vitro. Although it was acknowledged as a fundamental property of apoptosis more than 20 years ago little information is available on underlying mechanisms or its impact on cell death studies. Asynchronous death probably serves a number of beneficial functions in the organism; however, it greatly complicates acquisition, analysis, and interpretation of biochemical and molecular data dealing with programmed cell death. In most cases, only 5–10% of cells are in the active phase of apoptosis at any one time; therefore, standard biochemical assays designed to measure changes in a population of cells are often misleading or difficult to interpret relative to the events occurring in individual dying cells. A number of methods are being developed to deal with problems associated with asynchrony which should eventually make it possible to establish the sequence of cellular events responsible for programmed cell death/apoptosis.

REFERENCES

1. Martin DP, Schmidt RE, DiStefano PS, Lowry OH, Carter JG, Johnson EM. Inhibitors of protein synthesis and RNA synthesis prevent neuronal death caused by nerve growth factor deprivation. *J Cell Biol* 1988, **106**: 829–844.
2. Edwards SN, Tolkovsky AM. Characterization of apoptosis in cultured rat sympathetic neurons after nerve growth factor withdrawal. *J Cell Biol* 1994, **124**: 537–546.
3. Wyllie AH, Kerr JFR, Currie AR. Cell death: The significance of apoptosis. *Int Rev Cytol* 1980, **68**: 251–307.
4. Hamburger V. Cell death in the development of the lateral motor column of the chick embryo. *J Comp Neurol* 1975, 160: 535–546.
5. Oppenheim RW. Neuronal cell death and some related regressive phenomena during neurogenesis: A selective historical review and progress report, in *Studies in Developmental Neurobiology: Essays in Honor of Viktor Hamburger* (Cowan WM, ed.), Oxford University Press, New York, 1981, pp. 74–133.
6. Earnshaw WC. Nuclear changes in apoptosis. *Cur Opin Cell Biol* 1995, 7: 337–343.
7. Rukenstein A, Rydel R, Greene L. Multiple agents rescue PC12 cells from serum-free cell death by translation-transcription-independent mechanisms. *J Neurosci* 1991, **11**: 2552–2563.
8. Mesner P, Winters T, Green SH. NGF withdrawal-induced cell death in neuronal PC12 cells resembles that in sympathetic neurons. *J Cell Biol* 1992, **119**: 1669–1680.
9. Pittman RN, Wang S, DiBenedetto AJ, Mills JC. A system for characterizing cellular and molecular events in programmed neuronal cell death. *J Neurosci* 1993, **13**: 3669–3680.
10. Messam CA, Pittman RN. Dynamic morphological changes and increases in intracellular free calcium during apoptosis of PC12 cells (submitted).
11. Mills JC, Lee, VM-Y, Pittman RN. Dephosphorylation of t protein characterizes onset of the active phase of apoptosis (submitted).
12. Kerr JFR, Harmon B, Searle J. An electron-microscope study of cell deletion in the anuran tadpole tail during spontaneous metamorphosis with special reference to apoptosis of striated muscle fibres. *J Cell Sci* 1974, **14**: 571–585.

13. Chu-Wang IW, Oppenheim RW. Cell death of motoneurons in the chick embryo spinal cord. I. A light and electron microscopic study of naturally-occurring and induced cell loss during development. *J Comp Neurol* 1978, **177**: 33–58.
14. Clarke PGH. Developmental cell death: Morphological diversity and multiple mechanisms. *Anat Embryol* 1990, **181**: 195–213.
15. Evan G, Wyllie A, Gilbert C, Littlewood T, Land H, Brooks M, Waters C, Penn L, Hancock D. Induction of apoptosis in fibroblasts by c-Myc protein. *Cell* 1992, **69**: 119–128.
16. Fisher DE. Apoptosis in cancer therapy: Crossing the threshold. *Cell* 1994, **78**: 539–542.
17. DiBenedetto AJ, Pittman RN. Death in the balance. *Persp Dev Neurobiol* 1995, **3**: 109–117.
18. Xia Z, Dickens M, Raingeaud J, Davis RJ, Greenberg ME. Opposing effects of ERK and JNK-p38 MAP kinases on apoptosis. *Science* 1995, **270**: 1326–1331.
19. Wang S, Pittman RN. Altered protein binding to the octamer motif appears to be an early event in programmed neuronal cell death. *Proc Natl Acad Sci USA* 1993, **90**: 10,385–10,389.
20. Estus S, Zaks WJ, Freeman RS, Gruda M, Bravo R, Johnson EM. Altered gene expression in neurons during programmed cell death: Identification of c–jun as necessary for neuronal apoptosis. *J Cell Biol* 1994, **127**: 1717–1727.
21. Ham J, Babij C, Whitfield J, Pfarr CM, Lallemand D, Yaniv M, Rubin LL. A c-jun dominant negative mutant protects sympathetic neurons against programmed cell death. *Neuron* 1995, **14**: 927–939.
22. Greenlund LJS, Deckwerth TL, Johnson EM. Superoxide dismutase delays neuronal apoptosis: A role for reactive oxygen species in programmed neuronal death. *Neuron* 1995, **14**: 303–315.
23. Mills JC, Nelson D, Erecinska M, Pittman RN. Metabolic and energetic changes during apoptosis in neural cells. *J Neurochem* 1995, **65**: 1721–1730.
24. Mesner PW, Epting CL, Hegarty JL, Green SH. A timetable of events during programmed cell death induced by trophic factor withdrawal from neuronal PC12 cells. *J Neurosci* 1995, **15**: 7357–7366.
25. Smith JA, Martin L. Do cells cycle? *Proc Natl Acad Sci USA* 1973, **70**: 1263–1267.
26. Shields R, Smith JA. Cells regulate their proliferation through alterations in transition probability. *J Cell Physiol* 1977, **91**: 345–356.
27. Skehan P. Control models of cell cycle transit, exit, and arrest. *Biochem Cell Biol* 1988, **66**: 467–477.
28. Grasman J. A deterministic model of the cell cycle. *Bull Math Biol* 1990, **52**: 535–547.
29. Lazebnik YA, Cole S, Cooke CA, Nelson WG, Earnshaw WC. Nuclear events of apoptosis *in vitro* in cell-free mitotic extracts: a model system for analysis of the active phase of apoptosis. *J Cell Biol* 1993, **123**: 7–22.
30. Solary E, Bertrand R, Kohn KW, Pommier Y. Differential induction of apoptosis in undifferentiated and differentiated HL-60 cells by DNA topoisomerase I and II inhibitors. *Blood* 1993, **81**: 1359–1368.
31. Bertrand R, Solary E, O'Conner P, Kohn KW, Pommier Y. Induction of a common pathway of apoptosis by staurosporin. *Exp Cell Res* 1994, **211**: 314–321.
32. Lazebnik YA, Kaufmann SH, Desnoyers S, Poirier GG, Earnshaw WC. prICE, an ICE-like protease from apoptotic cells, cleaves poly (ADP-ribose) polymerase. *Nature* 1994, **371**: 346–347.
33. Casciola-Rosen LA, Miller DK, Anhalt GJ, Rosen A. Specific cleavage of the 70-kDa protein component of the U1 small nuclear ribonucleoprotein is a characteristic feature of apoptotic cell death. *J Biol Chem* 1994, **269**: 30,757–30,760.

3
Free Radicals and Mitochondria Dysfunction in Excitotoxicity and Neurodegenerative Disease

James A. Dykens

INTRODUCTION

The study of neurodegenerative diseases has begun converging with the previously disparate fields of free radical pathophysiology, excitotoxin-mediated neuronal death and mitochondrial physiology. This convergence has been prompted by increasing evidence that excitatory amino acids and free radical toxicity contribute to neuronal apoptosis *(1)* and to the etiology of chronic neurodegenerative diseases *(2–12)*. Moreover, a growing body of evidence implicates mitochondrial energetic and oxidative dysfunction, due to congenital genetic defects, and perhaps to radical-induced mutations and oxidative enzyme impairment, in several neurodegenerative disorders (reviewed by *[5,6,13–17]*).

The clinical manifestations of neurodegenerative diseases, oxidative pathology, excitotoxicity, and mitochondrial dysfunction have been the subject of recent comprehensive review *(12)* and are beyond the scope of this review. Rather, the author has relied on an historical perspective to summarize the evolution of the hypothesis that free radicals contribute to neurodegenerative disease, and has limited the scope to only a few processes capable of killing neurons, or of precipitating mitochondrial dysfunction, and that are therefore likely to contribute to neurodegenerative disease. In so doing, the author has clearly overlooked the myriad other ways in which radicals and mitochondrial dysfunction interrelate, and the reader is encouraged to consult the more comprehensive reviews of mitochondrial Ca^{2+} metabolism by Gunter and Gunter *(18)* and Gunter et al. *(19)*; of free radical pathophysiology by Halliwell and Gutteridge *(20)*; of oxidative pathology in the central nervous system by Siesjö et al. *(21)*, Götz et al. *(12)*, Bondy and LeBel *(22)*, and Halliwell *(23)*; of age related changes in mitochondrial function and oxidative equilibria by Benzi and Moretti *(24)*; and of the role oxidative and energetic pathology may play in the etiology of neurodegenerative disease *(5–7,15)*.

INTRACELLULAR AND FREE RADICAL REACTIVITY

Although the central focus here is on free radical reactivity and mitochondrial dysfunction in neurodegenerative disease, there are heuristic benefits from a broader perspective that encompasses cellular and mitochondrial pathology associated with ischemia and reoxygenation of aerobically poised tissues such as myocardium, as well as neuronal excitotoxicity of glutamate and its dicarboxylic analogs such as kainate

From: Cell Death and Diseases of the Nervous System
Edited by: V. E. Koliatsos and R. R. Ratan © Humana Press Inc., Totowa, NJ

and NMDA. These pathologies share in common several functional parallels with mitochondrial dysfunction in neurodegenerative disease, such as reductions in the ATP/ADP ratio that result when ATP utilization outpaces production during repeated or prolonged neuronal depolarizations that are characteristic of excitotoxin stimulation, or when availability of O_2 as the terminal electron acceptor in the mitochondrial electron transfer system is inadequate due to ischemia or hypoxia, or when genetic or radical-induced defects in electron transfer components impair the ability of mitochondria to generate ATP. Although the causes of ATP depletion are diverse, the functional responses of the mitochondria to accelerate oxidative phosphorylation, and for glycolytic carbon flux to increase, are similar.

It is also evident that increases in free cytosolic Ca^{2+} precede neuronal death following exposure to dicarboxylic excitotoxins *(25,26)*, and after transient hypoxia in both neurons and cardiac myocytes *(27)*. Such increases in free Ca^{2+} result in a host of responses, including activation of phospholipase A_2 with consequent arachidonate metabolism, activation of proteases, and alterations in numerous metabolic pathways. Depending on circumstances, such increases in free Ca^{2+} can also perturb oxidative phosphorylation, or even completely destroy mitochondrial function *(see below)* *(18,28–32)*.

Although this chapter focuses on the functional consequences of mitochondrial pathology, particularly in terms of free radical generation, it bears reiteration that cellular oxidative status results from an equilibrium between radical production and the sum total of the cellular (and extracellular) oxidative defenses, including enzymes such as superoxide dismutase (both cytosolic Cu^{2+}-Zn^{2+} and the mitochondrial Mn^{2+} morphs), catalase, and glutathione peroxidase-reductase, as well as antioxidants such as a-tocopherol, ascorbate, and urate. For example, between approximately 1–6% of the O_2 consumed by normally respiring mammalian mitochondria is univalently reduced to superoxide radical (O_2^-) by the electron transfer system instead of tetravalently to water by the terminal cytochrome oxidase. Although such low-level, chronic production of oxygen radicals reasonably contributes both to aging and to increased mutation rates in mitochondrial DNA *(33,34)*, this radical production can hardly be regarded as a potentially lethal acute oxidative threat, unless the antioxidative defenses are so impaired that even such mild radical production exceeds defenses and oxidative damage accrues. Of course, the antioxidant defenses are not absolutely efficient, and oxidative damage gradually accumulates as a function of age. More acutely, the gene defect in some familial cases of amyotrophic lateral sclerosis (ALS) that reduces Cu–Zn superoxide dismutase activity (aside from the likelihood of acquired deleterious function) *(35,36)* could well be responsible for increased oxidative stress, as reflected by increased levels of carbonyl proteins *(36)*, without invoking increases in radical production.

By the same token, however, a wealth of evidence indicates that mitochondrial dysfunction, with corresponding degradation of adenylate charge and increased free radical production, figures prominently in excitotoxicity and neurodegenerative disease despite normal, or even increased, levels of cellular antioxidant potential (reviewed by *[5,6,15]* [also *see* chapters by Olney and Ishimaru, O'Hearn and Molliver, Flanagan and Ratan, and Vornov]). Nevertheless, in assessing total antioxidant capacity, it bears re-emphasis that the appropriate antioxidant must also be appropriately situated if it is to function; cytosolic superoxide dismutase will do little to moderate intramitochondrial

O_2^- production, and intramembranous vitamin E will do little to forestall cytosolic radical reactivity. Moreover, the total antioxidant capacity is also a derivative of the overall oxidative environment, which in large measure is dictated by temporal effects. For example, although quite rapid, the reaction between O_2^- and SOD is about sixfold slower than the competing reaction of O_2^- with NO; when NO is available, formation of peroxynitrite anion will be the preferred reaction over dismutation, even in the presence of what appears to be ample activities of SOD. At physiologically relevant pH, the conjugate peroxynitrous acid spontaneously isomerizes into the less stable *trans* configuration, decomposition of which results in the production of highly reactive hydroxyl radicals and nitrogen dioxide Thus, activities of antioxidant enzymes, such as SOD, although indicative of antioxidant capacity, only imperfectly describe the oxidative environment in vivo where availability of other reactants, such as NO, will shift reaction equilibria and correspondingly alter the character of radical production.

It is also worth reiterating the basic tenet that radical reactivity characterized in vitro need not translate into physiologically relevant pathology in vivo. For example, it has long been known that because of orbital spin restrictions, molecular oxygen preferentially undergoes univalent reduction to yield superoxide radical, an anion capable of inactivating key regulatory metabolic enzymes *(37,38)*. Creatine kinase (CK), among other enzymes, is exquisitely susceptible to inactivation by low concentrations of either superoxide or hydroxyl radicals, and the data are convincing that such inhibition occurs in stenoxic tissues during hypoxia *(38,39)*. Although CK activity is demonstrably diminished in stenoxic tissues following hypoxia-reoxygenation, the reaction catalyzed by CK, i.e., maintaining the equilibrium between phosphocreatine and ATP, shows no impairment when followed in real time using ^{31}P-NMR techniques *(38)*. Indeed, because of a 3000 fold excess in CK activity over that necessary to maintain the equilibrium between the phosphagen and adenylate pools, even at maximal rates of ATP utilization there would be no functional deleterious consequence to the cell even if over 99% of the CK activity were oxidatively inactivated. Maintaining such excess CK activity is a common strategy in many species, leading to the reasonable inference that this is a likely reflection of oxidative stress as an evolutionary selective pressure.

The physiology and reactivity of nitric oxide (NO•) in vivo is similarly complicated by the intracellular environment. Like molecular oxygen, NO undergoes a series of step-wise reductions yielding species of varying reactivities *(40)*. Due primarily to the ready availability of thiol reductants, metals, and a plethora of electron acceptors in the intracellular environment, nitrogen monoxide (NO) is physiologically and functionally equivalent to a host of more, and sometimes less, reactive species, such as nitrosonium (NO^+), nitric oxide radical (NO•), and nitroxyl anion (NO^-), each of which in turn can undergo subsequent reactions (reviewed in ref. *41*). The net result is that, predicting the intracellular reactivity of a given radical and its potential contribution to pathology by extrapolation from its in vitro behavior may be misleading. Nevertheless, the data are clear that increased radical production and ensuring oxidative chemistry due to elevated P_{O2}, and/or to alterations in the equilibrium between radical production and antioxidant defenses, contributes to acute neuronal necrosis, excitotoxicity, apoptosis, and chronic neurodegenerative disease.

It seems counterintuitive to propose that free radicals, with patently high reactivity and inherently acute effects, should underlie late onset and/or chronic neurodegenerative diseases *(2,3)*. Even the most acute forms of excitotoxic neuronal death mediated by dicarboxylate analogs of glutamate such as kainate or NMDA, require hours or days to occur *(2,3,5,6)*, a duration that suggests a less debilitating toxicity for free radicals than is predicted by their extreme reactivity. Clearly, acute, lethal cellular damage, such as plasma membrane lipid peroxidation and consequent loss of integrity, can result from intense radical exposure or loss of oxidative defenses. However, cellular radical production only slightly exceeding total antioxidant defenses can also yield lethal long-term consequences for neurons and other cells. For example, several key metabolic regulatory enzymes including glutamine synthase, phosphofructokinase (PFK), creatine phosphokinase and lactate dehydrogenase are unusually susceptible to oxidative inactivation, showing significant losses of activity after O_2^- and ·OH exposures three orders of magnitude less than those first used to study enzyme inactivation by oxygen radicals *(37,38,42)*. Oxidative impairment of glutamine synthase will correspondingly elevate glutamate concentrations, potentially exacerbating excitotoxicity. Pyruvate dehydrogenase, but not lactate dehydrogenase, activity is reduced in canine cerebral cortex following transient hypoxia in vivo, and both this inactivation and protein oxidation are partially ameliorated by treatment with the antioxidant acetyl-L-carnitine *(43)*. Likewise, activities of several enzymes susceptible to oxidative inactivation including PFK, are reportedly reduced in postmortem brain from Alzheimer's patients and in gerbil brain following ischemia-reperfusion *(42,44)* (reviewed in *[7]*); but see *[45]*). Moreover, DNA repair and transcription mechanisms are damaged by radical exposure, imposing long-term deleterious consequences for the cell or mitochondria *(46)*.

Because transition metals that are both ubiquitous and obligatory for functioning of biological systems serve as catalysts in free radical cascades, even seemingly minor perturbations in metal metabolism can profoundly alter the cellular and mitochondrial oxidative milieu *(20,47–49)*. In addition to the well-known Fenton reactions where H_2O_2 undergoes heterolytic cleavage to yield reactive hydroxyl radical, metals play central roles in the interconversions of NO· into more biologically reactive forms *(41)*. More to the point of the oxidative etiology of late-onset chronic diseases, however, are the numerous observations that oxidative load increases with age, due to impairment of antioxidant defenses and/or increased radical production *(50–52)*.

Finally, it should be noted in this context that distinctions between acute radical damage capable of killing cells outright, gradual accumulation of oxidative damage over a lifetime, and the intermediate circumstance where an increase in oxidative stress presages and initiates apoptosis, are really points on a continuum of oxidative load and cellular responses. Physiological function depends on the integration of separate, but interrelated processes, and cellular response to a given stimuli is composed of a complex myriad of individual responses. For example, low-level oxidative stress increases as the cell ages, but mild, transient, increases induce apoptosis in young cells if a threshold in the equilibrium between oxidative and antioxidative defenses is exceeded *(53)*. Likewise, acute increases in oxidative load sufficient to damage cellular architecture can induce apoptosis, but if the oxidative damage exceeds the threshold where critical cellular integrity is breached or metabolic processes are destroyed beyond repair, necrosis ensues. Thus, the two variables of oxidative load and cellular response both have a wide spectrum of responses, with acute oxidative death and swift necrosis at one extreme and a gradual decline into senescence at the other. At intermediate oxidative loads, components of both necrosis and apoptosis likely co-occur, with one or the other dominating or pre-empting the other depending on circumstances (for other considerations on the apoptosis-necrosis continuum, *see* chapter by Clarke).

OXIDATIVE EXCITOTOXICITY

There is compelling evidence that increases in free cytosolic Ca^{2+} precede neuronal death following exposure to dicarboxylic excitatory neurotransmitters, such as glutamate, and numerous xenobiotic excitotoxins such as kainate, ibotenate, and NMDA *(25,26)*. Stimulation of postsynaptic NMDA-type glutamate receptors contributes to neuronal death due to hypoxia, hypoglycemia, and seizures *(29,42,54)* (*see* chapters by Vornov and Shin and Lee) and probably in chronic neurodegenerative disorders such as Parkinson's, Alzheimer's, and Huntington's diseases as well *(3,5–10,15)*.

The evidence is similarly persuasive that free radicals mediate excitotoxicity *(2,12,21,29,54,55)* (reviewed in refs. *23,56*, and *57*). Increases in neuronal free Ca^{2+} occur via both ligand- and voltage-gated mechanisms, as well as via metabotropic responses depending on which receptor is stimulated, yielding a variety of potentially damaging consequences *(31)*.

In light of simultaneous Ca^{2+} increases and ATP depletion *(58)*, the author proposed some years ago *(2)* that the metabolic pathology responsible for excitotoxin-induced neurodegeneration was analogous to the tissue damage that is observed in aerobically poised tissues, such as myocardium, following transient hypoxia. According to this proposal, Ca^{2+}-induced activation of a cytosolic serine protease causes the conversion of xanthine dehydrogenase into xanthine oxidase, a form that generates H_2O_2 and superoxide radicals (O_2^-). Although O_2^- is reactive, the proximate mediator of oxyradical toxicity is likely not O_2^-, but rather an ensuing flux of hydroxyl radicals ($^\bullet OH$) formed by either Fenton reactions that involve heterolytic H_2O_2 cleavage catalyzed by Fe^{2+} and other transition metals, or by decomposition of the peroxynitrite that is formed in reactions between O_2^- and nitric oxide (NO) (reviewed in ref. *20*). Hydroxyl radicals are the most oxidizing radical species found in biological systems; with rate constants on the order of 10^9–10^{10} $M^{-1} s^{-1}$, they react within five molecular radii of their production. Considering the ubiquity of both transition metals and NO synthetase in tissues including brain, $^\bullet OH$ is likely to be responsible for most oxyradical toxicity.

Despite much evidence for radical involvement in excitotoxicity, only very low xanthine dehydrogenase/oxidase activities have been detected in neurons *(2,59)*, thus raising doubts that xanthine oxidase is an important source of radicals under conditions of ATP depletion and Ca^{2+} mobilization. Nevertheless, the specific xanthine oxidase inhibitor allopurinol, and the antioxidant dimethylthiourea, ameliorate rat cortical necrosis in an acute middle cerebral artery occlusion model *(60)*, and the active allopurinol metabolite oxypurinol attenuates $^\bullet OH$ electron paramagnetic resonance (EPR) signals detected in rat cerebral cortex after ischemia-reoxygenation *(61)*. These data, plus evidence that allopurinol protects rat cerebellar granule cells in vitro *(2)* and rat olfactory cortex in vivo *(62)* from kainate toxicity, have supported the interpretation that radical production from xanthine oxidase contributes, at some level, to excitotoxicity, neuronal reoxygenation injury, and, perhaps, neurodegenerative diseases *(3,10,11,56)*.

In addition to their inhibitory effects on xanthine oxidase and other pyrrolases, both allopurinol and its metabolite oxypurinol can serve as effective •OH sinks *(63)*, suggesting that the ability of allopurinol to protect neurons against kainate toxicity could be due to •OH scavenging independent of xanthine oxidase inhibition. However, given the low aqueous solubility coefficients of both allopurinol and its metabolite, plus the low doses used, it is not likely that either one attains high enough concentrations in the cell to serve as effective mass action radical scavengers. Rather, it is more reasonable to ascribe the protective effect of allopurinol to its primary function, i.e., inhibition of adenylate catabolism and prevention of purine efflux. By forestalling degradation of the purine backbone required for the adenylate, allopurinol correspondingly facilitates ATP recharge and recovery upon reoxygenation in the case of transient hypoxia, or upon cessation of stimulation in the case of excitotoxicity.

In light of the extremely low xanthine dehydrogenase/oxidase activities in neurons, the extreme aerobic poise of the brain, and the Ca^{2+} sensitivity of mitochondrial function, the author has proposed *(64–66)* that radical production from mitochondrial electron transfer is a more likely candidate for oxidative stress during excitotoxicity, hypoxia, and chronic neurodegenerative diseases with sporadic occurrence such as Alzheimer's, ALS and Parkinson's. The connection(s) between perturbations in Ca^{2+} homeostasis and mitochondrial dysfunction in the case of excitotoxicity are fairly direct; acute increases in Ca^{2+} have been shown to elicit mitochondrial radical production *(64)* which in turn disrupts normal electron transport and collapses the transmembrane potential *(19,32)*. Indeed, Ca^{2+} influx and mitochondrial Ca^{2+} release upon NMDA receptor stimulation has already been shown to collapse the mitochondrial membrane potential via the permeability transition mechanism in cultured forebrain neurons *(67)*, and to elicit free radical production from cerebellar granule cells *(57)*. On the other hand, the link between mitochondrial dysfunction in late-onset neurodegenerative disease is more circuitously a function of mitochondrial radical production, oxidatively induced impairment of electron transfer components, oxidatively induced and inherited mutations in the mitochondrial genome, and the dynamics of mitochondrial populations in long-lived, terminally differentiated cells, like neurons. As such, the specifics of Ca^{2+} homeostasis, mitochondrial radical production, and the connection(s) between mitochondrial failure and neurodegenerative disease are the focus of the remainder this chapter.

MITOCHONDRIAL RADICAL PRODUCTION

Ubiquinone as Source of Free Radicals

Free radicals are produced by the mitochondrial electron transfer system in direct proportion to ambient P_{O2} and to the rate of O_2 utilization *(68–71)*. In addition, mitochondrial radical production is elicited not only by Ca^{2+} loading, and/or transient ischemia in aerobically poised cells such as myocardial myocytes *(48,72–77)* and neurons *(67)*, but also by tumor necrosis factor-α *(78)*.

From a thermodynamic perspective, any of the mitochondrial electron transfer components, all of which have lower reduction potentials than molecular O_2, are capable of reducing O_2 to O_2^- ($E^{o'}$ for electron transfer components range from -0.32 V for NADH–CoQ reductase to $+0.56$ V for cytochrome c oxidase, vs $+0.82$ V for O_2 reduction). Although

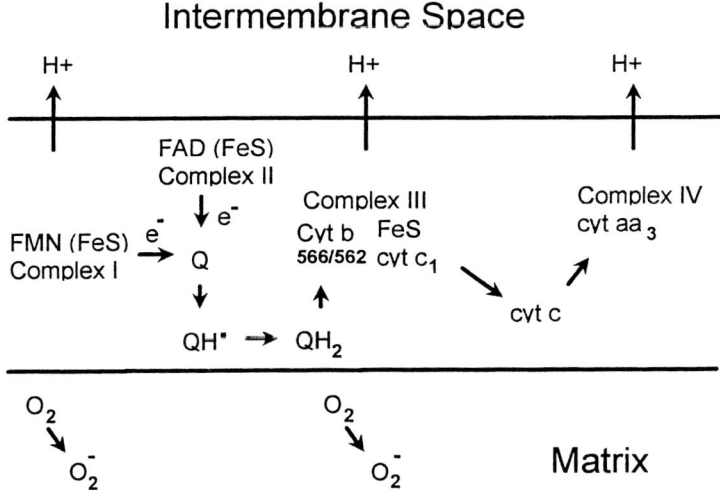

Fig. 1. Schematic of electron transfer components within inner mitochondrial membrane. Although only complex I and II (succinate dehydrogenase) are pictured, ubiquinone is also reduced by electron-transfer-flavoprotein dehydrogenase and glycerol-3-phosphate dehydrogenase. Two other transmembrane complexes, F_0F_1-ATPase (complex V) that couples potential energy of the electrochemical proton gradient to phosphorylation of ADP to ATP, and the adenine nucleotide translocase that exchanges matrix ATP for cytosolic ADP, are not shown. Electron flow through Complex III is likely via two catalytic sites, with oxidation occurring at Q_o (for proton output), and reduction at Q_i (for proton input). This double-occupancy model of the Q cycle is summarized by Ding et al. *(140)*. Although often depicted as a linear sequence, the various complexes and carriers are not present in equimolar amounts. Rather, electron transfer among them is dictated not only by reduction potentials, but also by random collisions, and hence by rates of lateral diffusion, within the inner membrane. When individual diffusion coefficients are considered, the functional ratio among the various components approaches unity *(116)*.

redox reactions within the normally functioning electron transfer system occur via paired electron transfers, orbital spin restrictions dictate that O_2 reduction proceed univalently, the first product of which is O_2^- (reviewed in ref. *20*). Most studies of mitochondrial O_2^- and H_2O_2 production assess the effect of various inhibitors of electron transfer based on the reasoning that "downstream" blockade will increase the number of "upstream" carriers in the reduced state, and correspondingly increase the probability of their undergoing autoxidation and redox cycling. For example, using rotenone and antimycin A, inhibitors of electron transfer that block electron transfer between ubiquinone (coenzyme Q) and cytochrome b, and between cytochrome b and c1, respectively, the early circumstantial data indicated that NADH-CoQ reductase (complex I) and ubiquinone were the two most likely sources for radical production (Fig. 1) *(68,79)*.

Although thermodynamics predicts that all the electron transfer components are capable of reducing O_2, it can not predict the kinetic feasibility of a given reaction in a complex biological system, a problem which has opened debate on whether any electron transfer component has physical access to molecular O_2, and hence the opportunity to actually generate O_2^- and H_2O_2. For example, little doubt remains that the univalently reduced radical form of Coenzyme Q, ubisemiquinone, is an intermediate step in both ubiquinone formation and reversible electron transfers in the electron transfer system *(80,81)*. Loschen *(69)* and others *(70–77)* concluded that the source of mitochondrial radical production was between succinate dehydrogenase and cytochrome b-c_1, and attention focused on the univalently reduced ubisemiquinone as a likely source of electrons for O_2 reduction. However, based on its hydrophobic nature and the aprotic milieu of its intramembrane location, Nohl has argued *(82,83)* that neither Coenzyme Q nor ubisemiquinone is able to interact with ambient molecular O_2, thereby eliminating it as a primary source for radical production. For example, intact mitochondria in isosmotic aprotic media (acetonitrile) show a stable EPR spectrum of membrane-bound ubisemiquinone despite the ready availability of molecular O_2 (reviewed in ref. *82*). Addition of water to the media, however, causes the ubisemiquinone signal to decrease as carbon-centered ·OH, and O_2^- EPR signals begin to increase *(83)*, indicating that transfer of electrons between the semiquinone and O_2 is occurring. Comparable studies during hypoxia-reoxygenation of mitochondria isolated from myocardium also show loss of the semiquinone signal, an effect coincident with the appearance of radical signals *(83)*. Such radical production is likely due to alterations within the membrane that permit ubisemiquinone to move closer to protons in the aqueous phase, and not to increased fluidity or permeability of the inner membrane to protons *(84)*. Regardless of whether the mechanism entails direct redox coupling or an indirect transfer of electrons (*see* ref. *85*), during normal respiration under physiologically relevant conditions, components of the mitochondrial electron transfer system between ubisemiquinone and cytochrome c_1 in complex III are capable of univalently reducing O_2 (Fig. 1). Moreover, production of oxyradicals from this region of the respiratory system is exacerbated by numerous pathologically relevant perturbations, such as transient hypoxia, exposure to O_2^- or ·OH, and changes in cytosolic Ca^{2+} that correspondingly increase mitochondrial Ca^{2+} *(see below)*.

Complex I as a Source of Mitochondrial Free Radicals

Complex I (NADH-Coenzyme Q reductase) has also been identified a site of electron leakage to molecular O_2 *(75)*. Complex I consists of at least 40 different protein subunits (only 7 of which are encoded in the mitochondrial genome) *(33,34,86,87)*, at least six, and perhaps eight, nonheme Fe–S centers, plus noncovalently bound FMN, and two species of bound ubiquinone. Rates of mitochondrial H_2O_2 and O_2^- production increase significantly when electron transfer from the NADH-dehydrogenase to Coenzyme Q is blocked with antimycin A, a pattern suggesting a site capable of oxidation within the complex that is accessible to molecular O_2 *(77)*. A flavin radical signal is detected using either EPR or potentiometric techniques upon complete reduction of complex I with NADH (reviewed in ref. *77*), and such a signal can only arise from single or, less likely, odd numbered polyvalent reductions, despite the obligatory initial two-electron redox nature of NADH *(88)*. Regardless of the mechanisms of internal electron trafficking within the complex, addition of NADH or succinate to rat liver submitochondrial particles induces production of hydroxyl and carbon-centered radicals when electron flow from complex I is impeded using antimycin A *(89)*, but also formation of organic radicals in the absence of such blockade *(90)*. It is intriguing to note that complex I activity is elevated in familial ALS patients *(36)*, yet it is diminished in Parkinson's patients *(91)*, perhaps as a consequence of oxidative inactivation (*see* refs. *36,64,92*).

CALCIUM REGULATION AND MITOCHONDRIAL RADICAL PRODUCTION

Mitochondria isolated from adult rat cerebral cortex and cerebellum produce a variety of radicals, including ˙OH- and carbon-centered species, when exposed to elevated Ca^{2+} concentrations typical of those occurring in the cytosol during excitotoxic receptor stimulation (64,65). The EPR data indicate that these signals are likely due to a mixture of hydroxyl and lipid peroxyl radical adducts (64) and that they are not due to hypoxia; no EPR signals are detected from mitochondria where P_{O2} was buffered by passing O_2 over the sample in an O_2-permeable Teflon tube, until Ca^{2+} and Na^+ were increased (64,65).

This radical production requires sustained Ca^{2+} exposure, and does not occur when Ca^{2+} is increased only briefly. Interestingly, morbidity of hippocampal neurons following glutamate exposure in vitro is also precipitated by a gradual sustained increase in cytosolic Ca^{2+} that occurs hours after an initial transient increase (93). When Ca^{2+} alone is increased, EPR signals from isolated neuronal mitochondria are 80% of those obtained when both Ca^{2+} and Na^+ are increased, a pattern suggesting that the mitochondrial radical production is likely elicited by Ca^{2+} alone and raising the question of how Ca^{2+}, a physiologically essential second messenger, could so profoundly disrupt equally essential mitochondrial function.

It appears that a continuous slow cycling of Ca^{2+} and Na^+ occurs across the mitochondrial inner membrane, and that normal alterations of intramitochondrial Ca^{2+} are associated with normal metabolic regulation (reviewed in refs. 19 and 30). In cardiac myocytes, fluctuating levels of mitochondrial free Ca^{2+} between 0.2–2 μM are responsible for increases in oxidative metabolism in response to increased activity, due primarily to allosteric regulation of enzymes (reviewed in ref. 94) and the glycerophosphate shuttle (18).

Bearing in mind caveats concerning activity vs concentration for bound ions, cytosolic Ca^{2+} concentration is typically 50–100 nM, depending on cell type. When Ca^{2+} levels reach 200–300 nM, mitochondria begin to accumulate Ca^{2+} as a function of the equilibrium between influx via a Ca^{2+} uniporter vs efflux via both Na^+-dependent and Na^+-independent carriers. Ca^{2+} influx via the uniporter is fast, passive and second order for Ca^{2+}, indicating initial potentiation of influx by Ca^{2+} (Fig. 2). Although the K_m of the uniporter for Ca^{2+} is substantially higher than typical cytosolic Ca^{2+}, evidence from studies using several tissues and different assay techniques indicates that, despite such a low affinity, intramitochondrial Ca^{2+} does fluctuate in parallel with changes in cytosolic Ca^{2+}. Indeed, it has been argued that such a low affinity for so rapid a uniport mechanism indicates that its primary function is to lower cytosolic Ca^{2+} when it is pathologically elevated due to ATP depletion and/or abnormal influx across the plasma membrane (18,19).

Ca^{2+} influx is completely dependent upon the negative transmembrane electrochemical potential ($\Delta\Psi$) established by electron transfer, and influx fails to occur in the absence of $\Delta\Psi$ even when an eightfold Ca^{2+} concentration gradient is imposed (95). A significant consequence of such dependence on $\Delta\Psi$ is that mitochondria release Ca^{2+} via the uniporter when the membrane potential is dissipated, as occurs with uncouplers like 2,4-dinitrophenol and carbonyl cyanide p-trifluoro-methoxyphenylhydrazone (FCCP). Important in the present context, in addition to Ca^{2+}, the uniporter can also transfer a host of other divalent cations, including Fe^{2+}. Because O_2^- increases free

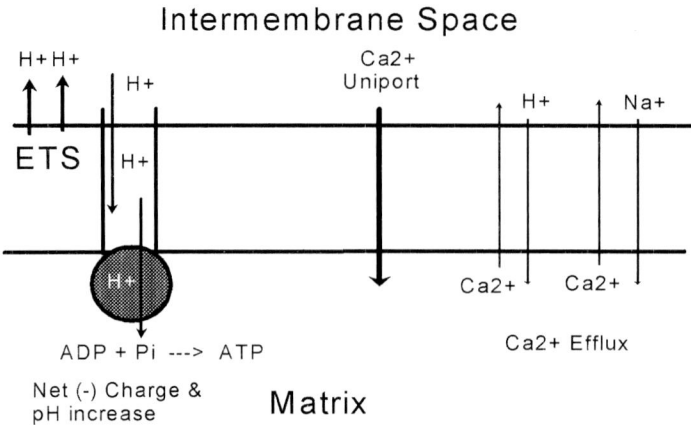

Fig. 2. Mitochondrial Ca^{2+} homeostasis reflects the equilibrium between rapid Ca^{2+} influx down an electrochemical gradient via a uniporter, versus active efflux via a Na^+-dependent, and Na^+-independent ion exchangers. The latter rely on favorable gradients across the inner membrane for Na^+ and H^+, respectively, for the energy to pump Ca^{2+} out. Although no direct hydrolysis of ATP is required to efflux Ca^{2+}, the energy needed to maintain a favorable Na^+ gradient, and the use of H^+ as a counter ion, both represent a loss of potential energy to the ATP pool. As such, even transiently elevated cytosolic Ca^{2+} results in expenditure of energy to regulate intramitochondrial Ca^{2+}.

Fe^{2+} availability by releasing it from protein carriers (20), such oxidative stress could result in increased mitochondrial Fe^{2+} uptake, thereby exacerbating intramitochondrial Fenton reactions.

Given the huge potential for rapid Ca^{2+} influx by the uniporter, regulation of intramitochondrial $[Ca^{2+}]$ falls to the efflux mechanisms, of which there are two. Both are active, but substantially slower than the uniporter; the V_{max} of the Ca^{2+} uniporter is three orders of magnitude higher than the Na^+-independent Ca^{2+} efflux mechanism, and some 100-fold faster than the Na^+-dependent mechanism (reviewed in ref. 19) (Fig. 3). Most cell types contain both the Na^+-independent and Na^+-dependent carriers, but the proportion of the faster Na^+-dependent protein increases in cells that normally experience rapid changes in cytosolic Ca^{2+} such as neurons and cardiac myocytes. Although initially considered passive ion exchangers, both efflux mechanisms pump Ca^{2+} out of mitochondria against an electrochemical gradient using energy derived from opposing favorable gradients for Na^+ (in the case of the Na^+-dependent mechanism) and H^+ (for the Na^+-independent mechanism). Use of the proton motive gradient to fuel Ca^{2+} efflux represents a loss of phosphorylation potential because the exchanged H^+ bypasses F_0F_1-ATPase. Likewise, activity of the inner membrane Na^+–H^+ exchanger that maintains a favorable mitochondrial Na^+ gradient for Ca^{2+} efflux is not associated with ATP hydrolysis (but *see* ref. 18).

To the extent that the H^+ gradient is used to maintain Ca^{2+} disequilibrium and therefore not for phosphorylation of ADP via F_0F_1-ATPase, Ca^{2+} regulation is an energy sink that is subject to rapid alterations. When cytosolic Ca^{2+} levels increase abnormally—whether due to opening of voltage gated NMDA channels, failure of plasma

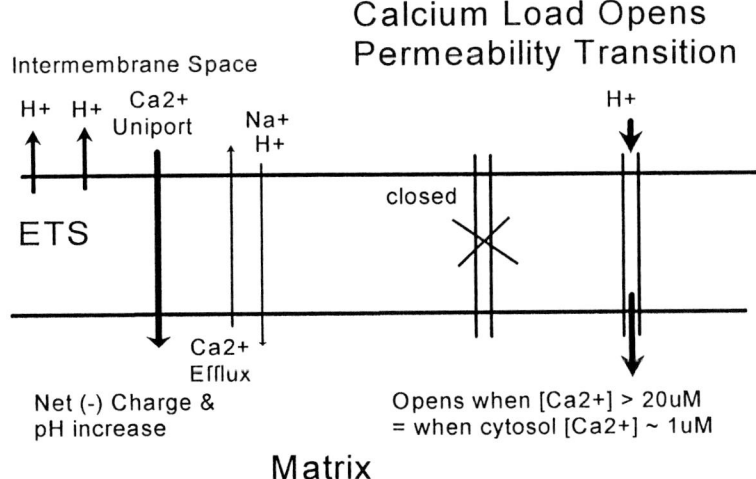

Fig. 3. When cytosolic Ca^{2+} exceeds approximately 1 μM, a proteinaceous pore in the inner membrane opens collapsing the transmembrane electrochemical gradient and blocking oxidative phosphorylation. Oxidants such as H_2O_2 potentiate opening of this permeability transition pore, while ATP serves to forestall its opening. As such, oxidative stress may imperil mitochondrial function by rendering it more susceptible to perturbations in Ca^{2+} homeostasis.

membrane ion-dependent ATPases from lack of ATP, or oxidative damage to the plasma membrane—the energy required to pump Ca^{2+} out of the mitochondria is lost to the adenylate pool in direct proportion to the Ca^{2+} load, and this can become a significant energetic drain (reviewed in ref. 96). As cells age and the potential for ATP production declines *(see below)*, Ca^{2+} loading that would have previously been minor may become increasingly difficult to compensate for, and the resulting increases in intramitochondrial Ca^{2+} can have lethal consequences because of irreversible mitochondrial collapse.

PERMEABILITY TRANSITION AND MITOCHONDRIAL FAILURE

Intramitochondrial Ca^{2+} sequestration and removal during pathological overloading of cytosolic Ca^{2+} results in futile energy dissipation, whereas acutely increased intramitochondrial Ca^{2+} levels (~25 μM) are capable of completely destroying the integrity of the inner mitochondrial membrane and eliciting a rapid increase in mitochondrial radical production that presages complete loss of mitochondrial function and cell death *(97,98,* reviewed in refs. *94* and *99)*. Such Ca^{2+} levels are not impossible; Ca^{2+}-activated fluorescent dyes such as fura 2 indicate that intramitochondrial Ca^{2+} is between 2- and 5-fold higher than cytosolic [Ca^{2+}] *(18,19*; but *see* ref. *99)*, and in respiring mitochondria, an intramitochondrial Ca^{2+} concentration of 25 μM could be established when external Ca^{2+} is 1 μM *(98,100)*.

This collapse of the mitochondrial transmembrane potential is due to the opening of a nonselective proteinaceous pore in the inner membrane, likely associated with adenylate translocase. This permeability transition (PT) also causes mitochondria to swell because of osmotic pressures. The ability of Ca^{2+} to open the permeability transition pore is dramatically synergized by a variety of physiologically pertinent "inducing agents," including hydroperoxide and free radicals *(101–103)*. Indeed, augmentation of mitochondrial antioxidant potential protects against Ca^{2+}-induced PT (reviewed in ref. *18*), although efficacy of some ostensible antioxidants may be due to more circuitous mechanisms such as membrane stabilization rather than repression of radicals *(104)*. In liver and myocardial mitochondria, the PT pore is selectively inhibited by the immunosupressive agent cyclosporin A; no comparable studies have been conducted in neuronal mitochondria *(67)*.

Loss of membrane potential causes mitochondria to release Ca^{2+}, thereby increasing the Ca^{2+} load on nearby mitochondria and setting up a chain reaction *(105)*. Of course, independent of the pathological sequelae of PT collapse—including increased radical production from uncoupled electron transfer—the ensuing loss of ATP *per se* is also potentially lethal to aerobically poised cells such as neurons *(106)*. Moreover, aside from the obvious problems caused by depletion of cellular energy, lack of ATP exacerbates PT collapse; exogenous ATP (but not ADP or AMP) prevents PT pore opening in rat cardiac mitochondria even when Ca^{2+} is 5 μM, a concentration normally sufficient to cause pore opening *(98)*. In any event, permeability transition collapse of the mitochondrial potential, the consequent uncoupling of electron transfer and increase in radical production, plus the loss of phosphorylation potential—all results of perturbations in Ca^{2+} regulation—carry both direct, and indirect, dire consequences for neurons *(107)*.

PHYSIOLOGICAL CONSEQUENCES OF RADICAL EXPOSURE

Severe radical exposures and excess Ca^{2+} loading rapidly destroy mitochondrial ATP generation and Ca^{2+} regulation *(96)*. However, less extreme Ca^{2+} or radical exposures are not without deleterious consequences to both electron transfer and subsequent mitochondrial free radical efflux *(105,107)*. For example, Fe-S centers in the electron transfer complexes are exquisitely susceptible to radical-mediated inactivation *(20,108)*, as are enzymes such as pyruvate dehydrogenase, which shows inactivation following ischemia-reoxygenation in canine cerebral cortex *(37,43,45)*. Moreover, mitochondrial transcription is extremely sensitive to oxidative inactivation by a variety of pro-oxidants *(46)*. As Vroegop et al. *(109)* point out, mitochondria are susceptible to different oxidants in different ways, with H_2O_2, cumin hydroperoxide and 6-hydroxydopamine exerting various toxic effects at different neuronal mitochondrial sites.

Exposure of mitochondria isolated from adult rat cerebral cortex or cerebellum to hydroxyl radicals (75 μmol H_2O_2 plus 1 mM Fe^{2+}-EDTA) in the presence of 2.5 μM Ca^{2+} impairs function of complex I, reducing respiration rates fueled by glutamate and malate by 62% *(64,65)*. Likewise, exposure to ˙OH decreases by half the EPR signals detected from rat cortical mitochondria with complex I substrates. Indeed, bovine heart mitochondrial complex I, NADH oxidase, succinate dehydrogenase, succinate oxidase and F_0F_1-ATPase are all exquisitely susceptible to inactivation by ˙OH (not surprising given its extreme reactivity), while complex I, NADH oxidase, and ATPase are also substantially inactivated by O_2^- despite the latter being a less reactive radical than ˙OH *(108)*. The adenine nucleotide translocase of the inner membrane is degraded under oxidative conditions that cause only mild peroxidation of inner membrane lipids *(110,111)*. In addition, many enzymes including mitochondrial ATPase *(112)* are inactivated or impaired when tyrosine residues are nitrated, a likely result of peroxynitrite undergoing heterolytic cleavage in a reaction catalyzed by transition metals *(40)*.

In addition to alterations of enzyme function by radical exposure, peroxidation of membrane lipids is another potentially deleterious consequence of radical production that overwhelms the antioxidant potential. Mitochondrial membrane lipid peroxidation, loss of structural integrity and functional impairment are all detected in rat hepatocytes after O_2^- and •OH exposures *(113)*. In rat brain mitochondria, transient ischemia-reoxygenation causes differential hydrolysis of mitochondrial membrane phospholipids *(114)*. Cyanide treatment in vivo also induces lipid hydroperoxides (conjugated dienes) in brain, but not in liver or heart, a pattern indicating that there are real tissue differences in susceptibility to oxidative stress (*see* EPR data below) *(115)*. Moreover, removal of exogenous Ca^{2+}, or pretreatment of cortical slices with the Ca^{2+} channel blocker diltiazem, diminishes CN-induced lipid peroxidation *(115)*, an effect indicating that perturbations in Ca^{2+} regulation contribute to the pathology observed. In addition to permeability issues (crucial to mitochondrial $\Delta\Psi$), peroxidation and consequent alterations in membrane fluidity likely alter diffusion coefficients of the electron transfer components, and thus could undermine coordinated electron flow that is dependent on lateral diffusion within the inner mitochondrial membrane *(116)*.

Oxidative susceptibility of complex I and repression of state 3 respiration have also been reported for isolated rat brain mitochondria exposed to the 1-methyl-4-phenylpyridinium (MPP^+) or •OH *(117)*. Inhibition of complex I by MPP^+ is blocked by glutathione, catalase and ascorbate, underscoring an oxidative etiology *(118)*. Impairment of oxidative phosphorylation by MPP^+ induces ATP depletion and compensatory lactate accumulation in vivo, and generates lesions in the rat striatum *(119)*. These lesions are prevented by MK-801, indicating that blockade of NMDA-receptor, and hence Ca^{2+} perturbations and ensuing oxidative mitochondrial dysfunction, likely contribute to neuronal death under these circumstances.

Hydroxyl exposure restricts electron entry via complex I, but when succinate is provided as substrate for electron entry via complex II, such treatment significantly increases respiration of neuronal mitochondria and, more importantly, also increases radical production by almost ninefold *(64,65)*. Comparable increases in radical production are also reported after t-butyl-hydroperoxide exposure, when isolated rat liver mitochondria are fueled with succinate *(120)*. These data support the study mentioned above showing that O_2^- or •OH exposure impairs complex I respiration of isolated liver mitochondria, yet has little effect on succinate respiration *(121)*. Similar results have also been reported for isolated renal mitochondria where exposure to •OH plus 30 μM Ca^{2+} significantly impairs complex I activity and uncouples oxidative phosphorylation, while it has little effect on electron entry via complex II *(28)*.

In a rat strain with congenitally elevated mitochondrial radical production, state 3 respiration, uncoupled respiration, respiratory control ratios, and mitochondrial membrane potential are all reduced *(122)*. Surprisingly, no differences between respiration supported with glutamate/malate or succinate are detected in mitochondria from these animals. Moreover, respiration with complex I substrates declines in synaptic and nonsynaptic mitochondria isolated from normal rat brain as the animals age from 3 to 28 mo, although succinate-fueled respiration shows no comparable reductions *(123)*. In any event, it is apparent not only that respiration proceeds at a high rate following combined Ca^{2+} and radical exposure when complex II substrates are provided, but also that such treatment increases subsequent free radical generation from mitochondria isolated from neurons as well as other cells.

NEURODEGENERATIVE DISEASES AS SYSTEMIC ILLNESSES

Impairment in complex I activity comparable to that induced by $^{\bullet}$OH exposure *(64)* is reported in mitochondria from substantia nigra and systemic tissues of Parkinson's disease patients *(16,91,92,124*; but *see* ref. *125)*, from platelets of Huntington's disease patients *(126)*, and from neocortex of Alzheimer's patients (*127*, reviewed in ref. *7)*. In addition, activities of complexes II and IV are reportedly reduced in the basal ganglia of Huntington's patients, although reductions in complex II-III, with no apparent changes in complex I or IV, are also reported (reviewed in ref. *15)*. Compared to age-matched controls, Alzheimer's patients show reductions in complex IV activity in platelets *(128)* and postmortem cortex *(129)*. Mitochondria isolated from skin fibroblasts of Alzheimer's patients sequester less Ca^{2+} than mitochondria from age matched normal controls, yet take up more Ca^{2+} following exposure to Fe^{2+} as a source of oxidative stress *(130)*.

Although complex IV activity is reduced in postmortem Alzheimer's brain tissue, oxygen utilization by mitochondria isolated from cerebral biopsies from Alzheimer's patients is elevated, compared to normal cortex, under conditions of low ADP availability *(45,131)*; (reviewed in refs. *7* and *10)*. These data suggest that electron transfer is partially uncoupled in Alzheimer's disease *(127)*, but the polarographic techniques used in these studies do not resolve *how* O_2 was reduced. Univalent reduction of O_2 to yield oxygen-centered radicals and/or peroxide, as detected using EPR spin traps *(64,65)*, rather than normal tetravalent reduction to water, could well be responsible for excess O_2 utilization by mitochondria in Alzheimer's disease (and could be occurring in other neurodegenerative diseases regardless of oxygen utilization rates). Such radical production could be germane to observations of decreased mitochondrial function and increased mutation rates and oxidative damage in mitochondrial DNA in the context of aging and neurodegeneration *(51,132,133)*.

Finally, increased reliance on substrate-level phosphorylation from both glycolysis and the Kreb's cycle as a compensatory response to impaired oxidative phosphorylation offers a plausible explanation for the increased $^{14}CO_2$ production reported in neocortical biopsies from Alzheimer's patients *(45,134)*, and for elevated brain lactate levels detected with MRI in Huntington's disease patients *(135)*.

A consistent, and crucial, observation in the above data is that mitochondrial defects in patients with neurodegenerative diseases are also found in systemic tissues, raising the possibility that such diseases are systemic in nature and that the CNS is just the primary site of frank pathology. The brain, rich in unsaturated lipids susceptible to peroxidation, is responsible for 20% of total O_2 consumption, yet represents only 2% of the total biomass. Such elevated aerobic respiration exerts an increased oxidative load, yet brain antioxidants such as superoxide dismutase and catalase are lower than in other, less aerobically poised tissues *(20,23,136)*. Moreover, most of the brain glutathione and SOD are located in glia, not in neurons *(137)*, rendering glia resistant to oxidative insults *(138)*.

Neuronal mitochondria are similar to myocardial mitochondria in that both show rapid fluctuations in $[Ca^{2+}]$ in response to changes in cytosolic Ca^{2+}, primarily because of an increased proportion of the faster Na^+-dependent Ca^{2+} efflux mechanism. Liver mitochondria, on the other hand, respond more slowly to changes in cytosolic Ca^{2+}

Free Radicals in Neurodegeneration

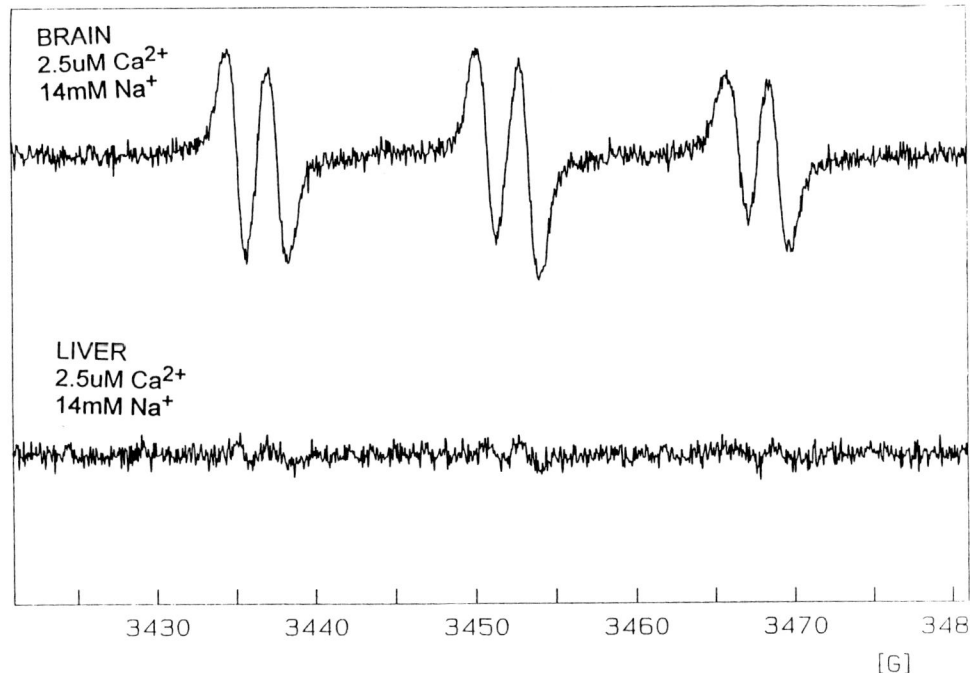

Fig. 4. Electron paramagnetic resonance (EPR) spectra from intact, coupled mitochondria isolated from adult rat liver in the presence of 10 mM succinate, 10 mM ADP, and 50 mM of the spin trap POBN. Minuscule POBN radical signals, just resolvable above noise, are detected in the presence or absence of 2.5 µM Ca^{2+} and 14 mM. However, under these same conditions, identically coupled mitochondria isolated from adult rat cerebellum generate a well-defined radical signal (a_N = 15.7 G, a_H = 2.5 G) indicative of a mixture of hydroxyl and carbon-centered radicals *(64,65)*. No EPR signals are observed from the combined buffers and substrates in the absence of mitochondria. EPR settings: receiver gain 5 × 10^6, scan width 45 G centered at 3465 G, modulation amplitude 1.013 G, time constant 1.26 ms, microwave power 40 mW, 20.5 s scan time with signal averaging over 16 scans. The spectra were not subjected to filtering.

(19). Such differences in response to prolonged Ca^{2+} exposure offers a testable mechanism for observed differences in radical production in brain versus liver mitochondria *(65,66)*; under conditions of increased Ca^{2+} and ADP, mitochondria isolated from either cerebral cortex or cerebellum generate free radicals *(64,65)*, whereas identical treatment fails to elicit radical production from identically isolated liver mitochondria (Fig. 4). The most parsimonious explanation for this disparity is that liver mitochondria contain more antioxidant capacity, thereby moderating radical production, and liver mitochondria do indeed contain more SOD activity than brain mitochondria.

CONCLUSION: A UNIFYING THEORY FOR AGE-RELATED, OXIDATIVE NEURODEGENERATION

If congenital mitochondrial defects are systemic in neurodegenerative diseases, neurons may be at particular risk, and the CNS consequently the primary site of pathology, in part because of lower antioxidant potential in the CNS, but in large measure because neuronal mitochondria generate free radicals under conditions where mitochondria in other tissues do not.

The oxidative pathophysiology of the brain is indeed different from other tissues. In addition to studies showing tissue differences in CN^- toxicity *(115)*, when rats are fed 36% of their daily caloric intake as ethanol, state 3 respiration declines, and O_2^- efflux increases with succinate as substrate in mitochondria isolated from brain, but not heart or liver *(139)*. Nevertheless, the evidence indicates that oxidative load from mitochondria increases with age, and the brain is the first organ to show oxidative degradation in aging *(50–52)*.

Neurons may be at a higher risk for oxidative injury because of a paucity of antioxidant defenses and because their mitochondria are more prone to generate free radicals when Ca^{2+} and ADP are elevated. But this does not explain why systemic genetic defects should result in neuronal oxidative pathology only late in life.

Acceleration of mitochondrial radical generation with age are reflected by increasing oxidative damage to nuclear and mitochondrial DNA; mutation rates in the mitochondrial genome are about 10-fold higher than in nuclear DNA. This is no doubt due to a variety of factors, such as: close proximity to the inner membrane site of radical production, lack of histones in mitochondrial DNA, poor DNA repair mechanisms, oxidative susceptibility of mitochondrial transcription mechanisms, and the absence of nonencoding regions so that oxidatively induced mutations are unavoidably transcribed *(33,34,46,52)*. Although mitochondria contain between 10–20 separate copies of DNA, accumulation of oxidatively induced mutations increases with age across the entire mitochondrial genome, not only in a few selected copies, so that the probability of expressing a defective protein increases correspondingly. Eventually, a point is reached where the mutational burden is such that defective proteins are the rule, not the exception. This results in reductions in the oxidative phosphorylation potential which, combined with decreased antioxidant defenses and diminished capacity to maintain Ca^{2+} homeostasis, plus increased oxidative load, puts the neuron at risk.

In addition to oxidatively induced mutations, congenital defects in electron transfer components become increasingly prominent over time. Such genetic defects need not be present in all copies of DNA in all mitochondria, but are more likely dispersed among subpopulations of mitochondria in the ovaplasm (*see* chapter by Flanigan and Ratan). At fertilization, the sperm typically contributes less than 1% of the total mitochondrial complement to the zygote, with the vast majority of the mitochondria inherited maternally from the ovum. Such patterns of inheritance neatly explain the non-Mendelian and sporadic distribution of many late-onset neurodegenerative diseases, including Alzheimer's, Parkinson's, and ALS.

In any event, after segregation into tissues during embryogenesis, the resulting mixed distribution of mitochondria with mutations and those without (a circumstance termed heteroplasmy) would remain stable in the absence of selective pressures within the cell. However, mitochondrial populations are not static, and highly coordinated communication between the nucleus and mitochondria results in mitochondrial reproduction via fission, with an organelle half-life in normal hepatocytes of approx 5 d (*de novo* mitochondrial synthesis does not occur). DNA replication precedes fission, and there is little evidence of DNA exchange between individual mitochondria.

Although the signal initiating mitochondrial reproduction remains unknown, it is likely a marker of mitochondrial senescence, probably related to reduced potential for oxidative phosphorylation or increased oxidative damage. Compared to normal mitochondria, those organelles expressing genetic mutations (regardless of whether congenital or oxidatively induced) are more likely to exhibit decreased oxidative phosphorylation and/or increased radical production, and are therefore more likely to be replaced. If mitochondria expressing mutations are replaced via reproductive fission more often than normal organelles, over time this reproductive "advantage" will cause an inexorable drift of the initially heteroplasmic mix of healthy and defective organelles toward homoplasmy of defective mitochondria, until a critical threshold is achieved where the cell is energetically or oxidatively imperiled.

The rapidity with which this threshold will be obtained is dictated by the nature of the mutations and by the proportion of mitochondria initially carrying them; clearly, the larger the percentage of mitochondria carrying defective copies of DNA, the sooner homoplasmy will be obtained and the earlier will be the onset of disease. Regardless, in the absence of genetic exchange among mitochondria, eventual achievement of homoplasmy becomes inevitable. Such gradual dominance of the mitochondrial population by energetically and/or oxidatively dysfunctional organelles offers a parsimonious explanation for the etiology of late-onset pathology in long-lived cells, such as terminally differentiated neurons. This, combined with tissue specific differences in mitochondrial radical production, may be the real mechanisms whereby systematically distributed mitochondrial defects cause sporadic neurodegenerative diseases.

ACKNOWLEDGMENTS

The author thanks M. N. Dykens for editorial comments, D. Gasparella for typing assistance, and Rajiv Ratan for his patience.

REFERENCES

1. Ratan RR, Murphy TH, Baraban JM. Oxidative stress induces apoptosis in embryonic cortical neurons. *J Neurochem* 1994, **62:** 376–379.
2. Dykens JA, Stern A, Trenkner E. Mechanism of kainate toxicity to cerebellar neurons *in vitro* is analogous to reperfusion tissue injury. *J Neurochem* 1987, **49:** 1223–1228.
3. Choi DW. Glutamate neurotoxicity and diseases of the nervous system. *Neuron* 1988, **1:** 623–634.
4. Blass JP, Gibson GE. The role of oxidative abnormalities in the pathophysiology of Alzheimer's disease. *Rev Neurol Paris* 1991, **147:** 513–525.
5. Beal MF. Role of excitotoxicity in neurological diseases. *Curr Opin Neurobiol* 1992a, **2:** 657–662.
6. Beal MF. Mechanisms of excitotoxicity in neurologic diseases. *FASEB J* 1992b, **6:** 3338–3344.
7. Beal MF, Hyman BT, Koroshetz W. Do defects in mitochondrial energy metabolism underlie the pathology of neurodegenerative diseases? *TINS* 1993a, **16:** 125–131.
8. Davis RE, Miller S, Hernstadt C, Ghosh SS, Fahy, E, Shinobu LA, Glasko D, Thal LJ, Beal MF, Nowell N, Parker WD. Mutations in mitochondrial cytochrome c oxidase genes segregate with late-onset Alzheimer disease *PNAS* 1997, **94:** 4526–4531.
9. Parker WD, Davis RE. Primary mitochondrial DNA as a causative event in Alzheimer's disease, in *Mitochondria and Free Radicals in Neurodegenerative Diseases* (Beal MF, Howell N, Bodis-Wollner I, eds.) Wiley-Liss, New York 1997, pp. 319–334.
10. Coyle JT, Puttfarken P. Oxidative stress, glutamate, and neurodegenerative disorders. *Science* 1993, **262:** 689–695.
11. Schulz JB, Beal MF. Mitochondrial dysfunction in movement disorders. *Curr Opin Neurol* 1994, **7:** 333–339.

12. Götz, ME, Kunig G, Riederer P, and Youdim MBH. Oxidative stress: free radical production in neural degeneration. *Pharmac Ther* 1994, **63:** 37–122.
13. Luft R. The development of mitochondrial medicine. *Proc Natl Acad Sci USA* 1994, **91:** 8731–8738.
14. Schapira AHV, Cooper JM. Mitochondrial function in neurodegeneration and aging. *Mutat Res* 1992, **275:** 133–143.
15. Browne SE, Beal MF. Oxidative damage and mitochondrial dysfunction in neurodegenerative diseases. *Biochem Soc Trans* 1994, **22:** 1002–1006.
16. Swerdlow RH, Parks JK, Miller SW, Tuttle JB, Trimmer PA, Sheehan JP, Bennett JP, Davis RE, Parker WD. Origin and functional consequences of the complex 1 defect in Parkinson's disease. *Ann Neurol* 1996, **40:** 663–671.
17. Chiueh CC, Gilbert CL, Colton CA. (eds.). "The Neurobiology of NO· and ·OH," Annals New York Acad Sci, vol 738, New York, 1995, 470 pp.
18. Gunter KK, Gunter TE. Transport of calcium by mitochondria. *J Bioenerg Biomembr* 1994, **26:** 471–485.
19. Gunter, TE, Gunter KK, Sheu SS, Gavin CE. Mitochondrial calcium transport: Physiological and pathological relevance. *Am J Physiol* 1994, **267:** 313–39.
20. Halliwell B, Gutteridge JMC. "Free Radicals in Biology and Medicine." Oxford University Press, 1989.
21. Siesjö BK, Agardh CD, Bengtsson F. Free radicals and brain damage. *Cerebrovasc Brain Metab Rev* 1989, **1:** 165–211.
22. Bondy SC, LeBel CP. The relationship between excitotoxicity and oxidative stress in the central nervous system. *Free Rad Biol Med* 1993, **14:** 633–642.
23. Halliwell B. Reactive oxygen species and the central nervous system. *J Neurochem* 1992, **59:** 1609–1623.
24. Benzi G, Moretti A. Age- and peroxidative stress-related modifications of the cerebral enzymic activities linked to mitochondria and the glutathione system. *Free Rad Biol Med* 1995, **19:** 77–101.
25. Murphy SN, Thayer SA, Miller RJ. The effects of excitatory amino acids on intracellular calcium in single mouse striatal neurons *in vitro*. *J Neurosci* 1987, **7:** 4145–4158.
26. MacDermott AB, Mayer KL, Westbrook GL, Smith SJ, Barker JL. NMDA-receptor activation increases cytoplasmic calcium concentration in cultured spinal cord neurons. *Nature* 1986, **321:** 519–522.
27. Siegmund B, Schlüter K-D, Piper HM. Calcium and the oxygen paradox. *Cardiovasc Res* 1993, **27:** 1778–1783.
28. Malis CD, Bonventre JV. Mechanism of calcium potentiation of oxygen free radical injury to renal mitochondria. *J Biol Chem* 1986, **261:** 14,201–14,208.
29. Cao W, Carney JM, Duchon A, Floyd RA, Chevion M. Oxygen free radical involvement in ischemia and reperfusion injury to brain. *Neurosci Lett* 1988, **88:** 233–238.
30. Gunter KK, Pfeiffer DR. Mechanisms by which mitochondria transport calcium. *Am J Physiol* 1991, **27:** C755–786.
31. Shier WT, Dubourdieu DJ. Evidence for two calcium-dependent steps and a sodium-dependent step in the mechanism of cell killing by calcium ions in the presence of ionophore A23187. *Am J Path* 1985, **120:** 304–315.
32. Hermes-Lima M. How do Ca^{2+} and 5-aminolevulinic acid-derived oxyradicals promote injury to isolated mitochondria? *Free Rad Biol Med* 1995, **19:** 381–390.
33. Wallace DC. Diseases of the mitochondrial DNA. *Ann Rev Biochem* 1992a, **61:** 1175–1212.
34. Wallace DC. Mitochondrial genetics: a paradigm for aging and degenerative disease? *Science* 1992b, **256:** 628–632.
35. Deng H-X et al. Amyotrophic lateral sclerosis and structural defects in Cu^+, Zn^+ superoxide dismutase. *Science* 1993, **261:** 1047–1051.

36. Bowling AC, Schulz JB, Brown RH Jr, Beal MF. Superoxide dismutase activity, oxidative damage, and mitochondrial energy metabolism in familial and sporadic amyotrophic lateral sclerosis. *J Neurochem* 1993a, **61**: 2322–2325.
37. Stadtman ER. Protein oxidation and aging. *Science* 1992, **257**: 1220–1224.
38. Dykens JA, Wiseman RW, Hardin CD. Phosphotransferase function is conserved during hypoxia-reoxygenation despite impairment of enzyme activity. *J Comp Physiol* 1996, **166**: 359–368
39. McCord JM, Russell WJ. Inactivation of creatine phosphokinase by superoxide during reperfusion injury. *Basic Life Sci* 1988, **49**: 869–873.
40. Beckman JS, Ischiropoulos H, Zhu L, van der Woerd M, Smith C, Chen J, Harrison J, Martin JC, Tsai M. Kinetics of superoxide dismutase and iron catalyzed nitration of phenolics by peroxynitrite. *Arch Biochem Biophys* 1992, **298**: 438–445.
41. Stamler JS, Singel DJ, Loscalzo J. Biochemistry of nitric oxide and its redo-activated forms. *Science* 1992, **258**: 1898–1902.
42. Oliver CN, Starke-Reed PE, Stadtman ER, Liu GJ, Carney JM, Floyd RA. Oxidative damage to brain proteins, loss of glutamine synthetase activity, and production of free radicals during ischemia/reperfusion-induced injury to gerbil brain. *Proc Natl Acad Sci USA* 1990, **87**: 5144–5147.
43. Bogaert YE, Rosenthal RE, Fiskum G. Postischemic inhibition of cerebral cortex pyruvate dehydrogenase. *Free Rad Biol Med* 1994, **16**: 811–820.
44. Gibson GE, Sheu KF, Blass JP, Baker A, Carlson KC, Harding B, Perrino P. Reduced activities of thiamine-dependent enzymes in the brains and peripheral tissues of patients with Alzheimer's disease. *Arch Neurol* 1988, **45**: 836–840.
45. Sims NR, Bowen DM, Neary D, Davison AN. Metabolic processes in Alzheimer's disease: adenine nucleotide content and production of $^{14}CO_2$ from [U-^{14}C]glucose *in vitro* in human neocortex. *J Neurochem* 1983, **41**, 1329–1334.
46. Kristal BS, Chen J, Yu BP. Sensitivity of mitochondrial transcription to different free radical species. *Free Rad Biol Med* 1994, **16**: 323–329.
47. LeBel CP, Ali SF, McKee M, Bondy SC. Organometal-induced increases in oxygen reactive species: The potential of 2', 7' dichlorofluorescin diacetate as an index of neurotoxic damage. *Toxicol Appl Pharmacol* 1990, **104**: 17–24.
48. Nohl H, Stolze K, Napetschnig S, Ishikawa T. Is oxidative stress primarily involved in reperfusion injury of the ischemic heart? *Free Rad Biol Med* 1991, **11**: 581–588.
49. Stohs SJ, Bagchi D. Oxidative mechanisms in the toxicity of metal ions. *Free Rad Biol Med* 1995, **18**: 321–336.
50. Sohal RS, Sohal BH. Hydrogen peroxide release by mitochondria increases during aging. *Mech Ageing Dev* 1991, **7**: 187–202.
51. Bowling AC, Mutisya EM, Walker LC, Price DL, Cork LC, Beal MF. Age-dependent impairment of mitochondrial function in primate brain. *J Neurochem* 1993b, **60**: 1964–1967.
52. Mecocci P, MacGarvey U, Kaufman AE, Koontz D, Shoffner JM, Wallace DC, Beal MF. Oxidative damage to mitochondrial DNA shows marked age-dependent increases in human brain. *Ann Neurol* 1993, **34**: 609–616.
53. Johnson EM Jr., Deckwerth TL, Deshmukh M. Neuronal death in developmental models: possible implications in neuropathology. *Brain Pathol* 1996, **6**: 397–409.
54. Schulz JB, Henshaw DR, Siwek D, Jenkins BG, Ferrante RJ, Cipolloni PB, Kowall NW, Rosen BR, Beal MF. Involvement of free radicals in excitotoxicity in vivo. *J Neurochem* 1995, **64**: 2239–2247.
55. Beal MF, Kowall NW, Ellison DW, Mazurek MF, Swartz KJ, Martin JB. Replication of the neurochemical characteristics of Huntington's disease by quinolinic acid. *Nature* 1986, **321**: 168–171.

56. Reynolds IJ, Hastings TG. Glutamate induces the production of reactive oxygen species in cultured forebrain neurons following NMDA receptor activation. *J Neurosci* 1995, **5:** 3318–3327.
57. Lafon-Cazal M, Pietri S, Culcasi M, Bockaert J. NMDA-dependent superoxide production and neurotoxicity. *Nature* 1993, **364:** 535–537.
58. Nicklas WK. Alteration by kainate of energy stores and neuronal-glia metabolism of glutamate *in vitro*, in *Excitotoxins* (Fuxe K, Roberts P, Schwarcz R, eds.), Plenum Press, New York, 1984, pp. 55–65.
59. Markley HG, Faillace LA, Mezey E. Xanthine oxidase activity in rat brain. *Biochem Biophys Acta* 1973, **309:** 23–31.
60. Martz D, Rayos G, Schielke, GP, Betz AL. Allopurinol and dimethylthiourea reduce brain infarction following middle artery occlusion in rats. *Stroke* 1989, **20:** 488–494.
61. Phillis JW, Sen S. Oxypurinol attenuates hydroxyl radical production during ischemia-reperfusion injury of the rat cerebral cortex; an ESR study. *Brain Res* 1993, **628:** 309–312.
62. Facchinetti F, Virgili M, Contestabile A, Barnabei O. Antagonists of the NMDA receptor and allopurinol protect the olfactory cortex but not the striatum after intracerebral injection of kainic acid. *Brain Res* 1992, **585:** 330–334.
63. Moorhouse PC, Grootveld M, Halliwell B, Quinlan JG, Gutteridge JM. Allopurinol and oxypurinol are hydroxyl radical scavengers. *Fed Eur Biochem* 1987, **213:** 23–28.
64. Dykens JA. Isolated cerebellar and cerebral mitochondria produce free radicals when exposed to elevated Ca^{2+} and Na^+: Implications for neurodegeneration. *J. Neurochemistry*, 1994, **63:** 584–591.
65. Dykens JA. Mitochondrial radical production and mechanisms of oxidative excitotoxicity, in *The Oxygen Paradox* (Davies KJA, Ursini F, eds.), Padova, CLEUP Press, 1995, pp. 453–468.
66. Dykens JA. Mitochondrial free radical production and the etiology of neurodegenerative disease, in *Neurodegenerative Diseases: Mitochondria and Free Radicals in Pathogenesis* (Beal MF, Bodis-Wollner I, and Howell N, eds.), John Wiley, 1997, 29–55.
67. White RJ, Reynolds IJ. Mitochondrial depolarization in glutamate-stimulated neurons: an early signal specific to excitotoxin exposure. *J Neurosci* 1996, **16:** 5688–5697.
68. Boveris A, Chance B. The mitochondrial generation of hydrogen peroxide. General properties and effect of hyperbaric oxygen. *Biochem J* 1973, **134:** 707–716.
69. Loschen G, Azzi A, Richter C, Flohe L. Superoxide radicals as precursors of mitochondrial hydrogen peroxide. *FEBS Lett* 1974, **42:** 68–72.
70. Sanders SP, Squire JL, Kuppusamy P, Harrison SJ, Bassett DJ, Gabrielson EW, Sylvester JT. Hyperoxic sheep pulmonary microvascular endothelial cells generate free radicals via mitochondrial electron transport. *J Clin Invest* 1993, **91:** 46–52.
71. Ueta H, Ogura R, Sugiyama M, Kagiyama A, Shin G. Spin trapping of cardiac submitochondrial particles isolated from ischemic and non-ischemic myocardium. *J Mol Cell Cardiol* 1990, **22:** 893–899.
72. Turrens JF, Beconi M, Barilla J, Chavez UB, McCord JM. Mitochondrial generation of oxygen radicals during reoxygenation of ischemic tissues. *Free Rad Res Commun* 1991, **12–13:** 681–689.
73. Paraidathathu T, de Groot H, Kehrer JP. Production of reactive oxygen by mitochondria from normoxic and hypoxic rat heart tissue. *Free Rad Biol Med* 1992, **13:** 289–297.
74. Paraidathathu T, Palamanda J, Kehrer JP. Modulation of rat heart mitochondrial function and the production of reactive oxygen by vitamin E deficiency. *Toxicology* 1994, **90:** 103–114.
75. Boveris A, Cadenas E. Production of superoxide radicals and hydrogen peroxide in mitochondria, in *Superoxide Dismutase*, Vol. II (Oberley LW, ed.), Boca Raton FL, CRC Press, 1982, pp. 15–30.

76. Ambrosio G, Zweier JL, Duilio C, Kuppusamy P, Santoro G, Glia PP, Tritto I, Cirillo P, Condorelli M, Chiariello M. Evidence that mitochondrial respiration is a source of potentially toxic oxygen free radicals in intact rabbit hearts subjected to ischemia and reflow. *J Biol Chem* 1993, **25**: 18,532–18,541.
77. Nohl H. Generation of superoxide radicals as byproduct of cellular respiration. *Ann Biol Clin Paris* 1994, **52**: 199–204.
78. Schulze-Osthoff K, Bakker AC, Vanhaesebroeck B, Beyaert R, Jacob WA, Fiers W. Cytotoxic activity of tumor necrosis factor is mediated by early damage of mitochondrial functions. Evidence for the involvement of mitochondrial radical generation. *J Biol Chem* 1992, **267**: 5317–5323.
79. Nohl H, Hegner D, Summer KH. The mechanisms of toxic action of hyperbaric oxygenation on the mitochondria of rat heart cells. *Biochem Pharm* 1981, **30**: 1753–1757.
80. De Jong AM, Albracht SP. Ubisemiquinones as obligatory intermediates in the electron transfer from NADH to ubiquinone. *Eur J Biochem* 1994, **222**: 975–982.
81. Miki T, Yu L, Yu CA. Characterization of ubisemiquinone radicals in succinate-ubiquinone reductase. *Arch Biochem Biophys* 1992, **293**: 61–66.
82. Nohl H, Stolze K. Ubisemiquinones of the mitochondrial respiratory chain do not interact with molecular oxygen. *Free Rad Res Commun* 1992, **16**: 409–419.
83. Nohl H. Ischemia/reperfusion impairs mitochondrial energy conservation and triggers O_2^- release as a byproduct of respiration. *Free Rad Res Commun* 1993, **18**: 127–137.
84. Nohl H, Gille L, Schönheit K, Liu Y. Conditions allowing redox-cycling ubisemiquinone in mitochondria to establish a direct redox couple with molecular oxygen. *Free Rad Biol Med* 1996, **20**: 207–213.
85. Beyer RE. An analysis of the role of coenzyme Q in free radical generation and as an antioxidant. *Biochem Cell Biol* 1992, **70**: 390–403.
86. Finel M. The proton-translocating NADH:ubiquinone oxidoreductase: a discussion of selected topics. *J Bioenerg Biomemnr* 1993, **25**: 357–366.
87. Degli-Espoti M, Ghelli A. The mechanism of proton and electron transport in mitochondrial complex I. *Biochim Biophys Acta* 1994, **1187**: 116–120.
88. Sled VD, Rudnitzky NI, Hatefi Y, Ohnishi T. Thermodynamic analysis of flavin in mitochondrial NADH. *Biochemistry* 1994, **33**: 10,069–10,075.
89. Giulivi C, Boveris A, Cadenas E. Hydroxyl radical generation during mitochondrial electron transfer and the formation of 8-hydroxydeoxyguanosine in mitochondrial DNA. *Arch Biochem Biophys* 1995, **316**: 909–916.
90. Fukushima T, Yamada K, Isobe A, Shiwaku K, Yamane Y. Mechanism of cytotoxicity of paraquat I. NADH oxidation and paraquat radical formation via complex I. *Exp Toxicol Pathol* 1993, **45**: 345–349.
91. Schapira AHV, Cooper JM, Dexter D, Clark JB, Jenner P, Marsden CD. Mitochondrial complex 1 deficiency in Parkinson's disease. *J Neurochem* 1990, **54**: 823–827.
92. Sanchez-Ramos JR, Övervik E, Ames BN. A marker of oxyradical-mediated DNA damage (8-hydroxy-2'deoxyguanosine) is increased in nigro-striatum of Parkinson's disease brain. *Neurodegen* 1994, **3**: 197–204.
93. Randall RD, Thayer SA. Glutamate-induced calcium transient triggers delayed calcium overload and neurotoxicity in rat hippocampal neurons. *J Neurosci* 1992, **12**: 1882–1895.
94. Crompton M, Andreeva L. On the involvement of a mitochondrial pore in reperfusion injury. *Basic Res Cardiol* 1993, **88**: 513–523.
95. Kapus A, Szaszi K, Kaldi K, Ligeti E, Fonyo A. Is the mitochondrial Ca uniporter a voltage-modulated transport pathway? *FEBS Lett* 1991, **282**: 61–64.
96. Ferrari R, Pedersini P, Bongrazio M, Gaia G, Bernocchi P, Di Lisa F, Visioli O. Mitochondrial energy production and cation control in myocardial ischemia and reperfusion. *Basic Res Cardiol* 1993, **88**: 495–512.

97. Chacon E, Acosta D. Mitochondrial regulation of superoxide by Ca2+: an alternate mechanism for the cardiotoxicity of doxorubicin. *Toxicol Appl Pharmacol* 1991, **107**: 117–128.
98. Duchen MR, McGuinness O, Brown LA, Crompton M. On the involvement of a cyclosporin A sensitive mitochondrial pore in myocardial reperfusion injury. *Cardiovasc Res* 1993, **27**: 1790–1794.
99. Mehendale HM, Roth RA, Gandolfi AJ, Klaunig JE, Lemasters, JJ, Curtis, LR. Novel mechanisms in chemically induced hepatotoxicity. *FASEB J* 1994, **8**: 1285–1295.
100. Crompton M. The role of Ca2+ in the function and dysfunction of heart mitochondria, in *Calcium and the Heart* (Langer GA, ed.), Raven Press, New York, 1990, pp. 167–197.
101. Novgorodov SA, Gudz TI, Kushnareva YE, Roginsky VA, Kudrjashov YB. Mechanism accounting for the induction of nonspecific permeability of the inner mitochondrial membrane by hydroperoxides. *Biochem Biophys Acta* 1991, **1058**: 242–248.
102. Takeyama N, Matusuo N, Tanaka T. Oxidative damage to mitochondria is mediated by the Ca(2+)-dependent inner-membrane permeability transition. *Biochem J* 1993, **294**: 719–725.
103. Guidox R, Lambelet P, Phoenix J. Effects of oxygen and antioxidants on the mitochondrial Ca-retention capacity. *Arch Biochem Biophys* 1993, **306**: 139–147.
104. Kora S, Sado M, Koike H, Terada H. Are free radicals involved in Ca(2+)-induced membrane damage of mitochondria? *J Pharmacobiodyn* 1992, **15**: 333–338.
105. Darley-Usmar VM, Stone D, Smith D, Martin JF. Mitochondria, oxygen and reperfusion damage. *Ann Med* 1991, **23**: 583–588.
106. Jurkowitz-Alexander MS, Altschuld RA, Hohl CM, Johnson JD, McDonald JS, Simmons TD, Horrocks LA. Cell swelling, blebbing and death are dependent on ATP depletion and independent of calcium during chemical hypoxia in a glial cell line (RPC-1). *J Neurochem* 1992, **59**: 344–352.
107. Marklund SL, Westman NG, Lundgren E, Roos G. Copper- and zinc-containing superoxide dismutase, manganese-containing superoxide dismutase, catalase, and glutathione peroxidase in normal and neoplastic human cell lines and normal human tissues. *Cancer Res* 1982, **42**: 1955–1961.
108. Zhang Y, Marcillat O, Giulivi C, Ernster L, Davies KJ. The oxidative inactivation of mitochondrial electron transport chain components and ATPase. *J Biol Chem* 1990, **265**: 16,330–16,336.
109. Vroegop SM, Decker DE, Buxser SE. Localization of damage induced by reactive oxygen species in cultured cells. *Free Rad Biol Med* 1995, **18**: 141–151.
110. Zwinzinski CW, Schmid HH. Peroxidative damage to cardiac mitochondria: Identification and purification of modified adenine nucleotide translocase. *Arch Biochem Biophys* 1992, **294**: 178–183.
111. Giron-Calle J, Zwinzinski CW, Schmid HH. Peroxidative damage to cardiac mitochondria. *Arch Biochem Biophys* 1994, **315**: 1–7.
112. Guierrieri F, Yagi T, Papa S. On the mechanism of H^+ translocation by mitochondrial H^+-ATPase. Studies with chemical modifier of tyrosine residues. *J Bioenerg Biomembr* 1984, **16**: 251–262.112.
113. Mehrotra S, Kakkar P, Viswanathan PN. Mitochondrial damage by active oxygen species in vitro. *Free Rad Biol Med* 1991, **10**: 277–285.
114. Sun D, Gilboe DD. Ischemia-induced changes in cerebral mitochondrial free fatty acids, phospholipids, and respiration in the rat. *J Neurochem* 1994, **62**: 1921–1928.
115. Ardelt BK, Borowitz JL, Maduh EU, Swain SL, Isom GE. Cyanide-induced lipid peroxidation in different organs: Subcellular distribution and hydroperoxide generation in neuronal cells. *Toxicology* 1994, **89**: 127–137.
116. Hackenbrock, CR. Lateral diffusion and electron transfer in the mitochondrial inner membrane. *Trends Biochem Sci* 1981, **6**: 151–154.

117. Bates TE, Heales SJ, Davies SE, Boakye P, Clark JB. Effects of 1-methyl-4-phenylpyridinium on isolated rat brain mitochondria: Evidence for a primary involvement of energy depletion. *J Neurochem* 1994, **63**: 640–648.
118. Cleeter MW, Cooper JM, Schapira AH. Irreversible inhibition of mitochondrial complex I by 1-methyl-4-phenylpyridinium: Evidence for free radical involvement. *J Neurochem* 1992, **58**: 786–789.
119. Storey E, Hyman BT, Jenkins B, Brouillet E. 1-Methyl-4-phenylpyridinium produces excitotoxic lesions in rat striatum as a result of impairment of oxidative metabolism. *J Neurochem* 1992, **58**: 1975–1978.
120. Kennedy CH, Church DF, Winston GW, Pryor WA. Tert-butyl hydroperoxide-induced radical production in rat liver mitochondria. *Free Rad Biol Med* 1992, **12**: 381–387.
121. Corbisier P, Raes M, Michiels C, Pigeolet E, Houbion A, Delaive E, Remacle J. Respiratory activity of isolated rat liver mitochondria following *in vitro* exposure to oxygen species. *Mech Ageing Dev* 1990, **51**: 249–263.
122. Salganik RI, Shabalina IG, Solovyova NA, Kolosova NG, Solovyov VN, Kolpakov AR. Impairment of respiratory functions in mitochondria of rats with an inherited hyper production of free radicals. *Biochem Biophys Res Commun* 1994, **205**: 180–185.
123. Harmon HJ, Nank S, Floyd RA. Age-dependent changes in rat mitochondria of synaptic and non-synaptic origins. *Mech Ageing & Devel* 1987, **38**: 167–177.
124. Parker WD, Boyson SJ, Parks JK. Abnormalities of the electron transport chain in idiopathic Parkinson's disease. *Ann Neurol* 1989, **26**: 719–723.
125. Di Monte DA, Sandy MS, Jewell SA, Adornato B, Tanner CM, Langston JW. Oxidative phosphorylation by intact muscle mitochondria in Parkinson's disease. *Neurodegen* 1993, **2**: 275–281.
126. Parker WD, Boyson SJ, Luder AS, Parks JK. Evidence for a defect in NADH:ubiquinone oxidoreductase (complex 1) in Huntington's disease. *Neurology* 1990, **40**: 1231–1234.
127. Sims NR, Finegan JM, Blass JP, Bowen D, Neary D. Mitochondrial function in brain tissue in primary degenerative dementia. *Brain Res* 1987b, **436**: 30–38.
128. Parker WD, Mahr NJ, Filley CM, Parke JK, Hughes D, Young DA, Cullum CM. Reduced platelet cytochrome c oxidase activity in Alzheimer's disease. *Neurology* 1994, **44**: 1086–1090.
129. Mutisya EM, Bowling AC, Beal MF. Cortical cytochrome oxidase activity is reduced in Alzheimer's disease. *J Neurochem* 1994, **63**: 2179–2184.
130. Kumar U, Dunlop DM, Richardson JS. Mitochondria from Alzheimer's fibroblasts show decreased uptake of calcium and increased sensitivity to free radicals. *Life Sci* 1994, **54**: 1855–1860.
131. Sims NR, Blass JP, Murphy C, Bowen DM, Neary D. Phosphofructokinase activity in the brain in Alzheimer's disease. *Ann Neurol* 1987a, **21**: 509–510.
132. Corral DM, Horton T, Lott MT, Shoffner JM, Beal MF, Wallace DC. Mitochondrial DNA deletions in human brain: regional variability and increase with advancing age. *Nat Genet* 1992, **2**: 324–329.
133. Ferrándiz ML, Martínez M, DeJuan E, Díez A, Bustos G, Miguel J. Impairment of mitochondrial oxidative phosphorylation in the brain of aged mice. *Brain Res* 1994, **644**: 335–338.
134. Pettegrew JW, Klunk WE, Panchalingham K, Kanfer JN, McClure RJ. Alterations of cerebral metabolism in probable Alzheimer's disease. *Neurobiol Aging* 1994, **15**: 117–132.
135. Jenkins BG, Koroshetz WJ, Beal MF, Rosen BR. Evidence for impairment of energy metabolism in vivo in Huntington's disease using localized ^{1}H NMR spectroscopy. *Neurology* 1993, **43**: 2689–2695.
136. Marklund, SL, Westman, NG, Lundgren, E, Roos G. Copper- and zinc-containing superoxide dismutase, maganese-containing superoxide dismutase, catalase, and glutathione peroxidase in normal and neoplastic human cell line and normal human tissues. *Cancer Res* 1982, **42**: 1955–1961.

137. Savolainen H. Superoxide dismutase and glutathione peroxidase activity in rat brain. *Res Commun Chem Pathol Pharmacol* 1978, **21:** 173–176.
138. Ben-Yoseph O, Boxer P, Ross BD. Oxidative stress in the central nervous system: monitoring the metabolic response using the pentose phosphate pathway. *Dev Neurosci* 1995, **16:** 328–336.
139. Ribiere C, Hininger I, Saffar-Boccara C, Sabourault D, Nordmann R. Mitochondrial respiratory activity and superoxide radical generation in the liver, brain and heart after chronic ethanol intake. *Biochem Pharmacol* 1994, **47:** 1827–1833.
140. Ding H, Robertson DE, Daldal F, Dutton PL. Cytochrome bc_1 complex [2Fe-2S] cluster and its interaction with ubiquinone and ubihydroquinone at the Qo site: a double-occupancy site model. *Biochemistry* 1992, **31:** 3144–3158.

4
Calcium and Cell Death

Marcel Leist and Pierluigi Nicotera

INTRODUCTION

The key role of Ca^{2+} ions (Ca^{2+}) for the function of excitable tissues was already described in 1882 in the classical experiments by Ringer. Since then, a plethora of cellular functions have been found to require Ca^{2+} signals or simply the maintenance of a set Ca^{2+} concentration. The major requirement for the signaling function of Ca^{2+} is the existence of a concentration gradient between the interstitial fluid and the interior of the cell. Whereas extracellular concentrations of free Ca^{2+} range between 1.3 and 1.8 mM, the free intracellular Ca^{2+} concentration ($[Ca^{2+}]_i$) is about 100 nM. Thus, cells are continuously faced with the problem of maintaining a concentration gradient of more than four orders of magnitude across the plasma membrane. Various physiological stimuli increase $[Ca^{2+}]_i$ transiently and thereby induce cellular responses. However, under pathological conditions, changes of $[Ca^{2+}]_i$ are generally more pronounced and sustained. Pronounced elevations of $[Ca^{2+}]_i$ activate hydrolytic enzymes, lead to exaggerated energy expenditure, impair energy production, initiate cytoskeletal degradation, and ultimately result in cell death. Such Ca^{2+}-induced cytotoxicity may play a major role in several neuropathological phenomena including chronic neurodegenerative diseases as well as acute neuronal losses (i.e., during the pathogenesis of stroke).

HISTORICAL PERSPECTIVE

The idea that Ca^{2+} may be cytotoxic dates back to Fleckenstein's suggestion in 1968 that excessive entry of Ca^{2+} into myocytes may be the underlying mechanism of cardiac pathology following ischemia (for review, *see ref. 1*). In 1979, it was shown that agonist overstimulation *(2)* or cytotoxic xenobiotics may cause lethal Ca^{2+} entry into cells *(3)*. Since the early 1980s, the role of Ca^{2+} in cell death was examined intensively, especially in hepatocytes, in the kidney and in the brain *(4–8)*. It became evident that cellular Ca^{2+} overload may involve multiple intra- and extracellular routes, most of which are also used for physiological signaling. In addition, Ca^{2+} is sequestered into separate intracellular pools that can contribute to Ca^{2+}-induced toxicity even in the absence of extracellular Ca^{2+}.

Along with the understanding of the role of Ca^{2+} as physiological regulator, it soon became clear that not only alterations of the normal Ca^{2+} homeostasis, but also changes

From: Cell Death and Diseases of the Nervous System
Edited by: V. E. Koliatsos and R. R. Ratan © Humana Press Inc., Totowa, NJ

in Ca^{2+} signaling would have adverse effects. These include alterations in cell growth, differentiation, and sensitivity to the activation of natural cell death, presumably occurring via apoptosis. In a large number of experimental paradigms, the following has now been shown.

1. Direct sustained elevation of $[Ca^{2+}]_i$, e.g., by exposure of cells to ionophores, causes cell death.
2. A $[Ca^{2+}]_i$ elevation precedes cell death induced by pathophysiological stimuli.
3. Prevention of $[Ca^{2+}]_i$ elevation during such experiments can inhibit cell death.
4. Alterations of Ca^{2+} signaling pathways (e.g., potentiation or inhibition of Ca^{2+} currents) can result in cytotoxicity.

Examples for such cases in the nervous system will be given below.

EXPERIMENTAL APPROACHES

The understanding of the role of Ca^{2+} in cell death has benefited greatly from advances in analytical methods to estimate Ca^{2+} concentrations in individual cells *(9)*. A major step has been the design and synthesis of indicators, that change their fluorescent properties upon Ca^{2+} binding. Such indicators can be loaded into cells in form of their more hydrophobic, nonfluorescent acetoxymethylesters. Subsequently, enzymatic cleavage releases the active compound that remains trapped inside the cell membrane. The use of dyes such as fura-2, that changes its excitation maximum upon Ca^{2+} binding, allows an estimation of the absolute $[Ca^{2+}]_i$ independently from the intracellular indicator concentration. The determination of $[Ca^{2+}]_i$ is facilitated by the so-called ratiometric approach, i.e., excitation at different wavelengths and calculation of $[Ca^{2+}]_i$ from the ratio of the intensities at a defined emission wavelength. This method has been successfully applied to the measurement of $[Ca^{2+}]_i$ in individual cells in suspension, by fluorescent-activated cell sorter (FACS) analysis or in individual adherent cells (e.g., neurons) by quantitative photometric or video-imaging techniques. In the latter case, subcellular fluctuations of $[Ca^{2+}]_i$ can be followed. Recently, the development of multiple wavelength confocal laser scanning microscopes has allowed the application of fluorescent imaging techniques to estimate $[Ca^{2+}]_i$ within individual neurons in brain slice preparations.

Other analytical methods that have proven to be extremely useful for a variety of applications include: The measurement of Ca^{2+} concentrations with Ca^{2+}-sensitive microelectrodes or the luminescent protein aequorin, the electrophysiological characterization of Ca^{2+} channels, especially by patch clamp techniques, and the use of the radioisotope Ca^{45} for the examination of Ca^{2+} fluxes, Ca^{2+} pools and redistribution phenomena.

Furthermore, the development of pharmacological agonists and antagonists has allowed the study of different pathways of Ca^{2+} trafficking and downstream targets known to be affected by Ca^{2+}. Most recently, molecular biological approaches have helped to characterize the proteins involved in intracellular Ca^{2+} binding, in Ca^{2+} transport across the membranes and those mediating Ca^{2+} signaling. For example, Ca^{2+} response elements have been identified and their activation has been investigated in neurons using appropriate reporter genes *(10)*.

REGULATION OF CA^{2+} CONCENTRATIONS

Within the cell Ca^{2+} is compartmentalized in various pools. At each of these locally restricted sites, Ca^{2+} can be either free or it may bind proteins as part of its effector function (e.g., calmodulin) or for storage (e.g., calbindins *[11]*). It is generally assumed that the localized or general alterations of $[Ca^{2+}]_i$ are more relevant to signaling and cytotoxicity than changes in the absolute amount of Ca^{2+}.

Calcium Influx

Ca^{2+} influx from the extracellular space following the steep concentration gradient of this ion may easily raise $[Ca^{2+}]_i$ and elicit toxicity. A detailed knowledge about the routes of entry is essential, since raised $[Ca^{2+}]_i$ derived from distinct sources may result in different cellular responses, even with identical average cytosolic concentrations *(10,12–14)*. This may be explained by Ca^{2+} compartmentalization even within the cytosol *(15)*, by significant and very localized Ca^{2+} gradients near the plasma membrane, or by the activation of parallel costimulatory pathways. Also, certain signals involve oscillations of $[Ca^{2+}]_i$ at defined frequencies rather than static changes.

The main source for Ca^{2+} entry into neurons are voltage- or ligand-gated Ca^{2+} channels. The voltage-dependent Ca^{2+} channels (VDCCs) open, when the neuronal plasma membrane hypopolarizes. VDCC can be subdivided into several groups *(16)*. N-type or P-type channels are involved in presynaptic Ca^{2+} entry, necessary for vesicle fusion and neurotransmitter release. They have therefore been implicated in distal neuronal death observed after epileptic attacks *(17)*, in global ischemia *(18)*, or after lesions of dopaminergic neurons by 6-hydroxydopamine or MPP^+ *(19,20)*. L-type channels are involved in post-synaptic depolarization and their direct contribution to neurotoxicity is well established *(21–23)*.

Among the ligand-activated Ca^{2+} channels, the *N*-methyl-D-aspartate (NMDA) receptor (NMDAR) is probably the most important for neuronal cell death *(24,25)*. The NMDAR is a heteromeric molecule belonging to the class of ionotropic glutamate receptors *(26)*. Upon agonist stimulation, the NMDAR channel opens to Na^+ and Ca^{2+}. Interestingly, this ligand-gated channel is also voltage controlled (potential-dependent Mg^{2+} block). The subunit composition and the resulting gating characteristics differ among different brain regions. Kainate or quisqualate receptors, i.e., other classes of ionotropic glutamate receptors involved in excitotoxicity *(17)*, may also function as Ca^{2+} channels *(27–29)*. This may be due to special subunit reorganization *(30)* or to an amino acid exchange reducing the specificity of the channel for Na^+ and allowing the influx of Ca^{2+}. Such amino acid exchange is due to post-transcriptional RNA editing of AMPA/kainate receptor subunits *(31)*.

Further possible routes of Ca^{2+} entry are gap junctions between neurons and glial cells *(32)* or the electrogenic Na^+/Ca^{2+} exchanger *(33,34)*, that is activated in conditions of intracellular Na^+ overload.

Calcium Sequestration/Export

Intracellular Ca^{2+} is sequestered into different organelles largely by energy-consuming processes (for review, *see* ref. 35). The uptake systems with the highest affinity are located in the endoplasmic reticulum. At high Ca^{2+} concentrations, potential-driven uptake into mitochondria may also be active. A third independent pool of Ca^{2+} is localized in the cell nucleus *(36–38)* and its role and regulation are still controversial. In addition to Ca^{2+} sequestration into organelles, excessive amounts of this cation can also be removed by extrusion via H^+ or Na^+ antiporters *(14,39–42)* and by a high-affinity ATP-dependent Ca^{2+} pump in the plasma membrane *(43)*. Continuously elevated $[Ca^{2+}]_i$ can therefore increase the cellular ATP consumption.

Intracellular Ca^{2+} Release and Translocation

$[Ca^{2+}]_i$ may rise not only because of influx from the extracellular space, but also due to inhibition of Ca^{2+} efflux, or release from intracellular stores *(5)*. In neurons, e.g., in excitotoxic injury, it has been demonstrated that a high percentage of the raised intracellular Ca^{2+} is derived from intracellular pools *(44,45)*. Therefore, local Ca^{2+} changes may regulate certain intracellular sites independently from Ca^{2+} influx through the plasma membrane. For example, a Ca^{2+} pool located in the nuclear envelope can be released by specific agonists in the perinuclear space *(36)* or possibly into the nucleus. This may regulate the entry of macromolecules through the nuclear pores *(46)* and contribute to elicit selective changes in the intranuclear free Ca^{2+} concentration.

The Ca^{2+} pool stored in the endoplasmic reticulum (ER) is released by stimulation of two classes of receptors: the IP3 receptors and the neuronal ryanodine receptor. Therefore, stimulation of cell membrane receptors not linked to Ca^{2+} channels (e.g., metabotropic glutamate receptors) can result in increased $[Ca^{2+}]_i$ due to the generation of second messengers. In addition to excessive Ca^{2+} entry through the plasma membrane, the ER Ca^{2+} pool plays a significant role in xenobiotic-induced toxicity and oxidant cellular injury. This pool is not static, but a steady state Ca^{2+} level is maintained by constant leakage of Ca^{2+} into the cytosol and reimport via an ATP-driven transporter, which is very sensitive to oxidant attack. Besides the ER, a further source for toxic increases of $[Ca^{2+}]_i$ are the mitochondria, which may release this ion upon depolarization, permeability transition (PT) or alterations in the respiratory chain. For example, mitochondrial Ca^{2+} release can contribute to nitric oxide (NO) cytotoxicity *(47)*.

Neuronal Set-Point Hypothesis

Increases of $[Ca^{2+}]_i$ above baseline do not necessarily result in cell death. For example, Ca^{2+}-influx through VDCC seems to be better tolerated by neurons than that occurring through NMDAR channels. There are even conditions, where an increased intracellular Ca^{2+} concentration can foster neuronal survival *(48)*, whereas Ca^{2+} depletion induces cell death *(49)*. The neuronal populations to be best characterized in this respect are peripheral sympathetic and sensory neurons. These neurons have intracellular Ca^{2+} concentrations ≤100 nM and are strictly dependent upon nerve growth factor (NGF) for survival, directly after isolation. Over the course of three weeks, $[Ca^{2+}]_i$ rises to about 250 nM. Concomitantly, the neurons lose their NGF requirement that prevents the onset of apoptosis. Artificially raising $[Ca^{2+}]_i$ at the beginning of the culture abolishes the NGF requirement. These findings have suggested that this neuronal population has a developmentally regulated Ca^{2+} setpoint, and that $[Ca^{2+}]_i$ determines the dependence on trophic factors, and controls neuronal survival *(50)*.

EFFECTOR SYSTEMS OF RAISED CA^{2+}

Unlike excessive $[Na^+]_i$ that may damage cells by a direct osmotic effect, increased $[Ca^{2+}]_i$ probably does not elicit neuronal death by itself. Rather, downstream reactions are activated. Some are the result of the continuous presence of elevated Ca^{2+}, others just require a transient $[Ca^{2+}]_i$ increase as a triggering signal. Their nature and contribution to Ca^{2+}-induced cell death have only been partially elucidated.

Nitric Oxide Synthase

Nitric oxide synthases (NOS) are cytochrome P450-related enzymes, that convert arginine to NO and citrulline. Different classes of isoenzymes exist in brain. Two constitutive forms are expressed in neurons (bNOS) or endothelial cells (eNOS), respectively, and are activated by Ca^{2+}/calmodulin following an increase in $[Ca^{2+}]_i$. Another isoform in microglia or astroglia (iNOS) is inducible by a variety of stimuli such as cytokines, and functions at basal Ca^{2+} concentrations. Activation of bNOS, following Ca^{2+} entry through the NMDAR has been implicated in excitotoxicity to cortical neuronal cultures (51) and in ischemia due to middle cerebral artery occlusion (52). The possible terminal cytotoxic mediator may be peroxynitrite ($ONOO^-$) formed from NO and O_2^- (see chapter by Dykens). Some neurons, especially cortical neurons expressing high levels of bNOS, seem to be resistant to NO toxicity, but may kill neighboring neurons because of their Ca^{2+}-induced NO production (53). In cerebellar granule cells (CGCs) elevated $[Ca^{2+}]_i$ causes both bNOS-activation and cytotoxicity. However, in these neurons, glutamate-triggered, Ca^{2+}-mediated cell death is independent of endogenous NO production (54). Rather, an inverted mechanism is more relevant in CGCs: exposure to NO donors leads to stimulation of NMDAR, likely because NO related species or $ONOO^-$ stimulate the release of endogenous agonists (Leist and Nicotera, unpublished results). This sort of autocrine stimulation eventually elicits apoptosis (55).

Hydrolytic Enzymes

High $[Ca^{2+}]_i$ can activate several hydrolytic enzymes. The main classes include proteases, nucleases, and lipases. Among proteases implicated as effectors of Ca^{2+}-elicited toxicity, calpains are Ca^{2+} activated cysteine proteases. They have been implicated in toxic cell death in the liver, and in excitotoxic neuronal death in the brain (27,56,57). In addition, there are nuclear Ca^{2+} activated proteases that may have a role in the execution phase of apoptosis (4,58).

Ca^{2+}-dependent nucleases are responsible for DNA degradation, which is frequently observed during apoptosis. Despite several attempts to identify and purify endonucleases involved in apoptosis, the nature of the enzyme(s) responsible for the typical oligonucleosomal DNA cleavage is still unclear.

Among lipases, the Ca^{2+}-dependent phospholipase A_2 (PLA_2) has been implicated in neurotoxicity. Its activation by Ca^{2+} results in the release of arachidonic acid and related polyunsaturated fatty acids, which are further metabolized by lipoxygenases or cyclooxygenases with concomitant generation of reactive oxygen species (ROS). In addition, PLA_2 activation generates lysophosphatids that alter membrane structures. This may facilitate Ca^{2+} influx and Ca^{2+} release from internal stores (56). In neurons, Ca^{2+} influx through NMDAR and PLA_2 activation have been closely linked (33,59). The release of arachidonic acid following activation of PLA_2 can inhibit glutamate uptake into both neurons and glial cells and may therefore prolong the excitotoxic action of this amino acid on its receptors (60).

Xanthine Dehydrogenase

Sustained elevations of $[Ca^{2+}]_i$, caused, for example, by ischemia, can promote the conversion of xanthine dehydrogenase to xanthine oxidase (i.e., during the catalytic cycle electrons are transferred to molecular oxygen instead of adenin-nicotin dinucleotides). Under low-energy conditions, where large parts of ATP are converted to hypoxanthine, this may result in massive generation of ROS. The activation of xanthine oxidase has been implicated in ischemic neuronal death in vivo and in kainate toxicity to CGCs in vitro *(61,62)*.

Mitochondria

Mitochondria play a dual role in Ca^{2+} toxicity (*see* chapters by Dykens, and Flanigan and Ratan). On one hand, they may reduce the cytosolic Ca^{2+} overload by sequestering the ion. On the other hand, following breakdown of the membrane-potential, they may release Ca^{2+} and generate ROS. Furthermore, compromised mitochondria are not only passively involved in cytotoxicity (i.e., because they do not provide the cell with sufficient ATP), but they may generate active signals involved in the execution of apoptosis. Thus, mitochondria may act as a deciding switch helping cells to recover or accelerating their demise *(63–66)*.

Ca^{2+} is sequestered into mitochondria mainly via a Ca^{2+} uniporter or via a $Ca^{2+}/2Na^+$ antiporter, under conditions of Na^+-overload. The uniporter is driven by the eletrochemical membrane-potential and has a high capacity, but a relatively low affinity. The lowest level at which brain mitochondria regulate $[Ca^{2+}]_i$ is 300 nM in the presence of spermine and may require even higher Ca^{2+} concentrations (1 µM) under unfavorable conditions. Thus, it has been assumed that Ca^{2+} is imported into mitochondria only during conditions of prolonged stimulation and overload, or when transient high Ca^{2+} concentrations are created at local sites close to mitochondria. In fact, studies in non-neuronal cells have shown that mitochondria can load Ca^{2+} during physiological agonist stimulation and may therefore contribute to lower elevated $[Ca^{2+}]_i$ *(67)*. Mitochondria have been shown to reduce considerably elevated $[Ca^{2+}]_i$ following excitotoxic glutamate-stimulation of neurons *(14,68)*, and mitochondrial Ca^{2+} deposits were found in CGCs, lethally challenged with NMDA *(69)* or in hippocampal neurons after stroke *(70)*.

Various mechanisms have been postulated to explain Ca^{2+} release from mitochondria *(64)*. Mitochondrial Ca^{2+} extrusion is an energy-requiring process (33 kJ/mol) linked to H^+ exchange. The net effect is the import of two H^+ in exchange for the extrusion of one Ca^{2+}. Mitochondrial Ca^{2+} release is stimulated by oxidative stress. Oxidation of NADH with subsequent mono-ADP-ribosylation of mitochondrial proteins or formation of cyclic ADP-ribose have been suggested as regulatory mechanisms. Such enhanced Ca^{2+} extrusion may be the basis of "Ca^{2+} cycling," i.e., continuous uptake and release of Ca^{2+} by mitochondria, which leads ultimately to the dissipation of the membrane potential and to mitochondrial failure (Fig. 1). The interaction of raised $[Ca^{2+}]_i$ and ROS may therefore lead to a vicious loop, since mitochondria, stressed as a consequence of NMDAR stimulation *(13,71)* and Ca^{2+} overload *(72)*, will produce increasing amounts of ROS and Ca^{2+} cycling, further damaging an already uncoupled respiratory chain.

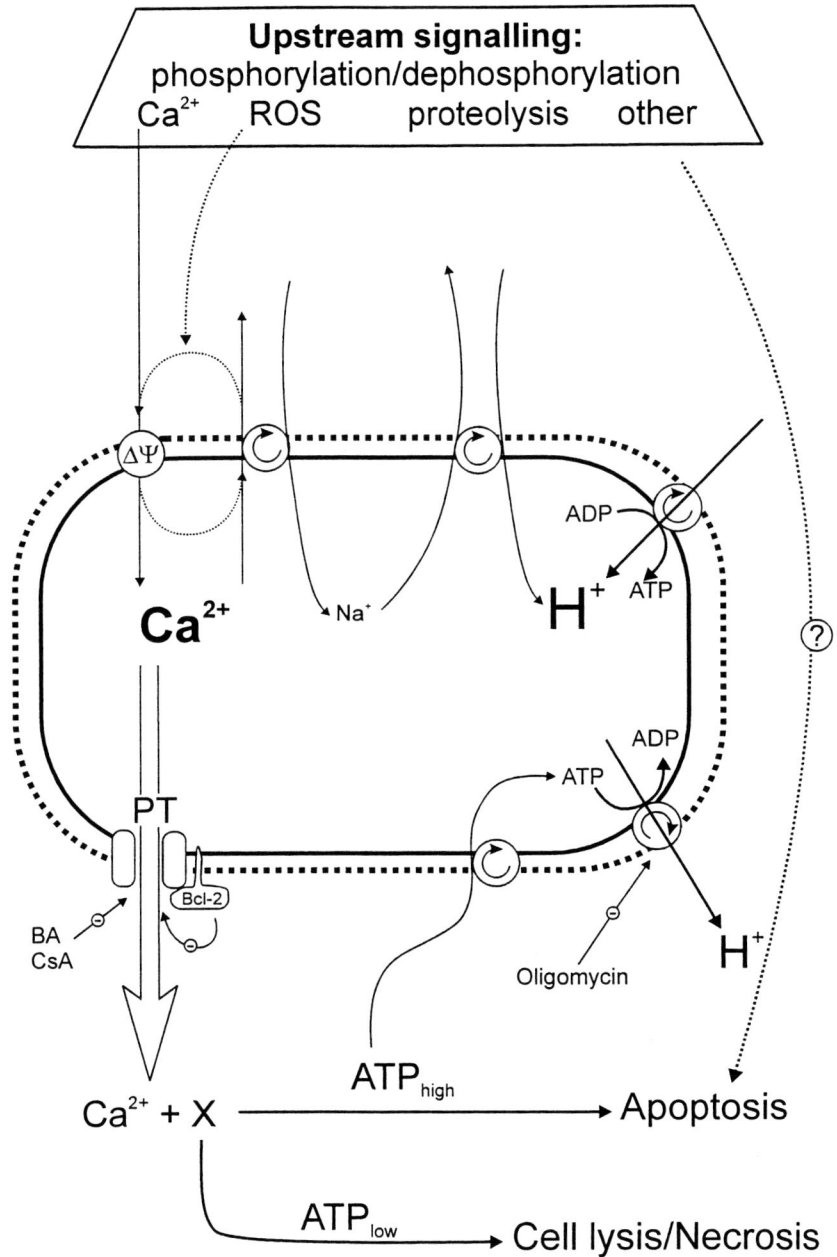

Fig. 1. A hypothetical role for mitochondria in apoptotic or necrotic cell death. Increased $[Ca^{2+}]_i$ or mitochondrial Ca^{2+} cycling may cause a decrease in membrane potential with subsequent permeability transition (PT). Availability of cytosolic ATP may buffer this effect. Permeability transition is associated with the release of mitochondrial factors (X), involved in further downstream events (i.e., nuclear degradation) in apoptosis. In isolated mitochondria, PT can be stimulated by atractyloside and prevented by bongkrekic acid (BA) or cyclosporine A (CsA). Possible sites of action for the interaction of antiapoptotic proteins (Bcl-2) or for agents that may inhibit PT by increasing mitochondrial transmembrane potential ($\Delta\psi$) such as oligomycin are also shown.

A mechanism for mitochondrial Ca^{2+} release fundamentally different from the one described above involves the PT of mitochondria (Fig. 1) *(64)*. The PT is associated with the opening of a pore in the inner mitochondrial membrane, which makes it completely permeable to ions and small molecules. Under such conditions, mitochondrial Ca^{2+} could be released without energy requirement. The PT is induced as a consequence of intracellular Ca^{2+} overload or oxidative stress and eventually results in the breakdown of mitochondrial membrane potential and swelling of the mitochondria. The PT may be a key switch responsible for the induction of apoptosis. It has been recently suggested that mitochondrial factors (e.g., cytochrome c), can be released during PT, and cause apoptotic-like changes in isolated nuclei *(63,65,66)*. In this context, it is important to note that energization of mitochondria and maintenance of their membrane potential does not necessarily require a functional respiratory chain. ATP may be imported from the cytosol via the ATP/ADP-translocator and then generate a membrane potential through the oligomycin-sensitive proton pump (Fig. 1). Consequently, also mitochondria unable to perform oxidative phosphorylation due to the lack of proteins coded by mitochondrial DNA are able to undergo PT and to induce nuclear apoptotic changes *(65)*.

Protein Phosphorylation

Persistent effects of Ca^{2+} on cellular functions and viability do not necessarily depend on the continuous elevation of $[Ca^{2+}]_i$. Rather, transient elevation of $[Ca^{2+}]_i$ may trigger persistent effects by altering the state of protein phosphorylation and gene transcription *(73)*.

At least four different effector systems, related to phosphorylation/dephosphorylation may be involved in Ca^{2+}-dependent neurotoxicity.

1. The isoforms type II and type IV of Ca^{2+}/calmodulin-dependent protein kinases (CaMK) are involved in Ca^{2+}-dependent transcriptional regulation. Type II CaMK is involved also in the activation of quisqualate receptors. In addition, this isoenzyme may act as a memory for transient elevations of $[Ca^{2+}]_i$, since its autophosphorylated form remains active even after the return of $[Ca^{2+}]_i$ to basal concentrations.
2. Ca^{2+}-sensitive adenylate cyclases especially of the type I are highly expressed in hippocampus and neocortex, and are activated upon glutamate stimulation.
3. A variety of Ca^{2+}-dependent protein kinase C (PKC) isoenzymes is expressed in brain. PKC potentiates the NMDA response *(19,74)*, may be involved in delayed Ca^{2+} influx *(75)*, and may thereby modify excitotoxic reactions.
4. Ca^{2+}-activated phosphatases, e.g., calcineurin, are involved in the regulation of various enzymes *(76)*, which play a role in neurotoxicity. Dephosphorylation of the NMDAR by calcineurin limits/shortens Ca^{2+} influx into the cell *(77–79)*.

This self-limiting effect on Ca^{2+} entry is blocked by pharmacological intervention with cyclosporine A or FK506. In addition, the extrusion of increased Na^+ following glutamate receptor activation, is dependent on the activation of Na^+/K^+-ATPase by calcineurin. Inhibition of Na^+ export with cyclosporine A may therefore preserve ATP (80). Furthermore, calcineurin dephosphorylates NOS and thereby increases this enzyme activity. The overall effect of calcineurin-inhibitors on ischemic/excitotoxic neuronal damage seems indeed to be beneficial *(80–82)* (Leist and Nicotera, unpublished observations).

The transcriptional effects of elevated $[Ca^{2+}]_i$ are mainly indirect and mediated by various signaling cascades, involving protein kinase A, MAP-kinases and CaMK.

Interestingly, the pathway of Ca^{2+} entry can determine the downstream signaling events. For instance, brain-derived neurotrophic factor (BDNF) is induced in cortical neurons only by Ca^{2+} entry through VDCC, but not through the NMDAR *(12)*. Ca^{2+} dependent immediate early genes such as *c-fos (83)* have been associated with excitotoxic neuronal death and developmental neuronal apoptosis *(84,85)*. There are different pathways leading to *fos* activation, which may result in different patterns of gene transcription. Ca^{2+} entry through VDCC induces phosphorylation of the transcription factor CREB that in turn activates *fos* expression via Ca^{2+}/cAMP-responsive elements (CaRE/CRE) in the promotor region. In the case of Ca^{2+} entry through the NMDAR activation of additional transcription factors such as Elk-1 or serum response factor (SRF) is instead required for *fos*-transcription (i.e., after Ca^{2+} entry via the NMDAR the phosphorylation of CREB is a necessary, but not sufficient condition *[10,12,73,86]*).

Transglutaminases

Tissue transglutaminase is a Ca^{2+}-dependent enzyme that catalyses protein crosslinking during apoptosis *(87)*. Transglutaminase activity has been causally associated with apoptosis of neuroblastoma cells *(88,89)* and may have an important role in the limitation of focal ischemic brain damage, by preventing neuronal secondary lysis and massive release of neurotransmitters (i.e., glutamate).

Cytoskeletal Components

Ca^{2+} can modify cytoskeletal organization and dynamics *(90)*. Increased $[Ca^{2+}]_i$ may either directly affect cytoskeletal protein organization, or it may change their phosphorylation/polymerization state. In addition, raised $[Ca^{2+}]_i$ may activate proteases cleaving major cytoskeletal constituents or ancillary proteins that anchor the cytoskeleton to the plasma membrane. Indeed, ischemic hippocampal damage involves proteolysis of fodrin and other cytoskeletal elements and it is reduced by calpain inhibitors *(57,91)*. In cultured cells, high local $[Ca^{2+}]_i$ can cause microtubule depolymerization *(15)*. In CGCs, NO-induced Ca^{2+} influx triggers the sequential depolymerization and degradation of microfilaments, nuclear lamins and microtubules *(55)*.

A consequence of Ca^{2+}-induced cytoskeletal alterations is the alteration of membrane receptors and channels. Microfilaments are involved in the desensitization of the NMDAR after stimulation, i.e., stabilization of F-actin with phalloidin prevents receptor desensitization *(92)*. Conversely, depolymerization of microfilaments with cytochalasin prevents the Ca^{2+} influx induced by Alzheimer's disease-related β-amyloid peptides *(93)*. Modifications of Ca^{2+} have also been observed after treatment of CGCs with the microtubule-depolymerizing agent colchicine *(94)*.

THE ROLE OF CA^{2+} IN DIFFERENT PARADIGMS OF NEUROTOXICITY

Heavy Metal Toxicity

Several metals can trigger selective neurotoxic processes, at least in part, by disturbing Ca^{2+} homeostasis *(95)*. Metals such as tin or mercury may directly cause an increase of $[Ca^{2+}]_i$ either by displacing Ca^{2+} bound to cellular macromolecules, or because they alter the function of proteins involved in Ca^{2+} import, export or sequestration. This may lead directly to acute or chronic neurodegeneration *(96)*. In addition, metals at low concentrations may have very subtle effects, not directly disturbing the Ca^{2+} homeostasis of resting neurons, but altering second messenger systems and channels in a way, that modifies Ca^{2+} responses towards physiological stimuli. This may lead to an increased or decreased Ca^{2+} influx upon stimulation, resulting in altered neurotransmitter release, alterations in cell differentiation, and eventually sensitization to apoptosis by otherwise nontoxic stimuli *(23,97)*.

HIV Coat Protein gp120

HIV infection and the subsequent AIDS syndrome are frequently associated with CNS defects, including dementia *(98)*. One of the pathogenic triggers may be the release of the viral gp120 protein *(99)*. This protein is known to cause neuronal apoptosis in vivo and in vitro *(100–102)*. The present experimental evidence suggests that neurotoxicity of gp120 is not direct, but involves secondary excitotoxic mechanisms resulting in Ca^{2+} overload *(103)*. Accordingly, NMDA and gp120 synergistically induce neurotoxicity *(104)*, whereas blockage of VDCC *(22)* or NMDAR channels by various pharmacological tools *(100,105)* prevents Ca^{2+} overload and gp120-induced cell death. The viral protein may in fact stimulate macrophages to produce terminal excitotoxic mediator acting on NMDAR *(106)*.

Excitotoxic Cell Death

Excitotoxicity is a phenomenon typically encountered in neurons or myocytes, following a stimulation that exceeds the physiologic range with respect to duration or intensity. Typical excitotoxic mediators are acetylcholine or—most important in the CNS—glutamate. A large variety of chronic neurodegenerative diseases seem to have an excitotoxic component *(107,108)* (see chapters by Dykens, Olney and Ishimaru, O'Hearn and Molliver, Fuller and Bohr, Vornov, Dietrich, and Bar-Peled and Rothstein). A causal contribution of excitotoxicity to neuronal damage has also been established in stroke or head trauma *(98,109,110)* as well as in acute poisoning phenomena, e.g., exposure to CO or MPP^+ *(20,111)*. Generally, excitotoxicity is induced by conditions favoring glutamate accumulation in the extracellular space, and it is enhanced by conditions (e.g., energy depletion) that hinder cellular protective mechanisms *(112)*. Typical conditions leading to increased extracellular glutamate concentrations *(113–117)* are depolarization of neurons, energy depletion due to hypoglycemia or hypoxia *(118–121)*, or defects in the glutamate reuptake systems *(60,122)*.

The contribution to neurotoxicity of synaptic activity, that is the release of glutamate and stimulation of its receptors on postsynaptic membranes, was deduced early from the observations that direct injection of glutamate was neurotoxic in vivo *(123,124)*, and that inhibition of neurotransmission by Mg^{2+} or glutamate antagonists protected neurons from hypoxic damage *(114,125)*. More recently, it has been shown that neurotransmitter release triggered by electrically stimulating hippocampal neurons with autosynapses was sufficient to stimulate glutamate-induced increases of $[Ca^{2+}]_i$ within the same cell *(78)*. Thus, there is a very close reciprocal relationship between increased extracellular glutamate concentrations and raised $[Ca^{2+}]_i$, which may finally result in neuronal death under excitotoxic conditions.

Glutamate can trigger Ca^{2+} influx by various routes. A Ca^{2+} channel is directly opened by stimulation of the NMDAR *(31)*. The key role of this channel in excitotoxicity is supported by pharmacological intervention studies *(98,126,127)*, and by the protective effect of NMDAR antisense RNA in an experimental stroke model *(128)*. However, experimental treatment with NMDAR antagonists does not achieve a full protection, especially in models of global ischemia as compared to focal, transient ischemia *(18,110,129,130)*. Additional mechanisms seem therefore to be involved especially in neuronal death in the ischemic core region.

Non-NMDA glutamate receptor subtypes (non-NMDAR) are also important mediators of excitotoxicity *(27–29)*. Non-NMDAR agonists cause Ca^{2+} influx into neurons *(131–133)* and neurotoxicity in vivo as well as in vitro *(62,75,134–137)*. The mechanisms may involve direct opening of Ca^{2+} channels *(30,138)*, the release of glutamate, which subsequently acts on NMDAR *(17,114–117,132)*, Ca^{2+} release from intracellular stores *(131)*, or excessive Na^+ influx, which subsequently triggers Ca^{2+} import. In addition, depolarization of neurons, following non-NMDAR stimulation releases the Mg^{2+} block of the NMDAR *(139)*.

Other mechanisms may also contribute indirectly to raise $[Ca^{2+}]_i$. Glutamate may, for example, increase oxidative stress in neurons by thiol depletion *(140)*. Neuronal depolarization following stimulation of glutamate receptors may activate VDCC *(21,22,141)*, and Ca^{2+} release from intracellular pools *(131,142)* can be triggered by the initial Ca^{2+} influx through the plasma membrane.

The key role of Ca^{2+} in excitotoxicity is suggested by three different lines of evidence *(8,126,143)*. First, there is an obvious increase in $[Ca^{2+}]_i$ in in vivo and in in vitro models of excitotoxic cell death. This has been observed in ischemic brain *(70,144)* or in brain slices exposed to NMDAR agonists or anoxia *(69,145)*. In addition, glutamate-stimulated Ca^{2+} influx has been shown directly in cultured neurons by the Ca^{45} technique *(146,147)*, and increased $[Ca^{2+}]_i$ after NMDAR stimulation has been observed repeatedly using fluorescent probes *(127,136,148–152)*. Additional evidence comes from in vivo microelectrode measurements, demonstrating an 80–90% decline of extracellular Ca^{2+} during ischemia and a corresponding increase of intracellular Ca^{2+} *(153,154)*. Second, decreasing Ca^{2+} entry into the neuron by removal of extracellular Ca^{2+} *(41,75,155,156)*, depletion of NMDAR *(128)*, pharmacological inhibition of glutamate receptor subtypes or of VDCCs *(21)* prevents death in many paradigms of excitotoxicity. Third, prevention of neurotoxicity by inhibition of Ca^{2+}-dependent downstream effects most strongly supports a causal role for Ca^{2+} in toxicity. Intracellular Ca^{2+} chelators can protect against ischemic damage in vivo and excitotoxic neuronal damage in vitro *(157)*. Also, inhibition of effectors of Ca^{2+} toxicity such as calmodulin-inhibition *(137)*, calcineurin *(53,80)*, or bNOS *(53)* protects neurons against the toxicity of excitatory amino acids.

Nevertheless, several issues concerning the role of Ca^{2+} in excitotoxicity remain to be resolved. Results in different experimental models are often not consistent. For example, only some studies show a quantitative correlation between increased $[Ca^{2+}]_i$ and excitotoxicity *(134,136,158)*, whereas others fail to do so in similar culture systems (for review, see ref. *126*). It is apparent that minor alterations in cell culture conditions seem to have a major influence on the final outcome of experiments. Additional confounding factors often derive from experimental approximations and pitfalls. For example, the measurement of average $[Ca^{2+}]_i$ ignores the fact that pronounced Ca^{2+} gradients exist within the cell *(159)* and that Ca^{2+} concentrations elicited by glutamate within the dendritic spines may exceed by far the average cytoplasmic concentrations.

The source of Ca^{2+} entry should be accounted for, since it may determine the cellular fate *(148)*. Changes of pH *(153)*, protein expression *(11)*, as well as many intercellular mediators have been shown to prevent alterations of neuronal Ca^{2+} homeostasis *(118,120)*. In several studies, the time component of the alterations in $[Ca^{2+}]_i$ is not sufficiently accounted for. Kinetics of Ca^{2+} elevation may be very complex and measuring $[Ca^{2+}]_i$ at just one defined time point may not yield sufficient information. It is important to consider that Ca^{2+} may modify its own homeostasis by influencing receptor activation/inactivation *(78,160)* by stimulating neurotransmitter release or the discharge of intracellular Ca^{2+}-induced Ca^{2+} release pools. The understanding of these phenomena, especially in vivo is still a major research task.

The reciprocal relationship between $[Ca^{2+}]_i$ and glutamate implies that Ca^{2+} or downstream cellular responses encountered during excitotoxic challenge would modify glutamate release. For instance, inhibition of presynaptic, N-type VDCC can reduce ischemic brain damage, possibly by inhibiting glutamate release *(18)*. Endogenous glutamate release following stimulation of cells, e.g., with NO or following injury has been repeatedly observed *(18,121,132,161–163)*. Accordingly, prevention of neurosecretion with tetanus toxin *(164)* or botulinum neurotoxin C (Leist and Nicotera, unpublished observations) has proven to be neuroprotective against hypoglycemia or NO challenge.

Alzheimer's Disease Associated Amyloid β Peptide

Alzheimer's disease is a neurodegenerative disorder with poorly defined pathogenetic mechanisms. The actions of amyloid-β (Aβ)-peptides derived form amyloid precursor proteins (APP) *(165)* have been implicated as a cause of neuropathological changes. The neurodegeneration observed post-mortem in subject with Alzheimer's disease shows apoptotic features *(166)*. Apoptosis is also evoked in vitro by exposure of neuronal cultures to Aβ *(167–169)*. The mechanism of Aβ toxicity in vitro may involve excessive Ca^{2+} entry into cells. Antigenic changes of cytoskeletal elements typical of Alzheimer's disease are mimicked by excessive Ca^{2+} influx into cultured hippocampal neurons *(170)*, and block of receptor-mediated Ca^{2+} entry by actin depolymerization prevents Aβ toxicity *(93)*. Interestingly, Aβ-induced disturbances of Ca^{2+} homeostasis in hippocampal neurons *(171)* is corrected by pretreatment of these cultures with various peptide mediators such as fibroblast growth factor *(172)* or tumor necrosis factor *(173)*. The toxicity of Aβ is strongly enhanced by conditions found during ischemic brain injury, i.e., raised concentrations of excitotoxic amino acids and hypoglycaemia. It may be speculated that the pathology of Alzheimer's disease is related to a prolonged, chronic form of excitotoxicity following the conversion of APP to Aβ. Accordingly, the unprocessed APP protein seems to have a dampening effect on $[Ca^{2+}]_i$, by opening neuronal K^+ channels, which results in hyperpolarization *(174)*. In agreement with these findings, soluble APP protects neurons from Aβ toxicity *(175,176)*.

Decision Point: Neuronal Apoptosis vs Necrosis

The mode of cell death has a large bearing on the fate of the tissue. Apoptosis and necrosis, in their classical definition, may be two fundamentally different modes of cell death *(177)* (for reservations, *see* chapter by Clarke). Besides the obvious morphological differences *(177)*, perhaps the most relevant distinction between the two types of cell death is the early preservation of membrane integrity, in apoptosis, whereas a rapid release of intracellular constituents occurs in the case of cell necrosis/lysis. The latter can lead to inflammatory reactions in the neighboring tissue. In contrast, apoptotic cells shrunken and with intact membranes can be removed by phagocytic cells. In fact, there is increasing evidence that classical apoptosis and necrosis represent the extreme ends of a wide range of possible morphological and biochemical deaths. Thus, apoptosis and necrosis may both be featured in vivo under pathological conditions or in vitro model systems as a function of the intensity of the insult or the availability of phagocytic cells.

The duration and extent of Ca^{2+} influx may determine whether neurons survive, die by apoptosis, or undergo necrotic lysis *(143)*. According to this paradigm, continuous, but moderate increases in $[Ca^{2+}]_i$ such as those produced by a sustained slow influx may cause apoptosis, whereas an exceedingly high influx rate would cause rapid cell lysis. For instance, stimulation of cortical neurons with high concentrations of NMDA results in necrosis, whereas exposure to lower concentrations causes apoptosis *(178)*. Correspondingly, neuronal death in experimental stroke models is necrotic in the ischemic core, but delayed and apoptotic in the less severely compromised penumbra or border regions *(179,180)*. The sensors that switch neurons towards one or the other fate may multiple. However, there is reason to believe that some would be located in the mitochondria *(181)* (Fig. 1). Ca^{2+} overload or other forms of cellular stress may elicit mitochondrial permeability transition (64) and a consequent release of Ca^{2+} *plus* a proteinaceous factor related to apoptotic death *(65,66)*. A complete deenergization of the cell, however, may not allow the ordered sequence of changes required for the apoptosis demise. In such a case, the combination of multiple random processes would result in rapid uncontrolled cell lysis/necrosis. Therefore, it seems likely that apoptosis ensues under conditions, where there remains sufficient energy production to execute an internal "death program" *(182,183)*. A common finding in apoptosis is, for example, that of morphologically intact mitochondria *(177,184)*, which may be energized by electron transport or by import of cytoplasmic ATP. Accordingly, ATP levels are maintained in PC12 cells, in CGCs, or hippocampal neurons undergoing apoptosis *(156,181,185)*.

CONCLUSIONS

Ca^{2+} is an ubiquitous intracellular messenger. Therefore, it is not surprising that in many instances of cell death, alterations of Ca^{2+} homeostasis are involved. In contrast to the role of Ca^{2+} as harbinger of cell death, its role as executor is still discussed. In various instances, a rise of the $[Ca^{2+}]_i$ may only parallel or follow cell death, without being causally involved. For example, Ca^{2+} influx may accompany influx of other ions such as Zn^{2+}, which has been recently suggested to be a major effector in an experimental paradigm of stroke *(186)*. Nevertheless, there is compelling evidence from a large number of different experimental models, that intracellular Ca^{2+} overload and the downstream processes that this event activates are the actual reason for cell demise, be it apoptotic or necrotic. Thus, the development of strategies to control intracellular Ca^{2+} homeostasis remains a useful approach to prevent neurotoxicity.

REFERENCES

1. Fleckenstein A. Calcium antagonism: history and prospects for a multifaceted pharmacodynamic principle, in *Calcium Antagonists and Cardiovascular Disease* (Opie LH, ed.), Raven Press, New York, 1984: pp. 9–28.

2. Leonard JP, Salpeter MM. Agonist-induced myopathy at the neuromuscular junction is mediated by calcium. *J Cell Biol* 1979, **82:** 811–819.
3. Schanne FAX, Kane AB, Young EE, Farber JL. Calcium dependence of toxic cell death. *Science* 1979, **206:** 700–702.
4. Nicotera P, Zhivotovsky B, Orrenius S. Nuclear Ca^{2+} transport and the role of Ca^{2+} in apoptosis. *Cell Calcium* 1994, **16:** 279–288.
5. Nicotera P, Bellomo G, Orrenius S. Calcium-mediated mechanisms in chemically induced cell death. *Annu Rev Pharmacol Toxicol* 1992, **32:** 449–470.
6. Trump BF, Berezesky IK. Calcium-mediated cell injury and cell death. *FASEB J* 1995, **9:** 219–228.
7. Siesjö BK. Cell damage in the brain: a speculative synthesis. *J Cereb Blood Flow Metab* 1981, **1:** 155–185.
8. Siesjö BK, Bengtsson F. Calcium fluxes, calcium antagonists, and calcium-related pathology in brain ischemia, hypoglycemia, and spreading depression: a unifying hypothesis. *J Cereb Blood Flow Metab* 1989, **9:** 127–140.
9. A practical guide to the study of calcium in living cells, in *Methods in Cell Biology* (Nuccitelli R, ed.) Academic, San Diego, 1994.
10. Ginty DD, Kornhauser JM, Thompson MA, Bading H, Mayo KE, Takahashi JS, Greenberg ME. Regulation of CREB phosphorylation in the suprachiasmatic nucleus by light and circadian clock. *Science* 1993, **260:** 238–241.
11. Mattson MP, Rychlik B, Chu C, Christakos S. Evidence for calcium-reducing and excito-protective roles for the calcium-binding protein Calbindin-D28k in cultured hippocampal neurons. *Neuron* 1991, **6:** 41–51.
12. Ghosh A, Carnahan J, Greenberg ME. Requirement for BDNF in activity-dependent survival of cortical neurons. *Science* 1994, **263:** 1618–1623.
13. Dugan LL, Sensi SL, Canzoniero LMT, Handran SD, Rothman SM, Lin T-S, Goldberg MP, Choi DW. Mitochondrial production of reactive oxygen species in cortical neurons following exposure to N-methyl-D-aspartate. *J Neurosci* 1995, **15:** 6377–6388.
14. White RJ, Reynolds IJ. Mitochondria and Na^+/Ca^{2+} exchange buffer glutamate-induced calcium loads in cultured cortical neurons. *J Neurosci* 1995, **15:** 1318–1328.
15. Shelanski ML. Intracellular ionic calcium and the cytoskeleton in living cells. *Ann NY Acad Sci* 1990, **568:** 121–124.
16. Miljanich GP, Ramachandran J. Antagonists of neuronal calcium channels: structure, function, and therapeutic implications. *Annu Rev Pharmacol Toxicol* 1995, **35:** 707–734.
17. Pollard H, Charriaut-Marlangue C, Centagrel A, Represa A, Robain O, Moreau J, Ben-Ari Y. Kainate-induced apoptotic cell death in hippocampal neurons. *Neuroscience* 1994, **63:** 7–18.
18. Valentino K, Newcomb R, Gadbois T, Singh T, Bowersox S, Bitner S, Justice A, Yamashiro D, Hoffman BB, Ciaranello R, Miljanich G, Ramachandran J. A selective N-type calcium channel antagonist protects against neuronal loss after global cerebral ischemia. *Proc Natl Acad Sci USA* 1993, **90:** 7894–7897.
19. Cooper AJ, Wooller S, Mitchell IJ. Elevated striatal Fos immunoreactivity following 6-hydrodopamine lesioning of the rat is mediated by excitatory amino acid transmission. *Neurosci Lett* 1995, **194:** 73–76.
20. Turski L, Bressler K, Rettig K-J, Löschmann P-A, Wachtel H. Protection of substantia nigra from MPP+ neurotoxicity by N-methyl-D-aspartate antagonists. *Nature* 1991, **349:** 414–418.
21. Sucher NJ, Lei SZ, Lipton SA. Calcium channel antagonists attenuate NMDA receptor-mediated neurotoxicity of retinal ganglion cells in culture. *Brain Res* 1991, **297:** 297–302.
22. Dreyer EB, Kaiser, PK, Offermann JT, Lipton SA. HIV-1 coat protein neurotoxicity prevented by calcium channel antagonists. *Science* 1990, **248:** 364–367.

23. Rossi AD, Viviani B, Zhivotovsky B, Manzo L, Orrenius S, Vahter M, Nicotera P. Inorganic mercury modifies Ca^{2+} signalling, triggers apoptosis and potentiates NMDA toxicity in neural cells. *Cell Death Different.* 1997, **4**: 317–324.
24. Rothman, S. M., Olney JW. Excitotoxicity and the NMDA receptor—still lethal after eight years. *Trends Neurosci* 1995, **18**: 57–58.
25. Choi DW. Calcium-mediated neurotoxicity: relationship to specific channel types and role in ischemic damage. *Trends Neurosci* 1988, **11**: 465–469.
26. Hollmann M, Heinemann S. Cloned glutamate receptors. *Annu Rev Neurosci* 1994, **17**: 31–108.
27. Brorson JR, Manzolillo PA, Miller RJ. Ca^{2+} entry via AMPA/kainate receptors and excitotoxicity in cultured cerebellar Purkinje cells. *J Neurosci* 1994, **14**: 187–197.
28. Holzwarth JA, Gibbons SJ, Brorson JR, Philipson LH, Miller RJ. Glutamate receptor agonists stimulate diverse calcium responses in different types of cultured rat cortical glial cells. *J Neurosci* 1994, **14**: 1879–1891.
29. Brorson JR, Manzolillo PA, Gibbons SJ, Miller RJ. AMPA-receptor desensitization predicts the selective vulnerability of cerebellar purkinje cells to excitotoxicity. *J Neurosci* 1995, **15**: 4515–4524.
30. Gu JG, Albuquerque C, Lee CJ, MacDermott AB. Synaptic strengthening through activation of Ca^{2+}-permeable AMPA-receptors. *Nature* 1996, **381**: 793–795.
31. Seeburg PH. The TINS/TiPS lecture. The molecular biology of mammalian glutamate receptor channels. *Trends Neurosci* 1993, **16**: 359–365.
32. Nedergaard M. Direct signaling from astrocytes to neurons in cultures of mammalian brain cells. *Science* 1994, **263**: 1768–1771.
33. Dumuis A, Sebben M, Fagni L, Prézeau L, Manzoni O, Cragoe EJ Jr, Bockaert J. Stimulation by glutamate receptors of arachidonic acid release depends on the $Na+/Ca^{2+}$ exchanger in neuronal cells. *Mol Pharmacol* 1993, **43**: 976–981.
34. Carini R, Bellomo G, Dianzini MU, Albano E. Evidence for a sodium-dependent calcium influx in isolated rat hepatocytes undergoing ATP depletion. *Biochem Biophys Res Comm* 1994, **202**: 360–366.
35. Orrenius S, Burkitt MJ, Kass GEN, Dypbukt JM, Nicotera P. Calcium ions and oxidative injury. *Ann Neurol* 1992, **32**: S33–S42.
36. Nicotera P, Orrenius S, Nilsson T, Berggren P-O. An inositol 1,4,5-triphosphate-sensitive Ca^{2+} pool in liver nuclei. *Proc Natl Acad Sci USA* 1990, **87**: 6858–6862.
37. Nicotera P, McConkey DJ, Jones DP, Orrenius S. ATP stimulates Ca^{2+} uptake and increases the free Ca^{2+} concentration in isolated rat liver nuclei. *Proc Natl Acad Sci USA* 1989, **86**: 453–457.
38. Gerasimenko OV, Gerasimenko JV, Tepikin AV, Petersen OH. ATP-dependent accumulation and inositol triphosphate- or cyclic ADP-ribose-mediated release of Ca^{2+} from the nuclear envelope. *Cell* 1995, **80**: 439–444.
39. Fujita K, Lazarovici P, Guroff G. Regulation of the differentiation of PC12 Pheochromocytoma Cells. *Environ Health Perspectives* 1989, **48**: 127–142.
40. Andreeva N, Khodorov B, Stelmashook E, Cragoe E Jr, Victorov I. Inhibition of Na^+/Ca^{2+} exchange enhances delayed neuronal death elicited by glutamate in cerebellar granule cell cultures. *Brain Res* 1991, **548**: 322–325.
41. Hartley DM, Choi DW. Delayed rescue of N-methyl-D-aspartate receptor-mediated neuronal injury in cortical culture. *J Pharmacol Exp Ther* 1989, **250**: 752–758.
42. Mattson MP, Guthrie PB, Kater SB. A role for Na+-dependent Ca^{2+} extrusion in protection against neuronal excitotoxicity. *FASEB J* 1989, **3**: 2519–2526.
43. Carafoli E. The Ca^{2+} pump of the plasma membrane. *Physiol Rev* 1991, **71**: 129–153.
44. Frandsen A, Schousboe A. Dantrolene prevents glutamate cytotoxicity and Ca^{2+} release from intracellular stores in cultured cerebral cortical neurons. *J Neurochem* 1991, **56**: 1075–1078.

45. Bouchelouche P, Belhage B, Frandsen A, Drejer J, Schousboe A. Glutamate receptor activation in cultured cerebellar granule cells increases cytosolic free Ca^{2+} by mobilization of cellular Ca2+ and activation of Ca^{2+} influx. *Exp Brain Res* 1989, **76**: 281–291.
46. Stehno-Bittel L, Perez-Terzic C, Clapham DE. Diffusion across the nuclear envelope inhibited by depletion of the nuclear Ca^{2+} store. *Science* 1995, **270**: 1835–1838.
47. Richter C, Gogvadze V, Schlapbach R, Schweizer M, Schlegel J. Nitric oxide kills hepatocytes by mobilizing mitochondrial calcium. *Biochem Biophys Res Commun* 1994, **205**: 1143–1150.
48. Galli C, Meucci O, Scorziello A, Werge TM, Calissano P, Schettini G. Apoptosis in cerebellar granule cells is blocked by high KCl, forskolin, and IGF-1 through distinct mechanisms of action: the involvement of intracellular calcium and RNA synthesis. *J Neurosci* 1995, 15: 1172–1179.
49. Kluck RM, McDougall CA, Harmon BV, Halliday JW. Calcium chelators induce apoptosis—evidence that raised intracellular ionised calcium is not essential for apoptosis. *Biochim Biophys Acta* 1994, **1223**: 247–254.
50. Johnson EM, Koike T, Franklin J. A "calcium set-point hypothesis" of neuronal dependence on neurotrophic factor. *Exp Neurol* 1992, **115**: 163–166.
51. Dawson VL, Dawson TM, London ED, Bredt DS, Snyder SH. Nitric oxide mediates glutamate neurotoxicity in primary cortical cultures. *Proc Natl Acad Sci USA* 1991, **88**: 6368–6371.
52. Huang Z, Huang PL, Panahian N, Dalkara T, Fishman MC, Moskowitz MA. Effects of cerebral ischemia in mice deficient in neuronal nitric oxide synthase. *Science* 1994, **265**: 1883–1885.
53. Dawson VL, Dawson TM, Bartley DA, Uhl GR, Snyder SH. Mechanisms of nitric oxide-mediated neurotoxicity in primary brain cultures. *J Neurosci* 1993, **13**: 2651–2661.
54. Lafon-Cazal M, Clucasi M, Gaven F, Pietri S, Bockaert J. Nitric oxide, superoxide and peroxynitrite: putative mediators of NMDA-induced cell death in cerebellar granule cells. *Neuropharmacology* 1993, **32**: 1259–1266.
55. Bonfoco E, Leist M, Zhivotovsky B, Orrenius S, Lipton SA, Nicotera P. Cytoskeletal breakdown and apoptosis elicited by NO-donors in cerebellar granule cells require NMDA-receptor activation. *J Neurochem* 1996, in press.
56. Traystman RJ, Kirsch JR, Koehler RC. Oxygen radical mechanisms of brain injury following ischemia and reperfusion. *J Appl Physiol* 1991, **71**: 1185–1195.
57. Siman R, Noszek JC. Excitatory amino acids activate calpain I and induce structural protein breakdown in vivo. *Neuron* 1988, 1: 279–287.
58. Clawson GA, Norbeck LL, Hatem CL, Rhodes C, Amiri P, McKerrow JH, Patierno SR, Fiskum G. Ca^{2+}-regulated serine protease associated with the nuclear scaffold. *Cell Growth & Differentiation* 1992, 3: 827–838.
59. Dumuis A, Sebben M, Haynes L, Pin J-P, Bockaert J. NMDA receptors activate the arachidonic acid cascade system in striatal neurons. *Nature* 1988, 336: 68–70.
60. Volterra A, Trotti D, Cassutti P, Tromba C, Salvaggio A, Melcangi RC, Racagni G. High sensitivity of glutamate uptake to extracellular free arachidonic acid levels in rat cortical synaptosomes and astrocytes. *J Neurochem* 1992, 59: 600–606.
61. Coyle JT, Puttfarcken P. Oxidative stress, glutamate, and neurodegenerative disorders. *Science* 1993, 262: 689–695.
62. Dykens JA, Stern A, Trenkner E. Mechanism of kainate toxicity to cerebellar neurons in vitro is analogous to reperfusion tissue injury. *J Neurochem* 1987, **49**: 1222–1228.
63. Liu X, Kim CN, Yang J, Jemmerson R, Wang X. Induction of apoptotic program in cell-free extracts: requirement for dATP and cytochrome c. *Cell* 1996, **86**: 147–157.
64. Gunter TE, Pfeiffer DR. Mechanisms by which mitochondria transport calcium. *Am J Physiol* 1990, **258**: C755–C786.
65. Zamzami N, Susin SA, Marchetti P, Hirsch T, Gómez-Monterrey I, Castedo M, Kroemer G. Mitochondrial control of nuclear apoptosis. *J Exp Med* 1996, **183**: 1533–44.

66. Newmeyer DD, Farschon DM, Reed JC. Cell-free apoptosis in xenopus egg extracts: inhibition by bcl-2 and requirement for an organelle fraction enriched in mitochondria. *Cell* 1994, **79**: 353–64.
67. Rutter GA, Theler JM, Murgia M, Wollheim CB, Pozzan T, Rizzuto R. Stimulated Ca^{2+} influx raises mitochondrial free Ca^{2+} to supramicromolar levels in a pancreatic beta-cell line. Possible role in glucose and agonist-induced insulin secretion. *J Biol Chem* 1993, **268**: 22,385–22,390.
68. Kiedrowski L, Costa E. Glutamate-induced destabilization of intracellular calcium concentration homeostasis in cultured cerebellar granule cells: role of mitochondria in calcium buffering. *Mol Pharmacol* 1995, **47**: 140–147.
69. Garthwaite G, Garthwaite J. Amino acid neurotoxicity: intracellular sites of calcium accumulation associated with the onset of irreversible damage to rat cerebellar neurones in vitro. *Neurosci Lett* 1986, **71**: 53–58.
70. Simon, R. P., Griffiths T, Evan MC, Swan JH, Meldrum BS. Calcium overload in selectively vulnerable neurons of the hippocampus during and after ischemia: an electron microscopic study in the rat. *J Cerebr Blood Flow Metabol* 1984, **4**: 350–361.
71. Reynolds IJ, Hastings TG. Glutamate induces the production of reactive oxygen species in cultured forebrain neurons following NMDA receptors activation. *J Neurosci* 1995, **15**: 3318–3327.
72. Dykens JA. Isolated cerebral and cerebellar mitochondria produce free radicals when exposed to elevated Ca^{2+} and Na+: implications for neurodegeneration. *J Neurochem* 1994, **63**: 584–591.
73. Ghosh A, Greenberg ME. Calcium signalling in neurons: molecular mechanisms and cellular consequences. *Science* 1995, **268**: 239–247.
74. Tingley WG, Roche KW, Thompson AK, Huganir RL. Regulation of NMDA receptor phosphorylation by alternative splicing of the C-terminal domain. *Nature* 11993, **364**: 70–73.
75. Manev H, Favaron M, Guidotti A, Costa E. Delayed increase of Ca^{2+} influx elicited by glutamate: role in neuronal death. *Mol Pharmacol* 1989, **36**: 106–112.
76. Snyder SH, Sabatini DM. Immunophilins and the nervous system. *Nature Med* 1995, **1**: 32–37.
77. Wang YT, Salter MW. Regulation of NMDA receptors by tyrosine kinases and phosphatases. *Nature* 1994, **369**: 233–235.
78. Tong G, Shepherd D, Jahr CE. Synaptic desensitization of NMDA receptors by calcineurin. *Science* 1995, **267**: 1510–1512.
79. Lieberman DN, Mody I. Regulation of NMDA channel function by endogenous Ca^{2+}-dependent phosphatase. *Nature* 1994, **369**: 235–239.
80. Marcaida G, Kosenko E, Minana M-D, Grisolía S, Felipo V. Glutamate induces a calcineurin-mediated dephosphorylation of Na+, K+ -ATPase that results in its activation in cerebellar neurons in culture. *J Neurochem* 1996, **66**: 99–104.
81. Sharkey J, Butcher SP. Immunophilins mediate the neuroprotective effects of FK506 in focal cerebral ischaemia. *Nature* 1994, **371**: 336–339.
82. Dawson TM, Steiner JP, Dawson VL, Dinerman JL, Uhl GR, Snyder SH. Immunosuppressant FK 506 enhances phosphorylation of nitric oxide synthase and protects against glutamate neurotoxicity. *Proc Natl Acad Sci USA* 1993, **90**: 9808–9812.
83. Morgan JI, Curran T. Role of ion flux in the control of c-fos expression. *Nature* 1986, **322**: 552–55.
84. Gorman AM, Scott MP, Rumsby PC, Meredith C, Griffiths R. Excitatory amino acid-induced cytotoxicity in primary cultures of mouse cerebellar granule cells correlates with elevated, sustained c-fos protooncogene expression. *Neurosci Lett* 1995, **191**: 116–120.
85. Smeyne RJ, Vendrell M, Hayward M, Baker SJ, Miao GG, Schilling K, Robertson LM, Curran T, Morgan JI. Continuous c-fos expression precedes programmed cell death in vivo. *Nature* 1993, **363**: 166–169.

86. Bading H, Ginty DD, Greenberg ME. Regulation of gene expression in hippocampal neurons by distinct calcium signaling pathways. *Science* 1993, **260**: 181–186.
87. Fesus L, Thomazy V, Falus A. Induction and activation of tissue transglutaminase during programmed cell death. *FEBS Lett* 1987, **224**: 104–108.
88. Piacentini M, Annicchiarico-Petruzzelli M, Oliverio S, Piredda L, Biedler JL, Melino G. Phenotype-specific "tissue" transglutaminase regulation in human neuroblastoma cells in response to retinoic acid: correlation with cell death by apoptosis. *Int J Cancer* 1992, **52**: 271–278.
89. Melino G, Annicchiarico-Petruzzelli M, Piredda L, Candi E, Gentile V, Davies PJA, Piacentini M. Tissue transglutaminase and apoptosis: sense and antisense transfection studies with human neuroblastoma cells. *Mol Cell Biol* 1994, **14**: 6584–6596.
90. Mattson MP, Barger SW, Begley JG, Mark RJ. Calcium, free radicals, and excitotoxic neuronal death in primary cell culture. *Meth Cell Biol* 1995, **46**: 187–216.
91. Lee KS, Frank S, Vanderklish P, Arai A, Lynch G. Inhibition of proteolysis protects hippocampal neurons from ischemia. *Proc Natl Acad Sci USA* 1991, **88**: 7233–7237.
92. Rosenmund C, Westbrook GL. Calcium-induced actin depolymerization reduces NMDA channel activity. *Neuron* 1993, **10**: 805–814.
93. Furukawa K, Mattson MP. Cytochalasins protect hippocampal neurons against amyloid beta-peptide toxicity: evidence that actin depolymerization suppresses Ca^{2+} influx. *J Neurochem* 1995, **65**: 1061–1068.
94. Bonfoco E, Ceccatelli S, Manzo L, Nicotera P. Colchicine induces apoptosis in cerebellar granule cells. *Exp Cell Res* 1995, **218**: 189.
95. Nicotera P, Rossi A. Molecular mechanisms of metal neurotoxicity. *J Trace Elem Electrolytes Health Dis* 1993, **7**: 254–256.
96. Viviani B, Rossi AD, Chow SC, Nicotera P. Organotin compounds induce Ca^{2+} overload and apoptosis in PC12 cells. *Neuro Toxicol* 1995, **16**: 19–26.
97. Rossi AD, Larsson O, Manzo L, Orrenius S, Vather M, Berggren P-O, Nicotera P. Modification of Ca^{2+} signaling by inorganic mercury in PC12 cells. *FASEB J* 1993, **7**: 1507–1514.
98. Lipton SA, Rosenberg PA. Excitatory amino acids as a final common pathway for neurologic disorders. *New Engl J Med* 1994, **330**: 613–622.
99. Brenneman DE, Westbrook GL, Fitzgerald SP, Ennist DL, Elkins KL, Ruff MR, Pert CB. Neuronal cell killing by the envelope protein of HIV and its prevention by vasoactive intestinal peptide. *Nature* 1988, **335**: 639–642.
100. Müller WEG, Schröder HC, Ushijima H, Dapper J, Bormann J: gp120 of HIV-1 induces apoptosis in rat cortical cell cultures: prevention by memantine. *Eur J Pharmacol* 1992, **226**: 209–214.
101. Bagetta G, Corasaniti T, Berliocchi L, Navarra M, Finazzi-Agró A, Nisticó G. HIV-1 gp120 produces DNA fragmentation in the cerebral cortex of rat. *Biochem Biophys Res Comm* 1995, **211**: 130–136.
102. Toggas SM, Masliah E, Rockenstein EM, Rall GF, Abraham CR, Mucke L. Central nervous system damage produced by expression of the HIV-1 coat protein gp120 in transgenic mice. *Nature* 1994, 367: 188–1193.
103. Lipton SA. Models of neuronal injury in AIDS: another role for the NMDA receptor? *TINS* 1992, **15**: 75–79.
104. Lipton SA, Sucher NJ, Kaiser PK, Dreyer EB. Synergistic effects of HIV coat protein and NMDA receptor-mediated neurotoxicity. *Neuron* 1991, **7**: 111–118.
105. Lipton SA: 7-Chlorokynurenate ameliorates neuronal injury mediated by HIV envelope protein gp120 in rodent retinal cultures. *Eur J Neurosci* 1992, **4**: 1411–1415.
106. Giulian D, Vaca K, Noonan CA. Secretion of neurotoxins by mononuclear phagocytes infected with HIV-1. *Science* 1990, **250**: 1593–1596.

107. Choi DW. Bench to bedside: the glutamate connection. *Science* 1992, **258**: 241–243.
108. Meldrum B, Garthwaite J. Excitatory amino acid neurotoxicity and neurodegenerative disease. *TiPS* 1990, **11**: 379–387.
109. Bullock R. Strategies for neuroprotection with glutamate antagonists. Extrapolating from evidence taken from the first stroke and head injury studies. *Ann N Y Acad Sci* 1995, **765**: 272–278.
110. Myseros JS, Bullock R. The rationale for glutamate antagonists in the treatment of traumatic brain injury. *Ann N Y Acad Sci* 1995, **765**: 262–271.
111. Ishimaru H, Katoh A, Suzuki H, Fukuta T, Kameyama T, Nabeshima T. Effects of N-methyl-D-aspartate receptor antagonists on carbon monoxide-induced brain damage in mice. *J Pharmacol Exp Therapeutics* 1992, **261**: 349–352.
112. Novelli A, Reilly JA, Lysko PG, Henneberry RC. Glutamate becomes neurotoxic via the N-methyl-D-aspartate receptor when intracellular energy levels are reduced. *Brain Res* 1988, **451**: 205–212.
113. Bullock R, Zauner A, Myseros JS, Marmarou A, Woodward JJ, Young HF. Evidence for prolonged release of excitatory amino acid in severe human head trauma: Relationship to clinical events. *Ann NY Acad Sci* 1995, **765**: 290–297.
114. Rothman S. Synaptic release of excitatory amino acid neurotransmitter mediates anoxic neuronal death. *J Neurosci* 1984, **4**: 1884–1891.
115. Sandberg M, Butcher SP, Hagberg H. Extracellular overflow of neuroactive amino acids during severe insulin-induced hypoglycemia: in vivo dialysis of the rat hippocampus. *J Neurochem* 1986, **47**: 178–184.
116. Drejer J, Beneviste H, Diemer NH, Schousboe A. Cellular origin of ischemia-induced glutamate release from brain tissue in vivo and in vitro. *J Neurochem* 1985, **45**: 145–151.
117. Beneviste H, Drejer J, Schousboe A, Diemer NH. Elevation of the extracellular concentration of glutamate and aspartate in rat hippocampus during transient cerebral ischemia monitored by intracerebral microdialysis. *J Neurochem* 1984, **43**: 1369–1374.
118. Cheng B, Mattson MP. NGF and bFGF protect rat hippocampal and human cortical neurons against hypoglycemic damage by stabilizing calcium homeostasis. *Neuron* 1991, **7**: 1031–1041.
119. Cheng B, Mattson MP. IGF-I and IGF-II protect cultured hippocampal and septal neurons against calcium-mediated hypoglycemic damage. *J Neurosci* 1992, **12**: 1558–1566.
120. Cheng B, Christakos S, Mattson MP. Tumor necrosis factors protect neurons against metabolic-excitotoxic insults and promote maintenance of calcium homeostasis. *Neuron* 1994, **12**: 139–153.
121. Wieloch T. Hypoglycemia-induced neuronal damage prevented by an N-methyl-D-aspartate antagonist. *Science* 1985, **230**: 681–683.
122. Rothstein JD, Dykes-Hoberg M, Pardo CA, Bristol LA, Jin L, Kuncl PW, Kanai Y, Hediger MA, Wang Y, Schielke JP, Welty DF. Knockout of glutamate transporters reveals a major role for astroglial transport in excitotoxicity and clearance of glutamate. *Neuron* 1996, **16**: 576–586.
123. Olney JW. Glutamate-induced retinal degeneration in neonatal mice. Electron microscopy of the acutely evolving lesion. *J Neuropathol Exp Neurol* 1969, **28**: 455–474.
124. Lucas DR, Newhouse JP. The toxic effect of sodium L-glutamate on the inner layers of the retina. *Arch Ophthalmol* 1957, **58**: 193–201.
125. Rothman SM. Synaptic activity mediates death of hypoxic neurons. *Science* 1983, **220**: 536–537.
126. Dubinsky JM. Examination of the role of calcium in neuronal death. *Ann NY Acad Sci* 1992, **679**: 34.
127. Michaels RL, Rothman SM. Glutamate neurotoxicity in vitro: antagonist pharmacology and intracellular calcium concentrations. *J Neurosci* 1990, **10**: 283–292.

128. Wahlestedt C, Golanov E, Yamamoto S, Yee F, Ericson H, Yoo H, Inturrisi CE, Reis DJ. Antisense oligodeoxynucleotides to NMDA-R1 receptor channel protect cortical neurons from excitotoxicity and reduce focal ischemia infarctions. *Nature* 1993, **363**: 260–263.
129. Buchan A, Li H, Pulsinelli WA. The N-methyl-D-aspartate antagonist, MK-801, fails to protect against neuronal damage caused by transient, severe forebrain ischemia in adult rats. *J Neurosci* 1991, **11**: 1049–1056.
130. Buchan A, Pulsinelli WA. Hypothermia but not the N-methyl-D-aspartate antagonist, MK-801, attenuates neuronal damage in gerbils subjected to transient global ischemia. *J Neurosci* 1990, **10**: 311–316.
131. Murphy SN, Miller RJ. Two distinct quisqualate receptors regulate Ca^{2+} homeostasts in hippocampal neurons in vitro. *Mol Pharmacol* 1989, **35**: 671–680.
132. Courtney MJ, Lambert JJ, Nicholls DG. The interactions between plasma membrane depolarization and glutamate receptor activation in the regulation of cytoplasmic free calcium in cultured cerebellar granule cells. *J Neurosci* 1990, **10**: 3873–3879.
133. Murphy SN, Miller RJ. Regulation of Ca^{2+} influx intro strital neurons by kainic acid. *J Pharmacol Exp Therap* 1989, **249**: 184–193.
134. Frandsen A, Drejer J, Schousboe A. Direct evidence that excitotoxicity in cultured neurons is mediated via N-methyl-D-aspartate (NMDA) as well as non-NMDA receptors. *J Neurochem* 1989, **53**: 297–299.
135. Cox JA, Felder CC, Henneberry RC. Differential expression of excitatory amino acid receptor subtypes in cultured cerebellar neurons. *Neuron* 1990, **4**: 941–947.
136. Milani D, Guidolin D, Facci L, Pozzan T, Buso M, Leon A, Skaper SD. Excitatory amino acid-induced alterations of cytoplasmic free Ca^{2+} in individual cerebellar granule neurons: role in neurotoxicity. *J Neurosci Res* 1991, **28**: 434–441.
137. Marcaida G, Minana M-D, Grisolía S, Felipo V. Lack of correlation between glutamate-induced depletion of ATP and neuronal death in primary cultures of cerebellum. *Brain Res* 1995, **695**: 146–150.
138. Mattson MP, Guthrie PB, Hayes BC, Kater SB. Roles for mitotic history in the generation and degeneration of hippocampal neuroarchitecture. *J Neurosci* 1989, 9: 1223–1232.
139. Monyer H, Sprengel R, Schoepfer R, Herb A, Higuchi M, Lomeli H, Burnashev N, Sakmann B, Seeburg PH. Heteromeric NMDA receptors: molecular and functional distinction of subtypes. *Science* 1992, **256**: 1217–1221.
140. Ratan RR, Murphy TH, Baraban JM. Oxidative stress induces apoptosis in embryonic cortical neurons. *J Neurochem* 1994, **62**: 376–379.
141. Bührle CP, Sonnhof U. The ionic mechanism of the excitatory action of glutamate upon the membranes of motoneurones of the frog. *Pflügers Arch* 1983, **396**: 154–62.
142. Verkhratsky A, Shmigol A. Calcium-induced calcium release in neurons. *Cell Calcium* 1996, **19**: 1–14.
143. Choi DW. Calcium: still center-stage in hypoxic-ischemic neuronal death. *Trends Neurosci* 1995, **18**: 58–60.
144. Dienel GA. Regional accumulation of calcium in postischemic rat brain. *J Neurochem* 1984, **43**: 913–925.
145. Kass IS, Lipton P. Calcium and long-term transmission damage following anoxia in dentate gyrus and CA1 regions of the rat hippocampal slice. *J Physiol* 1986, **378**: 313–334.
146. Mogensen HS, Hack N, Balázs R, Jorgensen OS. The survival of cultured mouse cerebellar granule cells is not dependent on elevated potassium-ion concentration. *Int J Dev Neurosci* 1994, **12**: 451–460.
147. Wroblewski JT, Nicoletti F, Costa E. Different coupling of excitatory amino acid receptors with Ca^{2+} channels in primary cultures of cerebellar granule cells. *Neuropharmacol* 1985, **24**: 919–921.

148. Tymianski M, Charlton MP, Carlen PL, Tator CH. Source specificity of early calcium neurotoxicity in cultured embryonic spinal neurons. *J Neurosci* 1993, **13**: 2085–2104.
149. De Erausquin GA, Hanev H, Guidotti A, Costa E, Brooker G. Gangliosides normalize distorted single-cell intracellular free Ca^{2+} dynamics after toxic doses of glutamate in cerebellar granule cells. *Proc Natl Acad Sci USA* 1990, **87**: 8017–8021.
150. Dubinsky JM, Rothman SM. Intracellular calcium concentrations during "chemical hypoxia" and excitotoxic neuronal injury. *J Neurosci* 1991, **11**: 2545–2551.
151. Murphy SN, Thayer SA, Miller RJ. The effects of excitatory amino acid on intracellular calcium in single mouse striatal neurons in vitro. *J Neurosci* 1987, **7**: 4145–4158.
152. Dubinsky JM. Intracellular calcium levels during the period of delayed excitotoxicity. *J Neurosci* 1993, **13**: 623–631.
153. Kristián T, Katsura K-I, Gidö G, Siesjö BK. The influence of pH on cellular calcium influx during ischemia. *Brain Res* 1994, **641**: 295–302.
154. Marciani MG, Louvel J, Heinemann U. Aspartate-induced changes in extracellular free calcium in 'in vitro' hippocampal slices of rats. *Brain Res* 1982, **238**: 272–277.
155. Garthwaite G, Garthwaite J. Neurotoxicity of excitatory amino acid receptor agonists in rat cerebellar slices: dependence on calcium concentration. *Neurosci Letters* 1986, **66**: 193–198.
156. Rothman SM, Thurston JH, Hauhart RE. Delayed neurotoxicity of excitatory amino acids in vitro. *Neurosci* 1987, **22**: 471–480.
157. Tymianski M, Wallace MC, Spigelman I, Uno M, Carlen PL, Tator CH, Charlton MP. Cell-permanent Ca^{2+} chelators reduce early excitotoxic and ischemic neuronal injury in vitro and in vivo. *Neuron* 1993, **11**: 221–235.
158. Randall RD, Thayer SA. Glutamate-induced calcium transient triggers delayed calcium overload and neurotoxicity in rat hippocampal neurons. *J Neurosci* 1992, **12**: 1882–1895.
159. Connor JA, Wadman WJ, Hockberger PE, Wong RKS. Sustained dentritic gradients of Ca^{2+} induced by excitatory amino acids in CA1 hippocampal neurons. *Science* 1988, **240**: 649–653.
160. Ehlers MD, Zhang S, Bernhardt JP, Huganir RL. Inactivation of NMDA receptors by direct interaction of calmodulin with the NR1 subunit. *Cell* 1996, **84**: 745–755.
161. Choi DW. Glutamate neurotoxicity and diseases of the nervous system. *Neuron* 1988, **1**: 623–634.
162. Levi G, Aloisi F, Ciotti MT, Gallo V. Autoradiographic localization and depolarization-induced release of acidic amino acids in differentiating cerebellar granule cell cultures. *Brain Res* 1984, **290**: 77–86.
163. Gallo V, Ciotti MT, Coletti A, Aloisi F, Levi G. Selective release of glutamate from cerebellar granule cells differentiating in culture. *Proc Natl Acad Sci USA* 1982, **79**: 7919–7923.
164. Monyer H, Giffard RG, Hartley DM, Dugan LL, Goldberg MP, Choi DW. Oxygen or glucose deprivation-induced neuronal injury in cortical cell cultures is reduced by tetanus toxin. *Neuron* 1992, **8**: 967–973.
165. Mattson MP, Barger SW, Cheng B, Lieberburg I, Smith-Swintowsky VL, Rydel RE. Beta-amyloid protein metabolites and loss of neuronal Ca^{2+} homeostasis in Alzheimer's disease. *Trends Neurosci* 1993, **16**: 409–414.
166. Smale G, Nichols NR, Brady DR, Finch CE, Horton WE Jr. Evidence for apoptotic cell death in alzheimer's disease. *Exp Neurol* 1995, **133**: 225–230.
167. Forloni G, Chiesa R, Smiroldo S, Verga L, Salmona M, Tagliavini F, Angeretti N. Apoptosis mediated neurotoxicity induced by chronic application of beta-amyloid fragment 25–35. *Neuroreport* 1993, **4**: 523–526.
168. Pike CJ, Burdick D, Walencewicz AJ, Glabe CG, Cotman CW. Neurodegeneration induced by beta-amyloid peptides in vitro: the role of peptide assembly state. *J Neurosci* 1993, **13**: 1676–1687.

169. Le W-D, Colom LV, Xie W-J, Smith RG, Alexianu M, Appel SH. Cell death induced by beta-amyloid 1-40 in MES 23.5 hybrid clone: the role of nitric oxide and NMDA-gated channel activation leading to apoptosis. *Brain Res* 1995, **686:** 49–60.
170. Mattson MP. Antigenic changes similar to those seen in neurofibrillary tangles are elicited by glutamate and Ca^{2+} influx in cultured hippocampal neurons. *Neuron* 1990, **2:** 105–117.
171. Mattson MP, Cheng B, Davis D, Bryant K, Lieberburg I, Rydel RE. Beta-amyloid peptides destabilize calcium homeostasis and render human cortical neurons vulnerable to excitotoxicity. *J Neurosci* 1992, **12:** 376–389.
172. Mattson MP, Tomaselli KJ, Rydel RE. Calcium-destabilizing and neurodegenerative effects of aggregated beta-amyloid peptide are attenuated by basic FGF. *Brain Res* 1993, **621:** 35–49.
173. Barger SW, Hörster D, Furukawa K, Goodman Y, Krieglstein J, Mattson MP. Tumor necrosis factor alpha and beta protect neurons against amyloid beta-peptide toxicity: evidence for involvement of a kappaB-binding factor and attenuation of peroxide and Ca^{2+} accumulation. *Proc Natl Acad Soc USA* 1995, **92:** 9328–9332.
174. Furukawa K, Barger SW, Blalock EM, Mattson MP. Activation of K^+ channels and suppression of neuronal activity by secreted beta-amyloid-precursor protein. *Nature* 1996, **379:** 74.
175. Goodman Y, Mattson MP. Secreted forms of beta-amyloid precursor protein protect hippocampal neurons against amyloid beta-peptide-induced oxidative injury. *Exp Neurol* 1994, **128:** 1–12.
176. Mattson MP, Cheng B, Culwell AR, Esch FS, Lieberburg I, Rydel RE. Evidence for excitoprotective and intraneuronal calcium-regulating roles for secreted forms of the beta-amyloid precursor protein. *Neuron* 1993, **10:** 243–254.
177. Wyllie AH, Kerr JF, Currie AR. Cell death: the significance of apoptosis. *Int Rev Cytol* 1980, **68:** 251–306.
178. Bonfoco E, Krainc D, Ankarcrona M., Nicotera, P, Lipton SA. Apoptosis and necrosis: two distinct events induced respectively by mild and intense insults with NMDA or nitric oxide/superoxide in cortical cell cultures. *Proc Natl Acad Sci USA* 1995, **92:** 72,162–72,166.
179. Li Y, Sharov VG, Jiang N, Zaloga C, Sabbah HN, Chopp M. Ultrastructural and light microscopic evidence of apoptosis after middle cerebral artery occlusion in the rat. *Am J Pathol* 1995, **146:** 1045–1051.
180. Charriaut-Marlangue C, Margaill I, Walsh RJ, Plotkine M, Ben-Ari Y. NG-nitro L-Arginine methylester (L-NAME) reduces cortical infarct and necrotic damage but not apoptotic cell loss. *Soc Neurosci Abstr* 1995, **21:** 998.
181. Ankarcrona M, Dypbukt JM, Bonfoco E, Zhivotovsky B, Orrenius S, Lipton SA, Nicotera P. Glutamate-induced neuronal death: a succession of necrosis or apoptosis depending on mitochondrial function. *Neuron* 1995, **15:** 961–973.
182. Chou CC, Lam CY, Yung BYM. Intracellular ATP is required for actinomycin D-induced apoptotic cell death in HeLa cells. *Cancer Lett* 1995, **96:** 181–187.
183. Hartley A, Stone JM, Heron C, Cooper JM, Schapira AHV. Complex I Inhibitors induce dose-dependent apoptosis in PC12 cells: relevance to Parkinson's disease. *J Neurochem* 1994, **63:** 1987–1990.
184. Hajos F, Garthwaite G, Garthwaite J. Reversible and irreversible neuronal damage caused by excitatory amino acid analogues in rat cerebellar slices. *Neurosci* 1986, **18:** 417–436.
185. Mills JC, Nelson D, Erecinska M, Pittman RN. Metabolic and energetic changes during apoptosis in neural cells. *J Neurochem* 1995, **65:** 1721–1730.
186. Koh J-Y, Suh SW, Gwang BJ, He YY, Hsu CY, Choi DW. The role of zinc in selective neuronal death after transient global cerebral ischemia. *Science* 1996, **272:** 1013–1016.

5
Proteases
Critical Mediators of Apoptosis

Anthony Rosen and Livia Casciola-Rosen

INTRODUCTION

Apoptosis, a prominent form of physiologic cell death, is assuming increased importance in human physiology and pathology (reviewed in ref. *1*). Unlike necrosis, an uncontrolled process in which acute cellular injury leads to rapid cell swelling and lysis, apoptosis follows an ordered and stereotyped sequence, which results in cell shrinkage, fragmentation into membrane-bound bodies, and rapid phagocytosis by surrounding cells *(2,3)*. Although the morphologic characteristics of apoptosis are readily recognizable, and internucleosomal DNA degradation is a frequent feature of apoptotic death *(4)*, the actual apoptotic mechanism has been difficult to define *(5)*.

PROTEASES AND THE APOPTOTIC MECHANISM

Several recent approaches have strongly implicated specific proteolysis as an essential element of the apoptotic mechanism. Genetic studies in *C. elegans* identified two genes (*ced-3* and *ced-4*) that were essential for somatic cells to undergo apoptosis *(6)* (*see* chapter by Royal and Driscoll). When *ced-3* was cloned, it was noted to be homologous to human interleukin 1β-converting enzyme (ICE/caspase-1 *[7]*), a novel cysteine protease initially identified as the enzyme responsible for processing of the interleukin-1 (IL-1) precursor into the active cytokine *(8)*. The CED-3 protein was subsequently demonstrated to be a cysteine protease, with its protease activity required for induction of apoptosis in *C. elegans (9)*. An extended family of mammalian ced-3 homologues (recently termed "caspases," for *c*ysteinyl *asp*artate-specific proteinases *(10) [see below]*)[1] has recently been identified, whose expression is able to induce apoptosis in a variety of different cells *(11–25)*. Activation of caspases is a uniform feature of apoptosis, and has been observed in several examples of apoptosis (e.g., *see* ref. *26*). Specific inhibition of the activity of CED-3 or its mammalian homologs by macromolecular (crmA or p35) or peptide-based (Ac-DEVD-CHO) inhibitors attenuates apoptosis both in vivo and in vitro *(27–36)*. Caspase-3 (CPP32) knockout mice have defective nervous system apoptosis during development *(see below)*, confirming an essential and nonredundant role for caspase-3 in the nervous system *(37)*. Taken

From: Cell Death and Diseases of the Nervous System
Edited by: V. E. Koliatsos and R. R. Ratan © Humana Press Inc., Totowa, NJ

together, these studies argue persuasively for a central and conserved role of this unique cysteine protease family in the apoptotic mechanism (some reservations regarding this position have been expressed by Clarke in Chapter 1). In mammals, there is great complexity within the system, with differing degrees of promiscuity among various proteases, and with multiple proteases having similar specificities, potentially imbuing the system with redundancy. This chapter will review the biology and enzymology of the caspases, and will pay particular attention to their activation, and their downstream substrates.

THE CASPASE FAMILY

ICE/caspase-1 was originally identified as the protease responsible for the processing of the interleukin 1β precursor (31 kDa) at Asp^{116}–Ala^{117} to generate the mature, biologically active cytokine (17.5 kDa) *(8,38)*. This protease is the prototype member of the caspase family, with many features defined for caspase-1 being representative of other family members. Because the enzymology and structure of caspase-1 have been defined in great detail *(39,40)*, we will initially discuss the features of this protease, and then use this framework to discuss other family members.

Caspase-1 is a cysteine protease, which is unique in its near-absolute specificity for aspartic acid in the P_1 position of both macromolecular and peptide substrates (reviewed in ref. *41*). The mature enzyme is a heterodimer of 10 kDa and 20 kDa, derived from a 45 kDa precursor by autocatalysis at Asp-X sites *(8)*. Although both subunits are essential for catalytic activity, the function of the prodomain remains unclear at this time, but is thought to be involved in the activation process. Early studies on ICE/caspase-1 demonstrated that the substrate P_1–P_4 amino acids are critical determinants of specific recognition by this protease *(8)*. The minimal peptide substrate for this protease (Ac-YVAD-NH-CH_3) was as efficiently cleaved by ICE/caspase-1 as was pro-interleukin 1β, indicating that most of the specificity information required by ICE resides in these P_1–P_4 amino acids. The optimal substrate has a large hydrophobic residue in P_4, and the obligatory aspartic acid in P_1. Solving the crystal structure of caspase-1 has defined those regions of the protease that determine this substrate specificity *(39,40)*. The mature protease appears to be a heterotetramer, composed of two heterodimers of p20 and p10 subunits. These two subunits are intimately associated, with both contributing key residues to the active site, forming a single catalytic domain. Pockets accommodating the P_1 and P_4 amino acids are clearly evident.

The pentapeptide around the catalytic Cys^{285} within caspase-1 (QACRG), which is identical to that in Ced-3, served as a useful sequence with which to search for additional members of this protease family. Subsequent cloning studies have identified nine additional family members *(11–25)*; these have recently been termed "caspases," due to their cysteine protease mechanism and their absolute requirement for Asp in the P_1 position *(10)*. All members of this family share the signature pentapeptide around the catalytic cysteine (QACXG), as well as those residues that form the unique S_1 subsite (Arg^{179}, Gln^{283}, Arg^{341}, and Ser^{347} in caspase-1), which stabilizes binding of the P_1 aspartic acid, and is responsible for the unique substrate specificity of these proteases. In contrast, those residues that accommodate the P_4 amino acid are considerably more variable in the different caspases *(36)*; the shapes and sizes of the resulting S_4 subsites (which appear to determine the specificity and promiscuity of these proteases) are therefore likely to differ for each protease *(42)*. Distinct P_4 specificities have previously been suggested, since caspase-1 prefers large hydrophobic amino acids in P_4, whereas caspase-3 has a near absolute requirement for aspartic acid at this position *(42,43)*. The molecular basis for these differences has recently been directly demonstrated by comparing the crystal structures of caspase-1 and caspase-3 *(42)*. Whereas the S_4 subsite in caspase-1 is a shallow depression that easily accommodates large hydrophobic amino acids, the equivalent site in caspase-3 is a narrow pocket, that contains several side-chains that serve to stabilize the P_4 aspartic acid *(42)*.

The critical and near-complete contribution of the P_1–P_4 residues to specificity of the caspases has led to the design of a panel of highly specific tetrapeptide aldehyde inhibitors for different members of this protease family *(8,36)*, which, in combination, have been useful as quantitative tools to identify caspase activities in apoptotic extracts in vitro *(43)*. For example, Ac-YVAD-CHO is a potent inhibitor of caspase-1 (K_i = 0.76 nM), but a poor inhibitor of caspase-3 (K_i > 10 μM), whereas Ac-DEVD-CHO is a potent inhibitor of caspases 3 and 7 (K_i = 0.2 nM and 2 nM, respectively), and a less potent inhibitor of caspase-1 (K_i = 17 nM). Because they are impermeable, these reagents have not been useful as probes of specific caspase activity in intact cell systems. Although more permeable tripeptide inhibitors have been designed, their specificity for different caspases in intact cells has not yet been well defined *(44)*.

A HIERARCHY OF CASPASES EXISTS IN APOPTOSIS

The caspases are all synthesized as inactive precursors which are cleaved at Asp-X sites during apoptosis to generate a large and small subunit in each case; these heterodimers constitute the active protease. Although some caspase precursors have the potential for initiating efficient initiating autoprocessing at Asp-X sites (particularly caspase-1), this does not appear to be a uniform feature of the family. This has focused attention on upstream activities that process precursor caspases at Asp-X sites. Recent studies have implicated several caspases (e.g., caspase 8 and 10) as well as granzyme B in the initiation of apoptosis induced by specific stimuli, by inducing the processing of caspases *(23,45–48)*. However, the upstream mechanisms responsible for caspase activation in response to a diverse range of apoptotic stimuli remain incompletely understood at the present time.

Of note, there are examples of caspase family members (e.g., caspase-3) which, although not capable of efficiently intitiating autoprocessing, are nevertheless able to process and activate their own precursor once they themselves are activated *(49)*. This creates the potential for autoamplification once a small amount of initial processing has been accomplished, and focuses attention on both those activities that may initiate caspase processing, as well as potential inhibitory activities which prevent autoamplification under nonapoptotic circumstances.

UPSTREAM ACTIVITIES IN THE APOPTOTIC PROTEOLYTIC CASCADE

Although multiple stimuli can induce the apoptotic process, most of these appear to converge upon a final common effector pathway, in which caspase-3 and similar homologues appear to play a central role. Several earlier studies suggested that proteases might also have a role in the transduction of the apoptotic signal upstream of the final common pathway, potentially by activating the downstream executioner proteases. Recent data strongly supports this concept, and three distinct examples of multitiered proteolysis during apoptosis have been clearly defined to date. It is predicted that several more examples of this type of mechanism will be identified in the future.

Involvement of Caspases in Fas- and TNF-Receptor Signaling

Recent studies have demonstrated that caspase-8 (and likely also caspase-10) are the most proximal components in the caspase pathway involved in transducing the death signal from the Fas- and TNF-receptors *(23,24)*. Caspase-8 has several unique features: It has an unusually long prodomain (containing two similar death effector domains), which becomes separated from the catalytic subunits during activation; it becomes physically associated with the cytoplasmic domain of the Fas receptor complex, after ligation of the Fas receptor; this association is not direct, but rather occurs via an interaction with the adapter protein, FADD/MORT1 through similar death effector domains that are found in both molecules. The actual mechanism by which the prodomain is removed during activation remains unclear at the present time. Several other studies have demonstrated that specific inhibition of the upstream activity results in failure to process the effector proteases caspase-3 and caspase-7, establishing that a multitiered caspase cascade exists downstream of Fas-ligation *(44,51)*.

Activation of Effector Caspases by Granzyme B

Granzyme B is a serine protease that is found in the granules of cytolytic lymphocytes. Granule-induced cytolysis is apoptotic in character, and is associated with the processing of the effector caspases to their mature forms. Significant data suggests that this processing is catalyzed by granzyme B, and potentially also by other granule proteases (reviewed in ref. *52*). Indeed, granzyme B, which is similar to caspases in its preference for aspartic acid in the P_1 position, has been demonstrated to cleave several caspases in vitro, with activation of their proteolytic function *(23,45–48)*. These observations reinforce the concept that upstream proteolytic activities are of critical importance in activating the effector proteases during apoptosis.

Involvement of the Proteasome in Activation of Effector Caspases

The role of upstream activating activities has been further emphasized in several recent studies, which have demonstrated that specific inhibitors of the proteasome prevent apoptosis in response to some stimuli *(53–55)*. Importantly, these inhibitors also abolished the cleavage of PARP, implying that the activity of the proteasome is upstream of the effector caspases during apoptosis. The exact target(s) of the proteasome remain unknown at the present time, but these might include the upstream activators of the caspases, or some of the caspases themselves.

EFFECTOR CASPASES MODIFY CRITICAL CELL STRUCTURE AND HOMEOSTATIC FUNCTIONS

Since protease activity is critical in mediating apoptosis, and affects a very limited subset of cellular proteins, identification of these downstream substrates, and assessing the effects of cleavage on their function, is essential. The studies performed to date have focused on identification of the cleaved substrates, and have only begun to address

Table 1
Cellular Proteins Cleaved During Apoptosis

Function	Substrate	Cleavage site (P_4–P_1')
Homeostasis repair	Poly(ADP-ribose) polymerase	DEVD-G
	U1-70 kDa	DGPD-G
	DNA-dependent protein kinase (DNA-PKcs)	DEVD-N
	SREBP	DEPD-S
	D4-GDI	DELD-S
	PKCδ	DMQD-N
	RB	DEAD-G
Unknown	Huntingtin	DXXD (multiple)
Structure	Lamin A	VEID-N
	Gas-2	SRVD-G
	Fodrin	?
	NuMA	?
	G-actin	LVVD-N/ELVD-G

the functional consequences of these cleavages. It is important to note that the caspases are highly specific endoproteinases, which do not result in the degradation of the protein, but rather generate discrete fragments that are not further processed (56). In many instances, cleavage separates key functional domains of the molecules, thus potentially liberating functional activities from regulation.

At the present time, only a limited number of substrates that are cleaved early during apoptosis have been defined; these can be divided into at least two distinct groups (see Table 1), which include proteins involved in cell structure and organization (57–61), and catalytic proteins active in several homeostatic pathways, including DNA repair (43,62–63), cell cycle control (64,65), splicing of precursor mRNA (66,67), and various signal transduction pathways (68,69). Huntingtin, a protein of unknown function that is mutated in Huntington's disease (see chapter by Ross, Becheu, and Koliatsos) is also a substrate for caspase-3, with cleavage most likely occurring at multiple DXXD sites (70). It is of interest that the majority of the homeostatic proteins share DXXD cleavage sites, whereas the structural proteins do not have DXXD sites (where these are known). Substrates cleaved at DXXD sites appear to be most efficiently cleaved by caspase-3 and caspase-7, whereas other caspases (e.g., caspase-6) appear to be responsible for cleavage of some of the substrates with structural functions. It is likely that the different groups of caspases, with different specificities, mechanisms of activation, and subcellular distributions, have evolved to cleave unique groups of substrates.

Interestingly, cleavage of downstream substrates by ICE-like proteases may have very different functional consequences. There is evidence that cleavage of some substrates abolishes their activity. For example, cleavage of DNA-PK$_{cs}$ (43) and G-actin (58) decreases kinase activity and polymerizing ability, respectively. In other instances, cleavage activates a previously dormant function. Examples of this include the cleavage by caspase-1 of the inactive caspase-3 precursor, thereby activating the proteolytic function of the latter (71); the fragment generated by cleavage of Gas2, which induces dramatic changes in the organization of the actin cytoskeleton and cell morphology *(60)*; and activation of kinase activity of the fragment generated by caspase-mediated cleavage of PKCδ, which produces several morphologic elements of the apoptotic phenotype. There is also considerable indirect evidence that suggests that protein cleavage during apoptosis generates fragments with proapoptotic potential. For example, caspase-3-mediated cleavage of PARP separates the *N*-terminal DNA-binding domain form the C-terminal catalytic portion of the molecule *(72)*. This *N*-terminal portion of PARP has profound effects on cell cycle control, and upon expression in HeLa cells, stimulates a greatly increased rate of cell doubling in response to DNA damage, with apoptosis occurring at an elevated rate *(73)*. This occurred even when the zinc finger required for DNA-binding was mutated, suggesting that the N-terminal domain of PARP interacts with other molecules in addition to DNA, thereby profoundly influencing cell cycle control *(73)*. It is not yet clear whether the *N*-terminal fragment released by caspase-3 during apoptosis, which is smaller than the domain used in the studies described above, exerts similar effects. Similarly, studies on the function of U1-70 kDa in mRNA splicing examined the functional effects of various domains of this molecule. U1-70 kDa has two extremely similar C-terminal domains that are highly enriched in arginine, serine, and aspartic acid (called "RSD domains") . Expression of a fragment containing both of these domains had a dominant negative effect on mRNA splicing *(74)*, predicting that expression of this domain would strongly inhibit the transcriptional homeostatic response. A crucial unanswered question is whether the single RSD domain generated by cleavage during apoptosis would act similarly.

APOPTOSIS: THE AGGREGATED PROCESSES FACILITATING LOSS OF INDEPENDENT EXISTENCE WITHIN TISSUES

The critical requirement of proteases both in the signaling and execution pathways during apoptosis continues to focus attention on the downstream substrates that are the final targets of these proteolytic cascades. The available data suggests that cleavage of a single downstream substrate is most unlikely to be responsible for all elements of the apoptotic phenotype, and that it is the concerted cleavage of substrates, abolishing some functions and freeing others from regulation, which generates apoptosis. Apoptosis might therefore be viewed as the combined processes leading to the loss of independent existence within tissues. The proteolytic alteration of critical cell structures, together with the removal of the cell's ability to respond in a homeostatic way to perturbing environmental and internal forces, are likely to be critical components of the apoptotic process, by creating the surface signals that initiate engulfment by neighboring cells, by generating the optimal cell structure that permits efficient engulfment, and by rendering the cell powerless to prevent these changes. Identification of the substrates cleaved, and the functions of cleavage, will continue to highlight those critical structural and homeostatic functions upon which continued independent existence as a cell depends, and will likely identify several potential targets for therapeutic intervention in diseases in which abnormal apoptosis plays a role.

REFERENCES

1. Thompson CB. Apoptosis in the pathogenesis and treatment of disease. Science 1995, **267:** 1456–1462.

2. Arends MJ, Wyllie AH. Apoptosis: Mechanisms and roles in pathology. *Int Rev Exp Pathol* 1991, **32**: 223–254.
3. Wyllie AH, Kerr JFR, Currie AR. Cell death: the significance of apoptosis. *Int Rev Cytol* 1980, **68**: 251–306.
4. Wyllie AH, Morris RG. Hormone-induced cell death. Purification and properties of thymocytes undergoing apoptosis after glucocorticoid treatment. *Am J Pathol* 1982, **109**: 78–87.
5. Steller H. Mechanisms and genes of cellular suicide. *Science* 1995, **267**: 1445–1449.
6. Ellis RE, Yuan J, Horvitz HR. Mechanisms and functions of cell death. *Ann Rev Cell Biol* 1991, **7**: 663–698.
7. Yuan J, Shaham S, Ledoux S, Ellis HM, Horvitz HR. The C. elegans cell death gene ced-3 encodes a protein similar to the mammalian interleukin-1β-converting enzyme. *Cell* 1993, **75**: 641–652.
8. Thornberry NA, Bull HG, Calaycay JR, Chapman KT, Howard AD, Kostura MJ, Miller DK, Molineaux SM, Weidner J, Aunins J, Elliston KO, Ayala JM, Casano FJ, Chin J, Ding GJF, Egger LA, Gaffney EP, Limjuco G, Palyha OC, Raju SM, Rolando AM, Salley JP, Yamin T, Lee TD, Shively JE, MacCross M, Mumford RA, Schmidt JA, Tocci MJ. A novel heterodimeric cysteine protease is required for interleukin-1β processing in monocytes. *Nature* 1992, **356**: 768–774.
9. Xue D, Shaham S, Horvitz HR. The *Caenorhabditis elegans* cell-death protein CED-3 is a cysteine protease with substrate specificities similar to those of the human CPP32 protease. *Genes Dev* 1996, **10**: 1073–1083.
10. Alnemri ES, Livingston DJ, Nicholson DW, Salvesen G, Thornberry NA, Wong WW, Yuan JY. Human ICE/CED-3 protease nomenclature. *Cell* 1996, **87**: 171.
11. Miura M, Zhu H, Rotello R, Hartwieg EA, Yuan J. Induction of apoptosis in fibroblasts by IL-1β-converting enzyme, a mammalian homolog of the *C. elegans* cell death gene ced-3. *Cell* 1993, **75**: 653–660.
12. Kamens J, Paskind M, Hugunin M, Talanian RV, Allen H, Banach D, Bump N, Hackett M, Johnston CG, Li P, Mankovich JA, Terranova M, Ghayur T. Identification and characterization of ICH-2, a novel member of the interleukin-1β-converting enzyme family of cysteine proteases. *J Biol Chem* 1995, **270**: 15,250–15,256.
13. Munday NA, Vaillancourt JP, Ali A, Casano FJ, Miller DK, Molineaux SM, Yamin T, Yu VL, Nicholson DW. Molecular cloning and pro-apoptotic activity of ICE$_{rel}$II and ICE$_{rel}$III, members of the ICE/CED-3 family of cysteine proteases. *J Biol Chem* 1995, **270**: 15,870–15,876.
14. Faucheu C, Diu A, Chan AWE, Blanchet A-M, Miossec C, Herve F, Collard-Dutilleul V, Gu Y, Aldape RA, Lippke JA, Rocher C, Su MSS, Livingston DJ, Hercend T, Lalanne JL. A novel human protease similar to the interleukin 1β converting enzyme induces apoptosis in transfected cells. *EMBO J* 1995, **14**: 1914–1922.
15. Kumar S, Kinoshita M, Noda M, Copeland NG, Jenkins NA. Induction of apoptosis by the mouse Nedd2 gene, which encodes a protein similar to the product of the *Caenorhabditis elegans* cell death gene ced-3 and the mammalian IL-1β-converting enzyme. *Genes Devel* 1994, **8**: 1613–1626.
16. Fernandes-Alnemri T, Litwack G, Alnemri ES. CPP32, a novel human apoptotic protein with homology to *Caenorhabditis elegans* cell death protein Ced-3 and mammalian interleukin-1β-converting enzyme. *J Biol Chem* 1994, **269**: 30,761–30,764.
17. Fernandes-Alnemri T, Takahashi A, Armstrong R, Krebs J, Fritz L, Tomaselli KJ, Wang L, Yu Z, Croce CM, Salveson G, Earnshaw WC, Litwack G, Alnemri ES. Mch3, a novel human apoptotic cysteine protease highly related to CPP32. *Cancer Res* 1995, **55**: 6045–6052.
18. Fernandes-Alnemri T, Litwack G, Alnemri ES. *Mch2*, a new member of the apoptotic *Ced-3/Ice* cysteine protease gene family. *Cancer Res* 1995, **55**: 2737–2742.
19. Lippke JA, Gu Y, Sarnecki C, Caron PR, Su MSS. Identification and characterization of CPP32/Mch2 homolog 1, a novel cysteine protease similar to CPP32. *J Biol Chem* 1996, **271**: 1825–1828.

20. Alnemri ES, Fernandes-Alnemri T, Litwack G. Cloning and expression of four novel isoforms of human interleukin-1β converting enzyme with different apoptotic activities. *J Biol Chem* 1995, **270**: 4312–4317.
21. Wang L, Miura M, Bergeron L, Zhu H, Yuan J. Ich-1, an ICE/ced-3-related gene, encodes both positive and negative regulators of programmed cell death. *Cell* 1994, **78**: 739–750.
22. Muzio M, Chinnaiyan AM, Kischkel FC, O'Rourke K, Shevchenko A, Ni J, Scaffidi C, Bretz JD, Zhang M, Gentz R, Mann M, Krammer PH, Peter ME, Dixit VM. FLICE, a novel FADD-homologous ICE/CED-3-like protease, is recruited to the CD95 (Fas/APO-1) death-inducing signaling complex. *Cell* 1996, **85**: 817–827.
23. Duan H, Orth K, Chinnaiyan AM, Poirier GG, Froelich CJ, He W, Dixit VM. ICE-LAP6, a novel member of the ICE/Ced-3 gene family, is activated by the cytotoxic T cell protease granzyme B. *J Biol Chem* 1996, **271**: 16,720–16,724.
24. Boldin MP, Goncharov TM, Goltsev YV, Wallach D. Involvement of MACH, a novel MORT1/FADD-interacting protease, in Fas/APO-1- and TNF receptor-induced cell death. *Cell* 1996, **85**: 803–815.
25. Fernandes-Alnemri T, Armstrong RC, Krebs J, Srinivasula SM, Wang L, Bullrich F, Fritz LC, Trapani JA, Tomaselli KJ, Litwack G, Alnemri ES. *In vitro* activation of CPP32 and Mch3 by Mch4, a novel human apoptotic cysteine protease containing two FADD-like domains. *Proc Natl Acad Sci USA* 1996, **93**: 7464–7469.
26. Schlegel J, Peters I, Orrenius S, Miller DK, Thornberry NA, Yamin TT, Nicholson DW. CPP32 apopain is a key interleukin 1β converting enzyme-like protease involved in Fas-mediated apoptosis. *J Biol Chem* 1996, **271**: 1841–1844.
27. Gagliardini V, Fernandez P, Lee RKK, Drexler HCA, Rotello RJ, Fishman MC, Yuan J. Prevention of vertebrate neuronal death by the crmA gene. *Science* 1994, **263**: 826–828.
28. Los M, Van de Craen M, Penning LC, Schenk H, Westendorp M, Baeuerle PA, Dröge W, Krammer PH, Fiers W, Schulze-Osthoff K. Requirement of an ICE/CED-3 protease for Fas/APO-1-mediated apoptosis. *Nature* 1995, **375**: 81–83.
29. Beidler DR, Tewari M, Friesen PD, Poirier G, Dixit VM. The baculovirus p35 protein inhibits Fas- and tumor necrosis factor-induced apoptosis. *J Biol Chem* 1995, **270**: 16,526–16,528.
30. Tewari M, Beidler DR, Dixit VM. CrmA-inhibitable cleavage of the 70-kDa protein component of the U1 small nuclear ribonucleoprotein during Fas- and tumor necrosis factor-induced apoptosis. *J Biol Chem* 1995, **270**: 18,738–18,741.
31. Tewari, M., W. G. Telford, R. A. Miller, V. M. Dixit. CrmA, a poxvirus-encoded serpin, inhibits cytotoxic T-lymphocyte-mediated apoptosis. *J Biol Chem* 1995, **270**: 22,705–22,708.
32. Bump NJ, Hackett M, Hugunin M, Seshagiri S, Brady K, Chen P, Ferenz C, Franklin S, Ghayur T, Li P, Licari P, Mankovich J, Shi LF, Greenberg AH, Miller LK, Wong WW. Inhibition of ICE family proteases by baculovirus antiapoptotic protein p35. *Science* 1995, **269**: 1885–1888.
33. Xue D, Horvitz HR. Inhibition of the Caenorhabditis elegans cell-death protease CED-3 by a CED-3 cleavage site in baculovirus p35 protein. *Nature* 1995, **377**: 248–251.
34. Clem RJ, Fechheimer M, Miller LK. Prevention of apoptosis by a baculovirus gene during infection of insect cells. *Science* 1991, **254**: 1388–1390.
35. Hay BA, Wassarman DA, Rubin GM. Drosophila homologs of baculovirus inhibitor of apoptosis proteins function to block cell death. *Cell* 1995, **83**: 1253–1262.
36. Nicholson DW, Ali A, Thornberry NA, Vaillancourt JP, Ding CK, Gallant M, Gareau Y, Griffin PR, Labelle M, Lazebnik YA, Munday NA, Raju SM, Smulson ME, Yamin T, Yu VL, Miller DK. Identification and inhibition of the ICE/CED-3 protease necessary for mammalian apoptosis. *Nature* 1995, **376**: 37–43.
37. Kuida K, Zheng TS, Na SQ, Kuan CY, Yang D, Karasuyama H, Rakic P, Flavell RA. Decreased apoptosis in the brain and premature lethality in CPP32-deficient mice. *Nature* 1996, **384**: 368–372.

38. Kostura MJ, Tocci MJ, Limjuco G, Chin J, Cameron P, Hillman AG, Chartrain NA, Schmidt JA. Identification of a monocyte specific pre-interleukin 1β convertase activity. *Proc Natl Acad Sci USA* 1989, **86**: 5227–5231.
39. Wilson KP, Black J, Thomson FJA, Kim EE, Griffith JP, Navia MA, Murcko MA, Chambers SP, Aldape RA, Raybuck SA, Livingston DJ. Structure and mechanism of interleukin-1β converting enzyme. *Nature* 1994, **370**: 270–275.
40. Walker NPC, Talanian RV, Brady KD, Dang LC, Bump NJ, Ferenz CR, Franklin S, Ghayur T, Hackett MC, Hammill LD, Herzog L, Hugunin M, Houy W, Mankovich JA, McGuiness L, Orlewicz E, Paskind M, Pratt CA, Reis P, Summani A, Terranova M, Welch JP, Xiong L, Moller A, Tracey DE, Kamen R, Wong WW. Crystal structure of the cysteine protease interleukin-1β-converting enzyme: A (p20/p10)2 homodimer. *Cell* 1994, **78**: 343–352.
41. Thornberry NA, Molineaux SM. Interleukin-1 beta converting enzyme: a novel cysteine protease required for IL-1 beta production and implicated in programmed cell death. *Protein Science* 1995, **4**: 3–12.
42. Rotonda J, Nicholson DW, Fazil KM, Gallant M, Gareau Y, Labelle M, Peterson EP, Rasper DM, Ruel R, Vaillancourt JP, Thornberry NA, Becker JW. The three-dimensional structure of apopain/CPP32, a key mediator of apoptosis. *Nature Struct Biol* 1996, **3**: 619–625.
43. Casciola-Rosen LA, Nicholson DW, Chong T, Rowan KR, Thornberry NA, Miller DK, Rosen A. Apopain/CPP32 cleaves proteins that are essential for cellular repair: A fundamental principle of apoptotic death. *J Exp Med* 1996, **183**: 1957–1964.
44. Slee EA, Zhu HJ, Chow SC, MacFarlane M, Nicholson DW, Cohen GM. Benzyloxycarbonyl-Val-Ala-Asp (OMe) fluoromethylketone (Z-VAD.FMK) inhibits apoptosis by blocking the processing of CPP32. *Biochem J* 1996, **315**: 21–24.
45. Darmon AJ, Nicholson DW, Bleackley RC. Activation of the apoptotic protease CPP32 by cytotoxic T-cell-derived granzyme B. *Nature* 1995, **377**: 446–448.
46. Martin SJ, Amarante-Mendes GP, Shi LF, Chuang TH, Casiano CA, O'Brien GA, Fitzgerald P, Tan EM, Bokoch GM, Greenberg AH, Green DR. The cytotoxic cell protease granzyme B initiates apoptosis in a cell-free system by proteolytic processing and activation of the ICE/CED-3 family protease, CPP32, via a novel two-step mechanism. *EMBO J* 1996, **15**: 2407–2416.
47. Chinnaiyan AM, Hanna WL, Orth K, Duan HJ, Poirier GG, Froelich CJ, Dixit VM. Cytotoxic T-cell-derived granzyme B activates the apoptotic protease ICE-LAP3. *Curr Biol* 1996, **6**: 897–899.
48. Darmon AJ, Ley TJ, Nicholson DW, Bleackley RC. Cleavage of CPP32 by granzyme B represents a critical role for granzyme B in the induction of target cell DNA fragmentation. *J Biol Chem* 1996, **271**: 21,709–21,712.
49. Liu XS, Kim CN, Pohl J, Wang XD. Purification and characterization of an interleukin-1β-converting enzyme family protease that activates cysteine protease P32 (CPP32). *J Biol Chem* 1996, **271**: 13,371–13,376.
50. Han ZY, Johnston C, Reeves WH, Carter T, Wyche JH, Hendrickson EA. Characterization of a Ku86 variant protein that results in altered DNA binding and diminished DNA-dependent protein kinase activity. *J Biol Chem* 1996, **271**: 14,098–14,104.
51. Orth K, O'Rourke K, Salvesen GS, Dixit VM. Molecular ordering of apoptotic mammalian CED-S/ICE-like proteases. *J Biol Chem* 1996, **271**: 20,977–20,980.
52. Henkart PA. ICE family proteases: Mediators of all apoptotic cell death. *Immunity* 1996, **4**: 195–201.
53. Fujita E, Mukasa T, Tsukahara T, Arahata K, Omura S, Momoi T. Enhancement of CPP32-like activity in the TNF-treated U937 cells by the proteasome inhibitors. *Biochem Biophys Res Commun* 1996, **224**: 74–79.
54. Grimm LM, Goldberg AL, Poirier GG, Schwartz LM, Osborne BA. Proteasomes play an essential role in thymocyte apoptosis. *EMBO J* 1996, **15**: 3835–3844.

55. Lin K, Baraban JM, Ratan RR. Inhibitors of the proteasome prevent Sindbis virus-induced apoptosis. 1997, in press.
56. Ashkenas J, Werb Z. Proteolysis and the biochemistry of life-or-death decisions. *J Exp Med* 1996, **183**: 1947–1951.
57. Lazebnik YA, Takahashi A, Moir RD, Goldman RD, Poirier GG, Kaufmann SH, Earnshaw WC. Studies of the lamin proteinase reveal multiple parallel biochemical pathways during apoptotic execution. *Proc Natl Acad Sci USA* 1995, **92**: 9042–9046.
58. Kayalar C, Ord T, Testa MP, Zhong L, Bredesen DE. Cleavage of actin by interleukin 1β-converting enzyme to reverse DNase I inhibition. *Proc Natl Acad Sci USA* 1996, **93**: 2234–2238.
59. Martin SJ, O'Brien GA, Nishioka WK, McGahon AJ, Mahboubi A, Saido TC, Green DR. Proteolysis of fodrin (non-erythroid spectrin) during apoptosis. *J Biol Chem* 1995, **270**: 6425–6428.
60. Brancolini C, Benedetti M, Schneider C. Microfilament reorganization during apoptosis: The role of Gas2, a possible substrate for ICE-like proteases. *EMBO J* 1995, **14**: 5179–5190.
61. Weaver VM, Carson CE, Walker PR, Chaly N, Lach B, Raymond Y, Brown DL, Sikorska M. Degradation of nuclear matrix and DNA cleavage in apoptotic thymocytes. *J Cell Sci* 1996, **109**: 45–56.
62. Kaufmann SH, Desnoyers S, Ottaviano Y, Davidson NE, Poirier GG. Specific proteolytic cleavage of poly-(ADP-ribose) polymerase: An early marker of chemotherapy-induced apoptosis. *Cancer Res* 1993, **53**: 3976–3985.
63. Casciola-Rosen LA, Anhalt GJ, Rosen A. DNA-dependent protein kinase is one of a subset of autoantigens specifically cleaved early during apoptosis. *J Exp Med* 1995, **182**: 1625–1634.
64. An B, Dou QP. Cleavage of retinoblastoma protein during apoptosis: An interleukin 1β-converting enzyme-like protease as candidate. *Cancer Res* 1996, **56**: 438–442.
65. Janicke RU, Walker PA, Lin XU, Porter AG. Specific cleavage of the retinoblastoma protein by an ICE-like protease in apoptosis. *EMBO J* 1997, in press.
66. Casciola-Rosen LA, Miller DK, Anhalt GJ, Rosen A. Specific cleavage of the 70-kDa protein component of the U1 small nuclear ribonucleoprotein is a characteristic biochemical feature of apoptotic cell death. *J Biol Chem* 1994, **269**: 30,757–30,760.
67. Waterhouse N, Kumar S, Song QH, Strike P, Sparrow L, Dreyfuss G, Alnemri ES, Litwack G, Lavin M, Watters D. Heteronuclear ribonucleoproteins C1 and C2, components of the spliceosome, are specific targets of interleukin 1β-converting enzyme-like proteases in apoptosis. *J Biol Chem* 1996, **271**: 29,335–29,341.
68. Emoto Y, Manome Y, Meinhardt G, Kisaki H, Kharbanda S, Robertson M, Ghayur T, Wong WW, Kamen R, Weichselbaum R, Kufe D. Proteolytic activation of protein kinase C δ by an ICE-like protease in apoptotic cells. *EMBO J* 1995, **14**: 6148–6156.
69. Na SQ, Chuang T, Cunningham A, Turi TG, Hanke JH, Bokoch GM, Danley DE. D4-GDI, a substrate of CPP32, is proteolyzed during Fas-induced apoptosis. *J Biol Chem* 1996, **271**: 11,209–11,213.
70. Goldberg YP, Nicholson DW, Rasper DM, Kalchman MA, Koide B, Graham RK, Bromm M, Kazemi-Esfarjani P, Thornberry NA, Vaillancourt JP, Hayden MR. Cleavage of huntingtin by apopain, a proapoptotic cysteine protease, is modulated by the polyglutamine tract. *Nature Genetics* 1996, **13**: 442–449.
71. Tewari M, Quan LT, O'Rourke K, Desnoyers S, Zeng Z, Beidler DR, Poirier GG, Salvesen GS, Dixit VM. Yama/CPP32β, a mammalian homolog of CED-3, is a CrmA-inhibitable protease that cleaves the death substrate poly(ADP-ribose) polymerase. *Cell* 1995, **81**: 801–809.
72. Lazebnik YA, Kaufmann SH, Desnoyers S, Poirier GG, Earnshaw WC. Cleavage of poly(ADP-ribose) polymerase by a proteinase with properties like ICE. *Nature* 1994, **371**: 346–347.

73. Schreiber V, Hunting D, Trucco C, Gowans B, Grunwald D, de Murcia G, Menissier de Murcia J. A dominant-negative mutant of human poly(ADP-ribose) polymerase affects cell recovery, apoptosis, and sister chromatid exchange following DNA damage. *Proc Natl Acad Sci USA* 1995, **92:** 4753–4757.
74. Romac JMJ, Keene JD. Overexpression of the arginine-rich carboxy-terminal region of U1 snRNP 70K inhibits both splicing and nucleocytoplasmic transport of mRNA. *Genes Dev* 1995, **9:** 1400–1410.

6
The Cell Cycle and Neuronal Cell Death

Robert S. Freeman

INTRODUCTION

Birth and death, the extremes of life, conjure images of antithetic and counteracting events necessary for maintaining a balance among living organisms. Their cellular counterparts, cell division and cell death, work in similar opposition to ensure proper development of an organism and to maintain homeostasis and normal function. A breakdown of either process can be detrimental and, in humans, may lead to birth defects, cancer, autoimmune disease, and neurological disorders. Although researchers have been well aware of the importance of proper cell division for decades, only recently has the significance of cell death attracted widespread appreciation. Consequently, we now have a relatively sophisticated understanding of the machinery that controls cell division; in comparison, we have only a rudimentary knowledge of the mechanisms that underlie cell death. But surprisingly, given our inclination to categorize birth and death as opposing and distinct forces, much of what we currently know about cell death has been fueled by research into cell cycle control. In this chapter, the hypothesis that the control of cell death and the cell cycle overlap will be discussed.

Several observations raise the possibility that the molecular mechanisms for control of cell division and cell death are related. Most apparent is the notion that dividing cells and dying cells share gross morphological and structural features. For completion of the cell cycle, chromosomes must be replicated, condensed, segregated, and decondensed. The mitotic spindle must be assembled and disassembled, the nuclear membrane must break down and be rebuilt. Dividing cells round up and become less adherent, and their membranes invaginate as cytokinesis begins. During apoptotic cell death, chromatin becomes condensed, marginated, and fragmented. The nuclear membrane irreversibly breaks down. The cell membrane rounds up, becomes less adherent, and blebs, eventually pinching off membrane-enclosed nuclear and cytosolic remnants called apoptotic bodies. Thus, transient and reversible structural changes involving chromatin condensation, cell rounding, and cytoskeletal rearrangements and that facilitate cell division, occur irreversibly and with destructive consequences during cell death.

Another tie between cell cycle control and cell death stems from the realization that tumor suppressor genes and oncogenes can function as regulators of cell death. The best characterized examples are the p53 tumor suppressor gene and the c-myc protooncogene. Induction of p53 leads to cell cycle arrest or to cell death; which fate is dependent on cell type, growth factor availability, and the expression of genes such as the retinoblastoma gene and E2F-1. Stress, hypoxia, and irradiation can all induce p53-mediated apoptosis *(1)*. Expression of c-myc is tightly linked to mitogen stimulation and is required for cell cycle progression. However, enforced c-myc expression in cells deprived of growth factors produces cells that fail to growth arrest and concomitantly undergo apoptosis *(2,3)*. The ability of genes such as c-myc and p53 to regulate apoptosis forms the basis of the hypothesis that cell death can be initiated by conflicting growth regulatory signals *(4)*.

A third link between the cell cycle and cell death is indicated by the direct involvement of cell cycle proteins in specific examples of cell death. Cyclin-dependent protein kinases (CDK), their regulatory subunits the cyclins, and certain CDK inhibitors and substrates have recently been implicated as regulators of apoptosis. In addition, several examples exist where the deregulated control of cell cycle proteins in proliferating cells ultimately leads to cell death, or so-called mitotic catastrophe *(5)*. But in neurons and other postmitotic cells, the notion that cell cycle proteins function as part of the cell death mechanism, or that entry into an abnormal or aborted cell cycle contributes to cell death, is less tenable. Accordingly, much of the following discussion will be based on studies performed in nonneuronal systems such as lymphocytes, fibroblasts, and various immortalized cell lines. Where data are available or reasonable inferences can be made, the involvement of cell cycle proteins in neuronal death will be considered.

To understand the potential involvement of cell cycle regulators in cell death, we must first consider their role in the normal cell cycle. For this discussion, the author refers the reader to several excellent review articles rather than cite all of the pertinent original studies.

A Cell Cycle Overview

Cell division requires the replication of a cell's genetic material and the partitioning of its nuclear and cytosolic contents into two daughter cells, each essentially identical to the parent. In most cases, this is accomplished as the cell cycles through four phases—G_1, S, G_2, and M. Elaborate feedback mechanisms called checkpoints regulate passage from one phase to the next *(6)*. Checkpoints work by temporarily halting the cell cycle if damage to DNA or the mitotic apparatus is detected. If the damage can be repaired, progression through the remaining stages of the cycle continues and no harm comes to the cell; if the damage is irreparable, cell death may result. The p53 protein is an example of an S-phase checkpoint controller that links cell cycle and cell death mechanisms when DNA damage repair is not possible *(7)*.

Cell cycle entry and early G_1 events are dependent upon the binding of extracellular mitogens such as growth factors to specific cell surface receptors. At a time in G_1 termed the restriction point, cell cycle progression becomes independent of growth factors. The restriction point is marked by the activation of G_1-CDK and the subsequent inactivation of the product of the retinoblastoma gene, pRb *(8)*. These events are followed closely by the increased transcription of genes necessary for DNA synthesis. DNA synthesis occurs during S-phase, generally the longest phase of the cell cycle. After the DNA is successfully replicated and the criteria of the S-phase checkpoint are satisfied, the cell enters G_2 where preparations for mitosis begin. In G_2, the proteins that drive a

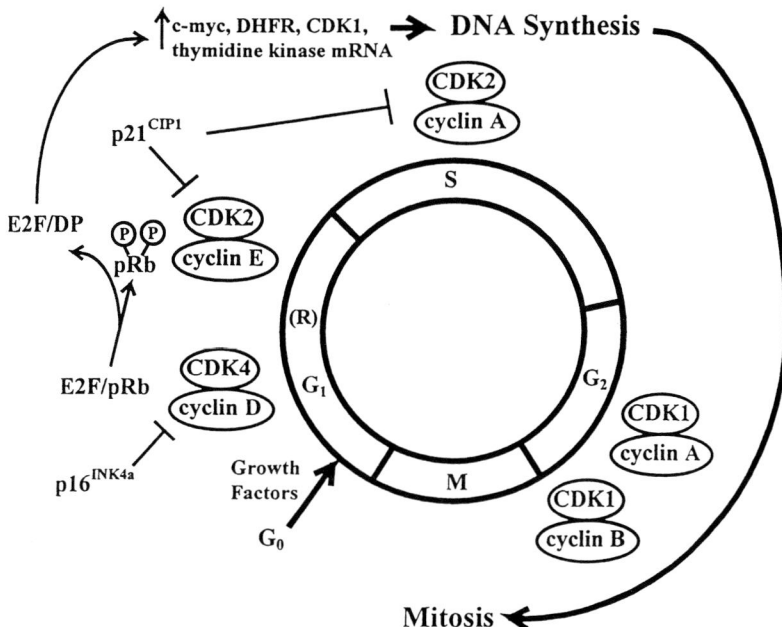

Fig. 1. Molecular components of the cell cycle machinery. Growth factors promote entry into the cell cycle and are required for progression through G_1 up to the restriction point (R). Phosphorylation of the retinoblastoma protein (pRb) by specific cyclin/cyclin-dependent kinase (CDK) complexes occurs at R and results in the release of E2F from pRb. E2F proteins associate with DP proteins to activate the transcription of genes necessary for S-phase and cell cycle progression such as c-myc, CDK1, dihydrofolate reductase (DHFR), and thymidine kinase. Specific protein inhibitors of CDK regulate orderly passage through the cell cycle and induce cell cycle arrest when DNA damage is detected. Additional cyclin/CDK complexes function during the S-, G_2-, and M-phases.

cell into and through the M-phase accumulate and assemble into an inactive complex. When activated at the G_2/M-phase border, this complex, called the mitosis promoting factor (MPF), initiates chromatin condensation, spindle formation, nuclear envelope breakdown, and chromosome segregation. If proper chromosome segregation occurs, MPF is abruptly destroyed, cytokinesis ensues, and the newly formed daughter cells either continue in the cell cycle (provided they receive sufficient mitogen) or withdraw from the cell cycle into a quiescent state.

Critical Components of the Cell Cycle Machinery

The workhorses of the cell cycle are the CDK, a family of cytosolic serine/threonine protein kinases that currently includes eight members. The various CDK are activated at major transitions in the cell cycle: near the restriction point in G_1, at the G_1/S border, and at the G_2/M border (Fig. 1). Their activation is tightly controlled by a complex series of events that includes the binding of regulatory proteins called cyclins and the subsequent phosphorylation of the CDK subunit by the CDK-activating kinase, itself a CDK *(9)*. CDK-cyclin complexes are inactive when bound to CDK inhibitory proteins or when specific threonine and tyrosine residues in the CDK catalytic domain are phosphorylated.

In mammalian cells, a succession of cyclins teams up with a succession of CDK as cells progress into G_1 and through mitosis *(10,11)*. The different CDK-cyclin complexes are thought to drive the cell cycle from one phase into the next with progression determined by the specific subset of proteins that are activated or inactivated as a result of phosphorylation by a particular CDK-cyclin complex. CDK4 or CDK6 (depending on the cell type), complexed with one of several D-type cyclins (cyclins D1–D3), is the first CDK to function in G_1. CDK4 activity is coupled to extracellular signals due to the dependency of cyclin D expression on growth factor supplies. Late in G_1, a second cyclin, cyclin E, forms complexes with a different CDK, CDK2. The kinase activity of this complex correlates with cyclin E protein levels which peak in late G_1 and decline after cells have entered S-phase. Microinjection of antibodies to cyclin E or cyclin D1, or antisense cyclin D1 cDNAs, arrests cells in G_1 and prevents entry into S-phase *(12–14)*. Ectopic expression of a dominant negative mutant of CDK2 similarly results in G_1 arrest *(15)*. In contrast, overexpression of either G_1 cyclin accelerates the cell cycle by shortening the G_1-phase *(13,16)*. Therefore, G_1 regulation and duration is closely tied to the G_1 cyclins and their affiliated CDK.

CDK2 kinase activity is found in a complex with cyclin A during S-phase and the integrity of this complex is required for the onset of DNA synthesis *(17)*. Cyclin A also binds and activates CDK1 (also called Cdc2), with peak activation of this complex occurring in G_2. Cyclin A is subsequently degraded during the G_2/M transition. Entry into M phase requires the activation of MPF, which consists of CDK1 in association with cyclin B. At the metaphase/anaphase transition, cyclin B is abruptly degraded by ubiquitin-dependent proteolysis. The destruction of cyclin B and resulting inactivation of MPF is necessary for cells to complete mitosis and may be part of a self-regulating negative feedback loop *(18)*.

As outlined above, CDK activity is influenced by binding to cyclins, by specific phosphorylation and dephosphorylation events, and by proteolysis of cyclins. An additional mechanism for regulating CDK activity involves a diverse family of proteins known as CDK inhibitors, or CKI *(19)*. Seven mammalian CKIs have been described so far and these fall into two groups based on similarities in sequence and function. One group that includes $p16^{INK4a}$, $p15^{INK4b}$, $p18^{INK4c}$, and $p19^{INK4d}$, specifically interacts with monomeric CDK4 and CDK6 thereby preventing the binding of D-type cyclins to these CDK. The second group consists of $p21^{CIP1}$, $p27^{KIP1}$, and $p57^{KIP2}$ and inhibits G_1 CDK as well as CDK2 by associating with preactivated CDK-cyclin complexes. The expression of many of the CKIs is regulated negatively by extracellular mitogens and positively by growth suppressing factors such as TGF-β and interferon-γ. Although CKI were originally thought to be specific for CDK, at least one, $p21^{CIP1}$, can interact with a second class of protein kinases. The stress-activated protein kinases (SAPK), also known as the c-Jun amino-terminal kinases (JNK), are activated in response to several cellular stresses including DNA damage, heat shock, tumor-necrosis factor, and neurotrophic factor withdrawal. $p21^{CIP1}$, but not $p16^{INK4a}$, can bind SAPKs/JNK and inhibit their kinase activity *(20)*.

Throughout the cell cycle cyclins are thought to target CDK to specific substrates. The best studied G_1-cyclin/CDK substrate is pRb, the protein encoded by the retinoblastoma tumor suppressor gene *(21)*. Prior to the restriction point in G_1, pRb exists mostly in a hypophosphorylated form. Hypophosphorylated pRb actively suppresses cell cycle progression, at least in part by binding and consequently inactivating members of the E2F/DP family of transcriptional regulators. When free from pRb, E2F-1 (either as a homodimer or as a heterodimer with a DP protein) contributes to the transcription of several genes that are important for DNA replication and cell division control *(22)*. In addition to E2F-1, hypophosphorylated pRb also serves as a cellular target for several DNA tumor virus proteins including adenovirus E1A protein, SV40 T antigen, and the E7 protein of human papillomavirus. To promote DNA synthesis and proliferation of infected cells, these viral oncoproteins bind and sequester hypophosphorylated pRb, thereby preventing pRb from performing its normal growth suppressing functions.

Close to or perhaps concurrent with the restriction point in the cell cycle, much of the pRb becomes hyperphosphorylated. Hyperphosphorylation occurs at more than a dozen distinct sites in the protein, many of which lie within sequences typical of those phosphorylated by CDK. CDK4 and CDK6, in association with D-type cyclins, are likely candidates for phosphorylating pRb. Late in G_1, CDK2/cyclin E complexes probably also participate in the hyperphosphorylation of pRb. As mentioned above, an important consequence of pRb hyperphosphorylation is the release and subsequent activation of the E2F-1 transcription factor.

MITOTIC CATASTROPHE

Mutations that affect cell cycle machinery often have lethal consequences in dividing cells *(5)*. Mitotic catastrophe is a term used to describe the cell death that occurs if a cell enters mitosis before completing DNA replication. Mitotic catastrophe is initiated by the uncoupling of M-phase activating events from S-phase checkpoint controls and is associated with premature activation of the cyclin and CDK proteins that govern the G_2/M transition. Although not originally described as apoptosis, this kind of cell death is characterized by rounded cell morphology and hypercondensation of chromatin, features typical of apoptosis. That premature or inappropriate activation of M-phase events by biochemical or genetic means can lead to cell death with features that resemble apoptosis is certain. Whether apoptosis generally involves deregulated cell cycle events, however, is less certain. What follows is a description of some of the more significant biochemical, pharmacological, and genetic data that suggest the involvement of cell cycle proteins in at least some instances of cell death.

Most physiologically appropriate cell deaths as well as certain pathologic cell deaths occur by apoptosis. Some of the hallmarks of apoptosis—cell rounding, loss of adhesion, disassembly of the nuclear lamina and nucleolus, and chromatin condensation—resemble similar events that occur during cell division *(23)*. In fact, a cell-free system useful for analyzing nuclear changes that accompany apoptosis was originally devised as a model for studying the nuclear events of mitosis *(24)*. In dividing cells, CDK-mediated phosphorylation of proteins such as the nuclear lamins, histone H1, nucleolin, and pp60[c-src] has been suggested as a mechanism by which many of these structural changes occur *(25)*. With the exception of the lamins, little is known about the state of these potential CDK substrates during cell death. The lamins assemble to form the scaffold that underlies the nuclear membrane. This scaffold breaks down during mitosis and apoptosis, although apparently by distinct mechanisms. In mitosis, phosphorylation of lamin polymers by CDK1 leads to their disassembly, although individual lamin subunits remain intact. In apoptotic cells, the lamin polymers also become phosphorylated and solubilized, but subsequently (or concurrently), monomeric lamins get cleaved, probably by an ICE-like protease *(26)*. Although breakdown of the nuclear lamina and lamin cleavage have been demonstrated in only a few cell death paradigms (none of which involve neurons) it seems likely that this will be a general feature of apoptotic cell death. Irreversible breakdown of the lamina may facilitate the hypercondensation of chromatin in the nucleus of apoptotic cells. Consistent with this, pharmacological inhibition of the lamin protease prevents the hypercondensation of chromatin into compact spheres that occurs late in apoptosis, but, interestingly, does not prevent the initial condensation of the chromatin around the perimeter of the nucleus nor its cleavage to a nucleosomal ladder *(26)*.

Studies on the cytotoxicity of antitumor drugs and other drugs that block the cell cycle have provided much correlative data supporting a role for S- and G_2/M-phase CDK and cyclins in the control of cell death. For example, treatment of Chinese hamster ovary cells with the anticancer drug cisplatin results in G_2 phase arrest, activation of CDK1, and apoptosis *(27)*. When caffeine is used to accelerate the activation of CDK1, the rate of cisplatin-induced apoptosis is also accelerated. CDK1 is also activated during DNA-damage-induced-apoptosis caused by topoisomerase inhibitors *(28)*. Moreover, both CDK1 and CDK2 are activated during apoptosis of S-phase-arrested HeLa cells induced by the protein kinase inhibitor staurosporine, the phosphatase inhibitor okadaic acid, or caffeine *(29)*. Staurosporine-induced apoptosis in leukemic T-cell lines is similarly accompanied by activation of CDK1 and CDK2 and growth arrest *(30)*. As a final example, CDK1 activation occurs coincident with apoptotic DNA fragmentation in fibroblasts treated with taxol, a microtubule stabilizer and antitumor drug *(31)*. These results seem to indicate that apoptosis is closely linked to the activation of cell cycle regulators, at least when cell death is caused by agents that directly target the cell cycle machinery.

Because the instances of cell death described above resemble mitotic catastrophe more than physiologic cell death, the involvement of CDK1 or CDK2 in apoptosis may be restricted to special cases of dividing cells. There is no doubt that such special cases are of great importance to cancer biologist, but their relevance to neuronal death is less evident. However, such studies do highlight the potential for CDK or CDK-like enzymes to play a role in other cell death mechanisms.

CELL CYCLE REGULATORS AND NEUROTROPHIC FACTOR DEPRIVATION

More significant links connecting cell cycle regulators and neuronal death come from studies of neurotrophic factor deprivation. Sympathetic neurons undergo RNA and protein synthesis-dependent apoptosis when deprived of nerve growth factor (NGF) *(32)*. Despite early declines in RNA synthesis and RNA abundance after NGF withdrawal, mRNA levels of cyclin D1 and of several immediate early genes associated with cell proliferation increase substantially *(33,34)*. Although the significance of increased cyclin D1 expression during apoptosis in sympathetic neurons remains unknown, a case for cyclin D1 involvement in neuronal death has been made using differentiated cells produced from the mouse NIE-115 neuroblastoma cell line. In these neuron-like cells, increases in cyclin D1 protein and CDK4 kinase activity accompany apoptosis induced by serum withdrawal *(35)*. Much of the cell death is blocked by overexpression of the cyclin D1/CDK4 inhibitor p16^{INK4a}. Ectopic expression of p16^{INK4a}, pRb, or a dominant-negative CDK4 protein also blocks apoptosis in fibroblasts caused by overexpression of cyclin D1.

Increased cyclin D1 levels are also observed during apoptosis in cerebellar granule cells that die in either of two neurological mutant mice, *lurcher* and *staggerer (36)*. In both mutants, granule cell death results from the absence of normal trophic support provided by their postsynaptic targets, the Purkinje cells. (In *lurcher*, Purkinje cells die first; in *staggerer*, they fail to mature properly.) In addition to increased cyclin D1 levels, granule cells in the internal granule layer have elevated levels of the proliferating cell nuclear antigen and apparently enter S-phase prior to cell death. These observations are particularly intriguing since the granule neurons of normal mice are postmitotic at the corresponding developmental stage.

Although these studies suggest a role for cyclin D1 in neuronal death induced by trophic factor deprivation, cyclin D1 does not appear to be a universal regulator of apoptosis. Mice lacking both copies of the cyclin D1 gene are viable and have defects in retina and breast development but do not exhibit any phenotypes yet attributable to defective cell death mechanisms *(37)*.

The M-phase-specific cyclin B, rather than cyclin D1, increases in abundance in differentiated rat PC12 cells induced to die by NGF withdrawal *(38)*. CDK1 kinase activity appears in these neuron-like cells concurrent with increased cyclin B. However, whether CDK1 or cyclin B is required for PC12 cell death has not been demonstrated. An alternative possibility is that the increased CDK1 activity could be derived from a small fraction of PC12 cells that may successfully reenter the cell cycle after NGF removal. Other data have established an intriguing correlation between the ability of various pharmacological agents to prevent cell death and their ability to inhibit the cell cycle. Various agents that block proliferation by inhibiting the G_1/S transition effectively inhibit the apoptosis of dividing PC12 cells caused by serum deprivation, and the death of differentiated PC12 cells or sympathetic neurons induced by NGF withdrawal *(39)*. In contrast, several S-, G_2-, and M-phase blockers fail to prevent cell death in these paradigms. Two CDK inhibitors, flavopiridol and olomoucine, also block apoptosis of NGF differentiated PC12 cells and sympathetic neurons deprived of NGF, but these drugs promote rather than prevent the death of dividing PC12 cells *(40)*. Although these are strong correlations, the exact targets of these compounds as well as their ultimate specificities remain unknown, as do the mechanisms by which they inhibit either proliferation or apoptosis.

Studies on the mechanism of neuronal death frequently employ embryonic or relatively immature neurons, or immortalized cell lines that have undergone several days of treatment to produce a differentiated neuron-like cell. Studies with such cell lines suggest that the differentiation state of a cell can affect its susceptibility to apoptotic stimuli. Taking this further, one may speculate that the extent of neuronal differentiation may also dictate the mechanism by which cells die. Thus, neurons from an adult animal may or may not die by the same mechanism as newly postmitotic neurons. If one function of neurotrophic factors is to keep newly born neurons in a postmitotic state, perhaps by actively suppressing the cell cycle machinery, then the removal of trophic factor soon after the final cell division may permit cells to successfully reenter the cell cycle. If the trophic factor is removed at a later stage, some cells may still resume the cell cycle while others may lose this capacity and undergo a fatally flawed cell cycle attempt or mitotic catastrophe. More mature neurons may lose altogether the ability to attempt cell cycle reentry and die by a mechanism unrelated or only partly related to cell cycle control.

A REQUIREMENT FOR CYCLINS AND CDK DURING CELL DEATH

Nonneuronal paradigms have, so far, proven most useful for providing evidence that the function of specific cyclins and CDK is required for apoptosis. For example, activation-induced death of T cell hybridomas is inhibited by cyclin B-specific antisense oligonucleotides (the same antisense oligonucleotides do not affect glucocorticoid induced death) *(41)*. Antisense oligonucleotides against a conserved sequence found in cyclins B, E, and A reduce the ability of HIV-1 Tat protein to induce apoptosis in lymphocyte cell lines *(42)*. Other reports show a CDK requirement for cell death. CDK1 activation occurs coincident with DNA fragmentation during apoptosis in lymphoma cells treated with the lymphocyte granule-derived protease fragmentin-2. Cells expressing a temperature sensitive mutant of CDK1 become less sensitive to fragmentin-2-induced apoptosis when CDK1 is inactivated *(43)*, although, as described above for cyclin B, apoptosis induced by other stimuli still occurs in the absence of CDK1 *(44)*. Surprisingly, however, in these same cells the level of apoptosis induced by DNA-damaging drugs actually increases when CDK1 is inactivated *(45)*.

Dominant negative mutants of CDK 1, 2, or 3 can also protect cells from apoptosis. HeLa cells transfected with any one of these mutants are more resistant to TNF-α or staurosporine-induced death than are control cells. The protection provided by dominant negative CDK is reversed by cotransfecting the wild type CDK or cyclin A. In addition, overexpression of the wild type CDK or cyclin A decreases the ability of the Bcl-2 protein to protect cells from TNF-α *(46)*. Further connections between CDK and Bcl-2 remain to be determined.

The c-myc protooncogene is an example of a mediator of both proliferation and cell death that may function by activating CDK. Normally, fibroblasts downregulate c-myc expression and growth arrest when depleted of growth factors. However if c-myc expression is kept artificially high, then growth factor withdrawal results in apoptosis and not growth arrest *(2,3)*. The c-Myc protein is a transcription factor that acts in conjunction with its partner, Max. One target of Myc/Max heterodimers is the cdc25A gene *(47)*. The cdc25 gene family encodes phosphatases that contribute to the activation of CDK by dephosphorylating specific threonine and tyrosine residues conserved among the CDK proteins. Increased transcription of the cdc25A gene may account for the activation of CDK2 and CDK4 and subsequent phosphorylation of pRb that occurs after induction of c-myc expression in growth arrested cells. Like c-myc, deregulated expression of cdc25A can induce apoptosis in cells depleted of growth factors and antisense depletion of cdc25A produces cells that are significantly impaired in their ability to undergo c-myc-induced apoptosis. A plausible, but still unexplored explanation for these results is that the activation of cdc25A expression by c-myc, in the absence of other growth promoting signals, leads to the activation of CDK2 and CDK4 and subsequent phosphorylation and inactivation of pRb. This could result in deregulation of E2F-1 complexes which, by a still unknown mechanism, may lead to apoptosis.

Studies on the differentiation of myoblasts into multinucleated myotubes have revealed a possible role for CDK inhibitors in protecting differentiating cells from programmed cell death. After mitogen withdrawal, cultured myocytes either terminally differentiate or undergo apoptosis. Those cells that differentiate become apoptosis resistant and this is correlated with the induction of p21^{CIP1} *(48)*. Myocytes transfected with p21^{CIP1} or p16^{INK4a} are protected from mitogen withdrawal-induced death whereas cells transfected with a mutant form of p21^{CIP1} lacking the CDK binding domain are not protected. p21^{CIP1} or p16^{INK4a} may directly block death by inhibiting a CDK or other protein kinase (SAPK/JNK?) that functions as an integral component of the cell death program. Alternatively, the CDK inhibitors may indirectly block death by actively promoting a differentiation pathway that is dominant over the cell death program.

While ectopic p21^{CIP1} expression can prevent cell death, lack of p21^{CIP1} expression can promote apoptosis. Neuroblastoma cells induced to differentiate with NGF upregulate p21^{CIP1} expression; blockage of p21^{CIP1} induction with antisense oligonucleotides causes cell death instead of differentiation *(49)*. In tumor cells the absence of p21^{CIP1} uncouples S/M phase controls by overriding a checkpoint that normally brings on growth arrest following DNA damage. Thus, tumor cells that lack p21^{CIP1} only transiently growth arrest after radiation- or drug-induced DNA damage and then undergo additional S phases without intervening mitoses. The resulting deregulated cell cycle ultimately ends in death by apoptosis *(50)*.

Importantly, no studies have implicated a particular CDK or cyclin as a general mediator of apoptosis. Even within a single cell type, different death-inducing stimuli activate pathways that differ in their dependence on specific cell cycle proteins. What these studies do suggest is that CDK have a role in many instances of apoptosis that occur after interruption or withdraw from the cell cycle.

pRb AND CELL DEATH

In addition to the cyclins, CDK, and CKI, the pRb protein regulates certain types of apoptosis. As outlined above, G_1-specific CDK phosphorylate pRb. Hyperphosphorylation of pRb decreases its affinity for E2F-1, and, E2F-1, once free from pRb, activates the transcription of a subset of genes important for DNA synthesis and cell cycle control *(22)*. pRb and related growth suppressing proteins are also the target of viral oncoproteins, such as the adenovirus E1A and human papilloma virus E7 proteins. E1A and E7 bind pRb, thereby preventing the formation of pRb/E2F-1 complexes. Without the G_1/S checkpoint normally provided by pRb, the cell is free to enter S phase and complete mitosis. Expression of E1A and E7 is associated not only with proliferation and oncogenic transformation, but also with cell death *(51,52)*. These observations together with data discussed next suggest that one function of pRb may be to suppress programmed cell death.

The first indication that pRb might play a role in neuronal survival came from analyses of transgenic mice lacking functional pRb *(53–55)*. Mice homozygous for loss of pRb function die before the 16th embryonic day with multiple defects including massive amounts of cell death throughout the nervous system, particularly in the hindbrain and dorsal root ganglia. Large numbers of mitotic figures are found outside of their normal neurogenic regions in both the central and peripheral nervous systems indicating deregulated proliferation or migration in these animals. Many of these ectopically dividing cells die shortly after entering S-phase *(56)*. In addition, fibroblasts from pRb knockout animals are more susceptible to apoptosis induced by antitumor agents or by growth factor deprivation than are fibroblasts containing one or no inactivated alleles *(57)*.

pRb may also have a role in establishing the neurotrophin responsiveness of certain peripheral neurons. Neurons from the dorsal root and trigeminal ganglia of mice lacking functional pRb have decreased survival and neurite outgrowth *in vitro*, even in the presence of the appropriate neurotrophins. This may be explained, at least partially, by the significantly reduced expression of the TrkA, TrkB, and p75^{NGFR} neurotrophin receptors in these ganglia *(56)*.

Additional support for an anti-apoptotic function for pRb comes from studies on human osteosarcoma SAOS-2 cells, which lack endogenous pRb expression. SAOS-2 cells undergo apoptosis in a dose- and time-dependent manner after exposure to ionizing radiation. Ectopic expression of pRb in these cells promotes viability and decreases apoptosis induced by radiation exposure. Cells expressing a mutant form of pRb that fails to complex with E2F-1 or E1A are not protected from apoptosis *(58)*. Overexpression of pRb also inhibits liver cell apoptosis caused by TGF-β1 and apoptosis of tumor cells caused by interferon-γ *(59,60)*. Hyperphosphorylation of pRb by a CDK could be one possible mechanism for inactivating pRb during cell death. An intriguing alternative mechanism for pRb inactivation is suggested by one report demonstrating cleavage of pRb by an ICE-related protease activity during DNA damage-induced death of leukemia cells *(61)*.

Data described above and elsewhere *(21)* suggest that inactivation of pRb can promote oncogenic transformation when conditions are favorable for proliferation, or apoptosis when the cell cycle is perturbed. How might inactivation of pRb enhance cell death? The phenotypes of cells that ectopically express E2F-1 and mice that lack E2F-1 may provide some clues to this question. Deregulated expression of E2F-1 is sufficient to drive quiescent cells all the way through G_1 and into S-phase. However, cells forced to enter S-phase by deregulated E2F-1 expression do not complete a normal cell cycle. Instead, they enter a pathway leading to apoptosis *(62–64)*. This pathway appears to be dependent on p53 function *(62,64,65)* and can be suppressed by coexpressing pRb *(62)*. Moreover, E2F-1-induced apoptosis is enhanced when E2F/pRb interactions are precluded by point mutations within E2F-1 *(66)*. In interleukin-3 (IL-3)-dependent myeloid cells, E2F-1 overexpression accelerates IL-3 withdrawal-induced death. Overexpression of the E2F-1 binding partner DP-1 has no effect on myeloid cell survival by itself, but when coexpressed with E2F-1, DP-1 leads to rapid cell death even in the presence of IL-3 *(67)*.

As cells move into S-phase, E2F-1 complexes with a CDK bound to cyclin A; this ultimately leads to suppression of E2F-1 DNA-binding activity. E2F-1 mutants that are defective in cyclin A/CDK binding cause S-phase delay and apoptosis in fibroblasts in the presence of serum *(68)*. Only mutant E2F-1 proteins with intact DNA binding and transactivation domains induce apoptosis. Mice that lack E2F-1 have been produced and found to exhibit a defect in T-lymphocyte development, apparently due to a defect in thymocyte apoptosis *(69,70)*. Surprisingly, these mice also develop a wide range of tumors and exhibit increased cell proliferation in several tissues suggesting that E2F-1 can function in vivo to regulate apoptosis and suppress cell proliferation.

Viral Oncogene Expression in Neurons Can Result in Cell Death

Activation of cellular protooncogenes or ectopic expression of viral oncogenes causes transformation in dividing stem cell populations that eventually can result in clonal expansion and neoplasia. But, unlike stem cells, terminally differentiated neurons can not successfully re-enter the cell cycle and do not give rise to tumors. Instead, the expression of certain oncogenes in neurons often results in cell death (71). This occurs in transgenic mice that express the viral oncogene SV40 T-antigen under the control of neuron-specific promoters. For instance, when T-antigen is directed to rod photoreceptors or horizontal cells of the retina, progressive neuronal death rather than tumor formation is the result (72,73). Similarly, cell death is the outcome when T-antigen is expressed in cerebellar Purkinje cells (74). In both retinal horizontal cells and Purkinje cells, T-antigen expression is turned on only after the neurons become postmitotic, and, in Purkinje cells, T-antigen induction causes ectopic DNA synthesis prior to DNA fragmentation and cell death. The analysis of mice expressing different mutant T-antigen proteins demonstrates that the pRb binding domain of T-antigen is both necessary and sufficient for Purkinje cell degeneration (75). Thus, in this example at least, an oncogene expressed in a postmitotic neuron can interfere with the normal growth suppressing machinery, causing an apparent reentry into the cell cycle and, ultimately, the death of the cell. Based on these observations, one might speculate that cells that have not full exited the cell cycle remain susceptible to oncogenic transformation whereas oncogene expression in newly postmitotic neurons leads to cell death.

The fate of neurons in which pRb function is missing or compromised may depend on the status of a second tumor suppressor gene, p53. p53 itself has been associated with apoptosis caused by DNA damage (76,77), hypoxia (78), c-myc expression (79), and growth factor withdrawal (80). A connection between pRb and p53 is suggested by studies on lens fiber cells in the developing ocular lens (81). Homozygous inactivation of pRb results in unchecked proliferation, impaired differentiation, and inappropriate apoptosis in lens fiber cells. However, ectopic apoptosis of fiber cells is almost totally suppressed in mouse embryos doubly null for pRb and p53. Inactivation of pRb with the pRb-binding E7 protein of human papillomavirus produces a result similar to targeted disruption of pRb, i.e., death of lens fiber cells (82,83). Once again, inactivation of p53 in addition to pRb (obtained by expressing in mice both the p53-binding E6 protein and the E7 protein) reduces the apoptosis of fiber cells and eventually gives rise to lens tumors in adult animals (83). Thus, neuronal death that occurs via inactivation of pRb may be dependent on p53 function.

CONCLUSIONS

The morphology of cells undergoing apoptosis is similar to cells undergoing both normal mitosis and an aberrant form of mitosis called mitotic catastrophe. During each of these processes, cells loosen substrate attachments, lose cell volume, condense their chromatin, and dismantle their nuclear membranes. These morphological similarities suggest that the underlying biochemical changes that accompany these processes may also be related. Consistent with this idea, the search for molecules that regulate cell death has turned up numerous proteins first characterized as cell cycle regulators.

Cyclins and CDK are activated in a wide range of cells undergoing apoptosis in response to an equally wide range of stimuli. Inactivation of pRb, a process mediated

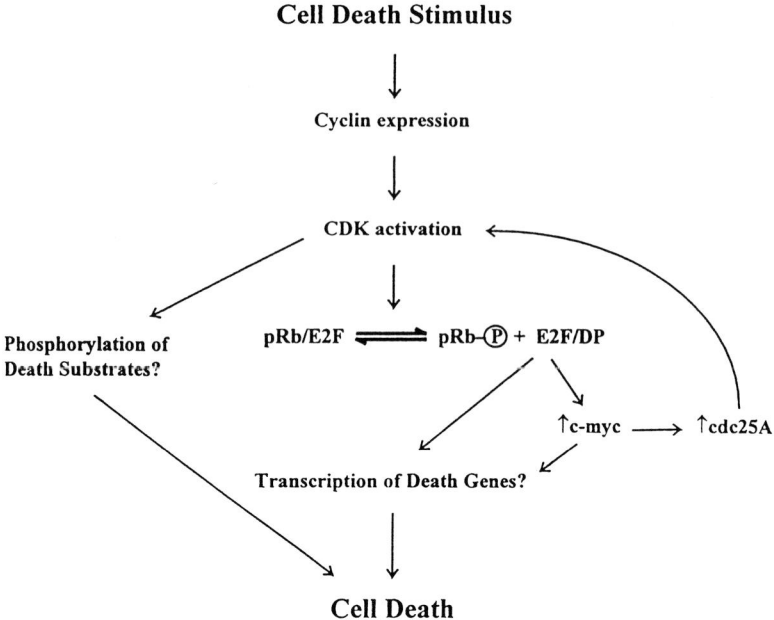

Fig. 2. A hypothetical model for cell death. Various cell death stimuli are associated with cyclin expression and CDK activation. CDK phosphorylate and regulate a variety of proteins during the cell cycle—some of these proteins (e.g., pRb, nuclear lamins), or a novel set of proteins, may be regulated by CDK during cell death. CDK-mediated phosphorylation of pRb leads to activation of E2F and subsequent transcription of genes such as *c-myc*. Myc, in turn, can induce the transcription of cdc25A mRNA. Cdc25A dephosphorylates inhibitory sites on CDK which promotes CDK activation and may propagate the cell death program. E2F and Myc may either positively or negatively regulate the transcription of death-associated or survival-associated genes.

by G_1-specific CDK in proliferating cells, is associated with the enhancement of cell death in neurons and other cells under growth suppressing conditions. E2F-1 activation, which follows pRb inactivation in the cell cycle, induces cell death in quiescent fibroblasts and immune cells, and the absence of E2F-1 results in defective thymocyte apoptosis in transgenic mice. Apoptosis induced by c-myc, a target of E2F-1 transcription factor activity, may be mediated by cdc25A, a target of c-myc transcription factor activity. cdc25A, in turn, functions by dephosphorylating inhibitory threonine and tyrosine residues on CDK leading to further CDK activation (Fig. 2). In this model, the role of pRb inactivation and subsequent E2F-1 activation may be restricted to perpetuating the cell death signal via a positive feedback mechanism that loops back on CDK activation. Potential death-mediating substrates for CDK phosphorylation remain to be defined but may include proteins involved in cytoskeletal rearrangement and chromatin condensation. Alternatively, E2F-1 may activate the transcription of yet unknown genes that are important for later steps in the death program, or, equally possible, inhibit the expression of genes that promote survival.

Are cell cycle regulators involved in *neuronal* cell death? Before this question can be definitively addressed, several other questions must first be answered. For example, are requisite components of the cell cycle machinery present in neurons and is their function actively suppressed by trophic factors or other survival promoting factors? Is the activation of CDK in *in vitro* models of neuronal death (which usually rely on immortalized or transformed cells or very young or immature primary neurons) an important part of the neuronal death program or a last ditch effort at cell division by a recently differentiated cell? Does apoptosis of mature neurons in adult animals involve mechanisms similar to those that occur in the developing nervous system? Although there is now a sizable body of evidence in support of a role for cell cycle proteins in cell death, exactly how this evidence relates to the death of neurons that accompanies nervous system development, diseases, and disorders remains to be seen.

ACKNOWLEDGMENTS

The author gratefully acknowledges support from the NIH (NS34400), the Paul Stark Endowment Fund at the University of Rochester, and a Markey Charitable Trust award to the University of Rochester.

REFERENCES

1. Ko LJ and Prives C. p53: puzzle and paradigm. *Genes Dev* 1996, **10**: 1054–1072.
2. Askew DS, Ashmun RA, Simmons BC, Cleveland JL. Constitutive c-myc expression in an IL-3-dependent myeloid cell line suppresses cell cycle arrest and accelerates apoptosis. *Oncogene* 1991, **6**: 1915–1922.
3. Evan GI, Wyllie AH, Gilbert CS, Littlewood TD, Land H, Brooks M, Waters CM, Penn LZ, Hancock DC. Induction of apoptosis in fibroblasts by c-myc protein. *Cell* 1992, **69**: 119–128.
4. Evan G, Harrington E, Fanidi A, Land H, Amati B, Bennett M. Integrated control of cell proliferation and cell death by the c-myc oncogene. *Phil Trans R Soc Lond B* 1994, **345**: 269–275.
5. Heald R, McLoughlin M, McKeon F. Human wee1 maintains mitotic timing by protecting the nucleus from cytoplasmically activated Cdc2 kinase. *Cell* 1993, **74**: 463–474.
6. Hartwell LH and Kastan MB. Cell cycle control and cancer. *Science* 1994, **266**: 1821–1828.
7. Enoch T and Norbury C. Cellular responses to DNA damage: cell-cycle checkpoints, apoptosis and the roles of p53 and ATM. *Trends Biochem Sci* 1995, **20**: 426–430.
8. Zetterberg A, Larsson O, Wilman KG. What is the restriction point? *Curr Opin Cell Biol* 1995, **7**: 835–842.
9. Morgan DO. Principles of CDK regulation. *Nature* 1995, **374**: 131–134.
10. Hunter T and Pines J. Cyclins and cancer. II. Cyclin D and CDK inhibitors come of age. *Cell* 1994, **79**: 573–582.
11. Sherr CJ. G_1 phase progression: cycling on cue. *Cell* 1994, **79**: 551–555.
12. Baldin V, Lukas J, Marcote MJ, Pagano M, Draetta G. Cyclin D1 is a nuclear protein required for cell cycle progression in G1. *Genes Dev* 1993, **7**: 812–821.
13. Quelle DE, Ashmun RA, Shurtleff SA, Kato J-Y, Bar-Sagi D, Roussel MF, Sherr CJ. Overexpression of mouse D-type cyclins accelerates G1 phase in rodent fibroblasts. *Genes Dev* 1993, **7**: 1559–1571.
14. Ohtsubo M, Theodoras AM, Schumacher J, Roberts JM, Pagano M. Human cyclin E, a nuclear protein essential for the G1-to-S phase transition. *Mol Cell Biol* 1995, **15**: 2612–2624.
15. van den Heuvel S and Harlow E. Distinct roles for cyclin-dependent kinases in cell cycle control. *Science* 1993, **262**: 2050–2054.
16. Ohtsubo M, Roberts J.M. Cyclin-dependent regulation of G1 in mammalian fibroblasts. *Science* 1993, **259**: 1908–1912.
17. Girard F, Strausfeld U, Fernandez A, Lamb NJC. Cyclin A is required for the onset of DNA replication in mammalian fibroblasts. *Cell* 1991, **67**: 1169–1179.

18. King RW, Jackson KP, Kirschner MW. Mitosis in transition. *Cell* 1994, **79**: 563–571.
19. Sherr CJ and Roberts JM. Inhibitors of mammalian G1 cyclin-dependent kinases. *Genes Dev* 1995, **9**: 1149–1163.
20. Shim J, Lee H, Park J, Kim H, Choi E. A non-enzymatic p21 protein inhibitor of stress-activated protein kinases. *Nature* 1996, **381**: 804–807.
21. Weinberg RA. The retinoblastoma protein and cell cycle control. *Cell* 1995, **81**: 323–330.
22. La Thangue NB. DRTF1/E2F: an expanding family of heterodimeric transcription factors implicated in cell-cycle control. *Trends Biochem Sci* 1994, **19**: 108–114.
23. Ucker DS. Death by suicide: one way to go in mammalian cellular development? *New Biol* 1991, **3**: 103–109.
24. Lazebnik YA, Cole S, Cooke CA, Nelson WG, Earnshaw WC. Nuclear events of apoptosis in vitro in cell-free mitotic extracts: a model system for analysis of the active phase of apoptosis. *J Cell Biol* 1993, **123**: 7–22.
25. Nigg EA. Targets of cyclin-dependent protein kinases. *Curr Opin Cell Biol* 1993, **5**: 187–193.
26. Lazebnik YA, Takahashi A, Moir RD, Goldman RD, Poirier GG, Kaufmann SH, Earnshaw WC. Studies of the lamin proteinase reveal multiple parallel biochemical pathways during apoptotic execution. *Proc Natl Acad Sci USA* 1995, **92**: 9042–9046.
27. Demarcq C, Bunch R.T, Creswell D, Eastman A. 1994, The role of cell cycle progression in cisplatin-induced apoptosis in chinese hamster ovary cells. *Cell Growth Differ* 1994, **5**: 983–993.
28. Shimizu T, O'Connor PM, Kohn KW, Pommier Y. Unscheduled activation of cyclin B1/Cdc2 kinase in human promyelocytic leukemia cell line HL60 cells undergoing apoptosis induced by DNA damage. *Cancer Res* 1995, **55**: 228–231.
29. Meikrantz W, Gisselbrecht S, Tam SW, Schlegel R. Activation of cyclin A-dependent protein kinases during apoptosis. *Proc Natl Acad Sci USA* 1994, **91**: 3754–3758.
30. Wang Q, Worland PJ, Clark JL, Carlson BA, Sausville EA. Apoptosis in 7-hydroxy-staurospurine-treated T lymphoblasts correlates with activation of cyclin-dependent kinases 1 and 2. *Cell Growth Differ* 1995, **6**: 927–936.
31. Donaldson KL, Goolsby G, Kiener PA, Wahl AF. Activation of p34^{cdc2} coincident with taxol-induced apoptosis. *Cell Growth Differ* 1994, **5**: 1041–1050.
32. Deckwerth TL and Johnson EM Jr. Temporal analysis of events associated with programmed cell death (apoptosis) of sympathetic neurons deprived of nerve growth factor (NGF). *J Cell Biol* 1993, **123**: 1207–1222.
33. Estus S, Zaks WJ, Freeman RS, Gruda M, Bravo R, Johnson EM Jr. Altered gene expression in neurons during programmed cell death: identification of c-jun as necessary for neuronal apoptosis. *J Cell Biol* 1994, **127**: 1717–1727.
34. Freeman RS, Estus S, Johnson EM Jr. Analysis of cell cycle-related gene expression in postmitotic neurons: selective induction of cyclin D1 during programmed cell death. *Neuron* 1994, **12**: 343–355.
35. Kranenburg O, van der Eb AJ, Zantema A. Cyclin D1 is an essential mediator of apoptotic neuronal cell death. *EMBO J* 1996, **15**: 46–54.
36. Herrup K and Busser JC. The induction of multiple cell cycle events precedes target-related neuronal death. *Development* 1995, **121**: 2385–2395.
37. Sicinski P, Donaher JL, Parker SB, Li T, Fazeli A, Gardner H, Haslam SZ, Bronson RT, Elledge SJ, Weinberg RA. Cyclin D1 provides a link between development and oncogenesis in the retina and breast. *Cell* 1995, **82**: 621–630.
38. Gao CY and Zelenka PS. Induction of cyclin B and H1 kinase activity in apoptotic PC12 cells. *Exp Cell Res* 1995, **219**: 612–618.
39. Farinelli SE and Greene LA. Cell cycle blockers mimosine, ciclopirox, and deferoxamine prevent the death of PC12 cells and postmitotic sympathetic neurons after removal of trophic support. *J Neurosci* 1996, **16**: 1150–1162.

40. Park DS, Farinelli SE, Greene LA. Inhibitors of cyclin-dependent kinases promote survival of post-mitotic neuronally differentiated PC12 cells and sympathetic neurons. *J Biol Chem* 1996, **271**: 8161–8169.
41. Fotedar R, Flatt J, Gupta S, Margolis RL, Fitzgerald P, Messier H, Fotedar A. Activation-induced T-cell death is cell cycle dependent and regulated by cyclin B. *Mol Cell Biol* 1995, **15**: 932–942.
42. Li CJ, Friedman DJ, Wang C, Metelev V, Pardee AB. Induction of apoptosis in uninfected lymphocytes by HIV-1 Tat protein. *Science* 1995, **268**: 429–431.
43. Shi L, Nishioka WK, Th'ng J, Bradbury EM, Litchfield DW, Greenberg AH. Premature p34cdc2 activation required for apoptosis. *Science* 1994, **263**: 1143–1145.
44. Martin SJ, McGahon AJ, Nishioka WK, La Face D. $p34^{cdc2}$ and apoptosis. *Science* 1995, **269**: 106–107.
45. Ongkeko W, Ferguson DJP, Harris AL, Norbury C. Inactivation of Cdc2 increases the level of apoptosis induced by DNA damage. *J Cell Sci* 1995, **108**: 2897–2904.
46. Meikrantz W and Schlegel R. Suppression of apoptosis by dominant negative mutants of cyclin-dependent protein kinases. *J Biol Chem* 1996, **271**: 10205–10209.
47. Galaktionov K, Chen X, Beach D. Cdc25 cell-cycle phosphatase as a target of c-myc. *Nature* 1996, **382**: 511–517.
48. Wang J, Walsh K. Resistance to apoptosis conferred by Cdk inhibitors during myocyte differentiation. *Science* 1996, **273**: 359–361.
49. Poluha W, Poluha DK, Chang B, Crosbie NE, Schonhoff CM, Kilpatrick DL, Ross AH. The cyclin-dependent kinase inhibitor $p21^{Waf1}$ is required for survival of differentiating neuroblastoma cells. *Mol Cell Biol* 1996, **16**: 1335–1341.
50. Waldman T, Lengauer C, Kinzler KW, Vogelstein B. Uncoupling of S phase and mitosis induced by anticancer agents in cells lacking p21. *Nature* 1996, **381**: 713–716.
51. Rao L, Debbas M, Sabbatini P, Hockenbery D, Korsmeyer S, White E. The adenovirus E1A proteins induce apoptosis, which is inhibited by the E1B 19-kDa and Bcl-2 proteins. *Proc Natl Acad Sci USA* 1992, **89**: 7742–7746.
52. White AE, Livanos EM, Tisty TD. Differential disruption of genomic integrity and cell cycle regulation in normal human fibroblasts by the HPV oncoproteins. *Genes Dev* 1994, **8**: 666–677.
53. Clarke AR, Maandag ER, van Roon M, van der Lugt NMT, van der Valk M, Hooper ML, Berns A, Riele HT. Requirement for a functional Rb-1 gene in murine development. *Nature* 1992, **359**: 328–330.
54. Jacks T, Fazeli A, Schmitt EM, Bronson RT, Goodell MA, Weinberg RA. Effects of an *Rb* mutation in the mouse. *Nature* 1992, **359**: 295–300.
55. Lee EY-HP, Chang C-Y, Hu N, Wang Y-CJ, Lai C-C, Herrup K, Lee W-H, Bradley A. Mice deficient for Rb are nonviable and show defects in neurogenesis and haematopoiesis. *Nature* 1992, **359**: 288–294.
56. Lee EY-HP, Hu N, Yuan S-SF, Cox LA, Bradley A, Lee W-H, Herrup K. Dual roles of the retinoblastoma protein in cell cycle regulation and neuron differentiation. *Genes Dev* 1994, **8**: 2008–2021.
57. Almasan A, Yin Y, Kelly RE, Lee EY-HP, Bradley A, Li W, Bertino JR, Wahl GM Deficiency of retinoblastoma protein leads to inappropriate S-phase entry, activation of E2F-responsive genes, and apoptosis. *Proc Natl Acad Sci USA* 1995, **92**: 5436–5440.
58. Haas-Koogan DA, Kogan SC, Levi D, Dazin P, T'Ang A, Fung Y, Israel MA. Inhibition of apoptosis by the retinoblastoma gene product. *EMBO J* 1995, **14**: 461–472.
59. Berry DE, Lu Y, Schmidt B, Fallon PG, O'Connell C, Hu S-X, Xu H-J, Blanck G. Retinoblastoma protein inhibits interferon-γ induced apoptosis. *Oncogene* 1996, **12**: 1809–1819.
60. Fan G, Ma X, Kren BT, Steer CJ. The retinoblastoma gene product inhibits TGF-β1 induced apoptosis in primary rat hepatocytes and human HuH-7 hepatoma cells. *Oncogene* 1996, **12**: 1909–1919.

61. An B and Dou QP. Cleavage of retinoblastoma protein during apoptosis: an interleukin 1β-converting enzyme-like protease as candidate. *Cancer Res* 1996, **56**: 438–442.
62. Qin X, Livingston DM, Kaelin WG Jr, Adams PD. Deregulated transcription factor E2F-1 expression leads to S-phase entry and p53-mediated apoptosis. *Proc Natl Acad Sci USA* 1994, **91**: 10,918–10,922.
63. Shan B and Lee W. Deregulated expression of E2F-1 induces S-phase entry and leads to apoptosis. *Mol Cell Biol* 1994, **14**: 8166–8173.
64. Kowalik TF, DeGregori J, Schwarz JK, Nevins JR. E2F-1 overexpression in quiescent fibroblasts leads to induction of cellular DNA synthesis and apoptosis. *J Virol* 1995, **69**: 2491–2500.
65. Wu X, Levine, AJ. p53 and E2F-1 cooperate to mediate apoptosis. *Proc Natl Acad Sci USA* 1994, **91**: 3602–3606.
66. Shan B, Durfee T, Lee W-H. Disruption of RB/E2F-1 interaction by single point mutations in E2F-1 enhances S-phase entry and apoptosis. *Proc Natl Acad Sci USA* 1996, **93**: 679–684.
67. Hiebert SW, Packman G, Strom DK, Haffner R, Oren M, Zambetti G, Cleveland JL. E2F-1:DP-1 induces p53 and overrides survival factors to trigger apoptosis. *Mol Cell Biol* 1995, **15**: 6864–6874.
68. Krek W, Xu G, Livingston DM. Cyclin A-kinase regulation of E2F-1 DNA binding function underlies suppression of an S phase checkpoint. *Cell* 1995, **83**: 1149–1158.
69. Field SJ, Tsai F, Kuo F, Zubiaga AM, Kaelin WG Jr, Livingston DM, Orkin SH, Greenberg ME. E2F-1 functions in mice to promote apoptosis and suppress proliferation. *Cell* 1996, **85**: 549–561.
70. Yamasaki L, Jacks T, Bronson R, Goillot E, Harlow E, Dyson NJ. Tumor induction and tissue atrophy in mice lacking E2F-1. *Cell* 1996, **85**: 537–548.
71. Heintz N. Cell death and the cell cycle: a relationship between transformation and neurodegeneration? *Trends Biochem Sci* 1993, **18**: 157–159.
72. Al-Ubaidi MR, Hollyfield JG, Overbeek PA, Baehr W. Photoreceptor degeneration induced by the expression of simian virus 40 large tumor antigen in the retina of transgenic mice. *Proc Natl Acad Sci USA* 1992, **89**: 1194–1198.
73. Hammang JP, Behringer RR, Baetge EE, Palmiter RD, Brinster RL, Messing A. Oncogene expression in retinal horizontal cells of transgenic mice results in a cascade of neurodegeneration. *Neuron* 1993, **10**: 1197–1209.
74. Feddersen, R.M, Ehlenfeldt, R, Yunis, WS, Clark, HB, Orr, HT. Disrupted cerebellar cortical development and progressive degeneration of Purkinje cells in SV40 T antigen transgenic mice. Neuron 1992, **9**: 955–966.
75. Feddersen RM, Clark HB, Yunis WS, Orr HT. In vivo viability of postmitotic Purkinje neurons requires pRb family member function. *Mol Cell Neurosci* 1995, **6**: 153–167.
76. Clarke AR, Purdie CA, Harrison DJ, Morris RG, Bird CC, Hooper ML, Wyllie AH. Thymocyte apoptosis induced by p53-dependent and independent pathways. *Nature* 1993, **362**: 849–852.
77. Lowe SW, Schmitt EM, Smith SW, Osborne BA, Jacks T. p53 is required for radiation-induced apoptosis in mouse thymocytes. *Nature* 1993, **362**: 847–849.
78. Graeber TG, Osmanian C, Jacks T, Housman DE, Koch CJ, Lowe SW, Giaccia AJ. Hypoxia-mediated selection of cells with diminished apoptotic potential in solid tumours. *Nature* 1996, **379**: 88–91.
79. Hermeking H and Eick D. Mediation of c-Myc-induced apoptosis by p53. Science 1994, **265**: 2091–2093.
80. Gottlieb E, Haffner R, von Ruden T, Wagner EF, Oren M. Down-regulation of wild-type p53 activity interferes with apoptosis of IL-3-dependent hematopoietic cells following IL-3 withdrawal. *EMBO J* 1994, **13**: 1368–1374.
81. Morgenbesser SD, Williams BO, Jacks T, DePinho RA. p53-dependent apoptosis produced by Rb-deficiency in the developing mouse lens. *Nature* 1994, **371**: 72–74.

82. Fromm L, Shawlot W, Gunning K, Butel JS, Overbeek PA. The retinoblastoma protein-binding region of simian virus 40 large T antigen alters cell cycle regulation in lenses of transgenic mice. *Mol Cell Biol* 1994, **14:** 6743–6754.
83. Pan H and Griep AE. Altered cell cycle regulation in the lens of HPV-16 E6 or E7 transgenic mice: implications for tumor suppressor gene function in development. *Genes Dev* 1994, **8:** 1285–1299.

II
Animal Models

7
Neuronal Cell Death in *C. elegans*

Dewey Royal and Monica Driscoll

INTRODUCTION

The biochemical mechanisms involved in the regulation and execution of developmental cell death are strikingly conserved from nematodes to humans. As a consequence, powerful genetic approaches to the dissection of the developmental cell death pathway in the nematode *Caenorhabditis elegans* (referred to as programmed cell death) have provided insights into rules that govern cellular life/death decisions, the process by which cells die, and the elimination of cell corpses that are relevant to understanding similar processes in higher organisms. As is the case for most metazoans, *C. elegans* neurons can also undergo a necrotic-like death when injured. Study of the basic biology of degenerative cell death in nematodes may thus also extend understanding of mechanisms of aberrant cell death in higher organisms. Here we first review the features of the nematode experimental system and then discuss current understanding of both developmental and degenerative cell death mechanisms in *C. elegans*.

THE *C. elegans* EXPERIMENTAL SYSTEM

Life Cycle and System Features

C. elegans is a small (1.3 mm), free-living soil nematode that feeds on *E. coli* in the laboratory. The predominant sexual form is the self-fertilizing hermaphrodite (XX), which gives rise to approx 300 offspring over a period of five days. *C. elegans* oocytes are fertilized internally and eggs are laid early in embryogenesis. After hatching, animals proceed through four larval stages (L1–L4) before reaching adulthood (sexual maturity). The reproductive life cycle is approx 3.5 d (20°C) and the life span of the animal is on the order of 15 d. Males (XO) can be produced spontaneously by nondisjunction of the X chromosome and are maintained in the laboratory by mating to hermaphrodites.

The cuticle of *C. elegans* is transparent, enabling the nucleus of every cell in the body to be visualized using Nomarski differential interference contrast optics microscopy. This feature facilitated the recording of the complete sequence of cell divisions that occur as the fertilized egg develops into the 959 somatic-celled adult hermaphrodite or the 1031 somatic-celled adult male *(1,2)*. The developmental plan is essentially invariable from animal to animal. A striking feature of the cell lineage is the generation of cells destined to undergo programmed cell death at specific times in development.

From: Cell Death and Diseases of the Nervous System
Edited by: V. E. Koliatsos and R. R. © Ratan. Humana Press Inc., Totowa, NJ

Serial section electron microscopy has yielded a description of the morphology and pattern of synaptic connectivity for each of the 302 neurons of the hermaphrodite *(3,4)*. Predictions about neuronal circuitry deduced from the wiring diagram can be tested by eliminating cells using a laser microbeam *(5–7)*.

Genetics and Molecular Biology

The key strength of the *C. elegans* model system resides in the extensive genetic analyses that can be conducted with this animal. The ability of *C. elegans* to reproduce by self-fertilization renders the production and recovery of mutants easy—homozygous mutants segregate as F2 progeny of mutagenized parents without any required genetic crossing. Mutant alleles are readily transferred by male matings so that complementation analysis and construction of double mutant strains is straightforward. Positions of hundreds of genes on the six *C. elegans* chromosomes have been determined. This genetic map has been aligned with the physical map of the genome (a collection of overlapping DNA clones that spans the six chromosomes). Considerable progress in sequence analysis of the *C. elegans* genome has been accomplished *(8,9)*, a project expected to be completed in 1998. Transgenic nematodes are constructed by injecting DNA into the hermaphrodite gonad where it is packaged into developing oocytes *(10,11)*. Transgenic animals harbor introduced DNA on extrachromosomal arrays that usually contain many gene copies. Vectors that facilitate identification of transformants *(10)*, construction of reporter gene fusions to *E. coli lacZ (12)* or the *A. victoria* green-fluorescent protein *(13)*, and the expression of genes in specific cell types are commonly used for molecular analyses.

ANALYSES OF PROGRAMMED CELL DEATH IN *C. elegans*

Programmed Cell Death in Development

Programmed cell death in the *C. elegans* hermaphrodite results in the elimination of 131 of the 1090 cells that are generated during development *(1,2)*. The majority of these cell deaths occur during embryogenesis (113 of the 671 embryonically generated cells die). Dying cells adopt a characteristic button-like appearance that is easily recognized in living animals using Nomarski differential interference contrast light microscopy (Fig. 1).

In contrast to situations such as vertebrate neuronal selection in which cells differentiate and send axons towards their target tissue before life/death decisions are made *(14)*, programmed cell death occurs in the *C. elegans* before cells differentiate into recognizable cell types. For deaths that have been examined at the ultrastructural level *(15)*, it has been noted that the cell destined to die can be distinguished almost immediately after its generation from the parent cell division—the ill-fated cell is smaller than its viable sister and is often surrounded by a cytoplasmic process that extends from a neighboring cell. Although most doomed cells do not differentiate, it has been possible to infer their developmental potential by analyzing nuclear morphology, the fates of surviving cells in homologous cell lineages, and the fates of cells that are spared from death by genetic mutations *(16–19)*. These studies suggest that most of the cells destined to undergo programmed cell death have neuronal potential. When rescued from the death fate (these cells have been called "undead" cells) at least some can differentiate and function *(16,17)*.

The morphological changes that transpire during developmental cell death in *C. elegans* have features of both invertebrate and vertebrate programmed cell deaths *(15)*. The

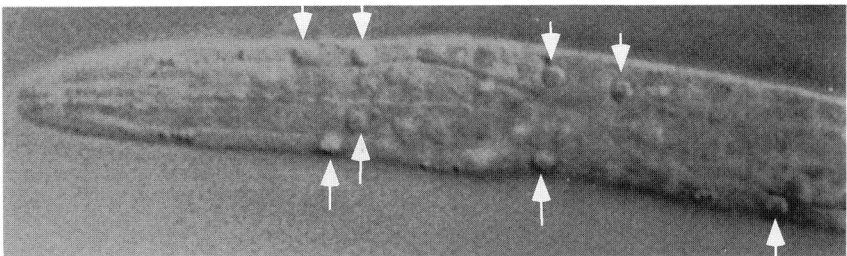

Fig. 1. Programmed cell death corpses appear as button-like structures that are readily apparent using Nomarski DIC optics. Cell death corpses are indicated by white arrows. Animal is a *ced-7*; *ced-5* double mutant in which programmed cell death corpses persist.

cytoplasm of the ill-fated cell becomes condensed and chromatin aggregates. Portions of the cell become split into membrane-bound fragments by the phagocytotic arms of a neighboring cell. Whorls of membrane and electron-dense material can be detected within vacuoles of the neighboring hypodermis, which engulfs all detectable remains of the dead cell. The time course of cell death is approximately one hour.

The Genetic Pathway for Programmed Cell Death

Hedgecock et al. exploited the fact that programmed cell death occurs in an invariable but easily visualized pattern in their screens for mutations that disrupt this pattern. Such screens identified at least 12 genes that function in programmed cell death *(16,20–22)*. Results of epistasis analyses have ordered these into a genetic pathway for programmed cell death (*see* Fig. 2). Five distinct steps of the cell death process are affected by these genes:

1. Control of the life/death decision in specific cells.
2. Global regulation of the death program.
3. Execution of cell death.
4. Engulfment of the dying cell.
5. Degradation of corpse DNA.

Cell-Specific Regulation of the Decision to Live or Die

Genetic analyses indicate that the decision of a cell to live or die can be controlled in a cell-specific manner *(22)*. Mutations in two genes, *ces-1* and *ces-2* (*c*ell death *s*pecification), affect the life/death decision of two cells called the NSM sisters *(22)* without influencing most other programmed cell deaths in the animal. The NSM neurons are the only two neurons flanking the pharynx that express serotonin *(23)*. Mutations in *ces-1* and *ces-2* were identified in screens for mutants with additional serotonergic neurons in this region (*see* Fig. 3). In a *ces-2* reduction-of-function *(lf)* mutant, the two NSM sisters do not die as they do in wild-type animals, but instead survive and differentiate into serotoninergic neurons *(22)*. Since cell survival is the loss-of-function phenotype, the wild-type function of *ces-2* is inferred to be to promote the deaths of the NSM sisters.

The situation with *ces-1* is a bit more complicated. A *ces-1* gain-of-function mutation (*gf*) prevents the deaths of the NSM sisters and the I2 sisters (two extra pair of serotoninergic neurons are produced). A *ces-1(lf)* mutation does not disrupt the deaths of the NSM and I2 sisters. Therefore, *ces-1* is inferred to be a negative regulator of the

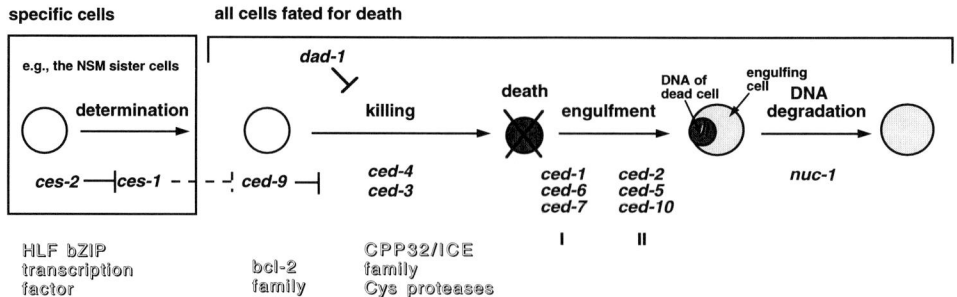

Fig. 2. The genetic pathway for programmed cell death in *C. elegans*. Mammalian counterparts of *C. elegans* genes are listed in outlined type below. Life/death fate decisions can be regulated in a cell-specific manner, as occurs for the *ces-1* and *ces-2* regulators of the NSM sister fate. bZIP transcription factor CES-2 negatively regulates *ces-1* which itself negatively regulates the death program in these two cells. Interestingly, the HLF transcription factor, which can negatively regulate mammalian apoptosis, is a member of the same PAR subfamily of bZIP proteins as CES-2. It is not certain whether the *ces-1/ces-2* pathway directly regulates *ced-9* activity, although it does act upstream of *ced-3*. *ced-9* negatively regulates the activities of two genes essential for all programmed cell deaths, *ced-3* and *ced-4*. Members of the Bcl-2 family related to CED-9 can either protect against or promote apoptosis. Several members of the CED-3/CPP32/ICE family of Cys-proteases have been implicated in apoptosis. *ced-4* transcripts are alternatively spliced to generate a death-promoting *ced-4S* product or a life-promoting *ced-4L* product. Although *ced-9* negatively regulates both these *ced-4* activities, their relationships relative to each other and relative to *ced-3* remain to be determined and, for clarity, these activities are not shown here. The *dad-1* gene can protect against programmed cell death, but its relationship to *ced-9*, *ced-3* and *ced-4* is not known. Two groups of genes (I: *ced-1*, *ced-6*, *ced-7* and II: *ced-2*, *ced-5*, *ced-10*) are needed for the engulfment of the corpse and one gene, *nuc-1*, is known to be needed for degradation of corpse DNA. This figure was adapted by permission from Horvitz et al. *(114)*.

deaths of the NSM and I2 sisters. Genetic analysis has placed *ces-2* upstream of *ces-1*; in a *ces-1(gf) ces-2(lf)* double mutant, the NSM sisters die. Thus, genetic evidence supports a model in which CES-2 negatively regulates the activity of CES-1, which in turn negatively regulates the programmed cell death of the NSM sisters (*see* Figs. 1 and 3).

Cloning of *ces-2* revealed that it encodes a basic region leucine zipper (bZIP) transcription factor that is similar in sequence and binding specificity to the proline and acid-rich (PAR) subfamily of bZIP proteins *(24)*. This suggests that life/death decisions can be specifically controlled at the level of transcription. Given the molecular identity of CES-2 and the deduced genetic order of activities of *ces-1* and *ces-2*, it is possible that CES-2 could bind the *ces-1* promoter and act to repress transcription of this negative regulator of cell death *(24)*. Alternatively, CES-2 could associate with CES-1 via a "zipper" interaction to modulate CES-1 activity. Cloning of the *ces-1* gene and identification of the downstream target genes will help elaborate how transcriptional regulation can influence individual life/death decisions.

At least some regulatory mechanisms controlling death of specific cells appear to be conserved between *C. elegans* and mammalians. The recently identified CES-2 transcription factor bears a remarkable similarity to the hepatic leukemia factor (HLF) BZIP transcription factor, which negatively regulates apoptotic cell death in lymphocytes *(25)*.

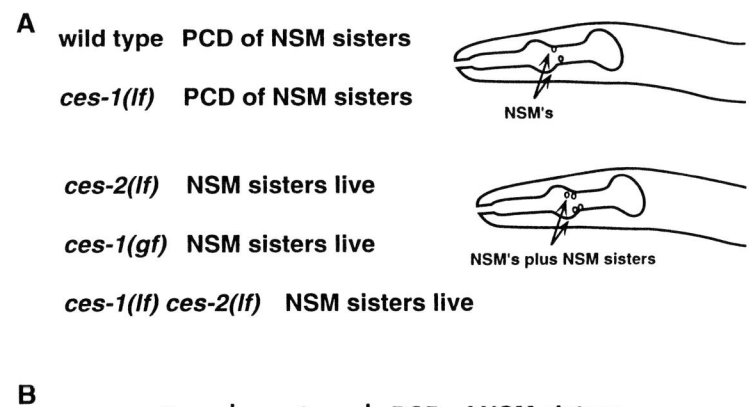

Fig. 3. *ces-1* and *ces-2* affect the cell fates of the NSM sisters. **(A)** In wild-type and *ces-1* loss-of-function mutants the sister cells of the NSM neurons undergo programmed cell death. In *ces-2* reduction-of-function, *ces-1* gain-of-function and the *ces-1* reduction-of-function *ces-2* loss-of-function double mutant the NSM sister cells live and produce serotonin. **(B)** A model that is consistent with these results proposes that *ces-2* negatively regulates *ces-1*. In the down regulated state *ces-1* cannot inhibit the programmed cell death of the NSM sister neurons.

CED-9, A General Negative Regulator of the Cell Death Program, is Similar in Sequence to the Mammalian Bcl-2 Family

The product of the *ced-9* gene (*ce*ll *d*eath abnormal) plays a critical role in keeping the death program turned off in most *C. elegans* cells. *ced-9* was first identified by an unusual gain-of-function mutation (allele *n1950*) that prevents programmed cell deaths of the 131 cells fated to die *(26)*. To generate loss-of-function *ced-9* alleles, tightly linked *cis*-dominant suppressors of *n1950* that were second site mutations which inactivate *ced-9* function, were isolated *(26)*. Such *ced-9(lf)* mutations are lethal because they cause extensive cell death, implying that the normal activity of *ced-9* is to prevent programmed cell death of a large number of cells. Because the *ced-9(gf)* mutation affects cells normally fated to die and *ced-9(lf)* mutations affect cells normally fated to live, it appears that *ced-9* is expressed in many, if not all, cell types. High levels of *ced-9* transcripts detected during embryogenesis (and low constitutive levels later in development) support broad expression of *ced-9 (27)*. It is not known how *ced-9* expression differs in cells fated to live versus cells fated to die.

ced-9 encodes a 280 amino acid protein that shares significant sequence similarity (23% identity) with the product of the mammalian *Bcl*-2 gene *(27)*. Bcl-2 generally promotes cell survival (reviewed in ref. *28*), although under some circumstances it can be involved in promoting cell death *(29)*. Expression of Bcl-2 in *C. elegans* using a heat-shock promoter partially blocks normal programmed cell death *(27,30)*, dramatically demonstrating that regulators of the death program are functionally conserved from nematodes to humans.

The mechanism by which CED-9 and Bcl-2 protect against cell death remains a key mystery in the cell death field. Bcl-2 is a member of a family of homologous proteins that can either inhibit (Bcl-2 and Bcl-x_L *[31]*) or promote (Bcl-x_s *[31]*, Bax *[32]*, Bak *[33]*, Bad *[34]* programmed cell death. The activities of Bcl-2 family members depend in part upon the relative levels of homo- or heterodimers that can form between them in different cell types (see refs. 35 and 36). In C. elegans, CED-9 is the only family member identified to date. Recent evidence suggest Bcl-2 and related anti-apaptotic Bcl-X_L, which are taught to associate with mitochondrial and endoplasmic reticulum (ER) membranes, can act as channels (37–39) and as protein docking/interaction sites for several proteins including CED-4 (see below and see refs 40–42), protein kinase Raf-1, calcineurin, H-Ras, P53-BP2, BAG-1, Nip-1, Nip-2, and Nip-3 (reviewed in ref. 43). One role of the multifunctional death-5regulatory protiens may be to regualte partitions of molecules such as cytochrome c and apoptosis-inducing factor (AIF) between the mitochondria and cytoplasm.

dad-1 *Can Negatively Regulate* C. elegans *Programmed Cell Death*

A temperature-sensitive mutation in the *dad-1* (*d*efender *a*gainst *d*eath) gene causes hamster cells to undergo apoptosis at the restrictive temperature *(48)*. A *C. elegans* homolog of the highly conserved *dad-1* gene has been cloned *(49)*. Both the *C. elegans* and the human *dad-1* genes can partially block programmed cell death when overexpressed under control of a heat-shock promoter in *C. elegans (49)*, suggesting that *dad-1* function in programmed cell death is conserved. Exactly how DAD-1 protects against programmed cell death remains obscure despite the fact that a yeast *dad-1* homolog is known to be a subunit of oligosaccharlytransferase *(50)*.

ced-3 *and* ced-4

Two *C. elegans* genes, *ced-3* and *ced-4*, have been identified as key executors of programmed cell death. Loss-of-function mutations in either of these genes prevent all programmed cell deaths in *C. elegans (16)*. Thus, the normal activities of *ced-3* and *ced-4* can be inferred to promote cell death. *ced-3* and *ced-4* mutants, which accommodate 131 extra cells in their bodies, are not markedly affected by the lack of all programmed cell death. They are fully viable, reproduce effectively, and exhibit only minor defects in their rates of development and chemotaxis *(16)* (M. Chalfie and C. Bargmann, personal communication).

ced-3 *Encodes a Cysteine Protease*

ced-3 encodes a protein related to the cytoplasmic cysteine protease interleukin 1β-converting enzyme (ICE) *(51)*, which processes a precursor protein to generate cytokine IL1β *(52,53)* (for a topical discussion of proteases as mediators of cell death, *see* chapter by Rosen and Casciola-Rosen). Similarity between CED-3 and ICE is highest (43%) over a 115-amino acid region that includes the ICE active site cysteine. CED-3 does function as a protease which, among other things, can autoprocess in vitro to generate subunits of the active protease (p13 and p17; *[54]*). CED-3-mediated proteolysis is dependent upon the presence of the active-site cysteine and is sensitive to inhibitors that specifically target CED-3/ICE family members *(54,55)*. Importantly, protease activity is critical to in vivo *ced-3* function. A *ced-3* allele engineered to encode a substitution for the active site Cys residue fails to complement a *ced-3* loss-of-function mutation *(56)* and the severity of the ced-3 phenotype is related to the level of *ced-3* protease activity assayed in vitro—the lower the protease activity, the more severe the mutant phenotype *(54)*. Thus, proteolysis is a critical component of the programmed cell death mechanism in the *C. elegans*.

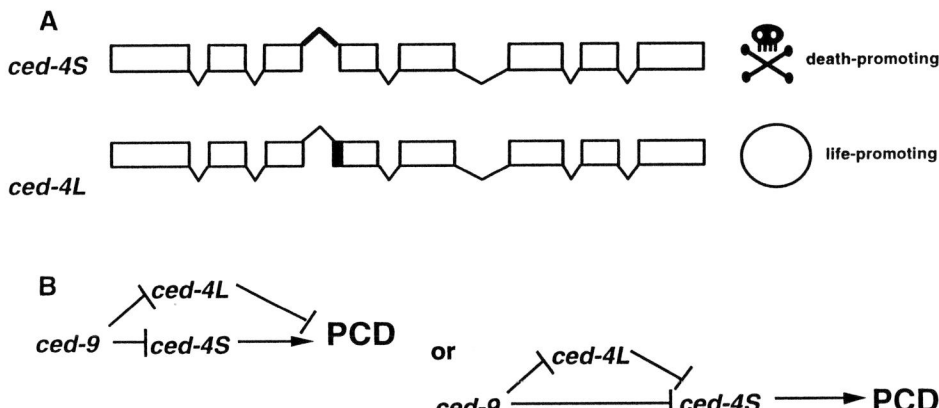

Fig. 4. *ced-4* encodes both death-promoting and death-protecting activities. **(A)** Depiction of two alternative splice forms, *ced-4S* and *ced-4L* (exons represented by boxes). *ced-4S* promotes cell death; *ced-4L*, which includes 25 extra amino acids at the beginning of exon 4 (dark box), promotes cell life. **(B)** Models for regulatory relationships between *ced-9*, *ced-4S* and *ced-4L*. *ced-9* inhibits both *ced4S* and *ced-4L* and the activity of *ced-4S* does not depend on *ced-4L*. In the first model, CED-4S and CED-4L act at the same point in the death pathway. In the second model CED-4L directly antagonizes the activity of CED-4S. This figure was adapted with permission from Petit et al. *(44)*.

CED-3 can induce apoptosis in mammalian cell culture *(45)*, supporting that the proteolysis is another aspect of the death program that is conserved from nematodes to humans. Consistent with this, several mammalian members of the CED-3/ICE family can induce apoptosis when expressed in cell culture (reviewed in ref. *46*). Of these, CED-3 falls into a subfamily closely related in sequence and in vitro substrate specificity to CPP32/Yama/apopain *(42)*, which is a good candidate for the true mammalian homolog of CED-3. Thus far only one *C. elegans* Cys-protease of this family has been identified.

Biochemical analyses indicate that CED-3 usually cleaves at Asp residues that are followed by small side-chain amino acids *(42)*. In vitro CED-3 substrates include human Il-1β, proteases hICH-1 *(47)* and hCPP32 *(48)*, poly(ADP-ribose) polymerase which is cleaved early in mammalian apoptosis *(49,50)*, and baculovirus p35 protein *(51)*, which inhibits programmed cell death in nematodes *(37)*, flies *(52)*, and mammals *(53)*. Inhibition of *C. elegans* programmed cell death by p35, which might act as a competitive inhibitor of the CED-3 protease, requires proteolytic cleavage of p35 *(51)*.

One of the most pressing questions in the cell death field concerns the identities of the substrates of the death proteases. Is there a key substrate whose cleavage activates a *C. elegans* death cascade? Candidate substrates include the death-protective CED-9 and DAD-1 proteins, the death executor CED-4, and the products of genes that act in corpse removal *(see below)*, but the substrate(s) remain to be identified.

ced-4 Encodes Death-Promoting and Death-Protective Proteins

In addition to the *ced-3* protease, the activity of the *ced-4* gene is needed for the execution of all *C. elegans* programmed cell deaths *(16)*. *ced-4* encodes two alternatively spliced transcripts referred to as *ced-4S* and *ced-4L* *(66)*. The shorter *ced-4S* transcript is 10–30 times more abundant than the *ced-4L* transcript; the *ced-4L* protein differs from *ced-4S* by the addition of only 25 internal amino acids (*see* Fig. 4A). The *ced-4* products are hydrophilic but are not strikingly similar to any known protein *(67)*, although they have limited similarity to the mammalian death effect or domain *(68)*. On the other hand, it is known that CED-4S and CED-4L have markedly different biological activities. Overexpression of *ced-4S* in transgenic worms showed that this splice variant encodes a protein that promotes cell death in a cell autonomous manner *(56)*. In contrast, *ced-4L* actually protects against cell death when ectopically expressed *(66)*.

The identification of alternative *ced-4* splice variants that encode opposing functions in cell death raises a number of intriguing questions regarding programmed cell death in the *C. elegans*. What are the signals that regulate alternative splicing of *ced-4*? What are the relative concentrations of *ced-4S* and *ced-4L* in cells fated to live? What are the relative concentrations of *ced-4S* and *ced-4L* in cells destined to die? How do the 25 additional amino acids inserted into CED-4S alter the cell fate decision? How do CED-4S and CED-4L interact with CED-3 and CED-9? Investigation of such questions will provide insight into the mysterious death mechanism and its regulation.

One thing that is clear from the analysis of *ced-4* is that alternative splicing is critical to the life/death fate. Interestingly, alternative splicing influences the life/death decision in mammals by affecting some Bcl-2 and ICE family members. Bcl-x can be alternatively spliced to generate Bcl-x_s, which promotes programmed cell death in culture, and Bcl-x_L, which protects against cell death *(31)*. CED-3/ICE family member Ich-1 can be alternatively spliced to produce a death-promoting product with proteolytic activity and a death-inhibiting product that does not encode an active protease *(59)*. Thus, ultimate understanding of cell death regulation in many cell types will require elaboration of splice site selection rules.

Ordering ced Activities into a Genetic Pathway

ced-9 vs *ced-3* and *ced-4*. The relationship between *ced-9* and *ced-3* and *ced-4* was determined by examining the phenotypes of double mutant strains. The lethality conferred by *ced-9(lf)* mutations is fully blocked by loss-of-function mutations in *ced-3* or *ced-4*, i.e., *ced-9(lf)*, *ced-3(lf)*, and *ced-4(lf) ced-9(lf)* strains are viable and do not exhibit programmed cell deaths *(26)*. This epistasis relationship suggests that *ced-3* and *ced-4* must be active for the *ced-9* gene product to function and positions *ced-9* upstream of *ced-3* and *ced-4* in a genetic pathway (*see* Fig. 2). In the simplest model, *ced-9* inhibits *ced-3* and *ced-4* killing activities (*ced-4S*) in cells destined to live. This model is consistent with the finding that killing by overexpression of either *ced-3* or *ced-4S* in the touch neurons is partially inhibited by the presence of a *ced-9(+)* allele in the genetic background *(56)*.

The simple model of the genetic pathway must now be modified to accommodate the recently identified death-promoting *ced-4S* and death-preventing *ced-4L* products. At present, not enough is understood to unambiguously determine the relationships between *ced-9*, *ced-4S*, and *ced-4L*. However, it is known that *ced-9*-mediated inhibition of *ced-4S* is not dependent on *ced-4L* activity because the *ced-9-ced-4S* relationship is unaffected in a mutant background that specifically eliminates *ced-4L (56)*. How does *ced-9* influence *ced-4L* activity? Constitutive expression of a *ced-4L* transgene can rescue the lethal phenotype of a *ced-9(lf)* mutant, suggesting that *ced-4L* acts downstream of, or in parallel to, *ced-9 (66)*. Thus, available genetic data are consistent with two alternative models for *ced-9*, *ced-4S*, and *ced-4L* interactions (*see* Fig. 4B).

Cell Death in C. elegans

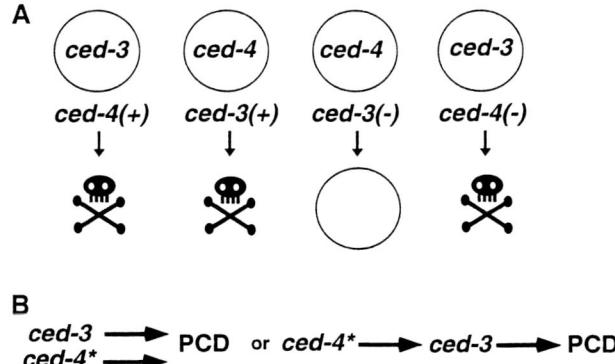

Fig. 5. Relationship between *ced-3* and *ced-4*. **(A)** Summary of studies of overexpression *of ced-3* and *ced-4* in the touch receptor neurons. Circles indicate cells and the gene over-expressed in these cells; underneath is indicated the genetic background for a particular experiment. Skull and crossbones icon indicates cell death. *ced-3* can kill in the absence of *ced-4*, but *ced-4* is not an effective killer in the absence of *ced-3*. **(B)** Models for *ced-3* and *ced-4* action in cell death. *ced-3* and *ced-4* could act in parallel to cause cell killing, but *ced-4* might not be an effective killer in the absence of *ced-3*. Alternatively, *ced-4* could act upstream of *ced-3* to potentiate is killing activity. Asterisk indicates that *ced-4* refers to *ced-4S* killing activity. The situation is complicated by the existence of *ced-4L*, which has a competing death-inhibiting activity. This figure was derived from Shaham and Horvitz *(56)*.

As noted above, the life/death outcome for an individual cell may depend upon alternative splicing or competition between CED-4S and CED-4L activities. How CED-9 influences this competition at the molecular level remains to be determined. CED-9 has recdntly been demonstrated to interact physically with CED-4S *(40–42)* which can also interact with CED-3 *(41)*. Thus, protien interactions may be centrlly important to regulation.

Interestingly, the mechanism of action of the *ced-9(gf)* mutation, *n1950* appears to be mediated through *ced-4L*. *ced-9(n1950)* encodes substitution G169E, which affects an invariant residue in the highly conserved BH1 domain of the CED-9/Bcl-2 family *(69)*. The equivalent change in bcl-2 (which affects a domain that mediates Bcl-2-Bax interaction *[34]*) inactivates the bcl-2 death-protective function in both mammalian cells *(34)* and in *C. elegans (69)*. Genetic studies revealed that *ced-4L* is needed for *ced-9(n1950)* to increase cell survival *(66)*. Thus, CED-9(G169E) appears to promote cell viability because it fails to inhibit survival-promoting CED-4L (while still inhibiting CED-4S). CED-9 could do this by associating with CED-4L, perhaps via the BH1 homology domain.

ced-3 *vs* ced-4

The phenotypes of *ced-3* and *ced-4* loss-of-function mutants are the same—all cells normally destined to undergo programmed cell death survive *(16)*. Thus, the order of *ced-3* and *ced-4* gene action in the death program cannot be inferred by constructing double mutants and examining epistasis relationships. Molecular studies suggest that *ced-3* and *ced-4S* do not influence each other at the transcriptional level since *ced-3* is expressed at wild type levels in a *ced-4* null background *(67)* and *ced-4* is expressed at wild-type levels in a *ced-3* mutant background (*[70]*, these experiments did not distinguish *ced-4S* and *ced-4L*).

Experiments in which *ced-3* or *ced-4S* were ectopically expressed at high levels in the well-characterized touch receptor neurons provide some insight into the order in which the *ced-3* and *ced-4S* gene products may function *(56)* (*see* Fig. 5). Overexpression of either *ced-3* or *ced-4S* promotes programmed cell death, confirming that these genes act cell autonomously to kill cells *(56)*. Expression of *ced-3* in a *ced-4(lf)* background causes killing at levels similar to that observed in the *ced-4(+)* background, indicating that ectopic killing by over-expressing *ced-3* occurs independently of *ced-4* function. In contrast, overexpression of *ced-4S* in *ced-3* mutant backgrounds causes significantly less cell killing than occurs in a *ced-3(+)* background. Two alternative models can explain these results as follows: *ced-4S* may act in parallel with *ced-3* to promote cell death but by itself might be a less effective killer than *ced-3* or *ced-4S* might function upstream of *ced-3* to increase CED-3 activity (*see* Fig. 5).

Although genetic and molecular investigations of *ced-9*, *ced-3*, and *ced-4* have extended our understanding of programmed cell death mechanisms, they have also opened up a number of fascinating questions. The determination of when and where *ced-9*, *ced-3*, and *ced-4* are expressed and how they physically interact to influence each others biochemical activities remain the subjects of future investigation.

Corpse Removal: The Undertaker Genes

Cells that have undergone programmed cell death are rapidly removed from the nematode body by a process that requires engulfment of the cell corpse by a neighboring cell *(15)*. Six genes are needed for efficient corpse removal *(20,21)*. In the *ced-1*, *ced-2*, *ced-5*, *ced-6*, *ced-7* and *ced-10* (the engulfment *ced* genes) mutant backgrounds, dead cells persist in the refractile "button-like" stage for hours (refer to Fig. 1). Because cells die efficiently in engulfment-deficient mutants, it is clear that phagocytosis is not the cause of cell death—it is simply involved in corpse removal.

None of the engulfment *ced* mutations completely block phagocytosis of all dying cells. Genetic analysis suggests that this weak expressivity occurs because at least two parallel pathways function in corpse phagocytosis *(21)*. Inactivation of one pathway by a loss-of-function mutation that disrupts one of its components does not dramatically block corpse removal because the other pathway can still be utilized for phagocytosis. Double mutant combinations among *ced-2*, *ced-5*, and *ced-10* exhibit the same degree of corpse persistence as single mutants in any one of these three genes, suggesting they function in the same pathway. Likewise, characterization of double and triple mutants among *ced-1*, *ced-6*, *ced-7* support that a similar relationship exists among these three genes. In contrast, double mutants including one member from each of these sets, (*ced-2*, *ced-5*, *ced-10*, and *ced-1*, *ced-6*, *ced-7*) exhibit significantly increased numbers of corpses as compared to those harboring single mutations, a result expected if at least two processes operate in parallel to mediate the engulfment of dead cells.

The mechanism by which engulfment *ced* genes contribute to corpse removal is not known, but insight into the way in which a dead cell is recognized and digested should result from molecular studies of these genes. Corpse removal requires some type of corpse recognition, extension of cytoplasmic processes to surround and engulf corpse parts, and finally, degradation of cell contents. The engulfment genes are most likely to participate in recognition and/or process extension since ultrastructural analysis has shown that engulfment itself does not occur in *ced-1*, *ced-2*, *ced-5*, *ced-6*, *ced-7*, and *ced-10* mutant backgrounds *(20,21)*.

A Nuclease Needed for Degradation of Corpse DNA

The product of the *nuc-1* gene (abnormal nuclease), is required for degradation of the DNA from cells that have undergone programmed cell death *(1)*. Without *nuc-1* activity, corpse DNA persists and can be visualized by DNA stains such as Feulgen. In a *nuc-1* mutant background, cell death and engulfment proceed normally so it is clear that the activity of the *nuc-1* gene is not required to initiate programmed cell death. Biochemical studies support that the *nuc-1* gene encodes or controls the activity of a Ca^{+2} Mg^{+2}-independent deoxyribonuclease which might be a lysosomal enzyme since it operates at a low pH optimum. The *nuc-1*-controlled nuclease is also required for DNA digestion in the gut *(1)*. Because only a few cells undergo programmed cell death at a given point in development and because individual *C. elegans* cells cannot readily be cultured or isolated for biochemical study, the question of whether NUC-1 mediates internucleosomal DNA degradation in a manner similar to that observed in mammalian apoptosis has not been answered.

In addition to *nuc-1* and the six identified engulfment genes, it is likely that the activities of many genes function in corpse degradation. Such genes might have thus far eluded genetic detection because they may not be essential for efficient corpse degradation, their activities may be redundantly encoded, or they might be essential for viability.

PATHOLOGIES INVOLVING CELL DEATH IN *C. elegans*

As is the case in higher organisms, where it is clear that misregulation of programmed cell death can result in severe developmental abnormalities, oncogenesis and neurodegeneration *(71,72)*, inappropriate programmed cell death in *C. elegans* can also result in pathology. In the most extreme case, *ced-9(lf)* mutations induce extensive cell death which is lethal to the animal *(26)*. In some instances, inappropriate cell death affects only a few cells with lethal consequences for the animal. For example, certain alleles of the *egl-1* gene (*egg-l*aying defective) cause the HSN neurons, which are required for egg expulsion, to undergo programmed cell death *(16,73)*. *egl-1* mutants eggs cannot lay eggs and consequent damage inflicted by larvae which develop internally cause the parents to die prematurely. How *egl-1* acts to influence the HSN life/death decision is not known. In contrast to the situation in higher organisms, the absence of programmed cell death in *C. elegans* does not markedly compromise animal health *(16)*. The 131 extra cells generated in *ced-3* or *ced-4* mutants do not undergo additional cell divisions to produce tumors and appear to be well tolerated.

AN ENGULFMENT ABNORMALITY CAN CAUSE CELL DEATH

The inappropriate activity of one group of engulfment ced genes can cause the death of specific cells. Semidominant mutations in two genes, *lin-24* and *lin-33* (lineage abnormal) cause Pn.p precursor cells to adopt abnormal morphologies and degenerate *(74)*. This cell death occurs independently of the *ced-3* and *ced-4* genes and thus is not executed via the normal programmed cell death pathway. *lin-24*- and *lin-33*-induced deaths are dependent upon the activities of the *ced-2*, *ced-5*, and *ced-10* engulfment genes. Neither the identities of the *lin-24* and *lin-33* proteins nor the mechanisms by which the Pn.p cells die when these genes are defective are

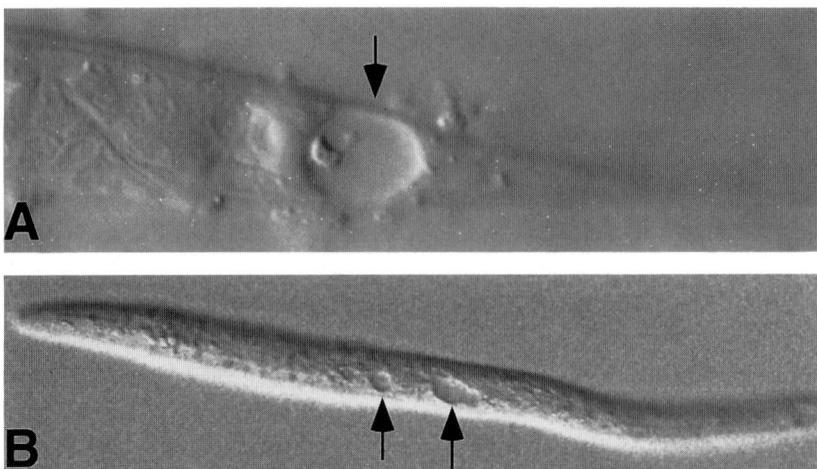

Fig. 6. Degenerative cell death. **(A)** A tail touch receptor undergoing neuronal degeneration in a *mec-4(d)* animal. The cell body is markedly swollen and the nucleus is highly distorted. **(B)** Dying neurons are easily recognized in living animals. Arrows point to vacuolar appearance of dying anterior touch receptor neurons.

known. One possibility is that in the *lin-24(sd)* and *lin-33(sd)* backgrounds, Pn.p cells express cell surface proteins that mark them as targets for engulfment by their neighbors.

INAPPROPRIATE ION CHANNEL ACTIVITY AND NEURODEGENERATION *C. elegans*

Unusual gain-of-function mutations in several *C. elegans* ion channel genes induce necrotic-like deaths of the neurons that express these channel genes. For example, dominant mutations in the *mec-4* gene (*mec*hanosensory abnormal; *mec-4(d)*) induce degeneration of six touch receptor neurons required for the sensation of gentle touch to the body *(75)*. In contrast, most *mec-4* mutations are recessive loss-of-function mutations that disrupt body touch sensitivity without affecting touch receptor ultrastructure or viability *(75,76; see* ref. *77)* for a detailed discussion of the normal function of MEC-4 in touch reception). Similarly, dominant mutations in *deg-1* (degeneration) induce death of a group of neurons that includes the PVC interneurons of the posterior touch sensory circuit *(78,79)*. (Loss-of-function mutations in *deg-1* appear wild-type in behavior, suggesting that the normal activity of the *deg-1* product may be redundantly encoded *[78]*.)

Morphology, Timing, and Ultrastructure of mec-4(d)- *and* deg-1(d)-*induced Neurodegeneration*

Although *mec-4(d)* and *deg-1(d)* mutations kill different groups of neurons, the morphological features of cell deaths they induce are the same *(80)*. The time course of degeneration depends upon the dosage of the toxic allele, but on average can take approximately 8 h *(80)*. When viewed under the light microscope, the nucleus and cell body of the affected cell first appear distorted and then the cell swells to several times its normal cell diameter (Fig. 6). Eventually the swollen cell disappears, often after shrinking but sometimes as a consequence of cell lysis. Interestingly, the swollen character of *mec-4(d)-* and *deg-1(d)*-induced deaths resembles the morphologies of mammalian cells undergoing necrotic cell death *(81,82)*.

At the ultrastructural level, cells dying as a consequence of *mec-4(d)* and *deg-1(d)* expression exhibit some remarkable features *(80)*. The first detectable abnormality apparent in an ill-fated cell is the formation of small tightly wrapped membrane whorls that seem to originate at the plasma membrane. These whorls are internalized and appear to coalesce into large electron-dense membranous structures. Large internal vacuoles form and distortion of the nucleus by these vacuoles is associated with chromatin clumping. Finally, organelles and cytoplasmic contents are degraded, usually leaving a membrane-enclosed shell. Degenerating touch neurons appear to be engulfed by the hypodermis *(80)* and corpses appear to be degraded via the activities of the engulfment *ced* genes (S. Chung and M. Driscoll, unpublished). The striking membranous inclusions suggest that intracellular trafficking may contribute to degeneration. Interestingly, in some mammalian degenerative conditions such as neuronal ceroid lipofuscinosis (Batten disease; the mnd mouse *[83,84]*) and that occurring in the wobbler mouse *(85)*, cells develop vacuoles and whorls (fingerprint bodies) that look similar to internalized structures in dying *C. elegans* neurons. This suggests that some degenerative processes may be similar in nematodes and mammals.

The touch receptor neurons in *mec-4(d)* mutants express terminally differentiated properties before they die *(85)* and the PVC neurons in *deg-1(d)* mutants differentiate and function before they degenerate *(78)*. *mec-4(d)-* and *deg-1(d)*-induced cell deaths have therefore sometimes been referred to as the nematode version of "late onset" neurodegeneration. Careful studies of the timing of *mec-4* expression relative to the onset of degeneration support that onset of neurodegeneration is correlated with the initial expression of the toxic gene product *(80)*.

mec-4(d)- and deg-1(d)-Induced Neurodegeneration Occur Autonomously and Independently of Programmed Cell Death Executors

Genetic mosaic analyses first indicated that *mec-4(d)* kills as a consequence of a toxic activity within the cells that die *(87)*. Ectopic expression of *mec-4(d)* can induce swelling and death of cells other than the touch receptor neurons, confirming the cell autonomy of *mec-4(d)* action *(88,89;* H. Singh, N. Tavernarakis and M. Driscoll, unpublished observations). The execution of degenerative cell death occurs by a mechanism that appears distinct from that utilized in programmed cell death. At the genetic level, it has been demonstrated that *ced-3(lf)* and *ced-4(lf)* mutations do not block *mec-4(d)-* and *deg-1(d)*-induced cell degeneration *(78)*. Likewise, *mec-4(d)* and *deg-1(d)* alleles do not disrupt programmed cell deaths (M. Chalfie, personal communication). There are also clear morphological distinctions between degenerative and programmed cell death. During programmed cell death, cells become compact and refractile whereas during *mec-4(d)-* and *deg-1(d)*-induced degenerative death cellular swelling occurs (compare Figs. 1 and 6). As mentioned above, mechanisms of corpse removal appear to be common to both programmed and degenerative cell death (S. Chung and M. Driscoll, unpublished observations).

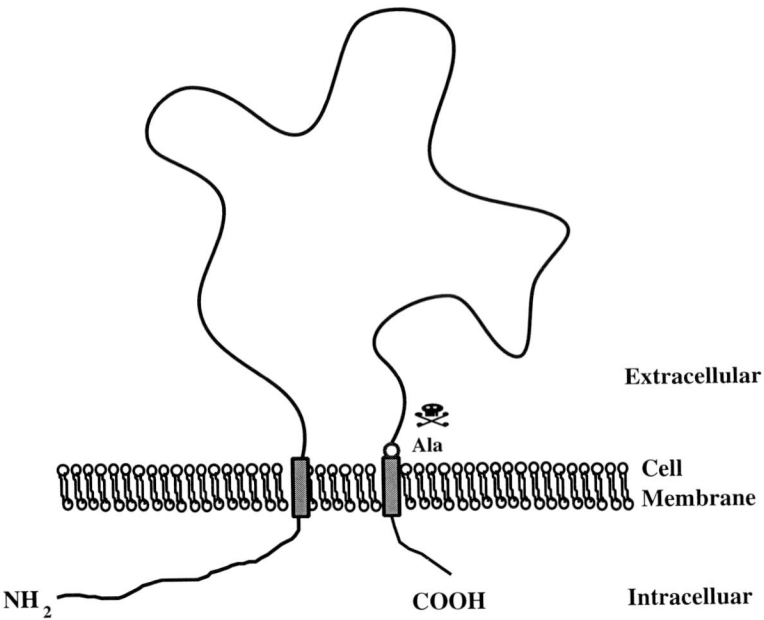

Fig. 7. General structure and transmembrane topology of *C. elegans* MEC-4. Two membrane-spanning domains (MSDI, MSDII) are grey boxes; indicated is Ala713, the site of large sidechain amino acids that induce neurodegeneration.

mec-4 and *deg-1* Encode Ion Channel Subunits of the DEG/ENaC Superfamily

mec-4 and *deg-1* encode proteins that are 51% identical *(78,79,90,91)*. These genes were the first identified members of the *C. elegans* "degenerin" family, so named because several members can mutate to forms that induce cell degeneration. Included in this family are *mec-10*, which can be engineered to encode toxic degeneration-inducing substitutions *(92)*; *unc-8 (89)*, which can mutate to a semidominant form that induces swelling and dysfunction of ventral cord *(93)*; and *unc-105*, which appears to be expressed in muscle and can mutate to a semi-dominant form that induces muscle hypercontraction *(94)*. Thus, a general feature of the degenerin gene family is that specific gain-of-function mutations have deleterious consequences for the cells in which they are expressed.

C. elegans degenerins share sequence similarity with subunits of the vertebrate amiloride-sensitive epithelial Na^+ channel *(95–97)*. α-, β-, and γ-ENaC (for epithelial Na^+ channel) are homologous subunits of the multimeric Na^+ channel that mediates Na^+ absorption in epithelia of the distal part of the kidney tubule, the urinary bladder, the distal colon and the lung (reviewed in ref. *98*). All members of this superfamily, called DEG/ENaC proteins, have two hydrophobic transmembrane domains (designated MSDI and MSDII; Fig. 7A). MSDII is predicted to form an amphipathic α-helix, and polar residues aligned along the hydrophilic face are postulated to project into the channel lumen to line the channel pore and influence channel conductance *(99,100)*. Although the activities of the nematode channels have not been directly assayed, the conservation of most primary sequence motifs between nematodes and mammalian family members and the finding that certain rat/nematode chimeras function in vivo in

C. elegans (99) and form channels in Xenopus oocytes *(100)*, support that the *C. elegans* degenerins function as ion channel subunits.

The degenerin family of *C. elegans* currently includes 15 members that have been characterized or predicted by the *C. elegans* Genome Sequencing Consortium. Given such a large *C. elegans* gene family, it is predicted that the mammalian ENaC family should likewise be large. Indeed, the number of identified members of the mammalian ENaC family has been continuously increasing. Two family members, FaNaC and MDEG, are expressed in the nervous system *(101,102)*. Because many *C. elegans* degenerins can mutate to toxic forms that induce neurodegeneration, the neuronally expressed mammalian family members are logical candidates for genes that can mutate to cause neurodegeneration in higher organisms. In this regard it is highly interesting that mammalian MDEG, engineered to encode an amino acid substitution analogous to the change in *mec-4(d)*, induces degeneration when expressed in *Xenopus* oocytes and embryonic hamster kidney cells *(102)*.

Death-Inducing Channel Mutations and Models for Initiation of Neurodegeneration

mec-4(d) and *deg-1(d)* alleles encode substitutions for a conserved alanine that is positioned extracellularly, adjacent to the predicted pore-lining domain MSDII (Fig. 8) (position 713 in MEC-4 *[90,91]*, position 707 in DEG-1 *[79]*). The size of the amino acid side chain at this position is correlated with toxicity—substitution of a small sidechain amino acid does not induce degeneration whereas replacement of Ala713 with a large side chain amino acid is toxic (Driscoll and Chalfie, 1991). This "rule" suggests that steric hindrance plays a role in the degeneration mechanism and supports the following working model for *mec-4(d)*-induced degeneration (Fig. 8). MEC-4 is postulated to be a subunit of a channel that, like other channels, can assume alternative open and closed conformations. In adopting the closed conformation, the sidechain of the amino acid at MEC-4 position 713 is proposed to come into close proximity to another part of the channel. Steric interference conferred by a bulky amino acid sidechain prevents such an approach, causing the channel to close less effectively. Increased cation influx results, initiating neurodegeneration. Although electrophysiological analyses have not yet been performed on this channel, it has been shown that an amino acid substitution for a likely channel pore-lining residue (a change that is predicted to disrupt ion influx) can prevent neurodegeneration when present *in cis* to the A713V substitution *(99)*, consistent with the working model of channel structure and toxicity (Fig. 8). Mutations affecting a second extracellular degenerin domain can also induce neurodegeneration that appears dependent upon the integrity of the channel pore domain *(79,89)*.

Mutations in other ion channel genes that increase channel activity cause vacuolar degeneration of *C. elegans* neurons. For example, a dominant mutation in the α7-like acetylcholine receptor gene *deg-3(d)* causes the degeneration of a defined group of neurons *(103)*. The dominant allele encodes substitution I293N within TMDII, which affects a site equivalent to that affected by the chicken α7-4 V251T substitution. The mutant chicken channel is defective in desensitization when expressed as a homomeric channel in *Xenopus* oocytes *(104)*, suggesting that the *C. elegans deg-3* channel might allow increased ion influx. Consistent with this hypothesis, some nicotinic antagonists partially suppress *deg-3(d)*-induced defects *(103)*.

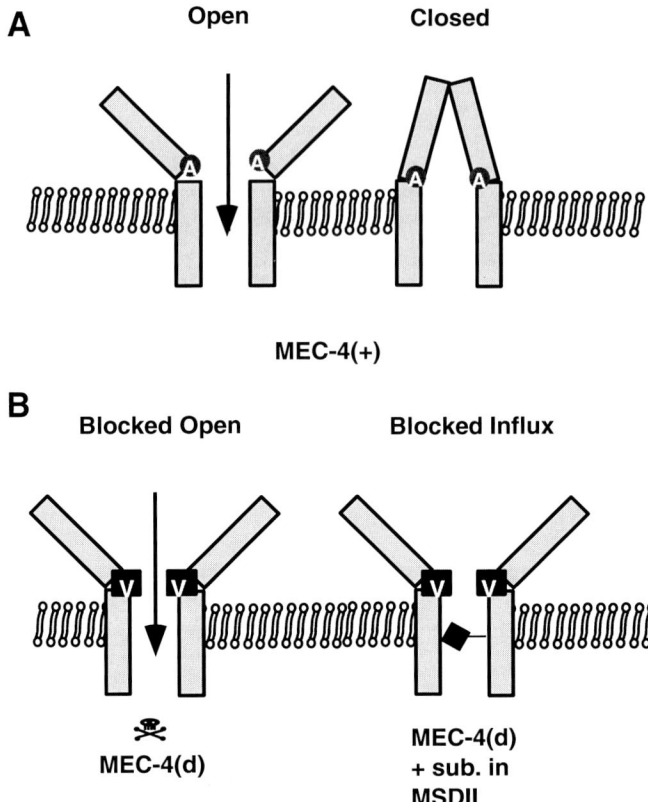

Fig. 8. Working model for the mechanism of action of the MEC-4(d) channel. **(A)** The normal MEC-4 channel can move freely between open and closed conformations because the sidechain of Ala713 can be accommodated in the channel complex. **(B)** Steric interference cause by a large amino acid sidechain at position 713 prevents the channel from closing efficiently **(left)**. The resulting elevation in ion uptake is toxic. The presence of a second amino acid substitution within the channel pore lining domain, which disrupts ion transport, can block degeneration **(right)**.

Parallels Between Neurodegenerative Cell Death in C. elegans and Higher Organisms—A Common Degenerative Death Mechanism?

Inappropriate channel activity is known to be causative for some mammalian neurodegenerative conditions. For example, it is interesting that the working model for the initiation of degenerative cell death in *C. elegans* is remarkably similar to events that initiate excitotoxic cell death in higher organisms. In excitotoxicity, glutamate receptor ion channels are hyperstimulated by the excitatory transmitter glutamate and the resultant elevated Na^+ and Ca^{+2} transport induces neuronal swelling and death *(105,106)*. Mammalian ion channel mutations can also induce neurodegeneration. In the weaver mutant mouse, altered gating and ion selectivity properties of the GIRK2 potassium channel are associated with vacuolar cell death in the cerebellum, dentate gyrus and olfactory bulb *(107,108)*.

It is noteworthy, however, that mutations in channel genes are not the sole means by which vacuolar neurodegeneration can be induced in *C. elegans*. Necrotic-like death of some *C. elegans* cells can be induced by expression of human β-amyloid peptide (1–42) *(109)*. Mutations in transcription factor *lin-26* cause hypodermal cells to become neuroblasts which swell and die *(110)*. Also, it has been noted that mutations that cause swelling and death can be isolated at a relatively high frequency, suggesting that there might be multiple types of genes that can mutate to induce necrotic-like death (A. Chisolm, R. Horvitz; M. Chalfie, personal communication). These observations suggest that degenerative cell death might be induced by a variety of cellular "injuries" and that a common death mechanism could operate to eliminate injured cells. Indeed, electron micrograph analyses of *mec-4(d)*-induced cell death revealed a reproducible sequence of cellular changes that transpire during degeneration, suggesting that specific regulated steps (rather than chaotic cellular destruction) are involved *(80)*. The peculiar internalized membranous whorls observed suggests degenerin-induced death could involve disrupted intracellular trafficking, an interesting implication given that disrupted trafficking has been implicated in Alzheimer's disease *(111)*, Huntington's disease *(112)*, and ALS *(113)*. Perhaps endocytotic responses provoked by diverse types of damage might be a common element of diverse degenerative conditions.

If specific genes enact different steps of the degenerative process, then such genes should be identifiable by mutation in *C. elegans*. Indeed, suppressor mutations in several genes that block *mec-4(d)*-induced degeneration have been isolated (D. Royal, M. Royal, K. Xu, E. Wu, R. Lints, M. Driscoll, unpublished). Although some suppressor mutations affect channel function (for example, mutations in *mec-6*, a candidate channel-associated protein *[78]*), others are expected to be more generally involved in the death process. Analysis of such genes should result in the description of a genetic pathway for degenerative cell death. Perhaps, as has proven to be the case for the analysis of *C. elegans* programmed cell death mechanisms, elaboration of an injury-induced death pathway in *C. elegans* may provide insight into neurodegenerative death mechanisms in higher organisms.

ACKNOWLEDGMENTS

The authors would like to thank Mary Anne Royal for help with the manuscript and figures and Nektarios Tavernarakis for help with the figures. This work was supported in part by the National Institutes of Health (NS34435) and the Amyotrophic Lateral Sclerosis Association. Dewey Royal was supported by a National Science Foundation Minority Postdoctoral Research Fellowship.

REFERENCES

1. Sulston J, Horvitz H. Post-embryonic cell lineages of the nematode, *Caenorhabditis elegans*. *Dev Biol* 1976, **56**: 110–156.
2. Sulston JE, Schierenberger E, White JG, Thomson JN. The embryonic cell lineage of the nematode *Caenorhabditis elegans*. *Dev Biol* 1983, **100**: 64–119.
3. White JG. The structure of the nervous system of *Caenorhabditis elegans*. *Phil Trans Royal Soc Lond B Biol Sci* 1986, **314**: 1–340.
4. White JG, Southgate E, Thomson JN, Brenner S. The structure of the ventral nerve cord of *Caenorhabditis elegans*. *Phil Trans R Soc Lond B Biol Sci* 1976, **275**: 327–348.

5. Chalfie M, Sulston JE, White JG, Southgate E, Thomson JN, Brenner S. The neural circuit for touch sensitivity in *Caenorhabditis elegans*. *J Neurosci* 1985, **5:** 956–964.
6. Bargmann C, Avery L. Laser killing of cells in *Caenorhabditis elegans*, in *Methods in Cell Biology. Caenorhabditis elegans*: Modern Biological Analysis of an Organism (Epstein HF, Shakes DC, eds.), Academic, San Diego, 1995, pp. 225–250.
7. Bargmann C, Horvitz H. Control of larval development by chemosensory neurons in *Caenorhabditis elegans*. *Science* 1991, **251:** 1243–1246.
8. Wilson R, Ainscough R, Anderson K, Baynes C, Berks M, Burton J, Connell M, Bonfield J, Copsey T, Cooper J. 2.2 Mb of contiguous nucleotide sequence from chromosome III of *C. elegans*. *Nature* 1994, **368:** 32–38.
9. Waterston R, Sulston J. The genome of *Caenorhabditis elegans*. *Proc Natl Acad Sci USA* 1995, **92:** 10836–10840.
10. Mello CC, Kramer JM, Stinchcomb D, Ambros V. Efficient gene transfer in *C. elegans*: extrachromosomal maintenance and integration of transforming sequences. *EMBO J* 1991, **10:** 3959–3970.
11. Fire A. Integrative transformation of *Caenorhabditis elegans*. *EMBO J* 1986, **5:** 2673–2680.
12. Fire A, Harrison SW, Dixon D. A modular set of lacZ fusion vectors for studying gene expression in *Caenorhabditis elegans*. *Gene* 1990, **93:** 189–198.
13. Chalfie M, Tu Y, Euskirchen G, Ward W, Prasher D. Green fluorescent protein as a marker for gene expression. *Science* 1994, **263:** 902–3.
14. Oppenheim RW. Cell death during the development of the nervous system. *Annu Rev Neurosci* 1991, **14:** 453–501.
15. Robertson AMG, Thomson JN. Morphology of programmed cell death in the ventral cord of *Caenorhabditis elegans*. *J Embryol Exp Morphol* 1982, **67:** 89–100.
16. Ellis HM, Horvitz HR. Genetic control of programmed cell death in the nematode *Caenorhabditis elegans*. *Cell* 1986, **44:** 817–829.
17. Avery L, Horvitz HR. A cell that dies during wild-type *C. elegans* development can function as a neuron in a *ced-3* mutant. *Cell* 1987, **51:** 1071–1078.
18. Horvitz HR, Ellis HM, Sternberg PW. Programmed cell death in nematode development. *Neuroscience Comment* 1982, **1:** 56–65.
19. White JG, Southgate E, Thomson JN. On the nature of undead cells in the nematode *Caenorhabditis elegans*. *Phil Trans R Soc Lond B* 1991, **331:** 263–271.
20. Hedgecock E, Sulston JE, Thomson JN. Mutations affecting programmed cell death in the nematode *Caenorhabditis elegans*. *Science* 1983, **220:** 1277–1280.
21. Ellis RE, Jacobson DM, Horvitz HR. Genes required for engulfment of cell corpses during programmed cell death in *Caenorhabditis elegans*. *Genetics* 1991, **129:** 79–94.
22. Ellis RE, Horvitz HR. Two *C. elegans* genes control the programmed deaths of specific cells in the pharynx. *Development* 1991, **112:** 591–603.
23. Horvitz HR, Chalfie M, Trent C, Subtan JE, Evans PD. Serotonin and octopamine in the nematode *C. elegas*. *Science* 1982, **216:** 1012–1014.
24. Metstein MM, Hengartner MO, Tsung N, Ellis RE, Horvitz HR. Transcriptional regulator of programmed cell death encoded by *Caenorhabditis elegans* gene *ces-2*. *Nature* 1996, **382:** 545–547.
25. Inada T, Inukari T, Yoshihara T, Seyschab H, Ashmun RA, Cannon CE, Laken SJ, Kastan MB, Look AT. Reversal of apoptosis by the leukemia-associated E2A-HLF chimeric transcription factor. *Nature* 1996, **382:** 541–544.
26. Hengartner MO, Ellis RE, Horvitz HR. *Caenorhabditis elegans* gene *ced-9* protects cells from programmed death. *Nature* 1992, **356:** 494–499.
27. Hengartner MO, Horvitz HR. *C. elegans* cell survival gene *ced-9* encodes a functional homologue of the mammalian proto-oncogene bcl-2. *Cell* 1994, **76:** 665–676.

28. Reed J. Bcl-2 and the regulation of programmed cell death. *J Cell Biol* 1994, **124**: 1–6.
29. Cortazzo M, Schor N. Potentiation of enediyne-induced apoptosis and differentiation by Bcl-2. *Cancer Res* 1996, **56**: 1199–1203.
30. Vaux DL, Weissman IL, Kim SK. Prevention of programmed cell death in *Caenorhabditis elegans* by human bcl-2. *Science* 1992, **258**: 1955–1957.
31. Boise LH, Gonzalez-Garcia M, Postema CE, Ding L, Lindsten T, Turka T, Mao X, Nunez G, Thompson CB. bcl-x, a bcl-2-related gene that functions as a dominant regulator of apoptotic cell death. *Cell* 1993, **74**: 597–608.
32. Oltvai ZN, Milliman CL, Korsmeyer SJ. Bcl-2 heterodimerizes in vivo with a conserved homologue, Bax, that accelerates programmed cell death. *Cell* 1993, **74**: 609–619.
33. Chittenden T, Harrinton EA, O'Connor R, Flemington C, Lutz RJ, Evan GI, Guild BC. Induction of apoptosis by the Bcl-2 homologue Bak. *Nature* 1995, **374**: 733–736.
34. Yang E, Zha J, Jockel J, Boise LH, Thompson CB Korsmeyer SJ. Bad, a heterodimeric partner for Bcl-XL and Bcl-2, displaces Bax and promotes cell death. *Cell* 1995, **80**: 285–291.
35. Farrow SN, Brown R. New members of the Bcl-2 family and their protein partners. *Curr Op Genet Dev* 1996, **6**: 45–49.
36. Oltvai ZN, Korsmeyer SJ. Checkpoints of dueling dimers foil death wishes. *Cell* 1994, 79, 189–192.
37. Muchmore SW, Sattler M, Liang H, Meadows RP, Harlan JE, Yoon HS, Nettesheim D, Chang BS, Thompson CB, Wong SL, Ng SW. X-ray and NMR structure of human Bcl-x_L, an inhibitor of programmed cell death. *Nature* 1996, **381**: 335–341.
38. Minn AJ, Velez P, Schendel SL, Liang H, Muchmore SW, Fesik SW, Fill M, Thompson CB. Bcl-x_L forms an ion channel in synthetic lipid membranes. *Nature* 1997, **385**: 353–357.
39. Schendel SL, Xie Z, Reed JC. Channel formation by anti-apoptoic protein, Bcl-2. *Proc Natl Acad Sci USA* 1997, **94**: 5113–5118.
40. Spector MS, Desnoyers S, Hoeppner DJ, Hentgartner MO. Interaction between the *C. elegans* cell-death regulators CED-9 and CED-4. *Nature* 1997, **385**: 653–656.
41. Chinnaiyan A, O'Rourke K, Dixit V. Interaction of CED-4 with CED-3 and CED-9: A molecular framework for cell death. *Science* 1997, **275**: 1122–1126.
42. Wu D, Wallen H, Nunez G. Interaction and regulation of subcellular localization of CED-4 by CED-9. *Science* 1997, **275**: 1126–1129.
43. Hacker G, Vaux DL. A sticky business. *Curr Biol* 1995, **5**: 622–624.
44. Petit PX, Susin SA, Kroemer G. Mitochondria and programmed cell death: Back to the future. *FEBS Lett* 1996, **396**: 7–13.
45. Susin SA. Bcl-2 inhibits the mitochondrial release of apoptogenic protease. *J Exp Med* 1996, **184**: 1331–1342.
46. Kluck RM, Bossy-Wetzel E, Newmeyer D. The release of cytochrome c from the mitochondria: A primary site for Bcl-2 regulaton of apoptosis. *Science* 1997, **275**: 1132–1136.
47. Tang J. Prevention of apoptosis by Bcl–2: Release of cytochrome c from mitochondria blocked. *Science* 1997, **275**: 1129–1132.
48. Nakashima T, Sekiguchi T, Kuraoka A, Fukushima K, Shibata Y, Komiyama S, Nishimoto T. Molecular cloning of a human cDNA encoding a novel protein, DAD1, whose defect causes apoptotic cell death in hamster BHK21 cells. *Mol Cell Biol* 1993, **13**: 6367–6374.
49. Sugimoto A, Friesen PD, Rothman JH. Baculovirus p35 prevents developmentally programmed cell death and rescues a *ced-9* mutant in the nematode *Caenorhabditis elegans*. *EMBO J* 1994, **13**: 2023–2028.
50. Silberstein S, Collins PG, Kellecher DJ, Gilmore R. The essential OST2 gene encodes the 16kD subunit of the yeast oligosaccaryltransferase, a highly conserved protein expressed in diverse eukaryotic organisms. *J Cell Biol* 1995, **131**: 371–383.

51. Yuan J, Shaham S, Ledoux S, Ellis HM, Horvitz HR. The C. elegans cell death gene ced-3 encodes a protein similar to mamailan interleukin-1b-converting enzyme. *Cell* 1993, **75**: 641–652.
52. Cerretti DP, Kozlosky CJ, Mosley B, Nelson N, Ness KV, Greenstreet TA, March CJ, Kronheim SR, Druck T, Cannizzaro LA, Huebner K, Black RA. Molecular cloning of the interleukin-1b converting enzyme. *Science* 1992, **256**: 97–100.
53. Thornberry NA, Bull H, Calaycay J, Chapman K, Howard A, Kostura M, Miller D, Molineaux S, Weidner J, Aunins J, Lee TD, Shively JE, MacCross M, Mumford RA, Schmidt JA, Tocci MJ. A novel heterodimeric cysteine protease is required for interleukin-1β-processing in monocytes. *Nature* 1992, **356**: 768–774.
54. Xue D, Shaham S, Horvitz HR. The *Caenorhabditis elegans* cell-death protein CED-3 is a cysteine protease with substrate specificities similar to those of the human CPP32 protease. *Genes Dev* 1996, **10**: 1073–1083.
55. Hugunin M, Quintal LJ, Mankovich JA, Ghayur T. Protease activity of *in vitro* transcribed and translated *Caenorhabditis elegans* cell death gene (*ced-3*) product. *J Biol Chem* 1996, **271**: 3517–3522.
56. Shaham S, Horvitz HR. Developing *Caenorhabditis elegans* neurons may contain both cell-death protective and killer activities. *Genes Dev* 1996, **10**: 578–591.
57. Miura M, Zhu H, Rotello R, Hartwieg EA, Yuan J. Induction of apoptosis in fibroblasts by IL-1b-converting enzyme, a mammalian homologue of the *C. elegans* cell death gene *ced-3*. *Cell* 1993, **75**: 653–660.
58. Martin SJ, Green DR. Protease activation during apoptosis: death by a thousand cuts? *Cell* 1995, **82**: 349–352.
59. Wang L, Miura M, Bergeron L, Zhu H, Yuan J. Ich-1, an Ice/*ced-3*-related gene, encodes both positive and negative regulators of programmed cell death. *Cell* 1994, **78**: 739–750.
60. Fernandes-Alnemri T, Litwack G, Alnemri ES. CPP32, a novel human apoptotic protein with homology to *Caenorhabditis elegans* cell death protein CED-3 and mammalian interleukin-1beta-converting enzyme. *J Biol Chem* 1994, **269**: 30,761–30,764.
61. Kaufmann SHD, Ottaviano S, Davidson Y, Poirier NE. Specific proteolytic cleavage of poly (ADP-ribose) polymerase: an early marker of chemotherapy-induced apoptosis. *Cancer Res* 1993, **53**: 3976–3985.
62. Tewari M, Quan L, O'Rourke K, Desnoyers S, Zeng Z, Beidler D, Poirier G, Salvesen G, Dixit V. Yama/CPP32b, a mammalian homologue of CED-3, is a crmA-inhibitable protease that cleaves the death substrate poly(ADP-ribose) polymerase. *Cell* 1995, **81**: 801–809.
63. Xue D, Horvitz HR. Inhibition of the *Caenorhabditis elegans* cell-death protease CED-3 by a CED-3 cleavage site in baculovirus p35 protein. *Nature* 1995, **377**: 248–251.
64. Hay BA, Wolff T, Rubin GM. Expression of baculovirus p35 prevents cell death in *Drosophila*. *Develop* 1994, **120**: 2121–2129.
65. Rabizadeh S, Bitler C, Butcher L, Bredesen D. Expression of the baculovirus p35 gene inhibits mammalian neural cell death. *J Neurochem* 1993, **61**: 2318–2321.
66. Shaham S, Horvitz HR. An alternatively spliced *C. elegans ced-4* RNA encodes a novel cell death inhibitor. *Cell* 1996, **86**: 201–208.
67. Yuan J, Horvitz HR. The *Caenorhabditis elegans* cell death *ced-4* encodes a novel protein and is expressed during the period of extensive programmed cell death. *Development* 1992, **116**: 309–320.
68. Baver MK, Wesselborg S, Schulze-Osthoff K. The Caenorhabditis elegans death protien Ced-y contains a motif with similarity to the mammalian 'death effector domain.' *FEBS Lett* 1997, **402**: 256–258.
69. Hengartner MO, Horvitz HR. Activation of *C. elegans* cell death protein CED-9 by an amino-acid substitution in a domain conserved in BCL-2. *Nature* 1994, **369**: 318–320.

70. Yuan J, Horvitz HR. The *Caenorhabditis elegans* genes *ced-3* and *ced-4* act cell autonomously to cause programmed cell death. *Devel Biol* 1990, **138**: 33–41.
71. Bredesen D. Neural apoptosis. *Ann Neurol* 1995, **38**: 839–851.
72. Thompson CB. Apoptosis in the pathogenesis and treatment of disease. *Science* 1995, **267**: 1456–1462.
73. Trent C, Tsung N, Horvitz HR. Egg-laying defective mutants of the nematode *C. elegans*. *Genetics* 1983, **104**: 619–647.
74. Kim S-C. Ph.D. thesis. MIT 1994.
75. Chalfie M, Au M. Genetic control of differentiation of the *Caenorhabditis elegans* touch receptor neurons. *Science* 1989, **243**: 1027–1033.
76. Chalfie M, Sulston J. Developmental genetics of the mechanosensory neurons of *Caenorhabditis elegans*. *Devel Biol* 1981, **82**: 358–370.
77. Driscoll M, Kaplan J. Mechanotransduction, in C. elegans II (Riddle D, ed.) Cold Spring Harbor Laboratory Press, Cold Spring Harbor, NY, 1996, pp. 645–677.
78. Chalfie M, Wolinsky E. The identification and suppression of inherited neurodegeneration in *Caenorhabditis elegans*. *Nature* 1990, **345**: 410–416.
79. Garcia-Anoveros J, Ma C, Chalfie M. Regulation of *Caenorhabditis elegans* degenerin proteins by a putative extracellular domain. *Curr Biol* 1995, **5**: 441–448.
80. Hall DH, Gu G, Garcia-Anoveros J, Gong L, Chalfie M, Driscoll M. Neuropathology of degenerative cell death in *C. elegans*. *J Neurosci* 1997, **17**: 1033–1045.
81. Kerr JFR, Wyllie AH, Currie AR. Apoptosis: a basic biological phenomenon with wide-ranging implications in tissue kinetics. *Br J Cancer* 1972, **26**: 239–257.
82. Wyllie AH, Kerr JFR, Currie AR. Cell death: the significance of apoptosis. *Int Rev Cytol* 1980, **68**: 251–306.
83. March PA, Wurzelmann S, Walkley SU. Morphological alterations in neocortical and cerebellar GABAergic neurons in a canine model of juvenile Batten disease. *Am J Med Genetics* 1995, **57**: 204–212.
84. Pardo CA, Rabin BA, Palmer DN, Price DL. Accumulation of the adenosine triphosphate synthase subunit C in the mnd mutant mouseA model for neuronal ceroid lipofuscinosis. *Am J Pathol* 1994, **144**: 829–835.
85. Andrews J. The fine structure of the cervical spinal cord, ventral root and brachial nerves in the wobbler (wr) mouse. *J Neuropathol Exp Neurol* 1975, **34**: 12–27.
86. Mitani S, Du H, Hall D, Driscoll M, Chalfie M. Combinatorial control of touch receptor neuron expression in *Caenorhabditis elegans*. *Develop* 1993, **119**: 773–83.
87. Herman RK. Mosaic analysis of two genes that affect nervous system structure on *Caenorhabditis elegans*. *Genetics* 1987, **116**: 377–388.
88. Maricq A, Peckol E, Driscoll M, Bargmann C. Mechanosensory signaling in *C. elegans* mediated by the GLR-1 glutamate receptor. *Nature* 1995, **378**: 78–81.
89. Tavernarakis N, Scheffler W, Wang S, Driscoll M. *unc-8*, DEG/ENaC superfamily member, encodes a subunit of a candidate mechanically gated channel that modulates *C. elegans* locomotion. *Neuron* 1997, **18**: 107–119.
90. Driscoll M, Chalfie M. The *mec-4* gene is a member of a family of *Caenorhabditis elegans* genes that can mutate to induce neuronal degeneration. *Nature* 1991, **349**: 588–593.
91. Lai C, Hong K, Kinnell M, Chalfie M, Driscoll M. Sequence and transmembrane topology of MEC-4, an ion channel subunit required for mechanotransduction in *Caenorhabditis elegans*. *J Cell Biol* 1996, **133**: 1071–1081.
92. Huang M, Chalfie M. Gene interactions affecting mechanosensory transduction in *Caenorhabditis elegans*. *Nature* 1994, **367**: 467–470.
93. Shreffler W, Magardino T, Shekdar K, Wolinsky E. The *unc-8* and *sup-40* genes regulate ion channel function in *Caenorhadbitis elegans* motorneurons. *Genetics* 1995, **139**: 1261–1272.

94. Liu J, Schrank B, Waterston R. Interaction between a putative mechanosensory membrane channel and a collagen. *Science* 1996, **273**: 361–364.
95. Chalfie M, Driscoll M, Huang M. Degenerin similarities. *Nature* 1993, **361**: 504.
96. Canessa C, Horisberger JD, Rossier BC. Epithelial sodium channel related to proteins involved in neurodegeneration. *Nature* 1993, **361**: 467–470.
97. Canessa CM, Schild L, Buell G, Thorens B, Gautschi I, Horisberger JD, Rossier BC. Amiloride-sensitive epithelial Na+ channel is made of three homologous subunits. *Nature* 1994, **367**: 412–413.
98. Palmer LGA. Epithelial Na+ channels: function and diversity. *Ann Rev Physiol* 1992, **54**: 51–66.
99. Hong K, Driscoll M. A transmembrane domain of the putative channel subunit MEC-4 influences mechanotransduction and neurodegeneration in *C. elegans*. *Nature* 1994, **367**: 470–473.
100. Waldmann R, Champigny G, Lazdunski M. Functional degenerin-containing chimeras identify residues essential for amiloride-sensitive Na+ channel function. *J Biol Chem* 1995, **270**: 11,735–11,737.
101. Lingueglia E, Champigny G, Lazdunski M, Barbry P. Cloning of the amiloride-sensitive FMRFamide peptide-gated sodium channel. *Nature* 1995, **378**: 730–733.
102. Waldmann R, Champigny G, Voilley N, Lauritzen I, Lazdunski M. The mammalian degenerin MDEG, an amiloride-sensitive cation channel activated by mutations causing neurodegeneration in *Caenorhabditis elegans*. *J Biol Chem* 1996, **271**: 10,433–10,436.
103. Treinin M, Chalfie M. A mutated acetylcholine receptor subunit causes neuronal degeneration in *C. elegans*. *Neuron* 1995, **14**: 871–877.
104. Galzi J, Devillers-Thiery A, Hussy N, Bertrand S, Changeux J, Bertrand D. Mutations in the channel domain of a neuronal nicotinic receptor convert ion selectivity from cationic to anionic. *Nature* 1992, **359**: 500–505.
105. Choi DW. Excitotoxic cell death. *J Neurobiol* 1992, **23**: 1261–1276.
106. Choi DW. Glutamate receptors and the induction of excitotoxic neuronal death. *Prog Brain Res* 1994, **100**: 47–51.
107. Slesinger PA, Patil N, Liao J, Jan YN, Jan YL, Cox DR. Functional effects of the mouse weaver mutation on G protein-gated inwardly rectifying K+ channels. *Neuron* 1996, **16**: 321–331.
108. Patil N, Cox D, Bhat D, Faham M, Myers R, Peterson A. A potassium channel mutation in weaver mice implicates membrane excitability in granule cell differentiation. *Nature Genetics* 1995, **11**: 126–129.
109. Link C. Expression of human beta-amyloid peptide in transgenic *Caenorhabditis elegans*. *PNAS* 1995, **92**: 9368–9372.
110. Labouesse M, Sookhareea S, Horvitz H. The *Caenorhabditis elegans* gene *lin-26* is required to specify the fates of hypodermal cells and encodes a presumptive zinc-finger transcription factor. *Dev* 1994, **120**: 2359–2368.
111. Sherrington R, Rogaev E, Liang Y, Rogaeva E, Levesque G, Ikeda M, Chi H, Lin C, Li G, Holman K, et al. Cloning of a gene bearing missense mutations in early-onset familial Alzheimer's disease. *Nature* 1995, **375**: 754–760.
112. DiFiglia M, Sapp E, Chase K, Schwarz C, Meloni A, Young C, Martin E, Vonsattel J, Carraway R, Reeves S. Huntingtin is a cytoplasmic protein associated with vesicles in human and rat brain neurons. *Neuron* 1995, **14**: 1075–1081.
113. Brown RHJ. Amyotrophic lateral sclerosis: recent insights from genetics and transgenic mice. *Cell* 1995, **80**: 687–692.
114. Horvitz H, Shaham S, Hengartner M. The genetics of programmed cell death in the nematode caenorhabditis elegans, in *Cold Spring Harbor Symposia on Quantitative Biology* Cold Spring Harbor Laboratory Press, Cold Spring Harbor, NY, 1994, pp. 377–385.

8
Cellular Interactions that Regulate Programmed Cell Death in the Developing Vertebrate Nervous System

Michael J. Burek and Ronald W. Oppenheim

INTRODUCTION

The development of individual neurons and the nervous system as a whole are characterized by progressive cellular events involving the gradual attainment of various phenotypes, connectives, and functions. In this context, it is somewhat paradoxical that one of the most fundamental features of the developing nervous system is the widespread death of large numbers of neurons. Characterizing both the cell–cell interactions and molecular mechanisms that regulate neuronal survival during developmental cell death is currently at the forefront of research in developmental neurobiology. First, understanding how neuronal death, like other developmental cellular events such as proliferation, migration, and axon growth confer structure and function to the developing nervous system will provide a deeper understanding of neural and behavioral development in general. Second, vigorously pursuing the cellular interactions and molecular mechanisms that regulate *developmental* neuronal death may ultimately result in therapeutic strategies for attenuating *pathological* neuronal death induced by aging, injury, or disease. In fact, such therapeutic approaches have already begun thanks, in part, to several decades of basic research on the role of target-derived signals in regulating the extent of cell death during development. Several putative target-derived neurotrophic factors such as nerve growth factor (NGF), ciliary neurotrophic factor (CNTF), brain-derived neurotrophic factor (BDNF), and glial cell line-derived neurotrophic factor (GDNF), which rescue neurons from developmental death, are also remarkably successful at preventing neuronal death in animal models of injury and neurodegenerative disease (e.g., refs. *1* and *2*). These neurotrophic factors, characterized originally within the framework of basic cell biological research, now for the first time offer the hope of slowing or arresting neuronal death induced by injury, disease, or aging in humans (*see* chapter by Koliatsos and Mocchetti for a topical discussion of trophic factors).

Because other chapters in this book provide a more in depth overview of the molecular pathways by which neurons undergo both normal and pathologically induced cell death, we will focus here on the cellular interactions that control the survival of neurons, as well as the possible biological significance of neuronal death during development. A critical first step in providing an overview of cell death in the nervous system is to define the phenomenon. We choose to use the term programmed cell death (PCD) to describe the spatially and temporally reproducible death of cells during embryonic, fetal, or early postnatal ages. Importantly, the term "programmed" emphasizes the stereotypical and species-specific nature of this cell death and in no way infers that this regressive phenomenon is genetically determined or inevitable. In fact, as we will discuss in this chapter, there is overwhelming evidence that the PCD of developing, and even some mature neuronal populations, is controlled by epigenetic signals derived primarily from cell–cell interactions. Our definition of PCD also generally excludes cell death as a consequence of injury and disease, while at the same time it makes no assumptions regarding possible biochemical or morphological pathways by which cells commit suicide. For example, PCD and apoptosis are not synonymous. Instead, apoptosis describes one cell death pathway (characterized by DNA fragmentation, nuclear condensation etc.) that, while utilized by many developing cells during PCD (*see* chapter by Clarke), is also activated in pathological conditions *(3–6)*.

PCD IS A UBIQUITOUS DEVELOPMENTAL PHENOMENON

Since massive cell death characterizes the ontogeny and turnover of virtually all other tissues and organs, and has been identified even among bacterial populations *(7)*, it is not surprising that neurons are also lost in large numbers during vertebrate development. As part of the attempt to elucidate the biological significance of neuronal PCD, a major effort in the field over the past several decades has been to address whether this regressive phenomenon is limited to specific regions of the nervous system, or involves only certain cell types. With only a few exceptions though, neuronal PCD has been described virtually everywhere in the nervous system it has been systematically investigated (*see* Table 1, Figs. 1 and 2). Motoneurons, sensory neurons, autonomic neurons, interneurons and projection neurons in the brain, spinal cord and peripheral nervous system have all been shown to die in large numbers during early periods in development. More recently PCD has even been described among populations of oligodendrocytes, Schwann cells and astrocytes. Although the magnitude of cell death varies from region to region, generally, one-half or more of an original cell population is eliminated by PCD. In some circumstances where cell death removes neurons that subserve transient developmental functions (e.g., Rohon–Beard sensory neurons in frogs), virtually all cells die. Because cells death occurs on such a massive scale, and PCD effects virtually all cell populations, including glia, this "regressive" phenomenon is likely to be a critical cellular event in the construction of the nervous system. Nevertheless, while several explanations have been offered to account for the evolution and possible biological significance of neuronal and non-neuronal PCD, so far no hypothesis has been put directly to the test. With the advent of molecular biological techniques that allow for the selective overexpression or deletion of molecules that govern neuronal survival, a virtual frontier of research opportunities are in store for those interested in testing whether altering PCD in the developing nervous system is functionally and/or behaviorally maladaptive.

NEURON-TARGET INTERACTIONS
CONTROL NEURONAL SURVIVAL DURING PCD

Many neuronal populations undergo PCD during or shortly after establishing synaptic connections with afferents and efferent targets (Fig. 3). This observation alone impli-

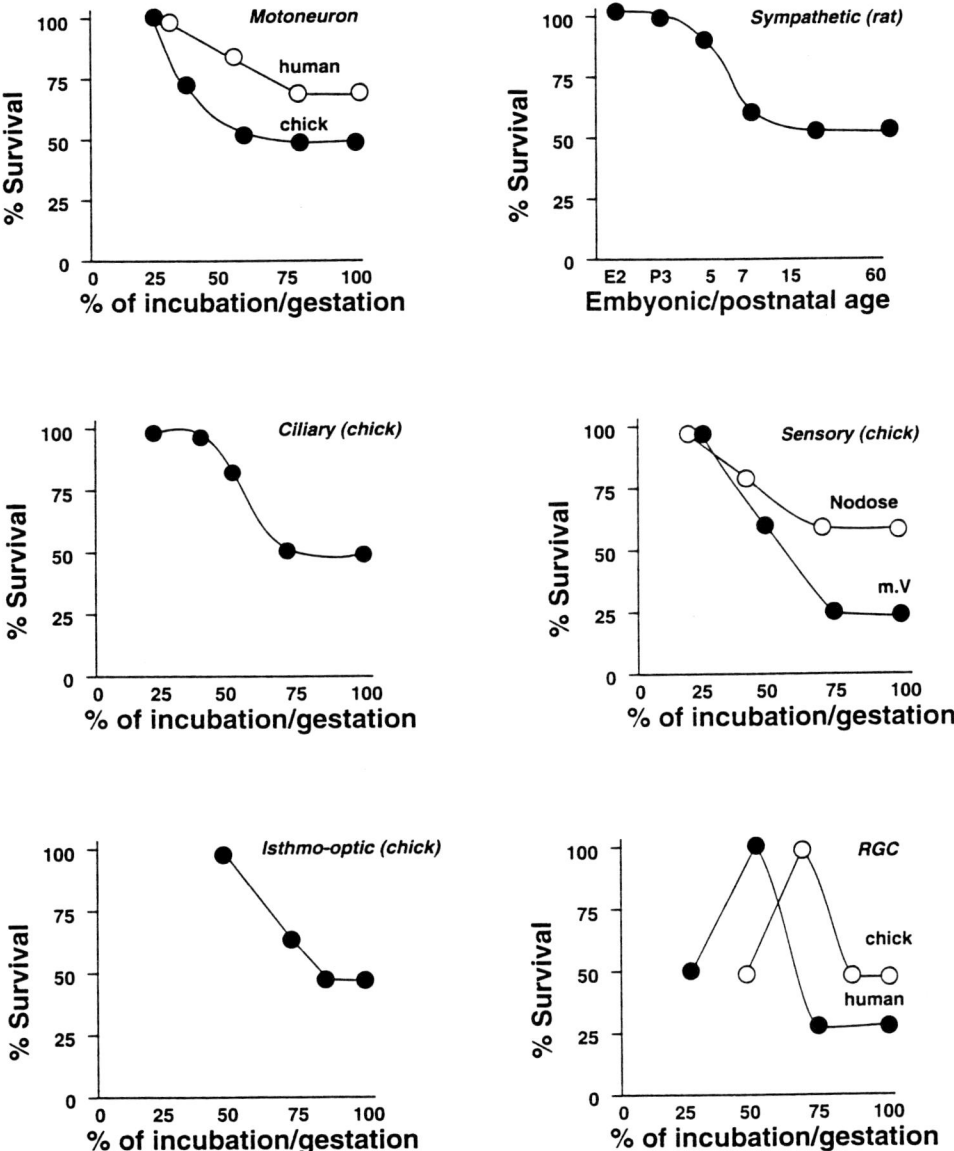

Fig. 1. The normal or "programmed" cell death of various central and peripheral neuronal populations in chick, rat, and human embryos. MV: mesencephalic nucleus of the fifth (trigeminal) nerve. RGC: retinal ganglion cells.

cates cellular interactions in the control of neuronal survival during PCD. Consistent with this, there are several examples where the extent of neuronal loss is regulated by interactions with postsynaptic targets. Generally, PCD is exacerbated when targets are partially or completely removed whereas more neurons survive the cell death period when target size is experimentally increased (*see* Fig. 4, *8*). In many, but not all instances, the number of presynaptic neurons that survive the cell death period is proportional to target size and/or the number of target neurons *(9,10)*. These observations suggest that neurons are initially overproduced and then compete for limited amounts

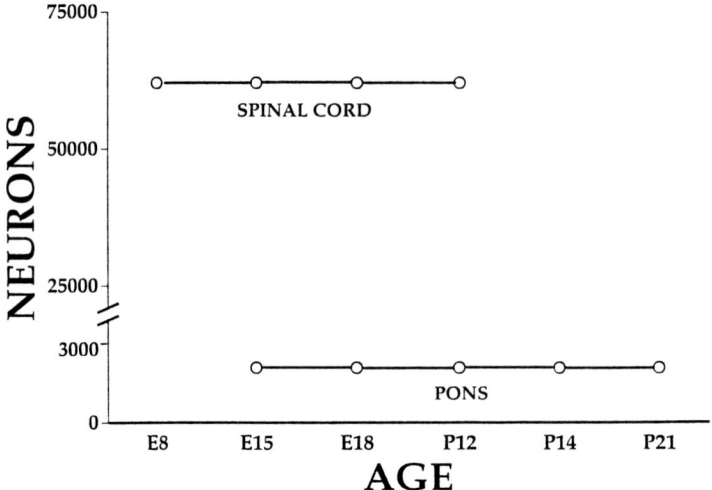

Fig. 2. The absence of PCD in two populations of avian neurons. Interneurons in the chick embryo spinal cord and pons remain constant throughout development (*see* refs. *16a* and *16b*).

of target-derived survival-promoting or "trophic" molecules that prevent their death. In this way, during early development, the number of presynaptic neurons is sculpted to match the number of neurons in their efferent targets and/or efferent target size.

Historically, experimental emphasis on neuron-target interactions as critical variables that regulate neuronal survival during PCD provided the framework for the characterization of the prototypical target-derived trophic factor, NGF, and the formulation of the neurotrophic theory. There are several criteria that make NGF a bona fide target-derived trophic factor for sensory and sympathetic neurons of the vertebrate nervous system. First, these neuronal populations are overproduced during development and subsequently undergo PCD in vivo. Second, although NGF is synthesized and secreted within the targets of sensory and sympathetic neurons there is not enough to maintain the viability of all presynaptic neurons. Third, sensory and sympathetic neurons have specific high affinity cell surface receptors *(11)* and the absence of these receptors results in the massive death of neurons *(12)*. Finally, eliminating endogenous NGF *(13,14)* or blocking its retrograde transport *(15)* kills sensory and sympathetic neurons, while exogenous NGF treatment *(16)* attenuates the PCD of these neuronal populations (Fig. 5). These experimental observations formed the basic tenets of the neurotrophic theory which states that developing neurons are initially overproduced and then later compete for limited amounts of population specific target-derived trophic factors that maintain their viability; the losers in this competitive process undergo PCD.

To what extent target-derived neurotrophic factors (versus other target-derived molecular signals) play a role in other forms of target-regulated neuronal PCD is not yet clear. However, putative target-derived neurotrophic factors related to NGF have recently been implicated in the early survival of some peripheral and central neuronal populations. While most sensory neurons depend on NGF for their survival certain subpopulations require other neurotrophins such as NT-3, BDNF, and NT-4/5. Exogenously administered neurotrophic factors also attenuate the PCD of central nervous

Table 1
A Partial List of Neuronal Populations That Undergo PCD During Vertebrate Development

Neuronal population	Animal
Motoneurons	
Spinal	Fish, frog, turtle, opposum, wallaby, mouse, rat, chicken, quail, monkey, human
Trochlear	Salamander, frog, chick, quail, duck, human
Abducens	Duck
Oculomotor	Duck, mouse, chicken
Facial	Mouse
Trigeminal	Chicken, mouse, rat
Electro-motor	Electric fish
Spinal ganglion	Frog, chicken, rat
Pineal ganglion	Chicken
Ciliary ganglion	Chicken
Sympathetic ganglion	Chicken, rat
Cochlear ganglion	Chicken
Vestibular ganglion	Chicken
Nodose ganglion	Chicken, quail
Trigeminal ganglion	Chicken, mouse, rat
Otocyst	Rat
Enteric neurons	Guinea pig
Lateral line	Frog
Sympathetic preganglionic cells	Chicken
Dorsal motor nucleus of vagus	Chicken
Mesencephalic nucleus of trigeminal	Chicken, hamster, frog
Rohon-Beard (sensory) neurons	Frog
Neuronal precursor cells of the adrenal medulla	Rat
Cochlear nuclei	Chicken, mouse
Inferior olive	Rat
Inferior colliculus	Rat
Isthmo-optic nucleus	Chicken, duck
Optic tectum	Hamster, chicken, rat, lizard
Ectomammilarly nucleus	Chicken
Visual nuclei	Tree shrew
Parabigeminal nucleus	Rat
Lateral geniculate nucleus	Mouse
Retina	Mouse, rat, rabbit, guinea pig, cat quokka and wallaby, frog, chicken, hamster, monkey, human
Cerebellum	Chicken, lizard, mouse, rat
Habenulae nucleus	Mouse
Thalamus	Lizard
Hippocampus	Mouse, rat
Corpus striatum	Lizard, rat
Cerebral hemispheres (forebrain)	Chicken, zebra finch
Olfactory cortex	Rat
Cerebral cortex	Rat, cat, mouse, hamster
Olfactory epithelium	Rat, mouse

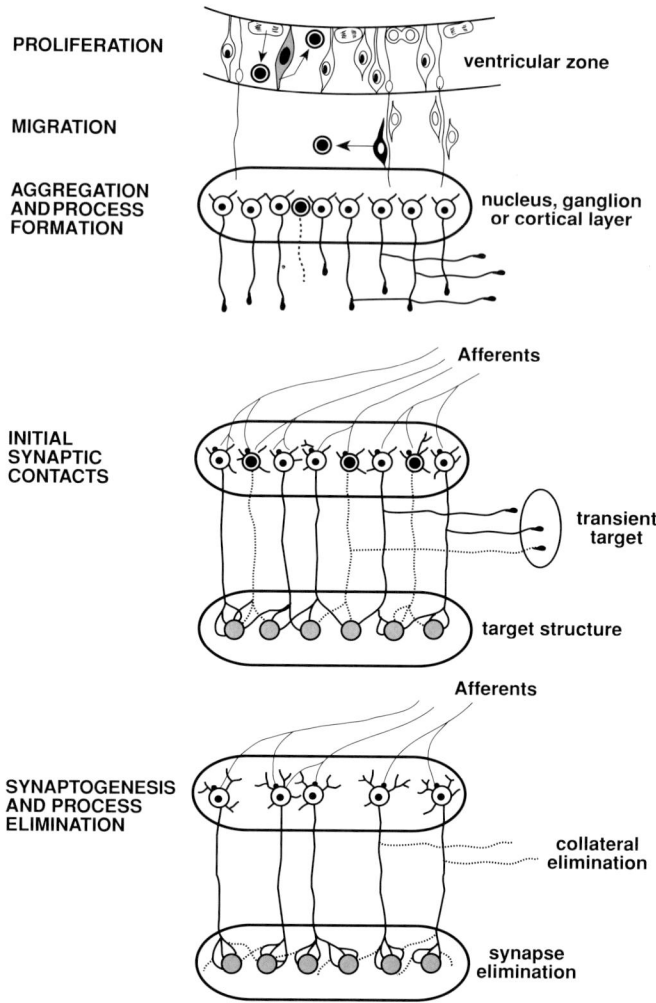

Fig. 3. Schematic illustration of several major stages in neuronal development when the PCD of neurons or neuronal precursors can occur. Round cells with large, dark nuclei represent cells undergoing PCD. (Adapted with permission from ref. *27a*.)

system populations including neurons in the isthmo-optic nucleus (ION) of the chick embryo, which respond to target-derived BDNF *(17,18)* and motoneurons of the vertebrate spinal cord which respond to BDNF *(19–22)*, NT-3 *(19)*, NT 4/5 *(19,23)*, IGF *(24,25)*, and GDNF *(19,26,27)*. Despite the survival-promoting effects of several neurotrophins and the widespread expression of neurotrophic factors and their receptors in the vertebrate central nervous system, none of the recently identified factors (with the exception of NGF) satisfy all of the criteria necessary to qualify them as target-derived survival factors. Only in the case of the avian ION has it even been demonstrated that trophic factor treatment (BDNF) restricted to the *target* region reduces the PCD of central neurons *(18)*. In this system though it still remains to be demonstrated that ION neurons require *endogenously* produced BDNF *(17)*. While further work is necessary to characterize the trophic signals that regulate target-dependent neuronal PCD in the central nervous system, it is likely that neurotrophic factors will continue to be implicated.

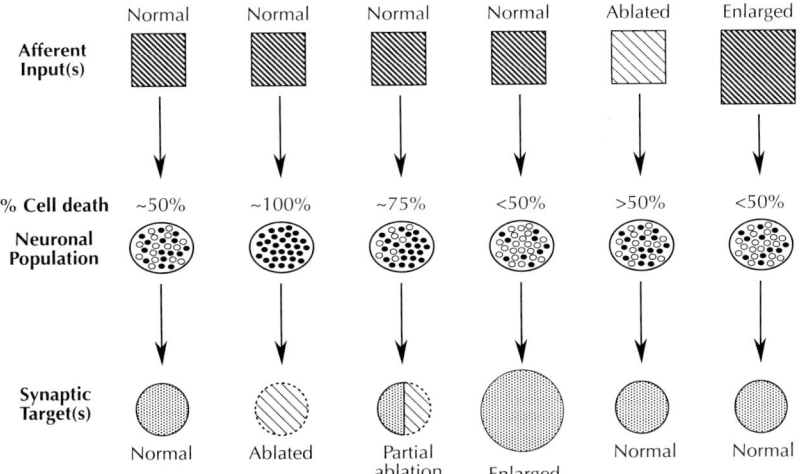

Fig. 4. Schematic illustration of the regulation of neuronal survival by targets and afferents during PCD. Solid black circles in the middle row (neuronal population) represent dying neurons. (Adapted by permission of the American Association for the Advancement of Science from ref. *28a.*)

If neurons compete for access to target-derived trophic factors or other target-derived trophic signals during PCD, what cellular attributes endow neurons with a competitive advantage over their counterparts at garnering such trophic support? There are several candidate biochemical and morphological characteristics (e.g., more synapses, more axonal branches) that could permit neurons better access to post-synaptic survival signals and therefore a better chance to escape extinction. The synaptic activation of post-synaptic cells is one well-characterized cellular process that has been extensively implicated in regulating neuronal survival during PCD *(8)*. The model system which best illustrates the role of synaptic activity in regulating PCD are motoneurons of the chick embryo spinal cord. Between embryonic d 6 and 11 approximately one-half of motoneurons die after these cells form synapses with their target muscles and commence synaptic transmission. The onset of such evoked synaptic activity in limb musculature results in spontaneous movements of the embryo and chronic blockade of this activity with nicotinic cholinergic antagonists rescues virtually all motoneurons from PCD *(28)*. Although the exact cellular and molecular mechanisms that link blockade of synaptic transmission at the neuromuscular junction with motoneuron survival are not well understood, two hypotheses have been generated (Fig. 6). First, the production hypothesis suggests that motoneurons, and perhaps neurons in general, compete for target-derived trophic factors, and that the production of these factors in targets is both limited and inversely proportional to post-synaptic activity. In contrast, the access hypothesis argues that the amount of trophic support in targets is neither limited nor activity-dependent, but that synaptic activity increases the ability of motoneurons to *access* target-derived trophic agents. For motoneurons, several lines of evidence come down on the side of the access hypothesis. First, muscle extracts from normally active and curare-paralyzed embryos are equally able to rescue motoneurons from PCD both in vitro and in vivo *(29,30)*. Second, activity blockade increases axonal branching and

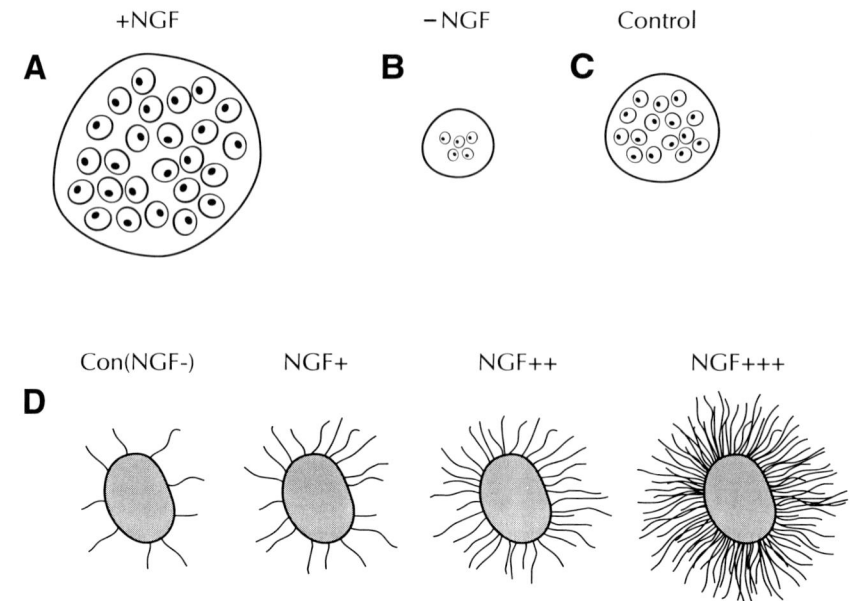

Fig. 5. Early studies characterizing the effects of NGF on sensory and sympathetic neurons showed that NGF could influence cell size, ganglion size and ganglion cell number **(A–C)**, as well as enhance the growth of neuronal processes in a dose-dependent manner **(D)**. In A, exogenous NGF was administered, whereas in B anti-NGF antibodies blocked the effects of endogenous NGF. In D, NGF- indicates the absence of NGF in explant cultures of ganglia, whereas +, ++, +++, indicates increasing amounts of exogenous NGF added to the cultures.

synapse formation suggesting a link between post-synaptic activation of nicotinic cholinergic receptors and an increase in the ability of pre-synaptic neurons to access target-derived neurotrophic support. Regardless of whether the access or production hypothesis (or both) is valid these observations illustrate the important role of targets and synaptic activity in the competitive interactions that decide the fate of presynaptic neurons during target-regulated PCD.

AFFERENTS AND THE PCD OF DEVELOPING NEURONS

Studies of neuronal cell death within the framework of the neurotrophic theory have historically emphasized neuron-target interactions in the competitive events that control how many neurons survive the PCD period. Accordingly, for several decades target-derived trophic factors were considered the primary, if not exclusive, source of trophic support for developing neurons. Only within the past 10 yr has it has become evident that afferent input is also critical for the survival of developing neurons, and may provide putative trophic signals that regulate the extent of neuronal PCD *(8,31)*. Just as removing or decreasing the size of targets exacerbates the target-regulated death of many presynaptic neuronal populations, experimental elimination of afferents just prior to the cell death period also reduces the number of surviving postsynaptic neurons (*see* Fig. 4) *(32–37)*. In one instance, it has been further shown that experimentally reducing the number of neurons providing afferent input (to the rat parabigeminal nucleus) proportionally decreases the number of postsynaptic neurons that survive PCD *(38)*. This

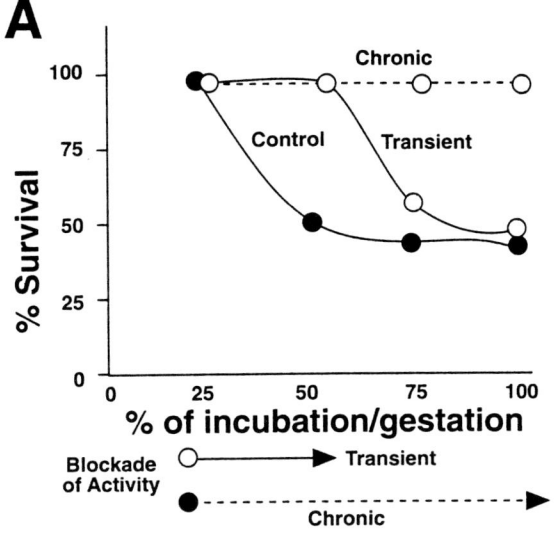
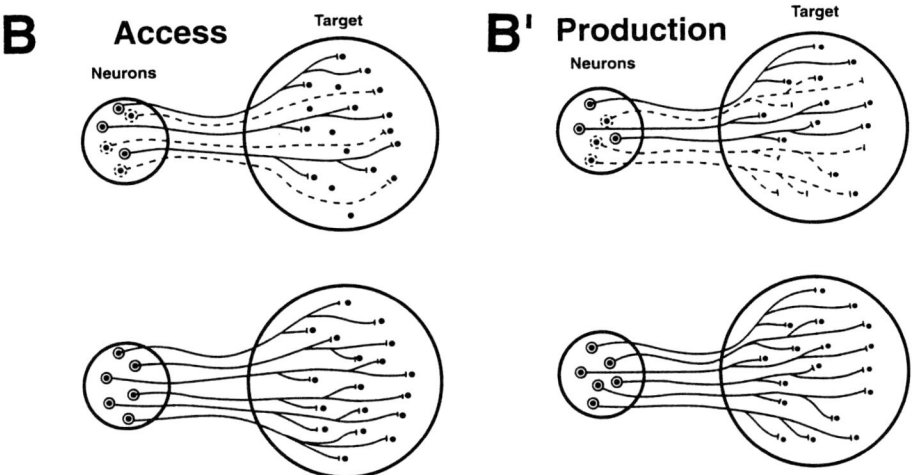

Fig. 6. Synaptic transmission between neurons and their targets **(A–B')** modulates the extent of neuronal PCD. In (A), spinal motoneurons in the chick embryo are rescued from normal PCD (control) by blocking neuromuscular transmission. Following transient blockade of activity (O– – – –O), rescued motoneurons undergo a delayed death as activity levels return to normal. Chronic activity blockade (O– – – –O) promotes the long term survival of motoneurons. In (B), two hypotheses for explaining the role of neuromuscular activity in regulating motoneuron survival are illustrated. Dying cells are indicated by dashed lines. On the left **(B)** activity blockade **(bottom)** is postulated to increase axonal branching and synapses over control levels **(top)** and thereby increase access to target-derived trophic factors (small black dots in target). By contrast, on the right (B') activity blockade **(bottom)** is thought to increase the production of trophic factors over control values **(top)** (A and B modified from Oppenheim, 1989 *Trends Neurosci* by permission of Elsevier Publishers).

suggests that, in addition to target-derived trophic factors, postsynaptic neurons compete with one another for access to limited amounts of afferent-derived trophic signals. This competitive process may adjust the number of postsynaptic neurons to the extent of their afferent projections, and in some cases to the number of neurons providing afferent input.

Although studies documenting enhanced neuronal death following deafferentation are consistent with a trophic role for afferent input this line of investigation alone cannot reveal whether afferents actually control the *number* of postsynaptic neurons that survive the cell death period. For instance, one simple explanation to account for deafferentation-induced cell death is that under such circumstances neurons are less competitive at accessing, retrogradely transporting, and/or responding to target-derived trophic factors. In addition, many neuronal populations deprived of afferents prior to the normal cell death period, (e.g., ciliary ganglion neurons *(34)*, motoneurons *(36)*, and isthmo-optic nucleus neurons in the chick embryo *(32,33)*, exhibit increased deafferentation-induced cell death only during the *latter portions* of the cell death period, when considerable neuron loss has already occurred and neurons are beginning to lose target-dependence. This suggests that the number of postsynaptic neurons that survive the cell death period may be relatively independent of the extent of their afferent input, but that afferents may nonetheless provide surviving neurons with an alternate source of trophic support.

The most effective way to address the role of afferents in neuronal PCD is to determine whether increasing afferent input, which presumably decreases neuronal competition for afferent-derived trophic signals, enhances the survival of postsynaptic neurons. Unfortunately, because of the technical difficulty associated with experimentally increasing afferent input to postsynaptic neuronal populations, only one study has demonstrated reduced cell death in these circumstances *(39)*. In contrast, a recent study by Sohal et al., 1992 provides the only compelling direct evidence *against* the hypothesis that afferents control postsynaptic neuron number by regulating the extent of PCD. In the developing chick embryo about one-half of motoneurons in the trochlear nucleus die between E13 and E18. These cells project to the superior oblique muscle and receive most of their afferent input from vestibular nuclei. To test the role of afferents in trochlear motoneuron PCD, these investigators surgically removed the hindbrain (the source of vestibular afferents) on E3 so that trochlear motoneurons would develop in the absence of virtually all of their normal afferent input. Although this manipulation reduced the initial number of precell death trochlear motoneurons, the *extent* of neuronal loss *during* PCD was virtually identical to that observed in normal embryos with intact trochlear afferents (about 50%). These observations suggest that, while afferents are an important source of trophic support for developing neurons, the percentage of neurons that survive the PCD period, at least for some neuronal populations, is independent of the extent of their afferent input.

Regardless of whether afferent input actually controls the number of neurons that survive the period of cell death, or merely provides neurons an alternate source of trophic support as they lose target dependence, recent work has focused on characterizing the possible molecular nature of afferent-derived trophic signals. There are several, not necessarily mutually exclusive scenarios, that could account for the trophic effects of afferent input on developing neurons. One possibility is that afferents release "anterograde" trophic agents or factors that rescue postsynaptic neurons, either by directly preventing their death or by augmenting their sensitivity to trophic support derived from other sources (e.g., targets, glia) *(38–42)*. Alternatively, presynaptic activity of afferents and/or evoked synaptic activity within postsynaptic neurons by afferent neurotransmission may provide neurons with important survival-promoting signals. Consistent with this, experimental blockade of afferent synaptic activity in vivo and in vitro induces the death of several neuronal populations *(41–45)*, while chronic depolarization in vitro maintains neuronal viability *(45)*. There are several activity-dependent cellular processes that could foster the survival of developing neurons. First, afferent activity could promote the presynaptic release of putative neurotrophic agents *(40)* or neurotransmitters *(46)* which directly prevent the death of postsynaptic neurons. Second, since depolarization-induced neuron survival in vitro depends on calcium influx through dihydropiridine-sensitive calcium channels *(45,47)* induced Ca^{2+} influx through ligand or voltage-gated channels within *postsynaptic* neurons could promote their survival. Such afferent-evoked increases in Ca^{2+} within postsynaptic neurons could up-regulate neurotrophic factors and/or their receptors *(48)* and neuron survival may either be a consequence of autocrine action *(49)*, or of an enhanced sensitivity to trophic support derived from other sources (e.g., target, glia). Finally, the activity-dependent regulation of cytoplasmic Ca^{2+} within postsynaptic neurons may directly regulate the molecular cascade implicated in PCD *(45,50)*.

As an initial step toward deciphering which or how many of these hypothesized mechanisms account for the trophic effects of afferent input, several investigators have addressed whether neuronal survival in vivo requires the *activity-dependent or activity-independent* release of trophic substances from afferent terminals (Fig. 7). One particularly well-suited model system to address this issue are neurons in the developing chick embryo optic tectum, which require retinal ganglion cell afferents for their early survival. Catsicas et al., 1992 *(41)* injected the eyes of embryonic day (E)16 chicks with tetrodotoxin (TTX) and colchicine to block action potentials and axoplasmic transport, respectively, within retinal ganglion cells. Intraocular injections of colchicine, but not TTX, transneuronally induced neuronal death in superficial layers of the tectum (Fig. 7A). This suggests that tectal neurons in these superficial layers rely on the *activity-independent* release of trophic substances from retinal ganglion cell afferents. These results are particularly exciting since neurotrophin-3 (NT-3), exogenously injected into the eyes of 16-d-old chick embryos, is anterogradely transported by retinal ganglion cell axons, released, and internalized by postsynaptic neurons in the tectum *(51)*. While these observations are consistent with the possibility that NT-3 is an "anterograde" trophic factor for tectal neurons, this hypothesis will remain speculative until the anterograde transport and release of *endogenous* neurotrophic factors from afferent terminals is directly linked to the *survival* of postsynaptic neurons (*see* also chapter by Koliatsos and Mocchetti).

Several studies suggest that, in addition to anterograde trophic factors, the *activity-dependent* release *(40,41)* of neurotransmitters and/or neuropeptides *(52)* from afferent terminals also promotes the survival of postsynaptic neurons (Fig. 7B). This is best illustrated in the developing nucleus magnocellularis (NM) (the chick homologue of the mammalian cochlear nucleus), a model system that has yielded tremendous insight into the cellular and molecular mechanisms by which afferents promote neuronal survival *(44)*. Although NM neurons exhibit little, if any, normal PCD, they depend on afferent input from the eighth nerve for their survival during late embryonic and early posthatch development. Removing the cochlea during this time induces a well characterized set of degenerative events culminating in the death of approximately 30% of NM neurons within several days *(53)*. The survival-promoting effects of eighth nerve

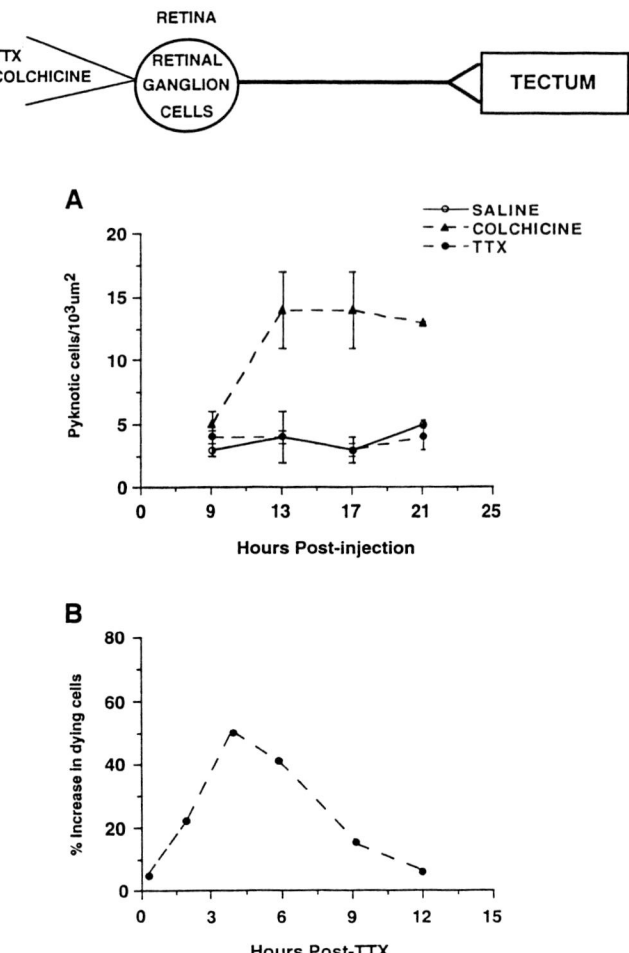

Fig. 7. In the developing chick retinotectal system, the survival of some tectal neurons reflects the activity-*independent* release of trophic substances (e.g., NT-3: see Von Bartheld et al., 1996) from retinal ganglion cell afferents. In **(A)**, intraocular injections of colchicine (which blocks action potentials), but not TTX (which silences electrical activity), increase the number of dying neurons in the stratum griseum et fibrosum superficiale (SGFS) layer of the chick optic tectum. (Adapted by permission of the Society of Neuroscience from ref. *54a*). By contrast, the activity-dependent release of trophic substances (e.g., glutamate) from retinal ganglion cell afferents maintains the viability of developing tectal neurons in the mammalian retino-tectal system. In **(B)**, there is a significant increase in the number of dying neurons in the rat superior colliculus when afferent electrical activity in retinal ganglion cells is blocked by intraocular injections of TTX. (Adapted by permission of the Society of Neuroscience from ref. *40*.)

afferents appear to stem from the activity-dependent release of trophic substances from these terminals since deafferentation-induced NM neuronal death can be mimicked by intralabyrinth administration of TTX *(54)*. Moreover, in an in vitro brain stem slice preparation, NM neurons survive when orthodromically stimulated (i.e., via eighth nerve afferents), but still undergo cell death resembling deafferentation when *antidromically* stimulated (i.e., without eighth nerve afferents) *(53)*.

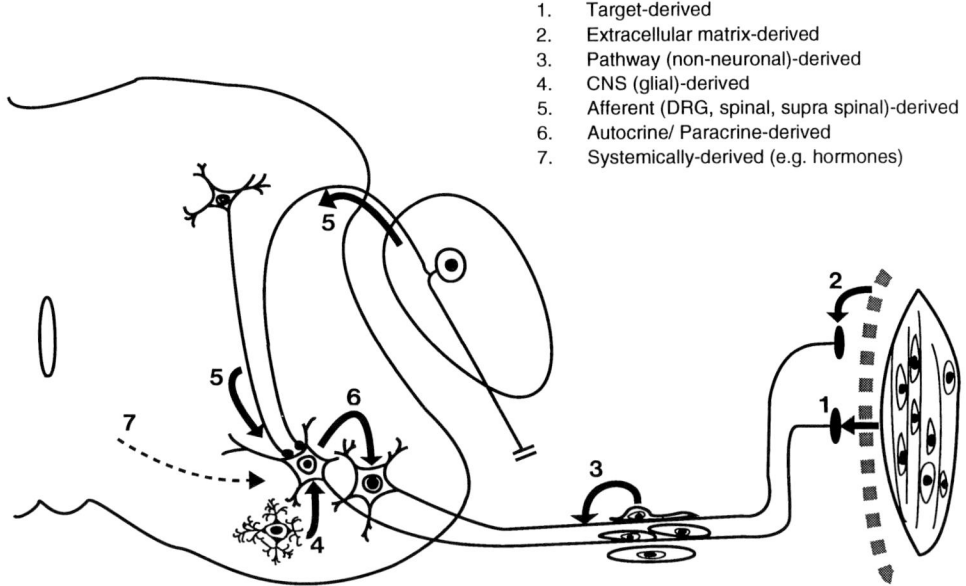

Fig. 8. Schematic illustration highlighting possible sources of trophic signals acting on developing neurons, using motoneurons of the spinal cord as an example.

If NM neurons require evoked activity via eighth nerve afferents, and yet these neurons still undergo cell death resembling deafferentation even when artificially depolarized, then what is the molecular nature of the eighth nerve-derived trophic signal? Recent evidence suggests that, by activating metabotropic glutamate receptors on NM neurons during synaptic transmission, eighth nerve afferents transneuronally promote the survival of NM neurons by buffering lethal increases in intracellular Ca^{2+}. Several observations support this hypothesis. First, activation of mGluRs in NM neurons attenuates voltage-dependent increases in Ca^{2+} influx *(55)*. Second, while NM neurons maintain normal intracellular Ca^{2+} following eighth nerve stimulation, levels dramatically increase when the eighth nerve is either not stimulated at all or stimulated while NM neurons are exposed to the mGluR receptor antagonist MCPG *(56)*. Finally, a marked elevation in Ca^{2+} within NM neurons is evident in brainstem slices within 1 h after the cochlea is removed *in ovo* and the calcium channel blocker flunarizine attenuates both this induced increase in Ca^{2+} in vitro as well as the death afferent-deprived NM neurons in vivo *(57)*. Taken together these observations imply that the activation of metabotropic glutamate receptors during synaptic transmission promotes the survival of postsynaptic NM neurons by preventing lethal accumulations of intracellular Ca^{2+}.

OTHER SOURCES OF TROPHIC SUPPORT FOR DEVELOPING NEURONS

Although the lion's share of attention has been devoted to the role of target and afferent-derived trophic signals in controlling neuronal survival during PCD, it is now evident that the survival of developing neurons, both during cell death and beyond, is critically dependent on other cell–cell-derived and even cell-intrinsic survival signals (Fig. 8). For instance, several glia-derived factors promote neuron survival in vitro *(58–60)* and non-neuronal cells, presumably glia, promote the early survival of sympathetic neuroblasts in vitro by providing these cells with NT-3 *(61)*. This hints at the possibility that glia-derived trophic factors regulate the survival of neurons before, during, and perhaps even after periods of PCD in vivo. Indirect support for a role of glia in regulating neuronal survival comes from a sexually dimorphic region of the developing songbird forebrain where enhanced gliogenesis in males, but not females, correlates with reduced PCD *(62)*. Neurons could also receive trophic signals from cellular interactions with other non-neuronal cells. Since extracellular matrix-derived molecular signals prevent the apoptotic death of mammary epithelial cells by suppressing the activity of the interleukin-one converting enzyme (ICE) family of proteases that have been implicated in cell death, similar trophic interactions could promote the survival of developing neurons *(63)*. Finally, neurons could receive trophic support from direct contact with other neuronal or non-neuronal cells, for instance, via exchange of neurotrophic factors through gap-junctions *(8)*.

Since steroid hormones promote neuronal survival both during PCD and even into adulthood they are also now considered an important source of trophic support for developing neurons. In vertebrates and invertebrates that undergo metamorphosis hormones regulate PCD to remove neurons that serve transient premetamorphic functions *(64,65)*. Also, during early critical periods in vertebrate sexual differentiation gonadal hormones rescue neurons from PCD to establish sex differences in cell number in several sexually dimorphic regions, including the superior cervical ganglion *(66)*, hippocampus *(67)*, the spinal nucleus of the bulbocavernosus *(68)*, and the telencephalon of song birds *(6)*. Finally, adrenal hormones control the survival of hippocampal granule neurons both during development and in adulthood *(69,70)*.

In addition to relying on a diverse array of cell-extrinsic survival signals (e.g., afferents, targets, hormones), developing neurons may also utilize cell-intrinsic molecular signals to prevent their demise. Since several neuronal populations in the developing nervous system express both neurotrophic factors and their complementary receptors it has been hypothesized that these cells stimulate their own survival via an autocrine loop of neurotrophic signaling *(71–74)*. Although there is no direct evidence yet that such autocrine trophic factor loops regulate the fate of developing neurons, *adult* sensory neurons require endogenously produced BDNF for their survival *(75)*. Nonetheless, it is tempting to speculate that autocrine regulation of neuronal survival could account for why some developing neuronal populations survive in the absence of cell-extrinsic survival signals prior to target innervation *(76,77)*, and why some forms of early neuronal PCD appear to be independent of target and afferent regulation *(78–81)*.

Collectively, these data indicate that the neurotrophic theory which states that neurons compete for limited amounts of population specific target-derived trophic factors simply cannot account for all instances of neuronal PCD. Afferents, hormones, glial effects, local trophic interactions and even cell-intrinsic molecular events (including neurotrophic signalling) may not only regulate the extent of neuronal PCD but may also provide trophic support for neurons before and after these regressive ontogenetic phases in development *(8,47)*. In addition, neurons change their trophic factor requirements through development *(78,80)*, respond to more than one putative neurotrophic factor, and a single neurotrophic factor can promote the survival of multiple neuronal populations *(47)*. Thus, with the exception of the prototypical target-derived trophic factor NGF, the neurotrophic theory may be a an oversimplification of the complex cellular interactions that control neuronal survival during PCD.

To complicate things further, neuronal survival during PCD and beyond may also reflect the *interaction* of various cell-extrinsic (and/or cell intrinsic) trophic signals. The hormonal regulation of PCD in vertebrates clearly illustrates this concept. In several sexually dimorphic regions, rather than *directly* targeting biochemical pathways that regulate a neuron's decision to live or die (e.g., PCD by apoptosis), gonadal steroids indirectly promote neuron survival by exploiting pre-existing trophic interactions between neurons and their afferents, efferent targets, and/or associative glia *(62,67,84)*. The best characterized example of this phenomenon is the neuromuscular system of the mammalian spinal nucleus of the bulbocavernosus (SNB). During the perinatal sexual differentiation period androgens attenuate the death of SNB motoneurons and their associated perineal muscles to establish permanent sex differences in copulatory behavior *(84)*. Interestingly, androgens *indirectly* promote SNB neuron survival by preventing the atrophy of target perineal muscles and thus preserving the ability of motoneurons to access putative target-derived trophic factors. In fact, since injections of ciliary neurotrophic factor (CNTF) into the perineal muscles of females may mimic the trophic effects of androgens on the SNB system, and because hormones regulate both neurotrophic factors and their receptors *(85–87)*, one possibility is that androgens save SNB motoneurons by augmenting ciliary neurotrophic factor (CNTF) signalling in their target perineal muscles *(88)*. According to this scenario, target-derived CNTF could stimulate SNB motoneuron survival directly, or prevent muscle involution and thereby promote cell survival *indirectly* by ensuring that SNB motoneurons access other target muscle-derived trophic agents.

WHY DO SOME NEURONS UNDERGO PCD?

Why would developing vertebrates devote precious resources to "overproduce" large numbers of neurons and other cells that are later eliminated by massive cell death? Since the development of virtually all tissues reflects the overproduction and subsequent death of cells, the nervous system may also have evolved this strategy because it is simply the most effective way to ensure optimal cell number. That is, even though proliferative events alone could regulate region specific differences in cell number, generating far more neurons than are ultimately needed may provide an "insurance policy" or buffer to compensate for the possible loss of neurons during the intervening cellular events between cell birth and neuronal differentiation (e.g., loss of cells that migrate incorrectly or cells with lethal mutations). Alternatively, the overproduction of neurons could be an inevitable outcome of the relative imprecision of cell cycle kinetics. For example, in the developing chick embryo at the end of the proliferative phase there are roughly 24,000 lumbar motoneurons; one-half of which die during the early cell death period. Assuming that the optimal number of lumbar motoneurons is 12,000 and that the lumbar motor nucleus is generated from 60 precursor cells that undergo a fixed number of symmetrical divisions, then after seven divisions there would be 7680 cells-too few for optimal innervation. However, after eight divisions of these same precursor cells there would be 15,360, cells—too many for optimal muscle innervation. In this regard, proliferative mechanisms alone are not able to precisely generate the number of motoneurons required for optimal muscle innervation. The major distinction between the two views is that in the first case,

Table 2
Some Possible Functions of PCD in the Developing Nervous System

Category	Examples
1. Systems matching	The creation of optimal levels of innervation between interconnected groups of neurons and between neurons and their non-neuronal targets.
2. Error correction	The death of neurons that make inappropriate synaptic connections.
3. Removal of cells that are no longer needed	Hormone-dependent dimorphisms in neuron death during sexual differentiation sculpt sex differences in physiology and behavior.
4. Removal of cells that serve transient developmental functions	The death of sensory, motor and CNS neurons that serve a transient physiological/behavioral function (e.g., Rohon-Beard cells in frogs).
5. Pattern formation and morphogenesis	The death of neural crest cells in specific segments of the hindbrain.
6. Removal of cells of an inappropriate phenotype	The death of neuronal precursors located in ectopic regions of the developing spinal cord.
7. Plasticity in the adult brain	Seasonal proliferation and PCD of projection neurons involved in canary song production provides the substrate for the seasonal learning of courtship song.

the overproduction and later death of cells is a strategy that may have been directly selected for because it is the only way to ensure optimal cell number. In the second case, the overproduction and subsequent death of cells is not selected for but is an inevitable outcome of the inability of proliferative mechanisms alone to precisely control neuron number.

Another possibility, paradoxically, is that the overproduction and wholesale death of large numbers of neurons is the most resource-efficient way to quantitatively adjust neuron number among highly interconnected neural regions. According to this scenario, in a set of several nuclei that develop initially independent of one another, cell proliferation, migration and differentiation need only be regulated or "programmed" in one or a few key regions. Subsequently, cellular interactions among interconnected regions (via changes in afferents and efferent targets) could indirectly regulate region-specific differences in the extent of PCD and therefore overall cell number. This so-called quantitative or systems matching hypothesis has some experimental support since several studies have shown a direct and linear correlation between the number of neurons in both afferent and efferent target populations and the number of neurons that survive the PCD period *(9,38,90)*.

In addition to controlling regional differences in neuron number, PCD is thought to serve other functions that are critical for the development of the nervous system (*see* Table 2). For instance, in invertebrates and vertebrates that undergo dramatic life history changes, PCD clearly removes cells that are not needed for postmetamorphosis behavior *(64,65)*. Also, in higher vertebrates, PCD eliminates neurons that serve transient developmental functions such as those that provide guidance cues for early axonal projections *(91,92)* or serve as temporary targets for early afferents *(93,94)*. PCD has also been implicated in removing cells with inappropriate phenotypes *(22,78)*, pattern formation, and formation of sexually dimorphic structures *(84)*.

PCD may also ensure the development of precise interconnectivity in the nervous system by eliminating neurons that have made incorrect or aberrant projections. For example, both mammalian retinal ganglion cells that project to inappropriate regions of the superior colliculus *(95,96)* and developing isthmo-optic nucleus (ION) of the chick embryo that project to the ipsilateral, rather than the contralateral retina *(88)* are selectively eliminated by PCD. Presumably, these neurons are at a competitive disadvantage and die because they fail to receive the appropriate target-derived trophic factors. However, such an error correction mechanism alone cannot account for all instances of PCD. Massive neuronal loss (>50%) has been documented both in neural regions where projection errors are nonexistent (e.g., spinal motoneurons), as well as in regions where such errors are relatively common *(97)*. Moreover, in some neural regions where cell death is thought to correct projection errors, the number of neurons that make targeting errors is very small relative to the actual number of neurons that undergo PCD. Finally, neurons that exhibit experimentally induced large-scale projection errors are not necessarily eliminated by PCD *(98)*. Despite these inconsistencies, sufficient indirect evidence still supports the hypothesis that, in some instances, PCD removes neurons that make erroneous projections *(8)*.

Although several plausible hypotheses regarding the biological significance of PCD have been generated to date, so far no hypothesis has been tested directly. If PCD really is critical for nervous system development then preventing cell death should be functionally and/or behaviorally maladaptive. One dramatic example of this approach comes from studies of the invertebrate *C. elegans* where mutations of the *ced-3* or *ced-4* genes (which code for cell death-inducing proteins) block all PCD (*see* chapter by Royal and Driscoll). Yet, remarkably, the anatomy, survival, and behavior of these organisms appears relatively normal *(99,100)*. This approach has recently been extended to the vertebrate nervous system where transgenic mice that overexpress trophic factors and/or genes that prevent cell death can now be used to directly test the hypothesis that PCD is required for nervous system development. For instance, transgenic mice that overexpress *bcl-2* (a gene that inhibits cell death), have hypertrophic brains and increased number of neurons in some regions of the nervous system (e.g., retinal ganglion cell layer, facial nucleus), but exhibit no obvious neurological impairments *(101–103)*. While these observations are consistent with the notion that some forms of PCD may not be critical for normal neural and behavioral development, there are caveats to these transgenic approaches. First, more systematic studies of the behavior of these transgenic animals will be obligatory and may reveal subtle functional deficits directly linked to increased cell number in specific neural regions. Second, while some cells that survive following mutations of the *ced-3* or *ced-4* genes in *C. elegans* actually function as neurons *(88)*, those cells that survive following transgenic manipulations in vertebrates may not necessarily have normal functional phenotypes. These neurons may simply be arrested at one junction in the cell death cascade and thus lack the expression and/or ability to respond to molecular signals that promote their differentiation. In support of this, while *bcl-2* overexpression protects motoneurons from cell death in a mutant mouse model of human motoneuron disease, it cannot prevent the degeneration of motoneuron axons *(104)*. This observation warrants some caution against overinterpreting the absence of obvious neurological deficits in transgenic mice that have increased neuronal numbers in specific regions of the nervous system *(101–103)*.

NOVEL FORMS OF PCD IN THE DEVELOPING NERVOUS SYSTEM

By far, the most common form of PCD in the developing nervous system, and consequently the most well-studied, involves developing neurons that are relatively differentiated and have well-established synaptic connections with afferents and efferent targets *(8)*. More recently however, studies have shown that PCD not only occurs much earlier in the development of the nervous system (e.g., PCD during neurulation) but that this "regressive" phenomenon also effects individual neurons much earlier in their life history, as soon as hours after withdrawal from the cell cycle *(78,79,105)*. The current view, therefore, is that PCD is not limited to any particular stage of development, but can eliminate neurons from the time they are precursor cells to long after the formation of synaptic connections and even into adulthood (*see* Fig. 3). Because of the historic emphasis on classic target-regulated neuronal death and the difficulty in characterizing and quantitating cell death among relatively immature neurons or neuronal precursors, relatively little is known regarding either the cellular and molecular mechanisms of this early PCD or its biological significance.

PCD that occurs at very early stages in nervous system development may shape the gross morphology of regions of the nervous system and/or create a permissive environment for the growth of axons *(8)*. However, it is much less clear why *individual neurons* die at early stages in their development. Vertebrate sensory neurons and olfactory cells die in large numbers as early as two hours after their production *(79,106)* and massive cell death occurs in proliferative zones of embryonic mouse cerebral cortex during neurogenesis *(78)*. This form of PCD may eliminate unwanted precursor cells (that could have lethal mutations) and/or cells with inappropriate phenotypes. Consistent with the latter possibility, pyknotic (degenerating) cells are evident in four distinct regions of the caudal neural tube of the developing chick embryo *(80)*. These dying cells are either recently postmitotic or even still within the cell cycle, and this PCD coincides with cellular differentiation and pattern formation of the neural tube along the dorsal-ventral axis. Moreover, perturbations that alter putative signals for dorsal-ventral differentiation also alter the number and distribution of dying cells *(80)*.

It is more difficult to generate plausible hypotheses that can account for why some developing neurons die in large numbers somewhat later in their development, just prior to the onset of classic target-regulated PCD. For example, motoneurons of the cervical spinal cord in the developing chick embryo die during their early differentiation, as these cells grow axons out to their target muscles *(81)*. Although cervical motoneurons are born between E2 and E4.5, and die in large numbers between E4 and E10, this PCD occurs in two distinct phases. A classic target-regulated neuronal death phase occurs between E6–E10 and is very similar to that which occurs in the lumbar and other regions of the spinal cord. Experimental increases in target size, blockade of synaptic activity, as well as known trophic factors and target-muscle derived extracts rescue cervical motoneurons from this later, more traditional form of target-regulated PCD. In contrast, these same manipulations have no effect on the earlier, massive death of cervical motoneurons between E4 and E5, implying that this novel form of PCD is target independent. Because early cervical motoneuron death is target-independent and occurs before the arrival of afferents, the survival of motoneurons during this regressive phase may be regulated by local cell-cell interactions (e.g., glia) or cell-intrinsic mechanisms (e.g., autocrine trophic factor loops). Although the biological significance of this early cervical motoneuron death is not yet clear, continuing to explore the underlying cellular and molecular mechanisms of this and other novel forms of PCD may yield new insight into the diverse cellular and molecular signals that promote neuronal survival during development.

GLIAL CELLS ALSO UNDERGO PCD IN THE DEVELOPING NERVOUS SYSTEM

Although there have been several anecdotal reports of glial cell death during development virtually none of these early studies systemically characterized the phenomenon, e.g., *(107,108)*. As a consequence, the relative prevalence of glial PCD, the underlying cellular and molecular mechanisms, as well as the possible adaptive function of glial cell death are only beginning to be elucidated *(109–112)*. Two recently developed model systems should shed light on the cellular and molecular mechanisms that control glial PCD (Fig. 9). Between embryonic days (E) 4 and 14 in the developing chick embryo, proliferating spinal cord Schwann cells (SCs) die in large numbers in the ventral roots; and over the first few postnatal weeks in the rat optic nerve, one-half of newly generated oligodendrocytes undergo PCD within 2–3 d after becoming postmitotic *(111)*. Axon-derived trophic signals regulate the survival of both Schwann cells and oligodendrocytes during their PCD *(109–112)*. For instance, far fewer retinal ganglion cells undergo PCD in transgenic mice that overexpress the anti-apoptotic gene Bcl-2, which results in an 80% increase in the number of axons within the optic nerve relative to normal animals. Remarkably, this manipulation *proportionately* increases the number of oligodendrocytes by increasing cell proliferation, but more importantly by attenuating PCD *(113)*. Using a similar line of investigation, *in ovo* administration of *N*-methyl-D-aspartate (NMDA) in doses that selectively kill motoneurons and their axons, dramatically increase the number of apoptotic SCs in the ventral roots of the lumbar spinal cord *(112)*. Conversely, manipulations that increase motoneuron survival, and thereby increase the number of motoneuron axons, reduce the *number* of dying SCs cells (Ciutat et al., unpublished observations).

Because axon number regulates the survival of myelinating glial cells, the normal PCD of these cells may reflect a competition for limited amounts of axon-derived trophic factors. In support of this, SC cell precursors are rescued from PCD in vitro by a survival factor present in neuron conditioned media *(114)* and Schwann cell death *induced* at developing neuromuscular junctions by muscle denervation is attenuated by the glial growth factor, a member of the recently identified neuregulin family of ligands *(115)*. Similarly, increasing levels of PDGF, IGF-1, CNTF, or NT-3 (which prevent the PCD of oligodendrocytes in vitro) in the developing optic nerve decreases the number of dying, and increases the number of healthy glial cells *(109–111)*. Moreover, transecting the optic nerve behind the eye during the PCD period increases the number of degenerating oligodendrocytes by fourfold *(116)* and this cell death is attenuated by exogenous IGF-1 or CNTF *(109)*. Since several exogenously administered trophic factors rescue myelinating glial cells from normal and induced cell death, the overproduction and death of SCs and oligodendrocytes may reflect a competition for limited amounts of axon-derived trophic factors. In this way, just as neuronal PCD adjusts the number of pre- and postsynaptic

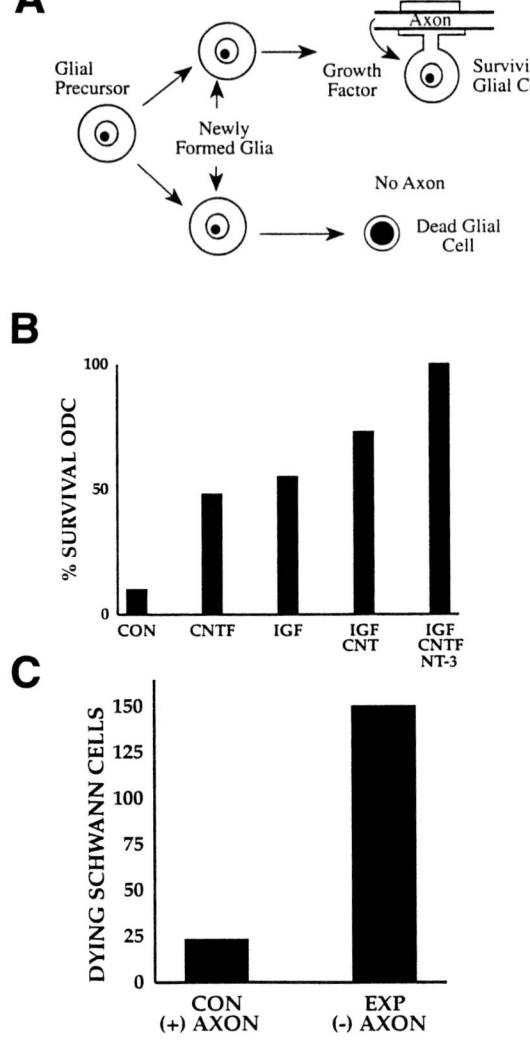

Fig. 9. The PCD of glial cells is regulated by axonally-derived signals and growth factors. Glial cells that fail to compete for these signals undergo PCD **(A)** and this regressive process may match glial cell number to the number of axons requiring myelination. The survival of cultured oligodendrocytes (ODC) from neonatal rat optic nerve is enhanced by multiple growth factors (A and **B**, modified from *[116]* by permission of The Company of Biologists). In **(C)**, the number of dying Schwann cells in the ventral root of the chick embryo is increased following the degeneration of motoneuron axons (modified from *[112]* by permission of The Society for Neuroscience).

neurons, the PCD of SCs and oligodendrocytes may control the ratio of glial cell number to the number of axons requiring myelination *(109,112)*. It will be imperative to test this hypothesis directly by determining whether manipulations that selectively arrest glial PCD, and therefore, alter the ratio of glial cell to axon number, disrupt the normal myelination and function of specific neuronal populations.

In addition to further characterizing the cellular and molecular mechanisms that govern the survival of myelinating glial cells, it will also be important to determine whether other non-neuronal cells such as astrocytes also die during early development. Although proliferation, but not PCD, regulates astrocyte number in the postnatal rat optic nerve *(108)*, large-scale astrocyte death has recently been described in the developing rat cerebellum *(117)*. During the first postnatal week, dying astrocytes are evident in both the white matter (WM) and internal granule cell layer of the developing cerebellum. Thus, cerebellar astrocytes, like neurons and myelinating glial cells, may be overproduced and then subsequently eliminated in order to precisely adjust cell number to the requirements of the surrounding tissue. Interestingly, oligodendrocyte precursor cells begin to migrate into the WM and myelinate axons at postnatal day 7, the time of peak astrocyte cell death, suggesting that astrocytes may provide transient trophic support for afferent and efferent axons in the developing WM. Once oligodendrocytes commence myelination and provide an alternate source of trophic support for WM axons, these cells may out-compete astrocytes for limited amounts of local survival factors resulting in large-scale astrocyte death occurs *(117)*.

Recent studies characterizing glial cell death represent a major advance in our understanding of PCD in the developing nervous system. First, by demonstrating that cell death in the nervous system is not restricted to neurons, these studies are consistent with the recent hypothesis that virtually all vertebrate cells are programmed to die unless they receive survival signals primarily derived from other cell–cell interactions *(118)*. Second, if cell death among glial cells is as prevalent and as massive as neuronal PCD, it may ultimately force us to abandon neuron-specific hypotheses to account for the loss of neurons during development. Third, because glia require survival signals from other cells (e.g., neurons) to prevent their demise, we need to consider the possibility that glial (and possibly neuronal) cell death associated with aging, injury or disease could reflect perturbations in such trophic cell–cell interactions.

PROLIFERATION AND PCD IN THE ADULT VERTEBRATE NERVOUS SYSTEM

Cellular turnover, characterized by the precise coregulation of cell proliferation and PCD, contributes to the homeostasis of virtually all cell types, tissues and organs. While neurogenesis and PCD interact to regulate neuron number during early nervous system development, these cellular events are generally over by the early postnatal period. However, there are some exceptions to this general principle. Some regions of the adult vertebrate nervous exhibit considerable neuronal turnover; older neurons that die are replaced by newly generated cells, while overall neuron number does not change. Two particularly well-studied examples of this phenomenon are mammalian olfactory receptor neurons (ORNs) *(119)* and neurons of the adult canary telencephalon *(79)*. New ORNs replace cells that die as a consequence of pathology, whereas in adult canaries seasonal changes in the turnover of specific cell populations may provide the neural substrate for seasonal modification of song behavior *(120,121)*. During spring (when circulating levels of testosterone (T) are highest), canaries sing frequent but stable courtship songs, and the turnover rate of projection neurons in one song-control region, the high vocal center (HVC), is relatively low. Conversely, in the fall, when T levels are low and song is being modified, older, pre-existing neurons die and large numbers of new neurons are incorporated into the HVC. In another example, depriving granule neurons in the dentate gyrus of adrenal hormones in adulthood induces their large-scale death as well as a compensatory increase in the proliferation of granule neuron precursors *(70,123)*. Finally, very recent work suggests that hippocampal neurons in chickadees are also replaced throughout adulthood *(124)*.

Because *adult* ORNs, HVC neurons, and hippocampal cells proliferate, migrate, differentiate and ultimately undergo PCD—i.e., are subjected to events normally restricted to early development—the characterization of the underlying molecular mechanisms may result in therapeutic strategies to *replace* neuronal populations compromised by injury, aging, and/or disease. With this in mind, several investigators have begun to elucidate the molecular interactions that coordinate cell production and cell death during normal and induced neuronal turnover in the adult brain. One hypothesis which has received recent attention is that differentiated neurons provide an inhibitory molecular signal that prevents the proliferation of their own precursor cells. In support of this, there is a temporal correlation between loss of ORNs by PCD and increased mitotic activity among ORN precursors in the olfactory epithelium *(119)*. Other evidence supports the alternative hypothesis that the genesis of precursor cells is stimulated by positive molecular signals derived from dying neurons. Following adrenalectomy, hormone-deprivation-induced granule neuron death in the adult rat dentate gyrus *precedes* a peak in the rate of neurogenesis among granule precursor cells *(69)*. Moreover, in adult canaries, seasonally or experimentally induced decreases in T are associated with a burst of HVC neuronal death followed *subsequently* by an increase in the incorporation of newly generated neurons *(120,121)*. In this latter case though, since there is no evidence of increased neurogenesis following manipulations of T *(120–122,125)*, dying neurons could provide molecular signals that promote the differentiation and/or survival of newly generated HVC neurons.

A COUNTDOWN FROM CELL BIRTH TO CELL DEATH

Traditionally, PCD has been viewed as a process whereby neurons compete for access to trophic support at a populational level and when, where and even what neurons survive or die relates only to the presence or absence of cellular attributes that place them at a competitive advantage or disadvantage (e.g., axonal branching, synapse number, etc.) In contrast to this view of cell death as a random, stochastic process, recent evidence suggests that this regressive phenomenon may, in one sense, be more highly stereotyped. Very recent work has revealed that developing neurons (and other cells) have a fixed interval with which to access trophic support *(126,127)*. As a consequence, cells die in orderly fashion according to when they were born; earliest-born cells are the first to die whereas later born cells are the last to go. For example, retinal ganglion cells, displaced amacrine cells, and even oligodendrocytes die within several days after their birth, regardless of when during the proliferative period these cells are produced *(109,126)*.

Why would developing cells have a fixed interval between their birth and PCD? One advantage to a cell death "countdown" is that it ensures that the fate of older cells has already been decided and therefore enhances or restricts competition for trophic support to cells born at approximately the same period in development. Since cells with similar phenotypes and trophic requirements are often generated during the same developmental time window this mechanism may have evolved to restrict *competition for trophic support to cells that share the same phenotype and/or trophic requirements*. This may be particularly important for neural structures such as the retina that have multiple cell types *(126)*. For instance, among the earliest born lumbar motoneurons of the chick embryo, the first to undergo cell death (Burek and Oppenheim, unpublished observations) project to the ventral muscle mass of the developing hindlimb (*see* ref. *127*). Conversely, later born cells project to the dorsal muscle mass of the hindlimb. Exogenous BDNF reduces lumbar motoneuron death only during the latter half of the cell death period *(128)*, presumably because only later-born motoneurons are responsive to this factor. These observations suggest that competition for target-derived trophic support occurs among motoneuronal subpopulations that share birthdates, efferent projection patterns, and trophic factor responsiveness.

What molecular mechanisms adjust the temporal window between neurogenesis and PCD? For neurons that undergo target-dependent PCD, one possibility is that efferent connections to targets provide a facilitatory feedback signal that "switches on" target-dependence. Yet, experimental observations support the alternative hypothesis that developing neurons have a cell-autonomous molecular program that regulates the onset of target dependence. Consistent with this hypothesis, prior to target innervation (i.e., before commencing dependence on target-derived BDNF), cranial sensory neurons survive in vitro in the absence of trophic support; also, sensory neurons whose axons take longer to grow to their targets in vivo, survive longer in vitro *(77)*. In addition, chick embryo motoneurons survive transiently in vitro in the absence of target-derived trophic agents and their target dependence can become established even in the complete absence of such factors *(127)*. These observations imply that developing neurons do not require target-derived signals to commence target dependence, but rather a *cell-intrinsic* molecular program gives neurons a limited temporal window in which to access trophic support. This mechanism may have evolved as the most efficient means of selectively eliminating neurons that have made targeting errors. That is, if neurons relied solely on target-derived signal(s) to initiate target dependence, cells that fail to grow axons to the appropriate locations would not be eliminated by cell death *(76)*. It will be interesting to determine whether molecular perturbations that maintain aberrantly projecting neurons in a target-independent state proves disruptive to subsequent events in neural and behavioral development.

If neurons survive transiently in the absence of cell-extrinsic trophic signals during the interval between their birth and the onset of target dependence, what molecular mechanisms underlie this phenomenon? This can be best addressed in the context of the calcium "set point" hypothesis for neuronal survival, which states that developing neurons have a relatively narrow optimal range of cytoplasmic Ca^{2+} that is compatible with their survival *(51)* (*see* chapter by Leist and Nicotera). Ca^{2+} levels within the "set-point" range may promote neuronal survival by controlling Ca^{2+}-sensitive protein kinase cascades or second messenger systems that, for instance, either upregulate the expression of genes that promote cell survival or down-regulate genes that initiate the cell death cascade *(45,47)*. Consistent with the "set-point" hypothesis for neuronal survival, depolarizing early sensory neurons (which increases cytosolic Ca^{2+}) prior to their onset of target-dependence enhances their overall survival in vitro, while pharmacological reductions of intracellular Ca^{2+} to sub-setpoint levels kills these cells *(129)*. Moreover, when both sensory and sympathetic neurons require target-derived NGF for their survival, these cells have relatively low levels of cytoplasmic Ca^{2+} *(130–132)*. Depolarizing both neuronal populations (using extracellular K^+) in vitro, increases cytoplasmic Ca^{2+} via dihydropyridine-sensitive L-type Ca^{2+} channels, and attenuates

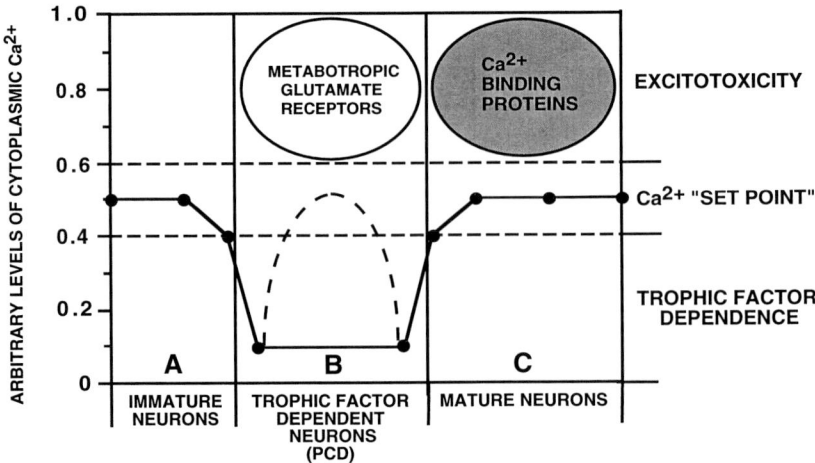

Fig. 10. Schematic illustration depicting how the regulation of cytoplasmic Ca^{2+} within neurons may account for age-dependent changes in their trophic requirements. Prior to target-innervation immature neurons **(A)** may promote their own viability by maintaining Ca^{2+} within a "set-point" range compatible with survival (Franklin and Johnson, 1992; Larment et al., 1992). In contrast, just prior to and during PCD **(B)** developing neurons may reduce levels of cytoplasmic Ca^{2+} to commence dependence on target-derived trophic factors (e.g., NGF). Afferents may promote neuronal survival during this time by depolarizing neurons, thereby increasing Ca^{2+} levels to the set-point range (—dashed lines). Co-activation of metabotropic glutamate receptors by afferent input may prevent activity-dependent increases in intracellular Ca^{2+} levels from becoming excitotoxic *(see text)*. Mature neurons **(C)** may increase Ca^{2+} back to pre-PCD levels to promote their own survival (Eichler et al., 1989, 1992; Koike and Tanaka, 1991), which could account for their decreased dependence on target-derived trophic factors and afferents (i.e., reduced susceptibility to axotomy and deafferentation: *see* Oppenheim, 1991; Snider et al., 1992). The augmented ability of mature neurons to buffer increases in Ca^{2+} (via increased expression of Ca^{2+} binding proteins) could also account for why these cells survive even in the absence of afferent input. Rises in intracellular Ca^{2+} in mature neurons induced by excess excitatory amino acids (e.g., glutamate) may result in cell death by excitotoxicity.

the death of sensory and sympathetic neurons following NGF withdrawal *(45,130,131)*. These results suggest that developing neurons are trophic-factor independent during axonal growth and early target innervation because they maintain Ca^{2+} levels within a survival promoting range, and that actively reducing cytoplasmic Ca^{2+} is the molecular trigger for the initiation of target-derived trophic factor dependence (*see* Fig. 10A,B).

DO MATURE NEURONS REQUIRE SURVIVAL SIGNALS TO MAINTAIN THEIR VIABILITY?

Neurons that undergo PCD in the *developing* vertebrate nervous system rely on trophic signals, primarily derived from cell–cell interactions such as afferents and targets in order to prevent their demise *(8,118)*. Since cell death during aging, neurodegenerative disease, and/or injury could stem from perturbations in some of the same cell–cell interactions operative during early development *(8,133)*, it is important to characterize whether and how the trophic requirements of neurons *change* throughout the lifespan of the organism. In this context, with a few exceptions (e.g., adult hippocampal granule neurons require circulating adrenal hormones for their survival *[70]*) developing neurons exhibit an age-related decline in their dependence on cell-extrinsic trophic signals such as targets and afferents *(8,134)*. Axotomizing or deafferenting neurons during early development induces rapid and massive neuronal death, whereas similar manipulations in adulthood generally cause relatively less or no cell loss *(8,134)*. This implies that mature neurons either receive alternate sources of cell-extrinsic neurotrophic support (e.g., glia) or lose their dependence on survival signals altogether.

Another possibility is that adult neurons do, in fact, require trophic molecular signals to prevent their demise, but such factors are derived from cell-intrinsic, rather than cell-extrinsic sources. Since, prior to target innervation, some developing neurons survive independently of exogenous trophic support by maintaining cytosolic Ca^{2+} within a "set-point" window compatible with cell survival, mature neurons may also adopt this strategy to foster their own survival (Fig. 10). Consistent with this hypothesis, both sympathetic and sensory neurons exhibit an age-dependent *increase* in intracellular Ca^{2+} levels which coincides with a decline in NGF dependence both in vitro and in vivo *(130–132)*. Preventing this age-dependent increase in cytoplasmic Ca^{2+} within sympathetic neurons *(132)* as well as experimentally lowering intracellular Ca^{2+} levels in *mature*, target-independent sensory and sympathetic neurons (to those levels characteristic of immature sensory neuron), makes these cells NGF-dependent *(135)*. The ability of mature neurons to prevent increases in intracellular Ca^{2+} (above the set-point window) may also account for their reduced trophic dependence on afferent input. While eighth nerve afferents promote the survival of chick nucleus magnocellularis neurons during *development* (by buffering potentially lethal increases in intracellular Ca^{2+}), *mature* NM neurons survive independent of afferent input. In rodents, this reduced susceptibility to deafferentation correlates with a dramatic increase in the cellular expression of cytoplasmic Ca^{2+}-binding proteins parvalbumin and calretinin within the rat cochlear nucleus *(136)*, suggestive of an increased ability of neurons to buffer Ca^{2+}. Cochlear neurons in adult guinea pigs also upregulate these same Ca^{2+} binding proteins following deafferentation *(136)*. Taken together, these observations imply that mature neurons exhibit reduced trophic dependence on targets and afferents (and perhaps other cell-extrinsic trophic signals), because they maintain intracellular Ca^{2+} levels within the "set-point" range. Consistent with this hypothesis, dramatic increases in intracellular Ca^{2+} that overwhelm the buffering capability of neurons has been implicated in exitotoxic cell death in the adult nervous system *(137)* (also *see* chapter by Olney and Ishimaru).

Part of the signal transduction mechanisms by which Ca^{2+} promotes the survival of mature (and developing) neurons may involve the upregulation of genes that arrest the PCD of developing neurons. This has recently been investigated in sympathetic neurons, which, when deprived of NGF, undergo PCD both in vivo and in vitro. Interestingly, mRNA levels for *bcl-2* in cultured sympathetic neurons decrease after NGF-deprivation and sympathetic neurons from *bcl-2*-deficient transgenic mice die more rapidly when NGF-deprived than cells from normal animals *(138)*. This implies that decreased levels of bcl-2 are associated with enhanced NGF dependence. Since sympathetic neurons gradually lose their dependence on target-derived NGF after the PCD period both in vivo and in vitro, one possibility is that these neurons upregulate *bcl-2* in order to acquire their trophic factor independence. However, no change in *bcl-2* mRNA or protein levels are evident in cultured sympathetic neurons during their transition to trophic factor independence, and cultured sympathetic neurons from *bcl-2* knockout mice exhibit a normal maturational decrease in NGF dependence *(138)*. These observations argue against the hypothesis that increased *bcl-2* expression is required for mature sympathetic neurons to develop trophic factor independence *(140)*. Nonetheless, either upregulation of *bcl-2* or related genes that arrest neuronal PCD (e.g., Bcl-x; 56), as well as the downregulation of genes that promote PCD (e.g., the *bcl-2* homolog *bak*; *[131]*), may stimulate the survival of adult neurons and account for their decreased dependence on cell-extrinsic sources of trophic support such as targets and afferents. Intracellular Ca^{2+} may provide the signal transduction cascades that influence the transcription of these cell death genes.

In addition to Ca^{2+}, another cell-intrinsic source of trophic support for mature neurons could be endogenously produced neurotrophic factors. An elegant study by Acheson et al. 1995 *(75)* provides strong support for this hypothesis. Although developing sensory neurons of the chick embryo die in vivo and in vitro when deprived of target-derived NGF, these cells survive for as long as one month when cultured from adult animals, even in the presence of anti-NGF antibodies *(141)*. Since *adult* sensory neurons in vivo express BDNF and the BDNF receptor trkB *(86,142)*, Acheson et al. reasoned that adult sensory neurons may promote their own survival via a BDNF autocrine loop. They tested this hypothesis directly by growing adult sensory neurons in single cell microwells in the presence of antisense oligonucleotides that reduced endogenous BDNF by 80%. Not only did this manipulation culminate in the rapid death of large numbers of sensory neurons, but these cells could be rescued by *exogenous* BDNF. More importantly though, anti-BDNF mRNA oligonucleotides do not kill adult sensory neurons harvested from ganglia of transgenic mice that lack the BDNF-sensitive subpopulation of sensory neurons. This rules out a nonspecific effect of anti-BDNF mRNA oligonucleotides and, together, these observations suggest that adult neurons become self-reliant, at least in part, by synthesizing and utilizing *endogenous* neurotrophic factors. Again, intracellular Ca^{2+} levels may be involved in regulating this cell-autonomous mechanism of neuronal survival.

DYSREGULATION OF NEURONAL PCD MAY UNDERLIE SOME FORMS OF PATHOLOGICAL NEURONAL DEATH—CONCLUSIONS

Both the regulation of cell survival during PCD, as well as the cessation of PCD late in development, are essential for sculpting final cell numbers in the nervous system. In this context, it is reasonable to speculate that at least some genetic or congenital neurological defects may stem from the inability of cell–cell survival signals, either extrinsic or intrinsic to cells, to appropriately suppress PCD. In support of this hypothesis, dysregulation of PCD has recently been implicated in the most severe form of the human neurodegenerative disease, spinal muscular atrophy (Werdnig–Hoffmann Disease) (*see* chapter by Rabin and Borchelt). SMA is caused by an autosomal recessive mutation characterized by the massive death of motoneurons during the perinatal period. This loss of motoneurons results in respiratory failure and death by one to three years of age. SMA has been mapped to a gene on chromosome 5 that is homologous to a baculovirus gene product (p35) which inhibits the normal PCD of mammalian cells *(143)*. The first two coding exons of the gene for this so-called "neuronal apoptosis inhibitory protein" (NAIP) are deleted in approximately 70% of Werdnig–Hoffman patients. This implies that failure to inhibit the normal PCD of motoneurons at the appropriate time during development in SMA patients may be responsible for increased cell loss.

Since pathological neuronal death may stem from the inability to suppress PCD, and both developmental PCD and neuronal survival in adulthood are regulated by neurotrophic factors, cell death caused by aging, injury, or neurodegenerative disease could reflect perturbations in neurotrophic factor signaling. However, while neurotrophic factors such as CNTF, BDNF, and GDNF are remarkably effective at rescuing neurons from induced death in animal models of injury and neurodegenerative disease, surprisingly, no pathological cell loss in the nervous system has yet been linked to defective neurotrophic interactions (see discussion of this point in chapter by Koliatsos and Mocchetti). This suggests that, rather than acting to compensate for a defect in trophic signalling, neurotrophic factors act as pharmacological agents to rescue neurons. Since increases in intracellular reactive oxygen species (ROS), which cause oxidative stress, are implicated as common molecular denominators in many forms of pathological cell death (144), the remarkable success of trophic factors in attenuating such induced-neuronal death in vivo may stem from their ability to increase cellular anti-oxidant defenses. Consistent with this notion, neurotrophic factors such as NGF (which rescue neurons from normal PCD) attenuate the oxidative stress-induced death of neurons in vitro by increasing antioxidant enzymes that detoxify ROS (e.g., superoxide dismutase) (145–148).

The possibility that neuronal death during development and disease share some common molecular elements provided the foundation for the recent demonstration that neurotrophic factors prevent injury-induced cell death. Another illustration of the power of this experimental approach are studies of the proto-oncogene bcl-2. Bcl-2 is very similar in structure to ced-9, a gene that negatively regulates PCD during C. elegans development, and overexpression of bcl-2 in immune cells prevents apoptosis induced by a variety of stimuli. Since human bcl-2, like ced-9, can prevent PCD in C. elegans, bcl-2 appears to be the vertebrate homolog of ced-9 (149). Consistent with this hypothesis, overexpression of bcl-2 rescues NGF-deprived sensory and sympathetic from PCD in vitro (122,150). Moreover, in transgenic mice that overexpress bcl-2, increased neuron number is evident in some regions, presumably due to reduced PCD (151). Remarkably, in these same transgenic mice, bcl-2 overexpression not only rescues motoneurons from axotomy (152) but also reduces cell death in the nervous system caused by ischemia (153,154). In fact, the pluripotent survival promoting effects of bcl-2, like neurotrophic factors, may stem from its ability to act as an antioxidant (155,157). These observations are groundbreaking because they suggest that theced-9/bcl-2 inhibition of PCD is a strongly conserved molecular pathway, not only across species (worms and humans), and cell types, but also regardless of the signals that actually trigger cell death (e.g., withdrawal of trophic factors versus ischemia). This makes a strong case for the notion that both normal and induced-neuronal death (and perhaps PCD in general) share common molecular steps (e.g., ROS), and reinforces the importance of basic research on both the cell–cell interactions and the molecular mechanisms that control the survival of developing neurons during PCD. This pursuit will not only provide new insight into the enigma of why cells die in massive numbers during normal development, but will continue to shed new light on ways to prevent pathological neuronal (and glial) cell death induced by injury, aging, and/or disease.

ACKNOWLEDGMENT

This work was supported by the NIH Grant NS20402 to R.W. Oppenheim and an N.R.S.A. Post-Doctoral Award (NS09838) to M.J. Burek.

REFERENCES

1. Gash DM, Zhang Z, Ovadia A, Cass W, Yi A, Simmerman L, Russell D, Martin D, Lapchak PA, Collins F, Hoffer BJ, Gerhardt GA. Functional recovery in parkinsonian monkeys treated with GDNF. *Nature* 1996, **380**: 252–255.
2. Sendtner M, Schmalbruch H, Stockli KA, Carroll P, Kreutzberg GW, Thoenen H. Ciliary neurotrophic factor prevents the degeneration of motor neurons in mouse mutant progressive motor neuropathy. *Nature* 1992, **358**: 502–504.
3. Anderson AJ, Su JH, Cotman CW. DNA damage and apoptosis in alzheimer's disease: colocalization with c-jun immunoreactivity, relationship to brain area, and effect of postmortem delay. *J Neurosci* 1996, **16**: 1710–1719.
4. Garcia-Valenzuela E, Gorczyca W, Darzynkiewicz Z, Sharma SC. Apoptosis in adult retinal ganglion cells after axotomy. *J Neurobiol* 1994, **25**: 431–438.
5. Nitatori T, Sato N, Waguri S, Karasawa Y, Araki H, Shibanai K, Kominami E, Uchiyama Y. Delayed neuronal death in the CA1 pyramidal cell layer of the gerbil hippocampus following transient ischemia is apoptosis. *J Neurosci* 1995, **15**: 1001–1011.
6. Portera-Cailliau C, Hedreen JC, Price D, Koliatsos VE. Evidence for apoptotic cell death in huntington disease and excitotoxic animal models. *J Neurosci* 1995, **15**: 3775–3787.
7. Yarmolinsky MB. Programmed cell death in bacterial populations. *Science* 1995, **267**: 836–837.
8. Oppenheim RW. Cell death during development of the nervous system. *Annu Rev Neurosci* 1991, **14**: 453–501.
9. Herrup K, Sunter K. Numerical matching during cerebellar development: quantitative analysis of granule cell death in staggerer mouse chimeras. *J Neurosci* 1987, **7**: 829–836.
10. Tanaka H, Landmesser L. Cell death of lumbosacral motoneurons in chick, quail, and chick-quail chimera embryos: a test of the quantitative matching hypothesis of neuronal cell death. *J Neurosci* 1986, **6**: 2889–2899.
11. Davies AM, Bandtlow C, Heumann R, Korsching S, Rohrer H, Thoenen H. Timing and site of nerve growth factor synthesis in developing skin in relation to innervation and expression of the receptor. *Nature* 1987, **326**: 353–358.
12. Smeyne RJ, Klein R, Schnapp A, Long LK, Bryant S, Lewin A, Lira SA, Barbacid M. Severe sensory and sympathetic neuropathy in mice carrying a disrupted trk/NGF receptor gene. *Nature* 1994, **368**: 246–249.
13. Johnson EM, Gorin PM, Brandeis LD, Pearson J. Dorsal root ganglion neurons are destroyed by exposure in utero to maternal antibody to nerve growth factor. *Science* 1980, **210**: 916–918.
14. Levi-Montalcini L. The nerve growth factor 35 years later. *Science* 1987, **237**: 1154.
15. Johnson EM. Destruction of the sympathetic nervous system in neonatal rats and hamsters by vinblastine: prevention by concomitant administration of nerve growth factor. *Brain Res* 1978, **141**: 105–118.
16. Oppenheim RW, Maderdrut JL, Wells DJ. Cell death of motoneurons in the chick embryo spinal cord. VI. Reduction of naturally occurring cell death in the thoracolumbar column of terni by nerve growth factor. *J Comp Neurol* 1982, **210**: 174–189.
16a. Armstrong RC, Clarke PGH. Neuronal death and the development of the pontine nuclei and inferior olive in the chick. Neuroscience 1979, **4**: 1635–1647.
16b. McKay SE, Oppenheim RW. Lack of evidence of cell death among avian spinal interneurons during normal development and following removal of targets and afferents. *J Neurosci* 1991, **22**: 721–733.

17. Clarke PGH. Neuronal death in development of the vertebrate central nervous system. *Sem in the Neurosci* 1994, **6:** 291–297.
18. Von Bartheld CS, Kinoshita Y, Prevette D, Yin QW, Oppenheim RW, Bothwell M. Positive and negative effects of neurotrophins on the isthmo-optic nucleus in chick embryos. *Neuron* 1994, **12:** 639–654.
19. Henderson CE, Camu W, Mettling C, Gouin A, Poulson K, Karihaloo M, Rullamas J, Evans T, McMahon SB, Armanini MP, Berkemeier L, Phillips HS, Rosenthal A. Neurotrophins promote motor neuron survival and are present in embryonic limb bud. *Nature* 1993, **363:** 266–270.
20. Koliatsos VE, Clatterbuck RE, Winslow JW, Cayouette MH, Price DL. Evidence that brain-derived neurotrophic factor is a trophic factor for motoneurons. *Neuron* 1993, **10:** 359–367.
21. Oppenheim RW, Yin Q-W, Prevette D, Yan Q. Brain-derived neurotrophic factor rescues developing avian motoneurons from cell death. *Nature* 1992, **360:** 755–757.
22. Yan Q, Elliott JL, Sun J, Matheson C, Zhang L, Mu X, Rex KL, Snider WD. Influences of neurotrophins on mammalian motoneurons *in vivo*. *J Neurobiol* 1993, **24:** 1555–1577.
23. Oppenheim RW, Prevette D, Haverkamp LJ, Houenou L, Qin-Wei Y, McManaman J. Biological studies of a putative avian muscle-derived neurotrophic factor that prevents naturally occurring motoneuron death in vivo. *J Neurobiol* 1993, **24:** 1065–1079.
24. Lewis ME, Neff NT, Contreras PC, Stong DB, Oppenheim RW, Grebow PE, Vaught JL. Insulin-like growth factor-I: potential for treatment of motor neuronal disorders. *Exp Neurol* 1993, **124:** 73–88.
25. Neff NT, Prevette D, Houenou LJ, Lewis ME, Glicksman MA, Qin-Wei Y, Oppenheim RW. Insulin-like growth factors: putative muscle-derived trophic agents that promote motoneuron survival. *J Neurobiol* 1993, **24:** 1578–1588.
26. Henderson CE, Phillips HS, Pollock RA, Davies AM, LeMeulle C, Armanini M, Simpson LC, Moffet B, Vandlen RA, Koliatsos VE, Rosenthal A. GDNF: a potent survival factor for motoneurons present in peripheral nerve and muscle. *Science* 1994, **266:** 970–972.
27. Oppenheim RW, Houenou LJ, Johnson JE, Lin L, Linxi L, Lo AL, Newsome AL, Prevette DM, Wang SW. Developing motor neurons rescued from programmed and axotomy-induced cell death by GDNF. *Nature* 1995, **373:** 344–346.
27a. Cowan, O'Leary. 1984 Medicine, Science, and Society: Celebrating the Harvard Medical School Bicentennial. Wiley, New York.
28. Pittman RH, Oppenheim RW. Neuromuscular blockade increases motoneurone survival during normal cell death in the chick embryo. *Nature* 1978, **271:** 364–366.
28a. Cowan WM, Fawcett JW, O'Leary DM, Stanfield BB. Regressive events in neurogenesis. *Science* **225:** 1258–1265.
29. Houenou LJ, McManaman JL, Prevette D, Oppenheim RW. Regulation of putative muscle-derived neurotrophic factors by muscle activity and innervation: in vivo and in vitro studies. *J Neurosci* 1991, **11:** 2829–2837.
30. Tanaka H. Chronic application of curare does not increase the level of motoneuron survival-promoting activity in limb muscle extracts during the naturally-occurring motoneuron cell death period. *Dev Biol* 1987, **124:** 347–357.
31. Linden R. The survival of developing neurons: a review of afferent control. *Neurosci* 1994, **58:** 671–682.
32. Clarke PGH. Neuronal death during development in the isthmo-optic nucleus of the chick: sustaining role of afferents from the tectum. *J Comp Neurol* 1985, **234:** 365–379.
33. Clarke PGH. Neuron death in the developing avian isthmo-optic nucleus, and its relation to the establishment of functional circuitry. *J Neurobiol* **23:** 1140–1158.
34. Furber S, Oppenheim RW, Prevette D. Naturally occurring neuron death in the ciliary ganglion of the chick embryo following removal of preganglionic input: evidence for the role of afferents in ganglion cell survival. *J Neurosci* 1987, **7:** 1816–1832.

35. Linden R, Pinon LGP. Dual control by targets and afferents of developmental neuronal death in the mammalian central nervous system: a study in the parabigeminal nucleus of the rat. *J Comp Neurol* 1987, **266:** 141–149.
36. Okado N, Oppenheim RW. Cell death of motoneurons in the chick embryo spinal cord. IX. The loss of motoneurons following removal of afferent inputs. *J Neurosci* 1984, **4:** 1639–1652.
37. Cunningham TJ, Huddleston C, Murray M. Modification of neuron numbers in the visual system of the rat. *J Comp Neurol* 1979, **184:** 423–424.
38. Linden R, Renteria AS. Afferent control of neuron numbers in the developing brain. *Dev Brain Res* 1988, **44:** 291–295.
39. Sohal GS, Hirano S, Kumaresan K, Ali MM. Influence of altered afferent input on the number of trochlear motor neurons during development. *J Neurobiol* 1992, **23:** 10–16.
40. Galli-Resta L, Ensini M, Fusco E, Gravina A, Margheritti B. Afferent spontaneous electrical activity promotes the survival of target cells in the developing retinotectal system of the rat. *J Neurosci* 1993, **13:** 243–250.
41. Catsicas M, Pequignot Y, Clarke PGH. Rapid onset of neuronal death induced by blockade of either axoplasmic transport or action potentials in afferent fibers during brain development. *J Neurosci* 1992, **12:** 4642–4660.
42. Yin Q-W, Johnson J, Prevette D, Oppenheim RW. Cell death of spinal motoneurons in the chick embryo following deafferentation: rescue effects of tissue extracts, soluble proteins, and neurotrophic agents. *J Neurosci* 1994, **14:** 7629–7640.
43. Maderdrut JL, Oppenheim RW, Prevette D. Enhancement of naturally occurring cell death in the sympathetic and parasympathetic ganglia of the chicken embryo following blockade of ganglionic transmission. *Brain Res* 1988, **444:** 189–194.
44. Rubel EW, Hyson RL, Durham D. Afferent regulation of neurons in the brain stem auditory system. *J Neurobiol* 1990, **21:** 169–196.
45. Franklin JL, Johnson EM. Suppression of programmed neuronal death by sustained elevation of cytoplasmic calcium. *Trends Neurosci* 1992, **15:** 501–508.
46. Lipton SA. Blockade of electrical activity promotes the death of mammalian retinal ganglion cells in culture. *Proc Natl Acad Sci USA* 1986, **83:** 9774–9778.
47. Johnson EM, Koike T, Franklin J. A "calcium set-point hypothesis" of neuronal dependence on neurotrophic factor. *Exp Neurol* 1992, **115:** 163–166.
48. Lu B, Yokoyama M, Dreyfus CF, Black IB. Depolarizing stimuli regulate nerve growth factor gene expression in cultured hippocampal neurons. *Proc Natl Acad Sci* 1991, **88:** 6289–6292.
49. Korsching S. The neurotrophic factor concept: a re-examination. *J Neurosci* 1993, **13:** 2739–2748.
50. Johnson EM, Deckwerth TL. Molecular mechanisms of developmental neuronal death. *Annu Rev Neurosci* 1993, **16:** 31–46.
51. Von Bartheld CS, Byers MR, Williams R, Bothwell M. Anterograde transport of neurotrophins and axodendritic transfer in the developing visual system. *Nature* 1996, **387:** 830–833.
52. Kaiser J, Lipton SA. VIP-mediated increase in cAMP prevents TTX-induced retinal ganglion cell death *in vitro*. *Neuron* 1990, **5:** 373–381.
53. Hyson RL, Rubel EW. Transneuronal regulation of protein synthesis in the brain stem auditory system of the chick requires synaptic activation. *J Neurosci* 1989, **9:** 2835–2845.
54. Born DE, Rubel EW. Afferent influences on brain stem auditory nuclei of the chicken: presynaptic action potentials regulate protein synthesis in nucleus magnocellularis neurons. *J Neurosci* 1988, **8:** 901–919.
54a. Catsicas M, Pequignot Y, Clarke PGH. Rapid onset of neuronal death induced by blockade of ether axoplasmic transport or action potentials in afferent bibers during brain development. 1992, *J Neurosci* **12:** 4642–4660.

55. Lahica EA, Rubsamen R, Zirpel L, Rubel EW. Glutamatergic inhibition of voltage-operated calcium channels in the avian cochlear nucleus. *J Neurosci* 1995, **15**: 1724–1734.
56. Zirpel L, Lahica EA, Lippe WR. Deafferentation increases the intracellular calcium of cochlear nucleus neurons in the embryonic chick. *J Neurophysiol* 1996, **74**: 1355–1357.
57. Hyde GE, Lahica EA, Rubel EW. Afferent influences on avian brainstem auditory system: flunarizine blocks calcium accumulation, mitochondrial biogenesis, and cell death following cochlea removal. *J Comp Neurol* (in press).
58. Eagleson KL, Raju TR, Bennett MR. Motoneuron survival is induced by immature astrocytes developing avian spinal cord. *Dev Brain Res* 1985, **17**: 95–104.
59. Engele J, Bohn MC. The neurotrophic effects of fibroblast growth factors on dopaminergic neurons *in vitro* are mediated by mesencephalic glia. *J Neurosci* 1991, **11**: 3070–3078.
60. Lindsay RM. Adult rat brain astrocytes support survival of both NGF-dependent and NGF-insensitive neurons. *Nature* 1979, **282**: 80–82.
61. Verdi JM, Groves AK, Farinas I, Jones K, Marchionni MA, Reichardt LF, Anderson DJ. A reciprocal cell-cell interaction mediated by NT-3 and neuregulins controls the early survival and development of sympathetic neuroblasts. *Neuron* 1996, **16**: 515–527.
62. Nordeen EJ, Nordeen KW. Sex difference among non-neuronal cells precedes sexually dimorphic neuron growth and survival in an avian song control nucleus. *J Neurobiol* 1996 (in press).
63. Boudreau N, Sympson CJ, Werb Z, Bissell MJ. Suppression of ICE and apoptosis in mammary epithelial cell by extracellular matrix. *Science* 1995, **267**: 891–893.
64. Kollros JJ. Transitions in the nervous system during amphibian metamorphosis, in *Metamorphosis: A Problem in Developmental Biology* (Gilbert LI, Frieden E, eds.), New York, Plenum, pp. 445–459.
65. Truman JW, Thorn RS, Robinow S. Programmed neuronal death in insect development. *J Neurobiol* 1992, **23**: 1295–1311.
66. Wright LL, Smolen AJ. The role of neuron death in the development of the gender difference in the number of neurons in the rat superior cervical ganglion. *Int J Dev Neurosci* **5**: 305–311.
67. Wimer RE, Wimer CC. On the development of strain and sex differences in granule cell number in the area dentata of house mice. *Dev Brain Res* 1988, **42**: 191–197.
68. Nordeen EJ, Nordeen KW, Sengelaub D.R, Arnold A.P. Androgens prevent normally occurring cell death in a sexually dimorphic spinal nucleus. *Science* 1985, **229**: 671–673
69. Gould E, McEwen BS. Neuronal birth and death. *Curr Opin Neurobiol* 1993, **3**: 676–682.
70. Sloviter RS, Dean E, Neubort S. Electron microscopic analysis of adrenalectomy-induced hippocampal granule cell degeneration in the rat: apoptosis in the adult central nervous system. *J Comp Neurol* 1993, **330**: 337–351.
71. Davies AM, Wright EM. Neurotrophin autocrine loops. *Curr Biol* 1995, **5**: 723–726.
72. Miranda RC, Sohrabji F, Toran-Allerand CD. Neuronal colocalization of mRNAs for neurotrophins and their receptors in the developing central nervous system suggests a potential for autocrine interactions. *Proc Natl Acad Sci* 1993, **90**: 6439–6443.
73. Wright EM, Vogel KS, Davies AM. Neurotrophic factors promote the maturation of developing sensory neurons before they become dependent on these factors for survival. *Neuron* 1992, **9**: 139–150.
74. Schecterson LC, Bothwell M. Novel roles for neurotrophins are suggested by BDNF and NT-3 mRNA expression in developing neurons. *Neuron* 1992, **9**: 449–463.
75. Acheson A, Conover JC, Fandl JP, DeChaiara TM, Rusell M, Thadani A, Squinto SP, Yancopoulos GD, Lindsay RM. A BDNF autocrine loop in adult sensory neurons prevents cell death. *Nature* 1995, **374**: 450–453.
76. Davies AM. Intrinsic programmes of growth and survival in developing vertebrate neurons. *Trends Neurosci* 1994, **17**: 195–199.

77. Vogel KS, Davies AM. The duration of neurotrophic factor independence in early sensory neurons is matched to the time course of target field innervation. *Neuron* 1991, **7:** 819–830.
78. Blaschke AJ, Staley K, Chun J. Widespread programmed cell death in proliferative and post-mitotic regions of the fetal cerebral cortex. *Development* 1996, **112:** 1165–1174.
79. Carr VM, Simpson SB. Rapid appearance of labelled degenerating cells in the dorsal root ganglia after exposure of chick embryos to tritiated thymidine. *Dev Brain Res* 1982, **2:** 157–62.
80. Homma S, Yaginuma H, Oppenheim RW. Programmed cell death during the earliest stages of spinal cord development in the chick embryo: a possible means of early phenotypic selection. *J Comp Neurol* 1994, **345:** 377–395.
81. Yaginuma H, Tomita M, Takashita N, McKay SE, Cardwell C, Yin QW, Oppenheim RW. A novel type of programmed neuronal death in the cervical spinal cord of the chick embryo. *J Neurosci* 1996, **16:** 3685–3703.
82. Barde YA, Edgar D, Thoenen H. Sensory neurons in culture: changing requirements for survival factors during embryonic development. *Proc Natl Acad Sci USA* 1980, **77:** 1199–1203.
83. Buchman VL, Davies AM. Different neurotrophins are expressed and act in a developmental sequence to promote the survival of embryonic sensory neurons. *Development* 1993, **118:** 989–1001.
84. Breedlove SM. Sexual dimorphism in the vertebrate nervous system. *J Neurosci* 1992, **12:** 4133–4142.
85. Lindholm D, Castren E, Hengerer B, Zafra B, Berninger B, Thoenen H. Differential regulation of nerve growth factor (NGF) synthesis in neurons and astrocytes by glucocorticoid hormones. *Eur J Neurosci* 1992, **4:** 404–410.
86. Lindholm D, Castren E, Berzaghi M, Blochl A, Thoenen H. Activity-dependent and hormonal regulation of neurotrophin mRNA levels in the brain: implications for neuronal plasticity. *J Neurobiol* 1994, **25:** 1362–1372.
87. Sohrabji F, Miranda RC, Toran-Allerand CD. Estrogen differentially regulates estrogen and nerve growth factor receptor mRNAs in adult sensory neurons. *J Neurosci* 1994, **14:** 459–471.
88. Forger NG, Roberts S, Wong V, Breedlove SM. Ciliary neurotrophic factor maintains motoneurons and their target muscles in developing rats. *J Neurosci* 1993, **13:** 4720–4726.
89. Oppenheim RW, Haverkamp LJ, Prevette D, McManaman JL, Appel SH. Reduction of naturally occurring motoneuron death in vivo by a target-derived neurotrophic factor. *Science* 1988, **240:** 919–922.
90. Wetts R, Herrup K. Direct correlation between Purkinje and granule cell number in the cerebellum of lurcher chimeras and wild-type mice. *Dev Brain Res* 1983, **10:** 41–47.
91. Kutch W, Bentley D. Programmed death of pioneer neurons in the grasshopper embryo. *Dev Biol* 1987, **123:** 517–525.
92. McConnell SK, Ghosh A, Shatz CJ. Subplate neurons pioneer the first axon pathway from the cerebral cortex. *Science* 1989, **245:** 978–982.
93. Lushkin MB, Shatz CJ. Studies of the earliest generated cells of the cat's visual cortex: cogeneration of subplate and marginal zones. *J Neurosci* 1985a, **5:** 1062–1075.
94. Lushkin MB, Shatz CJ. Neurogenesis of the cat's primary visual cortex. *J Comp Neurol* 1985b, **242:** 611–631.
95. Chalupa LM, Dreher B. High precision systems require high precision blueprints: a new view regarding the formation of connections in the mammalian visual system. *J Cognit Neurosci* 1991, **3:** 209–219.
96. O'Leary DDM, Fawcett JW, Cowan WM. Topographic targeting errors in the retinocollicular projections and their elimination by selective ganglion cell death. *J Neurosci* 1986, **6:** 3692–3705.

97. Catsicas S, Thanos S, Clarke PGH. Major role for neuronal death during brain development refinement of topographical connections. *Proc Natl Acad Sci* 1987, **84**: 8165–8168.
98. Landmesser L, O'Donovan MJ. The activation patterns of embryonic chick motoneurons projecting to inappropriate muscles. *J Physiol* 1984, **347**: 189–204.
93. Avery L, Horvitz HR. A cell that dies during wild-type C. elegans development can function as a neuron in a ced-3 mutant. *Cell* 1987, **51**: 1071–1078.
100. Ellis HM, Horvitz HR. Genetic control of programmed cell death in the nematode C. elegans. *Cell* 1986, **44**: 817–829.
101. Forss-Petter S, Danileson PE, Catsicas S, Battenberg E, Price J, Nerenberg M, Sutcliffe JG. Transgenic mice expressing b-galactosidase in mature neurons under neuron-specific enolase promoter control. *Neuron* 1990, **5**: 187–197.
102. Sadoul R, Catsicas S, Martinou J-C. Neuronal death: common mechanisms during development and disease. *Sem in the Neurosci* 1994, **6**: 343–346.
103. Veis DJ, Sorenson CM, Shutter JR, Korsemeyer SJ. Bcl-2-deficient mice demonstrate fulminant lymphoid apoptosis, polycystic kidneys, and hypopigmented hair. *Cell* 1993, **75**: 229–240.
104. Sagot Y, Dubois-Dauphin M, Tan SA, de Bilbao F, Aebischer P, Martinou J-C, Kato AC. Bcl-2 overexpression prevents motoneuron cell body loss but not axonal degeneration in a mouse model of a neurodegenerative disease. *J Neurosci* 1995, **15**: 7727–7733.
105. Horsburgh GM, Sefton AJ. Cellular degeneration and synaptogenesis in the developing retina of the rat. *J Comp Neurol* 1987, **263**: 553–566.
106. Carr VM, Farbman AI. The dynamics of cell death in the olfactory epithelium. *Exp Neurol* 1993, **124**: 308–314
107. Jackson KR, Duncan ID. Cell kinetics and cell death in the optic nerve of myelin deficient rat. *J Neurocytol* 1988, **17**: 657–670.
108. Knapp PE, Skoff RP, Redstone DW. Oligodendroglial cell death in jimpy mice: an explanation for the myelin deficit. *J Neurosci* 1986, **6**: 2813–2822.
109. Barres BA, Raff MC. Control of oligodendrocyte number in the developing rat optic nerve. *Neuron*, 1994, **12**: 935–942.
110. Barres BA, Hart IK, Coles HSR, Burne JF, Voyvodic JT, Richardson WD, Raff MC. Cell death and control of cell survival in the oligodendrocyte lineage. Cell 1992, **70**: 31–46.
111. Barres BA, Hart IK, Coles HSR, Burne JF, Voyvodic JT, Richardson WD, Raff MC. Cell death in the oligodendrocyte lineage. *J Neurobiol* 1992, **23**: 1221–1230.
112. Ciutat D, Caldero J, Oppenheim RW, Esquerda JE. Schwann cell apoptosis during normal development and after axonal degeneration induced by neurotoxins in the chick embryo. *J Neurosci* 1996, **16**: 3979–3990.
113. Burne JF, Staple JK, Raff MC. Glial cells are increased proportionately in transgenic optic nerves with increased numbers of axons. *J Neurosci* 1996, **16**: 2064–2073.
114. Jessen KR, Brennant A, Morgan L, Mirsky R, Kent A, Hasimoto Y, Cavrilovic J. The schwann cell precursor and its fate: a study of cell death and differentiation during gliogenesis in rat embryonic nerves. *Neuron* 1994, **12**: 509–527.
115. Trachtenberg JT, Thompson WJ. Schwann cell apoptosis at developing neuromuscular junctions is regulated by glial growth factor. *Nature* 1996, **379**: 174–177.
116. Barres BA, Scmid R, Sendtner M, Raff MC. Multiple extracellular signals are required for long-term oligodendrocyte survival. Development 1993, **118**: 283–295.
117. Krueger BK, Burne JF, Raff MC. Evidence for large-scale astrocyte death in developing cerebellum. *J Neurosci* 1995, **15**: 3366–3374.
118. Raff MC, Barres BA, Burne JF, Coles HS, Ishizaki Y, Jacobson MD. Social controls on cell survival and cell death. *Science* 1993, **262**: 695–700.

119. Calof AL, Hagiwara N, Holcomb JD, Mumm JS, Shou J. Neurogenesis and cell death in olfactory epithelium. *J Neurobiol* 1996, **30**: 67–81.
120. Doupe A. Songbirds and adult neurogenesis: a new role for hormones. *Proc Natl Acad Sci* 1994, **91**: 7836–7838.
121. Nordeen EJ, Nordeen KW. Hormonally-regulated neuron death in the avian brain. *Semi in the Neurosci* 1994, **6**: 299–306
122. Brown SD, Johnson F, Bottjer SW. Neurogenesis in adult canary telencephalon is independent of gonadal hormone levels. *J Neurosci* 1993, **13**: 2024–2032
123. Gould E, Cameron HE, Daniels DC, Woolley CS, McEwen BS. Adrenal hormones suppress cell division in the adult rat dentate gyrus. *J Neurosci* 1992, **12**: 3642–3650.
124. Barnea A, Nottebohm F. Recruitment and replacement of hippocampal neurons in young and adult chickadees: an addition to the theory of hippocampal learning. *Proc Natl Acad Sci* 1996, **93**: 714–718.
125. Rasika S, Nottebohm F, Alvarez-Buylla A. Testosterone increases the recruitment and/or survival of new high vocal center neurons in adult female canaries. *Proc Natl Acad Sci* 1994, **91**: 7854–7858.
126. Galli-Resta L, Ensini M. An intrinsic time limit between genesis and death of individual neurons in the developing retinal ganglion cell layer. *J Neurosci* 1996, **16**: 2318–2324.
127. Mettling C, Gouin A, Robinson M, M'Hamdi H, Camu W, Bloch-Gallego E, Buisson B, Tanaka H, Davies AM, Henderson CE. Survival of newly postmitotic motoneurons is transiently independent of exogenous trophic support. *J Neurosci* 1995, **15**: 3128–3137.
128. McKay SE, Herzog K-H, Garner A, Tucker R, Oppenheim RW, Large T. Expression of BDNF and trk B during the development of the neuromuscular system in the chick embryo. *Development* 1996, **122**: 715–724.
129. Larmet Y, Dolphin AC, Davies AM. Intracellular calcium regulates the survival of early sensory neurons before they become dependent on neurotrophic factors. *Neuron* 1992, **9**: 563–574.
130. Eichler ME, Rich KM. Death of sensory ganglion neurons after acute withdrawal of nerve growth factor in dissociated cultures. *Brain Res* 1989, **482**: 340–346.
131. Eichler ME, Dubinsky JM, Rich KM. Relationship of intracellular Ca2+ to dependence on nerve growth factor in DRG neurons in cell culture. *J Neurochem* 1992, **58**: 263–269.
132. Koike T, Tanaka S. Evidence that nerve growth factor dependence of sympathetic neurons for survival in vitro may be determined by levels of cytoplasmic free Ca2+. *Proc Natl Acad Sci* 1991, **88**: 3892–3896.
133. Appel SH. A unifying hypothesis for the cause of amyotrophic lateral sclerosis, Parkinsonism and Alzheimer's disease. *Ann Neurol* 1981, **10**: 499–501.
134. Snider WD, Elliott JL, Yan Q. Axotomy-induced neuronal death during development. *J Neurobiol* 1992, **23**: 1231–1246.
135. Tong JX, Eichler ME, Rich KM. Intracellular calcium levels influence apoptosis in mature sensory neurons after trophic factor deprivation. *Exp Neurol* 1996, **138**: 45–52.
136. Lohman C, Friauf E. Distribution of the calcium binding proteins parvalbumin and calretinin in the auditory brainstem of adult and developing rats. *J Comp Neurol* 1996, **367**: 90–109.
137. Choi DW. Calcium: still center stage in hypoxic-ischemic neuronal death. *Trends Neurosci* **18**: 58–60
138. Greenlund LJS, Korsemeyer SJ, Johnson EM. Role of Bcl-2 in the survival and function of developing and mature sympathetic neurons. *Neuron* 1995, **15**: 649–661.
139. Gonzalez-Garcia M, Garcia I, Ding L, O'Shea S, Boise LH, Thompson CB, Nunez G. Bcl-x is expressed in embryonic and postnatal neural tissues and functions to prevent neuronal cell death. *Proc Natl Acad Sci* 1995, **92**: 4304–4308.

140. Chittenden T, Harrington EA, O'Connor R, Flemington C, Lutz RJ, Evan GI, Guild BC. Induction of apoptosis by the Bcl-2 homologue Bak. *Nature* **374**: 733–736.
141. Lindsay RM. Nerve growth factors (NGF, BDNF) enhance axonal regeneration, but are not required for survival of adult sensory neurons. *J Neurosci* 1988, **8**: 2394–2405.
142. Ernfors P, Henschen A, Olson L, Persson H. Expression of nerve growth factor receptor mRNA is developmentally regulated and increased after axotomy in rat spinal cord motoneurons. *Neuron* 1989, **2**: 1605–1613.
143. Liston P, Roy N, Katsuyuki T, Lefebvre C, Baird S, Cherton-Horvat G, Farahani R, McLean M, Ikeda JE, MacKenzie A, Korneluk RG. Suppression of apoptosis in mammalian cells by NAIP and a related family of IAP genes. *Nature* 1996, **379**: 349–353.
144. Coyle JT, Puttfarcken P. Oxidative stress, glutamate and neurodegenerative disorders. *Science* 1993, **262**: 689–695.
145. Jackson GR, Apffel L, Werrbach-Perez K, Perez-Polo JR. Role of nerve growth factor in oxidant-anti-oxidant balance and neuronal injury. I. Stimulation of hydrogen peroxide resistance. *J Neurosci* Res 1990, **25**: 360–368.
146. Pan Z, Perez-Polo R. Role of nerve growth factor in oxidant homeostasis: glutathione metabolism. *J Neurochem* 1993, **61**: 1713–1721.
147. Sampath D, Jackson GR, Werrbach-Perez K, Perez-Polo J.R. Effects of nerve growth factor on glutathione peroxidase and catalase in PC12 cells. *J Neurochem* 1994, **62**: 2476–2479.
148. Spina MB, Squinto SP, Miller J, Lindsay RM, Hyman C. Brain-derived neurotrophic factor protects dopamine neurons against 6-hydroxydopamine and N-methyl-4-phenylpiridium ion toxicity: involvement of the glutathione system. *J Neurochem* 1992, **59**: 99–106.
149. Hengartner MO. Life and death decisions: ced-9 and programmed cell death in Caenorhabditis elegans. *Science* 1995, **270**: 931.
150. Allsopp TE, Wyatt S, Patterson HF, Davies AM. The proto-oncogene bcl-2 can selectively rescue neurotrophic factor dependent neurons from apoptosis. Cell 1993, **73**: 295–307.
151. McCaffery CA, Raju TR, Bennett MR. Effects of cultured astroglia on the survival of neonatal rat retinal ganglion cells in vitro. *Dev Biol* 1984, **104**: 441–448.
152. Dubois-Dauphin M, Frankowski H, Tsujimoto Y, Huarte J, Martinou JC. Neonatal motor neurons overexpressing the bcl-2 protooncogene in transgenic mice are protected from axotomy-induced cell death. *Proc Natl Acad Sci* 1994, **91**: 3309–3313.
153. Lawrence MS, Ho DY, Sun GH, Steinberg GK, Sapolsky RM. Overexpression of Bcl-2 with herpes simplex virus vectors protects CNS neurons against neurological insults in vitro and in vivo. *J Neurosci* 1996, **16**: 486–496.
154. Martinou JC, Dubois-Dauphin M, Syaple JK, Rodriguez I, Frankowsky H, Missotten M, Albertini P, Talabot D, Catsicas S, Pietra C, Huarte J. Overexpression of bcl-2 in transgenic mice protects neurons from naturally occurring cell death and experimental ischemia. *Neuron* 1994, **13**: 1–20.
155. Hockenbery DM, Nunez G, Milliman C, Schreiber RD, Korsemeyer SJ. Bcl-2 is an inner mitochondrial membrane protein that blocks programmed cell death. *Nature* 1990, **348**: 334–336.
156. Hockenbery DM, Oltvai ZN, Yin X-M, Milliman CL, Korsmeyer SJ. Bcl-2 functions in an antioxidant pathway to prevent apoptosis. *Cell* 1993, **75**: 241–251.
157. Kane DJ, Sarafian TA, Anton R, Hahn H, Gralla EB, Valentine JS, Ord T, Bredesen DE. Bcl-2 inhibition of neural death: decreased generation of reactive oxygen species. *Science* 1992, **262**: 1274–1277.

9
Axotomy-Induced Motor Neuron Death

Jeffrey L. Elliott and William D. Snider

INTRODUCTION

Transection of the axon (axotomy) represents a powerful paradigm for studying regeneration and death in neurons. Depending on the setting in which it is performed, axonal injury can lead to robust attempts by neurons to regenerate axons and reestablish functional contact with targets, or it can begin a complex process leading to neuronal death. Recent advances have begun to identify genes that execute or regulate a particular phenotype after axonal injury. These advances have been particularly noteworthy in the area of cell death regulation. In the last 5 yr, the importance of apoptosis as a principal mode of neuronal death has been recognized. Because axotomy can lead to neuronal death via apoptosis, it has emerged as powerful model for studying the molecular regulators and effectors of apoptosis in neurons.

This chapter will focus on our current understanding of axotomy-induced injury and death as it relates to motor neurons both during development and in adulthood. Motor neurons are a compelling neuronal population to study for at least two reasons. First, there is a significant clinical impetus to understand the processes by which motor neurons die in experimental paradigms, as it is possible that similar molecular mechanisms may operate in human motor neuron diseases such as amyotrophic lateral sclerosis (ALS) (*see* chapter by Rabin and Borchelt). Second, motor neurons are large cells frequently found in well-circumscribed nuclei with myelinated axons in readily accessible peripheral and cranial nerves. Thus, numbers of motor neurons and their axons can be easily and accurately determined without the uncertainties related to neuron counting that apply to other neuronal populations *(1)*.

MOTOR NEURON DEATH AFTER AXOTOMY IN THE NEONATAL PERIOD

The initial recognition that motor neurons in neonatal animals die after peripheral nerve injury was made over a century ago. However, cell death after axotomy was not systematically studied until 1946, when Romanes studied the effects of sciatic nerve transection in neonatal mice and found that the number of spinal motor neurons and ventral root axons rapidly declines after injury *(2)*. More recent investigations have confirmed this initial finding in neonatal lumbar spinal motor neurons, and have also

Table 1
Motor Neuron Growth Factors

Family	Factor	Receptor
Neurotrophin		
	BDNF	trkB
	NT-3	trkC
	NT-4	trkB
Neurocytokines		
	CNTF	CNTFRα, LIFR-B, gp130
	LIF	LIFR-β, gp130
	CT-1	LIFR-β, gp130, ? CT-1Rα
TGF-b		
	GDNF	GDNFR-α; c-Ret
	Neurturin	?
Insulin-like		
	IGF-I	IGF-1R
	IGF-II	

Because the survival promoting effects of bFGF and FGF-5 in vivo are controversial, these factors have not been included in this table (see refs. *16* and *72*).

extended it to cranial motor pools *(3–10)*. All of these studies agree with the observations of Schmalbruch, that when performed in the early postnatal period, axotomy results in the complete loss of motor neurons whose axons are transected by the procedure (Fig. 1A, B).

The profound loss of motor neurons following neonatal axotomy is most likely due to the fact that neurons during development are critically dependent for survival on target-derived trophic factors (*see* discussion of this point in the chapter by Burek and Oppenheim). Interruption of the axon leads to the deprivation of the critical target-derived substances and subsequent rapid neuronal death. In the 1950s, attempts by Hamburger and Levi-Montalcini to understand the nature of this peripheral influence on neuronal survival led to the purification of nerve growth factor (NGF) (*11*; for a review, *see* ref. *12*). However, NGF administration had only a minimal effect on motor neuron survival after neonatal axotomy, suggesting that motor neurons likely required different factor(s). In the past decade, multiple new classes of neurotrophic molecules have been identified, and several of these factors, indeed, have profound effects on motor neuron survival (*see* chapter by Koliatsos and Mocchetti).

GROWTH FACTORS PREVENT MOTOR NEURON DEATH AFTER NEONATAL AXOTOMY

Sendtner and colleagues were the first to demonstrate the usefulness of the neonatal axotomy paradigm in testing the ability of trophic factors to affect motor neuron survival in vivo *(4)*. Perhaps surprisingly, a large number of different trophic factors are capable of preventing motor neuron death after neonatal axotomy (Table 1). At least two members of the neurotrophin family of NGFs, brain-derived neurotrophic factor (BDNF) and neurotrophin-4 (NT-4), are powerful modulators of motor neuron survival during development and into adulthood (for a detailed review of neurotrophins, *see* ref. *13*). Application of BDNF or NT-4 to transected axons of postnatal day (PN) 1 facial or sciatic motor neurons (via gel foam pledglets soaked in factor) results in a survival rate of 50% when assayed at 1 wk postaxotomy *(5,6,14–18)*. Survival rates after BDNF administration improve to greater than 75% if axotomy is delayed until PN 2 or 3, although this finding likely reflects the exquisite age dependence effect on motor neuron survival *(see below)* as saline treated animals also demonstrate comparable improvements in relative motor neuron survival rates *(19,20)*.

That BDNF and NT-4 promote a similar degree of motor neuron survival postaxotomy is not unexpected, as both molecules produce effects via the same high-affinity trkB receptor *(21–23)*. Motor neurons robustly express trkB throughout the neonatal period and even into adulthood *(6,18)*, thus providing an explanation why motor neurons retain responsiveness to BDNF/NT-4 in maturity *(see below)*. Although neonatal motor neurons also express the NT-3 receptor trkC, only limited survival benefit from NT-3 administration after axotomy has been demonstrated *(6)*. These results suggest that NT-3 normally regulates pathways other than survival, at least for a majority of motor neurons. Interestingly, motor neurons also express the common neurotrophin receptor p75^{NGFR} transiently during development. NGF activation of p75^{NGFR} has been implicated in triggering apoptosis, which may explain why in some axotomy paradigms NGF administration has the paradoxical effect of decreasing motor neuron survival *(5,25)* *(see* also discussion of this point in chapter by Koliatsos and Mocchetti).

A member of the transforming growth factor-β superfamily, glial cell-line derived neurotrophic factor (GDNF), which initially had been identified as a survival promoting agent for dopaminergic neurons in culture *(26)*, promotes motor neuron survival in vitro, even at concentrations an order of magnitude lower than those required by the neurotrophins *(27)*. In vivo, administration of GDNF virtually completely prevented axotomy-induced death of motor neurons in the neonatal period in both mouse and rat with survival rates greater than 90% and approaching 100% *(7,9,27)*. Motor neurons are capable of retrogradely transporting GDNF back to the soma after its injection into muscle, indicating that motor neurons have specific receptor mediated uptake mechanisms for GDNF *(7)*. Recently, the functional receptor complex for GDNF has been identified as a GPI-linked α-GDNF subunit combined with the c-ret proto-oncogene *(28–31)*. *In situ* hybridization and protein studies demonstrate that both immature and adult spinal motor neurons express mRNA for GDNF receptors *(30)*. Another growth factor closely related to GDNF has recently been identified *(32)*, but whether this molecule, named neurturin, has similar effects on motor neuron survival after axotomy is unknown.

Various other classes of growth factors have also been shown to have significant survival promoting effects on motor neurons both in vitro and in vivo. Members of this interleukin-6 (IL-6) cytokine family include ciliary neurotrophic factor (CNTF), leukemia inhibitory factor (LIF), and the recently identified, cardiotrophin-1 (CT-1). CNTF, LIF, and CT-1 all promote a significant but incomplete motor neuron survival after neonatal facial or sciatic nerve transection *(4,16,20,33,34)*. Insulin-like growth factors (IGFs) I and II, acting via the IGF I receptor, are also capable of rescuing motor neurons from both naturally occurring and axotomy-induced death, although to a lessor extent than either BDNF/NT-4 or GDNF *(16,19,35)*.

The neonatal axotomy paradigm has revealed that motor neurons are capable of responding to a large and diverse group of growth factors in vivo. Which, if any, of these molecules is actually required for motor neuron survival remains unclear *(36)*. It is plausible that motor neuron properties including transmitter synthesis, glutamate receptor profiles, antioxidant enzyme expression, neurite outgrowth, or soma size might be preferentially regulated by distinct factors. For example, administration of GDNF, but not BDNF or CNTF, prevents the marked soma atrophy that follows neonatal axotomy *(6,7,9,27,37)*. This observation supports the view that differing motor neuron growth factors may regulate distinct aspects of motor neuron phenotype apart from survival.

Already the ability of certain growth factors to prevent motor neuron degeneration after axotomy has led to clinical trials of CNTF, BDNF, IGF-1, and GDNF in patients with ALS (*see* chapter by Hilt, Miller, and Malta). However, recent evidence suggests that the ability of growth factors promote motor neuron survival after axotomy diminishes over time. Continued administration of BDNF, CNTF, or CT-1 beyond the first week postaxotomy fails to prevent the subsequent degeneration of motor neurons in which 50% of motor neurons die between d 7 and 14 postaxotomy *(20,34)*. However, these experiments fail to address whether this continued motor neuron loss is due to a biologic phenomenon, such as the diminished ability of particular trophic factors to impact on survival pathways, or rather to technical considerations including impaired presentation/access of the growth factor to the motor neuron. This issue of long/term motor neuron rescue by trophic factors is of critical importance, if these agents are to be used rationally as treatments for human motor neuron disease.

Several lines of evidence suggest that motor neuron survival following neonatal axotomy may be dependent on the degree of motor neuron activity and afferent excitatory input. Administration of 2 mg/kg of the *N*-methyl-D-aspartate (NMDA) antagonist to neonatal rats markedly reduces the degree of spinal motor neuron death following a PN 0 sciatic nerve transection *(38,39)*. Conversely, administration of NMDA significantly increases the degree of motor neuron loss following a PN 7 axotomy *(40)*. Motor neurons robustly express NMDA receptors, which would allow the potential for NMDA-mediated excitotoxicity *(41)* in the setting of axotomy. The connection between axotomy-induced death by trophic factor deprivation and excitotoxicity is unclear (*see* ref. *42*). Currently, there is little data to indicate if motor neuron growth factors can modify gene expression of glutamate receptor profiles, calcium binding proteins, or other molecules potentially involved in excitotoxicity.

MOTOR NEURONS DIE BY APOPTOSIS AFTER NEONATAL AXOTOMY

Axotomized neonatal motor neurons undergo a stereotypical mode of death characteristic of apoptosis. Apoptotic death is characterized by certain morphological and biochemical features, including chromatin and cytoplasmic condensation, internucleosomal DNA fragmentation, apoptotic body formation, and eventual removal by phagocytic cells (for reviews, *see* refs. *43* and *44*) (*see* also chapter by Clarke in this volume). The temporal course of motor neuron death has recently been characterized in the rat facial nerve axotomy paradigm. When performed on PN 1, facial nerve transection at the level of the stylo-mastoid foramen does not lead to obvious morphological changes within facial motor nucleus during the first 24–48 h (*45*; Elliott and Snider, unpublished observations). Beginning at about 3 d after axotomy and peaking on d 4, fluores-

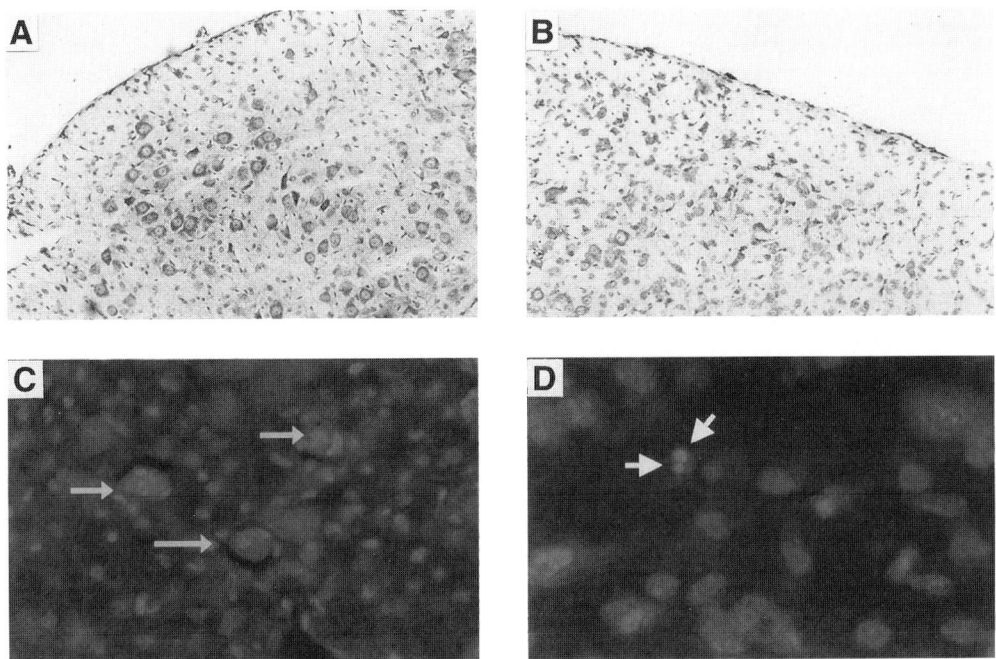

Fig. 1. Facial motor nucleus 1 wk after a PN 1 facial nerve transection in the rat contralateral (**A**) and ipsilateral (**B**) to the lesion. (**C,D**) Hoechst stain of facial motor neurons 4 d after a PN 1 facial nerve transection. Arrows indicate motor neurons with condensed and fragmented chromatin.

cent nuclear stains such as Hoechst highlight atrophic motor neurons with condensed and fragmented chromatin (Fig. 1C,D). Hoechst-positive neurons decline in number but still persist at d 5, commensurate with a decline in recognizable and countable motor neurons. By d 6–7, virtually all of the facial motor neurons have degenerated. Ultimately, the nucleus is replaced by glial and microglial elements (*46*; for a recent review, *see* ref. *47*). Only a small population of facial motor neurons remain, and these send axonal projections via the auricular branch of the nerve that was not subject to axotomy *(8)*.

Facial motor neurons degenerating after a PN 1 axotomy also express certain genes that have been associated with neuronal apoptosis in paradigms of growth factor deprivation in vitro. In a landmark study, Johnson and collaborators studied the death of sympathetic neurons in the setting of NGF deprivation in vitro, and determined that this particular death program requires the synthesis of new proteins to be successfully executed *(48)*. Using semiquantitative reverse transcriptase polymerase chain reaction (RT-PCR), they have subsequently identified messages for several transcription factors that are upregulated in dying superior cervical ganglion neurons, whereas levels of many other mRNAs declined *(49)*. Several of these genes are expressed at high levels in dying motor neurons. For example, 24 h after a PN 1 axotomy, virtually all facial motor neurons demonstrate a marked increase in *c-jun* expression, which then declines after 48 h. Nonaxotomized motor neurons do not normally express mRNA for *c-fos*. However, beginning at 48 h and peaking at 72–96 h

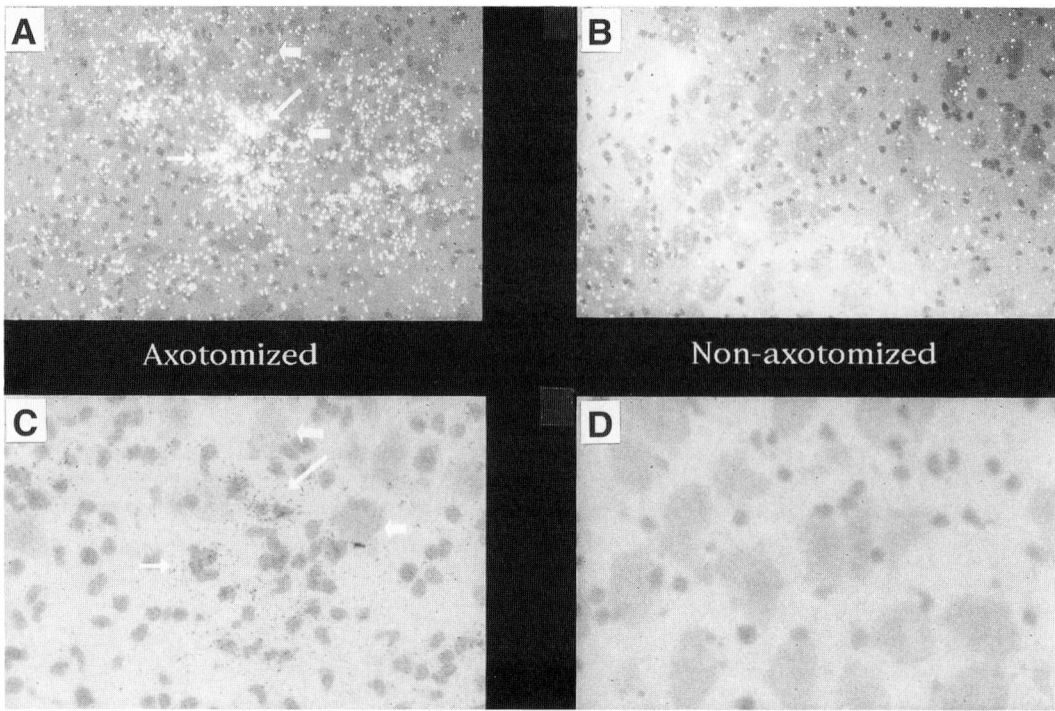

Fig. 2. Darkfield and brightfield views of c-fos mRNA expression in facial motor neurons 72 h after a PN 1 axotomy using a P^{33} oligonucleotide for *in situ* hybridization. Thin arrows indicate shrunken motor neurons with *c-fos* expression, whereas larger arrows highlight motor neurons without *c-fos* expression.

after a PN 1 axotomy, *c-fos* expression is upregulated. In contrast to *c-jun*, *c-fos* is expressed only in the subset of motor neurons that are shrunken (Fig. 2). After 96 h, the number of motor neurons expressing c-fos declines presumably due to death of the cells. For all other genes examined, including many bcl-2 and interleukin converting enzyme (ICE) homologs *(see below)*, levels of mRNA expression simply decline in motor neurons as they degenerate. These results indicate that motor neurons dying after axotomy in vivo exhibit a pattern of transcription factor expression that is remarkably similar to that described for sympathetic ganglion cells undergoing apoptosis in vitro.

Although *c-jun* and *c-fos* expression is strongly upregulated in dying motor neurons, the exact role of these genes in contributing to the cell death process is unclear. Because expression of these factors occurs in situations that do not result in cell death, the question arises whether expression of *c-jun* or *c-fos* is necessary but not sufficient for activating a cascade of events that ultimately lead to death. In support of this hypothesis, antibodies to *c-jun* or the entire *fos* family can block programmed cell death in sympathetic neurons in vitro, as can expression of a dominant *c-jun* negative *(49,50)*. However, lumbar motor neurons in neonatal transgenic mice lacking *c-fos* exhibit normal apoptotic death following sciatic nerve transection, indicating that programmed cell death can proceed without this specific transcription factors *(51)*.

GENETIC REGULATORS OF AXOTOMY-INDUCED MOTOR NEURON DEATH

Genetic approaches to the study of apoptosis have recently yielded remarkable insights into molecular mechanisms of motor neuron death after neonatal axotomy. Some of these studies have exploited the developmental pattern of cell death in the nematode, *C. elegans*, where the same 131 cells invariably die under a tightly genetically controlled program *(52)* (*see* chapter by Royal and Driscoll). Horvitz and coworkers using targeted mutational analysis and overexpression studies have been able to identify those genes that are critical in determining whether these cells live or die in *C. elegans*. They have identified a gene, *ced-9*, that acts as a death suppressor and a gene, *ced-3*, which is required for the successful loss of these 131 cells *(53)*.

Remarkably, the human proto-oncogene *bcl-2* first identified by Korsmeyer and collaborators was noted to have a strong sequence homology to the *ced-9* gene in *C. elegans* and thus potentially be an important regulator of cell survival in mammals *(53,54)*. In mammals, the *ced-9/bcl-2* gene family has evolved into a complex and large family of related molecules, including the death repressors *bcl-2*, *bcl-x_L*, or *mcl-1* and death promoters *bax*, *bcl-x_S*, *bad*, *bak*, or *bik* (for a review, *see* ref. 55). Proteins encoded by the previous genes share certain regions of high homology that are required for function as well as necessary for the selective formation of homo or heterodimers *(56,57)*. This observation has lead to the formulation of a "rheostat" model of apoptosis regulation, where relative levels of bax:bax homodimers or bcl-x or bcl-2:bax heterodimers determine cell progression along an apoptotic pathway *(58)*.

Certain members of the *ced-9/bcl-2* family can function as critical regulators of motor neuron survival following neonatal axotomy. *Bcl-2* is normally expressed in the motor neurons of neonatal wild type mice during their period of extreme vulnerability to axotomy-induced death *(59)*, indicating that endogenous levels of *bcl-2* are not sufficient to prevent this phenomenon. However, overexpression of *bcl-2* in transgenic mice is capable of inhibiting axotomy-induced death in both facial and spinal motor neurons *(8,60)*. The surviving motor neurons greatly atrophy, and it is unclear to what extent a functional motor neuron phenotype is preserved, although a recent report suggests that such motor neurons retain functional electrophysiologic properties *(61)*.

Targeted disruption of both *bcl-2* alleles leads to a small but significant loss of motor neurons in the postnatal period *(62)*. However, because *bcl-2* deficient animals are viable and exhibit only a mild neuronal phenotype it is likely that other *bcl-2* family members may actually play more pivotal roles in regulating apoptosis within the nervous system *(63)*. Neonatal motor neurons express a full repertoire of *bcl-2* family proteins including bcl-x, bax, bad, and bak *(64–66)*, and one or more of these molecules is likely to function as critical anti- or pro-apoptotic regulators within motor neurons. *Bcl-x* knockout mice are embryonic lethal with substantially increased telencephalic and dorsal root ganglia apoptosis at E 13 *(67)* demonstrating that *bcl-x* is a critical neuronal apoptosis regulator during early neural development. Preliminary evidence also suggests that *bcl-x* overexpression may prevent axotomy-induced death of facial motor neurons in the neonatal period *(68)*.

The essential role of the proapoptotic *bcl-2* family member, *bax*, in axotomy-induced motor neuron death has recently been demonstrated *(10)*. In the absence of *bax*, virtually all facial motor neurons survive a PN 1 facial nerve axotomy when quantified at

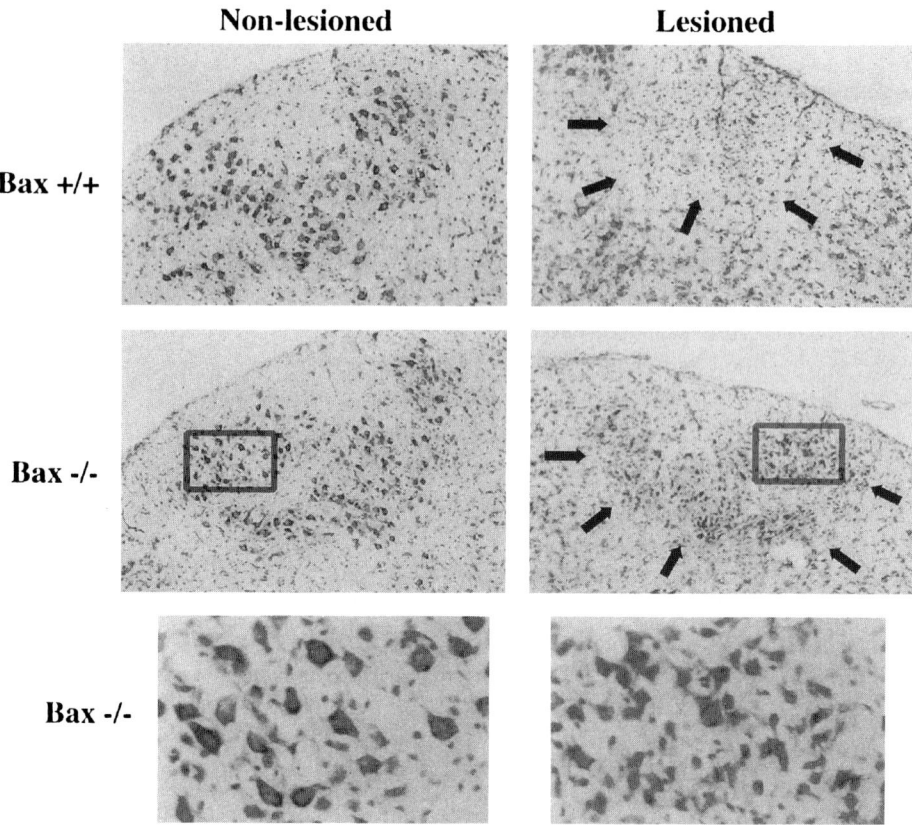

Fig. 3. Both lesioned and nonlesioned facial nucleus 1 wk after a PN 1 axotomy in *bax* wild type (+/+) and knockout (–/–) mice.

1 wk post-lesion (Fig. 3). Remarkably, motor neurons lacking *bax* survive a PN1 axotomy for extended time periods even up to 4 wk, the longest interval assessed. This prolonged motor neuron survival period in *bax* –/– mice contrasts with the limited temporal rescue afforded by trophic factor administration *(20,34)*. The motor neurons that survive axotomy in *bax* –/– mice do not appear normal, but show marked atrophy. This suggests that survival can be disassociated from other functions mediated by target-derived neurotrophic factors.

Interestingly, changes in mRNA levels for members of the *bcl-2* family in wild-type motor neurons dying after neonatal axotomy do not clearly correlate with predicted function. For *bcl-2, bcl-xl, bax, bad*, and *bak*, levels of mRNA gradually decline in degenerating axotomized motor neurons along a similar temporal profile (Elliott and Snider, unpublished observations). We did not observe increases in mRNA for apoptosis-promoting *bcl-2* family members, nor did we observe precipitous drops in mRNA levels for apoptotic inhibiting members of the *bcl-2* family in dying motor neurons. These results suggest that at least for the axotomy paradigm, *bcl-2* family function may not be regulated at the transcriptional level. This pattern may contrast with that observed in the cerebral ischemia model, where levels of bax protein appear to increase after vessel occlusion in the territories where death will occur *(64)*.

There exist at least ten known mammalian homologs for the *ced-3* gene, which is required for successful programmed cell death in the nematode *C. elegans* (for a review, *see* ref. *69*). Proteins encoded by these genes (including ICE, ICE rel-III, ICH-1/Nedd-2, ICH-2, CPP-32, Mch-2, Mch-3, and Flice) function as cysteine proteases with the ability to cleave proteins after specific aspartate residues, although the exact targets of these proteases as related to cell death are unclear. Peptide inhibitors of the ICE family proteases are capable of reducing naturally occurring cell death in chick embryonic motor neurons *(170)*, demonstrating that one or more ICE-like molecules are important effectors of apoptosis in motor neurons (for a comprehensive discussion of ICE family neurons, *see* chapter by Rosen and Casciola-Rosen). These results raise the possibility that axotomy-induced death in neonatal motor neurons may proceed via ICE-family mediated pathways, although which of the possible ICE-like molecules is important for motor neuron death is unknown.

AXOTOMY OF ADULT MOTOR NEURONS

Although immature motor neurons respond to axotomy by activating a programmed genetic pathway that leads to degeneration, motor neurons in adult animals normally respond to axotomy with a different phenotype, largely directed at axonal regeneration. Motor neuron survival following axotomy is critically dependent on the age of animal at the time of nerve transection. When the facial nerve is transected on PN 0, only 6% of facial motor neurons survive until PN 7, whereas if the axotomy is performed 1 d later, survival improves to about 12% *(5–7)*. Motor neuron survival rates following axotomy continue to increase through the first postnatal week reaching 24% when axotomy is performed on PN 2 *(8)*, 42% for a PN 7 axotomy *(71)*, and 67% for a PN 18 axotomy *(72)*. Survival rates for motor neurons following axotomy at PN 21, or beyond, plateau at levels comparable to those to found in adults, where the degree of motor neuron loss is generally small *(3,12,71)*. In contrast to the rapid loss of motor neurons that occurs over days following a neonatal axotomy, the small degree of adult motor neuron loss is gradual and evolves chronically over many weeks *(73)*. For spinal motor neurons, the decline in axotomy-induced motor neuron death is even more marked. Schmalbruch found that sciatic nerve transection performed in animals greater than 4 wk of age led to no significant change in the number of L4–L6 ventral root axons *(3)*.

The intriguing question as to what accounts for this resistance to axotomy-induced death at the molecular level remains unanswered. Certainly, the neurotrophic theory of neuronal development provides a conceptional framework for approaching this question, in that motor neurons may no longer be dependent on target-derived trophic factors for survival. However, the genetic mechanisms underlying this change during neuronal maturation remain unknown. Differences in expression/activity for certain bcl-2 and ICE family members, or other molecules related to programmed cell death, may be potential contributors in explaining this change in survival *(64)*, although current evidence does not clearly support such a role *(74)*.

After axotomy, the synthesis of certain molecules by motor neurons is markedly downregulated, whereas expression of other proteins is greatly increased. Table 2 summarizes many of these axotomy-induced changes *(75–93)*. An important point arising from these data is that motor neurons tend to upregulate the synthesis of proteins important for axonal regeneration such as cytoskeletal elements, growth associated proteins,

Table 2
Changes in Expression Following Axotomy in Adult Motor Neurons

Molecule	Change	Species	Ref.
Tα1-tubulin	Inc	M,R	75,76
IIβ-tubulin	Inc	M,R	76
Actin	Inc	R	76
GAP-43	Inc	M,R	76,77
Peripherin	Inc	R	78
Trk B (full length)	Inc	R	79,80
p75^{NGFR}	Inc	R	81,82
BDNF	Inc	R	79
MnSOD	Inc	R	102
NOS	Inc	R	83–85
c-jun	Inc	R	84
α-CGRP	Inc	R	77
βAPP	Inc	R	86
Galinin	Inc	R	87
Subs. P	Inc	R	87
VIP	Inc	R	87
Somatostatin	Inc	R	87
ChAT	Dec	R, M	77,82,88
MAPs	Dec	R	89,90
Neurofilaments	Dec	R	78,83
NMDAR1	Dec	R	91
NR2B	Dec	R	91
NR2D	Dec	R	91
CCK	Dec	R	77
Androgen receptor	Dec	R	92
α-subunit, I Na channel	Dec	R	93

and trophic factor receptors. In contrast, expression of proteins important for normal functioning such as neurotransmitter synthesis enzymes, glutamate receptors, or voltage-gated ion channels are significantly downregulated. The signals that regulate the activation or repression of motor neuron genes after axotomy are not clearly understood, although peripherally derived trophic factors are clear candidates.

The administration of appropriate growth factors can significantly alter the normal adult motor response to axotomy. Indeed, receptor expression patterns suggest that motor neurons remain responsive in maturity to the same molecules that promote survival during development *(6,30)*. Application of either BDNF, NT-4, and GDNF can attenuate the decline in ChAT activity that normally follows axotomy *(7,81,82;* but *see* ref. *37)*. In addition, BDNF and NT-4 are also capable of increasing the postaxotomy expression of p75^{NFGR} in adult motor neurons *(81,82)*.

The degree of motor neuron death following axotomy in adult animals can be markedly enhanced in a variety of settings. Ventral root avulsion combines a proximal axotomy with traction on the cell body and has been shown to cause a reproducible and massive loss of cervical and lumbar motor neurons in both adult rats and mice *(94–96)*.

Twenty percent of ipsilateral cervical motor neurons (at the appropriate segmental level) have degenerated by 7 d following a C7 avulsion, and this number increases to 70% by 3 wk *(79)*. Growth factors, most notably GDNF and BDNF, are capable of attenuating the severity of avulsion-induced motor neuron death by about 50% *(96,97)*. GDNF administration also prevents the considerable motor neuron soma atrophy that accompanies root avulsion *(96)*.

Nitric oxide synthase (NOS) has been implicated as one potential regulator of this phenomenon. After avulsion of their ventral roots, adult rat motor neurons upregulate expression of NOS with the concomitant increase in nitric oxide production *(84,97)*. Furthermore, inhibition of NOS reduces the degree of avulsion-induced motor neuron death *(94)*. Finally, BDNF administration, which promotes motor neuron survival after avulsion, also blocks expression of NOS in rat *(97)*. However, the exact importance of NOS in this paradigm remains unclear, as Oppenheim and collaborators did not observe a similar increase in NOS expression in murine motor neurons dying after avulsion *(96)*.

Motor neuron loss following axotomy in rats can be greatly accentuated using pharmacological manipulations of excitotoxic pathways. Brief treatment of adult rats with the NMDA antagonist MK-801 during the peri-axotomy period results in a 67% loss of facial motor neurons over an 8 wk time period *(73)*. This value compares to only a 19% loss in control animals. Semiquantitative *in situ* hybridization experiments demonstrated that the MK-801 administration also induced a highly significant upregulation of mRNA encoding NMDA R1 subunits. Thus, a combination of nerve injury in the setting of increased NMDAR1 receptor expression (and presumed activation) leads to a subacute death of motor neurons, suggesting that glutamate-induced excitotoxicity can lead to enhanced adult motor neuron death.

We and others have recently found that the degree of motor neuron loss following axotomy in adult mice is greater than has been reported in rats and appears to be highly strain dependent *(88,98*; Elliott and Snider, unpublished observations). For example, as early as 4 wk after facial axotomy in inbred strains such as balb/c, motor neuron loss has already reached 30% *(88*; *see* also ref. *98)*. In contrast, outbred strains such as CF-1 exhibit considerably less motor neuron loss (Elliott and Keller-Peck, unpublished observations). It should in theory be possible to map the genetic locus for susceptibility to axotomy-induced death as is being done for the ethanol susceptibility locus in the C57/bl6 strain *(99)*.

Motor neuron survival after axotomy in adult mice is also critically dependent on levels of the cytosolic antioxidant Cu/Zn superoxide dismutase (SOD1), for which mutations have been linked to one form of familial amyotrophic lateral sclerosis (FALS) *(100)*. Transgenic mice with targeted deletions of one or both SOD1 alleles have been generated and, remarkably, develop normal motor neuron number and function *(98)*. However, adult homozygous SOD1$^{-/-}$ mice exhibit a 66% increase in motor neuron loss following a facial nerve transection compared to wild-type littermates when evaluated at a 5-wk time point after the axotomy *(98)*. Interestingly, heterozygotes (SOD1$^{+/-}$) demonstrate an intermediate degree of axotomy-induced motor neuron loss. These experiments demonstrate that, although not essential during normal homeostasis, SOD1 activity is critical for motor neuron survival following the likely oxidative burdens generated by axotomy. In vitro experiments have shown that sympathetic ganglia neurons disconnected from their targets initially generate a transient burst of reactive oxygen species *(101)*. Whether reactive oxygen species are generated after axotomy, and what role, if any, they play in the death of motor neurons is unknown.

CONCLUSIONS

For over a century, axotomy has served as a useful tool for studying motor neuron death or regeneration. Neonatal peripheral nerve axotomy is now the preferred method for assessing responsiveness of motor neurons to growth factors in vivo. It now seems virtually certain that axotomy-induced motor neuron death in the neonatal period occurs by apoptosis. Again, the ease and convenience of the neonatal axotomy paradigm has been exploited in transgenic and pharmacological approaches to demonstrate powerful regulation of neuron survival by many of the bcl-2 and ICE homologs known to regulate apoptosis in nonneuronal systems. Whether these molecular regulators of neuronal apoptosis can influence the course of motor neuron degeneration in animal models of human motor neuron diseases is now under active investigation.

The implications of motor neuron axotomy in adult animals for studies of motor neuron disease remain less clear. Under certain conditions, adult axotomy can lead to a subacute or chronic degeneration of motor neurons. Although the molecular mechanisms of this degeneration have not been clarified, a demonstration of "chronic" motor neuron apoptosis would have implications for theories about how motor neurons die in ALS. Importantly, the extent of adult motor neuron loss after axotomy appears to be remarkably strain dependent in mice. It should be straightforward to localize and ultimately clone genes related to this differential susceptibility. Such genes would certainly be candidates for future investigations related to pathogenesis or therapy of ALS.

REFERENCES

1. Coggeshall RE, Lekan HA. Methods for determining numbers of cells and synapses: a case for more uniform standards of review. *J Comp Neurol* 1996, **364**: 6–15.
2. Romanes GJ. Motor localization and the effects of nerve injury on the ventral horn cells of the spinal cord. *J Anat* 1946, **42**: 117–131.
3. Schmalbruch H. Motoneuron death after sciatic nerve section in newborn rats. *J Comp Neurol* 1984, **224**: 252–258.
4. Sendtner M, Kreutzberg GW, Thoenen H. Ciliary neurotrophic factor prevents the degeneration of motor neurons after axotomy. *Nature* 1990, **345**: 440–441.
5. Sendtner M, Holtmann B, Kolbeck R, et al. Brain derived neurotrophic factor prevents the death of motoneurons in newborn rats after nerve transection. *Nature* 1992, **360**: 757–759.
6. Yan Q, Elliott JL, Matheson C, et al. Influences of neurotrophins on mammalian motoneurons in vivo. *J Neurobiol* 1993, **24**: 1555–1577.
7. Yan Q, Matheson C, Lopez OT. In vivo neurotrophic effects of GDNF on neonatal and adult facial motor neurons. *Nature* 1995, **373**: 341–344.
8. Dubois-Dauphin M, Frankowski H, Tsujimoto Y, et al. Neonatal motoneurons overexpressing the bcl-2 protooncogene in transgenic mice are protected from axotomy-induced cell death. *Proc Natl Acad Sci USA* 1994, **91**: 3309–3313.
9. Oppenheim RW, Houenou LJ, Johnson JE, Lin LF, Li L, et al. Developing motor neurons rescued from programmed and axotomy-induced cell death by GDNF. *Nature* 1995, **373**: 344–346.
10. Deckwerth TL, Elliott JL, Knudson CM, et al. Bax is required for neuronal death after trophic factor deprivation and during development. *Neuron* 1996, **17**: 401–411.
11. Cohen S, Levi-Montalcini R, Hamburger V. A nerve growth stimulating factor isolated from sarcomas 37 and 180. *Proc Natl Acad Sci USA* 1954, **40**: 1014–1018.

12. Snider WD, Elliott JL, Yan Q. Axotomy-induced neuronal death during development. *J Neurobiol* 1992, **23:** 1231–1246.
13. Lewin GR, Barde YA. Physiology of the neurotrophins. *Ann Rev Neurosci* 1996, **19:** 289–317.
14. Oppenheim RW, Yin QW, Prevette D, Yan Q. Brain-derived neurotrophic factor rescues developing avian motoneurons from axotomy-induced death. *Nature* 1992, **360:** 755–757.
15. Yan Q, Elliott J, Snider WD. Brain-derived neurotrophic factor rescues spinal motor neurons from axotomy-induced cell death. *Nature* 1992, **360:** 755–777.
16. Hughes RA, Sendtner M, Thoenen H. Members of several gene families influence survival of rat motoneurons in vitro and in vivo. *J Neurosci Res* 1993, **36:** 663–671.
17. Koliatsos VE, Clatterbuck RE, Winslow JW, et al. Evidence that brain-derived neurotrophic factor is a trophic factor for motor neurons in vivo. *Neuron* 1993, **10:** 359–367.
18. Koliatsos VE, Cayouette MH, Berkemeier LR, Clatterbuck et al. Neurotrophin 4/5 is a trophic factor for mammalian facial motor neurons. *Proc Natl Acad Sci USA* 1994, **91:** 3304–3308.
19. Li L, Oppenheim RW, Lei M, Houenou J. Neurotrophic agents prevent motoneuron death following sciatic nerve section in the neonatal mouse. *J Neurobiol* 1994, **25:** 759–766.
20. Vejsada R, Sagot Y, Kato AC. Quantitative comparison of the transient rescue effects of neurotrophic factors on axotomized motoneurons in vivo. *Eur J Neurosci* 1995, **7:** 108–115.
21. Klein R, Nanduri V, Jing S, et al. The trkB tyrosine kinase is a receptor for brain derived neurotrophic factor and neurotrophin-3. *Cell* 1991, **66:** 395–403.
22. Klein R, Lamballe F, Bryant S, Barbacid M. The trkB typrosine protein kinase is a receptor for neurotrophin-4. *Neuron* 1992, **8:** 947–956
23. Soppet D, Escandon E, Maragos J, et al. The neurotrophic factors brain-derived neurotrophic factor and neurotrophin-3 are ligands for the trkB tyrosine kinase receptor. *Cell* 1991, **65:** 895–903.
24. Merlio JP, Ernfors P, Jaber M, Persson H. Molecular cloning of rat trkC and distribution of cells expressing mRNAs for members of the trk family in the rat central nervous system. *Neuroscience* 1992, **51:** 513–532.
25. Frade JM, Rodrieguez-Tebor A, and Barde YA. Induction of cell death by endogenous nerve growth factor through its p75 receptor. *Nature* 1996, **383:** 166–168.
26. Lin FH, Doherty D, Lile J, et al. GDNF: a glial cell line derived neurotrophic factor for midbrain dopaminergic neurons. *Science* 1993, **260:** 1130–1132.
27. Henderson CE, Phillips HS, Pollock RA, et al. GDNF: a potent survival factor for motoneurons present in peripheral nerve and muscle. *Science* 1994, **266:** 1062–1064.
28. Durbec P, Marcos-Gutierrez CV, Kilkenny C, Grigoriou M, Wartiowaara K, et al. GDNF signalling through the Ret receptor tyrosine kinase. *Nature* 1996, **381:** 384.
29. Jing S, Wen D, Yu Y, Holst PL, Luo L, et al. GDNF-induced activation of the ret protein tyrosine kinase is mediated by GDNFR-α, a novel receptor for GDNF. *Cell* 1996, **85:** 1113–1124.
30. Treanor JJS, Goodman L, deSauvage F, Stone DM, Poulsen KT. Characterization of a multicomponent receptor for GDNF. *Nature* 1996, **382:** 80–83.
31. Trupp M, Arenas E, Fainzilber M, Nilson AS, Sieber BA, et al. Functional receptor for GDNF encoded by the c-ret proto-oncogene. *Nature* 1996, **381:** 785–789.
32. Kotzbauer PT, Lampe PA, Heuckeroth RO, et al. Neurturin, a relative of glial cell-line-derived neurotrophic factor. *Nature* 1996, **384:** 467–470.
33. Pennica D, Shaw KJ, Swanson TA, et al. Cardiotrophin-1: Biological activities and binding to the leukemia inhibitory factor receptor/gp130 signaling complex. *J Biol Chem* 1995, **270:** 10,915–10,922.
34. Pennica D, Arse V, Swanson TA, et al. Cardiotrophin-1, a cytokine present in embryonic muscle, supports long term survival of spinal motor neurons. *Neuron* 1996, **17:** 63–74.
35. Houenou LJ, Li L, Lo AC, et al. Naturally occurring and axotomy induced motor neuron death and its prevention by neurotrophic agents: a comparison between chick and mouse. *Prog Brain Res* 1994, **102:** 217–225.

36. Elliott JL and Snider WD. Motor neuron growth factors. Neurology 1996, **47(Suppl 2):** S47–S53.
37. Clatterbuck RE, Price DL, Koliatsos VE. Further characterization of the effects of brain-derived neurotrophic factor and ciliary neurotrophic on axotomized neonatal and adult mammalian motor neurons. *J Comp Neurol* 1994, **342:** 45–56.
38. Mentis GZ, Greensmith L, Vrbova G. Motoneurons destined to die are rescued by blocking n-methyl-D-aspartate receptors by MK-801. *Neuroscience* 1993, **54:** 283–285.
39. Dick J, Greensmith L, Vrbova G. Blocking of NMDA receptors during a critical stage of development reduces the effects of nerve injury at birth on muscles and motoneurons. *Neromusc Dis* 1995, **5:** 371–382.
40. Greensmith L, Hasan HI, Vrbova G. Nerve injury increases the susceptibility of motoneurons to *n*-methyl-D-aspartate induced neurotoxicty in the developing rat. *Neuroscience* 1994, **58:** 727–733.
41. Tolle TR, Berthelle A, Zieglgansberger W, et al. The differential expression of 16 NMDA and non-NMDA receptor subunits in the rat spinal cord and periaqueductal grey. *J Neurosci* 1993, **13:** 5009–5028.
42. Lowrie MB, Vrbova G. Dependence of post natal motoneurons on their targets: review and a hypothesis. *Trends Neurosci* 1992, **15:** 80–84.
43. Kerr JFR. Neglected opportunities in apoptosis research. *Trends Cell Biol* 1995, **5:** 55–57.
44. Bortner CD, Oldenburg NBE, Cidlowski JA. The role of DNA fragmentation in apoptosis. *Trends Cell Biol* 1995, **5:** 21–26.
45. Gerfen RW, Elliott JL, Johnson EJ, Snider WD. Genes associated with apoptosis are expressed by dying motoneurons after neonatal axotomy. *Soc Neurosci Abs* 1995, **21:** p. 2019.
46. Graeber MB, Kreutzberg GW. Delayed astrocyte reaction following facial nerve axotomy. *J Neurocytol* 1988, **17:** 209–220.
47. Moore S, Thanos S. The concept of microglia in relation to central nervous system disease and regeneration. *Prog Neurobiol* 1996, **48:** 441–460.
48. Martin DP, Schmidt RE, DiStefano DI, et al. Inhibitors of protein and RNA synthesis prevent neuronal death caused by nerve growth factor deprivation. *J Cell Biol* 1988, **106:** 829–844.
49. Estus S, Zaks WJ, Freeman RS, et al. Altered gene expression in neurons during programmed cell death; identification of c-jun as necessary for neuronal apoptosis. *J Cell Biol* 1994, **127:** 1717–1727.
50. Ham J, Babij C, Whitfield J, et al. A c-jun dominant negative mutant protects sympathetic neurons against programmed cell death. *Neuron* 1995, **14:** 927–939.
51. Roffler-Tarlov S, Gibson-Brown JJ, Tarlov E, Stolarov J, Chapman DL, et al. Programmed cell death in the absence of c-Fos and C-Jun. *Development* 1996, **122:** 1–9.
52. Ellis RE, Horvitz HR. Genetic control of programmed cell death in the nematode C. elegans. *Cell* 1986, **44:** 817–829.
53. Hengartner MO, Horvitz HR. C-elegans cell survival gene Ced-9 encodes a functional homolog of the mammalian proto-oncogene Bcl-2. *Cell* 1994, **76:** 665–676.
54. Bakhshi A, Jensen JP, Goldman P. et al. Cloning the chromosomal breakpoint of t(14;18) human lymphomas: clustering around JH on chromosome 14 and near a transcriptional unit on 18. *Cell* 1985, **41:** 889–906.
55. Korsmeyer SJ. Regulators of cell death. *Trends Genet* 1995, **11:** 101–105.
56. Oltvai ZN, Milliman CL, Korsmeyer SJ. Bcl-2 heterodimerizes in vivo with a conserved homolog, Bax, that accelerates programmed cell death. *Cell* 1993, **74:** 609–619.
57. Sedlak TW, Oltvai ZN, Yang E, Wang K, Boise LH, Thompson CB, Korsmeyer SJ. Multiple Bcl-2 family members demonstrate selective dimerizations with Bax. *Proc Natl Acad Sci USA* 1995, **92:** 7834–7838.
58. Oltvai ZN, Korsmeyer SJ. Checkpoints of dueling dimers foil death wishes. *Cell* 1994, **79:** 189–192.

59. Merry DE, Veis DJ, Hickey WF, Korsmeyer SJ. Bcl-2 protein expression is widespread in the developing nervous system and retained in the adult PNS. *Development* 1994, **120**: 301–311.
60. Farlie PG, Dringen R, Rees SM, Kannourakis G, Benard O. bcl-2 transgene expression can protect neurons against developmental and induced cell death. *Proc Natl Acad Sci USA* 1995, **92**: 4397–4401.
61. Alberi S, Raggenbass M, de Bilbao F, Dubois-Dauphain, M. Axotomized neonatal motoneurons overexpressing the bcl-2 proto-oncogene retain functional electrophysiological properties. *Proc Natl Acad Sci USA* 1996, **93**: 3978–3983.
62. Michaelidis TM, Sendter M, Cooper JD, et al. Inactivation of bcl-2 results in progressive degeneration of motoneurons, sympathetic and sensory neurons during early postnatal development. *Neuron* 1996, **17**: 75–89.
63. Veis DJ, Sorenson CM, Shutter JR, Korsmeyer SJ Bcl-2 deficient mice demonstrate fulminant lymphoid apoptosis, polycystic kidney disease and hypopigmented hair. *Cell* 1993, **75**: 229–240.
64. Krajewski S, Krajewska M, Shabaik A, et al. Immunohistochemical determination of in vivo distribution of Bax, a dominant inhibitor of Bcl-2. *Am J Pathol* 1994, **145**: 1323–1336.
65. Frankowski H, Misotten M, Fernandez P-A, et al. Function and expression of the Bcl-x gene in the developing and adult nervous system. *NeuroReport* 1995, **6**: 1917–1921.
66. Parsadanian A-Sh, Elliott JL, Snider WD Multiple ced-3 and ced-9 homologues are expressed in the murine nervous system. *Soc Neurosci Abs* 1995, **21**: p. 2019.
67. Motoyama N, Wang F, Roth KA, et al. Massive cell death of immature hematopoietic cells and neurons in bcl-x deficient mice. *Science* 1995, **267**: 1506–1510.
68. Parsadanian A-Sh, Keller-Peck C, Elliott JL, Snider WD. bcl-xl overexpression in transgenic mice prevents facial motor neurons against axotomy induced death. *Soc Neurosci Abs* 1996, **22**: p. 570.
69. Henkart PA. ICE family proteases: mediators of all apoptotic cell death. *Immunity* 1996, **4**: 195–201.
70. Milligan CE, Prevette D, Yaginuma H, et al. Peptide inhibitors of the ICE protease family arrest programmed cell death of motoneurons in vivo and in vitro. *Neuron* 1995, **15**: 385–393.
71. Snider WD, Thanedar S. Target deprivation of hypoglossal motoneurons during development and during maturity. *J Comp Neurol* 1989, **279**: 489–498.
72. Grothe C, Unsicker K. Basic fibroblast growth factor in the hypoglossal system; specific retrograde transport, trophic, and lesion related responses. *J Neurosci Res* 1992, **32**: 317–328.
73. Sanner C, Elliott JL, Snider WD. Upregulation of NMDAR1 mRNA induced by MK-801 is associated with massive death of axotomized motor neurons in adult rats. *Neurobiol Dis* 1994, **1**: 121–129.
74. Greenlund LJS, Korsmeyer SJ, Johnson EM Jr. Role of bcl-2 in the survival and function of developing and mature sensory neurons. *Neuron* 1995b, **15**: 649–661.
75. Gloster A, Wu W, Speelman A, Weiss S, et al. The Tα1-tubulin promoter specifies gene expression as a function of neuronal growth and regeneration in transgenic mice. *J Neurosci* 1994, **14**: 7319–7327.
76. Tetzlaff W, Alexander SW, Miller FD, Bisbey MA. Responses of facial and rubrospinal neurons to axotomy: changes in mRNA expression for cytoskeletal and GAP-43. *J Neurosci* 1991, **11**: 2528–2544.
77. Piehl F, Arvidsson U, Johnson H, Cullheim S, Dagerlind A, et al. GAP-43, aFGF, CCK, α and β CGRP in rat spinal motoneuons subjected to axotomy and/or dorsal root severance. *Eur J Neurosci* 1993, **5**: 1321–1333.
78. Troy CM, Muma NA, Greene LA, et al. Regulation of peripherin and neurofilament expression in regenerating rat motor neurons. *Brain Res* 1990, **529**: 232–238.
79. Kobayashi NR, Bedard AM, Hincke MT, Tetzlaff W. Increased expression of BDNF and trkB mRNA in rat facial motoneurons after axotomy. *Eur J Neurosci* 1996, **8**: 1018–1029.

80. Piehl F, Frisen J, Risling M, Hofelt T, Cullheim S. Increased trkB mRNA expression by axotomized motoneurones. *Neuroreport* 1994, **5:** 697–700.
81. Friedman B, Kleinfeld D, Ip NY, et al. BDNF and NT-4/5 exert neurotrophic influences on injured adult spinal motor neurons. *J Neurosci* 1995, **15:** 1044–1056.
82. Yan Q, Matheson C, Lopez OT, Miller JA. The biological responses of axotomized adult motoneurons to brain-derived neurotrophic factor. *J Neurosci* 1994, **14:** 5281–5291.
83. Muma NA, Hoffman PN, Slunt HH, et al. Alterations in levels of mRNA coding for neurofilament protein subunits during regeneration. *Exp Neurol* 1990, **107:** 230–235.
84. Wu W, Li Y, Schinco FP. Expression of c-jun and neuronal nitric oxide synthase in rat spinal motoneurons following axonal injury. *Neurosci Letts* 1994, **179:** 157–161.
85. Yu WHY. Nitric oxide synthase in motor neurons after axotomy. *J Histo Chem* 1994, **42:** 451–457
86. Sola C. Garcia-Ladona F, Sarasa M, Mengod G, Probst A, et al. BAPP gene expression is increased in the rat brain after motor neuron axotomy. *Eur J Neurosci* 1993, **5:** 795–808.
87. Zhang X, Verge VMK, Wisenfeld-Hallin Z, Piehl F, Hokfelt T. Expression of neuropeptides mRNAs in spinal cord after axotomy in the rat with a special reference to motoneurons and galanin. *Exp Brain Res* 1993, **93:** 450–461.
88. Kou SY, Chiu AY, Patterson PH. Differential regulation of motor neuron survival and choline acetyltransferase expression following axotomy. *J Neurobiol* 1995, **27:** 561–572.
89. Svensson M, Aldskogius H. The effect of axon injury on microtubule associated protein MAP2 mRNA in the hypoglossal nucleus of the adult rat. *Brain Res* 1992, **581:** 319–322.
90. Svensson M, Aldskogius H. The effect of axonal injury on microtubule associated proteins MAP2, 3 and 5 in the hypoglossal nucleus of the adult rat. *J Neurocytol* 1992, **21:** 222–231.
91. Piehl F, Tabar G, Cullheim S. Expression of NMDA receptor mRNAs in rat motoneurons is down regulated after axotomy. *Eur J Neurosci* 1995, **7:** 2101–2110.
92. Lubischer JL, Arnold AP. Axotomy transiently regulates androgen receptors in motor neurons of the spinal nucleus of the bulbocavernosus. *Brain Res* 1995, **694:** 61–68.
93. Iwahashi Y, Furuyama T, Ingaki S, Morita Y, Takagi H. Distinct regulation of sodium channel types I, II, and III following nerve transection. *Mol Brain Res* 1994, **22:** 341–345.
94. Wu W, Li L. Inhibition of nitric oxide synthase reduces motoneuron death due to spinal root avulsion. *Neurosci Lett* 1993, **153:** 121–124.
95. Koliatsos VE, Price WI, Pardo CA, Price DL. Ventral root avulsion: an experimental model of death of adult motor neurons. *J Comp Neurol* 1994, **342:** 35–44.
96. Li L, Wu W, Lin L-FH, Lei M, Oppenheim RW, Houenou LJ. Rescue to adult mouse motoneurons from injury-induced cell death by glial cell line-derived neurotrophic factor. *Proc Natl Acad Sci USA* 1995, **92:** 9771–9775.
97. Novikov L, Novikova L, Kellerth JO. Brain derived neurotrophic factor promotes survival and blocks nitric oxide synthase expression in adult rat spinal motoneurons after ventral root avulsion. *Neurosci Letts* 1995, **200:** 445–448.
98. Reaume AG, Elliott JL, Hoffman EK, Kowall NK, Ferrante RJ, et al. Motor neurons in Cu/Zn superoxide dismutase deficient mice develop normally but exhibit enhanced cell death after axonal injury. *Nat Genet* 1996, **13:** 43–47.
99. Melo JA, Shendure J, Pociask K, Silver LM. Identification of sex specific quantitative trait loci controlling alcohol preference in C57BL/6 mice. *Nature Genet* 1996, **13:** 147–153.
100. Rosen DR, Siddüqe T, Patterson D, et al. Mutations in the Cu/Zn superoxide dismutase gene are associated with familial amyotrophic lateral sclerosis. *Nature* 1993, **362:** 59–62.
101. Greenlund LJS, Deckwerth TL, Johnson EM Jr. Superoxide dismutase delays neuronal apoptosis; a role for reactive oxygen species in programmed cell death. *Neuron* 1995a, **14:** 303–315.
102. Yoneda T, Inagaki S, Hayashi Y, et al. Differential regulation of manganese and copper/zinc superoxide dismutase by the facial nerve transection. *Brain Res* 1992, **582:** 342–345.

10
Excitotoxic Cell Death

John W. Olney and Masahiko J. Ishimaru

INTRODUCTION

In recent years, the excitatory amino acids (EAAs), glutamate (Glu) and aspartate (Asp), have become recognized as the Jekyll/Hyde molecules of the central nervous system (CNS). These common acidic amino acids, which are naturally present in higher concentrations than any other amino acids in the CNS, serve vitally important metabolic, neurotrophic and neurotransmitter roles, but also harbor treacherous neurotoxic (excitotoxic) potential. Recent advances in understanding the excitotoxic properties of these compounds include the identification of several receptor subtypes that mediate Glu/Asp excitotoxicity, the generation of evidence potentially linking excitotoxins of both exogenous and endogenous origin to both acute and chronic neurodegenerative disorders and the development of antiexcitotoxic drugs for protecting against such disorders. Currently, in addition to classical excitotoxicity resulting from hyperactivation of EAA ionotropic receptors, new information is beginning to emerge pertaining to other forms of excitatory transmitter neurotoxicity. Surprisingly, some of the new forms are triggered not by hyperactivation but by hypoactivation of EAA receptors. Over the past several years we have witnessed an explosion in the molecular biology of Glu receptors; over 20 receptor subunits or subtypes having been cloned and sequenced within the past five years. In addition, several Glu transporter receptors have recently been cloned and are currently being studied for their potential role in neurodegenerative diseases. These important new developments are certain to accelerate progress in understanding the relationship between Glu receptor systems and both physiological and pathological processes in the mammalian CNS.

ORIGINS OF THE EXCITOTOXIC CONCEPT

It has only been within the past decade that Glu has become widely accepted as a major neurotransmitter in the vertebrate CNS, and only over that same period that the neurotoxic properties of Glu have generated significant interest among neuroscientists. In the preceding decades, it was shown *(1)* that Glu and several of its structural analogs depolarize (excite) CNS neurons, and that systemic administration of Glu to animals of various species causes acute degeneration of neurons in the retina *(2)* or several regions of brain that lack blood–brain barriers *(3,4)*. Only Glu analogs possessing the

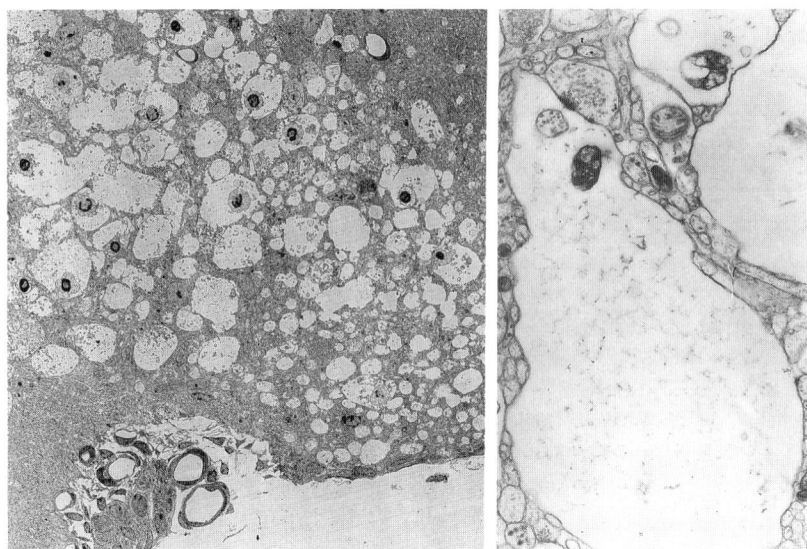

Fig. 1. On the left is an electron microscopic overview (×900) of the arcuate nucleus of the hypothalamus of an infant mouse 4 h after oral administration of glutamate. The cytopathological reaction is fulminatingly acute and consists of massive edematous swelling of neuronal cell bodies and dendrites, accompanied by clumping and pyknotic changes affecting nuclear chromatin. On the right is a magnified (×9000) view of a swollen dendrite that contains degenerating organelles and particulate debris. An axon terminal making synaptic contact with this dendrite has a normal appearance signifying that the degenerative reaction is confined to postsynaptic dendrosomal structures. Acute edematous degeneration of postsynaptic dendrosomal structures with sparing of presynaptic axons is a hallmark characteristic of classical excitotoxic neurodegeneration. (Portions of this illustration were adapted with permission from ref. 4.)

neuroexcitatory properties of Glu reproduced its neurotoxic effects, and these analogs displayed a parallel order of potencies for their excitatory and toxic actions (5). Moreover, ultrastructural studies (4) localized the apparent site of toxic action to postsynaptic dendrosomal membranes known to be responsive to the excitatory actions of Glu (Fig. 1). From these and related observations the excitotoxic concept was proposed (5,6), which holds that an excitatory mechanism and EAA synaptic receptors mediate Glu neurotoxicity.

Using pharmacological and electrophysiological methods, Watkins and colleagues identified three different EAA receptor subtypes that were differentially sensitive to specific agonists (NMDA [N-methyl-D-aspartate], AMPA [amino-3-hydroxy-5-methylisoxazole-4-propionic acid], and KA [kainic acid]), and they discovered antagonists that block the excitatory actions of EAA agonists at such receptors (7). Availability of specific antagonists permitted the excitotoxic hypothesis to be rigorously tested and confirmed. In these experiments (8) the first EAA antagonists that had been identified electrophysiologically (α-amino adipate, D-2-amino-5-phosphonovalerate), were found to protect neurons in the in vivo mouse hypothalamus against the neurotoxic actions of Glu or its more potent analog, NMDA. By in vitro methods (9,10), many EAA antagonist candidates were subsequently screened systematically and found to have antiexcitotoxic

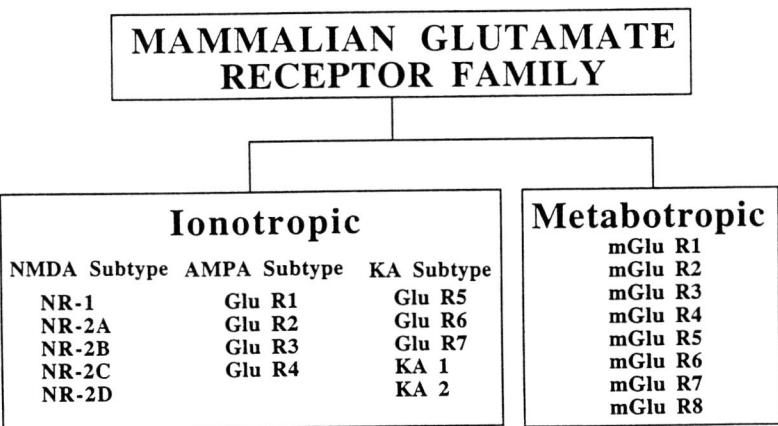

Fig. 2. This chart shows the major members of the mammalian glutamate receptor family that have been identified thus far. Within the ionotropic subfamily there are three subtypes, each of which has a number of subunits that can combine to form heteromeric assemblies. Within the metabotropic subfamily there are eight subtypes, each of which comprises a functional receptor by itself.

activities corresponding in potency and receptor specificity to their known antiexcitatory activities.

EXCITATORY AMINO ACID RECEPTOR SYSTEMS

Molecular biology techniques have recently revealed that the glutamate receptor family is comprised of two major subfamilies, ionotropic and metabotropic, and within these subfamilies there are additional categories and subcategories (Fig. 2). By the current classification scheme, the ionotropic receptor family continues to be subdivided into the three major categories (NMDA, AMPA, and KA) originally identified by Watkins et al. (7), but within each of these categories, multiple subunits have now been identified. It is believed that these subunits form heteromeric (probably pentomeric) assemblies, the exact number and types of which remain to be determined. Since the functional properties of a given ionophore assembly varies as a function of its subunit composition, nature appears to have provided for a large variety of functionally distinct ionophore complexes through which glutamate can mediate its myriad physiological functions and pathological misdeeds (11–13). Within the metabotropic family at least eight receptor subtypes have been identified, some of which achieve signal transduction through G protein coupling to a phosphoinositide second messenger system and others to an adenylate cyclase system (14). Neuroscientists are still identifying additional glutamate receptor subtypes and splice variants within these established subfamilies, and they are just embarking on the arduous task of characterizing each receptor subtype or variant in terms of its pharmacological and electrophysiological properties and in terms of the period in ontogeny when it may be most robustly expressed (15). In addition, molecular geneticists have begun the search for abnormalities in the human genome that could help explain neuropsychiatric disorders in terms of a disturbance in the makeup, expression or function of a particular glutamate receptor subtype (16,17).

Each of the major categories of glutamate receptors (NMDA, AMPA, KA, and metabotropic) is capable of causing neurotoxic effects culminating in nerve cell death. For two of these receptor systems (NMDA and metabotropic) either increased or decreased activity can have neurotoxic consequences. For every type of neurotoxicity associated with ionotropic receptors, an excitotoxic mechanism has been implicated. Whether an excitotoxic mechanism also mediates metabotropic receptor-linked neurotoxicity remains to be determined.

EAA RECEPTOR-LINKED NEUROTOXICITY SYNDROMES
Classical Excitotoxicity (Ionotropic Receptor Hyperactivity)

By "classical excitotoxicity" we mean the paradoxical property, shared by Glu and specific excitatory amino acid (EAA) analogs, of causing acute neuronal degeneration by excessive stimulation of postsynaptic EAA ionotropic receptors—ion channel-linked receptors through which Glu functions physiologically as a transmitter. It is well established that each of the three major ionotropic receptor subtypes (NMDA, AMPA, and KA) is capable of mediating classical excitotoxic reactions. It is believed that most of the neurons in the CNS possess one or more subtype of EAA ionotropic receptor on their dendritic or somal surfaces, which makes them vulnerable to excitotoxic degeneration, and which explains why the degenerative process conspicuously involves dendrosomal but not axonal portions of the neuron.

The NMDA receptor (depicted schematically in Fig. 3) has been studied more intensively than the other two ionotropic receptor subtypes and more is known about the factors that govern its operation. However, at the present time, a great deal remains unknown. For example, it is often maintained that Ca^{2+} influx or intracellular Ca^{2+} mobilization may be the single most important factor responsible for triggering the cascade of pathophysiological events leading to excitotoxic cell death (for a topical discussion of this subject, see chapters by Leist and Nicotera). However, in ion substitution experiments conducted in the chick retina, omission of Na^+ or Cl^- in the extracellular medium prevents excitotoxic neurodegeneration induced by NMDA, AMPA, or KA but omission of Ca^{2+} is not protective; in fact, it triggers excitotoxic neurodegeneration. Surprisingly, excitotoxic neurodegeneration triggered by Ca^{2+} omission is mediated by NMDA receptors and can be totally blocked by the powerful NMDA antagonist, MK-801 (18). Moreover, in many regions of the in vivo CNS and in some CNS in vitro preparations non-NMDA receptors are more efficient mediators of excitotoxic cell death than are NMDA receptors, despite the fact that most non-NMDA receptors have a substantially lower Ca^{2+} conductance than NMDA receptors. Thus, the role of Ca^{2+} in classical excitotoxicity remains unclear at the present time. A role for Na^+ as a primary triggering factor seems very likely since each of the ionotropic receptor subtypes has a high conductance for Na^+, and Na^+ omission blocks excitotoxicity mediated through each receptor subtype. Since preventing Cl^- influx blocks excitotoxic neurodegeneration through each Glu receptor subtype, Cl^- must also be considered an important factor, but one which has received very little attention. For example, the mechanism(s) and conduit(s) that mediate abnormal Cl^- uptake accompanying excitotoxic neurodegeneration remain largely unknown. Other potentially important factors that are thought to play a downsteam role include arachidonic acid, nitric oxide, oxidative free radicals, proteases, lipases and, eventually, nucleases, but it is not well understood precisely how these factors participate in the cascade.

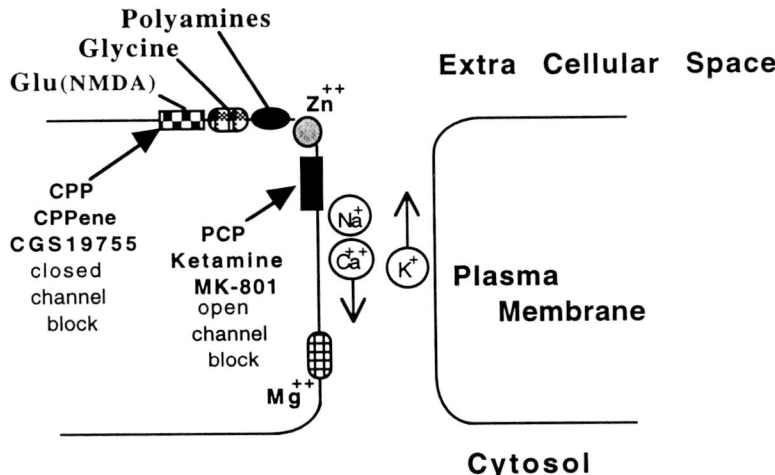

Fig. 3. A schematic depiction of the various components comprising the NMDA receptor-ionophore complex. Recent evidence suggests that transmitter Glu released from presynaptic axon terminals activates NMDA receptors on postsynaptic dendrosomal membranes resulting in opening of a Na^+/Ca^{2+} ion channel. Glycine, acting at strychnine-insensitive receptors and spermine or spermidine, acting at a polyamine site, facilitate opening of this channel, whereas PCP, Zn^{2+}, and Mg^{2+}, each acting at a separate site and presumably by separate mechanisms, are antagonists of channel function. Currently, it is recognized that NMDA receptor blockade results in psychosis in humans and cerebrocortical damage in rats regardless whether the blockade is induced by agents acting at the PCP site to perform an open channel block (PCP, ketamine, MK-801) or agents acting at the NMDA site to perform a closed channel block (CPP, CPPene, CGS 19755).

Based on early in vitro studies, it has become popular to think in terms of two separate forms of excitotoxic cell death, an acute form and a delayed form. Acute excitotoxic cell death has been depicted as occurring within minutes or a few hours, as being associated with uptake of Na^+, Cl^-, and water and resulting in rapid swelling of the neuron with cell death occurring hypothetically by lysis; this process has been studied primarily in the isolated chick embryo retina and has been monitored primarily by morphological methods. The delayed process has been studied primarily in cerebrocortical neuronal cultures and has been monitored primarily by biochemical measurement of lactic dehydrogenase (LDH) release as an indicator of cell death. This has been described as a Ca^{2+} dependent and NMDA receptor-mediated process leading to cell death as measured by LDH release occurring primarily in the 16–24 h time interval. Recently, Romano et al. *(19)* adapted the chick embryo retina model to allow incubations for up to 24 h and employed both LDH and histological analysis to monitor cell death. The results of this study do not support the concept of a delayed cell death process that is separate and distinct from acute excitotoxic cell death. Rather, by morphological (including ultrastructural) analysis the cells show signs of acute but quite advanced degeneration within the first hour or two after exposure to an excitotoxic agonist, but the release of LDH is a slowly evolving phenomenon that occurs on a delayed basis primarily in the 16–24 h interval after the degenerating neurons have shown morphological signs of advanced deterioration for many hours. Thus, at least in the chick retina, the LDH indicator system characteristically provides a delayed signal for detecting a cell death process which by morphological analysis would be described in more acute terms. Unfortunately, neither morphological analysis nor LDH assay provides a certain means of detecting precisely when the neuron becomes irreversibly committed to cell death. However, morphological (including ultrastructural) analysis is an indispensable component of the armamentarium for analyzing excitotoxic cell death, and the LDH assay is a useful method for quantitatively assessing the severity of the cell death response and the efficacy of various drugs in protecting against that response.

Neoclassical Excitotoxicity (NMDA Receptor Hypoactivity)

In recent years it has been found that blockade of NMDA receptors (NR) by NR antagonist drugs can cause an extensive neurodegenerative reaction in the adult rat brain. NR antagonist neurotoxicity was first described *(20)* as a reversible reaction, but it was soon learned that it can be irreversible depending on the length of time NR are maintained in a hypofunctional state. Treating an adult rat with a low dose of an NR antagonist induces brief NR hypofunction (NRH), which causes specific neurons in the posterior cingulate and retrosplenial (PC/RS) cortex to develop intracytoplasmic vacuoles that are conspicuous at 4 h but disappear by 24 h after treatment. However, administering an NR antagonist in high dosage or by continuous infusion for several days induces a prolonged NRH state that causes neurons in many cerebrocortical and limbic brain regions to degenerate unto death *(22,25)*. Pyramidal and multipolar neurons are preferentially affected and the full pattern of damage includes the posterior cingulate, retrosplenial, frontal, temporal, entorhinal, perirhinal, piriform and prefrontal cortices, the amygdala and hippocampus. At 4 h, the reaction in PC/RS cortex consists of intracytoplasmic vacuole formation, but in other brain regions a spongiform reaction featuring edematous swelling of spines on proximal dendrites is the most prominent cytopathological change *(22)*. At 24–48 h, the affected neurons become argyrophilic (de Olmos cupric silver method) and immunopositive for heat shock protein 72 kDa and they begin to display cytoskeletal abnormalities, including a conspicuous corkscrew deformity of their apical dendrites. In the 72–96 h interval, the dying neurons undergo fragmentation and elicit an inconspicuous glial response.

In a series of recent studies (reviewed in ref. *23*) we have found that several classes of drugs effectively block the PC/RS neurotoxic action of NR antagonists, including muscarinic receptor antagonists, $GABA_A$ receptor agonists, σ receptor antagonists, non-NMDA glutamate receptor antagonists, and α_2-adrenergic receptor agonists. These findings have provided new insight into the receptor mechanisms and neural circuitry (Fig. 4) involved in NRH neurodegeneration. Of considerable potential import, if our hypothesis *(see below)* proves correct that an NRH mechanism triggers neuronal degeneration in Alzheimer's disease and/or schizophrenia, these findings may lead to new therapeutic or prophylactic approaches to these disorders.

To fully appreciate how a deficit in NR activity can trigger a neurodegenerative syndrome, it is necessary to begin thinking of Glu in a new light, as an agent that performs major inhibitory functions. By tonically activating NR on GABAergic neurons (Fig. 4), Glu regulates inhibitory tone and ordinarily protects the brain against its own self destructive potential; removing this inhibitory mechanism from certain net-

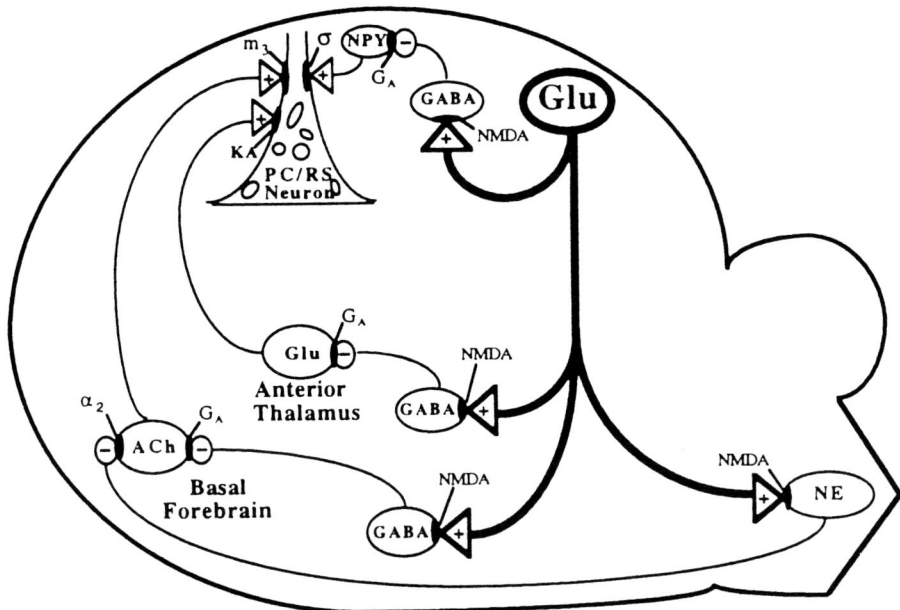

Fig. 4. Postulated circuitry to explain the neoclassical excitotoxic neurodegenerative reaction induced by NMDA antagonist drugs. Glu acting at NMDA receptors on GABAergic or noradrenergic neurons maintains tonic inhibition over three excitatory inputs to PC/RS neurons. All three excitatory inputs are subject to disinhibition when NMDA receptors are blocked. Removal of inhibition results in excessive activation of the PC/RS neuron through all three excitatory pathways as the proximal mechanism that triggers the neurodegenerative reaction.

works unleashes excitotoxic forces that proceed to wreak self destruction within the network. The excitotoxic forces are complex and include glutamatergic, cholinergic, and other less well defined components, but the key to understanding this neurotoxic process is to recognize that Glu, the master of all paradoxes, is not only an excitotoxic contributor to the pathological outcome, it is the driver of the inhibitory mechanism that normally holds the excitotoxic forces in check.

Neurodevelopmental Failure Associated with Metabotropic Receptor Hypoactivity

When the metabotropic receptor is activated physiologically by Glu, or experimentally by quisqualate or the more selective agonist, *trans*-ACPD [*trans*-(±)-1-amino-1,3-cyclopentane-dicarboxylic acid], it causes phosphoinositide hydrolysis (PiH) which triggers mobilization of intracellular Ca^{2+} stores and activation of protein kinase C (PKC), which in turn mediates phosphorylation of proteins that may be vitally important for both the function and survival of CNS neurons. It has been shown *(24)* that AP3 (2-amino-3-phosphonopropionate) blocks ACPD-stimulated PiH in hippocampal slices, and that daily sc administration of AP3 to infant rats for several consecutive days causes rats to develop with an almost complete absence of optic nerves *(21)*. The primary impact site of AP3 on the developing visual system is the retina, which shows gradual evolution of cytopathological changes over a 2–3 d period and total degeneration after 5–7 d *(26)*. In the brain, toxic changes were usually limited to regions that lack blood–

brain barriers, but in some animals there was a more generalized pattern of damage affecting neurons scattered throughout much of the neuraxis. Tentatively, these findings are being interpreted as follows: The immature retina lacks a blood–retina barrier so that the entire organ, following sc administration of AP3, is flooded with the compound, and this provides an in vivo testing ground that demonstrates what happens to CNS neurons if they are continuously exposed during critical periods of development to an agent that blocks the physiological action of Glu at its metabotropic receptor. As mentioned above, at this receptor, Glu stimulates a sequence of second messenger functions that may be vital for maturation and survival of the neuron.

RELEVANCE OF EAA NEUROTOXICITY TO HUMAN DISEASES

Acute CNS Injury

It is difficult to name an acute CNS injury syndrome that does not involve an excitotoxic mechanism. This is not to say that other mechanisms are not also involved in the pathophysiology of acute CNS injury, but excitotoxicity usually is of primary significance in triggering the cell death process. It is the classical type of excitotoxic mechanism that is believed to be involved in acute CNS injury syndromes and excitotoxins of both exogenous and endogenous origin have been implicated.

Caused by Exogenous Excitotoxin

Domoate encephalopathy provides an excellent example of an acute CNS injury syndrome caused by human exposure to an exogenous excitotoxin. In 1987, there was an outbreak of food poisoning in Canada that affected 145 people, some of whom died and were found at autopsy to have disseminated lesions in the CNS *(27,28)*. Some of the survivors sustained permanent brain damage and continued to show signs of cognitive deficits, especially impairment of memory, many months after the poisoning incident. All of the afflicted individuals had eaten mussels which were found to have exceedingly high concentrations (>800 µg/g) of domoate, an excitotoxic analog of Glu that interacts selectively and powerfully with the KA receptor *(29)*. It is known that KA, when administered systemically to adult rats, induces persistent seizures and a seizure-mediated brain damage syndrome, and it has been shown that domoic acid is approximately five times more powerful than KA in reproducing the same seizure–brain damage syndrome in adult rats *(29)*. Persistent seizure activity (status epilepticus) was a prominent feature of the clinical symptomatology observed in the Canadian poisoning incident, so it seems reasonable to conclude that the brain damage sustained by the victims of domoate poisoning was at least a partially a seizure-mediated damage. A disproportionately high percentage of severely affected individuals were elderly, which suggests that the KA receptor is a potentially sensitive mediator of excitotoxic neuropathology in the aged human brain. However, the situation is complicated by the fact that domoic acid and KA appear to act by a similar, if not identical, mechanism and much of the brain damage induced by KA in rats is seizure-mediated damage that can be prevented by pretreatment with an NMDA antagonist *(30)*. In other words, although the persistent seizure activity induced by either domoic acid or KA is triggered by activation of KA receptors, excessive seizure-mediated release of an excitotoxic transmitter (presumably Glu) at NMDA receptors is apparently responsible for most of the brain damage that ensues. Thus, available evidence supports the conclusion that both kainate and NMDA receptors are sensitive mediators of excitotoxic pathology in the aged human brain. Moreover, although the acute syndrome is triggered by ingestion of an exogenous excitotoxin, at least part of the resulting acute brain injury is caused by excessive release of endogenous excitotoxin (Glu).

Caused by Endogenous Excitotoxin

A number of major Acute CNS injury syndromes, including those associated with stroke *(31–34)*, trauma *(35,48)*, hypoglycemia *(37)* and status epilepticus *(38)*, are believed to be triggered by excessive activation of Glu receptors by endogenous excitotoxins, primarily Glu but probably also aspartate and possibly certain sulfur-containing amino acids. Excessive release and/or impaired uptake of endogenous excitotoxins occurs in each of these conditions and this leads to an extracellular accumulation of excitotoxins at ionotropic EAA receptors and consequent acute degeneration of neurons by a classical excitotoxic mechanism. In many of these conditions, a deficiency of energy available to the cell plays a key role. Under energy deficient conditions, Glu uptake mechanisms become defective, a condition resulting in an excessive amount of Glu accumulating at EAA receptors *(39)*. In addition, energy deficiency can cause failure of a physiological mechanism by which Mg^{2+} ordinarily blocks the NMDA receptor ion channel and suppresses ion flow through that channel. When the Mg^{2+} block fails, even normal physiological concentrations of Glu at the NMDA receptor cause pathologically excessive ion flow *(40)*. In head trauma, increased intracranial pressure is a major source of morbidity and mortality. An excitotoxic mechanism may play a major role in raising the intracranial pressure *(41)*. An excessive stimulation of EAA receptors causes the hyperstimulated neurons and the surrounding astroglia to accumulate intracellular fluid, the net effect of which is to transfer large amounts of water from the vascular to the intracellular compartment. Depending on the severity of the trauma, hundreds of thousands or millions of CNS cells will participate in this pathological process.

Chronic Neuropsychiatric Disorders

Evidence implicating both classical and neoclassical excitotoxic mechanisms and both exogenous and endogenous excitotoxins in chronic neuropsychiatric disorders is slowly accumulating. In the following sections, the status of evidence linking an excitotoxic mechanism to several chronic neuropsychiatric disorders is briefly summarized.

Neurolathyrism

Neurolathyrism is a crippling motor neuron disorder caused by chronic ingestion of *lathyrus sativus*, a legume containing β-N-oxalylamino-L-alanine (BOAA, also sometimes abbreviated ODAP), an acidic amino acid with powerful excitotoxic properties mediated through non-NMDA Glu receptors *(42,62)*. Spencer and coworkers *(43)* have demonstrated that the paralytic symptoms of neurolathyrism can be reproduced in monkeys maintained chronically on a diet enriched in BOAA. Evidence linking BOAA to neurolathyrism is of considerable interest: if chronic exposure to this exogenous excitotoxin acts selectively at non-NMDA Glu receptors and can cause motor neurons to degenerate slowly over a period of months or years, it may be possible for endogenous Glu acting at non-NMDA receptors to cause chronic degeneration of motor neurons in idiopathic motor neuron disorders such as amyotrophic lateral sclerosis. Moreover, should research on neurolathyrism reveal a mechanism by which ingested exogenous excitotoxins can cause degeneration of neurons in parts of the CNS that are thought to be inaccessible to such agents, it will be necessary to reevaluate the possibility that exogenous excitotoxins, including those used as food additives, might contribute (in concert with endogenous excitotoxins) to a variety of chronic neurodegenerative conditions.

Amyotrophic Lateral Sclerosis (ALS)

ALS provides an example of a motor neuron disorder in which endogenous Glu very likely plays a role, even in familial forms of the disorder in which a genetic defect not directly involving the Glu transmitter system may be the primary etiologic factor (*see* chapter by Bar-Peled and Rothstein). Several lines of evidence have implicated an excitotoxic mechanism in sporadic ALS, including evidence suggesting a metabolic defect impairing the glutamate transporter system *(44)*. In familial ALS, a genetic defect involving the enzyme superoxide dismutase-1 (SOD-1) has been discovered *(45)* and SOD-1 transgenic mice bearing this defect show both motor neuron degeneration and paralytic symptoms resembling those of ALS *(46,47)* (*see* chapter by Rabin and Borchelt). We have shown *(36)* that motor neurons in the rat lumbar spinal cord are selectively sensitive to the excitotoxic action of kainic acid (acting at non-NMDA Glu receptors) and that the excitotoxic changes induced in motor neurons by kainic acid are essentially identical to motor neuron degenerative changes observed by Gurney et al. *(46)* and Wong et al. *(47)* in SOD-1 transgenic mice. Consistent with these findings, it has recently been shown that Rilutek® (riluzole), an agent that suppresses glutamate neurotransmission, can retard the progression of clinical signs and symptoms in ALS patients *(49)*. Based on this evidence, riluzole was recently approved by the Food and Drug Administration as the first officially recognized effective treatment for ALS.

Huntington's Disease

Huntington's disease serves as an example of a chronic neurodegenerative disorder in which endogenous Glu (together with intracellular energy defects and possibly oxidative stress) may play a role even though a genetic defect not directly involving the Glu transmitter system may be the primary etiologic factor (*see* chapter by Ross, Becher, and Koliatsos). Many researchers have proposed that an excitotoxic mechanism may underlie the neuronal degeneration in Huntington's disease. For example, Beal *(50)* has proposed that a defect in energy metabolism may be responsible for unleashing the excitotoxic process in that various cellular energy poisons cause striatal neurons to degenerate in animal brain by a mechanism that is blocked by glutamate receptor antagonists. Recent discovery of the specific gene locus that is defective in Huntington's disease did not result immediately in clarification of the pathophysiological steps leading to neuronal degeneration. However, very recent evidence *(51)* suggests that the genetic defect results in abnormal expression of an enzyme that plays a key role in cellular energy metabolism. Thus, there is a strong possibility that neuronal degeneration in Huntington's disease can be explained in terms of a genetic defect that primarily disrupts intracellular energy metabolism and secondarily unleashes an excitotoxic process that is the critical mediator of cell death.

Alzheimer's Disease

In a recent publication *(42)*, we have advanced a new unified hypothesis to explain neurodegeneration in Alzheimer's disease (AD). This hypothesis posits an interaction between genetic factors (e.g., unfavorable ApoE genotype and β-amyloid-linked genetic mutations) and a new form of excitotoxicity (neoclassical excitotoxicity) that occurs spontaneously in the brain under conditions in which NMDA glutamate receptors are blocked or impaired. This is a complex hypothesis based on evidence that in the aging brain NMDA receptors are markedly impaired (hypofunctional) and on the paradoxical observation that, when NMDA receptors are rendered hypofunctional in the rat brain, this condition results in a widespread pattern of corticolimbic neuronal degeneration resembling the pattern seen in Alzheimer's disease *(22,42)*. We propose that the hypofunctional status of NMDA receptors in the aging brain sets the stage for widespread corticolimbic neurodegeneration that may or may not occur depending on certain auxilliary factors present in the AD brain but not the "normal" aging brain. The critical auxilliary factors may be genetically determined and may involve mechanisms that promote amyloidopathy which, in turn, promotes excitotoxic activation of NMDA receptors *(52)* and degeneration of NMDA receptor-expressing neurons. The end result would be a loss of NMDA receptors and a worsening of the hypofunctional status of the NMDA receptor system, i.e., the condition necessary to unleash widespread corticolimbic neurodegeneration. It is a strength of this hypothesis that it does not require abandoning other candidate hypotheses. In particular, currently popular genetic hypotheses that focus primarily on amyloid mechanisms are compatible with this hypothesis and would gain in explanatory power if combined with it.

Schizophrenia

Recently, we presented a new combined glutamate/dopamine hypothesis to explain the pathophysiology of schizophrenia *(23)*. This hypothesis, like the hypothesis pertaining to Alzheimer's disease, features NMDA receptor hypofunction (neoclassical excitotoxicity) as a key mechanism. However, in the schizophrenia hypothesis it is proposed that NMDA receptor hypofunction is a latent defect present at birth; it begins to manifest as a psychosis in early adulthood, whereas in AD it only becomes an operative mechanism late in life—at which time genetic factors peculiar to AD contribute to its neuropathological expression, and cause it to be expressed not as a psychosis but as a dementing syndrome. Consistent with this hypothesis, several recent studies have shown that elderly schizophrenics tend to have a decrease in psychotic symptomatology and an increase in AD-like dementia symptoms, but at autopsy these schizophrenic patients do not show the amyloid-linked (and genetically determined) neuropathological stigmata characteristic of AD.

EXCITOTOXICITY AND THE "APOPTOSIS VS NECROSIS" DICHOTOMY

As several authors have commented, e.g., Farber *(53)*, the apoptosis field has been subject to a great deal of confusion and controversy. In this section, we will present our analysis regarding some of the major sources of confusion in the apoptosis field and will describe our approach to the problem, which is aimed at reducing confusion and further clarifying the potential role of "apoptosis" in nerve cell death processes triggered by excitotoxic mechanisms (for further insights into this issue, *see* chapter by Clarke).

Historical Context

In the early 1970s, Kerr et al. *(54)* introduced the term "apoptosis" to refer to a cell death process that occurred in a variety of different circumstances, all having to do with controlled cell deletion *(54)*. Subsequently, on the basis of morphological (ultrastructural) criteria, these authors proposed that all cell death processes might fit into two broad categories termed "apoptosis" and "necrosis" *(55)*. In recent years, neuroscientists have become increasingly interested in understanding the role of apoptosis and genetically controlled cell death processes in the CNS, and in the possibility that such processes may be relevant to neurodegenerative diseases. Concurrently, as already mentioned, many neuroscientists have become interested in the role of excitotoxic mechanisms in neurodegenerative diseases. In very recent years, neuroscientists have begun addressing the question whether nerve cell death triggered by an excitotoxic stimulus entails mechanisms that can be considered apoptotic. Typically, consistent with the precedent set by Wyllie et al. *(55)*, workers on CNS apoptosis have framed the issue in terms of an "apoptosis vs necrosis" dichotomy and the findings reported thus far have been thoroughly confusing and contradictory; it has been claimed that excitotoxic neuronal degeneration is decidedly apoptotic, that it is decidedly necrotic, and that it may be both apoptotic and necrotic.

Sources of Confusion

Contributing to the above confusion are certain terminological and methodological problems that trouble the apoptosis field. It is inherently problematic to frame the issue in terms of an "apoptosis vs necrosis" dichotomy, because "necrosis" is an all-inclusive generic term that literally means the state of being "dead," and "apoptosis" presumably means one type of process that can lead to the state of being "dead." If apoptosis is one process that can lead to necrosis, it defies logic to think and talk in terms of "apoptosis versus necrosis." In addition, it may be overly simplistic to assume that all cell death processes can be subdivided into two mutually exclusive categories without any mechanisms operative in one category being also operative in the other. Also exceedingly problematic is the strong tendency in recent studies to conclude that excitotoxin-mediated cell death processes are either "apoptotic" or "necrotic" on the basis of recently developed DNA fragmentation methods that have never been validated as specific tests for distinguishing apoptotic from nonapoptotic cell death processes. Several authors have found that the TUNEL staining method is positive for cell death processes which, by ultrastructural criteria, are nonapoptotic as well as processes that ultrastructurally are apoptotic *(56–58)*. The same is true for the gel electrophoresis "laddering" test. In fact, in a study by Collins et al. *(59)* which was co-authored by Kerr who introduced the apoptosis concept, it is specifically cautioned that the laddering test was positive in their hands for a process considered necrotic as well as processes considered apoptotic.

A related source of confusion is the fact that, although almost all authorities in the field of apoptosis have acknowledged that ultrastructural analysis is the only reliable and definitive approach for identifying an apoptotic process, recent CNS studies usually have alluded to the Wyllie et al. *(55)* ultrastructural criteria but have not used ultrastructural methods themselves nor faithfully applied ultrastructural criteria to confirm their diagnosis of apoptosis. The problem is further compounded by the fact that, although Wyllie et al. *(55)* originally developed their ultrastructural criteria for identifying "apoptosis" from specific examples selected from a wide variety of tissues and a wide variety of conditions, none of the examples studied by them was selected from the

CNS. Thus, even if modern day workers on CNS apoptosis were faithfully following the ultrastructural guidelines that they cite, these guidelines do not even pertain to CNS cells. With increasing frequency, students of CNS apoptosis are finding evidence for "apoptosis" almost everywhere they look (cell culture and animal models of excitotoxicity, including ischemia and trauma, and human brain diseases, including Alzheimer's and Huntington's diseases). However, in diagnosing "apoptosis" they do not appear to be using any *ultrastructurally* established example of CNS "apoptosis" as a bona fide reference standard to corroborate their diagnosis.

Studies Aimed at Resolving the Confusion

Recently, we undertook studies aimed at clarifying the potential role of apoptotic mechanisms in excitotoxin-induced cell death processes. In an effort to avoid the pitfalls described above, we relied heavily on an ultrastructural approach but also employed other methods, including two types of DNA fragmentation analysis (TUNEL method of Gavrieli et al. *[60]*) and agarose gel electrophoresis assay for internucleosomal cleavage). We examined four cell death processes, one being a prototypic example of in vivo CNS apoptosis and the other three being different types of in vivo excitotoxic processes. Our primary goals were to clarify whether an apoptotic mechanism plays a role in any of the three excitotoxic processes and to determine whether any methods other than ultrastructural analysis are useful for distinguishing between apoptotic and nonapoptotic cell death processes. The neuropathological syndromes we have evaluated and our findings for each syndrome are detailed below.

Physiological Cell Death (PCD)

First we examined PCD, a cell death process that occurs spontaneously in the neonatal (postnatal d 1–7) rat brain *(4,56)*. In the neonatal rat brain there is a high concentration of cell profiles that are undergoing PCD in certain brain regions, e.g., regions surrounding the aqueduct of Sylvius and fourth ventricle, and such cells can also be found scattered throughout the neuraxis. These cells can be identified at the light microscopic level as presumably dying cells by virtue of their showing a positive reaction by the TUNEL method, which marks cells in which a specific pattern of DNA fragmentation is occurring. DNA extracts taken from regions where a high concentration of the TUNEL positive cells are located show a laddering pattern by agarose gel electrophoresis. By electron microscopic evaluation, these cells display a pattern of morphological changes that fulfill the criteria described by Wyllie et al. *(55)* as an apoptosis pattern (Fig. 5). The earliest changes are twofold: Conspicuous changes in the nuclear chromatin pattern consisting of the formation of one or several large electron-dense balls of clumped chromatin that usually have a remarkably regular circumferential contour conforming to the shape of a perfect sphere; and condensation of the cell body and nucleus. In a relatively early stage, the nuclear envelope disintegrates into fragments that float randomly about the cytoplasm. In the absence of a nuclear membrane, the cell becomes unpartitioned with nucleoplasmic contents freely intermingling with cytoplasmic contents. In some cases, the cell divides into apoptotic bodies consisting of one or more nuclear chromatin balls surrounded by a contingent of cytoplasm. These apoptotic bodies are phagocytosed by cells that appear to be resident glia. In the early stages when these conspicuous changes are occurring, the cytoplas-

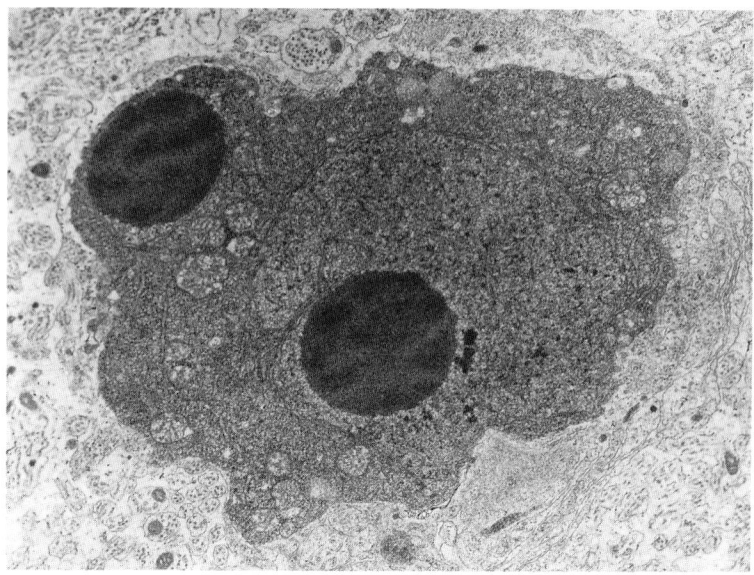

Fig. 5. An electron micrographic illustration of a neuron undergoing PCD in the periaqueductal region of the infant rat brain. This is a moderately advanced stage of PCD showing numerous changes consistent with an apoptotic process. The entire cell mass is condensed and there are two large spherical balls of clumped chromatin with many smaller aggregates of flocculent chromatin material dispersed throughout the matrix of the cell. Nucleoplasmic and cytoplasmic compartments have become confluent with one another due to discontinuity and fragmentation of the nuclear membrane. In the upper left corner it appears that a portion of the cell containing a chromatin ball surrounded by a contingent of condensed cytoplasm is being pinched off to form a separate apoptotic body that seems to be in the process of becoming membrane-bound, possibly by fragments of nuclear membrane that undergo reorganization to form a bilaminar membranous enclosure for the apoptotic body. Mitochondrial organelles do not appear to be entirely normal nor do they appear grossly abnormal. (×3000).

mic organelles (e.g., endoplasmic reticulum and mitochondria) do not show dramatic changes, although mitochondria typically display mild swelling and minor alterations in membrane structure. After being phagocytosed, all components of the ingested cell fragments show pleomorphic changes consistent with a biological degradation process.

Excitotoxic Cell Death-1 (ECD-1)

The first excitotoxic cell death process we examined was that which occurs very acutely in the neonatal rodent hypothalamus following sc administration of glutamate (4,56). The ECD-1 degenerative process is morphologically quite different from the PCD process, although the two processes do have one feature in common—both involve conspicuous clumping of nuclear chromatin. The earliest changes associated with the ECD-1 process consist of massive swelling of neuronal dendrites and cell bodies with accompanying changes in cytoplasmic organelles, including vacuolation of endoplasmic reticular membranes, initial condensation followed by massive swelling of mitochondria and disaggregation of polyribosomes and detachment, dispersal and dissolution of ribosomal particles (Fig. 6A). Changes in the nuclear chromatin become detect-

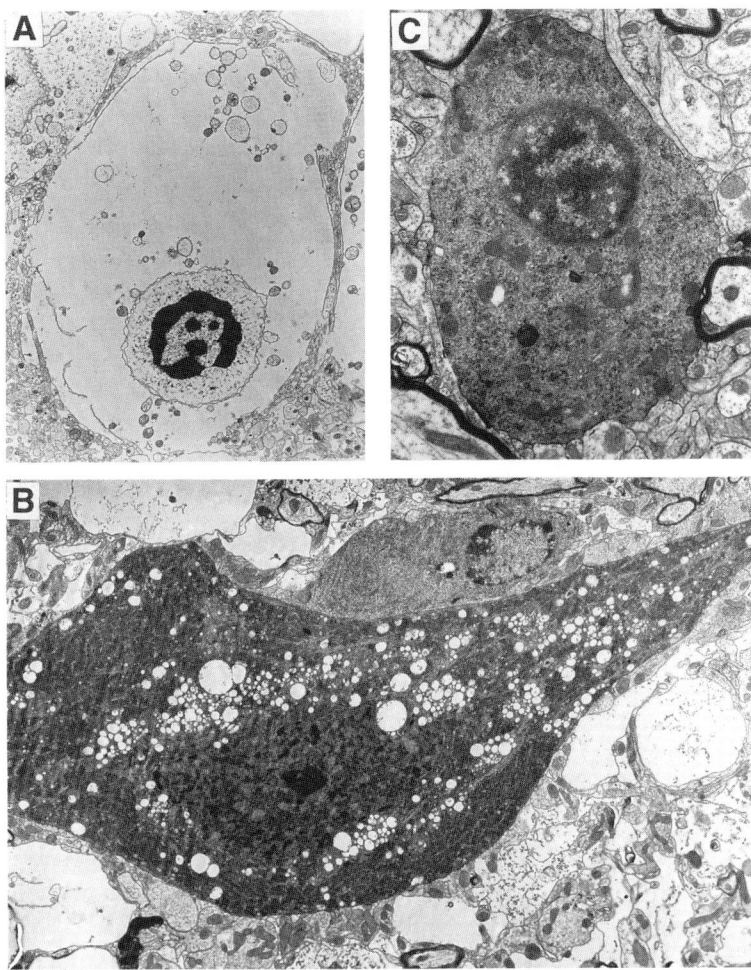

Fig. 6. Each of the three panels illustrates by electron microscopy at low magnification (×3000) a different type of ECD. **(A)** ECD-1 process in which cytoplasmic swelling and edematous degeneration of cytoplasmic organelles occurs first and this is followed by clumping and central coalescence of nuclear chromatin with the nuclear membrane becoming scalloped but remaining intact throughout the degenerative process. Multiple ruptures of the plasma membrane, apparently due to extreme swelling, is evident. **(B)** Spinal ventral horn motor neuron undergoing ECD-2 type of degeneration 4 h after application of kainic acid to the exposed dura of the lumbar spinal cord. This cell shows marked condensation of both the cytoplasm and nucleus accompanied by a striking display of clear cytoplasmic vacuoles that originate from both the endoplasmic reticulum and mitochondria. The condensation and vacuolation processes occur simultaneously as dual early manifestations of this type of degeneration. The nucleus does not show early changes other than condensation. **(C)** Pyramidal neurons in the posterior cingulate cortex showing a moderately late stage of the ECD-3 type of neurodegeneration 4 d following treatment of an adult rat with the NMDA antagonist, MK-801. In a much earlier stage of ECD-3 degeneration (4 h following treatment), this cell would have shown a vacuole reaction affecting cytoplasmic organelles, both mitochondria and endoplasmic reticulum, without any other changes. The vacuole reaction disappears by 24 h and the cell undergoes progressive condensation and nuclear changes of the type depicted here.

able only after cytoplasmic organelles have already undergone conspicuous disintegrative changes. Nuclear chromatin changes evolve gradually through a series of steps, beginning with the formation of small clumps of chromatin in a marginal clockface pattern. These marginal clumps become progressively larger and finally coalesce into a single large central chromatin mass that has an irregular circumferential contour and irregular shape in contrast to the perfectly smooth and spherical chromatin balls formed in the PCD process. The nuclear envelop assumes a crenulated appearance but remains intact and there is no evidence of nuclear chromatin material becoming dispersed into the cytoplasm or of the cell dividing into separate "bodies" containing a mixture of nuclear and cytoplasmic material. In late stages, the entire cell becomes condensed and assumes a roughly spherical shape as it is engulfed *in toto* (with little or no fragmentation) by a phagocytic cell. Microglial activation, evident by immunocytochemistry, is not a particularly conspicuous response. Hypothalamic neurons dying by the ECD-1 process show TUNEL positivity during the 8–24 h post treatment interval with peak positivity at 16 h. DNA extracts from tissue containing the degenerating hypothalamic neurons show laddering by gel electrophoresis in the 8–24 h post treatment interval.

Excitotoxic Cell Death-2 (ECD-2)

The second excitotoxic process examined was degeneration of ventral horn spinal motor neurons that occurs relatively acutely following local application of kainic acid to the exposed dura of the lumbar spinal cord *(36)*. ECD-2 is an acute cell death process but it evolves slightly more slowly and does not have the same morphological appearance as we have described above for the ECD-1 process induced by glutamate in the infant rodent hypothalamus. Instead of showing cytoplasmic swelling as the earliest sign of morphological change, motor neurons exposed to kainic acid show progressive condensation of both the cytoplasm and nucleus. Early changes also occur in cytoplasmic organelles consisting of conspicuous vacuoles being formed from endoplasmic reticular and Golgi membranes and mitochondria becoming initially condensed, then swollen (Fig. 6B). The rough endoplasmic reticular system does not disintegrate and the nuclear chromatin pattern does not show changes until a relatively late stage and clumping of nuclear chromatin is not as prominently displayed in the ECD-2 process as in ECD-1 or PCD.

As discussed by Ikonomidou et al. *(36)*, the factor that determines whether the neurodegenerative reaction will be of the ECD-1 or ECD-2 type is not primarily the type of excitotoxic stimulus, but rather the type of neuron. In general, parvocellular neurons throughout many regions of the CNS show the ECD-1 type of response to an excitotoxic stimulus and certain large pyramidal and multipolar neurons, including those in the neocortex, hippocampus and spinal cord, typically show the ECD-2 response. These two types of ECD responses are seen not only following exposure of CNS neurons to an excitotoxin agonist but also in acute brain injury syndromes in which endogenous glutamate is thought to trigger the neurodegenerative response (e.g., hypoxia/ischemia, status epilepticus, hypoglycemia and trauma).

Excitotoxic Cell Death-3 (ECD-3)

The third excitotoxic process examined was degeneration of retrosplenial cerebrocortical pyramidal neurons that occurs subacutely following intraperitoneal administration of MK-801 to the adult rat *(20,35,61)*. ECD-3 is a nerve cell death process that is triggered by any agent that induces a profound blockade of NMDA receptors, thereby releasing an excitotoxic process by a disinhibition principle. This is a special type of excitotoxicity (referred to above as neoclassical excitotoxicity) in which several excitatory inputs to the retrosplenial neuron are simultaneously disinhibited and thereby elevated to an abnormally high and sustained level of excitatory activity. The receptors on the retrosplenial neuron that are hyperstimulated in the ECD-3 syndrome are of at least three different types, muscarinic (m3 subtype), non-NMDA glutamatergic (kainic acid subtype), and sigma (subtype undetermined). Ultrastructurally, the earliest sign of morphological change associated with the ECD-3 process is conspicuous vacuole formation in the cytoplasm of the retrosplenial pyramidal neuron. Initially, the vacuoles are formed from endoplasmic reticulum but in a relatively early stage mitochondria show extreme swelling such that the swollen mitochondrial profiles contributes to the vacuolated appearance of the cytoplasm. The cytoplasmic vacuole formation reaches a peak at 4–12 h after MK-801 treatment and recedes in the 12–24 h period *(20)*. In the 24–48 h interval, either the cell recovers or becomes committed to cell death. Cells committed to die become progressively more condensed in both the cytoplasm and nucleus (Fig. 6C). The dendritic arbor becomes condensed and distorted into a corkscrew configuration in the 24–48 h interval and then undergoes a progressive fragmentation in the 48–96 h interval. Slowly, over a 2-wk period the dendritic fragments and shrunken cell bodies are phagocytosed and degraded by resident microglia. Nuclear chromatin changes are relatively late and consist of the formation of inconspicuous irregular masses of clumped chromatin. There is a transient microglial response to this type of cell death. The degenerating neurons do not show TUNEL positivity at any post-treatment interval from 4 to 7 d.

Our Interpretation of the Above Findings

Of the four cell death processes evaluated above, the only one that genuinely meets the ultrastructural criteria set forth by Wyllie et al. *(55)* for identifying apoptosis is the PCD process, a form of cell death that has for many decades been considered a prototypic example of a controlled cell death process. Each of the three examples of excitotoxin-triggered cell death is different from one another and each is fundamentally different from the PCD process, although each has one or more features in common with the PCD process that could cause the ECD process to be misdiagnosed as "apoptosis" by an investigator who is not using a systematic diagnostic approach, including ultrastructural evaluation as the primary means of establishing or confirming the diagnosis.

In general, all three of the ECD processes are fundamentally different from the PCD process and, therefore, fundamentally different from apoptosis as defined by Wyllie et al. *(55)*, in that none of them reproduces both the type and sequential pattern of morphological changes that characterize the PCD (apoptotic) process. For example, ECD-1 might easily be mistaken (and often has been in recent studies) as an apoptotic process because it shows conspicuous chromatin clumping which is visible by light microscopy and it yields a positive TUNEL staining response. However, the chromatin clumping by electron microscopic analysis does not become apparent until after conspicuous dendrosomal swelling and gross degenerative changes have occurred in the cytoplasmic organelle systems, and when it does occur it does not conform to the pattern of morphological changes that characterize the clumping of nuclear chromatin in the PCD process. It does not begin with the early formation of smoothly contoured spherical chromatin balls, but rather begins with the late formation of small marginal chromatin clumps that coalesce into a single large irregular mass of chromatin. The chromatin changes in the ECD-1 process are not accompanied by dissolution of the nuclear envelope and intermingling of nucleoplasmic and cytoplasmic contents, nor the formation of separate bodies comprised of chromatin balls together with a cytoplasmic contingent. Thus, ECD-1 is fundamentally very different from an apoptotic process and the fact that it gives a positive TUNEL staining response does not constitute evidence that it is an apoptotic process, but rather signifies that the TUNEL stain is not specific for apoptosis and is not a valid test for apoptosis unless it is accompanied by a systematic ultrastructural evaluation to ensure against a false positive diagnosis.

ECD-2 might be mistaken for an apoptotic process because it begins with condensation of the entire cell. However, this is accompanied not by early changes in nuclear chromatin but rather by early conspicuous vacuolization of endoplasmic reticulum and specific changes in mitochondria. Moreover, at no stage in the ECD-2 process does it involve the formation of smoothly contoured chromatin balls. Thus, ECD-2 is also fundamentally different from an apoptotic process and the fact that it involves early condensation of the cell body should not be considered evidence that it is an apoptotic process. Rather, it should be viewed as evidence that early condensation of the cell body *per se* is not an adequate criterion for diagnosing apoptosis. We have not determined whether cells degenerating by the ECD-2 process undergo a phase of TUNEL positivity, but clearly if the TUNEL test were positive this would be irrelevant, as we have already determined by more definitive ultrastructural analysis that the ECD-2 process is fundamentally different from apoptosis.

ECD-3 could be mistaken for an apoptotic process in that conspicuous condensation of the cell body occurs and the entire dendritic tree breaks up into many separate fragments that are phagocytosed separately from the cell body by resident microglia. However, the ECD-3 changes are fundamentally different from an apoptotic process in that the earliest changes are in the endoplasmic reticulum and mitochondria and, when cytoplasmic condensation occurs in a later interval, it is not accompanied by the formation of smoothly contoured nuclear chromatin balls or by any of the other changes characteristic of an apoptotic process. In addition, fragmentation of the dendritic tree followed by phagocytosis of individual fragments does not resemble the process by which the apoptotic cell divides into separate apoptotic bodies consisting of nuclear chromatin balls together with a surrounding mass of cytoplasm. Moreover, the ECD-3 cell death process evolves very slowly over a 2-wk period in contrast to the original description of "apoptosis" by Wyllie et al. *(55)* as an exceedingly acute process that evolves over a period of hours. Since ECD-3 does not show TUNEL positivity during any phase of the degenerative process, it is less likely than ECD-1 to be misdiagnosed as an apoptotic process

Conclusions Regarding the Apoptosis vs Necrosis Dichotomy

It is clear from the above analysis that we do not believe any of the three ECD examples qualifies as an apoptotic process. We believe that our position is entirely consistent with that of Kerr who originally coined the term "apoptosis" *(54)*, characterized the phenomenon in ultrastructural terms and more recently has cautioned against attempting to make an apoptosis diagnosis without first performing an ultrastructural analysis *(59)*.

Regarding the separate question of how the ECD processes relate to "necrosis," our position is difficult to compare with the position developed by Wyllie et al. *(55)*. The first basic problem is that, for reasons already given, we consider it unfortunate that Wyllie and colleagues chose the term "necrosis" to represent all forms of cell death other than apoptosis. Second, these authors provided a description of what they meant by "necrosis" based on various non-CNS tissues and indicated that they anticipated that this description would fit all nonapoptotic cell death processes, but none of our three CNS examples of ECD fits very well into their "necrosis" definition. Each comes closer to fitting their "necrosis" than their "apoptosis" definition but none conforms closely enough to "necrosis" to warrant the conclusion that "necrosis" is a useful basis for classifying nonapoptotic nerve cell death processes. Since currently we can specify at least three excitotoxic cell death processes which, by ultrastructural analysis, are different from one another, are different from "apoptosis" and are different from the Wyllie et al. *(55)* definition of "necrosis," it is likely that careful ultrastructural analysis of various other nerve cell death processes will reveal that there are yet many additional types of nerve cell death that will have to be recognized as separate entities for as long as we are limited to ultrastructural criteria as the only valid basis for categorizing cell death processes.

As a first step toward resolving the terminological confusion, we recommend allowing "necrosis" to mean what it traditionally has meant, i.e., a synonym for cell death, and quit thinking in terms of "apoptosis vs necrosis." Begin with the assumption that there are several, perhaps many, different processes that can lead to neuronal necrosis (death) and some of these may involve mechanisms that some day we may be able to define in ultrastructural, biochemical, time course and/or other terms as uniquely qualifying for "apoptosis." All nerve cell death processes that truly qualify by strict criteria should be classified as apoptotic neuronal death. All nerve cell death processes that do not qualify by strict criteria as apoptoic, should be classified as some other type of process that leads to neuronal death. In the interim period, while tools for clearly distinguishing different forms of cell death are being developed, each form should be referred to by a convenient term that communicates unambiguously what cell death process is being referred to. If only the agent or circumstance that triggers the cell death process is known, the cell death process should be named after that agent or circumstance. To the extent that the mechanism triggering the cell death process is known, the process should be referred to by that mechanism. Thus, a cell death process triggered by glutamate or any of its excitatory analogs would be referred to as excitotoxic because it is triggered by an excitotoxic mechanism. Calling excitotoxic cell death either "apoptosis" or "necrosis" should be discouraged since ultrastructural analysis does not corroborate the diagnosis of apoptosis and nothing is gained by calling it "necrosis" because this is less specific than "excitotoxic cell death." Moreover, as a practical measure, it is advisable to avoid the term "necrosis" simply because it has been rendered hopelessly ambiguous and meaningless by too many years of misuse.

None of the above should be construed as a lack of interest in the intriguing possibility that excitotoxic mechanisms might play a role in triggering or promoting cell death processes that may otherwise be actively and intrinsically controlled by gene-linked mechanisms. Much of the fervor in the field of CNS apoptosis appears to stem from an intense drive to relate the genome to human disease and cell death processes. We share this interest but believe that fervor must give way to sound science in the interests of truly advancing this important research goal.

CONCLUSIONS

In this chapter, we have reviewed the excitotoxic concept, described the known ways in which excitotoxicity can be expressed, considered the role of excitotoxic mechanisms in neuropsychiatric disorders, and expressed an unconventional viewpoint regarding the potential relationship between excitotoxic and apoptotic cell death mechanisms.

ACKNOWLEDGMENTS

This work was supported in part by NIMH Research Scientist Award MH 38894 and NIA Grant AG 11355, NIDA Grant DA 05072, NEI Grant EY 08089, and a NARSAD Established Investigator Award.

REFERENCES

1. Curtis DR, Watkins JC. The excitation and depression of spinal neurons by structurally related amino acids. *J Neurochem* 1960, **6:** 117–141.
2. Lucas DR, Newhouse JP. The toxic effect of sodium L-glutamate on the inner layers of the retina. *AMA Arch Ophthalmol* 1957, **58:** 193–201.
3. Olney JW. Brain lesions, obesity and other disturbances in mice treated with monosodium glutamate. *Science* 1969, **164:** 719–721.
4. Olney, JW. Glutamate-induced neuronal necrosis in the infant mouse hypothalamus: an electron microscopic study. *J Neuropathol Exp Neurol* 1971, **30:** 75–90.
5. Olney JW, Ho OL, Rhee V. Cytotoxic effects of acidic and sulphur-containing amino acids on the infant mouse central nervous system. *Exp Brain Res* 1971, **14:** 61–76.
6. Olney JW. Toxic effects of glutamate and related amino acids on the developing central nervous system, in *Heritable Disorders of Amino Acid Metabolism* (Nyhan WN, ed.), John Wiley, New York, 1974, pp. 501–512.
7. Watkins JC, Evans RH. Excitatory amino acid transmitters. *Ann Rev Pharmacol Toxicol* 1981, **21:** 165–204.
8. Olney JW, Labruyere J, Collins JF. D-Aminophosphonovalerate is 100-fold more powerful than d-alpha-aminoadipate in blocking N-methylaspartate neurotoxicity. *Brain Res* 1981, **221:** 207–210.
9. Olney JW, Price MT, Fuller TA, Labruyere J, Samson L, Carpenter M, Mahan K. The antiexcitotoxic effects of certain anesthetics, analgesics and sedative-hypnotics. *Neurosci Lett* 1986a, **68:** 29–34.
10. Olney J, Price M, Shahid Salles K, Labruyere J, Frierdich G. MK-801 powerfully protects against N-methyl aspartate neurotoxicity. *Eur J Pharmacol* 1987, **141:** 357–361.
11. Boulter J, Hollmann M, O'Shea-Greenfield A, Hartley M, Deneris E, Maron C, Heinemann S. Molecular cloning and functional expression of glutamate receptor subunit genes. *Science* 1990, **249:** 1033–1037.
12. Nakanishi S. Molecular diversity of glutamate receptors and implications for brain function. *Science* 1992, **258:** 597–602.

13. Monyer H, Sprengel R, Schoepfer R, Herb A, Higuchi M, Lomeli H, Burnashev N, Sakmann B, Seeburg PH. Heteromeric NMDA receptors: molecular and functional distinction of subtypes. *Science* 1992, **256**: 1217–1221.
14. Schoepp DS. Novel functions for subtypes of metabotropic glutamate receptors. *Neurochem Int* 1994, **24**: 439–449.
15. Monyer H, Seeburg PH, Wisden W. Glutamate-operated channels: developmentally early and mature forms arise by alternative splicing. *Neuron* 1991, **6**: 799–810.
16. Gregor P, Reeves RH, Jabs EW, Yang X, Dackowski W, Rochelle JM, Brown RH, Haines JL, O'Hara BF, Uhl GR, Seldin MF. Chromosomal localization of glutamate receptor genes: Relationship to familial amyotrophic lateral sclerosis and other neurological disorders of mice and humans. *Proc Natl Acad Sci USA* 1993, **90**: 3053–3057.
17. Hollmann M, Heinemann S. Cloned glutamate receptors. *Annu Rev Neurosci* 1994, **17**: 31–108.
18. Chen Q, Olney JW, Lukasiewicz P, Romano C. Mechanisms of kainate-induced excitotoxicity in retina: unique role of chloride. *Neurosci Abstr* 1996, **22**: 1278.
19. Romano C, Price MT, Olney JW. Delayed excitotoxic neurodegeneration induced by excitatory amino acid agonists in isolated retina. *J Neurochem* 1995, **65**: 59–67.
20. Olney JW, Labruyere J, Price, MT. Pathological changes induced in cerebrocortical neurons by phencyclidine and related drugs. *Science* 1989, **244**: 1360–1362.
21. Fix AS, Schoepp DD, Olney JW, Vestre WA, Griffey KI, Johnson JA, Tizzano JP. Neonatal exposure to D,L-2-amino-3-phosphonopropionate (D,L-AP3) produces lesions in the eye and optic nerve of adult rats. *Dev Brain Res* 1993, **75**: 223–233.
22. Corso TD, Sesma MA, Tenkova TI, Der TC, Wozniak DF, Farber NB, Olney JW. Multifocal brain damage induced by phencyclidine is augmented by pilocarpine. *Brain Res* 1997, **752**: 1–14.
23. Olney JW, Farber NB. Glutamate receptor dysfunction and schizophrenia. *Arch Gen Psychiatry* 1995, 52: 998–1007.
24. Schoepp, Johnson BG. Inhibition of excitatory amino acid-stimulated phosphoinositide hydrolisis in the neonatal rat hippocampus by 2-amino-3-phosphonopropionate. *J Neurochem* 1989, **53**: 273–278.
25. Fix AS, Horn JW, Wightman KA, Johnson CA, Long GG, Storts RW, Farber N, Wozniak DF, Olney JW. Neuronal vacuolization and necrosis induced by the noncompetitive N-methyl-D-aspartate (NMDA) antagonist MK(+)801 (Dizocilipine Maleate): A light and electron microscopic evaluation of the rat retrosplenial cortex. *Exp Neurol* 1993, **123**: 204–215.
26. Price MT, Romano C, Fix AS, Tizzano JP, Olney JW. Blockade of the second messenger functions of the glutamate metabotropic receptor is associated with degenerative changes in the retina and brain of immature rodents. *Neuropharmacology* 1995, **34**: 1069–1079.
27. Perl TM, Bedard L, Kosatsky T, Hockin JC, Todd ECD, Remis RS. An outbreak of toxic encephalopathy caused by eating mussels contaminated with domoic acid. *New Engl J Med* 1990, **322**: 1775–1780.
28. Teitelbaum JS, Zatorre RJ, Carpenter S, Gendron D, Evans AC, Gjedde A, Cashman NR. Neurologic sequelae of domoic acid intoxication due to the ingestion of contaminated mussels. *N Eng J Med* 1990, **322**: 1781–1787.
29. Stewart GR, Zorumski CF, Price MT. Domoic acid: A dementia-inducing excitotoxic food poison with kainic acid receptor specificity. *Exp Neurol* 1990, **110**: 127–138.
30. Clifford DB, Olney JW, Benz AM, Fuller TA, Zorumski CF. Ketamine, phencyclidine and MK-801 protect against kainic acid induced seizure-related brain damage. *Epilepsia* 1990, **31**: 382–390.
31. Rothman SM. Synaptic release of excitatory amino acid neurotransmitter mediates anoxic neuronal death. *J Neurosci* 1984, **4**: 1884–1891.
32. Rothman SM, Olney JW. Glutamate and the pathophysiology of hypoxic-ischemic brain damage. *Ann Neurol* 1986, **19**: 105–111.

33. McDonald JW, Silverstein FS, Johnston MV. MK-801 protects the neonatal brain from hypoxic-ischemic damage. *Eur J Pharmacol* 1987, **140:** 359–361.
34. Ikonomidou C, Price MT, Mosinger JL, Frierdich G, Labruyere J, Shahid Salles K, Olney JW. Hypobaric-ischemic conditions produce glutamate-like cytopathology in infant rat brain. *J Neurosci* 1989, **9:** 1693–1700.
35. Faden AI, Demediuk S, Panter S, Vink R. The role of excitatory amino acids and NMDA receptors in traumatic brain injury. *Science* 1989, **244:** 798–800.
36. Ikonomidou C, Qin YQ, Labruyere J, Olney JW. Motor neuron degeneration induced by excitotoxin agonists has features in common with that seen in the SOD-1 transgenic mouse model of amyotrophic lateral sclerosis. *J Neuropathol Exp Neurol* 1996a, **55:** 211–224.
37. Wieloch T. Hypoglycemia-induced neuronal damage prevented by an N-methyl-D-aspartate antagonist. *Science* 1985, **230:** 681–683.
38. Olney JW, Collins RC, Sloviter RS. Excitotoxic mechanisms of epileptic brain damage, in *Basic Mechanisms of the Epilepsies: Molecular and Cellular Approaches* (Delgado-Escueta AV, Ward AA, Woodbury DM, Porter RJ, eds.), Raven, New York, 1986b, pp: 857–877.
39. Benveniste H, Drejer J, Schousboe A. Elevation of the extracellular concentrations of glutamate and aspartate in rat hippocampus during transient cerebral ischemia monitored by intracerebral microdialysis. *J Neurochem* 1984, **43:** 1369–1374.
40. Henneberry RL, Novelli A, Cox JA, Lysko PG. Neurotoxicity at the N-methyl-D-aspartate receptor in energy-compromised neurons. An hypothesis for cell death in aging and disease. *Ann NY Acad Sci* 1989, **568:** 225–233.
41. Di X, Harpold T, Watson JC, Bullock MR. Excitotoxic damage in neurotrauma: fact or fiction. *Neurol Neurosci* 1996, **9:** 231–241.
42. Olney J, Wozniak D, Ishimaru M, Farber N. NMDA receptor dysfunction in Alzheimer's disease, in *Alzheimer Disease: From Molecular Biology to Therapy* (Becker R, Giacobini E, eds.), Birkhäuser, Boston, 1996, pp. 107–112.
43. Spencer PS, Schaumburg HH, Cohn DF, Seth PK. Lathyrism: a useful model of primary lateral sclerosis, in *Research Progress in Motor Neurone Disease* (Rose FC, ed.), Pitman, London, 1984, pp. 312–327.
44. Rothstein JD, Van Kammen M, Levey AI, Martin LJ, Kuncl RW. Selective loss of glial glutamate transporter GLT-1 in amyotrophic lateral sclerosis. *Annal Neurol* 1995, **38:** 73–84.
45. Rosen DR, Siddique T, Patterson D, et al. Mutations in the Cu/Zn superoxide dismutase gene are associated with familial amyotrophic lateral sclerosis. *Nature* 1993, **362:** 59–62.
46. Gurney M, Pu H, Chiu AY, et al. Motor neuron degeneration in mice that express a human Cu, Zn superoxide dismutase mutation. *Science* 1994, **264:** 1772–1775.
47. Wong PC, Pardo CA, Borchelt DR, Lee MK, Copeland NG, Jenkins NA, Sisoda SS, Cleveland DW, Price DL. An adverse property of a familial ALS-linked SOD1 mutation causes motor neuron disease characterized by vacuolar degeneration of mitochondria. *Neuron* 1995, **14:** 1105–1116.
48. Ikonomidou C, Qin Y, Labruyere J, Kirby C, Olney JW. Prevention of trauma-induced neurodegeneration in infant rat brain. *Peiatr Res* 1996, **39:** 1020–1027.
49. Bensimon TG, Lacomblez L, Meininger V. ALS/Riluzole Study Group. A controlled trial of riluzole in amyotrophic lateral sclerosis. *N Engl J Med* 1994, **330:** 585–591.
50. Beal MF. Role of excitotoxicity in human neurologic disease. *Curr Opinion Neurobiol* 1992, **2:** 657–662.
51. Burke JR, Enghild JJ, Martin ME, Jou YS, Myers RM, Roses AD, Vance JM, Strittmatter WJ. Huntingotn and DRPLA protiens selectively interact with the enzyme GAPDH. *Nature Med* 1996, **2:** 347–350.
52. Koh J, Yang LL, Cotman CW. Beta-amyloid protein increases the vulnerability of cultured cortical neurons to excitotoxin damage. *Brain Res* 1990, **533:** 315–320.
53. Farber E. Programmed cell death: Necrosis versus apoptosis. *Mod Pathol* 1994, **7:** 605–609.

54. Kerr JFR, Wyllie AH, Currie AR. Apoptosis: a basic biological phenomenon with wideranging implications in tissue kinetics. *Br J Cancer* 1972, **26**: 239–257.
55. Wyllie AH, Kerr JFR, Currie AR. Cell death: the significance of apoptosis. *Int Rev Cytol* 1980, **68**: 251–306.
56. Ishimaru M, Der TC, Tenkova TI, Sesma MA, Thurston JH, Olney JW. Three types of excitotoxicity evaluated for "apoptosis" signals. *Soc Neurosci Abst* 1995, **21**: p. 1584.
57. Charriaut-Marlangue C, Ben-Ari Y. A cautionary note on the use of the TUNEL stain to determine apoptosis. *Neuroreport* 1995, **7**: 61–64.
58. Grasl-Kraupp B, Ruttkay-Nedecky B, Koudelka H, Bukowska K, Bursch W, Shulte-Hermann R. In situ detection of fragmented DNA (TUNEL Assay) fails to discriminate among apoptosis, necrosis, and autolytic cell death: a cautionary note. *Hepatology* 1995, **21**: 1465–1468.
59. Collins RJ, Harmon BV, Gobé GC, Kerr JFR. Internucleosoma DNA cleavage should not be the sole criterion for identifying apaptosis. *Int J Radiat Biol* 1992, **61**: 451–453.
60. Gavrieli Y, Sherman Y, Ben-Sasson SA. Identification of programmed cell death in situ via specific labeling of nuclear DNA fragmentation. *J Cell Biol* 1992, **119**: 493–501
61. Olney JW, Labruyere J, Wang G, Wozniak DF, Price MT, Sesma MA. NMDA antagonist neurotoxicity: Mechanism and prevention. *Science* 1991, **254**: 1515–1518.
62. Olney JW, Misra CH, Rhee C. Brain and retinal damage from the lathyrus excitotoxin, b-N-oxalyl-L-ab-diaminopropionic scid (ODAP). *Nature* 1976, **264**: 659–661.

11
Neurotoxins and Neuronal Death
An Animal Model of Excitotoxicity

Elizabeth O'Hearn and Mark E. Molliver

INTRODUCTION

This chapter depicts a systems approach to analyze the manifestations and determinants of excitotoxic injury in the central nervous system (CNS) as illustrated by a recently described experimental model of excitotoxicity in cerebellum. In this model of neurotoxicity, administration of certain indole alkaloid derivatives (e.g., harmaline or ibogaine) leads to excitotoxic insult in a subset of neurons in cerebellar cortex. The resulting neuronal injury is highly selective for Purkinje cells and exhibits a unique, spatial pattern of degeneration. Complementary to data presented in other chapters of this volume focusing on the molecular and biochemical mechanisms of neuronal injury and death (*see* chapters by Dykens, Leist and Nicotera, Rosen and Casciola-Rosen, and Freeman), the present chapter takes a systems approach in order to characterize the role of neuronal circuitry in the production of selective neuronal injury (*see* also chapter by Olney and Ishimaru and, for a developmental systems perspective, *see* chapter by Burek and Oppenheim). The model of neurotoxicity described here exemplifies how a particular form of neuronal circuitry can determine the location of excitotoxic injury in the brain, the specific neurons affected and the spatial distribution of degenerating neurons. This model provides insights into the glial reactions induced by injury to a particular type of neuron, and it demonstrates that the pattern of termination of synaptic inputs to a neuron is an important determinant of neuronal vulnerability.

In the present model of drug-induced excitotoxicity, the compounds administered are not directly cytotoxic in themselves. The neuronal insult results from activation of intrinsic neuronal circuits that produce injury via the release of endogenous neurotransmitters (excitatory amino acids). Hence, the selective neuronal vulnerability observed in this excitotoxicity model derives from neuronal circuits that are natural components of the brain. Therefore, normal brain circuitry produces "built in" patterns of selective vulnerability that are likely to be important determinants of neurotoxicity in several forms of excitotoxic injury. The present model demonstrates that increased or prolonged stimulation of a normal brain circuit (e.g., the olivocerebellar projection) can lead to excitotoxic cell death. This model may closely simulate naturally occurring CNS insults that have been proposed to result in excitotoxic neuronal degeneration in

From: Cell Death and Diseases of the Nervous System
Edited by: V. E. Koliatsos and R. R. Ratan © Humana Press Inc., Totowa, NJ

man *(1,2)*. Since normal brain activity is not expected to produce neuronal damage, "protective mechanisms" are likely to be active in the form of neuronal circuits (e.g., inhibitory interneurons) that prevent injury caused by physiologic neuronal activity. Under special conditions of abnormally increased activity, such protective mechanisms may fail, increasing the potential for excitotoxic injury. Future research in excitotoxicity should characterize, in addition to mechanisms of injury, intrinsic protective factors that are based upon local neuronal circuitry.

AN EXPERIMENTAL MODEL FOR EXCITOTOXICITY IN CEREBELLUM

This experimental model of excitotoxicity is based upon the observation that administration of ibogaine, or the related drug harmaline, produces neuronal injury in the cerebellum of rats. The neurotoxicity is manifest by selective degeneration of Purkinje cells located primarily in the vermis of the cerebellum *(3,4)*. The injured neurons have a highly distinctive, spatial distribution in cerebellar cortex: degenerating Purkinje cells are aligned in the parasagittal plane and form discrete longitudinal bands that extend over long distances within the vermis. These longitudinal bands of degenerating neurons alternate with bands of neurons that appear morphologically intact. Complementing this highly ordered pattern of neuronal degeneration, activated glial cells, both microglia and astrocytes, form sagittally oriented radial stripes that are in register with the longitudinal bands in which Purkinje cells have degenerated.

IBOGAINE AND HARMALINE: BACKGROUND AND EFFECTS ON MOTOR BEHAVIOR

Both ibogaine and harmaline are indole alkaloids that are extracted from plants which grow in tropical rainforests, ibogaine from west central Africa (Gabon) *(5)* and harmaline from the Amazon basin of South America (Brazil) *(6)*. These drugs are CNS stimulants and hallucinogens and their most characteristic clinical effect is a prominent tremor. Both drugs have been used in religious ceremonies in the regions where the plants are endemic. Ibogaine came to public and scientific attention largely through claims that its ingestion may suppress the craving associated with drug addiction and that it may reduce signs of drug withdrawal in humans *(7–10)*. This drug is being evaluated for clinical use in the treatment of drug addiction and the Federal Drug Administration (FDA) has approved initial Phase I studies at the University of Miami. At the present time, the efficacy of ibogaine for the prevention of drug craving remains controversial. Most of the available human data is ancedotal, yet Glick and colleagues have reported suppression of morphine self-administration in rats *(11)* and a reduction in opiate withdrawal signs *(12)*. In contrast, other laboratories have found little effect of ibogaine on morphine or cocaine self-administration in animals *(13,14)*.

Although ibogaine and harmaline are potent hallucinogens, they are best known in the scientific literature for their tremorigenic properties. Ibogaine administration rapidly produces abnormal motor behavior characterized by tremor and ataxia in both mice *(15,16)* and rats *(4,12)*. The motor effects induced by ibogaine are the same as those produced by two chemically similar alkaloids, harmaline and ibogaline *(15,17)* (*see* Fig. 1). The best characterized of these drugs is harmaline, which produces a sustained 8-12 Hz generalized tremor in all species tested *(16–19)*. Ibogaline, like ibogaine, also produces a marked tremor at the same frequency (8–12 Hz) *(16,20)*. Based on the

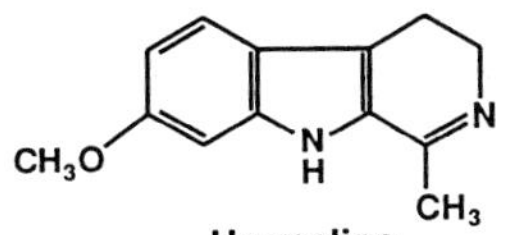

Neurotoxins and Neuronal Death

[Chemical structures of Harmaline, Ibogaine, and Ibogaline]

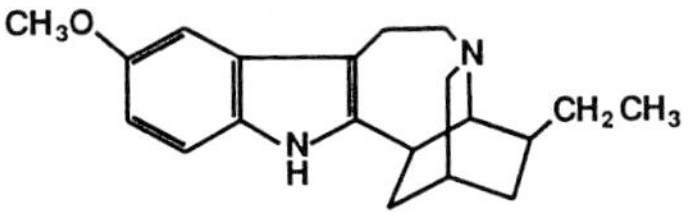

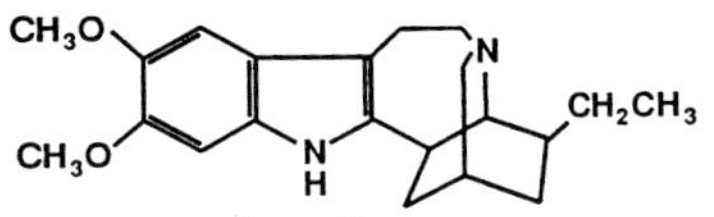

Fig. 1. Hallucinogenic indole alkaloids that are tremorigenic and produce Purkinje cell degeneration.

chemical similarities (Fig. 1) and nearly identical pharmacologic properties among these compounds, ibogaine is likely to share the same mechanism of action and produce the same physiologic effects, namely to increase the activity of neurons in the inferior olive.

Electrophysiologic studies of harmaline and ibogaline indicate that the tremor caused by these indole alkaloids results from drug-induced excitation of neurons in the inferior olivary nucleus. Harmaline produces sustained activation of inferior olivary neurons when given systemically *(18,21,22)*, when microinjected onto inferior olivary neurons *(23)*, or superfused on the inferior olive in brainstem slices *(24)*. Harmaline-treated animals exhibit rhythmic bursts of single unit activity at 8–12 Hz, which are synchronous throughout every level of the spino-cerebellar system, including inferior olive, Purkinje cells, deep cerebellar nuclei, vestibular nuclei, reticular formation, and motoneurons in spinal cord *(18,22,25)*. The rhythmic bursts of neuronal activity induced by harmaline are time-locked to the tremor, as shown by simultaneous electromyographic recording *(21,26)*. Moreover, interrupting all connections between the inferior olive and the cerebellum prevents the harmaline tremor without blocking activation of inferior olivary neurons *(21,22,27)*. The drug-induced tremorigenic activity, therefore, clearly arises in the inferior olive, and production of the tremor depends upon integrity of the olivocerebellar projection. Brainstem descending pathways are entrained by the rhythmic activity of the cerebellum, leading to repetitive excitation of motoneurons, which results in the tremor.

Neurotoxic Effects of Ibogaine

Evidence of neuronal injury following ibogaine administration in rats is almost entirely limited to the cerebellum. In most cases used to assess neurotoxicity, ibogaine was given as a single dose (100 mg/kg in water) by ip injection (unless otherwise noted). The entire brain was screened using sagittal sections stained by immunocytochemistry for numerous neuronal markers and with a Nissl method for cell bodies. One week after receiving ibogaine, no signs of neurotoxic injury were seen in forebrain or brainstem sections prepared for Nissl, serotonin, dopamine, GABA, or cytoskeletal proteins (neurofilament

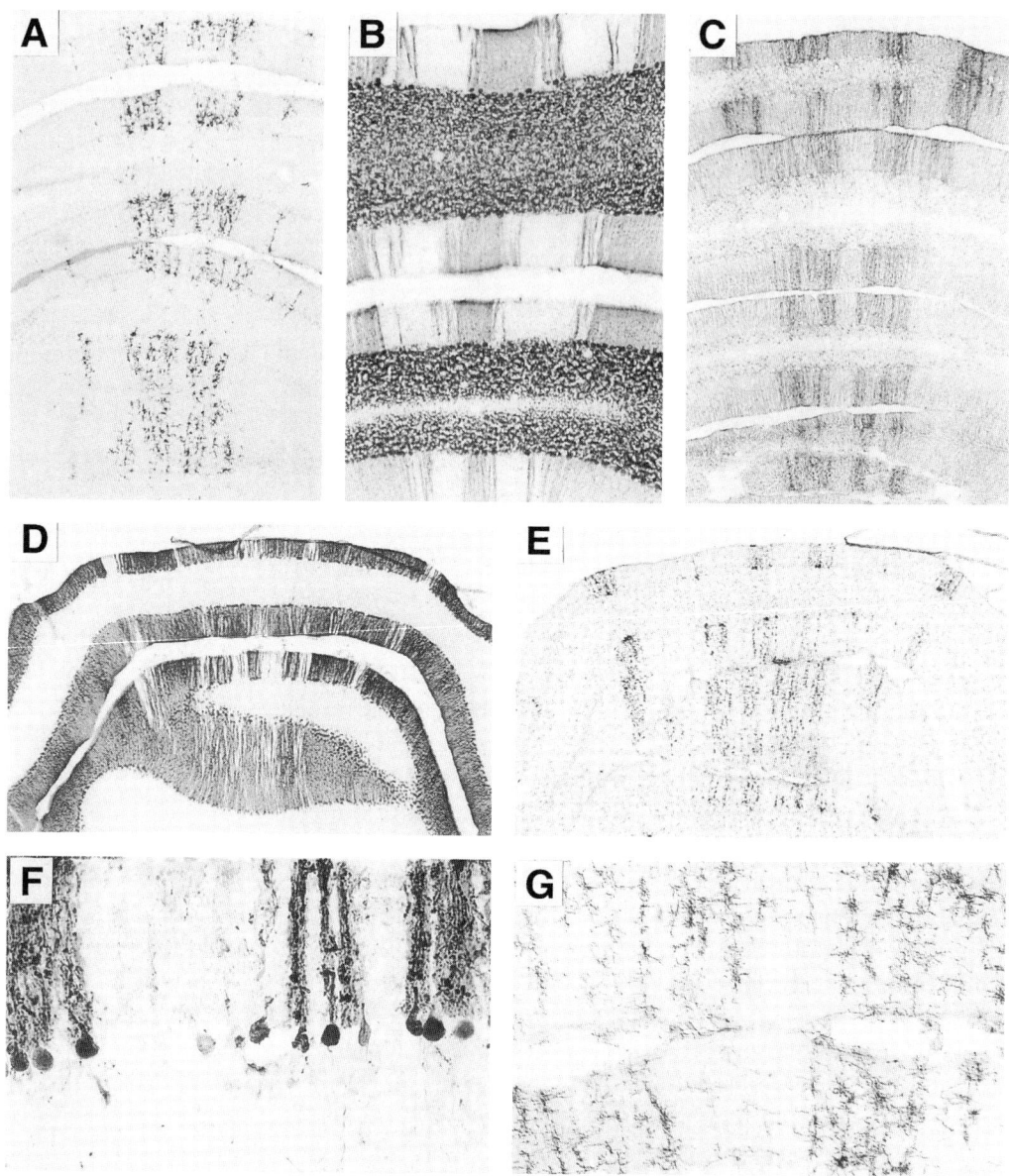

Fig. 2. Cerebellar degeneration following systemic administration of ibogaine. Neuronal degeneration in cerebellar cortex is highly selective for a subset of Purkinje cells. The damaged cells are grouped in discrete clusters, resulting in radial stripes that are devoid of Purkinje cells. The unstained zones in the molecular and Purkinje cell layers of cerebellar cortex (B,D,F) reflect the loss of Purkinje cell bodies and dendrites. Activated glial cells also form discrete radial bands that typically coincide with groups of degenerating Purkinje cells. **(A)** Following ibogaine treatment, activated microglial cells form symmetrical radial stripes in the vermis. These microglia are immunostained with an antibody (OX6) to the MHC II antigen. **(B)** Groups of Purkinje cells stained (brown) for CaM-kinase II remain intact, whereas pale bands indicate the regions where Purkinje cells are damaged or degenerating. In this Nissl counter-stained section, numerous small neurons (purple) are densely packed in the granule cell layer demonstrating that the granule cells are spared and remain intact. (The unstained horizontal strip in

proteins, MAP2), and there was no evidence of glial activation. In contrast, sections of cerebellar cortex reveal loss of staining in a subset of Purkinje cells (at 2–10 d after treatment) *(4)*. The loss of cellular staining was observed using markers for several neuronal proteins found in Purkinje cells including MAP2, calbindin, Ca^{2+}-calmodulin-dependent protein kinase II (CaM kinase II), and phosphorylated neurofilament proteins.

Purkinje cells are well visualized with antibodies to the Ca^{2+}-binding protein, calbindin, or to CaM-kinase II, which intensely stain the cell bodies, axons and dendrites. In cerebellum from normal rats, Purkinje cell bodies form an uninterrupted monolayer throughout the cerebellar cortex and their dendrites comprise a network of processes that extends continuously through the molecular layer. Following administration of ibogaine, small groups of Purkinje cell bodies and dendrites are damaged, yet no other neuron types exhibit altered morphology. At short intervals after treatment (1–2 d), a moderate number of irregular, shrunken and fragmented Purkinje cell bodies can be seen in the vermis. By 1 wk after drug administration, coronal sections of cerebellum reveal a striking abnormality *(4)*: several radial bands of cerebellar cortex are extremely pale, as they are no longer stained with neuronal markers, and these bands form discontinuities that reflect the loss of Purkinje cells (Figs. 2B and 2D). Between these pale bands of absent neurons, intact Purkinje cells remain densely stained, as in control rats. Followed through a series of coronal sections, degenerating or absent Purkinje cells are observed to form narrow rows from 1–8 Purkinje cells in width, aligned in the parasagittal plane. These longitudinal bands of neuronal loss span the Purkinje cell and molecular layers of cerebellar cortex (Fig. 2B,D,F). Within these pale bands, Purkinje cell bodies and dendrites are not detectable with antibodies to neuronal markers (e.g., calbindin or CaM-kinase II) at 1 wk survival. Adjacent Nissl-stained sections reveal unstained patches in which several Purkinje cell bodies are conspicuously absent or injured; these patches are in register with the gaps in calbindin and CaM-kinase II staining. The Purkinje cell loss induced by ibogaine is especially prominent in the vermis of the cerebellum, but some degeneration is found in the paravermis and, less commonly, in the cerebellar hemispheres and paraflocculus.

At short survivals (1–2 d), occasional thickened, dysmorphic Purkinje cell dendrites (stained for CaM-kinase II) arise from darkly-stained, pyknotic Purkinje cell bodies that are adjacent to large vacuoles (Fig. 2F). In some cases, dark, CaM-kinase-positive Purkinje cell bodies with a crescent shape form the edge of large vacuoles. Such vacuoles, not seen in control rats, indicate a degenerative process. Not uncommonly after ibogaine treatment, the

---

the lower folium is white matter.) **(C)** Groups of activated astrocytes that are stained for GFAP form darkly stained radial stripes that are vertically aligned across multiple folia. **(D,E)** Several radial stripes (unstained) of missing Purkinje cells (D) match (in an adjacent section) the deeply stained stripes that consist of activated microglia (E). Stain: (D) CaM-kinase II; (E) OX42 for CR3. **(F)** Bands of degenerating Purkinje cells in section stained for calbindin, at a higher magnification than in (D). When Purkinje cells degenerate, cell body and dendrites are no longer stained due to loss of specific neuronal proteins or cell death and eventual phagocytosis. Large cavities are associated with two degenerating cell bodies. Survival 5 d. **(G)** Groups of activated microglia aligned in narrow radial bands, at a higher magnification than in E. Survival 6 d. Stain: OX42. Treatment: Ibogaine 150 mg/kg ip in (A,F), 100 mg/kg ip in all other plates. All sections are in the coronal plane.

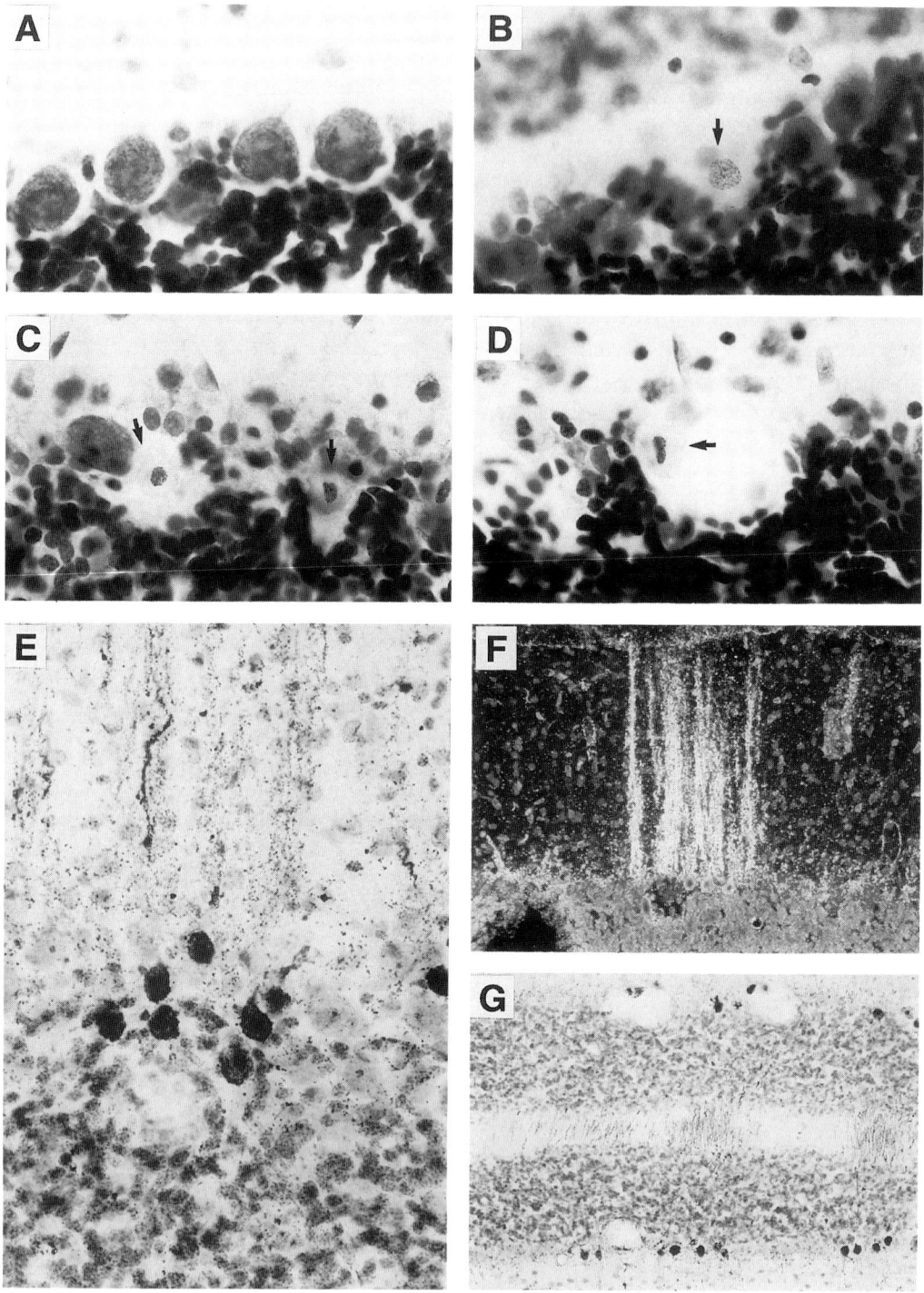

Fig. 3. Degeneration of Purkinje cells following ibogaine administration. (B–D) show early cytologic changes in injured neurons, as seen in Nissl-stained sections. (E–G) provide evidence of degeneration in silver-stained sections. **(A)** Intact Purkinje cell bodies contain moderately fine Nissl bodies dispersed through the cytoplasm. **(B–D)** Ibogaine-induced excitotoxic injury

descending axons from abnormal Purkinje cells have a spheroidal swelling along their course (seen with staining for calbindin or CaM-kinase II). These axonal enlargements found in the granule layer are reminiscent of axonal clubs that Cajal described along the axons of damaged Purkinje cells *(28)*. Similar dysmorphic axons, called "torpedoes" in neuropathology *(29)*, are commonly associated with degenerating Purkinje cells *(30)*.

Nissl-Stained Purkinje Cells

Purkinje cell bodies typically lie in a continuous tangential layer of cerebellar cortex as seen in Nissl-stained sections from untreated control rats. After treatment with ibogaine or harmaline, the Purkinje cell layer exhibits discontinuities marked by unstained or degenerating Purkinje cell bodies adjacent to zones with normal appearing cell bodies. Purkinje cell bodies that are aligned with those showing loss of CaM-kinase II immunoreactivity can be examined in cresyl violet stained sections. Adjacent to the zones that lack CaM-kinase II staining, Nissl-stained Purkinje cells exhibit several pathologic changes. In these zones of neuronal damage, many of the Purkinje cell bodies have disappeared by 1–2 wk after treatment, yet some of the cell bodies remain, but have abnormal staining and morphology. The early morphologic changes observed in Nissl-stained frozen sections are seen at 12 h to 2 d survivals (Figs. 3A–D, Fig. 4). Abnormal Purkinje cells are characterized by complete loss of Nissl body staining and a homogeneous, translucent cytoplasm, which has a uniformly pale, gelatinous appearance with a slight tinge of pale blue color. The nucleus also has a pale matrix, but chromatin is condensed into numerous punctate dots that are darkly stained and widely dispersed throughout the nucleus (Fig. 3B; Fig. 4A–D). These dots of chromatin are small and fairly uniform in size with an appearance similar to ground pepper, unlike the large balls of condensed chromatin seen in apoptosis *(31,32)* (*see* also descriptions by Clarke and Olney and Ishimaru in this volume). Commonly, there are 1–2 slightly larger and darker stained dots that may represent remnants of the nucleolus. The gelatinous achromatic cells of this type are first readily observed at 12–18 h survival, when several such cells are found laterally adjacent to each other within the bands of Purkinje cell injury. Initially, these gelatinous cells are large and ovoid with round but eccentric nuclei.

---

produces early cytologic changes (B—18 h) in affected cells that resemble chromatolysis. (Arrows indicate damaged Purkinje cell bodies.) Loss of Nissl body (rough ER) staining results in uniformly pale cytoplasm, with a translucent, gelatinous appearance. The nuclei of these cells are initially round with a pale matrix; chromatin is condensed into numerous small punctate dots that are dispersed throughout the nucleus. A larger dot is often seen, presumably the nucleolus. The chromatin dots are small and uniform in size with an appearance similar to ground pepper; they differ from the large balls of condensed chromatin seen in apoptosis *(31,32)*. By 36–48 h (C,D—42 h), the nuclei become somewhat shrunken, pyknotic and flattened, yet remain speckled with chromatin dots. At 2 d survivals, several of these gelatinous achromatic cell bodies lie within large cavities. The cavities rapidly disappear over the next few days, as do most of the damaged Purkinje cells. **(E,F,G)** The Gallyas silver method *(33,34)* for degeneration reveals positive, argyrophilic Purkinje cells 2 days after ibogaine treatment. Purkinje cell bodies are darkly stained (E,G), and silver-stained dendritic fragments form radial columns in the molecular layer above the somata. The alignment of degenerating Purkinje cells in radial stripes is well appreciated with dark field optics (F). Degenerating cell bodies are often situated at the edge of extracellular vacuoles or cavities (F,G). Argyrophilic axons arising from damaged Purkinje cells can be followed into the subcortical white matter (G).

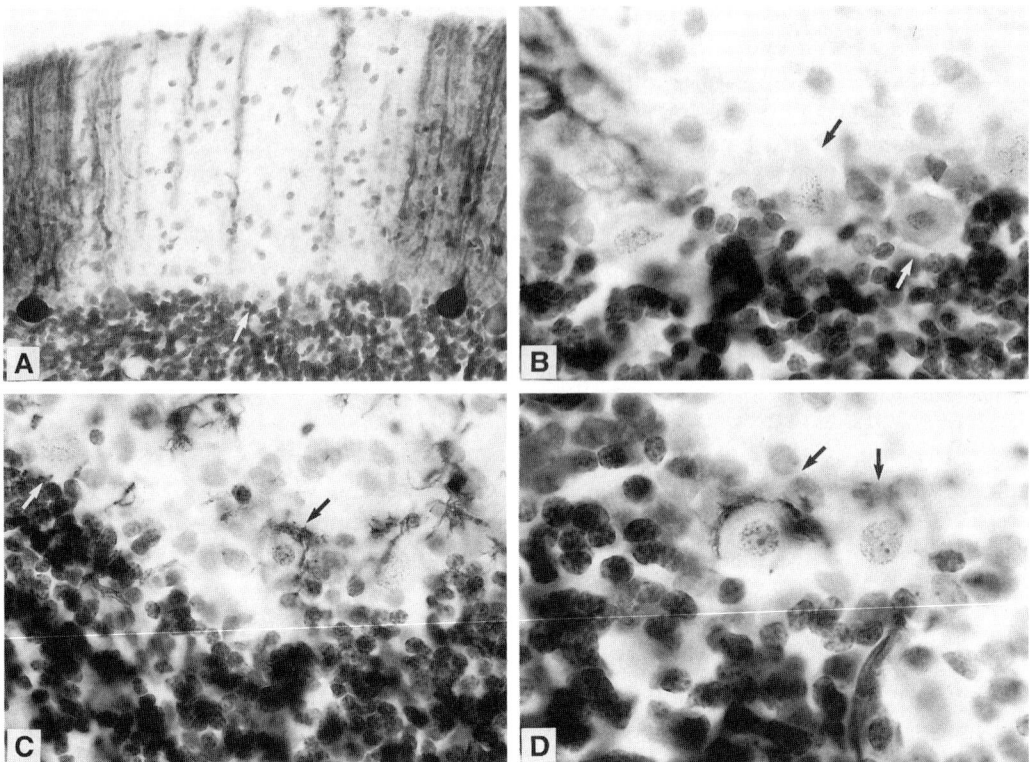

Fig. 4. Cytologic changes in injured Purkinje cells at 24 h after administration of ibogaine. Immunocytochemistry combined with Nissl-stain. **(A,B)** CaM-kinase II; **(C,D)** microglia stained for CR3 (antibody OX42). A band of absent CaM-kinase II reactivity (A) shows loss of stained Purkinje cell bodies and dendrites, yet Nissl-stain reveals a damaged Purkinje cell body with pale cytoplasm and granular nucleus (white arrow). Damaged Purkinje cell bodies (B) exhibit punctate dots of condensed chromatin (black arrow) and gelatinous, chromatolytic cytoplasm (black arrow). Damaged cell at white arrow has irregular contours and retains a reduced level of immunoreactivity for CaM-kinase II. (C,D) Microglial processes surround injured Purkinje cell bodies (arrows) that have pale cytoplasm and dots of condensed chromatin. Magnification (A) × 40 objective; (B,D) × 100; (C) × 63. (Narrow depth of field precludes seeing entire microglial cells.)

They are approximately the same size as normal Purkinje cells, and although some of the abnormal cells may be slightly enlarged, massively edematous Purkinje cells have not been observed. However, 1 d after treatment, the immediate environment of these Nissl-stained cells presents an unusual pathologic appearance: at 2 d survival, many of the injured, gelatinous cells appear to lie within a large cavity (Fig. 3C,D) or extracellular vacuole. The cell body and plasma membrane appear intact, although they are difficult to see (or to photograph) because the cytoplasm is translucent and barely stained. Over the first several days post treatment, many of these cells shrink in size, assume an irregular, crescentic or pleomorphic shape, and occupy an eccentric location at a margin of the cavity (Fig. 3D). At 48 h, the plasma membrane may become scalloped or exhibit irregular surface contours. During the first week after ibogaine treatment, most of the gelatinous cells disappear, presumably through cell death and phagocytosis *(see below)*. The large cavities are transient and have disappeared by 1 wk. However, at 1–2 wk survivals, a small number of individual, gelatinous Purkinje cells persist, but have become shrunken and irregular in shape with eccentric nuclei that are small and crescent shaped. Even at that late stage, the nuclear chromatin of these cells remains in the form of numerous small, dark, punctate dots that are evenly dispersed.

Silver Stain Reveals Degeneration of Cell Bodies, Axons, and Dendrites

Further evidence that the injured Purkinje cells are undergoing degeneration is based on the Gallyas reduced silver method *(33,34)* for degenerating neurons that reveals densely stained, argyrophilic Purkinje cells in the vermis 48 h after rats were treated with ibogaine *(4)* (Fig. 3E–G). Small clusters of silver-stained, degenerating Purkinje cells are aligned in thin parasagittal stripes with the same distribution as the zones which lack CaM-kinase II. Within these circumscribed zones, Purkinje cell bodies are opaque due to deposition of silver, and granular, silver-positive dendritic fragments form thin radial columns in the molecular layer above the somata (Fig. 3E,F). Silver stained, degenerating Purkinje cell bodies are often situated adjacent to extracellular vacuoles, similar to the cavities described above (Fig. 3E–G). Argyrophilic axons arising from damaged cells can be traced into the subcortical white matter (Fig. 3G) where clusters of degenerating axons form compact bundles. In the fastigial and lateral vestibular nuclei, silver-stained axon terminals surround neuronal cell bodies, consistent with these nuclei receiving direct synaptic projections from Purkinje cells in the vermis. The silver stain in these nuclei reflects degenerating axon terminals of injured Purkinje cells.

Glial Response to Purkinje Cell Degeneration

A striking morphologic feature of ibogaine-treated animals is the presence of highly activated microglial cells that are distributed in longitudinal stripes in cerebellar cortex *(3)*. Activated microglia are exquisitely sensitive indicators of neuronal injury and can be readily detected by immunocytochemistry using antisera directed against several different markers that are specific to microglial cells in the brain. Microglial markers that have been used most extensively in this model are antibodies against the complement receptor 3 (CR3, antiserum OX42) and against the MHC II antigen (antiserum OX6). Astrocytes were identified by staining for the glial intermediate filament protein, GFAP, which reveals astrocytic processes. In control brain, resting microglia (Fig. 5A) are readily identified by their expression of CR3; these cells have extensive, extremely fine processes and are widely dispersed throughout the CNS with a distribution that is characteristic for each brain region *(35)*. Resting microglia are not usually detectable with the OX6 antiserum, except for some cells in the white matter. In rats that received ibogaine, cerebellar cortex exhibits clusters of enlarged, darkly stained microglial cells located in the molecular and Purkinje cell layers. These activated microglia, found primarily in the vermis, form distinct radial bands with a location that matches that of damaged or degenerating Purkinje cells (Fig. 2A,E,G). Both of the microglial markers (OX6 and OX42) stain activated cells that have the same distribution and morphology (Fig. 2A,E). Activated astrocytes (Bergmann glia) are more densely stained with antibodies to GFAP than are resting astrocytes, and their radial processes are slightly

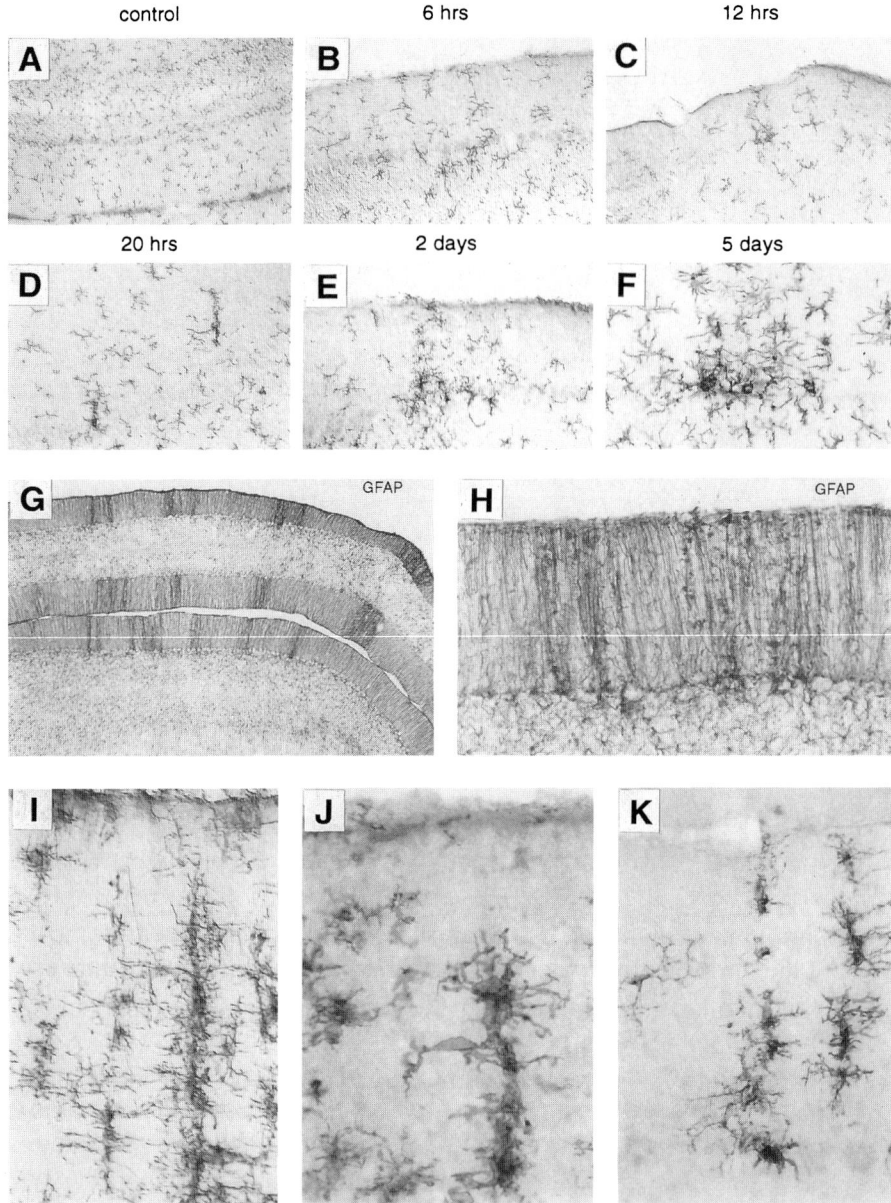

Fig. 5. Activated glial cells are associated with Purkinje cell degeneration. Early phases of microglial activation are seen in (A–F). Activated astrocytes are shown in (G) and (H). Later stages of microglial activation are seen in I–K. **(A)** Resting microglia are widely dispersed in cerebellum of control rat. (Sections in A-F are stained for CR3 with antibody OX42.) The early stages of microglial activation evolve over 6–48 h following drug treatment. **(B)** The earliest signs of microglial activation are detected at 6 h after ibogaine treatment by increased expression of CR3. The first activated microglial cells that exhibit increased staining for CR3 lie adjacent to Purkinje cell bodies. **(C,D)** At 12–20 h survival, additional activated microglia appear in the molecular layer, lying just above those in the Purkinje cell layer. **(E)** At 24–48 h, the number of radial groups of microglial cells increases and individual cells express greater CR3 immunoreactivity. The microglia responding to dendritic damage (molecular layer) become detectable

thickened (Fig. 5G,H). The astrocytes with increased GFAP-positivity form narrow, radially oriented stripes that are coextensive with the stripes of activated microglia in the Purkinje cell and molecular layers (Fig. 2C).

In contrast to resting microglia that have fine, delicate processes, activated microglial cells are more intensely immunoreactive and have enlarged cell bodies with short, thickened processes that appear to surround and engulf damaged Purkinje cells (Fig. 5E,F,J,K). Several activated microglial cells aggregate about individual neurons, forming radial arrays that appear as signposts for degenerating Purkinje cells (Fig. 5I–K). Typically, the soma of a damaged Purkinje cell is completely surrounded by densely stained microglia, and these perisomatic cells appear to be the most activated of the microglial cells (Fig. 4C,D). Several additional microglia, aligned directly above the Purkinje cell body, extend thickened processes that envelop the dendritic tree of the Purkinje cell. These radial microglial clusters outline an entire Purkinje cell with thickened, immunoreactive microglial processes (Fig. 5). Within a zone containing several damaged Purkinje cells, multiple radial clusters of activated microglia may appear nearly confluent. Adjacent groups of activated microglia can be traced from one section to the next, forming parasagittal bands that extend longitudinally through the vermis.

The microglial response seen in this model of cerebellar degeneration has features that are specific both to the region involved and to Purkinje cells in particular. The regional distribution of activated microglia is determined by the local pattern of neuronal injury. In the present example, several microglial cells become closely associated with each degenerating Purkinje cell, and their position, alignment and morphology reflect the particular neuron to which they are responding. Specific features of the microglial response vary with the particular CNS region and cell type that is injured. For example, cell damage in brain stem nuclei presents a quite different picture of activated microglia. A diffuse mass of activated microglia is present in the facial nucleus after axotomy *(36)* and a similar pattern is seen in the inferior olive following a neurotoxic lesion that kills most of the intrinsic olivary neurons *(see below)*. Yet a different picture of activated microglia is found in entorhinal cortex and hippocampus following lesions *(37,38)*.

---

later, starting at 1 to 2 d post-injury, suggesting that the first cues for neuronal injury arise from cell bodies. **(F)** Five days after the neuronal insult, intensely activated microglia engulf the Purkinje cell body and dendrites, and are involved in phagocytosis of cellular debris. Microglial cell bodies become more intensely reactive for CR3 and their cytoplasm increases in volume. **(G,H)** Radial bands of activated astrocytes express increased GFAP in slightly thickened astrocytic processes that ascend radially through the molecular layer. **(I–K)** Strongly activated microglia 1 wk after ibogaine administration are aligned in radial bands where they are associated with damaged Purkinje cell bodies and dendrites. Activated microglial cells have enlarged cell bodies with short, thickened processes that surround damaged Purkinje cells (J,K). Several activated microglia aggregate about individual neurons, forming radial arrays that appear as signposts for degenerating Purkinje cells (I). Typically, a damaged Purkinje cell body is surrounded by several densely stained microglia, which are the most activated of the microglial cells. Several additional microglia, directly above the Purkinje cell body, extend thickened processes that envelop the dendritic tree of the Purkinje cell. These microglial clusters outline an entire Purkinje cell with thickened, immunoreactive microglial processes.

Time Course of Microglial Activation

The first signs of activated microglia can be detected 6 h after ibogaine treatment by increased expression of CR3. Microglial cells that are clearly activated are found in the Purkinje cell layer of the vermis (Fig. 5B,C), but they are fewer in number and their processes are thinner than after longer survivals (e.g., 1 wk). The sequential process of microglial activation evolves over the period of 6–48 h following drug treatment. The first microglial cells that appear to be activated based upon increased CR3 expression lie adjacent to cell bodies of injured Purkinje cells that are encircled by microglial processes (Fig. 4C–D). At the early time points (from 6 to 12 h) a small number of activated cells is also seen in the granule layer, lying just below an activated microglial cell in the Purkinje cell layer. This pattern suggests that the earliest activated microglia are associated with cell bodies of injured Purkinje cells, in addition to some deeper ones along Purkinje cell axons, likely associated with the axonal "torpedoes" described above. At a 12 h survival, additional activated microglia appear in the molecular layer, lying just above those in the Purkinje cell layer (Fig. 5). At 24 and 48 h, the number of radial groups of microglial cells has increased and the individual cells express greater immunoreactivity for CR3. The microglia responding to dendritic damage become detectable only at 1–2 d post-injury, suggesting that injured neuronal cell bodies express the initial cues leading to microglial activation. From 2 to 7 d after neuronal insult, numerous, intensely activated microglia engulf the Purkinje cell body and are involved in phagocytosis and clearing of cellular debris, as suggested by Kreutzberg *(39)*. In the early stages of activation (6–20 h), the fine processes of affected microglia decrease in length and number, presumably due to retraction *(39,40)*. During that time, microglial cell bodies become more intensely reactive for CR3 and their cytoplasm increases in volume. Subsequently, from 24 to 48 h, the cells extend blunt, thickened processes that are closely associated with the affected neurons (Fig. 5).

Activated microglial cells can be detected at 6 h after ibogaine administration, which is earlier than the first appearance of silver-stained neurons and prior to that of activated astrocytes, since increased GFAP is not seen before 24–48 h. The initial signs of microglial activation in the present model precede even the first detectable morphologic changes found in Nissl-stained Purkinje cells. Microglial activation is therefore deemed to be the earliest and most sensitive morphologic sign of neuronal injury, suggesting that microglia, which constitute the immune system of the brain, are the first cells to respond to damaged neurons. Although they may have a protective or scavenger function, Kreutzberg and colleagues have proposed that microglia, as the resident macrophages of the CNS, may be involved in killing neurons by direct contact and phagocytosis or by release of toxic cytokines *(39,41,42)*. During the first month following the lesion induced by ibogaine, microglial activation remains more evident than astrocyte activation. The activation of astrocytes appears to be regulated by microglia, which can stimulate astrocytosis by release of IL-1 *(43,44)*. Astrocytes provide a less sensitive marker of neuronal injury than do microglia, particularly in the case of subtle lesions. Baseline resting levels of GFAP in the brain are fairly high and astrocyte activation produces a relatively small further increase; as staining of quiescent astrocytes is strong, the ability to detect small increases is limited. Moreover, the astrocytes associated with Purkinje cells (Bergmann glia) largely appear as fine radial processes rising through the molecular layer such that morphologic changes in these astrocytes are less evident than those in microglia, which have broad, blunt processes. Thus, microglial activation provides a far more dramatic and sensitive marker of neuronal injury, especially at early stages, than does activation of astrocytes. However, astrocyte and microglial markers differ in their persistence following very long survivals. Microglial markers of neuronal damage are maximally detectable at 2 wk after an insult, followed by a gradual decrease in expression of immune markers. Microglia slowly return to a resting state after phagocytosis is completed, such that only a subtle elevation in CR3 can be detected at 4–5 mo after the ibogaine lesion. At a survival of 1 yr following the injury, the increased astrocytic GFAP remains readily detectable, but before that time, microglia have returned to a quiescent state and leave no sign of previous activation. In addition to acutely isolating injured neurons, astrocytes are generally believed to form a "scar" that provides structural stability to the tissue after dead neurons have been removed. In contrast, microglia are active participants during the acute phases of neuronal degeneration and later become quiescent after their phagocytic housecleaning role is completed.

Does Nitric Oxide Contribute to Purkinje Cell Degeneration?

Nitric oxide (NO) has been proposed as a putative messenger of neuronal injury *(45–47)* and nitric oxide synthase (NOS), the enzyme responsible for NO synthesis *(48,49)* can be induced in neurons by CNS lesions in cerebellum *(50,51)* and in other regions *(52,53)*. Mature Purkinje cells do not express NOS under normal conditions, making these neurons ideal for studying injury-related NOS induction. In normal adult rats, Purkinje cells neither exhibit immunoreactivity for neuronal NOS (nNOS) nor are they positive for diaphorase (NADPH-d), a histochemical marker for NOS activity *(54)*. However, starting 3 d after ibogaine treatment, a small number of Purkinje cells display NADPH-d activity in the cell body and dendrites and have become immunoreactive for nNOS *(55)*. In ibogaine-treated rats, Purkinje cells that are nNOS-positive may appear morphologically normal (Fig. 6F), but some have irregularly shaped somata and dendrites (Fig. 6D,E,G). The NADPH-d and nNOS-positive Purkinje cells are situated predominantly in the vermis and consistently found near sagittal zones of neuronal degeneration, often adjacent to a zone of missing Purkinje cells. The labeled cells are generally surrounded by activated microglia (Fig. 6C,E). At longer survival times, the number of NADPH-d-positive Purkinje cells progressively increases over several days, reaching a maximum at 2 wk after ibogaine administration. Subsequently, they decrease in number, falling to only an occasional positive cell at 33 d after treatment *(55)*.

The main results of the NOS investigation show that ibogaine-induced neuronal damage leads to induction of nNOS in a subset of Purkinje cells that are typically located in the company of other degenerating Purkinje cells *(55)*. The long time course of this process demonstrates that NOS induction is delayed, reaching a maximum only at 2 wk after the initial insult. Given that mature Purkinje cells do not normally express NOS, the induction of this enzyme is interpreted as a delayed response to sublethal neuronal injury. The abnormal morphology observed in many diaphorase-positive Purkinje cells and their proximity to degenerating neurons after ibogaine treatment supports the hypothesis that neuronal damage is associated with induction of nNOS. In most studies of NOS induction due to injury, the principal neuronal insult has been axotomy *(52,56)*, which classically leads to a chromatolytic response to neuronal injury *(57)*. The induction of nNOS in the present excitotoxic injury model in which the cytologic changes

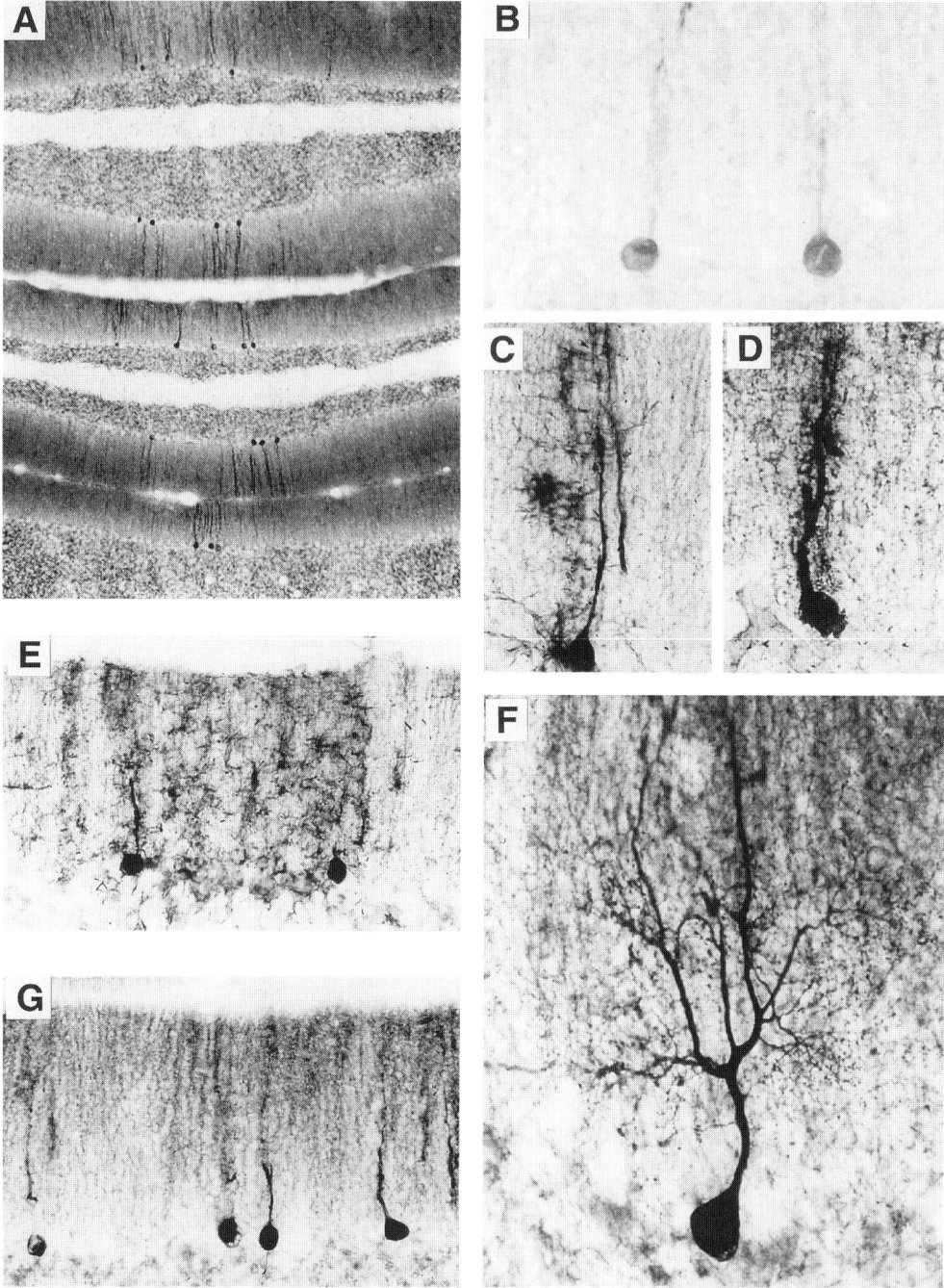

Fig. 6. Induction of nNOS in injured Purkinje cells. Only a small fraction of the damaged Purkinje cells express NOS. Normal, mature Purkinje cells do not exhibit immunoreactivity for neuronal NOS (nNOS) nor are they positive for diaphorase (NADPH-d), a histochemical marker for NOS activity *(54)*. Beginning 3 d after ibogaine treatment, a small number of Purkinje cells become positive for nNOS and for NADPH-d activity. The number of NADPH-d-positive Purkinje cells subsequently increases, reaching a maximum at 2 wk after ibogaine administration. The induction of this enzyme is interpreted as a delayed response to sublethal neuronal injury, seen only in those cells that survive the initial insult. **(A)** Several Purkinje cells express

mirror those of chromatolysis suggests an additional link between the mechanisms involved in excitotoxic injury and in axotomy-induced chromatolysis.

NOS induction is not observed on the first day after ibogaine treatment, when acute Purkinje cell damage is evident; rather, it is first detected after 3 d, when NOS-positive cells are still quite sparse. The number of positive cells increases much later (1–2 wk), when many of the damaged Purkinje cells are already necrotic or have been phagocytized. Based on this lengthy delay, NO is unlikely to be the initial mediator of damage in the acute phase of ibogaine-induced degeneration. Rather, the subsequent increase in nNOS-positive cells over 3–14 d is consistent with NOS induction being a response to neuronal injury, rather than a cause. The late appearance of nNOS may indicate the existence of a second, delayed phase in the expression of Purkinje cell injury well after the initial insult, and raises the possibility that additional neurons may undergo degeneration several weeks after ibogaine treatment. Whether the few gelatinous, chromatolytic Purkinje cell bodies present at 2 wk are the same ones that express nNOS has not been yet determined, nor is it known whether the nNOS-positive cells die or go on to recover. Yet the data show that there are multiple patterns of response to this excitotoxic insult, ranging from acute cell death to a prolonged evolution of changes similar to chromatolysis.

IS IBOGAINE-INDUCED PURKINJE CELL DEGENERATION MEDIATED BY THE OLIVOCEREBELLAR PROJECTION?

The present chapter began with the proposal that this form of cerebellar toxicity is put forth as a model of excitotoxic neuronal injury that is produced by intrinsic neuronal circuitry. The hypothesis that ibogaine or harmaline toxicity is dependent on and mediated by the projection from inferior olive to cerebellum has recently been subject to experimental verification *(58)*. To determine whether climbing fibers from the inferior olive mediate the neurotoxicity, the inferior olivary nucleus was chemically lesioned using a neurotoxic drug regimen that produces relatively selective degeneration of neurons in the inferior olive *(59–61)*. Six days after ablating the olivary neurons, the rats were injected with ibogaine and maintained for an additional week. The animals were perfused 1 wk after ibogaine treatment, and brain sections were prepared in order to analyze Purkinje cell damage and to verify the extent of olive ablation. The magnitude of Purkinje cell injury produced by ibogaine administered after ablation of

---

NADPH-diaphorase activity 7 d after ibogaine treatment. Diaphorase activity is a histochemical marker for NOS. **(B)** Two Purkinje cells are immuno-positive with an antibody for nNOS, the enzyme responsible for NO synthesis. **(C,D)** Diaphorase-positive Purkinje cells following ibogaine treatment. The cell in (C) is partially surrounded by activated microglia (double stain: diaphorase plus CR3). Cell in (D) with ragged margins is injured. **(E)** A radial band of activated microglial cells is associated with a diaphorase-positive Purkinje cell at each edge of the band. The activated microglia correspond to a zone where PKCs have degenerated. (NADPH-d and OX42. Surv. 13 d.) **(F)** A diaphorase-positive Purkinje cell body and its dendritic tree in the sagittal plane. Brown profiles in background are activated microglia. (NADPH-d and OX6. Surv. 12 d.) **(G)** Four NADPH-d positive PKCs. Two cells on left have irregular shape of soma. (NADPH-d. Surv. 12 d.) Plates (A,B,E,F) are reproduced from *[55]* with permission of Rapid Science Publishers.

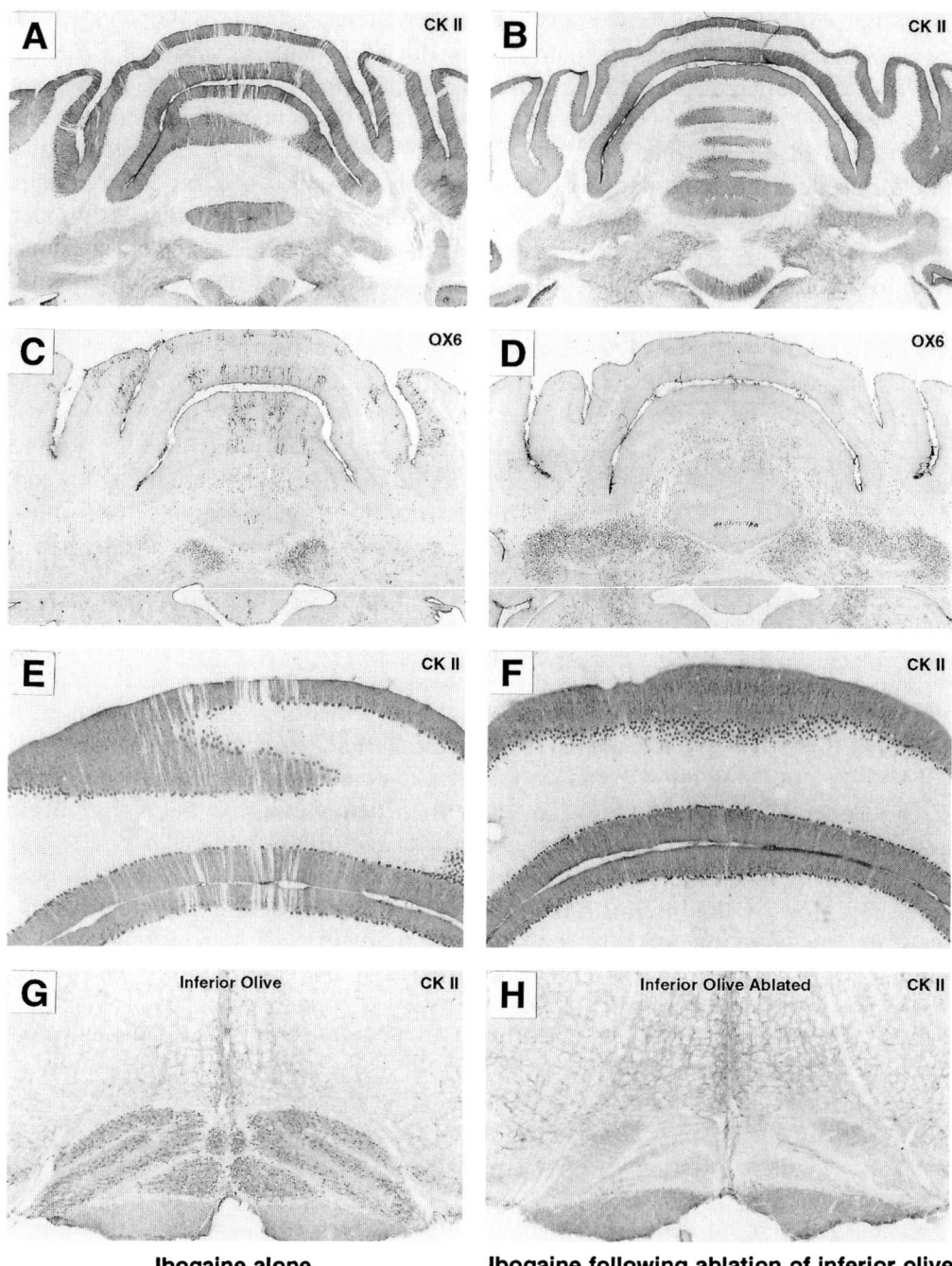

Fig. 7. Ibogaine-induced Purkinje cell degeneration is prevented by prior ablation of the inferior olivary nucleus. **(A,C,E)** Treatment with ibogaine alone (100 mg/kg) leads to radial bands of Purkinje cell loss at 1 wk survival (stain for Cam-kinase II) and microglial activation (MHC II) in corresponding bands. **(G)** Intact olivary neurons make up the inferior olivary nucleus (Cam-kinase II). **(B,D,F,H)** Figures in the right column follow olive ablation produced by the 3-AP regimen. Ibogaine given 1 wk after olive ablation produces almost no degeneration

the inferior olive was compared to the damage caused by ibogaine in normal rats, with intact inferior olivary nuclei. If the Purkinje cell degeneration induced by ibogaine is mediated by the olivocerebellar projection, then ablation of the olive prior to giving ibogaine should prevent the neuronal damage.

Ablation of the inferior olive was achieved by a neurotoxic lesion produced by giving systemic 3-acetylpyridine (3-AP). The 3-AP regimen consists of sequentially injected 3-AP, harmaline, and nicotinamide *(59)*. Nearly all inferior olivary neurons rapidly degenerate after rats are administered the 3-AP regimen. One to two weeks later, the site of the inferior olive contains densely-packed, small glial nuclei, and the large neuronal cell bodies that are present in the inferior olive of control rats have degenerated and are missing (Fig. 6G,H). Sections through the inferior olive stained with microglial markers reveal that the olivary neurons have been almost totally replaced by activated microglial cells packed in a gliotic zone that resembles the inferior olivary nucleus in location and shape *(58)*.

Prior ablation of the inferior olive almost completely prevents subsequent ibogaine-induced degeneration of Purkinje cells *(58)*. In contrast to the prominent neuronal damage in cerebellar cortex that follows treatment with ibogaine alone (Fig. 7A,E), the amount of Purkinje cell loss is markedly reduced when olive ablation precedes ibogaine administration (Fig. 7B,F). Radial arrays of degenerating Purkinje cells are nearly absent in the vermis of rats that received 3-AP plus ibogaine (compare Fig. 7E,F). In addition, compared with the effects of ibogaine alone, microglial activation was profoundly decreased in these rats (compare Fig. 7C,D). This neuroprotective effect of olive ablation has been found in every animal that was treated with the 3-AP regimen prior to receiving ibogaine.

Mechanism of Purkinje Cell Degeneration Induced by Ibogaine: A Model of Excitotoxicity

The demonstration that ablation of the inferior olivary nucleus prevents subsequent ibogaine-induced Purkinje cell degeneration indicates that ibogaine is not directly toxic to Purkinje cells. (The latter formulation assumes that exposure of Purkinje cells to the drug is identical regardless of whether the inferior olive is present or absent. However, there are brain areas where direct excitotoxicity of a drug is well documented, and yet ablation of a natural glutamatergic input offers some protection from the direct excitotoxic effects–*see* striatal excitotoxicity in the chapter by Ross, Becker, and Koliatsos) The protective effect of olive ablation supports the conclusion that ibogaine produces sustained activation of the olivocerebellar projection which then causes Purkinje cell degeneration by an excitotoxic mechanism. The experimental use of ibogaine to induce degeneration of Purkinje cells provides a useful in vivo model of excitotoxic neuronal injury. The excitotoxicity is mediated trans-synaptically through intrinsic olivocerebellar circuitry. The systemic administration of ibogaine or harmaline as a model system to investigate excitotoxic neuronal injury has several methodological advantages over other models. This experimental paradigm employs a

---

in cerebellar cortex (B,F) nor activation of microglia (D). (H) Following the 3-AP regimen, nearly all neurons in the inferior olive have degenerated and are consequently unstained. Comparison of right and left columns shows the neuroprotective effect of olive ablation. These data demonstrate that Purkinje cell degeneration induced by ibogaine is mediated by the olivocerebellar projection.

non-invasive procedure that does not damage the blood–brain barrier or involve surgical transection of neuronal processes or tissues. There is no artificial window that might facilitate invasion of macrophages or other inflammatory cells. Moreover, the site of ibogaine action (inferior olivary nucleus—medulla) is distant from the site of neuronal damage (Purkinje cells—cerebellum) such that drug administration does not artifactually alter cells in the target site, interfere with local blood flow, or provoke microglial activation. In addition, the neuronal insult caused by ibogaine is mediated by synaptic release of an endogenous neurotransmitter (presumably glutamate) at its natural postsynaptic target, in contrast to models of experimental excitotoxicity in which exogenous neurotoxins are injected directly into the region under investigation where they may well produce physical damage and nonspecific effects. These advantages of the present model of neurotoxicity should facilitate the analysis of neuronal injury, unobscured by confounding effects due to invasive procedures. In order to determine the influence of neuronal circuitry in producing excitotoxic cellular damage, it is important and advantageous to use an in vivo system where normal circuits and pathways are intact and functional.

Olivocerebellar Projection Confers Purkinje Cell Vulnerability

The olivocerebellar projection has several unusual features that determine the pattern of degeneration induced by ibogaine and that contribute to the vulnerability of Purkinje cells to excitotoxic insult. The longitudinal stripes of Purkinje cell degeneration caused by ibogaine and harmaline (Fig. 2) are consistent with the sagittal organization of the projection from the inferior olivary nucleus to cerebellar cortex. It is well established anatomically that the olivocerebellar projection has a precise topographic organization *(62–64)*. The neurons in the inferior olive give rise to axons that terminate as climbing fibers in cerebellar cortex where they form repeated synapses upon Purkinje cells *(65–67)*. Small groups of inferior olivary neurons that are coupled by gap junctions *(24,68,69)*, and thereby fire synchronously, project to a narrow band of Purkinje cells aligned in parasagittal rows. Hence, small zones of the inferior olive project to narrow longitudinal zones in cerebellar cortex *(64,70)*. Llinás and colleagues have shown that Purkinje cells aligned in rostro-caudal rows are activated synchronously by climbing fibers; in contrast, there is poor correlation of activity among Purkinje cells that lie lateral to each other *(71,72)*. Thus, the strict topographic organization of the olivocerebellar projection determines the pattern in which Purkinje cells in the same rostro-caudal row are excited, a factor that likely underlies the parasagittal bands of degeneration produced by ibogaine. Given that ibogaine and harmaline greatly increase rhythmic activity in the inferior olive, the finding of Purkinje cell damage in longitudinal stripes together with the sagittal organization of olivocerebellar circuitry support the proposal that, in this system, the organization of neuronal circuitry determines the spatial pattern of degeneration.

Although the distribution of degenerating cells is set by the topography of the olivocerebellar projection, the synaptic termination of climbing fibers establishes the vulnerability of individual neurons. Each Purkinje cell is innervated by a single climbing fiber, the branches of which ascend in close proximity to the Purkinje cell dendrites *(66,67)*. One climbing fiber forms repeated synaptic contacts upon the surface of a Purkinje cell, with the total number estimated to be at least several hundred synapses on each cell *(73)*. This specialized synaptic termination of climbing fibers provides the basis for their uniquely powerful excitatory action upon Purkinje cells *(74,75)*. The arrangement of climbing fiber axon terminals is referred to as a "distributed synapse" *(76)*, as each climbing fiber action potential results in nearly synchronous release of an excitatory transmitter (presumably glutamate *[77,78]*) at hundreds of synapses that are widely distributed over the Purkinje cell surface. According to Llinás and Nicholson *(76)*, the multiplicity of climbing fiber synapses distributed over the surface of the Purkinje cell is likely to form the most efficient, high priority type of synaptic contact, having a high safety factor for transmission and the largest possible depolarization. Based on the large number of these synapses per cell and the unique security of synaptic transmission, Eccles *(74)* described the excitation of Purkinje cells by a climbing fiber as the "most powerful and specific excitatory synapse yet discovered in the central nervous system." We propose that the distributed nature of this synaptic arrangement, combined with the great security of neurotransmission, confers upon Purkinje cells a high degree of vulnerability to excitotoxic injury. This enhanced vulnerability places the Purkinje cell at great risk and is likely to underlie the heightened susceptibility of Purkinje cells to a wide variety of insults *(79)*. By virtue of this unique pattern of neuronal circuitry, any event that produces sustained excitation of the inferior olive can result in Purkinje cell damage due to an excitotoxic mechanism. Excitotoxic injury of this type is proposed to result from repetitive glutamatergic excitation over the entire neuronal surface, leading to elevation of intracellular calcium levels throughout the Purkinje cell cytoplasm. Persistence of calcium overload can activate multiple Ca^{2+}-dependent enzymes that then damage neuronal constituents *(2,80–82)*.

It is likely that AMPA and metabotropic receptors mediate climbing fiber activation of Purkinje cells and that AMPA receptor activation is most responsible for the excitotoxic injury *(83)*. Activation of AMPA receptors leads to influx of sodium, that causes membrane depolarization followed by opening of voltage-sensitive calcium channels *(84,85)*. Sodium influx is likely to be accompanied by water leading to swelling of the Purkinje cell, which would contribute to neuronal damage by causing structural injury to the cell. In addition, some AMPA receptors also permit Ca^{2+} influx directly *(92)*. Metabotropic receptor stimulation leads to mobilization of intracellular calcium via the formation of IP3 *(87,88)*. Consequently, prolonged, rhythmic climbing fiber activation is likely to produce massive increases in intracellular calcium levels throughout the dendrites and cell body, leading to subsequent activation of Ca^{2+}-dependent enzymes that can initiate a proteolytic cascade leading to Purkinje cell degeneration *(86,89)*. Depending upon the particular experimental model and cell types involved, different glutamate receptors may be the main effectors of excitotoxicity *(83)*. Moreover, individual neuron types may exhibit differential vulnerability to particular glutamate agonists or may manifest different responses to injury due in part to the prevalence of glutamate receptor subtypes expressed by that cell. For example, incubation of cerebellar slices in glutamate agonists has revealed that NMDA causes selective degeneration of granule cells, but not Purkinje or Golgi cells; Purkinje cells are most vulnerable to AMPA, and Golgi cells to kainate *(83)*. In addition, particular cell types may express a different degenerative response to the same agonist: after slice incubation in AMPA, Golgi cells become pale and edematous whereas Purkinje cells are dark and vacuolated *(83)*. Thus, there is no single, universal response that is a sign of excitotoxicity; rather, the response profile of neurons to excitotoxic insult is influenced by multiple factors that are specific to individual cell types. The main factors include the pattern of synaptic input to particular neurons, the presence of local inhibitory interneurons that may be neuroprotective, and the profile of receptors expressed by the postsynaptic neuron. The local glial environment may also influence the neuronal response to insult, but as yet there have been few studies of differential glial responses to neuronal injury. We may speculate that regional differences may exist in the glial release of cytokines or trophic factors, and in the capacity for removal or metabolism of offending neurotransmitters or toxins. Further studies on the functional role of microglia and astrocytes should help to clarify these possibilities.

The saga of physiologic effects of ibogaine on the inferior olive that lead ultimately to sustained rhythmic release of glutamate at hundreds of synapses over the entire Purkinje cell surface provides an example of what appears to be the most vulnerable circuit in the brain to excitotoxic injury. This phenomenon should provide a useful model for studying the multiple factors that mediate and lead to excitotoxic injury in the CNS. It should also provide a system that may serve to test new pharmacologic approaches to ameliorate excitotoxic damage.

CONCLUSIONS

This chapter presents an experimental model of excitotoxicity based on the administration of drugs (ibogaine and harmaline) that produce trans-synaptic degeneration in the cerebellum, a region remote from the site of drug action, the inferior olive. These drugs cause sustained, rhythmic bursts of activity in olivary neurons. The vulnerability of Purkinje cells is postulated to result from the synaptic organization of the olivocerebellar projection, whereby a single climbing fiber forms hundreds of synaptic contacts that are distributed over the soma and dendrites of a Purkinje cell. Every climbing fiber impulse leads to simultaneous glutamate release at each synapse, producing depolarization and opening of ion channels over the entire neuronal surface. The resultant accumulation of intracellular Ca^{2+} is postulated to activate Ca^{2+}-sensitive proteolytic enzymes that produce destruction of organelles leading to cell death. Degenerating Purkinje cells are distributed in narrow bands, a pattern determined by the parasagittal projection of climbing fibers from the inferior olive to the cerebellar cortex. Activated microglial cells form similar longitudinal rows that are coextensive with the bands of degenerating neurons. The circuitry of this projection is the factor that determines this unique pattern of neuronal degeneration. The specialized synaptic arrangement of climbing fiber terminals confers great vulnerability upon Purkinje cells and may explain their high susceptibility to injury in many pathologic conditions.

This model system should prove useful to investigate excitotoxic neuronal injury since a predictable pattern of Purkinje cell degeneration is produced by a noninvasive procedure that does not alter the blood–brain barrier or cause direct physical damage. The cytologic features seen in this form of degeneration differ from most other examples of excitotoxic neuronal injury and further studies are in progress to better characterize the nature of this lesion. The injured cells exhibit unusual morphologic features by light microscopy. Purkinje cell cytoplasm appears chromatolytic in Nissl-stained frozen sections. Nuclear changes differ from those usually described in apoptosis (globular chromatin condensation) or in necrosis (dark pyknotic appearance). After ibogaine administration, the Purkinje nuclei become translucent, dotted with abundant small chromatin clumps that are widely dispersed through the nucleus. These changes do not match standard depictions of apoptosis or necrosis but are moderately close to the picture of excitotoxic cell death described by Clarke *(31)* and by Olney *(32)*. The present data suggest that this form of excitotoxicity has many features in common with axotomy-induced chromatolysis *(42,57,85)* although the same nuclear changes are not often described. Examples of granular clumping of chromatin have occasionally been reported in kainate-induced degeneration *(91)* and after AMPA treatment *(83)* in Golgi but not Purkinje cells. Moreover, the presence of barely stained Purkinje cells 1 d after ibogaine treatment, with both Nissl and immunocytochemical methods, underscores the need for caution in declaring that a neuron is dead or missing. Multiple neuronal markers and well-defined criteria will be needed to establish neuron loss. Hence, in agreement with Clarke (*see* his chapter in present volume), there is no universal set of morphologic criteria for the identification of apoptosis versus necrosis. Individual cases of cell death should be carefully assessed with regard to neuron type and nature of the insult. The several forms of excitotoxic neuronal death may involve different manifestations *(92)*, and detailed cytologic analysis of changes at both light and electron microscopic levels is needed to formulate a meaningful classification of the types of cell death.

ACKNOWLEDGMENTS

This study was supported by USPHS Grants DA 08692, DA 00225, and NO1DA-3-7301.

REFERENCES

1. Choi DW. Glutamate neurotoxicity and diseases of the nervous system. *Neuron* 1988, **1:** 623–634.
2. Olney JW. Neurotoxicity of excitatory amino acids, in *Kainic Acid as a Tool in Neurobiology* (McGeer EG, Olney JW, McGeer PO, eds.), Raven, New York, 1978, pp. 95–171.
3. O'Hearn E, Long DB, Molliver ME. Ibogaine induces glial activation in parasagittal zones of the cerebellum. *NeuroReport* 1993, **4:** 299–302.
4. O'Hearn E, Molliver ME. Degeneration of Purkinje cells in parasagittal zones of the cerebellar vermis after treatment with ibogaine or harmaline. *Neuroscience* 1993, **55:** 303–310.
5. Dhahir HI. A comparative study on the toxicity of ibogaine and serotonin. Indiana University PhD Thesis, 1971, 71: 1–151.
6. Liwszyc GE, Vuori E, Rasanen I, Issakainen J. Daime—a ritual herbal potion. *J Ethnopharmacol* 1992, **36:** 91–92.
7. Lotsof HS. Rapid method for interrupting the narcotic addiction syndrome. US Patent No. 4,499,096 (1985).
8. Lotsof HS. Rapid method for interrupting the cocaine and amphetamine abuse syndrome. US Patent No. 4,587,243 (1986).
9. Lotsof HS. Ibogaine in the treatment of chemical dependency disorders: Clinical perspectives. *Multidisciplinary Assoc Psychedelic Studies* 1995, **5:** 16–27.
10. Sheppard SG. A preliminary investigation of ibogaine: Case reports and recommendations for further study. *J Subst Abuse Treat* 1994, **11:** 379–385.
11. Glick SD, Rossman K, Steindorf S, Maisonneuve IM, Carlson JN. Effects and aftereffects of ibogaine on morphine self-administration in rats. *Eur J Pharmacol* 1991, **195:** 341–345.
12. Glick SD, Rossman K, Rao NC, Maisonneuve IM, Carlson JN. Effects of ibogaine on acute signs of morphine withdrawal in rats: Independence from tremor. *Neuropharmacology* 1992, **31:** 497–500.

13. Dworkin SI, Gleeson S, Meloni D, Koves TR, Martin TJ. Effects of ibogaine on responding maintained by food, cocaine and heroin reinforcement in rats. *Psychopharmacology* 1995, **117:** 257–261.
14. Sharpe LG, Jaffe JH. Ibogaine fails to reduce naloxone-precipitated withdrawal in the morphine-dependent rat. *NeuroReport* 1990, **1:** 17–19.
15. Singbartl G, Zetler G, Schlosser L. Structure-activity relationships of intracerebrally injected tremorigenic indole alkaloids. *Neuropharmacology* 1973, **12:** 239–244.
16. Zetler G, Singbartl G, Schlosser L. Cerebral pharmacokinetics of tremor-producing harmala and iboga alkaloids. *Pharmacology* 1972, **7:** 237–248.
17. Zetler G, Back G, Iven H. Pharmacokinetics in the rat of the hallucinogenic alkaloids harmine and harmaline. *Naunyn Schmiedebergs Arch Pharmacol* 1974, **285:** 273–292.
18. Lamarre Y, De Montigny C, Dumont M, Weiss M. Harmaline-induced rhythmic activity of cerebellar and lower brain stem neurons. *Brain Res* 1971, **32:** 246–250.
19. Poirier LJ, Sourkes TL, Bouvier G, Boucher R, Carabin S. Striatal amines, experimental tremor and the effect of harmaline in the monkey. *Brain* 1966, **89:** 37–52.
20. De Montigny C, Lamarre Y. Activity in the olivo-cerebello-bulbar system of the cat during ibogaline- and oxotremorine-induced tremor. *Brain Res* 1974, **82:** 369–373.
21. De Montigny C, Lamarre Y. Rhythmic activity induced by harmaline in the olivo-cerebello-bulbar system of the cat. *Brain Res* 1973, **53:** 81–95.
22. Llinás R, Volkind RA. The olivo-cerebellar system: Functional properties as revealed by harmaline-induced tremor. *Exp Brain Res* 1973, **18:** 69–87.
23. De Montigny C, Lamarre Y. Effects produced by local applications of harmaline in the inferior olive. *Can J Physiol Pharmacol* 1975, **53:** 845–849.
24. Llinás R, Yarom Y. Oscillatory properties of guinea-pig inferior olivary neurones and their pharmacological modulation: An in vitro study. *J Physiol* 1986, **376:** 163–182.
25. Lamarre Y, Weiss M. Harmaline-induced rhythmic activity of alpha and gamma motoneurons in the cat. *Brain Res* 1973, **63:** 430–434.
26. Milner TE, Cadoret G, Lessard L, Smith AM. EMG analysis of harmaline-induced tremor in normal and three strains of mutant mice with Purkinje cell degeneration and the role of the inferior olive. *J Neurophysiol* 1995, **73:** 2568–2577.
27. Lamarre Y, Mercier LA. Neurophysiological studies of harmaline-induced tremor in the cat. *Can J Physiol Pharmacol* 1971, **49:** 1049–1058.
28. Cajal SR. *Cajal's degeneration and regeneration of the nervous system.* Oxford University Press. NY, 1991, pp. 631–677.
29. Adams JH, Duchen LW. Greenfield's Neuropathology. Oxford University Press, New York, 1992.
30. Sotelo C. Axonal abnormalities in cerebellar Purkinje cells of the 'hyperspiny Purkinje cell' mutant mouse. *J Neurocytol* 1990, **19:** 737–755.
31. Clarke PGH. Apoptosis versus necrosis: How valid a dichotomy for neurons? in *Cell Death in Diseases of the Nervous System* (Koliatsos VE, Ratan RR, eds.), Humana Press, Totowa, NJ, 1997, this volume.
32. Olney JW, Ishimaru MJ. Excitotoxic cell death, in *Cell Death in Diseases of the Nervous System* (Koliatsos VE, Ratan RR, eds.), Humana Press, Totowa, NJ, 1997, this volume.
33. Gallyas F, Guldner FH, Zoltay G, Wolff JR. Golgi-like demonstration of "dark:" neurons with an argyrophil III method for experimental neuropathology. *Acta Neuropathol* 1990, **79:** 620–628.
34. Gallyas F, Wolff JR, Bottcher H, Záborszky L. A reliable and sensitive method to localize terminal degeneration and lysosomes in the central nervous system. *Stain Technol* 1980, **55:** 299–306.
35. Lawson LJ, Perry VH, Dri P, Gordon S. Heterogeneity in the distribution and morphology of microglia in the normal adult mouse brain. *Neuroscience* 1990, **39:** 151–170.

36. Graeber MB, Streit WJ, Kreutzberg GW. Axotomy of the rat facial nerve leads to increased CR3 complement receptor expression by activated microglial cells. *J Neurosci Res* 1988, **21**: 18–24.
37. Gehrmann J, Schoen SW, Kreutzberg GW. Lesion of the rat entorhinal cortex leads to a rapid microglial reaction in the dentate gyrus. A light and electron microscopical study. *Acta Neuropathol* 1991, **82**: 442–455.
38. Jorgensen MB, Finsen BR, Jensen MB, Castellano B, Diemer NH, Zimmer J. Microglial and astroglial reactions to ischemic and kainic acid-induced lesions of the adult rat hippocampus. *Exp Neurol* 1993, **120**: 70–88.
39. Kreutzberg GW. Microglia: A sensor for pathological events in the CNS. *Trends Neurosci* 1996, **19**: 312–318.
40. Andersson P-B, Perry VH, Gordon S. The kinetics and morphological characteristics of the macrophage-microglial response to kainic acid-induced neuronal degeneration. *Neuroscience* 1991, **42**: 201–214.
41. Gehrmann J, Matsumoto Y, Kreutzberg GW. Microglia: intrinsic immuneffector cell of the brain. Brain Res Rev 1995, **20**: 269–287.
42. Kreutzberg GW. Reaction of the neuronal cell body to axonal damage, in *The Axon. Structure, Function and Pathophysiology* (Waxman SG, Kocsis JD, Stys PK, eds.), Oxford University Press, New York, 1995, pp. 355–374.
43. Giulian D, Baker TJ. Peptides released by ameboid microglia regulate astroglial proliferation. *J Cell Biol* 1985, **101**: 2411–2415.
44. Giulian D, Young DG, Woodward J, Brown DC, Lachman LB. Interleukin-1 is an astroglial growth factor in the developing brain. *J Neurosci* 1988, **8**: 709–714.
45. Dawson VL, Dawson TM. Nitric oxide neurotoxicity. *J Chem Neuroanat* 1996, **10**: 179–190.
46. Dawson VL, Dawson TM, London ED, Bredt DS, Snyder SH. Nitric oxide mediates glutamate neurotoxicity in primary cortical cultures. *Proc Natl Acad Sci USA* 1991, **88**: 6368–6371.
47. Dawson VL, Kizushi VM, Huang PL, Snyder SH, Dawson TM. Resistance to neurotoxicity in cortical cultures from neuronal nitric oxide synthase-deficient mice. *J Neurosci* 1996, **16**: 2479–2487.
48. Bredt DS, Snyder SH. Nitric oxide, a novel neuronal messenger. *Neuron* 1992, **8**: 3–11.
49. Knowles RG, Palacios M, Palmer RM, Moncada S. Formation of nitric oxide from L-arginine in the central nervous system: a transduction mechanism for stimulation of the soluble guanylate cyclase. *Proc Natl Acad Sci USA* 1989, **86**: 5159–5162.
50. Chen, S, Aston-Jones, G. Cerebellar injury induces NADPH diaphorase in Purkinje and inferior olivary neurons in the rat. *Exp Neurol* 1994, **126**: 270–276.
51. Saxon DW, Beitz AJ. An experimental model for the non-invasive trans-synaptic induction of nitric oxide synthase in Purkinje cells of the rat cerebellum. *Neuroscience* 1996, **72**: 157–165.
52. Wu W, Liuzzi FJ, Schinco FP, Depto AS, Li Y, Mong JA, Dawson TM, Snyder SH. Neuronal nitric oxide synthase is induced in spinal neurons by traumatic injury. *Neuroscience* 1994, **61**: 719–726.
53. Herdegen T, Brecht S, Mayer B, Leah J, Kummer W, Bravo R, Zimmermann M. Long-lasting expression of JUN and KROX transcription factors and nitric oxide synthase in intrinsic neurons of the rat brain following axotomy. *J Neurosci* 1993, **13**: 4130–4145.
54. Vincent SR, Kimura H. Histochemical mapping of nitric oxide synthase in the rat brain. *Neuroscience* 1992, **46**: 755–784.
55. O'Hearn E, Zhang P, Molliver ME. Excitotoxic insult due to ibogaine leads to delayed induction of neuronal NOS in Purkinje cells. *NeuroReport* 1995, **6**: 1611–1616.
56. Verge VMK, Xu Z, Xu X-J, Wiesenfeld-Hallin Z, Hökfelt T. Marked increase in nitric oxide synthase mRNA in rat dorsal root ganglia after peripheral axotomy: *In situ* hybridization and functional studies. *Proc Natl Acad Sci USA* 1992, **89**: 11,617–11,621.

57. Lieberman AR. The axon reaction: A review of the principal features of perikaryal responses to axon injury. *Int Rev Neurobiol* 1971, **14:** 49–124.
58. O'Hearn E, Molliver ME. The olivocerebellar projection mediates ibogaine-induced degeneration of Purkinje cells: A model of indirect, trans-synaptic excitotoxicity. *J Neurosci* 1997, **17:** 8828–8841.
59. Llinás R, Walton K, Hillman DE, Sotelo C. Inferior olive: Its role in motor learning. *Science* 1975, **190:** 1230–1231.
60. Anderson WA, Flumerfelt BA. A light and electron microscopic study of the effects of 3-acetylpyridine intoxication on the inferior olivary complex and cerebellar cortex. *J Comp Neurol* 1980, **190:** 157–174.
61. Balaban CD. Central neurotoxic effects of intraperitoneally administered 3-acetylpyridine, harmaline and niacinamide in Sprague-Dawley and Long-Evans rats: a critical review of central 3-acetylpyridine neurotoxicity. *Brain Res* 1985, **356:** 21–42.
62. Groenewegen HJ, Voogd J. The parasagittal zonation within the olivocerebellar projection. I. Climbing fiber distribution in the vermis of cat cerebellum. *J Comp Neurol* 1977, **174:** 417–488.
63. Buisseret-Delmas C. Sagittal organization of the olivocerebellonuclear pathway in the rat. I. Connections with the nucleus fastigii and the nucleus vestibularis lateralis. *Neurosci Res* 1988, **5:** 475–493.
64. Oscarsson O. Spatial distribution of climbing and mossy fiber inputs into the cerebellar cortex. *Brain Res* 1976, **1:** 36–42.
65. Desclin JC. Histological evidence supporting the inferior olive as the major source of cerebellar climbing fibers in the rat. *Brain Res* 1974, **77:** 365–384.
66. Palay SL, Chan-Palay V. *Cerebellar Cortex, Cytology and Organization*. Springer-Verlag, Berlin, 1974, pp. 1–348.
67. Eccles JC, Ito M, Szentágothai J. *The Cerebellum as a Neuronal Machine*. Springer, Berlin, 1967, pp. 1–335.
68. Sotelo C, Llinás R, Baker R. Structural study of inferior olivary nucleus of the cat: morphological correlates of electrotonic coupling. *J Neurophysiol* 1974, **37:** 541–559.
69. Llinás R, Baker R, Sotelo C. Electrotonic coupling between neurons in cat inferior olive. *J Neurophysiol* 1974, **37:** 560–571.
70. Azizi SA, Woodward DJ. Inferior olivary nuclear complex of the rat: morphology and comments on the principles of organization within the olivocerebellar system. *J Comp Neurol* 1987, **263:** 467–484.
71. Sasaki K, Bower JM, Llinás R. Multiple purkinje cell recording in rodent cerebellar cortex. *Eur J Neurosci* 1989, **1:** 572–586.
72. Llinás R, Sasaki K. The functional organization of the olivo-cerebellar system as examined by multiple purkinje cell recordings. *Eur J Neurosci* 1989, **1:** 587–602.
73. Llinás R, Bloedel JR, Hillman DE. Functional characterization of neuronal circuitry of frog cerebellar cortex. *J Neurophysiol* 1969, **32:** 847–870.
74. Eccles JC, Llinás R, Sasaki K. The excitatory synaptic action of climbing fibres on the Purkinje cells of the cerebellum. *J Physiol* 1966, **182:** 268–296.
75. Eccles JC, Llinás R, Sasaki K. Intracellularly recorded responses of the cerebellar Purkinje cells. *Exp Brain Res* 1966, **1:** 161–183.
76. Llinás R, Nicholson C. Reversal properties of climbing fiber potential in cat Purkinje cells: an example of a distributed synapse. *J Neurophysiol* 1976, **39:** 311–323.
77. Zhang N, Ottersen OP. In search of the identity of the cerebellar climbing fiber transmitter: Immunocytochemical studies in rats. *Can J Neurol Sci* 1993, **20:** S36–S42.
78. Zhang N, Walberg F, Laake JH, Meldrum BS, Ottersen OP. Aspartate-like and glutamate-like immunoreactivities in the inferior olive and climbing fibre system: A light microscopic and semiquantitative electron microscopic study in rat and baboon (*Papio anubis*). *Neuroscience* 1990, **38:** 61–80.

79. Blackwood W, Corsellis JAN. *Greenfield's Neuropathology.* Edward Arnold, London, 1976.
80. Garthwaite G, Hajos F, Garthwaite J. Ionic requirements for neurotoxic effects of excitatory amino acid analogues in rat cerebellar slices. *Neuroscience* 1986, **18**: 437–447.
81. Choi, D.W. Ionic dependence of glutamate neurotoxicity. *J Neurosci* 1987, **7**: 369–379.
82. Siman R, Noszek JC. Excitatory amino acids activate calpain I and induce structural protein breakdown in vivo. *Neuron* 1988, **1**: 279–287.
83. Garthwaite G, Garthwaite J. AMPA neurotoxicity in rat cerebellar and hippocampal slices: Histological evidence for three mechanisms. *Eur J Neurosci* 1991, **3**: 715–728.
84. Kostyuk PG. Calcium channels in cellular membranes. *J Mol Neurosci* 1990, **2**: 123–141.
85. Bertolino M, Llinás R. The central role of voltage-activated and receptor-operated calcium channels in neuronal cells. *Annu Rev Pharmacol Toxicol* 1992, **32**: 399–421.
86. Brorson JR, Manzolillo PA, Miller RJ. Ca2+ entry via AMPA/KA receptors and excitotoxicity in cultured cerebellar Purkinje cells. *J Neurosci* 1994, **14**: 187–197.
87. Ross CA, Meldolesi J, Milner TA, Satoh T, Supattapone S, Snyder SH. Inositol 1,4,5-trisphosphate receptor localized to endoplasmic reticulum in cerebellar Purkinje neurons. *Nature* 1989, **339**: 468–470.
88. Ferris CD, Huganir RL, Supattapone S, Snyder SH. Purified inositol 1,4,5-trisphosphate receptor mediates calcium flux in reconstituted lipid vesicles. *Nature* 1989, **342**: 87–89.
89. Nicotera P, Bellomo G, Orrenius S. Calcium-mediated mechanisms in chemically induced cell death. *Annu Rev Pharmacol Toxicol* 1992, **32**: 449–470.
90. Koliatsos VE, Price DL. Axotomy as an experimental model of neuronal injury and cell death. *Brain Pathol* 1996, **6**: 447–465.
91. Ferrer I, Martin F, Reiriz J, Pérez-Navarro E, Alberch J, Macaya A, Planas AM. Both apoptosis and necrosis occur following intrastriatal administration of excitotoxins. *Acta Neuropathol. (Berl.)* 1995, **90**: 504–510.
92. Portera-Cailliau C, Price DL, Martin LJ. Non-NMDA and NMDA receptor-mediated excitotoxic neuronal deaths in adult brain are morphologically distinct: further evidence for an apoptosis-necrosis continuum. *J Comp Neurol* 1997, **378**: 88–104.

III
Nerve Cell Death in Human Diseases

12
DNA Repair and Neurological Diseases

Brian G. Fuller and Vilhelm Bohr

INTRODUCTION

The central nervous system (CNS) is composed of a heterogeneous population of cells with differing replicative and transcriptional programs. Current approaches to the analysis of DNA repair are beginning to address differences in repair pathways that correlate with patterns of gene expression and DNA replication. Early studies of DNA repair in the nervous system largely ignored the inherent heterogeneity in repair, allowing the assumption that repair in the bulk genome was representative of the total genome including active genes. In few organs was this more misleading than in the brain, where neurons can transcribe up to 30% of the expressed genome. Animal studies indicated continued accumulation of DNA damage in neurons and a reduction in repair capacity with advancing age. On the other hand, it was generally believed that the brain was relatively resistant to DNA damage.

The implementation of molecular techniques resulting in the discovery of gene-specific repair has revolutionized current views of DNA repair in the brain and of the neurologic diseases which can afflict DNA repair mutants. Since transcriptionally active genes are repaired preferentially, studies of repair in active genes may more clearly reflect cellular repair capacity in postmitotic cells. Therefore, analysis of changes in neuronal repair capacity with advancing age should be correlated with the evolution of transcriptional patterns and with changes in chromatin structure over the neuronal life span. Damage accumulation in the bulk genome may have limited significance if transcriptionally active genes are maintained.

In this chapter, the general aspects of DNA damage will be briefly reviewed followed by a short summary of the major DNA repair pathways in mammalian cells. Studies describing DNA repair in neuronal and glial tissue at the cellular and molecular level will be reviewed. Finally, autosomal recessive DNA repair syndromes will be presented. Theses syndromes illustrate the cellular hypersensitivity, genomic instability and neurologic disease which can accompany defective repair genes. The clinical features of these syndromes also suggest that DNA repair genes have a pleiotropic influence on development.

From: Cell Death and Diseases of the Nervous System
Edited by: V. E. Koliatsos and R. R. Ratan © Humana Press Inc., Totowa, NJ

FORMS OF DAMAGE

Cellular metabolism results in several types of DNA damage which are potentially mutagenic. Spontaneous depurinations and depyrimidinations resulting in apurinic/apyrimidinic (AP) sites occur an estimated 100,000 times/d/cell. Depurination of cytosine to uridine, or 5-methyl cytosine to thymidine, occurs several hundred times/d/cell. Cells in the brain metabolize an excess of 10^{11} oxygen molecules per d resulting in highly reactive oxygen species in roughly 2% of the reaction products. This results in approximately 10,000 oxidative lesions in DNA/cell/d *(1)*.

Oxidative DNA damage is perhaps the most relevant type of DNA damage in the brain. Studies show that more than 100 different lesions are formed in DNA after oxidative stress *(2)*. Rates of oxidative damage can vary depending on the cell type and location in the brain. The most frequently reported mutations from oxidative damage are C-T transitions. The 8-OH deoxy guanosine (8-OH-dG) and AP sites produced by oxidative damage can result in G-T transversions. Oxidative damage can also result in strand breaks and crosslinks. The relative biological importance of these lesions is not clear, but it is generally recognized that 8-OH-dG is an indicator of oxidative stress, and this has been the most extensively studied oxidative adduct.

Exogenous sources of DNA damage include irradiation, dietary and environmental carcinogens, and chemotherapeutic agents. The analysis of applied ultraviolet (UV) damage has been instrumental in the development of our present understanding of DNA repair. Although most mammalian cells including neurons and glia have the capacity to repair UV damage, it is likely that environmental UV irradiation produces negligible, if any, damage in the brain. Ionizing radiation from environmental and medical sources can produce a wide range of DNA damage including oxidative base damage, AP sites, DNA-protein crosslinks, single strand breaks and double strand breaks. Benzopyrenes and aflatoxins are metabolized to epoxide intermediates which then form DNA adducts and AP sites. However, the significance of dietary carcinogens to DNA damage in the nervous system is unclear. Chemotherapeutic agents such as mitomycin-C, nitrogen mustard, nitrosourea and platinum compounds produce strand breaks, nitrogen or oxygen alkylations, intrastrand adducts (IA) and less frequently, interstrand crosslinks (ICL) *(3)*.

The classic experimental DNA damaging agent has been UV irradiation, which forms two major photolesions in DNA, the cyclobutane dimer, and the 6–4 photoproduct. These lesions are directly linked to skin cancer, since patients with deficiencies in the repair of photolesions develop carcinomas in skin exposed to UV described below *(3)*. Other bulky lesions in DNA that have been widely studied include those made by carcinogens such as 4-nitroquinoline (4NQO), N-acetoxy acetylaminofluorene (NAAAF), and alkylating agents such as nitrogen mustard (HN_2).

DNA REPAIR MECHANISMS

There are several different repair pathways in mammalian cells. These are listed in Table 1. They include: one-step reactions, in which direct reversal of damage is accomplished by single enzyme mechanisms, single and multistep base excision mechanisms, and multistep reactions with pleiotropic specificities involving multiple protein components (i.e., nucleotide excision repair and double strand break repair).

Table 1
DNA Repair Pathways

Enzymatic pathway	Type of lesion repaired
Nucleotide Excision Repair (NER)	UV cyclobutane dimer
Bulk genome repair	Bulky adducts
Gene specific repair	
Preferential repair	
Strand specific repair	
Direct photoreversion	UV cyclobutane dimer
Methyltransferase	DNA alkylation
Base excision repair	Oxidative base damage
DNA alkylation	
Mismatch repair	Mismatched bases
Double strand break repair	Double strand breaks

Reversal of Damage

The single step process of damage reversal is exemplified by the removal of methyl groups from DNA by O^6-methyl-DNA-alkyltransferase, and by the reversal of pyrimidine dimers by bacterial photolyase. There is evidence of both types of enzyme activity in the mammalian brain.

Mismatch Repair

Mismatch repair can correct mispaired bases which result from cytosine deamination, replication errors or recombination. In *E. coli*, these mismatch bases are repaired by a set of enzymes, the MutS, MutL, and MutH proteins. The MutS protein recognizes the lesion and initiates the assembly of a repair complex containing all three proteins. The MutH protein incises at a GATC sequence in the unmethylated strand. Next, a MutS, MutL, and MutU dependent excision step removes a section of DNA containing the GATC site and the mismatch. The resulting single stranded gap is filled in by DNA polymerase III. Alternatively, GT and GA mispairs can be repaired by specific mismatch glycosylases. Homologues of the MutS and MutL genes have been identified in humans. Mismatch recognition in humans does not involve methylation of adenines as in *E. coli*, but likely involves recognition of strand incontinuities. Mutations in the MutS homologues hMSH2, hpMS1, hpMS2, and in the MutL homologue hMLH1 have been associated with HPNCC (hereditary nonpolyposis colon cancer), which may affect up to 1/200 people in the United States. Mutations in these genes have also been associated with development of glioblastoma multiforme in patients with Turcots syndrome. Recently, the GTBP/p160 protein has been shown to form a heterodimer with hMSH2. Mutations in GTBP/p160 prevent correction of GT mispairs and single displaced bases. Recent evidence also indicates that components of the mismatch repair pathway participate in transcription coupled repair in mammalian cells *(4)*.

Base Excision Repair

Base excision is a simple multistep process in which damaged bases are replaced. Glycosylases which recognize various forms of base damage cleave the N-glycosidic bond, thus releasing the damaged base moiety from its phosphoribose backbone. Endonuclease cleavage of the phosphodiester bonds at the resulting AP site is followed by replacement of the correct nucleotide by DNA polymerase and sealing of the 3' nick by DNA ligase. Mismatch glycosylases, and the DNA lyases which posses both glycosylase and endonuclease activity, also repair DNA via this pathway *(3)*. Base excision repair is viewed as an important mechanism for the repair of oxidative damage, which is perhaps the most relevant form of DNA damage in the CNS. More complex, bulky lesions are removed by the nucleotide excision repair pathways.

Nucleotide Excision Repair

Perhaps the most widely studied DNA repair pathway is nucleotide excision repair (NER). It is responsible for repair of bulky lesions which distort the double helix. These lesions include UV induced photoproducts and bulky adducts such as those derived from cisplatin and 4NQO. Understanding of the enzymology of this complex set of reactions was previously based on knowledge from work done in *E. coli*, but now the molecular events are being characterized in human cells. This has been facilitated by analysis of cell lines from patients with xeroderma pigmentosum (XP) and from analysis of excision repair defective Chinese hamster (CHO) mutant cell lines. Human genes isolated due to their ability to complement repair defective CHO mutants were initially called excision repair cross complementing (i.e., *ERCC 1–7)*. Where possible, they are now named after the XP group which they complement.

As will become evident throughout the rest of this chapter, transcription by RNA pol II has assumed increasing importance in our understanding of NER and its associated medical disorders. Initiation of transcription can be divided into three steps: assembly of the initiation complex, isomerization of the initiation sequence, and promoter clearance. During assembly, the TATA binding protein binds the TATA element at the promoter site to form the pre-initiation complex. This recruits DNA pol II and other basal transcription factors to bind to the pre-initiation complex. Subsequently, there is melting of DNA at the initiation site, followed by binding of the basal transcription factors TFIIE and TFIIH. This forms the activated initiation complex. In a process which requires hydrolysis of ATP, the RNA pol II core enzyme clears the promoter region and proceeds to elongate the primary transcript. TFIIH which is of particular relevance to NER, is believed to be required for the promoter clearance step of transcription initiation *(5)*.

TFIIH is a multiprotein complex composed of at least 8–10 polypeptides including XPB, XPD, Cdk7, Cyclin H, p62, p52, p44, and p34. It exhibits multiple enzyme activities including DNA dependent ATPase, bidirectional helicase and protein kinase activity. Dominant mutations in XPB and antibodies to XPD completely inhibit NER and transcription in vitro. An increasing number of DNA damage recognition and repair proteins which bind or interact with TFIIH are being identified. TFIIH represents a direct molecular link between DNA repair and transcription.

NER involves recognition, incision, displacement, polymerization, and finally, ligation. The recognition step involves XPA which forms a protein complex with replication protein A (RPA). TFIIH is then recruited to the lesion site possibly in association with the XPC protein and the XPG protein. The XPG protein incises 3' of the lesion and the XPF/ERCC1 protein incises 5' of the lesion. The resulting single stranded gap is filled in by DNA polymerase δ or E and sealed by DNA ligase. There appears to be a requirement for proliferating cell nuclear antigen (PCNA) in NER, possibly for the stimulation of polymerase activity or for increasing the catalytic turnover of the enzymes involved in excision. There are a number of recent reviews that discuss this pathway in detail and compare the pathways in bacteria and mammalian cells *(6,7)*.

Recent work has shown that the nucleotide excision repair pathways differ in different regions of the mammalian genome such that one pathway preferentially repairs active or essential genomic regions, versus those regions that are noncoding. This repair pathway has been called transcription coupled repair (TCR). In vitro cell free extract assays, which have been used with considerable success to determine aspects of the DNA repair enzymology, are all limited to studying inactive DNA. As of yet, there is no assay for in vitro repair *during* active transcription. Such an approach would allow optimal characterization of TCR. The preferential repair pathways for active genes may also demonstrate specificity for the transcribed strand. Further, there can be substantial differences in repair rates between various genes, and there can be variations within genes as well: certain codons are repaired better than others.

Gene-Specific DNA Repair

In the mid-1980s, a general method was developed to measure DNA damage and its repair in individual gene fragments *(8)*. This approach involves the generation of a strand break at the site of a lesion, followed by the resolution of single stranded DNA by electrophoresis and quantitative Southern analysis. Such experiments have been widely used, and are now adapted to the study of several DNA lesions. A number of different approaches were used to generate strand breaks at the sites of DNA lesions. These methods permit reprobing of the same biological sample of DNA for repair in different genes and in different strands of a gene. The limitations of this technique have been that fairly high doses of DNA damage are necessary, and that the experiments are complicated and time consuming. However, experiments using this technique have yielded much new insight into the fine structure of DNA repair and have demonstrated the strand bias of repair *(7,9)*. In the last few years techniques have been developed to measure DNA repair in genes at low doses of cellular damage, and to measure the repair at the nucleotide level *(9)*.

Early work on gene specific repair showed that essential or active genes were preferentially repaired when compared to inactive genomic regions or to the bulk of the genome *(8)*. The bulk of the genome is about 99% noncoding. When DNA repair experiments are done using the general bulk or total DNA of a cell, the results essentially provide a measurement of the events in the bulk of inactive DNA. In such experiments it is, thus, not possible to discern whether there is any preferential repair in active parts of the genome. Several results from the work on gene specific repair have suggested that repair in the essential or active genes reflects the cell's ability to survive DNA damage *(9)*. It is therefore likely that gene-specific DNA repair is a better measure of the biological importance of repair than are measures of repair in the overall genome. These observations become increasingly important when considering theories of CNS degeneration which are based on the accumulation of genetic damage.

A Cellular Hierarchy of Gene-Specific Repair

There are major differences in the repair efficiency of different genomic regions in mammalian cells. The repair efficiency can change with the transcriptional activity as

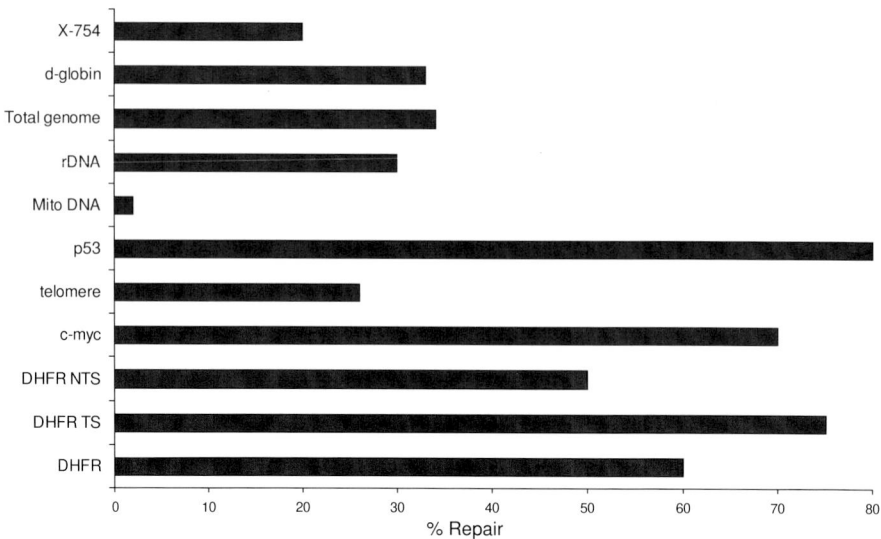

Fig. 1. Extent of repair at 8 h in mammalian cells. Assays were performed as described in ref. *9*. This demonstrates heterogeneity in repair rates among different genes. Rates of repair can be dictated by the transcriptional status in some genes, whereas in other genes nuclear matrix association or other factors may influence repair rates. X-754, a gene on the X chromosome; d globin, delta globin; rDNA, ribosomal DNA; Mito, mitochondrial; NTS, nontranscribed strand; TS, transcribed strand.

originally observed in the hamster and human metallothionein genes after modulation of transcriptional activity. However, repair efficiency can also be independent of the level of expression. A DNA repair hierarchy within the cell spans not only genes but also inactive regions of DNA. There are differences in repair rates among different genes. For example, the *p53* tumor suppressor gene, and another DNA damage-inducible gene, *gadd 153*, which are repaired faster than the housekeeping dihydrofolate reductase *(DHFR)* gene (M. K. Evans, unpublished data). Examples of different repair rates are shown in Fig. 1. While transcriptional rates can be important, the repair differences may relate to aspects of chromatin structure such as sequence accessibility, or to the level CpG methylation, or possibly to whether these regions are associated with the nuclear matrix. This repair heterogeneity extends to changes in DNA repair efficiency within a gene. In the hamster *DHFR* gene there is more efficient repair of the 5' end of the gene than of the 3' end. In a study on repair at the nucleotide level in the tumor suppressor gene p53, there were considerable differences in the repair of different codons *(10)*.

Strand Bias of Repair and the Connection to Transcription

Further study of gene-specific DNA repair led to the observation that the repair process can be biased towards the transcribed DNA strand in some genes *(9)*. The strand bias appears to be due to a direct molecular link between DNA repair and transcription, and the strand bias may be influenced by the state of differentiation of the cell. Terminally differentiated cells appear to lose this strand bias even though active sequences are preferentially repaired. This is discussed further below. Strand-specific repair was also demonstrated in *E. coli*, and the mfd gene product was identified as the transcription-repair coupling factor *(7,11)*. In mammalian cells this link may be the *ERCC6/CS-B* gene that is mutated in Cockayne's syndrome (reviewed below). Interestingly, there appears to be no additional function for the *NER* gene products other than their involvement in transcription and thus in repair *(11)*.

Double-Strand Break Repair

Spontaneous double-strand breaks (DSB) are believed to be relatively rare events within differentiated cells (less than 10/d/cell). Ionizing radiation from medical sources, particularly radiotherapy, is a principal cause of DSB in quantities high enough to affect cell survival. It is generally accepted that one unrepaired DSB is sufficient for lethality in eukaryotic cells. The kinetics of double strand break repair following ionizing radiation are biphasic. There is an initial fast component resulting in 50% of the repair being completed within 40 min, and a slower component such that greater than 90% of repair is complete within 4 h. The biphasic kinetics of double strand break repair have prompted the suggestion that there are two mechanisms for DSB repair: one involving rapid relegation of DNA ends, the other involving homologous recombination. Evidence supporting either mechanism in the repair of radiation-induced double strand breaks has yet to be firmly established in mammalian cells.

Biochemical similarities between immunoglobulin VDJ recombination and DSB repair have been established through identification of mutants defective in both mechanisms. X-ray-sensitive CHO mutants and murine severe combined immunodeficiency (SCID) cells are defective in VDJ recombination and DSB repair. Both are also defective in DNA-dependent protein kinase (DNA-PK) activity, implicating DNA-PK involvement in the repair of DSB and VDJ recombination. The DNA-PK holoenzyme is composed of a DNA binding subunit (Ku) and a catalytic subunit (p350). The Ku autoantigen is a heterodimer composed of a 70-kDa and an 80-kDa polypeptide. The human XRCC5 gene encodes the Ku80 protein and complements VDJ recombination and X-ray sensitivity in CHO xrs cell lines. V3 CHO cells and cells from scid mice are complemented by a fragment of the human chromosome 8, or by yeast artificial chromosomes which encode p350. Interestingly, the presence of Ku80 protein appears to impart stability to the Ku70 protein. Cells that lack the Ku80 protein have normal Ku70 mRNA expression, but Ku70 protein is undetectable.

Current models indicate that the Ku heterodimer binds to the free ends of DNA, which may form hairpin structures during VDJ recombination. Ku antigen bound to DNA recruits the p350 catalytic subunit to the site of the DSB. Once bound, p350 becomes an activated serine/threonine kinase. The molecular events following DNA-PK activation likely include the preparation of the free ends of DNA by endo- or exonucleases, alignment in regions of homology, polymerization to fill-in gaps, and ligation *(12)*. The substrates phosphorylated by activated DNA-PK have not been completely elucidated, but include p350 itself, Ku70, Ku80, p53, replication protein A (RPA), RNA polymerase II CTD, and various transcription factors including Fos, Jun, cMyc, and TFIID. These observations implicate DNA-PK involvement in DNA damage repair, recombination, and transcription. The p350 protein and its homologs including the ATM gene product (mutated in Ataxia Telangiectasia) comprise a group of phosphatidylinositol 3-kinase (PI-3 kinases) signal transduction proteins which regulate cell cycle check points in response to DNA damage *(13)*. Thus, it is likely that DNA-PK plays a central role in the detection of DSBs, the local activation of repair proteins at the site of a DSB, modulation of transcription, and the transduction of cellular signals indicating the presence of DNA damage. It will be of interest to determine if additional roles exist for DNA-PK in the normal physiology of neurons and glia.

DNA REPAIR IN THE MAMMALIAN BRAIN

Analysis of the fine structure of DNA repair in the human CNS has been limited. There has been historic interest in documenting the accumulation of DNA damage in terminally differentiated neurons as possible evidence for the genetic theory of aging. However, the general unavailability of human brain tissue for experimentation has resulted in the majority of studies being conducted in rodents. Since responses to DNA damage can vary considerably depending upon the gene, the organ, and the species affected, extrapolation of results from animal studies to humans should be approached cautiously.

A large literature exists describing clinical CNS injury resulting from irradiation and chemotherapeutic agents. However, in most cases the molecular pathways activated in response to DNA damaging agents are poorly characterized. Although animal studies indicate a reduction in overall DNA repair capacity with age, repair of transcriptionally active genes in terminally differentiated neurons remains efficient. Thus, future studies of DNA repair in the CNS should focus on the transcriptional response to DNA damage, the repair of specific lesions in genes of interest, and should emphasize a more detailed characterization of repair enzyme metabolism with regard to different regions of the brain.

Autoradiography

Experimental approaches to study DNA repair in the brain have included the disaggregation of brain tissue followed by separation of neuronal and glial cells by sucrose centrifugation. This allowed enrichment of the more heavily sedimenting neurons from the less pure populations of glial cells. The enriched populations were then utilized for enzyme purification or autoradiography. In an alternative technique, rodent brains are thinly sliced and prepared for autoradiography before or after exposure to DNA damaging agents *(14–16)*. Tritiated thymidine (^{3}[H]T) uptake as quantitated by scintillation counting or by autoradiography, has been used as a measure of DNA synthesis. The results of these studies are summarized here. Embryonic and newborn rodent neurons and glia demonstrate diffuse ^{3}[H]T uptake consistent with cell proliferation in the developing brain. DNA synthesis at this stage of development can be completely suppressed by hydroxyurea (HU) or the DNA polymerase inhibitors aphidicolin and di-deoxythymidine (d_2TTP). In HU-pretreated embryonic and newborn neurons, DNA damaging agents can stimulate ^{3}[H]T incorporation, allowing the quantitation of repair synthesis. In terminally differentiated adult rodent neurons, ^{3}[H]T uptake is resistant to HU, occurs spontaneously at low levels, and can be induced to higher levels by DNA damage *(15)*. Induction of unscheduled DNA synthesis (UDS) by UV damage is almost completely inhibited by d_2TTP, an inhibitor of DNA polymerase β (pol β). This is consistent with data indicating that pol β is the major polymerase present in adult neurons. As neurons age beyond adulthood they become less able to respond to UV damage with UDS, and spontaneous UDS decreases to low levels. This reduction in spontaneous UDS with age is also seen in glial cells. The highest levels of spontaneous and irradiation-induced UDS in adult mouse neurons can be observed in the cortex and caudate nucleus, and in Purkinje and hippocampal granule cells, respectively. In rats, UDS following N-acetoxy-N-acetyl-2-aminoflourene exposure appears most inducible in neurons of the thalamus and dentate gyrus *(15,16)*.

In a study of human neuroblastoma cells which were differentiated in vitro with nerve growth factor, differentiation was associated with reduced UDS in response to UV irradiation. The removal of benzopyrene and N'-methylguanosine was also decreased in differentiated cells. Levels of AP-endonuclease increased threefold in the differentiated state while DNA polymerase α and methyltransferase decreased, as might be expected in postmitotic cells. DNA pol β and uracil glycosylase levels remained unchanged *(17)*. Although it was concluded in that study that differentiation reduced nucleotide excision repair, the repair of actively transcribed genes was not evaluated. As discussed previously, repair of active genes may be a more accurate measure of repair capability.

In summary, autoradiography studies have demonstrated spontaneous and inducible DNA synthesis in neurons at all stages of development, indicating persistent DNA synthesis throughout the neuronal life span. Autoradiography studies have not provided detail regarding specific repair mechanisms, nor have they produced insights regarding gene specific or enzymatic changes in repair pathways during development.

DNA Repair Enzymes in the Brain

An incomplete list of enzymes of neuronal and glial origin involved in various aspects of DNA repair has been compiled in the literature. These studies have been reviewed by Kuenzle *(18)* and by Rao *(19)*, and will be summarized below. The majority of studies describe results obtained from rodent species and may therefore not reflect expression patterns in humans. Most of the repair enzymes identified in the brain have general roles in DNA metabolism. Excision repair proteins which have been implicated in the autosomal recessive syndromes such as XP, have only recently been described in neural tissue. Similarly, the endogenous role of *DNA-PK* or its homologues such as the *ATM* gene have yet to be defined in the brain.

Photolyase

Pyrimidine photoproducts of UV light can take the form of cyclobutane dimers (CPD) or 6–4 photoproducts. Photolyases exist which are specific for either form of damage. Recently, a homologue of the *Drosophila* 6–4 photolyase was isolated from a human brain cDNA library. Its function remains unknown, however it has been proposed that it may function in signal transduction much like the blue light photoreceptor in plants *(20)*.

Methyl Transferase

O^6 alkyl-guanine methyl transferase (MT) is present in the rodent and human brain throughout development. In mice, MT is expressed at high levels in neurogenic zones in the embryonic brain consistent with cell proliferation. In later stages of development, MT expression is maintained in neurons at low levels, but its expression persists in adult and aging neurons in the cerebellar granular layer, and in the hippocampus.

Expression of MT in mature glial cells is exceedingly low in comparison to differentiated neurons in adult mice. It has been postulated that MT activity in adult neurons participates in transcriptional regulation of genes through methylation (21). MT knockout mice die at embryonic d 7. The delayed removal of alkyl or ethyl groups from cerebral DNA in rats has been postulated as a cause for the high incidence of neural tumors in that species following treatment with nitrosourea alkylating agents. The cerebellum is the region of highest MT activity in the rat.

Brain MT activity in the human fetus is approximately one third of that present in other tissues such as lung or stomach, and it remains at a fairly constant level throughout life. In the adult human as in the fetus, MT activity is lowest in the brain, although MT activity in the human brain is 7–8 times greater than in the rat. The activity of MT in malignant brain tumors is similar to that of normal brain tissue, however meningiomas appear to express higher MT activity than normal brain tissue (21,22).

Uracyl Glycosylase

In the developing brain, uracyl glycosylase activity closely parallels DNA synthesis. Enzyme levels are highest in the perinatal period and subsequently fall to low levels within 3 wk of birth. Uracyl glycosylase activity assayed in human fetal brain tissue was similar to levels in other fetal tissues (22).

Poly-ADP ribose polymerase (PARP) - PARP has been isolated from adult bovine neurons and glia, and also from the rat in which activity peaks between postnatal d 4 and 7. PARP activity can be stimulated in rat cerebellar granular neurons in culture by exposure to toxic levels of glutamate. Inhibitors of PARP prevented neuronal death following glutamate-induced calcium influx, indicating that PARP may contribute to the neuronal cell death resulting from glutamate toxicity (23).

Endo- and Exonucleases

The first reported enzyme in brain tissue with a possible role in DNA repair was an endonuclease isolated from the lamb brain which was specific for single strand DNA. Its degradation products consisted of 5' oligonucleotides 5–14 residues in length. Its role or significance in DNA repair remains unclear. In rats, a 3'–5' exonuclease specific for single stranded DNA containing sites was isolated from neuronal cell nuclei, but not from liver nuclei. An AP endonuclease was also isolated from rat neocortex. In an in vitro reaction together with DNA pol β and *E. coli* DNA ligase, these endo and exonucleases were shown to remove AP sites from phage DNA, allowing subsequent gap filling and ligation. This most likely represents the first in vitro repair reaction carried out with purified components from nuclei of nondividing neuronal cells of brain origin (24).

The human equivalent of mouse APEX (AP endonuclease/exonuclease) is called HAP-1. HAP-1 is an AP endonuclease with 3' diesterase, 3' phosphatase and 3'–5' exonuclease activity. It participates in base excision repair. HAP-1 is identical to Ref-1, a redox protein which regulates the affinity of the transcriptional activator AP-1 for its DNA binding site. AP-1 is composed of the Fos–Jun heterodimer. Through reduction of a cysteine residue in the DNA binding domain of Jun, Ref-1 regulates transcriptional activity. Immunohistochemical stains for Ref-1 revealed localization in the nucleus of neurons throughout the adult rat brain. The highest levels of neuronal staining were in the dentate cells of the hippocampal formation, neurons of the piriform cortex, and Purkinje cells of the cerebellum. Intense staining was also identified in glial fibrillary acid protein-positive astrocytes, but not in microglia. Astrocytic staining was more intense than neuronal staining and it was strongest in the dendritic layer of the dentate gyrus, corpus callosum and superior colliculus *(25)*. Through regulation of transcriptional activation following oxidative stress, and through repair of oxidative damage in DNA, the Ref-1 protein may protect against oxidative damage in the cell.

DNA Polymerases

The activities of the five major DNA polymerases in mammalian cells have been identified in the brain. Of these, polymerase alpha (pol α) and polymerase beta (pol β) have been best characterized in neural and glial cell preparations *(18,19)*. During fetal development, pol α and pol β levels are high, reflecting cell proliferation in the developing nervous system. In the developing mouse, pol α peaks first in neurons then in glial cells. Polymerase alpha activity falls precipitously during the postnatal period to very low levels in neurons and glia. Pol β activity peaks at prenatal d 3 and again at postnatal d 30. It remains at a relatively high level in adult neurons and glial cells. A similar pattern of expression of pol α and pol δ activity has been described in rat neurons and glia. DNA pol β activity mimics that of pol α. DNA pol β accounts for 99% of the polymerase activity in the adult rodent brain. UDS in UV irradiated neuronal nuclei can be suppressed to below control levels by the pol β inhibitor d_2TTP. The significance of pol β activity and spontaneous UDS in adult and aging rodent neurons requires further study.

DNA Ligase

DNA ligase isolated from the brains of newborn and postnatal rats demonstrates peak activity at postnatal d 6 followed by a decline to low levels by d 20. Although neurons and glial fractions were not assayed separately, the cerebellum possessed higher DNA ligase activity than other regions of the brain *(26)*. In adult guinea pigs, neuronal nuclei appeared to have a sevenfold higher DNA ligase activity compared to glial nuclei *(27)*. The expression patterns of the various DNA ligases in the human brain should be further investigated.

ERCC1 and XPD

ERCC1 is involved in the targeting of DNA repair proteins to bulky lesions in DNA. It is expressed in the brain of rodents and humans. Its role in NER is reviewed above. In adult mice, *ERCC1* mRNA levels are the highest in the brain and lowest in the liver. Primary embryonic fibroblasts from *ERCC1* knockout mice were hypersensitive to UV killing and were defective in the incision of UV damaged DNA. Null mutant *ERCC1* mice are runted and die of liver failure in the fourth postnatal week. Although there were no detectable neurologic abnormalities in the these mice, brain nuclei stained positive for p53 *(28)*.

Dabholkar, Reed and colleagues demonstrated mRNA expression of *ERCC1*, *XPD* and *XPA* in malignant and nonmalignant brain tissue from patients undergoing resection of brain tumors. An RT-PCR technique was used to demonstrate concordant expression of all three mRNAs in nonmalignant tissue. High-grade malignant tumors more frequently expressed an alternatively spliced *ERCC1* transcript (in which exon VIII is deleted) than nonmalignant tissue or low grade malignant tumors *(29)*. In another study from the same group, the gene copy number of *ERCC1* and *ERCC2* in malignant gliomas was quantitated by radiodensitometry. Abnormal gene copy number was observed in 11/24 (46%) gliomas studied. Increases and decreases in gene copy number were observed for *ERCC1*, while decreases in gene copy number were observed for *XPD*. Both of these excision repair genes map to chromosome 19. Allelic loss at 19q is frequent in malignant gliomas, and both *ERCC1* and *XPD* localize to the region of frequent interstitial deletion at 19q13.

XPB

XPB is a component of TFIIH and is involved in the incision step of NER. XPB mutants can develop Cockaynes syndrome (reviewed below). *XPB* mRNA is ubiquitously expressed in the mouse brain during development. *In situ* hybridization of antisense *XPB* transcripts demonstrated diffuse expression of *XPB*, with the most intense staining in areas of highest cell density. *XPB* is expressed in both neurons and glial cells, and its expression remains fairly constant in the maturing and adult mouse brain *(30)*.

The haywire gene in *Drosophila* encodes a protein which is 66% homologous to the XPB gene product. Various mutant alleles of haywire produce sterility, reduced fertility, or recessive lethality. Transformation of haywire mutants with the wild type cDNA under control of the HSP 70 heat-shock promoter, rescued lethality and sterility. Analysis of different heat-shock regimens used to rescue lethality indicated that haywire activity was necessary during late larval to pupal development for production of viable adults. Flies treated with marginally effective heat-shock regimens were runted, had defects in coordination and motor function, and were unable to fly. Viable heterozygotes with mutant haywire alleles were hypersensitive to UV irradiation. Embryos suffering from semidominant maternal haywire mutations had CNS anomalies consisting of ventral nerve cord defects.

The neurologically defective phenotypes of haywire mutants have been postulated to result from disruption of the normal regulation of genes critical for CNS development. This is consistent with the role of *XPB* as a transcription factor, and with the development of neurologic disease in *XPB* mutants. Thus haywire mutants may serve as models in which to study the molecular basis for the neurodevelopmental defects observed in Cockayne's syndrome patients *(31)*.

XRCC1

The *XRCC1* gene product participates in the repair of X-ray damage. Messenger RNA expression of this gene has been documented in rodent and baboon tissues. In the baboon, *XRCC1* expression is highest in the testis, ovary, and brain (32).

Mismatch Repair in the Brain

Repair of GT and GU mismatches in extracts from adult rat cerebellar neurons has been recently described. GT mismatches were repaired more efficiently than GU mispairs. Other mismatches were not repaired. It remains to be determined whether the activity described was a result of a mismatch glycosylase, or a result of the so called "all type" mismatch repair (33).

Gene Specific Repair In The Brain

The fine structure of NER in PC-12 (rat pheochromocytoma) cells before and after terminal differentiation with nerve growth factor (NGF) has been analyzed. Global genomic repair of UV damage decreased from 15% to undetectable levels in terminally differentiated neuronal cells. Repair of the GAP-43 gene encoding, which is transiently induced by NGF, increased from 22–30% to 45–55% following exposure to NGF. Synapsin I which is expressed throughout the life span of neurons, was 70–80% repaired in both proliferating and differentiated cells, despite the very low level of repair in bulk DNA of differentiated cells. Strand specificity of repair was evaluated by hybridization of strand-specific mRNA. In the actively transcribed GAP-43 and synapsin I gene encoding, coding and noncoding strands were repaired at equal rates, demonstrating a lack strand specificity. The results of this study illustrate how measurements of global DNA repair may result in underestimations of repair in specific regions of the genome in neurons. The lack of strand specificity in terminally differentiated cells may reflected the requirement for the coding strand to direct repair of the transcribed template strand in postmitotic cells. Differentiated myoblasts display a similar lack of strand specificity *(34)*.

DNA REPAIR AND NEUROLOGIC DISEASE

The presence of neurologic disease in some, but not all patients with hypersensitivity to UV or X irradiation, lead to an intense effort to identify DNA repair defects in various neurologic diseases. Results of these studies have not clearly identified DNA repair defects in all cases. Moreover, DNA repair capacity in neural tissues was seldom assayed directly. Because of space limitations a discussion of these conditions will not be attempted. The repair syndromes discussed below share certain similarities with respect to the neurologic syndromes that accompany them. In general, there may be a relationship between neurologic disease and the degree of hypersensitivity to DNA damage. In addition, many of the mutant genes identified encode proteins which interact with transcriptional machinery, either as a direct component or through protein phosphorylation. In all of these syndromes somatic abnormalities can accompany neurologic disease. And finally, the patterns of presentation, and the pathologic findings in the brains of affected individuals are consistent with a defect early in CNS development.

Xeroderma Pigmentosum

XP is a rare autosomal recessive syndrome which manifests as clinical sun sensitivity, early onset freckling and a 1000-fold increased risk of skin cancer *(3,35)*. It is the most common and best studied nucleotide excision repair (NER) disorder. Approximately 1 in 200,00 people are affected in the US and Europe, while the incidence in Japan is 1 in 40,000. Patients with XP have defective mutations in certain genes encoding excision repair proteins. These proteins are responsible for the initial endonucleolytic cleavage of UV light-induced damage in cellular DNA. There are seven complementation groups (XP A, B, C, D, E, F, G); in addition there are patients with XP variant forms who manifest the clinical stigmata of XP, but who have normal or near normal NER. Some patients display characteristics of both XP and Cockayne's syndrome. Severely affected individuals with XP manifest short stature, arrested gonadal development and neurologic dysfunction *(35,36)*. Approximately 20% of patients with XP will develop degeneration of the central and peripheral nervous system with characteristic signs and symptoms *(3,36)*. Neuropathologic studies of patients with neurologic disease reveal primary neuronal loss among other changes *(37)*. The etiogenesis of XP neurologic disease remains unknown. Accumulation of DNA damage in the nervous system, and defective transcription have been proposed as alternate explanations for the neurologic degeneration associated with XP *(37,38)*.

The diagnosis of XP is suspected in patients with a history of profound sun sensitivity, photophobia, excessive freckling, or the early development of skin cancer. Defective NER as indicated by reduced unscheduled DNA synthesis (UDS) following UV irradiation in cultured fibroblasts confirms the diagnosis.

Most patients develop symptoms by age two, although the onset of symptoms can be delayed until adolescence. Sun burns, freckling, chronic actinic changes, and skin cancers are most prevalent in the face, head, neck, and other areas of sun-exposed skin. Characteristic ocular finding include: photophobia, conjuctivitis of the intrapalpebral fissures, ectropion keritits, as well as benign and malignant tumors of the lids, conjunctiva and corneal region. In addition to skin cancers, an excessive number of squamous cell carcinomas develop on the tip of the tongue. The neurologic abnormalities which arise in roughly 20% of patients are described in more detail below. Finally, an increased incidence of internal malignancies including brain tumors in XP patients has been reported, suggesting that certain DNA repair defects can lead to cancer *(36)*.

The seven complementation groups of XP were classically defined by the restoration of UDS (repair synthesis) following UV irradiation in cells fused from different affected individuals *(2)*. A lack of complementation of defective repair indicated that the fused cells belonged to the same complementation group. Through utilization of CHO mutants that resemble XP cells, it has been possible to clone most of the human repair genes *(6,7,38)*. Using in vitro repair assays in which purified components are added to cell extracts, it has been possible to reconstitute almost the entire incision process *(6)*. Recent work in the field implicates at least 25 proteins *(6,11)*. The different complementation groups of XP display different patterns of residual DNA repair. Until recently, our knowledge of the repair phenotype in these cells was based on repair measurements in the bulk genome, but information about the gene-specific repair patterns has recently become available. In XPA mutant cells, there is no repair in either the bulk genome or in active genes. XPC mutants represent an interesting repair phenotype that resembles the repair pattern in rodent cells: active genomic regions are repaired, but there is very little repair in the bulk of the genome *(39)*. The repair phenotype in cells from the premature aging condition Cockayne's syndrome demonstrate general genome repair which is normal, but the ability to preferentially repair active genes is lacking *(40)*.

The observation that neurologic disease is more common in certain complementation groups remains unexplained. However, a greater cellular hypersensitivity to UV killing in vitro has been correlated with the presence of neurologic disease *(41)*. A brief description of the molecular pathology of individual *XP* genes will follow. For a more detailed information one is referred to recent reviews *(3,6,7,11)*.

XPA

The *XPA* gene encodes a 273-amino acid protein which contains a DNA binding zinc finger domain. This protein is believed to participate in the initial recognition of DNA damage. It binds 6–4 UV photoproducts with much greater affinity than native DNA. XPA also interacts with the ERCC1/XPF complex as well as TFIIE and TFIIH. Mutations in the carboxy terminal domain of the *XPA* gene abolish interaction with

TFIIH and eliminate NER. It has been proposed that XPA recruits TFIIH to the site of UV damage in DNA. Genetic analysis of Japanese XPA patients reveal that the majority of patients posses a splicing mutation at intron 3 and or missense mutations at codons 116 or 228. The majority of XPA patients will develop neurologic disease. Homozygous mutations seem to a confer worse clinical prognosis than heterozygous mutations.

XPB

The *XPB* gene encodes a 782 amino acid protein with 3' helicase activity. The XPB protein is identical to the p89 subunit of TFIIH. It is believed that XPB protein along with XPD confer bidirectional helicase activity to TFIIH, facilitating the required local unwinding of DNA during promoter clearance, or at a sites of DNA damage. Splicing and missense mutations have been discovered in the *XPB* gene in the two kindreds which have been studied. Combined symptoms of XP and Cockaynes syndrome can be observed in XPB patients. Dominant mutations in XPB abolish NER and transcription in vitro.

XPC

XPC mutants express normal repair of the template strand of actively transcribed DNA, but are defective in the repair of nontemplate and bulk genomic DNA. *XPC* mutants comprise the largest group of XP patients. The development of primary brain tumors and classic neurologic disease has been documented in XPC patients. The XPC gene resides on chromosome 3, and it codes for a 160-kDa protein of unknown function.

XPD

The XPD protein, like XPB, has helicase activity and is a subunit of TFIIH. The *XPD* gene is located on chromosome 19 and it encodes a protein of 760 amino acids. Patients in the XPD group can have diverse presentations ranging from mild skin manifestations to more severe generalized abnormalities as seen with Trichothiodystrophy (described below). Neurologic disease develops in a minority of patients with XPD.

XPE

UV sensitivity in XPE cells in some patients is complemented by microinjection of XPE binding factor (XPE BF). This protein specifically binds damaged DNA and its expression is reduced in some XPE cell lines. A cDNA encoding XPE BF has been cloned, however identification of the mutations responsible for the XPE phenotype awaits a more thorough analysis of the *XPE* gene. No case of neurologic disease have been reported in XPE patients.

XPF

XPF cells are complemented by the human repair gene *ERCC4* and it appears to be the *XPF* gene. XPF and ERCC1 can form a complex with XPA and RPA. Addition of a purified ternary complex consisting of XPF/ERCC1 and XPA corrects the repair defect in XPA, ERCC1, ERCC4 and XPF mutants. The precise enzymatic role of XPF in NER has not been clearly defined. However the XPF/ERCC1 complex has been proposed to function as a 5' endonuclease during the later stages of the excision reaction. Neurologic disease is rare in XPF.

XPG

The *XPG* gene is located on chromosome 13. It encodes a 133-kDa protein with endonuclease activity. XPG is involved in the incision of DNA at the site of damage.

Diverse clinical phenotypes, ranging from those with mild sun sensitivity to more severely affected patients with neurologic disease or Cockayne's syndrome have been described among patients with XPG.

XP-Associated Neurologic Disease

Neurologic degeneration characteristic of XP is most common in complementation groups A and D. Patients in groups C, E, and F are generally free of neurologic disease. Individuals belonging to complementation group B have been described with the combined clinical features of XP and Cockayne's syndrome, and with features of Trichothiodystophy. Ninety-percent of the 500+ documented cases of XP neurologic disease have been among complementation groups A, C, and D. Patients in the other groups make up the remaining 10% of the cases. XPA accounts for the majority of cases in Japan, and is most frequently associated with the more severe form of early onset neurologic degeneration. Neurologic findings consistent with XP-associated neurologic disease can exist in patients prior to the development of clinically significant dysfunction. These very rare "asymptomatic" cases have been described among patients in complementation group C *(37)*.

While the majority of patients with neurologic disease belong to groups A and D, not all XPA or XPD patients will develop neurologic disease. No case of XP neurologic disease has been described in an XPE patient. Likewise, neurologic disease has only rarely been reported in XPC patients or in XP variant forms *(6,7)*. Recently, neurologic disease has been described in an XPF patient *(42)*. Somatic abnormalities such as gonadal hypoplasia and dwarfism are rare in XP, but when present are always associated with neurologic disease.

Perhaps the richest source of information describing XP-associated neurologic disease are the extensive studies of XPA patients from Japan *(37)*. Almost all reported Japanese patients with neurologic disease develop clinical symptoms before age seven. This pattern of presentation has been designated the "early-onset type" of XP neurologic disease, and has not only been documented in non-Japanese XPA patients, but also in patients with XPG. The constellation of neurologic findings described in these early onset patients established the classic presentation of XP neurologic disease *(37)*. These findings include: microcephaly, cerebral atrophy, intellectual impairment, cognitive deterioration, abnormal EEG activity, choreoathetosis, ataxia, dysmetria, dysarthria, extensor plantar response, sensorineural deafness, diminished or absent reflexes, peripheral sensory deficits, abnormal Romberg sign, nerve conduction abnormalities, and finally EMG and biopsy evidence of peripheral motor neuropathy. A comparison of clinical and neuropathologic features associated with defective DNA repair is presented in Table 2. Patients with clinical onset between 7 and 12 yr of age have been designated as the intermediate-onset type of juvenile XP neurologic disease. Patients belonging to complementation groups D and G have been described with this presentation. Clinical onset beyond age 12 has been designated as late onset, and has been described in XPA and XPD patients *(37)*. An adult onset form has been more recently described *(43)*. The usual course of XP-associated neurologic disease is continued deterioration with advancing age.

Kraemer described the clinical and neurologic findings reported for 830 published cases of XP *(36)*. Neurologic abnormalities were described in approx 18%. Although

Table 2
Neurologic Features of DNA Repair Syndromes

	Repair defect	Neurologic features	Neuroradiologic features	Neuropathology
Xeroderma pigmentosum	Excision repair	Mental retardation	Microcephaly Cerebral atrophy Sensorineural deafness	Cortical atrophy Neuronal loss Purkinje cell loss (white matter intact)
Cockayne's syndrome	Transcription coupled repair	Mental retardation	Microcephaly White matter atrophy Thickened calvarium Basal ganglia calcification Normal pressure hydrocephalus Sensorineural deafness	Demyelination Purkinje cell loss (cortical gray matter intact)
Trichothiodystrophy	Likely excision repair	Mental retardation	Microcephaly Sensorineural deafness Cortical atrophy Dysmyelination	Incomplete neuronal migration to the cortex Arrested neuronal development
Ataxia telangiectasia	X-ray damage Double strand break Repair ?	Cerebellar ataxia Mental retardation	Cerebellar atrophy	Cerebellar degeneration Purkinje cell loss Incomplete neuronal migration

the median age of onset of any symptoms and the initial episode of sensitivity to light both tended to occur at an earlier age in patients with neurologic disease, other non-neurologic aspects of XP were similar to those patients without neurologic disease. The most frequent cited neurologic abnormality was low IQ, reported in 80% of patients with neurologic disease. Abnormal motor function was noted in 30%, reflexes were abnormal in 20%, impaired hearing was reported in 18%, abnormal speech in 17%, EEG abnormalities in 13%, and microcephaly was reported in 24% of patients. Somatic abnormalities were reported in 23%, and delayed or absent secondary sexual characteristics were described in 12% of reported cases with XP neurologic disease *(36)*.

Neuropathologic descriptions of XP-associated neurologic disease have been reported relatively rarely. Gross findings include cerebral atrophy, low brain weight and ventricular dilatation. Focal atrophic changes can be observed in the cerebellum, hippocampus, substantia nigra and spinal cord. Microscopic findings include marked Purkinje cell loss in the cerebellum with increased lipofuscin in remaining Purkinje cells. A similar pattern of neuronal loss and lipofuscin deposition was also noted in the substantia nigra and in brain stem nuclei. Diffuse loss of neurons in the cerebral cortex with lipofuscin deposition in remaining neurons and associated axonal swelling underlie cerebral atrophy. Anterior horn cell loss and demyelination of the posterior columns can be present in the spinal cord. Loss of myelinated fibers in the peripheral nerves has also been described *(37)*.

XP/Cockayne's Syndrome

Individuals presenting with clinical features of Cockayne's syndrome but who also have the skin manifestations of XP including skin cancers, are said to have combined XP/Cockayne's syndrome *(35,36)*. Patients with this syndrome have been described in complementation groups B, D, and G. The presence of neurologic findings characteristic of Cockayne's syndrome such as cerebral white matter atrophy and intracranial calcifications, helps to distinguish XP/Cockayne's syndrome from with XP neurologic disease.

Cockayne's Syndrome

Cockayne's syndrome (CS) is a rare autosomal recessive disorder of which less than 300 cases have been reported. The initial description by Cockayne in 1936 was the finding of mental degeneration, short stature, premature aging and gonadal hypoplasia in two siblings *(44)*. Early-onset sun sensitivity in CS patients led to the discovery of UV hypersensitivity in cultured fibroblasts, however the increased incidence of skin cancer characteristic of XP was not observed in CS patients. Among the clinical features distinguishing CS from XP is the universal presence of neurologic degeneration in all CS patients. Of particular interest is the association of neurologic degeneration, developmental arrest, premature aging and the recently characterized molecular defect in transcription-coupled repair in these patients *(40,45)*.

The CS can be diagnosed when there is growth delay or dwarfism; neurodevelopmental arrest and/or later neurologic dysfunction; and one or more of the following: sun sensitivity, pigmented retinopathy or other characteristic ocular findings, sensorineural hearing loss, and dental caries *(46)*. Delayed recovery of RNA synthesis in cultured fibroblasts following UV irradiation confirms the diagnosis *(47)*.

Clinically, CS is a disease of progressive degeneration. Delayed psychomotor development is usually followed by mental retardation. Cachectic dwarfism, ataxia, retinopathy, deafness, and early death are features of this disease. Other characteristic features include: cataracts, progeria, sunken eyes with large ears and nose (characteristic facies), disproportionately long arms and legs relative to the short torso, short stature, type II lipoproteinemia, thickened calvarium, microcephaly, and normal pressure hydrocephalus. Prenatal growth delay, severe neurologic dysfunction, cataracts before age three and congenital structural anomalies of the eyes confer a particularly poor prognosis. Even in the absence of these poor prognostic features, CS patients rarely survive beyond adolescence (46). Characteristics which distinguish CS from XP include the absence of cutaneous, ocular or internal malignancies, and the presence of retinal degeneration and intracranial calcifications.

Neuropathologic analysis of brains from CS patients reveal a brain weighs less than 600 g. Gross pathologic features include: meningeal thickening, central white matter atrophy with tigroid demyelination, basal ganglia calcification, ventricular dilatation, and thickening of the corpus callosum. The cortical gray matter usually appears intact. The microscopic findings consists of patchy demyelination and reactive gliosis, Purkinje cell loss in the cerebellum, and perivascular calcifications. Lipofuscin accumulation and neurofibrillary tangles in neurons have also been described. Peripheral nerve and muscle biopsies reveal demyelination (48).

The CS cells display a characteristic delayed recovery of RNA synthesis following UV radiation (47). Assays of recovery of RNA synthesis have been utilized to define two complementation groups (CS-A, CS-B). A greater insight regarding the molecular pathology of CS was made possible by the discovery of a preferential repair of actively transcribed genes (8). It was subsequently shown that, while genomic repair of UV damage was normal in CS cells, preferential repair of the template strand of actively transcribed genes, so-called transcription-coupled repair (TCR) was defective (40).

It was further demonstrated that defective TCR and UV sensitivity were corrected in CS-B cells by the ERCC6 gene product, thus identifying mutations in *ERCC6* (renamed *CS-B*) as the cause of CS-B. The CS-B gene product is a 168-kDa peptide with helicase activity, a molecular feature which is characteristic of other transcriptional activators. The CS-B gene product has been proposed to be the human homologue of the *E. coli* transcription coupling repair protein, mfd (11).

The cloning of the gene defective in CS-A, several years after *ERCC6* was identified as *CS-B*, was a key development in the analysis of the molecular pathology of CS. Henning et al. (45) screened a human cDNA library for complementation of UV sensitivity in CS-A fibroblasts. A 2.2-kb cDNA fragment was isolated which localized to the centromeric region of human chromosome 5. This cDNA complemented CS-A cells, but not CS-B or XPC. Sequence analysis revealed that the encoded protein belonged to the WD repeat family of transcriptional activators. Analysis of PCR products from CS-A cells revealed internal deletions which were caused by mutations resulting from aberrant mRNA splicing. The most important observation was that the CS-A protein binds the CS-B protein as well as the p44 subunit of TFIIH, suggesting a dual role for CS-A in transcription and repair (45).

It has been proposed that CS-A and CS-B may be involved in optimizing chromatin structure to facilitate general transcription. Therefore mutations in CS-A or CS-B proteins which can result in defective transcription coupled repair, may also affect general transcription. Support for this hypothesis comes from recent work demonstrating defective basal transcription in CS-A and CS-B cellular extracts in vivo and in vitro *(49,50)*. The constellation of phenotypic features characteristic of CS may be explained by defective transcriptional regulation during somatic and nervous system development. Identification of the target gene(s) affected by the transcriptional dysfunction will further support the "transcription syndrome" hypothesis.

Trichothiodystrophy

The term trichothiodystrophy (TTD) has been applied to a very rare group of autosomal recessive neurocutaneous syndromes. The hallmark of these disorders is the presence of brittle hair resulting from reduced sulfur content. Icthyosis and impaired growth and intelligence are other common features of this syndrome *(51)*. Patients have been clinically grouped by an acronym (SPIBIDS) based on the presence of several or all of the following findings: osteosclerosis (S), photosensitivity (P), icthyosis (I), brittle hair (B), impaired intelligence (I), decreased fertility (D), and short stature (S). The majority of patients with TTD have impaired intelligence *(51)*. Among the many clinical similarities to CS, photosensitivity in TTD patients is not associated with carcinogenesis, and the underlying biochemical defect of TTD has been linked to transcription. It is becoming increasingly evident that the various presentations of CS, TTD, XPD, XPB, and XPG may represent a spectrum of phenotypes which result from mutations in the various proteins which comprise or interact with TFIIH *(7,38)*.

The TTD can be diagnosed by the presence of brittle hair with low cysteine and sulfur content. Examination of hair with polarizing light microscopy reveals characteristic alternating light and dark bands. Trichoschisis (transverse fractures through the hair shaft) and a decreased cuticular layer may be present and support the diagnosis *(49)*. Other clinical features of TTD include: osteosclerosis, oculocutaneous abnormalities, cataracts, dental diseases, immunodeficiency, skeletal anomalies, and other dysmorphic abnormalities.

Neurologic dysfunction in TTD is frequent, however its true incidence is difficult to estimate from the literature. The most frequently reported neurologic finding is mental retardation, which can be mild or severe. Microcephaly, delayed or arrested psychomotor development, dystonia, ataxia, pyramidal tract signs, sensorineural deafness, spasticity, abnormal reflexes, diminished muscle tone, peripheral neuropathy, and seizures are among the reported neurologic findings in TTD *(51)*.

Neuropathologic descriptions of brain tissue from TTD patients are scant. A brain biopsy from a patient with the related Pollits syndrome was consistent with arrest of fetal migration of neurons to the cortex *(52)*. Brain imaging reports have documented agenesis of the corpus callosum, cortical atrophy, periventricular leukomalacia, immature myelination, dysmyelination, intracranial calcifications and diffuse high signal intensity in the white matter on T2 weighted images *(53)*.

The photosensitivity observed in approx 50% of patients lead to the evaluation of NER in TTD, and to the demonstration of reduced NER in some but not all cell lines from photosensitive patients with TTD. Subsequent analysis has defined three complementation groups. The most extensively studied group is that in which photosensitivity is corrected by the XP-D gene product. The majority of TTD patients manifesting photosensitivity belong to that group.

An analysis of mutations in the cloned sequences of the *XPD* gene from three cases of TTD was recently reported by Takayama *(54)*. Several cell lines derived from donors with different phenotypes (i.e., XPD vs TTD) shared a common allele, yet differed in the second allele, suggesting that single allele mutations maybe responsible for phenotypic differences between XPD and TTD individuals. Mutations in the conserved helicase domains correlated with photosensitivity, while mutations or deletions in the carboxy terminal region of the protein correlated with TTD *(54)*

The TTD complementation group A (TTDA) was discovered when cells from a photosensitive patient were found not to belong to any of the known XP or TTD complementation groups. Although the gene for TTDA has not yet been identified, NER and unscheduled DNA synthesis were restored when purified TFIIH was added to TTDA cell extracts or microinjected in TTDA fibroblasts. This finding suggests that the TTDA protein may be a subunit of TFIIH *(38,55)*.

Recently, a third TTD complementation group has been described in which cells from a photosensitive patient with mild clinical features of TTD were shown to belong to XP complementation group B. As noted above, XPB (ERCC3) is a subunit of TFIIH, and purified TFIIH corrects the NER defect in XPB and TTD/XPB cells *(55)*. Similar to CS, many of the clinico-pathologic hallmarks of the TTD which are not easily explained by defective repair, have been proposed to be the result of subtle transcription defects *(7,38)*.

Ataxia Telangiectasia

Ataxia telangiectasia (AT) is an autosomal recessive disorder characterized by progressive cerebellar ataxia, oculocutaneous telangiectasia and immunodeficiency. AT cells demonstrate profound sensitivity to X-rays and radiomimetic agents *(3)*. Chromosome instability and defective cell cycle check points are the molecular consequences which predispose affected patients and heterozygote carriers to the development of cancer. The incidence of AT is 1 in 40,000 live births, while the estimate of AT heterozygotes is approx 1% of the US population *(54)*. Cancer, primarily of the lymphoreticular system, develops in approximately 10% of AT patients, almost all of whom are 20 yr or younger *(3)*. Relatives of AT patients who are heterozygotes for the mutated AT gene also have an increased risk of cancer, and it is estimated that 8.8% of Caucasian patients with breast cancer in the United States may be heterozygotes *(57)*.

The clinical diagnosis of AT is made when there is early onset of progressive cerebellar ataxia followed by the development of ocular telangiectasia at 4–6 yr of age. Clinical and laboratory evaluations which confirm the diagnosis include cerebellar atrophy on magnetic resonance or computed tomography scans, elevated alpha fetoprotein, and radioresistant DNA synthesis in cultured fibroblasts. Genetic screening techniques are being developed which should allow prenatal diagnosis and the identification of heterozygote carriers.

AT patients frequently suffer from chronic sinus, pulmonary and middle ear infections which result from their underlying immunodeficiency. Thymic atrophy or dysgenesis can accompany diminished humoral and cellular immunity. Cutaneous telangiectasias and other progeric skin changes including basal cell carcinomas develop in sun exposed skin. Endocrine disorders, gonadal hypoplasia and delayed sexual development are common.

In addition to ataxia, the neurologic findings in AT patients can include mental retardation, gaze abnormalities, drooling, nystagmus, dysarthria, choreoathetosis and abnormal reflexes. Arrested cognitive development occurs in the 30% of patients who survive to their second decade.

The most consistent neuropathologic finding of AT is degeneration of the cerebellar folia. Gross pathologic findings may also include degeneration of the olivary nuclei, posterior and lateral column degeneration, anterior horn cell atrophy, and occasional gliovascular nodules with modest gliosis and demyelination. Peripheral neuropathic changes may also be present. The extra-cerebellar changes involving the spinal cord, brainstem and other regions of the neuraxis are seen in later stages of the disease in older patients *(58)*.

Microscopic descriptions of the well characterized yet poorly understood neurologic degeneration indicate that Purkinje cells are the first cell population affected, followed by granular cell loss and thinning of the molecular layer of the cerebellar cortex *(58,59)*. In the later stages of AT, the cell populations supplying afferent signals to and receiving efferent signals from the cerebellum become atrophic (for other perspectives on AT, *see* chapter by Wood).

The molecular pathology of AT includes: X-ray sensitivity, radioresistant DNA synthesis, aberrant cell cycle check point function in response to X-rays, chromosome instability, and an increased rate of spontaneous sister chromatid exchange. Although a defect in repair of double strand breaks was difficult to demonstrate conclusively, the noted cellular phenotypic characteristics are consistent with a defective response to DNA strand breaks and aberrant regulation of recombinant events *(3)*. The molecular pathology underlying early cerebellar degeneration and the other CNS changes remains speculative.

The normal response to X-irradiation is a transient inhibition of DNA synthesis. AT cells lack this response and thus demonstrate radioresistant DNA synthesis. AT cells have been divided into four complementation groups based on their ability to complement radioresistant DNA synthesis. However, it was discovered through linkage analysis that three of the complementation groups mapped to a specific region on chromosome 11 *(60)*. This was supported by chromosome transfer studies which indicated that the complementing gene resided on chromosome 11. Through utilization of positional cloning techniques, a cDNA encoding a 5.9 kb fragment of the *ATM* (mutated in ataxia-telangiectasia) gene was isolated. This cDNA identified the *ATM* gene as a homolog of the yeast and mammalian phosphatidylinositol—3' kinases that are involved in signal transduction, cell cycle regulation, double strand break repair, and meiotic recombination *(59,60)*. *ATM* gene transcripts from 14 patients representing the four complementation groups were sequenced and found to contain mutations in the same regions of the *ATM* gene. This confirmed that the AT complementation resulted from intragenic rather than intergenic complementation as seen in XP, CS, and TTD. A more complete description of the range of mutations observed in patients awaits complete sequencing of the *ATM* gene. At present the physiologic substrate(s) of the *ATM* gene product and its biologic function are under investigation. However, a wide range of substrates, including the carboxy terminal domain of RNA pol II, and TFIID, have been identified for the *ATM* homologue, *p350* (reviewed above) *(13,62)*.

CONCLUSIONS

Our knowledge of DNA repair in the nervous system is developing rapidly, yet much remains to be learned. The unique properties of postmitotic neurons, including their high transcriptional activity, suggest that neuronal DNA repair will have distinct gene specificities. Similar patterns of repair heterogeneity might also extend to supporting glial cell populations which influence neuronal transmission. Finally, choroiod, ependymal, and vascular elements would in turn display repair specificities which reflect their transcriptional and cell cycle programming.

Anatomic barriers protect the brain from many environmental sources of DNA damage. Although oxidative damage may be the most prevalent type of DNA damage in the brain, its contribution to neuronal death has been difficult to quantify. Future approaches must focus on transcriptional responses to oxidative DNA damage, repair in vital genes, and the development of experimental systems in which gene specific repair can be correlated neuronal cell survival.

The pathways, which repair DNA damage in the mammalian brain, may have unique characteristics and unique gene specificities. The expression patterns of several repair enzymes in neurons suggest that they may function in processes other than repair. For example, it remains to be discovered whether repair activity is influenced by neural transmission. Experimental approaches to this question might involve the measurement of repair in genes which are the targets of excitation-transcription coupling, or analysis of signal transduction by repair enzymes such as DNA-PK, Ref-1, or the human 6–4 photolyase in neuronal cells.

Several repair pathways are directly linked to transcription, either through proteins with dual functions in repair and transcription, or through the ability of repair proteins to phosphorylate transcription factors. It remains a major challenge to relate the deficiencies in DNA repair with the neurological symptoms observed. The wide range of somatic and neurologic abnormalities, which can accompany mutations in repair genes, is consistent with disrupted transcriptional regulation during development. Future efforts should focus on identifying genes, which are aberrantly regulated during CNS development in DNA repair mutants.

REFERENCES

1. Ames BN, Shigenaga MK, Hagen TM Oxidants, antioxidants and degenerative diseases of aging. *Proc Natl Acad Sci USA* 1994, **90:** 7915–7922.
2. Dizdaroglu M. Characterization of free radical induced DNA damage to DNA by the combined use of enzymatic hydrolysis and gas chromatography-mass spectrometry. *J Chromatogr* 1986, **367:** 357–366.
3. Friedberg EC, Walker GC, Siede W. *DNA Repair and Mutagenesis*. ASM Press, Washington, D.C., 1995.
4. Modrich P, Lahue, R. Mismatch repair in replication fidelity, genetic recombination, and cancer biology. *Ann Rev Biochem* 1996, **65:** 101–133.
5. Goodrich JA, Tijan R. Transcription factors TFIIE and TFIIH and ATP hydrolysis direct promoter clearance by RNA polymerase II. *Cell* 1994, **77:** 145–156.
6. Wood R. DNA repair in eukaryotes. *Ann Rev Biochem* 1996, **65:** 135–138.
7. Friedberg EC. Relationship between DNA repair and transcription. *Ann Rev Biochem* 1996, **65:** 14–43.
8. Bohr VA, Smith CA, Okumoto DS, Hanawalt PC. DNA repair in an active gene: removal of pyrimidine dimers from the DHFR gene of CHO cells is much more efficient than in the genome overall. *Cell* 1985, **40:** 359–369.

9. Bohr VA. The fine structure of DNA repair and genomic instability. *Carcinogenesis* 1995, **16:** 2885–2895.
10. Tornaletti S, Pfeifer G. Slow repair of pyrimidine dimers at p53 mutational hot spots in skin cancer. *Science* 1994, **263:**1436–1438.
11. Sancar A. Nucleotide excision repair. *Ann Rev Bioch* 1994, **65:** 43–81.
12. Kirchgessner C, Brown JM. Radiosensitivity, DNA double strand break repair and VDJ recombination. in *DNA Repair Mechanisms: Impact on Human Disease and Cancer* (Vos JH, ed.), RG Landes, Austin, TX, 1995, pp. 349–373.
13. Lavin MF, Khana K, Beamish H, Spring K, Watters D, Shiloh D. Relationship of the ataxia-telangiectasia protein ATM to phosphoinositide 3-kinase. *TIBS* 1995, **20:** 382–383.
14. Korr H, Schultze B. Unscheduled DNA synthesis in various types of cells of the mouse brain in vivo. *Exp Brian Res* 1989, **74:** 573–578.
15. Desousa J, De Boni U, Cinader B. Age-related decrease in ultraviolet induced DNA repair in neurons but not in lymph node cells of inbred mice. *Mech. Aging Dev* 1986, **36:** 1–12.
16. Heytig C, van't Veer, L. Repair of ethylnitrosourea-induced DNA damage in the newborn rat. II. Localization of unscheduled DNA synthesis in the developing rat brain. *Carcinogenesis* 1981, **2:** 1173–1180.
17. Jensen L, Linn S. A reduced rate of bulky DNA adduct removal is coincident with differentiation of human neuroblastoma cells induced by nerve growth factor. *Mol Cell Biol* 1988, **8:** 3964–3968.
18. Kuenzle, CC. Enzymology of DNA replication and repair in the brain. *Brain Res Rev* 1985, **10:** 231–245.
19. Rao S. Genomic damage and its repair in the young and aging brain. *Mol Neurobiol* 1993, **7:** 23–48.
20. Todo T, Ryo H, Yamamoto K, Toh H, Inui T, Ayaki H, Nomura T, Ikenaga M. Similarity among the Drosophila (6-4) photolyase, a human photolyase homologue, and the DNA photolyase-blue-light photoreceptor family. *Science* 1996, **272:** 109–112.
21. Goto K, Numata M, Komura J, Ono Bestor T, Kondo H. Expression of DNA methyltransferase gene in mature and immature neurons as well as proliferating cells in mice. *Differentiation* 1994, **56:** 39–44.
22. Krokan H, Haugen A, Myrnes B, Guddal P Repair of premutagenic DNA lesions in human fetal tissues: evidence for low levels of O6-methylguanine -DNA methyltransferase and uracil-DNA glycosylase activity in some tissues. *Carcinogenesis* 1983, **4:** 1559–1564.
23. Bilen J, Itel ME, Niedergang C, Okazaki H, Mandel P. Poly(adenosine diphosphate ribose) polymerase activity in neuronal and glial nuclei from bovine cerebrum. *Neurochemical Res* 1981, **6:** 1253–1263.
24. Ivanov VA, Tretyak TM, Afonin YN. Excision of apurinic and/or apyrimidinic sites from DNA by nucleolytical enzymes from rat brain. *Eur J Biochem* 1988, **172:** 155–159.
25. Dragunow M. Ref-1 expression in adult mammalian neurons and astrocytes. *Neurosc Lett* 1995, **191:** 189–192.
26. Nakaya N, Sawasaki Y, Teraoka H, Nakajima, H, Tsukda K. Changes of DNA ligase in the developing rat brain. *J Biochem* 1977, **81:** 1575–1577.
27. Inoue N, Kato T. Nuclear DNA ligase and its action on chromatin DNA in neuronal, glial, and liver nuclei isolated from adult guinea pigs. *J Neurochem* 1980, **34:** 1574–1583.
28. McWhir J, Selfridge J, Harrison DJ, Squires S, Melton DW. Mice with DNA repair gene (ERCC-1) deficiency have elevated levels of p53, liver nuclear abnormalities and die before weaning. *Nature Genetics* 1993, **5:** 217–224.
29. Dabholkar MD, Berger MS, Vionnet JA, Egwuagu C, Silber JR, Yu JJ, Reed E. Malignant and nonmalignant brain tissues differ in their messenger RNA expression patterns for ERCC1 and ERCC2. *Cancer Res* 1995, **55:** 1261–1266.
30. Hubank M, Mayne L. Expression of the excision repair gene, ERCC3 (excision repair cross-complementing), during mouse development. *Dev. Brain Res* 1994, **81:** 66–76.

31. Mounkes LC, Jones RS, Liang B, Gelbart W, Fuller MT. A drosophila model for Xeroderma Pigmentosum and Cockaynes syndrome: haywire encodes the fly homologue of ERCC3, a human excision repair gene. *Cell* 1992, **71**: 925–937.
32. Zhou Z, Walter C. Expression of the DNA repair gene XRCC1 in baboon tissues. *Mut Res* 1995, **348**: 111–116.
33. Brooks PJ, Marietta C, Goldman D. DNA mismatch repair and methylation in adult brain neurons. *J Neurosci* 1996, **16**: 939–945.
34. Hanawalt PC, Gee P, Ho L, Hsu RK, Kane CJM. Genomic heterogeneity of DNA repair. *Ann NY Acad Sci* 1992, **633**: 17–25.
35. Cleaver JC, Kraemer K. Xeroderma Pigmentosum in (Scriver CR, Beaudet AL, Sly WS, Valle D, eds.) *Metabolic and Molecular Basis of Inherited Disease*. McGraw Hill, New York, 1995, pp. 4393–4419.
36. Kraemer KH, Lee MM, Scotto J. Xeroderma Pigmentosum, cutaneous, ocular and Neurologic abnormalities in 830 published cases. *Arch Derm* 1987, **123**: 241–249.
37. Robbins JH, et al. Neurologic disease in Xeroderma Pigmentosum. *Brain* 1991, **14**: 1335–1361.
38. Hoeijmakers JHJ. Nucleotide excision repair: Molecular and clinical implications, in DNA Repair Mechanisms: Impact on Human Disease and Cancer (Vos, J.H., ed.), RG Landes, Austin TX, 1995, pp. 125–150.
39. Evans MK, Robbins JH, Ganges MB, Tarone RE, Nairn RS, and Bohr VA. Gene-specific DNA repair in xeroderma pigmentosum complementation groups A, C, D, and F. Relation to cellular survival and clinical features. *J Biol Chem* 1993, **268**: 4839–4847.
40. Venema J, Mullenders LHF, Natarajan AT, Van Zeeland AA, Mayne LV. The genetic defect in Cockayne syndrome is associated with a defect in repair of UV-induced DNA damage in transcriptionally active DNA. *Proc Natl Acad Sci USA* 1990, **87**: 4707–4711.
41. Andrews AD, Barrett SF, Robbins, JH. Xeroderma Pigmentosum neurological abnormalities correlate with colony-forming ability after ultraviolet radiation. *Proc Natl Acad Sci USA* 1978, **75**: 1984–1988.
42. Moriwaki S, Nishigori C, Imamura S, Takahashi C, Fujimoto N, Takebe H. A case of xeroderma pigmentosum complementation group F with neurological abnormalities. *Br J Derm* 1993, **128**: 91–94.
43. Robbins JH, Brumback RA, Moshell AN. Clinically asymptomatic xeroderma pigmentosum neurological disease in an adult: evidence for a neurodegeneration in later life caused by defective DNA repair. *Eur Neurol* 1993, **33**: 188–90.
44. Cockayne EA. Dwarfism with retinal atrophy and deafness. *Arch Dis Child* 1936, **11**: 1–8.
45. Henning KA, Li L, Iyer N, McDaniel LD, Reagan MS, Legerski R, Schultz RA, Stefanini M, Lehmann AR, Mayne LV, Friedberg EC. The Cockayne syndrome group A gene encodes a WD repeat protein that interacts with CS-B protein and a subunit of RNA polymerase II TFIIH. *Cell* 1995, **82**: 555–564.
46. Nance MA, Berry SA. Cockayne syndrome: Review of 140 cases. *Am J Med Genet* 1992, **42**: 68–84.
47. Lehmann AR. Cockaynes syndrome and trichothiodystrophy: Defective repair without cancer. *Cancer Rev* 1987, **7**: 82–103.
48. Soffer D, Grotsky HW, Rapin I, Suzuki K. Cockayne syndrome: Unusual neuropathological findings and review of the literature. *Ann Neurol* 1979, **6**: 340–348.
49. Balajee AS, Ma A, Dianov GL, Friedberg EC, Bohr VA. Reduced RNA polymerase II transcription in intact and permeabilized Cockayne syndrome cells. *Proc Nat Acad Sci USA* 1997, **94**: 4306–4311.
50. Dianov GL, Houle JF, Iyer N, Bohr VA, Friedberg EC. Reduced RNA Polymerase II Transcription in Extracts of Cockayne Syndrome and Xeroderma Pigmentosum/Cockayne Syndrome Cells. Nucleic Acids Res 1997, (in press).

51. Itin PH, Pittelkow MR. Trichothiodystrophy: Review of sulfur-deficient brittle hair syndromes and association with the ectodermal dysplasias. *J Am Acad Derm* 1990, **22:** 705–717.
52. Coulter DL, Beals TF, Allen RJ. Neurotrichosis: Hair-shaft abnormalities associated with neurological diseases. *Dev Med Child Neurol* 1982, **5:** 634–44.
53. Chen E, Cleaver JE, Weber CA, Packman S, Barkovich AJ, Koch TK, Williams ML, Golabi M, Price VH. Trichothiodystrophy: Clinical spectrum, central nervous system imaging, and biochemical characterization of two siblings. *J Invest Dermatol* 1994, **103:** 154S–158S.
54. Takayama K, Salazar EP, Broughton BC, Lehmann AR, Sarasin A, Thompson LH, Weber CA. Defects in the DNA repair and transcription gene ERCC2 (XPD) in trichothiodystrophy. *Am J Hum Genet.* 1996, **58:** 263–270.
55. Vermeulen W, et al. (1994) Three unusual repair deficiencies associated with transcription factor BTF2(TFIIH): evidence for the existence of a transcription syndrome, in *Cold Spring Harbor Symposia on Quantitative Biology*, LIX, 1994, 317–329.
56. Swift M, Reitnauer PJ, Morrell D, Chase CL. Breast and other cancers in families with Ataxia-Telangiectasia. *N Engl J Med* 1987, **316:** 1289–94.
57. Swift M, Morrell D, Massey RB, Chase CL. Incidence of cancer in 161 families affected by Ataxia-Telangiectasia. *N Engl J Med* 1991, **325:** 1831–1836.
58. Sedgwick RP. Neurological abnormalities in ataxia-telangiectasia in *Ataxia-Telangiectasia—A Cellular and Molecular Link Between Cancer, Neuropathology, and Immune Deficiency* (Bridges BA, Harnden DG, eds.), Wiley, NY, 1982, pp. 23–35.
59. Vinters HV, Gatti RA, Rakic P. Sequence of cellular events in cerebellar ontogeny relevant to expression of neuronal abnormalities in ataxia-telangiectasia, in *Ataxia-Telangiectasia: Genetics, Neuropathology, and Immunology of a Degenerative Disease of Childhood* (Gatti, RA, Swift M, eds.) Alan R. Liss, NY, 1985, pp. 233–255.
60. Gatti RA, Berkel I, Boder E, Braedt G, Charmley P, Concannon P, Ersoy F, Foroud T, Jaspers NG, Lange K, et al. Localization of an ataxia-telangiectasia gene to chromosome 11q22-23. *Nature* 1988, **336:** 577–580.
61. Savitsky K, Bar-Shira A, Gilad S, Rotman G, Ziv Y, Vanagaite L, Tagle DA, Smith S, Uziel T, Sfez S, et al. A single ataxia telangiectasia gene with a product similar to PI-3 kinase. *Science* 1995, **268:** 1749–1753.
62. Hartley KO, Gell, D, Smith GC, Zhang H, Divecha N, Connelly MA, Admon A, Lees-Miller SP, Anderson CW, Jackson SP. DNA-dependent protein kinase catalytic subunit: A relative of phosphatidylinositol 3-kinase and the ataxia telangiectasia gene product. *Cell* 1995, **82:** 849–856.

13
Cell Death and the Mitochondrial Encephalomyopathies

Kevin M. Flanigan and Rajiv R. Ratan

INTRODUCTION

Mitochondria are intracellular organelles where energy production, via oxidative phosphorylation, takes place. Oxidative phosphorylation occurs at the inner mitochondrial membrane where electron transfer through the respiratory chain assemblies leads to efficient production of adenosine triphosphate (ATP), the major currency of energy within the cell *(1)*. The majority of mitochondrial respiratory proteins are encoded by genes in the nuclear genome. The mitochondria, however, contain their own unique DNA (mtDNA), in the form of a 16.6 kB circular molecule that encodes 37 genes, including those for 13 mitochondrial respiratory chain proteins, 22 transfer RNAs, and 2 ribosomal RNAs *(1)*.

Several human diseases are caused by mutations in mtDNA. These include point mutations in both tRNAs and in proteins of the respiratory chain, as well as deletions of portions of the mitochondrial genome *(2–4)*. As the sperm contributes essentially no mitochondria to the zygote, diseases associated with mtDNA mutations demonstrate a maternal inheritance, with the trait only passed through the maternal line. Mutations may also occur in nuclear genes encoding mitochondrial proteins; such mutations are associated with an autosomal pattern of disease inheritance *(5–7)*.

There are multiple mitochondria within a given cell (the number varies depending on the tissue), and within each mitochondria are multiple copies of the mtDNA molecule. The proportion of mutated mtDNA molecules can vary from tissue to tissue, a state termed heteroplasmy; alternatively, all mtDNA in an individual tissue may contain the same mutation (homoplasmy). A heteroplasmic state is achieved by the phenomenon of replicative segregation: when a cell divides, segregation of mitochondria occurs randomly, and daughter cells may receive unequal portions of mutant DNA molecules. This may occur with each mitotic division, with the result that differentiated tissues may vary in their degree of heteroplasmy *(8)*.

The human mitochondrial genome is completely sequenced, and mutations within that genome have been increasingly associated with disease phenotypes (*see* Tables 1 and 2). In spite of this growing genetic classification, the pathophysiology of those phenotypes remains obscure. Expression of disease may be related to the oxidative

From: Cell Death and Diseases of the Nervous System
Edited by: V. E. Koliatsos and R. R. Ratan © Humana Press Inc., Totowa, NJ

Table 1
Phenotypes Commonly Associated with Mitochondrial Point Mutations

Mutation, nucleotide	Gene	Phenotype	Reference
3243	tRNA$^{Leu(UUR)}$	MELAS	15
		PEO	16
		Diabetes/deafness	17
		MERRF/PEO	18
3271	tRNA$^{Leu(UUR)}$	MELAS	19
3303	tRNA$^{Leu(UUR)}$	Cardiomyopathy	20
3260	tRNA$^{Leu(UUR)}$	Cardiomyopathy/myopathy	21
4269	tRNA^{Ile}	Multisystem/cardiomyopathy	22
8344	tRNA^{Lys}	MERRF	23
8356	tRNA^{Lys}	MERRF/MELAS	24
15990	tRNA^{Pro}	Myopathy	25
8993	ATP synthase	NARP/Leigh	26,27
11778	ND4	LHON	28
4160	ND1	LHON	29
3460	ND1	LHON	30,31
14484	ND6	LHON	32
15257	Cytochrome b	LHON	33

MELAS, Mitochondrial encephalomyopathy, lactic acidosis, and stroke-like episodes; PEO, progressive external ophthalmoplegia; MERRF, myoclonic epilepsy and ragged-red fibers; NARP, neuropathy, ataxia, and retinitis pigmentosa; tRNA, transfer RNA; LHON, Leber hereditary optic neuropathy. Table modified from ref. 2.

demands of the tissue involved: the brain, a tissue with disproportionate oxidative energy demands, is commonly affected, along with other tissues with high oxidative demands (e.g., muscle). However, deficits in oxidative phosphorylation and energy production are not consistently observed in diseased tissues from patients with individual mitochondrial mutations (see below). Thus, the precise cellular factors that confer selective vulnerability to certain cell populations afflicted with mitochondrial mutations remains unknown.

Study of the contribution of mtDNA to cell biology and pathophysiology has been (and will continue to be) aided by the development of ρ^0 cells (9). These cells are cultured in ethidium bromide, resulting in the selective depletion of mtDNA. They lack a functional respiratory chain, and require pyruvate and uridine for growth; fusion with cytoplasts containing mitochondria results in complementation of the respiratory defect. Fusion of ρ^0 cells to enucleated cytoplasts containing mtDNA of interest allows examination of the contributions of mtDNA (vs nuclear DNA) to cellular function. A useful development in ρ^0 studies has been to use platelets, the anucleate derivatives of megakaryocytes, as mtDNA donors (10). Studies in ρ^0 fibroblasts have contributed to our understanding of diseases related to oxidative phosphorylation and aging (11), and the recent development of ρ^0 cells with neuronal properties promises to further clarify the pathophysiology of human neurologic disease (12).

In addition to evidence implicating abnormal mitochondria in a variety of degenerative diseases (13,14), a growing body of literature suggests that mitochondria play a

Table 2
Phenotypes Commonly Associated with Fixed Mitochondrial Deletions[a]

Syndrome	Phenotype	Reference

Kearns-Sayre syndrome (KSS)	Onset before age 20
	PEO
	Pigmentary retinopathy
	  plus one of the following:
	    heart block, cerebellar syndrome,
	    of CSF protein concentration >1 g/L
		34,25

PEO with ragged-red fibers	Ophthalmoplegia, ptosis, myopathy	36
Pearson marrow/pancreas syndrome	Refractory anemia, pancreatic dysfunction (rare survivors to adolescence develop KSS)	37,38

[a]A deletion of a single size, identical in all tissues, although the proportion of deleted vs wild type molecules may vary from tissue to tissue.

central role in apoptosis. Converging lines of evidence suggest a central role for mitochondrial function in the initiation and effector phases of apoptosis, and these studies promise to shed light on the pathophysiology of diseases related to mitochondrial DNA or function. The first part of this chapter will review this evidence, discussing the possible mechanisms by which mitochondrial function may play a role, and we will consider what this evidence may suggest for the pathophysiology of mitochondrial diseases associated with mitochondrial mutations (15–39).

APOPTOSIS, MITOCHONDRIA, AND THE EFFECTS OF THE PLURIOPOTENT ANTIAPOPTOTIC PROTEIN, BCL-2

Apoptosis has been historically defined by the features that distinguish it from necrosis. Some of these require involvement of the nucleus as part of the definition of apoptosis; nuclear changes, including chromatin condensation and endonuclease cleavage of nuclear DNA into 180–200 bp fragments, have been described as key morphological features of apoptosis. However, these nuclear changes are not the earliest events in apoptosis (39), and other features (including chromatin condensation and cytoplasmic contraction) may be seen in the absence of a nucleus (40). These observations have implied the existence of cytosolic effectors that do not need to be synthesized anew for apoptosis to occur.

Evidence for a constitutively expressed effector pathway of apoptosis has come from studies in *C. elegans* (*see* Chapter 7, this volume), in which at least three genes have been shown to be involved in the execution and prevention of apoptosis: *ced-3*, *ced 4*, and *ced-9*. Eukaryotic homologs have been identified for two of these genes. One gene, *ced-9*, encodes a protein that prevents cells from undergoing apoptosis (41) and is homologous to the mammalian *bcl-2* gene (42,43). The second, *ced-3*, encodes a cysteine protease related to the interleukin 1β-converting enzyme (ICE) family of proteases (now known as caspases [*see* Chapter 5, this volume]); activation of ced-3 in *C. elegans* is required to initiate apoptosis, and inhibitors of the caspase family have been shown to prevent apoptosis induced by multiple stimuli in phylogenetically unrelated organisms. Like ced-9 in *C. elegans*, bcl-2 has been shown to act upstream of activation of the caspase family (i.e., ced-3 and ICE family members) in preventing apoptosis; however, despite intensive study, the precise mechanism by which bcl-2 prevents cell death remains unclear.

Much evidence suggests that Bcl-2, in some paradigms, acts to prevent apoptosis through interactions with mitochondria. Subcellular localization studies of Bcl-2 have shown that the protein, through its carboxy-terminal hydrophobic domain, is located in the outer mitochondrial as well as outer nuclear and endoplasmic reticulum membranes *(44,45)*. Evidence that targeting of Bcl-2 to mitochondria may be of import to its antiapoptotic function comes from several observations.

In Vitro Apoptosis Requires Mitochondria

In a cell-free system, in which cytosolic lysate from cells committed to apoptosis induces changes consistent with apoptosis in isolated nuclei, Bcl-2 protein added to the lysate protects against nuclear apoptotic changes; notably, fractionation of the cytosolic lysate suggests that mitochondria play a central role, because a fraction heavily enriched in mitochondria is required for nuclear apoptotic changes *(46)*. This mitochondrial fraction acts to permit apoptosis by mechanisms other than providing ATP for energy-dependent nuclear apoptotic events, because a nonmitochondrial ATP-generating complex is present, as a part of the cell-free system, prior to the addition of the mitochondrial fraction.

One mechanism by which mitochondria might signal or trigger apoptosis is via generation of reactive oxygen species, but there is conflicting data on whether such species play a role. Bcl-2 was found to protect cells from γ irradiation-induced cell death (the immediate effect of such radiation is the generation of hydroxyl radical *[24]*), as well as from H_2O_2 and menadione-induced cell death *(47)*. However, ρ^0 cells—lacking a respiratory chain—undergo apoptosis when exposed to the protein kinase inhibitor staurosporine or when deprived of growth factors *(45)*, and cells cultured in anaerobic conditions cannot generate reactive oxygen species, yet enter apoptosis via pathways inhibited by Bcl-2 and the related protein Bcl-x_L *(48,49)*.

Subsequent studies in cell-free systems have suggested that mitochondria participate in the activation of proteases involved in apoptosis and in the induction of apoptotic nuclear changes by releasing intramitochondrial proteins. Two candidate proteins released from mitochondria that may play causal roles in the execution of apoptosis have been identified. Liu et al. demonstrated that cytochrome C, an essential member of the mitochondrial respiratory enzyme chain, is necessary for the induction of apoptosis by deoxyadenosine triphosphate in a cell-free system. Additionally, they showed that cytochrome C, a 15 kDa soluble protein normally resident in the intramitochondrial space, is released from the mitochondria into the cytoplasm in intact cells by the apoptosis-inducing agent, staurosporine *(50)*. By contrast, Susin et al. identified a 50 kDa protein, which they termed apoptosis inducing factor (AIF), released from isolated mitochondria by agents that disrupt the inner mitochondrial membrane potential *(51)*. Like cytochrome C, this protein causes isolated nuclei to undergo apoptotic changes such as chromatin condensation and internucleosomal DNA fragmentation. Of note, Bcl-2 hyperexpressed in the outer mitochondrial membrane blocks the release of AIF or cytochrome C from mitochondria *(50,51)*. Collectively, these studies suggest that Bcl-2 can prevent apoptotic changes by favoring the retention of proteins in the mitochondria. Further, they highlight the power of cell-free apoptotic systems for studying the role of normal as well as genetically abnormal mitochondria in apoptotic cell death.

Targeting Bcl-2 the Outer Mitochondrial Membrane Abrogates Apoptosis in Some Paradigms

Another strategy utilized to study the role of mitochondria in apoptosis is construction of fusion proteins containing mitochondrial targeting sequences to deliver overexpressed proteins such as Bcl-2 specifically to this organelle. Mutant Bcl-2 protein containing a mitochondrial targeting domain (in exchange for the native carboxy-terminal insertion sequence) are more effective in preventing apoptosis, in some paradigms, than mutant Bcl-2 targeted to the endoplasmic reticulum or Bcl-2 without an insertion sequence *(52,53)*. Although most studies have used the yeast protein Mas70p as a membrane-targeting domain, more specific targeting to the outer mitochondrial membrane can be achieved using a 27 amino acid sequence from the ActA protein. ActA is a listerial protein that is normally involved in the nucleation of the actin cytoskeleton by the listerial parasite; however, truncation of the amino terminal secretory signal sequence of ActA targets the rest of the molecule to the outer mitochondrial membrane *(54,55)*. Fusing ActA to Bcl-2 appears to give more specific mitochondrial staining than the use of the mitochondrial localization sequence Mas70p; therefore, this approach may be a useful strategy for targeting overexpressed proteins to the outer mitochondrial membrane in future studies of neuronal apoptosis.

MECHANISMS OF CELL LOSS IN MITOCHONDRIAL DISEASE

Altogether, these studies raise the hypothesis that inappropriate apoptosis may play a critical role in the clinical manifestations of mitochondrial disease. In fact, the pathophysiology of mitochondrially inherited disorders remains obscure. This is particularly true in the case of the central nervous system (CNS), from which patient samples cannot easily be obtained for in vitro studies of respiratory chain function. Indirect evidence for cell death in MELAS is suggested by the generalized cerebral atrophy commonly seen with longstanding disease *(56)*. Neuropathological studies in KSS, MELAS, and MERRF have revealed spongiform changes, primarily in the white matter, associated with gliosis and neuronal loss distributed asymmetrically through the gray matter and basal ganglia *(57,58)*. Abnormal mitochondria have been demonstrated in smooth muscle fibers of small intramuscular or cerebral blood vessels in many patients with MELAS *(59)*. In one report, accumulation of subendothelial mitochondria was associated with narrowing of the capillary lumen *(60)* leading the authors to suggest that the small-caliber cerebral vasculature is the primary site of pathophysiology in this syndrome, and that the neuronal dysfunction is a result of ischemia resulting from disordered autoregulation. Indeed, cerebral blood flow abnormalities have been demonstrated in some reports *(61,62)*. However, studies of brain biopsy tissue revealed morphologically abnormal mitochondria within neurons *(63)*. Furthermore, MELAS lesions are often widespread and confluent, an unlikely distribution if the lesions are due to small-vessel disease.

In contrast to brain, muscle is in many ways an ideal tissue for the study of pathophysiology. It is accessible, and cultured patient myoblasts can be fused with ρ^0 fibroblasts. Furthermore, muscle has a "warning system: fibers which contain abnormal accumulations of mitochondria are readily recognized as ragged-red fibers with histochemical staining. In spite of these useful features, the pathogenesis of muscle symptoms has not been fully elucidated, and some problems are inherent in the study of muscles in vitro. In syndromes associated with heteroplasmic mutations (e.g., the MELAS-associated A3243G mutation), the proportion of mutant DNA varies from tissue to tissue, and within a single muscle from fiber to fiber *(64)*; a similar tissue-specific (and, presumably, fiber-specific) defect in respiratory complex function may be present. This heterogeneity of mitochondrial functional defects implies that respiratory defects may not be discernible in muscle homogenates, where the vast majority of fibers may be unaffected. And in spite of the demonstration of respiratory chain defects in muscle tissue from patients with clinical myopathy, the molecular mechanism by which these defects results in clinical weakness is unclear.

Although the mechanism of cell loss in the mitochondrial encephalomyopathies has not been determined, the development of early, more specific markers for mitochondrial dysfunction and for apoptosis suggest that answers to this question are forthcoming.

Acknowledgment of a central role for mitochondria in apoptosis and necrosis has focused the attention on defining the precise role that mitochondria play as organelles mediating cell death. Several plausible schemes have been proposed, including disruption of mitochondrial calcium homeostasis, increased mitochondrial free radical production, and loss of mitochondrial ATP production. These topics have been covered in detail in other chapters of this book (*see* chapters by Dykens, and by Leist and Nicotera) and will not be reiterated here. Rather, approaches to monitoring specific physiologic functions of mitochondria will be discussed, and the attendant advantages and disadvantages of these approaches for identifying defects underlying mitochondrial diseases will be examined.

Mitochondrial Electron Transport and Mitochondrial Disease

The primary function of mitochondria is to generate ATP. Depletion of ATP is believed to lead to a loss of ion homeostasis and disruption of synaptic transmission as well as cell death. Efficient generation of ATP in neurons, as in other cells of the body, occurs as a result of transfer of electrons from nicotinamide adenine dinucleotide (reduced form) (NADH) or flavin adenine dinucleotide (reduced form) ($FADH_2$) to oxygen by a series of electron carriers at the inner mitochondrial membrane. NADH and $FADH_2$ are generated primarily in the citric acid cycle within the inner mitochondrial membrane. Except for the quinones, the electron carriers responsible for transferring electrons to oxygen in mitochondria are prosthetic groups of proteins. Each of these carriers is part of distinct multienzyme complexes, designated complexes I–IV. At three of four of these complexes, transfer of electrons is associated with pumping of protons out of the inner mitochondrial matrix. This pumping, combined with the poor permeability of the inner mitochondrial membrane to hydrogen ions, leads to the generation of a mitochondrial membrane potential (Ψ_m). ATP is synthesized as protons flow back into the mitochondrial matrix down their electrochemical gradient through proton channels.

Mitochondrial Electron Transport

Respiratory complex function can be dissected in vitro polarographically by measuring oxygen consumption and/or ATP production in the presence of substrates specific for each complex. Polargraphic measurements carried out on isolated mitochondria require a sufficient number of mitochondria, which can be difficult to obtain; an alternative is digitonin-permeabilizion of cells, which removes the permeability barrier of the plasma membrane, and reduces difficulty with yield *(65)*. In digitonin-permeabilized cells, complex I respiration is measured with malate and glutamate, or malate and pyruvate, as substrates; the corresponding dehydrogenases produce NADH, which is oxidized by the NADH:ubiquinone oxidoreductase of complex I, and electrons are transferred through complexes III and IV to oxygen. With succinate as a substrate, ubiquinone is reduced by the succinate:ubiquinone component of complex II, without complex I involvement. Respiration of complex IV (cytochrome c oxidase) can be measured by use of the substrates *N, N, N', N'*-tetramethylene-*p*-phenylenediamine (TMPD) and ascorbate, which reduce cytochrome c.

Enzyme activity, as opposed to respiration, can be measured spectrophotometrically, by measuring the rates of reduction of substrates for each enzyme in mitochondrial membranes exposed by sonication or freezing *(66)*. For example, rotenone-sensitive NADH:CoQ reductase activity represents isolated complex I enzyme function; rotenone-sensitive NADH:cytochrome c reductase activity is a measure of the combined activities of complex I and III *(67)*. Measurements of these activities are usually normalized against citrate synthase activity, a mitochondrial enzyme encoded by a nuclear gene; in this way, a control is provided for any differences in mitochondrial yield. Notably, some complexes require an intact membrane, and require freeze–thaw treatment of mitochondria rather than sonication or whole cell homogenization. This is particularly true with complex I, and implies a dependence on a high degree of organization of the subunits, as opposed to that seen with complex II *(67)*.

In some diseases associated with mtDNA mutations, a discordance has been noted between respiratory function, measured polarographically, and specific enzyme activity, measured by spectrophotometry. In Leber's hereditary optic neuropathy (LHON) patient muscle *(68)*, fibroblast *(69)*, and platelet *(70)* mitochondria, malate/glutamate or malate/pyruvate driven respiration is decreased, whereas complex I oxidoreductase activity shows no decline. This implies that polarography is the more sensitive test for determining the presence of a mitochondrial respiratory defect, and respiratory measurements are more likely to represent the true in vivo activity of respiratory chain complexes—perhaps because spectrophotometric assays do not couple enzyme activity to ADP phosphorylation *(68)*.

The reasons for the discrepancy between polargraphic and spectrophotometric results in LHON are unclear. One possibility is that in intact mitochondria (or digitonin-permeabilized cells) NADH may be directly delivered to the complex I binding site by the mitochondrial matrix NAD-linked dehydrogenases; these have been suggested to be closely associated with complex I *(71)*. The 11778 mtDNA mutation might interfere with the interaction between complex I and the NAD-linked dehydrogenases, resulting in decreased activity with an oxidizable substrate, but not in the presence of the NADH excess used in spectrophotometric assays *(69)*. Another possibility is that diffusion of the hydrophilic ubiquinone analogs used as substrates in enzyme assays may be a rate limiting step; if so, the assays would not measure the full activity of the wild type enzyme, and decreases of enzyme activity may be hard to detect.

Protein Translation

Although it might be expected that mutations in subunits of the respiratory chain would lead directly to altered respiratory chain activity, the connection between mitochondrial tRNA point mutations and respiratory chain dysfunction might seem less direct. Respiratory chain defects are apparently the result of more generalized defects of mitochondrial protein synthesis. In MERRF, mutations at nt 8344 in the mitochondrial tRNA^{Lys} result in decreased tRNA^{Lys} aminoacylation capacity, leading to impaired protein synthesis *(72)*. In MELAS, a mutation at nt 3243 in the mitochondrial tRNA^{Leu} gene results in impaired mitochondrial protein synthesis, although the mechanism of the impairment is less well characterized; proposed factors include defective mRNA processing, altered aminoacylation, incorrect conjugation of amino acids to the tRNA, and slowed protein synthesis due (at least in part) to altered transcripts binding abnormally to ribosomes *(73–75)*.

In both MERRF and MELAS, decreases in the activity of varying respiratory complexes have been described. In MERRF, deficiencies of complex II, complex III, complex IV, and combinations of complexes I and IV or III and IV have been described, although many of these reports predated molecular characterization of the gene mutation *(76)*. In MELAS, deficiencies in complexes I and IV have been described. However, recent work has determined that, although complex I activity is decreased in both disorders, complex IV activity is decreased only in MERRF and complex II activity is increased only in MELAS *(66)*. Polarographic studies demonstrated a decrease in glutamate/malate-driven (complex I) respiration, whereas no significant change was seen with succinate-driven (complex II) respiration; notably, these findings reached statistical significance only when respiration was uncoupled with FCCP (carbonyl cyanide-*p*-triflouromethoxyphenylhydrazone), resulting in maximal possible activity of the respiratory chain *(66)*.

ATP Production

In spite of the demonstration of respiratory complex defects, measurements of ATP have generally been normal in mitochondrial cytopathies, including cultured fibroblasts from patients with MELAS and with KSS *(77)*. A defect in ATP production is implied by the growth rate abnormalities seen when LHON fibroblasts are grown in galactose-containing medium *(69)*. As mammalian cells derive 45–60% of their ATP from glycolysis, even respiratory complex-deficient cells can grow in glucose-containing medium; galactose, however, is only slowly metabolized through the glycolytic pathway. Similarly, nonviability of fibroblasts in a galactose-containing medium has been utilized as a screening test for major oxidative defects *(78)*.

Mitochondrial Free Radicals

Mitochondrially derived reactive oxygen species are believed to play a central role in apoptosis and excitotoxicity (*see* chapter by Dykens). In mitochondrial diseases, free radicals may play a similar central role in pathogenesis. One mechanism by which, under basal conditions, mitochondrial radicals are believed to be generated, is by leakage of electrons from electron-rich intermediates onto oxygen. It is, therefore, reasonable to propose that inefficiencies in mitochondrial electron transfer resulting from mitochondrial mutations may lead to an accumulation of electron-rich intermediates and increased flux of mitochondrial free radicals. Such an increase in mitochondrial radicals, if not appropriately neutralized, could lead to cell damage and cell death, or to enhancement of cell damage already caused by the basic biochemical defect (*see* chapter by Dykens).

A model for mitochondrially derived free radical damage exists with tumor necrosis factor (TNF)-mediated cytotoxicity, in which altered mitochondrial electron transport and free radical production alter the gene-inductive and cytotoxic effects of TNF. In response to TNF exposure, superoxide radicals are generated in complex III by the univalent reduction of oxygen by the electron carrier ubisemiquinone. This conclusion is supported by the observation that inhibitors of complex I (rotenone) or complex II (thenoyltrifluoroacetone [TTFA])—which result in a decrease in the number of downstream electron carriers in the reduced state, capable of undergoing autoxidation— abrogate the cytotoxic effects of TNF *(79)*. Conversely, inhibition of complex III by antimycin-A (which acts to increase the number of electron carriers capable of forming superoxide) potentiates TNF-induced cell death *(79)*. Finally, ρ^0 cells (lacking a functional respiratory chain) are resistant to TNF-induced cytotoxicity *(80)*. These TNF studies suggest that mitochondrial radical production may be enhanced or inhibited depending on the nature of the genetic mitochondrial defect, and suggest a system for testing these effects. For example, a mutation in complex I that blocks electron transfer to complex III might be expected to inhibit TNF-induced death (as does rotenone). Conversely, mutations in complex III or IV, which may result in an accumulation of upstream electron carriers, might enhance TNF-induced death (as does antimycin-A). Thus, the effects of mitochondrial mutations on free radical generation are likely to be quite heterogeneous.

In mitochondrial encephalomyopathies, evidence supporting a pathophysiological role for "oxidative stress" (by definition, an imbalance in cellular oxidants and antioxidants in favor of oxidants) is limited. In one study, markers of "oxidative stress" were examined in the plasma of eleven patients with mitochondrial myopathy and progressive external ophthalmoplegia, six of whom had large-scale mtDNA deletions by Southern analysis. A statistically significant increase in plasma lipid peroxidation was observed, along with an associated decrease in levels of the antioxidant glutathione in plasma and erythrocytes *(81)*. However, whether or not these changes (or the patients' symptoms) improved was not addressed. Furthermore, whether these findings can be extrapolated to distinct mitochondrial disease phenotypes is unclear, and the precise meaning of these studies awaits further investigation.

An in vitro approach to determine the effect of mtDNA mutations on mitochondrial free radical production relies on the use of ρ^0 cells fused with cytoplasts derived from patients with mtDNA mutations (cybrids), in conjunction with the oxidant sensitive reporter, dihydrorhodamine 123 (DHR 123). DHR 123 is converted to rhodamine 123 by intracellular peroxide in the presence of trace metals, cytochrome C, or peroxidase, but is not converted by superoxide *(82)*. The rhodamine 123 produced is a positively charged, lipophilic species that accumulates in the matrix of mitochondria because of its inside-negative membrane potential; increased cellular oxidative load thus leads to increased rhodamine 123 mitochondrial labeling. Since DHR 123 staining is dependent on the mitochondrial membrane potential as well as the cellular oxidant load, control experiments are crucial, and must demonstrate that addition of rhodamine 123 to the cells under study results in mitochondrial staining which does not change with the experimental stimulus or variable. Studies have established that DHR 123 is trapped intracellularly for at least an hour with continued presence of reporter in the medium, and that DHR 123 is not sensitive to changes in other second messengers such as calcium or pH. In nerve cells, DHR 123 has been used effectively to monitor free radical burden in excitotoxic and growth factor deprivation paradigms *(83)*.

Sobreira et al. used DHR 123 in cybrids and hybrids from patients with mtDNA mutations (both point mutations and deletions). They found cellular fluorescence to be significantly reduced only in clones containing >99% mutant DNA, correlating with decreased oxygen consumption in these cells; fluorescence was not significantly changed in cybrids and hybrids with greater degrees of heteroplasmy *(84)*. Whereas these results are quite intriguing, it is not clear whether the changes observed are related to decreased oxidation of DHR 123 to rhodamine 123, or to decreased accumulation of rhodamine 123 into the mitochondria because the mitochondrial membrane potential has been dissipated by the mitochondrial genetic defect. Similar studies using DHR 123, newer generation oxidant reporters, and other in vivo markers of oxidative stress (*see* chapter by Ratan) promise to clarify the role of free radicals in mitochondrial encephalomyopathies.

Mitochondrial Transmembrane Potential

There is growing evidence that an early marker of apoptosis is loss of mitochondrial transmembrane potential ($\Delta\Psi_m$); loss of $\Delta\Psi_m$ precedes apoptosis associated with glucocorticoid-induced apoptosis of lymphocytes and TNF-induced apoptosis in U937 cells *(85)*. Mitochondrial transmembrane potential can be assayed using the dye JC-1 (5,5',6,6'-tetrachloro-1,1', 3,3'-tetraethylbenzimidazolocarbocyanine iodide). This lipophilic, positively charged dye displays green fluorescence in its monomeric form. Because of its charge, it accumulates in the negatively charged mitochondrial matrix. After the monomers exceed a critical concentration, the molecules form J-aggregates, which show red fluorescence *(86)*. The presence of J-aggregates require an intact electrochemical gradient, or mitochondrial transmembrane potential ($\Delta\Psi_m$) *(87)*; loss of the gradient (via protonophores or antimycin A) is associated with loss of red fluorescence. The JC-1 monomer, however, is not completely released from the matrix with loss of the electrochemical gradient; the amount of green or red fluorescence allows estimation of the energy state of mitochondria. Another dye that accumulates in the mitochondria in a energy-dependent fashion (correlating to the $\Delta\Psi_m$) is $DiOC_6(3)$ [3,3'dihexyloxacarbocyanine iodide] *(88)*. Methyltriphenylphosphonium (TPMP) also accumulates in mitochondria in a $\Delta\Psi_m$-dependent fashion, but is not fluorescent; uptake of [^{3}H]TPMP into the permeabilized or intact cells can be measured, providing an index of mitochondrial energization *(66)*. The advantage of fluorescent techniques is that individual mitochondria can be visualized, with change in energy state over time; in addition, in the case of JC-1, a ratio of energized to unenergized mitochondria (red to green fluorescence) is readily calculated.

Mitochondrial Calcium Homeostasis

Calcium regulation plays a central role in cell death, and the role of mitochondria in calcium regulation and signal transduction is increasingly recognized (*see* chapter by

Leist and Nicotera); therefore, an attractive hypothesis is that altered mitochondrial calcium regulation plays a role in cell death in the mitochondrial cytopathies. Calcium influx into the mitochondria matrix has been shown to disassociate electron transport from ATP production *(89,90)*. The mechanism by which an increase in intracellular Ca^{2+} ($[Ca^{2+}]_I$) results in loss of $\Delta\Psi_m$ is likely via the permeability transition pore (PTP). Activation of this membrane pore allows solutes and proteins of <1.5 kDa to freely enter the mitochondrial matrix, resulting in mitochondrial swelling and loss of membrane potential *(90)*. Activation occurs via many signals, including adenosine dinucleotides (particularly adenosine diphosphate [ADP]), marked increases in matrix and cytosolic Ca^{2+} (and, to a lesser extent, Mg^{2+}) concentrations, and some polyamines *(90)*. Enhanced permeability through the PTP is blocked by cyclosporin A. Of note, cyclosporin A abrogates the loss of $\Delta\Psi$ and apoptosis associated with glucocorticoid treatment of lymphocytes and TNF treatment of U937 cells, or necrotic cell death associated with exposure of neurons to N-methyl-D-aspartate (NMDA).

Studies in MELAS fibroblasts have revealed abnormalities in calcium homeostasis. These consist of elevated baseline $[Ca^{2+}]_i$ as well as abnormally sustained $[Ca^{2+}]_i$ in response to depolarization, in comparison to fibroblasts from normal individuals and from KSS patients *(77)*. Because plasma-membrane Ca^{2+} extrusion is an energy-dependent process *(93)*, low levels of ATP may have contributed to the sustained rise in $[Ca^{2+}]_i$; however, measurement of ATP revealed normal levels *(77)*. In the same experiments, analysis of membrane potential with JC-1 revealed a decreased proportion of energized mitochondria in MELAS fibroblasts as compared to controls *(77)*. This loss of membrane potential may not be by a PTP-dependent mechanism; in MELAS and MERRF fibroblasts, treatment with cyclosporin A did not alter the decrease in membrane potential as assayed by TPMP accumulation *(66)*. Future studies will clarify whether increases in cytoplasmic calcium in MELAS fibroblasts are correlated with increased sequestration of calcium into mitochondria.

CONCLUSIONS AND REMAINING QUESTIONS REGARDING THE PATHOPHYSIOLOGY OF MITOCHONDRIAL ENCEPHALOMYOPATHIES

Although the mitochondrial genome has been well characterized, much information regarding the pathophysiology of the mitochondrial encephalomyopathies remains to be learned. As outlined above, the growing sophistication of techniques may allow researchers to begin to answer outstanding questions.

What is the Nature of Cell Death?

Given the growing evidence that suggests that loss of mitochondrial transmembrane potential may act—directly and indirectly—as a trigger for, marker of commitment to, or as a positive modulator of programmed cell death under excitotoxic and other conditions *(94)*, it is tempting to speculate that mutations in mitochondrial protein genes, resulting in altered respiratory chain function, might predispose cells to apoptosis. This concept is only beginning to be thoroughly addressed. Although cellular dysfunction in diseases of mtDNA has often been presumed to be due to energy depletion with lack of ATP, in vitro studies do not entirely support this view. (Notably, lack of ATP production has been associated with necrotic, rather than apoptotic, death *[95]*.)

If Apoptosis Occurs, Is It Mediated Through Mitochondrial Proteins?

The recent description of mitochondrial proteins that serve as cytosolic effectors of apoptosis *(50,51)* confirms the central role mitochondria play in apoptosis. If apoptosis occurs in mitochondrial encephalomyopathies, what is the relationship between these effectors (cytochrome C and AIF) and mitochondrially encoded proteins?

What Is the Origin of the Respiratory Chain Defects?

In many cases, defects in oxidation (measured polarographically) do not correlate with activity of specific respiratory enzyme complexes (measured spectrophotometrically). This implies that some point mutations of the mitochondrial genome may alter the interaction of complexes with other components of the respiratory chain, rather than interfering with electron transport itself *(see above)*.

What Is the Nature of the Interaction of Nuclear and Mitochondrial Genes and Their Products?

The assembly of the mitochondrial respiratory chain requires the interaction of many nuclearly and mitochondrially encoded proteins. Further characterization of the protein–protein interactions of the subunits of the respiratory complexes will help clarify the pathophysiology of these disorders. The converse is true as well, as study of mitochondria carrying mutations have already helped to clarify organization of the respiratory chain; a role for ND4 in binding of rotenone was demonstrated in mitochondria from LHON patients carrying the 11778 mtDNA mutation, which showed resistance to the toxin *(70)*.

Clarification of the regulation of mitochondrial transcription and replication may reveal how the interaction of nuclear and mitochondrial genes affect human disease. Although the pathways leading to mitochondrial replication have not been completely characterized, there is evidence that mutations in nuclear genes may cause diseases with a "mitochondrial" phenotype due to depletion of total mtDNA *(96)*. Multiple depletion syndromes have been associated with low tissue levels of mtTFA, a promoter of mitochondrial transcription *(97,98)*. mtTFA activation has been shown to be dependent on a variety of other nuclear genes, including the nuclear respiratory factors NRF-1 and NRF-2 *(99)*, which are therefore candidate nuclear genes for mitochondrial syndromes. Autosomally inherited syndromes with "mitochondrial" phenotypes have been demonstrated to be due to mutations of nuclear genes encoding subunits of the respiratory chain *(5)*, and in some cases functional alterations in mitochondrial enzymes have been shown *(5,7)*. Finally, there is significant clinical and statistical genetic evidence for the presence within the nuclear genome of modifier genes, which alter the penetrance and the phenotype of mtDNA syndromes *(100–102)*, although this has been debated *(103)*; such genes await identification.

Is Cell Death Mediated Through Ca^{2+}-Dependent Pathways Similar to Those Seen in Excitotoxicity and Various Paradigms of Apoptosis?

Similarities in apoptosis and excitotoxic cell death have accentuated the role of mitochondrial calcium regulation in mechanisms of cell death, and recent studies explore these common mechanisms. For example, Bcl-2 expression in nerve cells causes increases mitochondrial energy-dependent Ca^{2+} uptake capacity, and confers resistance to Ca^{2+}-induced respiratory inhibition *(104)*. Interestingly, the increase in uptake capacity was seen only with nicotinamide adenine dinucleotide (NAD^+)-linked substrates, and not with the flavin adenine dinucleotide (FAD)-linked substrate succinate; this suggests a role for activity of complex I, and suggests an avenue of exploration for the mitochondrial encephalomyopathies. Further study will benefit from the application of other new techniques, such as study of the calcium photoprotein apoaequorin, which can be targeted to the mitochondria *(105)* (or the endoplasmic reticulum *[106]*) by adding a specific membrane-targeting sequence; visualization of the protein's bioluminescence would allow the characterization of intracompartmental calcium flux in response to various stimuli.

Can Insights Gained at the Cellular Level Lead to Novel Therapeutic Strategies?

Thus far, no treatment has proven to successfully slow the course of the mitochondrial encephalomyopathies *(107)*. Proof that these syndromes stem from apoptotic neuronal death may provide a rationale for treatment, as modifiers of the apoptotic program make their way into clinical trials.

ACKNOWLEDGMENT

K. Flanigan receives support from the Charles E. Culpeper Foundation.

REFERENCES

1. Cooper JM, Clark JB. The structural organization of the mitochondrial respiratory chain, in *Mitochondrial Disorders in Neurology* (Schapira AHV, DiMauro S, eds.), Butterworth-Heinemann, Oxford, 1994, pp. 1–30.
2. DiMauro S, Moraes CT. Mitochondrial encephalomyopathies. *Arch Neurol* 1993, **50:** 1197–1208.
3. Schon EA. Mitochondrial DNA and the genetics of mitochondrial disease, in *Mitochondrial Disorders in Neurology* (Schapira AHV, DiMauro S, eds.), Butterworth-Heinemann, London. 1994, pp. 31–48.
4. Johns DR. The other human genome: mitochondrial DNA and disease. *Nat Med* 1996. **2:** 1065–1068.
5. Bourgeron T, Rustin P, Chretien D, Birch Machin M, Bourgeois M, Viegas Pequignot E, Munnich A, Rotig A. Mutation of a nuclear succinate dehydrogenase gene results in mitochondrial respiratory chain deficiency. *Nat Genet* 1995, **11:** 144–149.
6. Drugge U, Holmberg M, Holmgren G, Almay BG, Linderholm H. Hereditary myopathy with lactic acidosis, succinate dehydrogenase and aconitase deficiency in northern Sweden: a genealogical study. *J Med Genet* 1995, **32:** 344–347.
7. Hall RE, Henriksson KG, Lewis SF, Haller RG, Kennaway NG. Mitochondrial myopathy with succinate dehydrogenase and aconitase deficiency. Abnormalities of several iron-sulfur proteins. *J Clin Invest* 1993, **92:** 2660–2666.
8. Wallace DC. Diseases of the mitochondrial DNA. *Annu Rev Biochem* 1992, **61:** 1175–1212.
9. King MP, Attardi G. Human cells lacking mtDNA: repopulation with exogenous mitochondria by complementation. *Science* 1989, **246:** 500–503.
10. Chomyn A, Lai ST, Shakeley R, Bresolin N, Scarlato G, Attardi G. Platelet-mediated transformation of mtDNA-less human cells: analysis of phenotypic variability among clones from normal individuals—and complementation behavior of the tRNALys mutation causing myoclonic epilepsy and ragged red fibers. *Am J Hum Genet* 1994, **54:** 966–974.

11. Laderman KA, Penny JR, Mazzucchelli F, Bresolin N, Scarlato G, Attardi G. Aging-dependent functional alterations of mitochondrial DNA (mtDNA) from human fibroblasts transferred into mtDNA-less cells. *J Biol Chem* 1996, **271**: 15,891–15,897.
12. Miller SW, Trimmer PA, Parker WD, Davis RE. Creation and characterization of mitochondrial DNA-depleted cell lines with "neuronal-like" properties. *J Neurochem* 1996, **67**: 1897–2907.
13. Beal MF. Does impairment of energy metabolism result in excitotoxic neuronal death in neurodegenerative illnesses? *Ann Neurol* 1992, 31: 119–130.
14. Beal MF. Aging, energy, and oxidative stress in neurodegenerative diseases. *Ann Neurol* 1995, **38**: 357–366.
15. Goto Y, Nonaka I, Horai S. A mutation in the tRNA(Leu)(UUR) gene associated with the MELAS subgroup of mitochondrial encephalomyopathies. *Nature* 1990, **348**: 651–653.
16. Moraes CT, Ciacci F, Silvestri G, Shanske S, Sciacco M, Hirano M, Schon EA, Bonilla E, DiMauro S. Atypical clinical presentations associated with the MELAS mutation at position 3243 of human mitochondrial DNA. *Neuromuscul Disord* 1993, **3**: 43–50.
17. van den Ouweland JM, Lemkes HH, Ruitenbeek W, Sandkuijl LA, de Vijlder MF, Struyvenberg PA, van de Kamp JJ, Maassen JA. Mutation in mitochondrial tRNA(Leu)(UUR) gene in a large pedigree with maternally transmitted type II diabetes mellitus and deafness. *Nat Genet* 1992, **1**: 368–371.
18. Verma A, Moraes CT, Shebert RT, Bradley WG. A MERRF/PEO overlap syndrome associated with the mitochondrial DNA 3243 mutation. *Neurology* 1996, **46**: 1334–1336.
19. Goto Y, Nonaka I, Horai S. A new mtDNA mutation associated with mitochondrial myopathy, encephalopathy, lactic acidosis and stroke-like episodes (MELAS). *Biochim Biophys Acta* 1991, **1097**: 238–240.
20. Silvestri G, Santorelli FM, Shanske S, Whitley CB, Schimmenti LA, Smith SA, DiMauro S. A new mtDNA mutation in the tRNA(Leu(UUR)) gene associated with maternally inherited cardiomyopathy. *Hum Mutat* 1994, **3**: 37–43.
21. Zeviani M, Gellera C, Antozzi C, Rimoldi M, Morandi L, Villani F, Tiranti V, DiDonato S. Maternally inherited myopathy and cardiomyopathy: association with mutation in mitochondrial DNA tRNA(Leu)(UUR). *Lancet* 1991, **338**: 143–147.
22. Taniike M, Fukushima H, Yanagihara I, Tsukamoto H, Tanaka J, Fujimura H, Nagai T, Sano T, Yamaoka K, Inui K, et al. Mitochondrial tRNA(Ile) mutation in fatal cardiomyopathy. *Biochem Biophys Res Commun* 1992, **186**: 47–53.
23. Shoffner JM, Lott MT, Lezza AM, Seibel P, Ballinger SW, Wallace DC. Myoclonic epilepsy and ragged-red fiber disease (MERRF) is associated with a mitochondrial DNA tRNA(Lys) mutation. *Cell* 1990, **61**: 931–937.
24. Silvestri G, Moraes CT, Shanske S, Oh SJ, DiMauro S. A new mtDNA mutation in the tRNA(Lys) gene associated with myoclonic epilepsy and ragged-red fibers (MERRF). *Am J Hum Genet* 1992, **51**: 1213–1217.
25. Moraes CT, Ciacci F, Bonilla E, Ionasescu V, Schon EA, DiMauro S. A mitochondrial tRNA anticodon swap associated with a muscle disease. *Nat Genet* 1993, **4**: 284–288.
26. Holt IJ, Harding AE, Petty RK, Morgan Hughes JA. A new mitochondrial disease associated with mitochondrial DNA heteroplasmy. Am J Hum Genet, 1990. 46(3): p. 428–33.
27. Tatuch Y, Christodoulou J, Feigenbaum A, Clarke JT, Wherret J, Smith C, Rudd N, Petrova Benedict R, Robinson BH. Heteroplasmic mtDNA mutation (T----G) at 8993 can cause Leigh disease when the percentage of abnormal mtDNA is high. *Am J Hum Genet* 1992, **50**: 852–858.
28. Wallace DC, Singh G, Lott MT, Hodge JA, Schurr TG, Lezza AM, Elsas LJD, Nikoskelainen EK. Mitochondrial DNA mutation associated with Leber's hereditary optic neuropathy. *Science* 1988, **242**: 1427–1430.
29. Howell N, Kubacka I, Xu M, McCullough DA. Leber hereditary optic neuropathy: involvement of the mitochondrial ND1 gene and evidence for an intragenic suppressor mutation. *Am J Hum Genet* 1991, **48**: 935–942.

30. Huoponen K, Vilkki J, Aula P, Nikoskelainen EK, Savontaus ML. A new mtDNA mutation associated with Leber hereditary optic neuroretinopathy. *Am J Hum Genet* 1991, **48**: 1147–1153.
31. Howell N, Bindoff LA, McCullough DA, Kubacka I, Poulton J, Mackey D, Taylor L, Turnbull DM. Leber hereditary optic neuropathy: identification of the same mitochondrial ND1 mutation in six pedigrees. *Am J Hum Genet* 1991, **49**: 939–950.
32. Johns DR, Neufeld MJ, Park RD. An ND-6 mitochondrial DNA mutation associated with Leber hereditary optic neuropathy. *Biochem Biophys Res Commun* 1992, **187**: 1551–1557.
33. Johns DR, Neufeld MJ. Cytochrome b mutations in Leber hereditary optic neuropathy. *Biochem Biophys Res Commun* 1991, **181**: 1358–1364.
34. Rowland LP, Blake DM, Hirano M, Di Mauro S, Schon EA, Hays AP, Devivo DC. Clinical syndromes associated with ragged red fibers. *Rev Neurol Paris* 1991, **147**: 467–473.
35. Moraes CT, DiMauro S, Zeviani M, Lombes A, Shanske S, Miranda AF, Nakase H, Bonilla E, Werneck LC, Servidei S, et al. Mitochondrial DNA deletions in progressive external ophthalmoplegia and Kearns-Sayre syndrome. *N Engl J Med* 1989, **320**: 1293–1299.
36. Holt IJ, Harding AE, Cooper JM, Schapira AH, Toscano A, Clark JB, Morgan Hughes JA. Mitochondrial myopathies: clinical and biochemical features of 30 patients with major deletions of muscle mitochondrial DNA. *Ann Neurol* 1989, **26**: 699–708.
37. Rotig A, Cormier V, Blanche S, Bonnefont JP, Ledeist F, Romero N, Schmitz J, Rustin P, Fischer A, Saudubray JM, et al. Pearson's marrow-pancreas syndrome. A multisystem mitochondrial disorder in infancy. *J Clin Invest* 1990, **86**: 1601–1608.
38. McShane MA, Hammans SR, Sweeney M, Holt IJ, Beattie TJ, Brett EM, Harding AE. Pearson syndrome and mitochondrial encephalomyopathy in a patient with a deletion of mtDNA. *Am J Hum Genet* 1991, **48**: 39–42.
39. Sun DY, Jiang S, Zheng LM, Ojcius DM, Young JD. Separate metabolic pathways leading to DNA fragmentation and apoptotic chromatin condensation. *J Exp Med* 1994, **179**: 559–568.
40. Schulze Osthoff K, Walczak H, Droge W, Krammer PH. Cell nucleus and DNA fragmentation are not required for apoptosis. *J Cell Biol* 1994, **127**: 15–20.
41. Hengartner MO, Ellis RE, Horvitz HR. Caenorhabditis elegans gene ced-9 protects cells from programmed cell death. *Nature* 1992, **356**: 494–499.
42. Vaux DL, Weissman IL, Kim SK. Prevention of programmed cell death in Caenorhabditis elegans by human bcl-2. *Science* 1992, **258**: 1955–1957.
43. Hengartner MO, Horvitz HR. C. elegans cell survival gene ced-9 encodes a functional homolog of the mammalian proto-oncogene bcl-2. *Cell* 1994, **76**: 665–676.
44. Nguyen M, Millar DG, Yong VW, Korsmeyer SJ, Shore GC, Targeting of Bcl-2 to the mitochondrial outer membrane by a COOH-terminal signal anchor sequence. *J Biol Chem* 1993, **268**: 25,265–25,268.
45. Jacobson MD, Burne JF, King MP, Miyashita T, Reed JC, Raff MC. Bcl-2 blocks apoptosis in cells lacking mitochondrial DNA. *Nature* 1993, **361**: 365–368.
46. Newmeyer DD, Farschon DM, Reed JC, Cell-free apoptosis in Xenopus egg extracts: Inhibition by Bcl-2 and requirement for an organelle fraction enriched in mitochondria. *Cell* 1994, **79**: 353–364.
47. Hockenbery DM, Oltvai ZN, Yin XM, Milliman CL, Korsmeyer SJ. Bcl-2 functions in an antioxidant pathway to prevent apoptosis. *Cell* 1993, **75**: 241–251.
48. Jacobson MD, Raff MC. Programmed cell death and Bcl-2 protection in very low oxygen. *Nature* 1995, **374**: 814–816.
49. Shimizu S, Eguchi Y, Kosaka H, Kamiike W, Matsuda H, Tsujimoto Y. Prevention of hypoxia-induced cell death by Bcl-2 and Bcl-xL. *Nature* 1995, **374**: 811–813.
50. Liu X, Kim CN, Yang J, Jemmerson R, Wang X. Induction of apoptotic program in cell-free extracts: requirement for dATP and cytochrome c. *Cell* 1996, **86**: 147–157.
51. Susin SA, Zamzani N, Castedo M, Hirsch T, Marchetti P, Macho A, Daugus E, Geuskens M, Kroemer G. Bcl-2 inhibits the mitochondrial release of an apoptogenic protease. *J Exp Med* 1996, **184**: 1331–1341.

52. Tanaka S, Saito K, Reed JC. Structure-function analysis of the Bcl-2 oncoprotein. Addition of a heterologous transmembrane domain to portions of the Bcl-2 beta protein restores function as a regulator of cell survival. *J Biol Chem* 1993, **268:** 10,920–10,926.
53. Nguyen M, Branton PE, Walton PA, Oltvai ZN, Korsmeyer SJ, Shore GC. Role of membrane anchor domain of Bcl-2 in suppression of apoptosis caused by E1B-defective adenovirus. *J Biol Chem* 1994, **269:** 16,521–16,524.
54. Pistor S, Chakraborty T, Niebuhr K, Domann E, Wehland J. The ActA protein of Listeria monocytogenes acts as a nucleator inducing reorganization of the actin cytoskeleton. *Embo J* 1994, **13:** 758–763.
55. Zhu W, Cowie A, Wasfy GW, Penn LZ, Leber B, Andrews DW. Bcl-2 mutants with restricted subcellular location reveal spatially distinct pathways for apoptosis in different cell types. *EMBO J* 1996, **15:** 4130–4141.
56. Koo B, Becker LE, Chuang S, Merante F, Robinson BH, MacGregor D, Tein I, Ho VB, McGreal DA, Wherrett JR, et al. Mitochondrial encephalomyopathy, lactic acidosis, stroke-like episodes (MELAS): clinical, radiological, pathological, and genetic observations. *Ann Neurol* 1993, **34:** 25–32.
57. McKelvie PA, Morley JB, Byrne E, Marzuki S. Mitochondrial encephalomyopathies: a correlation between neuropathological findings and defects in mitochondrial DNA. *J Neurol Sci* 1991, **102:** 51–60.
58. Sparaco M, Bonilla E, DiMauro S, Powers JM. Neuropathology of mitochondrial encephalomyopathies due to mitochondrial DNA defects. *J Neuropathol Exp Neurol* 1993, **52:** 1–10.
59. Ohama E, Ohara S, Ikuta F, Tanaka K, Nishizawa M, Miyatake T. Mitochondrial angiopathy in cerebral blood vessels of mitochondrial encephalomyopathy. *Acta Neuropathol Berl* 1987, **74:** 226–233.
60. Kishi M, Yamamura Y, Kurihara T, Fukuhara N, Tsuruta K, Matsukura S, Hayashi T, Nakagawa M, Kuriyama M. An autopsy case of mitochondrial encephalomyopathy: biochemical and electron microscopic studies of the brain. *J Neurol Sci* 1988, **86:** 31–40.
61. Clark JM, Marks MP, Adalsteinsson E, Spielman DM, Shuster D, Horoupian D, Albers GW. MELAS: Clinical and pathologic correlations with MRI, xenon/CT, and MR spectroscopy. *Neurology* 1996, **46:** 223–227.
62. Gropen TI, Prohovnik I, Tatemichi TK, Hirano M. Cerebral hyperemia in MELAS. *Stroke* 1994, **25:** 1873–1876.
63. Gilchrist JM, Sikirica M, Stopa E, Shanske S. Adult-onset MELAS. Evidence for involvement of neurons as well as cerebral vasculature in strokelike episodes. *Stroke* 1996, **27:** 1420–1423.
64. Moraes CT, Ricci E, Bonilla E, DiMauro S, Schon EA. The mitochondrial tRNA(Leu(UUR)) mutation in mitochondrial encephalomyopathy, lactic acidosis, and strokelike episodes (MELAS): genetic, biochemical, and morphological correlations in skeletal muscle. *Am J Hum Genet* 1992, **50:** 934–949.
65. Schapira AHV, Cooper JM. Biochemical and molecular features of deficiencies of Complexes I, II, and III, in *Mitochondrial Disorders in Neurology* (Schapira AHV, DiMauro S, eds.), Butterworth-Heinemann, Oxford, 1994, 75–90.
66. James AM, Wei YH, Pang CY, Murphy MP. Altered mitochondrial function in fibroblasts containing MELAS or MERRF mitochondrial DNA mutations. *Biochem J* 1996, **318:** 401–407.
67. Robinson BH. The use of tissue culture in the diagnosis of mitochondrial disease, in *Mitochondrial Disorders in Neurology* (Schapira AHV, DiMauro S, eds.), Butterworth-Heinemann, Oxford, 1994, pp. 166–180.
68. Larsson NG, Andersen O, Holme E, Oldfors A, Wahlstrom J. Leber's hereditary optic neuropathy and complex I deficiency in muscle. *Ann Neurol* 1991, **30:** 701–708.

69. Hofhaus G, Johns DR, Hurko O, Attardi G, Chomyn A. Respiration and growth defects in transmitochondrial cell lines carrying the 11778 mutation associated with Leber's hereditary optic neuropathy. *J Biol Chem* 1996, **271**: 13,155–13,1561.
70. Degli Esposti M, Carelli V, Ghelli A, Ratta M, Crimi M, Sangiorgi S, Montagna P, Lenaz G, Lugaresi E, Cortelli P. Functional alterations of the mitochondrially encoded ND4 subunit associated with Leber's hereditary optic neuropathy. *FEBS Lett* 1994, **352**: 375–379.
71. Sumegi B, Srere PA. Complex I binds several mitochondrial NAD-coupled dehydrogenases. *J Biol Chem* 1984, **259**: 15,040–15,045.
72. Enriquez JA, Chomyn A, Attardi G. MtDNA mutation in MERRF syndrome causes defective aminoacylation of tRNA(Lys) and premature translation termination. *Nat Genet* 1995, **10**: 47–55.
73. Schon EA, Koga Y, Davidson M, Moraes CT, King MP. The mitochondrial tRNA(Leu)(UUR)) mutation in MELAS: a model for pathogenesis. *Biochim Biophys Acta* 1992, **1101**: 206–209.
74. King MP, Koga Y, Davidson M, Schon EA. Defects in mitochondrial protein synthesis and respiratory chain activity segregate with the tRNA(Leu(UUR)) mutation associated with mitochondrial myopathy, encephalopathy, lactic acidosis, and strokelike episodes. *Mol Cell Biol* 1992, **12**: 480–490.
75. Hess JF, Parisi MA, Bennett JL, Clayton DA. Impairment of mitochondrial transcription termination by a point mutation associated with the MELAS subgroup of mitochondrial encephalomyopathies. *Nature* 1991, **351**: 236–239.
76. Byrne E, Trounce I, Marzuki S, Dennett X, Berkovic SF, Davis S, Tanaka M, Ozawa T. Functional respiratory chain studies in mitochondrial cytopathies. Support for mitochondrial DNA heteroplasmy in myoclonus epilepsy and ragged red fibers (MERRF) syndrome. *Acta Neuropathol Berl* 1991, **81**: 318–323.
77. Moudy AM, Handran SD, Goldberg MP, Ruffin N, Karl I, Kranz Eble P, DeVivo DC, Rothman SM. Abnormal calcium homeostasis and mitochondrial polarization in a human encephalomyopathy. *Proc Natl Acad Sci USA* 1995, **92**: 729–733.
78. Robinson BH, Petrova Benedict R, Buncic JR, Wallace DC. Nonviability of cells with oxidative defects in galactose medium: a screening test for affected patient fibroblasts. *Biochem Med Metab Biol* 1992, **48**: 122–126.
79. Schulze Osthoff K, Bakker AC, Vanhaesebroeck B, Beyaert R, Jacob WA, Fiers W. Cytotoxic activity of tumor necrosis factor is mediated by early damage of mitochondrial functions. Evidence for the involvement of mitochondrial radical generation. *J Biol Chem* 1992, **267**: 5317–5323.
80. Schulze Osthoff K, Beyaert R, Vandevoorde V, Haegeman G, Fiers W. Depletion of the mitochondrial electron transport abrogates the cytotoxic and gene-inductive effects of TNF. *Embo J* 1993, **12**: 3095–3104.
81. Piccolo G, Banfi P, Azan G, Rizzuto R, Bisson R, Sandona D, Bellomo G. Biological markers of oxidative stress in mitochondrial myopathies with progressive external ophthalmoplegia. *J Neurol Sci* 1991, **105**: 57–60.
82. Royall JA, Ischiropoulos H. Evaluation of 2',7'-dichlorofluorescin and dihydrorhodamine 123 as fluorescent probes for intracellular H_2O_2 in cultured endothelial cells. *Arch Biochem Biophys* 1993, **302**: 348–355.
83. Dugan LL, Sensi SL, Canzoniero LM, Handran SD, Rothman SM, Lin TS, Goldberg MP, Choi DW. Mitochondrial production of reactive oxygen species in cortical neurons following exposure to N-methyl-D-aspartate. *J Neurosci* 1995, **15**: 6377–6388.
84. Sobreira C, Davidson M, King MP, Miranda AF. Dihydrorhodamine 123 identifies impaired mitochondrial respiratory chain function in cultured cells harboring mitochondrial DNA mutations. *J Histochem Cytochem* 1996, **44**: 571–579.
85. Zamzami N, Marchetti P, Castedo M, Decaudin D, Macho A, Hirsch T, Susin S.A, Petit PX, Mignotte B, Kroemer G. Sequential reduction of mitochondrial transmembrane potential and generation of reactive oxygen species in early programmed cell death. *J Exp Med* 1995, **182**: 367–377.

86. Reers M, Smith TW, Chen LB. J-aggregate formation of a carbocyanine as a quantitative fluorescent indicator of membrane potential. *Biochemistry* 1991, **30:** 4480–4486.
87. Smiley ST, Reers M, Mottola Hartshorn C, Lin M, Chen A, Smith TW, Steele GD Jr, Chen LB. Intracellular heterogeneity in mitochondrial membrane potentials revealed by a J-aggregate-forming lipophilic cation JC-1. *Proc Natl Acad Sci USA* 1991, **88:** 3671–3675.
88. Petit PX, JE OC, Grunwald D, Brown SC. Analysis of the membrane potential of rat- and mouse-liver mitochondria by flow cytometry and possible applications. *Eur J Biochem* 1990, **194:** 389–397.
89. Zoratti M, Szabo I. Electrophysiology of the inner mitochondrial membrane. *J Bioenerg Biomembr* 1994, **26:** 543–553.
90. Bernardi P, Broekemeier KM, Pfeiffer DR. Recent progress on regulation of the mitochondrial permeability transition pore; a cyclosporin-sensitive pore in the inner mitochondrial membrane. *J Bioenerg Biomembr* 1994, **26:** 509–517.
91. White RJ, Reynolds IJ. Mitochondrial depolarization in glutamate- stimulated neurons: an early signal specific to excitotoxin exposure. *J Neurosci* 1996, **16:** 5688–5697.
92. Schinder AF, Olson EC, Spitzer NC, Montal M. Mitochondrial dysfunction is a primary event in glutamate neurotoxicity. *J Neurosci* 1996, **16:** 6125–6133.
93. Gunter KK, Gunter TE. Transport of calcium by mitochondria. *J Bioenerg Biomembr* 1994, **26:** 471–485.
94. Kroemer G, Petit P, Zamzami N, Vayssiere JL, Mignotte B. The biochemistry of programmed cell death. *Faseb J* 1995, **9:** 1277–1287.
95. Nicotera P, Thor H, Orrenius S, Cytosolic-free Ca2+ and cell killing in hepatoma 1c1c7 cells exposed to chemical anoxia. *Faseb J* 1989, **3:** 59–64.
96. Mariotti C, Uziel G, Carrara F, Mora M, Prelle A, Tiranti V, DiDonato S, Zeviani M. Early-onset encephalomyopathy associated with tissue-specific mitochondrial DNA depletion: a morphological, biochemical and molecular-genetic study. *J Neurol* 1995, **242:** 547–556.
97. Poulton J, Morten K, Freeman Emmerson C, Potter C, Sewry C, Dubowitz V, Kidd H, Stephenson J, Whitehouse W, Hansen FJ, et al. Deficiency of the human mitochondrial transcription factor h-mtTFA in infantile mitochondrial myopathy is associated with mtDNA depletion. *Hum Mol Genet* 1994, **3:** 1763–1769.
98. Larsson NG, Oldfors A, Holme E, Clayton DA. Low levels of mitochondrial transcription factor A in mitochondrial DNA depletion. *Biochem Biophys Res Commun* 1994, **200:** 1374–1381.
99. Virbasius JV, Scarpulla RC. Activation of the human mitochondrial transcription factor A gene by nuclear respiratory factors: a potential regulatory link between nuclear and mitochondrial gene expression in organelle biogenesis. *Proc Natl Acad Sci USA* 1994, **91:** 1309–1313.
100. Bu XD, Rotter JI. X chromosome-linked and mitochondrial gene control of Leber hereditary optic neuropathy: evidence from segregation analysis for dependence on X chromosome inactivation. *Proc Natl Acad Sci USA* 1991, **88:** 8198–8202.
101. Harding AE, Sweeney MG, Govan GG, Riordan Eva P. Pedigree analysis in Leber hereditary optic neuropathy families with a pathogenic mtDNA mutation. *Am J Hum Genet* 1995, **57:** 77–86.
102. Vilkki J, Ott J, Savontaus ML, Aula P, Nikoskelainen EK. Optic atrophy in Leber hereditary optic neuroretinopathy is probably determined by an X-chromosomal gene closely linked to DXS7. *Am J Hum Genet* 1991, **48:** 486–491.
103. Chalmers RM, Davis MB, Sweeney MG, Wood NW, Harding AE. Evidence against an X-linked visual susceptibility locus in Leber hereditary optic neuropathy. *Amer J Hum Genet* 1996, **59:** 103–108.

104. Murphy AN, Bredesen DE, Cortopassi G, Wang E, Fiskum G. Bcl-2 potentiates the maximal calcium uptake capacity of neural cell mitochondria. *Proc Natl Acad Sci USA* 1996, **93**: 9893–9898.
105. Rizzuto R, Simpson AW, Brini M, Pozzan T. Rapid changes of mitochondrial Ca2+ revealed by specifically targeted recombinant aequorin. *Nature* 1992, **358**: 325–327.
106. Kendall JM, Badminton MN, Sala Newby GB, Campbell AK, Rembold CM. Recombinant apoaequorin acting as a pseudo-luciferase reports micromolar changes in the endoplasmic reticulum free Ca2+ of intact cells. *Biochem J* 1996, **318**: 383–387.
107. Walker UA, Byrne E. The therapy of respiratory chain encephalomyopathy: a critical review of the past and current perspective. *Acta Neurol Scand* 1995, **92**: 273–280.

14
Viral Encephalitis

J. Marie Hardwick, David N. Irani, and Diane E. Griffin

INTRODUCTION

Encephalitis, strictly defined as inflammation of the brain, most commonly results from acute viral infection of the central nervous system (CNS). While many thousands of cases of viral encephalitis occur worldwide each year, infections caused by the arthropod-borne viruses (arboviruses) and herpes simplex virus (HSV) are the most life-threatening. The arboviruses that cause encephalitis in humans include members of the flavivirus family, the bunyavirus family, and the alphavirus genus of the togavirus family. The prototype alphavirus, Sindbis virus, causes an epidemic arthritis in humans and an acute encephalitis in mice. Since murine Sindbis virus infection exhibits many features that are similar to the human alphavirus-induced encephalitides, it serves as a useful animal model for studying these illnesses.

In vitro, Sindbis virus infection induces apoptosis in a variety of cell types. Cells which die by virus-induced apoptosis exhibit many morphologic and biochemical characteristics of apoptosis including membrane blebbing, severe chromatin condensation around the periphery of the nucleus, and activation of cellular endonucleases that cleave DNA into nucleosome-length fragments. These fragments are detectable as 180–200 base pair ladders when cellular DNA is separated on agarose gels. In vivo, many virus-infected tissues exhibit histologic evidence of apoptosis. This can occur to such a degree that the surviving cells are of insufficient numbers to engulf and degrade the apoptotic cells. As a result, the accumulation of apoptotic bodies that express foreign viral antigens likely contributes to the intense inflammatory response observed in many virus-infected tissues. In this way, virus-induced apoptosis may differ from the naturally occurring, noninflammatory apoptosis that maintains tissue homeostasis or that occurs during embryogenesis.

Both HSV and Sindbis virus specifically target neurons in the CNS, and lytic replication results in neuronal cell death. In addition, HSV and Sindbis virus can both establish long term infections of neurons, albeit through different mechanisms. HSV establishes a truly latent state in neuronal nuclei where only one virus-encoded RNA accumulates to significant levels and the transcription of other viral genes is suppressed. This RNA is referred to as the latency-associated transcript (LAT), and its precise role in the virus life cycle has not been clearly defined. In contrast, Sindbis virus lives out its entire life cycle in the cytoplasm of infected neurons and persists in a state of suppressed lytic replication rather than true latency.

From: Cell Death and Diseases of the Nervous System
Edited by: V. E. Koliatsos and R. R. Ratan © Humana Press Inc., Totowa, NJ

It has been postulated that apoptotic cell death is an important host defense mechanism for eliminating virus-infected cells from the body. That is, infected cells commit suicide prior to completion of the virus replication cycle, thereby blocking the spread of progeny virus to neighboring cells and tissues. This hypothesis is supported by the fact that some viruses such as adenovirus and baculovirus encode one or more anti-apoptotic genes that prolong cell survival during infection. In contrast, other viruses such as Sindbis appear to thrive in apoptotic cells, perhaps in part because they have short replication cycles that are typically completed prior to cell death. The apoptosis induced by viruses such as Sindbis may then serve an important mechanism of disease pathogenesis. This is particularly true in differentiated cell populations such as neurons that are not renewable. Neurotropic viruses such as Sindbis virus can, in fact, induce the drop-out of neurons in the brains and spinal cords of infected animals that leads to permanent hind limb paralysis and/or death *(1)*. This chapter will review some of the clinical features of viral encephalitis; then, using the paradigm of Sindbis virus infection, we will discuss the role of virus-induced neuronal apoptosis in the pathogenesis of encephalitis and integrate this information with what is known about the molecular mechanisms.

CLINICAL FEATURES OF VIRAL ENCEPHALITIS

Arboviruses

Many arthropod-borne viruses cause acute CNS infections (Table 1). The most important clinical syndrome that results from these infections is encephalomyelitis (inflammation of the brain and spinal cord), but less severe forms such as aseptic meningitis actually occur with greater frequency. Arboviruses replicate asymptomatically in both their reservoir hosts (e.g., birds) and their particular insect vectors (e.g., mosquitoes, ticks). Maintenance of these viruses in nature relies on a cycle of transmission between the reservoir and the vector. For all of these pathogens, humans are terminal hosts because infections are not of sufficient duration to allow for efficient transmission of the virus back to the vector. In many parts of the world, arbovirus encephalitis results in significant mortality and produces substantial long-term neurologic dysfunction in many of the survivors. Over 20 different arboviruses are known to cause encephalitis in humans, but specific pathogens are restricted geographically and seasonally to the habitat and propagation of their particular insect vectors. Many arboviruses are named for the location where they were originally isolated.

Arboviruses are transmitted to humans via the saliva of an infected mosquito or tick. Initial virus replication occurs locally at the site of the insect bite. A viremia can result and disseminate the virus through the bloodstream to the CNS *(2)*. Yet even the most neuroinvasive of the arboviruses produce encephalitis in only one of at least 100 infections. Once in the brain, however, neurons are the primary targets for these pathogens. Patients typically develop fever, headache, neck stiffness, altered level of consciousness, and seizures. Specific regions of the brain or spinal cord infected by each of these viruses may also cause unique clinical manifestations or neurologic sequelae. For instance, Japanese encephalitis virus (JEV) often infects neurons in the basal ganglia giving rise to a characteristic Parkinsonian-like syndrome *(3,4)*. Likewise, the tick-borne encephalitides can infect motor neurons of the cervical spinal cord resulting in weakness and atrophy in the muscles of the shoulder girdle *(5,6)*.

**Table 1
Arboviruses and Herpesviruses that Cause Encephalitis**

Encephalitic arboviruses
  Flaviviridae
    Mosquito-borne
      Japanese encephalitis virus (JEV)
      St. Louis encephalitis virus (SLE)
      Murray Valley encephalitis virus (MVE)
      West Nile encephalitis virus
    Tick-borne
      Kyasanur Forest encephalitis virus
      Louping-ill encephalitis virus
      Powassan encephalitis virus
      Negishi encephalitis virus
      Far Eastern encephalitis virus
      Central European encephalitis virus
  Bunyaviridae
    Mosquito-borne
      La Crosse virus
  Alphaviruses (Togaviridae)
    Mosquito-borne
      Eastern equine encephalitis virus (EEE)
      Western equine encephalitis virus (WEE)
      Venezuelan equine encephalitis virus (VEE)
      Sindbis virus (causes encephalitis in laboratory mice, but causes arthritis in humans)
Encephalitic herpesviruses
  Herpes simplex virus type 1 (HSV-1)
  Herpes simplex virus type 2 (HSV-2)
  Varicella-zoster virus (VZV)

Arboviruses also generally cause an age-dependent disease. Young children more commonly develop encephalitis, have a poorer prognosis, and are typically left with greater neurologic residua following infection than are older children or adults *(7–10)*. Susceptibility to neurologic disease may increase again in old age *(11)*. This age-dependent pattern is reminiscent of the situation in experimental Sindbis virus infections where newborn mice are more susceptible to fatal encephalitis than older mice *(12)*. The potential molecular explanations for age-dependent disease in animals are discussed below.

Flaviviruses

The flaviviruses are small, enveloped, positive-strand RNA viruses that are similar in size to the alphaviruses, but which have a distinct genomic organization and replication cycle *(13)*. Flavivirus encephalitis has been reported on all continents except Antarctica. Viruses are transmitted to humans either by mosquitoes or by ixodid ticks. The mosquito-borne flaviviruses that cause encephalitis are all members of the West Nile antigenic complex and are found worldwide. The largest number of cases of encephalitis occur in Asia, where JEV causes both endemic and epidemic disease. Sporadic outbreaks also occur in the Americas (St. Louis encephalitis), Australia (Murray Valley encephalitis), and Africa (West Nile encephalitis) *(14,15)*.

Japanese encephalitis (JE) is widely distributed throughout Asia including Japan, China, the Soviet Union, the Philippines, Southeast Asia, and India. The range of this disease, in fact, appears to be extending *(13)*. JE is by far the most important arbovirus-induced encephalitis in terms of worldwide morbidity and mortality, with tens of thousands of cases and thousands of deaths occurring annually *(14,17)*. In endemic areas such as southern Thailand, children are most commonly infected since older individuals typically have preexisting immunity. Here, cases occur throughout the year *(17)*. In epidemic areas such as China, however, all ages are equally susceptible and cases occur in outbreaks beginning in late summer *(17)*. JEV infection may be asymptomatic or manifested by fever alone, aseptic meningitis, or bona fide encephalitis. The onset of encephalitis rapidly follows a prodrome of headache, fever, chills, malaise, and nausea that typically lasts for 2–3 d. In children, abdominal symptoms may also be prominent. Encephalitis is characterized by sustained fever (usually > 104°F), meningismus, photophobia, confusion, and delirium that lasts for another 2–4 d. Neurological signs can also include a Parkinsonian picture of mask-like facies, rigidity, and involuntary movements, as well as altered consciousness, and focal or diffuse weakness. Tremor is present in 90% of patients *(3,4)*. Seizures are frequent in children, but occur in <10% of adults. Recovery or death rapidly ensue, but rare individuals show signs that the infection can linger for more than two weeks *(18)*. Permanent neurologic sequelae in survivors of the encephalitis are common.

St. Louis encephalitis is the most common flavivirus-induced encephalitis in the United States. It is widely distributed across the country. Outbreaks usually occur in August, September, and October—somewhat later than is common for the other arboviruses *(14,15)*. This virus has both urban (epidemic) and rural (endemic) cycles *(19)*. The risk of encephalitis following infection increases sharply with age *(20,21)* and with coexistent HIV infection *(22)*. After a nonspecific prodrome of fever and malaise that lasts several days, severe headache, nausea, and vomiting abruptly occur. This is then typically followed by disorientation, irritability, and stupor. Low serum sodium due to inappropriate secretion of antidiuretic hormone and elevated blood urea nitrogen are relatively common associated laboratory findings *(20,21,23)*. Mortality from the encephalitis is low in younger patients but is over 20% in the elderly *(14)*. Convalescence may be prolonged, and significant neurologic residua remain in 20% of survivors *(14)*.

The tick-borne flaviviruses form a separate antigenic group and are most prominent in Europe and Asia. Transmission to humans occurs by ingestion of infected goat milk or following a bite from an infected tick. Neurological disease results from six closely related viruses that can cause encephalitis: Kyasanur Forest, louping-ill, Powassan, Negishi, Far Eastern (Russian spring-summer) and Central European encephalitis viruses *(14)*. The Far Eastern and Central European strains of tick-borne encephalitis virus are endemic over a wide area of Europe and the Soviet Union and cause thousands of cases of encephalitis each year. These occur primarily in adults working or vacationing in wooded areas where ticks are plentiful. The onset of Far Eastern encephalitis is gradual with fever and headache progressing to meningismus, paralysis, and seizures over many days. Evidence of lower motor neuron disease with loss of muscle tone and reflexes can occur. Weakness that is limited to the upper extremities may reflect a selective

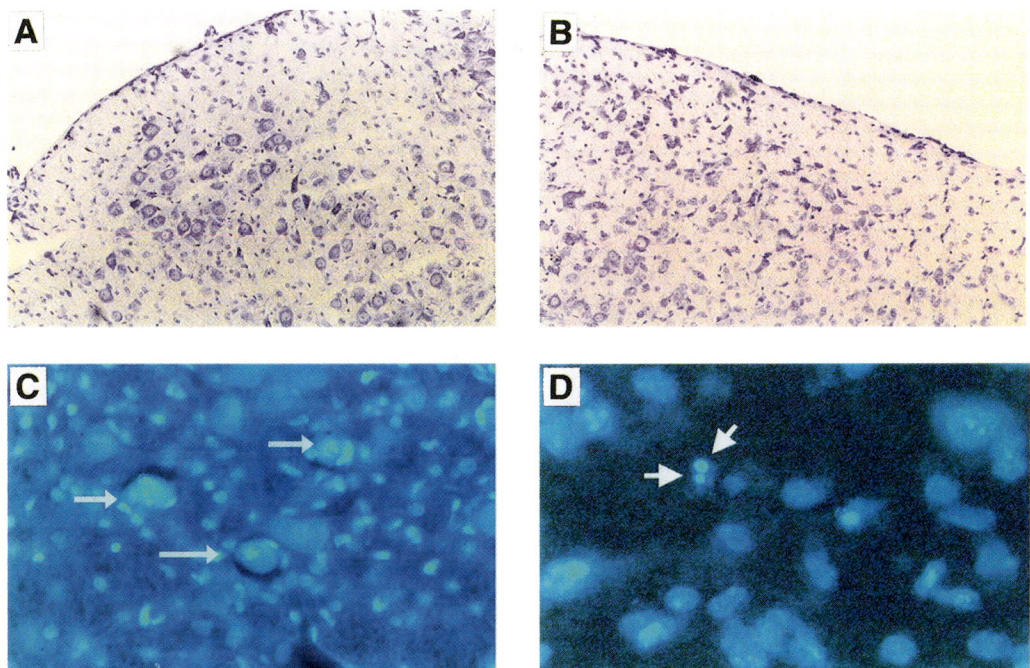

Plate 1 (Fig. 1A-D; see full caption on p. 185 and discussion in Chapter 9). Apoptosis in neonatal facial motor neurons after axotomy.

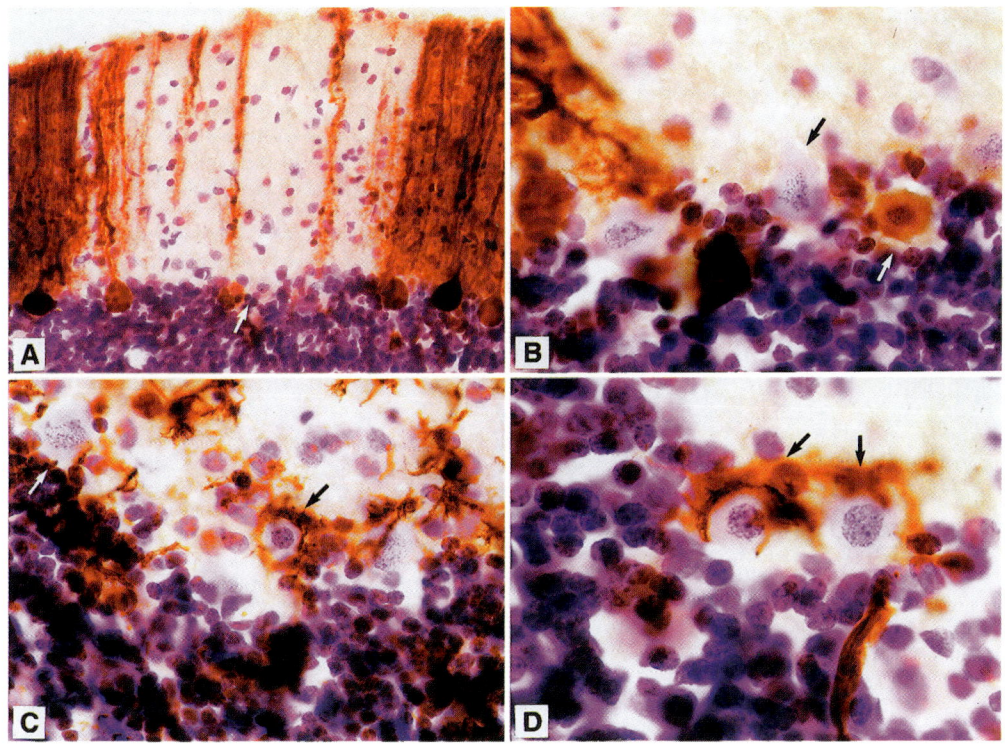

Plate 2 (Fig. 4A-D; see full caption on p. 228 and discussion in Chapter 11). Cytologic changes in injured Purkinje cells 24 h after ibogaine administration.

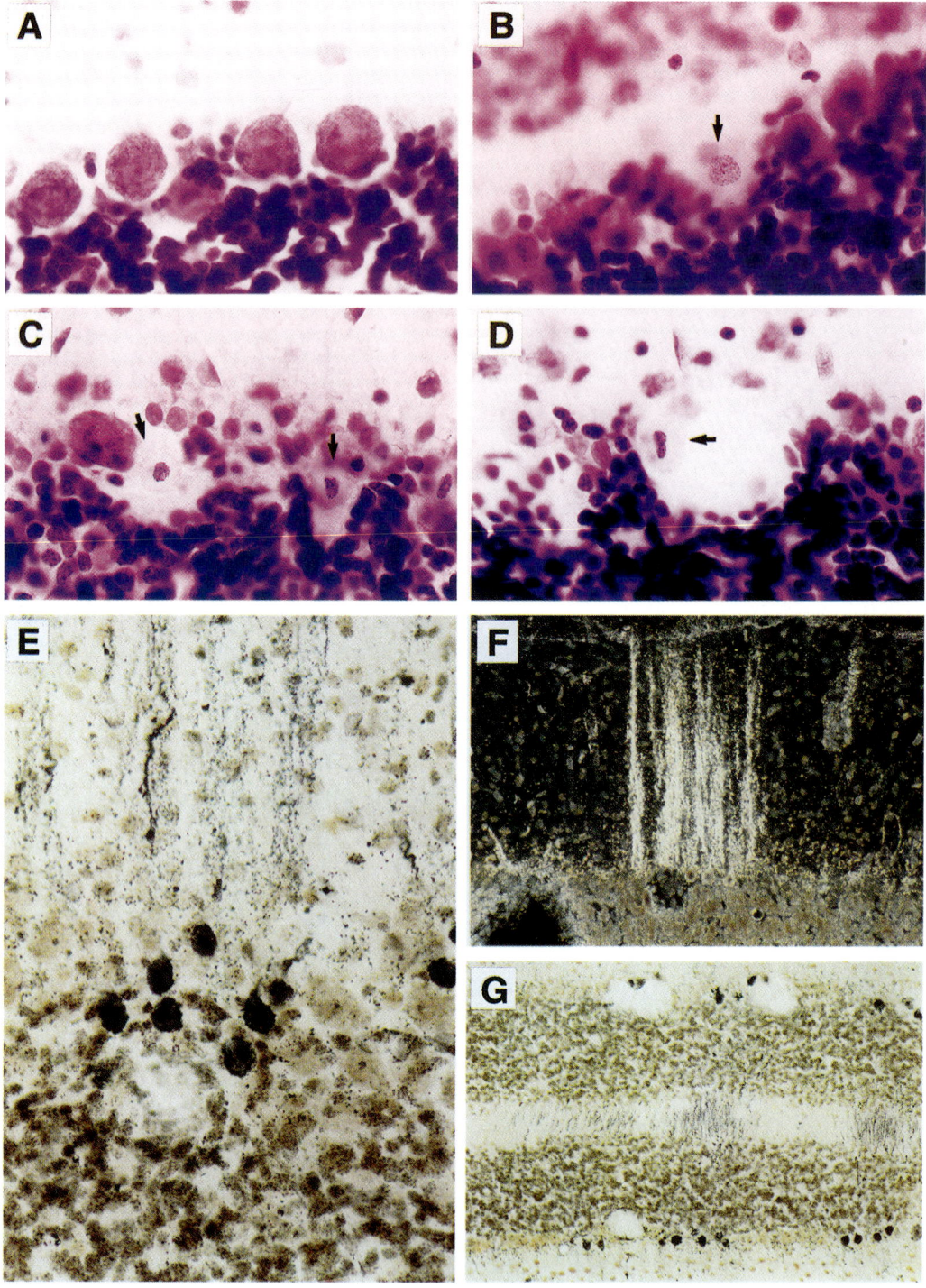

Plate 3 (Fig. 3A-G; see full caption on pp. 226–227 and discussion in Chapter 11). Degeneration of Purkinje cells following administration of the indole alkaloid ibogaine.

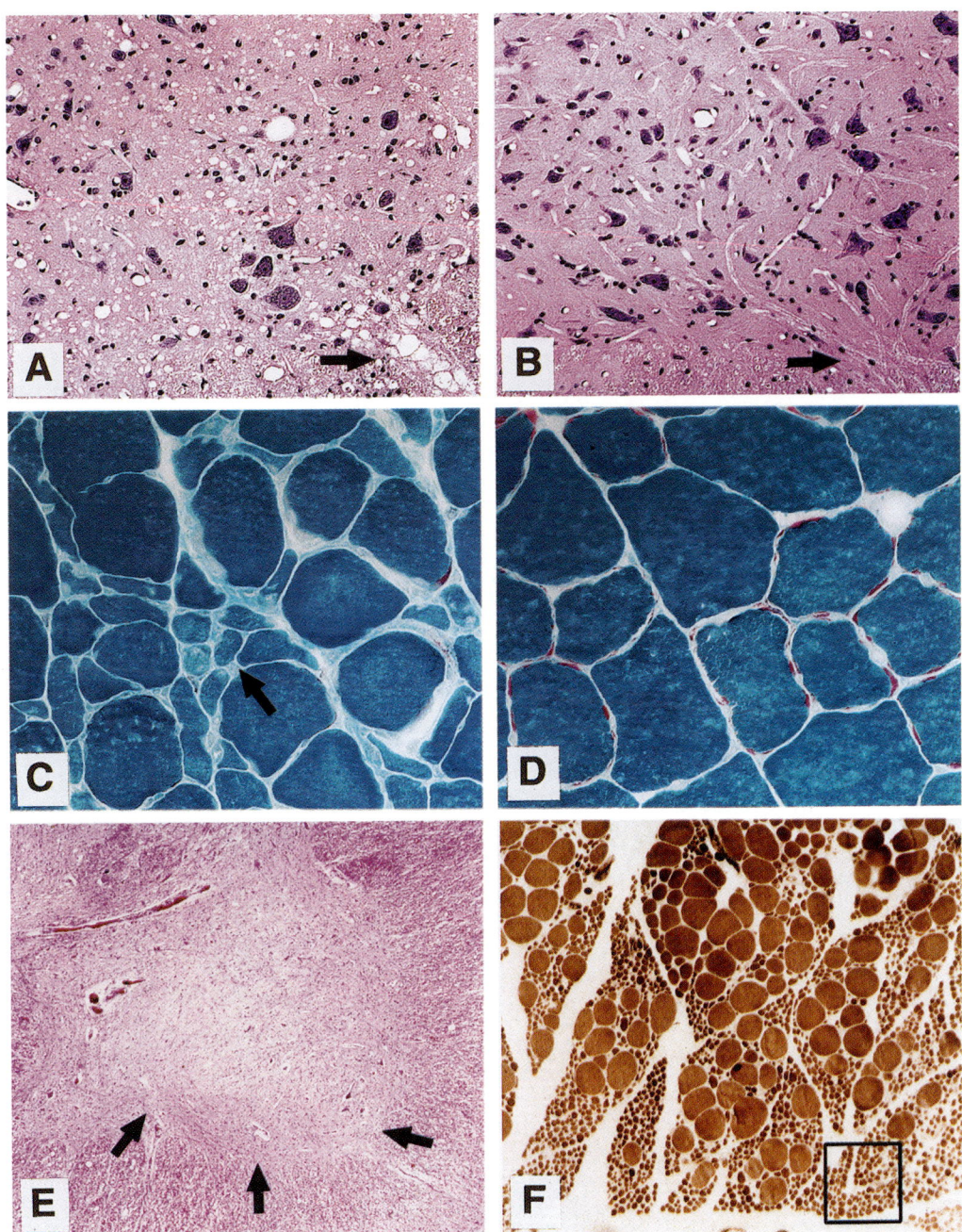

Plate 4 (Fig. 2A-F; see full caption on pp. 432–433 and discussion in Chapter 20). Similarities in spinal cord and muscle pathology between G37R transgenic mice and patients with amyotrophic lateral sclerosis.

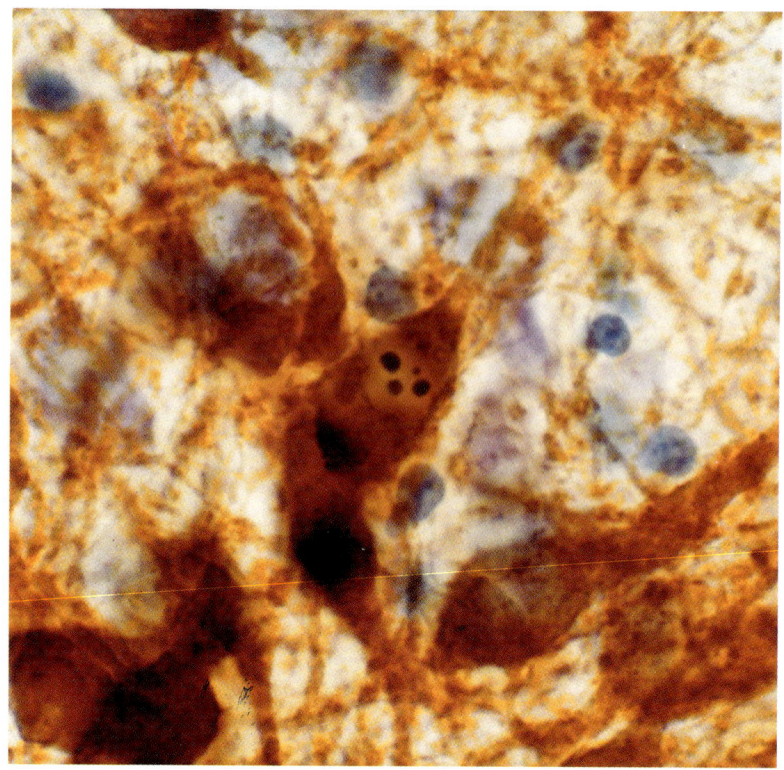

Plate 5 (Fig. 1; see full caption on p. 466 and discussion in Chapter 22). Apoptosis in dopaminergic neurons of the substantia nigra after striatal excitotoxic lesions.

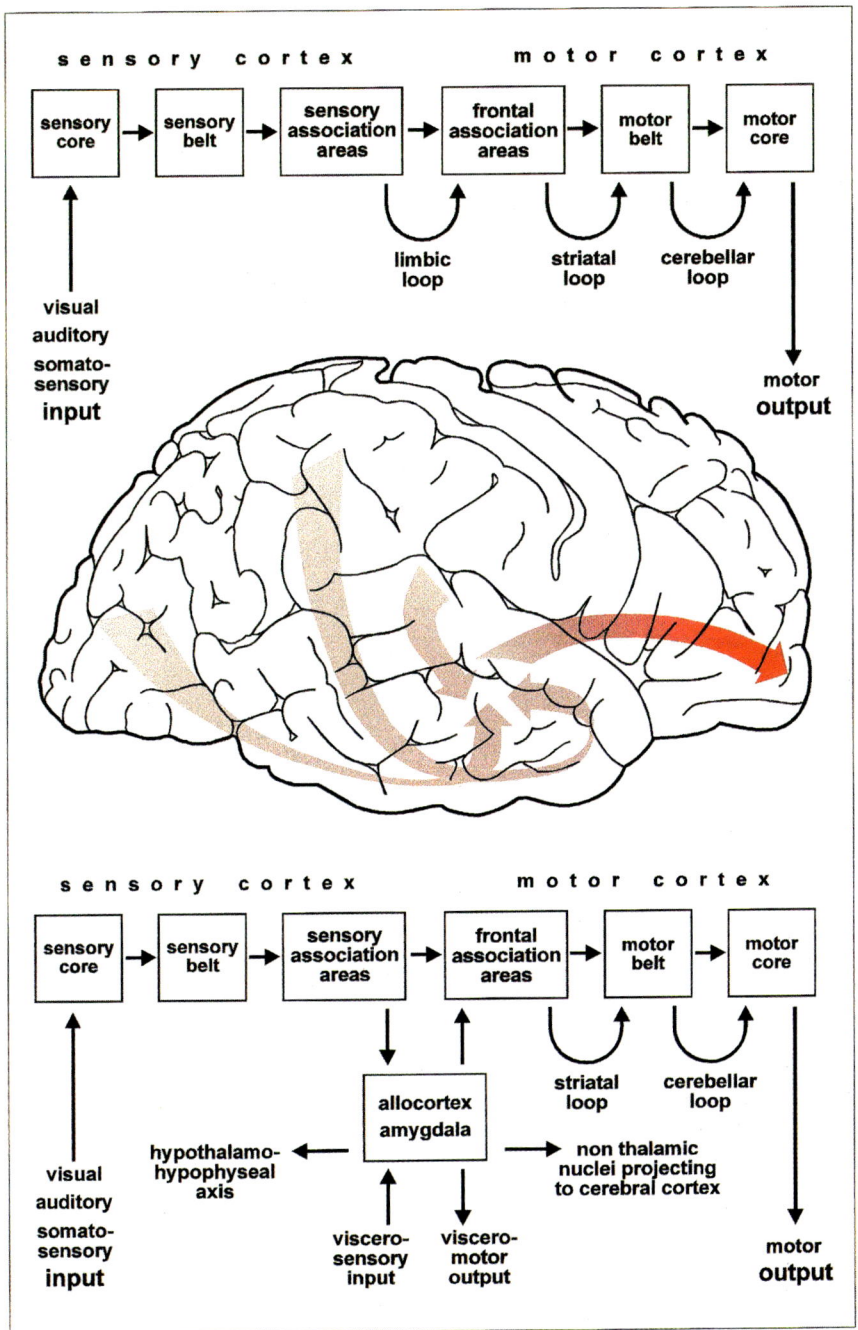

Plate 6 (Fig. 4; see full caption on p. 503 and discussion in Chapter 24). Patterns of information flow along associative pathways relate to the distribution of neuropathological changes in brains with Alzheimer's disease.

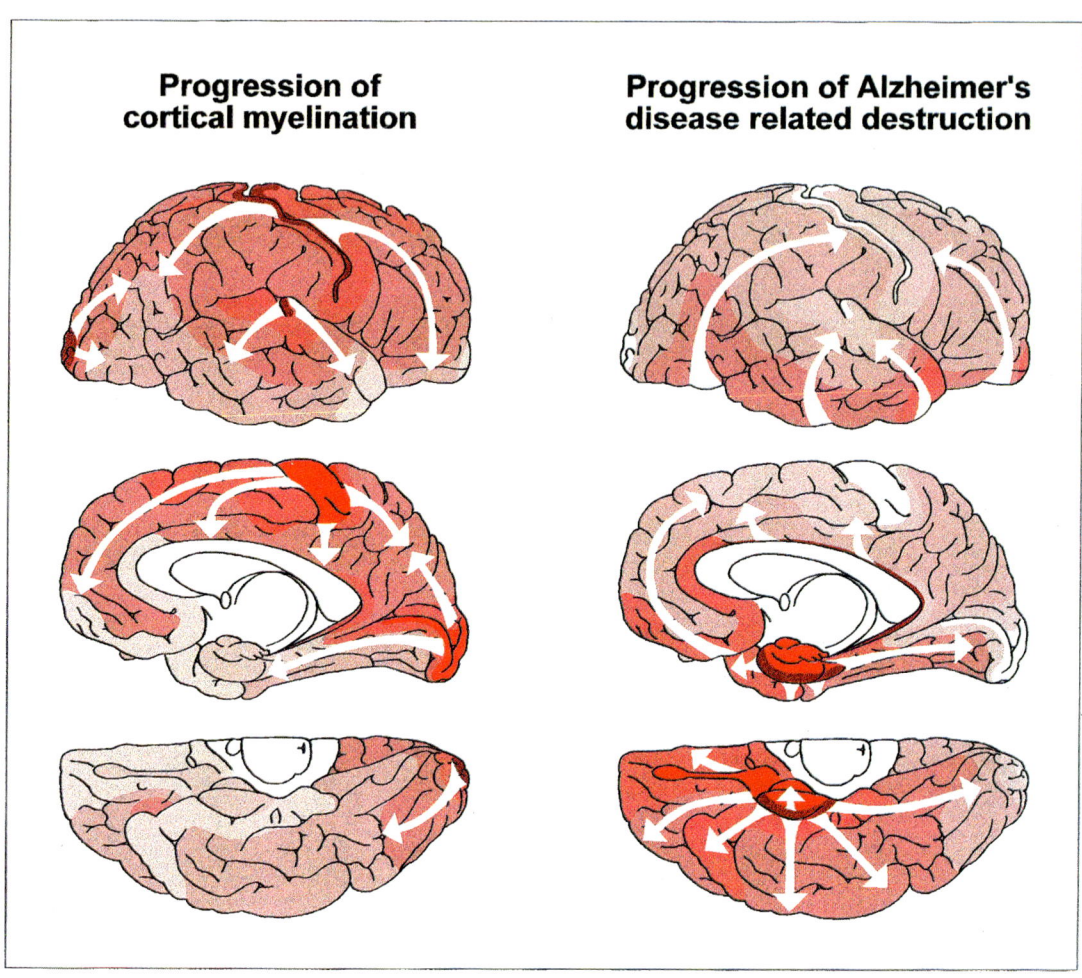

Plate 7 (Fig. 7; see full caption on p. 507 and discussion in Chapter 24). Myelination patterns in human brain relate to the progression of neuropathogy in Alzheimer's disease.

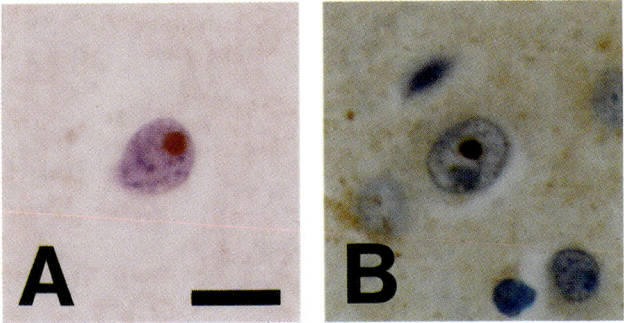

Plate 8 (Fig. 3; see full caption on p. 486 and discussion in Chapter 23). Intranuclear aggregates of huntingtin provide clues in the pathogenesis of Huntington's disease.

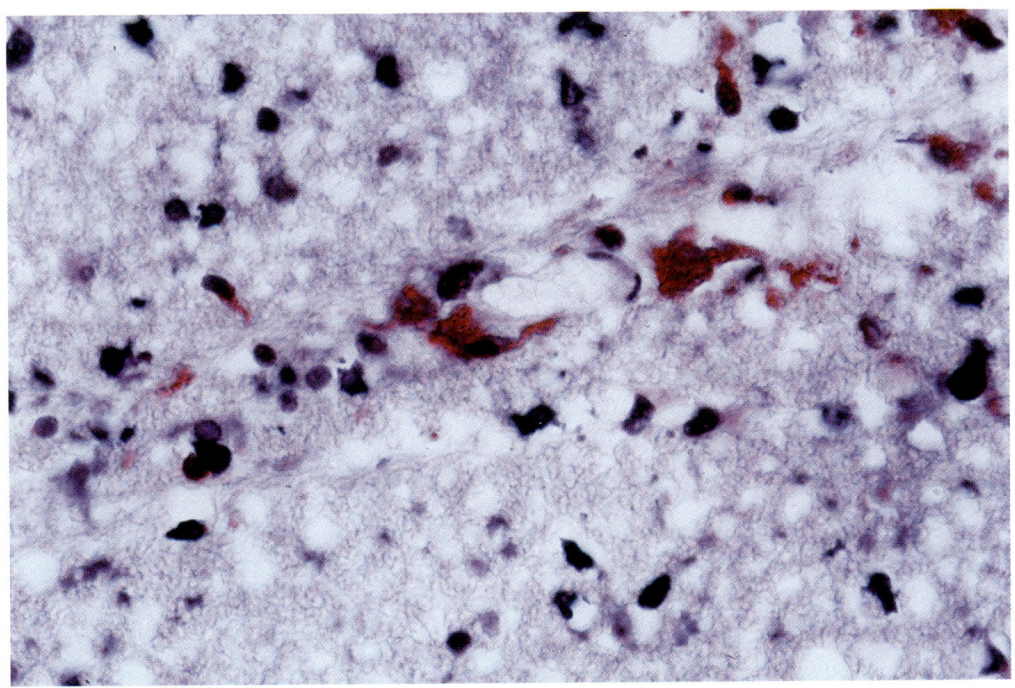

Plate 9 (Fig. 2; see full caption on p. 515 and discussion in Chapter 25). Double staining of Bax (cytoplasm) and TUNEL (nucleus) in the brain of a child with HIV encephalopathy.

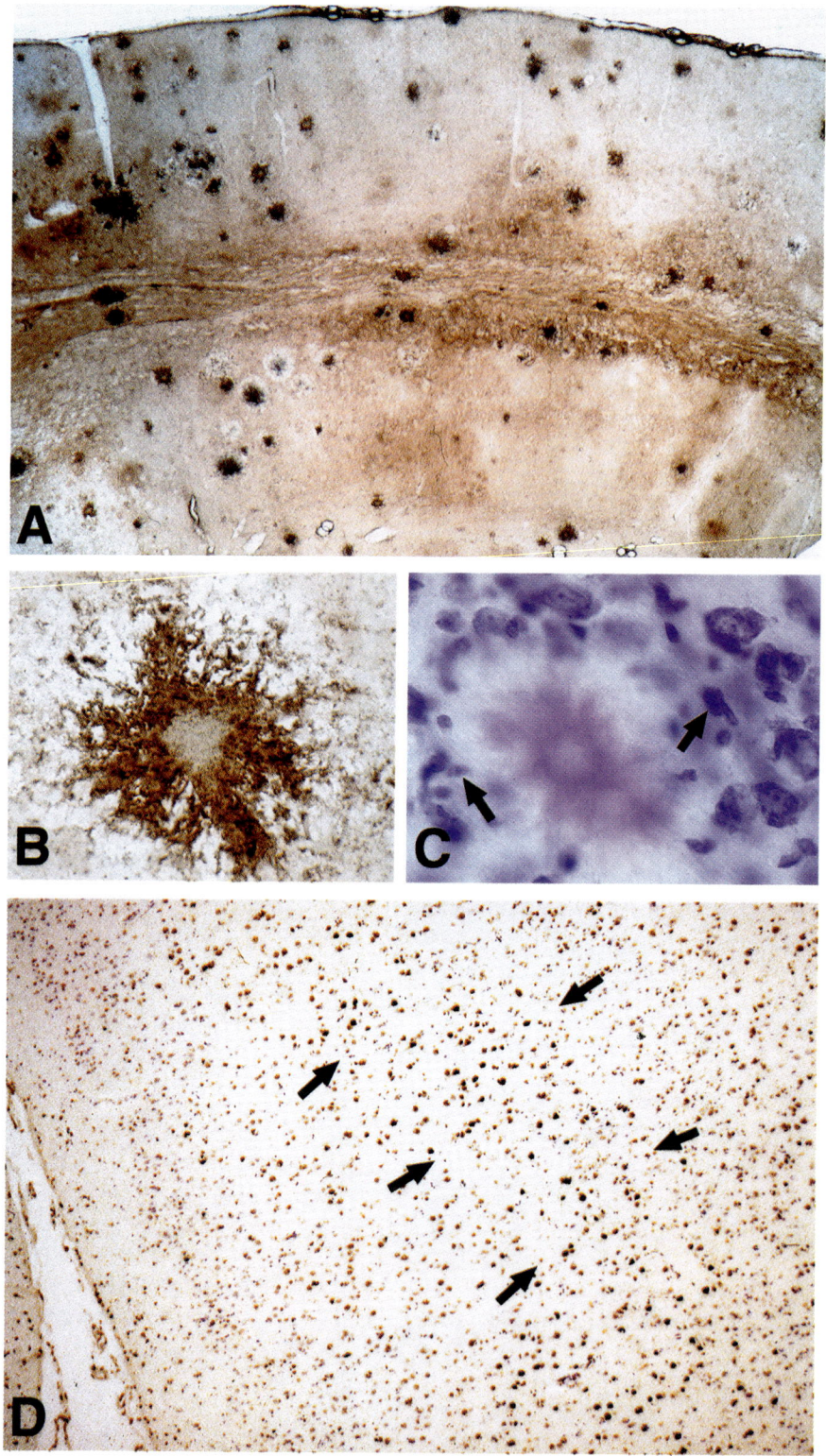

Plate 10 (Fig. 2A-D; see full caption on pp. 554–555 and discussion in Chapter 28). Partial imitation in APP transgenic mice, of the neuropathological features of Alzheimer's disease.

involvement of the motor neurons in the cervical spinal cord. The case fatality rate is approximately 20% and neurologic sequelae occur in 30–60% of survivors. Especially characteristic of this disease is a residual flaccid paralysis of the shoulder girdle and arms *(5,6,14)*. Rarely, chronic CNS infections may account for some cases of epilepsia partialis continua in the Soviet Union *(24)*.

Central European encephalitis is milder than the Far Eastern disease and exhibits a diphasic course in over half of cases. The first phase is a nonspecific, flu-like illness that lasts for approx 1 wk and that is followed by a 1–3-d remission. The neurologic phase then may manifest as aseptic meningitis or as an encephalitis that produces tremor, diplopia, altered mental status, and weakness. The case fatality rate is 1–5%, and approx 20% of survivors have mild neurologic residua *(5,14,25,26)*. As a group, the other members of the tick-borne complex produce many fewer cases of encephalitis and will not be discussed here.

Bunyaviruses

Bunyaviruses are a large family of enveloped, negative-strand or ambisense RNA viruses with more than 200 members. The most important bunyaviruses that cause encephalitis belong to the California serogroup which includes La Crosse virus, the most common cause of arbovirus encephalitis in the United States *(27)*. All of these viruses are transmitted by mosquitoes and have small mammals as their natural reservoirs. Both urban and rural cases have occurred because one important vector for these viruses, *Aedes triseriatus*, breeds in water that collects in such diverse containers as old tires and tree holes *(20,27)*. The introduction and spread of another container-breeding mosquito in the United States that serves as an efficient vector for these pathogens, *Aedes albopictus*, has increased the potential for greater spread of the California serogroup viruses as well as for yellow fever and dengue viruses *(28)*. Encephalitis due to LaCrosse virus occurs nearly exclusively among children *(29)*. The overall case fatality rate is less than 1%. During the acute illness, seizures occur in 50% of patients while focal weakness or other neurologic signs occur in 25%. Some neurologic residua (usually chronic seizure disorders) are noted in approximately 10% of survivors *(29)*.

Alphaviruses

The alphaviruses belong to the togavirus family of small, enveloped, positive-strand RNA viruses. In North and South America, Eastern and Western equine encephalitis viruses are the most important causes of arthropod-borne encephalitis *(15)*. Both are transmitted by mosquitoes. Birds are the primary reservoirs for these pathogens and horses and humans are terminal hosts *(2)*. Eastern equine encephalitis virus is endemic along the eastern and Gulf coasts of the United States, the Caribbean, and in South America. It causes small localized outbreaks of equine and human encephalitis during the summer *(15,30)*. Eastern equine encephalitis is a severe disease that is associated with a high case fatality rate (50–75%) in patients of all ages. The encephalitis is associated with fever, headache, altered consciousness, and seizures, and the majority of survivors have significant neurologic residua *(10,31)*. Western equine encephalitis virus produces epidemics of equine and human encephalitis in the western and midwestern United States *(15,30)*. Signs and symptoms are similar to those of Eastern encephalitis, but the disease is typically much less severe and the overall case fatality rate is lower (10%). Complications increase at the extremes of age *(32)*, and permanent sequelae are most likely to occur in infants and young children *(19,33,34,34a)*.

Herpes Simplex Virus (HSV)

The herpesviruses are large, enveloped, double-strand DNA viruses that are classified into three subfamilies: α, β, and γ. While approximately one hundred herpesviruses have been at least partially characterized, only eight of these pathogens are known to infect humans. However, all of the human herpesviruses, with the exception of the newly described Kaposi's sarcoma-associated herpesvirus (HHV-8), have been recognized to be at least occasional causes of CNS disease that ranges from a benign aseptic meningitis to a fatal encephalitis. The most severe infections are produced by HSV type-1 (HSV-1) and HSV type-2 (HSV-2); these are particularly important to recognize because they are the only forms of viral encephalitis that benefit from treatment with antiviral therapy.

Although HSV-1 and HSV-2 are regularly found latent within sensory ganglia, symptomatic CNS infection is unusual. Encephalitis occurs in neonates who become infected at the time of delivery from a mother with active genital herpes infection. As a result, neonatal encephalitis is almost always caused by HSV-2. In adults, CNS infection with HSV-2 is most often an aseptic meningitis which can be recurrent *(35a)*. Significant HSV-2 encephalitis occurs almost exclusively in immunocompromised adults *(36)*. While HSV-1 is occasionally associated with meningitis in adults *(37,38)*, it most frequently produces a severe, focal encephalitis. It may be, however, that the spectrum of neurologic disease associated with HSV-1 will broaden with improved diagnostic techniques.

Herpes simplex encephalitis (HSE) is the most common fatal, nonepidemic encephalitis in the Western World *(39,40)*. It occurs equally in both sexes and at all times of the year. In adults, well over 95% of cases are caused by HSV-1 *(37)*. Since virus isolated from a peripheral site may or may not be the same strain as found in the brain of that individual *(41)*, HSE may be a manifestation of primary or reactivated HSV infection *(37)*. There is no evidence for point source outbreaks of encephalitis due to more virulent strains of HSV *(42,43)*. The route by which the virus enters the CNS is not yet clear, but the consistency with which the infection localizes to the temporal lobe suggests that entry is by a neural route rather than from the blood.

HSE may be preceded by a nonspecific prodrome of fever and malaise. The onset of headache may be acute or subacute and is usually accompanied by fever, behavioral abnormalities, focal or generalized seizures, and a progressive deterioration in the level of consciousness. Focal temporal lobe signs including aphasias, superior quadrant visual field deficits, and a contralateral hemiparesis are the rule. The disease progresses to coma over days to several weeks without treatment resulting in a 70% overall mortality. An additional 20% of untreated patients are left with severe neurologic sequelae including speech and cognitive deficits, seizures, paralysis, and ataxia. Because the virus is difficult to culture from the cerebrospinal fluid (CSF) of patients with HSE, the diagnosis was dependent on isolating virus directly from brain tissue obtained by a brain biopsy. Presently, however, HSV-1 DNA can be routinely detected in the CSF of patients with suspected HSE by polymerase chain reaction (PCR). This less invasive procedure has been validated against the standard of direct brain biopsy and is now the diagnostic method of choice *(44,45)*. All patients with proven HSE should be treated with the antiviral drug acyclovir, and in combination with aggressive supportive care, mortality is now less than 25% *(39,46)*.

APOPTOSIS IN AN ALPHAVIRUS ENCEPHALITIS MODEL

Sindbis Virus Infection of Mice Serves as a Model for the Human Alphavirus-Induced Encephalitides

Sindbis virus replicates in the neurons and ependymal cells of mice following intracerebral inoculation *(1)*. The resulting encephalitis has many clinical and pathological features that resemble the human alphavirus-induced encephalitides, thus providing an animal model to study many aspects of these infections. Not only is the outcome from acute Sindbis virus infection age-dependent *(12)*, but different viral strains also produce varying degrees of neurologic morbidity and mortality in adult mice *(47)*. In immunocompetent animals, a mononuclear cell inflammatory response comprised of $CD4^+$ and CD8+ T cells, B cells, monocyte/macrophages, and a wide range of inflammatory cytokines develops in the brain within 3–4 days of infection *(48,49,50)*. The antiviral antibody response appears to be important in clearing avirulent strains of the virus from the CNS and promoting the recovery of animals from infection *(51)*. Paradoxically, CNS inflammation contributes to the mortality produced by neurovirulent strains since adult SCID mice which lack functional T and B cells do not succumb to these infections as readily as adult animals with normal immune function *(48)*. Yet, SCID mice exhibit the same age-dependent susceptibility that occurs in immunocompetent host strains (P. Tucker and M. Hardwick, personal observation), highlighting the interrelationship between other host factors and immune responses to disease survival. In all of the situations examined to date, survival of the host with acute Sindbis virus infection correlates with the survival of its infected neurons *(52)*. Because Sindbis virus can induce neuronal apoptosis, we will review virus-, cellular-, and immune-mediated factors that regulate this process.

Overview of Sindbis Virus Replication

Sindbis virus has a plus sense single strand RNA genome of 11.7 kb (Fig. 1) *(53)*. The 5' two-thirds of the genome encodes a polyprotein precursor which is cleaved into four nonstructural proteins (nsP1, nsP2, nsP3, and nsP4) by a cysteine protease encoded in the C-terminal half of nsP2 *(54–56)*. Other functions encoded by these nonstructural proteins include helicase, NTPase, methyltransferase, and RNA polymerase activities *(57–64)*. Together, these proteins make up the replication complex that is responsible for transcribing both the negative strand RNA (a replication intermediate complementary to the viral genome) as well as subsequent plus strands *(65–67)*.

A subgenomic 26S RNA, initiated from an internal promoter sequence, is translated into a second polyprotein containing the structural proteins that form the virus particle (Fig. 1) *(68)*. The N-terminal capsid protein is a chymotrypsin-like serine proteinase that is autocatalytically cleaved from the precursor polyprotein *(59,64,68,69)*. The remaining structural proteins, p62/E2, 6K, and E1, are derived through further proteolytic cleavage of the precursor via the action of cellular enzymes *(53,70)*. E2 is derived from the precursor, p62 *(71,72)*. E2 and E1 heterodimerize through noncovalent interactions to form the virion glycoprotein spikes. These proteins are responsible for viral binding to one of several potential cellular receptors and fusion of the viral envelope to the cell membrane *(73–75)*. Mutants which fail to cleave p62 into E2 make virions but are not infectious, presumably because the fusion activity of E1 is masked until after p62 cleavage *(76)*. p62/E2 has a transmembrane domain and a 33 amino acid cytoplasmic tail.

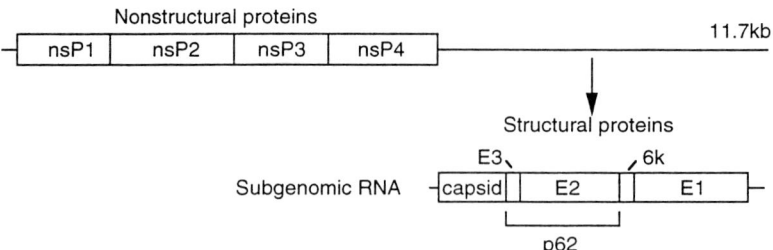

Fig. 1. Sindbis virus genome. The 11.7 kb genome function as a mRNA and is translated into the nonstructural proteins 1–4 (nsP1–4) shortly after infection. Subsequently, a subgenomic RNA is transcribed from the minus-sense RNA and translated into the structural proteins that form the virus particle. Both the nonstructural and structural proteins are translated as polyproteins that are subsequently processed by viral and cellular proteins.

Evidence suggests that a 10-residue segment within the cytoplasmic tail of E2, immediately following a five-residue highly charged stop transfer region, is responsible for oligomerization of E2 and binding to the capsid *(77–81)*. Studies with anti-idiotype antibodies have shown that the C-terminal 18 residues of E2 can associate with other viral nonstructural proteins (or perhaps cellular proteins) *(80)*. The tail is also modified by palmitoylation and phosphorylation, and both modifications appear to be important for completion of the viral life cycle *(82–84)*. Mutations in the E2 tail also have deleterious effects on virus growth and neurovirulence in infected mice *(85)*. Tyr400 and Leu402 of the E2 cytoplasmic tail are thought to bind in the pocket domain of the capsid protein and facilitate the assembly of virions *(86)*. In turn, a 68-amino acid segment of the Sindbis capsid protein is sufficient to bind the viral genomic RNA. The nucleotide sequences that mediate the interaction with the capsid lie between nucleotides 945 and 1076 (within the nsP1 coding sequence) *(87)*.

In vitro translation of the closely related Semliki Forest virus glycoproteins in the presence of microsomal membranes has delineated several processing events *(70)*. The p62/E2 cytoplasmic tail contains a hydrophobic stretch that serves as a signal peptide for the adjacent 6K protein. p62/E2 is thus cleaved from 6K cotranslationally by the cellular signal peptidase (short arrows, Fig. 2). Mutations near the peptidase site abolish this cleavage *(70,88)*. Following its cleavage from 6K, the p62/E2 tail apparently flips to the cytosolic side of the membrane where it mediates a direct interaction with viral capsids, a function essential to virion formation *(89)*. The C-terminus of 6K contains a transmembrane domain and a signal peptide for the adjacent E1 polypeptide. 6K is cleaved from E1 by a cellular signal peptidase *(35,70,90,91)*, is transported to the membrane, and incorporated into Sindbis virions *(92,93)*. Some evidence suggests that 6K may function as an ion channel similar to the influenza virus M2 protein *(94)*. In the absence of 6K, E2 and E1 heterodimerize and reach the plasma membrane, but virions are not assembled *(95)*. A hydrophobic region near the N-terminus of E1 has been implicated in fusion activity since mutations in this region alter this process *(96,97)*.

Following entry into the cell, most of the capsid proteins are dissociated from the viral genome and bind to ribosomes, possibly because viral RNA shares extensive sequence homology with 28S ribosomal RNA *(98–100)*. Sindbis virus infection then potently

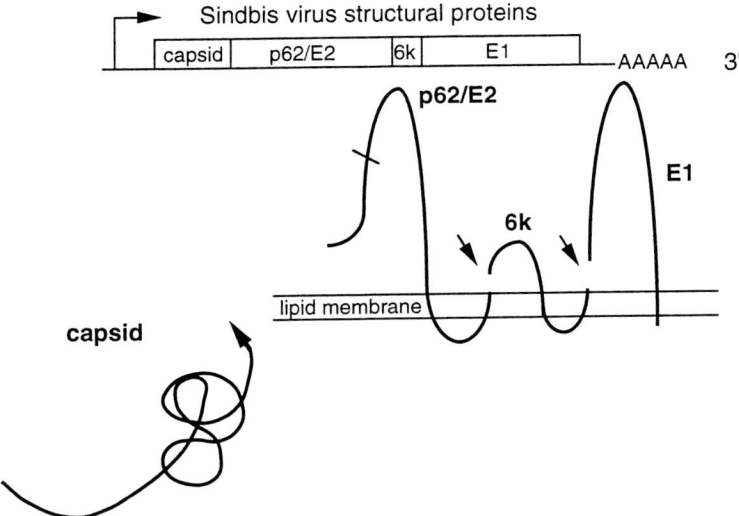

Fig. 2. Cleavage of a polyprotein precursor into viral structural proteins. The subgenomic RNA is translated into a polyprotein that is cleaved cotransitionally by the protease domain of the capsid protein. The remaining C-terminal portion is targeted to the emebrane via an uncleaved signal peptide in the N-terminus of p62. The cleavage of p62 to E2 (hatch mark) and cleavage between E2 and 6k, and between 6k and E1 (arrows) are accomplished by cellular proteases.

inhibits cellular protein synthesis. Although this has been associated with both capsid and glycoprotein synthesis, the mechanism that mediates this process is not known *(35,101)*. The viral RNA remains associated with intracellular vesicles where viral replication occurs *(102,103)*.

It is likely that alphavirus proteins have additional functions which have not yet been identified. For example, a poliovirus protease that is responsible for cleavage of the viral polyprotein also cleaves and inactivates a cellular translation initiation factor *(104)*. Other poliovirus proteins inactivate basal transcription factors, activate cellular kinases, and block vesicular transport from the endoplasmic reticulum to the golgi *(104)*. Although alphaviruses have no known nuclear functions, it is intriguing that the nsP2 protein which has protease activity appears to be translocated to the nucleus *(54,105)*.

Sindbis Virus-Induced Apoptosis Is Age-Dependent and Correlates with Neurovirulence

In contrast to adult mice, infection of newborn mice with the wild type Sindbis virus (strain AR339) results in fatal encephalitis *(12)*. By two weeks of age, however, 97% of animals recover from infection with AR339 *(106)*. Thus, Sindbis virus recapitulates the age-dependent susceptibility that is observed in human CNS infection by a wide range of arboviruses. Both *in situ* TUNEL staining on mouse tissue sections and the detection of fragmentation in DNA extracted from infected mouse brains have confirmed that Sindbis virus induces the apoptotic cell death of infected neurons in vivo *(52)*. In contrast, despite the abundance of virus, very little apoptosis was detected in the brains of infected two-week-old mice *(52)*. Thus, the induction of apoptosis by Sindbis virus in neurons of mouse brains and spinal cords correlates with mortality.

Neurovirulent strains of Sindbis virus have acquired the ability to kill older animals and induce more abundant apoptosis in vivo compared to less virulent strains *(52)*. Sequence comparisons of avirulent and virulent strains have identified a number of genetic determinants responsible for neurovirulence *(47,107)*. A particularly potent mutation is the change from a glutamine to a histidine at position 55 of the Sindbis virus E2 glycoprotein *(108–110)*. It thus appears that one or more viral determinants allows certain neurovirulent strains to induce apoptosis more efficiently. Not only do neurovirulent Sindbis viruses induce significantly more apoptosis in mouse brains and spinal cords, but they also replicate more efficiently at these sites. This suggests that the E2 glycoprotein can modulate viral replication efficiency. The mechanism by which a single amino acid mutation in a Sindbis virus structural protein modulates virus replication in mouse nervous system tissues and alters neurovirulence is not known. Accumulating evidence, however, suggests that E2 His55 has little effect on the ability of virus to bind to cells, but has a significant effect on early steps in viral replication that are specific for neuronal cells *(111,111a)*.

While amino acid changes in Sindbis virus glycoproteins clearly modulate neurovirulence, cell-derived factors also contribute to the eventual outcome of infection. Both virulent (E2 His55) and avirulent (E2 Gln55) strains of Sindbis virus replicate efficiently in the brains of newborn mice, resulting in 100% mortality. By the time mice reach two weeks of age, however, the avirulent strain shows a reduced ability to replicate in the CNS and has a low-mortality rate compared to the neurovirulent strain *(112)*. Host factors must therefore be responsible for protecting older mice from avirulent viruses. A similar phenomenon has also been observed in cell culture. While both the virulent and avirulent strains of Sindbis virus readily kill freshly explanted cultures of primary dorsal root ganglia, these neurons become increasing resistant to virus-induced cell death when infected with the avirulent strain (E2 Gln55) as they mature in culture over the next one to six weeks *(113)*. Just as both virus strains replicate efficiently in newborn mouse brains, these viruses are indistinguishable by many criteria when grown in non-neuronal BHK cell lines *(109)*. These results collectively suggest that age-dependent factor(s) develop in neurons which are responsible for suppressing virus-induced apoptosis and/or replication of avirulent viruses and which protect mice from a lethal infection.

How Does Sindbis Virus Induce Apoptosis?

Although some viral genes such as adenovirus E1A or apoptin of chicken anemia virus are known to induce apoptosis *(114)*, a specific Sindbis virus protein responsible for the induction of cell death has not been identified. Nevertheless, Sindbis virus glycoproteins enhance the cytopathic effects of the virus and implicate the p62/E2, 6K, and/or E1 transmembrane proteins in the induction of apoptosis *(115)*.

Although the molecular mechanisms by which Sindbis virus activates the cell death program are not well understood, several possible explanations exist. Sindbis virus is a potent inhibitor of host cell protein synthesis, in part mediated through activation of the protein kinase PKR (p68) which phosphorylates the alpha subunit of translation initiation factor eIF-2 *(116,117)*. Activation and autophosphorylation occur when PKR binds double strand RNAs which are found in viral replicative intermediates. Expression of PKR in a vaccinia virus vector where the PKR inhibitor E3L is deleted, induces apoptosis in HeLa cells while vaccinia containing an inactivated point mutant of PKR fails to induce cell death *(118)*. Thus, it is possible that Sindbis virus activates the cell death pathway by activating PKR and/or inhibiting host cell protein synthesis. The shutoff of host protein synthesis has in fact been postulated as a mechanism by which nonpermissive polioviruses can activate the death pathway in HeLa cells *(119)*. Furthermore, the observation that uninfected HeLa cells undergo apoptosis following treatment with metabolic inhibitors supports the hypothesis that viruses could trigger apoptosis in this manner *(119,120)*. One infers from these studies that HeLa cells require a protective protein to avoid activating the death pathway. In contrast, primary neurons and the BHK cell line, which both support Sindbis virus infection, are actually protected from apoptosis by metabolic inhibitors. This implies that these cells require the expression of new genes in order to activate the death pathway *(116,121,122)*. As a result, it seems unlikely that Sindbis virus would induce apoptosis through direct inhibition of cellular protein synthesis. Yet, these data do not eliminate the possibility that PKR plays a role in Sindbis virus-induced cell death since PKR may phosphorylate important targets that are distinct from those involved in translation regulation.

One potentially important target of PKR is IkB which regulates the function of the cellular transcription factor NFkB *(123)*. Like many other viruses, Sindbis virus activates NFkB within 1–2 h of infection by unknown mechanisms *(124)*. Upon activation, IkB is phosphorylated and degraded, thereby liberating NFkB from its cytoplasmic retainer. NFkB then translocates from the cytoplasm to the nucleus where it binds specific DNA sequences and activates gene transcription *(125)*. Treatment of cells with oligonucleotide decoys that bind NFkB and prevent activation of downstream genes protects AT-3 cells from Sindbis virus induced-apoptosis, indicating an essential role for NFkB in the death pathway activated by Sindbis virus in AT-3 cells *(124)*. The requirement for new gene expression, through the action of NFkB, to activate the death pathway seems contrary to the observation that Sindbis virus efficiently shuts off host protein synthesis. It is conceivable, however, that some cellular transcripts escape the inhibitory effects of the virus just as viral transcripts escape these effects.

Although NFkB activation has generally been considered a cell proliferation rather than a cell death signal, the induction of cell proliferation by a variety of proteins including adenovirus E1A and the cellular transcription factor E2F can also lead to cell death *(126,127)*. The identification of NFkB sites in the ICE protease promoter and demonstration that the p53 and TNF promoters are activated by NFkB is consistent with a role for NFkB in transcription-dependent induction of cell death *(128–130)*. However, cell type specific activities and distinct NFkB-related transcription factors make the role of NFkB in cell death a complex problem.

CELLULAR INHIBITORS OF APOPTOSIS MODULATE SINDBIS VIRUS ENCEPHALITIS

The Bcl-2 Family Modulates Apoptosis in the Nervous System

The *bcl-2* oncogene was first identified at *t*(14;18) translocations that occur in the majority of follicular B-cell lymphomas *(131)*. This translocation event results in the overexpression of *bcl-2* that allows B-cells to survive when they would normally die by apoptosis. Overexpression of Bcl-2 protects a wide variety of cell types, both in vivo and in vitro, from many death-inducing stimuli including serum and nerve growth factor withdrawal, treatment with calcium ionophores, glucose withdrawal, membrane peroxidation, glucocorticoid treatment, chemotherapeutic agents, and virus infection *(112,132–138)*. Transgenic mice overexpressing *bcl-2* in the B-cell lineage exhibit prolonged survival of responsive B cells and develop an autoimmune disease resembling systemic lupus erythematosus *(139,140)*. In *bcl-2*-deficient mice, the immune system starts to develop normally, but massive apoptosis in the spleen and thymus then occurs *(141–144)*. These animals also have defects in the small intestine and in epithelial cells which normally express substantial levels of *bcl-2* and have delayed growth and early mortality *(143)*. *bcl-2* is abundantly expressed in the nervous system during development and overexpression of *bcl-2* using the neuron-specific enolase or phosphoglycerate kinase promoters results in increased numbers of neurons and increased survival following axotomy or ischemia *(145,146)*. A growing number of *bcl-2* homologs have been identified and some of these, including *bcl-w* and *bcl-x_L*, are also abundantly expressed in brain *(147–149)*. *bcl-x_L*-deficient mice die during embryogenesis with severe apoptosis of neurons and hematopoietic cells *(150)*.

Although *bcl-2*-deficient mice have a less severe phenotype than the *bcl-x_L* knock-outs, embryonic retinal ganglion cells derived from *bcl-2*-deficient animals show impaired regeneration of retinal axons *(151)*. The ability to stimulate axonal regeneration appears to be independent of the prosurvival activity of *bcl-2* since the number of neurons remains constant. This finding is consistent with results obtained with a human neural-crest-derived cell line *(152)*, and suggests that *bcl-2* may have a role in neuronal maturation that extends beyond its ability to block apoptosis. *bcl-2* can also alter cell cycle progression since T-cells derived from *bcl-2*-deficient animals exhibit an accelerated cell cycle *(153)*. Conversely, increased expression of *bcl-2* in T-cells delays the transition from G_0 to S phase.

While some members of the Bcl-2 family protect cells from cell death, other Bcl-2-related proteins such as Bax and Bak actually promote cell death *(154–157)*. Furthermore, these death-inducing proteins heterodimerize with the death inhibitory family members. The domains of Bcl-2 and Bcl-x_L that mediate heterodimerization with Bax and Bak are short 15–20 amino acid regions that are highly conserved among Bcl-2 family members, designated BH1 and BH2 *(157,158)* (*see* Fig. 3). Based on the structure of Bcl-x_L, BH1 and BH2 consist of helix-loop-helix motifs that are in close proximity in the three-dimensional structure *(160)*. The domain of Bax required for heterodimerization with Bcl-2 and Bcl-x_L is a third, more N-terminal homology domain known as BH3 *(161)*. A 46-amino acid peptide containing the BH3 homology domain of Bak is not only sufficient to bind Bcl-x_L but also induces cell death *(161)*.

A role for *bax* in neuronal cell death during development is suggested by the observation that neurons from *bax*-deficient mice survive axotomy *(162)* (*see* chapter by Elliott and Snider). Cultured neurons derived from *bax*-deficient mice also survive withdrawal of nerve growth factor. In the absence of growth factors, however, these neurons undergo atrophy and neurite degeneration, while replacement of growth factors stimulates hypertrophy and neurite outgrowth *(162)*.

How Do Bcl-2 Family Members Regulate Cell Death?

The mechanism by which Bcl-2 family members block apoptosis is not known. Because the protective proteins form heterodimers with the death-promoting members of the family such as Bax and Bak, Bcl-2 may sequester Bax and Bak thereby blocking

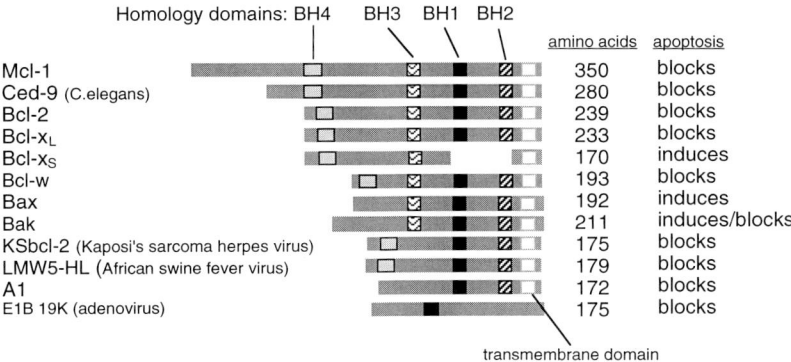

Fig. 3. Bcl-2 protein family. All listed family members are mammalian except where indicated.

their death-promoting activity. Support for this hypothesis comes from the observation that Bax alone induces cell death *(156,163,164)*. In this scenario, the death-promoting members of the Bcl-2 family trigger downstream events that lead to cell death and the role of the protective members of the family is to heterodimerize with Bax and Bak and intercept the death program.

An alternative hypothesis is that protective Bcl-2-related proteins are capable of blocking cell death by a mechanism other than inhibiting Bax and Bak *(159,161)*. This hypothesis therefore suggests that Bcl-2 and Bcl-x_L mediate downstream events that promote cell survival. Support for this hypothesis comes from the observation that mutations in the anti-apoptotic protein Bcl-x_L, that render it unable to bind to Bax and Bak, do not significantly impair its ability to block apoptosis *(157)*. In addition, a viral homolog of Bcl-2, KSbcl-2, that is encoded by the human herpes virus Kaposi's sarcoma-associated virus, also fails to bind Bax and Bak, but is a potent inhibitor of cell death *(165)*. It is possible that this viral homolog can escape the negative regulatory effects of Bax and/or Bak.

These two hypotheses are not mutually exclusive. That is, both the death-inducing and the death-inhibiting members of the Bcl-2 family may have independent activities. Recent evidence described below is consistent with this notion. Nevertheless, heterodimerization between the pro- and antideath family members is likely to modulate their individual activities.

The identification of target proteins through which Bcl-2 and Bax mediate their effects may help to elucidate their mechanisms of action. The observation that Bcl-2 interacts with R-ras suggests its involvement in a signal transduction pathway *(166)*. Ras, a 21 kDa GTP-binding protein, has been implicated in neuronal differentiation and protection of neural cells from apoptosis induced by nerve growth factor (NGF) withdrawal *(167)*. A dominant negative mutant of Ras (Ha Ras^{Asn17}) also inhibits Sindbis virus-induced cell death in PC12 cells *(168)*. The recent demonstration that Bcl-x_L binds directly to CED-4, a *C. elegans* protein that in turn binds, and presumably regulates, the function of death-inducing proteases known as caspases *(169–171)*, is a significant advance in the field (*see* chapter by Rosen and Casciola-Rosen). This interaction potentially places Bcl-2 family members in a protein complex with the caspases that are thought to be part of a common downstream pathway in cell death *(172)*. Thus, Bcl-2 may inhibit caspase activity through a mammalian homolog of CED-4.

The possible localization of Bcl-2 proteins to nuclear pores and the ability of Bcl-2 to block nuclear localization of p53 has raised the possibility that Bcl-2 could serve as a gate keeper for the nucleus *(173,174)*. The observation that Bcl-2 blocks the activation of NFkB in Sindbis virus-infected cells is consistent with this possibility *(124)*. Others have suggested that Bcl-2 regulates intracellular Ca^{2+} concentrations *(175,176)*. This is an attractive hypothesis because the endonucleases involved in DNA laddering are Ca^{2+}-dependent and because calcium ionophores can induce apoptosis in thymocytes and neural cells *(133,176,177)*. While death stimuli such as IL-3 withdrawal and glucocorticoids can deplete calcium stores in the endoplasmic reticulum *(175,178)*, cell death can also occur without mobilization of calcium.

The localization of Bcl-2 primarily to the outer mitochondrial membrane has focused much attention on this particular organelle. Early studies predicted that Bcl-2 may function in an antioxidant pathway to block apoptosis *(179)*. However, cells grown under anaerobic conditions where the production of reactive oxygen species is unlikely to occur still die by an apoptotic cascade which can be blocked by Bcl-2 *(156,180)*. In addition, cells lacking mitochondrial DNA, thereby lacking a functional respiratory chain, still die by apoptosis that can also be inhibited by Bcl-2 *(181)*. Thirteen proteins of the multisubunit enzyme complexes of the electron-transport chain are encoded by mitochondrial DNA. Therefore, cells lacking mitochondrial DNA cannot carry out oxidative phosphorylation and must rely exclusively on glycolysis (*see* also chapters by Dykens, and by Flanigan and Ratan). The recent finding that dATP-induced release of cytochrome c from mitochondria contributes to apoptosis has prompted the suggestion that Bcl-2 could potentially mediate the translocation of cytochrome c *(182–184)*. Bcl-2 has also been shown to preserve the mitochondrial transmembrane potential which is lost early in the death process *(185,186)*. It has been speculated that Bcl-2 may accomplish this task, or that proteins such as Bax may facilitate the loss of mitochondrial membrane potentials by forming pores in the membrane similar to bacterial colicins *(187)*. The structure of Bcl-x_L was recently reported and found to have striking similarity to the B fragment of Diphtheria toxin which facilitates translocation of the toxin A fragment across membranes *(160)*. The colicins and the δ-endotoxins of *Bacillus thuringiensis* are believed to exert their cytocidal activities by destroying membrane potential through pore formation. Similar to these toxins, Bcl-x_L is composed of two central alpha helices (predicted to span the membrane) surrounded by (five) amphipathic alpha helices that shield the hydrophobic core until it is loaded into the membrane.

Regulators of Apoptosis Modulate Sindbis Virus Infection

The molecular mechanisms for the age-dependent susceptibility to virus infections are not well understood. Ubol and Griffin *(75)* reported that a putative Sindbis virus receptor is more abundant in mouse brains at 16 h compared to 96 h after birth. Thus, it is conceivable that an amino acid change in the Sindbis virus E2 glycoprotein that enhances the ability to grow in mature brains (e.g., Glu to His at position 55) could alter the efficiency of binding to this vanishing receptor. Another possibility is that the developmental regulation of anti-apoptotic genes could modulate the age-dependent outcome of infection *(112)*. The anti-apoptotic Bcl-2 and Bcl-x_L proteins are normally expressed in neurons of the brain and spinal cord. Thus, cellular genes that protect neurons from undergoing apoptosis may block virus-induced apoptosis as well. A potential role for cellular anti-apoptotic genes in suppressing virus-induced apoptosis was explored in vitro by infecting a cell line stably expressing the Bcl-2 protein. AT-3 cells overexpressing Bcl-2 are protected from cell death following infection with Sindbis virus (E2 Gln55), while the AT3Neo control cells all die *(113)*. However, more neurovirulent viruses with a His at E2 position 55 overcome the protective effects of Bcl-2 and kill AT3Bcl-2 cells *(108)*. Thus, the ability of neurovirulent Sindbis viruses to kill animals correlates with their ability to kill cells overexpressing Bcl-2. Furthermore, replication of the more avirulent virus was diminished in AT3Bcl-2 cells compared with the virulent strain, consistent with that observed in infected mice *(52,108,113)*. Taken together, these results suggest the possibility that inhibitors of apoptotic cell death may alter the outcome of virus infection in animals (Fig. 4).

Based on the finding that cell lines expressing *bcl-2* are resistant to Sindbis virus-induced apoptosis, a Sindbis virus vector was generated in which *bcl-2* was inserted into the viral genome (Fig. 5). Recombinant viruses carrying a copy of the *bcl-2* gene under the control of a Sindbis virus promoter can infect a variety of cell types but are impaired in their ability to kill cells *(159)*. In contrast, control viruses with a stop codon in the *bcl-2* open reading frame or viruses in which the *bcl-2* gene was inserted in reverse orientation killed cells efficiently. These analyses have been extended to include a wide variety of viral and cellular genes that regulate cell death including iap-related proteins and caspase inhibitors *(188)* (V. Nava, R. Clem, and M. Hardwick, personal observations). This system permits rapid analysis of wild-type and mutant genes without the necessity of generating stable cell lines *(159)*.

The Sindbis virus vector system has also been used in vivo. To determine if *bcl-2* could protect mice from a lethal Sindbis virus infection, animals were infected with a Sindbis virus vector carrying the *bcl-2* gene. Those mice expressing human Bcl-2 protein in neurons were protected from fatal encephalitis *(189)*. Furthermore, exogenous Bcl-2 suppressed Sindbis virus replication in mouse brains *(189)*. However, to determine if endogenous *bcl-2* protects mice from Sindbis virus infection, it will be necessary to assess the outcome of infection in *bcl-2* knock-out mice.

Virus-Induced Apoptosis vs Viral Persistence

The ability of cellular anti-apoptotic genes to block virus-induced cell death supports the hypothesis that failure to activate the death pathway following virus infection leads to a persistent infection *(113)*. Such a mechanism could explain the persistence of avirulent Sindbis virus strains in neurons of mouse brains for extended periods of time *(51,190,191)*. In support of this possibility, AT-3 cells overexpressing Bcl-2 were found to be capable of sustaining a low-grade Sindbis virus infection over many months of passaging *(113)*. Similar mechanisms may exist for other viruses as well (Fig. 4).

IMMUNE-MEDIATED REGULATION OF APOPTOSIS DURING SINDBIS VIRUS ENCEPHALITIS

Antiviral Antibody Downregulates Sindbis Virus Infection in Neurons

In vivo, Sindbis virus replicates predominantly within neurons following intracerebral inoculation. Immunologically normal mice are able to clear avirulent strains of the virus from infected neurons within 7–8 d. Although cytotoxic T-cells have long been

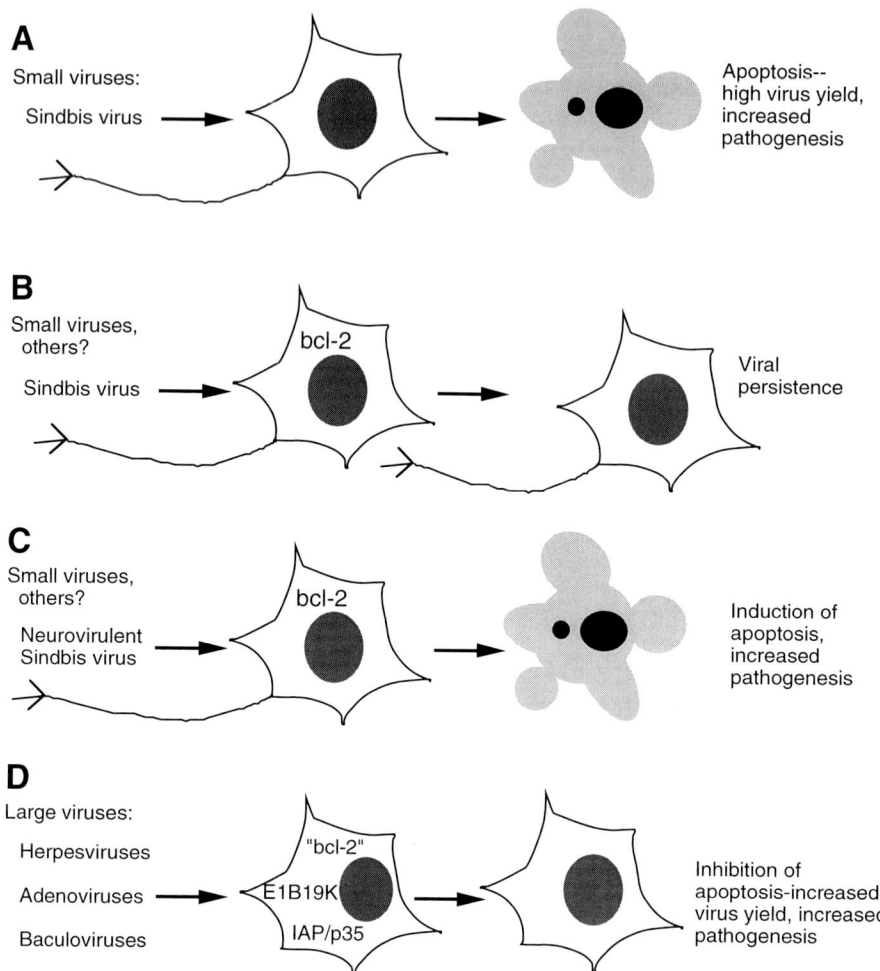

Fig. 4. Induction of apoptosis in response to a virus infection can be a protective mechanism or it can contribute to disease.

considered the primary mechanism whereby viruses are cleared from infected cells, Sindbis virus is instead cleared from infected neurons by specific antibodies against the E2 glycoprotein both in vitro and in vivo *(51)*. In SCID mice that are unable to mount a specific immune response on their own, Sindbis virus establishes a persistent infection in neurons without being cleared. The adoptive transfer of specific anti-E2 monoclonal antibodies, but not antibodies to other viral proteins or immune T-cells, clears infectious virus from the brain *(51)*. This phenomenon can be modeled in neuron cultures where treatment of persistently infected dorsal root ganglion cells with the same anti-E2 antibody significantly downregulates virus replication *(51)*. Although the mechanism by which antiviral antibodies performs this function is not understood, antibody treatment delays virus-induced shutoff of host protein synthesis and restores the ability of cells to respond to antiviral defenses such as alpha-interferon *(192,193)*. In fact, the virus is not completely eradicated from infected neurons but instead is forced into a more quiescent state which can be reactivated in mouse brains if antiviral

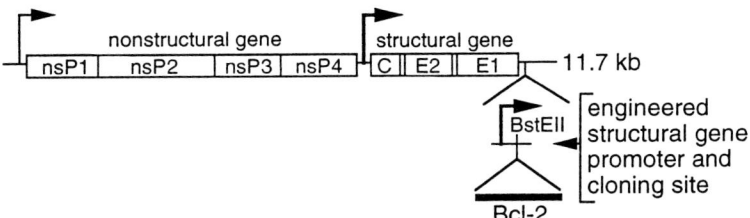

Fig. 5. Sindbis virus vector. A second copy of the viral subgenomic promoter and a unique cloning site (BstEII) are inserted downstream of the coding sequences and upstream of the 3' regulatory region. The Bcl-2 coding sequences are inserted into the BstEII site and incorporated into infectious particles.

antibody titers fall below a certain threshold *(191)*. Nevertheless, the continued presence of these antibodies inhibits viral replication, and preexisting antibody can fully protect animals from neurovirulent Sindbis virus infection and is likely to block virus-induced neuronal apoptosis in this situation *(190)*.

Cell-Mediated Immune Responses Are Involved in the Pathogenesis of Neurovirulent Sindbis Virus Infection

Since adult SCID mice survive neurovirulent Sindbis virus infections more readily than immunocompetent animals, T- and/or B-cells may play a detrimental role in the pathogenesis of this disease *(48)*. The persistent infection that develops in the brains of these animals occurs without obvious neuropathological changes *(51)*. These observations have recently been extended in adult mice that are genetically deficient in CD8+ T-cells who die more slowly following neurovirulent Sindbis virus infection than do strain-matched controls (J. Rowell and D. Griffin, personal observations). Although the precise role that cellular immunity plays in these situations remain undefined, it is presumed that less neuronal apoptosis underlies the survival advantage seen in these immunodeficient animals.

In other studies that were initiated to explore the role of the immune response against nonstructural viral proteins, it was observed that animals who were immunized with recombinant vaccinia viruses expressing the nonstructural portion of the Sindbis virus genome were protected against a subsequent challenge with a neurovirulent viral strain *(194)*. This protection was abrogated if T-cells were deleted before challenge with the virus but after the immunization, suggesting that they were the primary effector of the beneficial effect. Although T-cell-derived product(s) that correlate with protection have not been identified, DNA isolated from the CNS of immunized animals clearly has less fragmentation than DNA from controls suggesting that protection is associated with inhibition of virus-induced neuronal apoptosis (D. Griffin, personal observation).

Innate Immunity in Neurovirulent Sindbis Encephalitis

In addition to acquired T- and B-cell responses, innate immunity can sometimes exert a protective effect during neurovirulent Sindbis virus infection. When examined in adult animals of two closely related strains, BALB/cJ and BALB/cByJ, mortality following infection differs quite dramatically. While more than 80% of BALB/cJ mice eventually succumb to infection, less than 20% of the BALB/cByJ animals die *(195)*. If both strains of animals are treated with a competitive inhibitor of nitric oxide synthetase (NOS), the resistant animals now develop a fatal disease and exhibit more apoptosis *(195)*. Since the virus replicates equally in the brains of the two hosts and NOS inhibition has no effect on this process in either strain, nitric oxide (NO) may instead confer protection against apoptosis in infected neurons. This conclusion was confirmed by treating a neuronal cell line with drugs that produce NO before infecting them with Sindbis virus in vitro *(195)*, and has since been supported by studies examining the amount of DNA fragmentation in the brain tissue of these two animal strains (P. Tucker and S. Wesselingh, personal communication). Because macrophages and microglia are important producers of NO within the CNS and are activated during Sindbis virus encephalitis *(195)*, this innate immunity appears particularly important to the survival of certain host strains following neurovirulent infection. Pharmacologic inhibition of another macrophage-derived product, tumor necrosis factor-alpha, likewise renders BALB/cByJ susceptible to this disease (P. Tucker and D. Griffin, personal communication). Collectively, these data support the hypothesis that innate immune responses can modulate Sindbis virus-induced neuronal apoptosis as a means of protecting animals from an otherwise lethal disease.

REGULATION OF APOPTOSIS BY HERPES SIMPLEX VIRUS (HSV)

Because Sindbis virus and probably a number of other small viruses replicate efficiently in cells that have activated the apoptotic death pathway, Sindbis virus does not need to encode an inhibitor of cell death. However, endogenous host cell inhibitors of apoptosis may modulate Sindbis virus infections in individuals as discussed in detail above. In contrast, viruses with large genomes that require longer times to complete their replication cycles are likely to require inhibitors of apoptosis for efficient progeny virus production. Gamma herpesviruses encode homologs of cellular Bcl-2, adenoviruses encode two distinct apoptosis inhibitors (E1b 19K and 55K), and baculoviruses encode IAP proteins and a caspase inhibitor called P35 *(196–199)* (Fig. 4). Although the specific role of Bcl-2 homologs in gamma herpesviruses is not known, deletion of the adenovirus or baculovirus apoptosis inhibitors severely impairs production of progeny virus presumably because the cell dies by apoptosis prior to completion of viral replication. Based on these findings, one might predict that herpes simplex virus would also encode anti-apoptotic gene(s) that could delay death of the host cell during lytic replication *(200)*. This is consistent with the finding that HSV-infected cells do not exhibit evidence of apoptosis *(201,202)*. Furthermore, induction of apoptosis in HEp-2 cells by treatment with sorbitol can be inhibited by infection with HSV, suggesting that HSV encodes a gene that protects HEp-2 cells *(201)*. The 175-kDa immediate early transcription factor encoded by HSV, ICP4, is a potential candidate for an apoptosis inhibitor. Mutant viruses lacking ICP4 induce apoptosis in Vero cells whereas the wild type virus fails to induce apoptosis *(202)*. Whether ICP4 will turn out to mimic a cellular inhibitor of apoptosis, like several anti-apoptotic genes encoded by other viruses, remains to be determined. It is possible that large DNA viruses will harbor many types of genes that modulate the cell death pathway and lie awaiting discovery. Likewise, the viral factors that signal apoptosis are not known for most viruses.

During HSV latency, only one viral gene is transcribed, the latency-associated transcript, LAT. The primary transcription product of LAT is an 8.3 kb RNA which gives rise to several stable LATs including a 2 kb RNA *(203,204)*. This 2 kb LAT RNA is localized to the nucleus, is not polyadenylated, and appears to be a stable splicing product as indicated by its lariat-like structure *(205)*. Several lines of evidence suggest that LAT is important for reactivation from latency *(206,207)*. Although LAT RNA contains open reading frames, there is no convincing evidence to date that LAT is translated into protein. However, the related bovine herpesvirus-1 (BHV-1) latency-related RNA appears to encode a 41 kDa protein *(208)*. Because this 41 kDa protein inhibits cell cycle progression and can be coprecipitated with cyclin A, which is induced in infected cells, it has been suggested that this latency protein of BHV-1 may play a role in inhibition of apoptosis during infection of neurons *(209)*. A role for HSV LAT RNAs in inhibition of apoptosis would be consistent with the observation that LAT is important for efficient reactivation of the virus from its latent state, and for the induction of a disease state. However, this hypothesis remains to be tested. It is possible that the latency products from different viruses will function by different mechanisms, thus complicating the study of these transcripts and their potential protein products. Nevertheless, the conservation of latency RNAs among α herpesviruses insures an important role for these RNAs and/or proteins in the biology of the virus, providing fertile territory for future investigations.

CONCLUSIONS

Viral encephalitis is not uncommon in humans and infections produced by pathogens, such as the arboviruses, can lead to substantial neurologic morbidity and mortality. In an animal model of arboviral encephalitis where mice are inoculated with Sindbis virus, infected neurons undergo apoptosis in proportion to the mortality incurred by the disease. Virus, host cell, and immune factors all influence the survival of infected neurons. As a result, our understanding of the pathogenesis of these infections has been clarified. More fundamentally, however, this model highlights the critical role that neuronal apoptosis plays in neurological disease.

REFERENCES

1. Jackson AC, Moench TR, Griffin DE, Johnson RT. The pathogenesis of spinal cord involvement in the encephalomyelitis of mice caused by neuroadapted Sindbis virus infection. *Lab Invest* 1987, **56**: 418–423.
2. Griffin DE. Alphavirus pathogenesis and immunity, in *The Togaviridae and Flaviviridae* (Schlesinger S, Schlesinger MJ, eds.), Plenum Publ., New York, 1986, 209–249.
3. Sabin AB. Epidemic encephalitis in military personnel. *J Amer Med Assoc* 1947, **133**: 281–293.
4. Dickerson RB, Newton JR, Hansen JE. Diagnosis and immediate prognosis of Japanese B encephalitis. *Am J Med* 1952, **12**: 277–288.
5. Smorodintsev AA. Tick-borne spring-summer encephalitis. *Prog Med Virol* 1958, **1**: 210–248.
6. Grascenkov NI. Tick-borne encephalitis in the USSR. *Bull Wrld Hlth Org* 1964, **30**: 187–196.
7. Hurwitz ES, Schell W, Nelson D, Washburn J, LaVenture M. Surveillance for California encephalitis group virus illness in Wisconsin and Minnesota, 1978. *Am J Trop Med Hyg* 1983, **32**: 595–601.
8. Longshore WA, Stevens IM, Hollister AC, Gittelsohn A, Lennette EH. Epidemiologic observations on acute infectious encephalitis in California with special reference to the 1952 outbreak. *Amer J Hyg* 1956, **63**: 69–86.

9. Earnest MP, Goolishian HA, Calverley JR, Hayes RO, Hill HR. Neurologic, intellectual and psychologic sequelae following western encephalitis: A followup study of 35 cases. *Neurol* 1971, **21:** 969–974.
10. Przelomiski MM, O'Rourke E, Grady GF, Berardi VP, Markley HG. Eastern equine encephalitis in Massachusetts: A report of 16 cases, 1970–1984. *Neurol* 1988, **38:** 736–739.
11. Monath TP, Tsai TF. St. Louis encephalitis: Lessons from the last decade. *Amer J Trop Med Hyg* 1987, **37 (suppl):** 40S–59S.
12. Taylor RM, Hurlbut HS, Work TH, Kingston JR, Frothingham TE. Sindbis virus: a newly recognized arthropod-transmitted virus. *Amer J Trop Med Hyg* 1955, **4:** 844–862.
13. Schlesinger MJ, Schlesinger S. Formation and assembly of alphavirus glycoproteins, in *The Togaviridae and Flaviviridae* (Schlesinger S, Schlesinger MJ, eds.), Plenum Press, New York, 1986, 121–148.
14. Monath TP. Flaviviruses, in *Fields Virology*, 2nd edn. (Fields BN, ed.), Raven Press, New York, 1990, 763–814.
15. Calisher CH. Medically important arboviruses of the United States and Canada. *Clin Microbiol Revs* 1994, **7:** 89–116.
16. Paul WS, Moore PS, Karabatsos N, Flood SP, Yamada S, Jackson T, Tsai TF. Outbreak of Japanese encephalitis on the island of Saipan, 1990. *J Infect Dis* 1993, **167:** 1053–1058.
17. Umenai T, Krzysko R, Bektimirov TA, Assaad FA. Japanese encephalitis: Current worldwide status. *Bull World Hlth Org* 1985, **63:** 625–631.
18. Ravi V, Desai AS, Shenoy PK, Satishchandra P, Chandramuki A, Gourie-Devi M. Persistence of Japanese encephalitis virus in the human nervous system. *J Med Virol* 1993, **40:** 326–329.
19. Kokernot RH, Hayes J, Will RL, Tempelis CH, Chan DHM, Radivojivic B. Arbovirus studies in the Ohio-Mississippi basin, 1964–1967. II. St. Louis encephalitis virus. *Amer J Trop Med Hyg* 1969, **18:** 750–761.
20. Tsai TF, Canfield MA, Reed CM, Flannery VL, Sullivan KH, Reeve GR, Bailey RE, Poland JD. Epidemiological aspects of a St. Louis encephalitis outbreak in Harris County, Texas, 1986. *J Infect Dis* 1988, **157:** 351–356.
21. Southern PM, Smith JW, Luby JP, Barnett JA, Sanford JP. Clinical and laboratory features of epidemic St. Louis encephalitis. *Ann Int Med* 1969, **71:** 681–690.
22. Okhuysen PC, Crane JK, Pappas J. St. Louis encephalitis in patients with human immunodeficiency virus infection. *Clin Infect Dis* 1993, **17:** 140–141.
23. White MG, Carter NW, Rector FC, Seldin DW. Pathophysiology of epidemic St. Louis encephalitis. I. Inappropriate secretion of antidiuretic hormone. *Ann Int Med* 1969, **71:** 691–702.
24. Asher DM. Chronic encephalitis, in *Search for the Cause of Multiple Sclerosis and Other Chronic Diseases of the Central Nervous System* (Boese A, Weinheim, eds.), Verlag Chemie; 1980, 272–279.
25. Cruse RP, Rothner AD, Erenberg G, Calisher CH. Central European tick borne encephalitis: An Ohio case with a history of foreign travel. *Amer J Dis Child* 1979, **133:** 1070–1071.
26. Blaskovic D. The public health importance of tick borne encephalitis in Europe. Bull Wld Hlth Org 36 suppl, 1967, **1:** 5–13.
27. Gonzalez-Scarano F, Nathanson N. Bunyaviruses, in *Fields Virology*, 2nd edn. (Fields BN, ed.), Raven Press, New York, 1990, 1195–1228.
28. Centers for Disease Control. Aedes albopictus infestation—United States, Brazil. *MMWR* 1986, **35:** 493–495.
29. Centers for Disease Control. LaCrosse encephalitis in West Virginia. *MMWR* 1988, **37:** 79–82.
30. Markoff L. Alphaviruses, in *Principles and Practice of Infectious Diseases*, 4th edn. (Mandell GL, Dolin K, Bennett JE, eds.), Churchill Livingstone, New York, 1995, 1455–1459.
31. Farber S, Hill A, Connerly ML, Dingle JH. Encephalitis in infants and children caused by the virus of the Eastern variety of equine encephalitis. *J Amer Med Assoc* 1940, **114:** 1725–1731.

32. Centers for Disease Control. Arboviral infections of the central nervous system—United States, 1987. *MMWR* 1988, **37:** 506–515.
33. Lennette EH, Longshore WA. Western equine and St. Louis encephalitis in man, California, 1945–1950. *Calif Med* 1951, **75:** 189.
34. Kokernot RH, Shinefield HR, Longshore WA. The 1952 outbreak of encephalitis in California. *Calif Med* 1953, **79:** 73–77.
34a. Somekh E, Glode MP. Multiple intracranial calcifications after western equine encephalitis. *Ped Infect Dis J* 1991, **10:** 408–409.
35. Schlesinger MJ, London SD, Ryan C. An in-frame insertion into the Sindbis virus 6K gene leads to defective proteolytic processing of the virus glycoproteins, a trans-dominant negative inhibition of normal virus formation, and interference in virus shut off of host-cell protein synthesis. *Virology* 1993, **193:** 424–432.
35a. Schlesinger Y, Tebas P, Gaudreault-Keener M, Buller RS, Storch GA. Herpes simplex virus type 2 meningitis in the absence of genital lesions: improved recognition with use of the polymerase chain reaction. *Clin Infect Dis* 1995, **20:** 842–848.
36. Linneman CC, First MR, Alvira MM, Alexander JW, Schiff GM. Herpesvirus hominis type 2 meningoencephalitis following renal transplantation. *Amer J Med* 1976, **61:** 703–708.
37. Nahmias AJ, Whitley RJ, Visintine AN, Takei Y, Alford Jr. CA, Collaborative Antiviral Study Group. Herpes simplex virus encephalitis: Laboratory evaluations and their diagnostic significance. *J Infect Dis* 1982, **145:** 829–836.
38. Heller M, Dix RD, Baringer JR, Schachter J, Conte JE Jr. Herpetic proctitis and meningitis: Recovery of two strains of herpes simplex type 1 from cerebrospinal fluid. *J Infect Dis* 1982, **146:** 584–588.
39. Whitley RJ, Lakeman F. Herpes simplex virus infection of the central nervous system: therapeutic and diagnostic considerations. *Clin Infect Dis* 1995, **20:** 414–420.
40. Olson LC, Buescher EL, Artenstein MS. Herpesvirus infections of the human central nervous system. *N Engl J Med* 1967, **24:** 1271–1277.
41. Whitley R, Lakeman AD, Nahmias A, Roizman B. DNA restriction enzyme analysis of herpes simplex virus isolates obtained from patients with encephalitis. *N Engl J Med* 1982, **307:** 1060–1062.
42. Hammer SM, Buchman TG, D'Angelo LJ, Karchmer AW, Roizman B, Hirsch MD. Temporal cluster of herpes simplex encephalitis: Investigation by restriction endonuclease cleavage of viral DNA. *J Infect Dis* 1980, **141:** 436–440.
43. Landry ML, Berkovits IV, Summers WP, Booss J, Hsiung GD, Summers WC. Herpes simplex encephalitis: Analysis of a cluster of cases by restriction endonuclease mapping of virus isolates. *Neurol* 1983, **33:** 831–835.
44. Cinque P, Cleator GM, Weber T, Monteyne P, Sindic CJ, van Loon AM. The role of laboratory investigation in the diagnosis and management of patients with suspected herpes simplex encephalitis: a consensus report. *J Neurol Neurosurg Psych* 1996, **61:** 339–345.
45. Lakeman FD, Whitley RJ. Diagnosis of herpes simplex encephalitis: application of polymerase chain reaction to cerebrospinal fluid from brain-biopsied patients and correlation with disease. *J Infect Dis* 1995, **171:** 857–863.
46. Whitley RJ. Neonatal herpes simplex virus infections. *J Med Virol* Suppl 1993, **1:** 13–21.
47. Griffin DE, Hahn CS, Jackson AC, Lustig S, Strauss EG, Strauss JH. The basis of Sindbis virus neurovirulence, in *Cell Biology of Virus Entry, Replication, and Pathogenesis* (Compans RW, Helenius A, Oldstone MBA, eds.), Alan R. Liss, New York, 1989, 387–396.
48. Wesselingh SL, Levine B, Fox RJ, Choi S, Griffin DE. Intracerebral cytokine mRNA expression during fatal and nonfatal alphavirus encephalitis suggests a predominant type 2 T cell response. *J Immunol* 1994, **152:** 1289–1297.
49. Griffin DE, Hess JL. Cells with natural killer activity in the cerebrospinal fluid of normal mice and athymic nude mice with acute Sindbis virus encephalitis. *J Immunol* 1986, **136:** 1841–1845.

50. Moench TR, Griffin DE. Immunocytochemical identification and quantitation of mononuclear cells in cerebrospinal fluid, meninges, brain during acute viral encephalitis. *J Exp Med* 1984, **159**: 77–88.
51. Levine B, Hardwick JM, Trapp BD, Crawford TO, Bollinger RC, Griffin DE. Antibody-mediated clearance of alphavirus infection from neurons. *Science* 1991, **254**: 856–860.
52. Lewis J, Wesselingh SL, Griffin DE, Hardwick JM. Sindbis virus-induced apoptosis in mouse brains correlates with neurovirulence. *J Virol* 1996, **70**: 1828–1835.
53. Schlesinger S, Schlesinger MJ. *The Togaviridae and Flaviviridae*, Plenum, New York, 1986.
54. Strauss EG, De Groot RJ, Levinson R, Strauss JH. Identification of the active site residues in the nsP2 proteinase of Sindbis virus. *Virology* 1992, **191**: 932–940.
55. Hsieh P, Rosner MR, Robbins PW. Host-dependent variation of asparagine-linked oligosaccharides at individual glycosylation sites of Sindbis virus glycoproteins. *J Biol Chem* 1983, **258**: 2548–2554.
56. Lemm JA, Rice CM. Roles of nonstructural polyproteins and cleavage products in regulating Sindbis virus RNA replication and transcription. *J Virol* 1993, **67**: 1916–1926.
57. Wengler G1, Wengler G2. The carboxy-terminal part of the NS 3 protein of the West Nile flavivirus can be isolated as a soluble protein after proteolytic cleavage and represents an RNA-stimulated NTPase. *Virology* 1991, **184**: 707–715.
58. Gorbalenya AE, Koonin EV. Viral proteins containing the purine NTP-binding sequence pattern. *Nucl Acids Res* 1989, **17**: 8413–8441.
59. Strauss EG, Strauss JH. Structure and replication of the alphavirus genome, in *The Togaviridae and Flaviviridae* (Schlesinger S, Schlesinger MJ, eds.), Plenum Press, New York, 1986, 35–90.
60. Mi S, Durbin R, Huang HV, Rice CM, Stollar V. Association of the Sindbis virus RNA methyltransferase activity with the nonstructural protein nsP1. *Virology* 1989, **170**: 385–391.
61. Wang Y-F, Sawicki SG, Sawicki DL. Sindbis virus nsP1 functions in negative-strand RNA synthesis. *J Virol* 1991, **65**: 985–988.
62. Li G, LaStarza M, Hardy WR, Strauss JH, Rice CM. Phosphorylation of Sindbis nsP3 in vivo and in vitro. *Virology* 1990, **179**: 416–427.
63. Poch O, Sauvaget I, Delarue M, Tordo N. Identification of four conserved motifs among the RNA-dependent polymerase encoding elements. *EMBO J* 1989, **8**: 3867–3874.
64. Hahn YS, Grakoui A, Rice CM, Strauss EG, Strauss JH. Mapping of RNA- temperature-sensitive mutants of Sindbis virus: complementation group F mutants have lesions in nsP4. *J Virol* 1989, **63**: 1194–1202.
65. Barton DJ, Sawicki SG, Sawicki DL. Solubilization and immunoprecipitation of alphavirus replication complexes. *J Virol* 1991, **65**: 1496–1506.
66. Kuhn RJ, Griffin DE, Zhang H, Niesters HGM, Strauss JH. Attenuation of Sindbis virus neurovirulence by using defined mutations in nontranslated regions of the genome RNA. *J Virol* 1992, **66**: 7121–7127.
67. Pogue GP, Cao X-Q, Singh NK, Nakhasi HL. 5' sequences of rubella virus RNA stimulate translation of chimeric RNAs and specifically interact with two host-encoded proteins. *J Virol* 1993, **67**: 7106–7117.
68. Simmons DT, Strauss JH. Translation of Sindbis virus 26 S RNA and 49 S RNA in lysates of rabbit reticulocytes. *J Mol Biol* 1974, **86**: 397–409.
69. Choi H-K, Tong L, Minor W, Dumas P, Boege U, Rossmann MG, Wengler G. Structure of Sindbis virus core protein reveals a chymotrypsin-like serine proteinase and the organization of the virion. *Nature* 1991, **354**: 37–43.
70. Liljestrom P, Garoff H. Internally located cleavable signal sequences direct the formation of Semliki Forest virus membrane proteins from a polyprotein precursor. *J Virol* 1991, **65**: 147–154.

71. Ding M, Schlesinger MJ. Evidence that Sindbis virus NSP2 is an autoprotease which processes the virus nonstructural polyprotein. *Virology* 1989, **171**: 280–284.
72. Mayne JT, Bell JR, Strauss EG, Strauss JH. Pattern of glycosylation of sindbis virus envelope proteins synthesized in hamster and chicken cells. *Virology* 1985, **142**: 121–133.
73. Wahlberg JM, Bron R, Wilschut J, Garoff H. Membrane fusion of Semliki Forest virus involves homotrimers of the fusion protein. *J Virol* 1992, **66**: 7309–7318.
74. Wang K-S, Kuhn RJ, Strauss EG, Ou S, Strauss JH. High-affinity laminin receptor is a receptor for Sindbis virus in mammalian cells. *J Virol* 1992, **66**: 4992–5001.
75. Ubol S, Griffin DE. Identification of a putative alphavirus receptor on mouse neural cells. *J Virol* 1991, **65**: 6913–6921.
76. Presley JF, Polo JM, Johnston RE, Brown DT. Proteolytic processing of the Sindbis virus membrane protein precursor PE2 is nonessential for growth in vertebrate cells but is required for efficient growth in invertebrate cells. *J Virol* 1991, **65**: 1905–1909.
77. Metsikko K, Garoff H. Oligomers of the cytoplasmic domain of the p62/E2 membrane protein of Semliki Forest virus bind to the nucleocapsid in vitro. *J Virol* 1990, **64**: 4678–4683.
78. Suomalainen M, Garoff H. Alphavirus spike-nucleocapsid interaction and network antibodies. *J Virol* 1992, **66**: 5106–5109.
79. Zhao H, Garoff H. Role of cell surface spikes in alphavirus budding. *J Virol* 1992, **66**: 7089–7095.
80. Suomalainen M, Liljestrom P, Garoff H. Spike protein-nucleocapsid interactions drive the budding of alphaviruses. *J Virol* 1992, **66**: 4737–4747.
81. Vaux DJT, Helenius A, Mellman I. Spike-nucleocapsid interaction in Semliki Forest virus reconstructed using network antibodies. *Nature* 1988, **336**: 36–42.
82. Gaedigk-Nitschko K, Schlesinger MJ. The Sindbis virus 6K protein can be detected in virions and is acylated with fatty acids. *Virology* 1990, **175**: 274–281.
83. Ivanova L, Schlesinger MJ. Site-directed mutations in the Sindbis virus E2 glycoprotein identify palmitoylation sites and affect virus budding. *J Virol* 1993, **67**: 2546–2551.
84. Liu N, Brown DT. Phosphorylation and dephosphorylation events play critical roles in Sindbis virus maturation. *Virology* 1993, **196**: 703–711.
85. Levine B, Jiang HH, Kleeman L, Yang G. Effect of E2 envelope glycoprotein cytoplasmic domain mutations on Sindbis virus pathogenesis. *J Virol* 1996, **70**: 1255–1260.
86. Lee S, Owen KE, Choi H-K, Lee H, Lu G, Wengler G, Brown DT, Rossmann MG, Kuhn RJ. Identification of a protein binding site on the surface of the alphavirus nucleocapsid and its implication in virus assembly. Structure 1996, **4**: 531–541.
87. Weiss B, Geigenmuller-Gnirke U, Schlesinger S. Interactions between Sindbis virus RNAs and a 68 amino acid derivative of the viral capsid protein further defines the capsid binding site. Nucleic Acids Res 1994, **22**: 780–786.
88. Gaedigk-Nitschko K, Schlesinger MJ. Site-directed mutations in Sindbis virus E2 glycoprotein's cytoplasmic domain and the 6K protein lead to similar defects in virus assembly and budding. *Virology* 1991, **183**: 206–214.
89. Liu N, Brown DT. Transient translocation of the cytoplasmic (Endo) domain of a type I membrane glycoprotein into cellular membranes. *J Cell Biol* 1993, **120**: 877–883.
90. Melancon P, Garoff H. Reinitiation of translocation in the Semliki Forest virus structural polyprotein: identification of the signal for the E1 glycoprotein. *EMBO J* 1986, **5**: 1551–1560.
91. Hashimoto K, Erdel S, Keranen S, Saraste J, Kaariainen L. 1981, Evidence for a separate signal sequence for the carboxy-terminal envelope glycoprotein E1 of Semliki Forest virus. *J Virol* **38**: 34–40.
92. Lusa S, Garoff H, Liljestrom P. Fate of the 6K membrane protein of Semliki Forest virus during virus assembly. *Virology* 1991, **185**: 843–846.
93. Gaedigk-Nitschko K, Ding M, Levy MA, Schlesinger MJ. Site-directed mutations in the Sindbis virus 6K protein reveal sites for fatty acylation and the underacylated protein affects virus release and virion structure. *Virology* 1990, **175**: 282–291.

94. Sanz MA, Perez L, Carrasco L. Semliki Forest virus 6K protein modifies membrane permeability after inducible expression in Escherichia coli cells. *J Biol Chem* 1994, **269:** 12,106–12,110.
95. Liljestrom P, Lusa S, Huylebroeck D, Garoff H. In vitro mutagenesis of a full-length cDNA clone of Semliki Forest virus: the small 6,000-molecular-weight membrane protein modulates virus release. *J Virol* 1991, **65:** 4107–4113.
96. Levy-Mintz P, Kielian M. Mutagenesis of the putative fusion domain of the Semliki Forest virus spike protein. *J Virol* 1991, **65:** 4292–4300.
97. Boggs WM, Hahn CS, Strauss EG, Strauss JH, Griffin DE. Low pH-dependent Sindbis virus-induced fusion of BHK cells: Differences between strains correlate with amino acid changes in the E1 glycoprotein. *Virology* 1989, **169:** 485–488.
98. Singh I, Helenius A. Role of ribosomes in Semliki Forest virus nucleocapsid uncoating. *J Virol* 1992, **66:** 7049–7058.
99. Singh I, Helenius A. Nucleocapsid uncoating during entry of enveloped animal RNA viruses into cells. *Virology* 1992, **3:** 511–518.
100. Geigenmuller-Gnirke U, Nitschko H, Schlesinger S. Deletion analysis of the capsid protein of Sindbis virus: Identification of the RNA binding region. *J Virol* 1993, **67:** 1620–1626.
101. Elgizoli M, Dai Y, Kempf C, Koblet H, Michel MR. Semliki Forest virus capsid protein acts as a pleiotropic regulator of host cellular protein synthesis. *J Virol* 1989, **63:** 2921–2928.
102. Peranen J, Kaariainen L. Biogenesis of type I cytopathic vacuoles in Semliki Forest virus-infected BHK cells. *J Virol* 1991, **65:** 1623–1627.
103. Froshauer S, Kartenbeck J, Helenius A. Alphavirus RNA replicase is located on the cytoplasmic surface of endosomes and lysosomes. *J Cell Biol* 1988, **107:** 2075–2086.
104. Rikkonen M, Peranen J, Kaariainen L. Nuclear and nucleolar targeting signals of Semliki Forest virus nonstructural protein nsP2. *Virology* 1992, **189:** 462–473.
105. Hambidge SJ, Sarnow P. Early events in poliovirus-infected cells. *Virology* 1992, **3:** 501–510.
106. Griffin DE, Levine B, Tyor WR, Tucker PC, Hardwick JM. Age-dependent susceptibility to fatal encephalitis: alphavirus infection of neurons. *Arch Virol* 1994a, **9:** 31–39.
107. Lustig S, Jackson AC, Hahn CS, Griffin DE, Strauss EG, Strauss JH. Molecular basis of Sindbis virus neurovirulence in mice. *J Virol* 1988, **62:** 2329–2336.
108. Ubol S, Tucker PC, Griffin DE, Hardwick JM. Neurovirulent strains of alphavirus induce apoptosis in bcl-2-expressing cells; Role of a single amino acid change in the E2 glycoprotein. *Proc Natl Acad Sci USA* 1994, **91:** 5202–5206.
109. Tucker PC, Strauss EG, Kuhn RJ, Strauss JH, Griffin DE. Viral determinants of age-dependent virulence of Sindbis virus for mice. *J Virol* 1993, **67:** 4605–4610.
110. Levine B, Griffin DE. Molecular analysis of neurovirulent strains of Sindbis virus that evolve during persistent infection of scid mice. *J Virol* 1993, **67:** 6872–6875.
111. Dropulic LK, Hardwick JM, Griffin DE. A single amino acid change in the E2 glycoprotein of Sindbis virus confers neurovirulence by altering an early step of virus replication. *J Virol* 1997, **71:** 6100–6105.
111a. Tucker, PC, Lee CH, Bui N, Martinie D, Griffin DE. Amino acid changes in the Sindbis virus glycoprotein that increase neurovirulence improve entry into the neuroblastoma cells. *J Virol* 1997, **71:** 6106–6112.
112. Griffin DE, Levine B, Ubol S, Hardwick JM. The effects of alphavirus infection on neurons. *Annals Neurol* 1994, **35:** S23–S27.
113. Levine B, Huang Q, Isaacs JT, Reed JC, Griffin DE, Hardwick JM. Conversion of lytic to persistent alphavirus infection by the bcl-2 cellular oncogene. *Nature* 1993, **361:** 739–742.
114. Zhuang S-M, Shvarts A, van Ormondt H, Jochemsen AG, van der EB AJ, Mathieu HM, Noteborn HM. Apoptin, a protein derived from chicken anemia virus, induces p53-independent apoptosis in human osteosarcoma cells. *Can Res* 1995, **55:** 486–489.

115. Frolov I, Schlesinger S. A comparison of the effects of Sindbis virus and Sindbis virus replicons on host cell protein synthesis and cytopathogenicity in BHK cells. *J Virol* 1994, **68**: 1721–1727.
116. Saito S. Enhancement of the interferon-induced double-stranded RNA-dependent protein kinase activity by Sindbis virus infection and heat-shock stress. *Microbiol Immunol* 1990, **34**: 859–870.
117. Lewis J, Jagus R, Hardwick JM. Unpublished data, 1997.
118. Lee SB, Esteban M. The interferon-induced double-stranded RNA-activated protein kinase induces apoptosis. *Virology* 1994, **199**: 491–496.
119. Tolskaya EA, Romanova LI, Kolesnikova MS, Ivasnnikova TA, Smirnova EA, Raikhlin NT, Agol VI. Apoptosis-inducing and apoptosis-preventing functions of poliovirus. *J Virol* 1995, **69**: 1181–1189.
120. Clem RJ, Miller LK. Control of programmed cell death by the baculovirus genes p35 and iap. *Mol Cell Biol* 1994, **14**: 5212–5222.
121. Martin DP, Schmidt RE, DiStefano PS, Lowry OH, Carter JG, Johnson EM Jr. Inhibitors of protein synthesis and RNA synthesis prevent neuronal death caused by nerve growth factor deprivation. *J Cell Biol* 1988, **106**: 829–844.
122. Estus S, Zaks WJ, Freeman RS, Gruda M, Bravo R, Johnson EM Jr. Altered gene expression in neurons during programmed cell death—identification of c-Jun as necessary for neuronal apoptosis. *J Cell Biol* 1994, **127**: 1717–1727.
123. Proud CG. PKR: a new name and new roles. *TIBS* 1995, **20**: 241–246.
124. Lin K-I, Lee S-H, Narayanan R, Baraban JM, Hardwick JM, Ratan RR. Thiol agents and Bcl-2 identify an alphavirus-induced apoptotic pathway that requires activation of the transcription factor NF-kappa B. *J Cell Biol* 1995, **131**: 1149–1161.
125. Beg AA, Baldwin AS Jr. The IkB proteins: multifunctional regulators of Rel/NF-kB transcription factors. *Genes Dev* 1993, **7**: 2064–2070.
126. Rao L, Debbas M, Sabbatini P, Hockenbery D, Korsmeyer S, White E. The adenovirus E1A proteins induce apoptosis, which is inhibited by the E1B 19-kDa and Bcl-2 proteins. *Proc Natl Acad Sci USA* 1992, **89**: 7742–7746.
127. Qin X-Q, Livingston DM, Kaelin WG Jr, Adams PD. Deregulated transcription factor E2F-1 expression leads to S-phase entry and p53-mediated apoptosis. *Proc Natl Acad Sci USA* 1994, **91**: 10918–10922.
128. Casano FJ, Rolando AM, Mudgett JS, Molineaux SM. The structure and complete nucleotide sequence of the murine gene encoding interleukin-1-beta converting enzyme (ICE). *Genomics* 1994, **20**: 474–481.
129. Wu H, Lozano G. NF-kappa B activation of p53. A potential mechanism for suppressing cell growth in response to stress. *J Biol Chem* 1994, **269**: 20,067–20,074.
130. Trede NS, Tsytsykova AV, Chatila T, Goldfeld AE, Geha RS. Transcriptional activation of the human TNF-alpha promoter by superantigen in human monocytic cells: role of NF-kappa B. *J Immunol* 1995, **155**: 902–908.
131. Tsujimoto Y, Bashir MM, Givol I, Cossman J, Jaffe E, Croce CM. DNA rearrangements in human follicular lymphoma can involve the 5' or the 3' region of the bcl-2 gene. *Proc Natl Acad Sci USA* 1987, **84**: 1329–1331.
132. Miyashita T, Reed JC. bcl-2 gene transfer increases relative resistance of S49.1 and WEH17.2 lymphoid cells to cell death and DNA fragmentation induced by glucocorticoids and multiple chemotherapeutic drugs. *Can Res* 1992, **52**: 5407–5411.
133. Zhong L-T, Sarafian T, Kane DJ, Charles AC, Mah SP, Edwards RH, Bedesen DE. Bcl-2 inhibits death of central neural cells induced by multiple agents. *Proc Natl Acad Sci USA* 1993, **90**: 4533–4537.
134. Mah SP, Zhong LT, Liu Y, Roghani A, Edwards RH, Bredesen DE. The protooncogene bcl-2 inhibits apoptosis in PC12 cells. *J Neurochem* 1993, **60**: 1183–1186.
135. Garcia I, Martinou I, Tsujimoto Y, Martinou J-C. Prevention of programmed cell death of sympathetic neurons by the bcl-2 proto-oncogene. *Science* 1992, **258**: 302–304.

136. Allsopp TE, Wyatt S, Paterson HF, Davies AM. The proto-oncogene bcl-2 can selectively rescue neurotrophic factor-dependent neurons from apoptosis. *Cell* 1993, **73**: 295–307.
137. Dubois-Dauphin M, Frankowski H, Tsujimoto Y, Huarte J, Martinou JC. Neonatal motoneurons overexpressing the bcl-2 protooncogene in transgenic mice are protected from axotomy-induced cell death. *Proc Natl Acad Sci USA* 1994, **91**: 3309–3313.
138. Hinshaw VS, Olsen CW, Dybdahl-Sissoko N, Evans D. Apoptosis: a mechanism of cell killing by influenza A and B viruses. *J Virol* 1994, **68**: 3667–3673.
139. Núñez G, Hocknebery D, McDonnell TJ, Sorensen CM, Korsmeyer SJ. Bcl-2 maintains B cell memory. *Nature* 1991, **353**: 71–73.
140. Strasser A, Whittingham S, Vaux DL, Bath ML, Adams JM, Cory S, Harris AW. Enforced BCL2 expression in B-lymphoid cells prolongs antibody responses and elicits autoimmune disease. *Proc Natl Acad Sci USA* 1991, **83**: 8661–8665.
141. Nakayama K-I, Nakayama K, Negishi I, Kuida K, Shinkai Y, Louie MC, Fields LE, Lucas PJ, Stewart V, Alt FW, Loh DY. Disappearance of the lymphoid system in bcl-2 homozygous mutant chimeric mice. *Science* 1993, **261**: 1584–1588.
142. Veis DJ, Sorenson CM, Shutter SR, Korsmeyer SJ. Bcl-2-deficient mice demonstrate fulminant lymphoid apoptosis, polycystic kidneys, and hypopigmented hair. *Cell* 1993, **75**: 229–240.
143. Kamada S, Shimono A, Shinto Y, Tsujimura T, Takahashi T, Noda T, Kitamura Y, Kondoh H, Tsujimoto Y. bcl-2 deficiency in mice leads to pleiotropic abnormalities: accelerated lymphoid cell death in thymus and spleen, polycystic kidney, hair hypopigmentation, and distorted small intestine. *Can Res* 1995, **55**: 354–359.
144. Nakayama K, Nakayama K-I, Negishi I, Kuida K, Sawa H, Loh DY. Targeted disruption of Bcl-2αβ in mice: Occurrence of gray hair, polycystic kidney disease, and lymphocytopenia. *Proc Natl Acad Sci USA* 1994, **91**: 3700–3704.
145. Martinou J-C, Dubois-Dauphin M, Staple JK, Rodriguez I, Frankowski H, Missotten M, Albertini P, Dominique T, Catsicas S, Pietra C, Haurte J. Overexpression of Bcl-2 in transgenic mice protects neurons from naturally occurring cell death and experimental ischemia. *Neuron* 1994, **13**: 1017–1030.
146. Dubois-Dauphin M, Frankowski H, Tsujimoto Y, Huarte J, Martinou JC. Neonatal motorneurons overexpressing the bcl-2 protooncogene in transgenic mice are protected from axotomy-induced cell death. *Proc Natl Acad Sci USA* 1994, **91**: 3309–3313.
147. Boise LH, González-Garcia M, Postema CE, Ding L, Lindsten T, Turka LA, Mao X, Nuñez G, Thompson CB. Bcl-x, a bcl-2-related gene that functions as a dominant regulator of apoptotic cell death. *Cell* 1993, **74**: 597–608.
148. Gibson L, Holmgreen SP, Huang DCS, Bernand O, Copeland NG, Jenkins NA, Sutherland GR, Baker E, Adams JM, Cory S. Bcl-w, a novel member of the Bcl-2 family, promotes cell survival. *Oncogene* 1996, **13**: 665–675.
149. Frankowski H, Missotten M, Fernandez P-A, Martinou I, Michel P, Sadoul R, Martinou J-C. Function and expression of the Bcl-x gene in the developing and adult nervous system. *NeuroReport* 1995, **6**: 1917–1921.
150. Motoyama N, Wang F, Roth KA, Sawa H, Nakayama K-I, Nakayama K, Negishi I, Senju S, Zhang Q, Fujii S, Loh DY. Massive cell death of immature hematopoietic cells and neurons in Bcl-x-deficient mice. *Science* 1995, **267**: 1506–1510.
151. Chen DF, Schneider GE, Martinou J-C, Tonegawa S. Bcl-2 promotes regeneration of severed axons in mammalian CNS. *Nature* 1997, **385**: 434–439.
152. Zhang K-Z, Westberg JA, Holtta E, Andersson LC. Bcl-2 regulates neural differentiation. *Proc Natl Acad Sci USA* 1996, **93**: 4504–4508.
153. Linette GP, Li Y, Roth K, Korsmeyer SJ. Cross talk between cell death and cell cycle progression: Bcl-2 regulates NFAT-mediated activation. *Proc Natl Acad Sci USA* 1996, **93**: 9545–9552.

154. Oltvai ZN, Milliman CL, Korsmeyer SJ. Bcl-2 heterodimerizes in vivo with a conserved homolog, bax, that accelerates programmed cell death. *Cell* 1993, **74**: 609–619.
155. Farrow SN, White JHM, Martinou I, Raven T, Pun K-T, Grinham CJ, Martinou J-C, Brown R. Cloning of a *bcl-2* homologue by interaction with adenovirus E1B 19K. *Nature* 1995, **374**: 731–733.
156. Chittenden T, Harrington EA, O'Connor R, Flemington C, Lutz RJ, Evan GI, Guild BC. Induction of apoptosis by the Bcl-2 homologue Bak. *Nature* 1995b, **374**: 733–736.
157. Kiefer MC, Brauer MJ, Powers VC, Wu JJ, Umansky SR, Tomei LD, Barr PJ. Modulation of apoptosis by the widely distributed Bcl-2 homologue Bak. *Nature* 1995, **374**: 736–739.
158. Yin X-M, Oltvai ZN, Korsmeyer SJ. BH1 and BH2 domains of Bcl-2 are required for inhibition of apoptosis and heterodimerization with Bax. *Nature* 1994, **369**: 321–323.
159. Cheng EH-Y, Levine B, Boise LH, Thompson CB, Hardwick JM. Bax-independent inhibition of apoptosis by Bcl-xL. *Nature* 1996, **379**: 554–556.
160. Muchmore SW, Sattler M, Liang H, Meadows RP, Harlan JE, Yoon HS, Nettesheim D, Chang BS, Thompson CB, Wong S-L, Ng S-C, Fesik SW. X-ray and NMR structure of human Bcl-xL, an inhibitor of programmed cell death. *Nature* 1996, **381**: 335–341.
161. Chittenden T, Flemington C, Houghton AB, Ebb RG, Gallo GJ, Elangovan B, Chinnadurai G, Lutz RJ. A conserved domain in Bak, distinct from BH1 and BH2, mediates cell death and protein binding functions. *EMBO J* 1995a, **14**: 5589–5596.
162. Deckwerth TL, Elliott JL, Knudson CM, Johnson EM Jr, Snider WD, Korsmeyer SJ. Bax is required for neuronal death after trophic factor deprivation and during development. *Neuron* 1996, **17**: 401–411.
163. Han J, Sabbatini P, Perez D, Rao L, Modha D, White E. The E1B 19K protein blocks apoptosis by interacting with and inhibiting the p53-inducible and death-promoting Bax protein. *Genes Develop* 1996, **10**: 461–477.
164. Zha H, Aime-Sempe C, Sato T, Reed JC. Proapoptotic protein Bax heterodimerizes with Bcl-2 and homodimerizes with Bax via a novel domain (BH3) distinct from BH1 and BH2. *J Biol Chem* 1996, **271**: 7440–7444.
165. Cheng EH-Y, Nicholas J, Bellows DS, Hayward GS, Guo H-G, Reitz MS, Hardwick JM. A Bcl-2 homolog encoded by Kaposi's sarcoma-associated virus, human herpesvirus 8, inhibits apoptosis but does not heterodimerize with Bax or Bak. *Proc Natl Acad Sci USA* 1997, **94**: 690–694.
166. Fernandez-Sarabia MJ, Bischoff JR. Bcl-2 associates with that ras-related protein R-ras p23. *Nature* 1993, **366**: 274–275.
167. Ferrari G, Greene LA. Proliferative inhibition by dominant negative ras rescues naive and neuronally differentiated PC12 cells from apoptotic death. *EMBO J* 1994, **13**: 5922–5928.
168. Joe AK, Ferrari G, Jiang HH, Liang XH, Levine B. Dominant inhibitory ras delays Sindbis virus-induced apoptosis in neuronal cells. *J Virol* 1996, **70**: 7744–7751.
169. Chinnaiyan AM, O'Rourke K, Lane BR, Dixit VM. Interaction of CED-4 with CED-3 and CED-9: a molecular framework for cell death. Science 1997, **275**: 1122–1126.
170. Wu D, Wallen HD, Nunez G. Interaction and regulation of subcellular localization of CED-4 by CED-9. *Science* 1997, **275**: 1126–1129.
171. Spector MS, Desnoyers S, Hoeppner DJ, Hengartner MO. Interaction between the *C. elegans* cell-death regulators CED-9 and CED-4. *Nature* 1997, **385**: 653–655.
172. Fraser A, Evan G. A license to kill. *Cell* 1996, **85**: 781–784.
173. de Jong D, Prins FA, Mason DY, Reed JC, van Ommen GB, Kluin PM. Subcellular localization of the bcl-2 in malignant and normal lymphoid cells. *Can Res* 1994, **54**: 256–260.
174. Ryan JJ, Prochownik E, Gottlieb CA, Apel IJ, Merino R, Nunez G, Clarke MF. c-myc and bcl-2 modulate p53 function by altering p53 subcellular trafficking during the cell cycle. *Proc Natl Acad Sci USA* 1994, **91**: 5878–5882.

175. Lam M, Dubyak G, Chen L, Nunez G, Miesfeld RL, Distelhorst CW. Evidence that Bcl-2 represses apoptosis by regulating endoplasmic reticulum-associated Ca2+ fluxes. *Proc Natl Acad Sci USA* 1994, **91:** 6569–6573.
176. McConkey DJ, Orrenius S. The role of calcium in the regulation of apoptosis. *J Leukocyte Biol* 1996, **59:** 775–783.
177. Durant S, Homo F, Duval D. Calcium and A23187-induced cytolysis of mouse thymocytes. *Biochem Biophysic Res Comm* 1980, **93:** 385–391.
178. Distelhorst CW, Lam M, McCormick TS. Bcl-2 inhibits hydrogen peroxide-induced ER Ca2+ pool depletion. *Oncogene* 1996, **12:** 2051–2055.
179. Hockenbery DM, Oltvai ZN, Yin X-M, Milliman CL, Korsmeyer SJ. Bcl-2 functions in an antioxidant pathway to prevent apoptosis. *Cell* 1993, **75:** 241–251.
180. Shimizu S, Eguchi Y, Kosaka H, Kamiike W, Matsuda H, Tsujimoto Y. Prevention of hypoxia-induced cell death by Bcl-2 and Bcl-xL. *Nature* 1995, **374:** 811–813.
181. Jacobson MD, Burne JF, King MP, Miyashita T, Reed JC, Raff MC. Bcl-2 blocks apoptosis in cells lacking mitochondrial DNA. *Nature* 1993, **361:** 365–369.
182. Liu X, Kin CN, Yang J, Jemmerson R, Wang X. Induction of apoptotic program in cell-free extracts: requirement for dATP and cytochrome c. *Cell* 1996, **86:** 147–157.
183. Yang J, Liu X, Bhalla K, Kim CN, Ibrado AM, Cai J, Peng T-I, Jones DP, Wang X. Prevention of apoptosis by Bcl-2: release of cytocrome c from mitochondria blocked. *Science* 1997, **275:** 1129–1132.
184. Kluck RM, Bossy-Wetzel E, Green DR, Newmeyer DD. The release of cytochrome c from mitochondria: a primary site for Bcl-2 regulation of apoptosis. *Science* 1997, **275:** 1132–1136.
185. Zamzami N, Susin SA, Marchetti P, Hirsch T, Gomez-Monterrey I, Castedo M, Kroemer G. Mitochondrial control of nuclear apoptosis. *J Exp Med* 1996, **183:** 1533–1544.
186. Zamzami N, Marchetti P, Castedo M, Decaudin D, Macho A, Hirsch T, Susin SA, Petit PX, Mignotte B, Kroemer G. Sequential reduction of mitochondrial transmembrane potential and generation of reactive oxygen species in early programmed cell death. *J Exp Med* 1995, **182:** 367–377.
187. Minn AJ, Velez P, Schendel SL, Liang H, Muchmore SW, Fesik SW, Fill M, Thompson CB. Bcl-xL forms an ion channel in synthetic lipid membranes. *Nature* 1997, **385:** 353–357.
188. Duckett CS, Nava VE, Gedrich RW, Clem RJ, Van Dongen JL, Gilfillan MC, Shiels H, Hardwick JM, Thompson CB. A conserved family of apoptosis inhibitors related to the baculovirus *iap* gene. *EMBO J* 1996, **15:** 2685–2694.
189. Levine B, Goldman JE, Jiang HH, Griffin DE, Hardwick JM. Bcl-2 protects mice against fatal alphavirus encephalitis. *Proc Natl Acad Sci USA* 1996, **93:** 4810–4815.
190. Tyor WR, Wesselingh SL, Levine B, Griffin DE. Longterm intraparenchymal immunoglobulin secretion after acute viral encephalitis in mice. *J Immunol* 1992, **149:** 4016–4020.
191. Levine B, Griffin DE. Persistence of viral RNA in mouse brains after recovery from acute alphavirus encephalitis. *J Virol* 1992, **66:** 6429–6435.
192. Despres P, Griffin JW, Griffin DE. Effects of anti-E2 monoclonal antibody on Sindbis virus replication in AT3 cells expressing *bcl-2*. *J Virol* 1995, **69:** 7006–7014.
193. Despres P, Griffin JW, Griffin DE. Antiviral activity of alpha interferon in Sindbis virus-infected cells is restored by anti-E2 monoclonal antibody treatment. *J Virol* 1995, **69:** 7345–7348.
194. Gorrell MD, Lemm JA, Rice CM, Griffin DE. Immunization with nonstructural proteins promotes functional recovery of alphavirus-infected neurons. *J Virol* 1997, **71:** 3415–3419.
195. Tucker PC, Griffin DE, Choi S, Bui N, Wesselingh S. Inhibition of nitric oxide synthesis increases mortality in Sindbis virus encephalitis. *J Virol* 1996, **70:** 3972–3977.
196. Hardwick JM. Virus-induced apoptosis. *Advan Pharmacol* 1997, (in press).
197. Clem RJ, Hardwick JM, Miller LK. Anti-apoptotic genes of baculoviruses. *Cell Death Different* 1996, **3:** 9–16.

198. Shen Y, Shenk TE. Viruses and apoptosis. *Curr Biol* 1995, **5**: 105–111.
199. White E. Function of the adenovirus E1B oncogene in infected and transformed cells. *Sem Virol* 1994, **5**: 341–348.
200. Chou J, Roizman B. The gamma-1-34.5 gene of herpes simplex virus 1 precludes neuroblastoma cells from triggering total shutoff of protein synthesis characteristic of programed cell death in neuronal cells. *Proc Natl Acad Sci USA* 1992, **89**: 3266–3270.
201. Koyama AH, Miwa Y. Suppression of apoptotic DNA fragmentation in herpes simplex virus type-1-infected cells. *J Virol* 1997, **71**: 2567–2571.
202. Leopardi R, Roizman B. The herpes simplex virus major regulatory protein ICP4 blocks apoptosis induced by the virus or by hyperthermia. *Proc Natl Acad Sci USA* 1996, **93**: 9583–9587.
203. Stevens JG. Human herpesviruses: a consideration of the latent state. *Microbiol Rev* 1989, **53**: 318–332.
204. Devi-Rao GB, Goodart SA, Hecht LM, Rochford R, Rice MK, Wagner EK. Relationship between polyadenylated and nonpolyadenylated herpes simplex virus type 1 latency-associated transcripts. *J Virol* 1991, **65**: 2179–2190.
205. Wu T-T, Su Y-H, Block TM, Taylor JM. Evidence that two latency-associated transcripts of herpes simplex virus type 1 are nonlinear. *J Virol* 1996, **70**: 5962–5967.
206. Perng G-C, Ghiasi H, Slanina SM, Nesburn AB, Wechsler SL. The spontaneous reactivation function of the herpes simplex virus type 1 LAT gene resides completely within the first 1.5 kilobases of the 8.3-kilobase primary transcript. *J Virol* 1996, **70**: 976–984.
207. Bloom DC, Hill JM, Devi-Rao G, Wagner EK, Feldman LT, Stevens JG. A 348-base-pair region in the latency-associated transcript facilitates herpes simplex virus type 1 reactivation. *J Virol* 1996, **70**: 2449–2459.
208. Hossain A, Schang LM, Jones C. Identification of gene products encoded by the latency-related gene of bovine herpesvirus 1. *J Virol* 1995, **69**: 5345–5352.
209. Schang LM, Hossain A, Jones C. The latency-related gene of bovine herpesvirus 1 encodes a product which inhibits cell cycle progression. *J Virol* 1996, **70**: 3807–3814.

15
Mechanisms of Cell Injury in Prion Diseases

David R. Borchelt

INTRODUCTION

Transmissible spongiform encephalopathies (TSE) are associated with the progressive destruction of neural systems involved in cognition and movement. Gene targeting and transgenic strategies have provided strong evidence for the central role of the prion protein in the transmission and pathogenesis of TSEs, now termed prion diseases, and have begun to be used to examine the mechanisms by which the formation of conformationally altered prion proteins cause cell dysfunction and death.

Prion diseases present with a great variety of neurological syndromes, ranging from dementia and insomnia to ataxia. Brain pathology usually includes spongiform degeneration and gliosis and, less frequently, amyloid deposits composed of prion protein (PrP). All prion diseases exhibit at least one of the following characteristics: missense mutations in the PrP gene (inherited disease), accumulation of protease resistant PrP, and/or transmissibility to laboratory hosts.

In humans, prion diseases include Creutzfeldt-Jakob disease (CJD), Gerstmann-Straussler-Scheinker disease (GSS), fatal familial insomnia (FFI), and kuru. Zoonotic prion diseases occur in sheep (scrapie), mink, and mule deer. Over the last few years, the disease has appeared in cattle in the United Kingdom (bovine spongiform encephalopathy [BSE]) and there has been increased concern that the consumption of products from affected cattle have led to an increase in the incidence of human disease in that country *(1–3)*. In contrast to the above concern with BSE, scrapie (ovine prion disease) has been endemic in sheep for centuries, yet there have never been clear examples of humans developing disease as a result of consuming ovine products.

CJD, the most common form of human prion disease, occurs sporadically in the population. Less frequently, CJD occurs in families as a dominantly inherited disease. Both sporadic and familial CJD can be transmitted to nonhuman primates and/or laboratory rodents. There have also been rare documented cases of accidental transmission of CJD through contaminated human growth hormone preparations (before the advent of recombinant hormone) and tissue transplantation. GSS and FFI appear to be exclusively familial, dominantly inherited, prion diseases. Kuru, transmitted among the Fore tribes of New Guinea (via ritualistic cannibalism), was the first human TSE to be experimentally transmitted *(4)* and has disappeared as a result of the cessation of cannibalistic practices.

WHAT ARE PRIONS?

The term "prion" was first applied to partially purified preparations of infectious material from the brains of laboratory hamsters infected with ovine scrapie. These infectious preparations contained little or no nucleic acid, and moreover, infectivity was undiminished by treatments that destroy nucleic acids. Importantly, treatments that damage (including strong denaturants) or destroy proteins diminished infectivity *(5)*. The most abundant protein in these preparations was a protease-resistant 27–30 kDa molecule, termed PrP 27-30 *(6,7)*, which could not be separated from infectivity. Additional studies indicated that PrP27-30 is generated during the purification process by partial proteolysis of a 33–35 kDa precursor protein, termed the scrapie prion protein (PrP^{Sc}) *(8)*. Peptide sequence analysis of purified PrP27-30 led to the identification and molecular cloning of the PrP gene and the realization that PrP is encoded by genes that are normal constituents of the mammalian genome *(8,9)*. In the normal Syrian hamster brain, the product of the PrP gene (termed cellular prion protein or PrP^C) is degraded by relatively low concentrations of protease. Thus, initially, the prion-associated form of the PrP (PrP^{Sc}) was defined by protease resistance, whereas the normal form of PrP (PrP^C) was defined by protease sensitivity. A large body of data, some of which is described in the succeeding sections of this chapter, suggest that PrP^{Sc} may be the only component of prions.

PRION PROTEINS PLAY A CENTRAL ROLE IN THE PROPAGATION AND PATHOGENESIS OF PRION DISEASE

The function of PrP^C and the role of PrP^C in scrapie propagation has, until recently, been enigmatic and it continues to be controversial. Several groups have used gene targeting strategies to generate mice with disrupted, nonfunctional, prion protein genes. Elegant studies by Büeler and colleagues convincingly demonstrated that mice devoid of PrP^C (PrP0/0 or PrP knockout mice) do not succumb to prion infection and, more importantly, do not propagate infectivity (Fig. 1) *(10)*. These data firmly establish that the host susceptibility to scrapie infection and the propagation of scrapie infectivity requires the expression of PrP^C.

Although highly conserved, mammalian prion protein genes show some sequence diversity. This diversity apparently underlies a phenomenon termed the species barrier, which is the observation that prions isolated from a specific host (e.g., Syrian hamster) often show low infectivity in hosts of a different species (e.g., mouse). Interestingly, in cases in which Syrian hamster prions do infect a mouse, the sick mice generate prions that are infectious only in mice and the prions obtained from these mice contain mouse PrPSc.

Several complementary studies in transgenic mice have established that amino acid sequence differences in PrP^C underlie the species barrier, and that the prion proteins play a critical role in disease transmission and pathogenesis. For example, transgenic mice expressing Syrian hamster prion proteins (but not nontransgenic littermates) are susceptible to prions isolated from Syrian hamsters (Fig. 2) *(11)*. When infected, these transgenic mice produce Syrian hamster PrP^{Sc} and prions that are infectious to hamsters but not mice. When these transgenic animals are infected with mouse prions, only mouse PrP^{Sc} is produced, as well as prions that can infect only mice (Fig. 3). Thus, Syrian hamster prions, containing Syrian hamster PrP^{Sc}, interact only with Syrian ham-

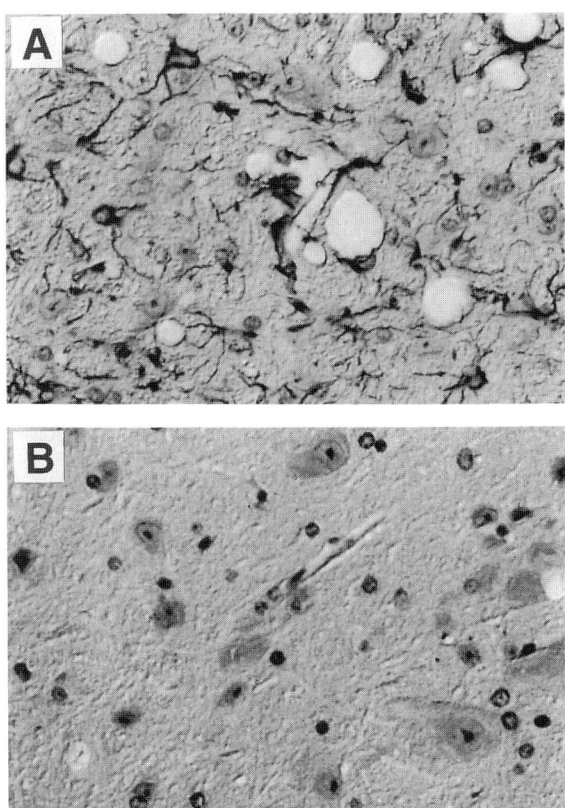

Fig. 1. Mice with targeted deletion of the PrP gene are not susceptible to prion infectivity. This figure shows the degree of neurodegeneration in a wild-type mouse (**A**) and PrP knockout mouse (**B**) inoculated with mouse scrapie prions. (Reprinted with permission from *ref. 10.*)

ster PrP^C to produce a Syrian hamster prion and vice versa. If the production of PrP^{Sc} were simply a by-product of prion infection, then transgenic animals expressing Syrian hamster PrP^C would have produced both hamster and mouse PrP^{Sc}. Moreover, if Syrian hamster PrP^C was merely serving as a receptor for the infecting prions, then transgenic animals would have produced mouse prions, regardless of the origin of the infecting agent.

A more dramatic example of the role of PrP^C in host specificity involved transgenic mice expressing chimeric hamster/mouse PrP^C. Following inoculation of these animals with mouse prions, the chimeric prion protein generated a chimeric mouse/hamster PrP^{Sc} and prions that were infectious in both mice and Syrian hamsters (Fig. 4) *(12)*.

Collectively, these studies demonstrate that the host repertoire of cellular prion proteins determines susceptibility to infectious prions. Hosts with PrP^C molecules isologous to the PrP^{Sc} in the infecting prion are more susceptible than hosts with heterologous PrP^C genotypes. Moreover, the genotype of the host PrP^C molecule encrypts the host specificity of the newly synthesized prion, which now contains PrP^{Sc} derived from newly converted host PrP^C.

HOW DO PRIONS REPLICATE?

The simplest interpretation of the biochemical and transgenic studies described in the previous section is that direct interactions between the PrP^{Sc} in the infecting prions

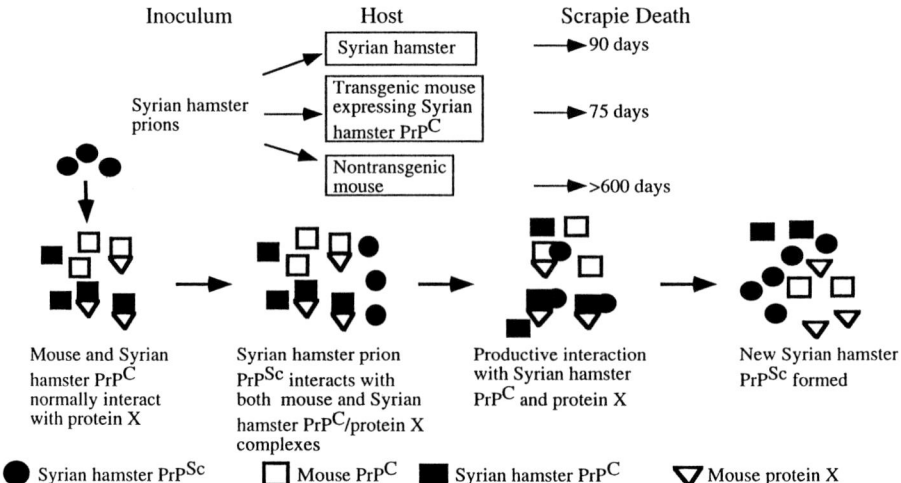

Fig. 2. Transgenic mice expressing Syrian hamster PrP are susceptible to Syrian hamster scrapie prions. Mice harboring a cosmid clone of the Syrian hamster prion protein gene develop neurologic illness, resembling rodent scrapie, and die by 75 d of age, whereas only a subset of nontransgenic mice develop disease after a very protracted incubation period. The molecular basis for these results appears to involve direct interactions between Syrian hamster PrP^{Sc} and the Syrian hamster transgene product (PrP^{C}). The conversion of PrP^{C} into PrP^{Sc} appears to involve a second cellular protein termed protein X *(31)*. (Data on Syrian hamster PrP transgenics (line Tg81) taken from *ref. 11.*)

and host PrP^{C} catalyze the formation of new PrP^{Sc}. Investigations of PrP^{Sc} biosynthesis in scrapie-infected neuroblastoma cells *(13–16)* indicated that the distinctive properties of PrP^{Sc} (protease resistance, detergent insolubility, and fibrillar organization into amyloid) are acquired post-translationally. However, biochemical studies of PrP^{C} and PrP^{Sc} composition suggested that these two isoforms of PrP do not significantly differ in the type or extent of post-translational modification; both PrP^{C} and PrP^{Sc} are sialoglycoproteins that possess C-terminal glycolipid anchor modifications *(17)*.

The biophysical characterization of PrP^{C} and PrP^{Sc} structures revealed that a greater percentage of PrP^{C} is folded into α-helical structure than PrP^{Sc} *(18)*. To date, the conformational changes that occur as PrP^{C} is converted into PrP^{Sc} are not fully understood. Computational studies predict that PrP^{C} contains four α-helical domains *(19)*. However, NMR spectroscopy of bacterially produced recombinant PrP polypeptides, encompassing residues 90–231 and 121–231, suggested that only three α-helical domains are present *(20–22)*. Unfortunately, at present, these investigations of PrP^{C} structure are hampered by the lack of a biological assay for functionally folded recombinant PrP^{C}. This limitation might be obviated, however, if it becomes possible to convert, in vitro, recombinant PrP^{C} into an infectious PrPSc-like molecule.

When first suggested, the notion that an infectious disease could be propagated via a specific protein conformation was a novel concept. However, it is now clear that similar phenomena occur in yeast. Two phenotypes of yeast, termed psi+ and URE3, are inherited in a nonmendelian fashion and can be transmitted to naive yeast via cytoplasmic mixing without nuclear fusion. Deletion of the genes encoding the polypeptides linked to psi+ (Sup35p) and URE3 (Ure2p) eliminates the ability of the yeast to propa-

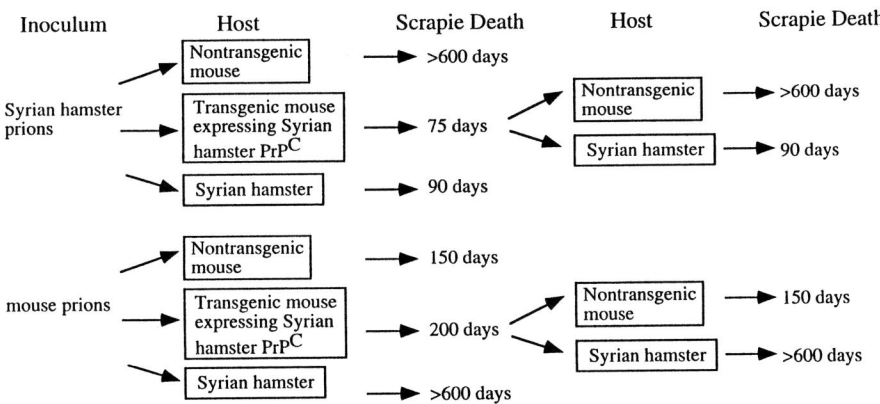

Fig. 3. Transmission of rodent scrapie in transgenic mice expressing Syrian hamster PrP^C demonstrates that PrP^{Sc} in the inoculum interacts with isologous host PrP^C in the propagation of prions and the pathogenesis of disease. (The figure is based on data taken from *ref. 11*.)

gate the psi+ and URE3 phenotypes *(23–26)*. In normal yeast, Ure2p is a soluble cytoplasmic protein, but in URE3 yeast, the protein is aggregated and relatively protease resistant *(27)*. Similarly, in yeast exhibiting the psi+ phenotype, Sup35p polypeptides aggregate and become protease resistant, whereas in normal yeast, Sup35p is a soluble cytoplasmic polypeptide, homologous to polypeptide chain releasing factor 3 *(28,29)*. Moreover, Sup35p-psi+ polypeptides interact with and conformationally alter the structure of newly synthesized Sup35p proteins *(29)*. Interestingly, the propagation of psi+ depends on the expression of the chaperone protein HSP104. Overexpression or inactivation of HSP104 can cure the psi+ phenotype, suggesting that protein refolding is an important event in the propagation of the psi+ phenotype *(28–30)*. Notably, the propagation of PrP^{Sc} in scrapie-infected mouse neuroblastoma N2a cells is disrupted by chemical chaperones *(16)*. Thus, the phenomenon of propagating protein conformations is not restricted to mammalian prions but can occur in highly divergent polypeptides.

Because of the many parallels between the propagation of the psi+ and URE3 phenotypes in yeast and prions in mammals, the observation that the propagation of psi+ requires the activity of HSP104 is noteworthy. Studies in transgenic mice harboring human PrP genes encoding a mutation (P102L) linked to GSS have suggested that other proteins may participate in the generation of PrP^{Sc}. Mice expressing human PrP-P102L do not develop disease, whereas chimeric human/mouse PrP genes encoding the same mutation cause neurologic disease. These data were interpreted to suggest that the chimeric molecule carries information necessary for association with other murine cellular proteins (protein X), which provide critical functions in generating PrP^{Sc} and prions (Fig. 5) *(31)*. Similarly, when the murine PrP-P102L gene is expressed on the background of wild-type endogenous PrP, disease develops between 150 and 300 d of age *(32)*. However, when the PrP-P102L gene is expressed on the background of PrP0/0 mice, then the onset of disease is highly synchronous, occurring at 145 d of age *(33)*. Collectively, these data argue that PrP^C normally associates with another cellular protein, which plays a role in the generation of the prion conformation.

In summary, data from transgenic mice (and yeast) argue that the conformation of PrP^C is irreversibly altered via a direct interaction with PrP^{Sc}. This process appears to

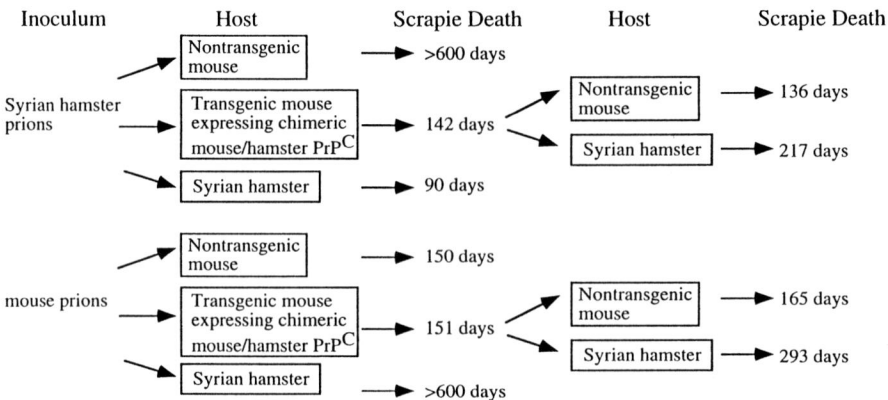

Fig. 4. Transmission of rodent scrapie in transgenic mice expressing a chimeric mouse/hamster PrP^C demonstrates that the host specificity of newly synthesized prions is dictated by the PrP genotype of the host. Mice expressing the chimeric mouse/hamster PrP^C can generate prions that are infectious to both hamsters and mice. (The figure is based on data taken from *ref. 12.*)

be essential to prion replication. It seems highly likely that other cellular proteins also participate in the formation of PrP^{Sc}, although it has been possible to demonstrate the conversion (albeit inefficient) of immunopurified PrP^C into a protease-resistant PrP^{Sc}-like molecule in cell-free systems containing essentially only PrP^C and PrP^{Sc} *(34–36)*. The biophysics of the conversion process remain ill-defined and will likely not be resolved until investigators are able to synthetically manufacture a PrP^C template, catalyze its conversion into a PrP^{Sc}-like protein in vitro, and demonstrate its infectivity.

TRANSGENIC ANIMALS PROVIDE A MECHANISM TO ASSESS PRION INFECTIVITY

Transgenic mice expressing wild-type human prion proteins in the presence or absence of endogenous mouse PrP^C have provided a tool to investigate the replication of human prions. Two separate studies, using essentially the same transgenic mice expressing human PrP^C and the same PrP0/0 mice have demonstrated that the expression of human PrP^C in mice increases susceptibility to human prions *(37,38)*. Notably, however, in one of these studies, only transgenic mice expressing high levels of human PrP^C (eight times the endogenous levels of mouse PrP^C) in the absence of endogenous mouse PrP^C were good hosts for human CJD and GSS *(37)*. In contrast, in the parallel study, deletion of the endogenous mouse PrP gene increased susceptibility to human prions, but was not a prerequisite for infection.

To examine the potential for BSE transmission to human populations, transgenic mice expressing human PrP^C in the presence or absence of endogenous mouse PrP^C were challenged with homogenates of brains from cattle dying of BSE *(38)*. Mice expressing both human and mouse PrP^C showed no enhanced susceptibility to BSE compared to nontransgenic controls, and mice expressing a human PrP transgene in PrP0/0 mice were free of disease up to 264 d of age. These data argue that the wave of fear for BSE transmission to humans may not be warranted.

Because of the diverse clinical presentation of prion diseases, and the diverse neuropathological findings, prion diseases can be difficult to diagnose *(39)*. It is likely that

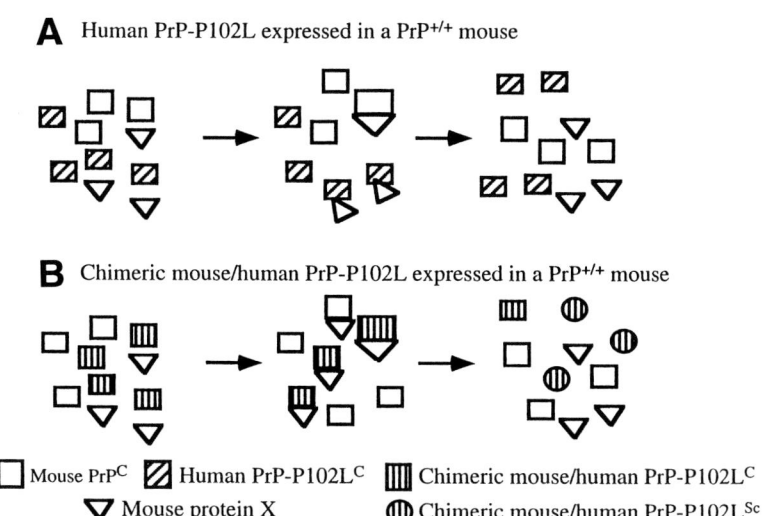

Fig. 5. The difference in the ability of human PrP-P102L (**A**) and a chimeric mouse/human PrP-P102L (**B**) to induce disease in transgenic mice appears to be due to the requirement of a second cellular protein (protein X) to catalyze the conversion of PrP-P102L^C into a PrP-P102L^{Sc}-like molecule. (The figure is based on data taken from *ref. 31*.)

some cases of prion diseases go undiagnosed because of an absence of hallmark features (i.e., PrP amyloid, spongiform pathology, protease-resistant PrP^{Sc}). Because human biological products are still used in the treatment of various diseases, these occult prion sources can unwittingly infect patients via transplantation, blood products, or endocrine preparations. For example, before the advent of recombinant growth hormone, a number of cases of early-onset CJD could be traced to the administration of growth hormone prepared from cadavers *(39)*. The development of transgenic mice that are readily susceptible to human prions may provide a means to identify prion diseases with unusual presentations and to monitor the safety of tissues and biological preparations derived from humans. A similar strategy may be employed to monitor cattle and beef products consumed by humans.

GENETICS OF FAMILIAL PRION DISEASES

Although the vast majority of cases of prion disease occur sporadically in the population (e.g., CJD), a subset of cases are inherited in an autosomal dominant fashion. To date, more than 20 different missense mutations in PrP (Fig. 6) have been linked to familial disease (for a review, *see* ref. *[40]*). Although the majority of familial cases present with clinical signs that are similar to classic sporadic CJD (rapidly progressing dementia being the most prominent clinical feature) a few cases show novel phenotypes, sufficiently novel to have been given distinctive names (for reviews, *see* refs. *[39 and 41]*). For example, GSS, caused by a mutation of codon 102 (Pro to Leu), usually presents initially with ataxia, followed by a protracted clinical course that eventually leads to dementia and death. In GSS, amyloid deposits composed of PrP are usually observed, but in other forms of familial prion disease, these structures occur far less frequently *(39)*.

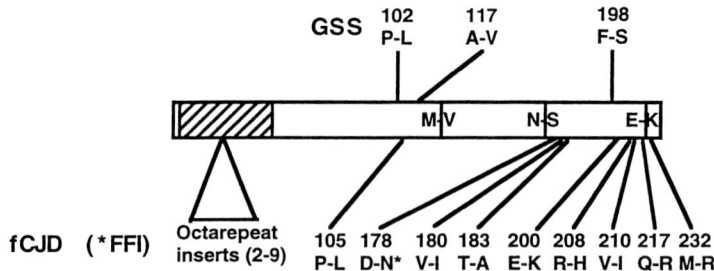

Fig. 6. Mutations in PrP that are linked to familial prion disease. Mutations linked to GSS are shown on top of the box depicting the PrP open reading frame, whereas mutations linked to familial CJD are shown underneath. Polymorphic amino acid changes are shown inside the box. The cross-hatched area represents the N-terminal segment of PrP, which contains polymorphic octarepeat motifs. (Data on familial mutations are taken from *ref. 40.*)

In addition to the mutations linked to disease, there are a number of polymorphic variants that appear to influence disease phenotype (Fig. 6). Among these polymorphisms, the variation of codon 129 (Met to Val) is the most common. A striking example of the influence of this variant on phenotype is found in individuals with FFI. Mutations at codon 178 (Asp to Asn) on PrP alleles with Met at codon 129 cause FFI, whereas the same mutation at codon 178 on alleles with Val at codon 129 cause familial CJD *(42)*. Similarly, classical GSS is associated with mutations at codon 102 on an allele with Met at codon 129, whereas a new variant of GSS has been described that is associated with a codon 102 mutation on an allele with Val at codon 129 *(43)*.

The polymorphism at codon 129 has also been shown to modify both susceptibility and clinical presentation of sporadic CJD. Homozygosity for methionine at codon 129 appears to increase the risk of developing sporadic CJD by 3.4-fold *(44)*. Parameters such as frequency of PrP-immunoreactive plaques, PrP^{Sc} accumulation, spongiosis, and astrogliosis are also influenced, in part, by the polymorphism at codon 129 *(45–48)*.

Thus, in the familial prion diseases, diverse neurologic manifestations result from disease-linked mutations in PrP under the modulatory influence of polymorphic variations in PrP sequence.

TRANSGENIC MODELS OF INHERITED PRION DISEASES

Initial efforts to model familial human prion diseases in transgenic mice were considerate of the species barrier and opted to introduce mutations linked to GSS (codon 102 Pro to Leu) into murine prion protein genes. Mice expressing a murine prion protein encoding the ProLeu substitution at codon 102 (at levels four times higher than endogenous PrP^C) developed neurologic illness with pathological changes similar to the human disease *(32)*. More importantly, nervous tissues from these animals harbored prions that were infectious to other transgenic mice expressing low levels of the mutant mouse PrP transgene (these latter mice did not spontaneously develop disease) *(49)*. Nontransgenic mice, however, were not susceptible to these mutant prions. These data strongly suggest that the mutant prions behaved essentially as distinct species, infecting preferentially animals harboring PrP molecules of isologous amino acid sequence (i.e., mutant mouse PrP).

Thus, in a controlled setting, the expression of mutated prion proteins caused neurologic disease that could be transmitted to other animals of the same genotype. These data provide the strongest possible evidence that mutations in PrP increase the propensity of PrP to adopt a PrP^{Sc}-like conformation, and that the dissemination of this conformation via protein:protein interactions propagates disease.

Data from transgenic mice expressing mutant PrP have also demonstrated that PrP^{Sc} can adopt conformations that do not impart high resistance to proteolytic degradation. Originally, PrP^{Sc} was defined as that form of PrP associated with scrapie infectivity *(5)*. In experimentally infected hamsters, PrP^{Sc} was biochemically defined by resistance to protease digestion, aggregation in the presence of detergents, and the propensity to form amyloid fibrils upon protease digestion in the presence of detergents *(8,50)*. However, in mice expressing mutant murine PrP-P102L, prions are produced in the absence of demonstrable protease-resistant PrP *(32)*. Likewise, mice expressing chimeric human/mouse PrP-P102L and dying of neurologic disease generate infectious prions in the absence of protease-resistant PrP^{Sc} *(31)*. These data indicate that the property of protease resistance is not a prerequisite for the adoption of a PrP^{Sc} conformation. Rather, the defining feature of the PrP^{Sc} conformation is its propensity to propagate via self-association and cause neurologic illness.

MECHANISMS OF CELL INJURY

The mechanisms by which the accumulation of PrP^{Sc} damages cells and leads to clinical disease are not well defined. PrP^{Sc} may be directly toxic to specific populations of neurons. In culture, synthetic fragments of PrP (amino acids 106–126) are toxic to hippocampal neurons *(51)*. These peptides include a central hydrophobic domain and, like Alzheimer amyloid peptides, self-assemble into filaments that resemble the amyloid fibrils. Interestingly, cells derived from PrP0/0 mice appear to be resistant to these toxic PrP peptides *(52)*. However, deposits of amyloid are not an essential component of prion diseases; a significant percentage of CJD cases lack amyloid deposits *(39)*. Moreover, the relevance of the culture studies with synthetic PrP fragments is uncertain, because most of the accumulating PrP^{Sc} in the brains of patients with prion diseases appears to encompass the full-length polypeptide.

In studies of cultured cells infected with prions and producing PrP^{Sc}, investigators have reported disturbances in bradykinin-receptor-mediated Ca^{2+} responses *(53)*. More recently, these disturbances have been attributed to diminished membrane fluidity, which leads to diminutions in receptor affinity for the ligand *(54)*. Whether these phenotypes result directly from an activity of PrP^{Sc}, however, is presently unknown.

Additional insight into the issue of PrP^{Sc} and amyloid toxicity has come from innovative studies in which normal murine brain tissue was grafted into the brains of PrP0/0 animals prior to inoculation with prions *(55)*. Despite the presence of PrP^{Sc} and amyloid deposits outside the graft, spongiform degeneration and cell death was restricted to the graft area, a pattern suggesting that extrinsically deposited PrP^{Sc} is not toxic.

One potential mechanism of injury may involve a loss of functional PrP^{C} as the host protein is converted into a "nonfunctional" PrP^{Sc}. Notably, in yeast, the psi+ and URE3 phenotypes appear to result from the loss of functional Sup35p and Ure2p, respectively. However, mice with targeted deletion of the endogenous PrP gene develop normally and show none of the neuropathological phenotypes associated with prion disease up to two years of age *(10)*. Thus, it was initially thought that the behavioral/neuropathological abnormalities that occur in the prion diseases do not arise from acute reductions of functional PrP^C.

Counter to this notion are studies of PrP null animals reporting that targeted deletion of the entire open reading frame of PrP caused Purkinje cell degeneration, leading to ataxia at approx 70 wk of age *(56)*. Whether these unusual phenotypes can be reversed by the interbreeding of PrP transgenes has not been evaluated; thus, it is uncertain whether the loss of Purkinje cells is due solely to the lack of PrP^C.

Although the dramatic phenotype described above appears to be unique to one line of PrP null mice, a study of two independently derived lines of PrP0/0 mice *(57)* reported altered circadian rhythms and sleep disturbances *(58)*. Importantly, these abnormalities could be reversed by the interbreeding of murine PrP transgenes. A subsequent study of sleep patterns in one of the above lines of PrP null mice found evidence of increased waking periods and an altered rate of transition between sleep and waking periods *(59)*. Thus, it is possible that a portion of the phenotype seen in prion disease could be due to the acute loss of functional PrP^C.

Additional studies of PrP0/0 animals that otherwise appear normal have revealed defects in synaptic inhibition and long-term potentiation *(60)*, abnormalities that could be reversed by breeding these animals to transgenic mice expressing human PrP *(61)*. Although these studies strongly suggested that the observed abnormalities in PrP0/0 mice were the result of the loss of normal PrP^C, a subsequent study, using the same PrP0/0 animals, could not confirm the original findings *(62)* and the role of PrP^C in synaptic transmission remains somewhat controversial.

Although the phenotypes of the PrP0/0 mice partially overlap with one or more symptoms of human prion disease (such as alterations in sleep patterns), the loss of functional PrP^C does not account for all the clinical and neuropathological phenotypes of prion diseases. PrP0/0 mice do not exhibit pathology typical of prion diseases (i.e., gliosis, spongiform degeneration). The inoculation of transgenic mice expressing Syrian hamster PrP^C causes spongiform degeneration in the host, but only Syrian hamster PrP^C is converted to PrPs, leaving the cellular pool of functional mouse PrP^C intact *(11,63)*. When PrP^C is overexpressed in transgenic animals, the incubation time to illness after inoculation with prions is shortened *(11,63)*. Thus, although the phenotypes of PrP0/0 mice imitate a subset of the clinical features of human prion diseases, the weight of evidence indicates that loss of functional PrP^C does not cause the entire spectrum of disease phenotypes.

Studies of disease progression in hamsters and mice have demonstrated that clinical disease does not manifest until levels of PrP^{Sc} accumulate to near peak levels. Interestingly, studies in transgenic mice expressing high levels of Syrian hamster PrP have demonstrated that the time of onset of disease correlates with the level of expression (higher expression leads to earlier onset) *(11,63)*. However, earlier onset does not appear to result from a proportionally higher level of PrP^{Sc} accumulation *(11)*. In wild-type mice, the onset of clinical symptoms usually begins within weeks of maximal accumulation of PrP^{Sc} and prion infectivity. However, in mice harboring a single targeted PrP allele (PrP+/0), the onset of disease occurs many months after maximal accumulation of prion infectivity and PrP^{Sc}. Intriguingly, the maximal levels of prion titer and PrP^{Sc} in PrP+/0 mice are not markedly lower than those in wild-type mice *(64)*. From these data, it appears that the rate of PrP^{Sc} generation and degradation (note that the vast majority of PrP^{Sc} in prion inoculum is rapidly cleared *[10]*) reaches an equilibrium that prevents the perpetual increase in the levels of PrP^{Sc}, and that cell injury results not from a direct toxicity of PrP^{Sc} or the acute loss of PrP^C, but rather from the actual process of creating PrP^{Sc}. In this context, one possible mechanism of cell injury could involve a disruption of cellular chaperone activities, leading to increased levels of malfolded and inactive proteins. Immunocytochemical studies of CJD have disclosed increased ubiquitin immunoreactivity in brain regions that show PrP^{Sc} deposition and spongiosis *(65)*. Cultured cells producing PrP^{Sc} show abnormal distributions of heat shock proteins and respond aberrantly to stress stimuli *(66)*. Moreover, astroglia increase expression of stress response proteins after prion infection *(67)*. The requirement for heat shock chaperone activities in the propagation of yeast analogs of prions (psi and URE3) further implicates molecular chaperone activities in prion replication; it suggests that prion replication may injure cells by increasing the burden on cellular chaperone activities, thus causing increased levels of malfolded and inactive polypeptides.

GENERAL MECHANISMS OF PRION DISEASES

The processes that lead to clinical manifestations of prion disease and the eventual death of afflicted individuals are poorly understood. The degree of neuronal pathology can vary substantially among different cases of prion disease *(39)*. Neuronal abnormalities, when present, include spongiform changes, PrP^{Sc} accumulation, synaptic loss, and cell death *(68,69)*. In scrapie, and scrapie-infected mice, dying neurons show internucleosomal DNA fragmentation typical of apoptotic cell death *(70,71)*. However, whether neuronal death is an invariant feature of human prion diseases is controversial and synaptic loss appears to be observed more consistently than cell death *(68)*.

In addition to neuronal abnormalities, the disease is characterized by a strong glial reaction. PrP is expressed in both neurons and astroglia *(72)*, and electron microscopic examinations of PrP^{Sc}-immunostained preparations have demonstrated PrP^{Sc} immunoreactivity in both types of cells *(73)*. To examine the types of cells involved in prion replication and the pathogenesis of disease, transgenic mice expressing Syrian hamster PrP transgenes under the influence of promoter elements from neuron-specific enolase were challenged with Syrian hamster prions *(74)*. The mice with the neuronal expression of Syrian hamster PrP were susceptible to prions derived from hamsters, a result demonstrating that astroglial cell replication of prions does not play a substantial role in the propagation or pathogenesis of prion diseases.

Although it appears that pathogenic processes occurring in neurons underlie prion diseases, the temporal and spatial sequences of dysfunction at the level of neural systems are not well understood. In experimental models of prion disease, a number of factors (e.g., route of inoculation, strain of prion, and PrP genotype of the host) converge to modulate the onset, clinical presentation, and progression of disease. In general, direct exposure of the nervous system to prions results in a much more rapid onset of disease; intracerebellar inoculation is associated with the shortest incubation period of disease *(75)*. In the CNS, prions appear to spread along anatomical pathways. For example, inoculation into the eye leads to synaptic loss and neuronal death in the lat-

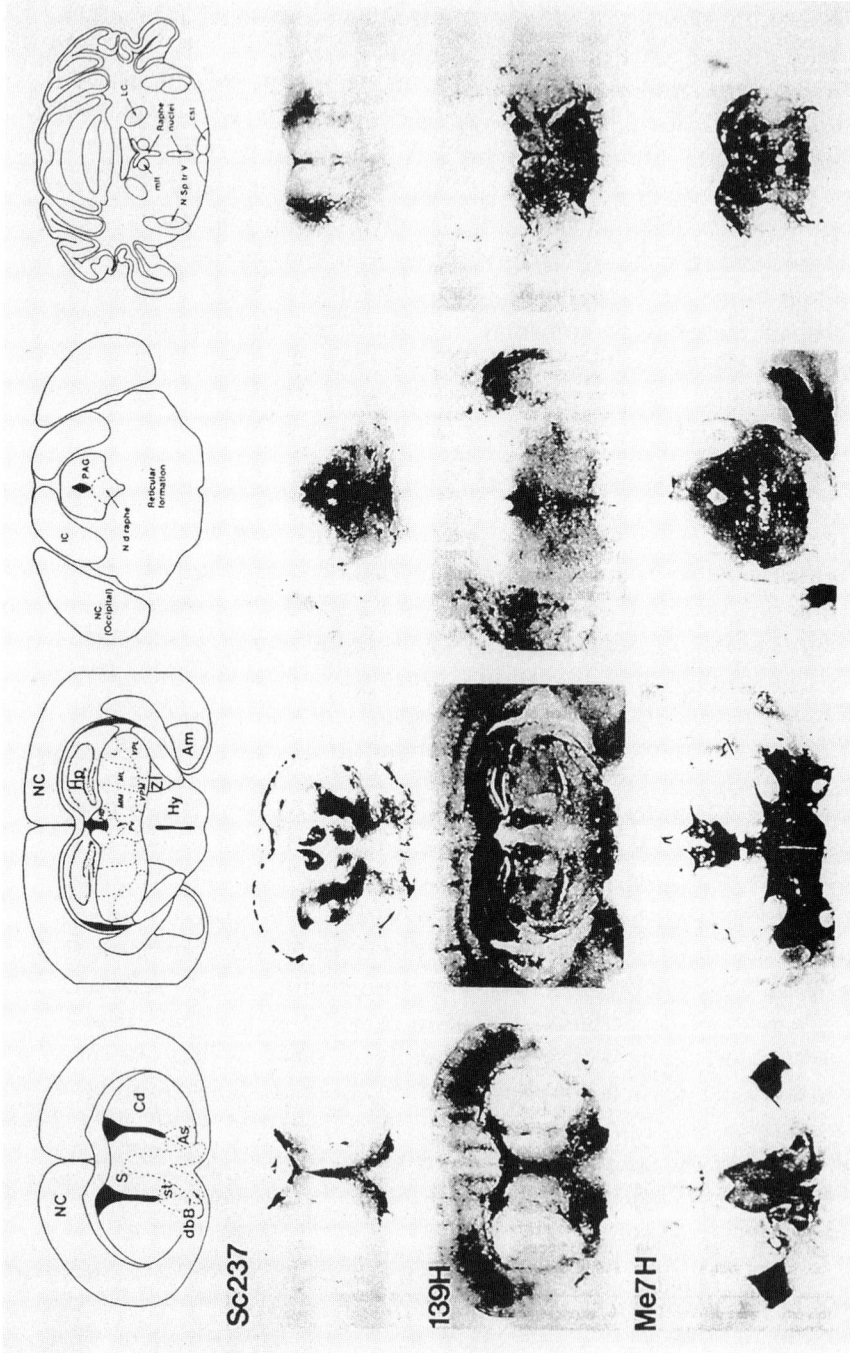

Fig. 7. The pattern of PrP^{Sc} deposition in transgenic mice expressing Syrian hamster PrP^C varies dramatically, depending upon the strain of prion inoculum. This figure displays histoblots of PrP^{Sc} (83) made from transgenic mice, expressing Syrian hamster PrP^C, that were infected with three different strains of Syrian hamster scrapie. PrP^{Sc} is specifically detected by limited proteolysis of the tissue (after imprinting on nitrocellulose membrane), followed by denaturation in guanidinium and immunostaining (83). Note that each strain displays a characteristic anatomical pattern of deposition. All animals were inoculated into the thalamus at weanling age (3–4 wk). (Used with permission from ref. 81.)

eral geniculate nucleus well before the appearance of other clinical symptoms *(76)*. However, the anterograde spread of prion infectivity along the optic nerve appears to be much slower (>100 d) than the rate of anterograde transport of PrP^C, which moves in the fast component *(77)*. The rate of axonal transport of PrP^{Sc} itself is not known. Because the anterograde transport of prions in the optic nerve requires the expression of PrP^C *(78)*, it is reasonable to assume that the propagation of prion infectivity from the optic nerve to neurons of the lateral geniculate requires the de novo synthesis of new prions and PrP^{Sc}. A similar mechanism for the propagation of prion infectivity has been suggested in other CNS systems *(79)*. In humans with sporadic CJD and in non-experimental zoonotic prion diseases, the rate of spread from one neural system to another undoubtedly influences the clinical presentation of disease. Presumably, these processes would play a less significant role in the dominantly-inherited familial prion diseases as the mutant PrP polypeptide is expressed in throughout the brain.

Investigations of the experimentally transmitted prion diseases have demonstrated that different experimentally derived isolates (or strains) of prions are associated with diverse clinical and pathological phenotypes (for a review, *see* Carp and Rubenstein *[80]*). For example, prions of different strain history, transmitted in a single host species over multiple generations, give rise to remarkably diverse patterns of PrP^{Sc} deposition (Fig. 7) *(81)*. These strain-specific patterns are heritable through multiple passages and thus appear to result from the encryption of strain-specific information in the prion. Notably, PrP^{Sc} molecules isolated from animals experimentally infected with different strains of prions show different conformations *(36,82)*. Thus, although alternative hypotheses have not been excluded, it appears that strain information is encrypted in the secondary structure of PrP^{Sc}.

The molecular basis for the selective accumulation of PrP^{Sc} in specific cell populations inoculated with different prion strains is not well understood. One possibility is that prion strains "home" to specific populations of neurons, which for some reason are more susceptible to a given strain. Alternatively, there may be variations in the abilities of neuronal populations to support the propagation of specific PrP^{Sc} conformations.

If strain information is encrypted in the secondary structure of PrP^{Sc}, then strain information is likely to play an important role in modulating the pathogenesis of both sporadic and familial human prion diseases. For example, each familial mutation could create a novel strain. In the case of sporadic CJD, the amino acid polymorphisms at codon 129 (and other positions) could alter the structure of the protein and create alternate strains. This notion is supported by a number of studies, demonstrating variations in PrP-immunoreactive plaque frequency, PrP^{Sc} accumulation, spongiform changes, and astrogliosis in patients harboring different codon 129 polymorphisms *(45–48)*. Thus, the diversity in the clinical and pathological features of human prion diseases, all of which appear to result from a conformational change in PrP, may be due, at least in part, to strain-related information carried in different PrP^{Sc} conformations.

CONCLUSION

The study of prion diseases, once a collection of enigmatic syndromes and disorders, has been revolutionized by the application of recombinant DNA technology and transgenesis. Biochemical, biophysical, and genetic studies all point to PrP^{Sc} as the central player in the pathogenesis of these diseases. Although it is not yet clear how the propagation of the "scrapie" conformation in PrP leads to cell dysfunction and the death of the organism, the molecular tools that are now available should hasten the resolution of these important questions.

ACKNOWLEDGMENTS

The author is grateful to Dr. Vassilis Koliatsos for critical review of this manuscript. He is supported by the Alzheimer's Association, the Devilbiss Fund, and the National Institutes of Health under grants AG0514 and NS2047.

REFERENCES

1. Collinge J, Sidle KCL, Meads J, Ironside J, Hill AF. Molecular analysis of prion strain variation and the aetiology of "new variant" CJD. *Nature* 1996, **383**: 685–690.
2. Aylin P, Rooney C, Drever F, Coleman M. Increasing mortality from Creutzfeldt-Jakob disease in England and Wales since 1979: ascertainment bias from increase in post-mortems? *Popul Trends* 1996, **85**: 34–38.
3. Nathanson N, Wilesmith J, Griot C. Bovine spongiform encephalopathy (BSE): Causes and consequences of a common source epidemic. *Am J Epidemiol* 1997, **145**: 959–969.
4. Gajdusek DC, Gibbs CJ, Jr., Alpers M. Experimental transmission of a kuru–like syndrome to chimpanzees. *Nature* 1966, **209**: 794–797.
5. Prusiner SB. Novel proteinaceous infectious particles cause scrapie. *Science* 1982, **216**: 136–144.
6. Prusiner SB, McKinley MP, Groth DF, Bowman KA, Mock NI, Cochran SP, Masiarz FR. Scrapie agent contains a hydrophobic protein. *Proc Natl Acad Sci USA* 1981, **78**: 6675–6679.
7. Prusiner SB, Groth DF, Bolton DC, Kent SB, Hood LE. Purification and structural studies of a major scrapie prion protein. *Cell* 1984, **38**: 127–134.
8. Oesch B, Westaway D, Wälchli M, McKinley MP, Kent SBH, Abersold R, Barry RA, Tempst P, Teplow DB, Hood LE, Prusiner SB, Weissmann C. A cellular gene encodes scrapie PrP 27-30 protein. *Cell* 1985, **40**: 735–746.
9. Basler K, Oesch B, Scott M, Westaway D, Wächli M, Groth DF, McKinley MP, Prusiner SB, Weissmann C. Scrapie and cellular PrP isoforms are encoded by the same chromosomal gene. *Cell* 1986, **46**: 417–428.
10. Büeler H, Aguzzi A, Sailer A, Greiner R-A, Autenried P, Aguet M, Weissmann C. Mice devoid of PrP are resistant to scrapie. *Cell* 1993, **73**: 1339–1347.
11. Scott M, Foster D, Mirenda C, Serban D, Coufal F, Wälchli M, Torchia M, Groth D, Carlson G, DeArmond SJ, Westaway D, Prusiner SB. Transgenic mice expressing hamster prion protein produce species-specific scrapie infectivity and amyloid plaques. *Cell* 1989, **59**: 847–857.
12. Scott M, Groth D, Foster D, Torchia M, Yang SL, DeArmond SJ, Prusiner SB. Propagation of prions with artificial properties in transgenic mice expressing chimeric PrP genes. *Cell* 1993, **73**: 979–988.
13. Borchelt DR, Scott M, Taraboulos A, Stahl N, Prusiner SB. Scrapie and cellular prion proteins differ in their kinetics of synthesis and topology in cultured cells. *J Cell Biol* 1990, **110**: 743–752.
14. Taraboulos A, Raeber AJ, Borchelt DR, Serban D, Prusiner SB. Synthesis and trafficking of prion proteins in cultured cells. *Mol Biol Cell* 1992, **3**: 851–863.
15. Caughey B, Raymond GJ. The scrapie-associated form of PrP is made from a cell surface precursor that is both protease- and phospholipase-sensitive. *J Biol Chem* 1991, **266**: 18217–18223.
16. Tatzelt J, Prusiner SB, Welch WJ. Chemical chaperones interfere with the formation of scrapie prion protein. *EMBO J* 1996, **15**: 6363–6373.

17. Stahl N, Baldwin MA, Teplow DB, Hood L, Gibson BW, Burlingame AL, Prusiner SB. Structural studies of the scrapie prion protein using mass spectrometry and amino acid sequencing. *Biochemistry* 1993, **32**: 1993–2002.
18. Pan KM, Baldwin M, Nguyen J, Gasset M, Serban A, Groth D, Mehlhorn I, Huang Z, Fletterick RJ, Cohen FE, Prusiner SB. Conversion of {\symbol}-helices into {\symbol}-sheets features in the formation of the scrapie prion proteins. *Proc Natl Acad Sci USA* 1993, **90**: 10,962–10,966.
19. Huang Z, Gabriel JM, Baldwin MA, Fletterick RJ, Prusiner SB, Cohen FE. Proposed three-dimensional structure for the cellular prion protein. *Proc Natl Acad Sci USA* 1994, **91**: 7139–7143.
20. Riek R, Hornemann S, Wider G, Billeter M, Glockshuber R, Wüthrich K. NMR structure of the mouse prion protein domain PrP(121-231). *Nature* 1996, **382**: 180–182.
21. Hornemann S, Glockshuben R, Autonomous and reversible folding of soluble amino-terminally truncated segment of the mouse prion protein. *J Mol Biol* 1996, **261**: 614–619.
22. Mehlhorn I, Groth D, Stockel J, Moffat B, Reilly D., Yansura D, Willett WS, Baldwin M, Fletterick R, Cohen FE, Vandlen R, Henner D, Prusiner SB. High-level expression and characterization of a purified 142-residue polypeptide of the prion protein. *Biochemistry* 1996, **35**: 5528–5537.
23. Ter-Avanesyan MD, Dagkesamanskaya AR, Kushnirov VV, Smirnov VN. The SUP35 omnipotent suppressor gene is involved in the maintenance of the non-Mendelian determinant [psi+] in the yeast saccharomyces cerevisiae. *Genetics* 1994, **137**: 671–676.
24. Wickner RB. [URE3] as an altered URE2 protein: evidence for a prion analog in saccharomyces cerevisiae. *Science* 1994, **264**: 566–569.
25. Wickner RB, Masison DC, Edskes HK. [PSI] and [URE3] as yeast prions. *Yeast* 1995, **11**: 1671–1685.
26. Lindquist S. Mad cows meet psi-chotic yeast: The expansion of the prion hypothesis. *Cell* 1997, **89**: 495–498.
27. Masison DC, Wickner RB. Prion-inducing domain of yeast Ure2p and protease resistance of Ure2p in prion-containing cells. *Science* 1995, **270**: 93–95.
28. Paushkin SV, Kushnirov VV, Smirnov VN, Ter-Avanesyan MD. Propagation of the yeast prion-like [psi+] determinant is mediated by oligomerization of the SUP35-encoded polypeptide chain release factor. *EMBO J* 1996, **15**: 3127–3134.
29. Patino MM, Liu JJ, Glover JR, Lindquist S. Support for the prion hypothesis for inheritance of a phenotypic trait in yeast. *Science* 1996, **273**: 622–626.
30. Chernoff YO, Lindquist SL, Ono B, Inge-Vechtomov SG, Liebman SW. Role of the chaperone protein Hsp104 in propagation of the yeast prion-like factor [psi+]. Science 1995, **268**: 880–884.
31. Telling GC, Scott M, Mastrianni J, Gabizon R, Torchia M, Cohen FE, DeArmond SJ, Prusiner SB. Prion propagation in mice expressing human and chimeric PrP transgenes implicates the interaction of cellular PrP with another protein. *Cell* 1995, **83**: 79–90.
32. Hsiao KK, Scott M, Foster D, Groth DF, DeArmond SJ, Prusiner SB. Spontaneous neurodegeneration in transgenic mice with mutant prion protein. *Science* 1990, **250**: 1587–1590.
33. Telling GC, Haga T, Torchia M, Tremblay P, DeArmond SJ, Prusiner SB. Interactions between wild-type and mutant prion proteins modulate neurodegeneration in transgenic mice. *Genes Dev* 1996, **10**: 1736–1750.
34. Kocisko DA, Come JH, Priola SA, Chesebro B, Raymond GJ, Lansbury PT, Caughey B. Cell-free formation of protease-resistant prion protein. *Nature* 1994, **370**: 471–474.
35. Kocisko DA, Priola SA, Raymond GJ, Chesebro B, Lansbury PT, Jr., Caughey B. Species specificity in the cell-free conversion of prion protein to protease-resistant forms: A model for the scrapie species barrier. *Proc Natl Acad Sci USA* 1995, **92**: 3923–3927.
36. Bessen RA, Kocisko DA, Raymond GJ, Nandan S, Lansbury PT, Caughey B. Non-genetic propagation of strain-specific properties of scrapie prion protein. *Nature* 1995, **375**: 698–700.

37. Telling GC, Scott M, Hsiao KK, Foster D, Yang SL, Torchia M, Sidle KCL, Collinge J, DeArmond SJ, Prusiner SB. Transmission of Creutzfeldt-Jakob disease from humans to transgenic mice expressing chimeric human–mouse prion protein. *Proc Natl Acad Sci USA* 1994, **91**: 9936–9940.
38. Collinge J, Palmer MS, Sidle KCL, Hill AF, Gowland I, Meads J, Asante E, Bradley R, Doey LJ, Lantos PL. Unaltered susceptibility to BSE in transgenic mice expressing human prion protein. *Nature* 1995, **378**: 779–783.
39. Prusiner SB, Hsiao KK. Human prion diseases. *Ann Neurol* 1994, **35**: 385–395.
40. Prusiner SB. Prion diseases and the BSE crisis. Science, in press
41. Tateishi J, Kitamoto T. Inherited prion diseases and transmission to rodents. *Brain Pathol* 1995, **5**: 53–59.
42. Monari L, Chen SG, Brown P, Parchi P, Petersen RB, Mikol J, Gray F, Cortelli P, Montagna P, Ghetti B, Goldfarb LG, Gajdusek DC, Lugaresi E, Gambetti P, Autilio-Gambetti L. Fatal familial insomnia and familial Creutzfeldt-Jakob disease: different prion proteins determined by a DNA polymorphism. *Proc Natl Acad Sci USA* 1994, **91**: 2839–2842.
43. Young K, Clark HB, Piccardo P, Dlouhy SR, Ghetti B. Gerstmann-Straussler-Scheinker disease with the PRNP P102L mutation and valine at codon 129. *Mol Brain Res* 1997, **44**: 147–150.
44. LaPlanche JL, Delasnerie-Laupretre N, Brandel JP, Chatelain J, Beaudry P, Alperovitch A, Launay JM. Molecular genetics of prion diseases in France. *French Research Group on Epidemiology of Human Spongiform Encephalopathies Neurology* 1994, **44**: 2347–2351.
45. Parchi P, Castellani R, Capellari S, Ghetti B, Young K, Chen SG, Farlow M, Dickson DW, Sima AAF, Trojanowski JQ, Petersen RB, Gambetti P. Molecular basis of phenotypic variability in sporadic Creutzfeldt-Jakob disease. *Ann Neurol* 1996, **39**: 767–778.
46. Schulzschaeffer WJ, Giese A, Windl O, Kretzschmar HA. Polymorphism at codon 129 of the prion protein gene determines cerebellar pathology in Creutzfeldt-Jakob disease. *Clin Neuropathol* 1996, **15**: 353–357.
47. MacDonald ST, Sutherland K, Ironside JW. Prion protein genotype and pathological phenotype studies in sporadic Creutzfeldt-Jakob disease. *Neuropathol Appl Neurobiol* 1996, **22**: 285–292.
48. Pickering-Brown SM, Mann DM, Owen F, Ironside JW, de Silva R, Roberts DA, Balderson DJ, Cooper PN. Allelic variations in apolipoprotein E and prion protein genotype related to plaque formation and age of onset in sporadic Creutzfeldt-Jakob disease. *Neurosci Lett* 1995, **187**: 127–129.
49. Hsiao KK, Groth D, Scott M, Yang S-L, Serban H, Rapp D, Foster D, Torchia M, DeArmond SJ, Prusiner SB. Serial transmission in rodents in neurodegeneration from transgenic mice expressing mutant prion proteins. *Proc Natl Acad Sci USA* 1994, **91**: 9126–9130.
50. McKinley MP, Meyer RK, Kenaga L, Rahbar F, Cotter R, Serban A, Prusiner SB. Scrapie prion rod formation in vitro requires both detergent extraction and limited proteolysis. *J Virol* 1991, **65**: 1340–1351.
51. Forloni G, Angeretti N, Chiesa R, Monzani E, Salmona M, Bugiani O, Tagliavini F. Neurotoxicity of a prion protein fragment. *Nature* 1993, **362**: 543–546.
52. Brown DR, Herms J, Kretzschmar HA. Mouse cortical cells lacking cellular PrP survive in culture with a neurotoxic PrP fragment. *Neuroreport* 1994, **5**: 2057–2060.
53. Kristensson K, Feuerstein B, Taraboulos A, Hyun WC, Prusiner SB, DeArmond SJ. Scrapie prions alter receptor-mediated calcium responses in cultured cells. *Neurology* 1993, **43**: 2335–2341.
54. Wong K, Qiu Y, Hyun W, Nixon R, VanCleff J, Sanchez-Salazar J, Prusiner SB, DeArmond SJ. Decreased receptor-mediated calcium response in prion–infected cells correlates with decreased membrane fluidity and IP3 release. *Neurology* 1996, **47**: 741–750.
55. Brandner S, Isenmann S, Raeber A, Fischer M, Sailer A, Kobayashi Y, Marino S, Weissmann C, Aguzzi A. Normal host prion protein necessary for scrapie-induced neurotoxicity. *Nature* 1996, **379**: 339–343.

56. Sakaguchi S, Katamine S, Nishida N, Moriuchi R, Shigematsu K, Sugimoto T, Nakatani A, Kataoka Y, Houtani T, Shirabe S, Okada H, Hasegawa S, Miyamoto T, Noda T. Loss of cerebellar Purkinje cells in aged mice homozygous for a disrupted PrP gene. *Nature* 1996, **380**: 528–531.
57. Manson JC, Clarke AR, Hooper ML, Aitchison L, McConnell I, Hope J. 129/Ola mice carrying a null mutation in PrP that abolishes mRNA production are developmentally normal. *Mol Neurobiol* 1994, **8**: 121–127.
58. Tobler I, Gaus SE, Deboer T, Achermann P, Fischer M, Rülicke T, Moser M, Oesch B, McBride PA, Manson JC. Altered circadian activity rhythms and sleep in mice devoid of prion protein. *Nature* 1996, **380**: 639–642.
59. Tobler I, Deboer T, Fischer M. Sleep and sleep regulation in normal and prion protein-deficient mice. *J Neurosci* 1997, **17**: 1869–1879.
60. Collinge J, Whittington MA, Sidle KCL, Smith CJ, Palmer MS, Clarke AR, Jefferys JGR. Prion protein is necessary for normal synaptic function. *Nature* 1994, **370**: 295–297.
61. Whittington MA, Sidle KCL, Gowland I, Meads J, Hill AF, Palmer MS, Jefferys JGR, Collinge J. Rescue of neurophysiological phenotype seen in PrP null mice by transgene encoding human prion protein. *Nature Genetics* 1995, **9**: 197–201.
62. Lledo P-M, Tremblay P, DeArmond SJ, Prusiner SB, Nicoll RA. Mice deficient for prion protein exhibit normal neuronal excitability and synaptic transmission in the hippocampus. *Proc Natl Acad Sci USA* 1996, **93**: 2403–2407.
63. Prusiner SB, Scott M, Foster D, Pan K–M, Groth D, Mirenda C, Torchia M, Yang S-L, Serban D, Carlson GA, Hoppe PC, Westaway D, DeArmond SJ. Transgenic studies implicate interactions between homologous PrP isoforms in scrape prion replication. *Cell* 1990, **63**: 673–686.
64. Bueler H, Raeber A, Sailer A, Fischer M, Aguzzi A, Weissmann C. High prion and PrPSc levels but delayed onset of disease in scrapie-inoculated mice heterozygous for a disrupted PrP gene. *Mol Med* 1994, **1**: 19–30.
65. Ironside JW, McCardle L, Hayward PA, Bell JE. Ubiquitin immunocytochemistry in human spongiform encephalopathies. *Neuropathol Appl Neurobiol* 1993, **19**: 134–140.
66. Tatzelt J, Zuo J, Voellmy R, Scott M, Hartl U, Prusiner SB, Welch WJ. Scrapie prions selectively modify the stress response in neuroblastoma cells. *Proc Natl Acad Sci USA* 1995, **92**: 2944–2948.
67. Diedrich JF, Carp RI, Haase AT. Increased expression of heat shock protein, transferrin, and {\symbol} 2-microglobulin in astrocytes during scrapie. *Microb Pathog* 1993, **15**: 1–6.
68. Clinton J, Forsyth C, Royston MC, Roberts GW. Synaptic degeneration is the primary neuropathological feature in prion disease: A preliminary study. *Neuroreport* 1993, **4**: 65–68.
69. DeArmond SJ, Prusiner SB. Etiology and pathogenesis of prion diseases. *Am J Pathol* 1995, **146**: 785–811.
70. Fairbairn DW, Carnahan KG, Thwaits RN, Grigsby RV, Holyoak GR, O'Neill KL. Detection of apoptosis induced DNA cleavage in scrapie-infected sheep brain. *FEMS Microbiol Lett* 1994, **115**: 341–346.
71. Giese A, Groschup MH, Hess B, Kretzschmar HA. Neuronal cell death in scrapie-infected mice is due to apoptosis. *Brain Pathol* 1995, **5**: 213–221.
72. Moser M, Colello RJ, Pott U, Oesch B. Developmental expression of the prion protein gene in glial cells. *Neuron* 1995, **14**: 509–517.
73. Diedrich JF, Bendheim PE, Kim YS, Carp RI, Haase AT. Scrapie-associated prion protein accumulates in astrocytes during scrapie infection. *Proc Natl Acad Sci USA* 1991, **88**: 375–379.
74. Race RE, Priola SA, Bessen RA, Ernst D, Dockter J, Rall GF, Mucke L, Chesebro B, Oldstone MBA. Neuron-specific expression of a hamster prion protein minigene in transgenic mice induces susceptibility to hamster scrapie agent. *Neuron* 1995, **15**: 1183–1191.

75. Casaccia-Bonnefil P, Kascsak RJ, Fersko R, Callahan S, Carp RI. Brain regional distribution of prion protein PrP27-30 in mice stereotaxically microinjected with different strains of scrapie. *J Infect Dis* 1993, **167**: 7–12.
76. Jeffrey M, Fraser JR, Halliday WG, Fowler N, Goodsir CM, Brown DA. Early unsuspected neuron and axon terminal loss in scrapie-infected mice revealed by morphometry and immunocytochemistry. *Neuropathol Appl Neurobiol* 1995, **21**: 41–49.
77. Borchelt DR, Koliatsos VE, Guarnieri M, Pardo CA, Sisodia SS, Price DL. Rapid anterograde axonal transport of the cellular prion glycoprotein in the peripheral and central nervous systems. *J Biol Chem* 1994, **269**: 14,711–14,714.
78. Brandner S, Raeber A, Sailer A, Blättler T, Fischer M, Weissmann C, Aguzzi A. Normal host prion protein (PrPc) is required for scrapie spread within the central nervous system. *Proc Natl Acad Sci USA* 1996, **93**: 13,148–13,151.
79. Jendroska K, Heinzel FP, Rotchia M, Stowring L, Kretzschmar HA, Kon A, Stern A, Prusiner SB, DeArmond SJ. Proteinase-resistant prion protein accumulation in Syrian hamster brain correlates with regional pathology and scrapie infectivity. *Neurology* 1991, **41**: 1482–1490.
80. Carp RI, Rubenstein R. Diversity and significance of scrapie strains. *Semin Virology* 1991, **2**: 203–213.
81. DeArmond SJ, Yang S-L, Lee A, Bowler R, Taraboulos A, Groth D, Prusiner SB. Three scrapie prion isolates exhibit different accumulation patterns of the prion protein scrapie isoform. *Proc Natl Acad Sci USA* 1993, **90**: 6449–6453.
82. Telling GC, Parchi P, DeArmond SJ, Cortelli P, Montagna P, Gabizon R, Mastrianni J, Lugaresi E, Gambetti P, Prusiner SB. Evidence for the conformation of the pathologic isoform of the prion protein enciphering and propagating prion diversity. *Science* 1996, **274**: 2079–2082.
83. Taraboulos A, Jendroska K, Serban D, Yang S-L, DeArmond SJ. Regional mapping of prion protins in brain. *Proc Natl Acad Sci USA* 1992, **89**: 7620–7624.

16
Blood Flow, Energy Failure, and Vulnerability to Stroke

James J. Vornov

INTRODUCTION

In recent years, the study of mechanisms of ischemic neuronal injury has focused on the cascade of intracellular events triggered by ischemic conditions, emphasizing the role of glutamate receptor activation. The focus on intracellular mechanisms has led to an approach of identifying individual mechanisms of injury in simple systems and then examining their role in more intact systems. Rarely, however, are these mechanisms placed into their proper context as events that occur during reduced blood flow, interacting with cellular energy status. Simple models of ischemic injury based on one or two of these basic mechanisms do not reflect what is known about the dynamics of blood flow and metabolism in ischemia. For example, it has been suggested that at the point of energy failure, there is massive release of glutamate into the extracellular space, triggering a cascade that resembles glutamate toxicity. Although glutamate receptors certainly play an important role in mediating ischemic neuronal injury, blockade of glutamate receptors has no effect in some models. In other models, glutamate toxicity can be observed in the absence of massive glutamate release. Thus, whereas glutamate receptors are often critical mediators of ischemic neuronal injury, energy failure and reperfusion are fundamental factors as well.

This chapter will attempt to place the mechanisms of neuronal injury in the context of blood flow and energy metabolism. In general, the process of energy failure resulting from ischemia can be seen in three distinct phases, i.e., the onset of ischemia, when complete energy failure can still be avoided; the time during complete energy failure; and the recovery from energy failure. The mechanisms of injury in cerebral ischemia can be viewed first as involving a group of factors that make cells more or less likely to reach the state of energy failure. A second set of mechanisms involve a group of factors that make cells more or less likely to remain alive and fully recover in the long term. Those factors determine obligatory time periods ("windows") within which therapeutic interventions must be made in order to avoid irreversible cell injury (Table 1).

GLOBAL ISCHEMIA

Ischemic damage to the central nervous system (CNS) is caused by reduced blood flow. The severity and duration of blood flow reduction determines the amount of brain injury. Complete absence of flow causes injury in a matter of minutes. The brain can

Table 1
Minimum and Maximum Durations of Therapeutic Windows

Ischemia phase	Minimum window	Maximum window
Onset	1 min Upon decapitation, ATP falls to less than 20% of normal.	3h, longer? In most animal models of ischemia, reperfusion fails to reduce injury at all unless begun within 3 h. The potential effects of chronic mild ischemia or episodic ischemia are not known.
Energy failure	5 min 7 min is about the minimum duration of ischemia needed for damage. Energy failure and recovery occur within a minute, so 5 min of energy failure may be required for injury.	1 to 6 h, longer? Under normal conditions, reperfusion after 30 min results in widespread damage, with some regions still resistant, e.g., brainstem. With severe hypothermia, many hours can be withstood.
Recovery	Nonexistent If severe energy failure persists long enough, even the most rapid recovery will not prevent injury.	6 h In some models, manipulations must be done near the onset of reperfusion for maximal effect, but effects can still be seen at several hours. The process of metabolic recovery probably blends with more delayed mechanisms.
Delayed death	Nonexistent With severe, prolonged energy failure, damage may be complete	Several hours to years Active events that lead to cell death begin as energy is restored, but a delay of hours is generally found. Cellular dropout could continue for years due to disuse or growth factor effects.

withstand mild reductions of blood flow for hours. The simpler case of absence of flow will be considered first, because with uniform absence of blood flow, vulnerability can only be determined by the intrinsic characteristics of brain regions and cell types.

Typically, complete absence of flow is the result of cardiac arrest, when the electrical and pumping activity of the heart stops. Very brief periods of complete ischemia result only in rapid loss of consciousness, which is called a faint or syncope. No permanent injury occurs. On the other hand, unless flow is restored within minutes, death results, in part because ischemia occurs throughout the entire animal, including the heart. Thus, selective injury to vulnerable regions of the brain occurs only when the duration of global ischemia lies between the extremes of simple syncope and death.

The simplest model of global ischemia is decapitation. Many early neurochemical studies of cerebral metabolic rates were performed using this method, owing to the fact that after decapitation the brain becomes a closed system. Phosphocreatine is the most labile high energy phosphate, dropping by 70% in the first 3 s after decapitation of a mouse. Adenosine triphosphate (ATP) falls more slowly, but requires only 1 min to fall by 80%. Compared to rodents, larger animals, such as dogs, may have a slower rate of ATP depletion, requiring about 5 min for ATP to drop by 80%.

In general, the flow during global ischemia is so uniformly low that the only variable is time. In complete global ischemia, then, there is a simple relationship between the duration of ischemia and the amount of injury. The onset of ischemic biochemical changes is uniform. Without reflow, death will occur.

If energy failure is due to consumption of available substrates, then resistance of the brain to injury should be increased by lowering metabolic rate. The earliest studies of cerebral metabolism demonstrated that hypothermia and anesthesia lowered cerebral oxygen use *(1,2)*. Subsequently, it was demonstrated that measures which decrease cellular energy demand slow the depletion of high energy phosphates by 30% to 40% under conditions of no blood flow *(3)*. Thus, hypothermia and anesthesia were the first recognized neuroprotective strategies *(4–6)* and remain the most widely used clinically. With the rapid loss of energy, these strategies clearly must be initiated before the onset of ischemia, a condition that limits their usefulness to specific settings such as the surgical suite.

SELECTIVE HIPPOCAMPAL VULNERABILITY TO ISCHEMIA—INSIGHTS FROM ORGANOTYPIC CULTURE

It is well established that hippocampus is the most vulnerable region of the brain to ischemia *(7)*. The development of rodent forebrain models of ischemia facilitated the study of hippocampal vulnerability. The incomplete circle of Willis of gerbils makes it possible to simply occlude the carotids bilaterally and achieve very low levels of flow throughout the forebrain. The sparing of blood flow to the brainstem helps increase the rate of survival with longer duration of ischemia. Electrocautery of the vertebral vessels allows a similar model of forebrain ischemia to be performed in the rat.

These global ischemia models demonstrated a reproducible differential vulnerability within the hippocampus. The subregion of the hippocampus called CA1 is the most vulnerable. As the duration of global ischemia is increased, the CA3 region becomes involved. With even longer durations, the dentate gyrus is also involved.

Our laboratory has had a long-term interest in determining why the CA1 sector of the hippocampus is most vulnerable to ischemia. The balance of information points to an intrinsic vulnerability of these neurons to energy depletion itself. This conclusion arises from our investigations of the intrinsic vulnerability of the hippocampus using a preparation called organotypic culture, in which a slice of the hippocampus is grown in long-term culture, preserving the characteristic regional differentiation *(8–10)*.

Newell et al. *(11)* first demonstrated that the selective vulnerability of the hippocampus seen in vivo was reproduced in organotypic culture in a simple model of ischemia. We found similar selective vulnerability using metabolic inhibition to simulate ischemic injury and demonstrated that neurons could be protected by *N*-methyl-D-aspartate (NMDA) receptor blockade during early recovery from metabolic inhibition *(12)*.

Even though we found that simulated ischemic injury was dependent on NMDA receptor activation, it was not the same as selective vulnerability of organotypic cultures to direct NMDA toxicity. Brief exposure of cultures to NMDA causes injury in the CA1 and dentate regions of the culture *(13)*, i.e., regions of the hippocampus with the highest density of NMDA receptors. The dentate is vulnerable to NMDA toxicity, but is selectively resistant to ischemic injury, both in vivo and in our tissue culture model. Interestingly, hypoglycemic injury in organotypic culture does correspond to the selective vulnerability to NMDA agonists *(14)*, perhaps because aspartate is released during hypoglycemia and aspartate is a more selective NMDA receptor antagonist.

Another important observation that has emerged from our studies in organotypic culture is the demonstration that injury can be caused by brief, selective activation of amino-3-hydroxy-5-methyl-4-isoxazole propionic acid (AMPA)/kainate receptors *(13,15)*. This is different from the situation in primary dissociated cultures and may be due to the more differentiated state of the organotypic cultures. The well established selective vulnerability of the CA3 region of the hippocampus can be reproduced in culture. It requires exposures of more than 24 h at low agonist concentration and may be due to the selective localization of high affinity kainate receptors in the CA3 region. At higher agonist concentrations, receptor activation may be less specific. When cultures are exposed to high concentrations of glutamate, toxicity results from the coactivation of NMDA and AMPA/kainate receptors *(15)*.

Work in our laboratory has centered on establishing whether the selective vulnerability of the hippocampus is determined by vulnerability to glutamate toxicity. Our most recent experiments (Fig. 1) suggest that the selective vulnerability of CA1 is a result of an intrinsic vulnerability to energy failure. Partial inhibition of mitochondrial function by low concentrations of potassium cyanide for several hours in the presence of glucose selectively damages the CA1 region in organotypic cultures. This injury cannot be prevented by glutamate receptor blockade. Other mitochondrial inhibitors such as azide and aminooxalic acid yield similar results. Similarly, exposure to the superoxide generating compound paraquat causes selective damage to the CA1 region. This effect may be due to selective vulnerability to superoxide. The nitric oxide (NO) generator, sodium nitroprusside, also selectively damages CAI. However, previous observations suggest that free radicals can contribute to mitochondrial inhibition (*see* chapter by Dykens). Thus, vulnerability to superoxide and NO may also be a reflection of vulnerability to mitochondrial dysfunction.

FOCAL ISCHEMIA

Although studies of hippocampal vulnerability provide basic insights into the mechanisms of ischemic injury, animal models of focal ischemia due to arterial occlusion resemble more closely the clinical situation in which stroke occurs. When a single large artery is occluded, the vascular anatomy results in a pattern of lowered perfusion within the territory of the blood vessel that is not uniform. The lowest flow occurs in the part of the vascular territory that depends exclusively on the vessel for its blood supply. The flow in such a region may stop entirely, producing conditions just like global ischemia within that limited area. Just as in global ischemia, severe injury will occur rapidly unless flow is rapidly restored. This region of most severe ischemia is called the "ischemic core."

Most interesting to the stroke investigator is the territory in which flow is reduced, but not stopped altogether. It is this region that can be rescued by a number of therapeutic approaches and is called the "penumbra." Although there have been a number of attempts to define the penumbra exactly, for now it is sufficient to recognize that it is a region of reduced blood flow which may eventually become infarcted if the arterial

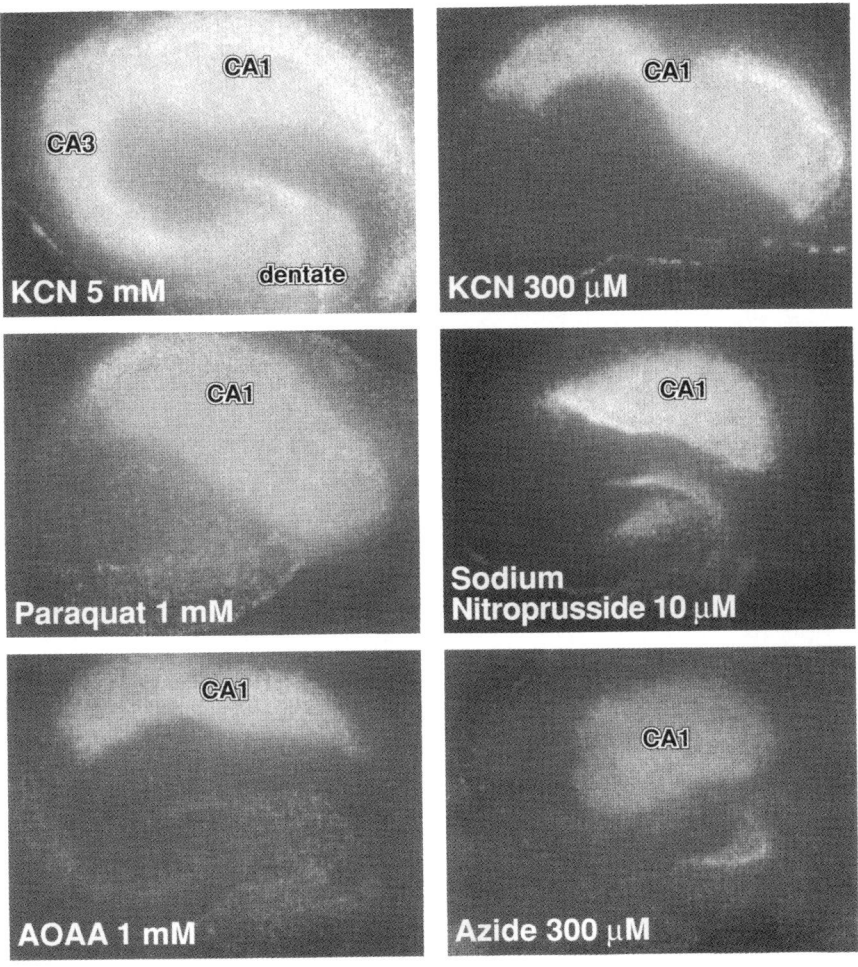

Fig. 1. Selective vulnerability of the hippocampus in organotypic culture. Low power fluorescent video images of the hippompal cultures are shown the day after a 6 h exposure to various oxidants and mitochondrial inhibitors. Propidium iodide was present throughout the experiment so that the bright areas denote injury to neurons within the culture. At the upper left, the injury caused by a high concentration of potassium cyanide (KCN) is shown to illustrate the anatomy of the cultures. In the remaining panels, doses of toxins are illustrated that selectively injure CA1. Note that while sodium nitroprusside (middle right) can liberate cyanide, it was toxic at concentrations less than 1/10 of the effective dose of cyanide (upper left), suggesting that the liberation of nitric oxide is responsible. Aminooxalic acid (AOAA) and azide are both inhibitors of oxidative metabolism.

occlusion is prolonged. The flow is low enough to eventually cause infarction, but not low enough to do so within minutes.

Focal ischemia can also occur during a severe decrease in blood pressure. For example, massive blood loss can lower blood pressure and cause a state of diffusely, low blood flow throughout the brain. Because the blood supply to the brain is not uniform, blood flow varies regionally. The areas on the border between two different main arteries have lower blood flow than those in the central region of a single artery's distribution. These border zones are called "watershed" regions and are more vulnerable to ischemic injury from severe hypotension because of focal reduction in blood flow.

THE SEQUENCE OF ISCHEMIC FAILURE

The classic studies of Brierley and coworkers in global ischemia showed that neural function was lost at levels of blood flow above those necessary to cause infarction *(16,17)*. Symon and coworkers then established that, in focal ischemia, there were distinct thresholds for electrical changes, biochemical changes and cellular injury *(18)*. Subsequent studies have examined a large number of electrical and biochemical events of varying severity that occur with the onset of ischemia.

These thresholds are most easily demonstrated soon after the onset of ischemia *(19)*. The most severely ischemic region has the same rapid and severe loss of cellular homeostasis that occurs with cardiac arrest. Areas with slightly higher flow rates maintain ATP levels, but become acidotic. At higher flow, glucose metabolism and pH are preserved, but electrophysiological activity is halted. At the mildest areas of disturbance, protein synthesis is inhibited, with the cells selectively expressing genes for stress proteins, like the heat shock proteins.

As time progresses after the permanent occlusion of a vessel, the ischemic core and the less affected penumbra change dynamically. Shortly after occlusion of the MCA in the rat, the ATP depletion that marks the ischemic core involves most of the striatum but only a small region of the immediately adjacent cortex *(20,21)*. As the duration of arterial occlusion increases, the severely involved area of the cortex expands. In the rat, it requires about 3 h for the region of ATP depletion to become maximal. This distribution corresponds with the territory of infarction as assessed at 24 h by standard biochemical or histological methods.

There is a simple explanation for the observation that reperfusion has no effect once focal ischemia exceeds a certain duration. The region of infarction expands over time as energy failure involves more and more of the vessel's territory. Eventually, all of the region at risk has arrived at the state of energy failure and remained in that state for long enough to cause irreversible injury. It makes no difference whether or not flow is ever restored.

Several characteristic biochemical changes occur in the penumbra between the onset of ischemia and energy failure, which may be important as markers of brain tissue that can still be rescued from infarction. The specific changes may be clues to the mechanisms that drive the tissue to eventual energy failure and infarction. Protein synthesis is inhibited in the penumbra, which is probably a sign of cellular stress. Similarly, changes in gene expression suggest cell stress, since immediate early genes are induced, including the heat shock proteins *(22)*.

One such characteristic change is an early increase in 2-deoxyglucose uptake in the penumbra *(23)*. Eventually, the 2-deoxyglucose uptake in these regions ceases, presumably as the tissue progresses to the point of energy failure. The significance of these biochemical changes are not fully understood, but are interpreted as denoting a period of hypermetabolism prior to energy failure. Glucose uptake might increase simply because of lack of available oxygen for a more efficient aerobic metabolism. Alternatively, the changes may be a response to ongoing physiologic events related to injury, for example, second messengers generated by the waves of depolarization that have been observed in the penumbra.

When NMDA receptors are blocked, the glucose hypermetabolism, the inhibition of protein synthesis and the repeated depolarizations are all blocked (24,25). The simple interpretation of these observations is that receptor activation in ischemic tissue is responsible for the cellular stress and energy demand, hastening the progression to energy failure.

An alternative explanation of the biochemical alterations in the penumbra is that these metabolic and genetic changes are a response by cells to increase their resistance to ischemia under conditions of low substrate supply. Turning off protein synthesis would conserve energy. It may also spare amino acids for metabolic uses, including the synthesis of the antioxidant tripeptide glutathione (26). The glucose hypermetabolism could also be an adaptive response to mitochondrial failure under low oxygen conditions.

ION FLUX AS THE CAUSE OF ENERGY FAILURE

Strategies to protect the penumbra can be viewed simply. Under conditions of no flow, tissue becomes nonviable after a brief period of time. Under conditions of low flow, the progression to energy failure just takes longer. If the state of energy failure persists for more than a short time, infarction occurs. In global ischemia, the state is reached very quickly. In focal ischemia models, the state is reached in minutes within the core, but takes hours in the region surrounding the core.

The most powerful protective strategy would be to prevent the tissue from ever reaching the stage of energy failure. This might be done by restoring blood flow before the state of ATP depletion occurs. Alternatively, a therapy might act by increasing the resistance of tissue to substrate deprivation, preventing the tissue from ever reaching a point of energy failure, even though blood flow remains low.

In cardiac ischemia, medical therapy is directed at preventing energy failure by reducing the work of the heart as a pump. This translates to a very simple view of ischemia as an imbalance between supply and demand. At very low flow rates, even basic demand is enough to cause full failure, although as discussed above severe metabolic depression induced by hypothermia and anesthesia can lower demand enough to increase resistance to ischemia. In the intermediate rates that we are most interested in, the tissue cannot survive indefinitely with normal energy use. If energy demand can be reduced, perhaps even slightly, more tissue will survive.

What drives the CNS tissue to energy failure? As the heart is a pump, the brain is an information processing system. The major energetic stress for neurons is synaptic signaling, which involves movement of ions across the plasma membrane and generation of second messenger molecules. Sodium and calcium are the major ions involved in synaptic transmission.

The strong protective effects of excitatory neurotransmitter blockade follows simply from viewing protection as a delay in the onset of energy failure in low-flow areas. This idea is supported by the observation that, when NMDA receptors are blocked, the glucose hypermetabolism, the inhibition of protein synthesis and the repeated depolarizations in the penumbra are all prevented (24,25). This blockade may be due to prevention of altered metabolic conditions by decreasing energy demands, or may be more directly due to the blockade of second messengers.

MECHANISMS OF ENERGY FAILURE—TISSUE CULTURE MODELS

Neuronal tissue culture has emerged as an important tool in the study of ischemic neuronal injury. As a simplified, accessible system, many more variables can be controlled. It is possible to examine many more conditions in a much briefer period of time with less variability than in animals. In addition, the homogeneous nature of dissociated tissue culture preparations makes biochemical measurements easier to interpret. Finally, the accessibility of tissue culture makes measurements possible that cannot be made in animal models.

Since the essence of ischemia is deprivation of oxygen and metabolic substrate, most tissue culture models of ischemia have been performed by simple removal of oxygen and glucose. These substrate deprivation preparations seem to model the onset of focal ischemia best. Injury by depletion of oxygen and glucose is relatively slow compared to complete global ischemia, taking an hour or more. There is always restoration of metabolism after a set interval. The same neuroprotective agents that work in focal ischemia tend to work in substrate deprivation models- especially glutamate receptor antagonists (27). These treatments increase resistance to the substrate deprivation. Although a control culture can survive only 60 min of substrate deprivation, blockade of NMDA receptors extends the duration of tolerability to 90 min. However, if substrate deprivation is extended beyond this period, the culture fails to survive.

When the duration of ischemia is increased to the point that NMDA receptor activation is no longer effective as a protective measure, blockade of other pathways of sodium entry become effective as added measures. AMPA/kainate-type glutamate receptors and voltage-sensitive channels can be blocked with tetrodotoxin (TTX). Extracellular acidosis adds additional protection, perhaps through blockade of sodium-hydrogen ion exchange.

There is good evidence that sodium movement is closely tied to cellular energy demand. The sodium-potassium ATPase is the major consumer of energy in the neuron during activity and is dependent on sodium influx into the cell (28). Voltage-sensitive and ligand-gated ion channel activity are, of course, closely related to synaptic activity. However, the most studied ligand-gated ion channels are the excitatory neurotransmitter receptors. The protective effects of excitatory neurotransmitter antagonists and sodium channel blockers imply that preventing sodium entry into neurons prevents injury, perhaps by delaying energy failure. In addition, whereas most sodium channels are rapidly inactivating and thus allow only small amounts of sodium to enter the cell, other, more slowly inactivating channels may allow a slower sodium leak into cells (29).

Another route for sodium entry is through the activity of ion exchangers. There are a number of sodium-dependent transport activities which may be activated during ischemia and contribute to sodium entry. Calcium extrusion from the cell occurs both by exchange for sodium and more directly by specific calcium ATPases. Available evidence suggests, however, that calcium accumulation occurs late, after energy failure (30).

The participation of this large number of ionic mechanisms is easily reconciled by the simple view that preventing ion entry into cultured neurons during substrate deprivation preserves neuronal energy and prevents ATP depletion. In animal models of focal ischemia, preventing infarction is clearly correlated with preventing ATP depletion and metabolic failure. Interestingly, such ATP measurements have not been reported in tissue culture.

EVENTS ASSOCIATED WITH ENERGY FAILURE

The events that occur at the time of energy failure have been studied best in models of global ischemia. A persistent depolarization occurs which may be similar to the transient depolarizations discussed earlier. Depolarization gives rise to prominent ionic shifts, including decreases in extracellular sodium and calcium concentrations and increases in extracellular potassium *(30)*.

There is also a marked intracellular and extracellular acidosis. Some of the acidosis is caused by accumulation of metabolic acids as blood flow slows to levels that cannot remove the protons. ATP hydrolysis also generates acid. The severity of this acidosis has been linked to injury, since exaggeration of acidosis by hyperglycemia worsens ischemic injury *(31,32)*.

Extracellular glutamate concentration rises dramatically at the time of energy failure *(33)*. Some of this release may be through normal calcium-dependent synaptic release mechanisms. Glutamate accumulation may be enhanced by failure of glutamate uptake via sodium-dependent transport. It has also been suggested that loss of membrane potential causes glutamate transporters to run backwards *(34)*.

Remarkably, there is little structural change in neurons at the time of energy failure. High-resolution electron microscopy shows disaggregation of ribosomes that is the ultrastructural correlate of the inhibition of protein synthesis discussed above. There is little structural evidence of acute cellular swelling, even as large ion shifts occur *(35)*.

ATP depletion may cause other biochemical changes. For example, as ATP is depleted, kinases can no longer function to phosphorylate proteins, whereas phosphatases remain active. This may have important functional effects on proteins like the NMDA receptors, which require intact phosphorylation mechanisms.

These severe biochemical changes are irreversible unless blood flow is quickly restored, energy metabolism resumed, and the transmembrane gradient reestablished.

REPERFUSION AND RECOVERY

Once flow reoccurs, the most important task for the recovering cell is to restore ionic and energetic homeostasis. For the most part, this process has been studied acutely in large animal models of global ischemia. NMR spectroscopic studies have shown that the speed of ATP and pH recovery predict cell recovery *(36,37)*. Without rapid recovery, infarction will follow.

In focal models of ischemia, reperfusion mechanisms will participate in the recovery of the ischemic core. The same factors that dictate recovery in global ischemia may be operating in these focal ischemic areas. On the other hand, less severely affected areas, which have not progressed to the point of energy failure at the time of reperfusion, may be protected if they are prevented from ever reaching energy failure, and not via reperfusion following energy failure.

The biochemical and cellular events following reperfusion have generally been studied in models of global ischemia. In these global ischemia models, energy depletion is sudden, uniform, and severe. Upon reperfusion, energy recovery is then rapid and complete. Surprisingly, even though energy has been restored, neurons may still be injured by mechanisms independent of the recovery of energy state. Such neuronal injury appears to occur selectively and in delayed fashion in the hippocampus following global ischemia. To what extent the mechanisms involved in the delayed death of hippocampal neurons following global ischemia and rapid recovery of energy homeostasis are operative in other brain regions subjected to similar insults remains unclear.

ISCHEMIC GLUTAMATE RELEASE AND EXCITOTOXICITY

The demonstration that neuronal injury could be reduced by antagonists of excitatory neurotransmission during recovery from global ischemia *(39)* has been a major factor driving neuroprotection research. These studies provided convincing evidence that ischemic neuronal injury was reversible well after reperfusion, allowing for a "therapeutic window." Global ischemia represents only one type of clinical ischemia, affecting a region of the brain which, while critical, represents a small portion of the human brain. These studies did, however, bring attention to a class of drugs which may be of wider use in the treatment of ischemia.

Unfortunately, early experiments demonstrating neuroprotection in global ischemia led to a highly simplified view of ischemic injury, centered around glutamate toxicity. It seemed logical to conclude that ischemic injury occurs when energy failure results in massive release of glutamate into the extracellular space. The large number of studies investigating the ionic and biochemical mechanisms of acute glutamate toxicity tend to reinforce this impression.

Several subsequent pieces of data strongly suggested that this view is inaccurate *(33,40)*. First, the protective effects of glutamate receptor antagonists can be observed long after the acute ischemic glutamate release has ended. Second, the classic signs of glutamate toxicity are not observed early in ischemia. Finally, protective effects of glutamate receptor antagonists are not universally observed. Under some conditions, the protective effects of MK-801 were demonstrated to be due to hypothermia rather than receptor blockade *(41)*.

An examination of the morphology of ischemic injury and glutamate toxicity shows strong contrasts between these two types of injury, suggesting that their mechanisms are distinct. For example, in tissue culture, glutamate or selective agonists at either NMDA or AMPA/kainate receptors causes acute swelling *(42)*. A similar swelling, most marked in dendrites, is observed in acutely prepared brain slices exposed to glutamate agonists *(43,44)*. The changes observed in animals may be slower, but show the same typical swelling reaction. These swollen dendrites are also seen under conditions of excessive synaptic activation such as sustained electrical stimulation of the perforant path *(45)*.

By contrast, the hallmark of early ischemic injury is a more subtle change in mitochondrial morphology, a polyribosome disaggregation that correlates with inhibition of protein synthesis and a subsequent cellular condensation *(35,46–48)*. The typical eosinophilic appearance of injured neurons within an infarct is generally absent from excitotoxic injury caused by glutamate agonists or excessive synaptic activation.

MECHANISMS OF REPERFUSION INJURY—TISSUE CULTURE MODELS

To study the physiological events of ischemia in our laboratory, we have adapted a technique more widely used in nonneuronal systems, in which oxidative metabolism is blocked by the mitochondrial inhibitor, potassium cyanide, and glycolysis is blocked by pharmacological concentrations of 2-deoxyglucose (2-DG). There are two major advantages to metabolic inhibition over methods that employ substrate deprivation, the latter of which is achieved by using chambers to induce hypoxia and cessation of mitochondrial function and by removing glucose from the media. First, with metabolic inhibition, the cultures remain accessible; the atmosphere does not have to be controlled and the cultures can remain under continuous observation for techniques such as ratio fluorescence imaging of ion-sensitive dyes for pH or intracellular calcium measurements.

Second, metabolic inhibition appears to be able to model more accurately than substrate deprivation some of the adverse events that occur very early after the reperfusion or recovery phase, in vivo, a phenomenon known as "reperfusion injury." In this type of injury, there is little morphological change in the appearance of neurons after metabolic inhibition. Rather, neurons slowly lose membrane integrity only after many hours of metabolic restoration; no evidence of acute, massive glutamate receptor activation is observed.

In our model, the protective effects of glutamate receptor blockade are found in the recovery phase, corresponding to the delayed cell death we observe. The NMDA receptor antagonist MK-801 has the same effects when present only during recovery as when present during exposure and recovery. Blockade of NMDA receptors only during metabolic inhibition does not protect against injury, whereas, blockade of NMDA receptors during recovery is necessary and sufficient to prevent injury.

One mechanism of injury that may be specific for early reperfusion is damage associated with pH recovery. In cardiac myocytes, Bond et al. demonstrated that cellular injury began when pH returned to normal after simulated ischemia. The recovery of pH and the onset of injury is dependent on the activity of sodium-hydrogen exchangers.

Our laboratory has now demonstrated that the same pH recovery injury occurs in cultured cortical neurons *(50)*. Neuronal injury is greatly decreased when pH recovery after metabolic inhibition is slowed by drugs that block the sodium-hydrogen exchanger. In both neurons and cardiac myocytes it is not yet clear whether the injury is caused by sodium entering in exchange for hydrogen ions, or whether maintenance of low pH inhibits some pH-sensitive mechanisms of injury. While it is known that inhibitors of the exchanger can alter pH recovery in rats after global ischemia, their potential protective effects have not yet been examined.

CALCIUM-MEDIATED INJURY DURING RECOVERY

In unpublished experiments, we have demonstrated that the injury during recovery is calcium dependent. There is substantial protection when extracellular calcium is removed during recovery, but no protection when calcium is removed only during metabolic inhibition. However, at the time the NMDA antagonists can protect against injury, intracellular calcium levels are normal, as measured by calcium-sensitive dyes. Even though free calcium is not elevated, the dependence of injury on extracellular calcium during recovery reinforces the principle that calcium entry into cells during energy stress is a principal determinant of injury.

There are a number of parallels between our model of ischemia and brief exposure to glutamate. It is possible to partially protect neurons during recovery from brief exposure to glutamate. Our calcium measurements are quite similar to some measurements of intracellular calcium in cultured neurons after brief exposure to glutamate, in which intracellular calcium was low, at the same time that addition of NMDA receptor antagonists can prevent injury *(51)*. Even though intracellular calcium may be low, there is calcium entry that mediates injury *(52,53)*. Interestingly, in acute slices, toxic glutamate depletes ATP *(54,55)*. Many of the other ionic changes associated with ischemia, including intracellular acidosis, occur with brief glutamate exposure *(56,57)*.

It is interesting, then, to view brief glutamate exposure as a model of reperfusion, in which neurons are challenged to recover from a brief ionic and energetic insult. Persistent glutamate receptor activation during recovery without apparent excitotoxicity is an important component of injury, preventing recovery. As detailed above, in slower models of ischemia, such as substrate deprivation, glutamate may also play a role in slowly driving neurons to energy failure. In our metabolic inhibition model, glutamate receptor activation may not be necessary in order for neurons to reach energy failure, as in the ischemic core.

There is a potential strong link between calcium accumulation and energy recovery, since mitochondria are a major buffer for intracellular free calcium *(58)*. Mitochondria may use their metabolic gradients to buffer calcium rather than produce ATP to restore cellular homeostasis. Enough mitochondrial calcium accumulation may inhibit electron transport itself *(59,60)* (*see* chapter by Dykens).

These results suggest that metabolic stresses on the recovering cell appear to be somewhat similar to those in the onset period. Sodium, calcium and protons may have accumulated intracellularly during energy failure. The pumps that restore transmembrane gradients, i.e., sodium and calcium ATPases, will be major consumers of ATP. In addition, calcium and hydrogen ions are moved out of the cell through exchangers that use the sodium gradient. Activity of these exchangers will bring in more sodium, further stressing the energy recovery of the cell. Of course, activation of glutamate receptors and voltage-sensitive ion channels will also bring in more sodium and calcium, adding to energetic stress. All of these mechanisms are simple extensions of those discussed during the onset of ischemia.

OXIDATIVE STRESS AND APOPTOSIS

Some mechanisms may be unique to the reperfusion phase. One obvious candidate is oxidative stress caused by the sudden resupply of oxygen to ischemic tissue *(61)*. Oxidative stress in cerebral ischemia may also be produced as a consequence of excitotoxicity and second messenger activation *(62)*. This raises the possibility that oxidative stress may occur during the onset of ischemia, and be exaggerated upon the onset of reperfusion *(63)*. Oxidative stress may contribute to failure of energy recovery once reperfusion occurs *(36)*. Mitochondrial injury and calcium accumulation may also contribute to generation of oxidative stress *(64,65)* (*see* chapters by Dykens and Leist and Nicotera).

Some mechanisms of cellular injury in ischemia models are very delayed. In the gerbil model of global ischemia in which CA1 is selectively injured, high energy phosphates and electrical activity recover rapidly *(38)*. Neurons degenerate only after 3 d or more. Whereas the CA1 region appears morphologically intact before degeneration, some of the signs of cellular stress that have been reported in the penumbra of focal ischemia have been demonstrated, including expression of stress genes *(22)*, glucose hypermetabolism *(66)* and inhibition of protein synthesis. In addition, there is an ongoing accumulation of calcium as measured by radiotracer techniques *(67)*.

Delayed degeneration resembles the process of apoptosis which is associated with programmed cell death in development and normal cell turnover. Some of the biochemical and histological features of apoptosis have been reported in both global and focal ischemia (for a different view, *see* chapter by Olney and Ishimaru in this volume). It may be that cells that survive the initial challenge of energetic recovery succumb to persistent stress during the days after reperfusion. Since apoptosis can be induced by oxidative stress *(68)* and oxidative stress may be caused by mitochondrial dysfunction, there may be a strong association between calcium entry into mitochondria, oxidative stress and the induction of apoptosis in neurons that survive the acute phase of reperfusion.

Quantitatively, the numbers of cells that undergo apoptosis may vary greatly depending on the time course and severity of ischemia. In permanent focal ischemia, apoptotic death may be a small component limited to the periphery of the infarction *(69)*. Apoptosis may be maximal when reperfusion is rapid, allowing recovery of the energy status.

CONCLUSIONS: IMPLICATIONS FOR CLINICAL ISCHEMIA

In animal models of cerebral ischemia, there are powerful methods to measure blood flow and multiple biochemical parameters at well-defined times after the onset of ischemia and reperfusion. Cerebral ischemia in patients is much more difficult to study. As a result, there is no way for the clinician to know whether large parts of the brain are in a low flow situation that could be rescued simply by reflow, in a situation of energy failure in which reflow would have to be accompanied by biochemical protection to aid recovery, or already in a state of inevitable infarction.

It is instructive to compare the situation faced in cerebral ischemia with that found in cardiac ischemia, i.e., patients presenting for medical treatment because of severe chest pain and symptoms of failure of the heart to pump, like shortness of breath and low blood pressure.

Stroke presents with the various clinical syndromes of arterial occlusion. This is analogous to the chest pain used in cardiac ischemia. However, the clinician attempting to diagnose cardiac ischemia has a simple, yet powerful tool—the EKG. The EKG simply records the electrical activity of the heart. The synchronous nature of cardiac depolarization and repolarization allows physiological assessment of the heart with electrical means.

No such simple, noninvasive technique, exists in cerebral ischemia. In the course of this review, several biochemical markers have been mentioned—ATP, glucose hypermetabolism, pH, calcium entry, protein synthesis, glutamate receptor activation. Techniques such as SPECT, PET, or NMR spectroscopy may some day allow the neurologist to follow the process of energy failure and recovery.

This knowledge may be critical because the optimal treatment may depend on the current state of the brain at risk. Simple rescue from energy failure by reperfusion or receptor blockade may work well soon after the onset of symptoms. Once therapies involve reperfusion, other approaches may be required in addition to receptor blockade, including prevention of apoptotic cell death.

Clearly, advances in the clinical therapy of cerebral ischemia will depend on knowledge gained from the multiple experimental systems discussed in this chapter. Each approach has its unique strengths and limitations. It is this author's hope that the simple approach to the mechanisms of neuronal vulnerability presented here will assist investigators in understanding how mechanisms contribute to the clinical syndrome of stroke and guide more focused work on developing clinical strategies to minimize death and disability in patients.

REFERENCES

1. Gatfield PD, Lowry OH, Schulz DW, Passonneau JV. Regional energy reserves in mouse brain and changes with ischaemia and anaesthesia. *J Neurochem* 1966, **13**: 185–195.
2. King LJ, Schoepfle GM, Lowry OH, Passonneau JV, Wilson S. Effects of electrical stimulation on metabolites in brain of decapitated mice. *J Neurochem* 1967, **14**: 613–618.
3. Michenfelder JD, Theye RA. The effects of anesthesia and hypothermia on canine cerebral ATP and lactate during anoxia produced by decapitation. *Anesthesiology* 1970, **33**: 430–439.
4. Laptook AR, Corbett RJ, Burns D, Sterett R. Neonatal ischemic neuroprotection by modest hypothermia is associated with attenuated brain acidosis. *Stroke* 1995, **26**: 1240–1246.
5. Verhaegen M, Iaizzo PA, Todd MM. A comparison of the effects of hypothermia, pentobarbital, and isoflurane on cerebral energy stores at the time of ischemic depolarization *Anesthesiology* 1995, **82**: 1209–1215.
6. Yager JY, Asselin J. Effect of mild hypothermia on cerebral energy metabolism during the evolution of hypoxic-ischemic brain damage in the immature rat. *Stroke* 1996, **27**: 919–925.
7. Schmidt-Kastner R, Freund TF. Selective vulnerability of the hippocampus in brain ischemia. *Neuroscience* 1991, **40**: 599–636.
8. Gahwiler BH. Organotypic monolayer cultures of nervous tissue. *J Neurosci Methods* 1981, **4**: 329–342.
9. Stoppini L, Buchs PA, Muller D. A simple method for organotypic cultures of nervous tissue. *J Neurosci Methods* 1991, **37**: 173–182.
10. Zimmer J, Gahwiler BH. Cellular and connective organization of slice cultures of the rat hippocampus and fascia dentate. *J Comp Neurol* 1984, **228**: 432–446.
11. Newell DW, Malouf AT, Franck JE, Glutamate-mediated selective vunerability to ischemia is present in organotypic cultures of hippocampus. *Neurosci Lett* 1990, **116**: 325–330.
12. Vornov JJ, Tasker RC, Coyle JT. Delayed protection by MK-801 and tetrodotoxin in a rat organotypic hippocampal culture model of ischemia. *Stroke* 1994, **25**: 457–464.
13. Vornov JJ, Tasker RC, Coyle JT. Direct observation of the agonist- specific regional vulnerability to glutamate, NM DA, and kainate neurotoxicity in organotypic hippocampal cultures. *Experimental Neurology* 1991, **114**: 11–22.
14. Tasker RC, Coyle JT, Vornov JJ. The regional vulnerability to hypoglycemia-induced neurotoxicity in organotypic hippocampal culture: protection by early tetrodotoxin or delayed MK-801. *J Neurosci* 1992, **12**: 4298–4308.
15. Vornov JJ, Tasker RC, Park J. Neurotoxicity of acute glutamate transport blockade depends on coactivation of both NMDA and AMPA/Kainate receptors in organotypic hippocampal cultures. *Exp Neurol* 1995, **133**: 7–17.
16. Brierley JB, Brown AW, Meldrum BS. The nature and time course of the neuronal alterations resulting from oligaemia and hypoglycaemia in the brain of Macaca mulatta. *Brain Res* 1971, **25**: 483–499.
17. Salford LG, Plum F, Brierley JB. Graded hypoxia-oligemia in rat brain. II. Neuropathological alterations and their implications. *Arch Neurol* 1973, **29**: 234–238.
18. Branston NM, Symon L, Crockard HA, Pasztor E. Relationship between the cortical evoked potential and local cortical blood flow following acute middle cerebral artery occlusion in the baboon. *Exp Neurol* 1974, **45**: 195–208.
19. Hossmann KA. Viability thresholds and the penumbra of focal ischemia. *Ann Neurol* 1994b, **36**: 557–565.
20. Mies G, Ishimaru S, Xie Y, Seo K, Hossmann KA. Ischemic thresholds of cerebral protein synthesis and energy state following middle cerebral artery occlusion in rat. *J Cereb Blood Flow Metab* 1991, **11**: 753–761.
21. Mies G, Paschen W, Hossmann KA. Cerebral blood flow, glucose utilization, regional glucose, and ATP content during the maturation period of delayed ischemic injury in gerbil brain. *J Cereb Blood Flow Metab* 1990, **10**: 638–645.

22. Nowak T Jr, Jacowicz M. The heat shock/stress response in focal cerebral ischemia. *Brain Pathol* 1994, **4:** 67–76.
23. Shiraishi K, Sharp FR, Simon RP. Sequential metabolic changes in rat brain following middle cerebral artery occlusion: a 2-deoxyglucose study. *J Cereb Blood Flow Metab* 1989, **9:** 765–773.
24. Mies G, Kohno K, Hossmann KA. MK-801, a glutamate antagonist, lowers flow threshold for inhibition of protein synthesis after middle cerebral artery occlusion of rat. *Neurosci Lett* 1993, **155:** 65–68.
25. Simon R, Shiraishi K. N-methyl-D-aspartate antagonist reduces stroke size and regional glucose metabolism. *Ann Neurol* 1990, **27:** 606–611.
26. Ratan RR, Murphy TH, Baraban JM. Macromolecular synthesis inhibitors prevent oxidative stress-induced apoptosis in embryonic cortical neurons by shunting cysteine from protein synthesis to glutathione. *J Neurosci* 1994a, **14:** 4385–4392.
27. Choi DW. NMDA receptors and AMPA/kainate receptors mediate parallel injury in cerebral cortical cultures subjected to oxygen-glucose deprivation. *Prog Brain Res* 1993, **96:** 137–143.
28. Peng L, Hertz L. Potassium-induced stimulation of oxidative metabolism of glucose in cultures of intact cerebellar granule cells but not in corresponding cells with dendritic degeneration. *Brain Res* 1993, **629:** 331–334.
29. Taylor CP. Na+ currents that fail to inactivate. *Trends Neurosci* 1993, **16:** 455–460.
30. Siesjo BK. Pathophysiology and treatment of focal cerebral ischemia. Part I: Pathophysiology. *J Neurosurg* 1992, **77:** 169–184.
31. Siesjo BK, Ekholm A, Katsura K, Theander S. Acid-base changes during complete brain ischemia. *Stroke* 1990, **21:** 194–199.
33. Obrenovitch TP, Richards DA. Extracellular neurotransmitter changes in cerebral ischaemia. *Cerebrovasc Brain Metab Rev* 1995, **7:** 1–54.
34. Szatkowskik M, Attwell D. Triggering and execution of neuronal death in brain ischaemia: two phases of glutamate release by different mechanisms. *Trends Neurosci* 1994, **17:** 359–365.
35. Deshpande J, Bergstedt K, Linden T, Kalimo H, Wieloch T. Ultrastructural changes in the hippocampal CA1 region following transient cerebral ischemia: evidence against programmed cell death. *Exp Brain Res* 1992, **88:** 91–105.
36. Kim H, Koehler RC, Hurn PD, Hall ED, Traystman RJ. Amelioration of impaired cerebral metabolism after severe acidotic ischemia by tirilazad posttreatment in dogs. *Stroke* 1996, **27:** 114–121.
37. Nishijima MK, Koehler RC, Hurn PD, Eleff SM, Norris S, Jacobus WE, Traystman RJ. Postischemic recovery rate of cerebral ATP, phosphocreatine, pH, and evoked potentials. *Am J Physiol* 1989, **257:** H1860–1870.
38. Arai H, Passonneau JV, Lust WD. Energy metabolism in delayed neuronal death of CA1 neurons of the hippocampus following transient ischemia in the gerbil. *Met Brain Dis* 1986, **1:** 263–278.
39. Gill R, Foster AC, Woodruff GN. MK-801 is neuroprotective in gerbils when administered during the post-ischaemic period. *Neuroscience* 1988, **25:** 847–855.
40. Hossmann KA. Glutamate-mediated injury in focal cerebral ischemia: the excitotoxin hypothesis revised. *Brain Pathol* 1994a, **4:** 23–36.
41. Buchan R, Pulsinelli WA. Hypothermia but not the N-methyl-D- aspartate antagonist, MK801, attenuates neuronal damage in gerbils subjected to transient global ischemia. *J Neurosci* 1990, **10:** 311–316.
42. Rothman SM. The neurotoxicity of excitatory amino acids is produced by passive chloride influx. *J Neurosci* 1985, **5:** 1483–1489.
43. Ellren K, Lehmann A. Calcium dependency of N-methyl-D-aspartate toxicity in slices from the immature rat hippocampus. *Neuroscience* 1989, **32:** 371–379.

44. Garthwaite J, Garthwaite G. Mechanisms of excitatory amino acid neurotoxicity in rat brain slices. *Adv Exp Med Biol* 1990, **268**: 505–518.
45. Olney JW, deGubareff T, Sloviter RS. "Epileptic" brain damage in rats induced by sustained electrical stimulation of the perforant path. II. Ultrastructural analysis of acute hippocampal pathology. *Brain Res Bull* 1983, **10**: 699–712.
46. Dietrich WD, Busto R, Yoshida S, Ginsberg MD. Histopathological and hemodynamic consequences of complete versus incomplete ischemia in the rat. *J Cereb Blood Flow Metab* 1987, **7**: 300–308.
47. Kalimo H, Paljarvi L, Vapalahti M. The early ultrastructural alterations in the rabbit cerebral and cerebellar cortex after compression ischaemia. *Neuropathol Appl Neurobiol* 1979, **5**: 211–223.
48. Simon RP, Griffiths T, Evans MC, Swan JH, Meldrum BS. Calcium overload in selectively vulnerable neurons of the hippocampus during and after ischemia: an electron microscopy study in the rat. *J Cereb Blood Flow Metab* 1984, **4**: 350–361.
49. Vornov JJ. Toxic NMDA receptor activation occurs during recovery in a tissue culture model of ischemia. *J Neurochem* 1995, **65**: 1681–1691.
50. Vornov JJ, Thomas AG, Jo D. Protective effects of extracellular acidosis and blockade of sodium/hydrogen ion exchange during recovery from metabolic inhibition in neuronal tissue culture. *J Neurochemistry*, 1996, **67**: 2379–2388.
51. Randall RD, Thayer SA. Glutamate-induced calcium transient triggers delayed calcium overload and neurotoxicity in rat hippocampal neurons. *J Neurosci* 1992, **12**: 1882–1895.
52. Hartley DM, Choi DW. Delayed rescue of N-methyl-D-aspartate receptor-mediated neuronal injury in cortical culture. *J Pharm Exp Ther* 1989, **250**: 752–758.
53. Hartley DM, Kurth MC, Bjerkness L, Weiss JH, Choi DW. Glutamate receptor-induced 45Ca2+ accumulation in cortical cell culture correlates with subsequent neuronal degeneration. *J Neurosci* 1993, **13**: 1993–2000.
54. Djuricic B, Rohn G, Paschen W, Hossmann KA. Protein synthesis in the hippocampal slice: transient inhibition by glutamate and lasting inhibition by ischemia. *Metab Brain Dis* 1994, **9**: 235–247.
55. Whittingham TS, Assaf H, Selman WR, Ratcheson RA, Lust WD. Glutamate-induced energetic stress in hippocampal slices: evidence against NMDA and glutamate uptake as mediators. *Metab Brain Dis* 1992, **7**: 77–92.
56. Hartley Z, Dubinsky JM. Changes in intracellular pH associated with glutamate excitotoxicity. *J Neurosci* 1993, **13**: 4690–4699.
57. Irwin RP, Lin SZ, Long RT, Paul SM. N-methyl-D-aspartate induces a rapid, reversible, and calcium-dependent intracellular acidosis in cultured fetal rat hippocampal neurons. *J Neurosci* 1994, **14**: 1352–1357.
58. Thayer SA, Miller RJ. Regulation of the intracellular free calcium concentration in single rat dorsal root ganglion neurones in vitro. *J Physiol* 1990, **425**: 85–115.
59. Duchen MR. Ca(2+)-dependent changes in the mitochondrial energetics in single dissociated mouse sensory neurons. *Biochem J* 1992, **283**: 41–50.
60. Duchen MR, Biscoe TJ. Relative mitochondrial membrane potential and [Ca2+]i in type I cells isolated from the rabbit carotid body. *J Physiol* 1992, **450**: 33–61.
61. Chan PH. Role of oxidants in ischemic brain damage. *Stroke* 1996, **27**: 1124–1129.
62. Coyle JT, Puttfarcken P. Oxidative stress, glutamate, and neurodegenerative disorders. *Science* 1993, **262**: 689–695.
63. Dugan LL, Lin TS, He YY, Hsu CY, Choi DW. Detection of free radicals by microdialysis/spin trapping EPR following focal cerebral ischemia-reperfusion and a cautionary note on the stability of 5,5-dimethyl-1-pyrroline N-oxide (DMPO). *Free Radic Res* 1995, **23**: 27–32.
64. Chacon E, Acosta D. Mitochondrial regulation of superoxide by Ca2+: an alternate mechanism for the cardiotoxicity of doxorubicin. *Toxicol Appl Pharmacol* 1991, **107**: 117–128.

65. Dawson TL, Gores GJ, Nieminen AL, Herman B, Lemasters JJ. Mitochondria as a source of reactive oxygen species during reductive stress in rat hepatocytes. *Am J Physiol* 1993, **264**: C961–C967.
66. Pulsinelli WA, Levy DE, Duffy TE. Regional cerebral blood flow and glucose metabolism following transient forebrain ischemia. *Ann Neurol* 1982, **11**: 499–502.
67. Araki T, Inoue T, Kato H, Kogure K, Murakami M. Neuronal damage and calcium accumulation following transient cerebral ischemia in the rat. *Mol Chem Neuropathol* 1990, **12**: 203–213.
68. Ratan RR, Murphy TH, Baraban JM. Oxidative stress induces apoptosis in embryonic cortical neurons. *J Neurochem* 1994b, **62**: 376–379.
69. Johnson EM JR, Greenlund LJ, Akins PT, Hsu CY. Neuronal apoptosis: current understanding of molecular mechanisms and potential role in ischemic brain injury. *J Neurotrauma* 1995, **12**: 843–852.

17
Epilepsy and Cell Death

Cheolsu Shin and Ki-Hyeong Lee

INTRODUCTION

Epilepsy is one of the most common neurological disorders afflicting mankind. It is a syndrome of unpredictable spontaneous recurrence of seizures. Its prevalence ranges from 1.5 to 19.5 per 1000 depending on the geographical area and the ethnic group *(1)*. Despite all the misconceptions and prejudicial history since the first description of this malady in Hippocrates' "On the Sacred Disease" 2500 yr ago, there have been dramatic improvements in diagnosis and treatment thanks to the modern brain imaging technology such as magnetic resonance image (MRI), positron emission tomography (PET), and single photon emission tomography (SPECT) as well as various surgical approaches to medically intractable epilepsies. Compared with these marked improvements in diagnostic and treatment modalities, the understanding of the basic pathophysiologic mechanisms of epilepsy remains rather elusive.

One of the most difficult problems encountered in elucidating the pathophysiologic mechanisms of human epilepsy is that the cause cannot be precisely dissected out from the effect. The pathological substrates found in the epileptic brain may contain both the primary causative lesion and the secondary changes resulting from the seizures themselves or the attendant hypoxia. This is especially important in understanding the pathology of complex partial epilepsy, the most common form of acquired epilepsy, because the hippocampus is frequently the epileptic focus and is also one of the most vulnerable areas to hypoxia which may be associated with the seizure activity. Another difficulty is the need for appropriate control specimen for comparison with the epileptic tissue. To get around these problems in the interpretation of human brain pathology, many animal models of seizures and epilepsy have been developed.

Epilepsy is usually classified by the clinical and electroencephalographic manifestations of the seizures. Seizures are defined as the episodic dysfunction of the central nervous system due to abnormal and excessive synchronized electrical activities in a group of neurons. About 40% of patients have primary generalized seizures (absence, tonic-clonic, myoclonic) while 60% have partial seizures (simple or complex). Of the partial seizures, complex partial seizures involving the temporal lobes have been of great interest to the researchers, not only because it accounts for roughly one-half of all partial seizures and one-third of all epilepsies *(1)*, but also comprises most of the medi-

From: Cell Death and Diseases of the Nervous System
Edited by: V. E. Koliatsos and R. R. Ratan © Humana Press Inc., Totowa, NJ

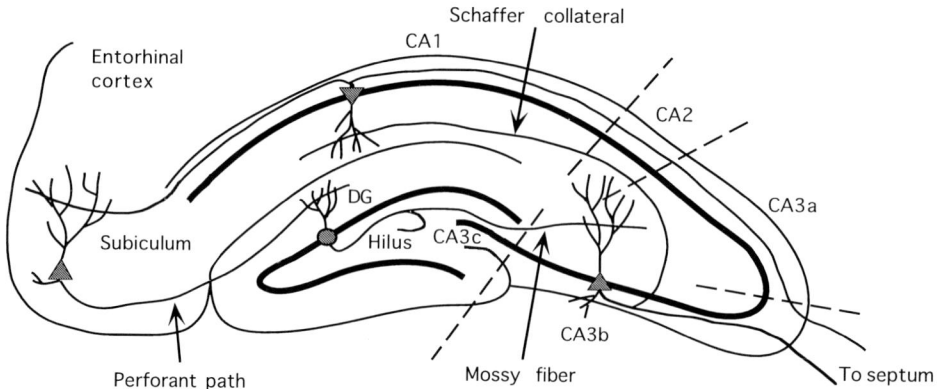

Fig. 1. Schematic diagram of the hippocampal formation and its synaptic connections. "DG," Dentate gyrus.

cally intractable cases that are amenable to surgical treatment. The availability of the specimen from surgical resection of epileptogenic temporal lobe has been a great impetus for more studies in the human epileptic brain, providing further clinical and pathological correlations.

In this chapter, we will review the historical evolution of concepts and theories regarding mechanisms of epileptic cell death and its implication on epileptogenesis in the human brain. Then, several animal models of human epilepsy will be described. The issue of the mode of cell death implicated in epilepsy will be addressed in detail. Finally, the cell death pathways believed to play a critical role in epileptogenesis will be discussed along with its clinical relevance and future directions. We will begin with a review of essential hippocampal anatomy.

ANATOMICAL CONSIDERATIONS

The hippocampal formation is easily identified on the mesial surface of the temporal lobes and extends along the floor of the temporal horn of the lateral ventricle. Three major subdivisions are described as the dentate gyrus, the hippocampus proper, and the subicular complex which is part of the parahippocampal gyrus contiguous with the hippocampal fields. Ammon's horn, or cornu ammonis (thus CA), is a classic anatomic term describing the hippocampus proper. It is the term used most widely to describe the subfields of the hippocampus: CA1, CA2, CA3, and CA4 (dentate hilus). CA1 is also called Sommer's sector.

On transverse section, the hippocampus is divided into four subfields: CA1, CA2, CA3, and dentate hilus (Fig. 1). CA2 is between CA1 and CA3 but is not well demarcated. The area of CA3 adjacent to the CA2 region are termed CA3a. CA3b is approximately the middle third and CA3c is the terminal portion between the superior and inferior blades of the dentate gyrus. The main cells of the hippocampus are the pyramidal cells with both their basal and apical dendrites. The axons of the pyramidal cells branch into collaterals that project within and outside the hippocampus. The collateral axons of the CA3 pyramidal cells which synapse onto the apical dendrites of the CA1 pyramidal cells are called Schaffer collaterals. The second major cell type is the basket cell which contains GAD (glutamic acid decarboxylase), the γ-aminobutyric acid (GABA) synthesizing enzyme, and exerts an inhibitory action on the pyramidal cells.

The dentate gyrus is a three-layered structure consisting of the stratum moleculare, stratum granulosum, and the polymorphic layer (from outside to inside). The dentate granule cells are the main cell type of the dentate gyrus. Mossy cells and other interneurons are found in the polymorphic layer in the dentate hilus (CA4). The axons of granule cells, referred to as mossy fibers, contain zinc and therefore stain with Timm's reagent. They make excitatory synaptic contacts with the proximal portion of the apical dendrites of CA3 pyramidal cells in the stratum lucidum. Mossy fibers also innervate the soma and dendrites of polymorphic interneurons (basket cells and mossy cells) in the dentate hilus. The dendrites of granule cells branch into the molecular layer where the perforant pathway (the main hippocampal afferent system originating from the entorhinal cortex), makes synaptic contacts. The granule cells then project by way of the mossy fibers to the CA3 pyramidal cells, and CA3 pyramidal cells send Schaffer collaterals to CA1 pyramidal cells. CA1 pyramidal cells then project to the subicular complex, which in turn projects to the entorhinal cortex. The aminoacid glutamate is the major neurotransmitter in these excitatory synaptic pathways (Fig. 1).

The dentate granule cells are glutamatergic excitatory neurons and contain dynorphin and cholecystokinin (CCK) in the terminal boutons of their axons (mossy fibers). Inhibitory interneurons containing GAD exist throughout the hippocampus. Basket cells are GABAergic inhibitory interneurons and are also immunoreactive for CCK. Somatostatin (SS) and Neuropeptide Y (NPY)-immunoreactive interneurons are found in the hilar region and stratum oriens of the CA1 subfields. The mossy cells are found in the dentate hilar region and have excitatory synapses onto the GABA interneurons (basket cells); this setting provides tonic baseline inhibition and also mediates phasic feedback inhibition to the granule cells.

Although glutamate receptors are ubiquitously found in cortex, there are some notable differences in the distribution of receptor subtypes *(2)*. In neocortex, NMDA receptors are particularly dense within the outer two layers, whereas high-affinity kainate receptors predominate in layers 5 and 6. In the hippocampus, high-affinity kainate receptors are found in high concentration around the mossy fiber terminals (CA3 stratum lucidum). In contrast, NMDA receptors are absent in the CA3 stratum lucidum, but are very dense in the basal and apical dendritic fields of CA1 (stratum oriens and radiatum, respectively). AMPA receptors like the NMDA receptors are in very high in density over the CA1 pyramidal cell bodies and dendritic layers.

HISTORICAL ASPECTS

The term "epileptic brain damage" has traditionally been used to refer to the selective pattern of damage found in patients with chronic epilepsy. More recently, it is also used to refer to the similar pattern of damage described in animals or humans after prolonged seizures. These two pathological patterns are similar but not identical, as the former necessarily includes lesions from the original insult, such as trauma or ischemia. Whether and how these two similar patterns are causally related is of course the critical question facing the field of experimental epilepsy.

Since Bouchet and Cazauvieilh *(3)* first noted that one or both Ammon's horns in the brains of some epileptic patients were small and "hard," the association of epilepsy with selective neuronal death and structural alteration in the hippocampus has been recognized by many neuropathologists. Margerison and Corsellis *(4)* found that one or both hippocampi were affected in 50–60% of cases of chronic occult epilepsy. There have been two contrasting theories about the implication of this brain pathology. The first one states that the development of hippocampal lesions precede the onset of recurrent seizures and, therefore, are themselves epileptogenic. This view was immediately criticized because in many patients with epilepsy no demonstrable abnormality could be found in the hippocampus or elsewhere in the brain. The second theory proposed that this lesion was merely the byproduct of repeated seizures. Thus, Spielmeyer found neuronal and glial abnormalities in the hippocampus or cerebellum in 80% of 127 epileptic brains and speculated that the vascular spasm occurring at the onset of each epileptic attack may result in subsequent brain damage *(5)*. Penfield and his colleagues *(6)*, on the other hand, suggested that the initial damage to hippocampus happens during birth by deformation of the skull and herniation of the medial temporal gyri through the incisure of the tentorium, leading to compromised blood supply to the temporal cortex via the posterior cerebral artery. This condition would result in ischemic injury to the mesial temporal gray matter with the resulting hippocampal lesion, "incisural sclerosis," becoming the epileptogenic focus. Although these theories proved not to be valid, they brought attention to the possibility of initial detrimental events early in life, whether ischemic or not, as causal mechanism of epilepsy later in life. Febrile convulsions and prolonged seizures early in life were reported to be associated with later development of epilepsy, even though later population based-studies showed no clear relationship between febrile convulsions and complex partial seizures. It now appears that a subpopulation of febrile convulsion cases may be more susceptible to development of epilepsy *(7)*. It remains to be seen whether this subpopulation of patients with febrile convulsion already has "epileptic brain damage" or will develop the damage as a result of the convulsions.

In pathological studies, hippocampal neuronal death and gliosis were the prominent findings *(8)*. Microscopically, the affected neurons showed the typical "ischemic nerve cell damage" *(5)*, which was originally described after hypoxic or ischemic neuronal insults. In this type of cell death, the cell body becomes scarcely visible and shows a shrunken and eosinophilic cytoplasm with a small condensed dark nucleus. The similarity of microscopic findings between the epileptic neuronal damage and ischemic cell death had been attributed to the fact that cerebral anoxia/ischemia contributed to the pathology after prolonged seizures. However, in the experimental model of status epilepticus, the damage following status epilepticus correlates with the excessive neuronal activity itself, but not with systemic factors such as hypoxia and hypotension *(9)*. Therefore, it is now accepted that excitotoxicity is likely the major pathophysiologic mechanism of epileptic brain damage.

The exact molecular mechanism of epileptic neuronal death and its relevance to the development of epilepsy, i.e., epileptogenesis, have recently been intensively studied. The critical questions can be summarized as follows.

1. Do recurrent epileptic seizures cause selective neuronal damage?
2. If so, what are the mechanism and mode of epileptic cell death?
3. What are the critical steps involved in epileptic cell death?
4. Does this selective neuronal deaths result in epilepsy?

DO RECURRENT EPILEPTIC SEIZURES CAUSE SELECTIVE NEURONAL DAMAGE?

Pathology of Human Temporal Lobe Epilepsy

Severe neuronal loss was found in the hippocampus of temporal lobe epilepsy patients who experienced prolonged seizures during early childhood *(10)*. Sustained seizures in early life are often followed by cognitive impairment *(11)*, cerebral atrophy and histological lesions *(12)*. In the epileptic human hippocampus, severe status epilepticus is associated with selective injury to the prosubiculum, CA1, and CA3 *(13)*.

Over the years, many terms have been used to describe the specific pattern of sclerosis found in the temporal lobe of patients with complex partial epilepsies: incisural sclerosis, pararhinal sclerosis, and mesial temporal sclerosis, as well as a subtype called end-folium sclerosis. Because the histologic abnormality found in the epileptic temporal lobe extends into the amygdala, subiculum, and parahippocampal gyrus, the most appropriate term seems to be mesial temporal sclerosis (MTS).

Most of the neuropathology of mesial temporal sclerosis is concentrated in the Ammon's horn. There is a loss of pyramidal neurons in the hippocampus with gliosis and shrinkage of the neuropil. Neuronal death is marked in the most susceptible fields, CA1 and dentate hilus (CA4), with less severe damage in the dentate gyrus and CA3 and minimal damage in CA2. With neuronal loss and reactive gliosis, the volume of the hippocampus is reduced, thus resulting in atrophic and sclerotic hippocampus. Milder sclerosis may involve only the dentate hilus region which has been termed end-folium sclerosis by Margerison and Corsellis *(4)*.

Immunohistochemically, somatostatin- and neuropeptide Y-immunoreactive interneurons and mossy cells are selectively lost *(14)*, whereas GAD-immunoreactive (GABAergic) interneurons are relatively preserved *(15)* in the dentate hilus. The previous selective pattern of neuronal damage is very similar to kainate-induced damage *(16)* and electrically induced status epilepticus *(17)*, thus providing abnormalities support for the usefulness of these animal models. Also, in the epileptic human brain, the selective loss of pyramidal neurons and interneurons in the hippocampus is accompanied by sprouting of the mossy fibers to innervate the inner molecular layer of the dentate gyrus *(18)*. These aberrant synaptic connections have been presumed to be an anatomical substrate contributing to hyperexcitability *(19)*.

To assess electrophysiological abnormalities, the excised hippocampus can be sliced in thin sections and perfused with oxygenated artificial CSF, a medium allowing temporary survival of the neurons in vitro. Masukawa et al. *(20)* reported a significant increase in the synaptic hyperexcitability of dentate granule neurons in cases of mesial temporal sclerosis compared to cases with lesions outside the hippocampus. Others have reported an increased tendency for burst firing and the presence of the late NMDA receptor-mediated component of excitatory postsynaptic potential *(21)*. These electrophysiologic findings seem to be consistent with the results of experimental studies on animals, a coincidence supporting the validity of these animal models *(22)*.

Changes in neurotransmitter receptors may also be linked to the synaptic hyperexcitability of neurons. Since receptor proteins do not have a rapid turnover and are well preserved in surgical specimens and postmortem brains, receptor binding studies using radioactive ligands are feasible in these tissues. GABA and glutamate are the principal inhibitory and excitatory neurotransmitters and benzodiazepines augment GABA receptor function. Their respective receptor systems have been examined in MTS using in vitro receptor autoradiography. Although the density of benzodiazepine receptors was decreased in MTS, when corrected for the loss of neurons, the receptor density per surviving neuron appeared to be unchanged *(23)*. Results from studies of the glutamate receptor system with various ligands for subtypes, NMDA and AMPA/quisqualate, have not been consistent. Increases as well as decreases in the binding of these ligands were found when corrected for neuronal density in MTS. Various investigators also had conflicting findings in the anatomical distribution of these ligands, making interpretation of all these results difficult *(24,25)*. Difference in the selection of controls may be one of the many explanations for these conflicting data.

Models of Experimentally Induced Epilepsy

The experimental models of status epilepticus (SE) have many advantages in investigating the damage induced by epileptic seizures. In addition to the reproducibility and ability to control various parameters, the direct toxic effect on neurons from the chemoconvulsant can be separated from the epileptic damage by blocking the seizure activity with anticonvulsants.

Chemoconvulsant-Induced Status Epilepticus

In this group, anti-GABAergic, cholinergic, or glutamatergic drugs are usually administered systemically to induce SE. Bicuculline, a $GABA_A$ receptor antagonist, induces SE with resulting neuronal damage in cerebral cortex and hippocampus *(26,27)*. Allylglycine, which is believed to inhibit the GAD and suppress GABA synthesis, also induces SE. In this model of SE, cell death was found in interneurons in the CA1, CA3, and dentate hilus of hippocampus. The cholinomimetic drug pilocarpine with/without lithium can induce SE with neuronal damage in neocortex, thalamic and amygdala nuclei, and the hippocampus *(28)*. Glutamatergic excitotoxicity likely underlies these damages, since the epileptic discharge and neuronal damage can be blocked by the NMDA antagonists CPP and MK-801 *(29)*.

The glutamate receptor subtype agonist and neurotoxin kainic acid (KA) can induce, by either systemic or intracranial injection, limbic SE in rodents which is clinically similar to the human complex partial seizures. Damaged areas induced by direct intracerebral injection into amygdala include the ipsilateral amygdala, hippocampal formation, and the contralateral hippocampus *(30)*. The primary neuronal damage is mainly found in hippocampal CA3 and less in CA1. This pattern of selective cell damage correlates with high-affinity KA receptor distribution in the hippocampus *(31)*.

Electrical Stimulation-Induced SE

Sustained electrical stimulation of the perforant pathway with trains of pulses through an implanted electrode in an anesthetized rat produces neuronal damage morphologically similar to that of KA-induced SE *(17)*. Distinct populations of dentate hilar neurons and CA3 and CA1 pyramidal cells are injured. This selective neuronal damage is accompanied by the loss of paired pulse inhibition, which normally represents recurrent inhibition of the dentate granule neurons. Recurrent inhibition in the granule cell layer is believed to be mediated by inhibitory interneurons in the dentate hilus *(32)*. Similar to human epileptic hippocampus, there is no significant loss of inhibitory GABA-containing basket cells in the dentate hilus *(15)*. In contrast, there is nearly complete loss of SS-immunoreactive interneurons and mossy cells. Sloviter suggested that the selective vulnerability of different cell populations may be related to relative differences in the concentration of calcium buffering proteins such as Calbindin-D_{28k} or parvalbumin *(33)*. By using calcium-binding protein (Calbindin-D_{28k}) and parvalbumin immunohistochemistry, he showed a positive correlation between selective death of hippocampal neurons and lack of calcium buffering proteins. In support of this idea, these vulnerable dentate hilar mossy cells were found to become resistant to the damages induced by perforant pathway stimulation when they were perfused with the calcium chelator BAPTA *(34)* (for a therapeutic use of these compounds, *see* chapter by Tymianski). However, this selective loss of neurons may not be explained by calcium buffering alone, since calcium binding protein immunoreactivity does not always correlate with neuronal vulnerability *(35)*. Other factors contributing to selective neuronal damage may include a lower threshold for activation (and therefore damage) of these vulnerable neurons by perforant path stimulation compared to a more resistant population, such as dentate granule cells *(36)*. The NMDA receptor/channel antagonist MK-801 reduces the damage to SS-immunoreactive interneurons. These findings suggest that excessive intracellular calcium mediated by NMDA receptor/channel play a major role in neuronal cell death induced by sustained electrical stimulation (for a topical discussion of calcium as mediator of cell death, *see* chapter by Leist and Nicotera in this volume).

WHAT IS THE MECHANISM OF CELL DEATH INDUCED BY THE EPILEPTIC SEIZURE?

Excitotoxicity appears to mediate neuronal death induced by epileptic seizure for the following reasons. First, the pattern of selective damage has a close relationship with the network across which seizure activity is propagated. For example, in rodents, focal neocortical seizure activity can induce thalamic damage *(37)*, and limbic seizures produce damage in a wide range of nuclei involved in limbic seizure activity. Second, the cytopathological changes occurring in vulnerable hippocampal or cortical neurons are quite similar to the changes induced by glutamate, KA, or other excitotoxic agents. In experimentally induced SE, the affected neurons show focal dendritic swellings containing distended mitochondria and increased calcium; nuclei with condensed chromatin; and perikarya with dilated Golgi apparatus and endoplasmic reticulum and swollen mitochondria, progressing to dense condensation of the cytoplasm with intense vacuolation *(26,27)*. Similar pathological changes are seen in susceptible dentate hilar neurons and CA3 pyramidal neurons by sustained electrical stimulation of the perforant pathway *(22,38)*. Third, NMDA antagonists can protect against the acute neuronal damage induced by epileptic seizures in various animal models *(39)*.

Admittedly, there are no consistent correlations between the anatomic distribution of glutamate receptor subtypes and selective neuronal loss (either in SE or the chronic epileptic brain), except for some correlation of the selective neuronal loss in KA-induced SE to the distribution of KA receptors *(31)*. In fact, as noted above, difference in calcium-buffering capacity in select populations of cells may be more responsible for specificity of damage than the rather ubiquitous distribution of glutamate receptors in these excitatory pathways *(33)*, as the initial critical event leading to cell death in excitotoxicity is likely the excessive intracellular free calcium. More insight may be forthcoming, as cloning of glutamate receptors provides powerful molecular tools to study the roles of various subunits.

WHAT MAY BE THE CRITICAL STEPS IN EPILEPTIC CELL DEATH?

Excessive electrical discharges induce massive release of excitatory neurotransmitters that activate receptors, open ionic channels, and presumably induce an excessive increase in intracellular calcium. However, the biochemical and molecular mechanisms that are subsequently activated leading to neuronal death are not yet fully understood. Possibilities include many calcium-mediated intracellular processes such as the calcium/calmodulin system with activation of CaM Kinase II, the c-AMP pathway, the nitric oxide and free radical pathways, and other protein kinases, proteases and nucleases.

Further downstream in these intracellular signal transduction pathways is a group of inducible immediate early genes (IEG) such as c-*fos* or c-*jun* which are sometimes called "third messengers." The protein products of these genes regulate the expression of other target genes by acting as DNA-binding transcription factors *(40)*. For the past few years, the evidence has accumulated that these IEG may be playing a key role in programmed cell death in several different settings. Smeyne et al. *(41)* have shown that prolonged c-*fos* expression in neurons precedes programmed cell death in vivo and in vitro. c-*jun* expression has been shown to precede oligonucleosomal-sized DNA fragmentation and cell death produced by chemotherapeutic drugs in astrocytoma cell lines *(42)*. Colotta et al. *(43)* reported that they could reduce cell death in NGF-deprived lymphoid cell lines by blocking *c-fos* and *c-jun* expression with antisense oligonucleotide strategies.

Evidence has accumulated that IEG may play a critical role in SE and hypoxia/ischemia-induced neuronal death. Morgan et al. *(44)* noted that seizures induce the expression of IEG such as c-*fos* in most hippocampal neurons. It has been also demonstrated that dentate granule cells and hippocampal pyramidal neurons express c-*fos* mRNA in the course of kindled seizures *(45)*. Studies with transgenic mice with β-galactosidase expression regulated by c-fos promoter shows that with repeated seizure activities induced by systemic administration of kainate, β-galactosidase/c-*fos* activity is induced in the hippocampus with a delayed time course in some hippocampal neurons destined to die, as well as immediately after the seizures *(41)*. In contrast, transient cerebral ischemia was reported not to be accompanied by prolonged expression of c-*fos* mRNA or protein in vulnerable regions of the gerbil hippocampus *(46)*. Dragunow et al. *(47)* reported that a severe hypoxia/ischemia insult suppresses IEG expression, whereas a moderate insult induces delayed expression of c-*jun* leading to apoptotic cell death. It is presumed that, with severe hypoxia/ischemia insults, neurons die rapidly by necrosis, so that protein synthesis is inhibited. It was recently shown that moderate hypoxia-ischemia causes delayed apoptotic neuronal death, whereas severe insult produces necrotic cell death *(48)*. Also, in a model of SE-induced brain injury, c-*jun* and c-*fos* were shown to be expressed for prolonged periods in dying hippocampal neurons, whereas neurons that survive SE showed only seizure- induced transient expression (dentate granule cells) or no expression (CA2 pyramidal cells) *(49)*. Expression of these IEG was shown to be mediated mainly through NMDA receptors. Even though these experiments provided possible correlation between IEG and delayed neuronal death in

SE, the causal relationship has not yet been established. The use of antisense oligonucleotide and transgenic mouse strategies may be useful to provide further support for this hypothesis.

WHAT IS THE MODE OF CELL DEATH ASSOCIATED WITH EPILEPTIC DAMAGE?

Apoptosis and necrosis represent two prominent modes of cell death. Traditionally, it was suggested that physiologically appropriate cell death is due to apoptosis and that pathological mechanisms involve necrosis. Acute excitotoxicity has been found to be associated with necrosis *(50)* characterized by cell swelling followed by rupture. However, there are several features in delayed excitotoxicity that could be attributed to apoptosis.

1. The selective and scattered pattern of neuronal death found in delayed ischemic or epileptic brain damage is more consistent with apoptosis than necrosis *(51)*.
2. The delayed time course of neuronal death after ischemia or epileptic insult is more likely to be due to active, programmed cell death, rather than passive, accidental cell death.
3. There appears to be a correlation between prolonged increase of immediate early gene expression (c-*fos*, c-*jun*, etc.) and neuronal damage after ischemia *(48,52)* and SE *(41,49,53)* which suggests that certain set of genes may be activated by these transcription factors in an orderly fashion during excitotoxic cell damage. In addition, there have been some reports that excitotoxic damages are blocked by inhibitors of macromolecular synthesis (cycloheximide or actinomycin D) *(54,55)*.

There are at least three criteria utilized to determine whether apoptosis or programmed cell death occurs in a given system: dependence on *de novo* protein synthesis, DNA fragmentation into oligonucleosomal-sized fragments, and cell shrinkage, condensation, and fragmentation of chromatin (morphological criteria) *(40)* (*see* chapters by Clarke and by Olney and Ishimaru).

In animal models of epilepsy, whether neuronal death depends on RNA and/or protein synthesis remains unclear. In KA-induced SE in rats, cycloheximide has been shown to prevent cell death as well as c-*fos* expression *(54)*. However, this finding does not prove that the cell death is an active, programmed event requiring *de novo* protein and RNA synthesis, as cycloheximide may have other effects. Cycloheximide can attenuate seizures induced in rats by lithium and pilocarpine *(56)*.

Contrary to the presumed protective action of cycloheximide in KA-induced SE, much evidence to the contrary exists in other circumstances of cell death. For example, protein synthesis inhibitors have been shown not to have any protective effect on glutamate neurotoxicity in vivo and in vitro *(57,58)*. In human tumor cell lines, cycloheximide and actinomycin D can even exacerbate or induce apoptosis, presumably because they inhibit the ongoing synthesis of necessary proteins such as growth factors *(59)*. Therefore, the dependence of cell death on RNA/protein synthesis may not be an absolute criterion for either apoptosis or necrosis in epileptic brain damage.

DNA fragmentation into multiples of 180 base-pair oligonucleosomes on gel electrophoresis has been considered by many a biochemical hallmark common to all forms of apoptotic cell death. Glutamate infusion into the hippocampus of rats produces DNA fragmentation *(55)*. DNA fragmentation is also detected in KA-induced SE *(60)*. On the contrary, Finiels et al. *(61)* could not detect DNA fragmentation either by gel electrophoresis or by *in situ* detection methods (TUNEL staining) in glutamate agonist-treated cortical cell cultures, despite the fact that the dying cells showed typical apoptotic morphology and neuronal death was inhibited by macromolecular synthesis inhibitors. Furthermore, Ignatowicz et al. *(62)* could not detect DNA fragmentation in KA-injected rat brain, although they did not observe the epileptic seizures. Finally, glutamate-induced cell death in cerebellar granule cell cultures was neither accompanied by DNA fragmentation nor blocked by protein synthesis inhibitors *(58)*.

Although internucleosomal DNA fragmentation has been considered a major biochemical feature of apoptosis, it is now known that key morphological features of apoptosis such as chromatin condensation may occur without DNA fragmentation *(63)*. Recent studies indicate that a certain type of calcium-activated endonuclease can cause single strand breaks, an effect rendering the internucleosomal fragmentation of DNA undetectable on ordinary gel electrophoresis *(64)*. In some cell systems, the chromosomal DNA may be partially degraded into large size fragments by endonucleases at the beginning of the apoptotic process, but not proceed all the way to the DNA laddering *(65)*. To further complicate the picture, there is a report that necrotic cell death can be accompanied by internucleosomal DNA cleavage in some tumor cell lines *(66)*.

Thus, the only reliable method for conclusively determining apoptosis may be the morphologic criteria through light microscopic and, especially, electron microscopic (EM) examination. As described previously, the microscopic findings on affected cells in epileptic hippocampus are quite similar to those seen in ischemic cell death, presumably because both ischemia and epilepsy may involve similar excitotoxic mechanisms, although ischemic cell death traditionally had been considered "necrotic."

In a classic EM study of acute hippocampal pathology induced by sustained electrical stimulation of the perforant path *(38)*, typical profiles of apoptotic cells were found, as well as pathological features of excitotoxicity, such as swelling of dendrites but sparing of the axons. According to this study, cell bodies of affected CA1 and CA3 pyramidal neurons and various interneurons in the hilus typically showed two distinct type of changes; "dark" cell and "nondark" cell changes. The "dark" cell type of degenerating neurons showed shrunken, homogeneously condensed cytoplasm, scattered vacuolar spaces enclosed in the membranes of endoplasmic reticulum, and a compacted nucleus. It was often impossible to observe the cytoplasmic detail because of extreme cell condensation. On the other hand, "nondark" degenerating neurons had swollen endoplasmic reticulum, sometimes with membrane-bound vacuoles (probably autophagic vacuoles), whereas the mitochondria appeared condensed. In both types of degeneration, the nucleus showed coarse chromatin clumping with an intact nuclear envelope in early stages. Similar changes were described in the hippocampus of other experimental epilepsy models. These ultrastructural findings are quite reminiscent of the apoptotic type II cells originally classified by Clarke *(67)* (also *see* his chapter in this volume).

Recently, Pollard et al. *(60)* reported that CA3 pyramidal neurons showed apoptotic morphology by EM and DNA fragmentation on gel electrophoresis 18 h after the injection of KA into the rat amygdala. According to their finding, dying cells in CA3 showed three different types of apoptotic morphology based on the classification of Clarke *(67)*, as well as necrosis. Sloviter and colleagues reported that, after repetitive perforant path stimulation, CA3 pyramidal cells and dentate hilar neurons show typically necrotic degeneration, whereas dentate granule cells simultaneously exhibit morphological features which closely resemble apoptosis *(68)*.

Considering all the evidence described, it is likely that both apoptosis and necrosis may occur in the same neuronal population in association with the epileptic damage. In our own as well as others' experience *(69)*, the mode of neuronal death may be determined by the severity of the insult, different environmental factors (e.g., the existence of growth factors, different synaptic connections) and the susceptibility of each neuron (glutamate receptor subtypes/subunits, threshold of activation, calcium-buffering capacity). When the initial insult is overwhelming, most cells will undergo necrosis. With a mild insult, no death or perhaps only apoptosis may occur. When there is a moderate insult, both necrosis and apoptosis may initially occur. When a critical subpopulation of neurons such as the mossy cells in the dentate hilus are thus damaged, the resulting loss of inhibition may cause a hyperexcitable state. Aberrant synaptic connection may then occur reinforcing hyperexcitability. As a result, even physiological synaptic activity may cause excessive excitatory stimulation on remaining neurons and, unless checked by an adaptive increase in inhibition, a vicious cycle of further neuronal damage will develop, leading eventually to the formation of a spontaneously active epileptic focus.

DOES EPILEPTIC CELL DEATH CAUSE HYPEREXCITABILITY?

As described above, experimental work in animals has shown that repeated prolonged seizures or SE is sufficient to cause hippocampal sclerosis through excitotoxic mechanisms without the involvement of hypoxia and/or hypoperfusion *(9)*. In humans, surgical removal of sclerotic hippocampus via temporal lobectomy results in dramatic improvement or "cure" of the epileptic condition. This suggests that the sclerotic hippocampus or its surrounding region within the resection margin somehow cause the epilepsy. Some of these patients have a history of complicated febrile seizures or SE in their early life, with later development of temporal lobe epilepsy *(8,10)*. One possibility is that a perinatal or intrauterine injury (perhaps subclinical) may have already caused a certain degree of hippocampal sclerosis and these individuals were then more susceptible to complicated febrile convulsions or SE which may have added further insult to the hippocampus and the surrounding area. In the absence of definite known perinatal insult, however, it seems equally plausible that seizures by themselves can cause hippocampal sclerosis, and that, once developed, the hippocampal sclerosis can cause epilepsy.

What, then, would be the process by which neuronal death and gliosis with accompanying functional and morphologic rearrangements lead to epileptogenesis in the hippocampus and the eventual emergence of temporal lobe epilepsy? SE in animals causes a loss of recurrent, GABA-mediated inhibition of dentate granule cells *(22)*. Detailed immunohistochemical studies of the sclerotic hippocampus in experimental models of partial epilepsy as well as from humans have shown that GABAergic interneurons are mostly well preserved *(15)*, but that mossy cells and SS/NPY-immunoreactive interneurons are lost in the dentate hilus *(70)*. Mossy cells (not to be confused with mossy fibers which are the axons of the dentate granule cells) are the most numerous neuron type in the hilus of the dentate gyrus, receiving synaptic inputs from both the dentate granule cells (via mossy fibers) and the entorhinal cortex (through the perforant path, the major afferent pathway to the hippocampus). Mossy cells project both ipsilaterally and contralaterally to the molecular layer of the dentate gyrus, where the dendrites of the granule cells are located *(32)*. Electrophysiologically, they are activated at a lower threshold than the dentate granule cells and appear to excite their targets *(36)*. Intense synaptic activation such as with repetitive seizures or SE can damage mossy cells through NMDA receptor-mediated excitotoxicity leading to a excessive rise of intracellular calcium *(34)*.

Preservation of GABAergic neurons in the face of functional loss of GABA-mediated recurrent inhibition of the dentate granule cells after intense seizures appeared paradoxical, leading to the "dormant basket cell" hypothesis (Fig. 2A,B) *(17,71)*. The hypothesis is that the seizure-induced cell death of the most susceptible mossy cells in the hilus removes a tonic excitatory input on the GABAergic inhibitory basket cells in the dentate gyrus, resulting in disinhibition. Thus, a partial loss of this inhibition combined with excitation from otherwise physiologic stimuli could lead to excessive firing of the granule cells, more mossy cell death, further loss of GABAergic inhibition, and ultimately the emergence of an epileptic condition even many years after the initial injury. Similar anatomic and synaptic arrangements are also found in the Ammon's horn (CA3/CA1 region) of the hippocampus proper *(72)*.

An alternative to the "dormant basket cell" hypothesis is "the mossy fiber sprouting" hypothesis (Fig. 2C,D). This hypothesis explains the observed hyperexcitability of dentate granule cells as a consequence of a pathologic synaptic rearrangement in which the excitatory granule cells innervate themselves, resulting in a recurrent excitatory circuit. According to this theory, death of neurons in the dentate hilus that normally project to the dendrites of the granule cells results in loss of synaptic contacts; these contacts are replaced by the mossy fibers of the granule cells themselves, presumably via sprouting. A dramatic increase in this projection in the molecular layer of the dentate gyrus has been demonstrated by Timm's staining after KA-induced seizures, in kindling models of epilepsy, and in surgical specimens from medically intractable epileptic patients *(18,19)*. The usual paired pulse inhibition (physiological indicator of recurrent GABAergic inhibition) was eliminated, whereas there was anatomical evidence of intense sprouting, both facts supporting the argument for the emergence of functional recurrent excitatory synapses *(73)*. However, the functional consequences of these aberrant projections remain controversial as some evidence suggests that the sprouting may actually be promoting inhibition, perhaps in response to the abnormal hyperexcitability.

It is possible that both "dormant basket cell" and "mossy fiber sprouting" hypotheses may be correct and work cooperatively to promote epileptogenesis (Fig. 3). It is interesting to note that in a fluid percussion model of closed-head injury, similar damage is observed in the dentate hilus *(74)*. This may be one explanation of mesial temporal sclerosis found in post-traumatic epilepsy. Once the process is initiated with any form of insult, even in a limited region of the hippocampus, a vicious cycle could follow with dormant basket cells and recurrent excitatory sprouting and the damage could spread throughout the hippocampus. This process of self-perpetuating hyperexcitability then inevitably will influence synaptically related structures. If the local protective inhibitory system is overcome by the constant hyperexcitable afferent input, CA3/CA1 hippocampus, subiculum and entorhinal cortex could sequentially be recruited into the seizure network, each with its own intrinsic hyperexcitability. When

Dormant Basket Cell Hypothesis

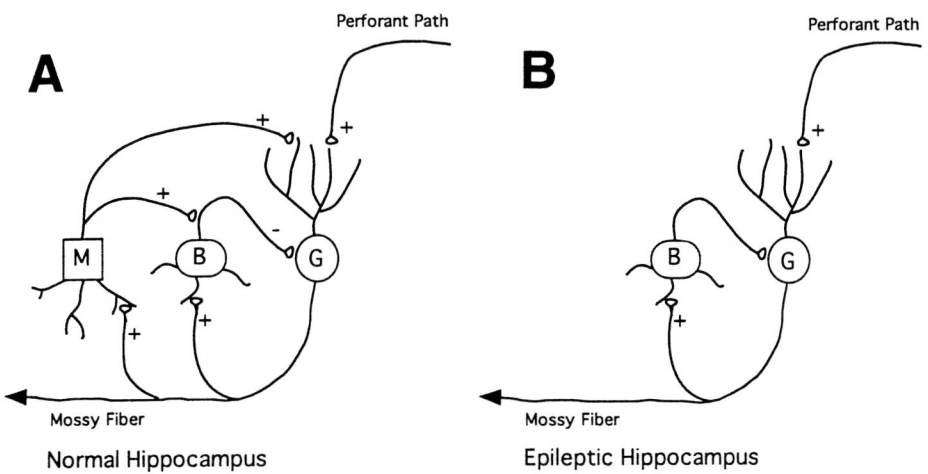

Mossy Fiber Sprouting Hypothesis

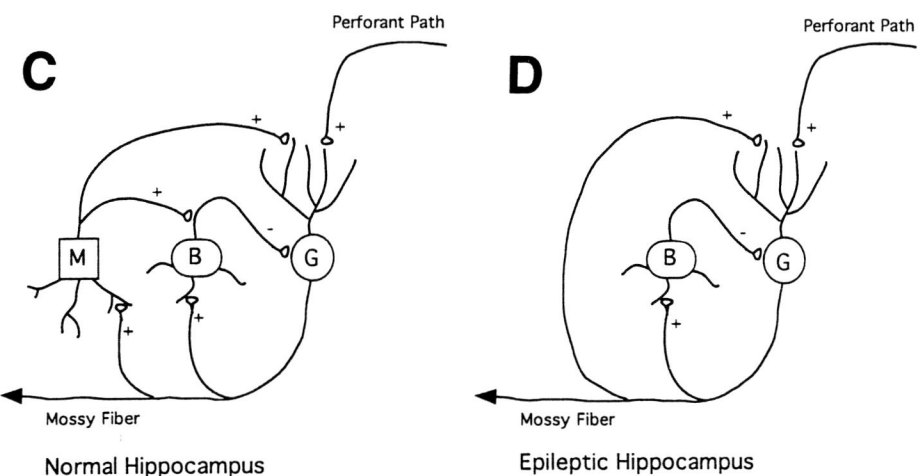

Fig. 2. Schematic diagram of the "dormant basket cell" and "mossy fiber sprouting" hypotheses. M, Mossy cells; B, basket cells; G, granule cells; +, excitatory synapse; –, inhibitory synapse.

this synaptically affiliated network matures enough to coordinate synchronization of individual hyperexcitability, then spontaneous seizures may occur whenever the brain's inhibitory control of the seizure network falls below a certain threshold.

CONCLUSIONS

Selective neuronal death likely plays an important role in the process of epileptogenesis in humans. Further insight into the mechanisms involved may provide a window of opportunity for the prevention of epilepsy. Whether intermittent brief seizure activity in an already epileptic patient results in ongoing neuronal damage is less clear. If so, this may provide an explanation for the behavioral and neurocognitive dysfunc-

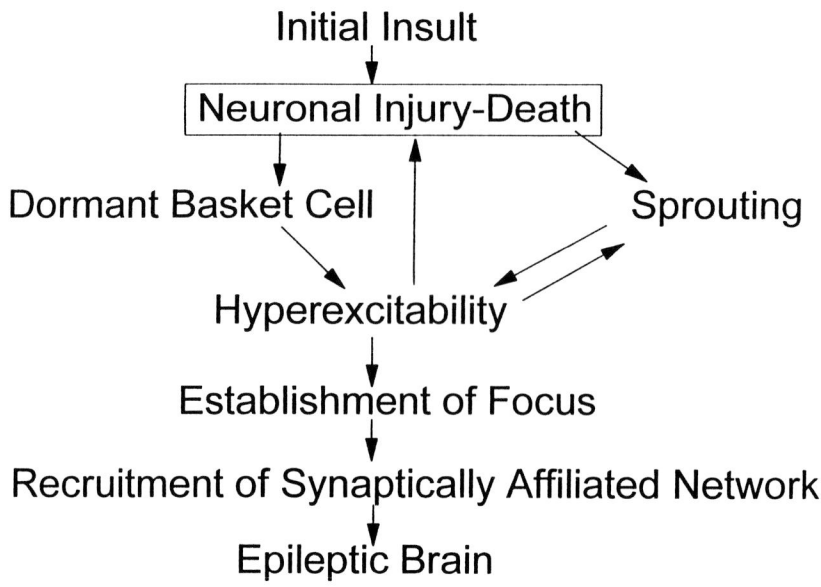

Fig. 3. Current hypothesis on the role of cell death in epileptogenesis.

tion found in epileptic patients in addition to the antiepileptic drug toxicity. Progression of the epileptic process toward a medically intractable situation could also be explained. On the other hand, clinically recognizable seizures may not be necessary for the pathological process to progress. Better understanding of the mechanisms of neuronal injury and death leading to the development of epilepsy and progression of the epileptic process would lead to formulation of novel therapeutic strategies at various stages of the epileptogenic process.

REFERENCES

1. Hauser WA, Hesdorffer DC. *Epilepsy: Frequency, Causes, and Consequences*, Demos Publications, 1990
2. Cotman CW, Monaghan DT, Ottersen J, Storm-Mathisen J. Anatomical organization of excitatory amino acid receptors and their pathways. *Trends Neurosci* 1987, **10:** 273–280.
3. Bouchet C, Cazauvieilh JB. De l'épilepsie considérée dans ses rapports avec l'aliénation mentale. *Archives generales de Medicine (Paris)* 1825, **9:** 510–542.
4. Margerison JH, Corsellis JAN. Epilepsy and the temporal lobes. A clinical, electroencephalographic and neuropathological study of the brain in epilepsy, with particular reference to the temporal lobes. *Brain* 1966, **89:** 499–530.
5. Spielmeyer W. Die Pathogenese des epileptischen Krampfes. *Zeitschrift für die gesamte Neurologie und Psychiatrie* 1927, **109:** 501–520.
6. Penfield W, Flanigin H. Surgical therapy of temporal lobe seizures. *Arch Neurol Psych* 1950, **64:** 491–500.
7. Berg AT, Shinnar S. Unprovoked seizures in children with febrile seizures: Short term outcome. *Neurology* 1996, **47:** 562–568.
8. Bruton CJ. *The Neuropathology of Temporal Lobe Epilepsy*, Oxford Press, NY, 1988.
9. Meldrum BD, Vigouroux RS, Brierly JB. Systemic factors and epileptic brain damage. *Arch Neurol* 1973, **29:** 82–87.

10. Sagar HJ, Oxbury JM. Hippocampal neuron loss in temporal lobe epilepsy: correlation with early childhood convulsions. *Ann Neurol* 1987, **22**: 334–340.
11. Aicardi J, Chevrie JJ. Consequences of status epilepticus in infants and children, in *Status epilepticus: Mechanisms of brain damage and treatment* (Delgado-Escueta AV, Wasterlain CG, Treiman DM, Porter RJ, eds.), Raven Press, NY, 1983, pp. 115–125.
12. Corsellis JAN, Brouton CJ. Neuropathology of status epilepticus in humans, in *Status epilepticus: Mechanisms of brain damage and treatment* (Delgado-Escueta AV, Wasterlain CG, Treiman DM, Porter RJ, eds.), Raven Press, NY, 1983, pp. 129–139.
13. Degiorgio CM, Tomiyasu U, Gott PS, Treiman DM. Hippocampal pyramidal cell loss in human status epilepticus. *Epilepsia* 1992, **33**: 23–27.
14. De Lanerolle NC, Kim JH, Robbins RJ, Spencer DD. Hippocampal interneuron loss and plasticity in human temporal lobe epilepsy. *Brain Res* 1989, **495**: 387–395.
15. Babb TL, Pretorius JK, Kupfer WR, Crandall PH. Glutamate decarboxylase-immunoreactive neurons are preserved in human epileptic hippocampus. *J Neurosci* 1989, **9**: 2562–2574.
16. Sperk G, Marksteiner J, Gruber B, Bellmann R, Mahata M, Ortler M. Functional changes in neuropeptide Y- and somatostatin-containing neurons induced by limbic seizures in the rat. *Neurosci* 1992, **50**: 831–846.
17. Sloviter RS. Permanently altered hippocampal structure, excitability, and inhibition after experimental status epilepticus in the rat: the 'dormant basket cell' hypothesis and its possible relevance to temporal lobe epilepsy. *Hippocampus* 1991, **1**: 41–46.
18. Sutula T, Cascino G, Cavazos J, Parada I, Ramirez L. Mossy fiber synaptic reorganization in the epileptic human temporal lobe. *Ann Neurol* 1989, **26**: 321–330.
19. Cavazos JE, Golarai G, Sutula TP. Mossy fiber synaptic reorganization induced by kindling: time course of development, progression, and permanence. *J Neurosci* 1991, **11**: 2795–2803.
20. Masukawa LM, Higashiona M, Kim JH, Spencer DD. Epileptiform discharges evoked in hippocampal brain slices from epileptic patients. *Brain Res* 1989, **493**: 168–174.
21. Urban L, Aitken PG, Freidman A, Somjen GG. An NMDA-mediated component of excitatory synaptic input to dentate granule cells in 'epileptic' human hippocampus studied in vitro. *Brain Res* 1990, **515**: 319–322.
22. Sloviter RS. "Epileptic" brain damage in rats induced by sustained electrical stimulation of the perforant path. I. Acute electrophysiological and light microscopic studies. *Brain Res Bull* 1983, **10**: 675–697.
23. Johnson EW, De Lanerolle NC, Kim JH, Sundaresan S, Spencer DD, Mattson RH, Joghbi SS, Baldwin RM, Hoffer PB, Seibul JP, Innis RB. Central and peripheral benzodiazepine receptors: opposite changes in human epileptogenic tissue. *Neurol* 1992, **42**: 811–815.
24. Hosford DA, Crain BJ, Cao Z, Bonhaus DW, Friedman AH, Okazaki MM, Nadler JV, Mcnamara JO. Increased AMPA-sensitive quisqualate receptor binding and reduced NMDA receptor binding in epileptic human hippocampus. *J Neurosci* 1991, **11**: 428–434.
25. Geddes JW, Cahan LD, Cooper SM, Kim RC, Choi BH, Cotman CW. Altered distribution of excitatory amino acid receptors in temporal lobe epilepsy. *Exp Neurol* 1990, **108**: 214–220.
26. Evans M, Griffiths T, Meldrum BS. Early changes in the rat hippocampus following seizures induced by bicuculline or L-allylglycine: a light and electron microscope study. *Neuropathol Appl Neurobiol* 1983, **9**: 39–52.
27. Griffiths T, Evans MC, Meldrum BS. Intracellular calcium accumulation in rat hippocampus during seizures induced by bicuculline or L-allylglycine. *Neurosci* 1983, **10**: 385–395.
28. Olney JW, De Gubareff T, Labruyere J. Seizure-related brain damage induced by cholinergic agents. *Nature* 1983, **301**: 520–522.
29. Sparenborg S, Brenneck LH, Jaax NK, Braitman DJ. Dizocilpine (MK-801) arrests status epilepticus and prevents brain damage induced by soman. *Neuropharmacol* 1992, **31**: 357–368.

30. Ben-ari Y, Tremblay E, Ottersen OP, Naquet R. Evidence suggesting secondary epileptogenic lesions after kainic acid: pretreatment with diazepam reduces distant but not local brain damage. *Brain Res* 1979, **165**: 362–365.
31. Unnerstall JR, Wamsley JK. Autoradiographic localization of high-affinity [^{3}H] kainic acid binding sites in the rat forebrain. *Eur J Pharmacol* 1983, **86**: 361–371.
32. Ribak CE, Seress L, Amaral DG. The development, ultrastructure and synaptic connections of the mossy cells of the dentate gyrus. *J Neurocytol* 1985, **14**: 835–857.
33. Sloviter RS. Calcium-binding protein (Calbindin-D$_{28k}$) and parvalbumin immunocytochemistry: Localization in the rat hippocampus with specific reference to the selective vulnerability of hippocampal neurons to seizure activity. *J Comp Neurol* 1989, **280**: 183–196.
34. Scharfman HE, Schwartzkroin PA. Protection of dentate hilar cells from prolonged stimulation by intracellular calcium chelation. *Science* 1989, **246**: 257–260.
35. Freund TF, Ylinen A, Miettinen R, Pitkanen A, Lahtinen H, Baimbridge KG, Riekkinen PJ. Pattern of neuronal death in the rat hippocampus after status epilepticus. Relationship to calcium binding protein content and ischemic vulnerability. *Brain Res Bull* 1991, **28**: 27–38.
36. Scharfman HE, Schwartzkroin PA. Responses of cells of the rat fascia dentata to prolonged stimulation of the perforant path: sensitivity of hilar cells and changes in granule cell excitability. *Neuroscience* 1990, **35**: 491–504.
37. Collins RC, Olney JW. Focal cortical seizures cause distant thalamic lesions. *Science* 1982, **218**: 177–179.
38. Olney JW, Degubareff T, Sloviter RS. "Epileptic" brain damage in rats induced by sustained electrical stimulation of the perforant path. II. Ultrastructural analysis of acute hippocampal pathology. *Brain Res Bull* 1983, **10**: 699–712.
39. Fariello RG, Golden GT, Smith GG, Reyes PF. Potentiation of kainic acid epileptogenicity and sparing from neuronal damage by an NMDA receptor antagonist. *Epilepsy Res* 1989, **3**: 206–213.
40. Dragunow M, Preston K. The role of inducible transcription factors in apoptotic nerve cell death. *Brain Res Rev* 1995, **21**: 1–28.
41. Smeyne RJ, Vendrell M, Hayward M, Baker SJ, Miao GG, Schilling K, Robertson LM, Curran T, Morgan JI. Continuous c-fos expression precedes programmed cell death in vivo. *Nature* 1993, **363**: 166–169.
42. Manome Y, Datta R, Fine HA. Early response gene induction following DNA damage in astrocytoma cell lines. *Biochem Pharmacol* 1993, **45**: 1677–1684.
43. Colotta F, Polentarutti N, Sironi M, Mantovani A. Expression and involvement of c-fos and c-jun proto-oncogenes in programmed cell death induced by growth factor deprivation in lymphoid cell lines. *J Bio Chem* 1992, **267**: 18,278–18,283.
44. Morgan JI, Curran T. Proto-oncogene transcription factors and epilepsy. *Trends Pharm Sci* 1991, **12**: 343–349.
45. Shin C, Mcnamara JO, Morgan JI, Curran T, Cohen DR. Induction of *c-fos* mRNA expression by afterdischarges in the hippocampus of naive and kindled rats. *J Neurochem* 1990, **55**: 1050–1055.
46. Kiessling M, Stumm G, Xie Y, Herdegen T, Aguzzi A, Bravo R, Gass P. Differential transcription and translation of immediate early genes in the gerbil hippocampus after transient global ischemia. *J Cereb Blood Flow Metab* 1993, **13**: 914–924.
47. Dragunow M, Beilharz E, Sirimanne E, Lawlor P, Williams C, Bravo R, Gluckman P. Immediate early gene protein expression in neurons undergoing delayed death, but not necrosis, following hypoxic-ischemic injury to the young rat brain. *Mol Brain Res* 1994, **25**: 19–23.
48. Beilharz EJ, Williams Ce, Dragunow M, Sirimanne ES, Gluckman PD. Mechanisms of delayed cell death following hypoxic-ischemic injury in the immature rat: evidence for apoptosis during selective neuronal loss. *Mol Brain Res* 1995, **29**: 1–14.

49. Dragunow M, Young D, Hughes P, Macgibbon G, Lawlor P, Singleton K, Sirimanne E, Beilharz E, Gluckman P. Is c-jun involved in nerve cell death following status epilepticus and hypoxic-ischaemic brain injury? *Mol Brain Res* 1993, **18**: 347–352.
50. Coyle JT, Puttfarcken P. Oxidative stress, glutamate and neurodegenerative disorders. *Science* 1993, **262**: 689–695.
51. Schreiber SS, Baudry M. Selective vulnerability in the hippocampus—a role for gene expression? *Trends Neurosci* 1995, **18**: 446–451.
52. Neumann-Haefelin T, Wiebner C, Vogel P, Back T, Hossmann KA. Differential expression of the immediate early genes *c-fos, c-jun, jun*B, and *NGFI*-B in the rat brain following transient forebrain ischemia. *J Cereb Blood Flow Metab* 1994, **14**: 206–216.
53. Sonnenberg JL, Mitchelmore C, Macgregor-Leon PF, Hempstead J, Morgan JI, Curan T. Glutamate receptor agonists increase the expression of Fos, Fra, and AP-1 DNA binding activity in the mammalian brain. *J Neurosci Res* 1989, **24**: 72–80.
54. Schreiber SS, Tocco G, Najm I, Thompson RF, Baudry M. Cycloheximide prevents kainate-induced neuronal death and c-fos expression in adult rat brain. *J Mol Neurosci* 1993, **4**: 149–159.
55. Kure S, Tominaga T, Tada K, Narisawa K. Glutamate triggers internucleosomal DNA cleavage in neuronal cells. *Biochem Biophy Res Commun* 1991, **179**: 39–45.
56. Williams MB, Jope RS. Protein synthesis inhibitors attenuate seizures induced by lithium plus pilocarpine. *Exp Neurol* 1994, **129**: 169–173.
57. Leppin C, Finiels-Marlier F, Crawley JN, Montpied P, Paul SM. Failure of a protein synthesis inhibitor to modify glutamate receptor-mediated neurotoxicity in vivo. *Brain Res* 1992, **581**: 168–170.
58. Dessi F, Charriatu-Marlangue C, Khrestchatisky M, Ben-Ari Y. Glutamate-induced neuronal death is not a programmed cell death in cerebellar culture. *J Neurochem* 1993, **60**: 1953–1955.
59. Martin DP, Schmidt RE, Distefano PS, Lory Oh, Carter JG, Johnson EM. Inhibitors of protein synthesis and RNA synthesis prevent neuronal death caused by nerve growth factor deprivation. *J Cell Biol* 1988, **106**: 829–844
60. Pollard H, Charriaut-Marlangue C, Cantagrel S, Represa A, Robain O, Moreau J, Ben-Ari Y. Kainate-induced apoptotic cell death in hippocampal neurons. *Neuroscience* 1994, **63**: 7–18.
61. Finiels F, Robert JJ, Samolyk ML, Privat A, Mallet J, Revah F. Induction of neuronal apoptosis by excitotoxins associated with long-lasting increase of 12-0-tetradecanoylphorbol 13-acetate-responsive element-binding activity. *J Neurochem* 1995, **65**: 1027–1034.
62. Ignatowicz E, Vezzani AM, Rizzi M, D'Incalci M. Nerve cell death induced *in vivo* by kainic acid and quinolinic acid does not involve apoptosis. *NeuroReport* 1991, **2**: 651–654.
63. Sun DY, Jiang S, Zheng LM, Ojcius DM, Young JDE. Separate metabolic pathways leading to DNA fragmentation and apoptotic chromatin condensation. *J Exp Med* 1994, **179**: 559–568.
64. Trump BG, Berezesky IK. Calcium-mediated cell injury and cell death. *FASEB J* 1995, **9**: 219–228.
65. Oberhammer F, Wilson JE, Dive C, Morris ID, Hickman JA, Wakeling AE, Walker PR, Sikorska M. Apoptotic death in epithelial cells: cleavage of DNA to 300 and/or 50 kb fragments prior to or in the absence of internucleosomal fragmentation. *EMBO J* 1993, **12**: 3679–3684.
66. Collins RJ, Harmon BV, Gobe GC, Kerr JFR. Internucleosomal DNA cleavage should not be the sole criterion for identifying apoptosis. *Int J Radiat Biol* 1992, **61**: 451–453.
67. Clarke PGH. Developmental cell death: morphological diversity and multiple mechanisms. *Anat Embryol* 1990, **181**: 195–213.
68. Sloviter RS, Dean E, Sollas AL, Goodman JH. Apoptosis and necrosis induced in different hippocampal neuron population by repetitive perforant path stimulation in the rat. *J Comp Neurol* 1996, **366**: 516–533.

69. Bonfoco E, Krainc D, Ankarcrona M, Nicotera P, Lipton SA. Apoptosis and necrosis: Two different events induced, respectively, by mild and intense insults with N-methyl-D-aspartate or nitric oxide/superoxide in cortical cell cultures. *Proc Natl Acad Sci USA* 1995, **92:** 7162–7166.
70. Houser CR. GABA neurons in seizure disorder: A review of immunocytochemical studies. *Neurochem Res* 1991, **16:** 295–308.
71. Sloviter RS. The functional organization of the hippocampal dentate gyrus and its relevance to the pathogenesis of temporal lobe epilepsy. *Ann Neurol* 1994, **35:** 640–654.
72. Schwartzkroin PA, Scharfman HE, Sloviter RS. Similarities in circuitry between Ammon's horn and dentate gyrus: local interactions and parallel processing, in *The Hippocampal Region as a Model for Studying Brain Structure and Function* (Storm-Mathisen J, Zimmer J, Ottersen OP, eds.), Elsevier, NY, 1990, pp. 269–286.
73. Tauck DL, Nadler JV. Evidence of functional mossy fiber sprouting in hippocampal formation of kainic acid-treated rats. *J Neurosci* 1985, **5:** 1016–1022.
74. Lowenstein DH, Thomas MJ, Smith DH, Mcintosh TK. Selective vulnerability of dentate hilar interneurons following traumatic brain injury: a potential mechanistic link between head trauma and disorders of the hippocampus. *J Neurosci* 1991, **12:** 4846–4853.

18
Trauma to the Nervous System

W. Dalton Dietrich

INTRODUCTION

The mortality and morbidity associated with traumatic brain injury (TBI) is considered to be one of the major public health problems in Western industrialized countries. About 200 cases per 100,000 indicates that about 500,000 persons sustain a head injury each year in the United States *(1)*. Although the incidence of vehicular TBI has decreased 43% from 1979 to 1992, the incidence related to firearms has increased *(2)*. Traumatic brain injury is broadly classified into mild, moderate and severe categories based on the Glasgow coma scale (GCS) *(1,3–5)*. The mild category includes the group with a GCS score of 13–15, the moderate score of 9–12, and the severe score of 3–8.

Although most traumatic head injuries are considered mild *(5)*, a significant proportion of these mild injuries result in specific disabilities. This is especially true in children who face increased demands on their cognitive abilities as they mature.

Recent advances have been made in defining and quantifying the types of brain damage that occur in closed head injury and in understanding the pathogenesis. Approaches to classifying brain injury emphasize primary and secondary damage as well as focal versus diffuse damage. To investigate the pathophysiology of human brain injury, experimental models of TBI have been established. Although no experimental model completely mimics the human condition, individual models have been shown to produce many features of human brain injury. Therapeutic strategies directed at specific pathogenic mechanisms have also been initiated in both experimental and clinical settings. This chapter reviews the pathological consequences of TBI and summarizes the many clinical and experimental strategies used to investigate and treat this condition.

NEUROPATHOLOGY OF HEAD INJURY

Several methods are used to categorize the structural changes associated with brain trauma *(5–133)*. A classification based on primary and secondary damage has advantages in grouping potentially treatable and nontreatable injury mechanisms. Another method involves classifying brain damage as focal versus diffuse *(8)*. In a review by Ribus and Jane *(120)* on traumatic contusions and intracerebral hematomas, events during or after the initial 24-h period were emphasized. The classification based on

From: Cell Death and Diseases of the Nervous System
Edited by: V. E. Koliatsos and R. R. Ratan © Humana Press Inc., Totowa, NJ

primary and secondary injury mechanisms is most advantageous in discussing the pathobiology of TBI. However, due to the ability of computerized tomography (CT) and magnetic resonance imaging (MRI) to identify focal lesions in the early postinjury period, the most helpful classification in the clinical setting appears to be that of focal versus diffuse brain damage. In addition to its advantages in terms of prognosis, such a classification can be helpful in assessing potential therapeutic interventions. Thus, in discussing the pathological presentation of head-injured patients, focal versus diffuse injury will be emphasized.

Focal Brain Injury

Cerebral Contusion

Cerebral contusions have long been considered the hallmark of head injury and are caused mainly by contact between the brain surface and bony protuberances on the base of the skull. Most but not all contusions are hemorrhagic. The pattern and distribution of cortical contusions can be important in determining the mechanism of injury. They frequently occur at the frontal and temporal lobes or in the inferior surfaces of the frontal and temporal lobes. Although contusions are most severe at the crest of gyri, they may also extend into the subcortical white matter. In the acute injury state, contusions are hemorrhagic and swollen, but with time they become shrunken brown scars. As reported in ref. *10*, parasagittal contusions tend to be most conspicuous in patients with diffuse axonal injury. Cortical contusions may give rise to focal neurologic manifestations and/or seizures.

Intracranial Hematoma

Hematomas constitute mass-occupying lesions and may occur above or underneath the dura mater *(7)*. Hemorrhage of the subarachnoid space (SAH) is the most common pathologic consequence of head trauma. Its distribution is most prominent over the convexity of the hemisphere and can be diffuse in severe head injury. These lesions are caused by marked displacement and herniation of brain structures and are a common cause of clinical deterioration and death in patients. The fracture of the squamous portion of the temporal bone and laceration of the middle meningeal artery frequently leads to epidural hemorrhages. These lesions can accumulate at a rapid rate and may also cause an acute mass effect and a fatal outcome.

Subdural hematomas originate from cortical contusions and tears of bridging vessels. The majority of subdural hematomas occur over the frontal, parietal, or temporal convexity. These lesions commonly accumulate at a slower rate than epidural hematomas and therefore are less likely to produce shifts in brain mass. Subdural hematomas are the most common type of intracranial injury found in infants subjected to trauma *(11)*.

Multiple petechial hemorrhages are particularly prominent in the white matter of the frontal and temporal lobes and in the brainstem and are seen principally in patients dying within a few hours after injury. This type of damage is referred to as "diffuse vascular injury" and involves arterial, venous, or capillary beds. Histopathologic analysis demonstrates that these lesions are small and perivascular in distribution.

Other types of focal brain damage include infection, direct damage to the pituitary stalk leading to infarction in the anterior lobe of the pituitary gland, avulsion of cranial nerves and pontomedullary tears. Tears at the junction between the pons and medulla may occur with severe hyperextension of the head and are commonly associated with instant death.

Diffuse Brain Damage

Diffuse Axonal Injury

In 1973, Mitchell and Adams reported that primary damage to the brain stem in nonmissile head injury did not occur in isolation but was associated with diffuse damage to white matter. Adams et al. *(8)* reviewed 45 cases and coined the term diffuse axonal injury (DAI) for this type of white matter damage. Severe cases of DAI have three distinctive features including a focal lesion of the corpus callosum, lesions in the dorsolateral quadrants of the brainstem, or microscopic evidence of widespread damage to axons *(7)*. Severe DAI not accompanied by an intracranial mass lesion occurs in almost 50% of patients with a severe head injury, and is the most common cause of a vegetative state and severe disability until death *(12)*. In mild, moderate or severe brain injury, axonal damage is a consistent feature with the distribution and number of damaged axons increasing with injury severity *(10,13–15)*. To date, it has been assumed that DAI leads to disconnection of various brain areas, a condition which translates into much of the morbidity seen with head injury *(16)*.

Evolution of Axonal Lesions

Depending on patient survival, DAI can take on several structural forms *(6,17,18)*. In cases where patients only survive for short periods of time, large numbers of axonal retraction balls may be observed throughout the white matter of the cerebral hemispheres, cerebellum and brain stem. In patients who survive for intermediate periods of time (wk), large clusters of microglia within the white matter, cerebellum and brain stem are frequently observed. Finally, in patients who survive vegetatively for long periods (mo), long tract degeneration of the Wallerian type is seen throughout the cerebral hemispheres, brain stem, and spinal cord.

Immunocytochemical techniques targeting neurofilament subunits have recently been used to assess the evolution of axonal lesions in head-injured patients *(18–20)*. Using ultrastructural techniques, Christman et al. *(18)* reported that TBI causes a disorganization of the neurofilaments and the axolemma as early as 6 h after TBI. At this stage, focally enlarged immunoreactive axons with axolemmal infolding or disordered neurofilaments are seen. By 30 and 60 h survival time, further accumulation of neurofilaments and organelles lead to the formation of distinct axonal swellings. These data implicate neurofilamentous disruption as a pivotal event in axonal injury.

Amyloid precursor protein (APP) is a membrane-bound glycoprotein that is transported along the axon and is the precursor of the β-amyloid protein seen in senile plaques *(21,22)*. Recently, axonal β-APP immunoreactivity has been used as a marker of DAI in head-injured patients *(23,24)*. As early as 3 h after trauma, β-amyloid-immunoreactive axons appear swollen, but after 24 h they are grossly swollen and resemble axonal retraction balls. Single-stained axons or clusters of stained axons are detected in the corpus callosum or the adjacent corona radiata of head-injured patients.

Brain Swelling and Secondary Complications

Brain swelling frequently occurs with TBI and may be localized or generalized. A localized swelling of the brain adjacent to a contusion is frequently seen in patients who sustain a nonmissile head injury. The swelling is thought to be a consequence of vasogenic edema brought on by damage to the blood–brain barrier (BBB) in areas adjacent to the damaged site *(10)*. Diffuse brain swelling of one cerebral hemisphere or both hemispheres can also occur. Unilateral brain swelling has been seen most often in patients with an acute subdural hematoma *(25)*. Diffuse swelling of both hemispheres tends to occur in younger patients and is associated with a reduction in ventricular volume *(26,27)*. Although the pathogenesis of this type of brain swelling is not clearly understood, the swelling may be contributed to a loss of vasomotor tone leading to vasodilatation, increase in blood volume, BBB dysfunction, and the genesis of vasogenic edema.

Posttraumatic hydrocephalus is another complication of TBI. In cases where there is a large amount of blood in the subarachnoid space, an impairment of cerebrospinal fluid (CSF) absorption via the arachnoid granulations can lead to ventricular enlargement *(6)*. In other cases of DAI or severe ischemia, ventricular enlargement can be the result of a loss of brain tissue.

Brain damage secondary to raised intracranial pressure is another complication of head injury. In a series by Adams and Graham *(28)*, 125 of 151 fatal head-injured patients experienced increased intracranial pressure. Secondary damage to the brain stem in the form of a midline hemorrhage and/or infarction with ischemic damage in territories supplied by the posterior cerebral arteries has also been reported. Brain damage secondary to an intracranial expanding lesion is a common cause of deterioration and coma in head-injured patients.

EXPERIMENTAL MODELS OF TRAUMATIC BRAIN INJURY

Severe closed head injury produces a range of cerebral lesions that may be divided into four general categories:

1. Diffuse axonal injury (DAI).
2. Vascular lesions, including subdural hematoma.
3. Contusion.
4. Neuronal degeneration within selectively vulnerable regions.

Various experimental models have been developed to critically investigate the pathophysiology and treatment of brain injury. In a recent review of animal models of head injury, Gennarelli *(29)* classified models of head injury according to the method of producing injury.

Fluid-percussion (F-P) and rigid indentation models are characterized as percussion concussion whereas inertial injury models and impact acceleration models are included under acceleration concussion. Finally, in an attempt to investigate the effects of mechanical deformation on specific cell types, in vitro models of stretch-induced injury have been recently developed.

Percussion Concussion

The central and lateral (parasagittal) F-P model are characterized by brief behavioral unresponsiveness, metabolic alterations, changes in cerebral blood flow and BBB permeability, and behavioral deficits. The central F-P model tends to have variable and relatively small contusions in the vicinity of the fluid pulse and scattered, mostly brain stem axonal damage *(30–32)*. Lateral and parasagittal F-P models are characterized by a cortical contusion which tends to be lateral (remote) to the impact site *(33–36)*. Evidence for axonal damage is seen scattered throughout the white matter tracts within the ipsilateral cerebral hemisphere as well as tissue tears at gray-white matter interfaces *(35–39)*. Hippocampal damage is pronounced in the lateral F-P injury model. However, fewer brain stem abnormalities are seen with lateral F-P injury compared to central. Thus, the parasagittal F-P injury model results in a range of pathologies including contusion, widespread axonal injury and selective neuronal necrosis *(36–40)*. Another important feature of F-P models is that varying severities of injury (mild, moderate, and severe) can be studied in a reproducible fashion. This is an important factor, because the degree of neuroprotection with a particular agent or treatment protocol would be expected to vary with injury severity.

Coma of variable durations can be produced in the rigid percussion models with a variable sized contusion in the parasagittal cortex beneath the impact *(29)*. In contrast to the central rigid concussive model where axonal damage has been described to be infrequent, lateral rigid percussion produces axonal damage scattered throughout the hemisphere and along the periphery of the indented hemisphere.

Acceleration Concussion

Inertial acceleration models can produce relatively pure acute subdural hematomas and diffuse axonal injury *(16)*. Tissue tear hemorrhages in the central white matter and gliding contusions in the parasagittal gray-white junctions also tend to occur. These models are characterized by a variable period of coma and widespread axonal damage throughout the white matter and hemispheres as well as the upper brain stem and cerebellum. Impact acceleration models have recently been demonstrated to produce prolonged coma and widespread axonal damage *(41,42)*. These models are typically associated with variable and somewhat uncontrolled skull fractures.

In terms of which are the more appropriate models, it is important to note that cerebral contusion is the most common sequela of head injury *(29,43)*. Acceleration and concussion models both reproduce this phenomenon. Diffuse brain injury is considered a phenomenon with greater relevance for patients who die or remain vegetative *(25)*. Recent data indicate that the lateral and parasagittal F-P models produce widespread axonal perturbations *(38–40)*. As discussed by Gennarelli *(29)*, human head injury is never as pure as an experimental model. Thus, the human injury condition may not be addressed in its entirety with a single animal model. However, once a particular feature of human brain injury is produced in an experimental model, mechanisms underlying the genesis of the injury can be critically investigated.

In Vitro Models

A shortcoming of animal models of brain injury is that such models preclude a critical assessment of individual cell responses to trauma. In animal experiments, for example, the initial cellular response to injury may be a consequence of both primary and secondary events initiated by a complex cascade of cellular interactions (Fig. 1). Thus, to critically investigate the consequences of injury on a specific cell type in the absence of confounding cellular and systemic factors, several in vitro models have been developed in conjunction with cell culture technology *(44–49)*. Models range from "scratching" the culture with a pipet tip to stretching cultured cells (the latter in order to induce deformation).

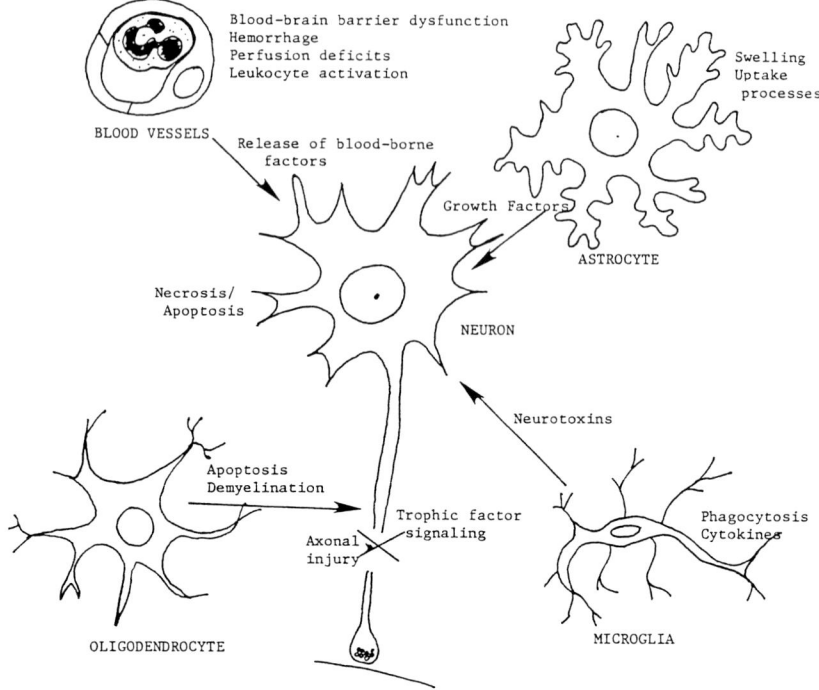

Fig. 1. Cellular interactions after trauma to the nervous system.

Using these approaches, investigators have shown that trauma induces astrocytic hyperplasia, hypertrophy, and increased glial fibrillary acidic protein (GFAP) content without the contribution of other cell types *(44)*. Goldberg and colleagues *(45)* reported in astrocytes that the rise in intracellular calcium could be blocked by the *N*-methyl-D-aspartate (NMDA) antagonist MK-801 or by the amino-3-hydroxy-5-methylisoxazole-4-propionic acid (AMPA)/kainate antagonist CNQX. Using the F-P model of injury, Hariri et al. *(50)* showed that traumatized astrocytes produce interleukin-6. Recently, Ellis et al. *(48)* have described the light and electron microscopic changes in astrocytes following different degrees of astrocytic stretching. Murphy and Horrocks *(51)* showed that neuroblastoma and oligodendroglioma cell lines produced membrane fatty acid and lactate dehydrogenase release after a 20-atm pressure pulse. Using neonatal cortical neurons, Tavalin and colleagues *(49)* reported a stretched-induced delayed depolarization that was dependent on NMDA receptor activation. In vitro experimental approaches like these should provide important information concerning mechanisms underlying cellular responses to trauma and the role of specific cell types in the pathophysiology of brain trauma.

NEURONAL DAMAGE AFTER TRAUMATIC BRAIN INJURY

Experimental Findings

The neuropathologic sequelae of experimental head injury have been characterized using various models of brain injury. Sutton and colleagues *(52)* utilized a cortical contusion device to investigate the primary and secondary neuronal responses to this insult. At 6 h after TBI, the contused cortical tissue appeared edematous, pyknotic neurons were present and loss of neuronal Nissl staining was apparent at the injury site. By 8 d, the cortical cavitation was surrounded by thin regions of necrotic tissue and/or glial scars. The temporal profile of neuronal perturbations after lateral F-P brain injury has also been assessed. In the study by Cortez et al. *(33)*, from 1–6 h after trauma, abnormal neurons were seen adjacent to the hemorrhagic site, in the pyramidal cell layer of the hippocampus and within the granule cell layer of the dentate gyrus. Acidophilic neurons were noted in the posterior, lateral dorsal, ventroposterior and medial thalamic nuclei and superior colliculi. Also in the lateral F-P model, Hicks et al. *(53)* reported neuronal loss within the ipsilateral parietoccipital cortex and hippocampal CA3 region after lateral F-P injury.

Light and electron microscopic techniques have recently been used to assess the acute and chronic consequences of parasagittal F-P brain injury *(35–37)*. At 1 h after moderate injury, dark shrunken neurons indicative of irreversible damage were seen in cortical layers overlying the gliding contusion that displayed BBB breakdown. Ultrastructural studies demonstrated that early BBB breakdown resulted from mechanical damage of small venules within the lateral external capsule. At more posterior brain levels (i.e., –4.8 mm behind bregma) a focal area of CA3 pyramidal cell necrosis was also associated with extravasated horseradish peroxidase (HRP) from vessels within the cistern separating the hippocampus and thalamus. In contrast, acute neuronal damage within the dentate hilus, lateral thalamus and anterior CA3 (–2.8 mm behind bregma) occurred in the absence of regional HRP extravasation. Thus, while some acute neuronal pathology is associated with local vascular perturbations, it also appears that other mechanisms, including circuit-specific excitotoxic mechanisms, may also participate. The acuteness of this pathology would be expected to limit the potential for therapeutic interventions directed against this specific neuronal response to trauma.

In the same F-P model, more chronic patterns of neuronal injury have also been documented *(54)*. At 3 d after TBI, scattered necrotic neurons are present throughout the frontoparietal cerebral cortex remote from the impact site. In addition, selective neuronal necrosis is seen in the CA3 and CA4 hippocampal sectors, the dentate hilus and lateral thalamus ipsilateral to the trauma. A well demarcated contusion overlying the lateral external capsule is also seen at this survival period. Finally, at 8 wk after TBI, a thinning of the cerebral cortex is apparent in histological sections, with enlargement of the ipsilateral lateral ventricle *(54)*.

With the continued investigation of experimental models of TBI, it has become increasingly apparent that DAI exists as a spectrum of injury involving widespread areas of the brain *(10,15)*. Yaghmai and Povlishock *(32)* have characterized patterns of reactive axonal change using neurofilament antibodies after midline F-P brain injury and controlled cortical impact injury. Within 1–2 h after injury, reactive axonal change was most conspicuous in anatomical sites which included the pontomedullary junction, cerebellar peduncles, vestibular and red nuclei, cranial nerves within the brain stem, the internal and external capsule, as well as the corpus callosum. Typically, reactive axons were most numerous within the brain stem, with a limited involvement of the subcortical white matter or the corpus callosum. This pattern of axonal damage is in contrast to the human condition, where callosal and subcortical white matter axonal damage predominates. Interestingly, recent data indicate that moderate F-P brain injury in rats can lead to both reversible and irreversible axonal perturbations, a finding that may explain some of the transient behavioral consequences of TBI *(39)*.

Selective Neuronal Vulnerability

Damage to the hippocampus is commonly reported in autopsy studies of head-injured patients *(55)*. In an acceleration model of brain injury in nonhuman primates, Kotapka et al. *(56)* reported CA1 hippocampal damage in 59% of their animals. In this TBI model, damage to CA1 was not the result of secondary global ischemic insult, marked elevation of intracranial pressure or seizures. In this regard, CA1 damage is not reported in other TBI models including cortical contusion and F-P injury. However, in a study by Jenkins et al. *(57)*, midline F-P injury followed by a sublethal global ischemic insult led to CA1 vulnerability. It will be important to determine whether any secondary injury process contributes to CA1 damage in acceleration models.

The dentate gyrus and CA3 region of the hippocampus have been reported to be selectively damaged in F-P models *(36,58)*. Lowenstein et al. *(58)* reported bilateral damage to dentate hilus neurons after lateral F-P injury. In an ultrastructural study of parasagittal F-P injury, neuronal damage within the dentate hilus and CA3 pyramidal neurons was reported at 1 h after injury. Interestingly, CA3 neuronal pathology was associated with local BBB damage *(35)*. Thus, vascular perturbations may account for some of the hippocampal damage produced by TBI.

Thalamic damage after brain injury has been reported in clinical and experimental studies *(54,59–61)*. In human TBI, Ross et al. *(59)* documented the selective loss of neurons within the thalamic reticular nucleus in severely head-injured patients. These investigators suggested that the loss of inhibitory thalamic reticular neurons might underlie some forms of attention deficits following head injury. In a study by Anderson et al. *(60)*, magnetic resonance scans of TBI patients were analyzed to examine relationships between injury severity, lesion volume, ventricle-to-brain ratio and thalamic volume. In their study, patients with moderate-to-severe injuries had smaller thalamic volumes and greater ventricle-to-brain ratios than patients with mild-to-moderate injuries. The decreased thalamic volumes suggested that subcortical brain structures may be susceptible to transneuronal degeneration following cortical lesions. Experimentally, thalamic atropy has been recently reported 8 wk after parasagittal F-P injury in rats *(54)*. In that study, focal damage to several thalamic nuclei may have resulted from progressive retrograde degeneration following axonal damage and/or neuronal necrosis.

Ventricular Dilation

A number of clinical studies have documented the morphological consequences of TBI in patients surviving for relatively long periods. Ventricular expansion not associated with hydrocephalus and/or increased intracranial pressure is felt to be a sensitive indicator of the degree of structural damage and an indirect measurement of white matter atrophy *(62,63)*. Anderson and Bigler *(61)* have correlated ventricular size with memory disturbances. Importantly, the group of patients with the highest ventricular volumes demonstrated significantly lower memory scores. In a recent experimental study, the chronic histopathological consequences of F-P brain injury were assessed *(54)*. At 8 wk after moderate parasagittal F-P injury, the most remarkable changes were a thinning of the ipsilateral cortical mantle and a significant increase in ventricular volume. Quantitation of tissue volumes showed significant decreases in cortical, hippocampal and thalamic volumes after TBI. Although decreased tissue volumes would be expected to reflect neuronal dropout after TBI, regional neuronal cell counts are required to critically assess this possibility. Nevertheless, in clinical and experimental studies of TBI, ventricular volume measurements after chronic survival periods may be a sensitive morphological indicator of progressive neurodegenerative processes and a predictor of therapeutic efficacy.

PATHOGENESIS OF TRAUMATIC BRAIN INJURY

Primary Injury Mechanisms

According to Holbourn and Edin *(64)*, two major types of forces are responsible for brain injury, one localized at the impact site and a second characterized by rotational forces. Depending on the force and/or location of the primary impact, head trauma can produce acute damage to blood vessels as well as axonal projections. Contact phenomena generate superficial or contusional hemorrhages through coup and contrecoup mechanisms *(16)*. Direct injury is commonly superficial; the coup-contrecoup hemorrhages may be adjacent or central *(65)*.

Axonal shearing is a common lesion of the cerebral white matter and occurs particularly in acceleration/deceleration injury. Interestingly, only recently has morphological evidence for axonal shearing become available *(19,66)*. In 1993, Maxwell and colleagues reported ultrastructural evidence for tearing or shearing of axons at 20 and 35 min after trauma using nonhuman primates exposed to lateral acceleration of the head. These studies indicated that axonal separation occurred without prior formation of axonal swelling and that resealing could occur at later periods. Thus, perturbations of the axolemma leading to the accumulation of cytoskeletal components and organelles, may represent a secondary injury process.

Shearing strains may also damage blood vessels and cause petechial hemorrhages, deep intracerebral hematomas and brain swelling. In one experimental study, mechanical damage to small venules, resulting in focal BBB breakdown was reported at 1 h after moderate F-P injury *(35)*. In that injury model, vascular damage lead to the formation of a gliding contusion within the external capsule *(36)*. Interestingly, contusion volume at 3 d correlated with the frequency of selective neuronal necrosis within the overlying cerebral cortex. These experimental findings demonstrate a spatial relationship between trauma-induced vascular damage and patterns of acute neuronal vulnerability. Interestingly, experimental studies of brain ischemia have also reported similar vascular/neuronal relationships in terms of acute pathology *(67)*. Early vascular damage following acute brain injury may increase neuronal vulnerability by several potential mechanisms including the genesis of posttraumatic perfusion deficits or the extravasation of neurotoxic blood-borne substances *(68)*.

Secondary Injury Mechanisms

In many head-injured patients, the extent of neurological recovery might depend on the contribution of posttraumatic secondary insults *(69–73)*. In the clinical setting, such secondary consequences of the injury include hypotension, hypoxia, hyperglycemia, anemia, sepsis and hyperthermia. Experimental evidence also indicates an increased susceptibility of the posttraumatic brain to secondary insults. For example, following mild F-P brain injury, CA1 hippocampal vulnerability is enhanced with imposed secondary ischemia *(57)*. An important area of research regarding the treatment of brain injury involves the characterization of secondary injury processes which may be targeted for intensive care management and for pharmacotherapy.

Cerebral Hypoxia/Ischemia

A high incidence of hypoxic/ischemic brain damage is seen in patients who die as a result of nonmissile head injury *(55)*. In a study by Adams et al. *(25)*, evidence for hypoxic damage was observed in 65 of the 151 cases of fatal brain injury. Hypoxic damage in the form of hemorrhagic infarction and diffuse necrosis of neurons most frequently occurred in arterial boundary zones between the major cerebral arteries, particularly the anterior and middle cerebral artery territories. Hypoxic damage is common in patients who have experienced an episode of intracranial hypertension. A significant correlation between hypoxic brain damage and nonmissile head-injured patients in the presence of arterial spasm has also been reported *(74)*. In an experimental study, Tanno et al. *(75)* reported that posttraumatic hypoxia aggravated the BBB consequences to F-P brain injury.

Posttraumatic hemodynamic impairments are also believed to be an important secondary injury mechanism following brain trauma. Various clinical and experimental studies have reported significant reductions in local cerebral blood flow (lCBF) after TBI *(40,76–79)*. In one experimental study of moderate F-P brain injury, lCBF reductions 30 min after TBI ranged from 40 to 80% of control, and occurred throughout the traumatized hemisphere *(40)*. Mild to moderate reductions in lCBF may result from the release of vasoactive substances from damaged tissue *(80)* or possibly as a secondary consequence of reductions in neuronal activity or metabolism *(35,81–82)*. In contrast, recent data from studies where severe F-P injury was investigated have shown lCBF reductions reaching ischemic levels (i.e., < 20 mL/g/min) 30 min after TBI *(83)*. Interestingly, evidence for subarachnoid and intracerebral hemorrhage as well as local platelet accumulation was associated with ischemic sites within the cerebral cortex. The severity of the initial impact is therefore a critical factor in determining the hemodynamic and pathological consequences of experimental TBI. Future studies are required to investigate the various mechanisms underlying these hemodynamic responses to TBI so that therapeutic strategies can be developed to treat this potentially important secondary injury mechanism.

Hypotension

Hypotension is present in 15–20% of patients who suffer TBI *(71,84)*. Marmarou et al. *(85)* reported that, in severely injured patients, outcome was significantly correlated with the proportion of mean arterial pressure recordings <80 mmHg. Hemorrhagic hypotension after F-P injury in rats is associated with a more severe depletion of high-energy phosphates than either TBI or hypotension alone *(88)*. In addition, mild hemorrhagic hypotension produces significant deficits in cerebral oxygen availability and neurological function after experimental F-P brain injury *(79)*.

The increased sensitivity of the posttraumatic brain to moderate levels of hypotension may occur from deficits in the autoregulation of cerebral blood flow *(87)*. Overgaard and Tweed *(88)* reported that blood flow autoregulation was impaired in all patients in the first 24 h and in most patients in the first 4 d after trauma. DeWitt et al. *(79)* reported in cats after F-P brain injury that global and regional autoregulation are absent after TBI. Hypotensive periods that may potentially occur during surgical procedures and/or anesthesia may be hazardous to the head-injured patient.

Reperfusion Injury

In many types of brain injury, organ reperfusion has been shown to exacerbate tissue injury *(89,90)*. In models of cerebral ischemia, for example, there is evidence for inflammatory processes influencing the pathological outcome *(91,92)*. In models of TBI, the temporal profile of polymorphonuclear leukocyte accumulation has recently been examined *(93,94)*. In the cortical impact model, Clark et al. *(94)* reported significant increases in myeloperoxidase activity (an indicator of neutrophil accumulation) in the traumatized hemisphere 1 and 2 d after injury. In the lateral F-P model, Soares et al. *(95)* reported that neutrophils could be identified at 4, 24, and 48 h after TBI. Studies by Uhl et al. *(96)* reported that neutropenia induced by vinblastin treatment did not affect the development of brain edema or lesion size but did reduce cerebral blood flow 24 h postinjury. Studies are needed to determine whether attenuating neutrophil infiltration after TBI leads to long-term structural and/or functional protection.

Recent data suggest that treatment with cell adhesion molecule (CAM) antibodies selectively reduce apoptosis after transient occlusion of the middle cerebral artery *(97)*. In that study, treatment with both anti-ICAM-1 and anti-CD11b antibodies both reduced numbers of apoptotic cells. In general, in CNS trauma, there is little evidence for apoptotic cell death. In one study, Raghupathi and McIntosh *(98)* investigated the effects of F-P injury on the expression of *bcl-2*. These investigators reported a decrease in bcl-2 immunoreactivity that appeared to precede neuronal death. Most recently, Bresnahan et al. *(99)* have reported evidence for apoptosis of oligodendroglia in long tracts undergoing Wallerian degeneration after spinal cord injury in monkeys. It was suggested that demyelination of long tracts after SCI may be due, in part, to apoptotic death of oligodendrocytes associated with dying axons. The role of apoptosis in the pathophysiology of TBI and spinal cord injury is important in that novel therapeutic strategies directed at this injury mechanism may prevent delayed degeneration.

Hyperthermia

Many patients have fever after TBI *(100,101)*, and recent clinical data indicate that brain temperature following TBI may be higher than core temperature *(102)*. Posttraumatic hyperthermia (39°C) induced artificially 24 h after trauma has recently been shown to increase mortality rate and aggravate the histopathological outcome in a F-P injury model *(37)*. Interestingly, in addition to increasing the frequency of neuronal necrosis, posttraumatic hyperthermia also increased the frequency of severely swollen myelinated axons. Thus, in the clinical setting, post-traumatic hyperthermia may represent a clinically manageable secondary injury mechanism.

Because delayed posttraumatic hyperthermia has been shown to have detrimental effects on outcome, aggressive attempts to reduce brain temperature in head-injured patients appear to be indicated. Temperature is known to affect multiple biological processes *(103,104)*, and undetected brain temperature elevations might abolish the beneficial effect of a therapeutic agent. This is one example on how secondary injury mechanisms can complicate data interpretation in clinical trials for TBI.

THERAPEUTIC INTERVENTIONS

Over the past several years, the pathophysiology of TBI has been investigated using a variety of animal models. New therapies driven by findings from these models have been initiated. Several reviews have recently summarized many of the agents that have been investigated in TBI *(105,106)*. Agents that provide protection in experimental models of TBI include opiate receptor antagonists *(34,107)*, NMDA receptor antagonists *(107–109)*, thyroid releasing hormone (TRH) and TRH analogs *(105,106)* and mild and moderate hypothermia *(36,110,111)*.

In reviewing this literature, it is obvious that the problem of neuronal injury after TBI involves many common final pathways that are also present in other types of brain injury including cerebral hypoxia/ischemia (*see* chapters by Olney and Ishimaru, and Vornov). However, cell injury after TBI is also unique in other ways and may require novel approaches. The present discussion will be limited to the major therapeutic strategies that are being currently assessed experimentally and clinically.

Glutamate Antagonists

Excitatory amino acid neurotransmitters have been implicated in the pathophysiology of TBI *(107–109,112–115)*. Microdialysis has documented elevated levels of extracellular amino acids after TBI *(114–117)* while NMDA receptor antagonists, including phenylcyclidine (PCP), ketamine, dextrorphin and dizocilpine (MK-801) have been reported to provide protection against brain trauma *(107–109)*. Hayes et al. *(109)* demonstrated that pretreatment with PCP attenuated some of the behavioral deficits associated with F-P brain injury in rats. Studies using the noncompetitive NMDA antagonist MK-801 reported improved neurologic function with postinjury treatment *(109)*. Postinjury administration of the competitive antagonist CPP has also been shown to limit the neurological deficit associated with lateral F-P injury *(108)*.

Bullock and Fujisawa *(113)* have reviewed the role of glutamate antagonists for the treatment of CNS injury. Clinical trials of the competitive NMDA antagonists CGS 19755 and D-CPP-ene have been initiated in stroke patients. On the other hand, the noncompetitive NMDA antagonist MK-801 has been withdrawn from clinical use due to side effects. Clinical studies are required to determine whether NMDA receptor blockade is of benefit for TBI patients. For a more general discussion of the pharmacologic approach using glutamate antagonists, *see* chapter by Bar-Peled and Rothstein.

Radical Scavengers

Excessive production of oxygen free radicals have been implicated in the pathophysiology of TBI *(118–121)*. Experimental F-P injury increases the production of oxygen radicals, and free radical scavengers have been reported to be protective in brain injury models *(121)*. Using 2,3 DHBA as a sensitive indicator of hydroxyl radical production, Globus et al. *(114)* reported significant elevations in extracellular hydroxyl radical production after parasagittal F-P injury using microdialysis. The free radical scavenger SOD has been shown to reduce BBB opening and the genesis of brain edema after TBI, while transgenic mice overexpressing the human copper/zinc SOD have been reported to be protected against brain trauma *(122)*. Recently, clinical trials using intravenously administered pegylated SOD have been initiated for the treatment of TBI. For a general discussion of the role of free radicals on neuronal injury and the design of therapeutics based on free radical scavenging, *see* chapter by Dykens.

Neurotrophic Factors

DAI is a unique problem associated with brain trauma. DAI, which causes circuit disruption, may not only produce immediate functional consequences, but also affect trophic signaling between neuronal populations, possibly resulting in delayed neuronal death. For a general discussion of the role of free radicals and neuronal aging and the design of therapeutics based on free radical scavenging, *see* chapter by Dykens. Thus, the addition of trophic factors after TBI may help maintain neuronal survival as well as promoting circuit reorganization and recovery. Neurotrophins have been shown to be neuroprotective in in vitro and in vivo studies of neuronal injury *(123,124)*. Basic fibroblast growth factor (bFGF) has been reported to protect neuronal cell cultures from a variety of potential lethal injuries, including hypoxia and ischemia *(123)*. Most recently, bFGF has been shown to be neuroprotective in a F-P model of TBI *(125)*. In that study, intravenous bFGF infusion initiated 30 min after trauma significantly reduced contusion volume and decreased the number of cortical necrotic neurons 3 d after injury. If experimental studies continue to show a benefit of bFGF on neuronal injury and behavioral outcome in models of TBI, clinical trials may be initiated in the future. It is pertinent to add here that recent experimental studies in focal cerebral ischemic have demonstrated that bFGF administered 24 h after the ischemic insult leads to accelerated behavioral recovery without significantly decreasing infarct volume *(126)*. For a more topical discussion on the role of trophic factors in the treatment of neurological disease, including TBI, *see* chapter by Koliatsos and Mochetti.

Therapeutic Hypothermia

Experimental Findings

Numerous studies have demonstrated that, while mild hypothermia (30–34°C) is neuroprotective in models of cerebral ischemia and TBI, mild hyperthermia (39°C) worsens the outcome *(37,103,104)*. In a parasagittal F-P model, posttraumatic hypothermia (30–34°C) has been shown to improve the histopathological outcome *(36)*. In one study, a 3-h period of posttraumatic hypothermia (30°C) initiated 30 min after TBI significantly reduced the contusion volume and the number of necrotic cortical neurons at 3 d *(36)*. Microdialysis studies have also reported that posttraumatic hypothermia reduces the acute surge in levels of extracellular glutamate and hydroxyl radicals after F-P injury *(114)*. These early indicators of neuroprotection correlate with an improved behavioral outcome of posttraumatic hypothermia in terms of sensorimotor tasks and cognitive function *(110,111,112)*. Regarding chronic outcome measures, a recent study has reported that posttraumatic hypothermia decreases the degree of cortical atrophy and ventricular enlargement 8 wk after TBI *(54)*. The ability of a given therapeutic intervention to provide long-term protection should be assessed prior to the initiation of a clinical trial.

Clinical Observations

While deep hypothermia (i.e., <27°C) as a means of cerebral protection in the neurosurgical setting has met with various problems *(110)*, the use of more moderate levels of hypothermia (>32°C) has been reported to improve outcome in a limited number of clinical studies *(100,128–130)*. Clifton et al. *(128)* reported that, in 46 severely injured patients (GCS 4–7), systemic hypothermia (32–33°C) begun within 6 h of injury and continued for 48 h resulted in no cardiac or coagulopathy-related complications and a lower seizure incidence. Importantly, hypothermia improved recovery and reduced disability in this study. In other studies, therapeutic hypothermia was reported to be effective in attenuating intracranial hypertension *(129)* while not affecting the incidence of delayed intracerebral hemorrhage *(130)*.

In the related field of ischemia, recent studies have indicated that the duration of cooling in the postinjury setting is an important variable in determining the extent of neuroprotection *(131)*. In severely head-injured patients, prolonged periods of hypothermia may be necessary to protect the severely traumatized brain from both primary and secondary mechanisms of injury. In this regard, Hayashi et al. *(132)* have recently reported cerebral thermo-dysfunction in head-injured patients several days after injury. In their clinical studies, brain cooling for up to 7 d resulted in improved outcome.

SUMMARY AND FUTURE DIRECTIONS

TBI has been shown in clinical and experimental studies to initiate a complex cascade of neuronal perturbations. The occurrence of DAI, vascular lesions, contusions and neuronal degeneration all represent important components of the injury process. Various pharmacological agents have been shown to be neuroprotective or provide improved behavioral outcome in established models of TBI. It is clear that therapies should be directed at both the early vascular consequences of head injury as well as neuronal vulnerability. Presently, there is a need for quantitative histopathological assessment of these lesions to determine which specific pathogenic mechanisms or cellular events are affected by specific therapies. Finally, chronic outcome measures including histopathological assessment and neurobehavioral testing are required to determine whether acute treatment protocols offer chronic or only temporary protection.

To date, few experimental studies have targeted DAI for therapeutic intervention. Because DAI is considered to be an important factor in the generation of the morbidity and mortality associated with TBI, these investigations are of the utmost importance. Recently, hypothermia has been reintroduced into the clinical arena and appears to provide some benefit to the head-injured patient. Thus, it will be important to assess the therapeutic window for posttraumatic hypothermia on DAI. Likewise, the effects of secondary injury mechanisms including ischemia, hypotension and hyperthermia on DAI require investigation. Because many of these same issues are relevant to spinal cord injury (where long tracts are damaged), investigations using reproducible models of spinal cord trauma should also lead to novel therapeutic strategies directed at both neuroprotection and repair.

In future years, the combination of pharmacological agents with mild hypothermia, may prove useful in the treatment of TBI. For example, recent cerebral ischemia studies with postischemic hypothermia indicate that the CA1 region of hippocampus undergoes a delayed excitotoxic injury *(131)*. If similar processes occur after CNS trauma, therapeutic agents or brain cooling may have to be administered in the acute as well as in the chronic postinjury setting. However, although hypothermia may protect neurons from acute injury, it might also retard naturally occurring recovery processes *(133)*.

Recent studies in TBI have demonstrated neuroprotection with the trophic factor bFGF. Thus, the combination of moderate hypothermia and neurotrophic factor treatment might lead to acute neuronal protection as well as promote circuit reorganization and functional recovery after TBI.

ACKNOWLEDGMENTS

This work was supported by USPHS Grants NS-30291 and NS27127. The author would like to thank H. Valkowitz for typing the chapter.

REFERENCES

1. Frankowski RF. Descriptive epidemiological studies of head injury in the United States 1974–1984. *Adv Psychom Med* 1986, **16**: 153.
2. Sosin DM, Sniezek JE, Waxweiler RJ. Trends in death associated with traumatic brain injury, 1979 through 1992: Success and failure. *JAMA* 1995, **273**: 1778.
3. Rimel RW, Ciordani B, Barth JT. Disability caused by minor head injury. *Neurosurgery* 1981, **9**: 221–228.
4. Teasdale G, Jennett B. Assessment of coma and impaired consciousness. A practical scale. *Lancet* 1974, **2**: 81–84.
5. Kraus JF, Black MA, Hessol N. The incidence of acute brain injury and serious impairment in a defined population. *Am J Epidemiol* 1984, **119**: 186–201.
6. Adams JH, Graham DI, Gennarelli TA. Contemporary neuropathological considerations regarding brain damage in head injury, in *Central Nervous System Trauma Status Report* (Becker DP, Povlishock JT, eds.), pp. 65–77, 1985.
7. Graham DI. Neuropathology of head injury, in *Neurotrauma* (Narayan RK, Wilberger JE, Povlishock JT eds.), McGraw-Hill, New York, 1996, pp. 43–59.
8. Adams JH, Graham DI, Murray LS, Scott G. Diffuse axonal injury due to nonmissile head injury in humans: An analysis of 45 cases. *Ann Neurol* 1982, **12**: 557–563.
9. Ribus GC, Jane JA. Traumatic contusions and intracerebral hematomas. *J Neurotrauma* 1992, **9**: S265–S278.
10. Adams JH, Doyle D, Ford I, Graham DI, McLellan D. Diffuse axonal injury in head injury: definition, diagnosis and grading. *Histopathology* 1989, **15**: 49–59.
11. Leestma TG, Klauber MR, Marshall LF. Outcome from head injury related to patients age. A longitudinal prospective study of adult and pediatric head injury. *J Neurosurg* 1988, **68**: 409–416.
12. McLellan DR, Adams JH, Graham DI. The structural basis of the vegetative state and prolonged coma after non-missile head injury, in *Le Coma Traumatique* (Papo I, Cohadon F, Massarotti M eds.), Liviana Editrice, Padova, 1986. p. 165.
13. Strich SJ. Shearing of nerve fibers as a cause of brain damage due to head injury. A pathological study of twenty cases. *Lancet* **2**: 443–448, 1961.
14. Peerless S, Rewcastle N. Shear injuries of the brain. *Can Med Assoc J* 1967, **96**: 577–582.
15. Pilz P. Axonal injury in head injury. *Acta Neurochir* 1983, **32**(suppl):119–123.
16. Gennarelli TA, Thibault LE, Adams JH, Graham DI, Thompson CJ, Marcincin RP. Diffuse axonal injury and traumatic coma in the primate. *Ann Neurol* 1982, **12**: 564–574.
17. Jane JA, Steward O, Gennarelli T. Axonal degeneration induced by experimental noninvasive minor head injury. *J Neurosurg* 1985, **62**: 96–100.
18. Christman CW, Grady MS, Walker SA, Holloway KL, Povlishock JT. Ultrastructural studies of diffuse axonal injury in humans. *J Neurotrauma* 1994, **11**: 173–186.
19. Povlishock JT. Traumatically induced axonal injury: pathogenesis and pathobiological implications. *Brain Pathol* 1992, **2**: 1–12.
20. Grady MS, McLaughlin MR, Christman CW, Valadka AB, Fligner CL, Povlishock JT. The use of antibodies targeted against the neurofilament subunits for the detection of diffuse axonal injury in humans. *J Neuropathol Exp Neurol* 1993, **52**: 143–152.
21. Koo EH, Sisodia SS, Archer DR, Martin LJ, Weidemann A, Beyruther K, Fischer P, Masters CL, Price DL. Precursor of amyloid protein in Alzheimer's disease undergoes fast antegrade axonal transport. *Proc Natl Acad Sci* 1990, **87**: 1561–1565.
22. Lewen A, Li GL, Nilsson P, Olsson Y, Hillered L. Traumatic brain injury in rat produces changes of beta-amyloid precursor protein immunoreactivity. *Neuro Report* 1995, **6**: 357–360.

23. Gentleman SM, Nash MJ, Sweeting CJ, Graham DI, Roberts GW. Beta-amyloid precursor protein (BAPP) as a marker for axonal injury after head injury. *Neurosci Lett* 1993, **160**: 139–144.
24. Sherriff FE, Bridges LR, Sivaloganathan S. Early detection of axonal injury after human head injury using immunocytochemistry for beta-amyloid precursor protein. *Acta Neuropathol* 1994, **87**: 55–62.
25. Adams JH, Graham DI, Scott G, Parker LS, Doyle D. Brain damage in fatal non-missile head injury. *J Clinical Pathol* 1980, **33**: 1132–1145.
26. Zimmerman RA, Bilaniuk LT, Gennarelli TA. Computed tomography of shearing injuries of the cerebral white matter. *Radiology* 1978, **127**: 393–396.
27. Snoek J, Jennett B, Adams JH, Graham DI, Doyle D. Computerized tomography after severe head injury in patients without acute intracranial hematoma. *J Neurol, Neurosurg and Psychiat* 1979, **42**: 215–222.
28. Adams JH and Graham DI. The pathology of blunt head injuries, in *Scientific Foundations of Neurology* (Critchley M, O'Leary JL, Jennett B, eds.), Heinemann, London, 1972, p. 478.
29. Gennarelli TA. Animal models of human head injury. *J Neurotrauma* 1994, **11**: 357–368.
30. Sullivan HG, Martinez AJ, Becker DP, Miller JD, Wist A, Griffith RL. The fluid percussion model of mechanical brain injury in the cat. *J Neurosurg* 1976, **45**: 520–534.
31. Dixon CE, Lyeth BG, Povlishock JT, Findling RL, Hamm RJ, Marmarou A, Young HF, Hayes RL. A fluid percussion model of experimental brain injury in the rat. *J Neurosurg* 1987, **67**: 110–119.
32. Yaghmai A, Povlishock J. Traumatically induced reactive change as visualized through the use of monoclonal antibodies targeted to neurofilament subunits. *J Neuropathol Exp Neurol* 1992, **51**: 158–176.
33. Cortez SC, McIntosh TK, Noble LJ. Experimental fluid percussion brain injury: vascular disruption and neuronal and glial alterations. *Brain Res* 1989, **482**: 272–282.
34. McIntosh TK, Vink R, Noble L, Yamakami I, Fernyak S, Soares H, Faden AI. Traumatic brain injury in the rat: characterization of a lateral fluid-percussion model. *Neuroscience* 1989, **28**: 233–244.
35. Dietrich WD, Alonso O, Halley M. Early microvascular and neuronal consequences of traumatic brain injury: A light and electron microscopic study in rats. *J Neurotrauma* 1994, **11**: 289–301.
36. Dietrich WD, Alonso O, Busto R, Globus MY-T, Ginsberg MD. Post-traumatic brain hypothermia reduces histopathological damage following concussive brain injury in the rat. *Acta Neuropathol* 1994, **87**: 250–258.
37. Dietrich WD, Alonso O, Halley M, Busto R. Delayed posttraumatic brain hyperthermia worsens outcome after fluid percussion brain injury: A light and electron microscopic study in rats. *Neurosurgery* 1996, **38**: 533–541.
38. Pierce JES, Trojanowski JQ, Graham DI, Smith DH, McIntosh TK. Immunohistochemical characterization of alterations in the distribution of amyloid precursor proteins and beta-amyloid peptide after experimental brain injury in the rat. *J Neurosci* 1996, **16**: 1083–1090.
39. Bramlett HM, Kraydieh S, Green EJ, Dietrich WD. Temporal and regional patterns of axonal damage following traumatic brain injury: a beta-amyloid precursor protein immunocytochemical study in rats. *J Neuropathol Exp Neurol* 1997, **56(10)**: 1132–1141.
40. Dietrich WD, Alonso O, Busto R, Prado R, Dewanjee S, Dewanjee MK, Ginsberg MD. Widespread hemodynamic depression and focal platelet accumulation after fluid percussion brain injury: A double-label autoradiographic study in rats. *J Cereb Blood Flow Met* 1996, **16**: 481–489.
41. Marmarou A, Foda MA, Brink WV, Kita H, Demetriadou K. A new model of diffuse brain injury in rats. Part 1: Pathophysiology and biomechanics. *J Neurosurg* 1994, **80**: 291–300.

42. Foda MA, Marmarou A. A new model of diffuse brain injury in rats. Part II: Morphological characterization. *J Neurosurg* 1994, **80**: 301–313.
43. Alberico AM, Ward JD, Chio SC, Marmarou A, Young HF. Outcome after severe head injury. Relationship to mass lesion, diffuse injury, and ICP course in pediatric and adult patients. *J Neurosurg* 1987, **67**: 648–656.
44. Yu ACH, Lee YL, Eng LF. Astrocytes in culture. 1. The model and the effect of antisense oligonucleotides on glial fibrillary acidic protein synthesis. *J Neurosci Res* 1993, **34**: 295–303.
45. Goldberg MP, Regan R, Choi DW. Mechanical trauma initiates propagation of elevated cytosolic calcium in cultured neurons and glia. *Neurology* 1991, **41**(Suppl. 1): 390.
46. Shepard SR, Ghajar JBG, Gianuzzi R, Kupferman S, Hariri RJ. Fluid percussion barotrauma chamber: A new in vitro model for traumatic brain injury. *J Surg Res* 1991, **51**: 417–424.
47. Ellis EF, McKinney JS, Liang S, Tavalin SJ, Satin LS. Biochemical and electrophysiologic studies of stretch-induced cells in culture. *J Cereb Blood Flow Metab* 1995, **15** (Suppl 10): S30.
48. Ellis EF, McKinney JS, Willoughby KA, Liang S, Povlishock JT. A new model for rapid stretch-induced injury in culture: Characterization of the model using astrocytes. *J Neurotrauma* 1995, **12**: 325–339.
49. Tavalin SJ, Ellis EF, Satin LS. Mechanical perturbation of cultured cortical neurons reveals a stretch-induced delayed depolarization. *J Neurophysiol* 1995, **74(6)**: 2767–2773.
50. Hariri RJ, Chang VA, Barie PS, Wang RS, Sharif SF, Ghajar JBG. Traumatic injury induces interleukin-6 production by human astrocytes. *Brain Res* 1994, **636**: 139–142.
51. Murphy EJ, Horrocks LA. A model for compression trauma: pressure-induced injury in cell cultures. *J Neurotrauma* 1993, **10**: 431–444.
52. Sutton RL, Lescaudron L, Stein DG. Unilateral cortical contusion injury in the rat: Vascular disruption and temporal development of cortcial necrosis. *J Neurotrauma* 1993, **10**: 135–149.
53. Hicks RR, Smith DH, Lowenstein DH, Marie RS, McIntosh TK. Mild experimental brain injury in the rat induces cognitive deficits associated with regional neuronal loss in the hippocampus. *J Neurotrauma* 1993, **10**: 405–414.
54. Bramlett HM, Dietrich WD, Green EJ, Busto R. Chronic histopathological consequences of fluid-percussion brain injury in rats: effects of posttraumatic hypothermia. *Acta Neuropathol* 1997, **93(2)**: 190–199.
55. Graham DI, Adams JH. Ischaemic brain damage in fatal head injuries. *Lancet* **1**: 265–266, 1971
56. Kotapka MJ, Gennarclli TA, Graham DI, Adams JH, Thibault LE, Ross DT, Ford I. Selective vulnerability of hippocampal neurons in acceleration-induced experimental head injury. *J Neurotrauma* 1991, **8**: 247–258.
57. Jenkins LW, Moszynski K, Lyeth BG, Lewelt W, DeWitt DS, Allen A, Dixon CE, Povlishock JT, Majewski TJ, Clifton GL, Young HF, Becker DP, Hayes RL. Increased vulnerability of the mildly traumatized brain to cerebral ischemia: The use of controlled secondary ischemia as a research tool to identify common or different mechanisms contributing to mechanical and ischemic brain injury. *Brain Res* 1989, **477**: 211–224.
58. Lowenstein DH, Thomas MJ, Smith DH, McIntosh TK. Selective vulnerability of dentate hilar neurons following traumatic brain injury: A potential mechanistic link between head trauma and disorders of the hippocampus. *J Neurosci* 1992, **12**: 4846–4853.
59. Ross DT, Graham DI, Adams JH. Selective loss of neurons from the thalamic reticular nucleus following severe human head injury. *J Neurotrauma* 1993, **10**: 151–165.
60. Anderson CV, Wood D-MG, Bigler ED, Blatter DD. Lesion volume, injury severity, and thalamic integrity following head injury. *J Neurotrauma* 1996, **13**: 35–40.
61. Anderson CV, Bigler ED. Ventricular dilation, cortical atrophy, and neuropsychological outcome following traumatic brain injury. *J Neuropsychiat Clinical Neurosci* 1995, **7**: 42–48.

62. MacNamara SE, Bigler ED, Blatter D. Magnetic resonance indentified ventricular dilation in traumatic brain injury: comparison of pre- and post-injury scan and postinjury results. *Archives of Clinical Neuropsychol* 1992, **7**: 275–284.
63. Johnson SC, Bigler ED, Burr RB, Blatter DD. White matter atrophy, ventricular dilation and intellectual functioning following traumatic brain injury. *Neurosychology* 1994, **8**: 307–315.
64. Holbourn AHS, Edin MA. Mechanisms of head injuries. *Lancet* 1943, **II**: 438–441.
65. Courville CB, Blomquist OA. Traumatic intracerebral hemorrhage with particular reference to its pathogenesis and its relation to "Delayed traumatic apoplexy." *Arch Surg* 1940, **41**: 1–28.
66. Maxwell WL, Watt C, Graham DI, Gennarelli TA. Ultrastructural evidence of axonal shearing as a result of lateral acceleration of the head in non-human primates. *Acta Neuropathol* 1993, **86**: 136–144.
67. Dietrich WD, Halley M, Valdes I, Busto R. Interrelationships between increased vascular permeability and acute neuronal damage following temperature-controlled brain ischemia in rats. *Acta Neuropathol* 1991, **81**: 615–625.
68. Povlishock JT, Dietrich WD. The blood–brain barrier in brain injury: An overview, in *The Role of Neurotransmitters in Brain Injury* (Globus MY-T, Dietrich WD, eds.), Plenum, NY, 1992, pp. 265–269.
69. Chesnut RM. Secondary brain insults after head injury: clinical perspectives. *New Horizons* 1995, **3**: 366–375.
70. Hovda DA, Becker DB, Katayama Y. Secondary injury and acidosis. *J Neurotrauma* 1992, **9** (Suppl 1): S47–S60.
71. Miller JD. Head injury and brain ischemia: Implication for therapy. *Br J Anaesth* 1985, **57**: 120–129.
72. Young W. Secondary CNS injury. *J Neurotrauma* 1988, **5**: 219–221.
73. DeWitt DS, Jenkins LW, Prough DS. Enhanced vulnerability to secondary ischemic insults after experimental traumatic brain injury. *New Horizons* 1995, **3**: 376–383.
74. MacPerson P, Graham DI. Correlation between angiographic findings and the ischemia of head injury. *J Neurol, Neurosurg and Psychiat* 1978, **41**: 122–127.
75. Tanno H, Nockels RP, Pitts LH, Noble LJ. Breakdown of the blood–brain barrier after fluid percussion brain injury in the rat: Part 2: Effect of hypoxia on permeability to plasma proteins. *J Neurotrauma* 1992, **9**: 335–347.
76. Bouma GJ, Muizelaar P, Choi SC, Newlon PG, Young HF. Cerebral circulation and metabolism after severe traumatic brain injury: the elusive role of ischemia. *J Neurosurg* 1991, **75**: 685–693.
77. DeWitt DS, Jenkins LW, Wei EP. The effects of fluid-percussion brain injury on regional cerebral blood flow and pial arteriolar diameter. *J Neurosurg* 1986, **64**: 787–794.
78. DeWitt DS, Prough DS, Tayor CT, Whitley JM. Reduced cerebral blood flow, oxygen delivery, and electroencephalographic activity after traumatic brain injury and mild hemorrhage in cats. *J Neurosurg* 1992, **76**: 812–821.
79. DeWitt DS, Prough DS, Taylor CL, Whitley JM, Deal DD, Vines SM. Regional cerebrovascular responses to progressive hypotension after traumatic brain injury in cats. *Am J Physiol* 1992, **263**: H1276–H1284.
80. Busto R, Dietrich WD, Globus MY-T, Alonso O, Ginsberg MD. Extracellular release of serotonin following fluid-percussion brain injury in rats. *J Neurotrauma* 1997, **14(1)**: 35–42.
81. Hovda DA, Yoshino A, Kawamata T, Katayama Y, Becker DB. Diffuse prolonged depression of cerebral oxidative metabolism following concussive brain injury in the rat: a cytochrome oxidase histochemistry study. *Brain Res* 1991, **567**: 1–10.
82. Yoshino A, Hovda DA, Kawamata T, Katayama Y, Becker DB. Dynamic changes in local cerebral glucose utilization following cerebral concussion in rats: evidence of a hyper- and subsequent hypometabolic state. *Brain Res* 1991, **561**: 106–119.

83. Dietrich WD, Alonso OF, Busto R, Loor J, Dewanjee MK, Ginsberg MD. Cerebral ischemia following traumatic brain injury in rats. *Neurosurgery* (in press).
84. Luerssen TG, Klauber MR, Marshall LF: Outcome from head injury related to patient age. A longitudinal prospective study of adult and pediatric head injury. *J Neurosurg* 1988, **68**: 409–416.
85. Marmarou A, Anderson RL, Ward JD, Choi SC, Young HF, Eisenberg HM, Foulkes MA, Marshall LF, Jane JA. Increased intracranial pressure in head injury and influence of blood volume. *J Neurotrauma* 1992, **9(1)**: S327–S332.
86. Ishige N, Pitts LH, Berry I, Carlson SG, Nishimura MC, Moseley ME, Weinstein PR: The effect of hypoxia on traumatic head injury in rats: alterations in neurologic function, brain edema, and cerebral blood flow. *J Cereb Blood Flow Metabol* 1987, **7**: 759–767.
87. Lewelt W, Jenkins LW, Miller JD. Autoregulation of cerebral blood flow after experimental fluid percussion injury of the brain. *J Neurosurg* 1980, **53**: 500–511.
88. Overgaard J, Tweed WA. Cerebral circulation after head injury: I. Cerebral blood flow and its regulation after closed head injury with emphasis on clinical correlations. *J Neurosurg* 1974, **41**: 531–541.
89. Dietrich WD. Morphological manifestations of reperfusion injury in brain. *Ann NY Acad Sci* 1994, **723**: 15–23.
90. Kochanek PM, Hallenbeck JM. Polymorphonuclear leukocytes and monocytes/macrophages in the pathogenesis of cerebral ischemia and stroke. *Stroke* 1992, **23**: 1367–1379.
91. del Zoppo GJ, Schmid-Schonbein GW, Mori E, Copeland BR, Chang CM. Polymorphonuclear leukocytes occlude capillaries following middle cerebral artery occlusion and reperfusion in baboons. *Stroke* 1991, **22**: 1276–1283.
92. Chopp M, Zhang RL, Chen H, Ning J, Rusche JR. Postischemic administration of an anti-Mac-1 antibody reduces ischemic damage after transient middle cerebral artery occlusion in rats. *Stroke* 1993, **25**: 869–876.
93. Schoettle RJ, Kochanek PM, Magargee MJ, Uhl MW, Nemoto EM. Early polymorphonuclear leukocyte accumulation correlates with the development of posttraumatic cerebral edema in rats. *J Neurotrauma* 1990, **7**: 207–217.
94. Clark RSB, Schiding JK, Kaczorowski SL, Marion DW, Kochanek PM. Neutrophil accumulation after traumatic brain injury in rats: comparison of weight drop and controlled cortical impact models. *J Neurotrauma* 1994, **11**: 499–506.
95. Soares HD, Hicks RR, Smith D, McIntosh TK. Inflammatory leukocyte recruitment and diffuse neuronal degeneration are separate pathological processes resulting from traumatic brain injury. *J Neuroscience* 1995, **15**: 8223–8233.
96. Uhl MW, Biagas KV, Grundl PD, Barmada MA, Schiding JK, Nemoto EM, Kochanek PM. Effects of neutropenia on edema, histology, and cerebral blood flow after traumatic brain injury in rats. *J Neurotrauma* 1994, **11**: 303–315.
97. Chopp M, Li Y, Jiang N, Zhang RL, Prostak J. Antibodies against adhesion molecules reduce apoptosis after transient middle cerebral artery occlusion in rat brain. *J Cereb Blood Flow Metabol* 1996, **16**: 578–584.
98. Raghupathi R, McIntosh TK. Programmed cell death in traumatic brain injury: Temporal alterations in bcl-2 immunoreactivity. *J Neurotrauma* 1995, **12**: 966.
99. Bresnahan JC, Shuman SL, Beattie MS: Evidence for apoptosis of oligodendroglia in long tracts undergoing wallerian degeneration after spinal cord injury (SCI) in monkeys. *Soc Neurosci Abstr* 1996, **22**: 1185.
100. Hayashi N, Hirayama T, Ohata M. The computed cerebral hypothermia management technique to the critical head injury patients. *Advances in Neurotrauma Res* 1993, **5**: 61–64.
101. Rousseaux P, Scherpered B, Bernard MH, Graftieaux JP, Guyot JF. Fever and cerebral vasospasm in ruptured intracranial aneurysms. *Surg Neurol* 1980, **14**: 459–465.

102. Sternau L, Thompson C, Dietrich WD, Busto R, Globus MY-T, Ginsberg MD. Intracranial tempertuare: observations in human brain. *J Cereb Blood Flow Metab* 1991, **11(2):** S123.
103. Dietrich WD. Nonpharmacologic strategies—hypothermia, in *Neurotrauma* (Narayan RK, Wilberger JE, Povlishock JT, eds.) McGraw-Hill, NY, 1996, pp. 1491–1506.
104. Dietrich WD, Busto R, Globus MY-T, Ginsberg MD. Brain damage and temperature: Cellular and molecular mechanisms, in *Cellular and Molecular Mechanisms of Ischemic Brain Damage* (B. Siesjo, T. Wieloch, eds.), *Advances in Neurology* Lippincott-Raven Publishers, Philadelphia, PA, 1996, **71:** 177–198.
105. McIntosh, TK. Novel pharmacologic therapies in the treatment of experimental brain injury: A review. *J Neurotrauma* 1993, **10:** 215–261.
106. Faden AI. Pharmacological treatment approaches for brain and spinal cord trauma, in *Neurotrauma*, (Narayan RK, Wilberger JE, Povlishock JT, eds.) McGraw-Hill, NY, 1996c, pp. 1479–1490.
107. Hayes RL, Jenkins LW, Lyeth BG. Neurotransmitter-mediated mechanisms of traumatic brain injury: acetylcholine and excitatory amino acids. *J Neurotrauma* 1992, **9:** S173–S187.
108. Faden AI, Demediuk P, Panter SS, Vink R: The role of excitatory amino acids and NMDA-receptors in traumatic brain injury. *Science* 1989, **244:** 798–800.
109. McIntosh TK, Vink R, Soarces H, Hayes R, Simon R. Effects of the N-methyl-D-aspartate receptor blocker MK-801 on neurologic function after experimental brain injury. *J Neurotrauma* 1989, **6:** 247–259.
110. Clifton GL, Jiang JY, Lyeth BG, Jenkins LW, Hamm RJ, Hayes RL: Marked protection by moderate hypothermia after experimental traumatic brain injury. *J Cereb Blood Flow Metab* 1991, **11:** 114–121.
111. Lyeth BG, Jiang JY, Shanliang L. Behavioral protection by moderate hypothermia initiated after experimental traumatic brain injury. *J Neurotrauma* 1993, **10:** 57–64.
112. Baker A, Moulton RJ, MacMillian VH, Shedden PM. Excitatory amino acids in cerebrospinal fluid following traumatic brain injury in humans. *J Neurosurg* 1993, **79:** 369–372.
113. Bullock R, Fujisawa H. The role of glutamate antagonists for treatment of CNS injury. *J Neurotrauma* 1992, **9**(Suppl 2): S443–S462.
114. Globus MY-T, Alonso O, Dietrich WD, Busto R, Ginsberg MD. Glutamate release and free radical production following brain injury: effects of posttraumatic hypothermia. *J Neurochem* 1995, **65:** 1704–1711.
115. Katayama Y, Becker DP, Tamura T, Hovda DA. Massive increases in extracellular potassium and the indiscriminate release of glutamate following concussive brain injury. *J Neurosurg* 1990, **73:** 889–900.
116. Palmer, AM, Marion, DW, Botsceller, ML, Redd, EE. Therapeutic hypothermia is cytoprotective without attenuating the traumatic brain injury-induced elevations in interstitial concentrations of aspartate and glutamate. *J Neurotrauma* 1993, **10:** 363–372.
117. Nilsson P, Hillered L, Ponton U. Changes in cortical extracellular levels of energy-related metabolites and amino acids following concussive brain injury in rats. *J Cereb Blood Flow Metab* 1990, **10:** 631–637.
118. Chan PH, Schmidley JW, Fishman RA, Longar SM. Brain injury, edema, and vascular permeability changes induced by oxygen-derived free radicals. *Neurology* 1984, **34:** 315–320.
119. Chan PH, Longar L, Fishman RA. Protective effects of liposome-entrapped superoxide dismutase on posttraumatic brain edema. *Ann Neurol* 1987, **21:** 540–547.
120. Kontos HA, Povlishock JT. Oxygen radicals in brain. *CNS Trauma* 1986, **3:** 257–263.
121. Kontos HA. Oxygen radicals in CNS damage. *Chem Biol Interact* 1989, **72:** 229–255.

122. Chan PH, Epstein CJ, Kinouchi H, Kamii H, Chen SF, Carlson E, Gafni J, Yang G, Reola L. Neuroprotective role of CuZn-superoxide dismutase in ischemic brain damage, in *Cellular And Molecular Mechanisms of Ischemic Brain Damage* (Siesjo BK, Wieloch T, eds.), Lippincott-Raven, Philadelphia, 1996, pp. 271–280.
123. Mattson MP, Scheff SW. Endogenous neuroprotection factors and traumatic brain injury: mechanisms of action and implications for therapy. *J Neurotrauma* 1994, **11**: 3–33.
124. Mocchetti I, Wrathall JR. Neurotrophic factors in central nervous system trauma. *J Neurotrauma* 1995, **12**: 853–870.
125. Dietrich WD, Alonso O, Busto R, Finklestein SP. Posttreatment with intravenous basic fibroblast growth factor reduces histopathological damage following fluid-percussion brain injury in rats. *J Neurotrauma* 1996, **13**: 309–316.
126. Kawamata T, Alexis NE, Dietrich WD, Finklestein SP. Intracisternal basic fibroblast growth factor (bFGF) enhances behavioral recovery following focal cerebral infarction in the rat. *J Cereb Blood Flow Metab* 1996, **16**: 542–547.
127. Bramlett H, Green EJ, Dietrich WD, Busto R, Globus MY-T, Ginsberg MD. Post-traumatic brain hypothermia provides protection from sensorimotor and cognitive behavioral deficits. *J Neurotrauma* 1995, **12**: 289–298.
128. Clifton GL, Allen S, Barrodale P, Plenger P, Berry J, Koch S, Fletcher J, Hayes RL, Choi SG. A phase II study of moderate hypothermia in severe brain injury. *J Neurotrauma* 1993, **10**: 263–271.
129. Marion DW, Obrist WD, Carlier PM, Penrod LE, Darby JM. The use of moderate therepeutic hypothermia for patients with severe head injuries. A preliminary report. *J Neurosurg* 1993, **79**: 354–362.
130. Resnick DK, Marion DW, Darby JM. The effect of hypothermia on the incidence of delayed traumatic intracerebral hemorrhage. *Neurosurg* 1994, **34**: 252–256.
131. Dietrich WD, Lin B, Globus MY-T, Green EJ, Ginsberg MD, Busto R. Effect of delayed MK-801 (dizocilpine) treatment with and without immediate postischemic hypothermia on chronic neuronal survival after global ischemia in rats. *J Cereb Blood Flow Metab* 1995, **15**: 960–968.
132. Hayashi N. The prevention of cerebral thermo-pooling, ICP elevation, and free radicals by control of brain tissue temperature in severely brain injured patient. *J Neurochem* 1996, **66**(Suppl 2): 66.
133. Goss JA, Styren SD, Miller PD, Kochanek PM, Palmer AM, Marion DW, DeKosky ST. Hypothermia attenuates the normal increase in interleukin 1-beta RNA and nerve growth factor following traumatic brain injury in the rat. *J Neurotrauma* 1995, **12**: 159–167.

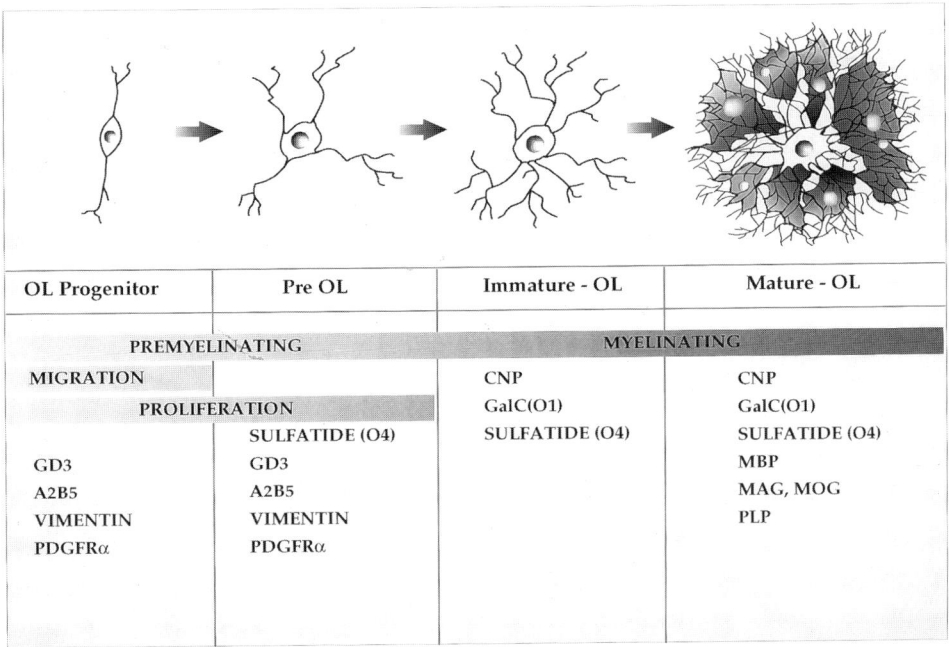

Fig. 1. Maturation of OL lineage cells is depicted. Four principal stages of development are shown from the OL-progenitor to the mature OL, together with the corresponding morphologic features and potential for proliferation, migration, and myelination. The expression of antigenic markers, which in combination define each stage of OL maturation, are shown.

mitted to the OL lineage has been identified which originate in the subependymal zone of the lateral ventricles (8). These cells are bipolar, proliferative and migratory. This population is generated over a short period during late gestation and early postnatal development in the rat and can be identified in vitro by immunoreactivity for the antibodies A2B5 ganglioside (24,25), GD3 ganglioside, and vimentin (20). These three markers are not specific for OLs but are useful, in combination, to define this stage of OL development. Reactivity to A2B5 occurs in both OLs and neurons, and to GD3 in both OLs and microglia. Vimentin is a marker for proliferating OLs and astrocytes, as well as postmitotic astrocytes.

Next in ontogenetic progression is the preOL, a multipolar mitotically active cell, identified by reactivity to the O4 MAb, which recognizes the cell surface glycolipids sulfatide and seminolipid (5,8,10,12,26). Immunoreactivity to the O4 MAb persists in subsequent stages. This transitional stage gives rise to immature OLs that are postmitotic and have a more extensively branching morphology. This stage appears to correlate with commitment to terminal progression into a mature OL. Labeling of such immature OLs with O1 is correlated with cell surface expression of several antigens, including galactocerebroside (5). Markers of mature OLs include CNS myelin proteins, myelin basic protein (MBP), proteolipid protein (PLP), myelin associated glycoprotein (MAG), myelin/oligodendrocyte glycoprotein (MOG), and 2'-3'-cyclic nucleotide 3'-phosphohydrolase (CNP)] and carbonic anhydrase II (CAII), a myelin-associated enzyme involved in membrane compaction (reviewed in ref. 4).

METHODS FOR ASSESSMENT OF OL DEATH

Anatomical Approaches to Visualize OLs

The recent development of techniques to visualize OLs by immunocytochemistry or *in situ* hybridization has permitted more precise characterization of the morphological and stage-specific features of OL death. These markers, when used in combination, can specifically identify a single stage of OL development. However, the use of fixatives, when attempting to visualize OL precursors, may result in artifactual staining of cross-reacting nonoligodendrogial cells or the loss of OL epitopes. For this reason, when employing the A2B5, O4, and O1 antibodies, it is advisable to immunocytochemically characterize live cells *(5)* or to perform appropriate fixation controls to confirm that the staining of fixed cells is equivalent to that of live cells.

The in vivo characterization of the distribution of OL markers has been largely restricted to analysis of populations of mature OLs, because of lack of suitable techniques to visualize OL precursors. Most studies have identified OL precursors histologically on the basis of imprecise morphological criteria. However, OL precursors cannot be distinguished from other neuroglial precursors with absolute certainty. Attempts to visualize OL precursors by combining morphological criteria with localization of ferritin, transferrin or iron lack stage-specificity. Moreover, as discussed below, these markers of iron metabolism are not specific to OLs and have been localized to multiple CNS cell types, including microglia, astrocytes and neurons. Immunocytochemical identification of OL precursors using antibodies directed against surface markers of OL precursors (e.g., the A2B5, O4, and O1 monoclonal antibodies) has been hampered either by limited or nonspecific staining arising from fixation artifacts. Approaches using 300–500-μm-thick slices of fresh tissue have generally been unacceptable for histological studies *(15)*. Some investigators have succeeded with acrolein-based fixatives for fetal tissues *(27)*, but this has worked inconsistently for perfused animal tissues or autopsy specimens (Back and Kinney, personal observations). For visualization of the O4 and O1 monoclonal antibodies in human autopsy tissues, we have recently developed a simple paraformaldehyde-based method (28), which achieves a cellular localization consistent with that obtained for OL precursors in the rodent forebrain *(17)*. Antibodies direceted against DM-20, a spliced product of proteolipid protein, have detectd premyelinating OLs in the developing rat brain before the onset of active myelination (28a).

In Vitro Cell Survival Assays

In vitro assessment of OL death has largely relied upon routine morphometric approaches (reviewed in ref. *29*). Two commonly employed assays of OL viability are exclusion of trypan blue by live cells and the MTT survival assay, which histochemically visualizes the formazan reaction product produced by mitochondrial cleavage of a tetrazolium ring substrate. Fluorescent chromatin-binding compounds have been used to detect cell death, including apoptosis. The bisbenzamidazole derivative Hoechst 33342 stains both live and dead fixed cells. Apoptotic cells are distinguished by the presence of a condensed or fragmented nucleus. Propidium iodide binds reversibly to DNA of dead cells. However, this approach has the limitation that samples must be counted immediately, because of leakage of dye into viable cells permeabilized by the fixation process. Propidium iodide labeling of cryostat sections of tissue has also been employed in vivo to count pyknotic nuclei of apparent OLs in rat optic nerve *(30)*. Apoptotic nuclei can also be visualized by the TUNEL method for *in situ* end labeling of 3'-end nicks of DNA. The results of this technique must be interpreted cautiously, because cells dying via necrosis can also be labeled. It is imperative, therefore, to confirm the occurrence of apoptosis by complementary ultrastructural or molecular approaches.

A limitation of all the above approaches is lack of specificity to OLs, and, thus, care must be taken if contaminating cells are present in the samples. Because these techniques are morphologically based, they are also time consuming and rely on subjective morphological criteria to arrive at cell counts. Biochemical assays of cell death, including solubilized MTT, lactate dehydrogenase or chromium release from dead cells, have not been widely used, because of problems of low specific signal or high background from other cell types in mixed glial cultures.

SELECTION OF IN VITRO MODEL SYSTEMS TO STUDY OL DEATH

The development of methods for the in vitro culture of OLs represents the single most important advance in the characterization of mechanisms governing the survival and differentiation of OLs. However, inherent to tissue culture are a number of potential pitfalls, as well as questions of biological relevance. Among the questions to be considered in assessing the experimental design of in vitro studies of OL death are the following:

1. Are the OLs under study a pure population or a combination of several stages of differentiation?
2. Are multiple cell surface markers employed to rigorously define the stage(s) under study or are morphological criteria used?
3. What is the effect of the density and age of the cultures upon OL survival and are these parameters consistent in all studies?
4. What is the degree of contamination of cultures by nonoligodendrocytes, usually astrocytes and microglia, and how much does this factor vary among experiments?
5. Does the culture system contain nonoligodendroglia which might affect the OL survival response?
6. What growth medium is employed and is it chemically defined or supplemented with serum?
7. If serum is employed, what type is employed and how might it affect the final concentrations of components of the medium (e.g., electrolytes, vitamins, hormones, iron)?
8. Does the growth medium contain certain components (e.g. glucose, iron, antibiotics) at concentrations which might adversely affect the question under study?

Culture of OLs was initially achieved with mixed glial culture systems, which consisted predominantly of OLs, but also contained contaminating astrocytes and microglia [reviewed in ref. *31*]. This approach offers the advantages of a stable culture system with generally high initial yield and survival. One limitation is the presence of contaminating glial cells. Astrocytes in vitro, for example, secrete a variety of substances which may influence OL survival or differentiation (*see*, for example, refs. *32* and *33*). Among such factors which may be trophic or toxic to OLs are insulin-like growth factors (IGFs), platelet-derived growth factor (PDGFs), neutrophin-3 (NT-3), ciliary neurotrophic factor (CNTF), tumor necrosis factor (TNF), interleukin-1 (IL-1), leukemia inhibitory factor (LIF), and interleukin-6 (IL-6). It should be noted that astrocyte factors produced in vitro are not necessarily also produced in vivo. Messenger RNA for PDGFα, for example, was localized only to neurons in developing rat brain, and thus the in vitro release of PDGF by astrocytes may be nonphysiological *(34)*. Another potential limitation of mixed glial cultures systems is that they may be maintained in serum-supplemented medium. Exposure to serum poses a number of potential problems due to the fact that its composition is not defined and may vary depending upon the source. An important consideration for studies of OL death is the iron content of the serum and whether the medium is supplemented with apo- or holo-transferrin, the former being iron-free. As discussed below, OLs are one of the primary CNS cell types that store and utilize iron and, under conditions of oxidative stress, these iron stores represent a source of potential toxicity. The final concentration of iron in growth medium can vary considerably with the serum composition. Calf serum, for example, is often iron supplemented to compensate for the anemia induced in the calves to be used in the production of veal.

In order to circumvent the confounding variables of mixed glial systems, a number of protocols have been developed which rely upon the selectivity of certain monoclonal antibodies for specific stages in the OL lineage *(30,35)*. Using this approach, termed immunopanning, it has been possible to isolate greatly enriched, nearly pure populations of OLs at a defined stage of maturation. In addition, protocols have been developed to promote the survival and differentiation of these OL cultures in chemically-defined media (i.e., serum-free) that have been supplemented with selected combinations of growth factors. This approach has been particularly valuable to define the actions of OL trophic factors. Barres and colleagues have argued that even minor contamination of OL cultures by astrocytes or other cell types might yield invalid conclusions regarding the actions of OL trophic factors. By testing the actions of individual trophic factors on single cell OL microcultures, they demonstrated, for example, that IGFs are potent survival factors for O2-A progenitor cells. Under these culture conditions, IGFs did not appear to act as mitogens, in contrast to results obtained from mixed culture systems *(36)*. One caveat of the microculture technique is that cell proliferation may be very slow and should be assessed over the course of several days.

APPROACHES TO ESTABLISH BIOLOGICAL RELEVANCE

Optic Nerve

Several investigators have found the optic nerve a useful model system to study OL survival and differentiation, because of its ready accessibility for manipulations, including nerve transection. Its large size makes it useful for quantitative measurements of cell death. In addition, sufficient numbers of OLs may be isolated for complementary in vitro studies. One question still to be addressed is whether the findings in the optic nerve can be generalized to OLs in other CNS regions.

Raff et al. have employed the optic nerve extensively as a source of OLs for in vitro characterization of OL trophic factors, as well as for in vivo studies of the actions of selected trophic factors on OL survival or proliferation. One novel approach was to implant COS cells, which released exogenous PDGF, into the vicinity of the optic nerve *(36)*. The effect of PDGF on OL precursors was consistent with its role as a survival factor but not as a potent mitogen in the developing optic nerve. A limitation of this approach is that it examines the effect of a nonphysiological elevation of the level of the growth factor on OL survival. A complementary approach is to implant hybridoma cells into the subarachnoid space in the vicinity of the nerve, in order to locally release a neutralizing monoclonal antibody directed against a bioactive molecule of interest. Using this technique to deliver an NT-3 antibody, Barres et al. observed a decrease in the number of OLs and OL precursors, consistent with a role of NT-3 in promoting cell survival *(37)*. Transgenic mouse models offer another novel approach to study mechanisms of OL death. A transgenic mouse that overexpressed the human *bcl*-2 gene was found to have an increased number of retinal ganglion axons as well as a proportional increase in OLs in the optic nerve *(38)*.

One important question readily addressed in the optic nerve is the relationship of axon number or functional integrity to OL survival. A variety of studies have examined the effects of axonal degeneration, following optic nerve transection, on nerve myelination, OL survival, or the response of other neuroglia. This approach has permitted a detailed ultrastructural analysis of the fate of OLs in the setting of active nerve demyelination. Intriguing recent observations were that many apparently dedifferentiated OLs and a small population of myelinating OLs persisted several weeks after nerve transection *(39)*. The fact that these cells did not appear to be dependent upon axons for their continued survival raises important questions regarding the mechanisms of OL survival following CNS injury *(40)*.

Mutant Mouse Models

Several strains of mutant mice display abnormal patterns of myelination and a characteristic phenotype notable for weakness and impaired motor performance, including the shiverer, quaking, and jimpy mice. The underlying defects in these mutants include either altered structure (e.g., myelin basic protein in the shiverer mouse) or altered expression (e.g., some strains of the jimpy mouse) of a myelin-related protein. Although most of these mutants display minimal loss of OLs, one exception is the animal with a point mutation in the protelolipid protein (PLP) gene (e.g., the jimpy mouse). Such animals display an increased proportion of OLs undergoing apoptosis relative to controls *(41,41a)*. These mutants are important animal models for Pelizaeus–Merzbacher disease, an X-chromosome-linked neurodegenerative demyelinating disease of childhood (Table 1). These mice have a marked reduction in myelin with sheaths of abnormal periodicity, develop seizures and die prematurely. A causal role for the misfolded jimpy PLP protein in triggering OL death is supported by the observation that PLP-transgenic/wild-type mice still show a severalfold increase in the number of pyknotic nuclei of apparent OLs despite complementation of the jimpy mutation with an autosomal PLP transgene *(42,43)*. Although the transgenic mice show functional expression of PLP, and compacted myelin sheaths of normal periodicity are present in surviving OLs, severe dysmyelination occurs in these mice supporting a dominant-negative effect of the mutant PLP protein. Expression of the PLP gene may be associated with secretion of a factor that promotes survival of OLs, a fact consistent with the observation that immature OLs persist at sites of demyelination *(44)*. Death of OLs derived for the jimpy mouse correlates with the onset of expression of the abnormal PLP/DM-20 myelin-associated proteins *(44b)*. Interestingly, in vitro studies in which the synthesis of PLP was blocked by antisense oligonucleotides resulted in prolonged survival of OL progenitors that failed to differentiate or elaborate membrane sheets *(44a)*. These studies suggest that prolonged OL survival and abnormal myelination may be associated with decreased expression of PLP, whereas abnormal PLP protein or PLP overexpression may be associated with a more severe phenotype that results in enhanced OL death.

One possible explanation for OL death in mutants expressing abnormal myelin-associated proteins is the aberrant accumulation of protein within the endoplasmic reticulum (ER) *(41a)*. COS-7 cells transfected with mutated PLP were associated with impaired intracellular transport of the protein *(45)*. Interestingly, transgenic mice, in which the class I major histocompatibility (MHC) gene is coupled to an MBP promoter, also show high levels of the MHC gene product in the ER *(46)*. These mice develop extensive demyelination and a marked reduction in OL number, display intense hindlimb shivering and die prematurely due to tonic seizures. There are several possible mechanisms by which impaired protein transport may occur in these mutants. The transporter may not bind the mutant protein. In the case of MHC class I proteins, the appropriate transporter may not be present in the cell, in view of the fact that OLs do not normally express class I MHC *(46)*.

EAE Models for Multiple Sclerosis (MS)

Experimental autoimmune encephalitis (EAE) has served as the primary animal model for the autoimmune-mediated demyelination that is believed to underlie MS *(47–49)*. When a susceptible species is immunized with CNS tissue or myelin in Freund's complete adjuvant, a CD4+ T-cell-mediated inflammatory response is triggered and directed against myelin-associated proteins. Depending upon the method of immunization, both acute and chronic forms of EAE can be induced, which share some of the features of the chronic progressive and relapsing-remitting forms of MS. The development of neurological symptoms ensues from lymphocytic infiltrates whose activation is mediated via antigen-presenting macrophages. EAE can be prevented by blocking the actions of either the T cells or macrophages. Moreover, acute EAE is characterized by minimal active remyelination after acute myelin destruction, a pattern supporting a role for the ongoing inflammatory process in the lack of CNS recovery. An important insight from EAE is the view that OL death may be a secondary event that ensues from the inflammatory processes that accompany immune-mediated destruction of myelin. This model has permitted a detailed molecular analysis of a number of cytokines released by immune mediators, including tumor necrosis factor (TNF) and interferon-γ (INF-γ), which are cytotoxic to OLs (*see* next section).

Intriguingly, recent studies indicate that populations of OL precursors may persist at sites of demyelination, but that they apparently fail to differentiate into the mature myelin-producing cells. In a model of antibody-induced demyelination, an early enhanced proliferation of OL progenitors was observed in response to demyelination, but it was not sustained as remyelination ensued *(49a)*. Interestingly, chronic MS lesions also contained significant numbers of OL precursors that showed no evidence of mitosis *(49b)*. These studies suggest that OL precursors may be impaired in their ability to regenerate following demyelination. Although OL progenitors in the adult CNS express receptors for PDGF-AA and bFGF *(49c)* (*see* below), it is unknown whether these growth factors could stimulate OL proliferation at sites of demyelination.

MECHANISMS OF OLIGODENDROCYTE DEATH

Characterization of the mechanisms of OL death, until recently, had focused on the mature oligodendrocyte, largely with the aim of characterizing the mechanisms underlying adult demyelinating disorders, including multiple sclerosis (MS). A variety of studies now indicate that the biology of OL precursors differs from that of the mature OL (reviewed in ref. 4). Hence, it remains to be determined whether the mechanisms that mediate death of developing OLs differ from those occurring in mature OLs. Studies of OL death have centered around three main areas to be reviewed below: the effect upon OL survival of cellular mediators, including cytokines and growth factors; the role of glutamate on the survival and differentiation of OL precursors; and the vulnerability of OL precursors to free radical-mediated toxicity triggered by intracellular depletion of glutathione.

Cytotoxic Cytokines

An active area of research into the pathogenesis of a number of human white matter disorders involves studies investigating a role for direct oligodendroglial injury by cytotoxic cytokines. In vivo studies, as well as human pathological studies, have shown that white matter injury can occur in the setting of a marked increase in reactive cell types—initially, macrophages and, later, reactive astrocytes, which may release a number of cytokines toxic to OLs. In the premature infant, for example, release of cytotoxic cytokines in response to circulating bacterial endotoxin remains an unresolved potential mechanism for periventricular white matter injury and subsequent cerebral palsy *(50)*. Perinatal white matter injury may arise in the setting of sepsis via overproduction or inappropriate release of cytokines as part of an acute or chronic inflammatory process. Systemic exposure to endotoxin can induce selective necrosis of cerebral white matter in some neonatal animals (reviewed in ref. *50*). In adrenoleulodystrophy, an X-linked disorder in which very long chain fatty acids accumulate in the brain and adrenal cortex, boys begin to rapidly deteriorate when demyelination triggers a concomitant intense inflammatory reaction by macrophages and reactive astrocytes at sites of active myelin destruction (reviewed in ref. *47*). An autoimmune process appears to mediate the demyelination in multiple sclerosis. In this condition, a myelin-specific autoantibody response is believed to be sustained, in part, by a variety of cytokincs that appear to also sustain the activation of clones of antimyelinogenic T-lymphocytes, as well as have direct cytotoxic effects on OLs (reviewed in ref. *51*).

Sites of acute white matter injury are characterized histologically by an initial tissue reaction in which infiltrates of macrophages/microglia predominate. During the subacute phase, hypertrophic astrocytes increase in number in and around the injury site. Focal or diffuse reactive astrocytosis is central to the subsequent process of tissue remodeling following injury. Among the cell mediators that stimulate reactive astrocytosis are a number of cytokines and growth factors that have trophic or toxic effects upon OLs in vitro. Microglia induce reactive astrocytosis via release of TNF, IL-1, IL-6, and INF-γ reviewed in ref. *52*. Three members of the IL-6 family of cytokines, including IL-6, LIF and CNTF, enhance the in vitro survival of OLs *(30,57,61)*. A number of other blood-borne mediators of astrogliosis, including PDGF, steroids, and IGF, also promote the survival of OLs (*see* below). In vitro studies limited to mature OLs, have shown that TNF/lymphotoxin *(53–59)* or INF-γ *(60)* have toxic effects upon OLs in culture. TNF mediates a slow form of OL injury occurring over 48 to 96 h, a temporal pattern consistent with apoptosis. Toxic effects of lymphotoxin occur more rapidly and at lower doses than with TNF *(56)*. The toxicity of TNF and INF-γ has not, however, been reproducible in all culture systems. Recently, Agresti et al. failed to find toxicity of either cytokine in primary cultures enriched in either OL precursors or immature OLs, but found reversible antiproliferative effects of INF-γ which were synergistically enhanced by TNF-α *(59b)*. The disparate toxicity of these cytokines in various systems may be partly explained by stage-dependent effects of these cytokines as supported by the observation that INF-γ also reversibly blocked the differentiation of OL precursors *(59b)*. Recently, studies with transgenic mice that overexpress TNF in a subpopulation of CNS neurons, have provided in vivo support for TNF-mediated OL toxicity. In one such study, the mice spontaneously developed a chronic inflammatory form of CNS demyelination associated with progressive paralysis, tremors, and premature death *(62)*. Pathologically, the CNS displayed a number of features also seen in EAE and MS. These included the presence of CD4+ and CD8+ T cells in the brain and meninges, extensive demyelination in the medulla and cervical cord, and a pronounced reactive astrocytosis and microgliosis. It remains to be determined whether the demyelination was a direct result of TNF overproduction or resulted from secondary sequelae of CNS inflammation triggered by TNF. Support for the latter comes from recent studies of transgenic mice that overproduce IL-3 in astrocytes in the CNS *(63)*. These mice also develop primary CNS demyelination in association with chronic CNS inflammation characterized by a robust activation and expansion of macrophages and microglia.

Future study will thus be required to determine the mechanisms by which dysregulation of cytokine-mediated pathways may promote OL death and demyelination. The variable degree of OL toxicity observed in vitro with some cytokines, including TNF and INF-γ, may reflect a requirement for cooperation among multiple cell types to initiate the actions of a given cytokine. It appears likely that activation of microglia is required for the subsequent expression of astrocytic cytokines *(64,65)*. Microglia, but not astrocytes, for example, are potently stimulated by bacterial endotoxin to produce IL-1β which secondarily stimulates astrocytic expression of both TNF-α and IL-6 *(66)*. Moreover, cytokines may promote the production of other cytotoxic mediators, such as nitric oxide, leading to secondary necrotic injury to OLs *(67)*.

Growth Factors and Regulation of Oligodendrocyte Survival

Both neurons and astrocytes produce growth factors that affect the survival of OLs (reviewed in ref. *68*). Two pathological processes that may occur in the setting of white matter injury and impair the normal release of growth factors trophic to OLs are axonal injury and reactive astrocytosis. The mechanisms by which axonal injury and reactive astrocytosis may contribute to disruption of myelination in some white matter disorders is currently unknown. In part, an alteration in the cellular mechanisms that mediate the production and release of growth factors for OL survival, proliferation or differentiation may be involved. The identification of these growth factors represents an emerging field of great potential importance to the understanding of the pathogenesis of white matter disorders.

The fate of each stage in the OL developmental lineage is regulated by a competition for selected growth factors, such that the mechanisms that promote OL survival are closely linked to those that promote OL death *(68,69)*. When a cell fails to receive the growth factors necessary for its survival, an internal program is activated resulting in the elimination of the cell by a process that involves apoptosis. Hence, the nature of white matter injury may be influenced by such factors as the timing of injury, the OL stages present, and the OL trophic factors whose availability may be disrupted. For this discussion, the effect of OL trophic factors will be considered in terms of the stages of OL development upon which they exert their actions.

Among the growth factors that regulate the survival of CNS precursors are bFGF, NT-3, PDGF, and IGF-1. Basic FGF inhibits the differentiation and promotes the proliferation of OL precursor cells in vitro *(70–73)*. Withdrawal of bFGF in vitro also increases death of rat forebrain OL precursors via apoptosis *(74)*. Notably, however, bFGF can trigger apoptotic death of mature OLs, perhaps by triggering reentry into the cell cycle *(75,75a)*. NT-3 derives from both astroycytes and neurons and promotes the survival and proliferation of OL precursors in vitro *(30,37)*. In vivo studies, in which anti-NT-3 monoclonal antibodies were delivered by hybridoma cells implanted in the vicinity of the optic nerve, also supported a role for NT-3 in stimulating the proliferation of OL precursors *(37)*.

The IGFs derive from astrocytes *(76,77)* and their primary action on OL precursors may be to promote their short-term survival *(36)*. Insulin, IGF-1 receptors, and IGF binding proteins localize to OL precursors in vitro and in vivo *(76,78,78a)*. IGF-l mRNA localizes to the subventricular zone at the time of appearance of OL progenitors *(79)* and autocrine expression of IGF-1 by OL precursors is suggested by detection of IGF-1 mRNA in OL precursors in vitro *(80)*. An increase in astrocytic IGF-1 gene expression has been observed in models of experimentally induced myelin regeneration in adult rats in association with transient OL expression of IGF receptors *(76)* or expression of myelin basic protein mRNA *(77)*. It remains to be determined whether release of IGF-1 from reactive astrocytes may represent a potential mechanism to promote survival of OL precursors during the process of tissue remodeling following injury.

PDGF derives largely from neurons *(34,81)*, is a potent mitogenic *(82)*, chemotactic *(22)*, and survival factor *(36)* for OL precursors and blocks their differentiation to immature OLs *(82–84)*. In vitro, PDGF potentiates the migratory and proliferative effects of bFGF on O2A progenitor cells *(85)*. In combination with PDGF, NT-3 promotes the proliferation of OL precursors in vitro *(30)*. In vivo, PDGF functions largely as a survival factor. Implants of the COS-7 cell line, which were transfected with a PDGF plasmid expression vector, release exogenous PDGF into the vicinity of the optic nerve and cause a twofold increase in the number of immature OLs *(36)*. The increase is consistent with enhanced OL survival but not an increase in cell proliferation, since the number of mitotic figures does not change. In the developing rat forebrain, Ellison and de Vellis localized the PDGFα receptor exclusively to OL precursors of two distinct lineage stages—the OL progenitor and the preOL (Fig. 1) and failed to find receptor expression on immature OLs or other CNS cells *(17)*. Taken together, these in vivo studies support a key role for PDGF in promoting the survival of proliferative OL precursor populations. It remains to be determined in vivo whether focal axonal injury might affect the survival of proliferative OL precursors by disrupting the release or actions of factors, including PDGF and NT-3.

In addition to OL precursors, postmitotic immature and mature myelinating OLs also populate the cerebral white matter around the time of birth in the rat. In vitro studies have shown that the survival of OL in these later stages is promoted by at least three classes of trophic factors: insulin and IGF *(30,86)*; neurotrophins, principally NT-3 *(37)*; and members of the interleukin-6 family, IL-6, CNTF, and LIF *(30,57,61)*. Exogenous NT-3 *(37)* and CNTF *(30)* each promote the in vivo survival of GC-positive OLs in the developing optic nerve. In addition to a role in OL survival, IGF-1 is an important inducer of CNS myelination in vivo. IGF-1 knockout mice display a reduction in size of white matter tracts associated with decreased numbers of OLs and myelinated axons *(87)*. Similarly, transgenic mice that ectopically express IGF-binding protein, which inhibits the action of IGF-1, show a decrease in the number of myelinated axons and of callosal and cortical OLs (88). Transgenic mice that overexpress IGF-1 show an increase in OL number, myelin content and expression of myelin protein genes *(88,89)*.

In summary, a number of trophic factors, including PDGF, NT-3, IGF-1 and CNTF, have been identified through in vitro and animal studies to play a role in OL survival. Withdrawal of these factors can result in cell death via apoptosis. At present, little is known about the actions of these factors during human cerebral white matter development. It is possible that focal axonal injury or diffuse gliosis may contribute to the pathogenesis of a variety of white matter disorders (e.g., PVL, Table 1) by disrupting the release or altering the balance of trophic factors required for OL survival and maturation. Future studies are needed to determine what role these and other trophic factors may play acutely and chronically to mitigate or promote OL death or disrupted myelination.

Role of Glutamate Receptors in OL Differentiation and Death

The actions of glutamate on OL survival and differentiation appear to be developmentally regulated through both receptor and nonreceptor-mediated mechanisms. Several recent in vitro studies have begun to clarify the role of receptor-mediated actions of glutamate on the differentiation and survival of OL progenitors (reviewed in ref. *90)*. OL progenitors principally express the kainate and AMPA classes of glutamate receptors. There is no compelling evidence for functionally active N-methyl-D-aspartate (NMDA) receptor expression at any stage of OL development. Glutamate has been proposed to block the differentiation of O2-A progenitors through an AMPA receptor-mediated mechanism involving potassium influx via a delayed-rectifier potassium channel *(91)*. These cells were relatively resistant to a 24 hour exposure to 1 mM kainate, which caused appearance of less than 20% apoptotic cells. Similarly, mixed populations of OL progenitors and GC-positive cells were only partially killed by a 24 hour exposure to 1 mM kainate *(92)*. Maturation of oligodendrocytes is associated with increased vulnerability to AMPA-kainate receptor agonists or oxygen–glucose deprivation via a mechanism of receptor-mediated excittotoxicity blocked by a selective AMPA receptor antagonist, but not by growh factors or antioxidants *(92a)*. Future studies are needed to clarify the role of stage-specific factors in determining the relative sensitivity of OLs to glutamate-mediated toxicity. Changing patterns of glutamate receptor and ion channel expression may significantly alter the relative responsiveness of OLs to the effects of glutamate as a survival and differentiation factor.

Role of Glutamate Uptake Mechanisms in OL Differentiation and Death

Precedent for nonreceptor-mediated glutamate cytotoxicity was first reported in immature cortical neurons that were shown to be killed in vitro via a glutamate uptake-mediated mechanism *(93,94)*. In contrast to mature cortical neurons, which undergo glutamate receptor-mediated toxicity *(95,96)*, immature cortical neurons are vulnerable to an apoptotic form of death that is triggered by uptake of glutamate or homocysteate, a glutamate analog which blocks cystine uptake via a glutamate–cystine antiporter (97,98). Glutamate uptake causes a depletion of intracellular cysteine, a precursor for glutathione. Glutamate toxicity is accompanied by a depletion of intracellular glutathione. Buthionine sulfoximine, an inhibitor of glutathione synthesis is also neurotoxic.

The authors have been studying mechanisms of death of OL precursors with the aim of understanding the pathogenetic mechanisms that may give rise to loss of OLs in cerebral white matter injury in the premature infant. This injury has a predilection for the periventricular white matter of the cerebral hemispheres and is termed periventricular leukomalacia (PVL) *(50)*. PVL is believed to be the antecedent injury for some forms of cerebral palsy (*see* Overview). The authors' initial studies of the effect of glutamate on the survival of galactocerebroside-positive OLs in culture showed that glutamate exposure killed these cells with an EC_{50} of 200 μM *(99)*. The mechanism of glutamate toxicity was found to be similar to that of immature cortical neurons, occurring via a nonreceptor-mediated mechanism, and differed from the receptor-mediated toxicity of glutamate in mature cortical cultures. Unlike mature neurons, where cellular toxicity is evident within minutes, toxic changes were not observed in immature OLs until hours after the initial exposure to glutamate. Neither the NMDA receptor antagonist, MK-801, nor CNQX, a competitive antagonist of non-NMDA receptors, protected OLs from glutamate toxicity. Rather, the toxicity of glutamate appeared to be mediated via both sodium-dependent and independent transport systems and was completely prevented by inhibition of glutamate uptake by D,L-threo-β-hydroxyaspartate, a blocker of sodium-dependent, high-affinity transport. The novel nature of this form of glutamate-triggered OL death was supported by the observation that death was preceded by intracellular depletion of glutathione, a key scavenger of oxygen free radicals *(99)*. A role for free radical toxicity in triggering the death of immature OLs was supported by the cytoprotection rendered by a number of structurally distinct free radical scavengers, including ascorbate, α-tocopherol and idebenone, despite glutamate uptake into the cell and despite glutathione depletion *(99)*. Other studies also support an increased vulnerability of mature OLs to cellular injury mediated by free radical toxicity *(100,101)*.

Cystine Deprivation and the Death of OL Precursors Via Oxidative Stress

Further insight into the mechanism by which oxidative stress mediates OL death came from the observation that the toxicity of glutamate could be enhanced by exposing OLs to glutamate under cystine-free conditions *(33,99)*. Cystine is the oxidized form of the amino acid cysteine and the principal in vitro form of this crucial amino acid. Cystine was found to protect against glutamate-toxicity in a dose-dependent manner with an EC_{50} of about 50 µM. Like glutamate exposure, cystine deprivation-induced death occurred in cultures dominated by immature OLs. These studies suggested that a form of oxidative stress caused the death of immature OLs by inducing intracellular cystine deficiency through the action of a glutamate–cystine exchanger, with uptake of glutamate accompanied by cystine efflux. Studies of OLs preloaded with radiolabeled cystine, in fact, showed an increased efflux of the radiolabel from OLs exposed to glutamate, a pattern supporting the presence of a glutamate–cystine exchanger in immature OLs.

The basis for the critical dependence of immature OLs on cystine appears to derive from the fact that cystine is not only required for protein synthesis but is also a precursor for the tripeptide glutathione, a key scavenger of intracellular free radicals. We found that when OLs were deprived of cystine, cell death was preceded by a steady decline in intracellular glutathione *(33)*. Independently, it was found that the free radical scavenger, *N*-acetylcysteine, also protected OLs from glutamate toxicity by supplying a precursor for glutathione synthesis *(58)*.

These studies indicated that the death of immature OLs was linked to a mechanism that is not uniquely triggered by glutamate. Rather, OL death might be induced by a variety of conditions that cause a depletion of intracellular glutathione, thereby rendering cells susceptible to injury from oxidative stress. The role for free radical toxicity in triggering the death of immature OLs depleted of glutathione by cystine deprivation was further supported by the cytoprotection provided by a number of structurally distinct free radical scavengers, including ascorbate, α-tocopherol and idebenone *(33)*. In a related study, hypoxia and agents that subjected mixed glial cultures to an increase in reactive oxygen species resulted in a selective loss of OL precursors *(102)*. OL precursors were also found to have a greater rise in photochemically generated intracellular reactive oxygen species compared to astrocytes, an event which correlated with a lower level of glutathione and a higher iron content in the OLs *(103)*.

We have recently identified that preOLs, but not mature OLs cultured in vivo demonstate a critical dependence on glutathone for survival *(136)*. Glutathione depletion of preOLs resulted in a marked rise in intracellular free radical generation that could be completely blocked by antioxidants, including vitamin E and a cell-permeable form of glutathione, glutathione ester. Mature OLs were more resistant to death induced by glutathione depletion despite the fact that they had a lower basal glutathione level than preOLs and pharmacologic agents more potently depleted glutathione in mature OLs than in preOLs. Moreover, mature OLs did not show evidence of free readical generation when their gluthione level was reduced to the same level that committed preOLs to die. Hence, oligodendroglial maturation was associated with decreased susceptibility to free-radical-mediated oxidative stress.

A Role for Iron in Free-Radical-Mediated OL Death?

It is currently unknown which free radical species may be toxic to OLs, but it may be significant that the OL is one of the primary iron reservoirs within the rodent and human brain *(104–109)*. In support of a role for iron in triggering free-radical OL injury, we have observed that both glutamate- (Oka and Volpe, unpublished observations) and cystine deprivation-induced death of OL precursors was blocked by pretreatment with the iron chelator desferoxamine *(33)*. In a related study, a photochemically-induced rise in ROS could also be prevented in OLs by desferoxamine *(103)*. These observations suggest that the hydroxyl radical is an oxygen species which may trigger OL death. The hydroxyl radical is highly toxic to cells, and is generated when hydrogen peroxide reacts with ferrous ions via the Fenton reaction *(110)*. The vulnerability of the immature brain to hydrogen peroxide and hydroxyl radical production is supported by the demonstration of accentuated brain injury after hypoxia-ischemia in copper/zinc superoxide dismutase (SOD) transgenic (overexpressing) mice in the perinatal period but not in the adult *(111)*. During hypoxia-ischemia, enhanced levels of hydrogen peroxide might occur in these mice as a result of a threefold higher level of expression of SOD.

Although a variety of intracellular trace metals might be a source of oxygen radicals, a number of observations indicate that OLs play a central role in CNS iron metabolism and, thus, may be at risk for iron-mediated oxygen radical toxicity. Within the adult rodent *(104,109,112,113)* and human CNS *(105,107,108,114,115)*, ferric iron, ferritin, and transferrin localize primarily to OLs, as well as to some restricted populations of neurons, microglia, and astrocytes. Myelin deficient rats, in which OLs fail to mature, show significant reductions in the distribution of iron, transferrin, and the transferrin receptor in the CNS *(113,116,117)*.

During white matter development, iron utilization and the expression of iron binding proteins are highly regulated. In the neonatal rat, rapid brain growth in the first two weeks of life is accompanied by a commensurate increase in the rate of iron uptake which peaks by d 15 of life *(118)*. The peak utilization of iron appears to occur in the early perinatal period with the onset of myelinogenesis. Iron-positive cells are localized predominantly in the subcortical white matter as early as postnatal d 3 in the rat and between 14 and 28 d of age reach an adult pattern in which clusters of iron-positive cells localize near blood vessels *(119)*. Moreover, OLs are enriched in heavy-chain (H) ferritin early in development; H-ferritin is associated with high iron utilization and low iron storage *(109)*. During human white matter development, ferritin-positive OLs, identified by morphological criteria, increase markedly in the cerebral white matter, from deep to superficial regions, from about 30 wk of gestation *(108)*. Prenatally, the distribution of transferin is most prominent in the choroid plexus, probably reflecting a role for this protein in the CSF uptake and transfer of iron into the CNS via receptor-mediated transport across cell membranes *(120,121)*. In contrast to ferritin, there is limited transferrin expression in white matter until near term, after which transferrin localizes most heavily to white matter tracts that are beginning to myelinate *(105,112,114,122)*. Studies in the developing rat optic nerve found that expression of the transferrin receptor precedes the appearance of staining for transferrin, myelin-basic protein, and galactocerebroside *(123)*.

Taken together, these observations support an important role for iron and transferrin during OL development and active myelination of white matter and suggest that a variety of reactive oxygen species (ROS) may be normally produced as a by-product of iron-requiring metabolic processes which are activated near the onset of myelination. Under physiological conditions, for example, ferritin can function as an iron donor for microsomal lipid peroxidation *(124)*. Moreover, the timing of expression of iron regulatory proteins may crucially influence the developmental vulnerability of white matter to injury. Several recent intriguing studies have begun to elucidate the functional importance of these proteins under physiological and pathophysiological conditions.

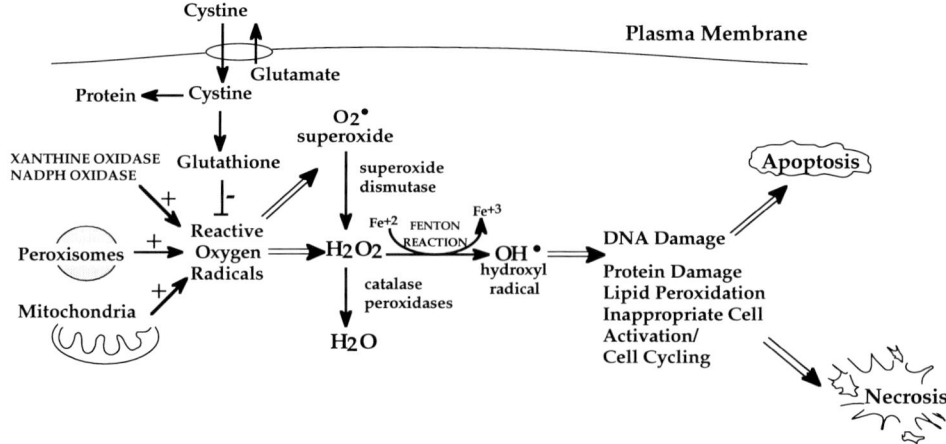

Hypothetical pathway for reactive oxygen radical mediated oligodendrocyte death.

Fig. 2. A proposed scheme by which glutathione depletion leads to oxidative stress-induced OL death.

tions. Ferritin can function as an antioxidant which sequesters intracellular iron stores to prevent iron-induced toxicity (125). Furthermore, under hypoxic or acidotic conditions, OLs in culture show both a reversible inhibition of MBP synthesis and a marked induction in the ferritin H chain (126,127). Ferritin induction appears to occur in response to increased free intracellular iron or iron-mediated free radical generation under acidotic conditions (128,129). Thus, it appears that iron stores are tightly regulated under both normal and hypoxic conditions, possibly reflecting the essential but potentially deleterious effects of this ion on OL survival. It is an intriguing possibility that, in the setting of white matter injury, regulation of iron might fail, if the death of ferritin-rich OLs were to release excessive iron into the surrounding white matter, thus, potentially leading to a self-perpetuation of free-radical mediated cell death.

A Role for Antioxidant Enzyme Expression in OL Vulnerability to Death?

An elevation in a variety of ROS is a well established sequela of ischemia (110,130,131). It remains to be determined whether a rise in ROS could result in cell death by exceeding the capacity of antioxidant enzymes in the OL to handle ROS. The vulnerability of OL precursors to oxidative stress may in part be due to a lack or imbalance in expression of antioxidant enzymes early in development.

Little is known in vitro about the developmental expression in OLs of the antioxidant enzymes catalase, SOD, and glutathione peroxidase, which are the primary enzymes required by most cell types to inactivate ROS. An imbalance in the expression of SOD and catalase has been observed in cultured mouse epidermal cells (132) but similar studies have not been done in OLs. Exogenous catalase will protect against OL death induced in vitro by oxygen radicals (101) and by catecholamine exposure (133). Recent studies suggest that OLs at several stages have lower levels of some antioxidant enzymes than do other types of glia. Mitochondrial manganese SOD was not immunocytochemically detected in either apparent OL progenitors or mature OLs, but was strongly visualized in both astrocytes and microglia (133a). Glutathione peroxidase levels were lower in OLs at several stages compared to astrocytes (133b).

In human brain, limited data suggest a progressive maturation of antioxidant systems near the time of birth as myelination begins. A temporal and regional gradient is observed for catalase and SOD immunoreactive glial cells, which are first visualized in deep cerebral white matter by 32 post conceptional weeks *(134, 134a)*. At term, all layers of the cerebral white matter contain glia that stain for catalase. It may also be significant that whereas SOD demonstrates a steady increase in activity in the early postnatal rat brain, catalase levels show more fluctuation *(135)*. In the rat, catalase levels are markedly higher than adult levels at birth but steadily decline until 11 wk of age, after which catalase gradually increases to adult levels by 40 wk.

Mechanisms of ROS-Mediated OL Death: Apoptosis vs Necrosis?

As shown schematically in Fig. 2, death of OL precursors can be induced in vitro by a number of conditions that cause a depletion of intracellular glutathione, thereby rendering OLs susceptible to injury from a form of oxidative stress. Cytoprotection from the toxicity of excess glutamate or cystine deprivation can be prevented by several structurally distinct free radical scavengers, including α-tocopherol, ascorbate, and idebenone, despite glutamate uptake into the cell and despite glutathione depletion *(33,99)*. It is currently unknown which ROS may be the most toxic to OLs, but it may be significant that the OL is the main CNS cell type that contains iron, transferrin, and ferritin (109). During development a variety of ROS may be normally produced as a by-product of iron-requiring metabolic processes that are activated near the onset of myelination. In particular, the hydroxyl radical, which is generated when hydrogen peroxide reacts with ferrous ions via the Fenton reaction, is highly toxic to cells *(110)*. Little is known in OLs about the developmental expression of antioxidant enzymes, such as catalase, SOD, and glutathione peroxidase, which are required by most cell types to inactivate ROS. One possible scenario is that the selective vulnerability of OL precursors to oxidative stress may relate to a developmental variation in the expression of antioxidant enzymes, which may be upregulated during the course of OL development to prevent toxicity from ROS generated during myelinogensis.

There are two potential forms of cell death—necrosis or apoptosis, both of which may result from OL injury secondary to ROS. We have observed that, under conditions that promote free radical toxicity, immature OLs undergo a form of death with a number of features consistent with apoptosis *(136)*. These cells slowly exhibit morphological features of toxicity that culminate in death hours after exposure to cystine-free conditions. At the light microscopic level, the OLs show evidence of nuclear condensation when stained with the nuclear dye bisbenzamide. Nuclei also show evidence of *in situ* DNA fragmentation with the TUNEL method. Ultrastructurally, these cells display chromatin condensation, fragmentation of the nucleolus, vacuolation of the cytoplasm, and preservation of the plasma and nuclear membranes *(136)*.

These findings are consistent with several converging lines of evidence that indicate that a number of triggers exist during normal CNS development to cause OL precursors or mature OLs to die via a mechanism involving apoptosis. In addition, a form of immune-mediated necrotic cell death can be induced in mature rat OLs by exposure to complement, even in the absence of cell-specific antibody, via direct interaction of complement with myelin to form pores that permit the influx of calcium *(137–139)*. Activated microglia in vitro also cause a form of necrotic injury to OLs that is mediated via nitric oxide *(64,67)*. These latter forms of OL necrosis may be pertinent to OL loss in the adult nervous system in immune-mediated conditions such as multiple sclerosis.

A number of cellular mediators, including cytokines and selected growth factors, have been shown to trigger OL death via a mechanism involving apoptosis. Cytotoxic effects of TNFα and TNFβ in mature OLs appear to be mediated via apoptosis *(56–59,140–142)*. INF-γ has been found to induce apoptosis of OLs in some *(60)*, but not all *(56)*, systems and the toxicty of INF-γ can be reversed by LIF *(60)*. A covalent dimer of IL-2, generated in vitro by the action of a nerve-derived transglutaminase *(143)*, triggers death of mature OLs via an apoptotic mechanism directly involving p53 *(144)*.

One mechanism by which cytokines may exert their toxicity is via stimulation of sphingomyelin hydrolysis to generate the lipid intermediate ceramide. Both sphingomyelinase *(144a)* and interleukin 1β *(144b)* induced ceramide production. Several studies demonstrated direct toxicity of ceramide to OLs that is maturation-dependent and caused death via apoptosis. C_2-ceramide was more toxic to O2A progenitors or CG-4 cells than to mature O1-positive OLs *(144b,144c)*. Cultured neurons and astrocytes were resistant to C_2-ceramide at doses toxic OLs suggesting that ceramide may be selectively toxic to OLs in the CNS *(144c)*.

A phenomenon with several features of apoptosis has also been shown to be involved in the determination of OL number during rat optic nerve development *(36,145)*. OL death is triggered by the withdrawal of growth factors, including PDGF, and IGF-1, required for OL survival. In view of the fact that the *bcl-2* gene has been shown to block cell death in a variety of cell types and cell lines, bcl-2 overexpression might be predicted to promote increased OL survival during optic nerve development. Using a transgenic mouse that overexpresses human bcl-2, an increase in OL number in the optic nerve was demonstrated to parallel an increase in axon number *(38)*. However, when placed in culture, OLs from the transgenic animals were not more protected from apoptosis than controls. This observation raises important questions concerning the pathway by which apoptotic OL death occurs in the optic nerve and the role of bc1-2 in this process. In vitro, withdrawal of growth factors, including bFGF, triggers death of OL precursors via apoptosis *(74)*. Interestingly, bFGF, which induces proliferation of OL progenitors, causes apoptosis in mature OLs *(75)*.

The conditions under which nerve growth factor (NGF) induces the survival or apoptosis of mature OLs remains contraversial. Mature OLs may express both functional TrkA tyrosine kinase and p75 neurotrophin receptors *(146–150)* and levels of TrkA may be upregulated during some culture conditions *(146)*. Death of mature rat OLs can be induced by NGF binding to p75 via a pathway that involves c-jun kinase (JNK) activation and an increase in intracellular ceramide *(147)*. Upregulation of TrkA blocks death in mature OLs susceptible to p75-mediated apoptosis and correlates with a suppression of c-jun kinase activity *(148)*. However, in human adult OLs p75, but not TrkA expression, is detected under a variety of culture conditions, but NGF exposure fails to induce OL death. These studies suggest that interspecies differences as well as culture conditions, but NGF exposure fails o induce OL death. These studies suggest that interspecies differences as well as culture conditions may influence NGF-mediated OL survival.

Taken together, data from both in vitro and in vivo studies indicate that both necrotic and apoptotic mechanisms of OL death can occur. Apoptosis appears to be central to normal white matter development by ensuring the correct number and distribution of OLs for axonal myelination. In the mature CNS, OL death may occur by either apoptosis or necrosis.

CONCLUSIONS

The study of OL death has greatly advanced in the last decade largely through in vitro studies that have permitted a detailed molecular approach. Future studies are required to characterize the molecular pathways, triggered by such disparate conditions as trophic factor withdrawal, cytotoxic cytokine or excitatory amino acid agonist exposure, or intracellular accumulation of ROS, that lead to OL death. During the course of a given white matter disease, multiple processes may contribute to OL death via both necrosis and apoptosis. Hence, consideration of the mechanisms of OL death in any given white matter disease must be viewed within the context of the pathogenetic process. Whereas adult white matter disorders involve the mature OL, developmental disorders must identify the predominant OL stages present at a given epoch in brain development that are vulnerable to injury. In both cases, the success of future therapeutic interventions to prevent OL death will depend both upon identification of the timing of insult and the resultant form of death. OL rescue from apoptosis, for example, may be most likely to be successful when injury occurs over a short time period to healthy cells (e.g., ischemia-reperfusion perinatal brain injury) as opposed to conditions in which the cells undergo a chronic insult (e.g., neurometabolic conditions such as adrenoleukodystrophy).

Key areas for future study of human white matter disorders include in vitro and in vivo approaches as follows:

1. Validation of in vitro mechanisms of OL death in appropriate animal models or human tissues, including identification of the timing of OL loss and whether OL death has occurred via necrosis or apoptosis;
2. Characterization of substances released by reactive cell types (e.g., lymphocytes, macrophages, reactive astrocytes) that may be cytotoxic to OLs and identification of the signal transduction pathways activated by these substances which trigger cell death programs;
3. Characterization of the relationship of focal or diffuse axonal injury to OL loss or dedifferentiation, and, particularly, both identification of the impact of axonal injury on OL trophic support and determination of whether injury has resulted in OL loss or dedifferentiation to a premyelinogenic stage;
4. Characterization of the potential for functional recovery or OL plasticity and identification of the factors which may regulate this response to injury; and
5. Formulation of therapeutic interventions tailored to the appropriate mechanisms of OL death with such agents as scavengers of free radicals, cytokine antagonists, inhibitors of apoptosis, or growth factor agonists.

ACKNOWLEDGMENTS

This work was supported by a Grass Foundation Morrison Fellowship, a Reynolds Rich Smith Fellowship, a Charles A. Janeway Child Health Research Center Award Fellowship (HD27805), a Hearst Fund Award and NINDS K08 NS01855 to SAB and NINDS P20NS32570 and Mental Retardation Center grant NICHD P30HD18655 to JJV. The authors thank Drs. Hannah Kinney, Julie Ellison, and Paul Rosenberg for many helpful discussions.

REFERENCES

1. Volpe JJ. *Neurology of the Newborn*, W. B. Saunders, Philadelphia, 1995.
2. Kettenmann H, Ransom BR. *Neurologia*, Oxford University Press, NY, 1995.
3. Wekerle H, Cusner ML. Immunopathogenesis of demyelinating diseases: Introduction. *Brain Pathology* 1996, **6**: 229–230.
4. Pfeiffer SE, Warrington AK, Bansal R. The oligodendrocyte and its many cellular processes. *Trends Cell Biol* 1993, **3**: 191–197.
5. Bansal R, Warrington A, Gard A, Rauscht B, Pfeiffer S. Multiple and novel specificities of monoclonal antibodies O1, O4, and R-mAb used in the analysis of oligodendrocyte development. *J Neurosci Res* 1989, **24**: 548–557.
6. Gard AL, Pfeiffer SE. Oligodendrocyte progenitors isolated directly from developing telencephalon at a specific phenotypic stage. myelinogenic potential in a defined environment. *Development* 1989, **106**: 119–132.
7. Armstrong R, Friedrich VL, Jr, Holmes KV, Dubois-Dalcq M. In vitro analysis of the oligodendrocyte lineage in mice during demyelination and remyelination. *J Cell Biol* 1990, **111**: 1183–1195.
8. Gard AL, Pfeiffer SE. Two proliferative stages of the oligodendrocyte lineage (A2B5+04- and 04GalC+) under different mitogenic control. *Neuron* 1990, **5**: 615–625.
9. Raff MC, Mirsky R, Fields KL, et al. Galactocerebroside is a specific cell surface antigenic marker for oligodendrocytes in culture. Nature 1978, 274:813–816.
10. Schachner M, Kim SK, Zehnle R. Developmental expression in central and peripheral nervous system of oligodendrocyte cell surface antigens (O antigens) recognized by monoclonal antibodies. *Dev Biol* 1981, **83**: 328–338.
11. Sommer I, Schachner M. Monoclonal antibodies (O1 to O4) to oligodendrocyte cell surfaces: An immunocytological study in the central nervous system. *Dev. Biol.* 1981, **83**: 311–327.
12. Sommer I, Schachner M. Cells that are O4-antigen positive and O1-negative differentiate into O1 antigen-positive oligodendrocytes. *Neurosci Lett* 1982, **29**: 183–188.
13. Reynolds R, Wilkins GP. Development of macroglial cells in rat cerebellum II. An *in situ* immunohistochemical study of oligodendroglial lineage from precursor to mature myelinating cell. *Development* 1988, **102**: 409–425.
14. Hasegawa M, Houdou S, Mito T, Takshima S, Asanuma K, Ohno T. Development of myelination in the human fetal and infant cerebrum: a myelin basic protein immunohistochemical study. *Brain Dev* 1992, **14**: 1–6.
15. Warrington AK, Pfeiffer SE. Proliferation and differentiation of O4+ oligodendrocytes in postnatal rat cerebellum: analysis in fixed tissue slices using anti-glycolipid antibodies. *J Neurosci Res* 1992, 338–353.
16. Bodhireddy SR, Lyman WD, Rashbaum WK, Weidenheim KM. Immunohistochemical detection of myelin basic protein is a sensitive marker of myelination in second semester human fetal spinal cord. *J Neuropath Exp Neurol* 1994, **53**: 144–149.
17. Ellison JA, De Vellis J. Platelet-derived growth factor receptor is expressed by cells in the early oligodendrocyte lineage. *J Neurosci Res* 1994, **37**: 116–128.
18. Goyne GE, Warrington AK, Devito JA, Pfeiffer SE. Oligodendrocyte precursor quantitation and localization in perinatal brain using a retrospective bioassay. *J Neurosci Res* 1994, **14**: 5365–5372.
19. Noble M, Murray K. Purified astrocytes promote the in vitro division of a bipotential glial progenitor cell. *EMBO J* 1984, **3**: 2243-2247.
20. Raff MC, Williams BP, Miller RH. The in vitro differentiation of a bipotential glial progenitor cell. *EMBO J* 1984, **3**: 1857–1864.
21. Aloisi F, Agresti C, D'urso D, Levi G. Differentiation of bipotential glial precursors into oligodendrocytes is promoted by interaction with type-1 astrocytes in cerebellar cultures. *Proc Natl Acad Sci USA* 1988, **85**: 6167–6171.

22. Armstrong RC, Harvath L, Dubois-Dalcq ME. Type 1 astrocytes and oligodendrocyte-type 2 astrocyte glial progenitors migrate toward distinct molecules. *J Neurosci Res* 1990, **27**: 400–407.
23. Wren D, Wolswijk G, Noble M. In vitro analysis of the origin and maintenance of 0-2^{adult} progenitor cells. *J Cell Biol* 1992, **116**: 167–176.
24. Fredman P, Magnani JL, Nirenberg M, Ginsburg V. Monoclonal antibody A2B5 reacts with many gangliosides in neuronal tissue. *Arch Biochem Biophys* 1984, **33**: 661–666.
25. Dubois C, Manuguerra J-C, Hauttecoeur B, Maze J. Monoclonal antibody A2B5, which detects cell surface antigens, binds to ganglioside GT_3 (II^3(NeuAc)$_3$LacCer) and to its 9-O-acetylated derivative. *J Biol Chem* 1990, **265**: 2797–2803.
26. Armstrong RC, Dorn HJ, Kufta CV, Friedman E, Dubois-Dalcq ME. Pre-oligodendrocytes from adult human CNS. *J Neurosci* 1992, **12**: 1538–1547.
27. Rivkin MJ, Flax J, Mozell R, Osathanondh R, Volpe JJ, Villa-Komaroff L. Oligodendroglial development in human fetal cerebrum. *Ann Neurol* 1995, **38**: 92–101.
28. Back SA, Volpe JJ, Kinney HH. Inmunocytochemical characterization of oligodendrocyte development in human cerebral white matter. *Soc Neurosci Abstr* 1996, **20**: 1722.
28a. Trapp BD, Nishiyama A, Cheng D, Macklin W. Differentiation and death of premyelinating oligodendrocytes in developing rodent brain *J Cell Biol* 1997, **137(2)**: 459–468.
29. McGahon AJ, Martin SJ, Bissonnette RP, et al. The end of the (cell) line: Methods for the study of apoptosis *in vitro*. Methods in Cell Biology. Academic Press, New York, 1995, vol 46, pp. 153–185.
30. Barres B, Schmid R, Sendnter M, Raff MC. Multiple extracellular signals are required for long-term oligodendrocyte survival. *Development* 1993, **118**: 283–295.
31. Levison SW, Mccarthy KD. Astroglia in culture, in *Culturing Nerve Cells* (Banker G, Goslin K, eds.) 1991, MIT Press, Cambridge, Massachusetts, pp. 309–336.
32. Gard AL, Burrell MR, Pfeiffer SE, Rudge JS, Williams WCN. Astroglial control of oligodendrocyte survival mediated by PDGF and leukemia inhibitory factor-like protein. *Development* 1995, **121**: 2187–2197.
33. Yonezawa M, Back SA, Gan X, Rosenberg PA, Volpe JJ. Cystine deprivation induces oligodendroglial death: Rescue by free radical scavengers and by a diffusible glial factor. *J Neurochem* 1996, **67**: 566–573.
34. Ellison JA, Scully SS, De Vellis J. Evidence for neuronal regulation of oligodendrocyte development: Cellular localization of platelet-derived growth factor receptor and A-chain mRNA during cerebral cortex development in the rat. *J Neurosci Res* 1996, **45**: 28–39.
35. Gard AL, Pfeiffer SE, Williams WCM. Immunopanning and developmental stage-specific primary culture of oligodendrocyte progenitors (04+GalC-) directly from postnatal rodent cerebrum. *Neuroprotocols* 1993, **2**: 209–218.
36. Barres BA, Hart IK, Coles HSR, et al. Cell death and control of cell survival in the oligodendrocyte lineage. *Cell* 1992, **70**: 31–46.
37. Barres BA, Raff MC, Gaese F, Bartke I, Dechant G, Barde Y-A. A crucial role for neurotrophin-3 in oligodendrocyte development. *Nature* 1994, **367**: 371–375.
38. Burne JF, Staple JK, Raff MC. Glial cells are increased proportionally in transgenic nerves with increased numbers of axons. *J Neurosci* 1995, **16**: 2064–2073.
39. Butt A, Kirvell S. Glial cells in transected optic nerves of immature rats. I. An analysis of individual cells by intracellular dye-injection. *J Neurocytology* 1996, **25**: 365–380.
40. Butt A, Kirvell S. Glial cells in transected optic nerves of immature rats. II: An immunohistochemical study. *J Neurocytology* 1996, **25**: 381–392.
41. Skoff RP. Programmed cell death in the dysmyelinating mutants. *Brain Pathol* 1995, **5**: 283–288.
41a. Gow A, Southwood CM, Lazzarini RA. Disrupted proteolipid protein trafficking results in oligodendrocyte apoptosis in an animal model of Pelizaeus-Merzbacher disease. *J Cell Biol* 1998, **140**: 925–934.

42. Schneider A, Griffiths IR, Readhead C, Nave K-A. Dominant-negative action of the jimpy mutation in mice complemented with an autosomal transgene for myelin proteolipid protein. *Proc Natl Acad Sci USA* 1995, **92:** 4447–4451.
43. Griffiths IR, Schneider A, Anderson J, Nave KA. Transgenic and natural mouse models of proteolipid protein (PLP)-related dysmyelination and demyelination. *Brain Pathol* 1995, **5:** 275–281.
44. Nakao J, Yamada M, Kagawa T, et al. Expression of proteolipid protein gene is directly associated with secretion of a factor influencing oligodendrocyte development. *J Neurochem* 1995, **64:** 2396–2403.
44a. Williams II WC, Gard AL. In vitro death of jimpy oligodendrocytes: correlation with onset of DM20/PLP expression and resistance to oligodendrogliotrophic factors. *J Neurosci Res* 1997, **50:** 177–189.
44b. Yang X, Skoff RP. Proteolipid protein regulates the survival and differentiation of oligodendrocytes. *J Neurosci* 1997, **17(6):** 2056–2070.
45. Gow A, Friedrich VL, Lazzarini RA. Many naturally occurring mutations of myelin proteolipid protein impair its intracellular transport. *J Neurosci Res* 1994, **37:** 574–583.
46. Turnley AM, Morahan G. Dysmyelination in class I MHC transgenic mice. *Micr Res Tech* 1996, **32:** 286–294.
47. Bradl M, Linington C. Animal models of demyelination. *Brain Pathol* 1996, **6:** 303–311.
48. Miller DJ, Asakura K, Rodriguez M. Central nervous system remyelination clinical application of basic neuroscience principles. *Brain Pathol* 1996, **6:** 331–344.
49. Steinman L. A few autoreactive cells in an autoimmune infiltrate control a vast population of nonspecific cells: A tale of smart bombs and the infantry. *Proc Natl Acad Sci USA* 1996, **93:** 2253–2256.
49a. Keirstead HS, Levine JM, Blakemore WF. Response of the oligodendrocyte progenitor cell population (defined by NG2 labelling) to demyelination of the adult spinal cord. *Glia* 1998, **22:** 161–170.
49b. Wolswijk G. Chronic stage multiple sclerosis lesions contain a relatively quiescent population of oligodendrocyte precursor cells. *J Neurosci* 1998, **18(2):** 601–609.
49c. Redwine JM, Blinder KL, Armstrong RC. In situ expression of fibroblast growth factor receptors by oligodendrocyte progenitors and oligodendrocytes in adult mouse central nervous system. *J Neurosci Res* 1997, **50:** 229–237.
50. Back SA, Volpe JJ. Cellular and molecular pathogenesis of periventricular white matter injury. *MRRD Research Reviews* 1997, **3:** 96–107.
51. Brosnan CF, Raine CS. Mechanisms of immune injury in multiple sclerosis. *Brain Pathol* 1996, **6:** 243–257.
52. Norenberg MD. Astrocyte response to CNS injury. *J Neuropath Exp Neurol* 1994, **53:** 213–220.
53. Robbins DS, Shirazi Y, Drysdale B-E, Lifrerman A, Shin HS, Shin ML. Production of cytotoxic factor for oligodendrocytes by stimulated astrocytes. *J Immunol* 1987, **139:** 2593–2597.
54. Selmaj K, Raine CS. Tumor necrosis factor mediates myelin and oligodendrocyte damage in vitro. *Ann Neurol* 1988, **23:** 339–346.
55. Merrill JE. Effects of interleukin-1 and tumor necrosis factor-α on astrocytes, microglia, oligodendrocytes, and glial precursors in vitro. *Dev Neurosci* 1991, **13:** 130–137.
56. Selmaj K, Raine CS, Farooq M, Norton WT, Brosnan CF. Cytokine cytotoxicity against oligodendrocytes: Apoptosis induced by lymphotoxin. *J Immunol* 1991, **147:** 1522–1529.
57. Louis J-C, Magal E, Takayama S, Varon S. CNTF protection of oligodendrocytes against natural and tumor necrosis factor-induced death. *Science* 1993, **259:** 689–692.
58. Mayer M, Noble M. N-acetyl-L-cysteine is a pluripotent protector against cell death and enhancer of trophic factor-mediated cell survival in vitro. *Proc Natl Acad Sci USA* 1994, **91:** 7496–7500.

59a. Prabhakar S, D'souza S, Antel JP, Mclaurin J, Schipper Hm, Wang E. Phenotypic and cell cycle properties of human oligodendrocytes in vitro. *Brain Res* 1995, **672**: 159–169.

59b. Agresti C, D'Urso D, Levi G. Reversible inhibitory effects of interferon-γ and tumor necrosis factor-α on oligodendroglial lineage cell proliferation and differentiation in vitro. *Eur J Neurosci* 1996, **8**: 1106–1116.

60. Vartanian T, Li Y, Zhao M, Stefanson K. Interferon-γ-induced oligodendrocyte cell death: implications for the pathogenesis of multiple sclerosis. *Mol Med* 1995, **1**: 732–743.

61. Kahn MA, De Vellis J. Regulation of an oligodendrocyte progenitor cell line by the interleukin-6 family of cytokines. *GLIA* 1994, **12**: 87–98.

62. Probert L, Akassoglou K, Pasparakis M, Kontogeorgos G, Kollias G. Spontaneous inflammatory demyelinating disease in transgenic mice showing central nervous system-specific expression of tumor necrosis factor-α. *Proc Natl Acad Sci USA* 1995, **92**: 11294–11298.

63. Chiang C-S, Powell HC, Gold LH, Samimi A, Campbell L. Macrophage/microglial-mediated primary demyelination ,and motor disease induced by the central nervous system production of Interleukin-3 in transgenic mice. *J Clin Invest* 1996, **97**: 1512–1524.

64. Merrill JE, Ignarro LJ, Sherman MP, Melinek J, Lane TE. Microglial cell cytotoxicity of oligodendrocytes is mediated through nitric oxide. *J Immunol* 1993, **151**:2 132–2141.

65. Mitrovic B, Martin FC Charles AC, et al. Neurotransmitters and cytokines in CNS pathology. *Prog Br Res* 1994, **103**: 319–330.

66. Lee SC, Liu W, Dickson DW, Bronsan CF, Berman JW. Cytokine production by human fetal microglia and astrocytes: Differential induction by lipopolysaccharide and IL-1β. *J Neurosurg* 1993, **150**: 2659–2667.

67. Mitrovic B, Ignarro LJ, Vinters HV, et al. Nitric oxide induces necrotic but not apoptotic death in oligodendrocytes. *Neuroscience* 1995, **65**: 531–539.

68. Collarini EJ, Pringle N, Mudhar H, et al. Growth factors and transcription factors in oligodendrocyte development. *J Cell Sci Supp* 1991, **15**: 117–123.

69. Barres BA, Raff MC. Control of oligodendrocyte number in the developing rat optic nerve. *Neuron* 1994, **12**: 935–942.

70. Bogler O, Wren D, Barnett SC, Land H, Noble M. Cooperation between two growth factors promotes extended self-renewal and inhibits differentiation of oligodendrocyte-type-2 astrocyte(O-2A) progenitor cells. *Proc Natl Acad Sci USA* 1990, **87**: 6368–6372.

71. Mckinnon RR, Matsui T, Dubois-Dalcq M, Aaronson Sa. FGF modulates the PDGF-driven pathway of oligodendrocyte development. *Neuron* 1990, **5**: 603–614.

72. Wolswijk G, Noble M. Cooperation between PDGF and FGF converts slowly dividing O-2^{adult} progenitor cells to rapidly dividing cells with characteristics of O-2A^{perinatal} progenitor cells. *J Cell Biol* 1992, **118**: 889–900.

73. Bansal R, Pfeiffer SE. Inhibition of protein and lipid sulfation in oligodendrocytes blocks biological responses to FGF-2 and retards cytoarchitechtural maturation, but not developmental lineage progression. *Dev Biol* 1994, **162**: 511–524.

74. Yasuda T, Grinspan J, Stern J, Fransceschini B, Bannerman P, Pleasure D. Apoptosis occurs in the oligodendroglial lineage, and is prevented by basic fibroblast growth factor. *J Neurosci Res* 1996, **40**: 306–317.

75. Muir DA, Compston DAS. Growth factor stimulation tri,ggers apoptotic cell death in mature oligodendrocytes. *J Neurosci Res* 1996, **44**: 1–11.

75a. Bansal R, Pfeiffer SE. FGF-2 converts mature oligodendrocytes to a novel phenotype. *J Neurosci Res* 1997, **50**: 215–228.

76. Komoly S, Hudson DL, Dewebster H, Bondy CA. Insulin-like growth factor I. Gene expression is induced in astrocytes during experimental demyelination. *Proc Natl Acad Sci USA* 1992, **89**: 1894–1898.

77. Yao DL, West NR, Bondy CA, et al. Cryogenic spinal cord injury induces astrocytic gene expression of insulin-like growth factor I and insulin-like growth factor binding protein 2 during remyelination. *J Neurosci Res* 1995, **40**: 647–659.

78. Baron-Van Evercooren A, Olichon-Berthe C, Kowalski A, Visciano G, Obberghen EV. Expression of IGF-1 and insuliin receptor genes in the rat central nervous sytem: a developmental, regional and cellular analysis. *J Neurosci Res* 1991, **28**: 244–253.
78a. Mewar R, McMorris R. Expression of insulin-like growth factor-binding protein messenger RNAs in developing rat oligodendrocytes and astrocytes. *J Neurosci Res* 1997, **50**: 721–728.
79. Bartlett WP, Li X-S, Williams M. Expression of IGF-1 mRNA in the murine subventricular zone during postnatal development. *Mol Brain Res* 1992, **12**: 285–291.
80. Shinar Y, McMorris FA. Developing oligodendroglia cells express mRNA for insulin-like growth factor-1, a regulator of oligodendrocyte development. *J Neurosci Res* 1995, **42**: 516–527.
81. Yeh H-J, Ruit K, Wang Y-X, Parks W, Snider W, Deuel T. PDGF A-chain gene is expressed by mammalian neurons during development and in maturity. *Cell* 1991, **64**: 209–216.
82. Raff MC, Lillen LE, Richardson WD, Burne JF, Noble MD. Platelet-derived growth factor from astrocytes drives the clock that times oligodendrocyte development in culture. *Nature* 1988, **333**: 562–565.
83. Noble M, Murray K, Stroobant P, Waterfield MD, Rddle P. Platelet-derived growth factor promotes division and motility and inhibits premature differentiation of the oligodendrocyte/type-2 astrocyte progenitor cell. *Nature* 1988, **333**: 560–562.
84. Hart IK, Richardson WD, Bolsover SR, Raff MC. PDGF and intracellular signalling in the timing of oligodendrocyte differentiatdion. *J Cell Biol* 1989, **109**: 3411–3417.
85. Wolswijk G, Noble M. Identification of an adult-specific glial progenitor cell. *Development* 1989, **105**: 387–400.
86. McMorris FA, Mozell RL, Carson MJ, Shinar Y, Meyer RD, Marchetti N. Regulation of oligodendrocyte development and central nervous system myelination by insulin-like growth factors. *Ann NY Acad Sci* 1993, **692**: 321–324.
87. Beck KD, Powell-Braxton L, Widmer HR, Valverde J, Hefti F. Igf1 gene disruption results in reduced brain size, CNS hypomyelination, and loss of hippocampal granule and striatal parvalbumin-containing neurons. *Neuron* 1995, **14**: 717–730.
88. Carson MJ, Behringer RR, Brinster RL, McMorris FA. Insulin-like growth factor I increases brain growth and central nervous system myelination in transgenic mice. *Neuron* 1993, **10**: 729–740.
89. Ye P, Carson J, D'Arcole AJ. In vivo actions of insulin-like growth factor-1 (IGF-1) on brain myelination: Studies of IGF-1 and IGF binding protein-1 (IGFBP-1) transgenic mice. *J Neurosci* 1995, **15**: 7344–7356.
90. Gallo V, Russell JT. Excitatory amino acid receptors in glia: different subtypes for distinct function? *J Neurosci Res* 1995, **42**: 1–8.
91. Gallo V, Zhou JM, Mcbain CJ, Wright P, Knutson PL, Armstrong RC. Oligodendrocyte progenitor cell proliferation and lineage progression are regulated by glutamate receptor-mediated K+ channel block. *J Neurosci* 1996, **16**: 2659–2670.
92. Yoshioka A, Hardy M, Younkin DP, Grinspan J, Stern JL, Pleasure D. a-Amino-3-hydroxy-5-methyl4-isoxazolepropionate (AMPA) receptors mediate excitotoxicity in the oligodendroglial lineage. *J Neurochem* 1995, **64**: 2442–2448.
92a. McDonald JW, Althomsons SP, Hyrc KL, Choi DW, Goldberg ME. Oligodendrocytes from forebrain are highly vulnerable to AMPA/kainate receptormediated excitotoxicity. *Nature Med* 1998, **1(3)**: 291–297.
93. Murphy T, Miyamoto M, Sastre A, Schnaar R, Coyle J. Glutamate toxicity in a neuronal cell line involves inhibition of cystine transport leading to oxidative stress. *Neuron* 1989, **2**: 1547–1558.
94. Murphy TH, Schnaar RL, Coyle JT. Immature cortical neurons are uniquely sensitive to glutamate toxicity by inhibition of cystine uptake. *FASEB J.* 1990, **4**: 1624–1633.
95. Choi DW. Cerebral hypoxia: some new approaches and unanswered questions. *J Neurosci* 1990, **10**: 2493–2501.

96. Lipton SA, Rosenberg PA. Mechanisms of disease: Excitatory amino acids as a final common pathway for neurologic disorders. *N Engl J Med* 1994, **330**: 613–622.
97. Ratan RR, Murphy TH, Baraban JM. Macromolecular synthesis inhibitors prevent oxidative stress-induced apoptosis in embryonic cortical neurons by shunting cysteine from protein synthesis to glutathione. *J Neurosci* 1994, **14**: 4385–4392.
98. Ratan RR, Murphy TH, Baraban JM. Oxidative stress induces apoptosis in embryonic cortical neurons. *J Neurochem* 1994, **62**: 376–379.
99. Oka A, Belliveau MJ, Rosenberg PA, Volpe JJ. Vulnerability of oligodendroglia to glutamate: pharmacology, mechanisms, and prevention. *J Neurosci* 1993, **13(4)**: 1441–1453.
100. Griot C, Vandevelde M, Richard A, Peterhans ERS. Selective degeneration of oligodendrocytes mediated by reactive oxygen species. *Free Rad Res Comms* 1990 **11**: 181–193.
101. Kim YS, Kim SU. Oligodendroglial cell death induced by oxygen radicals and its protection by catalase. *J Neurosci Res* 1991, **29**: 100–1045.
102. Husain J, Juurlink BHJ. Oligodendroglial precursor cell susceptibility to hypoxia is related to poor ability to cope with reactive oxygen species. *Brain Res* 1995, **698**: 86–94.
103. Thorburne SK, Juurlink BHJ. Low glutathione and high iron govern the susceptibility of oligodendroglial precursors to oxidative stress. *J Neurochem* 1996, **67**: 1014–1022.
104. Hill JM, Switzer RC. The regional distribution ancl cellular localization of iron in the rat brain. *Neuroscience* 1984, **11**: 595–603.
105. Dwork AJ, Schon EA, Herbert J. Nonidentical distribution of transferrin and ferric iron in human brain. *Neuroscience* 1988, **27**: 333–345.
106. Gerber MR, Connor JR. Do oligodendrocytes mediate iron regulation in the human brain. *Ann Neurol* 1989, **26**: 95–98.
107. Ozawa H, Nishida A, Mito T, Takasima S. Development of ferritin-containing cells in the pons and cerebellum of the human brain. *Brain Devel* 1994, **16**: 92–95.
108. Iida K, Takashima S, Ueda K. Immunohistochennical study of myelination and oligodendrocyte in infants with periventricular leukomalacia. *Pediatr Neurol* 1995, **13**: 296–304.
109. Connor JR, Menzies SL. Relationship of iron to oligodendrocytes and myelination. *GLIA* 1996, **17**: 83–93.
110. Halliwell B, Gutteridge JC. Role of free radicals and catclytic metal ions in human disease: An overview, in *Methods in Enzymology*, (Packer AN, ed.), Academic Press, San Diego 1990: 1–85. vol 186.
111. Ditelberg JS, Sheldon RA, Epstein CJ, Ferriero DM. Brain injury after perinatal hypoxia-ischemia is exacerbated in copper/zinc superoxide dismutase transgenic mice. *Pediatr Res* 1996, **39**: 204–208.
112. Connor JR, Fine RE. Development of transferrin-positive oligodendrocytes in the rat central nervous system. *J Neurosci Res* 1987, **17**: 51–59.
113. Connor JR, Menzes SL. Altered distribution of iron in the central nervous system of myelin deficient rats. *Neuroscience* 1990, **34**:
114. Connor JR, Menzes SL, St. Martin SM, Mufson EJ. Cellular distribution of transferrin, ferritin, and iron in normal and aged human brains. *J Neurosci Res* 1990, **27**: 595–611.
115. Morris CM, Candy JM, Bloxham CA, Edwardson JA. Immunocytochemical localisation of transferrin in the human brain. *Acta Anat (Baser)* 1992, **143**: 14–18.
116. Connor JR, Phillips TM, Lakshman MR, Barron KD, Fine RE, Csiza CK. Regional variations in the levels of transferrin in the CNS of normal and myelin-deficient rats. *J Neurochem* 1987, **49**: 1523–1529.
117. Roskams AJ, Connor JR. Transferrin receptor expression in myelin deficient (md) rats. *J Neurosci Res* 1992, **31**: 421–427.
118. Taylor EM, Morgan EH. Developmental changes in transferrin and iron uptake by brain in the rat. *Brain Res Dev Brain Res* 1990, **55**: 35–42.
119. Connor JR, Pavlick G, Karli D, Menzes SL, Palmer C. A histochemical study of iron-positive cells in the developing rat brain. *J Comp Neurol* 1995, **355**: 111–123.

120. Crickton Rr, Charloteaux-Wauters M. Iron transport and storage. *Eur J Biochem* 1987, 164:485-506.
121. Gocht A, Keith AB, Candy JM, Morris CM. Iron uptake in the brain of the myelin-deficient rat. *Neurosci Lett* 1993, **154:** 187–190.
122. Mollgard K, Stagaard M, Saunders NR. Cellular distribution of transferrin immunoreactivity in the developing rat brain. *Neurosci Lett* 1987, **78:** 35–40.
123. Lin HH, Connnor JR. The development of the transferrin-transferrin receptor system in relation to astroyctes, MBP and galactocerebroside in normal and myelin-deficient rat optic nerves. *Brain Res* 1989, **49:** 281–293.
124. Koster JF. Ferritin, a physiological iron donor for microsomal lipid peroxidation. *FEBS Lett* 1986, **199:** 85–88.
125. Balla G, Jacob HS, Balla J, et al. Ferritin: A cytoprotective antioxidant strategem of endothelium. *J Biol Chem* 1992, **267:** 18148–18153.
126. Qi Y, Dawson G. Hypoxia induces synthesis of a novel 22-kDa protein in neonatal rat oligodendrocytes. *J Neurochem* 1992, **59:** 1709–1716.
127. Qi Y, Dawson G. Hypoxia specifically and reversibly induces the synthesis of ferritin in oligodendrocytes and human oligodendrogliomas. *J Neurochem* 1994, **63:** 1485–1490.
128. Rehncrona S, Nelsen Hauge H, Siejsö BK. Enhancement of iron-catalyzed free radical formation by acidosis in brain homogenates: difference in effect by lactic acid and CO_2. *J Cereb Blood Flow Metab* 1989, **9:** 68–70.
129. Qi Y, Jamindar T, Dawson G. Hypoxia alters iron homeostasis and induces ferritin synthesis in oligodendrocytes. *J Neurochem* 1995, **64:** 2458–2464.
130. Chan PE. Oxygen radicals in focal cerebral ischemia. *Brain Path* 1992, **4:**59–65.
131. Rangan U, Bulkey GB. Prospects for treatment of free-radical-mediated tissue injury. *Brit Med Bull* 1993, **49:** 700–718.
132. Amstad P, Peskin A, Shah G, et al. The balance between Cu, Zn-superoxide dismutase and catalase affects the sensitivity of mouse epidermal cells to oxidative stress. *Biochemistry* 1991, **30:** 9305–9313.
133. Noble PG, Antel JP, Yong VW. Astrocytes and catalase prevent the toxicity of catecholamines to oligodendrocytes. *Brain Res* 1994, **663:** 831–890.
133a. Pinteaux E, Perraut M, Tholey G. Distribution of mitochondrial manganese superoxide dismutase among rat glial cells in culture. *Glia* 1998, **22:** 408–414.
133b. Juurlink BHJ, Thorburne SK, Hertz L. Peroxide-scavenging deficit underlies oligodendrocyte susceptibility to oxidative stress. *Glia* 1998, **22:** 371–378.
134. Houdou S, Kuruta H, Hasegawa M, et al. Developmental immunohistochemistry of catalase in the human brain. *Brain Res 1991,* **556:** 267–270.
134a Takashima S, Kuruta H, Mito T. Immunohistochemistry of superoxide dismutase-1 in developing humanbrain. *Brain Dev* 190, **12:** 211–213.
135. Del Maestro R, McDonald WG. Distribution of superoxide dismutase, glutathione peroxidase and catalase in developing rat brain. *Mech Ageing Devel* 1987, **41:** 29–38.
136. Back SA, Gan X, Rosenberg PA, Volpe JJ. Maturation-dependent vulnerability of oligodendrocytes to oxidative stress-induced death caused by glutathione depletion. *J Neurosci* 1998, in press.
137. Wren DR, Noble M. Oligodendrocytes and oligodendrocyte/type-2 astrocyte progenitor cells of adult rats are specifically susceptible to the lytic effects of complement in absence of antibody. *Proc Natl Acad Sci USA* 1989, **86:** 9025.
138. Scolding NJ, Morgan BP, Campbell AK, Compston DAS. Complement mediated serum cytotoxicity against oligodendrocytes: a comparison with other cells of the oligodendrocyte- type 2 astrocyte lineage. *J Neurol Sci.* 1990, **97:** 155–162.
139. Scolding NJ, Morgan BP, Campbell AK, Compton DAS. The role of calcium in oligodendrocyte injury and repair. *Neurosci Lett* 1992, **135:** 95–98.

140. D'Souza S, Alinauskas K, McCrea E, Goodyear C, Antel JP. Differential susceptibility of human CNS-derived cell populations to TNF-dependent and independent immune-mediated injury. *J Neurosci* 1995, **15**: 7293–7300.

141. McLaurin J, D'Souza S, Stewart J, et al. Effect of tumor necrosis factor alpha and beta on human oligodendrocytes and neurons in culture. *Int J Dev Neurosci* 1995, **13**: 369–381.

142. Wilt SG, Milward E, Zhou JM, et al. In vitro evidence for a dual role of tumor necrosis factor-alpha in human immunodefeciency virus type 1 encephalopathy. *Ann Neurol* 1995, **37**: 381–394.

142a. Hisahara S, Shoji S, Okano H, Miura M. ICE/CEO-3 family executes oligodendrocyte apoptosis by tumor necrosis factor. *J Neurochem* 1997, **69**:10–20.

143. Eitan S, Schwartz MD. A transglutaminase that converts interleukin-2 into a factor cytotoxic to oligodendroctyes. *Science* 1993, **261**: 106–108.

144. Eizenberg O, Faber-Elman A, Gottlieb E, Oren M, Rotter V, Schwartz MO. Direct involvment of p53 in programmed cell death of oligodendrocytes. *EMBO J* 1995, **14**: 1136–1144.

144a. Larocca JN, Farooq M, Norton WT. Induction of oligodendrocyte apoptosis by C2-Ceramide. *Neurochem Res* 1997, **22(4)**: 529–534.

144b. Brogi A, Strazza M, Melli M, Costantino-Ceccarini E. Induction of intracellular ceramide by interleukin-1β in oligodendrocytes. *J Cell Biochem* 1996, **66**: 523–541.

144c. Cassaccia-Bonnefil P, Aibel L, Chao MV. Central glial and neuronal populations display differential sensitivity to ceramide-dependent cell death. *J Neurosci Res* 1996, **43**: 382–389.

145. Raff MC. Programmed cell death and the control of cell survival: Lessons from the nervous system. *Science* 1993 **262**: 695–700.

146. Althaus HH, Hempel R, Kloppner S, Engel J, Schmidt-Schultz T, Kruska L, Heumann I. Nerve growth factor signal transduction in mature pig oligodendrocytes. *J Neurosci Res* 1997, **50**: 729–742.

147. Cassaccia-Bonnefil P, Carter BD, Dobrowsky RT, Chao MV. Death of oligodendrocytes mediated by the interaction of nerve growth factor with its receptor p75. *Nature* 1996, **383**:716–719.

148. Cohen R, Marmur R, Norton WT Mehler MF, Kessler JA. Nerve growth factor and neurotrophin-3 differentially regulate the proliferation and survival of developing rat brain oligodendrocytes. *J Neurosci* 1996, **16**: 6433–6442.

149. Yoon SO, Casaccia-Bonnefil P, Carter BD, Chao MV. Competitive signalling between TrkA and p75 nerve growth factor receptors determines cell survival. *J Neurosci* 1998. **18(9)**: 3273–3281.

150. Ladjwala U, Lachance C, Simoneau SJJ, Bhakar A, Barker PA, Antel JP. p75 neurotrophin receptor expression on adult human oligodendrocytes: signalling without cell death in response to NGF. *J Neurosci* 1998, **18(4)**: 1297–1304.

20
Motor Neuron Disease

Bruce A. Rabin and David R. Borchelt

INTRODUCTION

Motor neuron disease (MND) refers to the group of degenerative disorders characterized by progressive weakness and atrophy of skeletal muscle due to the selective dysfunction and degeneration of upper and/or lower motor neurons. In adults, the most common of these disorders is amyotrophic lateral sclerosis (ALS, i.e., Lou Gehrig's disease), which results from the degeneration of large motor neurons in the anterior horn of the spinal cord, brain stem nuclei, and motor cortex. The most common form of MND in children is spinal muscular atrophy (SMA), a distinct form of MND characterized by the dysfunction and loss of anterior horn cells of the spinal cord and lower brainstem nuclei without associated upper motor neuron findings. This chapter will focus on recent advances in understanding the molecular events leading to dysfunction and death of motor neurons in these two disorders.

ALS

ALS is a fatal form of motor neuron disease characterized by clinical and pathologic features associated with the degeneration of both upper and lower motor neurons. Approximately 90% of ALS cases are sporadic (SALS) and 10% are familial (FALS) *(1)*. Although the causes of sporadic ALS are unknown, recent theories have focused on the role of excitotoxicity *(2)* (*see* also chapter by Rothstein), oxidative stress *(3)* (*see* also chapter by Dykens), alterations in the cytoskeleton *(4)*, and autoimmunity *(5)*. The recent identification of superoxide dismutase-1 (SOD1) mutations in a subset of autosomal dominant FALS patients (ALS1) has provided important insight into potential mechanisms of disease pathogenesis and will be discussed in detail below.

Neuropathology of ALS

The spinal cord from patients with ALS may appear grossly normal or only mildly atrophic. Histochemical stains for myelin reveal degeneration of the descending crossed and uncrossed corticospinal tracts and a profound loss of motor neurons is often seen at the level of the cervical and lumbar enlargements. Motor neurons in various stages of degeneration may also be seen, particularly in rapidly progressive cases. Early findings include chromatolysis and the presence of axonal spheroids composed of randomly oriented 10 nm filaments. Ultrastructural analysis has shown that spheroids may be composed of membrane-bound structures such as mitochondria, vesicles, and fragmented endoplasmic reticulum.

Ubiquitin-positive immunocytochemical staining of anterior horn cells is a prominent and characteristic feature of ALS, as is the accumulation of phosphorylated neurofilaments (NF), fragmentation of the Golgi apparatus, and the presence of Bunina bodies (eosinophilic cytoplasmic inclusions) in many of the remaining motor neurons. Reactive gliosis is also frequently seen. In studies of roots and nerves, particularly the phrenic nerve (comprised of long, unbranched axons), a variety of other axonal abnormalities have been observed: axonal atrophy (reduction in axonal circumference), dying back, reduced numbers of large myelinated fibers, decreased axonal transport, and Wallerian degeneration *(6)*. Axonal terminals at neuromuscular junctions are small or absent and skeletal muscles show grouped atrophy and fiber-type grouping.

In some cases of FALS, intracytoplasmic inclusions contain SOD1, ubiquitin, and NFs. Perikaryal inclusions are enriched in randomly oriented NF, whereas swollen neurites often show numerous NF oriented parallel to the long axis of the process. Axonal abnormalities in FALS share many of the features described above for SALS.

Although ALS is characterized by motor neuron signs and pathologic changes are prominent in the cell bodies and axons of motor neurons, nonmotor neurons may also demonstrate significant pathologic changes. Posterior column degeneration, loss of neurons in Clarke's column, and degeneration of neurons in the forebrain are well described. Neurons in the dorsal root ganglion (DRG) and cranial nerves III, IV, and VI may also be affected.

One explanation for the appearance of these abnormalities in nonmotor systems is that there is a continuum of neuronal vulnerability in ALS, with motor neurons most sensitive. Rapidly progressive MND would be associated primarily with motor neuron degeneration and less dramatic progression may be associated with pathology extending into nonmotor systems. This hypothesis is supported by the observation that patients maintained for prolonged periods on respirators show widespread degeneration of nonmotor neural systems.

Despite the presence of pathologic abnormalities outside the motor system, however, these clinical and pathologic manifestations are generally less pronounced than those found in the corticospinal tracts and anterior horn cells of ALS patients. The underlying mechanism(s) of the selective cell death of motor neurons is unknown and remains a central question in understanding the pathophysiologic basis for this disease.

SOD1

Approximately 15–20% of FALS is caused by autosomal dominant mutations within the copper–zinc superoxide mutase (SOD1) gene on chromosome 21 *(7)*. In fact, over 45 different SOD1 missense mutations have been discovered in early-onset families with FALS. The major known biochemical function of this homodimeric copper and zinc-containing metalloprotein is to act as a free radical scavenger to reduce the concentration of intracellular superoxide molecules by catalyzing the formation of hydrogen peroxide and oxygen from superoxide anions. This scavenging activity of SOD1 is thought to protect various cellular constituents (e.g., nucleic acids, proteins, lipids) that would otherwise be damaged by free radicals (for a detailed discussion of these cellular

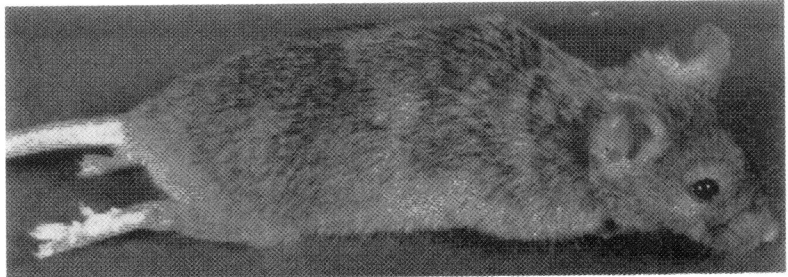

Fig. 1. A SOD1 mutant transgenic mouse demonstrating severe weakness of the limbs.

events, see chapter by Dykens). Thus, it was hypothesized that loss of motor neurons in FALS patients with SOD1 mutations is a direct result of reduced SOD1 activity. Indeed, measurements of SOD1 activity in red blood samples from affected patients initially showed reduced levels of activity in vitro, despite the fact that a normal SOD1 allele is present in each of these patients (8).

In addition, bacteria, yeast, and Drosophila with deletions of the SOD1 gene have increased sensitivity to superoxide anions, as demonstrated by reduced growth rates and survival times. In mammalian nerve cell cultures, reduction of SOD1 activity accelerates cell death, whereas increased levels of SOD1 activity is protective (9–12). However, in vitro measurement of SOD1 activity has shown that most FALS-associated mutant SOD1 enzymes possess normal or near-normal levels of enzyme activity and stability (13). Transformation of these FALS-associated mutant SOD1 genes into SOD1-deleted yeast, unable to grow on selective media, restores them to the wild type phenotype (12). These data demonstrate that SOD1 mutations do not uniformly eliminate or reduce SOD1 activity and imply that the mechanism of cell death in SOD1-associated FALS cannot be due to reduced SOD1 activity.

To directly test whether motor neuron death can result from the complete absence of SOD1 enzymatic activity, mice homozygous for a SOD1 gene deletion were analyzed (14). These SOD1-null mice developed normally, and, although motor neurons in these animals were more susceptible to retrograde degeneration after facial axotomy, they did not develop MND.

Examinations of transgenic mice expressing human SOD1 with either G93A, G37R, or G85R mutations have provided definitive proof that the mechanism(s) by which FALS mutations in SOD1 cause MND is unrelated to alterations in SOD1 activity levels (15–17). For example, at 4 mo of age, transgenic mice expressing G37R and G93A human SOD1 show reduced spontaneous movements, difficulty moving their hindlimbs, muscle wasting, and electromyographic (EMG) patterns consistent with active denervation (Fig. 1). Eventually, the forelimbs become weak and the hindlimbs completely paralyzed. These mice have significantly elevated levels of SOD1 protein and in vitro activity measurements indicate parallel increases in activity levels (15,17).

In mice expressing the G37R mutation, vacuoles, conspicuous in the neuropil, are present in dendrites > axons > cell bodies (Fig. 2). These vacuoles may be associated with swollen mitochondria in various stages of degeneration. Mitochondria in adjacent nonneuronal cells appear unaffected. Vacuoles in axons and dendrites show altered patterns of NF and MAP2 immunoreactivity, respectively. Cell bodies and dendrites

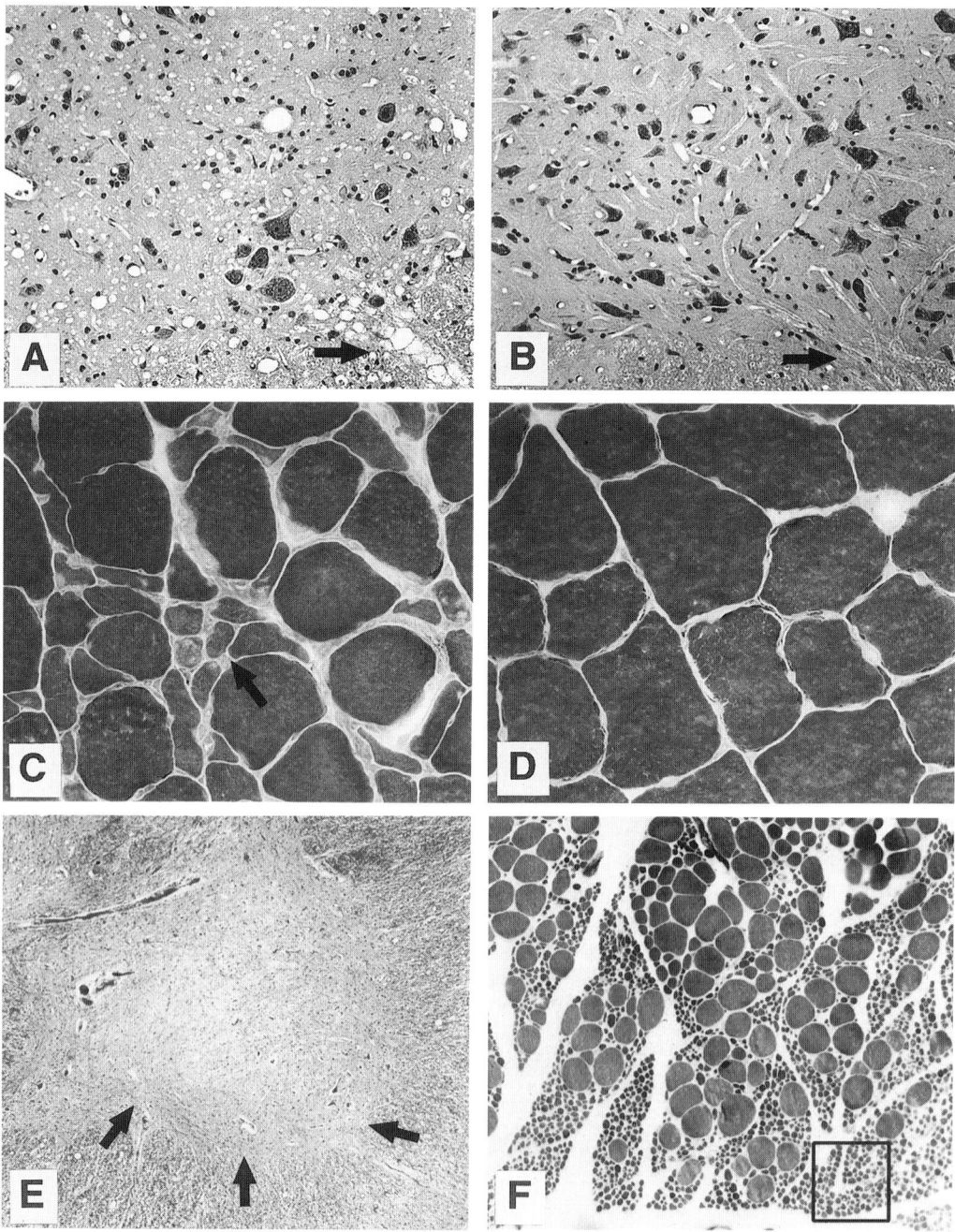

Fig. 2. (A) A section through the cervical spinal cord of a G37R transgenic mouse at 20 wk of age demonstrating loss of motor neurons and severe axonal pathology at the ventral root exit zone (arrow). **(B)** A section through the cervical spinal cord of a control mouse of the same age as shown in (A), showing a normal number of large anterior horn cells and a normal appearance of the ventral root exit zone (arrow). **(C)** This picture shows a section through the striated muscle of a G37R transgenic mouse at 11 wk of age stained with Gomori trichrome, demonstrating grouped atrophy of muscle fibers (arrow). **(D)** A section through the muscle of a control animal stained with Gomori trichrome, showing the normal appearance of striated muscle.

also stain with ubiquitin and phosphorylated NF antibodies. Astrogliosis is evident and often present in proximity to abnormal neurons. Motor axons in intraspinal segments and roots, which show abundant SOD1 immunoreactivity, exhibit a series of abnormalities including vacuoles, cytoskeletal alterations, Wallerian degeneration, reductions in the number of end plates, and muscle fiber type grouping and grouped atrophy. G37R transgenic mice with the highest levels of transgene product have similar abnormalities in other neuronal populations, including those in DRG, dorsal horn, brainstem nuclei, cerebellar nuclei, several thalamic nuclei, and piriform cortex *(17)*.

Mice expressing high levels (five times the endogenous levels) of G93A mutant human SOD1 have a very similar disease *(15,18)*. At 3 mo of age, these mice develop hindlimb weakness, a coarse coat, poor grooming, and reduced stride length. By 5 mo of age, a majority of the mice are moribund. Pathological analysis demonstrates reduced numbers of motor neurons with some containing Bielschowsky-positive fibrils interpreted to be phosphorylated NF. Vacuolation in cell bodies > dendrites is also seen. Variably sized axonal swellings (containing NF, internexin, and peripherin) are prominent. There is severe loss of myelinated axons in the ventral roots and peripheral nerves, as well as cervical and lumbar motor neurons *(18–20)*.

Transgenic mice expressing a murine SOD1 with the G85R mutation also develop MND *(16)*. At 3–4 mo, the founder mice developed generalized weakness that progresses within 72 h to total immobility. Several mice developed a spastic paralysis involving the hindlimbs. Large spinal motor neurons were decreased in number, and remaining neurons exhibited pyknosis and karyorrhexis associated with dystrophic neurites and swollen fragmented processes *(16)*. Degenerative changes were also seen in motor nerve cranial nuclei and a few neurons in layer V of neocortex.

Finally, transgenic mice expressing the human G85R SOD1 mutation develop a rapidly progressing MND *(24)*. In these animals, neuropathological abnormalities were evident in both neurons and glia of the spinal cord. Both nerve cell types contained ubiquitin and SOD1 immunoreactive inclusions that increased in number as the disease progressed.

Importantly, none of the SOD1 transgenic models show reduced SOD1 levels. In fact, most have elevated SOD1 levels. Moreover, mice overpressing wild type human SOD1 do not develop disease. Together, these results demonstrate that the development of motor neuron degeneration in transgenic SOD1 mice expressing mutant SOD1 protein is a direct result of the SOD1 mutations and that it is not due to decreased SOD1 activity or the presence of a human SOD1 protein.

If reduced enzyme activity level is not responsible for disease pathogenesis in SOD1 patients and transgenic animals, why do motor neurons die? It has been proposed that FALS mutant SOD1 molecules acquire novel toxic properties *(17,21,22)*. Although the disease-linked SOD1 missense mutations are distributed throughout the protein, Beckman and coworkers have suggested that FALS-linked mutations alter the active site to allow increased contact between bound copper and peroxynitrite, catalyzing the

---

(E) A section through the spinal cord of a patient with ALS, demonstrating the profound loss of large motor neurons in the ventral horn. **(F)** Neurogenic atrophy of striated muscle from a patient with ALS shows grouped atrophy (box), resembling muscle pathology in transgenic mice **(C)**. (A–D) were adapted with permission from ref. *17*.

nitration of protein tyrosine residues *(23)*. Nitrotyrosine formation on proteins important in the biology of motor neurons would then result in dysfunction and death of these cells. In support of this hypothesis, mice expressing the human G85R SOD1 mutation show elevated nitration of the glutamate transporters *(24)*.

An alternative hypothesis suggests that motor neuron degeneration in SOD1 patients may result from increased production of hydroperoxides. A4V and G93A mutations, but not free copper or the SOD1 apoenzyme, catalyze the oxidation of a spin-trap compound (5,5-dimethyl-1-pyroline *N*-oxide) at rates higher than those occurring with wild type SOD *(3)*. Chelating agents (e.g., diethyl-dithylidocarbonate and D,L-penicillamine) abolish this reaction in vitro. Moreover, in a cell culture model of SOD1 toxicity, apoptotic neural cell death (which is accelerated by these FALS SOD1 mutations) is delayed by treatment with copper chelators. However, examinations of FALS transgenic mice have not revealed significant levels of oxidative damage. Thus, whether or not motor neuron injury results from increased copper-mediated peroxidase reactions remains unclear.

Neurofilaments

Although abnormal accumulation of neurofilament proteins in perikarya and proximal axons are pathological hallmarks of ALS, their role in *causing* motor neuron cell death has been uncertain. Several lines of transgenic mice were constructed to examine whether accumulation of normal neurofilament or expression of neurofilament mutations could directly lead to motor neuron death.

Interestingly, mice with high levels of NF-L expression have weakness, denervation, and muscle atrophy and die within 3–4 wk of birth. Neurofilament accumulations are present in perikarya and large axonal swellings *(25)*. Overexpression of human NF-H also causes a MND-like disease, characterized by onset of weakness at 4–5 mo of age associated with abnormalities in axonal transport of neurofilament, tubulin, and actin *(26,27)*. Mice overexpressing NF-M developed neurofilamentous accumulations in cell bodies and proximal axons, but did not develop clinical signs. Therefore, overexpression of NF-L and NF-H in mice produces a disease characterized by progressive weakness, muscle atrophy, and axonal degeneration. However, the hallmark feature of ALS, motor neuron cell death, is not present in these animals.

In contrast with the phenotype produced by overexpression of wild type neurofilament genes, the overexpression of mutated NF-L in a transgenic system does result in motor neuron death. A point mutation was introduced into a conserved region of the mouse NF-L gene (leucine at position 394 was changed to proline) and expressed in mice as a transgene *(28)*. The expression of this gene in the background of a normal mouse NF-L gene produces a disease characterized by reduced motor activity, muscle weakness, grouped atrophy of muscle, and, importantly, motor neuron death. Axonal spheroids, another characteristic feature of ALS, were prominent in the ventral root exit zones and anterior horn regions of the spinal cord *(28)*.

These studies clearly demonstrate the ability of altered neurofilament expression to cause motor neuron dysfunction and death in animal models. Interestingly, screening of sporadic ALS patients for mutations in neurofilament genes has revealed 5/356 patients with deletions within the KSP repeat domain of NF-H *(24)*, whereas neurofilament mutations have not been identified in normal or FALS patients. Thus, although the neurofilament transgenic mice indicate that neurofilaments may play a role in motor neuron cell death, mutations within their genes are unlikely to be the direct agent responsible for disease onset in the majority of ALS patients.

Excitotoxicity

In 1990, it was observed that the excitatory amino acid neurotransmitter, glutamate, was elevated in the spinal fluid of patients with ALS *(29)*. Because the normal clearance of glutamate in the synaptic cleft occurs via specific glutamate transporters, it was hypothesized that abnormal extracellular accumulation of glutamate was due to dysfunctional glutamate transporter proteins. In fact, when glutamate transport was measured in postmortem brains and spinal cords of patients with ALS, Alzheimer's disease, Huntington's disease, and SMA brains, a significant reduction was demonstrated only in ALS motor cortex and spinal cord *(30)*.

Subsequently, it was shown that the astroglial glutamate transporter GLT-1 is reduced specifically in motor cortex and spinal cord of some ALS patients, despite the proliferation of astrocytes and glia in affected regions of ALS patients *(31)*. It was also shown that chronic glutamate uptake inhibition in an organotypic spinal cord culture results in the slow degeneration of motor neurons *(32)*.

Additional evidence for the role of glutamate in ALS pathogenesis is suggested by drugs acting on the glutamate pathway *(33)*. Both riluzole and gabapentin protect against motor neuron degeneration in a model of chronic glutamate toxicity and prolong the survival of G93A transgenic mice. Importantly, riluzole was approved in 1996 by the Food and Drug Administration for use in patients with ALS after two placebo-controlled multicenter trials demonstrated a statistically significant benefit on survival times. These data support the hypothesis that dysfunction and death of motor neurons in ALS may involve glutamate toxicity (for a more topical discussion of this subject, *see* chapter by Rothstein).

Trophic Factors

While there is no direct evidence that a deficiency or mutation within one of the neurotrophic factors results in MND, there is considerable interest in whether or not they can be used for preventing neuronal cell loss in affected patients. One well studied model of motor neuron loss is axotomy. Axotomy of neonatal motor neurons causes profound retrograde degeneration, with nearly complete loss of all neurons whose axons have been severed (*see* chapter by Elliott and Snider in this volume). Degeneration of these cells can be ameliorated by the presence of one of many of neurotrophic peptides, including brain-derived neurotrophic factor (BDNF), neurotrophin-4/5 (NT-4/5), ciliary neurotrophic factor (CNTF), glial-derived neurotrophic factor (GDNF), and cardiotrophin-1 (CT-1). To directly test whether a reduction in one of the neurotrophic factors is responsible for, or contributes to, motor neuron disease, gene ablation studies in transgenic mice of BDNF, NT-3, NT-4, CNTF, and leukemia-inhibiting factor (LIF) were undertaken. Mutations in these genes have failed to consistently produce mice with motor signs. However, disruption of the LIFβ-R or α portions of the CNTF receptor results in loss of half of the spinal cord neurons and GDNF knockout mice also have fewer spinal motor neurons, suggesting some role for supporting development or maintenance of neurons in vivo. The SOD1 and neurofilament transgenic animal models of ALS described above would be ideal for testing the in vivo neuroprotective abilities of these growth factors, despite the fact that pathogenetic mechanisms in ALS are unlikely to be related to growth factor deprivation (for a comprehenisve discussion of motor neuron trophic factors, *see* chapter by Koliatsos and Mocchetti; for discussion of recent clinical trials of trophic factors, *see* chapter by Hilt et al.).

Neurotoxins

The systemic administration of β,β-iminodipropionitile (IDPN) to rodents causes a MND phenotype manifested as limb weakness. Arrays of abnormally oriented NF accumulate in spheroids and, to a lesser extent, in cell bodies. Swollen axons are filled with SOD1 immunoreactivity. The spheroids are denuded of myelin and distal axons become atrophic, resembling the alterations seen in ALS and FALS. IDPN is thought to impair the slow transport of NF proteins, perhaps because the toxin dissociates NF from microtubules. Administration of aluminum salts also results in a disease characterized by accumulations of NF in perikarya and proximal processes of nerve cells, including motor neurons. As in the IDPN-induced disease, the transport of slow component proteins, particularly NF, is disrupted.

SMA

Childhood SMA is one of the most common lethal neuromuscular diseases of infants and children. It is inherited as an autosomal recessive disease with markedly varied severity. The most severe form (Werdnig-Hoffmann disease; SMA type I) is uniformly fatal before age 1–2 yr. Types II and III are less severe, with increasing ability for independent sitting and assisted ambulation. Although severity is variable, an international collaborative workshop has developed useful clinical inclusion criteria: all children with SMA have widespread symmetric weakness, more prominent proximally than distally, with legs affected more than the arms, weakness of the trunk muscles, and absence of upper motor neuron and sensory findings. There is associated clinical or paraclinical (electrophysiologic or pathologic) evidence of denervation.

Pathology

Examination of autopsy material characteristically shows profound loss of motor neurons in the spinal cord and brainstem nuclei. Chromatolytic and ballooned neurons are found in the ventral horn, Clarke's column, DRG, and thalamus. Immunocytochemical staining often shows phosphorylated neurofilaments and ubiquitin staining in the perikarya, whereas ultrastructural analysis demonstrates accumulation of mitochondria and other vesicular and membranous structures in the soma of degenerating neurons (reviewed in ref. *34*).

Genetics

Although there is clinical variability in disease severity (even within a family), all three forms of childhood SMA are linked to the chromosome 5q11.3–13.1 region *(35,36)*. The genomic organization of this region is polymorphic and quite complex, with numerous repeated DNA elements occurring in different orientations on chromosomes from different individuals. In 1995, two candidate genes were described within this region, survival motor neuron (SMN) *(37)* and neuronal apoptosis inhibitor protein (NAIP) *(38)*.

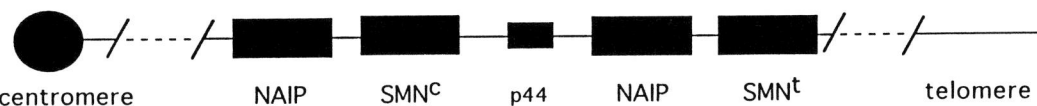

Fig. 3. Organization of the SMA-determining region of chromosome 5q11–13 in most people. There is some variability of individual gene copy number and orientation, presumably due to complex genetic rearrangements, including gene conversion.

SMN occurs within a 500 bp inverted repeat and itself occurs as a tandemly repeated gene (Fig. 3). Mutations are found in the telomeric copy, referred to as SMN or SMN^t, in 100% of affected patients. Partial or complete deletions are detected in 95% of patients, with remaining patients having other mutation types, including point mutations, frameshift mutations, and premature terminations. The centromeric copy, referred to as SMN^c or $^cBCD541$, differs from the telomeric version by five nucleotides—one of these changes is in the coding region and does not lead to an amino acid change, three are in intronic sequences, and one is in the 3' UTR. SMN^c has not been found deleted in any affected patient, although it is deleted in 4% of normal controls. This suggests that homozygous deletion of the centromeric and telomeric copies of SMN may be nonviable and likely results in an embryonic lethal event. If this explanation is correct, than SMN^c must be able to partially substitute for SMN^t.

Although the coding sequences are identical in the two genes, SMN^c undergoes alternative splicing, with only small amounts of the full-length transcript made. Thus, it may be the relative amount of full length SMN^c transcript (and native SMN protein made from it) which determines disease severity. In this model, the smaller transcripts made from SMN^c (particularly those lacking exons 7 and 8) produce proteins unable to rescue patients with deleted SMN^t. This hypothesis predicts that patients with higher levels of full-length SMN^c transcripts would have less severe disease.

SMN is highly conserved between mice and humans, with 82% identity *(39–41)*. Unlike human SMN, there is no alternative splicing of the mouse homolog *(39)*. Interestingly, the mouse contains only one copy of SMN *(39–41)*, which is also closely linked to NAIP on mouse chromosome 13 *(40,41)*. SMN is expressed widely in humans and mice. *In situ* hybridization of adult mouse brain and spinal cord demonstrates expression in large motor neurons, as well as most other cells. Similarly, almost all neurons in the brain express SMN, with highest levels seen in the hippocampus and cerebellum *(39)*. The presence of a single copy of SMN in the mouse will greatly simplify functional analysis of this protein and help to clarify the role of the centromeric and telomeric human genes.

The other candidate gene within this region is NAIP. NAIP is a large gene, spanning 70 kB of genomic DNA and encoding a 1232 aa protein. This gene has been given its name because of homology of exons 5–7 and 9–11 with two baculovius genes-inhibitors of apoptosis (iap), one from *Cydia pomonella* granulosis virus (33% over 189 amino acids) and the other from *Orgyia pseudotsugata* nuclear polyhydrosis virus (33% over 180 amino acids). These baculovirus genes function to inhibit the cellular response to viral infection (apoptosis). Notable is the absence in NAIP of the zinc-finger motif found in the baculovirus genes and proto-oncogenes involved in regulating apoptosis *(42)*. However, in vitro assays of antiapoptotic activity have demonstrated the ability of NAIP to function as an inhibitor of apoptosis *(43)*.

Variable numbers of truncated and deleted versions of the NAIP gene are found in normal controls and affected patients. However, SMA chromosomes are deleted in NAIP exons 5–6 at a much higher frequency than non-SMA chromosomes (67% type 1, 42% types 2 and 3 vs 3% normals) *(38)*. Although there is no direct genotype–phenotype correlation with regard to correlation of severity of disease with extent of deletion within NAIP, it does appear that more severely affected individuals have deleted exons 5–7 both in SMN and NAIP. Given the complexity of the NAIP gene region, it would be difficult to assess if patients who are severely affected but do not carry deletions in NAIP have point mutations within the coding region of NAIP.

The discovery of several asymptomatic carriers of mutations in either SMN or NAIP raises the possibility that neither SMN nor NAIP is the responsible gene. Almost all asymptomatic individuals carrying the SMN mutations, however, are relatives of patients with less severe forms of disease. In contrast, asymptomatic carriers of NAIP exon 5–7 deletions are more heterogeneous and may be related to SMA patients with SMA types I, II, or III. To explain these findings, it has been suggested that either a modifying gene is present (NAIP pseudogenes?) or that these individuals will have a delayed onset of symptoms (adult SMA). Indeed, some patients with adult SMA (SMA type IV), have SMN and/or NAIP deletions.

Although the homology of NAIP to iap genes is suggestive of a functional role in regulating apoptosis, its primary role in SMA pathogenesis is unlikely, given the genetic findings described above. Apoptosis does, however, play an important role during development of the adult nervous system, with neurons of all types, including motor neurons, undergoing apoptosis as a mechanism of cellular pruning (*see* chapter by Burek and Oppenheim for a topical discussion of this issue). It has thus been proposed that motor neuron loss in SMA is the result of continued or reactivated apoptotic pathways in these cells *(38)*. However, there has been no convincing demonstration to date that apoptosis in the CNS of SMA patients has a role in disease pathogenesis.

RNA Processing

Initial searches of genome databases revealed no significant homology of the SMN protein with any known protein and its function has remained unclear. An important clue has come from studies utilizing the yeast 2-hybrid system to identify proteins with which SMN interacts *(44)*. These studies demonstrated the ability of the SMN protein to interact with itself, fibrillarin, and heteronuclear ribonuclear protein (hnRNP) U. Fibrillarin is a small nucleolar RNA binding protein that contains an RNP motif and RGG box. In yeast, fibrillarin is required for pre-rRNA processing and is essential for viability. HnRNP U is a member of the hnRNP proteins that bind nuclear RNAs (precursor mRNA and nuclear mRNA) and is believed to a play a role in the transport and processing of mRNA. SMN was found to interact specifically with the RGG box-containing C-terminus region of hnRNP U. The interaction with these two proteins appears specific, as another RGG-containing protein, hnRNP A1, does not interact with SMN *(44)*.

A more recent search of the genome databases has revealed an intriguing C-terminal dodecapeptide which spans exons 6 and 7 of SMN: this dodecapeptide shows homology to proteins in mice, yeast, and *C. elegans (45)*. The conserved region contains the motif: Tyr-x-x-Gly-Tyr-x-x-Gly-Tyr-x-x-Gly. Three of six proteins found to contain this "YG box" are known RNA binding proteins: human initiation factor 4b, shrimp hnRNP A/B, and *Drosophila* hnRNP40. Most SMN mutations are deletions involving at least exon 7 (C-terminal domain). However, three missense mutations are known and all three lie within the conserved YG box *(45)*. This pattern suggests strongly that this region is crucial for SMN function.

Antibodies to SMN have been generated that identify a protein of molecular weight 38–40 kDa on Western blots. Immunocytochemistry of HeLa cells with these antibodies demonstrates several small dots in the nucleus of cultured cells. These nuclear dots are closely associated with coiled bodies, but are distinct entities and have been termed "gem bodies" for Gemini of coiled bodies *(44)*. Coiled bodies are found within nuclei and are thought to function in the metabolism of small nuclear ribonuclear particles (snRNPs). Double immunostaining with antibodies to SMN and snRNPs showed snRNP staining of coiled bodies, but no staining of gems *(44)*. The number of gems in a nucleus is variable, but usually between 2 and 6 are present in any one cell and they measure 0.1–1.0 µm in diameter.

Function of SMN

When SMN was initially identified in 1995, it was a unique protein with no known function. The data currently available now strongly suggest an important role in RNA metabolism. How this would lead to selective vulnerability of motor neurons remains unanswered. Does SMN have any additional functions? Does it interact with a set of motor neuron-specific RNA binding proteins? What role does expression of the centromeric copy of SMN play? Model systems currently being developed will greatly aid in our understanding of the normal biology of the SMN protein and consequent disease pathogenesis that result from mutations within this gene.

NATURALLY OCCURRING ANIMAL MODELS OF MOTOR NEURON DISEASE

As demonstrated by the transgenic neurofilament and SOD1 animal models described above, animal models that recapitulate all or part of human diseases provide unique opportunities to examine and manipulate the factors responsible for cell death and degeneration in MND. The ideal animal model would show identical disease progression to humans, share a common underlying pathogenic mechanism, and demonstrate identical pathological findings. Transgenic strategies have thus far targeted known pathogenic mechanisms (axonal transport and SOD1 metabolism) and examined the subsequent effects. Naturally occurring animal models of MND provide insight into other possible mechanisms of pathogenesis. Although there are many naturally occurring animal models, only a few fulfill the criteria of replicating specific aspects of human motor neuron disease with little or no involvement of other portions of the nervous system. Despite their weaknesses and limitations, both transgenic and naturally occurring animal models will prove crucial as new therapeutic strategies are developed to slow motor neuron cell death.

Motor Neuron Degeneration

The motor neuron degeneration (mnd) mouse was isolated as a spontaneous mutation with progressive muscle weakness *(46)*. Death usually occurs before 1 yr of age. Spinal cord motor neurons of mnd mice demonstrate presymptomatic accumulation of ubiquitin-positive material and autofluorescent inclusions in the cell body. Phosphorylated neurofilaments appear later. Although these findings resemble some of the pathologic changes seen in ALS, these mice have a much more severe systemic disease characterized by the accumulation of lipofuscin-like material in the cytoplasm of virtually all nerve cells, as well as most other cell types. There is also accumulation of ATP synthase subunit c in the cytoplasm of these cells. This constellation of findings most closely resembles those seen in the autosomal recessive neuronal disease ceroid lipofuscinosis *(47,48)*.

Hereditary Canine Spinal Muscular Atrophy

Hereditary canine spinal muscular atrophy (HCSMA) is an autosomal dominant disease of Brittany spaniels characterized by progressive weakness and atrophy of skeletal muscle with relative sparing of eye movements and sphincter function. Homozygous animals have a rapidly progressive disease, with tetraplegia by age 16 wk; heterozygous animals have either a subacute or chronic disease. Pathologic changes are present in the ventral horn and brainstem; there is no evidence of upper motor neuron involvement. All dogs with HCSMA develop neurofilamentous swellings in proximal axons of motor neurons with associated reduction in axonal diameter and perikaryal size. Axonal internodes are widely separated, accounting for some of the weakness observed in these animals; however, EMG studies and muscle biopsies are consistent with denervation. Despite the severe weakness and obvious axonal swellings, motor neuron numbers are only slightly reduced.

Progressive Motor Neuronopathy

The progressive motor neuronopathy (pmn) mouse was originally identified as a spontaneous autosomal recessive mutation on mouse chromosome 13, characterized by progressive weakness beginning in the hindlimbs and progressing to quadriplegia and death by 6 wk of age due to respiratory failure *(49)*. Although this mutation maps to the same chromosome as SMN, recombination events occur between them, demonstrating that they are separate genes. The disease is due to a dying-back process with distal axonal degeneration. Proximal axons and cell bodies of motor neurons are relatively spared. Thus, the disease appears to resemble more closely an axonal neuropathy, such as Charcot-Marie-Tooth, type 2 disease, than MND.

Wobbler

The wobbler (wr) mutation maps to mouse chromosome 11 and is transmitted as an autosomal recessive trait. His disease is characterized by progressive weakness beginning in the distal forelimbs and progressing to involve proximal muscles. Before the onset of weakness, however, there is vacuolar degeneration of neurons in the ventral root, thought to be associated with dilatation of the rough endoplasmic reticulum and Golgi apparatus, but perhaps also involving mitochondria. Once the disease becomes evident, there is degeneration of large myelinated fibers in the cervical ventral roots, associated with axonal vacuolation, neurofilamentous accumulation, and myelin ovoids.

CONCLUSIONS

Cell death in motor neuron disease is a complicated process and likely involves many, if not most, of the pathways known to cause cell death in model systems—

excitotoxicity, peroxidation, nitration, apoptosis, and abnormalities of axonal transport. However, the role each of these processes play in initiating and propagating motor neuron death is currently uncertain. Understanding this process, at the molecular and systems level, will be crucial to developing rational and effective therapies for patients affected by these devastating diseases.

ACKNOWLEDGMENTS

BA Rabin is supported by grants from the NIH (K08NS01781) and the Muscular Dystrophy Association. DR Borchelt is supported by Grants from the NIH (AG05146) and the Amyotrophic Lateral Sclerosis Association.

REFERENCES

1. Emery A, Holloway S. Familial motor neuron disease, in *Human Motor Neuron Disease* (Rowland LP, ed.), Raven Press, New York, 1982, pp. 139–147.
2. Rothstein J. Excitotoxicity hypothesis. *Neurology* 1996, **47:** S19–S26.
3. Wiedau-Pazos M, Goto JJ, Rabizadeh S, Gralla EB, Roe JA, Lee MK, Valentine JS, and Bredesen DE. Altered reactivity of superoxide dismutase in familial amyotrophic lateral sclerosis. *Science* 1996, **271:** 515–518.
4. Cleveland D, Bruijn LI, Wong PC, Marszalek JR, Vechio JD, Lee MK, Xu X-S, Borchelt DR, Sisodia SS, Price DL. Mechanisms of selective motor neuron death in transgenic mouse models of motor neuron disease. Neurology. 1996, **47:** S54–S62.
5. Drachman D, Kuncl R. Amyotrophic lateral sclerosis. *Ann Neurol* 1989; **26:** 269–274.
6. Brady S. Motor neurons and neurofilaments in sickness and health. *Cell* 1993, **73:** 1–3.
7. Rosen D, Siddique T, Patterson D, Filewicz DA, Sapp P, Hentati A, Donaldson D, et al. Mutations in the Cu/Zn superoxide dismutase gene are associated with familial amyotrophic lateral sclerosis. *Nature* 1993, **362:** 59–62.
8. Deng HX, Hentati A, Tainer JA, Iqbal Z, Cayabyab A, Hung W-Y, Getzoff ED, Hu P, Herzfeldt B, Roos, RP. Amyotrophic lateral sclerosis and structural defects in Cu,Zn superoxide dismutase. *Science* 1993, **261:** 1047–1051.
9. Rothstein J, Bristol LA, Hosler B, Brown RH Jr, Kuncl RW. Chronic inhibition of superoxide dismutase produces apoptotic death of spinal neurons. *Proc Natl Acad Sci USA* 1994, **91:** 4155–4159.
10. Troy C, Shelanski ML. Down-regulation of copper/zinc superoxide dismutase causes apoptotic death in PC12 neuronal cells. *Proc Natl Acad Sci USA* 1994, **91:** 6384–6387.
11. Greenlund L, Deckwerth TL, Johnson EM Jr. Superoxide dismutase delays neuronal apoptosis: a role for reactive oxygen species in programmed cell death. *Neuron* 1995, **14:** 303–315.
12. Rabizadeh S, Butler Gralla E, Borchelt DR, Gwinn R, Valentine JS, Sisodia S, Wong P, Lee M, Hahn H, Bredesen DE. Mutations associated with amyotrophic lateral sclerosis convert superoxide dismutase from an antiapoptotic gene to a proapoptotic gene: studies in yeast and neural cells. *Proc Natl Acad Sci USA* 1995, **92:** 3024–3028.
13. Borchelt D, Lee MK, Slunt HH, Guarnieri M, Xu Z-S, Wong PC, Brown RH Jr, Price DL, Sisodia SS, Cleveland DW. Superoxide dismutase 1 with mutations linked to familial amyotrophic lateral sclerosis possesses significant activity. *Proc Natl Acad Sci USA* 1994, **91:** 8292–8296.
14. Reaume A, Elliott JL, Hoffman EK, Kowall NW, Ferrante RJ, Siwek DF, Wilcox HM, Flood DG, Beal MF, Brown, RH Jr. Motor neurons in Cu/Zn superoxide dismutase-deficient mice develop normally but exhibit enhanced cell death after axonal injury. *Nat Genet* 1996, **13:** 43–47.

15. Gurney M, Pu H, Chiu AY, Dal Canto MC, Polchow CY, Alexander DD, Caliendo J, Hentati A, Kwon YW, Deng H-X, Chen W, Zhai P, Sufit RI, Siddique T. Motor neuron degeneration in mice that express a human Cu,Zn superoxide dismutase mutation. *Science* 1994, **264:** 1772–1775.
16. Ripps M, Huntley GW, Hof PR, Morrison JH, Gordon JW. Transgenic mice expressing an altered murine superoxide dismutase gene provide an animal model of amyotrophic lateral sclerosis. *Proc Natl Acad Sci USA* 1995, **92:** 689–693.
17. Wong P, Pardo CA, Borchelt DR, Lee MK, Copeland NG, Jenkins NA, Sisodia SS, Cleveland DW, Price DL. An adverse property of a familial ALS-linked SOD1 mutation causes motor neuron disease characterized by vacuolar degeneration of mitochondria. *Neuron* 1995, **14:** 1105–1116.
18. Dal Canto M, Gurney ME. The development of central nervous system pathology in a murine transgenic model of human amyotrophic lateral sclerosis. *Am J Pathol* 1994, **145:** 1–9.
19. Dal Canto M, Gurney ME. Neuropathological changes in two lines of mice carrying a transgene for mutant human Cu,Zn SOD1 and in mice overexpressing wild-type human SOD: a model of familial amyotrophic lateral sclerosis (FALS). *Brain Res* 1995, **676:** 25–40.
20. Tu P-H, Raju P, Robinson KA, Gurney ME, Trojanowski JQ, Lee VM-Y. Transgenic mice carrying a human mutant superoxide dismutase transgene develop neuronal cytoskeletal pathology resembling human amyotrophic lateral sclerosis. *Proc Natl Acad Sci USA* 1996, **92:** 954–958.
21. Brown RJ. Amyotrophic lateral sclerosis: recent insights from genetics and transgenic mice. *Cell* 1995, **80:** 687–692.
22. Wong P, Borchelt DR. Motor neuron disease caused by mutations in superoxide dismutase 1. *Curr Opin Neurol* 1995, **8:** 294–301.
23. Beckman JS, Carson M, Smith CD, Koppenol WH. ALS, SOD, and peroxynitrite. *Nature* 1993, **364:** p. 584.
24. Bruijn L, Becher MW, Lee MK, Anderson KL, Jenkins NA, Copeland NG, Sisodia SS, Rothstein JD, Borchelt DR, Price DL, Cleveland DW. ALS-linked SOD1 mutant G85R mediates damage to astrocytes and promotes rapidly progressive disease with SOD1-containing inclusions. *Neuron* 1997, **18:** 327–338.
25. Xu L, Cork LC, Griffin JW, Cleveland DW. Increased expression of neurofilament subunit NF-L produces morphological alterations that resemble the pathology of human motor neuron disease. *Cell* 1993, **73:** 23–33.
26. Cote F, Collard JF, Julien JP. Progressive neuronopathy in transgenic mice expressing the human neurofilament heavy gene: a mouse model of human amyotrophic lateral sclerosis. *Cell* 1993, **73:** 35–46.
27. Collard J-F, Cote F, Julien JP. Defective axonal transport in a transgenic mouse model of amyotrophic lateral sclerosis. *Nature* 1995, **375:** 61–64.
28. Lee M, Marszalek JR, Cleveland DW. A mutant neurofilament subunit causes massive, selective motor neuron death: implications for the pathogenesis of human motor neuron disease. *Nature* 1994, **13:** 975–988.
29. Rothstein J, Tsai G, Kuncl RW, Clawson L, Cornblath DR, Drachman DB, Pestronk A, Stauch BL, Coyle JT. Abnormal excitatory amino acid metabolism in amyotrophic lateral sclerosis. *Ann Neurol* 1990, **28:** 18–25.
30. Rothstein J, Martin LJ, Kuncl RW. Decreased glutamate transport by the brain and spinal cord in amyotrophic lateral sclerosis. *New Engl J Med* 1992, **22:** 1464–1468.
31. Rothstein J, Van Kammen M, Levey AI, Martin L, Kuncl RW. Selective loss of glial glutamate transporter GLT-1 in amyotrophic lateral sclerosis. *Ann Neurol* 1995, **38:** 73–84.
32. Rothstein J, Jin L, Dykes-Hoberg M, Kuncl RW. Chronic glutamate uptake inhibition produces a model of slow neurotoxicity. *Proc Natl Acad Sci USA* 1993, **90:** 6591–6595.
33. Bensimon G, Lancomblez L, Meninger V, ALS Riluzole Study Group. A controlled trial of riluzole in amyotrophic lateral sclerosis. *N Engl J Med* 1994, **330:** 585–591.

34. Crawford T, Pardo CA. The neurobiology of childhood spinal muscular atrophy. *Neurobiol Dis* 1996, **3:** 97–110.
35. Gilliam T, Brzustowicz LM, Castilla LH, Lehner T, Penchaszadeh GK, Daniels RJ, Byth BC, Knowles J, Hisiop JE, Shapira Y, Dubowitz V, Munstat TL, Ott J, Davies KE. Genetic homogeneity between acute and chronic forms of spinal muscular atrophy. *Nature* 1990, **345:** 823–825.
36. Melki J, Sheth P, Abdelhak S, Burlet P, Bachelot MF, Lathrop MG, Frezal J, Munnich A. Mapping of acute (type 1) spinal muscular atrophy to chromosome 5q12-q14. *Lancet* 1990, **336:** 271–273.
37. Lefebvre S, Burglen L, Reboullet S, Clermont O, Burlet P, Violiet L, Benichou B, Cruaud C, Millasseau P, Zeviani M, Le Paslier D, Frezal J, Cohen D, Weissenbach J, Munnich A, Melki J. Identification and characterization of a spinal muscular atrophy determining gene. *Cell* 1995, **80:** 155–165.
38. Roy N, Mahadevan MS, McLean M, Shutler G, Yaraghi Z, Farahani R, Baird S, Besner-Johnston A, Lefebvre C, Kang X, Salih M, Aubry H, Tamai K, Guan X, Ioannou P, Crawford TO, de Jong PJ, Surh L, Ikeda J, Korneluk G, MacKenzie A. The gene for neuronal apoptosis inhibitor protein is partially deleted in individuals with spinal muscular atrophy. *Cell* 1995, **80:** 167–178.
39. Bergin A, Kim G, Price DL, Sisodia SS, Lee MK, Rabin BA. Identification and characterization of a mouse homologue of the Spinal Muscular Atrophy-determining gene, survival motor neuron. *Gene* 1997, in press.
40. Viollet L BS, Bueno Brunialti AL, Lefebvre S, Burlet P, Clermont O, Cruaud C, Guenet J, Munnich A, Melki J. cDNA isolation, expression, and chromosomal localization of the mouse survival motor neuron gene (SMN). *Genomics* 1997, **40:** 185–188.
41. Didonato C, Chen X-N, Noya D, Korenberg JR, Nadeau JH, and Simard LR. Cloning, characterization, and copy number of the murine survival motor neuron gene: homolog of the spinal muscular atrophy-determining gene. *Gen Res* 1997, **7:** 339–352.
42. Clem R, Miller LK. Induction and inhibition of apoptosis by insect viruses, in *Apoptosis II: The Molecular Basis of Apoptosis in Disease* (Tomei LD, Lope, FO, eds.), Cold Spring Harbor Laboratory Press, Plainview, NY, 1994, pp. 89–110.
43. Liston P, Roy N, Tamai K, Lefebvre C, Baird S, Cherton-Horvat G, Farahani R, McLean M, Ikeda J, MacKenzie A, Korneluk RG. Suppression of apoptosis in mammalian cells by NAIP and a related family of IAP genes. *Nature* 1996, **379:** 349–353.
44. Liu Q, Dreyfus G. A novel structure containing the survival of motor neurons protein. *EMBO* 1996, **15:** 3555–3565.
45. Talbot K, Ponting CP, Theodosiou AM, NR Rodrigues, Surtees R, Mountford R, Davies KE. Missense mutation clustering in the survival motor neuron gene: a role for the conserved tyrosine and glycine rich region of the protein in RNA metabolism? *Hum Mol Gen* 1997, **6:** 497–500.
46. Messer A, Flaherty LA. Autosomal dominance in a late-onset motor neuron disease in the mouse. *J Neurogenet* 1986, **3:** 345–355.
47. Bronson RT, Lake BD, Cook S, Taylor S, Davisson MT. Motor neuron degeneration of mice is a model of neuronal ceroid lipofuscinosis (Batten's Disease). *Ann Neurol* 1993, **33:** 381–385.
48. Pardo C, Rabin BA, Palmer DN, Price DL. Accumulation of the Adenosine Triphosphate Synthase Subunit C in the mnd Mutant Mouse. *Am J Path* 1994, **144:** 829–835.
49. Schmalbruch H, Skovgaard Jensen HJ, Bjaerg M, Kamienicka Z, Kurland L. A new mouse mutant with progressive motor neuronopathy. *J Neuopathol Expt Med* 1991, **50:** 192–204.

21
Cerebellar Degenerations

Katherine A. Wood

INTRODUCTION

Study of neuronal cell loss now requires an appreciation of the mode of cell death. In neurons, several types of cell death have been described based on the morphology of the dying cell *(1)* such as necrosis, apoptosis, autophagic cell death, and cytoplasmic cell death (*see* chapter by Clarke). Among the previous types of cell death, apoptosis and necrosis appear to be distinct at both the biochemical and molecular level. Necrosis may be considered "accidental death" and usually involves rapid and massive disruption to cellular homeostasis. Once initiated, necrosis is essentially irreversible and treatment for human disorders involving necrosis must either prevent the insult from occurring, or prevent damage due to secondary effects, e.g., inflammation. Apoptosis, on the other hand, may offer far more opportunity for intervention: apoptosis is an active process involving the coordination of gene cascades over a period of time. Studies on the molecular events of apoptosis have identified gene products that are an absolute requirement for execution of the apoptotic program in certain cell types and in response to a specific initiating stimulus. The pathology of many human disorders appears to involve dysregulation of apoptosis which, in the healthy individual, is tightly controlled during development and in adulthood. Inappropriate apoptosis may lead to neurodegenerative disease, whereas the failure to remove unwanted cells by apoptosis can lead to tumor development. In this chapter, the importance of cell death in cerebellar disorders will be considered, with a particular emphasis on the relevance of apoptosis to neurodegenerative cerebellar disorders. Reference is made to some animal and cell culture models that help to improve our current understanding of physiological and pathological cell death in the human cerebellum.

NORMAL CEREBELLAR CYTOARCHITECTURE AND DEVELOPMENT

The Adult Cerebellum

The adult cerebellum consists of three cortical layers: an inner granule cell layer of densely packed interneurons, the majority being granule cells; the Purkinje cell layer consisting of a single row of the bodies of Purkinje cells; the molecular layer which contains a few interneurons but mainly is a zone of intense synaptic activity, consisting of the parallel axons of granule cells and the extensively branched dendritic trees of the Purkinje cells (Fig. 1).

From: Cell Death and Diseases of the Nervous System
Edited by: V. E. Koliatsos and R. R. Ratan. © Humana Press Inc., Totowa, NJ

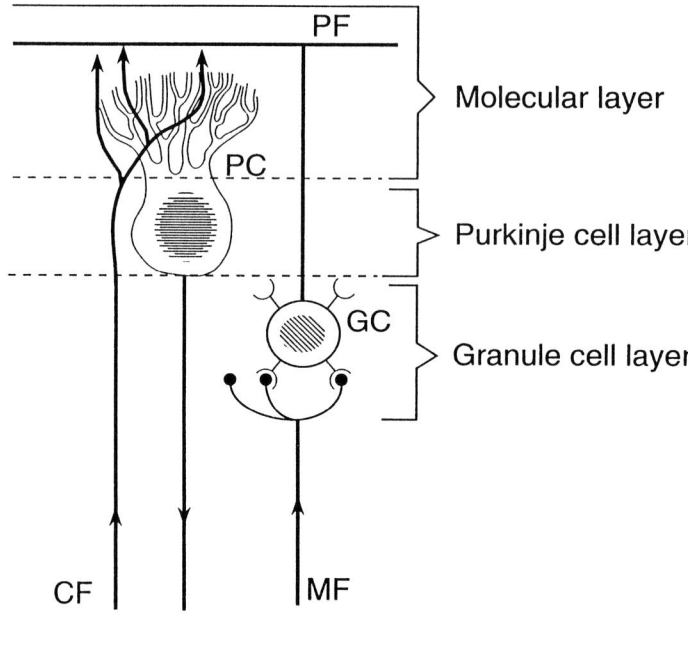

Fig. 1. The three cortical layers of the adult cerebellum.

The five types of cerebellar neurons are arranged in a complex but regular pattern of circuitry. The afferent mossy fibers, originating in all areas/nuclei innervating cerebellum except inferior olive, contact the dendrites of granule cells and the afferent climbing fibers, originating in the inferior olivary complex, traverse the granule cell layer and contact the dendrites of the Purkinje cells. The Purkinje cell to climbing fiber ratio is one to one. The parallel axons of the granule cells interact with the transverse dendrites of Purkinje cells in a way such that each Purkinje cell can receive stimuli from many different granule cells and each granule cell can contact several Purkinje cells. The only axons to leave the cerebellum are the Purkinje cell axons that cross the granule cell layer and terminate mainly in the deep cerebellar nuclei. A remarkably constant relationship exists between the Purkinje cells and the granule cell neurons; for each Purkinje neuron (about 15 million in humans), there are about 450 granule cells. The number of granule cells of Purkinje cells is fixed early in embryogenesis and the final number appears to be adjusted accordingly. Of all brain structures, cerebellum demonstrates the closest relationship between cell number and cell weight among mammals and the ratio is remarkably well conserved. In addition, the ratio of Purkinje cells to granule cells increases in the course of phylogenesis and brain evolution *(2)*.

Human Cerebellar Development

The development of the human cerebellum begins very early after conception, yet is one of the last subdivisions of the brain to achieve adult cytoarchitecture *(3)*. The cerebellum and pons arise from the rostral rhombencephalon. Outgrowth from the paired alar plates on either side of the rhombencephalic midline begins at between 34 and 42 d of gestation and ultimately meet and fuse at the midline beginning at about two months. The vermis develops as the primitive hemispheres meet at the midline and the cerebellar hemispheres develop laterally. The flocculus and superior cerebellar node is apparent by 44 d of gestation, and by 9 wk the flocculonodular fissure that separates the bulk of the cerebellum (cerebellar body) from the flocculonodular lobe is apparent. Interconnections between the flocculonodular elements and the vestibular system develop and mature. By 12 wk of development the primary horizontal fissure that transverses the superior cerebellar structures is apparent and connections between the vermis and superior cerebellar hemispheres with other brain stem structures are established. The primary and secondary fissures develop during the following weeks. Interestingly, the primordial cerebellum is laid down in a series of rhombomeres and each of these may act as a restriction unit preventing ingrowth from adjacent units *(4,5)*.

The neurons of the cerebellum arise from different sources *(6)*. The Purkinje cells and the deep cerebellar neurons arise from the germinal matrix lining the primitive fourth ventricle and radiate laterally. A cell layer present by wk 9 of gestation forms the single cell layer characteristic of Purkinje cells between 16 and 28 wk and the deep cerebellar nuclei begin to develop at about week 8. Up to about 10 wk of gestation, the generation of cerebellar cortical neurons takes place by mitosis of the ventricular epithelium. The granule neurons arise from primitive cells of the pial surface that migrate over the surface of the cerebellum beginning wk 11–13 and are identifiable as an external granule cell layer (EGL) by wk 27. At about wk 30, cells begin to migrate downwards into the cerebellum and this migration continues postnatally. An EGL is evident up to about 7 mo postnatal and there is a fivefold increase in internal granule cell number during the first two postnatal years with the maximum rate of growth between 240 and 650 d after conception.

PHYSIOLOGICAL CELL DEATH

Historically, the development of the cerebellum has been studied using histology, cytology, cell counting *(7,8)* and incorporation of radiolabeled thymidine *(9,10)* to identify proliferative cells. The recent advances in techniques to study cell death has exposed potential flaws in these methods. A mitotic figure is easily confused with an apoptotic figure due to the condensation of chromatin in both processes; nucleotides and their analogs such as bromodeoxyuridine are incorporated into the newly synthesized strands of DNA during mitosis but also during apoptosis when DNA damage and repair enzymes are induced; apoptotic cells are cleared rapidly in vivo, thus the true cell death is easily underestimated or not realized. Therefore, in vivo studies during development should also include biochemical studies of apoptosis in conjunction with the classical approaches used.

Until only recently, it was believed that there was a single phase of cell death that occurred in the developing cerebellum. Recent advances in the detection of apoptosis have revealed that the rodent cerebellum is subject to two significant phases of cell death that occur normally during development *(11)*. A phase of apoptotic cell death begins shortly after birth and occurs mainly in the EGL. Since apoptosis is more prominent in the folds of the folia, apoptotic cell death may also play a role in cerebellar foliation. Interestingly, Bcl-2 protein expression is very low in the outer germinal layer of the cerebellum, a pattern coinciding with high numbers of apoptotic cell death. Bcl-

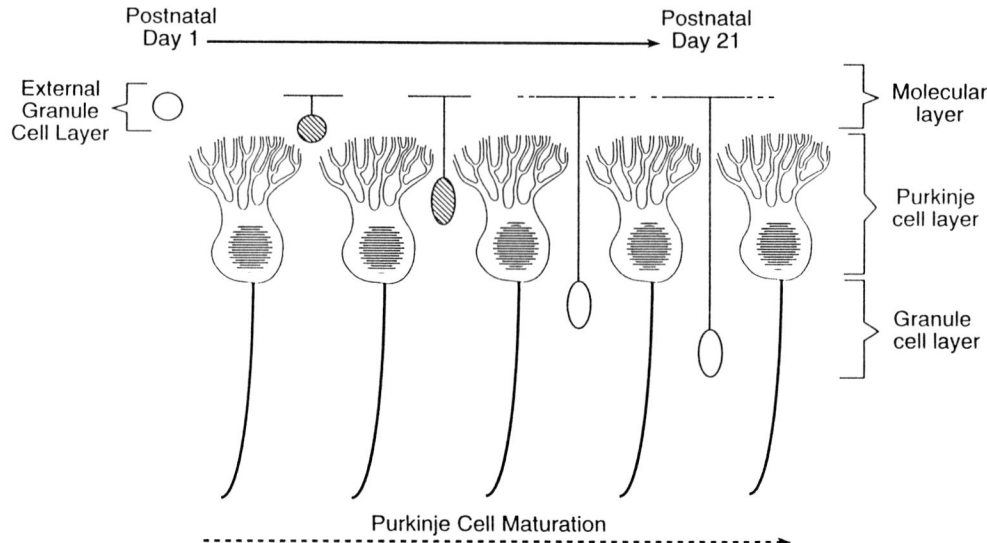

Fig. 2. Cell death in the EGL. The pattern suggests that synaptogenesis, growth factor release and/or growth factor receptor upregulation seem to be regulating the neuronal cell number at this early stage.

2 functions as a death repressor molecule and, during development, Bcl-2 expression is typically confined to zones of cell survival *(12)*. In addition to apoptotic cell death in the EGL, apoptosis is also observed in a few granule cells that have migrated to the internal granule cell layer. These cells rarely incorporate thymidine and are considered postmitotic. The first phase of cell death in the EGL ends by postnatal d 11 in the mouse, at which time levels of Bcl-2 begin to rise, and coincides with the arrival of mossy fibers and the maturation of Purkinje cells; this pattern suggests that synaptogenesis, growth factor release and/or growth factor receptor upregulation seem to be regulating the neuronal cell number at this early stage (Fig. 2).

Proliferating granule cells are particularly sensitive to apoptosis. At this stage of development, toxins and other noxious agents, including radiation, can cause profound disturbances to the cerebellum. The period of cerebellar growth that occurs during the first 10 postnatal days in rodents is equivalent to the third trimester of human development, a time of enhanced neuronal susceptibility to noxious substances.

A second phase of cell death occurs between the postnatal wk 3 and 5 in cells of the internal granule cell layer and this is believed to effect the final ratio of granule cells: Purkinje cells. During this time, Bcl-2 levels in the cerebellar cortex remain elevated *(13)*. This second phase of cell death has the hallmark of excitotoxic cell death in response to excess glutamate and there is no biochemical evidence of apoptosis. In primary culture of rat granule cells, high concentrations of glutamate cause slow neurotoxic cell death and perturbation of intracellular calcium homeostasis is believed to be responsible for this type of cell death *(14)*. The reduction in granule cell number that occurs during the postnatal weeks 3–5 in mice may reflect excitotoxic death of neurons that have excessive synaptic input, since glutamate is the major neurotransmitter. Another example of appropriate target matching includes the elimination, by postnatal d15, of all but one of the climbing fibers that innervate the Purkinje cell shortly after birth. Multiple innervation of the Purkinje cells is retained in the adult when secondary main input from the granule cell is absent, as it is found in some mutant mice or following radiation. It is believed that NMDA receptor activation is critical for the regression of these functional synapses *(15)*.

The two phases of physiological cell death in the developing cerebellum may act as a fine-tuning mechanism to retain those cells that are receiving both the signals for survival, such as synaptic input and growth factors, as well as the appropriate amount of survival factors, i.e., there is a qualitative and quantitative regulation of cerebellar neuron survival. In summary, during the protracted development of the cerebellum, there are two major phases of cell death, occurring by different mechanisms, that contribute to determining the final numbers and the temporal and spatial organization of neuronal populations in the adult.

SYSTEMS FOR STUDYING CEREBELLAR CELL DEATH
Cell Culture Models for Cell Death

Primary culture of cerebellar neuron and cerebellar slice cultures have offered alternative avenues for studying cell death in the cerebellum. It is possible to maintain Purkinje cells in culture for several weeks either in mixed culture or with a feeder layer of astrocytes. Rat Purkinje neurons are typically obtained from rats at embryonic d 17–19 after the cells have undergone a final mitotic division. Cerebellar slice culture can be maintained in vitro for only a short period for electrophysiological analyses.

The most common cell type studied in vitro are the granule cells which are usually established from rat cerebella at postnatal d 9. With the use of mitotic inhibitors, enriched granule cell cultures (greater than 95%) can be maintained for up to 3 wk in vitro. The granule cells develop processes, forming an extensive network between cells. Although the true differentiation state of granule cells in culture is not known, granule cells in primary cultures established from the cerebella of rodents during the first postnatal week are inherently susceptible to apoptosis *(16)*. With increasing time in culture, granule cell loss occurs through apoptotic cell death. Culture media must contain low concentrations of glutamate or depolarizing concentrations of potassium for granule cell survival in the absence of neurotrophic factors *(17,18)*. The mechanism by which high potassium concentration promotes granule neuron survival is unclear; however, there is evidence that depolarization mediates the influx of calcium and raises intracellular calcium levels. The survival-promoting ability of potassium can be blocked with intracellular calcium chelators and death induced by lowering the potassium concentration is preceded by a reduction in intracellular calcium concentration *(19)*.

Granule cell survival under depolarizing conditions can be promoted by activation of muscarinic cholinergic receptors. These receptors are functionally coupled to G proteins and activation of G_s has been shown recently to block apoptosis due to nondepolarizing concentrations of potassium *(20)*. Under physiological concentrations of potassium, granule cells undergo apoptosis, but can be rescued with agents that act as survival factors, including insulin-like growth factor I (IGF I) *(18)*. IGF-I is secreted by Purkinje cells and the receptor is expressed at high levels within the EGL, a pattern that may reflect a mechanism for regulation of target matching. Several other agents have been shown to enhance the survival of granule cells in culture, including brain-derived neurotrophic factor (BDNF), neurotrophin-3 (NT-3), and pigment epithelium-derived factor. The BDNF receptor TrkB is expressed in the cerebellum early in development and BDNF and NT-3 are present in the immature cerebellum. Because triiodothyronine enhances the expression of NT-3, it seems likely that some assets may act indirectly through regulation of the production of other factors in the cerebellum.

Animal Models For Developmental Disorders

Staggerer Mouse

Mice homozygous for the staggerer gene develop progressive ataxia during the first two weeks after birth. The disorder is due to a cell-autonomous defect in the Purkinje cell leading to reduced Purkinje cell number and a subsequent decrease in granule cell proliferation *(21)*. It is not clear whether the alteration in granule cell number is due to reduced cell proliferation and/or increased cell death, although the staggerer cerebellum is smaller than normal even before granule cell genesis has begun. Recently, the staggerer gene was identified as a mutated orphan nuclear hormone receptor, RORα, of the class that can bind to DNA and activate transcription as a monomer in the absence of ligand *(22)*. It has been proposed that the mutant RORα, that lacks the ligand binding domain, fails to interact with the thyroid hormone signaling pathway to induce Purkinje cell differentiation. In man, hypothyroidism is associated with reduced Purkinje cell dendrite arborization and reduced granule cell number. Recently, it has been shown that grafting the cerebella of hypothyroid rats postnatal d 3 with cell lines expressing high levels of NT-3 prevented death of neurons in the EGL and promoted Purkinje cell differentiation *(23)*.

Weaver Mouse

A second murine model is provided by the weaver mouse. Mice homozygous for the weaver gene show extensive neuropathology in the EGL of the cerebellum and in the nigrostriatal dopamine projection system. The adult weaver mouse is depleted of granule cells and Purkinje cells. The cell death within the cerebellum takes place before postnatal d 21 *(24)*. Cultured weaver granule cells survive poorly on laminin substratum and have a poor resting membrane potential compared to normal granule cells. A G-protein activated potassium channel has recently been proposed as the candidate weaver gene *(25)*. The weaver neurons that exhibit the greatest susceptibility are those exiting the cell cycle, i.e., at a time when they are most vulnerable to conflicting growth and apoptotic signals. Apoptosis has been implicated as the mechanism of granule cell death in these mutant mice.

Reeler and Lurcher Mouse

The reeler mouse exhibits an autosomal recessive disorder characterized by malpositioning of Purkinje cells and granule cells. The internal granular layer is significantly thinned and most granule cells are found external to the Purkinje cells, comprising an ectopic EGL. In cell culture systems, no difference has been found in migration or proliferation between normal and reeler granule cells, a pattern suggesting that the reeler defect is a consequence of a disruption early in the differentiation of granule cells that leads to fewer precursor cells *(26)*.

The lurcher mouse is an autosomal semidominant mutation that is lethal in homozygous mice. Mice heterozygous for lurcher show massive Purkinje cell death and a concomitant loss of granule cells. Purkinje cell abnormalities are apparent from postnatal d 8, i.e., at a time preceding cell death. Electron microscopic analysis of the degenerating Purkinje cells revealed chromatin clumping against the inner surface of the nuclear membrane, a morphology consistent with apoptosis (27). More recently, it was demonstrated that Purkinje cells in the lurcher express the sulfated glycoprotein-2 (SPG-2) and the Kv3.3b potassium channel just prior to death (28). SPG-2 mRNA is induced in certain cell types undergoing apoptosis, whereas the Kv3.3b potassium channel is associated with terminal Purkinje cell differentiation. These data implicate a contribution, to neuronal apoptosis, of signals that can be handled appropriately only by terminally differentiated cells.

Of note, positional cloning has been used to identify the gene responsible for Purkinje cell loss and ataxia in the lurcher heterozygote mouse. Mutation in one lurcher allele of the mouse delta2 glutamate teceptor gene (GluR delta2) was idenified. The mutation, involving a G-to-A transition, changes a highly conserved alanine to a threonine residue in transmembrane domain III of GluR delta2, and leads to a very high membrane conductance and depolarized resting potential in affected Purkinje cells. How these electrophysiological changes lead to apoptotic cell death in Purkinje cells of the lurcher heterozygote remains unclear (29).

Purkinje Cell Degeneration Mouse

The Purkinje cell degeneration (PCD) mice are unusual in that there is a rapid degeneration of a restricted neuronal population beginning at a relatively advanced age. Mice homozygous for PCD begin to show degeneration and loss of Purkinje cells at postnatal d 18. Purkinje cell degeneration apparently results from a direct effect of the mutant gene on the Purkinje cells (29). Cell loss also occurs in discrete populations of thalamic neurons beginning at the postnatal wk 8 and these neurons display some of the classic morphological features of apoptosis (30). Other murine models include the wobbler and shaker mice but, to date, the genetic defects in these mutants remain unknown.

HUMAN CEREBELLAR DISEASE

Historically, the classification of human cerebellar disorders has been based upon clinico-pathological findings whereas detailed histological examination of autopsy tissue is less common. Ataxia occurs in many cerebellar disorders and serves as a relatively reliable diagnostic sign for cerebellar dysfunction, although all of the signs and symptoms of cerebellar disease may be due to damage outside the cerebellum, such as the brain stem or spinal cord. Cerebellar ataxia is a disruption of coordinated muscle activity whereas strength and sensation are essentially unaffected. Other signs of cerebellar disease are coarse intention tremor, dysmetria, impairment in rate and rhythm of movements, hypotonia, and nystagmus.

The protracted nature of human cerebellar development allows for disruption of cerebellar formation from very early in gestation up to 10 yr of age. Some defects that are a consequence of disrupted embryogenesis are not evident until later in life. The complex local circuitry, the regulation of neuron-to-target matching, and interdependence among cerebellar neurons for survival makes cerebellar neurons vulnerable to cell death as a consequence of a primary defect elsewhere in the system. Defects in the cerebellum associated with reduced cell number may arise from disruption in neurogenesis, migration or synaptogenesis during development and are distinct from cerebellar neuron death due to toxic substances. Developmental cerebellar defects may also be a consequence of environmental agents when antenatal or perinatal exposure targets a susceptible population; examples include alcohol and radiation. Similarly, primary metabolic disorders causing dysfunction in organs outside the central nervous system may expose the developing cerebellum to toxic substances.

A second group of disorders involving cell loss affects the cerebellum once the adult cytoarchitecture and cell–cell interactions have been established. Environmental toxins such as alcohol, metabolic disease, and paraneoplastic disease fall into this category. The pattern of susceptibility of cerebellar neurons to damage is often age dependent and the susceptible population may be quite different in adulthood compared to development. Neuronal loss in the cerebellum also occurs during aging and this regressive event is accompanied by a remodeling of surrounding surviving structures *(31)*.

Paraneoplastic Disease

Paraneoplastic neurological syndromes consist of a variety of clinical neurological disorders that include paraneoplastic cerebellar degeneration (PNCD). This type of degeneration is a rare complication of malignancies, most frequently seen in ovarian and breast cancer, small cell lung cancer and Hodgkin's disease. Paraneoplastic disease is one of the most devastating disorders in clinical neurology and can antedate discovery of the neoplasm for up to several years. PNCD is clinically characterized by the subacute development of pancerebellar dysfunction *(32)* and the most striking neuropathological finding is severe diffuse loss of Purkinje cells throughout the cerebellar cortex, often with a mild to moderate loss of neurons in the granular layer, peridentatal demyelination and neuronal loss with gliosis in the deep cerebellar nuclei.

Speculation on the etiology of the disease includes viral infection and neurotoxins secreted by tumor cells. However, unfortunately, the most popular theory for paraneoplastic disorders is the development of autoimmunity against a Purkinje cell-specific antigen. Purkinje cells can bind IgG nonspecifically, a problem which has hampered identification of antineuronal antibodies specific for Purkinje cells.

In 1965, a specific antineuronal antibody was detected in the serum of patients with small cell lung cancer and sensory neuropathy and in 1979 the first antibody against the cytoplasm of Purkinje cells was found in a patient with cerebellar degeneration and Hodgkin's disease. An antibody discovered in 1983, now termed anti-Yo, binds to ribosomes, endoplasmic reticulum and Golgi of Purkinje cells. Several antigens that react with anti-Yo have now been cloned and these antigens are uniquely expressed in Purkinje cells and some neuroectodermal lineages. The antigens have a leucine-zipper motif and contain repeats of a hexapeptide. These antigens are only expressed in tumors of patients with anti-Yo positive PNCD (accounting for about 40% of the PNCD patients). The major epitope of anti-Yo has been localized at the site of the leucine-zipper motif. Proteins with a leucine-zipper motif are able to form "coiled coil" homodimers or heterodimers and several such proteins function to bind DNA and regulate transcription. The antibodies found in PNCD patients may therefore be DNA-binding proteins and may have the capacity to interfere in gene transcription after entry into the cytosol *(33)*.

Purkinje cells have the ability to take up macromolecules from the CSF and transport them through the dendrites to the perikaryon. Injection of eosinophil-derived neurotoxin into the CSF of guinea pigs causes specific ablation of Purkinje cells, as does delivery of other toxic ribonucleases. In addition, the delivery of nonspecific immunotoxins causes massive cell death within the Purkinje cell population. The morphology of these dying Purkinje cells is reminiscent of apoptosis (R. J. Youle, personal communication). In PNCD the anti-Yo antibodies are against intracellular targets; therefore, the antibody may be nonspecifically taken up from the CSF and delivered, perhaps through receptor-mediated endocytosis, to the cytosol of Purkinje cells.

PNCD is most commonly associated with cancers of the breast, ovary and female genital tract and the occurrence of polyclonal IgG anti-Purkinje cell antibodies are almost entirely confined to these patients. Some of the antigens identified by these antibodies have been cloned and sequenced *(34)* but their specific normal cellular function remains unknown. In experimental models of PNCD, direct injection of purified antibodies from patients with PNCD into the lateral ventricles of rats has no neuropathological effect. The presence of distinct antineuronal antibodies in PNCD patients, the restricted distribution of autoantibody reactivity and the absence of autoantibodies in some patients suggests that PNCD is the final common clinical expression of several pathophysiological disorders. It remains a real possibility that apoptosis plays an important role in the pathophysiology of this disorder.

Medications, Metals, Solvents, and Toxins

There are many agents that have been reported to cause ablation of specific cell populations within the cerebellum. The differentiation state of cerebellar neurons dictates both the sensitivity to toxic insults and the type of cell death (i.e., occurrence of apoptosis). In some instances, it has been shown that inhibition of protein synthesis can limit the damage due to certain insults, a pattern consistent with apoptosis as the mechanism of cell death.

Many drugs, solvents and metals that reach the brain can cause devastating neurologic damage. Although the cerebellum may be affected, symptoms may be masked due to damage to the cerebral cortex. Purkinje cells are commonly damaged after exposure to neurotoxins and, in adult life, they may be among the most sensitive to toxins brain cells. A wide range of environmental toxins are known to cause ataxia associated with altered Purkinje cell morphology or death. Examples include toluene, carbon monoxide, DDT, and bismuth. Detailed histology and biochemical data of the degenerating cells is not available to date; however, there are descriptions of pyknotic Purkinje cells and Purkinje cells with condensed chromatin, i.e., morphologies suggestive of apoptosis.

Phenytoin

Phenytoin, used to treat epilepsy, can cause acute and chronic damage to the central nervous system. Phenytoin-induced cerebellar atrophy is associated with a diffuse loss of Purkinje cells and a reduction in granule cell number. Phenytoin has recently been demonstrated to cause apoptotic cell death of granule cells in culture *(35)*.

Methylazoxymethanol

The cycads are a group of plants containing toxins called cyosins that are carcinogenic and cause cerebellar ataxia. Cycasin (methylazoxymethanol-β-glycoside) intoxication causes neuropathophysiological changes similar to those seen in the weaver mouse. Cycasin is cleaved at a beta-glucosidase linkage to yield the cycasin aglycone, methylazoxymethanol (MAM). Cycasin is thus toxic when administered orally, whereas MAM is toxic when delivered parenterally. MAM is relatively unstable and studies on the toxic effects of the cycasins has been undertaken using the water-soluble, stable MAM acetate, which can be administered orally.

MAM causes methylation of purine bases in both DNA and RNA and is thus a potent carcinogen. Direct DNA damage is one of the classic inducers of apoptosis in certain cell types (see chapter by Fuller and Bohr). It is postulated that, after DNA damage reaches a certain level, effector molecules such as p53 halt cell cycle progression; subsequently, if the damage is significant, repair is initiated or apoptosis is induced. Apoptosis thus provides a molecular mechanism for ridding multicellular organisms of cells that have sustained damage and may ultimately become neoplastic. In the cerebellum, proliferative cells in the germinal layers are susceptible to apoptosis and MAM is well known to cause significant granule cell death when administered to rodents shortly after birth.

Interestingly, it has recently been shown that MAM induces granule cell loss by killing an early precursor population through a p53-independent and apoptosis-independent pathway. In contrast, radiation induces direct granule cell death by a p53-dependent and apoptosis-dependent pathway. These results suggest that DNA damaging agents can induce granule cell loss by p53-dependent apoptosis or by p53-independent mechanisms.

Ibogaine

Ibogaine is an indole alkaloid isolated from the African shrub Tabernanthe iboga *(36)*. It has been used in Africa as a stimulant, appetite suppressant and is claimed to have psychotomimetic properties. Ibogaine has been reported to have considerable efficacy in reducing narcotic and cocaine dependence. More detailed studies have revealed that the drug causes degeneration of Purkinje cells and activates astrocytes and microglia in longitudinal bands corresponding to terminal domains of climbing fibers *(37)*. (For a more topical discussion of the subject of ibogaine-induced cerebellar degeneration, see chapter by O'Hearn and Molliver.)

Antineoplastic Treatments

Several antineoplastic agents have documented cerebellar toxicity. Fluorouracil is a halogenated pyrimidine that blocks DNA synthesis and can cause cerebellar ataxia with symptoms appearing 2 wk after the beginning of therapy. The neuropathology associated with this syndrome has not been characterized yet.

Cytosine arabinoside (ARA-C) is a DNA analog and interferes with DNA synthesis. Loss of Purkinje cells has been demonstrated in humans following ARA-C treatment *(38)*. In suckling mice, ARA-C has been shown to cause extensive death of undifferentiated cells in the EGL. In animals allowed to survive, there was partial regeneration of the germinal granule layer, heterotopic granule cells in the molecular layer, and irregularly arranged Purkinje cells; the morphology of the degenerating granule cells is reminiscent of apoptosis *(39)*.

The consequences of radiation delivered to the craniospinal region to treat CNS tumors has been examined in detail *(40)*. Cerebellar granule cells in the EGL are particularly sensitive to radiation *(41)*. In rats and mice during postnatal d 1–3, exposure to low doses of radiation cause a dose-dependent depletion of granule cells. However, the surviving cells have the proliferative capacity to reconstitute the entire EGL and at one mo after radiation there is no gross difference between the irradiated and age-matched controls *(42)*. When radiation is administered at later time points (postnatal d 7–10), the EGL is regenerated but an ectopic layer of granule cells is retained in the adult and there are reports of alterations in Purkinje cell dendrites. The radiation-induced death of cerebellar granule cells is an apoptotic and p53-dependent mechanism, unlike the physiological apoptosis that occurs during development and in response to MAM and methyl-mercury *(11)*.

Heavy Metals

Several heavy metals are known to be neurotoxic to the cerebellum including lead, mercury, thallium, manganese, and bismuth. Lead poisoning causes encephalopathy and the cerebellum is most severely affected. In adult humans, there is loss of Purkinje cells and a mild gliosis is reported. In children, even low-level exposure causes developmental delay. Although vascular changes are implicated, synaptogenesis is also affected. One theory for the molecular mechanism of delayed synaptogenesis involves the cell surface glycoprotein NCAM. NCAM becomes desialylated at the onset of synapse formation and lead exposure blocks its conversion from the sialylated form. The mechanism of cell loss due to lead toxicity has not been investigated.

Industrial mercury poisoning can result in tremors, but primarily causes renal toxicity due to the ready uptake of inorganic mercury by the kidneys. Methyl-mercury, which is not cleared by the kidneys and therefore shows increased uptake by the brain, is found in fish harvested from waters high in industrial effluents and in fungicide-coated seed grains inadvertently used in flour production. Methyl-mercury specifically targets the cerebellar granule cells die in large numbers through a nonapoptotic mechanism; there is also some reduction in Purkinje cell number. *In utero* exposure to mercury leads to severe neurodegeneration and disrupted migration of neuronal populations. The biochemical mechanism for mercury toxicity is not clear. Mercury binds to sulfhydryl moieties and to amino acids but the granule cell-specific toxicity may be due to oxidative or free radical-generated damage *(43)*.

Alcohol

The effect of alcohol on the cerebellum is age dependent. In neonatal rodents, a reduction in Purkinje or granule cell number has been reported *(44,45)*. Alcohol exposure in adult rodents causes morphological and functional changes in the cerebellum. In humans, a generalized cerebellar atrophy is seen with Purkinje cell loss. It is not yet clear if granule cell loss is secondary to Purkinje cell dysfunction or if alcohol has a direct toxic effect on the nerve cells. In general, the effects of alcohol consumption on cerebellum could be due to ethanol toxicity, acetaldehyde toxicity, nutritional deficiencies (as, e.g., in Wernicke's encephalopathy), or an unidentified toxin. (One putative mechanism of alcohol-induced cell loss involves NMDA receptors.) Indeed, NMDA receptor activation can prevent alcohol-induced death of cerebellar granule cells in culture. Moreover, NMDA is neurotrophic for granule cells in culture and alcohol inhibits NMDA receptor function. Other neurotrophic agents can also protect against alcohol-induced cell death, including basic fibroblast growth factor. Overall, the mechanism of alcohol-induced cerebellar degeneration and the means to prevent/treat it remain unclear.

Degenerative Disorders

Clarification of this group of overlapping, autosomal dominant hereditary ataxias will likely be based in a precise genetic and molecular characterization. Currently, classification of the degenerative disorders of the cerebellum is problematic due to the many overlapping clinical signs and the variability in the age of onset, severity and progression. Many of the degenerative disorders are evident early in life and may be considered developmental disorders. Certain degenerative diseases exhibit a later time of onset.

Spinocerebellar Ataxia Type 1

A pathogenic model based on findings in spinocerebellar ataxia type 1 (SCA1) and other neurodegerative disorders has been proposed *(47)*. SCA1 is characterized by ataxia, dysarthria, ophthalmoparesis, and varying degrees of neuropathy. Symptoms typically appear during the third or fourth decade; however, in subsequent generations, the severity often increases associated with an earlier age of onset, a phenomenon termed anticipation (*see* also discuaaion of this phenomenon in chapter by Ross, Becker, and Koliatsos). The SCA1 gene was identified by linkage analysis. The disease-causing mutation was identified as an expansion of a CAG trinucleotide repeat that lies within the coding region of a novel protein termed ataxin-1. Normal alleles contain 6–39 repeats with 1–3 CAT interrupts, whereas SCA1 alleles contain 40 to 81 perfect CAG repeats.

In addition to SCA1 there are four other known neurodegenerative disorders involving expansion of CAG repeats that encode for polyglutamine tracts within proteins. These are spinobulbar muscular atrophy (SBMA), dentatorubropallidolysian atrophy, Machado–Joseph disease, and Huntington disease. The current hypothesis is that these CAG expansions result in a gain of function. The mutated protein may interact with new target molecules or become transactivators. Expansion of CAG repeats may be the mutational mechanism for many of the neurodegenerative diseases, most particularly the dominantly inherited ataxias. For an in-depth discussion of these diseases, *see* chapter by Ross, Becker, and Koliatsos.

Ataxia Telangiectasia

Ataxia telangectasia (AT) is an autosomal recessive disease with a pleiotropic phenotype, characterized by cerebellar degeneration, chromosome instability, radiation sensitivity, immunodeficiency, and predisposition to cancer. Recently, a gene that is mutated in AT was identified (ATM) *(59)*. The ATM gene has similarity to several yeast and mammalian phosphatidylinositol-3' (PI-3) kinases which are involved in cell cycle control, mitogenic signal transduction and meiotic recombination. Although the complete sequence of the gene is not yet known, the sequence similarities of ATM suggest involvement of the gene product in cell cycle control, signal transduction, and cellular response to DNA damage. PI-3 kinases mediate cellular responses to mitogenic growth factors, to differentiation signals and to insulin. Of particular interest is that PI-3 kinase is required for the survival-promoting effects of nerve growth factor in rat pheochromocytoma cells. The loss of function of a PI-3-like activity in neurons may therefore lead to unscheduled apoptosis in the susceptible neuron population in AT patients. Cells from AT patients are more sensitive to apoptosis induced by agents that damage DNA. The primary neuropathology of AT is Purkinje cell loss. It is not entirely clear when the neuropathology is first evident, but there is indication that abnormalities in Purkinje cells are evident at a very young age. PI-3 kinase plays an important role in differentiation signals and in regulation of DNA damage and repair; therefore, the Purkinje cell loss associated with AT may be a consequence of apoptosis, perhaps induced by cumulative DNA damage or by conflicting signals for differentiation and cell cycling in the postmitotic neuron. Additional insight into the function of the ATM gene products must await complete cloning and sequencing.

CONCLUSIONS

During normal cerebellar development, cell death plays an important role in determining the size, structure, and relationships among the different neuronal populations. Apoptosis has been implicated as a major mechanism of cell loss during development of the nervous system. In the cerebellum, early postnatal, p53-independent apoptotic cell death is responsible, in part, for regulating granule cell number and perhaps foliation. Other types of cell death, such as excitotoxic cell death, also contribute to the regulation of the phenotype of the final complement of cerebellar neurons. There is now convincing evidence from animal model systems that dysregulation of the mechanisms regulating cell death can lead to developmental abnormalities.

In the development of therapies for disorders involving the cerebellum, a consideration of the mechanism of cell death is important. Apoptosis mediates selective loss of various cell types during normal development, but has also been postulated to mediate the pathological death of cells induced by various toxic insults or in the course of autoimmune disease and in neurodegenerative diseases. Apoptosis can be blocked, or at least delayed, under certain circumstances and thus offers the opportunity for intervention. For example, following ingestion of poisons, treatment to inhibit apoptosis in the short term may allow the body to metabolize and thus clear the drug from the system while preventing neuronal loss. Provided that the cell is not committed to undergo apoptosis, but has been held in the initiation stage of apoptosis, removing the initiating stimulus may allow full recovery of the affected population of neurons.

REFERENCES

1. Tomei LD, Cope FO. in *Apoptosis: The Molecular Basis of Cell Death* (Inglis J, Witkowski JA, eds.), Cold Spring Harbor Laboratory Press, New York, 1991.
2. Lange W. *Cell Tiss Res* 1975, **157**: 115–124.
3. Liechtenberg R. in *Handbook of Cerebellar Diseases*, vol. 16 (Koller WC, ed.), Marcel Dekker, New York, 1993.
4. Kandel ER, Schwartz JH. Principles of Neural Science. Elsevier Science, New York, 1985
5. Ramon Y, Cajal, S. Histology of the Nervous System, vol II. Oxford University Press, New York, 1995.
6. Barr ML, Kiernan JA. The Human Nervous System. An Anatomical Viewpoint, J. B. Lippencott, Philadelphia, PA, 1993.
7. Altman J. *J Comp Neurol* 1972, **145**: 353–398.
8. Altman J. *J Comp Neur* 1972, **145**: 399–464.
9. Fujita S, Shimada M, Nakamura T. *J Comp Neurol* 1966, **128**: 191–208.
10. Miale IL, Sidman RL. *Exp Neurol* 1961, **4**: 277–296.
11. Wood KA, Dipasquale B, Youle RJ. *Neuron* 1993, **11**: 621–632.
12. Novack DV, Korsmeyer, SJ. *Am J Pathol* 1994, **145**: 61–73.
13. Merry DE, Veis DJ, Hickey WF, Korsmeyer SJ. *Develop* 1994, **120**: 301–311.

14. Dessi F, Charriaut-Marlangue C, Khrestchatisky M, Ben-Ari Y. *J Neurochem* 1993, **60:** 1953–1955.
15. Rabacchi S, Bailly Y, Delhaye-Bouchaud N, Mariani J. *Science* 1992, **256:** 1823–1825.
16. Dipasquale B, Marini AM, Youle RJ. *Biochem Biophys Res Commun* 1991, **181:** 1442–1448.
17. Yan G-M, Weller M, Wood KA, Paul SM. *Brain Res* 1994, **656:** 43–51.
18. D'Mello SR, Galli C, Ciotti T, Calissano P. *Proc Natl Acad Sci USA* 1993, **90:** 10,989–10,993.
19. Collins F, Schmidt MF, Guthrie PB, Kater SB. *J Neurosci* 1991, **11:** 2582–2587.
20. Yan G-M, Lin S-Z, Irwin RP, Paul SM. *J Neurochem* 1995, **65:** 2425–2431.
21. Landis DMD, Sidman RL. *J Comp Neur* 1978, **179:** 831–864.
22. Hamilton BA, Frankel WN, Kerrebrock AW, Hawkins TL, FitzHugh W, Kusumi K, Russell LB, Mueller KL, Berkel VV, Birren BW, Kruglyak L, Lander ES. *Nature* 1996, **379:** 736–739.
23. Neveu I, Arenas E. *J Cell Biol* 1996, **133:** 631–646.
24. Roffler-Tarlov S, Martin B, Graybiel AM, Kauer JS. *J Neurosci* 1996, **16:** 1819–1826.
25. Patil N, Cox DR, Bhat D, Faham M, Myers RM, Peterson AS. *Nature Genet* 1995, **11:** 126–129.
26. Nagata I, Terashima T. *Int J Devel Neuroscience* 1994, **12:** 387–395.
27. Dumesnil-Bousez N, Sotelo C. *J Neurocytol* 1992, **21:** 506–529.
28. Norman DJ, Feng L, Cheng SS, Gubbay J, Chan E, Heintz N. *Develop* 1995, **121:** 1183–1193.
29. Zuo J, DeJager PL, Takahashi KI, Jiang W, Linden DW, Heinz N. *Nature* 1977, **388:** 769–773.
30. Mullen RJ. *Nature* 1977, **270:** 245–247.
31. O'Gorman S, Sidman RL. *J Comp Neurol* 1985, **234:** 298–316.
32. Strata P, Rossi F. *Neurochem Int* 1994, **25:** 85–91.
33. Dropcho EJ. *Sem Neurol* 1994, **14:** 179–187.
34. Moll JWB, Vecht CJ. *Clin Neurol Neurosurg* 1995, **97:** 71–81.
35. Fathallah-Shaykh H, Wolf S, Wong E, Posner JB, Furneaux H. *Proc Natl Acad Sci USA* 1991, **88:** 3451–3454.
36. Yan GM, Irwin RP, Lin SZ, Weller M, Wood KA, Paul SM. *J Pharmacol Exp Ther* 1995, **274:** 983–996.
37. O'Hearn E, Long DB, Molliver ME. *NeuroReport* 1993, **4:** 299–302.
38. O'Hearn E, Molliver ME. *Neurosci* 1993, **55:** 303–310.
39. Winkelman MD, Hines JD. *Ann Neurol* 1983, 520–527.
40. Shimada M, Wakaizumi S, Kasubuchi Y, Kusonoki T. *Arch Neurol* 1975, **32:** 555–559.
41. Bloom HJG, Wallace ENK, Henk JM. *Am J Roentgenol* 1969, **105:** 43–62.
42. Ferrer I, Serrano T, Rivera R, Olive M, Zujar MJ, Graus F. *Acta Neuropathol* 1993, **86:** 491–500.
43. Altman J, Anderson WJ, Wright KA. *Anat Rec* 1969, **163:** 453–472.
44. Sager PR, Doherty RA, Rodier PM. *Exp Neurol* 1982, **77:** 179–193.
45. Bauer-Moffett C, Altman J. *Brain Res* 1977, **119:** 249–268.
46. Bauer-Moffett C, Altman J. *Exp Neurol* 1975, **48:** 378–382.
47. Pantazis NJ, Dohrman DP, Luo J, Thomas JD, Goodlett CR, West JR. *Alcohol Clin Exp Res* 1995, **19:** 846–853.
48. Zoghbi HY. *Clin Neurosci* 1995, **3:** 5–11.
49. Savitsky K, Bar-Shira A, Gilad S, Rotman G, Ziv Y, Vanagaite L, Tagle DA, Smith S, Uziel T, Sfez S, Ashkenazi M, Pecker I, Frydman M, Harnik R, Patanjali SR, Simmons A, Clines GA, Sartiel A, Gatti RA, Chessa L, Sanal O, Lavin MF, Jaspers NGJ, Taylor AMR, Arlett CF, Miki T, Weissman SM, Lovett M, Collins FS, Shiloh Y. *Science* 1995, **268:** 1749–1753.

22
Parkinson's Disease

Robert E. Burke

INTRODUCTION

Parkinson's disease (PD) is a progressive, degenerative neurologic disease which usually occurs in late mid-life and presents clinically with motor impairment. There are four principal motor signs on which the diagnosis is based: tremor at rest, rigidity, bradykinesia (or slowing of movement) and postural instability. In recent years, it has become apparent that a decline in cognitive function and depression occur as well. PD is a major public health problem. Its prevalence has been estimated to be 200–300 affected individuals per 100,000 population, an estimate which would suggest that approximately 500,000 individuals in the United States are affected. The prevalence of the disorder increases with age, with estimates of up to 800 affected per 100,000 in the 65 yr and older group. Therefore, it is anticipated that the magnitude of PD as a public health problem will increase many fold as we enter the next century, when the proportion of Americans older than 65 is expected to double.

PD has been a model among neurologic diseases for the development of rational approaches to symptomatic treatment based on knowledge of the morphologic and biochemical pathology (for a definition of rational or "biological" therapeutics, *see* chapter by Koliatsos and Morchetti). The first successful treatment of PD with levodopa by Cotzias in 1967, and the subsequent successful therapies with direct-acting dopamine agonists, were solidly based on such knowledge. It had been known for many years that the disorder is characterized by loss of neurons in the substantia nigra pars compacta (SNpc). In the late 1950s, Carlsson showed that levodopa was capable of reversing reserpine-induced bradykinesia in animals, and that dopamine was measurable in brain. In 1960, Ehringer and Hornykiewicz showed that dopamine is deficient in the brains of PD patients. With the development, at that time, of histochemical fluorescence techniques to visualize central monoaminergic pathways, it became clear that loss of SNpc neurons could account for depletion of striatal dopamine in PD patients, thus providing a clear rationale for attempts to replace brain dopamine. Whereas the pathology of PD includes not only the SNpc but also the locus ceruleus, the nucleus basalis of Meynert and many other structures, it is now believed that loss of striatal dopamine secondary to the degeneration of SNpc neurons can alone account for all of the major motor manifestations. This concept has been supported by observation that the neurotoxin N-methyl-4-phenyl-1,2,3,6-tetrahydropyridine (MPTP) is selective for dopamine neurons of the SNpc in humans and other primates, and yet can produce the full spectrum of motor signs seen in PD *(1)*.

In spite of the success afforded by levodopa and dopamine agonist therapies, it is clear that these symptomatic treatments do not offer adequate control. After several years of exposure to levodopa, the large majority of patients suffer loss of response and therapy-related complications, including dyskinesias and chaotic fluctuations in response. Efforts to slow the progression of disease by use of the irreversible MAO B inhibitor deprenyl initially appeared to offer promise by delaying the need to start treatment with levodopa. However, more recent data suggests that the ability of deprenyl to delay the need for levodopa may have been mediated by a subtle ability to relieve Parkinson symptoms. While important new therapies for the control of parkinsonian symptoms continue to be developed, including surgical and transplantation approaches, it is clear that our efforts to prevent the progression of the disease must be guided by a better understanding of its fundamental pathogenesis.

PATHOGENESIS OF PD: GENERAL CONSIDERATIONS

Most recent analyses of the pathogenesis of PD have given broad consideration to the possible roles played by aging, environmental factors and genetic factors. The possible role of aging is suggested by the usual occurrence of PD in late middle age and by marked increases in its prevalence at older ages *(2)*. The possible contribution of age to the expression of the disease is supported by early studies showing a loss of striatal dopamine with age *(3)* and a loss of cells in the SN *(4)*. While the gradual loss of striatal dopaminergic markers *(5,6)* and SN neurons *(7)* with age has been confirmed in more recent studies, the pattern and timing of these losses differs from that which occurs in PD, indicating that aging is not likely to play a direct role in the degenerative process. For example, a marker for dopaminergic terminals, the binding of α-dihydrotetrabenazine to the vesicular monoamine transporter, decreases with age, but with a different time course than occurs in PD, in which loss occurs exponentially *(6)*. The loss of SN neurons in aging is linear and predominates in the dorsal tier, whereas in PD it is exponential and predominates in the lateral ventral tier *(7)*. In addition to these observations, it has been shown that the SN in PD contains numerous reactive microglia, which are much less frequent in age-matched control brains, indicating an active destructive process which is not present in the normal aged brain *(8,9)*. Thus, the precise relationship of aging to the pathogenesis of PD is unclear.

Consideration of a role for environmental factors in the cause of PD was given major impetus with the discovery, in 1983, that exposure to MPTP is capable of inducing Parkinsonism in humans *(1)*. The role of environmental factors was given additional weight by initial results of twin studies, as discussed below, which appeared to exclude any important role for genetic factors. The possible role of environmental factors has been addressed by a number of epidemiologic studies which have been well reviewed by others *(10,11)*. Many of these studies have shown associations between rural residence, well-water drinking or herbicide/pesticide exposure and the risk of developing PD *(10)*. However, the precise role played by any specific compounds has remained elusive.

The possible contribution of genetic factors to the pathogenesis of PD has been extensively reviewed by Duvoisin *(12,13)*. A study conducted by Ward and others in

1983 evaluated 65 twin pairs, 43 of which were monozygotic and included a proband with clinically typical PD, and found concordance in only one twin pair. The study concluded that genetic factors were unlikely to play a significant role. However, the data was subsequently reassessed, and it was concluded that the twin studies were not incompatible with a substantial genetic component *(14)*. The original twin study did not adequately address the problem of preclinical PD. More recent follow-up has shown that some originally unaffected members of twin pairs have gone on to develop PD *(12)*. In addition, the possible occurrence of preclinical parkinsonism has been addressed by positron emission tomographic study of striatal [^{18}F]-fluro-DOPA uptake. One study has shown a concordance of 45% for dopaminergic dysfunction in monozygotic twin pairs *(15)*. The possible contribution of genetic factors has also received support in more recent years by the description of a number of large pedigrees including histologically proven cases of PD *(16)*. Thus, it is now believed that genetic factors are likely to play an important role. Nevertheless, linkage studies in familial PD, and assessments of candidate genes have been unrevealing thus far *(13)*.

PATHOGENESIS OF PD: THE FREE RADICAL AND EXCITOTOXICITY HYPOTHESES

The concept that free radical-mediated injury may underlie the neuronal degeneration which occurs in PD has been, and continues to be, the leading hypothesis for its pathogenesis. The free radical theory has been the subject of many excellent recent reviews *(17,18)*, so it will be outlined here only briefly. This theory is also referred to as the "oxidant stress" hypothesis or the "endogenous toxin" hypothesis. In their recent review, Fahn and Cohen point out that the free radical hypothesis is appealing because four aspects of the neurochemistry of dopamine neurons and their local environment within the SN make the concept plausible. First, a major degradative pathway for dopamine is its oxidative deamination by monoamine oxidases A and B. This process results in the enzymatic production of H_2O_2, which, while itself not a free radical, can nevertheless react nonenzymatically with ferrous or cupric ions via Fenton-type reactions to form highly reactive hydroxyl radicals. Second, dopamine can react nonenzymatically with oxygen to form quinones and semiquinones, with the production of superoxide, hydrogen peroxide and hydroxyl radicals. Third, the SN, particularly the SN pars reticulata, is rich in iron which, as mentioned above, may in its ferrous state catalyze the formation of hydroxyl radicals from H_2O_2. A fourth neurochemical feature of the SN which may contribute to free radical-related mechanisms is the presence of neuromelanin, which is formed from the auto-oxidation of dopamine. This auto-oxidation generates toxic quinones and reactive oxygen species. In addition, the presence of neuromelanin in the cell may alter the ability of metal ions to participate in the production of reactive oxygen species *(19)*.

The possibility that dopamine neurons may undergo free radical-mediated injury in PD has received support from both the 6-hydroxydopamine and MPTP animal models of dopamine neuron injury. The 6-hydroxydopamine reacts with oxygen to produce superoxide anion radical, hydrogen peroxide and hydroxyl radical *(20,21)*. MPTP acts via its active product MPP$^+$, which is selectively taken up into dopamine neurons via the dopamine transporter, and inhibits complex I activity in mitochondria. Inhibition of complex I not only interferes with ATP synthesis, but also results in augmented production of superoxide anion radical. The possible role of superoxide radical in MPTP toxicity has received direct support by the demonstration by Przedborski et al. that transgenic mice with high Cu/Zn superoxide dismutase activity are resistant to MPTP *(22)*.

The free radical hypothesis of PD has also received support from studies of human postmortem brain. Free radicals can cause injury to cells by damaging DNA, proteins and lipids of the cell membrane (*see* chapters by Dykens and Flanigan and Ratan). There is evidence from postmortem studies for free radical-induced modification of each of these classes of molecules. Dexter and coworkers *(23)* have shown that in PD there is a reduction in levels of polyunsaturated fatty acids in the brain (which provide an index of the amount of substrate available for lipid peroxidation), and an increase in levels of malondialdehyde, an intermediate in the lipid peroxidation process. The increase in malondialdehyde was regionally specific for the SN. These workers have more recently confirmed evidence for abnormal lipid peroxidation in PD by identifying a 10-fold increase in cholesterol lipid hydroperoxide, an early marker in the lipid peroxidation process *(24)*. Free radicals are also capable of directly damaging DNA. Sanchez-Ramos et al. have shown that regional concentrations of 8-hydroxy-deoxyguanosine, as an index of oxyradical-mediated DNA damage, are increased in the caudate and SN of PD patients *(25)*. Relatively less attention has been given to the possibilities of oxygen-mediated damage to proteins, or of advanced glycosylation changes to proteins in PD *(26)*. The possibility that such protein changes may also occur in the brain in PD is supported by the recent demonstration that protein adducts of 4-hydroxy-2-nonenal, a cytotoxic product of lipid peroxidation, can be identified by immunohistochemistry in the SN of many patients with PD *(27)*.

Postmortem studies have also revealed neurochemical features that may predispose the PD brain to oxidative damage. Reduced glutathione is an important endogenous antioxidant, and it has been reported as reduced in the SN in PD *(28)*. More recently, Jenner et al. have confirmed low levels of reduced glutathione in the SN of PD patients, and have shown that the alteration is disease-specific *(29)*. Interestingly, they have also shown that reductions are observed in patients with incidental Lewy body disease, which may be a preclinical form of PD *(30)*. This finding would suggest that the reduced levels of glutathione may be a fundamental and primary abnormality in PD, rather than a secondary change.

A number of postmortem studies have also suggested that abnormalities of iron metabolism may underlie the neurodegeneration of PD. Iron metabolism is of particular interest in relation to the free radical hypothesis of PD because it normally is found in high concentrations in SN, and is capable of catalyzing free radical formation. Dexter et al. reported increased levels of iron in the SNpc of PD patients *(31)*. This observation took on a potentially greater significance when this group subsequently reported decreased levels of ferritin in PD brains *(32)*, as ferritin normally sequesters iron in an unreactive state. However, more recently, it has become apparent that increased iron levels may be observed in many neurodegenerative diseases of the basal ganglia *(33)*, so the specificity of changes in iron levels in PD is less clear. Nevertheless, the possible relationship of altered iron metabolism to the pathogenesis of PD remains of interest, based on the recent finding of a higher density of lactoferrin receptors on neurons and microvessels of patients with PD *(34)*. This finding suggests that lactoferrin receptors, which regulate intraneuronal iron content, may be overly expressed in vulnerable dopaminergic neurons in PD.

Another postmortem finding in PD patients which is compatible with the free radical hypothesis is that of a deficiency in mitochondrial complex I. Such a defect could either result in the abnormal production of free radicals, or be the result of free radical injury *(35)*. This defect takes on particular interest in light of the observation that MPP^+, the toxic oxidative product of MPTP, inhibits complex I *(36)*. The defect in complex I in PD patients has been demonstrated by Schapira to result in a mean 37% decrease in enzyme activity *(35)*. This decrease appears to be both regionally specific for the SN, and disease specific, among basal ganglia disorders, for PD.

Thus, the free radical hypothesis receives indirect support from a large number of separate lines of evidence, and, as stated above, it remains the foremost and most widely tested hypothesis of neural degeneration in PD. Nevertheless, it remains only a hypothesis, and it has its shortcomings *(37)*. For example, there is no specific aspect of the free radical hypothesis as it is currently posed to account for the relative vulnerability of ventral tier dopaminergic neurons in PD. In addition, it must be remembered that noncatecholamine neuronal groups, such as the nucleus basalis (which is cholinergic) also degenerate in PD, and aspects of the free radical hypothesis which are dependent on catecholamine metabolism are not relevant to the degeneration of these structures.

An alternate hypothesis for the pathogenesis of PD is that excitotoxic mechanisms may play a role. Excitotoxic mechanisms, possibly acting in concert with oxidative stress *(38)* or impaired energy metabolism *(39)* have been postulated to underlie a number of chronic, neuron phenotype-specific neurodegenerative disorders. However, there is minimal specific evidence implicating excitotoxicity in the pathogenesis of PD. One often-quoted study by Turski et al. *(40)* demonstrated an ability of NMDA antagonists to protect the substantia nigra from direct injection of MPP^+. However, the specificity of the direct intracerebral injection model, as opposed to the systemic injection of MPTP, has been questioned *(41)*, and, in any event, the finding could not be replicated *(41)*. The number of NMDA receptor binding sites in the SN is low *(42)*. Among human neurologic diseases, and animal models thereof, which are induced by energy failure on the basis of hypoxia, mitochondrial toxins or inherited defects in mitochondrial metabolism, and for many of which there is reasonable evidence that excitotoxicity may play a role, it is the striatum among the basal ganglia, not the SN, which shows relative vulnerability. In spite of these considerations, it is important to recognize that there is too little evidence to claim or refute a role for excitotoxicity in PD. There is no question that, in vitro, glutamate can be toxic for dopaminergic neurons *(43)*.

If excitotoxicity is considered in its broadest possible context, that of abnormal, sustained neuronal depolarization leading to cell death, then recent findings in the mouse mutant Weaver may be considered relevant. In this mutant, there is a late postnatal spontaneous degeneration of dopamine neurons of the SN, particularly those of the ventral tier. These dopamine neurons are analogous to the ventral lateral group which is most vulnerable to degeneration in PD. It has recently been shown that the mutation in the *weaver* results in a glycine to serine substitution in the pore forming region of a G-protein coupled inward rectifier potassium channel (Girk2) *(44)*. The functional effect of this mutation depends on the relative expression of *girk2* in the cell in relation to other Girk subunits with which Girk2 can form multimeric channels *(45)*.

In the SN of the adult rat, *girk2*, but not *girk1*, is expressed at high levels, a pattern suggesting that Girk2 homomultimers may exist. Such channels in the mutant lose their selectivity for K^+, and may develop an abnormal receptor-activated Na^+ current, leading to chronic depolarization and a form of excitotoxic cell death *(45)*. The possible role of mutations in *girk2* or other potassium channels has not been assessed in familial forms of PD. It would seem worthwhile to assess these genes as possible candidates.

PROGRAMMED CELL DEATH IN DOPAMINE NEURONS, ANIMAL MODELS OF PARKINSONISM, AND PD

The concept that programmed cell death, as a genetically regulated cell death process, may underlie the neuron-specific degenerations of later life has gathered great attention in recent years, as this volume attests. One appeal of the programmed cell death hypothesis is that it provides a conceptual framework in which to understand the high degree of specificity of these degenerations for particular neuronal phenotypes. Thus, while in relation to PD it may be difficult to understand how an environmental factor might act via free radical or excitotoxic mechanisms to cause the degeneration of a particular class of neurons, it is not difficult to envision disturbed cell death programs which are uniquely regulated within specific neuronal phenotypes leading to the death of those particular cells. During development, the magnitude of programmed cell death within specific neuronal phenotypes is highly regulated by the availability of unique neurotrophic factors and other molecules *(46) (see* chapters by Burek and Oppenheim and by Koliatsos and Mocchetti). It seems plausible to theorize that either an inherited genetic abnormality or an acquired somatic mutation due to an environmental factor or the aging process might lead to the disturbance of the programs which regulate cell death within a specific class of neurons. In this formulation, the cell-type specific degenerations of later life might be thought of as the converse of oncogenesis; in the latter, inherited or acquired genetic abnormalities lead to mitosis; in the former, they lead to programmed cell death. It is now known that apoptosis and mitosis are intimately related possible outcomes of the cell cycle *(see* chapter by Freeman).

It is important to emphasize that the programmed cell death hypothesis should not be thought of as entirely unrelated to concepts of free radical and excitotoxin-mediated cell death. While traditional concepts of free radical and excitotoxin injury have centered on the ability of toxic molecules to directly injure cellular constituents without any participation of the host cell's own genetic programs, it is now clear that in both in vitro and in vivo paradigms of free radical and excitotoxin-mediated injury programmed cell death may occur, as discussed elsewhere in this volume. It is also apparent that, in some settings, programmed cell death may be carried out by the controlled production of free radicals. Thus, the concept of programmed cell death should be thought of as closely related to mechanisms of free radical- or excitotoxin-mediated injury.

In specific relation to PD, it is important first to consider the evidence that programmed cell death does in fact occur within dopaminergic neurons. Our interest in this possibility was stimulated by the observation that an excitotoxic lesion to the target striatum during development led to a reduced adult number of nigro-striatal dopaminergic neurons *(47,48)*. This decrease occurred in spite of the axon-sparing nature of the lesion *(48,49)* and in the absence of any direct injury to the nigra. This observation suggested that the nigro-striatal dopaminergic system may be target dependent during development, and this possibility was supported by many earlier observations made in vitro which had demonstrated the ability of striatal components to support the viability and differentiation of developing dopaminergic neurons *(50–53)*. These observations of target effects on the developing dopaminergic nigro-striatal system suggested that it may develop according to the principles of classical neurotrophic theory *(46,54)*.

As predicted by classic theory, a natural cell death event does occur in the SNpc, with typical light microscopic morphology of apoptosis, demonstrated both by Nissl stain and suppressed silver staining *(55)*. More recently, we have used a double labeling technique to identify apoptotic natural cell death in phenotypically-defined dopaminergic neurons *(56)*. Natural cell death in these neurons has a bimodal time course. There is an initial, major peak which begins on embryonic d 20, and largely abates by postnatal day (P) 8. There is a second, minor peak of natural cell death on P 14. The neural basis of these two peaks is unknown. It does not appear to be due to a different time course of cell death in different anterior-to-posterior locations, because the same bimodal pattern is observed in all planes. Whether it is based on differences in dopamine neuron subtypes or ontogenetic ages remains to be determined. The presence of a major cell death event within the first postnatal week is in keeping with the demonstration by Tepper and colleagues that there is a decrement in the number of TH-positive neurons in SN postnatally, particularly in the first postnatal week *(57)*. Thus, natural cell death does occur within the dopaminergic neurons of the SN, with the morphology of apoptosis.

We have examined whether this death event in SNpc can be regulated during development. Our previous observation that early injury to striatum results in a diminished number of mature SNpc neurons would predict, in accord with the classical neurotrophic theory, that such injury would increase cell death. This does occur; excitotoxic injury to the striatum on P 7 results in an eightfold increase in the number of apoptotic profiles identified by the silver stain technique at 24 h postlesion *(58)*. These profiles are morphologically identical to those observed during natural cell death, and meet ultrastructural and DNA 3' end-labeling criteria for apoptosis. Within SNpc, induction of cell death is identified within phenotypically defined dopaminergic neurons (Fig. 1). We have recently examined the developmental dependence of early target injury to induce apoptotic cell death in SN, and have found that while a lesion on PND 14 shows a level of induction equal to that observed on P 7, lesions on P 21 and 28 do not result in a significant induction of cell death (Kelly and Burke, submitted). In adult animals, extensive lesion of the striatum with ibotenate does not result in any induction of cell death discernible by silver stain in SNpc *(59)*. This result is compatible with an observation by Lundberg et al that a similar striatal lesion resulted in atrophy, but no loss, of neurons in SNpc, prelabeled with the retrograde tracer Fluoro-Gold *(60)*. Krammer had similarly shown very little effect of such lesions on SNpc *(61)*. Thus, there appears to be a developmental "window" during which time the level of natural cell death in SNpc is sensitive to target size. Similar observations have been made for peripherally projecting neural systems with dependence on NGF *(62)*.

For the sake of relevance to PD, the preceding discussion focused strictly on events within the SNpc. We will mention, only briefly, that apoptotic natural cell death also occurs in SNpr *(63)* and that it is also induced by early striatal lesion *(58)* with a 2-wk developmental window of dependence for induction of death (Kelly and Burke, sub-

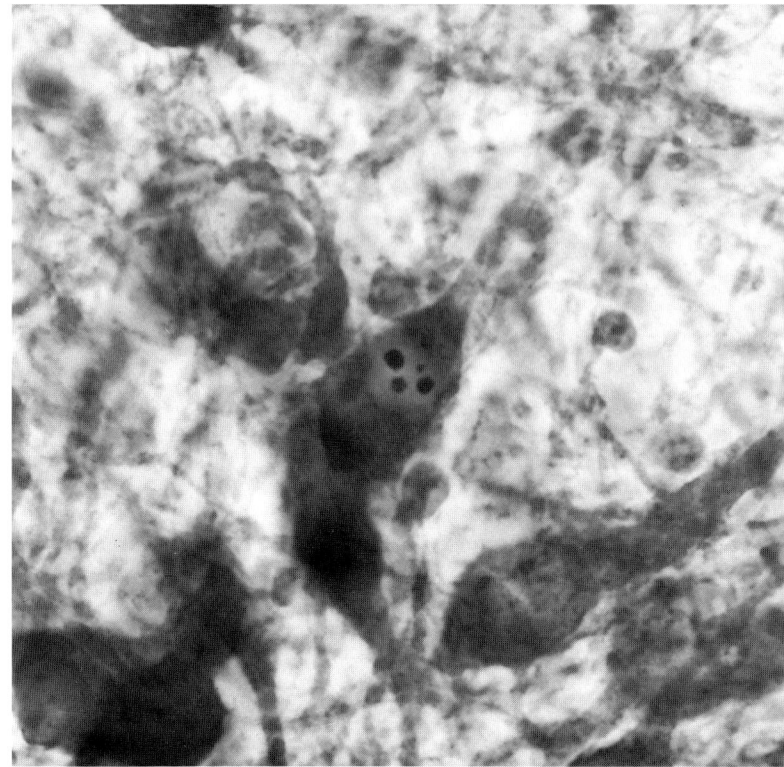

Fig. 1. Immunoperoxidase stain for TH, with Nissl counterstain, of the SNpc at 24 h after a striatal excitotoxic lesion with quinolinic acid, performed on postnatal d 7. The brown cytoplasmic reaction product identifies TH-positive dopaminergic neurons. The Nissl counterstain identifies four basophilic, rounded chromatin clumps, typical for apoptosis, within the nucleus of a single dopamine neuron in the center of the micrograph. (Reprinted with permission from Macaya et al. 1994.)

mitted). However, in SNpr, unlike SNpc, an induction of cell death is observed following striatal injury in adult animals *(59,61,64,65)*. This cell death is not apoptotic *(59)*, so how it relates to the programmed cell death observed during development, if at all, is unknown. The induced apoptotic cell death observed during development is likely to have a different neural basis in SNpr as compared to SNpc, because the striatum does not represent a major target of SNpr; in fact, the inverse is true, so induced death in SNpr may represent an instance of developmental cell death due to a loss of afferent input *(66)*.

Knowing that apoptotic cell death can occur in dopaminergic neurons, and that it is inducible, we have examined whether it occurs in known models of Parkinsonism. If early target support is necessary for the viability of dopaminergic neurons, early developmental destruction of their terminals with the selective neurotoxin 6-hydroxydopamine (6-OHDA) should interfere with attainment of support, and result in induced death. We have recently found that intrastriatal injection of 6-OHDA, with destruction of dopaminergic terminals, does result in induced apoptotic cell death in dopaminergic neurons of the SN *(67)*. Interestingly, this ability of intrastriatal 6-OHDA to induce apoptotic death, like early target injury, is developmentally dependent, with a major induction of death on P 7 and 14, but only a minor effect at later postnatal times. Interestingly, at later postnatal times, such as P 42, intrastriatal 6-OHDA induces two different morphologies of cell death in the SNpc, apoptotic and nonapoptotic. This result suggests that either the toxin induces cell death by two different mechanisms, or that the same fundamental mechanism, that of programmed cell death, induces an apoptotic morphology in less mature animals, and a nonapoptotic morphology in more mature animals. There are precedents for such an occurrence in studies of natural cell death *(68,69)*.

While we have emphasized the interpretation that 6-OHDA induces apoptosis in developing animals by interfering with target support, it is also theoretically possible that the toxin itself acts directly to induce apoptosis. 6-OHDA has been shown in vitro to induce apoptosis in PC12 cells *(70)*. It is known that intrastriatal injection of 6-OHDA can induce death in dopaminergic neurons of the SNpc in adult animals *(71)*, and if target support is not critical for viability in adults, as we have argued, we must postulate that this death is mediated by a toxic effect. Presumably, if such a direct toxic effect can occur in adults, it can occur in neonates as well. However, in the one setting in adults where 6-OHDA is quite likely to be exerting a direct toxic effect, i.e., that following direct injection into SN, apoptosis is not observed *(72)*. Nor has it yet been observed following intrastriatal injection in adults *(73)*. Nevertheless, we must keep an open mind to the possibility that 6-OHDA can induce apoptosis in immature animals by virtue of its direct intracellular toxicity as well as by its ability to interfere with target-derived support.

Other important animal models of parkinsonism, in rodents and primates, are induced by MPTP, or its active oxidized product, MPP$^+$. Is there evidence for apoptotic cell death in these models? A number of investigators have shown, in a variety of systems, that MPP$^+$ can induce apoptosis in vitro. Dipasquale and colleagues first showed that MPP+ can induce apoptotic morphology and DNA fragmentation in postnatal cerebellar granule cells in culture *(74)*. Subsequently, others have shown that MPP$^+$ appears to induce apoptosis in embryonic mesencephalic cultures *(75)*, in both differentiated *(76)* and undifferentiated *(77)* PC12 cells and in a human neuroblastoma cell line *(78)*. However, in the only in vivo study to date of MPTP in a mouse model, apoptotic cell death was not observed in SN by either silver staining or DNA end-labeling techniques at any time point following injection *(79)*. In view of the demonstration by Hartley and co-workers *(77)* that low doses of MPP$^+$ induce apoptosis in PC12 cells, but high doses produce necrosis, Jackson-Lewis and colleagues *(79)* examined the effects of injection of 80, 60, and 40 mg/kg MPTP. The lowest dose did not produce neural degeneration identifiable by silver stain, and the higher doses produced only a nonapoptotic morphology of cell death. Thus, there is no evidence at this time that MPTP induces apoptotic neuron death in vivo. It is important to recognize, however, that apoptosis is only one possible morphology of programmed cell death *(80,81)* (*see* also chapter by Clarke), so that the inability to identify it in this model does not exclude a possible role for programmed cell death, defined in its broadest sense as a death process which is mediated by genetic programs relatively independently of preceding cellular events. Such a possibility is suggested by the recent demonstration that MPTP treatment of mice induces an increase in *bax* mRNA, and Bax immunoreactivity in SN neurons (82). Bax, a protein homolog of Bcl-2, is capable of heterodimerizing with Bcl-2 and accelerating apoptotic cell death *(83,84)*.

As mentioned earlier, the Weaver mouse is the only genetic model of spontaneous degeneration of dopaminergic neurons of the SN, so it is of particular interest. In Weaver mouse, cerebellar granule cells degenerate, as well as dopaminergic neurons. Several investigators have shown that death of cerebellar granule cells in the Weaver is apoptotic by morphologic and DNA end-labeling criteria *(85–88)*. We, therefore, sought to identify the nature of cell death in the Weaver SN. We have found that while apoptotic cell death occurs postnatally in weaver SN, it also does so with equal magnitude in heterozygote and wild type littermates *(89)*. This apoptotic death therefore represents a natural cell death event, similar to that observed in the rat *(55)*. Unique to Weaver is the occurrence of a nonapoptotic morphology of cell death which occurs later in the postnatal period than the apoptotic natural cell death. This nonapoptotic form of cell death does not show positive DNA end-labeling. On ultrastructural analysis it is characterized by irregular, nonapoptotic, electron-dense nuclear aggregates, and pronounced cytoplasmic changes which include extensive vacuole formation and the formation of stacks of endoplasmic reticulum. While this may represent a variant of a cytoplasmic form of programmed cell death (such as Type 3b as defined by Clarke *[80]*), it may also represent a pathologic form of cell death mediated via an as yet unknown mechanism set forth by the *girk2* mutation in the *Weaver*.

Before considering evidence that apoptosis may play a direct role in of PD, it is important to define briefly terms and methodologic standards for identification of apoptosis. The term "apoptosis" is used by many investigators as synonymous with "programmed cell death." It is our opinion that, strictly speaking, the term "apoptosis" refers to one very well defined morphology of programmed cell death. This is the way the term was originally used by Kerr et al. *(90)*, and it is important to maintain that more restricted usage, because it is very clear that there are other morphologies of programmed cell death which are non-apoptotic *(80,81)*. The distinction between the two terms is also important for the reason that, operationally, while apoptosis can be strictly and definitively identified by morphologic criteria, as discussed below, the broad concept of programmed cell death, as a genetically regulated cell process leading to death, is much more difficult to identify definitively with any single morphologic or biochemical assessment. Apoptotic cell death is most definitively identified by ultrastructural studies *(91)* (*see* similar views expressed in the chapters by Clarke and by Olney and Ishimaru). One of the most distinctive features of apoptotic cell death is the formation of homogeneous, electron dense chromatin clumps with sharply delineated edges, either against the inner surface of the nuclear membrane, or rounded within the nucleus *(91)*. It is critically important to recognize that not all intranuclear electron dense aggregates are to be equated with apoptotic cell death. Examples of pleiomorphic, irregular, heterogeneous electron dense aggregates with poorly defined boundaries have been observed in many instances of apparently necrotic cell death *(91)*. While not as definitive as ultrastructure, demonstration at the light microscope level of intensely basophilic, rounded, intranuclear and sharply demarcated chromatin aggregates, by use of basic dyes for the staining of Nissl substance, is highly suggestive of apoptosis *(92)*. In numerous instances, such profiles have been confirmed by ultrastructural studies to conform to the classic morphology of apoptosis. Another widely used technique is that of *in situ* labeling of DNA free 3' ends, generated by fragmentation of genomic DNA during apoptosis *(93)*. However, this technique can also label free 3' ends generated by necrotic cell death *(94)*, so it must be used with strict attention to the presence of an apoptotic morphology *(95)* as well as a positive reaction product.

There have been recent reports of possible evidence for apoptotic cell death in the SN in the postmortem PD brain. A recent preliminary report has identified possible apoptotic profiles by electron microscopy in the SN of three patients with PD *(96)*. In this report, electron dense intranuclear aggregates were identified, but they did not have the characteristic features of apoptotic chromatin clumps, i.e., a uniform increase in electron density with sharply delineated edges *(91)*. In addition, observations were not made in controls, so the pathologic significance of the findings is unclear. Nevertheless, while apoptotic morphology was not convincingly demonstrated, some of the morphologic findings, such as relative preservation of mitochondria and cellular membranes, may be compatible with other morphologies of programmed cell death. Thus, the possibility that programmed cell death plays a direct role in neuron death in PD remains an important consideration which will undoubtedly be the focus of much attention in the years ahead.

Even if programmed cell death does not play a direct role in the neuron destruction of PD, it may be highly relevant to other aspects of the pathogenesis and treatment. First, the likelihood of developing PD may depend on an individual's endowment of dopaminergic neurons in the SN. It appears that the mature number of dopaminergic neurons is likely to be regulated during development by target-derived and other factors which control the magnitude of a natural cell death event which occurs among these neurons. Genetic and environmental factors which affect the regulation of this death event may, therefore, influence the likelihood of developing PD.

A major new approach to the treatment of PD is that of transplantation of fetal mesencephalic dopaminergic neurons *(97–99)*. While preliminary clinical results support transplantation as a promising approach, it is nevertheless clear that very few transplanted dopaminergic neurons survive *(100)*. While in the past it has been assumed that the death of implanted cells is mediated by exogenous factors such as tissue handling, local trauma, or the lack of a vascular supply, it has recently been shown that implanted mesencephalic tissue undergoes apoptotic cell death *(101)*. Interestingly, the time course of this cell death would suggest that it occurs at about the time when dopaminergic neurons of the SNpc undergo their major initial phase of natural cell death *(55,56)*. These findings would suggest that, if the factors which regulate natural cell death in the SN can be identified, they could be utilized to augment cell survival in transplants.

Programmed cell death may also be relevant to the effects of chronic levodopa treatment of PD. The subject is quite controversial, but there is some clinical evidence to question whether levodopa may aggravate some aspects of PD *(102)*. On theoretical grounds, if the metabolism of dopamine leads to oxidative stress, as discussed earlier, then augmentation of dopamine availability by ingestion of levodopa may worsen the underlying disease process. It has recently been shown in vitro that either dopamine or levodopa is capable of inducing apoptosis. Ziv and colleagues have shown that dopamine induces apoptosis in cultured embryonic chick sympathetic neurons *(103)*. Walkinshaw and Waters have demonstrated that levodopa, in micromolar concentrations, induces apoptosis in PC12 cells, and that this effect is inhibited by antioxidants *(104)*. As of yet, there have been no in vivo studies of possible levodopa toxicity, utilizing methods to identify apoptosis.

CONCLUSIONS

We now know that apoptotic programmed cell death can occur in dopamine neurons of the SN, that it is inducible, and that it occurs in some living models of parkinsonism. Whether or not programmed cell death plays a direct role in the degeneration of dopamine neurons in PD is a question which will warrant vigorous investigation in the years ahead. Whether or not programmed cell death plays a direct and primary role in the neuronal degeneration associated with PD, it is important to recognize that it has the potential to play an important downstream or effector role in processes that may primarily be initiated by oxidative stress or excitotoxic mechanisms. It is also important to recognize that, aside from any role that programmed cell death may play in the pathogenesis of PD, it almost certainly plays a role in processes that regulate developmental control of the mature number of dopamine neurons. These same mechanisms may also regulate the survival of implanted tissue, and the success of the transplantation approach. Thus, investigation of the molecular and cellular basis of programmed cell death in neural tissues should be an important research priority for PD, as well as other neurodegenerative disorders, in the years ahead.

NOTE ADDED IN PROOF

Several important recent observations have been made regarding the pathogenesis of PD and a possible role for programmed cell death. Polymeropoulos et al. *(105)* have shown that one form of familial PD is due to a mutation in α-synulein. Tatton and Kish *(106)* have shown that apoptotic cell death is observed in an MPTP model, induced by small doses of toxin, given over a prolonged period of time. Tompkins et al. *(107)* have demonstrated positive in situ end-labeling, in conjunction with apoptotic morphology, in cases of Lewy body-associated disorders.

ACKNOWLEDGMENTS

The author is grateful to Pat White for diligent secretarial assistance. He is also grateful to Dr. Serge Przedborski for helpful comments and suggestions on the manuscript. This work was supported by NS 26836 and by the Parkinson's Disease Foundation.

REFERENCES

1. Langston JW, Ballard PA, Tetrud JW, Irwin I. Chronic parkinsonism in humans due to a product of meperidine-analog synthesis. *Science* 1983, **219**: 979–980.
2. Mayeux R, Denaro J, Hemenegildo N, Marder K, Tang M-X, Cote LJ, Stern Y. A population-based investigation of Parkinson's disease with and without dementia. *Arch Neurol* 1992, **49**: 492–497.
3. Carlsson A, Winblad B. Influence of age and time interval between death and autopsy on dopamine and 3-methoxytyramine levels in human basal ganglia. *J Neural Trans* 1976, **38**: 271–276.
4. Mcgeer PL, Mcgeer EG, Suzuki JS. Aging and extrapyramidal function. *Arch Neurol* 1977, **34**: 33–35.
5. Kish SJ, Shannak K, Rajput A, Deck Jhn, Hornykiewicz O. Aging produces a specific pattern of striatal dopamine loss implications for the etiology of idiopathic Parkinsons Disease. *J Neurochem* 1992, **58**: 642–648.
6. Scherman D, Desnos C, Darchen F, Pollak P, Javoy-Agid F, Agid Y. Striatal dopamine deficiency in Parkinson's disease: Role of aging. *Ann Neurol* 1989, **26**: 551–557.

7. Fearnley M, Lees AJ. Ageing and Parkinson's disease: substantia nigra regional selectivity. *Brain* 1991, **114**: 228–301.
8. McGeer PL, Itagaki S, Akiyama H, McGeer EG. Rate of cell death in parkinsonism indicates active neuropathological process. *Ann Neurol* 1988, **24**: 547–576.
9. McGeer PL, Itagaki S, Boyes BE, McGeer EG. Reactive microglia are positive for HLA-DR in the substantia nigra of Parkinson's and Alzheimer's disease brains. *Neurology* 1988, **38**: 1285–1291.
10. Tanner CM. Epidemiology of Parkinson's disease. *Neurologic Clinics* 1992, **10**: 317–329.
11. Langston JW, Koller WC, Giron LT. Etiology of Parkinson's disease, in *The Scientific Basis for the Treatment of Parkinson's Disease*, (Olanow CW, Lieberman AN, eds), 1st edition, The Parthenon Publishing Group, Inc., Park Ridge, NJ, 1992, pp. 33–58.
12. Duvoisin RC. The genetics of Parkinson's disease: A review, in *Advances in Neurology, Volume 60: Parkinson's Disease from Basic Research to Treatment*, (Narabayashi H, Nagatsu T, Yanagisawa N, Mizuno Y.) Raven Press, NY, 1993, pp. 306–315.
13. Duvoisin RC. Recent advances in the genetics of Parkinson's disease, in *Advances in Neurology, Volume 69: Parkinson's Disease*, (Battistin L, Scarlato G, Caraceni T, Ruggieri S., eds), Lippincott-Raven Publishers, Philadelphia, PA, 1996, pp. 33–40.
14. Johnson Wg, Hodge Se, Duvoisin R. Twin studies and the genetics of Parkinson's disease — a reappraisal. *Mov Disord* 1990, **5**:187–194.
15. Burn DJ, Mark MH, Playford ED, Maraganore DM, Zimmerman TR, Duvoisin RC, Harding AE, Marsden CD, Brooks DJ. Parkinson's Disease in twins studied with F 18 DOPA and positron emission tomography. *Neurology* 1992, **42**:1894–1900.
16. Golbe LI, Di Iorio G, Bonavita V, Miller DC, Duvoisin RC. Autosomal dominant Parkinson's disease. *Ann Neurol* 1990, **27**: 276–282.
17. Olanow CW. An introduction to the free radical hypothesis in Parkinson's disease. *Ann Neurol* 1992, 32:S2-S9.
18. Fahn S, Cohen G. The Oxidant Stress Hypothesis In Parkinson's Disease. Evidence supporting it. *Ann Neurol* 1992, **32**: 804–812.
19. Swartz HM, Sarna T, Zecca L. Modulation by neuromelanin of the availability and reactivity of metal ions. *Ann Neurol* 1992, 32:S69-S75.
20. Cohen G, Heikkila RE. The generation of hydrogen peroxide, superoxide radical, and hydroxyl radical by 6-hydroxydopamine, dialuric acid, and related cytotoxic agents. *J Biol Chem* 1974, **249**: 2447–2452.
21. Heikkila RE, Cohen G. 6-Hydroxydopamine. evidence for superoxide radical as an oxidative intermediate. *Science* 1973, **181**: 456–457.
22. Przedborski S, Kostic V, Jackson-Lewis V, Naini AB, Simonetti S, Fahn S, Carlson E, Epstein CJ, Cadet JL. Transgenic mice with increased Cu/Zn-superoxide dismutase activity are resistant to N-methyl-4-phenyl-1,2,3,6-tetrahydropyridine-induced neurotoxicity. *J Neurosci* 1992, **12**: 1658–1667.
23. Dexter DT, Carter CJ, Wells FR, Javoy-Agid F, Agid Y, Lees A, Jenner P, Marsden CD. Basal lipid peroxidation in substantia nigra is increased in Parkinson's disease. *J Neurochem* 1989, **52**: 381–389.
24. Dexter DT, Holley AE, Flitter WD, Slater TF, Wells FR, Daniel SE, Lees AJ, Jenner P, Marsden CD. Increased levels of lipid hydroperoxides in the parkinsonian substantia nigra: An HPLC and ESR study. *Mov Disord* 1994, **9**: 92–97.
25. Sanchez-Ramos JR, Overvik E, Ames BN. A marker of oxyradical-mediated DNA damage (8-hydroxy-2'deoxyguanosine) is increased in nigro striatum of Parkinson's disease brain. *Neurodegen* 1994, 3:197–204.
26. Smith MA, Sayre LM, Monnier VM, Perry G. Radical ageing in Alzheimer's disease. *Trends Neurosci* 1995, **18**: 172–176.
27. Yoritaka A, Hattori N, Uchida K, Tanaka M, Stadtman ER. Immunohistochemical detection of 4-hydroxynonenal protein adducts in Parkinson's disease. *Proc Natl Acad Sci USA* 1996, 93:2696–2701.

28. Perry TL, Godin DV, Hansen S. Parkinson's disease: a disorder due to nigral glutathione deficiency? *Neurosci Lett* 1982, **33**: 305–310.
29. Jenner P, Dexter DT, Sian J, Schapira Ahv, Marsden Cd. Oxidative stress as a cause of nigral cell death in Parkinson's disease and incidental Lewy body disease. *Ann Neurol* 1992, **32**: S82–S87.
30. Gibb WRG, Lees AJ. The relevance of the Lewy body to the pathogenesis of idiopathic Parkinson's disease. *J Neurol* 1988, **51**: 745–752.
31. Dexter DT, Wells FR, Lees AJ, Agid F, Agid Y, Jenner P, Marsden CD. Increased nigral iron content and alterations in other metal ions occurring in brain in Parkinson's disease. *J Neurochem* 1989, **52**: 1830–1836.
32. Dexter DT, Carayon A, Vidailhet M, Ruberg M, Agid F, Agid Y, Lees AJ, Wells FR, Jenner P, Marsden CD. Decreased ferritin levels in brain in Parkinson's disease. *J Neurochem* 1990, **55**: 16–20.
33. Dexter DT, Carayon A, Javoy-Agid F, Agid Y, Wells FR, Daniel SE, Lees AJ, Jenner P, Marsden CD. Alterations in the levels of iron, ferritin and other trace metals in Parkinson's disease and other neurodegenerative diseases affecting the basal ganglia. *Brain* 1991, **114**: 1953–1975.
34. Faucheux BA, Nillesse N, Damier P, Spik G, Mouattprigent A, Pierce A, Leveugle B, Kubis N, Hauw JJ, Agid Y, Hirsch EC. Expression of lactoferrin receptors is increased in the mesencephalon of patients with Parkinson's disease. *Proc Natl Acad Sci USA* 1995, **92**, 9603–9607.
35. Schapira AHV, Mann VM, Cooper JM, Krige D, Jenner PJ, Marsden CD. Mitochondrial function in Parkinson's disease. *Ann Neurol* 1992, **32**, S116-S124.
36. Nicklas WJ, Vyas I, Heikkila RE. Inhibition of NADH-linked oxidation in brain mitochondria by MPP^+, a metabolite of the neurotoxin MPTP. *Life Sci* 1985, **36**: 2503–2508.
37. Calne Db. The free radical hypothesis in idiopathic parkinsonism: Evidence against it. *Ann Neurol* 1992, 32:799–803.
38. Coyle JT, Puttfarcken P. Oxidative stress, glutamate, and neurodegenerative disorders. *Science* 1993, **262**: 689–695.
39. Albin RL, Greenamyre JT. Alternative excitotoxic hypotheses. *Neurology* 1992, **42**: 733–738.
40. Turski L, Bressler K, Rettig K-J, Loschmann PA, Wachtel H. Protection of substantia nigra from MPP+ neurotoxicity by *N*-methyl-D-aspartate antagonists. *Nature* 1991, **349**: 414–418.
41. Sonsalla PK, Zeevalk GD, Manzino L, Giovanni A, Nicklas WJ. MK-801 fails to protect against the dopaminergic neuropathology produced by systemic 1-methyl-4-phenyl-1,2,3,6-tetrahydropyridine in mice or intranigral 1-methyl-4- phenylpyridinium in rats. *J Neurochem* 1992, **58**: 1979–1982.
42. Difazio MC, Hollingsworth Z, Young AB, Penney JB, Jr. Glutamate receptors in the substantia nigra of Parkinson's disease brains. *Neurology* 1992, **42**: 402–406.
43. Casper D, Blum M. Epidermal growth factor and basic fibroblast growth factor protect dopaminergic neurons from glutamate toxicity in culture. *J Neurochem* 1995, **65**: 1016–1026.
44. Patil N, Cox Dr, Bhat D, Faham M, Myers RM, Peterson AS. A potassium channel mutation in weaver mice implicates membrane excitability in granule cell differentiation. *Nature Genetics* 1995, **11**: 126–129.
45. Slesinger PA, Patil N, Liao J, Jan YN, Jan LY, Cox DR. Functional effects of the mouse *weaver* mutation on G protein-gated inwardly rectifying K^+ channels. *Neuron* 1996, 16:321–331.
46. Barde YA. Trophic factors and neuronal survival. *Neuron* 1989, **2**: 1525–1534.
47. Burke RE, Macaya A, Devivo D, Kenyon N, Janec EM. Neonatal hypoxic-ischemic or excitotoxic striatal injury results in a decreased adult number of substantia nigra neurons. *Neuroscience* 1992, 50:559–569.

48. Macaya A, Burke RE. Effect of striatal lesion with quinolinate on the development of substantia nigra dopaminergic neurons: a quantitative morphological analysis. *Dev Neurosci* 1992, **14**: 362–368.
49. Schwarcz R, Whetsell WO, Mangano RM. Quinolinic acid: an endogenous metabolite that produces axon-sparing lesions in the rat brain. *Science* 1983, **219**: 316–318.
50. Prochiantz A, Di Porzio U, Kato A, Berger B, Glowinski J. *In vitro* maturation of mesencephalic dopaminergic neurons from mouse embryos is enhanced in presence of their striatal target cells. *Proc Natl Acad Sci USA* 1979, **76**: 5387–5391.
51. Hemmendinger LM, Garber BB, Hoffmann PC, Heller A. Target neuron-specific process formation by embryonic mesencephalic dopamine neurons *in vitro*. *Proc Natl Acad Sci USA* 1981, **78**: 1264–1268.
52. Hoffmann PC, Hemmendinger LM, Kotake C, Heller A. Enhanced dopamine cell survival in reaggregates containing target cells. *Brain Res* 1983, **274**: 275–281.
53. Tomozawa Y, Appel SH. Soluble striatal extracts enhance development of mesencephalic dopaminergic neurons *in vitro*. *Brain Res* 1986, **399**: 111–124.
54. Clarke PGH. Neuronal death in the development of the veterbrate nervous system. *Trends Neurosci* 1985, 345–349.
55. Janec E, Burke RE. Naturally occurring cell death during postnatal development of the substantia nigra of the rat. *Mol Cell Neurosci* 1993, **4**: 30–35.
56. Oo TF, Burke RE. Time course of developmental cell death in phenotypically-defined dopaminergic neurons of the substantia nigra. *Soc Neurosci Abstr* 1996, **22**: in press
57. Tepper JM, Damlama M, Trent F. Postnatal changes in the distribution and morphology of rat substantia nigra dopaminergic neurons. *Neuroscience* 1994, **60**: 469–477.
58. Macaya A, Munell F, Gubits RM, Burke RE. Apoptosis in substantia nigra following developmental striatal excitotoxic injury. *Proc Natl Acad Sci USA* 1994, **91**: 8117–8121.
59. Stefanis L, Burke RE. Transneuronal degeneration in substantia nigra pars reticulata following striatal excitotoxic injury in adult rat: Time course, distribution, and morphology of cell death. *Neuroscience* 1996, in press
60. Lundberg C, Wictorin K, Bjorklund A. Retrograde degenerative changes in the substantia nigra pars compacta following an excitotoxic lesion of the striatum. *Brain Res* 1994, **644**: 205–212.
61. Krammer EB. Anterograde and transsynaptic degeneration "en cascade" in basal ganglia induced by intrastriatal injection of kainic acid: an animal analogue of Huntington's disease. *Brain Res* 1980, **196**: 209–221.
62. Lindsay RM. Brain-derived neurotrophic factor: an NGF-related neurotrophin, in *Neurotrophic Factors*, (Loughlin SE, Fallon JH, eds.), Academic Press, San Diego, CA, 1993: pp.257–284.
63. Szeto A, Oo T, Burke RE. Naturally occurring cell death during postnatal development of the substantia nigra pars reticulata of rat. *Mov Disord* 1994, **9, Suppl 1**: 106 (Abstract).
64. Saji M, Reis DJ. Delayed transneuronal death of substantia nigra neurons prevented by gamma-aminobutyric acid agonist. *Science* 1987, **235**: 66–68.
65. Pasinetti GM, Morgan DG, Finch CE. Disappearance of GAD-mRNA and tyrosine hydroxylase in substantia nigra following striatal ibotenic acid lesions: evidence for transneuronal regression. *Exp Neurol* 1991, **112**: 131–139.
66. Linden R. The survival of developing neurons: A review of afferent control. *Neuroscience* 1994, **58**: 671–682.
67. Marti MJ, James CJ, Oo TF, Kelly WJ, Burke RE. Destruction of striatal dopaminergic terminals by injection of 6-hydroxydopamine induces apoptotic cell death in dopaminergic neurons of the substantia nigra during development. *Mov Disord* 1996, **11(Suppl 1)**: 44.
68. Pilar G, Landmesser L. Ultrastructural differences during embryonic cell death in normal and peripherally deprived ciliary ganglia. *J Cell Biol* 1976, **68**: 339–356.

69. Cunningham TJ. Naturally occurring neuron death and its regulation by developing neural pathways. *Int Rev Cytol* 1982, **74:** 163–186.
70. Walkinshaw G, Waters CM. Neurotoxin induced cell death in neuronal PC12 cells is mediated by induction of apoptosis. *Neuroscience* 1994, **63:** 975–987.
71. Sauer H, Oertel WH. Progressive degeneration of nigrostriatal dopamine neurons following intrastriatal terminal lesions with 6 hydroxydopamine a combined retrograde tracing and immunocytochemical study in the rat. *Neuroscience* 1994, 59:401–415.
72. Jeon BS, Jackson-Lewis V, Burke RE. 6-Hydroxydopamine lesion of the rat substantia nigra: time course and morphology of cell death. *Neurodegen* 1995, **4:** 131–137.
73. Ichitani Y, Okamura H, Matsumoto Y, Nagatsu I, Ibata Y. Degeneration of the nigral dopamine neurons after 6-hydroxydopamine injection into the rat striatum. *Brain Res* 1991, **549:** 350–353.
74. Dipasquale B, Marini AM, Youle RJ. Apoptosis and DNA degradation induced by 1-methyl-4-phenylpyridinium in neurons. *Biochem Biophys Res Comm* 1991, **181:** 1442–1448.
75. Mochizuki H, Nakamura N, Nishi K, Mizuno Y. Apoptosis is induced by 1 Methyl 4 Phenylpyridinium ion (MPP(+)) in ventral mesencephalic striatal coculture in rat. *Neurosci Lett* 1994, **170:** 191–194.
76. Mutoh T, Tokuda A, Marini AM, Fujiki N. 1 Methyl 4 Phenylpyridinum kills differentiated PC12 cells with a concomitant change in protein phosphorylation. *Brain Res* 1994, **661:** 51–55.
77. Hartley A, Stone JM, Heron C, Cooper JM, Schapira AHV. Complex I inhibitors induce dose dependent apoptosis in PC12 cells relevance to Parkinsons Disease. *J Neurochem* 1994, **63:** 1987–1990.
78. Itano Y, Nomura Y. 1-Methyl-4-phenyl-pyridinium ion (MPP(+)) causes DNA fragmentation and increases the Bcl-2 expression in human neuroblastoma, SH-SY5Y cells, through different mechanisms. *Brain Res* 1995, **704:** 240–245.
79. Jackson-Lewis V, Jakowec M, Burke RE, Przedborski S. Time course and morphology of dopaminergic neuronal death caused by the neurotoxin 1-methyl-4-phenyl-1,2,3,6,-tetrahydropyridine. *Neurodegen* 1995, **4:** 257–269.
80. Clarke PGH. Developmental cell death: morphological diversity and multiple mechanisms. *Anat Embryol* 1990, **181:** 195–213.
81. Schwartz LM, Smith SW, Jones MEE, Osborne Ba. Do all programmed cell deaths occur via apoptosis?. *Proc Natl Acad Sci USA* 1993, **90:** 980–984.
82. Hassouna I, Wickert H, Zimmermann M, Gillardon F. Increase in bax expression in substantia-nigra following 1-methyl-4-phenyl-1,2,3,6- tetrahydropyridine (MPTP) treatment of mice. *Neurosci Lett* 1996, **204:** 85–88.
83. Oltvai ZN, Milliman CL, Korsmeyer SJ. Bcl-2 heterodimerizes *in vivo* with a conserved homolog, Bax, that accelerates programed cell death. *Cell* 1993, 74:609–619.
84. Korsmeyer SJ, Shutter JR, Veis DJ, Merry DE, Oltvai ZN. Bcl-2/Bax: a rheostat that regulates an anti-oxidant pathway and cell death. *Sem Cancer Biol* 1993, **4:** 327–332.
85. Smeyne RJ, Goldowitz D. Development and death of external granular layer cells in the weaver mouse cerebellum: a quantitative study. *J Neurosci* 1989, **9:** 1608–1620.
86. Migheli A, Attanasio A, Lee W-H, Bayer SA, Ghetti B. Detection of apoptosis in weaver cerebellum by electron microscopic in situ end-labeling of fragmented DNA. *Neurosci Lett* 1995, **199:** 53–56.
87. Wullner U, Loschmann PA, Weller M, Klockgether T. Apoptotic cell death in the cerebellum of mutant weaver and lurcher mice. *Neurosci Lett* 1995, **200:** 109–112.
88. Harrison Smw, Roffler-Tarlov S. Apoptotic and non-apoptotic cell death in the mouse mutant weaver. *Soc Neurosci Abstr* 1995, **424:** 16
89. Oo TF, Blazeski R, Harrison SMW, Henchcliffe C, Mason CA, Roffler-Tarlov S, Burke RE. Neuron death in the substantia nigra of weaver mouse occurs late in development and is not apoptotic. *J Neurosci* 1996, in press

90. Wyllie AH, Kerr JF, Currie AR. Cell death: the significance of apoptosis. *Int Rev Cytol* 1980, **68:** 251–306.
91. Kerr JFR, Gobe GC, Winterford CM, Harmon BV. Anatomical methods in cell death, in *Methods in Cell Biology: Cell Death*, (Schwartz LM, Osborne BA, eds.), Academic Press, NY, 1995: pp. 1–27.
92. Clarke PGH, Oppenheim RW. Neuron death in vertebrate development: *in vivo* methods, in *Methods in Cell Biology: Cell Death*, (Schwartz LM, Osborne BA, eds.), Academic Press, NY, 1995: pp. 277–321.
93. Gavrieli Y, Sherman Y, Ben-Sasson SA. Identification of programmed cell death in situ via specific labeling of nuclear DNA fragmentation. *J Cell Biol* 1992, **119:** 493–501.
94. Grasl-Kraupp B, Ruttkay-Ndicky B, Koudelka H, Bukowska K, Bursch W, Schulte-Hermann R. In situ detection of fragmented DNA (TUNEL assay) fails to discriminate among apoptosis, necrosis, and autolytic cell death: A cautionary note. *Hepatology* 1995, **21:** 1465–1468.
95. Oo TF, Henchcliffe C, Burke RE. Apoptosis in substantia nigra following developmental hypoxic-ischemic injury. *Neuroscience* 1995, **69:** 893–901.
96. Anglade P, Vyas S, Javoy-Agid F, Herrero MT, Michel PP, Marquez M, Mouatt-Prigent A, Ruberg M, Hirsch EC, Agid Y. Apoptotic degeneration of nigral dopaminergic neurons in Parkinson's disease. *Soc Neurosci Abstr* 1995, **21:** 1250
97. Lindvall O, Rehncrona S, Brundin P, Gustavii B, Astedt B, Widner H, Lindholm T, Bjorklund A, Leenders KL, Rothwell JC, Frackowiak R, Marsden CD, Johnels B, Steg G, Freedman R, Hoffer BJ, Seiger A, Bygdeman M, Stromberg I, Olson L. Human fetal dopamine neurons grafted into the striatum in two patients with severe Parkinson's disease. *Arch Neurol* 1989, **46:** 615–631.
98. Lindvall O, Brundin P, Widner H, Rehncrona S, Gustavii B, Frackowiak R, Leenders KL, Sawle G, Rothwell JC, Marsden CD, Bjorklund A. Grafts of fetal dopamine neurons survive and improve motor function in Parkinson's disease. *Science* 1990, **247:** 574–577.
99. Lindvall O, Sawle G, Widner H, Rothwell JC, Bjorklund A, Brooks D, Brundin P, Frackowiak R, Marsden CD, Odin P, Rehncrona S. Evidence for long-term survival and function of dopaminergic grafts in progressive Parkinson's disease. *Ann Neurol* 1994, **35:** 172–180.
100. Yurek DM, Sladek JR. Dopamine cell replacement: Parkinson's disease. *Ann Rev Neurosci* 1990, **13:** 415–440.
101. Mahalik TJ, Hahn WE, Clayton GH, Owens GP. Programmed cell death in developing grafts of fetal substantia nigra. *Exp Neurol* 1994, **129:** 27–36.
102. Fahn S. Controversies in the therapy of Parkinson's disease, in *Advances In Neurology, Volume 69: Parkinson's Disease*, (Battistin L, Scarlato G, Caraceni T, Ruggieri S., eds.), 1st edition, Lippincott-Raven Publishers, Philadelphia, PA, 1996: pp. 477–486.
103. Ziv I, Melamed E, Nardi N, Luria D, Achiron A, Offen D, Barzilai A. Dopamine induces apoptosis like cell death in cultured chick sympathetic neurons a possible novel pathogenetic mechanism in Parkinsons disease. *Neurosci Lett* 1994, **170:** 136–140.
104. Walkinshaw G, Waters CM. Induction of apoptosis in catecholaminergic PC12 cells by L-dopa. *J Clin Invest* 1995, **95:** 2458–464.
105. Polymeropoulos MH, Lavedan C, Leroy E, Ide SE, Dehejia A, Dutra A, Pike B, Root H, Rubenstein J, Boyer R, Stenroos ES, Chandrasekharappa S, Athanassiadou A, Papapetropoulos T, Johnson WG, Lazzarini AM, Duvoisin RC, Golbe LI, Nussbaum RL. Mutation in the a-synuclein gene identified in families with parkinson's disease. *Science* 1997, **276:** 2045–2047.
106. Tatton NA, Kish SJ. In situ detection of apoptotic nuclei in the substantia nigra compacta of 1-methyl-4-phenyl-1,2,3,6-tetrahydropyridine-treated mice using terminal deoxynucleotidyl transferase labeling and acridine orange. *Neuroscience* 1997, **77:** 1037–1048.
107. Tompkins MM, Basgall EJ, Zamrini E, Hill WD. Apoptotic-like changes in Lewy-body-associated disorders and normal aging in substantia nigral neurons. *Am J Path* 1997, **150:** 119–131.

23
Huntington's Disease and DRPLA:
Two Glutamine Repeat Diseases

Christopher A. Ross, Mark W. Becher, and Vassilis E. Koliatsos

HD AND DRPLA AS GLUTAMINE REPEAT DISEASES

Huntington's disease (HD) and Dentato-Rubro-Pallido-Luysian Atrophy (DRPLA) are two in a growing list of neurodegenerative diseases caused by the expansion of CAG repeats encoding glutamine in various genes. Other diseases known to be caused by glutamine repeat expansion include spinal cerebellar ataxia (types 1, 2, 3 and 7) and spinal and bulbar muscular atrophies (SBMA) *(1)*. Taken together, these conditions represent a subset of a more general class of diseases thought to result from expansion of trinucleotide repeats either in coding or in noncoding regions of the DNA and include conditions such as Fragile X syndrome, Friedreich's ataxia and myotonic dystrophy. Glutamine repeat diseases are the only trinucleotide expansion diseases in which triplet repeat expansion occurs within the coding DNA region. It is more than certain that the list of these diseases, all of which target excitable tissues in the human body (nervous system and muscles), will expand in the coming years.

A central feature of the glutamine repeat diseases — indeed, of all neurodegenerative diseases (*see* chapters by Braak and Braak and Burke) — is the remarkable degree of *selective vulnerability* of specific populations of neurons (Fig. 1). Each one of these diseases is characterized by a distinct pattern of neurodegeneration, although affected regions in the CNS overlap considerably *(1,2–7)*. Thus, a group of brain regions appear to be consistently vulnerable to the presence of glutamine repeats in the various genes implicated in the glutamine repeat diseases. These regions include the neostriatum, globus pallidus, the subthalamic and red nuclei, basis pontis, cerebellar Purkinje cells and dentate nucleus, and motor neurons in the brainstem and spinal cord. By contrast, brain regions affected in other neurodegenerative disorders remain free of pathology in every known glutamine repeat disease. For example, the hippocampus and the nucleus basalis of Meynert, which are affected in Alzheimer's disease, are intact in glutamine repeat disorders. The reason for this pattern of distinct, but overlapping, areas of vulnerability in this family of diseases remains unknown.

All glutamine repeat diseases are believed to result from pathogenetic mechanisms involving "gain of function" of proteins encoded by the disease genes. Although these

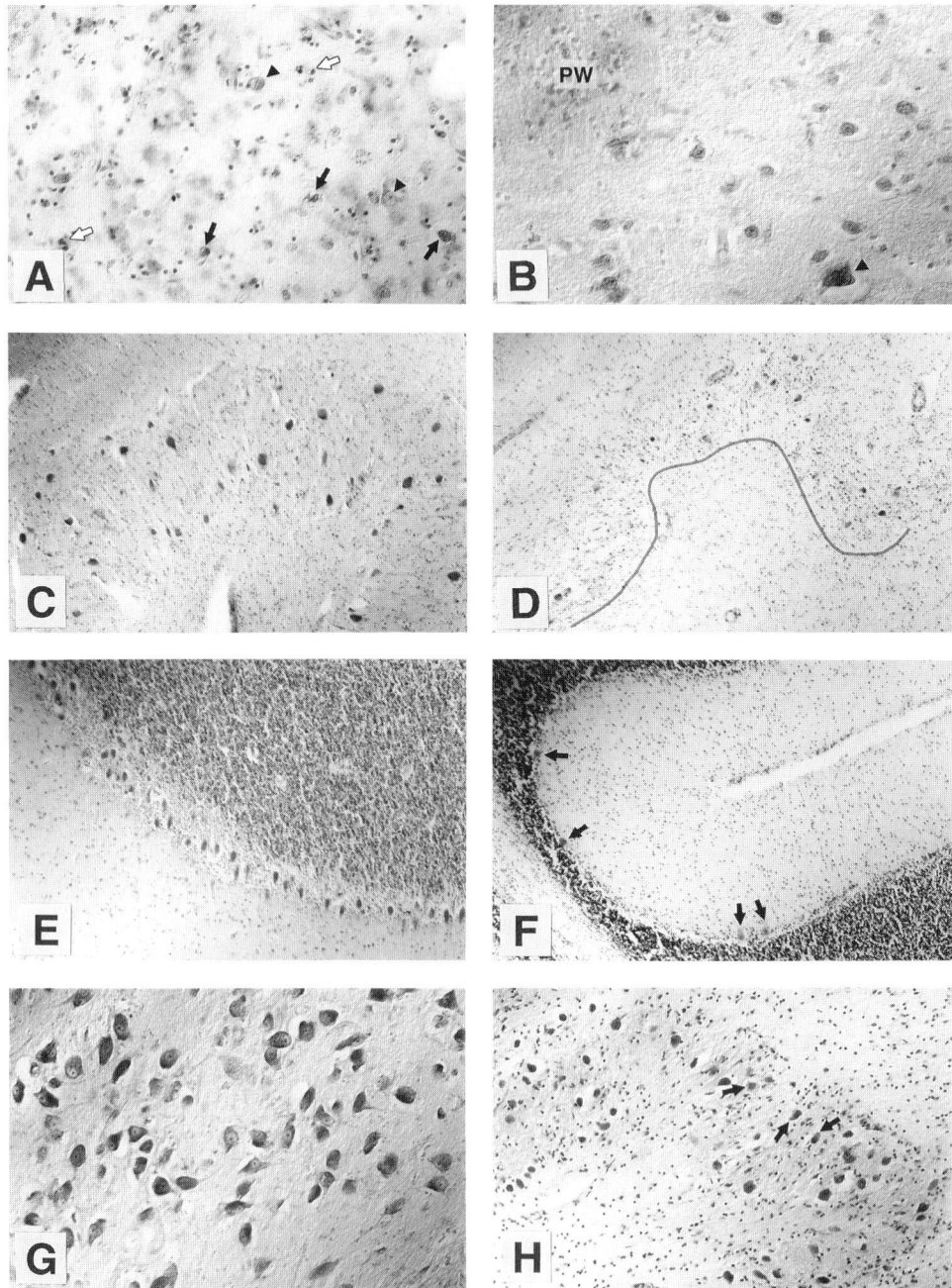

Fig. 1. Illustration of the principle of selective vulnerability that is so evident in the cases of HD (left-hand panels: A,C,E,G) and DRPLA (right-hand panels: B,D,F,H) but characterizes, to varying degrees, all neurodegenerative diseases. This figure is composed so as to document differing patterns of neuropathology between adult-onset HD and adult-onset DRPLA across same brain areas (A-B: neostriatum; C-D: cerebellar dentate nucleus; E-F: cerebellar cortex) and neuropathological distinctions between glutamine repeat and other neurodegenerative diseases, such as AD (G). All sections have been stained with dUTP-nick-end-labeling (TUNEL) and counterstained with cresyl violet. A-B. Although profound degeneration is *(continued)*

genes are expressed widely inside and outside the nervous system, pathology appears to be restricted to the central nervous system. How genes expressed so widely cause selective pathology remains a central mystery. In this chapter we will focus on two model repeat disorders, HD and DRPLA. We will examine their neuropathological differences and similarities; review possible mechanisms leading to neuronal injury and death in these diseases; and, finally, discuss recent developments in our understanding of protein-protein interactions in HD and DRPLA.

SELECTIVE VULNERABILITY OF NEURONS IN HD AND DRPLA

Generalities

In general, we can envision two, non-mutually exclusive, mechanisms through which the products of glutamine repeat genes can cause neuronal degeneration. First, the glutamine repeat expansion itself may cause cell injury, with the protein containing the glutamine repeat acting as a modifier of the degree of the injury. This viewpoint can explain the commonalities among the glutamine repeat diseases and implies that, when the glutamine repeat is long enough in any of the proteins involved in these diseases, the common toxic effect predominates and leads to a severe and more or less uniform neurodegeneration. Second, the expanded glutamine repeat may alter the protein involved in a particular disease in a specific way and, depending on the role of this protein in cells, the resulting conformational change may activate a different cascade of injury. This scenario places emphasis on the differences that exist among the proteins involved in various glutamine repeat diseases and predicts that, no matter how long the glutamine repeat is, each disease will be featured by a distinct pattern of pathology. As we mentioned above, the complexity of HD and DRPLA suggests that both viewpoints discussed above may be valid.

Clinical presentation and neuropathology differ between juvenile- and adult-onset HD and DRPLA. Thus, mutations in a single gene can cause variable pathogenetic processes, leading to various clinical and pathological phenotypes. The study of juvenile HD and DRPLA may provide clues to solving the dilemma of a common patho-

---

Fig. 1. *(continued)* seen in the sensorimotor area of putamen in the brain of a patient with Vonsattel stage 3 (intermediate) HD (A), the same brain region appears intact in a patient with advanced DRPLA (B). Neuropathological changes in (A) consist of TUNEL staining of nuclei belonging to medium-size nerve cells (black arrows; rightmost arrow points to a medium spiny neuron, the principal projection neuron of the neostriatum) or small-size cells representing microglia or oligondendrocytes (white arrows). Arrowheads in A indicate normal appearing medium spiny neurons. Normal-appearing cells in B include many medium spiny neurons scattered throughout the channel and a large striatal interneuron at the bottom (arrowhead). PW: pencil of Wilson, the characteristic neostriatal white matter bundle. C-D. A healthy-appearing cerebellar dentate nucleus in HD with normal cellularity (C) is contrasted with the profound degeneration of this nucleus in DRPLA, consishng of cell loss and atrophy of remaining neurons (D). E-F. The Purkinje cell layer of cerebellum appears intact in HD (E), whereas advanced loss of Purkinje cells is seen in DRPLA (F). Remaining Purkinje neurons in (F) are demarcated with arrows. G. Although the nucleus basalis of Meynert degenerates consistently in AD and commonly in PD, it appears intact in subjects with HD. H. Early stages of degeneration of the inferior olive in a case of DRPLA with TUNEL staining of many neurons (arrows).

genesis versus distinct pathogenetic mechanisms for these diseases. In other words, if there were a single unitary mechanism of glutamine repeat toxicity, juvenile cases (which express themselves when the glutamine repeats are sufficiently long) would tend to have identical patterns of pathology in all diseases. By contrast, if the pathogenetic mechanism were distinct for different diseases, one would expect distinct neuropathological patterns in the juvenile cases among various diseases.

Adult HD and DRPLA: Differences and Similarities

Despite remarkable similarities in their clinical presentation, adult-onset HD and DRPLA have different patterns of neuropathology (Fig. 1). HD is characterized by loss of neostriatal projection neurons accompanied by reactive astrocytosis and a less severe, but prominent degeneration of the globus pallidus *(8–11)*. In HD, variable degrees of neurodegeneration have been reported in the claustrum, subthalamic nucleus, amygdala, neocortex, pons, olivary complex, and cerebellar Purkinje cells *(11–14)*. Pathologic changes in these areas consist of astrocytosis with or without mild neuronal loss. Rarely, HD is characterized by marked loss of Purkinje cells *(14)*. Minimal, if any, neuronal loss is evident in the dentate nucleus of cerebellum.

Adult DRPLA is characterized by degeneration of the cerebellar dentate nucleus and globus pallidus with disruption of the dentatofugal and pallidofugal systems *(15–21)*. Neuronal loss in the dentate can be quite severe with astrocytosis and distortion of the dentate ribbon. Variable degrees of neuronal loss are found in the globus pallidus, red nucleus, and subthalamic nucleus (corpus Luysi). Some reports have stressed a preferential involvement of the external over the internal segment of globus pallidus, but this pattern is not consistent. White matter degeneration is a prominent feature of DRPLA, especially in the hilus of the dentate nucleus and adjacent deep cerebellar white matter. Focal white matter spongiosis can be found in the corona radiata *(15,18,21,22)*. Reactive astrocytosis without identifiable neuronal loss can be found in thalamus *(15,23)*, striatum *(18,20)*, and basis pontis *(17)*. Neuroaxonal dystrophy (axonal swellings and neuronal loss) is seen in the nucleus gracilis *(15,16,22,23)* and a variety of degenerative changes has been reported in the olivary complex *(15,18,23)*. Thus, areas of overlap in neuropathology between adult HD and DRPLA are degenerative changes in the globus pallidus and reactive changes in the subthalamic nucleus and thalamus.

Juvenile HD and DRPLA: Comparison with Adult-onset Disease

Juvenile HD and DRPLA are often different from their adult counterparts. In terms of clinical presentation, juvenile-onset HD frequently has rigidity rather than chorea and seizures are common. Juvenile-onset DRPLA is also frequently characterized by seizures, but without rigidity. Regarding neuropathology, juvenile variants of HD and DRPLA show more widespread brain changes than typical adult cases.

Pediatric-onset HD is a relatively rare entity, estimated to cover less than 1–2% of all HD cases *(24–27)*. Similar to the adult HD, caudate and putamen show marked neuronal loss and prominent astrocytosis but, in addition, most pediatric cases show cerebellar atrophy with loss of both Purkinje and granular cells. Globus pallidus appears markedly atrophic, although it has been reported to contain the normal number of neurons *(25)*. Astrocytosis appears to be greater in the external than in the internal pallidal segment. Cases of juvenile HD have also been reported to show astrocytosis or

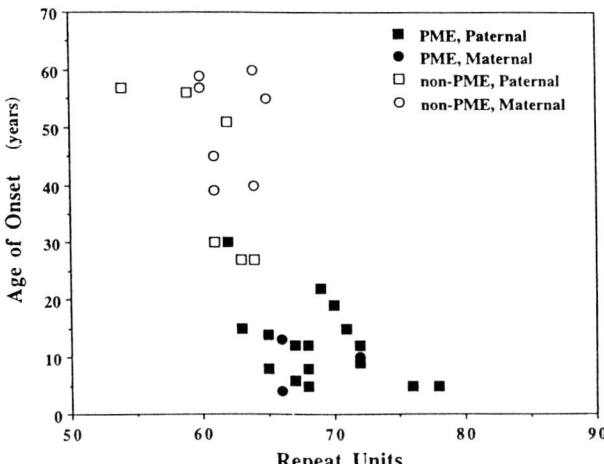

Fig. 2. Numbers of CAG triplet repeats correlate with age of onset, parental origins and clinical phenotypes in glutamine repeat diseases, in this case DRPLA. Higher repeat numbers are associated with earlier-onset disease and with a progressive myoclonus epilepsy (PME) phenotype. Most patients with the PME phenotype inherit their expanded alleles from their affected fathers. Reproduced with permission from (32).

variable loss of neurons in the dentate nucleus of the cerebellum, thalamus, hippocampus and lateral vestibular nuclei (25). Many cases show marked loss of neurons in neocortex, including focal laminar necrosis. Interestingly, numerous axonal spheroids have been noted in the nucleus gracilis in juvenile HD cases, a finding typical of the DRPLA cases originally reported as the Haw River syndrome in the United States (23,25,28) and of other adult and juvenile cases of DRPLA (15,16).

DRPLA has been thoroughly studied in the Japanese population, where the differences between juvenile- and adult-onset cases were first recognized in the course of research efforts to establish clinical subtypes of DRPLA (17,29,30). Although the pattern of neurodegeneration in adult DRPLA is quite variable, it appears that juvenile DRPLA is characterized by a more widespread loss of neurons and astrocytosis than adult DRPLA. In addition to degeneration of the dentate nucleus, neuropathological changes in juvenile DRPLA have been described in the striatum, thalamus and cerebral cortex (18,20,29).

A common denominator in juvenile HD and DRPLA is the larger length of the CAG triplet repeats than in adult-onset cases (31–34) (Fig. 2). Thus, the neuropathological differences between juvenile and adult onset cases may be correlated, at least in part, with the triplet repeat length, operating through an unknown pathogenetic mechanism (32,33,35). The significantly greater severity of the juvenile-onset cases provides a further opportunity to explore similarities and differences in the pathogenesis of HD and DRPLA. The neuropathological patterns discussed in the previous two passages of the present review are consistent with both etiopathogenetic scenarios presented in the introduction. The juvenile cases of both HD and DRPLA show degeneration of Purkinje cells (like SCA1), a pattern consistent with a common pathogenetic mechanism resulting from glutamine repeats. However, the neostriatum is relatively spared in juvenile DRPLA, and the dentate nucleus of the cerebellum is relatively spared in juvenile HD, a pattern indicating that there must also be important differences in pathogenesis.

MECHANISMS OF CELL INJURY AND DEATH

Long before the identification of the HD gene *(36)*, studies using animal models had implicated several mechanisms of cell death in HD. Excitotoxicity and metabolic toxicity represent two of the most thoroughly developed hypotheses of neuronal injury in HD.

Excitotoxicity

The result of "excessive" neuronal activation by excitatory neurotransmitters (acidic amino acids such as glutamate; *see* chapter by Olney), excitotoxicity was the first model of striatal cell death proposed for HD. Infusions of glutamate or ionotropic glutamate receptor agonists into the rat striatum reproduce many of the neuropathological changes seen in HD *(37,38)*. Striatal lesions caused by *N*-methyl-D-aspartate (NMDA) receptor agonists such as quinolinic acid are the best experimental replica of the neuropathology seen in HD, including gliosis and selective loss of medium spiny [γ-aminobutyric acid (GABA)ergic] projection neurons with relative sparing of large somatostatin, neuropeptide Y, and cholinergic neurons *(39–41)*. Quinolinic acid-lesioned medium spiny neurons show morphological changes in their processes and increases in calbindin D28K (a Ca^{2+} binding protein) immunoreactivity similar to those seen in HD brains *(42)*. These findings suggest a possible role for NMDA receptor-mediated excitotoxicity in HD, which is consistent with the normal biology of the striatum, i.e., the fact that striatal neurons receive dense excitatory (glutamatergic) inputs from cortex. In fact, ablation of giutamatergic corticostriatal projections prior to treatment with glutamate agonists attenuates striatal excitotoxicity *(38)*. In addition, a recent magnetic resonance spectroscopy study has suggested that glutamate is increased in the striatum of patients with early HD *(43)*.

One puzzle regarding the role of NMDA receptors in cell death in HD is the limited distribution of neuropathological changes compared to that of NMDA receptors *(44–47)*. NMDA receptors are equally dense in neocortex, striatum, and hippocampus, whereas pathology in HD is relatively selective for the striatum, with the hippocampus remaining essentially unaffected. Furthemmore, within cortex, degenerative changes in HD are more prominent in deep layers, whereas NMDA receptors are more abundant in superficial layers. Therefore, excitotoxicity mediated via ionotropic glutamate receptors cannot be the sole factor determining cell injury and death in HD.

The exact mechanisms of cell death resulting from excessive activation of ionotropic glutamate receptors are not known. In vitro models show two phases of neuronal cell death following exposure to toxic levels of excitatory amino acids or their agonists: an acute phase in which cells swell and die by osmotic Iysis, and a delayed phase, requiring hours to days and more closely resembling in vivo excitotoxicity *(48,49)*. Prevention of the acute phase does not abort the delayed phase. The delayed phase requires at least a transient increase in intracellular Ca^{2+} channels. Increased intracellular Ca^{2+} leads to the activation of a number of cellular processes that may be involved in cell death, including the activation of protein kineses, proteases, and phospholipases and the synthesis of nitric oxide (*see* chapter by Tymianski). Excitotoxicity can cause cell death by either necrosis or apoptosis, although a combined *in* vivo model has been proposed, in which apoptotic mechanisms predominate within the first 24 hours and are then succeeded by necrotic mechanisms *(50)*. Besides time, a critical determinant of the mode of cell death may be the dose of excitotoxic injury, as demonstrated by differences in the morphology of injured neurons between the center and the periphery of the injection site of an excitotoxic compound *(50,51)*.

Metabolic Toxicity

A more recent model than excitotoxicity, metabolic toxicity involves the subacute disruption of energy generation within neurons by mitochondrial toxins, such as malonate and 3-nitropropionic acid (3NPA) and produces a pattern of pathology very similar to that of HD *(52,54)*. Ingestion of 3NPA via contaminated food produces selective basal ganglia lesions in humans *(55)*. Systemic injection of this toxin in rats causes selective loss of striatal medium spiny neurons. The neurotoxic effects of mitochondrial toxins can be prevented by prior decortication or, *in vitro*, by antagonists of excitatory amino acids. Lesions produced by malonate are blocked by coenzyme Q10 and nicotinamide, an effect consistent with a mechanism involving energy depletion *(56)*. Chronic 3NPA treatment in baboons causes selective striatal lesions and replicates the cognitive and motoric deficits of HD *(57)*. These results suggest that a mild impairment of mitochondrial energy metabolism may selectively increase the vulnerability of medium spiny neurons to excitotoxicity *(58,59)*.

Recent evidence from studies of patients with HD is consistent with energy defects in neurons. Magnetic resonance spectroscopy can detect evidence of altered energy metabolism in HD patients *(60,61)*. Furthermore, recent studies have shown neurochemical defects in several mitochondrial enzyme complexes in postmortem brain tissues from patients with HD *(62)*. An interesting possibility that can help explain the heterogeneity of pathology is that disruption of energy generation within cells at different points can cause distinct patterns of pathology. As discussed above, 3NPA and malonate, both of which are inhibitors of complex 11 in the mitochondrial electron transport chain, cause selective neuronal degeneration in the striatum. Iodoacetic, an inhibitor of glyceraldehyde-3-phosphate dehydrogenase (GAPDH), causes striatal lesions that resemble those of excitotoxic compounds *(63)*. The previous two examples illustrate how inhibition of metabolism at disparate key points can cause similar lesions in identical brain areas. By contrast, cyanide, which is an electron transport chain inhibitor selective for complex IV, has been reported to cause a different pattern of injury, with brain imaging studies suggesting lesions in the globus pallidus, putamen, substantia nigra, subthalamic nucleus, and cerebellum *(64,65)*. In general, metabolic inhibitors tend to give rise to subcortical/cerebellar rather than cortical/hippocampal lesions, a pattern reminiscent of triplet repeat diseases.

It should be stressed that the excitotoxic and metabolic models of HD should not be viewed as separate mechanisms of neuronal injury. For example, free radicals have been implicated in both excitotoxic and metabolic models *(66–69)*. Lesions produced by intrastriatal injections of glutamate agonists in rats are attenuated by pretreatment of the animals with free radical spin trap compounds. One free radical that may be involved in excitotoxicity is nitric oxide, the product of the enzyme nitric oxide synthase (NOS) *(69–71)*. In cell cultures, nitric oxide is a mediator of NMDA toxicity *(71)*. Treatment with NOS inhibitors ameliorates excitotoxic injury in several models

(69). In mice with deletions of neuronal NOS, neurons are resistant to both excitotoxicity *(72)* and metabolic toxicity *(73)*.

GENETIC AND MOLECULAR DETERMINANTS OF NEURONAL INJURY AND DEATH IN HD AND DRPLA

Huntingtin and Atrophin-1

A central problem in the pathogenesis of HD and DRPLA is how a gene that is widely expressed causes such a restricted pathology. This question is relevant to other glutamine repeat diseases and, in general, to all neurodegenerative diseases, including Alzheimer's disease and Amyotrophic Lateral Sclerosis (ALS). As described above, selective vulnerability of neurons to generalized toxic insults may play a role, but is unlikely to provide the full explanation. Thus, for each disease, the particular features of the protein encoded by the disease gene and the interactions of this protein with other proteins are likely to be critical. There is some hope that transgenic animal models *(74)* may help clarify these processes in the future. The HD gene encodes a protein of 3144 amino acids (huntingtin) with the predicted molecular mass of 348 kDa. This protein has no significant homology to other known proteins, including proteins encoded by other genes with CAG repeat expansions *(6)*. Huntingtin has no hydrophobic domains suggestive of membrane spanning regions. It does have a repeating motif termed the "HEAT repeat", a loosely conserved repeating unit that is also present in a number of other cytoplasmic proteins. The function of this motif is unknown, but it is believed to be related to a structural role for these proteins in cells *(75)*.

Antibodies against peptides corresponding to the predicted 17-amino-acid portion of huntingtin on the N-terminal side of the glutamine repeat, as well as antibodies directed against other regions of the protein, recognize specifically a 350-kDa protein on Western blots, a finding showing that the CAG repeat is translated *(76–84)*. The use of these antibodies has shown that the huntingtin protein, like the mRNA, is localized in a variety of tissues throughout the body and in the brain *(85–88)*.

Immunocytochemical experiments using well-characterized antisera indicate that, in the brain, huntingtin is expressed widely and at all stages of development. Within neurons, huntingtin is localized in all major compartments, including perikarya, dendrites, axons, and terminals; huntingtin is not present in nuclei and, based on ultrastructural studies, is not localized in mitochondria *(77,78,82,83,89)*. Cell fractionation experiments are consistent with immunohistochemical studies in showing that huntingtin is absent in brain fractions enriched in nuclei *(82)* as well as in nuclear fractions from lymphoblastoid cell lines *(81)*. Huntingtin is also absent in brain fractions enriched in mitochondria *(82)*. Thus, contrary to initial speculations, huntingtin does not appear to have a direct involvement in functions related either to the cell nucleus or to mitochondria. Elegant experiments using colloidal gold-labeled secondary antibodies to detect immunoreactivity at the electron microscopic level *(77)* have shown huntingtin immnoreactivity around synaptic vesicles in nerve terminals as well as in the cytoplasm of postsynaptic dendrites. Gutekunst et al. *(78)* showed that, in Purkinje cell dendrites, a portion of huntingtin immunoreactivity appeared to be associated with microtubules.

The data reviewed in the previous paragraph suggest that normal huntingtin is involved in important functions in a variety of cell types, perhaps in a "housekeeping" mode. One possible role is the intracellular trafficking of vesicles, including at least some synaptic vesicles in neurons; such a role may include interactions with the neuronal cytoskeleton. Gutekunst et al. *(78)* suggested a role of huntingtin in the anchoring of vesicles or organelles to microtubules.

An important question regarding the role of huntingtin in HD is whether the expanded HD gene can be translated, and, if so, whether the properties of the expanded protein might differ from those of normal huntingtin. Several groups have detected translation product of the expanded allele *(76,78–84,90,91)* by SDS polyacrylamide gel electrophoresis (SDS-PAGE) with either a low percentage of total acrylamide or the use of crosslinker. The gel separation of the expanded from the normal protein appears to be greater than expected from a simple increase in molecular weight, a pattern suggesting that the expanded polyglutamine may alter the structure of the protein such that its electrophoretic properties are modified.

It has been suggested that expansion of the polyglutamine tract does not alter significantly the subcellular localization of huntingtin *(81,83,90,92)*. Ongoing studies on the subcellular localization of huntingtin in both normal and patient tissues using immunofluorescence in conjunction with confocal microscopy *(93,54)* demonstrate that huntingtin is enriched in the Golgi apparatus and the trans-Golgi network. In these studies, huntingtin appears to be co-localized with the transferrin receptor and with coated vesicles in the peripheral cytoplasm. There is no striking difference in huntingtin localization between control lymphoblasts and lymphoblasts from a patient homozygous for the CAG expansion, a pattern suggesting that, at least in these peripheral cells, the mutation does not alter the subcellular localization of the protein. COST cells transfected with mouse huntingtin that bears an expanded glutamine repeat derived from an HD patient show the same pattern of localization of the expressed protein as cells expressing normal-length huntingtin; this pattern is featured by an absence of nuclear staining *(83)*. Electron microscopic observations suggest that, in HD, mutant huntingtin may accumulate in the endosomal/ lysosomal compartment of brain neurons.

Contrary to the observations reviewed above that suggest a trivial cellular disposition of huntingtin that does not change with CAG expansion, investigators working with mice transgenic for exon 1 of the human HD gene carrying 115-116 CAG repeats have recently observed with the use of electron microscopy the presence of intranuclear inclusions in neurons in several brain areas, including neostriatum, prior to the development of a neurological phenotype *(95)*. These inclusions contain huntingtin, are ubiquitinated and appear to be succeeded by a pronounced indentation of the nuclear membrane and an increase in the density of nuclear pores. These inclusions are strikingly similar to nuclear structures observed with electron microscopy in biopsy samples from cortex and caudate nucleus of patients with HD twenty years ago *(96)* (Fig. 3).

Less is known about the DRPLA gene product (atrophin-1). Like the HD gene, the open reading frame of the DRPLA gene has no significant homology to other genes except for simple sequence repeats *(97–99)*. However, atrophin-1, unlike huntingtin, does appear to be a member of a gene family with at least one other member. This other gene, termed atrophin-related protein or ARP1, shows homology to atrophin-1 in both the N-terminus and C-terminus, but does not have a glutamine repeat *(100)*. The function of atrophin-1 is unknown. The DRPLA gene is widely expressed, with a predominantly cytoplasmic localization of atrophin-1 in neurons *(97,99,101–103)*. Like

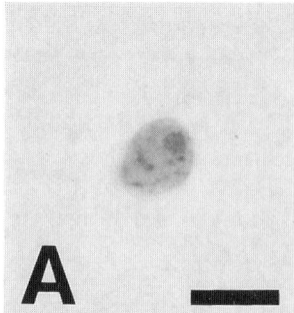 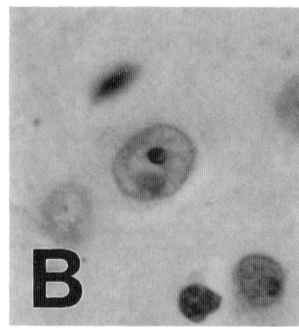

Fig. 3. Intranuclear inclusions in Huntington's Disease. Distinct inclusions which may have pathogenic significance can be found within neuronal nuclei in Huntington's disease patients using antibodies to an N-terminal epitope of huntingtin (A; AP78) or ubiquitin (B). These lesions suggest that mutant huntingtin with an expanded triplet repeat may be cleaved and the N-terminal portion, which contains the CAG triplet repeat, becomes ubiquinated and aggregates in the nucleus of vulnerable neurons. Bar = 10 μm.

huntingtin, the expanded atrophin-1 can be distinguished from the normal protein on SDS-PAGE from material from DRPLA patients *(103)*.

Glutamine Repeats

The role of glutamine repeats in proteins is unknown, but may involve interactions with other proteins. Glutamine-rich regions have been described in the factor interaction domain of transcription factors, although it is not clear whether the function of glutamine repeats is related to that of glutamine-rich regions. For example, in the androgen receptor, the expanding glutamine repeat is located in a region considered to be important for interactions with other cell type transcription factors *(104)*. In assays of transcriptional activity, the length of glutamine repeats influences transcription, presumably via interactions with other transcription factors *(105)*.

Several hypotheses regarding the role of glutamine repeats in protein-protein interactions have been proposed. Green *(106)* suggested that proteins with long glutamine repeats might become substrates for transglutaminase activity, resulting in cross-linked products involving an ε-γ glutamyl lysine isopeptide. The cross-linked protein can then be degraded by proteolysis, but, according to this hypothesis, the isodipeptide cannot be degraded and may exert toxic effects in cells.

Another possibility involves noncovalent protein interactions. Perutz and co-workers *(107–109)* have proposed that glutamine repeats can form a "polar zipper," involving a pleated sheet with hydrogen bonds linking the main peptide chain with the side chain amides. This hydrogen bonding can occur either between two proteins with glutamine repeats or by self-association within one glutamine repeat-containing protein. Lengthening of the glutamine repeat would presumably increase the stability of the association.

How the previous interactions cause toxic effects within cells is unclear. One model suggests that glutamine repeats can precipitate within cells as insoluble pleated sheets *(110)*. This hypothesis is based on the appearance, in gels, of an aggregate of a truncated glutamine repeat-containing peptide fragment and is analogous to the mechanism proposed for prion diseases (*see* chapter by Borchelt).

Protein-Protein Interactions

The models described above do not explain the selective vulnerability of neurons in HD, DRPLA, and the other glutamine repeat diseases. A possible mechanism is that the glutamine repeat-containing protein interacts with other proteins that have a more restricted pattern of expression. Many groups have begun to search for such associated proteins.

Consistent with the possibility that huntingtin associates with other proteins is the recent demonstration that huntingtin runs in a gel filtration column with a broad peak of >1000 kDa. Calmodulin is one protein with which huntingtin appears to interact indirectly *(111)*. Several other proteins interact directly with huntingtin. One such protein is GAPDH *(112,113)*. This interaction was initially demonstrated for both huntingtin and atrophin-1 using a polyglutamine amnity column *(112)*. This study suggested that the interaction was dependent on the length of the glutamine repeat. However, it is not clear that this dependence also exists in vivo. More recently, it was shown that GAPDH also interacts with ataxin 1 (the protein product of the SCA1 gene) and the androgen receptor (the protein product of the SBMA gene). However, these interactions do not appear to be altered by the length of the glutamine repeats *(114)*. It was also suggested that the huntingtin-GAPDH interaction might inhibit GAPDH enzyme activity *(115)*, although evidence for this effect has been thus far missing. Nevertheless, the huntingtin-GAPDH interaction is provocative for two reasons: First, GAPDH has been recently implicated in cell death *(116)*; second, inhibition of GAPDH can itself cause excitotoxic cell death of striatal neurons *(63)*. It is conceivable that the interaction between proteins containing glutamine repeats and GAPDH may be involved in some form of metabolic toxicity.

Huntingtin also interacts with a protein identified using the yeast two-hybrid system, termed huntingtin interactor protein 1 (HIP1) *(117,118)*. HIP1 is homologous to the previously identified yeast gene product Sla2p *(119)*. Mutations in this gene cause disturbances in endocytosis. HIP1 is also homologous to a portion of talin *(120)*, a molecule involved in cell membrane-cytoskeleton interactions. The interactions between huntingtin and HIP1 may be dependent on the length of the glutamine repeat; there is a striking decrease in the strength of the interaction when the glutamine repeat in huntingtin is expanded *(117)*. This effect is consistent with the possibility that the huntingtin-HIP1 interaction plays a role in the pathogenesis of HD. In addition, the huntingtin-HIP1 interaction is restricted to the brain, a pattern consistent with the brain-specific pathology of HD. It is still unclear how this interaction may be relevant to the regional brain pathology in HD. Biochemical evidence suggests that HIP1 is associated with the cytoskeleton and is consistent with the sequence homology of HIP1 to other cytoskeleton-associated proteins. In concert, the interactions of huntingtin with other proteins are compatible with the view that huntingtin is involved in the trafficking of vesicles, especially with regard to endocytosis, and interacts with the cytoskeleton.

Another candidate protein was identified using the yeast two-hybrid system *(121)* using the N-terminus of the HD protein with an expanded glutamine repeat as the target. This protein, termed "huntingtin-associated protein-1" (HAP-1), interacts more strongly with huntingtin when the latter bears an expanded glutamine repeat. Unlike huntingtin, the expression of HAP-1 is restricted to the brain; thus, HAP-1 could be implicated in the brain-specific pathology of HD. However, HAP-1 is not expressed selectively in vulnerable neurons. Moreover, it is not enriched in the striatum. Although in situ hybridization studies *(122)* indicate a pattern of HAP-1 expression similar to that of neuronal NOS, attempts to find a direct association between HAP-1 and NOS have been unsuccessful. HAP-1 does not interact with either huntingtin or atrophin-1. Unfortunately, HAP-1 (like huntingtin) has no strong homology to other proteins.

Recent experiments have used the yeast two-hybrid system to find proteins with which HAP-1 interacts. HAP-1 appears to dimerize, perhaps due to coiled-coil motifs. In addition, the N-temminus of HAP-1 associates with a protein we have termed "Duo" because of its homology to a recently identified protein called "Trio" *(123)*. Duo, like Trio, contains spectrin-like repeats. The cDNA coding for human Duo appears to be the human homologue of a rat cDNA called P-CIP10, identified on the basis of its binding to the C-terminal domain of the integral membrane protein Peptidylglycine α-Amidating Monooxygenase (PAM) found in secretory and endocytic vesicles *(124)*. Duo also contains a Rac1 GEF domain and a plecstrin homology (PH) domain with over 98% amino acid identity to the Rac1 GEF domain of Trio which has been demonstrated to be functionally active *(123)*.

These data are suggestive of a role of Duo in regulating Ras-like signal transduction and raise the possibility that huntingtin is also indirectly involved in this activity. Rac1 has been recently implicated in the activation of NADPH oxidase, an effect leading to the generation of superoxide radicals *(125)*. If this cascade is active in neurons, it may provide an indirect way by which huntingtin can activate free radical production and thereby cause neuronal injury and death.

Huntingtin also interacts with an E2 ubiquitin-conjugating (hE2-25k) enzyme and is also ubiquitinated itself, though neither the ubiquitinization nor the interaction appear to depend on the length of the glutamine repeat *(126)*. Antibodies to the E2 enzyme cross-react with a band which appears to be selectively expressed in brain regions affected in HD, a pattern suggesting that an unidentified protein could be involved in the selective pathology seen in HD. However, this possibility requires further investigation.

Another pathogenetic hypothesis centered on huntingtin involves proteolysis. Huntingtin can be cleaved by apopain (caspase 3) *(127,128)* an enzyme belonging to a family of proteases that have been implicated in apoptosis in a variety of cell systems (*see* chapter by Rosen and Casciola-Rosen). However, it is unclear whether proteolysis is a cause or an effect of cell death in HD.

In DRPLA, preliminary experiments have recently identified two putative interactor proteins for atrophin-1. Both have WW domains *(129)* which may interact with PY motifs (XPPXY) in atrophin-1. Expression of one of these proteins is restricted to the brain, while the other is expressed throughout the body. The strength of the interaction of these proteins with atrophin-1 does not appear to depend on the length of the glutamine repeat in atrophin-1. Neither of these proteins interacts with huntingtin. One of them bears homology to a family of E3 ubiquitin-protein ligases *(130)*, including the yeast Rsp5 protein. Strikingly, *rsp5* mutants have a very similar phenotype to *sla2* mutants in that both strains are deficient in the endocytosis and degradation of the Fur4 uracil pemmease *(131)*. This similarity suggests a possible link between huntingtin, atrophin-1 and endocytosis.

CONCLUSIONS

Two themes relevant to understanding the regional pathology of HD, DRPLA, and the other glutamine repeat diseases appear to be emerging. The first is the concept of

HD: POSSIBLE PATHOGENESIS

Triplet Repeat Expansion
↓
Abnormal Protein Product
↓
Altered Protein Interactions
HAP1, HIP1/Sla2p, hE2-25k, GAPDH, etc.
↓
Altered Cellular Processes
Proteolysis (Apopain/Caspase 3)
? Endocytosis, cytoskeletal interactions
? Rac1 activity/neuritogenesis
↓
Mechanisms of Cell Death
Excitotoxicity, metabolic stress, apoptosis, oxidative stress

Fig. 4. Hypothesized sequence of events in the pathogenesis of HD based on the present state of our knowledge. *See* text for details.

intrinsic cellular vulnerability to metabolic poisons. Different metabolic poisons give rise to slightly different patterns of pathology, which bear some similarities to the different glutamine repeat diseases. However, how the presence of the expanded glutamine repeats in the proteins encoded by the genes causing these diseases leads to metabolic vulnerability is uncertain. As discussed above, the proposal that the glutamine repeat expansions themselves are sufficient to cause pathology appears to be over-simplified. The properties of the proteins encoded by the disease genes must also be considered.

The other theme is that of specific protein interactions. It is evident that huntingtin and atrophin-1 interact with a number of cellular proteins. At least some of these interactions appear to depend on the length of the glutamine repeat and thus, to play a role in the pathogenesis of the disease. The growing evidence that huntingtin is associated with intracellular vesicles (possibly endosomes) and cytoskeletal elements is consistent with evidence for a similar role for the huntingtin-interacting proteins. An outline of a proposed pathogenetic cascade based on current data is shown in Fig. 4.

It should be kept in mind that the gain of function properties of proteins associated with the glutamine repeat diseases may be unrelated to their normal function. Nevertheless, the location of the proteins within cells is likely to be relevant. Huntingtin and atrophin-1 interact with different proteins. Atrophin-1 does not interact with HAP1 or huntingtin and the latter does not interact with the two newly identified atrophin-1 interactors. Therefore, protein-protein interactions may help explain the differences in cellular pathology in the different glutamine repeat diseases. Future work with models of glutamine repeat diseases at various levels may help clarify the relevance of the emerging biochemistry of these proteins to the pathogenesis of these diseases.

ACKNOWLEDGMENTS

Supported by NIH NS16375. Chapter was adapted from a review that appeared in Brain Pathology, vol 7 pp 1003–1016, 1997.

REFERENCES

1. LaSpada AR, Paulson HL, Fischbeck KH (1994) Trinucleotide repeat expansion in neurological disease. *Ann Neurol* **36**: 814–822.

2. Bates G (1996) Expanded glutamines and neurodegeneration—a gain of insight. *BioEssays* **18**: 175–178.
3. Housman D (1995) Gain of glutamines, gain of function? *Nature Genet* **10**: 3–4.
4. MacDonald ME, Gusella JF (1997) Huntington's disease: Translating a CAG repeat into a pathogenic mechanism. *Current Opin Neurobiol*, in press.
5. Nasir J, Goldberg YP, Hayden MR (1996) Huntington's disease: New insights into the relationship between CAG expansion and disease. *Hum Mol Genet* **5**: 1431–1435.
6. Ross CA (1995) When more is less: Pathogenesis of glutamine repeat neurodegenerative diseases. *Neuron* **15**: 493–496.
7. Warren ST, Nelson DL (1993) Trinucleotide repeat expansions in neurological disease. *Curr Opin Neurobiol* **3**: 752–759.
8. Albin RL (1995) Selective neurodegeneration in Huntington's disease. *Ann Neurol* **38**:835–836.
9. Oppenheimer DR, Esiri MM (1992) Diseases of the basal ganglia, cerebellum and motor neurons. In: *Greenfield's Neuropathology*, Adams JH, Duchen LW (eds.), pp 988–1045. Oxford University Press: New York.
10. Sharp AH, Ross CA (1996) Neurobiology of Huntington's disease. *Neurobiol Dis* **3**: 3–15.
11. Vonsattel J-P, Myers RH, Stevens TJ, Ferrante RJ, Bird ED, Richardson EP, Jr. (1985) Neuropathological classification of Huntington's disease. *J Neuropathol Exp Neurol* **44**: 559–577.
12. Hedreen JC, Peyser CE, Folstein SE, Ross CA (1991) Neuronal loss in layers V and VI of cerebral cortex in Huntington's disease. *Neurosci Lett* **133**:257–261.
13. Jackson M, Gentleman S, Lennox G, Ward L, Gray T, Randall K, Morrell K, Lowe J (1995) The cortical neuritic pathology of Huntington's disease. *Neuropathol Appl Neurobiol* **21**: 18–26.
14. Rodda RA (1981) Cerebellar atrophy in Huntington's disease. *J Neurol. Sci* **50**: 147–157.
15. Becher MW, Rubinsztein DC, Leggo J, Wagster MV, Stine OC, Ranen NG, Franz ML, Abbott MH, Sherr M, MacMillan JC, Barron L, Porteous M, Harper PS, Ross CA (1997) Dentatorubral and pallidoluysian atrophy (DRPLA): Clinical and neuropathological findings in genetically confirmed North American and European pedigrees. *Mov Disord* **12**: 519–530.
16. Goto I, Tobimatsu S, Ohta M, Hosokawa S, Shibasaki H, Kuroiwa Y (1982) Dentatorubro-pallidoluysian degeneration: Clinical, neuro-ophthalmologic, biochemical, and pathologic studies on autosomal dominant form. *Neurology* **32**:1395–99.
17. Iizuka R, Hirayama K, Maehara K (1984) Dentato-rubro-pallido-luysian atrophy: A clinico-pathological study. *J Neurol Neurosurg Psych* **47**: 1288–1298.
18. Naito H, Oyanagi S (1982) Familial myoclonus epilepsy and choreoathetosis: Hereditary dentatorubralpallidoluysian atrophy. *Neurology* **32**: 798–807.
19. Smith JK, Gonda VE, Malamud N (1958) Unusual form of cerebellar ataxia. Combined dentato-rubral and pallido-luysian degeneration. *Neurology* **8**: 205–209.
20. Takahashi H, Ohama E, Naito H, Takeda S, Nakashima S, Makifuchi T, Ikuta F (1988) Hereditary dentatorubral-pallidoluysian atrophy: Clinical and pathologic variants in a family. *Neurology* **38**: 1065–1070.
21. Warner TT, Lennox GG, Janota I, Harding AE (1994) Autosomal-dominant Dentatorubro-pallidoluysian atrophy in the United Kingdom. *Mov Disord* **9**: 289–296.
22. Warner TT, Williams LD, Walker RWH, et al. (1995) A clinical and molecular genetic study of dentatorubropallidoluysian atrophy in four European families. *Ann Neurol* **37**: 452–459.
23. Farmer TW, Wingfield MS, Lynch SA, et al. (1989) Ataxia, chorea, seizures and dementia. Pathologic features of a newly defined familial disorder. *Arch Neurol* **46**: 774–779.
24. Byers RK, Dodge JA (1967) Huntington's chorea in children. Report of four cases. *Neurol* **17**: 587–596.

25. Byers RK, Gilles FH, Gung C (1973) Huntington's disease in children: Neuropathologic study of four cases. *Neurol* **23**: 561.
26. Jervis GA (1963) Huntington's chorea in childhood. *Arch Neurol* **9**: 244–257.
27. Nelson JS (1995) Diseases of the basal ganglia. In: *Pediatric Neuropathology*, Ducket S (ed). Chapter 9, pp 212–214. Williams & Wilkins: Baltimore.
28. Burke JR, Wingfield MS, Lewis KE, Roses AD, Lee JE, Hulette C, Pericak-Vance MA, Vance JM (1994) The Haw River Syndrome: Dentato-rubropallidoluysian atrophy (DRPLA) in an African-American family. *Nature Genet* **7**: 521–524.
29. Tomoda A, Ikezawa M, Ohtani Y, Miike T, Kumamoto T (1991) Progressive myoclonus epilepsy: Dentato-rubro-pallido-luysian atrophy (DRPLA) in childhood. *Brain Dev* **13**: 266–269.
30. Uyama E, Kondo I, Uchino M, et al. (1995) Dentatorubral-pallidoluysian atrophy (DRPLA): Clinical, genetic, and neuroradiologic studies in a family. *J Neurol Sci* **130**: 146–153.
31. Duyao M, Ambrose C, Myers R, Novelletto A, Persichetti F, Frontali M, Folstein S, Ross C, Franz M, Abbott M, Gray J, Conneally P, Young A, Penney J, Hollingsworth Z, Shoulson I, Bonilla E, Alvir J, Bickham-Conde J, Cha J-H, Dure L, Gomez F, Ramos M, Sanchez-Ramos J, Snodgrass S, deYoung M, Wexler N, Moscowitz C, Penchaszadeh G, MacFarlane H, Anderson M, Jenkins B, Srinidhi J, Barnes G, Gusella J, MacDonald M (1993) Trinucleotide repeat length instability and age of onset in Huntington's disease. *Nat Genet* **4**: 387–392.
32. Ikeuchi T, Koide R, Tanaka H, et al. (1995) Dentatorubral-Pallidoluysian Atrophy: Clinical features are closely related to unstable expansions of trinucleotide (CAG) repeat. *Ann Neurol* **37**: 769–775.
33. Komure O, Sano A, Nishino N, Yamauchi N, Ueno S, Kondoh K, Sano N, Takahashi M, Murayama N, Kondo I, Nagafuchi S, Yamada M, Kanazawa I (1995) DNA analysis in hereditary dentatorubral-pallidoluysian atrophy: Correlation between CAG repeat length and phenotypic variation and the molecular basis of anticipation. *Neurology* **45**: 143–149.
34. Stine OC, Pleasant N, Franz ML, Abbott MH, Folstein SE, Ross CA (1993) Correlation between the onset age of Huntington's disease and length of trinucleotide repeat in IT-15. *Hum Mol Genet* **2**: 1547–1549.
35. Snell RG, MacMillan JC, Cheadle JP, et al. (1993) Relationship between trinucleotide repeat expansion and phenotypic variation in Huntington's disease. *Nat Genet* **4**: 393–397.
36. Huntington's Disease Collaborative Research Group (1993) A novel gene containing a trinucleotide repeat that is expanded and unstable on Huntington's disease chromosomes. *Cell* **72**: 971–983.
37. Coyle JT, Schwarcz R (1976) Lesion of striatal neurones with kainic acid provides a model for Huntington's chorea. *Nature* **263**: 244–246.
38. McGeer EG, McGeer PL, Singh K (1978) Kainate-induced degeneration of neostriatal neurons: Dependency upon corticostriatal tract. *Brain Res* **139**: 381–383.
39. Beal MF, Kowall NW, Ellison DW, Mazurek MF, Swartz KJ, Martin JB (1986) Replication of the neurochemical characteristics of Huntington's disease by quinolinic acid. *Nature* **321**: 168–171.
40. Beal MF, Brouillet E, Jenkins B, Ferrante RJ, Kowall NW, Miller JM, Storey E, Srivastava R, Rosen BR, Hyman BT (1993) Neurochemical and histologic characterization of striatal excitotoxic lesions produced by the mitochondrial toxin 3-nitropropionic acid. *J Neurosci* **13**: 4181–4192.
41. Young AB, Greenamyre JT, Hollingsworth Z, Albin R, D'Amato C, Shoulson I, Penney JB (1988) NMDA receptor losses in putamen from patients with Huntington's disease. *Science* **241**: 981–983.

42. Huang Q, Zhou D, Sapp E, Aizawa H, Ge P, Bird ED, Vonsattel J-P, DiFiglia M (1995) Quinolinic acid-induced increases in calbindin D_{28k} immunoactivity in rat striatal neurons in vivo and in vitro mimic the pattern seen in Huntington's disease. *Neuroscience* **65**: 397–407.
43. Taylor-Robinson SD, Weeks RA, Bryant DJ, Sargentoni J, Marcus CD, Harding AE, Brooks DJ (1996) Proton magnetic resonance spectroscopy in Huntington's disease: Evidence in favour of the glutamate excitotoxic theory? *Mov Disord* **11**:167–173.
44. Cotman CW, Monaghan DT, Ottersen OP, Storm-Mathisen J (1987) Anatomical organization of excitatory amino acid receptors and their pathways. *Trends Neurosci* **10**: 273–280.
45. Monaghan DT, Cotman CW (1985) Distribution of N-methyl-D-aspartate-sensitive L-[^{3}H]glutamate-binding sites in rat brain. *J Neurosci* **5**: 2909–2919.
46. Monaghan DT, Bridges RJ, Cotman CW (1989) The excitatory amino acid receptors: Their classes, pharmacology, and distinct properties in the function of the central nervous system. *Annu Rev Pharmacol Toxicol* **29**: 365–402.
47. Young AB, Fagg GE (1990) Excitatory amino acid receptors in the brain: Membrane binding and receptor autoradiographic approaches. *Trends Pharmacol Sci* **11**: 126–133.
48. Choi DW (1988) Glutamate neurotoxicity and diseases of the nervous system. *Neuron* **1**: 623–634.
49. Choi DW (1992) Bench to bedside: The glutamate connection. *Science* **258**: 241–243.
50. Portera-Cailliau C, Hedreen JC, Price DL, Koliatsos VE (1995) Evidence for apoptotic cell death in Huntington disease and excitotoxic animal models. *J Neurosci* **15**: 3775–3787.
51. Wilcox BJ, Applegate MD and Koliatsos VE (1995) Nerve growth factor prevents apoptotic cell death in injured central cholinergic neurons. *J Comp Neurol* **359**: 573–585.
52. Beal MF, Brouillet E, Jenkins B, Henshaw R, Rosen B, Hyman BT (1993) Age-dependent striatal excitotoxic lesions produced by the endogenous mitochondrial inhibitor malonate. *J Neurochem* **61**: 1147–1150.
53. Beal MF, Hyman BT, Koroshetz W (1993) Do defects in mitochondrial energy metabolism underlie the pathology of neurodegenerative diseases? *Trends Neurosci* **16**: 125–131.
54. Brouillet E, Jenkins BG, Hyman BT, Ferrante RJ, Kowall NW, Srivastaba R, Roy DS, Rosen BR, Beal MF (1993) Age-dependent vulnerability of the striatum to the mitochondrial toxin 3-nitropropionic acid. *J Neurochem* **60**: 356–369.
55. Ludolph AC, He F, Spencer PS, Hammerstad J, Sabri M (1991) 3-Nitropropionic acid-exogenous animal neurotoxin and possible human striatal toxin. *Can J Neurol Sci* **18**: 492–498.
56. Beal MF, Henshaw DR, Jenkins BG, Rosen BR, Schulz JB (1994) Coenzyme Q_{10} and nicotinamide block striatal lesions produced by the mitochondrial toxin malonate. *Ann Neurol* **36**: 882–888.
57. Palfi S, Ferrante RJ, Brouillet E, Beal MF, Dolan R, Guyot MC, Peschanski M, Hantraye P (1996) Chronic 3-nitropropionic acid treatment in baboons replicates the cognitive and motor deficits of Huntington's disease. *J Neurosci* **16**: 3019–3025.
58. Albin RL, Greenamyre JT (1992) Alternative excitotoxic hypotheses. *Neurology* **42**: 733–738.
59. Greene JC, Greenamyre JT (1995) Manipulation of membrane potential modulates malonate-induced striatal excitotoxicity *in vivo*. *Soc Neurosci Abst* **21**: 1039.
60. Jenkins B, Koroshetz W, Beal MF, Rosen B (1993) Evidence for an energy metabolism defect in Huntington's disease using localized proton spectroscopy. *Neurology* **43**: 2689–2695.
61. Jenkins BG, Brouillet E, Chen Y-C, Storey E, Schulz JB, Kirschner P, Beal MF, Rosen BR (1996) Non-invasive neurochemical analysis of focal excitotoxic lesions in models of neurodegenerative illness using spectroscopic imaging. *J Cereb Blood Flow Metab* **16**: 450–461.

62. Gu M, Gash MT, Mann VM, Javoy-Agid F, Cooper JM, Schapira AHV (1996) Mitochondrial defect in Huntington's disease caudate nucleus. *Ann Neurol* **39**: 385–389.
63. Matthews RT, Ferrante RJ, Jenkins BG, Browne SE, Goetz K, Berger S, Chen IY-C, Beal MF (1997) Iodacetate produces striatal excitotoxic lesions. *J Neurochem*, in press.
64. Rosenow F, Herholz K, Lanfermann H, Weuthen G, Ebner R, Kessler J, Ghaemi M, Heiss W-D (1995) Neurological sequelae of cyanide intoxication—the patterns of clinical, magnetic resonance imaging, and positron emission tomography findings. *Ann Neurol* **38**: 825–828.
65. Uitti RJ, Rajput AH, Ashenhurst EM, Rozdilsky B (1985) Cyanide-induced parkinsonism: A clinicopathologic report. *Neurology* **35**: 921–925.
66. Coyle JT, Puttfarcken P (1993) Oxidative stress, glutamate, and neurodegenerative disorders. *Science* **262**: 689–695.
67. Dugan LL, Sensi SL, Canzoniero LMT, Handran SD, Rothman SM, Lin TS, Goldberg MP, Choi DW (1995) Mitochondrial production of reactive oxygen species in cortical neurons following exposure to N-methyl-D-aspartate. *J Neurosci* **15**: 6377–6388.
68. Schulz JB, Henshaw DR, Siwek D, Jenkins BG, Ferrante RJ, Cipolloni PB, Kowall NW, Rosen BR, Beal MF (1995) Involvement of free radicals in excitotoxicity in vivo. *J Neurochem* **64**: 2239–2247.
69. Schulz JB, Matthews RT, Jenkins BG, Ferrante RJ, Siwek D, Henshaw DR, Cipolloni PB, Mecocci P, Kowall NW, Rosen BR, Beal MF (1995) Blockade of neuronal nitric oxide synthase protects against excitotoxicity in vivo. *J Neurosci* **15**: 8419–8429.
70. Bredt DS, Snyder SH (1992) Nitric oxide, a novel neuronal messenger. *Neuron* **8**: 3–11.
71. Dawson VL, Dawson TM, Bartley DA, Ulh GR, Snyder SH (1993) Mechanisms of nitric oxide-mediated neurotoxicity in primary brain cultures. *J Neurosci* **13**: 2651–2661.
72. Dawson VL, Kizushi VM, Huang PL, Snyder SH, Dawson TM (1996) Resistance to neurotoxicity in cortical cultures from neuronal nitric oxide synthase-deficient mice. *J Neurosci* **16**: 2479–2487.
73. Schulz JB, Huang PL, Matthews RT, Passov D, Fishman MC, Beal MF (1996) Striatal malonate lesions are attenuated in neuronal nitric oxide synthase knockout mice. *J Neurochem* **67**: 430–433.
74. Mangiarini L, Sathasivam K, Seller M, Cozens B, Harper A, Hetherington C, Lawton M, Trottier Y, Lehrach H, Davies SW, Bates GP (1996) Exon 1 of the HD gene with an expanded CAG repeat is sufficient to cause a progressive neurological phenotype in transgenic mice. *Cell* **87**: 493–506.
75. Andrade MA, Bork P (1995) HEAT repeats in the Huntington's disease protein. *Nature Genet* **11**: 115–116.
76. Aronin N, Chase K, Young C, Sapp E, Schwartz C, Matta N, Kornreich R, Landwehrmeyer B, Bird E, Beal MF, Vonsattel J-P, Smith T, Carraway R, Boyee FM, Young AB, Penney JB, DiFiglia M (1995) CAG expansion affects the expression of mutant huntingtin in the Huntington's disease brain. *Neuron* **15**: 1193–1201.
77. DiFiglia M, Sapp E, Chase K, Schwarz C, Meloni A, Young C, Martin E, Vonsattel J-P, Carraway R, Reeves SA, Boyce FM, Aronin N (1995) Huntingtin is a cytoplasmic protein associated with vesicles in human and rat brain neurons. *Neuron* **14**: 1075–1081.
78. Gutekunst C-A, Levy AI, Heilman CJ, Whaley WL, Yi H, Nash NR, Rees HD, Madden JJ, Hersch SM (1995) Identification and localization of huntingtin in brain and human lymphoblastoid cell lines with anti-fusion protein antibodies. *Proc Natl Acad Sci USA* **92**: 8710–8714.
79. Ide K, Nukina N, Masuda N, Goto J, Kanazawa I (1995) Abnormal gene product identified in Huntington's disease lymphocytes and brain. *Biochem Biophys Res Comm* **209**: 1119–1125.
80. Jou Y-S, Myers RM (1995) Evidence from antibody studies that the CAG repeat in the Huntington's disease gene is expressed in the protein. *Hum Mol Genet* **4**: 465–469.

81. Persichetti F, Ambrose CM, Ge P, McNeil SM, Srinidhi J, Anderson MA, Jenkins B, Barnes GT, Duyao MP, Kanaley L, Wexler NS, Myers RH, Bird ED, Vonsattel J-P, MacDonald ME, Gusella JF (1995) Normal and expanded Huntington's disease gene alleles produce distinguishable proteins due to translation across the CAG repeat. *Mol Med* **1**: 374–383.
82. Sharp AH, Loev SJ, Schilling G, Li S-H, Li X-J, Bao J, Wagster MV, Kotzuk JA, Steiner JP, Lo A, Hedreen J, Sisodia S, Snyder SH, Dawson TM, Ryugo DK, Ross CA (1995) Widespread expression of the Huntington's disease gene (IT-15) protein product. *Neuron* **14**: 1065–1074.
83. Trottier Y, Devys D, Imbert G, Saudou F, An I, Lutz Y, Wever C, Agid Y, Hirsch EC, Mandel J-L (1995) Cellular localization of the Huntington's disease protein and discrimination of the normal and mutated form. *Nature Genet* **10**: 104–110.
84. Trottier Y, Lutz Y, Stevanin G, Imbert G, Devys D, Cancel G, Saudou F, Weber C, David G, Tora L, Agid Y, Brice A, Mondel J-L (1995) Polyglutamine expansion as a pathological epitope in Huntington's disease and four dominant cerebellar ataxias. *Nature* **378**: 403–406.
85. Ambrose CM, Duyao MP, Barnes G, Bates GP, Lin CS, Srinidhi J, Baxendale S, Hummerich H, Lehrach H, Altherr M, Wasmuth J, Buckler A, Church D, Housman D, Berks M, Micklem G, Durbin R, Dodge A, Read A, Gusella J, MacDonald ME (1994) Structure and expression of the Huntington's disease gene: Evidence against simple inactivation due to an expanded CAG repeat. *Som Cell and Mol Genet* **20**: 27–38.
86. Landwehrmeyer GB, McNeil SM, Dure LS, Ge P, Aizawa H, Huang Q, Ambrose CM, Duyao MP, Bird ED, Bonilla E, de Young M, Avila-Gonzales AJ, Wexler NS, DiFiglia M, Gusella JF, MacDonald MD, Penney JB, Young AB, Vonsattel J-P (1995) Huntington's disease gene: Regional and cellular expression in brain of normal and affected individuals. *Ann Neurol* **37**: 218–230.
87. Li S-H, Schilling G, Young III WS, Margolis RL, Stine OC, Wagster MV, Abbott MH, Franz ML, Ranen NG, Folstein SE, Hedreen JC, Ross CA (1993) Huntington's disease is widely expressed in human and rat tissues. *Neuron* **11**: 985–993.
88. Strong TV, Tagle DA, Valdes JM, Elmer LW, Boehm K, Swaroop M, Kaatz KW, Collins FS, Albin RL (1993) Widespread expression of the human and rat Huntington's disease gene in brain and nonneuronal tissues. *Nature Genet* **5**: 259–265.
89. Bhide PG, Day M, Sapp E, Schwarz C, Sheth A, Kim J, Young AB, Penney J, Golden J, Aronin N, DiFiglia M (1996) Expression of normal and mutant huntingtin in the developing brain. *J Neurosci* **16**: 5523–5535.
90. Perisichetti F, Srinidhi J, Kanaley L, Ge P, Myers RH, D'Arrigo K, Barnes GT, MacDonald ME, Vonsattel J-P, Gusella JF, Bird ED (1996) Huntington's disease CAG trinucleotide repeats in pathologically confirmed post-mortem brains. *Neurobiol Dis* **1**: 159–166.
91. Schilling G, Sharp AH, Loev SJ, Wagster MV, Li S-H, Stine OC, Ross CA (1995) Expression of the Huntington's disease (IT15) protein product in HD patients. *Hum Mol Genet* **4**: 1365–1371.
92. Wood JD, MacMillan JC, Harper PS, Lowenstein PR, Jones AL (1996) Partial characterisation of murine huntingtin and apparent variations in the subcellular localisation of huntingtin in human, mouse and rat brain. *Hum Mol Genet* **5**: 481–487.
93. Sapp E, Schwarz C, Chase K, Bhide P, Young AB, Penney J, Vonsattel JP, Aronin N, DiFiglia M (1996) Altered neuronal expression and intracellular trafficking of huntingtin in the Huntington's disease brain. *Soc Neurosci Abst* **22**: 226.
94. Velier J, Schwarz C, Young C, Fallon J, Hyman B, Martin EJ, Hughes S, Vallee R, Aronin N, DiFiglia M (1996) Wild-type and mutant huntingtin localize to the Golgi complex and to vesicles in the peripheral cytoplasm in fibroblasts of control and HD patients. *Soc Neurosci Abst* **22**: 226.
95. Davies SW, Turmaine M, Cozens BA, DiFiglia M, Sharp AH, Ross CA, Scherzinger E, Wanker EE, Mangiarini L, Bates GP (1997) Formation of Neuronal Intranuclear Inclusions Underlies the Neurological Dysfunction in Mice Transgenic for the HD Mutation. *Cell* **90**: 537–548.

96. Roizin, L, Stellar, S, and Liu, JC (1979) Neuronal nuclear-cytoplasmic changes in Huntington's chorea: electron microscopic investigations. In: Advances in Neurology Volume 23, Huntington's Disease, T.N. Chase, N.S. Wexler and A. Barbeau, eds (New York, Raven Press), 95–122.
97. Margolis RL, Li S-H, Young WS, Wagster MV, Stine OC, Kidwai AS, Ashworth RG, Ross CA (1996) DRPLA gene (Atrophin-1) sequence and mRNA expression in human brain. *Mol Brain Res* **36**: 219–226.
98. Nagafuchi S, Yanagisawa H, Sato K, Shirayama T, Ohsaki E, Bundo M, Takeda T, Tadokoro K, Kondo I, Murayama N, Tanaka Y, Kikushima H, Umino K, Kurosawa H, Furukawa T, Nihei K, Inoue T, Sano A, Komure O, Takahashi M, Yoshizawa T, Kanazawa I, Yamada M (1994) Dentatorubral and pallidoluysian atrophy expansion of an unstable CAG trinucleotide on chromosome 12p. *Nature Genet* **6**: 14–18.
99. Onodera O, Oyake M, Takano H, Ikeuchi T, Igarashi S, Tsuji S (1995) Molecular cloning of a full-length cDNA for dentatorubral-pallidoluysian atrophy and regional expressions of the expanded alleles in the CNS. *Am J Hum Genet* **57**: 1050–1060.
100. Khan FA, Margolis RL, Loev SL, Sharp AH, Li S-H, Ross CA (1996) cDNA cloning and characterization of an atrophin-1 (DRPLA disease gene)-related protein. *Neurobiol Dis* **3**: 121–128.
101. Loev SJ, Margolis RL, Young WS, Li S-H, Schilling G, Ashworth RG, Ross CA (1995) Cloning and expression of the rat atrophin-1 (DRPLA disease gene) homologue. *Neurobiol Disease* **2**: 129–138.
102. Nagafuchi S, Yanagisawa H, Ohsaki E, Shirayama T, Tadokoro K, Inoue T, Yamada M (1994) Structure and expression of the gene responsible for the triplet repeat disorder, dentatorubral and pallidoluysian atrophy (DRPLA). *Nature Genet* **8**: 177–182.
103. Yazawa I, Nukina N, Hashida H, Goto J, Yamada M, Kanazawa I (1995) Abnormal gene product identified in hereditary dentatorubral-pallidoluysian atrophy (DRPLA) brain. *Nature Genet* **10**: 99–103.
104. Adler AJ, Danielsen M, Robins DM (1992) Androgen-specific gene activation via a consensus glucocorticoid response element is determined by interaction with nonreceptor factors. *Proc Natl Acad Sci USA* **89**: 11660–11663.
105. Gerber H-P, Seipel K, Georgiev O, Höfferer M, Hug M, Rusconi S, Schaffner W (1994) Transcriptional activation modulated by homopolymeric glutamine and proline stretches. *Science* **263**: 808–811.
106. Green H (1993) Human genetic diseases due to codon reiteration: Relationship to an evolutionary mechanism. *Cell* **74**: 955–956.
107. Perutz M (1994) Polar zippers: Their role in human disease. In: *Protein Science* **3**: 1629–1637, Cambridge University Press.
108. Perutz M, Johnson T, Suzuki M, Finch JT (1994) Glutamine repeats as polar zippers: Their possible role in inherited neurodegenerative diseases. *Proc Natl Acad Sci USA* **91**: 5355–5358.
109. Stott K, Blackburn JM, Butler PJG, Perutz M (1995) Incorporation of glutamine repeats makes protein oligomerize: Implications for neurodegenerative diseases. *Proc Natl Acad Sci USA* **92**: 6509–6513.
110. Ikeda H, Yamaguchi M, Sugai S, Aze Y, Narumiya S, Kakizuka A (1996) Expanded polyglutamine in the Machado-Joseph disease protein induces cell death *in vitro* and *in vivo*. *Nature Genet* **13**: 196–202.
111. Bao J, Sharp AH, Wagster MV, Becher M, Schilling G, Ross CA, Dawson VL, Dawson TM (1996) Expansion of poly-glutamine repeat in huntingtin leads to abnormal protein interactions involving calmodulin. *Proc Natl Acad Sci USA* **93**: 5037–5042.
112. Burke JR, Enghild JJ, Martin ME, Jou YS, Myers RM, Roses AD, Vance JM, Strittmatter WJ (1996) Huntingtin and DRPLA proteins selectively interact with the enzyme GAPDH. *Nature Med* **2**: 347–350.
113. Sirover MA (1996) Emerging new functions of the glycolytic protein, glyceraldehyde-3-phosphate dehydrogenase, in mammalian cells. *Life Sciences* **58**: 2271–2277.

114. Koshy B, Matilla T, Burright EN, Merry DE, Fischbeck KH, Orr HT, Zoghbi HY (1996) Spinocerebellar ataxia type-1 and spinobulbar muscular atrophy gene products interact with glyceraldehyde-3-phosphate dehydrogenase. *Hum Mol Genet* **5:** 1311–1318.
115. Roses AD (1996) From genes to mechanisms to therapies: Lessons to be learned from neurological disorders. *Nature Med* **2:** 267–269.
116. Ishitani R, Sunaga K, Hirano A, Saunders P, Katsube N, Chuang D-M (1996) Evidence that glyceraldehyde-3-phosphate dehydrogenase is involved in age-induced apoptosis in mature cerebellar neurons in culture. *J Neurochem* **66:** 928–935.
117. Kalchman MA, Koide HB, McCutcheon K, Graham RK, Nichol K, Nishiyama K, Kazemi-Esfarjani P, Lynn FC, Wellington C, Metzler M, Goldberg YP, Kanazawa I, Gietz RD, Hayden MR (1997) HIP1, a human homolog of S. cerevisiae Sla2p, interacts with membrane-associated huntingtin in the brain. *Nature Genet*, in press.
118. Wanker EE, Rovira C, Scherzinger E, Hasenbank R, Wälter S, Tait D, Colicelli J, Lehrach H (1997) HIP-1: A huntingtin interacting protein isolated by the yeast two-hybrid system. *Hum Mol Genet* **6:** 487–495.
119. Holtzman DA, Yang S, Drubin DG (1993) Synthetic-lethal interactions identify two novel genes, *SLA1* and *SLA2*, that control membrane cytoskeleton assembly in *Saccharomyces cerevisiae*. *J Cell Biol* **122:** 635–644.
120. Rees DJG, Ades SE, Singer SJ, Hynes RO (1990) Sequence and domain structure of talin. *Nature* **347:** 685–689.
121. Li X-J, Li S-H, Sharp AH, Nucifora FC Jr, Schilling G, Lanahan A, Worley P, Snyder SH, Ross CA (1995) A Huntingtin-associated protein enriched in brain with implications for pathology. *Nature* **378:** 398–402.
122. Li X-J, Sharp AH, Li S-H, Dawson TM, Snyder SH, Ross CA (1996) Huntingtin associated protein (HAP1): Discrete neuronal localizations in the brain resemble neuronal nitric oxide synthase. *Proc Natl Acad Sci USA* **93:** 4839–4844.
123. Debant A, Serra-Pages C, Seipel K, O'Brien S, Tang M, Park S-H, Streuli M (1996) The multi domain protein Trio binds the LAR transmembrane tyrosine phosphatase, contains a protein kinase domain, and has separate rac-specific and rho-specific guanine nucleotide exchange factor domains. *Proc Natl Acad Sci USA* **93:** 5466–5471.
124. Alam MR, Caldwell BD, Johnson RC, Darlington DN, Mains RE, Eipper BA (1996) Novel proteins that interact with the COOH-terminal cytosolic routing determinants of an integral membrane peptide-processing enzyme. *J Biological Chem* **271:** 28,636–28,640.
125. Irani K, Xia Y, Zweier JL, Sollott SJ, Der CJ, Fearon ER, Sundaresan M, Finkel T, Goldschmidt-Clermont PJ (1997) Mitogenic signaling mediated by oxidants in ras-transformed fibroblasts. *Science* **275:** 1649–1652.
126. Kalchman MA, Graham RK, Xia G, Koide HB, Hodgson JG, Graham KC, Goldberg YP, Gietz RD, Pickart CM, Hayden MR (1996) Huntingtin is ubiquitinated and interacts with a specific ubiquitin-conjugating enzyme. *J Biol Chem* **271:** 19385–19394.
127. Goldberg YP, Nicholson DW, Rasper DM, Kalchman MA, Koide HB, Graham RK, Bromm M, Kazemi-Esfarjani P, Thornberry NA, Vaillancourt JP, Hayden MR (1996) Cleavage of huntingtin by apopain, a proapoptotic cysteine protease, is modulated by the polyglutamine tract. *Nature Genet* **13:** 442–449.
128. Rosen A (1996) Huntingtin: New marker along the road to death? *Nature Genet* **13:** 380–382.
129. Bork P, Sudol M (1994) The WW domain: A signaling site in dystrophin? *Trends Biochem Sci* **19:** 531–533.
130. Huibregste JM, Scheffner M, Beaudenon S, Howley PM (1995) A family of proteins structurally and functionally related to the E6-AP ubiquitin-protein ligase. *Proc Natl Acad Sci USA* **92:** 2563–257.
131. Hein C, Springae J-V, Volland C, Haguenauer-Tsapis R, Andre B (1995) *NPL1*, an essential gene involved in induced degradation of Gap1 and Fur4 permeases, encodes the Rsp5 ubiquitin-protein ligase. *Mol Microbiol* **18:** 77–87.

24
Cortical Destruction and Cell Death in Alzheimer's Disease

Heiko Braak and Eva Braak

CORTICAL CONSTITUENTS AND AD-RELATED LESIONS

Anatomy

The neuronal constituents of the cerebral cortex are comprised of many types of pyramidal neurons and a heterogeneous group of nonpyramidal nerve cell types *(5–8)*. Most types of pyramidal cells generate an axon which enters the white matter and terminates in other cortical areas or subcortical structures. The axon and frequently its collateral branches are myelinated distal to the initial segment. Most nonpyramidal cells, by contrast, generate an axon that branches profusely in the vicinity of the parent soma. Functional maturity of most cortical projection cells is achieved only after myelination of the axon and establishment of bidirectional synaptic contacts to local circuit neurons. Projection cells, oligodendroglial cells, and local circuit neurons are the essential components of a cortical functional unit (Fig. 1).

In the brain of the adult human, the somata of pyramidal cells usually contain small, evenly distributed lipofuscin pigment granules, while nonpyramidal cells are either free of pigment or tightly filled with it (Fig. 1). The lipofuscin pigment pattern is especially valuable as a marker for differentiation of cortical neuronal types if autopsy material must be examined *(9)*.

In homotypic neocortical association areas, the number of pyramidal neurons in the supragranular layers (I-III) is roughly equivalent to that in the deep layers (IV-VI). The nonpyramidal cells are loosely distributed throughout layers II-VI. Pyramidal cells make up about 85%, and nonpyramidal cells about 15% of the neuronal constituents of the cortex *(10)*. The cortex displays a variable content in myelinated fibers. Radiant bundles consist of fibers perpendicular to the cortical surface, while myelinated lines are formed of fibers parallel to it. The inner and outer lines of Baillarger, located in layers Vb and IV, mainly represent myelinated axon collaterals of pyramidal cells. Upper portions of the third layer (sublayer IIIab) show a tendency to separate from the deeper portions, suggesting the future emergence of a new layer. Its affiliated myelinated plexus is the line of Kaes Bechterew which is a particularly late-maturing cortical component seen in only a few areas.

From: Cell Death and Diseases of the Nervous System
Edited by: V. E. Koliatsos and R. R. Ratan © Humana Press, Inc., Totowa, NJ

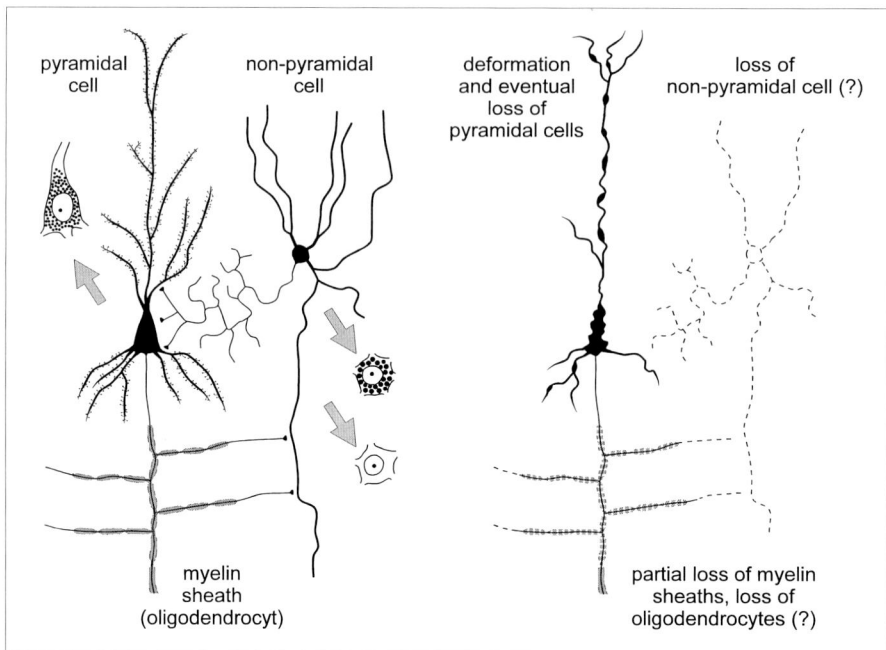

Fig. 1. *Left side*: Cortical functional units consist of projection neurons (pyramidal cells), related oligodendrocytes, and local circuit neurons (nonpyramidal cells). Pyramidal cells generally contain small, evenly distributed lipofuscin granules, while nonpyramidal cells are either free of pigment or filled with coarse granules (arrows). In general, the pyramidal cells have spiny dendrites and a myelinated axon which enters the white matter. Most nonpyramidal cells, by contrast, have smooth dendrites and a short, locally branching axon. Functional maturity is achieved only after myelination of the pyramidal cell axon and establishment of bidirectional synaptic contacts between the two neuronal types. *Right side*: AD-related destruction possibly includes premature dysfunction of specific oligodendrocytes, induction of intraneuronal cytoskeletal changes, appearance of neurofibrillary tangles and neuropil threads, and partial loss of varicose dendrites. A loss of local circuit neurons may occur secondary to the disappearance of projection cells.

Pathology

Deposition of virtually insoluble proteins in both extracellular and intraneuronal locations is among the most conspicuous cortical changes in AD. The extracellular precipitations consist mainly of specific Aβ-amyloid proteins, while abnormally phosphorylated and cross-linked tau proteins dominate the intraneuronal neurofibrillary tangles (NFTs) and neuropil threads (NTs, Fig. 2) *(11–18)*.

Aβ Amyloid

Depositions of Aβ amyloid are frequently, though not inevitably, found in the aging human brain. Initially, a few patches occur in basal regions of the neocortex. The number of deposits gradually increases, and eventually the entire cortex and adjacent portions of the underlying white matter become filled with plaque-like precipitations. An inverse relationship is observed between the degree of cortical myelination and the

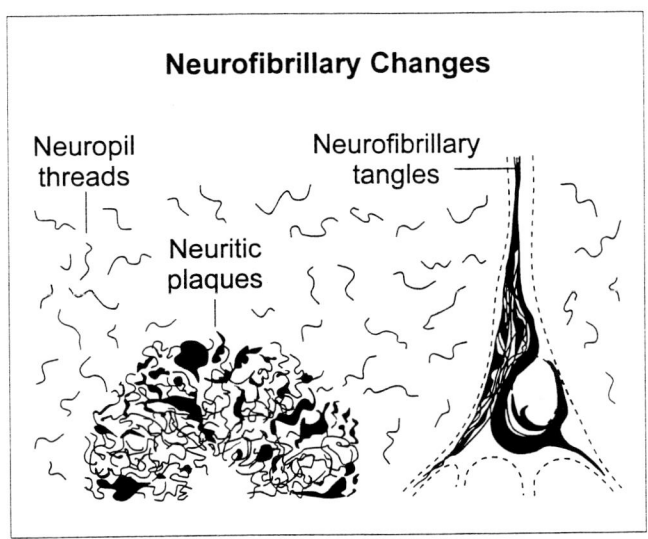

Fig. 2. Neurofibrillary tangles, neuropil threads, and dystrophic neurites of neuritic plaques are different forms of the argyrophilic changes in the neuronal cytoskeleton. Neurofibrillary tangles develop within nerve cell bodies, while neuropil threads occur in distal portions of dendritic processes. Only a few types of cortical pyramidal cells are prone to development of the changes.

density of Aβ depositions, with cortical areas rich in myelin displaying much sparser depositions than poorly myelinated fields.

There is no consistent relationship between the intensity of amyloid deposits and the severity of cortical dysfunction. In the initial stages of AD, high densities of NFTs and NTs are often observed, although no portions of entangled neurons are in contact with Aβ deposits *(2,15)*. This observation contradicts the theory that Aβ amyloid exerts a noxious influence upon sensitive nerve cells, thereby inducing the intraneuronal pathology. Whether Aβ deposits are capable of initiating other pathologic changes is not yet known. Recent findings indicate that Aβ deposits may contribute to neuronal degeneration and cell loss unrelated to cytoskeletal abnormalities *(19)*.

Neurofibrillary changes

Abnormalities of the neuronal cytoskeleton develop only in specific types of pyramidal neurons. Cells furnishing long ipsilateral cortico–cortical connections are particularly prone to accumulate the material, while local circuit neurons with short axons resist the development of these changes *(20–22,3)*. The evolution of cytoskeletal abnormalities is best examined in early stages of AD and in the brains of comparatively young individuals. Under these conditions, cytoskeletal pathology can be studied in the absence of Aβ deposits, vascular changes, or other overt pathologic lesions. Specific antibodies which react with the abnormally phosphorylated tau protein (AT8: *[23]*) decorate the altered pyramidal cells with all their neuronal processes, a picture closely resembling the result of a successful Golgi impregnation (Fig. 3). Such AT8-immunoreactive neurons exhibit no obvious destruction of their neurites. Since AT8-posi-

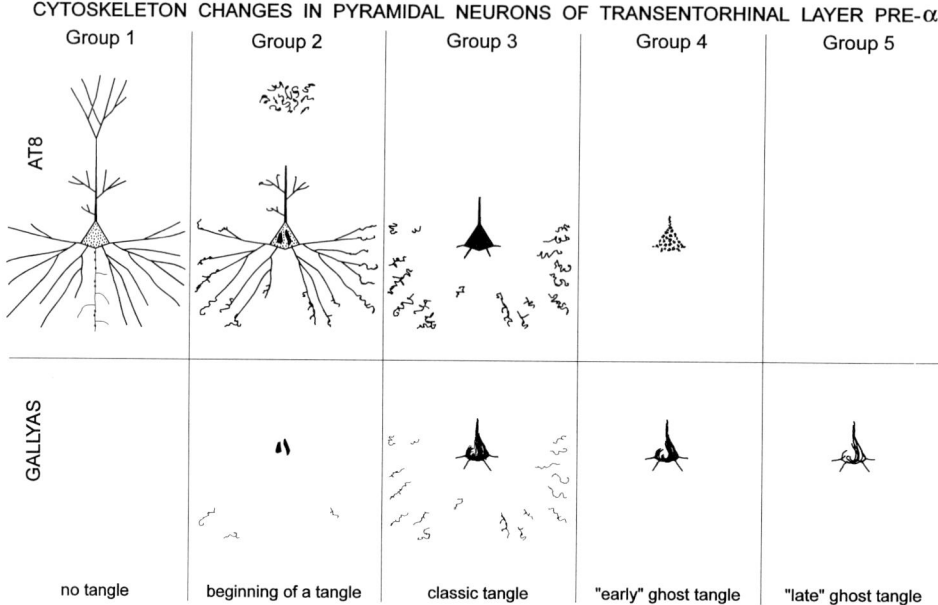

Fig. 3. Progression of AD-related changes in susceptible cortical pyramidal cells. Sequential changes in the AT8 immunoreaction are displayed for comparison with a corresponding pattern seen in Gallyas silver-stained sections (with permission from [23]).

tive phosphorylated tau proteins appear transiently during certain phases of cell division, it may be conjectured that the nonmitotic AT8-positive pyramidal cells have partially lost control of their cell cycle and exhibit an abortive initiation of the cell division process (24–26). These cells contain the soluble abnormal tau protein; aggregation to solid, argyrophilic NFTs or NTs has not yet taken place. It is not presently known just how long such cells are able to maintain the seemingly unaltered structure of their neurites. Sooner or later lesions appear (Fig. 2). The most conspicuous changes occur in the distal segment of the dendrites, which become dilated, follow a tortuous course, and develop short appendages. The altered, distal dendritic segments probably loose their connection to the proximal portion. Cells at this stage of destruction begin formation of slender argyrophilic NTs within the changed dendritic segments and, subsequently, a stout NFT appears within the soma (23).

First traces of the abnormal cytoskeletal material are generally seen in close association with the intraneuronal lipofuscin deposits (a juxtaposition suggesting, perhaps, a facilitation of crosslinking reactions). In some neuronal types, the NFT extends into the proximal dendrites, while in others it remains confined to the cell body. The NFT never extends into the proximal axon. Changes in the cellular processes cause severe disturbances in neuronal function long before the eventual disappearance of the cell bodies of entangled cortical projection cells. The circumstances of death of these neurons are not yet fully understood (27,19). Their mode of dying appears to be different from events usually referred to as apoptosis or necrosis. Neither blebbing, chromatin condensations, nor the appearance of abundant macrophages is generally observed. A possible explanation is that the stepwise reduction of the cellular processes takes a long time. Normal pyramidal cells show a centrally placed nucleus and well-developed Nissl material. In tangle-bearing neurons, by contrast, the staining capacity of the basophilic material is decreased and the nucleus is located eccentrically. After deterioration of the parent cell soma, the pathologic cytoskeletal material is converted into an extraneuronal "ghost" tangle *(28,23)*. During this process, the NFT becomes less densely twisted and gradually loses its specific argyrophilia. Cortical pyramidal cells may bear a tangle for years. "Ghost" tangles occur only in areas and layers with particularly early destruction and only in relatively late stages of the disease. As time passes, even the "ghost" tangles disappear from the tissue. The lipofuscin granules of entangled and dying cortical projection cells then also become extraneuronal, and specific staining of these components can be used to indicate the former position of the lost nerve cells *(29)*.

Very little information is currently available concerning AD-related changes of nonpyramidal cells *(30,31)*. Many types of local circuit neurons are unusually vulnerable, and disappear rapidly from the cortical tissue *(32,10)*. Some are sensitive to disconnection of their afferents *(33)*, while others show a marked sensitivity to hypoxia or noxious influences *(34)*. Cortical local circuit neurons that contain calcium-binding proteins seem to resist the development of neurofibrillary changes *(35,21,22)*. A loss of such cells secondary to the disappearance of projection cells is, however, likely (Fig. 1, right half).

Significance of Neuronal Loss in AD

It is questionable whether assessment of neuronal loss in AD provides any relevant information regarding the functional capabilities of the brain. In general, the pyramidal cells are sturdy constituents of the cortex. They survive for a long time despite marked alterations of their cytoskeleton and their cellular processes *(2,3)*. However, the mere preservation of such severely altered neurons should not lead to the assumption that their functional capabilities remain unimpaired (Fig. 1, right side). A mere quantitative assessment of the number of cortical nerve cells provides, thus, very little information about the severity of AD-related destruction. Moreover, a mild neuronal loss confined to specific areas, layers, and cell types of the cortex, as is typically seen in AD, almost certainly escapes recognition in global counts *(4)*.

CORTICAL SUBDIVISIONS AND AD-RELATED LESIONAL PATTERN

Anatomy

For a better understanding of the nonrandom pattern of AD-related pathology, recall that the human cerebral cortex consists of an extensive neocortex (proneocortex and mature neocortex: 95%) and a small allocortex (periallocortex and allocortex: 5%) *(5,8)*. The allocortex is located chiefly in the anteromedial portions of the temporal lobe and includes limbic system centers such as the hippocampal formation, the presubicular region, and the entorhinal region. Closely related is the subcortical nuclear complex of the amygdala *(36)*. Large territories within both the neocortex and the allocortex are a late development, both phylogenetically and ontogenetically.

The parietal, occipital, and temporal neocortices are each comprised of a primary core field, a belt region, and related association areas. Somatosensory, visual, and auditory information proceeds through the respective core and belt fields to a variety of association areas, from which the data is conveyed via long cortic-cortical pathways to the prefrontal cortex. Tracts generated in this highest organizational level of the brain then transmit the data via the frontal belt (premotor areas) to the primary motor area. The major routes of transport are the striatal loop and the cerebellar loop, which participate in the regulation of cortical output (Fig. 4).

Part of the main stream of data from various sensory association areas to the prefrontal cortex branches off and converges upon periallocortex, allocortex, and amygdala. Neocortical information is thus the dominant source of input to the human limbic system (Fig. 4). The transentorhinal region and the lateral nucleus of the amygdala are the major ports of entry for this highly processed data. In Fig. 5, the limbic loop is shown in more detail. The gray arrow emphasizes the strategic position of the limbic loop between the sensory association areas on the left and the prefrontal cortex on the right. The hippocampal formation, the entorhinal region, and the amygdala are densely interconnected, and the three together represent the highest level of organization of the limbic system. Part of the hippocampal, entorhinal, and amygdalar efferents terminate in the ventral striatum. The data are then transferred via the ventral pallidum and the magnocellular portion of the mediodorsal thalamic nucleus to the prefrontal cortex, in particular to its medial and orbitofrontal portions. All areas of the limbic loop play a significant role in the maintenance of emotional stability, learning skills, and memory functions *(37–39)*.

Pathology

The cortical and subcortical components of the limbic loop are affected the most by the neurofibrillary changes which gradually appear in the course of AD (Fig. 5). The changes involve only specific architectonic units, and spread in a predictable manner across other cortical areas and subcortical nuclei *(40,41,37,42,43,2,3,44–51)*. Specific projection cells in the periallocortical transentorhinal region, located in the depths of the rhinal sulcus, are the first cortical neurons to show the changes (clinically silent stage I). Stage II cases exhibit numerous transentorhinal NFTs/NTs and additional ones in the entorhinal region proper (Fig. 5). Cortical destruction at stage II slightly impedes the transmission of neocortical information—via the entorhinal region—to the hippocampal formation, but without exceeding the threshold above which initial clinical symptoms appear. Some individuals develop initial NFTs/NTs at a surprisingly young age *(52)*. Obviously, advanced age is no prerequisite for the development of the intraneuronal pathology. This observation casts doubt on theories which seek to explain the changes as a consequence of noxious influences generally expected to take effect in old age (peroxidative stress, mitochondrial dysfunction, or imbalance of glucose metabolism) (for a more topical discussion of these problems, *see* chapters by Dykens and by Flanigan and Ratan). This does not rule out the possibility that peroxidative stress may contribute to the changes in advanced stages of the disease or influence the pace of the pathological process *(53–57)*. However, it is not likely to be a primary factor in the pathogenesis of the initial AD-related lesions. In cases at the limbic stages III or IV, cortical destruction is already severe, but is limited to a few allocortical regions and adjacent areas. The key feature of these stages is the striking destruction of the entorhinal layers responsible for the data transfer from the neocortex to the hippocampus and vice versa. Initially, the hippocampal formation itself is only mildly involved; at stage IV, however, the destructive process spreads out markedly from the

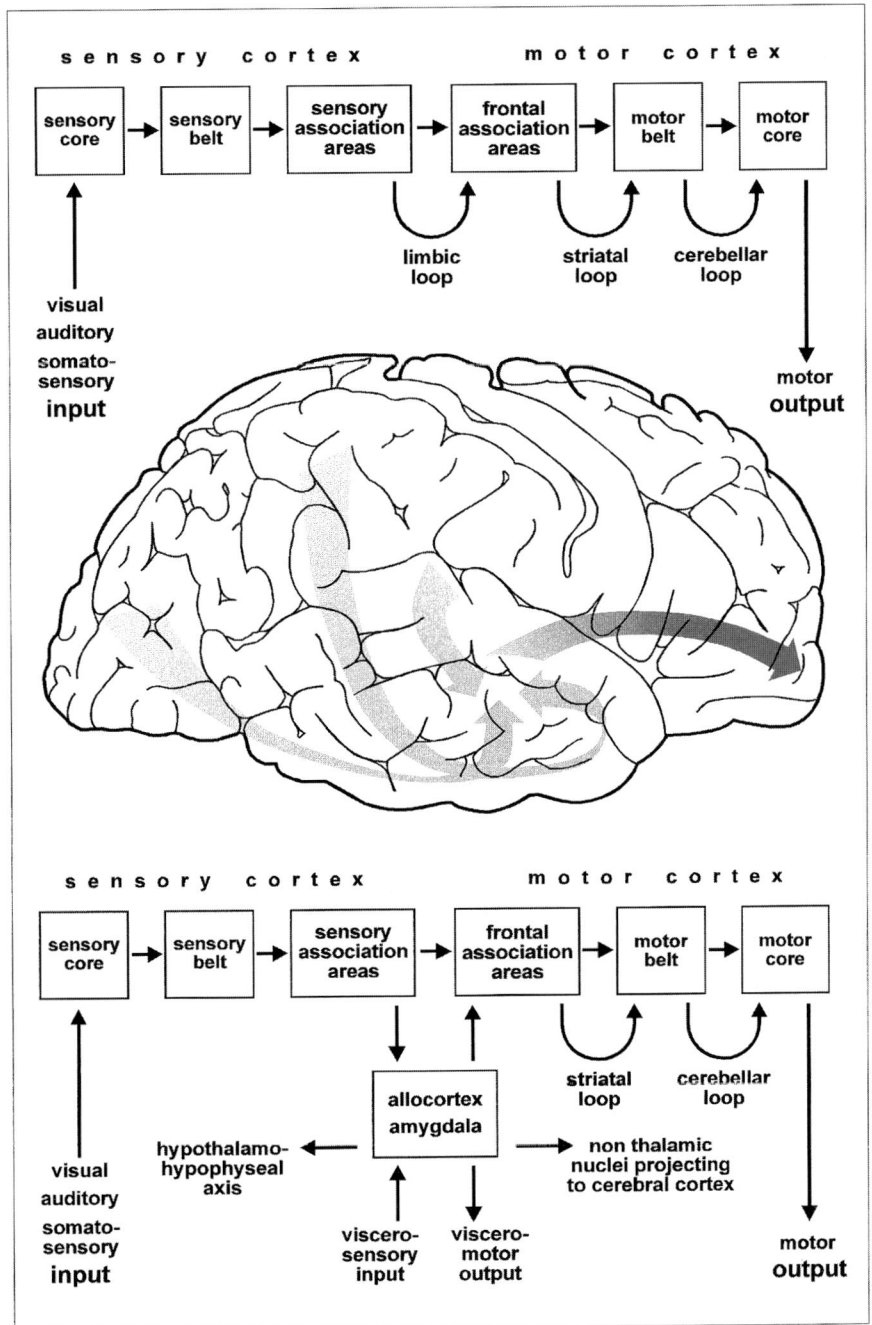

Fig. 4. Somatosensory, visual, and auditory information proceeds through core and belt fields of the neocortex to a variety of association areas. The data are then transported via long cortico–cortical pathways to the prefrontal association cortex. From here, the data are transferred—preferably via the striatal and cerebellar loops—to the primary motor field. Part of the stream of data from the sensory association areas to the prefrontal cortex branches off and converges upon the entorhinal region and the amygdala (afferent arm of the limbic loop). Projections from the entorhinal region, the amygdala, and the hippocampus exert important influence upon the prefrontal cortex (efferent arm of the limbic loop). Reproduced with permission from *(64)*.

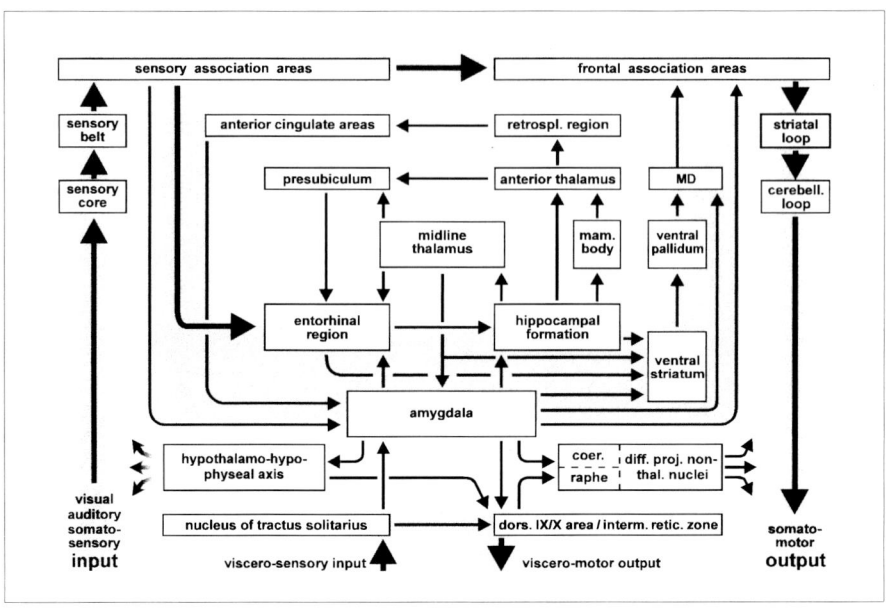

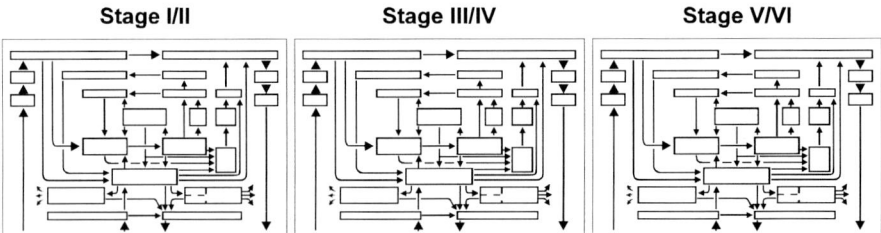

Fig. 5. *Upper half*: Schematic representation of neocortical and limbic input to the entorhinal region; data transfer between entorhinal region, amygdala, hippocampal formation, and organization of major limbic circuit connections. cerebell.; cerebellar; coer.; locus coeruleus; diff. proj. nonthal. nuclei; nonthalamic nuclei diffusely projecting to the cerebral cortex; dors. IX/X area; dorsal glossopharyngeal and vagal area; interm. retic. zone; intermediate reticular zone; mam.; mamillary nuclei; MD; mediodorsal nuclei of the thalamus; retrospl.; retrosplenial. Reproduced with permission from ref. *(64)*. *Lower half*: Distribution pattern of neurofibrillary changes during the course of Alzheimer's disease. Six stages (I-VI) can be distinguished. Stages I–II show alterations virtually confined to the transentorhinal and entorhinal regions; a key characteristic of stages III–IV is the severe involvement of the entorhinal region, hippocampus, and adjoining neocortical areas, whereas stages V–VI are marked by devastating destruction of virtually all neocortical association areas. Increasing density of shading indicates increasing severity of the changes.

entorhinal region into the amygdala, the hippocampus, and, especially, into the adjoining association areas of the basal temporal neocortex (Fig. 5). The clinical protocols of many individuals at stage III or IV record an impairment of cognitive functions and presence of subtle personality changes. In others, the appearance of symptoms is still camouflaged by individual reserve capacities which compensate for the local destruction. Because of the occasional expression of initial symptoms and the characteristic brain lesions, stage III or IV cases are considered to represent incipient AD. The final neocortical stages show large numbers of NFTs and NTs in virtually every subdivision of the cerebral cortex. A key feature of stage V is the extremely severe destruction of neocortical association areas, leaving only the primary motor field, the primary sensory areas, and their belt regions uninvolved or mildly affected. At stage VI, the pathological process extends into the primary areas (Figs. 5–7). Individuals featured by brain destruction at stages V or VI are all demented; these stages correspond to fully developed AD *(58–62)*.

CORTICAL MYELINATION AND AD-RELATED LESIONAL PATTERN

The consistency of the pattern of lesions which gradually develop in the course of AD is still an enigma. Figs. 6 and 7 illustrate that the sequence and pattern of destruction are strikingly similar to the progress of cortical myelination in early development, but in reverse order *(63)*.

The human cortex exhibits late and prolonged myelination lasting well into adulthood *(64a)*. The first traces of myelin appear in the neocortical core areas and then myelination continues through the border fields into the association areas. Temporal association areas close to the transentorhinal region and the transentorhinal region itself show a particularly late myelination; this results in a remarkably sparse myelin content in these areas in the adult brain. This region is precisely the site of the first neurofibrillary changes. Neurofibrillary alterations extend gradually into the neocortex, successively reaching association areas of the third and the second temporal gyri. Involvement of the auditory core field and other core areas is observed only in the end stages of AD (Figs. 6,7).

Regressive changes in components of the brain often repeat the process of their maturation, albeit in reverse order. The similarities between cortical myelination and development of AD-related changes can be explained by postulating premature dysfunction of late-maturing cortical oligodendrocytes. Factors released by oligodendrocytes exert an important influence on neighboring nerve cells *(65–67)*. Stability of neurons increases with myelination of their axons, and disease-related instability may be the result of decreasing influence of oligodendrocytes. Recent findings point to a loss of oligodendrocytes in AD. Presently, it is unclear whether this loss precedes the development of intraneuronal changes or is secondary to the disappearance of cortical projection neurons *(19)*. One speculative scenario is that a deficiency in oligodendrocyte factors caused by dysfunction or premature death of a specific subpopulation of these cells results in alterations of the cytoskeleton in a specific subpopulation of cortical projection neurons *(63)*.

CONCLUSIONS

A hallmark of Alzheimer's disease is area-, lamina-, and cell-type-specific involvement of cortical nerve cells. Initially, the changes are confined to a few cortical areas of the limbic system. They then spread in a predictable manner across adjoining areas, eventually encroaching upon the primary fields of the neocortex.

Only a small subset of cortical pyramidal cells is prone to develop the changes. Initially, soluble and abnormally phosphorylated tau-proteins appear in all portions of such cells. Later, the cellular processes undergo marked changes. Eventually, the material is converted into insoluble, argyrophilic neuropil threads within the dendrites and

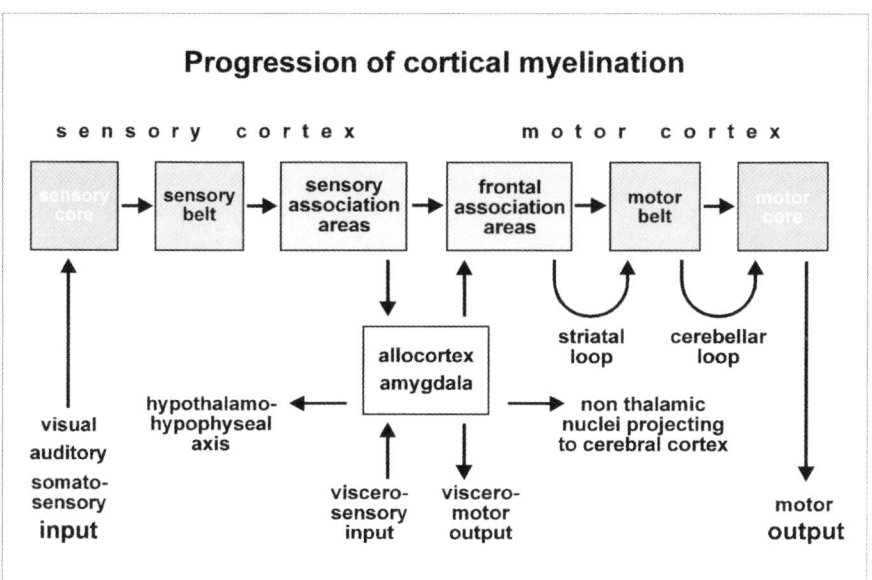

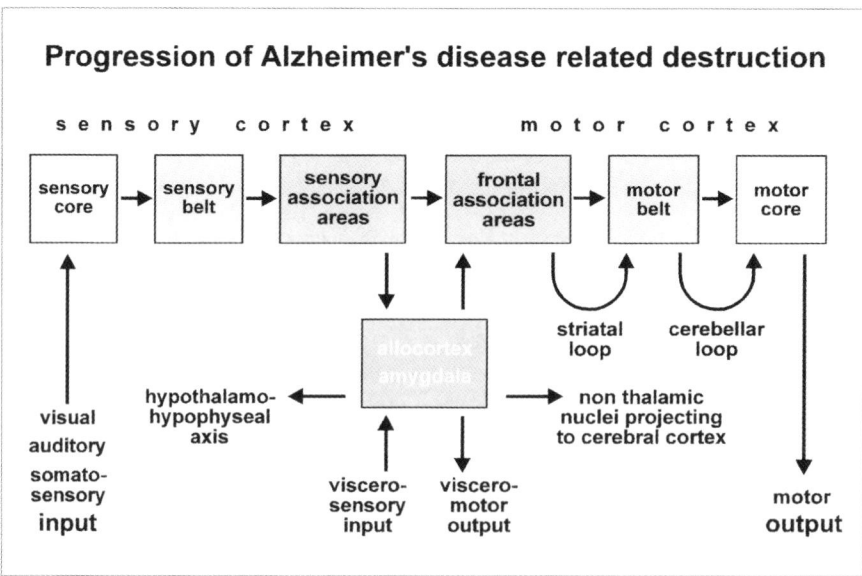

Fig. 6. *Upper half:* Cortical myelination begins in neocortical core fields and progresses via belt areas to the related association areas (indicated by different color intensity of the boxes). Lower half: AD-related cortical destruction begins in the transentorhinal region, from which the changes extend into adjacent areas, eventually reaching the neocortical core fields. Note that the sequence of destruction inversely recapitulates that of cortical myelination. Reproduced with permission from *(63)*.

neurofibrillary tangles within the soma. These cytoskeletal alterations are tolerated for a long time. However, entangled neurons eventually die and leave behind a "ghost" or "tombstone" tangle and small aggregates of neuronal lipofuscin deposits.

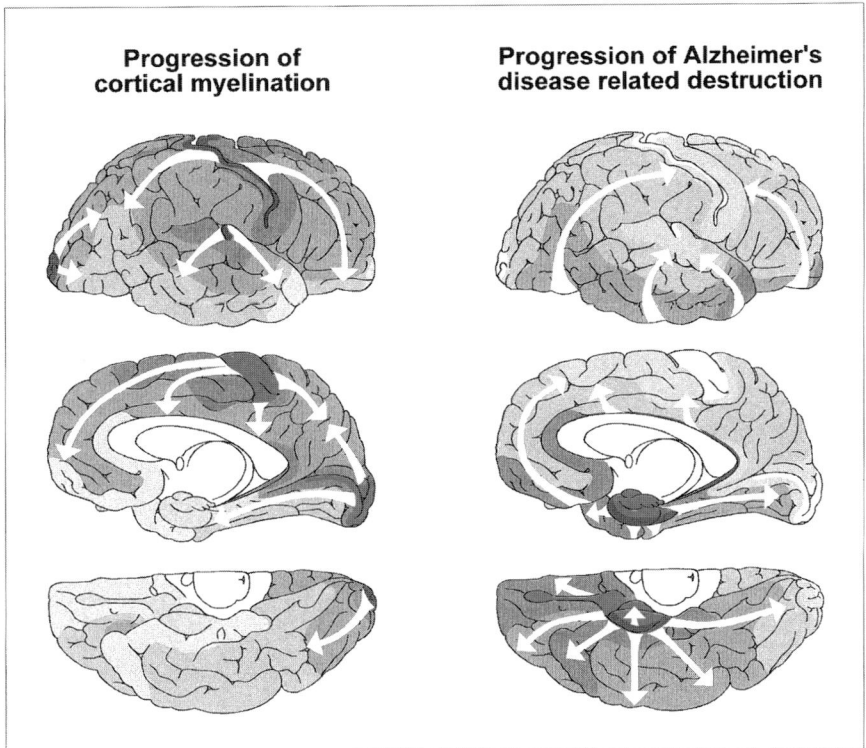

Fig. 7. Drawings of a right hemisphere showing the outward progression of myelination from the neocortical core fields into the association areas on the left side and the progression of AD-related neurofibrillary changes from the transentorhinal and entorhinal regions via neocortical association fields and belt areas into the core fields on the right side. Note the inverse pattern of the two processes. Reproduced with permission from *(63)*.

Specific pyramidal cells of the cerebral cortex show particularly late myelination of their axons. Dysfunction or loss of late maturing oligodendroglial cells may be an additional component of the brain destruction related to Alzheimer's disease. A loss of nerve cells without the formation of neurofibrillary changes is likely to occur as well and probably affects mainly local circuit neurons closely connected with entangled projection cells.

ACKNOWLEDGMENTS

This study was supported by grants from the Deutsche Forschungsgemeinschaft, the Bundesministerium für Forschung und Technologie, and Degussa, Hanau. The skillful assistance of Ms Szasz (drawings) is gratefully acknowledged.

REFERENCES

1. Corey-Bloom J, Galasko D, Thal LJ. Clinical features and natural history of Alzheimer's disease, in *Neurodegenerative Diseases*, (Calne DP, ed.), Saunders, Philadelphia, PA, 1994, 631–645.
2. Braak H, Braak E Neuropathological stageing of Alzheimer-related changes. *Acta Neuropathol* 1991, **82:** 239–259.

3. Braak H, Braak E. Pathology of Alzheimer's disease, in *Neurodegenerative Diseases*, (Calne DB, ed.), Saunders, Philadelphia, PA, 1994, 585–613
4. Regeur L, Jensen GB, Pakkenberg H, Evans SM, Pakkenberg B. No global neocortical nerve cell loss in brains from patients with senile dementia of Alzheimer's type. *Neurobiol Aging* 1994, **15:** 347–352.
5. Braak H. *Architectonics of the Human Telencephalic Cortex*, Springer, Berlin, Germany, 1980.
6. Fairén A, DeFelipe J, Regidor J. Nonpyramidal neurons: General account, in *Cerebral Cortex*, vol 1: *Cellular Organization of the Cerebral Cortex*. (Peters A, Jones EG, eds.) Plenum, NY, 1984, 201–253.
7. Feldman ML. Morphology of the neocortical pyramidal neurons, in *Cerebral Cortex*, vol 1: *Cellular Organization of the Cerebral Cortex*. (Peters A, Jones EG, eds.) Plenum, NY, 1984, 123–200.
8. Zilles K. Cortex, in *The Human Nervous System*, (Paxinos G, ed.), Academic, NY, 1990.
9. Braak H, Braak E. Architectonics as seen by lipofuscin stains, in in *Cerebral Cortex*, vol 1: *Cellular Organization of the Cerebral Cortex*. Plenum, NY, 1984, 59–104.
10. Braak H, Braak E. Ratio of pyramidal cells versus non-pyramidal cells in the human frontal isocortex and changes in ratio with ageing and Alzheimer's disease. *Progr Brain Res* 1986, **70:** 185–212.
11. Beyreuther K, Masters CL. Amyloid precursor protein (APP) and βA4 amyloid in the etiology of Alzheimer's disease: precursor product relationships in the derangement of neuronal function. *Brain Pathol* 1991, **1:** 241–252.
12. Goedert M, Spillantini MG, Crowther RA. (1991) Tau proteins and neurofibrillary degeneration. *Brain Pathol* 1991, **1:** 279–286.
13. Probst A, Langui D, Ulrich J (1991) Alzheimer's disease: A description of the structural lesions. *Brain Pathol* 1991, **1:** 229–239.
14. Dickson DW, Ksiezak-Reding H, Liu WK, Davies P, Crowe A, Yen S HC. (1992) Immunocytochemistry of neurofibrillary tangles with antibodies to subregions of tau protein: Identification of hidden and cleaved tau epitopes and a new phosphorylation site. *Acta Neuropathol* 1992, **84:** 596–605.
15. Goedert M Tau protein and the neurofibrillary pathology of Alzheimer's disease.*Trends Neurosci* 1993, **16:** 460–465.
16. Iqbal K, Alonso AC, Gong CX, Khatoon S, Singh TJ, Grundke-Iqbal I. Mechanism of neurofibrillary degeneration in Alzheimer's disease. *Molecular Neurobiol* 1994, **9:** 119–123.
17. Selkoe DJ. Alzheimer's disease: A central role for amyloid. *J Neuropathol Exp Neurol* 1994, **53:** 438–447.
18. Trojanowski JQ, Shin RW, Schmidt ML, Lee VMY. Relationship between plaques, tangles, and dystrophic processes in Alzheimer's disease. *Neurobiol Aging* 1995, **16:** 335–340.
19. Lassmann H, Bancher C, Breitschopf H, Wegiel J, Bobinski M, Jellinger K, Wisniewski HM. Cell death in Alzheimer's disease evaluated by DNA fragmentation in situ. *Acta Neuropathol* 1995, **89:** 35–41.
20. Lewis DA, Campbell MJ, Terry RD, Morrison JH. Laminar and regional distributions of neurofibrillary tangles and neuritic plaques in Alzheimer's disease: A quantitative study of visual and auditory cortices. *J Neurosci* 1987, **7:** 1799–1808.
21. Hof PR, Cox K, Young WG, Celio MR, Rogers J, Morrison JH. Parvalbumin-immunoreactive neurons in the neocortex are resistant to degeneration in Alzheimer's disease. *J Neuropathol Exp Neurol* 1991, **50:** 451–462.
22. Hof PR, Nimchinsky EA, Celio MR, Bouras C, Morrison JH. Calretinin-immunoreactive neocortical interneurons are unaffected in Alzheimer's disease. *Neurosci Lett* 1993, **152:** 145–149.
23. Braak E, Braak H, Mandelkow EM. A sequence of cytoskeleton changes related to the formation of neurofibrillary tangles and neuropil threads. *Acta Neuropathol* 1994, **87:** 554–567.
24. Lee S, Christakos S, Small MB. Apoptosis and signal transduction: clues to a molecular mechanism. *Curr Opin Cell Biol* 1993, **5:** 286–291

25. Preuss U, Döring F, Illenberger S, Mandelkow EM. Cell cycle-dependent phosphorylation and microtubule binding of tau protein stably transfected into chinese hamster ovary cells. *Mol Biol Cell* 1995, **6**: 1397–1410.
26. Vincent I, Rosado M, Davies P. Mitotic mechanisms in Alzheimer's disease? *J Cell Biol* 1996, **132**: 413–425.
27. Dickson DW. Apoptosis in the brain—physiology and pathology. *Am J Pathol* 1995, **146**: 1040–1044.
28. Bancher C, Brunner C, Lassmann H, Budka H, Jellinger K, Wiche G, Seitelberger F, Grundke-Iqbal I, Wisniewski HM. Accumulation of abnormally phosphorylated τ precedes the formation of neurofibrillary tangles in Alzheimer's disease. *Brain Res* 1989, **477**: 90–99.
29. Braak H, Braak E. Allocortical involvement in Huntington's disease. *Neuropathol Appl Neurobiol* 1992, **18**: 539–547.
30. Unger JW, Lange W. NADPH-diaphorase-positive cell populations in the human amygdala and temporal cortex: Neuroanatomy, peptidergic characteristics and aspects of aging and Alzheimer's disease. *Acta Neuropathol* 1992, **83**: 636–646.
31. Brion JP, Resibois A. A subset of calretinin-positive neurons are abnormal in Alzheimer's disease. *Acta Neuropathol* 1994, **88**: 33–43.
32. Braak H, Braak E. Golgi preparations as a tool in neuropathology with particular reference to investigations of the human telencephalic cortex. *Progr Neurobiol* 1995b, **25**: 93–139.
33. Fonseca M, Soriano E, Ferrer I, Martinez A, Tunon T. Chandelier cell axons identified by parvalbumin-immunoreactivity in the normal human temporal cortex and in Alzheimer's disease. *Neuroscience* 1993, **55**: 1107–1116.
34. Sloper JJ, Johnson P, Powell TPS. Selective degeneration of interneurons in the motor cortex of infant monkeys following controlled hypoxia: a possible cause of epilepsy. *Brain Res* 1980, **198**: 204–209.
35. Hof PR, Morrison JH. Neocortical neuronal subpopulations labeled by a monoclonal antibody to calbindin exhibit differential vulnerability in Alzheimer's disease. *Exp Neurol* 1991, **111**: 293–301.
36. Amaral DG, Price JL, Pitkänen A, Carmichael ST. Anatomical organization of the primate amygdaloid complex, in *The Amygdala: Neurobiological Aspects of Emotion, Memory, and Mental Dysfunction*, (Aggleton JP, ed.), Wiley-Liss, NY, 1992, pp. 1–66.
37. Hyman BT, van Hoesen GW, Damasio AR. Memory-related neural systems in Alzheimer's disease: An anatomic study. *Neurology* 1990, **40**: 1721–1730.
38. Damasio AR, Damasio H. Disorders of higher brain function, in *Comprehensive Neurology* (Rosenberg RN, ed.), Raven, NY, 1991, pp. 639–657.
39. Zola-Morgan S, Squire LR. Neuroanatomy of memory. *Ann Rev Neurosci* 1993, **16**: 547–563.
40. Kemper TL. Senile dementia: A focal disease in the temporal lobe, in *Senile Dementia: A Biomedical Approach* (Nandy E, ed.), Elsevier, Amsterdam, 1978, pp. 105–113.
41. Hyman BT, van Hoesen GW, Damasio AR, Barnes CL. Alzheimer's disease: Cell-specific pathology isolates the hippocampal formation. *Science* 1984, **225**: 1168–1170.
42. van Hoesen GW, Hyman BT. Hippocampal formation: anatomy and the patterns of pathology in Alzheimer's disease. *Progr Brain Res* 1990, **83**: 445–457.
43. Braak H, Braak E. On areas of transition between entorhinal allocortex and temporal isocortex in the human brain. Normal morphology and lamina-specific pathology in Alzheimer's disease. *Acta Neuropathol* 1985a, **68**: 325–332.
44. Arnold SE, Hyman BT, Flory J, Damasio AR, van Hoesen GW. The topographical and neuroanatomical distribution of neurofibrillary tangles and neuritic plaques in the cerebral cortex of patients with Alzheimer's disease. *Cerebral Cortex* 1991, **1**: 103–116.
45. Fewster PH, Griffin-Brooks S, MacGregor J, Ojalvo-Rose E, Ball MJ. A topographical pathway by which histopathological lesions disseminate through the brain of patients with Alzheimer's disease. *Dementia* 1991, **2**: 121–132.
46. Price JL, Davis PB, Morris JC, White DL. The distribution of tangles, plaques and related immunohistochemical markers in healthy aging and Alzheimer's disease. *Neurobiol Aging* 1991, **12**: 295–312.

47. van Hoesen GW, Hyman BT, Damasio AR. Entorhinal cortex pathology in Alzheimer's disease. *Hippocampus* 1991, **1:** 1–8.
48. Arriagada PV, Marzloff L, Hyman BT. Distribution of Alzheimer-type pathologic changes in nondemented elderly individuals matches the pattern in Alzheimer's disease. *Neurology* 1991, **42:** 1681–1688.
49. van Hoesen GW, Solodkin A. Some modular features of temporal cortex in humans as revealed by pathological changes in Alzheimer's disease. *Cerebral Cortex* 1993, **3:** 465–475.
50. Hyman BT, Gomez-Isla T. Alzheimer's disease is a laminar, regional, and neural system specific diasease, not a global brain disease. *Neurobiol Aging* 1994, **15:** 353–354.
51. Solodkin A, van Hoesen GW. Entorhinal cortex modules of the human brain. *J Comp Neurol* 1996, **365:** 610–627.
52. Braak H, Braak E, Bohl J, Reintjes R. Age, neurofibrillary changes, Aβ-amyloid and the onset of Alzheimer's disease. *Neurosci Lett* 1996, **210:** 87–90.
53. Volicer L, Crino PB. Involvement of free radicals in dementia of the Alzheimer's type: a hypothesis. *Neurobiol Aging* 1990, **11:** 567–571.
54. Pappolla MA, Omar RA, Kim KS, Robakis NK. Immunohistochemical evidence of antioxidant stress in Alzheimer's disease. *Amer J Pathol* 1992, **140:** 621–628.
55. Balazs L, Leon M. Evidence of an oxidative challenge in the Alzheimer's brain. *Neurochem Res* 1994, **19:** 1131–1137.
56. Benzi G, Moretti A. Are reactive oxygen species involved in Alzheimer's disease? *Neurobiol Aging* 1995, **16:** 661–674.
57. Choi BH. Oxidative stress and Alzheimer's disease. *Neurobiol Aging* 1995, **16:** 675–678.
58. Jellinger K, Braak H, Braak E, Fischer P. Alzheimer lesions in the entorhinal region and isocortex in Parkinson's and Alzheimer's diseases. *Ann New York Acad Sci* 1991, **640:** 203–209.
59. Bancher C, Braak H, Fischer P, Jellinger KA. Neuropathological staging of Alzheimer lesions and intellectual status in Alzheimer's and Parkinson's disease. *Neurosci Lett* 1991, **162:** 179–182
60. Braak H, Duyckaerts C, Braak E, Piette F. Neuropathological staging of Alzheimer-related changes correlates with psychometrically assessed intellectual status, in *Alzheimer's Disease: Advances in Clinical and Basic Research, Third International Conference on Alzheimer's Disease and Related Disorders,* (Corian B, Iqbal K, Nicolini M, Winblad B, Wisniewski H, Zatta P, eds.), Wiley, Chichester, 1993, pp. 131–137.
61. Duyckaerts C, He Y, Seilhean D, Delaère P, Piette F, Braak H, Hauw JJ. Diagnosis and staging of Alzheimer's disease in a prospective study involving aged individuals. *Neurobiol Aging* 1994, **15** (Suppl 1): 140–141.
62. Duyckaerts C, Delaère P, He Y, Camilleri S, Braak H, Piette F, Hauw JJ. The relative merits of tau- and amyloid markers in the neuropathology of Alzheimer's disease, in *Treating Alzheimer's and Other Dementias,* (Bergener M, Finkel SI, eds.), Springer, NY, 1995, pp. 81–89.
63. Braak H, Braak E. Development of Alzheimer-related neurofibrillary changes in the neocortex inversely recapitulates cortical myelogenesis. *Acta Neuropathol* 1996, **92:** 197–201.
64. Braak H, Braak E. Aspects of cortical destruction in Alzheimer's disease, in *Connections, Cognition and Alzheimer's Disease* (Hyman B, Duyckaerts C, Christen Y, eds.), Springer, Berlin, Germany, 1996, pp. 1–16.
64a. Yakovlev PI, Lecours AR. The myelogenetic cycles of regional maturation of the brain, in *Regional Development of the Brain in Early Life,* (Minkowksi A, ed.), Blackwell, Oxford, 1967, pp. 3–70.
65. Vaughan DW. The structure of neuroglial cells, in *Cerebral Cortex,* vol 2: *Functional Properties of Cortical Cells,* (Jones EG, Peters A, eds.), Plenum, NY, 1984, 285–329.
66. Schwab ME. Myelin-associated inhibitors of neurite growth and regeneration in the CNS. *Trends Neurosci* 1990, **13:** 452–456.
67. Kapfhammer JP, Schwab ME. Inverse patterns of myelination and GAP-43 expression in the adult CNS: Neurite growth inhibitors as regulators of neuronal plasticity. *J Comp Neurol* 1994, **340:** 194–206.

25
HIV-1 Infection of the CNS
Evidence for Apoptosis of Nerve Cells

Harris A. Gelbard

INTRODUCTION

The neurologic dysfunction associated with human immunodeficiency virus type 1 (HIV-1) infection of children and adults has become a problem of pandemic proportions. At present, there are approx 21.8 million individuals worldwide infected with HIV-1, and, of this group, at least one million are children. By the yr 2000, approx 6 million pregnant women and 5–10 million children are expected to have HIV-1 infection *(1)*. Furthermore, HIV-1-associated dementia is the most frequent cause of this type of neurologic disease in young adults in the USA *(2)*. HIV-1-associated dementia is estimated to occur at an annual rate of 7% in people with AIDS *(3)*. Because antiretroviral therapies and treatment for opportunistic infections have lengthened the survival time, but not eradicated the virus from the central nervous system (CNS), it is likely that the incidence of neurologic disease, including HIV-1 dementia, will increase.

To better understand how HIV-1 infection of the CNS results in selective dysfunction of the CNS and ultimately neural cell death, it is helpful to review the neurologic dysfunction that can accompany infection. The American Academy of Neurology has defined the clinical features of HIV-1 associated dementia (HIV-D; formerly AIDS dementia complex, ADC) as cognitive impairment, behavioral changes, and impairment of motor skills *(4)*. This dementia has the clinical features of a subcortical dementia; that is, primary abnormalities in language or seizures are uncommon *(5)*. People with HIV-1 associated dementia usually have rapid progression of their symptoms, with a mean survival time of ≤6 mo *(6)*. McArthur and colleagues have found that progressive neurologic deficits do not occur during the latent phases of HIV-1 infection, but rather occur after the onset of severe immunodeficiency. Indeed, only 0.4% of people had HIV-1 associated dementia during the early phase of HIV-1 infection, but the prevalence increased to 16% in patients with clinical AIDS *(6,7)*. McArthur et al. have recently predicted that 20–30% of all individuals with AIDS would develop dementia *(3)*.

HIV-1 infection of the central nervous system in children results in neurological abnormalities in 40–90% of patients. Symptoms include developmental delays, cognitive deficits, and motor abnormalities. Cognitive deficits may manifest themselves in older children as attention deficits *(8)*. In contrast to adults, the neurological abnormalities seen in children are largely due to a primary HIV-1 encephalopathy (HIVE). In one series, only 5% of children had opportunistic infections of the CNS *(9)*. Thus, neuropathologic studies of HIV-1 infection in the developing nervous system may provide the most insights into the effects of virus on brain tissue, since it is unlikely that opportunistic infections of brain may be present and confound our understanding of how the HIV-1 virus causes neurologic disease in the brain.

NEUROPATHOLOGY OF HIV-1 ENCEPHALITIS

The neuropathological hallmarks associated with HIVE, in the absence of opportunistic infections of brain, include multinucleated giant cells, microglial activation, astrogliosis, and myelin pallor (decreased staining for myelin). HIV-1 productively infects brain-resident macrophages and microglia *(10,11)*. This disease complex is most striking in infected children *(11)*. No evidence exists for productive or latent infection of endothelial cells or oligodendrocytes *(12)*.

HIV-1 can infect astrocytes in children and adults in a "restricted" fashion, that is, only regulatory gene products (Tat, nef) are made without production of progeny virus *(13–15)*. It is presently unknown how widespread this phenomenon is in patients with HIV-1 encephalitis, or what is its pathophysiological significance.

Productive infection of neurons with HIV-has never been demonstrated. One report has demonstrated latent proviral HIV-1 infection of neurons in brain tissue from a single patient with severe HIV-1-associated dementia of 5 yr duration using the technique of *in situ* polymerase chain reaction (ISPCR) *(16)*. A more recent report, also using ISPCR, failed to confirm this finding in brain tissue from patients with HIV-1- associated dementia *(12)*. Nevertheless, there is neuronal loss in discrete areas of the retina, neocortex, and in subcortical brain regions, including putamen, substantia nigra, and cerebellum. There is also loss of synaptic density and vacuolation of dendritic spines in affected areas *(17–21)*. Neuronal loss has been estimated between 18–38% in various cortical regions *(22,23)*. However, quantitative analysis of neuronal loss from post-mortem studies of patients with HIV-1 and neurologic dysfunction has not demonstrated a clear correlation between the magnitude of neuronal loss and neurologic disease *(24)*.

These neuropathologic findings suggest the following: Productive, cytolytic infection of neurons by HIV-1 is highly unlikely. Thus, changes in neuronal architecture and number are likely to be mediated by either viral gene products or cellular metabolites produced by productively infected macrophages or microglia present in focal inflammatory infiltrates. In support of this, we and others have demonstrated that HIV-1-infected macrophages, when antigenically stimulated in vitro, produce pro-inflammatory mediators of astrocyte and neuronal dysfunction and ultimately, neuronal death by apoptosis *(25–29)*. Although there appears to be a lack of correlation between neurologic deficits and neuronal density *(24)*, the same HIV-1 gene products or cellular metabolites that may be responsible for neuronal demise may also mediate neuronal dysfunction prior to death.

A recent study has demonstrated the that relative number of activated microglia and macrophages correlates with the magnitude of neurologic dysfunction *(30)*. Thus, it seems likely that interactions between HIV-1-infected macrophages and microglia and other neural cells are necessary to initiate pathological changes in neurons, but require continued infiltration of uninfected macrophages to produce neurologic symptoms.

These observations also pose several key questions beyond the scope of this chapter: what is the basis for the distribution of HIV-1-infected macrophages and microglia in brain parenchyma? What role, if any, do restrictively infected astrocytes play in the progression of neurologic symptoms? Rather this chapter will address what the fates of nerve cells are after HIV-1 infection of the CNS.

HIV-1 belongs to the group of pathogenic retroviruses known as lentiviruses. The idea that apoptosis may play a role in the pathogenesis of HIV-1 was first suggested in 1990. The hypothesis was that lentiviral infection caused cell loss and subsequent tissue atrophy in the brain and the immune system by inappropriate expression of genes involved in programmed cell death (PCD) *(31)*. In particular, it was suggested that HIV subverted normal inter- and intracellular signalling to induce pathological PCD of CD4+ lymphocytes. This has been subsequently demonstrated by a number of laboratories *(32,33)*. At the intracellular level, CD4+ and CD8+ lymphocytes from patients with both asymptomatic and symptomatic HIV-1 infection have increased levels of reactive oxygen species and a decreased mitochondrial transmembrane potential, which appears to be an early, irreversible step in the apoptotic pathway *(34)*.

Although in vitro models of HIV-1 neuropathogenesis established that HIV-1-infected macrophages and monocytoid cell lines secreted soluble neurotoxins that induced neuronal death in primary rodent and human neuronal cultures *(35–37)*, the question of whether neuronal apoptosis occurred remained unanswered. An in vitro study by Talley et al. *(38)* demonstrated that the pro-inflammatory cytokine tumor necrosis factor alpha (TNFα), present in high levels in conditioned media from activated HIV-1-infected macrophages, induced apoptosis in human neuronal cells *(38)*. TNF-induced neuronal apoptosis was blocked by overexpression of the anti-apoptosis gene product Bcl-2 and the cowpox virus gene product crmA, implicating the involvement of interleukin 1β-converting enzyme in the apoptotic pathway *(38,39)*.

EVIDENCE FOR APOPTOSIS IN THE CNS

Despite hypothesizing, in 1990, that neuronal loss documented in postmortem studies of patients with HIV-1 infection was due to apoptosis *(17,18)*, technical limitations with archival formalin-fixed postmortem brain tissue prevented answering the question of whether neurons indeed died by apoptosis. However, with the introduction of a technique to identify free 3'-OH ends of newly cleaved DNA *in situ (40)*, coupled with the light microscopic identification of some of the morphologic hallmarks of apoptosis (i.e., chromatin condensation), determination of whether apoptosis of nerve cells occurred in brain tissue from patients with HIV-1 infection became feasible. Using the TUNEL (terminal deoxynucleotidyl dUTP nick end labeling) technique, we demonstrated that apoptotic neurons were present in the cerebral cortex and basal ganglia of children that had HIVE and progressive encephalopathy (HIVE/PE) *(41)* (Fig. 1). Double-labeling immunocytochemistry for TUNEL and the HIV-1 p24 antigen revealed a spatial association between apoptotic neurons and perivascular inflammatory cell infiltrates containing HIV-1-infected macrophages and multinucleated giant cells *(41)*. Quantitative morphometric analysis of apoptotic neurons present in TUNEL-stained tissue sections through the basal ganglia of children with HIVE/PE revealed a highly significant (12-fold) increase in apoptotic neurons relative to children that were seronegative for HIV-1 infection, and a 3-fold increase in apoptotic neurons relative to

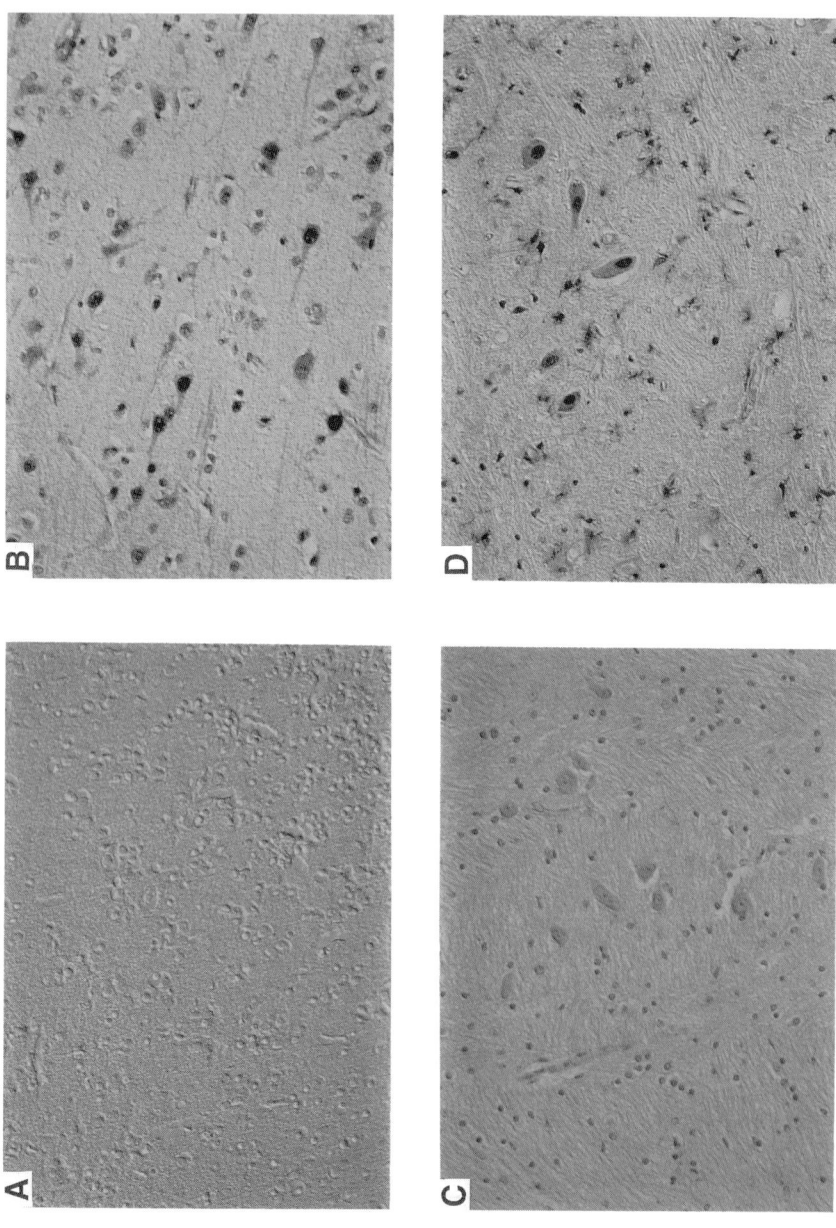

Fig. 1. Cerebral cortex (**A**) and basal ganglia (**C**) from a HIV-1 seronegative patient and cerebral cortex (**B**) and basal ganglia (**D**) from a pediatric patient with HIVE/PE immunostained with antisera to TUNEL reagent. No TUNEL staining is seen in cerebral cortex in (A). In (C), methyl green counterstain is used to identify cells with intact DNA; again, no TUNEL staining is seen. In (B) and (D) note frequent nuclear staining of neurons and both nuclear and cytoplasmic staining of microglia, as well as nuclear staining of macrophages. Original magnification is ×40, chromagen is Ni-enhanced DAB, photomicrograph is taken with a blue filter.

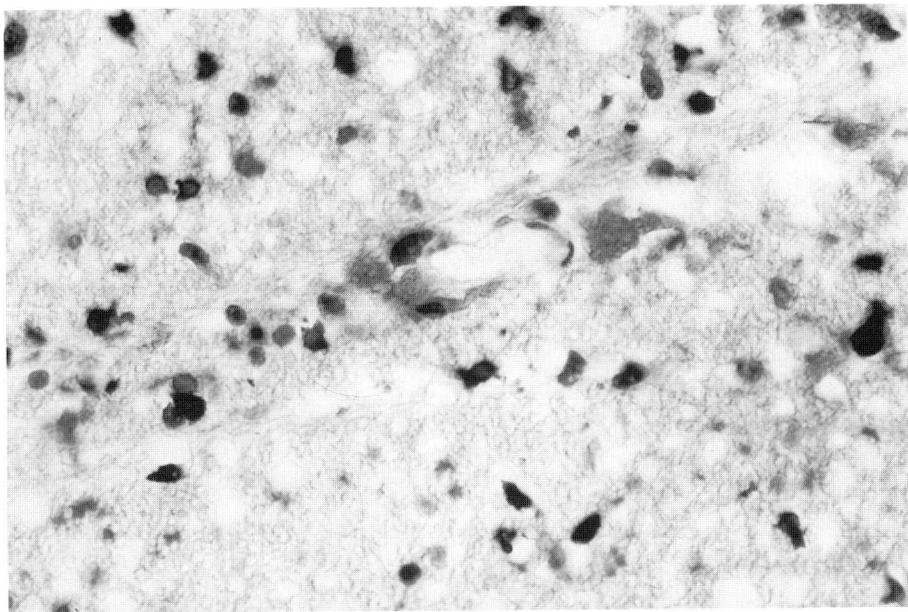

Fig. 2. Double-staining of cytoplasmic Bax and nuclear TUNEL in a high power field (×60) through the basal ganglia of a pediatric patient with HIVE/PE. Note the perivascular distribution of macrophages intensely immunoreactive for bax (red color), with many macrophages also showing TUNEL-positive nuclei. Chromagen is New Fuchsin for Bax and VIP for TUNEL. Because bax and p24 are both cytoplasmic, double staining of macrophages (or other cell types) to demonstrate HIV-1-infected cells is not possible; however, adjacent tissue sections of perivascular inflammatory infiltrates have ~10–20% p24-positive macrophages (data not shown).

children that had HIV-1 infection without the neuropathologic features of HIVE, or a premortem diagnosis of PE *(42)*.

In contrast, apoptotic neurons were infrequently observed in 3/9 cases of patients that were HIV-1 seronegative, ranging in age from the first postnatal month of life to 16.5 yr (Fig. 1). These data suggested that neuronal apoptosis was unlikely to be associated with postnatal development, but instead may be the end result of a neuropathologic process such as HIVE. Neuronal apoptosis in the CNS of adult patients with HIV-1 infection was confirmed in four separate reports *(43–46)*. One report suggested that neuronal apoptosis may occur prior to the onset of dementia *(46)*.

EXPRESSION OF CELL DEATH GENES IN HIVE

Four studies noted the present of apoptotic macrophages and microglia *(41,43,44,46)* in the brains of patients with HIVE, but only one report *(43)* noted the presence of apoptotic astrocytes in 2/7 brains of patients with HIVE. This finding was only observed with a DNA polymerase technique, not the TUNEL technique. To further investigate the fate of glial cells in the brains of pediatric patients with HIVE, we have recently analyzed the *in situ* expression of pro- (Bax) and anti-apoptosis (Bcl-2) gene products in cerebral cortex and basal ganglia *(47)*. Markedly elevated numbers of microglia and macrophages immunoreactive for Bax were present in basal ganglia and cerebral cortex of children with HIVE/PE (Fig. 2), in comparison to HIV-1 infected children or children who were seronegative for HIV-1. Similar findings were observed in an adult brain with HIVE.

In contrast, patients with HIVE/PE, but not HIV-1, or seronegative controls, had increased expression of *bcl-2* and *bcl-x* in reactive astrocytes in cortex and basal ganglia. The findings of TUNEL staining, increased bax expression, and decreased or absent *bcl-2* expression in brain-resident macrophages and microglia suggest that these cells are more prone to undergo apoptosis in patients with HIVE. In contrast, astrocytes may be resistant to apoptosis. This may represent a cellular mechanism to limit microglial activation and the spread of productive HIV-1 infection in the CNS of children with HIVE.

However, TUNEL-positive neurons in basal ganglia and cerebral cortex of children with HIVE did not have cytoplasmic expression of bax with double label immunocytochemistry. Neurons in tissue sections from basal ganglia and cerebral cortex of children with HIVE were not immunoreactive for bax alone. Possible explanations include fixation artifacts and suboptimal processing conditions secondary to delays in postmortem fixation. However, other nerve cells (i.e., microglia and brain-resident macrophages) were intensely immunoreactive for bax in brains of patients with HIVE. A more likely explanation would be that neurons in the brains of patients with HIVE may undergo apoptosis secondary to dysregulation of other pro-apoptotic genes including *bak*, *bcl-x_s*, and *bad (48)*. Resolution of this question is presently unfeasible because antibodies suitable for immunocytochemical studies in archival formalin-fixed tissue are unavailable.

Several studies have examined the effects of simian immunodeficiency virus infection on neuronal fate in the CNS. One report demonstrated that hippocampal neuronal atrophy occurs in rhesus macaques as early as 3 mo following SIV inoculation *(49)*. In younger macaques, there was a significant association between a reduction in neuronal density and duration of infection. A more recent study, using a neurovirulent strain of SIV, demonstrated a spatial association between apoptotic neurons and perivascular inflammatory cell infiltrates containing SIV-infected macrophages and multinucleated giant cells *(50)*. These findings are consonant with previous reports in patients with HIVE *(41,43,44)*. However, this report also noted glial cell apoptosis, in contrast to several of the human studies *(41,44)*.

Neuronal apoptosis has been demonstrated in a SCID mouse model of HIV-1 encephalitis. Here HIV-1-infected monocytes were stereotactically injected into brain parenchyma, resulting in microglial activation, astrogliosis, and TUNEL-stained neurons *(51)*. These findings were specific for HIV-1-infected monocytes, since stereotactically injected uninfected monocytes were not spatially associated with TUNEL-stained neurons. Preliminary studies have demonstrated neurobehavioral deficits in mice with intraparenchymal injections of HIV-1-infected monocytes (H. Gendelman, personal communication).

CONCLUSIONS

Thus, the available evidences suggest that neurons in CNS with a lentiviral infection (HIV-1, SIV) undergo apoptosis as opposed to necrosis. One of the questions of paramount importance that remains to be answered is why there is a discrepancy between neuronal loss and severity of neurologic dysfunction in adult patients with HIV-1 infection *(24)*. Hopefully, in vivo models of HIVE such as the SCID mouse model *(51)* may provide answers to whether many of the HIV-1-induced neurotoxins of macrophage and microglial origin induce neuronal dysfunction prior to neuronal apoptosis. By quantifying behavioral and electrophysiologic abnormalities in the HIV-1-infected SCID mouse brain and comparing these indices with the quantitative assessment of apoptotic neurons, it may be possible to answer whether neurologic disease is simply the result of loss of neurons or neuronal dysfunction prior to death by apoptosis.

REFERENCES

1. Scarlatti G. Paediatric HIV infection. *Lancet* 1996, **348**: 863–868.
2. Janssen RS, Nwanyanwu OC, Selik RM, Stehr-Green JK. 1992. Epidemiology of human immunodeficiency virus encephalopathy in the United States. *Neurology* 42: 1472–1476.
3. McArthur JC, Hoover DR, Bacellar H, Miller EN, Cohen BA, Becker JT, Graham NMH, McArthur JH, Selnes OA, Jacobson LP, Visscher BR, Concha M, Saah A. Dementia in AIDS patients: Incidence and risk factors. *Neurology* 1993, **43**: 2245–2252.
4. Janssen RS, Cornblath DR, Epstein LG, Foa RP, McArthur JC, Price RW, Asbury AK, Beckett A, Benson DF, Bridge TP, Leventhal CM, Satz P, Saykin AJ, Sidtis JJ, Tross S. Nomenclature and research case definitions for neurological manifestations of human immunodeficiency virus type-1 (HIV-1) infection. Report of a Working Group of the American Academy of Neurology AIDS Task Force. *Neurol* 1991, **41**: 778–785.
5. Navia BA, Jordan BD, Price RW. The AIDS dementia complex: I. Clinical features. *Ann Neurol* 1986, **19**: 517–524.
6. McArthur JC. Neurologic manifestations of AIDS, *Medicine* 1987; **66**: 407–437.
7. McArthur JC, Cohen BA, Selnes OA, Kumar AJ, Cooper K, McArthur JH, Soucy G, Cornblath DR, Chmiel JS, Wang MC, et al. Low prevalence of neurological and neuropsychological abnormalities in otherwise healthy HIV-1-infected individuals: results from the multicenter AIDS Cohort Study. *Ann Neurol* 1989, **26**: 601–611.
8. Cohen SE, Mundy T, Karassik B, Lieb L, Ludwig DD, Ward J. Neuropsychological functioning in human immunodeficiency virus type 1 seropositive children infected through neonatal blood transfusion. *Pediatrics* 1991, **88**: 58–68.
9. Civitello LA, Brouwers P, Pizzo PA. Neurological manifestations in 120 children with symptomatic human immunodeficiency virus infection. *Ann Neurol* 1993, **34**: 481.
10. Sharer LR, Epstein LG, Cho E-S, Joshi VV, Meyenhofer MF, Rankin LF, Petito CK. Pathologic features of AIDS encephalopathy in children: Evidence for LAV/HTLV-III infection of brain. *Hum Pathol* 1986, **17**: 271–284.
11. Sharer LR. Pathology of HIV-1 infection of the central nervous system (Review). *J Neuropath. Exp Neurol* 1992, **51**: 3–11.
12. Takahashi K, Wesselingh SL, Griffin DE, McArthur JC, Johnson RT, Glass JD. Localization of HIV-1 in human brain using polymerase chain reaction/in situ hybridization and immunohistochemistry. *Ann Neurol* 1993, **39**: 705–711.
13. Saito Y, Sharer LR, Epstein LG, Michaels J, Mintz M, Louder M, Golding K, Cvetkovich TA, Blumberg BM. Overexpression of nef as a marker for restricted HIV-1 infection of astrocytes in postmortem pediatric central nervous tissues. *Neurol* 1994, **44**: 474–480.
14. Tornatore C, Chandra R, Berger JR, Major EO. HIV-1 infection of subcortical astrocytes in the pediatric central nervous system. *Neurol* 1994, **44**: 481–487.
15. Ranki A, Nyberg M, Ovod V, Matti H, Elovaara I, Raininko R, Haapasalo H, Krohn K. Abundant expression of HIV Nef and Rev proteins in brain astrocytes in vivo is associated with dementia. *AIDS* 1995, **9**: 1001–1008.
16. Nuovo GJ, Galtery F, MacConnell P, Braun A. In situ detection of polymerase chain reaction-amplified HIV-1 nucleic acids and tumor necrosis factor-a RNA in the central nervous system. *Am J Pathol* 1994, **144**: 659–666.
17. Everall IP, Luthbert PJ, Lantos PL. Neuronal loss in the frontal cortex in HIV infection. *Lancet* 1991, **337**: 1119–1121.

18. Ketzler S, Weis S, Haug H, Budka H. Loss of neurons in the frontal cortex in AIDS brains. *Acta Neuropath* 1990, **80:** 92–94.
19. Tenhula WN, Xu SZ, Madigan MC, Heller K, Freeman WF, Sadun AA. Morphometric comparisons of optic nerve axon loss in acquired immunodeficiency syndrome. *Am J Ophthalmol* 1992, **15:** 14–20.
20. Wiley CA, Schrier RD, Nelson JA, Lampert PW, Oldstone MBA. Cellular localization of human immunodeficiency virus infection within the brains of acquired immune deficiency syndrome patients. *Proc Natl Acad Sci USA* 1986, **83:** 7089–7093.
21. Wiley CA, Masliah E, Morey M, Lemere C, DeTeresa R, Grafe M, Hansen L, Terry R. Neocortical damage during HIV infection. *Annals Neurolo.* 1991, **29:** 651–657.
22. Simpson DM, Tagliati M. Neurologic manifestations of HIV infection. *Ann Intern Med* 1994, **121:** 769–785.
23. Price RW. Understanding the AIDS dementia complex (ADC), in *HIV, AIDS, and the Brain* (Price RW, Perry SW, eds.), Raven Press, NY, 1994, pp. 1–45.
24. Everall IP, Glass JD, McArthur J, Spargo E, Lantos P. Neuronal density in the superior frontal and temporal gyri does not correlate with the degree of human immunodeficiency virus-associated dementia. *Acta Neuropathol* 1994, **88:** 538–544.
25. Fine SM, Angel RA, Perry SW, Epstein LG, Dewhurst S, Gelbard HA. Tumor necrosis factor a inhibits glutamate uptake by primary human astrocytes: implications for pathogenesis of HIV-1 dementia. *J Biol Chem* 1996, **271:** 15,303–15,306.
26. Gelbard HA, Dzenko K, DiLoreto D, del Cerro C, del Cerro M, Epstein LG. Neurotoxic effects of tumor necrosis factor in primary human neuronal cultures are mediated by activation of the glutamate AMPA receptor subtype: implications for AIDS neuropathogenesis. *Dev Neurosci* 1993, **15:** 417–422.
27. Gelbard HA, Nottet HSLM, Swindells S, Jett M, Dzenko KA, Genis P, White R, Wang L, Choi Y-B, Zhang D, Lipton SA, Tourtellotte WW, Epstein LG, Gendelman HE. Platelet activating factor: A candidate HIV-1-induced neurotoxin. *J Virol* 1994, **68:** 4628–4635.
28. Genis P, Jett M, Bernton EW, Gelbard HA, Dzenko K, Keane R, Resnick L, Volsky DJ, Epstein LG, Gendelman HE. Cytokines and arachidonic acid metabolites produced during HIV-infected macrophageastroglial interactions: Implications for the neuropathogenesis of HIV disease. *J Exp Med* 1992, **176:** 1703–1718.
29. Nottet SLM, Jett M, Flanagan CR, Zhai Q-H, Peridsy Y, Rizzino A, Bernton EW, Genis P, Baldwin T, Schwartz J, LaBenz CJ, Gendelman HE. A regulatory role for astrocytes in HIV-1 encephalitis. *J Immunol* 1995. **154:** 3567–3581.
30. Glass JD, Fedor H, Wesselingh SL, McArthur JC. Immunocytochemical quantitation of human immunodeficiency virus in the brain: correlations with dementia. *Ann Neurol* 1995, **38:** 755–762.
31. Ameisen JC, Capron A. Cell dysfunction and depletion in AIDS: the programmed cell death hypothesis. *Immunol Today* 1990, **12:** 102–104.
32. Terai C, Kornbluth RS. Pauza CD, Richman DD, Carson DA. Apoptosis as a mechanism of cell death in cultured T lymphocytes acutely infected with HIV-1. *J. Clin Invest* 1991, **87:** 1710–1715.
33. Groux H, Monte D, Bourez JM, Capron A, Ameisen JC. Activation-induced death by apoptosis in CD4+ T cells from HIV-infected asymptomatic individuals. *J Exp Med* 1992, **175:** 331–340.
34. Macho A, Castedo M, Marchetti P, Aguilar JJ, Decaudin D, Zamzami N, Girard PM, Uriel J, Kroemer G. Mitochondrial dysfunctions in circulating T lymphocytes from human immunodeficiency virus-1 carriers. *Blood* 1995, **7:** 2481–2487.
35. Giulian D, Vaca K, Noonan CA. Secretion of neurotoxins by mononuclear phagocytes infected with HIV-1. *Science* 1991, **250:** 1593–1596.
36. Gelbard HA, Dzenko K, DiLoreto D, del Cerro C, del Cerro M, Epstein LG. Neurotoxic effects of tumor necrosis factor in primary human neuronal cultures are mediated by acti-

vation of the glutamate AMPA receptor subtype: implications for AIDS neuropathogenesis. *Dev Neurosci* 1993, **15**: 417–422.
37. Gelbard HA, Nottet HSLM, Swindells S, Jett M, Dzenko KA, Genis P, White R, Wang L, Choi Y-B, Zhang D, Lipton SA, Tourtellotte WW, Epstein LG, Gendelman HE. Platelet activating factor: A candidate HIV-1-induced neurotoxin. *J Virol* 1994, **68**: 4628–4635.
38. Talley A, Dewhurst S, Perry S, Gummuluru S, Dollard SC, Fine SM, New D, Epstein LG, Gendelman HE, Gelbard HA. Tumor necrosis factor alpha induces apoptosis in human neuronal cells: protection by the antioxidant N-acetylcysteine and the genes *bcl-2* and *crmA*. *Mol Cell Biol* 1995, **15**: 2359–2366.
39. Miura M, Zhu H, Rotello R, Hartweig E, Yuan J. Induction of apoptosis in fibroblasts by IL-1 beta-converting enzyme, a mammalian homolog of the C. elegans cell death gene ced-3. *Cell* 1993, **75**: 653–660.
40. Gavrieli Y, Sherman Y, Ben-Sasson SA. Identification of programmed cell death in situ via specific labeling of nuclear DNA fragmentation. *J Cell Biol* 1992, **119**: 493–501.
41. Gelbard HA, James H, Sharer L, Perry SW, Saito Y, Kazee AM, Blumberg BM, Epstein LG. Identification of apoptotic neurons in post-mortem brain tissue with HIV-1 encephalitis and progressive encephalopathy. *Neuropathol Appl Neurobiol* 1995, **21**: 208–217.
42. Gelbard HA, Boustany R-M, Schor NF. Apoptosis in development and disease of the nervous system: Part 2: Apoptosis in childhood neurologic disease. *Ped Neurol* (in press).
43. Petito CK, Roberts B. Evidence of apoptotic cell death in HIV encephalitis. *Am J Pathol* 1995, **146**: 1121–1130.
44. Adle-Biassette H, Levy Y, Colombel M, Poron F, Natchev S, Keohane C, Gray F. Neuronal apoptosis in HIV infection in adults. *Neuropathol Appl Neurobiol* 1995, **21**: 218–227.
45. An SF, Giometto B, Scaravilli T, Tavolato B, Gray F, Scaravilli F. Programmed cell death in brains of HIV-1 positive AIDS and pre-AIDS individuals. *J Neurovirol* 1996, **2**: 24.
46. Neuen-Jacob, E, Arendt, G, von Giesen H-J, Wechsler W. Neuronal cell apoptosis in the basal ganglia occurs early in the course of HIV encephalitis and may precede the clinical signs of HIV-1 associated dementia. *Neuropathol Appl Neurobiol* 1996, **22:Suppl. 1**: 16–17.
47. Sharer LR, Krajewski S, James H, Ross J, Blumberg BM, Epstein LG, Dewhurst S, Reed JC, Gelbard HA. Bax is a marker for apoptotic microglia in pediatric patients with HIV-1 encephalitis. *J Neurovirol* 1996, **2**: 27.
48. Oltvai ZN, Korsmeyer SJ. Checkpoints of dueling dimers foil death wishes. *Cell* 1994, **79**: 189–192.
49. Luthert PJ, Montgomery MM, Dean AF, Cook RW, Baskerville A, Lantos PL. Hippocampal neuronal atrophy occurs in rhesus macaques following infection with simian immunodeficiency virus. *Neuropathol Appl Neurobiol* 1995, **21**: 529–534.
50. Adamson DC, Dawson TM, Zinc MC, Clements JE, Dawson VL. Neurovirulent simian immunodeficiency virus infection induces neuronal, endothelial, and glial apoptosis. *Mol Med* 1996, **2(4)**: 417–428.
51. Persidsky Y, Limoges J, McComb R, Bock P, Baldwin T, Tyor W, Patil A, Nottet HSLM, Epstein L, Gelbard H, Flanagan E, Reinhard SJ, Gendelman HE. A quantitative analysis of brain immunopathology in SCID mice with HIV-1 encephalitis. *Am J Pathol* 1996, **149**: 1027–1053.

26
Neurotoxicity of Drugs of Abuse

Jean Lud Cadet

INTRODUCTION

The amphetamines are drugs of abuse that are presently seeing a significant resurgence. In California, it has been suggested that methamphetamine (METH) will surpass cocaine as the preferential drug of abuse. This might be related to the relative ease of its synthesis and the focus of government agencies on cocaine. Unlike cocaine, the amphetamines are, however, known causes of major neurotoxic damage to mammalian monoaminergic systems *(1–3)* For example, METH depletes dopamine (DA) and its metabolites *(1,2,4,5)*, depletes DA uptake sites *(1,6)* and causes marked decreases in tyrosine hydroxylase (TH) activity in the nigrostriatal DA system. METH can also cause significant alterations in the serotonin (5-HT) system depending on the doses of the drug used in the experiments. In contrast to the effects of METH, another closely related analog, methyldioxymethamphetamine (MDMA, "ecstasy") exerts most of its effects on the serotoninergic system of most mammals (including humans) except in mice, where MDMA affects mainly the dopaminergic system *(2,7)*.

Although the toxic effects of these drugs have been known for many years, the detailed cellular and molecular events that lead to these changes have not been fully elucidated. A role for oxygen-based radicals and of excitotoxic damage has, however, been suggested (*see* chapter by Dykens). Thus, the purpose of this chapter is to briefly review the evidence from our laboratory, and those of others, which documents a role for superoxide radicals and for nitric oxide in the neurotoxic effects of METH and MDMA.

SUPEROXIDE RADICALS ARE MEDIATORS OF THE TOXIC EFFECTS OF METH AND MDMA

In order to examine the role of superoxide in the neurotoxic effects of these drugs, we have used transgenic (Tg) mice that express the normal human copper, zinc (Cu/Zn)-superoxide dismutase (SOD) gene (developed by Epstein et al. *[24]*). These mice have significantly increase activity of the Cu/Zn SOD enzyme in various brain regions, with heterozygous and homozygous transgenic mice having mean increases of about 2.5- and 5.7-fold, respectively. We reasoned that if superoxide radicals played a role in the toxic efects of these drugs, then SOD Tg mice should be protected against the toxic effects of both METH and MDMA. Indeed, when administered to wild-type mice, both METH and MDMA cause marked reductions in various markers of the nigrostriatal DA pathway in a dose-dependent fashion. However, Tg mice overexpressing SOD were significantly protected against the effects of these drugs in a gene-dosage dependent fashion, with the homozygous mice showing the greatest protection (7).

These results indicate that superoxide radicals are indeed some of the culprits involved in the toxic effects of METH and MDMA on the DA systems of rodents. We have envisioned the following scenario for the production of superoxide radicals by these drugs: After entering the brain, METH and MDMA are known to cause increased release of DA in the striatum (8). DA oxidation within the terminals and in the synaptic cleft could lead to the generation of oxygen-based radicals (9,10) which could then lead to the degeneration of DA terminals. These ideas are consonant with the previous demonstration that inhibition of SOD by diethyldithiocarbamate potentiates the deleterious effects of METH (11).

NITRIC OXIDE PARTICIPATES IN THE TOXIC EFFECTS OF METH AND MDMA IN VITRO

In addition to superoxide radicals, the production of nitric oxide (NO) might play an important role in neurodegenerative processes that occur after exposure to certain toxins (12,13). This occurs through the conversion of L-arginine to L-citrulline in a reaction that is catalyzed by nitric oxide synthase (NOS) (14,15). Reactions of superoxide radicals with NO can produce peroxynitrite so that the combination of these toxins can all play a role in neurodegeneration (16,17).

In order to test if NO production is involved in the toxic affects of METH and MDMA, we used in vitro models of fetal mesencephalic (DA) and raphe serotoninergic (5-HT) cells, respectively. These experiments showed that blocking of NO production by various NOS inhibitors significantly attenuated the toxic effects of these two drugs, in a dose-dependent fashion (18). These data are consistent with results showing that glutamate receptor antagonists can attenuate the toxicity of METH and suggest that METH might be causing the production of NO in vivo. This idea remains to be fully evaluated. Figure 1 shows a working model of the participation of superoxide radicals and nitric oxide in the toxic effects of METH and MDMA.

THE NEUROTOXIC EFFECTS OF METH AND MDMA INVOLVE ACTIVATION OF POLY-ADP-RIBOSE POLYMERASE

Oxygen-based radicals and NO can cause DNA damage (19,20). This DNA damage is associated with the activation of poly-ADP-ribose polymerase (PARP), an enzyme that catalyzes the conjugation of ADP-ribose to proteins. This process uses and depletes nicotinamide adenine dinucleotide (NAD) and can lead to ATP depletion. A net consequence of these reactions can be cellular demise if NAD is not replenished. NO is thought to work by a similar mechanism (21). Thus, given that both NO and superoxide are involved in the toxic effects of both METH and MDMA, we reasoned that activation of PARP might also be a step in the amphetamine-induced cell death cascade.

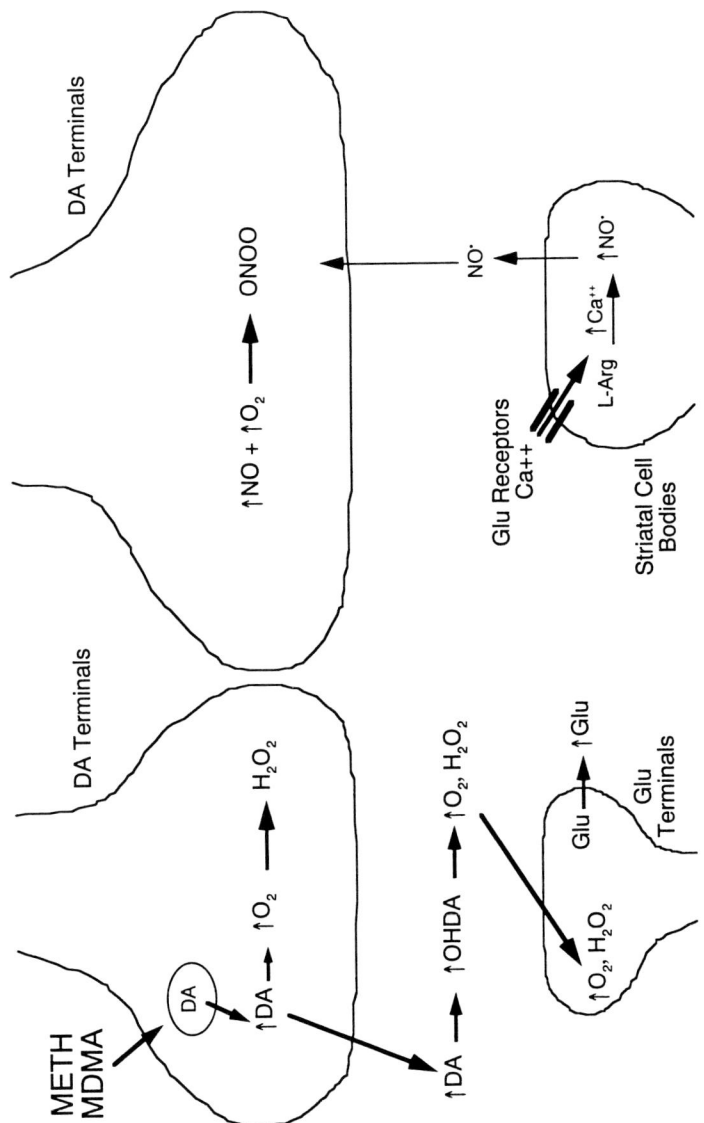

Fig. 1. Involvement of O_2^- and $NO^{\cdot}$ in METH- and MDMA-induced neurotoxicity.

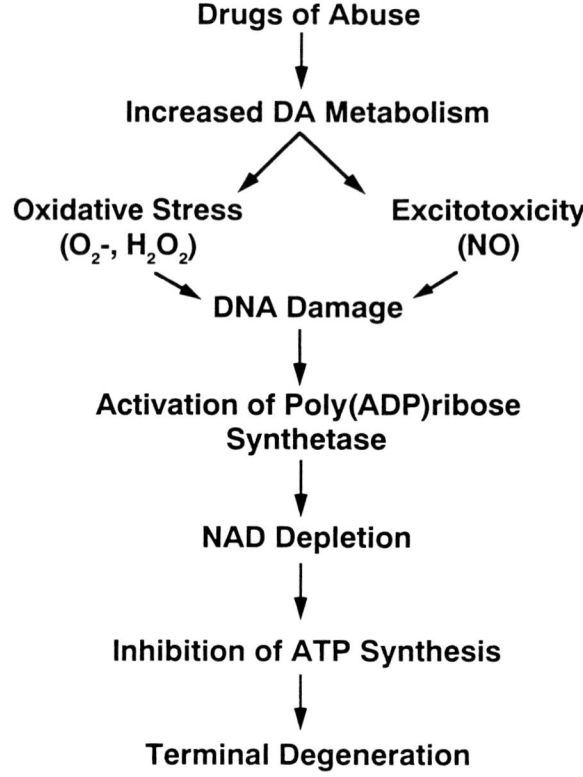

Fig. 2. Involvement of oxidative pathways in METH- and MDMA-induced neurotoxicity.

This idea suggested that inhibition of PARP might provide protection against the toxic effects of both METH and MDMA (Fig. 2). Benzamide or nicotinamide, inhibitors of PARP activity, caused dose-dependent inhibition of the METH- and MDMA-induced damage to cells in culture (18). Interestingly, the protective effects of the PARP inhibitors were of greater magnitude than those caused by NOS inbitors, suggesting that the effects of oxygen-based radicals and of NO might be converging through the PARP pathway. Figure 3 represents a working model of the cascade leading to toxin-induced degeneration of the nigrostriatal DA system after administration of these drugs of abuse.

CONCLUSION

Although the toxic effects of drugs of abuse have been known for a long time, the specific cellular and molecular events that lead to the demise of either DA or 5-TH systems remain to be fully elucidated. The studies briefly reviewed herein indicate that this cascade involves superoxide and NO radicals and the activation of the DNA repair enzyme PARP. Depletion of NAD and of ATP stores resulting from PARP activation might therefore play an important role in the acute neurotoxic effects of these drugs of abuse. More recent studies from this laboratory have also indicated that METH-induced cell death involves a process that is reminiscent of apoptosis since exposure to METH causes DNA strand breaks, cytoplasmic blebbing, and fragmented nuclei in vitro (Cadet et al., 1997). This process was inhibited by the anti-apoptotic proto-

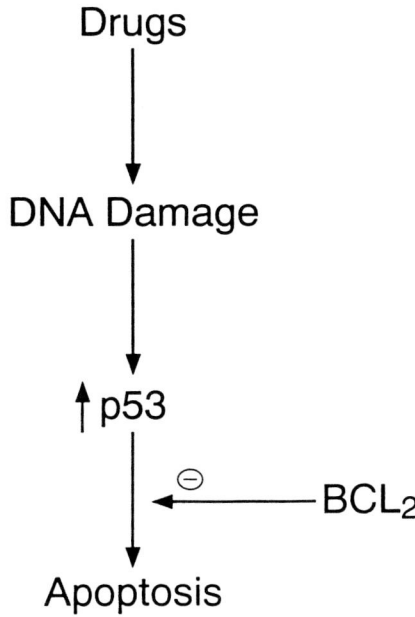

Fig. 3. Involvement of p53 and Bcl2 in METH- and MDMA-induced neurotoxicity.

oncogene, *bcl-2*. Thus, these results suggest that neurodegenerative processes in monoaminergic system might involve a process that is similar to the cell death program. These ideas are actively being investigated in this laboratory and are supported by recent demonstrations that p53 knockout mice are protected against the toxic effects of METH (Hirata and Cadet, 1997). Such studies promise to identify specific death genes that are activated after exposure to METH or MDMA.

REFERENCES

1. Ricaurte GA, Schuster CR, Seiden LS. Long-term effects of repeated methylamphetamine administration on dopamine and serotonin neurons in the rat brain: regional study. *Brain Res* 1980, **193**: 153–163.
2. O'Callaghan JP, Miller DB. Neurotoxicity profiles of substituted amphetamines in the C57BL/6J mouse. *J Pharmacol Exp Ther* 1994, **270**: 741–751.
3. Steranka LR, Sanders-Bush E. Long-term effects of continuous exposure to amphetamine in brain dopamine concentration and synaptosomal uptake in mice. *Eur J Pharmacol* 1980, **65**: 439–443.
4. Cadet JL., Ladenheim B, Baum I, Carlson E, Epstein C. CuZn-superoxide dismutase (CuZnSOD) transgenic mice show resistance to the lethal effects of methylenedioxyamphetamine (MDA) and methylenedioxymethamphetaine (MDMA). *Brain Res* 1994a, **655**: 259–262.
5. Wagner GC, Ricaurte GA, Seiden LS, Schuster CR, Miller RJ, Westley J. Long-lasting depletions of striatal dopamine and loss of dopamine uptake sites following repeated administration of methamphetamine. *Brain Res* 1980, **181**: 151–160.
6. Nakayama M, Koyama T, Yamashita I. Long-lasting decreases in dopamine uptake sites following repeated administration of methamphetamine in the rat striatum. *Brain Res* 1992, **601**: 209–212.

7. Cadet JL, Ladenheim B, Hirata H, Rothman RB, Ali RB, Carlson E, Epstein C, Moran TH. Superoxide radicals mediate the biochemical effects of methylendioxymethamphetamine (MDMA): evidence from using CuZn-superoxide transgenic mice. *Synapse* 1995, **21:** 169–175.
8. Marshall JF, O'Dell SJ, Weihmuller FB. Dopamine-glutamate interactions in methaphetamine-induced neurotoxicity. *J Neural Transm* 1993, **91:** 241–254.
9. Cadet JL. A unifying hypothesis of movement and madness: involvement of free radicals in disorders of the isodendritic core. *Med Hypotheses* 1988, **27:** 87–94.
10. Cohen G, Heikkila RE. The generation of hydrogen peroxide, superoxide radical and hydroxyl radical by hydroxydopamine, dialuric acid, and related cytotoxic agents. *J Biol Chem* 1974, **249:** 2447–2452.
11. De Vito MJ, Wagner GC Methamphetamine-induced neuronal damage: A possible role for free radicals. Neuropharmacology 1989, **28:** 1145–1150.
12. Coyle JT, Puttfarcken P. Oxidative stress, glutamate, and neurodegenerative disorders. *Science* 1993, **262:** 689–695.
13. Dawson VL, Dawson TM, Bartley DA, Uhl GR, Snyder SH Mechanisms of nitric oxide-mediated neurotoxicity in primary brain cultures. *J Neurosci* 1993, **13:** 2651–2661.
14. Moncada S, Palmer RMJ, Higgs EA. Nitric oxide: physiology, pathophysiology and pharmacology. *Pharmacol Rev* 1991, **43:** 109–142.
15. Nathan C. Nitric oxide as a secretory product of mammalian cells. *FASEB J* 1992, **6:** 3051–3064.
16. Beckman JS. The double-edged role of nitric oxide in brain and superoxide-mediated injury. *J Develop Physiol* 1991, **15:** 53–59.
17. Radi R, Beckman JS, Bush KM, Freeman BA. Peroxynitrite-induced membrane lipid peroxidation: The cytotoxic potential of superoxide and nitric oxide. *Arch Biochem Biophys* 1991, **288:** 481–487.
18. Sheng P, Cerutti C, Cadet JL. Methamphetamine (METH) causes reactive gliosis in vitro: attenuation by the ADP-ribosylation (ADPR) inhibitor, benzamide. *Life Sciences* 1994, **55:** 51–54.
19. Schraufstatter IU, Hinshaw DB, Hyslop PA, Spragg RG, Cochrane CG. Oxidant injury of cells: DNA strand-breaks activate polyadenosine diphosphate-ribose polymerase and lead to depletion of nicotinamide adenine dinucleotide. *J Clin Invest* 1986a, **77:** 1312–1320.
20. Schraufstatter JU, Hyslop PA, Minshaw DB, Spragg RG, Sklar LA, Cochrane CG. Hydrogen peroxide-induced injury of cells and its prevention by inhibitors of poly (ADP-ribose) polymerase. *Proc Natl Acad Sci USA* 1986b, **83:** 4908–4912.
21. Zhang J, Dawson VL, Dawson TM, Snyder SH. Nitric oxide activation of poly(ADP-Ribose) synthetase in neurotoxicity. *Science* 1994, **263:** 687–689.
22. Hirata H, Cadet JL. P53 knockout mice are protected against the longterm effects of methamphetamine on dopaminergic terminals and cell bodies. *J Neurochem* 1997, **69:** 780–790.
23. Cadet JL, Ordonez SV, Ordonez JV. Methamphetamine induces apoptosis in immortalized neural cells: protection by the proto-oncogene, bcl-2. *Synapse* 1997, **25:** 176–184.
24. Epstein CJ, Avraham KB, Lovett M, Smith S, Elroy-Stein O, Rotman G, et al. Transgenic mice with increased CuZn-superoxide dismutase activity: Animal model of dosage effects in Down Syndrome. *Proc Natl Acad Sci USA* 1997, **84:** 8044–8048.

27
Schizophrenia

Steven E. Arnold

INTRODUCTION

Schizophrenia is a severe mental illness that affects approx 1% of men and women worldwide. It is a disease with profound public health and economic implications, let alone the devastation it brings to afflicted individuals and their families. The first symptoms usually occur in late adolescence or early adulthood and last throughout life. While the manifestations may be diverse, the syndrome is typically characterized by disorganized thought processes, hallucinations and delusions, emotional and social withdrawal, and deterioration in independent functioning. Cognitive impairment has been increasingly well recognized as an important component of the disease.

Clinical findings in schizophrenia have suggested both neurodevelopmental and neurodegenerative changes; therefore, it is not unreasonable to predict a role for abnormal cell death in the pathophysiology of the disease at either end of the lifespan. However, contemporary neuropathological investigations are only beginning to define the extent to which this might occur or to elucidate its nature, if it indeed occurs.

EARLY NEUROPATHOLOGICAL STUDIES OF SCHIZOPHRENIA

For well over a century, pathologists have been examining the brains of patients with schizophrenia. Numerous findings have been reported and yet the defining lesion(s) of the disorder so far remain elusive. The first identifiable neuropathologic examination was in 1871 by E. Hecker, who noted ventricular enlargement and darkening of the cortex in a patient with symptoms of hebephrenia (1). Subsequently, Alois Alzheimer and a number of other European neuropathologists emphasized abnormalities in the cerebral cortex with descriptions of cell loss in specific layers of the cortex, disorganization of neuronal processes, fatty degeneration, and occasionally, accompanying gliosis (2). However, as further investigations ensued, findings became more diverse, inconsistent, and conflicted in both the locations and types of pathology described. Many of the prominent neuropathologists of the early 20th century working in the field convened to present their findings at the First International Congress of Neuropathology in Rome in 1952. The reported abnormalities ranged widely and included patchy cell loss in the cortex and diencephalon, specific neuronal degenerative changes, meningeal and cortical endarteritis, fibrinoid and exudative deposits around blood vessels, gliosis and glial degenerative changes, demyelinization, and lamina-specific neuron loss in the cortex thought to be due to hypoplasia. In addition, some investigators reported negative results or findings that could best be accounted for as artifactual *(3)*.

In view of such conflicting opinions, as well as the predominance of psychoanalytic theories in psychiatry during that era, neuropathologic research in schizophrenia entered a virtual moratorium for several decades. We now recognize that many of the inconsistencies in the early studies can be attributed to the methodological limitations of the time. Accuracy of psychiatric diagnoses was one of the principal limitations. Without the benefit of rigorous clinical and research diagnostic criteria that we now employ, it is likely that many of the samples used in early studies included patients with mood and other psychiatric disorders, and a host of schizophrenia-like organic psychoses. Often, little clinical information was known about the cases, and meaningful clinicopathologic correlation was not feasible. Furthermore, many studies had limited numbers of control cases, and these were generally unmatched for such factors as age, concomitant illnesses, and postmortem interval. Tissue handling procedures were neither uniform nor well described, and this likely produced artifacts. Another limitation was the capriciousness and lack of sensitivity and specificity of many of the classical histologic stains that were used. Last, the early neuropathological descriptions were mostly qualitative and, as such, lacked the sensitivity and specificity of currently available quantitative methods for determining subtle brain abnormalities.

With better attention to these methodological considerations, as well as recent technological advances in the neurosciences and an improved understanding of neural systems underlying mental processes, there has been renewed interest in identifying the neurobiological substrates of schizophrenia.

PATHOPHYSIOLOGIC MODELS: CLINICAL EVIDENCE OF BOTH ABNORMAL DEVELOPMENT AND DEGENERATION

Development

There have been two principal hypotheses regarding the etiology and pathophysiology of schizophrenia and there is clinical evidence to support both. The predominant model currently is that schizophrenia results from abnormalities in brain development. These abnormalities could be due to genetic, gestational, and/or perinatal factors. According to this hypothesis, developmental abnormalities ultimately result in a static encephalopathy manifesting as psychotic disorder after maturational factors allow its full expression in adolescence.

Some proponents of the developmental hypothesis have argued that the heritability of schizophrenia supports its neurodevelopmental nature. Extensive research of families, twins, and adoptees have demonstrated that there is a large genetic contribution to schizophrenia. While the lifetime prevalence for schizophrenia in the general population is around 1%, children of one schizophrenic parent have about a 10% risk and, if both parents are schizophrenic, the risk is almost 50% *(4)*. Particularly compelling are data from monozygotic twins reared apart which have a 58% concordance rate for the illness; this rate is slightly higher even than that of the control cohort of monozygotic twins reared together. It has been estimated that up to 85% of the variance in risk for the disorder is accounted for by genes *(5)*.

Even though schizophrenia is highly heritable, this in itself does not necessarily mean it is a neurodevelopmental disorder. Obviously, there are many genetic disorders that do result in a static encephalopathy (e.g., Fragile X syndrome). However, many others are purely neurodegenerative (e.g., Huntington disease), whereas still others may have both neurodevelopmental and neurodegenerative components (e.g., Down syndrome).

A history of gestational and perinatal complications has also been found to be common in patients with schizophrenia and this has figured prominently in discussions of neurodevelopmental factors in the etiology of the disease. The increased risk in the offsprings of mothers who are in the mid trimester of pregnancy during outbreaks of influenza is especially provocative *(6)*. This is when migration of nascent neurons to the cerebral cortex is very active in humans and, thus, infection at that time could serve to disrupt the normal pattern of cortical lamination and connectivity. Other factors that have been associated with an increased risk of schizophrenia are maternal starvation (especially in the first trimester), Rhesus and ABO blood type incompatibility, and birth complications associated with hypoxic injury *(7,8)*. Indeed, Cannon et al. have recently shown that the risk of later development of schizophrenia is proportional to the occurrence and severity of anoxic birth injury in individuals both with and without genetic loading for the disorder *(9)*. Again, this suggests that an environmental insult during development predisposes to schizophrenia.

The developmental model of schizophrenia is also supported by findings of reduced head circumference (which reflects decreased brain growth) at birth in "preschizophrenics" and in adults with schizophrenia, as well as other minor physical anomalies (e.g., low set ears, palate height) that are also often seen in association with abnormal CNS development *(10,11)*. Other evidence for abnormal development comes from retrospective analyses of early childhood development using such items as school records and home movies. These find that motor, language, and social development is often delayed among schizophrenics compared to normal and sibling controls *(12,13)*. Finally, one of the most consistent neurobiological findings in schizophrenia has been ventriculomegaly seen in CT and MRI studies *(14)*. The degree of ventricular enlargement correlates with poor premorbid adjustment and has been found at the first onset of symptoms, before chronic treatment. Altogether, these diverse findings provide strong circumstantial evidence for the role of abnormal neurodevelopment in schizophrenia.

Degeneration

The other major model for the pathophysiology of schizophrenia is that it results from a neurodegenerative process. For this, too, there is substantial supporting clinical evidence. While long-term longitudinal follow-up studies of schizophrenia extending into late life are relatively sparse, the few existing ones indicate a significant heterogeneity in the outcome *(15–17)*. Some patients show stabilization or even improvement, but some exhibit progressive deterioration. In a cross-sectional study of severely ill patients who required chronic hospitalization, Davidson et al. found a slow but steady decline in cognitive measures with each decade, culminating in severe dementia for the majority of patients *(18)*. In a demographically similar population of elderly, chronically hospitalized patients with schizophrenia, we found that at least two thirds met criteria for an additional diagnosis of dementia *(19)*. Psychometric studies with these patients revealed that the neuropsychological profile was similar to that seen in patients with Alzheimer disease (AD) *(20)*. Surprisingly, diagnostic neuropathological investigations in these patients failed to find abundant AD lesions or other lesions known to cause dementia *(21,22)*.

Psychophysiological studies of olfaction have revealed deficits in olfactory discrimination, with a severity correlating with the duration of the illness independent of age *(23)*. Again, this is suggestive of a neurodegenerative process. Another approach to clinically investigate the presence of degeneration in schizophrenia is with longitudinal neuroimaging. Although CT studies have not consistently revealed an increase in ventricular size over time *(24)*, two recent brain MRI volumetric studies have reported progressive temporal lobe atrophy *(25,26)*. More extensive data will be necessary to document and characterize changes over time, but accumulating clinical evidence does indicate that there may be an ongoing neurodegenerative process in addition to neurodevelopmental abnormalities in schizophrenia.

POSTMORTEM FINDINGS

Schizophrenia is a particularly vexing disease from a neuropathologist's perspective. While a clinical diagnosis can be made reliably, the disorder is still a syndrome that is almost certainly heterogeneous in nature in terms of polygenic inheritance, environmental contributors, and neurobiological signatures and no pathognomonic lesions in the brain have been yet identified with any consistency. Furthermore, while at this point in time there is reason to think that abnormal cell death does play a role in the schizophrenic disease process, there is still little consistent neuropathologic data to support or refute this position or to define the extent and timing of this loss in development or maturity.

Diagnostic Neuropathology

Contemporary diagnostic examinations of brains from schizophrenic patients have been relatively few. Yet such examinations are critical in order to distinguish any core pathologic features of schizophrenia, determine the frequency with which known neurologic disease entities might masquerade as schizophrenia, and account for pathology that could confound the interpretation of true schizophrenia-related findings. In 1982, Stevens examined the brains of 28 patients with schizophrenia with standard histologic stains *(27)*. She reported a number of abnormalities including neuron loss and areas of infarction in the globus pallidus in five patients, cerebellar Purkinje cell loss in 13, cerebellar white matter gliosis in five, and fibrillary gliosis in periventricular, periaqueductal, or basal forebrain regions bilaterally in 21, which she believed to be a distinctive finding. In another study which used a prospectively accrued, reliably diagnosed sample of schizophrenics, Bruton et al. found that 5 of 56 schizophrenics had developed other neurologic diseases after the initial diagnosis of schizophrenia, including stroke, epilepsy, multiple sclerosis, and Friedreich's ataxia *(28)*. These cases and three others who had received frontal leukotomies for the treatment of their schizophrenia were excluded from further analysis. In the remaining cases, the authors noted a higher frequency of "unexpected pathology" among the schizophrenics compared to controls. Various focal lesions were observed in 21 of the remaining schizophrenics, compared to 12 of 56 normal controls. There were no differences among samples for amyloid plaque deposition, neurofibrillary degeneration, or cerebrovascular disease. Similar to Stevens' study, an increase in fibrillary gliosis was observed in periventricular structures and white matter in addition to portions of cortex; however, this was usually found in brains in which the other focal pathology was present. When a "homogeneous" group was assessed after excluding those cases with other focal pathology, there was no significant increase in gliosis.

Diagnostic neuropathologic investigations have also been conducted in elderly, chronically hospitalized patient populations in whom severe dementia is prevalent. Among 26 patients, we found a sufficient neuropathological explanation for the dementia in only three (one with plaque-only AD, one with the Lewy body variant of AD, and one with adult polyglucosan body disease) and assorted minor abnormalities (e.g., lacunar infarct, meningioma, acoustic neuroma) in seven others *(19)*. In a series of 101 such patients, Golier et al. *(29)* similarly found 10 with AD, 15 with vascular lesions, two with Parkinson's disease, and eight with other abnormalities. Thus, in these end stage, deteriorated schizophrenics, there appears to be a modestly increased frequency of abnormal neuropathologic findings compared to what would be expected in a community-based sample. However these findings are extremely variable, and cannot be confidently associated with the schizophrenic illness.

Macroscopic Findings

Planimetric and volumetric methods have been applied in gross anatomic analyses of the brain in schizophrenia. Overall, these studies find that brains from patients with schizophrenia are shorter in length, weigh less, have enlarged ventricles, and have decreased cortical thickness compared to normal controls. However, not all investigators have been able to confirm these findings (for reviews, *see* refs. *30* and *31*). Studies focusing on selective brain areas have reported smaller hippocampi, smaller parahippocampal gyri, decreased parahippocampal cortical thickness, decreased sylvian fissure length and planum temporale volume on the left, decreased volumes of various nuclei of the basal ganglia, decreased volume of the mediodorsal nucleus of the thalamus, smaller cerebellar vermis, and decreased volumes of the substantia nigra and periventricular gray matter. Again, these abnormalities have not been confirmed in all instances while others still await replication. It should be noted that volumetric abnormalities do not in themselves address the issue of etiology in schizophrenia as either developmental or degenerative processes may account for them. Nonetheless, in at least some patients with schizophrenia, the sizes of various structures that are instrumental for complex behaviors are reduced and this would suggest that the normal complement of neurons is not present.

Are There Fewer Neurons in Schizophrenia?

In exploring the role of cell death in schizophrenia, quantitation of neurons is of obvious importance. Furthermore, given the nature of schizophrenia as a disease caused by remote or very protracted processes, the methodology used to discern subtle differences in cell density or number is especially critical. Most studies of schizophrenia have assessed neuron density on a per-unit-of-area basis. This approach is problematic, because loss of neurons may or may not result in decreased neuron density. Commensurate shrinkage of the neuropil, which is a component of the reference volume for the density measurement, would tend to hide any true reduction in overall neuron number. In diseases in which there is massive and fairly rapid neuron loss (e.g., AD), this reduction in neuron number may result in decreased density measurements. However, if the process were either remote or protracted, then the density of neurons in a given region could be near normal while tissue volume and total number of neurons are smaller. Determining the reference volume of a particular region of interest in schizophrenia (e.g., dorsolateral prefrontal cortex) is an arduous task and is rarely undertaken. Furthermore, most studies have used cell counting methods that do not account for under- or over-counting due to size, orientation, shape, or splitting of cells. Failure to account for these variables allows for substantial bias that could introduce variances of as much as 40% from the true counts. These concerns have been addressed by recently popularized stereological methods (32). To date, there have been three studies of schizophrenia that have employed them.

Most studies of neuron quantity in schizophrenia have focused on temporal and frontal cortices. Within the temporal lobe, decreased neuron density has been reported in the superior temporal gyrus, entorhinal cortex, and hippocampus (33–36). However, there have been almost as many studies that have failed to confirm these findings (37–39). Of particular note is the study of Heckers et al., which employed stereological methods to estimate total numbers of neurons in different regions of the hippocampus and entorhinal cortex in a small number of patients with schizophrenia and failed to find a difference from controls (40). Findings have been even more controversial in the frontal lobe with reports of decreased pyramidal and interneuron density in the anterior cingulate, decreased or unchanged neuron density in primary motor cortex and decreased, unchanged, or increased neuron densities in dorsolateral prefrontal cortex (37,41–43). It is difficult to make sense of the conflicting reports at this stage. Concerning the finding of an increased neuron density in frontal cortex, one explanation that has been suggested is that the neurons themselves are smaller and thus have a commensurably decreased dendritic arbor and a decreased neuropil compartment associated with them.

Abnormalities of neuron numbers have also been reported in various subcortical structures. Using stereological methods, Pakkenberg et al. found reduced numbers of neurons in the mediodorsal nucleus of the thalamus and the nucleus accumbens, while counts in the globus pallidus and the basolateral nucleus of the amygdala were unchanged (44,45). Other studies of neuron density in subcortical nuclei have reported decreases in neuron density in the substantia nigra and the Purkinje cell layer of the cerebellum, unchanged neuron densities in the nucleus basalis of Meynert and locus ceruleus, and increased density in the pedunculopontine nucleus (46–50). Most of these investigations await replication.

Studies of Neurodegeneration and Neural Injury

Diagnostic neuropathologic investigations of the brains of patients with schizophrenia have not demonstrated any noteworthy increases in the frequency of neurodegenerative disorders or other recognized explanations of the illness. However, it is still possible that remote injury, or accumulations of disease-associated or age-related neurodegenerative lesions could play a role in the pathophysiology of schizophrenia. There are a variety of common alterations in the cellular and molecular composition of the brain that occur as a result of a neurodegenerative disease or neural injury. Some of these are relatively disease specific while others are more general indicators of injury. Among the disease-specific markers that have been assessed in postmortem studies of schizophrenia are neurofibrillary tangles, amyloid plaques, Lewy bodies and Pick bodies. Quantitative studies of each of these markers in clinically well-characterized patients failed to find any significant differences between patients and normal controls *(21,22,51–53)*. In addition, postmortem studies searching for numerous CNS viruses have been negative for cytomegalovirus, Epstein-Barr virus, herpes simplex virus type I, human herpes virus type VI, varicella-zoster, mumps, measles, rubella, picorna virus, and influenza A virus, human immunodeficiency virus, influenza A, Borna disease virus, and bovine diarrhea virus *(54)*.

Among the general indicators of neural injury, astrocytosis has been the most extensively studied, but findings remain controversial. Stevens *(27)* and Bruton et al. *(28)* both reported increased fibrillary gliosis in periventricular and other subcortical regions using the conventional Holzer stain. In contrast, there have been numerous quantitative studies of gliosis in multiple brain regions using Nissl stains and glial fibrillary acidic protein (GFAP) immunohistochemistry which have found no differences in the number of astrocytes in schizophrenics compared to normal controls *(45,55–57)*; however, a recent study revealed that elderly, severely deteriorated schizophrenics have increased astrocytosis compared to those without concurrent dementia *(58)*.

The largely negative findings from these quantitative studies of astrocytosis have figured prominently in discussions of developmental vs postmaturational pathophysiologies of schizophrenia. Reactive astrocytosis is characterized by an upregulation of GFAP, enlargement and proliferation of astrocytes, and extension of astrocytic processes *(59)*. After the acute phase, GFAP levels decrease and astrocyte somata shrink, while fine fibrillary processes may persist as a chronic marker of injury. This reaction is orchestrated by a range of intrinsic as well as extrinsic immunologic factors that are developmentally regulated. Injury that occurs during fetal development (or as a result of programmed cell death) does not induce reactive gliosis and thus leaves no trace of the remote injury. In the studies of astrocytosis in schizophrenia, special attention has been paid to the ventromedial temporal lobe to elucidate the morphometric and various cytoarchitectural abnormalities that have been described there (see below). The negative findings have been interpreted as consistent with the hypothesis that the cytoarchitectural perturbation in schizophrenia occurs during brain development; otherwise, it should be accompanied by astrocytosis. What this interpretation fails to account for is that a discrete injury could have occurred postmaturationally, but because it was not active at the time of nerve cell death, GFAP levels might have returned to near normal and evidence of the astrocytic reaction might not be detectable.

In an attempt to examine the neurodegenerative hypothesis of schizophrenia in a relatively comprehensive manner, we have recently examined eight molecular markers of neurodegeneration and neural injury in a well-characterized, prospectively accrued cohort of elderly, chronically hospitalized individuals with schizophrenia compared to normal elderly controls, and "positive" controls with neuropathologically verified AD *(22)*. Antemortem clinical rating scales indicated that most of the schizophrenia patients had severe psychiatric illness and dementia. Furthermore, the brain regions and laminae selected for examination were those known to be especially vulnerable to the accumulation of neurodegenerative pathology (e.g., layer II–III of entorhinal cortex, layer V of orbitofrontal cortex). Thus, if there is any prominent, active or accumulated evidence of neurodegenerative disease or neural injury in the cerebral cortex in schizophrenia, it should have been detected with the previous approach. The immunohistochemical studies assessed hyperphosphorylated tau for neurofibrillary tangles, Lewy bodies, the 1–40 Aβ peptide for amyloid deposition, ubiquitin for a variety of intracellular inclusions and as a marker of impending cell death, GFAP and vimentin to assess reactive astrocytosis, and the CD68 antigen for microglia. In addition, Tdt d-UTP digoxigenin nick end labeling was conducted in a subset of patients and controls to assess for apoptotic bodies and dying cells. The densities of some markers were mildly increased in the schizophrenic patients compared to normal controls but they were at much lower levels than those observed in the AD control group, and none of the differences between schizophrenics and controls were statistically significant. Based on these findings, we concluded that the clinically deteriorated state of schizophrenia in late life is not explained by any robust neurodegenerative process or significant ongoing neural injury in the cerebral cortex. However, remote, occult, or very indolent processes could not be excluded.

Postmortem Evidence of Neurodevelopmental Abnormalities

One explanation for the deterioration that may be observed in schizophrenia over the lifespan is that there is an interaction between normal involutional processes and a static encephalopathy that is due to developmental abnormalities. It is interesting that, despite the substantial clinical evidence suggesting developmental aberrations in schizophrenia and the predominance of this hypothesis in the field, there have been relatively few postmortem studies that have directly sought evidence for it. However, it is not easy to detect residual evidence of developmental abnormalities that would have occurred almost a lifetime before the examination. As discussed above, one of the primary pieces of evidence used to support developmental theories is the lack of gliosis or other markers of postmaturational brain injury. However, this is essentially a negative result. The other major approach that has been taken is to examine cytoarchitecture as a reflection of developmental events.

Several cytoarchitectonic analyses have reported abnormal arrangement of neurons in the hippocampus and disturbed laminar distribution of neurons in the entorhinal cortex. Kovelman and Scheibel, and later Conrad et al., measured the alignment of pyramidal neurons in the hippocampus and found that there was significantly more variability in the axes of orientation of pyramidal neurons in schizophrenics than in controls *(38,60)*. The findings were interpreted as consistent with a developmentally based disarray in migration. However, attempts at replicating these findings have not been successful *(37,39,61,62)*.

Cytoarchitectural disturbances have also been observed in the entorhinal cortex with poor laminar differentiation, poorly developed clusters of neurons in layer II, apparent heterotopic displacement of layer II-type neurons deep into layer III, and decreased densities of neurons, especially in superficial laminae *(36,63,64)*. However, one recent attempt to qualitatively and quantitatively replicate these observations of cytoarchitectural disorganization was unsuccessful, although the decreased neuron density that was previously reported by other investigators was verified *(65)*.

If the findings of cytoarchitectural disturbances in limbic cortices of patients with schizophrenia are ultimately confirmed, they would indicate that neuronal migration and/or differentiation during brain development are abnormal. In humans, neuronal migration to superficial layers of limbic and cortical regions occurs chiefly within the second trimester of pregnancy. Therefore, any genetic, environmental or other gestational insult that causes heterotopic displacement of neurons would have to occur at or before this time. Indeed, the clinical associations mentioned earlier between schizophrenia and malnutrition in the first trimester or between schizophrenia and influenza infection during the second trimester of pregnancy are concordant with this model.

In order to further explore aspects of neuronal migration in schizophrenia, Akbarian et al. examined the distribution of immunohistochemically distinct interstitial neurons in the white matter of prefrontal and temporal lobe *(66)*. Interstitial white matter neurons are considered to be remnants of the cortical subplate which escaped programmed cell death. These investigators reported a maldistribution characterized by a decreased density of three types of interstitial white matter neurons—microtubule-associated protein-2 immunoreactive neurons, neurons immunoreactive for poorly phosphorylated neurofilament proteins, and NADPH-diaphorase staining neurons—in superficial portions of the white matter adjacent to the cortex, while densities of these neurons were increased in deeper portions. Akbarian et al. interpreted this finding as consistent with aberrant development in the cortical subplate with either faulty migration or altered programmed cell death of subplate neurons.

One promising experimental model that explores the neurodevelopmental hypothesis of schizophrenia is derived from the human postmortem findings of decreased neuron density and abnormal cytoarchitecture in the entorhinal cortex. Lipska et al. conducted a series of experiments in which ibotenic acid lesions were placed in the ventral hippocampus of neonatal and young adult rats *(67,68)*. Compared to sham-operated rats, the neonatally lesioned rats appeared normal when tested at a time shortly before puberty under various conditions (e.g., environmental stress, dopaminergic manipulation with amphetamines, apomorphine, and antipsychotic medication). However, after puberty, lesioned rats were hyperresponsive to conditions of environmental stress and dopaminergic challenges. In contrast to animals that received injections of ibotenic acid as neonates, animals that were lesioned as young adults showed a different pattern of behavioral abnormalities and responses to dopaminergic manipulation, with no delay in onset. Recently, it was demonstrated that the neonatally lesioned animals also exhibit a postpubertal onset of abnormal prepulse inhibition of the acoustic startle response *(69)*. This is a psychophysiologic sensorimotor gating response that is related to attention and information processing and which is disturbed in schizophrenic patients. Together, these findings demonstrate the importance of the timing of the lesion in relation to its manifestations and, to the extent that these animals demonstrate similarities with schizophrenics in terms of behavior, psychophysiologic parameters, and responses to pharmacologic challenges, it is a heuristically useful model for the disorder.

The postmortem findings indicating abnormal brain development in schizophrenia are still controversial and must be considered preliminary. Nevertheless, they constitute a promising start to the difficult task of elucidating developmental aberrations in human postmortem tissue decades after the events took place.

GLUTAMATE RECEPTOR DYSFUNCTION IN SCHIZOPHRENIA: EFFORTS TOWARDS A UNIFYING HYPOTHESIS OF DEVELOPMENT AND DEGENERATION IN SCHIZOPHRENIA

An integrative hypothesis model that is beginning to gain increasing attention is the glutamate receptor dysfunction model of schizophrenia (70,71). It has long been recognized that a schizophrenia-like psychosis can be induced by phencyclidine (PCP), a non-competitive NMDA receptor blocker. Not only does PCP cause hallucinations and delusions, but it also causes an associated apathetic state and a type of formal thought disorder, both of which are more distinctive features of schizophrenia. These features are not induced by other psychotogenic drugs (e.g., amphetamines). Another interesting clinical feature that is pertinent to the developmental model of schizophrenia is that ketamine, an anesthetic agent which binds to the PCP site of the NMDA receptor, rarely induces psychosis in children, while psychosis occurs upon emergence from ketamine anesthesia in over 50% of young and middle-aged adults. This age dependency of the response to glutamate antagonists is also seen in rats. Fetal and neonatal rat pups are immune to the neurotoxic effects of NMDA antagonists, but become gradually susceptible between puberty and adulthood.

Postmortem studies of glutamatergic mechanisms in schizophrenia have yielded complex and confusing results to date. Decreased glutamatergic synaptic activity has been suggested by reports of increased glutamate uptake and kainic acid receptor binding in the prefrontal cortex of patients with schizophrenia (72), whereas kainic acid receptor binding has been reported to be decreased in putamen, hippocampus and parahippocampal gyrus (73). A reduction in glutamate uptake and an increase in NMDA receptor binding has been observed in basal ganglia (74). There is also a reduction in the density of the non-NMDA GluR1 and GluR2 receptors and receptor transcripts in the hippocampus and parahippocampal gyrus (75). Increased PCP receptor binding has been reported in cortex and hippocampus (76). Finally, alterations in the brain levels of aspartate, glutamate, N-acetylaspartylglutamate and N-acetyl-a-linked acidic dipeptidase have been observed, which were most prominent in the prefrontal and hippocampal regions (77). Obviously, these postmortem findings to date are diverse, if not contradictory in character and topography, and the precise nature of glutamatergic dysfunction in schizophrenia is yet to be determined.

Derangements of inhibitory transmitter systems have also been invoked in schizophrenia and are compatible with the glutamate model. Benes et al. have reported decreased numbers of GABAergic interneurons in limbic cortex and increased $GABA_A$ receptor binding in superficial laminae of cingulate cortex (42,78), while Akbarian et al. found decreased messenger RNA levels for glutamic acid decarboxylase in prefrontal cortex (79). These findings indicate that inhibitory GABA mechanisms may be compromised in schizophrenia. Without this inhibitory counterbalance, excitatory mechanisms could be expected to have an excessive effect on brain function.

In an animal model that explores glutamatergic excitotoxicity in relation to other proposed neuropathologic features of schizophrenia, Bardgett et al. have assessed the consequences of a ventricular injection of kainic acid in rats (80). They observed a dose-dependent loss of neurons that was regionally specific for the ventral and dorsal hippocampus, thalamus, and piriform cortex; an increase in dopamine receptor binding in the nucleus accumbens that probably represents a neurochemical response to denervation of limbic cortical afferents after their destruction; and behavioral hyperactivity. Thus, by creating a diffusely excitotoxic state, the investigators were able to induce specific limbic neuropathology and neurochemical alterations similar to that seen in human postmortem studies.

In addition to its role in neurotransmission throughout life, glutamate has many other diverse effects at different timepoints and plays critical roles as a cellular signaling mechanism during development and in the course of degenerative processes *(81)*. In development, glutamate stimulates astroglia to release taurine which helps guide migration and viability of nascent neurons. It modulates dendritic outgrowth and synaptogenesis in highly selective ways. NMDA receptors, which are linked to excitotoxic cell death (see chapter by Olney and Ishimaru) become sensitive to glutamate at specific times during development, when synaptogenesis is most robust. At other points of the lifespan, glutamate plays a well-recognized role in disease-related cell death in ischemic brain injury, epilepsy, and AD. It has also been proposed as a factor in cell death occurring as part of normal aging. For all these reasons, glutamate dysfunction in the brain is especially well suited as a model for many of the clinical and neurobiological features of schizophrenia. Among these features are the clinical profile of schizophrenic psychosis, the neuroanatomic findings of cell loss and migrational disturbances in limbic regions without evidence of persistent astrocytosis, secondary effects on dopamine systems, the delayed onset of psychotic symptoms until early adulthood, and the slowly progressive cognitive deterioration observed in some patients over the lifespan.

CONCLUSIONS

It is curious that the neuropathological signatures of schizophrenia remain elusive despite over a 100 yr of investigation in a disease with such profound abnormalities in perception, cognition and personality. It is even uncertain whether there is abnormal cell death at all in the disorder, let alone what the timing or mechanisms for that cell death might be. Some of this uncertainty could be due to the heterogeneity of clinical phenotypes included in the diagnosis of schizophrenia which, in turn, may correspond to a heterogeneity in the neurobiological substrates. Another possibility is that some forms of schizophrenia could arise from remote or extremely protracted neuropathological processes that interact with normal neurodevelopmental events or aging. While subtle, these processes could severely perturb the anatomy and physiology of important anatomic substrates of complex behaviors and give rise to psychosis. Mounting clinical evidence as well as recent findings from postmortem investigations and experimental animal models are increasingly providing support for this notion.

REFERENCES

1. Hecker E. Die hebephrenie. *Arch Pathol Anat Physiol Klin Med* 1871, **52**: 394–401.
2. Alzheimer A. Beitrage zur pathologischen Anatomie der Dementia Praecox. *Allg Z Psychiatr Psychischgericht Med* 1913, **70**: 810–840.
3. *Proceedings of the First International Congress of Neuropathology*, vol 1. (Rosenberg and Sellier), Turin, Italy, 1952.
4. Gottesman II. *Schizophrenia Genesis*. W.H. Freeman, New York, 1991.

5. McGuffin P, Anderson P, Owen M, et al. The strength of the genetic effect: is there room for an environmental influence in the aetiology of schizophrenia? *Br J Psychiatry* 1994, **164:** 593–599.
6. Mednick SA, Machon RA, Huttunen MO, Bonnett D. Adult schizophrenia following prenatal exposure to an influenza epidemic. *Arch Gen Psychiatry* 1988, **45:** 189–92.
7. Susser E, Neugebauer R, Hoek HW, Brown AS, Lin S, Labovitz D, Gorman JM. Schizophrenia after prenatal famine. *Arch Gen Psychiatry* 1996, **53:** 25–31.
8. Hollister JM, Laing P, Mednick SA. Rhesus incompatibility as a risk factor for schizophrenia in male adults. *Arch Gen Psychiatry* 1996, **53:** 19–24.
9. Cannon TD, Megginson-Hollister J, Bearden CE, Hadley T. A prospective cohort study of genetic and perinatal influences in the etiology of schizophrenia. *NEJM* 1996, submitted.
10. McNeil TF, Cantor-Graae E, Nordstrom LG, Rosenlund T. Head circumference in "preschizophrenics" and control neonates. *Br J Psychiatry* 1993, **162:** 517–523.
11. O'Callaghan E, Larkin C, Kinsella A, Waddington JL. Familial, obstetric, and other clinical correlates of minor physical anomalies in schizophrenia. *Am J Psychiatry* 1991, **148:** 479–483.
12. Walker E. Developmentally moderated expressions of the neuropathology underlying schizophrenia. *Schizophr Bull* 1994, **20:** 453–480.
13. Jones P, Rodgers B, Murray R, Marmot M. Child development risk factors for adult onset schizophrenia in the British 1946 birth cohort. *Lancet* 1994, **344:** 1398–402.
14. Liddle PF. Brain imaging, in *Schizophrenia,* (Hirsch SR, Weinberger DR, eds.), Blackwell, London, England, 1995, pp. 295–323.
15. Ciompi L. Catamnestic long-term study on the course of life and aging of schziophrenics. *Schizophr Bull* 1980, **6:** 606–618.
16. Winokur G, Pfohl B, Tsuang M. A 40-year follow-up of hebephrenic-catatonic schizophrenia, in *Schizophrenia and Aging*, (Miller N, Cohen G, eds.), Guilford, New York, 1987, pp. 52–60.
17. Lindstrom LH. Clinical and biological markers for outcome in schizophrenia. *Neuropsychopharmacology* 1996, **14:** 23S–26S.
18. Davidson M, Harvey PD, Powchik P, Parella M, White L, Knobler HY, Losonczy MF, Keefe RSE, Katz S, Frecska E. Severity of symptoms in chronically institutionalized geriatric schizophrenic patients. *Am J Psychiatry* 1995, **152:** 197–207.
19. Arnold SE, Gur RE, Shapiro RM, Fisher KR, Moberg PJ, Gibney MR, Gur RC, Blackwell P, Trojanowski JQ. Prospective clinicopathological studies of schizophrenia: Accrual and assessment. *Am J Psychiatry* 1995, **152:** 731–737.
20. Moberg PJ, Mahr R, Gibney M, Arnold SE, Shapiro R, Kumar A, Gottlieb G, Gur RE. Neuropsychological functioning in elderly patients with schizophrenia and Alzheimer's disease. *JINS* 1995, **1:** 132.
21. Arnold SE, Franz BR, Trojanowski JQ. Elderly patients with schizophrenia exhibit infrequent neurodegenerative lesions. *Neurobiol Aging* 1994, **15:** 299–303.
22. Arnold SE, Trojanowski JQ, Gur RE, Han L-Y, Franz BR, Ruscheinsky DD. Investigations of neurodegneration and neural injury in the brains of elderly patients with schizophrenia. *Arch Gen Psychiatry* 1996, submitted.
23. Moberg PJ, Doty RL, Mahr RN, Arnold SE, Turetsky BI, McKeown DA, Gur RC, Gur RE. Olfactory deficits in young and elderly patients with schizophrenia. *Am J Psychiatry* 1996, submitted.
24. Vita A, Sacchetti E, Valvassori G, Cazullo CL. Brain morphology in schizophrenia: a 2- to 5-year CT follow-up study. *Acta Psychiatr Scand* 1988, **78:** 618–621.
25. DeLisi LE, Tew W, Xie SH, Hoff AL, Sakuma M, Kushner M, Lee G, Shedlack K, Smith AM, Grimson R. A prospective follow-up study of brain morphology and cognition in first-episode schizophrenic patients - preliminary findings. *Biol Psychiatry* 1995, **38:** 349–360.
26. Gur RE, Cowell P, Turetsky BI, Gallacher F, Cannon T, Gur RC. A follow-up study of neuroanatomical, clinical, and neurobehavioral measures in schizophrenia. *Arch Gen Psychiatry* 1996, submitted.

27. Stevens JR. Neuropathology of schizophrenia. *Arch Gen Psychiatry* 1982, **39**: 1131–1139.
28. Bruton CJ, Crow TJ, Frith CD, Johnstone EC, Owens DGC, Roberts GW. Schizophrenia and the brain: a prospective clinico-neuropathological study. *Psychol Med* 1990, **20**: 285–304.
29. Golier JA, Davidson M, Haroutunian V, Powchik P, Purohit D, Perl D, Davis KL. Neuropathological study of 101 elderly schizophrenics: Preliminary findings. *Schizophr Res* 1995, **15**: 120.
30. Bogerts B. Recent advances in the neuropathology of schizophrenia. *Schizophr Bull* 1993, **19**: 431–445.
31. Arnold SE, Trojanowski JQ. Recent advances in defining the neuropathology of schizophrenia. *Acta Neuropathol* 1996, in press.
32. West MJ. New stereological methods for counting neurons. *Neurobiol Aging* 1993, **14**: 287–293.
33. Holinger DP, Rosen GD, Galaburda AM. Decreased neuronal density in supragranular layer of area Tpt of the superior temproal gyrus of schizophrenics. *Soc Neurosci Abstr* 1995, **21**: 238.
34. Falkai P, Bogerts B. Cell loss in the hippocampus of schizophrenics. *Eur Arch Psychiatry Neurol Sci* 1986, **236**: 154–161.
35. Jeste DV, Lohr JB. Hippocampal pathologic findings in schizophrenia: a morphometric study. *Arch Gen Psychiatry* 1989, **46**: 1019–1024.
36. Falkai P, Bogerts B, Rozumek M. Limbic pathology in schizophrenia: the entorhinal region—a morphometric study. *Biol Psychiatry* 1988, **24**: 515–521.
37. Arnold SE, Franz BR, Gur RC, Gur RE, Shapiro RM, Moberg PJ, Trojanowski JQ. Smaller neuron size in schizophrenia in hippocampal subfields that mediate cortical-hippocampal interactions. *Am J Psychiatry* 1995, **152**: 738–748.
38. Kovelman JA, Scheibel AB. A neurohistological correlate of schizophrenia. *Biol Psychiatry* 1984, **19**: 1601–1621.
39. Benes FM, Sorensen I, Bird ED. Reduced neuronal size in posterior hippocampus of schizophrenic patients. *Schizophr Bull* 1991, **17**: 597–608.
40. Heckers S, Heinsen H, Geiger B, Beckmann H. Hippocampal neuron number in schizophrenia. A stereological study. *Arch Gen Psychiatry* 1991, **48**: 1002–1008.
41. Benes FM, Davidson J, Bird ED. Quantitative cytoarchtectural studies of the cerebral cortex of schizophrenics. *Arch Gen Psychiatry* 1986, **43**: 31–35.
42. Benes FM, McSparren J, Bird ED, SanGiovanni JP, Vincent SL. Deficits in small interneurons in prefrontal and cingulate cortices of schizophrenic and schizoaffective patients. *Arch Gen Psychiatry* 1991, **48**: 996–1001.
43. Selemon LD, Rajkowska G, Goldman-Rakic PS. Abnormally high neuronal density in the schizophrenic cortex: A morphometric analysis of prefrontal area 9 and occipital area 17. *Arch Gen Psychiatry* 1995, **52**: 805–818.
44. Pakkenberg B, Gundersen JG. Total number of neurons and glial cells in human brain nuclei estimated bythe disector and the fractionator. *J Microscopy* 1988, **150**: 1–20.
45. Pakkenberg B. Pronounced reduction of total neuron number in mediodorsal thalamic nucleus and nucleus accumbens in schizophrenia. *Arch Gen Psychiatry* 1990, **47**: 1023–1028.
46. Bogerts B, Hantsch J, Herzer M. A morphometric study of the dopamine-containing cell groups in the mesencephalon of normals, Parkinson patients, and schizophrenics. *Biol Psychiatry* 1983, **18**: 951–969.
47. Reyes MG, Gordon A. Cerebellar vermis in schizophrenia. *Lancet* 1981, **2**: 700–701.
48. Lohr JB, Jeste DV. Locus ceruleus morphometry in aging and schizophrenia. *Acta Psychiatr Scand* 1988, **77**: 689–697.
49. Karson CN, Garcia-Rill E, Biedermann J, Mrak RE, Husain MM, Skinner RD. The brain stem reticular formation in schizophrenia. *Psychiatry Res* 1991, **40**: 31–48.
50. Garcia-Rill E, Biedermann JA, Chambers T, Skinner RD, Mrak RE, Husain M, Karson CN. Mesopontine neurons in schizophrenia. *Neuroscience* 1995, **66**: 321–335.

51. el-Mallakh RS, Kirch DG, Shelton R, Fan KJ, Pezeshkpour G, Kanhouwa S, Wyatt RJ, Kleinman JE. The nucleus basalis of Meynert, senile plaques, and intellectual impairment in schizophrenia. *J Neuropsychiatry Clin Neurosci* 1991, **3**: 383–386.
52. Purohit DP, Davidson M, Perl DP, Powchik P, Haroutunian VH, Bierer LM, McCrystal J, Losonsczy M, Davis KL. Severe cognitive impairment in elderly schizophrenic patients: A clinicopathological study. *Biol Psychiatry* 1993, **33**: 255–260.
53. Powchik P, Davidson M, Nemeroff CB, Haroutunian V, Purohit DP, Losonczy M, Bissette G, Perl D, Ghanbari H, Miller B, Davis K. Alzheimer's-disease-related protein in geriatric schizophrenic patients with cognitive impairment. *Am J Psychiatry* 1993, **150**: 1726–1727.
54. Taller AM, Asher DM, Pomeroy KL, Eldadah BA, Godec MS, Falkai PG, Bogert B, Kleinman JE, Stevens JR, Torrey EF. Search for viral nucleic acid sequences in brain tissues of patients with schizophrenia using nested polymerase chain reaction. *Arch Gen Psychiatry* 1996, **53**: 32–40.
55. Roberts GW, Colter N, Lofthouse R, Johnstone EC, Crow TJ. Is there gliosis in schizophrenia? Investigation of the temporal lobe. *Biol Psychiatry* 1987, **22**: 1459–1468.
56. Stevens CD, Altshuler LL, Bogerts B, Falkai P. Quantitative study of gliosis in schizophrenia and Huntington's chorea. *Biol Psychiatry* 1988, **24**: 697–700.
57. Crow TJ, Ball J, Bloom SR, Brown R, Bruton CJ, Colter N, Frith CD, Johnstone EC, Owens DG, Roberts GW. Schizophrenia as an anomaly of development of cerebral asymmetry. A postmortem study and a proposal concerning the genetic basis of the disease. *Arch Gen Psychiatry* 1989, **46**: 1145–1150.
58. Arnold SE, Franz BR, Trojanowski JQ, Moberg PJ, Gur RE. Glial fibrillary acidic protein immunoreactive astrocytosis in elderly patients with schizophrenia and dementia. *Acta Neuropathol* 1996, **91**: 269–277.
59. Norton WT, Aquino DA, Hozumi I, Chiu F-C, Brosnan CF. Quantitative aspects of reactive gliosis: A review. *Neurochem Res* 1992, **17**: 877–885.
60. Conrad AJ, Abebe T, Austin R, Forsythe S, Scheibel AB. Hippocampal pyramidal cell disarray in schizophrenia as a bilateral phenomenon. *Arch Gen Psychiatry* 1991, **48**: 413–417.
61. Altshuler LL, Conrad A, Kovelman JA, Scheibel A. Hippocampal pyramidal cell orientation in schizophrenia. *Arch Gen Psychiatry* 1987, 44: 1094–1098.
62. Christison GW, Casanova MF, Weinberger DR, Rawlings R, Kleinman JE. A quantitative investigation of hippocampal pyramidal cell size, shape, and variability of orientation in schizophrenia. *Arch Gen Psychiatry* 1989, **46**: 1027–1032.
63. Jakob H, Beckmann H. Prenatal developmental disturbances in the limbic allocortex in schizophrenics. *J Neural Transm* 1986, **65**: 303–326.
64. Arnold SE, Hyman BT, Hoesen GWV, Damasio AR. Some cytoarchitectural abnormalities of the entorhinal cortex in schizophrenia. *Arch Gen Psychiatry* 1991, 48: 625–632.
65. Krimer LS, Herman MM, Saunders RC, Boyd JC, Kleinman JE, Hyde TM, Weinberger DR. Qualitative and quantitative analysis of the entorhinal cortex cytoarchitectural organization in schizophrenia. *Soc Neurosci Abstr* 1995, **21**: 239.
66. Akbarian S, Kim JJ, Potkin SG, Hetrick WP, Bunney WE, Jones EG. Maldistribuiton of interstitial neurons in prefrontal white matter of the the brains of schizophrenic patients. *Arch Gen Psychiatry* 1996, **53**: 425–436.
67. Lipska BK, Jaskiw GE, Weinberger DR. Postpubertal emergence of hyperresponsiveness to stress and to amphetamine after neonatal excitotoxic hippocampal damage: A potential animal model of schizophrenia. *Neuropsychopharmacology* 1993, **9**: 67–75.
68. Lipska BK, Weinberger DR. Behavioral effects of subchronic treatment with haloperidol or clozapine in rats with neonatal excitotoxic hippocampal damage. *Neuropsychopharmacology* 1994, **10**: 199–205.
69. Lipska BK, Swerdlow NR, Geyer MA, Jaskiw GE, Braff DL, Weinberger DR. Neonatal excitotoxic hippocampal damage in rats causes post-pubertal changes in prepulse inhibition of startle and its disruption by apomorphine. *Psychopharmacology* 1995, **122**: 35–43.

70. Olney JW, Farber NB. Glutamate receptor dysfunction and schizophrenia. *Arch Gen Psychiatry* 1995, **52**: 998–1007.
71. Kim JS, Kornhuber HH, Schmid-Burgk W, Holzmuller B. Low cerebrospinal fluid glutamate in schizophrenic patients and a new hypothesis on schizophrenia. *Neurosci Lett* 1980, **30**: 379–382.
72. Deakin JFW, Slater P, Simpson MDC, Gilchrist AC, Skan WJ, Royston MC, Reynolds GP, Cross AJ. Frontal cortical and left temporal glutamtergic dysfunction in schizophrenia. *J Neurochem* 1989, **52**: 1781–1786.
73. Kerwin R, Patel S, Meldrum B. Quantitative autoradiographic analysis of glutamate binding sites in the hippocampal formation in normal and schizophrenic brain post mortem. *Neuroscience* 1990, **39**: 25–32.
74. Simpson MDC, Slater P, Royston MC, Deakin JFW. Regionally selective deficits in uptake sites for glutamate and gamma-amino butyric acid in the basal ganglia in schizophrenia. *Psychiatry Res* 1992, **42**: 273–282
75. Eastwood SL, McDonald B, Burnet PW, Beckwith JP, Kerwin RW, Harrison PJ. Decreased expression of mRNAs encoding non-NMDA glutamate receptors GluR1 and GluR2 in medial temporal lobe neurons in schizophrenia. *Brain Res Molec Brain Res* 1995, **29**: 211–223.
76. Simpson MDC, Royston MC, Slater P, Deakin JFW. Phencyclidine and sigma receptor abnormalities in schizophrenic post-mortem brain. *Schizophr Res* 1990, **3**: 32.
77. Tsai G, Passani LA, Slusher BS, Carter R, Baer L, Kleinman JE, Coyle JT. Abnormal excitatory neurotransmitter metabolism in schizophrenic brains. *Arch Gen Psychiatry* 1995, **52**: 829–836.
78. Benes FM, Sorensen I, Vincent SL, Bird ED, Sathi M. Increased density of glutamate-immunoreactive vertical processes in superficial laminae in cingulate cortex of schizophrenic brain. *Cereb Cortex* 1992, **2**: 503–512.
79. Akbarian S, Kim JJ, potkin SG, Hagman JO, Tafazzoli A, Bunney WE, Jones EG. Gene expression for glutamic acid decarboxylase is reduced without loss of neurons in prefrontal cortex of schizophrenics. *Arch Gen Psychiatry* 1995, **52**: 258–266.
80. Bardgett ME, Jackson JL, Taylor GT, Csernansky JG. Kainic acid decreases hippocampal neuron number and increasesdopamine receptor binding in the nucleus accumbens: an animal model of schizophrenia. *Behav Brain Res* 1995, **70**: 153–164.
81. Mattson MP. Cellular signaling mechanisms common to the development and degeneration of neuroarchitecture. A review. *Mech Ageing Dev* 1989, **50**: 103–157.

IV
Approaches in Treating Nerve Cell Death

28
Trophic Factors as Therapeutic Agents for Diseases Characterized by Neuronal Death

Vassilis E. Koliatsos and Italo Mocchetti

TROPHIC FACTORS AS BIOLOGICAL THERAPIES

The unprecedented growth, over the last 20 yr, of information on the structure and function of the mammalian nervous system has generated high hopes for a more efficient management and treatment of nervous and mental disorders. A critical dimension of this development is the realization that the mature nervous system is much more capable of surviving and adapting to injury that once believed *(1)*. Although there are multiple manifestations of the innate capability of the adult nervous system to adjust to injury *(2)*, a remarkable indication of its regenerative potential is the presence, well within the adult life of mammalian species, of cells in the telencephalon that can differentiate into any type of neuron when given the proper instructive signals *(3)*.

The clinical implications of recently developed concepts on the adaptive potential of the nervous system are broad and may include possibilities such as the use of electronic chips attached to the nervous system for targeted depolarization of nerve or muscle cells (e.g., in cases of spinal cord injury *[4]*). However, the greatest hope lies in our ability to develop substances that can encourage the existing potential of the nervous system and can then be used as drugs. In other words, the highest challenge for modern clinical neurobiology is the development of pharmaceuticals based on a more fundamental understanding of what constitutes neural injury and how the nervous system responds in certain cases to prevent permanent damage.

The search for drugs that can address mechanisms of neural injury introduces the general concept of the so-called *biological therapies*, which are defined by two principal features: they presuppose some knowledge of pathogenetic mechanisms and address specific steps in the pathogenetic cascade; and, ideally, they use physiological compounds employed by the organism for reasons related to the aims of the therapeutic intervention. Both of these conditions insure a maximal approximation of the therapeutic intervention to physiological processes.

In sharp contrast to biological therapeutics, the aim of traditional organic pharmacology is to identify any type of molecule that can alleviate the symptoms of a disease and is reasonably safe, regardless of the mechanism of action. The original compounds in the pharmacopoeia were discovered by serendipity, and daughter drugs have been produced by analogical design, either on the basis of a structural homology to the pre-

From: Cell Death and Diseases of the Nervous System
Edited by: V. E. Koliatsos and R. R. Ratan © Humana Press Inc., Totowa, NJ

vious compounds or on the basis of a resemblance in their in vivo effects *(5)*. The design of biological pharmaceuticals is based on mechanisms of disease.

Trophic factors represent an example par excellence of a biological therapy, primarily because they are naturally implemented in mammalian organisms to promote the survival of cells, including neurons, during developmental cell death *(6)* (Fig. 1A). Equally important, trophic factors are naturally employed to induce and maintain the differentiation of surviving neurons to their mature phenotypes *(7)* (Fig. 1A,B). Because loss of differentiation and death of neurons are hallmarks of many diseases of the nervous system, especially neurodegenerative disorders, trophic factors are natural candi-

---

Fig. 1. *(opposite page)* A sample of preparations illustrating the foundations of the classical neurobiology of trophic peptides. **(A)** The ability of trophic factors to modulate cell death and to induce differentiation of neurons during development is best exemplified in the experiments of Levi-Montalcini. On the left, newborn mice were treated for three weeks with an antibody that blocked the action of NGF. The resulting deprivation of the peripheral sympathetic system (Sym) of NGF results in a dramatic loss of sympathetic neurons. This figure depicts low magnification whole mounts and high-magnification cross-sectional views of the upper sympathetic chain with the superior cervical and stellate ganglia. The extreme left whole mount and the upper cross section are preparations from control mice; the whole mount next to the one on the extreme left and the cross section at the bottom are taken from animals treated with the NGF antibody. On the right, NGF was added to an explant culture of a chick DRG shown untreated on top. This treatment results in a massive elaboration of processes from the periphery of the ganglion in the form of a characteristic halo. This observation was developed into a quantitative bioassay for NGF that is still in use. **(B)** Trophic factors continue to promote or stimulate the differentiation of neurons during the adult life of animals. These preparations were taken from the caudate nucleus of two macaques, one of which was treated with vehicle solution (V) and the other treated with human recombinant NGF (T) delivered into the ipsilateral lateral ventricle. Preparations were stained with a monoclonal antibody against ChAT to visualize cholinergic interneurons of the neostriatum (i.e., the only neostriatal neurons that respond to NGF). Note enlargement of the cell body and processes and the enrichment of the cholinergic neuropil, representing local axonal branches and terminals of cholinergic neurons, in "T". **(C,D)** These two panels illustrate the ability of trophic factors to restore pathology in developing and adult neurons following various types of lesions. (C) illustrates the ability of NGF to restore the phenotype of axotomized neurons in young adult animals, whereas (D) illustrates the ability of GDNF to prevent the retrograde death of early postnatal motor neurons following nerve section. In (C), the two sections were taken from the medial septal nucleus of two macaque monkeys that underwent unilateral transection of the fornix. The animal on top (V) was subsequently treated with vehicle solution, whereas the animal on the bottom (T) was treated with NGF in the lateral ventricle. Both preparations were stained for ChAT immunocytochemistry. The vehicle-treated monkey shows a near-complete disappearance of cholinergic phenotype in neurons of the medial septum, whereas the NGF-treated septum (bottom) shows a striking preservation of cholinergic phenotype. Asterisks indicate the lesioned side. Vertical lines indicate the medial plane. In (D), two sections were taken from the facial nucleus of 6-d-old rats subjected to a unilateral transection of the facial nerve immediately after birth. The animal on top was treated with a vehicle solution delivered into the stump of the transected facial nerve, whereas the animal on the bottom was treated with the cytokine GDNF, following the same procedures. Sections were stained with cresyl violet. Note the extensive degeneration of axotomized motor neurons on the top panel (many surviving neurons appear shrunken, as indicated by arrows). GDNF affords a complete protection from axotomy-induced death of neonatal facial motor neurons (bottom panel).

Trophic Factors as Therapeutic Agents

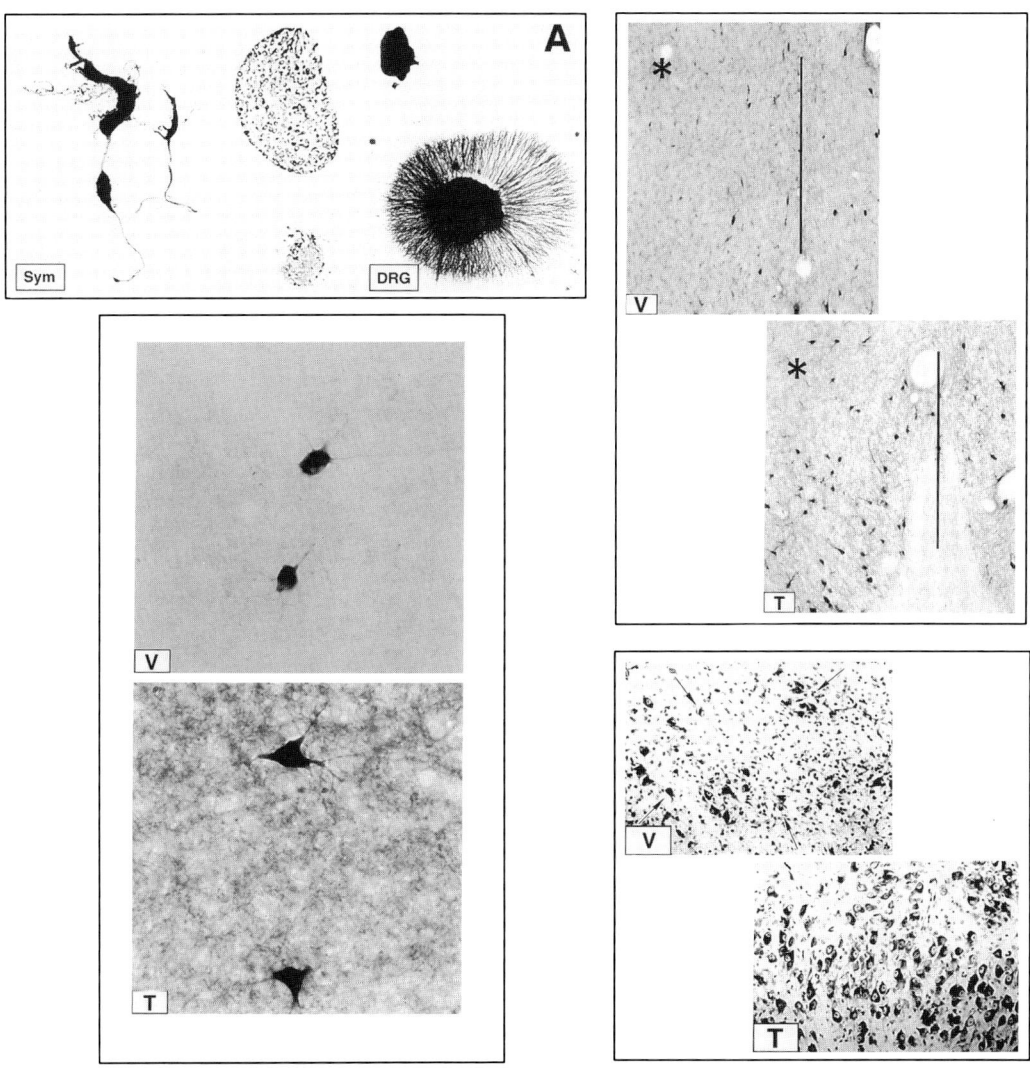

Fig. 1

dates for the treatment of these illnesses. In addition, their efficacy in heterogeneous lesion paradigms (Fig. 1C,D; *see below*) indicates that trophic factors can intervene in downstream steps in the pathogenetic cascade regardless of specific etiology or the nature of early ("upstream") events. On the other hand, trophic molecules influence neurons by binding to specific membrane-associated receptors, a condition enabling them to target specific populations of neurons without direct interference with ubiquitous second messenger systems. These features are expected to minimize potential side effects stemming from nonspecific activation of signal transduction in many brain areas and organs outside the brain.

It should be emphasized that trophic therapies are not considered for neurological diseases because it is generally accepted that these factors or their receptors are reduced in the brains of patients suffering from these illnesses *(8–12)*. Of course, the availability of trophic factors to target neurons may be disrupted in the presence of axonal and synaptic pathology, which commonly characterizes neurodegenerative disorders *(13)*. However, the principal aim of the trophic approach is neuronal regeneration, not the replacement of a missing substance. Although investigators have attempted sophisticated analyses of altered trophic factor availability/transduction using body fluids from living patients *(14)*, postmortem nervous tissues *(12,15)*, or biopsies from peripheral tissues *(16)*, findings derived from such studies may reflect secondary effects of the main pathological process or of cell death on trophic cascades, rather than primary pathogenetic events. This limitation is especially problematic for factors whose expression and transduction are regulated by neuronal activity, especially neurotrophins (*see* the Activity-Related Changes in Trophic Factor Expression and Utilization section). On the other hand, the presence of trophic factor receptors in the brains of patients with neurological diseases is a necessary requirement for the successful binding/transduction of trophic factors, if these molecules are to be used as drugs. For example, in the brains of patients with Alzheimer's disease (AD), there is evidence for the persistence of trophic factor receptors (e.g., $p75^{NGFR}$ *[10,11]* and trkA and trkB [Koliatsos, unpublished observations]) at various levels of expression. Although some of these receptors may be reduced in the brains of patients with AD *(10)*, it is encouraging that at least some trophic factor receptors (e.g., nerve growth factor [NGF] receptors) are upregulated when exposed to the corresponding trophic molecule (unlike neurotransmitter receptors), thereby maximizing the benefit of the exogenous factor *(17,18)*.

In its broad formulation, neurotrophism involves the entire set of interdependencies among connected neurons *(19)* (*see* also chapter by Burek and Oppenheim). If, for example, neuron A innervates neuron B, then A supports B, and B sustains A. Most investigators, however, work on the basis of a narrower definition, which only considers the retrograde support of neuron A by neuron B. According to this definition, neuron B releases trophic factors that influence neuron A as follows: they promote the survival and induce and maintain its differentiation; their effects are longterm (with a single quantum of released trophic factors causing sustained effects); and their effects on survival and differentiation continue when neuron A is injured, regardless of the specific type of injury.

The function of trophic factors as retrograde signals is perhaps an acceptable simplification for the biological actions of the NGF family of peptides and has inspired the construction of a powerful theory for the formation of the nervous system *(19)* (for a more detailed discussion of this theory, *see* chapter by Burek and Oppenheim). However, there is growing evidence that the anterograde support of cells, including neurons and myocytes, may also be an important trophic mechanism. Such support can be mediated via a variety of signals, including neurotransmitters and neuromodulators *(20)*. An important anterograde "trophic" signal for myocytes is agrin, which is transported by motor axons to synaptic endings in muscle *(21)* and then binds to postjunctional membranes to induce the synaptic clustering of several muscle surface proteins, including acetylcholine nicotinic receptors *(22)*. In the developing avian visual system, neurotrophin-3 (NT-3) can be anterogradely transported, released at the terminal, and utilized by second-order neurons *(23)*. The fibroblast growth factors (FGF) *(24)* and, possibly, brain derived neurotrophic factor (BDNF) *(25)* can be anterogradely transported in the mammalian central nervous system (CNS). Despite accumulating evidence for the anterograde transport of trophic factors and their possible utilization by second-order neurons, trophic effects on survival and phenotype of neurons have not yet been demonstrated in the anterograde fashion.

NEURONAL CELL DEATH AS A THERAPEUTIC TARGET: "DISTAL" AND "PROXIMAL" INTERVENTIONS

Neuronal death is a common outcome in many diseases of the nervous system, from neurodegenerative diseases, such as AD (*see* chapter by Braak and Braak in this volume) and stroke (*see* chapter by Vornov in this volume) to, probably, severe psychosocial trauma *(26)*. Neurodegenerative diseases, the most complex neurological conditions associated with death of neurons and nonneuronal cells, were the first to be considered as subjects of trophic therapies (*see* the Clinical Potential of Trophic Peptides section). In these conditions, there is usually a considerable period of time between the action of a (usually unknown) causative agent, to the establishment of pathological processes within neurons and glia, and, finally, to cell death (i.e., a long "incubation" period); the latter steps are also associated with the symptomatic phase of the illness. There are two advantages in using neurodegenerative disorders as an example to discuss various therapeutic approaches in treating cell death: first, these complex disorders were historically the first to be considered for a trophic treatment; and second, because morbid processes associated with neurodegenerative disease take a fairly long time to develop, we can distinguish among various steps or degrees of cell injury, representing distinct targets for intervention (causes → initial mechanisms → secondary mechanisms → cell death). In such a context, the trophic therapies can be placed in perspective.

Because it would be extremely difficult to discuss the previous interventions in a generic fashion covering all neurodegenerative conditions, we will use AD as the prototypical neurodegenerative disorder. We have previously presented the advantages of trophic approaches for AD in more topical discussions of the subject *(13,27)*.

The current approaches to treat AD aim at promoting the function of remaining synapses either by affording structural support or by stimulating transmitter release and binding. These approaches can lead to amelioration of symptoms that depend on a group of synapses specified by brain system or by type of neurotransmission. The majority of drugs in current use or in clinical trials for AD increase central cholinergic neurotransmission, a strategy consistent with the early universal involvement of forebrain cholinergic systems in AD and the major implications of brain cholinergic transmission for working memory *(28–30)*. These drugs maximize the binding of available acetylcholine to the postsynaptic receptor via acetylcholinesterase inhibition, an approach that can be improved with the increase of their half-life (as in the case of the recently released donepezil) and the design of molecules with predilection for the central cholinergic synapse or with specificities for molecular forms of acetylcholinesterase persisting through advanced stages of AD *(31,32)*. Cholinomimetic strategies can produce striking effects in brain function *(33)* and improve symptoms in a subset of patients with AD *(34,35)* but are unlikely to deal with ongoing pathology or to combat cell death. Therefore, their effects are limited to *short-term functional benefits*. The current development of direct cholinergic agonists or glutamate agonists (ampakines, to enhance corticocortical signaling) is essentially a variation of the previous approach.

Interventions that target cell death in AD by eliminating its distal causes (i.e., etiologies) are at the other end of the therapeutic continuum as compared to cholinomimetic agents. To the extent that we accept amyloidogenesis as a necessary and, perhaps, common pathogenetic step for a variety of causes of AD, an *etiopathogenetic intervention* for this illness would be to prevent the formation of senile plaques. These pathological structures interfere with neural circuitry and may precipitate neuronal death via a number of mechanisms, including neurotropism, direct toxicity, or macrophage activation. The core of the plaque is comprised of β-amyloid protein (Aβ), a 4-kDa fragment of a larger precursor molecule, the amyloid precursor protein (APP) *(36)*. Aβ spans through the transmembrane and the adjacent extracellular domain of APP and self-assembles to form amyloid fibrils in vitro *(37)*. APP is normally cleaved by an unknown protease (APP secretase) within the extracellular region of Aβ *(38)* to generate nonamyloidogenic secretory products. Although we do not have direct evidence for events that lead to Aβ formation and amyloidogenesis in the brains of patients with AD, the fact that APP mutations associated with familial AD cluster around the normal cleavage sites of Aβ *(39–42)* suggests strongly that the normal processing of APP is inhibited in AD. Therefore, strategies that facilitate a secretase-type cleavage of APP—either by enhancing the function of the enzyme or by inhibiting the enzyme to reduce APP turnover—appear as reasonable approaches to treatment *(43)*. However, although several candidates have been proposed (reviewed in ref. *43*), the nature of the secretase and the biochemical details of its action remain unknown. Another possible target is the intracellular Aβ pool, which comprises a portion of APP that undergoes recycling within the cytoplasm and never reaches the neuronal membrane *(44)*, as well as the reinternalized membrane-associated APP *(45)*. Lysosomes play a major role in this processing, but the specific pathways and enzymes involved remain unknown. In a related approach, attention is shifted from plaque formation *per se* to cytokine (i.e., interleukin-1 [IL-1]) cascades set forth by the recruitment of antigen-presenting cells in the vicinity of plaques *(46)*.

A different strategy to prevent cell death is to modify biological steps downstream of etiopathogenetic processes (i.e., when neurons have already become afflicted with the initial injury but before they are committed to cell death) *(47)*. This strategy would *assist neurons in confronting the injury* regardless of its nature in AD, thus, preventing the degenerative process from advancing irreversibly to cell death. The significance of this strategy becomes apparent if we consider the kinetics of the degenerative process—unless we can prevent cell death, the precipitous loss of nerve cells will ultimately overcome any therapeutic effort aimed at a gain of function (e.g., with anticholinesterases). The need to protect neurons from dying invites the consideration of agents that interfere with the fundamental mechanisms of cell death. Trophic factors are the longest known and best understood agents for this task.

NEUROBIOLOGY OF TROPHIC FACTORS: BASIC TENETS
Molecules
Neurotrophins

The prototypical trophic factor NGF was discovered in the mid 1950s by Stanley Cohen and Rita Levi-Montalcini *(48,49)*, purified and characterized by Angeletti and colleagues around 1970 *(50,51)*, and obtained in crystal form just a few years ago *(52)*. Until Barde and coworkers *(53)* undertook the labor-intensive task to purify BDNF from kilograms of pig brains, NGF represented a type of biochemical oddity. The purification of BDNF allowed the determination of a partial amino acid sequence, leading to the construction of probes and the ultimate cloning of the BDNF gene *(54)*, which was found to have considerable homology to NGF and predicted an NGF family of related peptides. Within 3 yr after the cloning of BDNF, two additional related peptides were identified with methods imported from molecular genetics: NT-3 *(55–59)* and neurotrophin-4/5 (NT-4/5) *(60–62)*. Members of the NGF family, the so-called *neurotrophins*, share a 50–60% homology at the amino acid level and have significant structural similarities *(63)*. Neurotrophins share important biological features with NGF, i.e., powerful effects on the survival and differentiation of neurons and selectivity for restricted populations of neurons that express distinct neurotrophin receptors (*see* the Mechanisms section).

Cytokines

Cytokines can also exert trophic effects on the nervous system. Unlike neurotrophins, cytokines do not comprise a single gene family but are heterologous molecules including FGF, transforming growth factors β (TGFβ), and leukemia inhibitory factor (LIF)-like peptides. These factors are characterized by their ability to influence extremely variable types of cells and their tendency to have fundamentally different effects on different populations of target cells, ranging from the maintenance of the pluripotentiality of embryonic stem cells to induction of the differentiation of sympathetic neuronal precursors *(64)*.

The current consideration of cytokines as neuroactive substances is largely the result of the cloning of the ciliary neurotrophic factor (CNTF) gene *(65)*, which encodes a peptide initially described as an activity that supports the survival of chick ciliary ganglion neurons *(66)*. Unlike neurotrophins, CNTF lacks a signal peptide sequence obligatory for processing in the secretory pathway and appears late in development (i.e., after the first postnatal week in the mammalian peripheral nervous system [PNS]) *(67)*. This pattern of expression is incompatible with a role in developmental cell death and may explain why CNTF null mice have a relatively unimpressive phenotype *(68)*.

Based more on a partial similarity of effects in the nervous and hematopoietic systems and much less on amino acid sequence homology, CNTF has been classified in a group of cytokines together with LIF, oncostatin M (OM), and the interleukins (especially IL-6). Despite only a 15–20% primary sequence identity, these molecules are predicted to have similar tertiary structures containing four amphipathic helices *(69,70)* and share many common transduction elements (*see* the Mechanisms section). Although, for some time, CNTF enjoyed the exclusive status of a "neural cytokine," based on its presumed selective effects in the nervous system *(71)*, it is now becoming apparent that CNTF shares all the generic features of cytokines, including various effects in many organ systems and a cachectin-like effect on muscle and bone marrow *(72)*. Under certain conditions, CNTF (like LIF) may promote cell death in mammalian neurons in vitro *(73)*. A recent addition to the CNTF–LIF family of factors has been cardiotrophin 1 (CT-1) *(74)*, expressed in a number of peripheral tissues and with potent in vitro effects on myocytes; this cytokine has been recently found to promote the survival of mammalian motor neurons in vitro and in vivo *(75)*.

Cytokines other than CNTF and its relatives have been the subject of investigation for several years. For example, the heparin-binding growth factors acidic and basic FGF (or FGF-1 and FGF-2) had already been cloned by the mid 1980s *(76)*; their multiple effects in the nervous system have been characterized extensively in the past ten years *(77)*. Although the profound effects of FGF in nonneural tissues and, specifically, their angiogenic properties *(78)* have raised some concerns about the use of these molecules as selective and nontoxic promoters of neuronal viability, the pleiomorphic effects of these molecules may make them reasonable candidates for conditions with complex damage to the nervous system and associated tissues, such as spinal cord injury *(see below)*.

Insulin-Like Growth Factors (IGFs)

Other interesting molecules are the insulin-like growth factors (IGFs), IGF-I and IGF-II. Although these substances were thought to be produced in the liver in response to growth hormone and to mediate the effects of this hormone (somatomedins), only IGF-I appears to fulfill the criteria of a somatomedin, whereas IGF-II is only minimally dependent on growth hormone *(79)*. IGF are synthesized in many tissues in developing and adult animals. It has been shown that IGF-I has profound anabolic and thymoleptic effects in adult men but these effects are weaker than those exerted by growth hormone in doses resulting in comparable concentrations of plasma IGF-I *(80)*. IGF-I induces sprouting of motor neuron terminals with likely functional implications in animal models of spinal muscular atrophies *(81)*, but evidence for a direct regenerative effect on motor neurons is limited *(82,83)*.

TGFβ Family

More recently, there has been an increased interest in the TGFβ superfamily, which contains >25 members including TGFβ 1–5, activins, bone morphogenetic proteins, glial cell-derived neurotrophic factor (GDNF), and neurturin. Except for a few cases, these peptides and their receptors are not highly expressed in the mammalian brain *(84,85)*. GDNF was purified from the supernatant of the B49 glial cell line and has been shown to promote the survival of mesencephalic dopaminergic neurons in vitro *(86)*. Several in vivo studies have shown significant effects of this factor on motor neurons *(87–89)* and dopaminergic neurons in the substantia nigra *(90,91)*. GDNF is an impressive molecule with unprecedented potency in all cell systems where it exerts its effects. For embryonic motor neurons in vitro, GDNF is active at concentrations as low as 0.2 pg/mL, making it the most potent survival factor for motor neuron cultures maintained for short-term periods *(87)*. At least some members of the TGFβ superfamily, including the TGFβ 1–3 and activin A, share some of the neurotrophic effects of GDNF in vitro *(92)*. Neurturin, the most recently identified member of the TGFβ family *(93)*, has a surprising 42% amino acid homology with GDNF, placing it closer to GDNF than any other TGFβ-like protein. Although neurturin is a very potent survival factor for sympathetic neurons in vitro, it is still unclear if it shares some of the most clinically relevant biological effects of GDNF, especially on motor and mesencephalic dopaminergic neurons.

Mechanisms

Neurotrophin Receptors

An essential step in the physiological characterization of trophic factors (i.e., signal transduction), the identification of their receptors, has been the subject of intense and prolific investigations during the last decade. Because of their historical precedence, neurotrophins have enjoyed the lion's share in these investigations. The earliest neurotrophin receptor to be identified is a 75-kDa protein that binds all neurotrophins with the same low affinity *(94,95)*. This so-called p75^{NGFR} is expressed in restricted populations of central and peripheral neurons. Interestingly, p75^{NGFR}

bears significant homologies with several receptors that mediate cell death, such as the two receptors for tumor necrosis factor and the fas antigen receptor, which transduce signals that lead to apoptotic cell death in human cell lines *(96)*. Despite a dramatic shift in attention from p75^{NGFR} to the high-affinity neurotrophin receptors *(see below)* in the past 5 yr, there is currently a renewed interest in p75^{NGFR}, starting with the realization that the small cytoplasmic domain of p75^{NGFR} can initiate (signal) ceramide cascades *(97)*. Under certain conditions, this type of signaling can promote, rather than inhibit, processes related to cell death. For example, unoccupied p75^{NGFR} mediates cell death in PC12 cells *(98)*. NGF binding to p75^{NGFR} on Schwann cells activates the cell death mediator (transcription factor) nuclear factor kappa B (NFκB), an effect selective for NGF and not seen with BDNF or NT-3 *(99)*. In unpublished experiments, using primary dorsal root ganglia (DRG) sensory neurons, we have found that blocking p75^{NGFR} delays neuronal death following deprivation of NGF (Koliatsos and Ehlers, personal observations). The previous in vitro observations have been recently corroborated by evidence that, during the development of the chick retina, NGF causes the death of retinal neurons that express p75^{NGFR} but not trkA *(100)*. Research in p75^{NGFR} null mice, which initially revealed only defects in skin perception of temperature *(101)*, has recently shown that these mice have larger numbers of basal forebrain and striatal cholinergic neurons, an observation consistent with the fact that p75^{NGFR} mediates the developmental death of these neurons, presumably upon binding to NGF *(102)*.

A second group of neurotrophin receptors belongs to the tyrosine kinase superfamily, which also includes the receptors for insulin and epidermal growth factor. These receptors comprise the trk family, named after *t*ropomyosin *r*eceptor *k*inase. (The first such receptor, trkA, was originally detected as an oncogene, whose tumorigenic activity resulted from a genetic rearrangement event in which nonmuscle tropomyosin sequences were aberrantly fused onto the transmembrane and cytoplasmic domains of trkA *[103,104]*.) Trk receptors bind neurotrophins with high affinity, and there is a differential preference of various members of this family for different neurotrophins. The first identified member, trkA, has a selective affinity for NGF. TrkB is preferentially a receptor for BDNF and NT-4/5, and trkC prefers NT-3 *(105)*. Despite an earlier hypothesis proposing the formation of heterodimers between p75^{NGFR} and trk, there is no consistent evidence for coprecipitation of the two types of receptors, although p75^{NGFR} may influence the binding of neurotrophins to trk receptors through unknown mechanisms *(106,107)*.

A majority of investigators agree that trk receptors are the principal neurotrophin transducers, and results from trk null mice, in general, correlate well with results from neurotrophin null mice, revealing a closely kept trk-neurotrophin rapport, at least during development *(see below)*. However, some trkB and trkC isoforms lack the catalytic kinase domain, and, to increase further the level of complexity, 2–3 kinase-containing trkC isoforms contain amino acid inserts within the kinase region that block catalytic activity. The physiological role of noncatalytic trk receptors remains elusive.

TrkA mRNA is expressed selectively in neurons known to respond to NGF, such as basal forebrain cholinergic neurons, striatal cholinergic interneurons, peripheral sympathetic, and medium-to-small sized sensory neurons that mediate pain and temperature. TrkA null mice show loss of small DRG neurons responsible for pain and temperature but have minimal pathology in the brain *(108)*; this pattern is very consistent with NGF null mice *(109)*.

The expression of trkB and trkC follows patterns very different from the selective expression of trkA and p75^{NGFR}. Catalytic trkB is expressed widely in the PNS and in most areas of the telencephalon and diencephalon, including thalamus, neocortex, and hippocampus. Expression in telencephalon is higher than in diencephalon (Koliatsos, personal observations). An additional curiosity in the expression of trkB in the mammalian brain is the abundance of the truncated forms of the receptor, which, as mentioned above, cannot transduce neurotrophin signals *(110)*. The majority of expressed trkB in the mature mammalian nervous system is noncatalytic, a condition that raises questions concerning the efficacy of trkB-transducible trophic factors in adult animals. TrkB nulls, with loss of motor and neural crest- and placode-derived sensory neurons, do not show significant brain pathology (like trkA nulls) *(111)*. BDNF nulls show decreased numbers of peripheral sensory neurons, but motor neurons appear normal *(112,113)*. The discrepancy between the degree of involvement of the CNS and PNS after the elimination of neurotrophins and neurotrophin receptor genes suggests some redundancy of trophic transduction in the brain and spinal cord.

TrkC has the same widespread distribution in the brain as trkB but without any predilection for the telencephalon. NT-3 and trkC nulls show a mutually consistent pattern of proprioceptive and motoric defects with atrophy of large sensory neurons but without changes in the brain *(114,115)*.

Cytokine Receptors

Although much more needs to be known concerning the mechanisms of transduction of neural cytokines, a general schema is emerging—that three members of the group (i.e., CNTF, IL-6, and LIF) share at least two receptor components, the LIF receptor β and gp130, a transduction unit for IL-6 *(116)*. CNTF also requires a binding receptor (CNTFRα) *(117)* and two receptor-associated kinase units on the cytoplasmic side (JAK/TYK kinases) *(118)*. Although it was postulated initially that CNTF was the principal cytokine for the nervous system based on the selective expression of CNTFRα in neural tissues *(71)*, CNTF can also act on nonneural tissues, raising questions about the selectivity of CNTFRα or its role in CNTF transduction. Interestingly, CNTFRα nulls are unable to feed because of facial muscle weakness and die within the first postnatal day *(119)*. Motor neurons are significantly reduced in the brainstem and spinal cord. Thus, mice that lack CNTFRα differ from CNTF nulls especially with regard to numbers of motor neurons, suggesting that CNTFR may serve as a receptor for at least one other CNTF-like trophic peptide.

Other Trophic Factor Receptors

A GPI-anchored glycoprotein has been recently cloned as the binding, but not transducing, receptor for GDNF *(120,121)*. GDNF transduction appears to be mediated via the product of the protooncogene c-ret. This receptor tyrosine kinase forms complexes with GDNFR and GDNF that can be isolated by immunoprecipitation; in addition, GDNF can phosphorylate ret in vitro *(120,122,123)*. GDNF transduction via *c-ret* suggests a broader function of this trophic factor because loss-of-function mutations of c-ret have been implicated in Hirschsprung's disease *(124,125)* and gain-of-function mutations have been linked to the multiple endocrine neoplasia syndromes 2A and 2B *(126)*. GDNF nulls are similar to ret nulls because they display complete renal agenesis and lack of an enteric nervous system *(127–130)*. However, GDNF nulls also show loss of neurons in the brainstem and spinal cord and in dorsal root, sympathetic, and certain parasympathetic ganglia, an effect consistent with the observed neurotrophic actions of this peptide *(128,130)*. Monoaminergic neurons of the brainstem appear intact in GDNF null animals *(128,130)*. CT-1 appears to be acting via a GPI-linked receptor subunit, like CNTF and GDNF, but does not bind to CNTFRα. No further information exists on mechanisms of CT-1 transduction at the present time *(75)*.

Models

The development of our ideas for a potential role of trophic factors as drugs for human neurological illness has been based largely on the ability of these molecules to ameliorate structural and functional abnormalities induced in animals by a variety of lesions or as a result of naturally occurring disease. Inducible lesions can be accomplished by neurotoxins, surgical interventions, or alterations of the genotype either via the introduction of altered genes (transgenes) or the elimination of genes of interest via gene targeting (to generate "null" mice). Regardless of the preferred method, the success of any of these interventions depends on their ability to generate structural and functional disruptions that: are selective for systems or populations of neurons involved in a particular neurological illness; and can recapitulate at least part of the neuropathology typical of the disease to be modeled (i.e., amyloid deposition in the form of β-pleated fibrils in AD or Lewy bodies in Parkinson's disease). The former requirement above has been much more difficult to achieve than the latter. However, with the careful application of some of the previous strategies, investigators have been able to recapitulate fairly consistently at least two generic features of neurodegenerative diseases (i.e., *dedifferentiation* of neurons and *cell death* in neural systems implicated in illnesses such as AD, Parkinson's disease, motor neuron disease, and sensory neuropathies).

Axotomy

Axotomy has been a particularly popular model to study the responses of nerve cells to injury and ways to modify/ameliorate some of the consequences of injury. Usually delivered at a distance from the cell body, axotomy lesions avoid complications introduced by direct injury of the perikarya and disruption of the blood–brain barrier (BBB). Moreover, axotomy often recapitulates crucial pathological features of neurodegenerative diseases, including alterations in the cytology of nerve cells, changes in the expression of specific genes (including downregulation of transmitter markers and perturbations in cytoskeletal constituents), and, in some cases, cell death *(131)*. Another advantage of axotomy is the fact that it allows the affected neuron to either regenerate or make adaptive changes prior to degeneration (i.e., axotomy provides a window of time within which to observe the responses of lesioned neurons). Axotomy models have been used successfully to generate pathology in the PNS (i.e., motor neurons and sensory and sympathetic neurons) and CNS (i.e., thalamic, retinal ganglion, corticospinal and rubrospinal neurons, neurons in the inferior olive, and cholinergic neurons in the basal forebrain) (*see* chapter by Elliott and Snider, this volume) (reviewed by Lieberman *[132]*; Snider et al. *[133]*; Koliatsos and Price *[131]*).

Neurotoxins

In general, neurotoxic lesions lead to a much more rapid degeneration of afflicted neurons than axotomy. Neurotoxins have been especially successful in lesioning central monoamine neurons (i.e., systems classically resistant to the effects of axotomy because of profuse regenerative sprouting of axons/terminals). These neurotoxins include toxic derivatives of transmitters (e.g., 6-OH dopamine for dopamine systems), by-products of opiate biosynthesis (e.g., 1-methyl-4-phenyl-1,2,3,6-tetrahydropyridine [MPTP]) for dopamine systems and N-[2-chloroethyl]-*N*-methyl-2-bromobenzylamine [DSP-4] for noradrenergic terminals and neurons in locus coeruleus *[134]*, and amphetamine derivatives (e.g., methamphetamine, p-chloroamphetamine, 3,4-methylenedioxyamphetamine [MDA], 3,4-methylenedioxymethamphetamine [MDMA], or fenfluramine) for serotonin and dopamine systems (*see* chapters by Burke and Cadet, this volume). Excitatory amino acids (i.e., structural analogs of glutamate), such as kainic acid, ibotenic acid, and the endogenous tryptophan metabolite quinolinic acid have been used in large (excitotoxic) doses as generic neuronal toxins in all areas that they are injected; their most important contribution to the neurobiology of disease has been the replication of neuropathology that occurs in HD when these compounds are injected into the neostriatum *(135)* and the replication of neuropathology seen in anoxic-ischemic injury (*see* Chapters by Olney and Ishimaru, Vornov, and by Ross, Becker, and Koliatsos).

Genetics

Lately, neuronal injury has been generated with the introduction, in the mammalian nervous system, of altered genes (transgenes) bearing either mutations associated with dominantly inherited neurological disorders or loss-of-function mutations (the latter for the generation of "null" mice). Transgenes are delivered directly into the pronucleus of fertilized zygotes *(136)* or into mouse blastocysts via genetically modified embryonic stem cells *(137,138)*; newer approaches include the use of large regulatory sequences in very large pieces of DNA delivered into the germ line of mice via yeast artificial chromosomes or cosmids *(139,140)*. In these DNA sequences, the upstream introduction, of specific promoters makes it possible to target the expression of a transgene in a tissue- and cell type-specific manner or at determined stages of development *(141)*. Major contributions for the neurobiology of human disease have been the generation of: mice with brain amyloid deposits akin to AD following the introduction of familial AD-linked APP mutations (Fig. 2) *(142,143)*; mice that express mutant superoxide dismutase 1 and manifest phenotypes similar to familial amyotrophic lateral sclerosis (FALS) *(144)*; and ataxic mice that express high levels of a transgene with expanded triplet repeats causing spinocerebellar ataxia type 1 (ataxin-1) *(145)*. Con-

Fig. 2. *(opposite page)* A major task in the experimental therapeutics involving trophic factors (i.e., preclinical trial of these molecules in a way that predicts their potential clinical utility) is their application in carefully selected animal models of disease. Recently developed transgenic models of disease provide the best approximation to some of the illnesses for which trophic therapies are considered, such as AD. **(A,B)** This transgenic mouse bears a double APP mutation, $Lys^{670} \rightarrow Asn$, $Met^{671} \rightarrow Leu$ (Tg 2576), responsible for early-onset AD in a large Swedish family *(143)*. (A) shows the magnitude of plaque burden in a low-magnification view of the neocortex (top) and hippocampus (bottom) of a 12-mo-old transgenic mouse; the unstained halo present around several amyloid deposits is filled with abnormal neurites. (B) is a high-power magnification of one of the amyloid deposits seen in the same animal as in (A); this amyloid deposit shows a fibrillar pattern of staining. Sections in (A) and (B) are stained with an antibody recognizing human Aβ that detects the amyloid core of senile plaques. **(C)** Section through the cerebral cortex of a Tg 2576 mouse stained with cresyl violet and showing an amyloid plaque at the same magnification as (B). The congophilic core of the plaque is

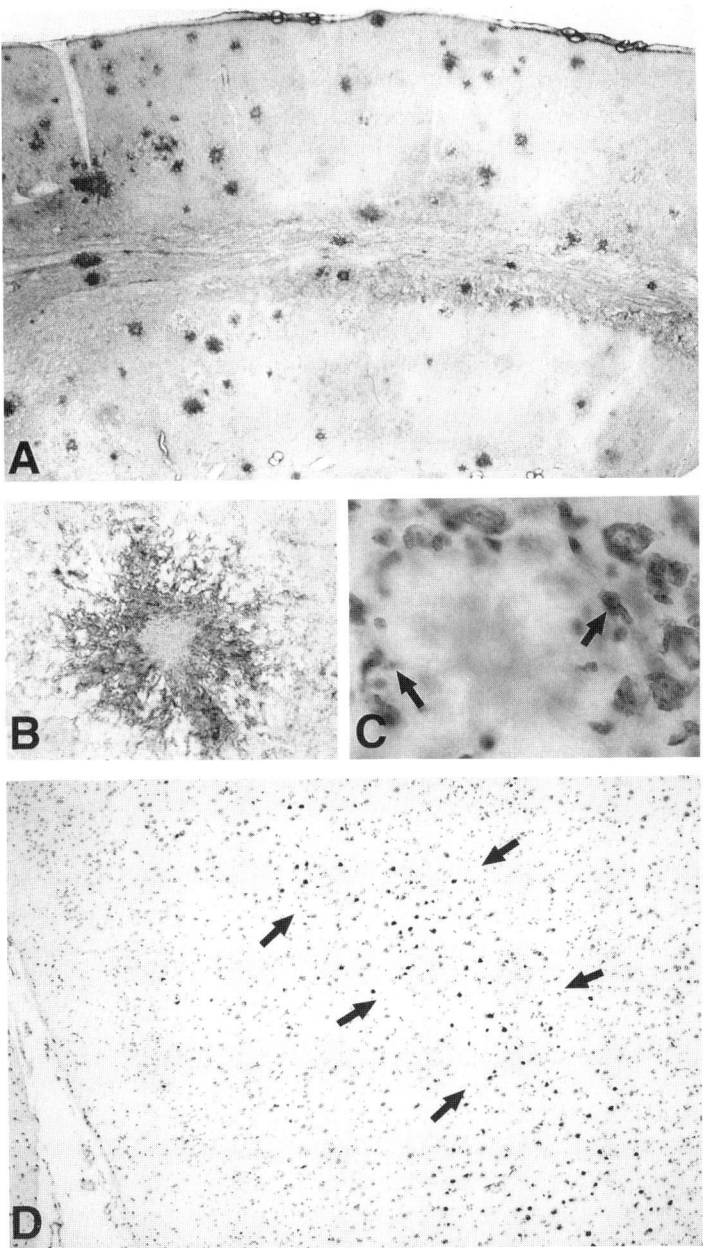

metachromatic for cresyl violet (pink against a purple pattern of staining). Small cellular profiles in satellite positions around the plaque (arrows) belong to microglia; lesion is surrounded by normal cortical neurons. **(D)** Massive degeneration of cortical neurons projecting in associative pathways (indicated here in upper layer III of temporal neocortex with TUNEL stain) is a central feature of AD. Although, thus far, transgenic approaches have reproduced in experimental animals with pathology akin to AD, there is no evidence that amyloidogenesis in these animals in the form of senile plaques causes death of neurons. However, these models continue to be analyzed, and it is hoped that some evidence of cell death may emerge from these studies. (A), (B), and (C) courtesy of Dr. Lee J. Martin.

cerning the use of null (and, to some extent, transgenic) mice as physiological preparations (i.e., to study function), an important caveat is that the expression of a specific gene is eliminated (or altered) during development (i.e., a period of time with significant molecular redundancy). This limitation may explain why, in many cases, null mice have failed to show dramatic phenotypes even when the knocked-out gene was of proven importance for the neurobiology of the adult animal as shown from other lines of investigation. Antisense strategies, in which the expression of a certain gene is suppressed abruptly in adult animals *(146)*, or "second generation" transgenic approaches, in which a specific transgene is targeted to a particular region only during the adult life of the animal using Cre/lox recombination systems *(147)* provide a more physiological alternative to traditional knockout methodologies. Replication-defective viral vectors that carry altered genes to be injected into specific sites in the nervous system may also provide a powerful alternative strategy.

Animals with spontaneous mutations that affect the nervous system or animals that undergo age-associated neurodegenerative changes have been also used widely as models for human neurological illnesses. Such animals include the Wobbler *(148,149)* and progressive motor neuronopathy (pmn) *(150)* mice (to study alterations in motor neuron anatomy and physiology akin to human motor neuron disease) and aged rodents and nonhuman primates *(151)* (used as models of human aging and even AD).

Effects of Trophic Factors on Animal Models

As we stated in previous topical reviews *(152)*, several criteria predict the responsivity of specific populations of neurons to a particular trophic factors including: the expression of the trophic factor in projection targets or the local environment of the putative responders; the expression by the putative responders of binding and transducing receptors for the trophic factor; the retrograde transport or internalization of the trophic factor by the putative responders under high-affinity conditions (i.e., successful competition of the internalization of radiolabeled trophic factor by an excess of cold peptide); physiological effects of the factor on putative responders, such as regulation of genes pertinent to neurotransmission and the neuronal cytoskeleton, or intermediate signals, such as immediate early genes (preferably with picomolar or nanomolar concentrations of the factor); and pharmacological effects of the factor in preventing lesion-induced injury, preferably including scenarios in which the injury results in cell death. An additional criterion could be the expression (or, better, the upregulation) of the trophic factor in the environment of putative responders during the period of naturally occurring cell death.

The previous criteria have been satisfied to various degrees for different factors in various populations of neurons. From the extensive literature that pertains to the (partial) satisfaction of the previous criteria, we have selected below studies that show in vivo effects on injured neurons because these studies have the greatest clinical relevance (*see* the Neurodegenerative Diseases section). Many relevant animal models have been treated with trophic factors, and specific effects of these molecules in preventing death or protecting the phenotype of injured cells have been documented in the following populations of neurons: basal forebrain cholinergic neurons, principally in models using fimbria-fornix lesions *(153–158)*, or aged rodents *(159,160)*; neostriatal cholinergic neurons *(161,162)*; axotomized thalamic neurons undergoing retrograde degeneration *(163)*; lateral geniculate neurons degenerating as a result of monocular deprivation *(164)*; mesencephalic dopaminergic neurons injured with the neurotoxins 6-hydroxy dopamine (6-OHDA) and MPTP *(90,91,165,166)* or undergoing axotomy-induced injury *(167)*; noradrenergic neurons in locus coeruleus after excitotoxic lesions *(168)*; neonatal motor neurons undergoing axotomy-induced retrograde degeneration in the brainstem *(87,89,169,223)* and spinal cord *(75)*; adult motor neurons undergoing axotomy-induced phenotypic injury *(170,171)* or degenerating secondary to unknown genetic injury in Wobbler *(172,173)* and pmn *(174)* mice; peripheral sensory neurons undergoing axotomy-induced dedifferentiation *(175)* or axotomy-induced retrograde degeneration in development *(176,177)* and adult life *(178)*; and peripheral sympathetic neurons undergoing axotomy-induced phenotypic injury *(179)* or guanethidine- *(180)* and 6-OHDA *(181)* induced degeneration.

In some animal models, trophic effects have not been demonstrated directly on the cell bodies of lesioned or degenerating neurons, but, rather, on their processes or in functions dependent on the integrity of lesioned neurons. Important examples in the first category include the NGF-induced sprouting of central cholinergic axons *(182)* and peripheral sensory *(183)* and sympathetic *(184)* axons or the NGF-mediated regeneration of injured sensory axons *(185)* and the NT-3-mediated prevention of large sensory fiber degeneration in a metabolic/nutritional model of peripheral neuropathy *(186)*. Of particular theoretical importance is the BDNF-mediated sprouting of serotoninergic axons in mammalian cortex in a rat model of *p*-chloroamphetamine-induced degeneration *(187)*. Although, in general, trophic effects demonstrable on cell bodies/dendrites have their counterparts on axons/terminals in normal neurons or in relatively simple models of disease *(182,184)*, in more complex settings (such as genetically determined disorders in animals), the effects of a particular factor on the two neuronal compartments may be dissimilar. GDNF, for example, has been shown to protect degenerating motor neurons but has no effects on motor axons in the pmn mouse *(188)*.

In some models, investigators have placed special emphasis on the functional ("clinical") effects of trophic peptides. Such examples include NGF-mediated amelioration of sensory deficits in a variety of models of peripheral neuropathy, including streptozotocin-induced diabetic neuropathy *(189)*, antineoplastic agent-induced toxic neuropathy *(190,191)*, and pyridoxine-induced large fiber sensory neuropathy *(186)* in rats. Such in vivo effects of trophic peptides not only on the structural/biochemical aspects of neurons, but also on functions mediated by these neurons will certainly continue to accumulate in various combinations. From the standpoint of experimental therapeutics, it is essential to know both the specific effects of trophic factors on animal models and the relevance of these effects to the neuropathology and clinical features of the human disease represented by the animal model. The latter issue rarely receives the attention it deserves, a problem that has had a negative impact on the success of recent clinical trials (see the section Clinical Potential).

CLINICAL POTENTIAL OF TROPHIC PEPTIDES

Neurodegenerative Diseases

Since the presentation of Appel's hypothesis that developmental mechanisms involving trophic factor deprivation may play a role in neurological diseases of late life *(192)* and Hefti's enthusiastic suggestion of the relevance of NGF for AD *(193)*, neurodegenerative disorders have been considered as the prototypical clinical targets for trophic therapies. As discussed above, the universal occurrence, in neurodegenerative diseases, of neuronal death preceded by a long phase of dedifferentiation, combined with the physiological role of trophic factors in promoting the survival and differentiation of neurons, adds further credence to the previous suggestions. At present, clinical trials are planned or under way for testing neurotrophic factors in a variety of neurodegenerative diseases, including NGF and NT-3 in peripheral neuropathy, GDNF in Parkinson's disease and ALS and IGF-1 in peripheral neuropathy, ALS, and postpolio syndrome.

However, the recent negative experience with the completed trials of CNTF and BDNF for ALS (see chapter by Hilt, Miller, and Malta, this volume) necessitates a reexamination of the conditions under which trophic factor trials for neurodegenerative diseases are initiated. In the case of CNTF in ALS, the trial not only showed no efficacy on several measures of muscle strength, but subjects in the drug group showed increased mortality and developed a clinical syndrome characterized by fever and cachexia. Interestingly, CNTF was given in doses ~30 times smaller than those used in animal models. These rather disappointing effects were not surprising to many members of the trophic community having first-hand knowledge of this molecule. To name only a few concerns stemming directly from animal experiments, the effects of CNTF in simple models of motor neuron injury (194) have not been reproduced by other groups (195); more complex models of motor neuron disease where CNTF was tried (172) are not exactly representative of human ALS; and CNTF-induced cachexia has been observed very frequently in preclinical in vivo studies (87,196). The previous problems are representative of the kinds of concerns that should delay or arrest further advance of a trophic molecule into clinical trials until more information is available. In addition, all of these issues are not only of theoretical or historical importance because molecules with even more potent preclinical effects than CNTF (e.g., GDNF) may share with CNTF a cytokine-related side effect profile, including weight loss and cachexia (Koliatsos, personal observations).

In the case of BDNF, delivered in doses on average higher (20 times) than those used in animal models, although preliminary analyses showed an effect on forced vital capacity in patients with ALS (i.e., a smaller decline in forced vital capacity in subjects treated with BDNF than those taking placebo), no effects on other measures of muscle strength or on survival were seen. At the end of the trial, the result on forced vital capacity was shown to be nonsignificant (see chapter by Hilt, Miller, and Malta, this volume). As in the case of CNTF, warning signs had already emerged in the preclinical experience with this trophic factor. For example, it is unclear that BDNF is as effective in mature as in developing motor neurons (195); concerns have been expressed on the degree of penetration of BDNF into the nervous parenchyma, a problem attributed to the avid binding of this peptide to noncatalytic trkB receptors (89). More importantly, the issue of the effective retrograde transport of this peptide through an injured neuromuscular system has not been addressed adequately using the appropriate animal model of FALS transgenic mice.

Clinical trials are very costly, and the clinical and psychological stakes of involving patients with desperate clinical conditions are enormous. Perhaps more importantly in the long run, a premature disappointment with an otherwise promising therapeutic approach risks an annulment, in the minds of the scientific community and the public, of justified hopes and the need for further experimentation. Lessons from the disappointing clinical trials with CNTF and BDNF must be absorbed promptly and used to guide further clinical handling of these molecules. Some points to be considered have been presented above in the discussion of the two clinical trials. An additional question to be considered is what constitutes an adequate model of disease in which to try a trophic therapy at the preclinical stage (Fig. 2). Clearly, the in vitro effects of a trophic peptide cannot be accepted as sufficient justification for clinical use of the molecule. Simple lesion models need to be carefully reviewed and their relevance established for specific therapeutic tasks in human disease. Models should be used to test side effects and effective dose ranges (for example it is worrisome that, because of side effects, the largest dose of NGF used in clinical trials of peripheral sensory neuropathy is 10^{-4} of the dose shown to be effective in animal models). Finally, animal models should also be used for studies of the responses of the whole animal to the trophic peptide, including the possibility of the generation of antibodies that neutralize the delivered peptide *(197)*.

Stroke

Acidic amino acids naturally employed as excitatory transmitters (e.g., glutamate) or structurally related compounds of environmental origin can, under certain conditions, mediate cell death (*see* chapter by Olney and Ishimaru) and appear to play a major role in neuronal death associated with CNS trauma, hypoxia-ischemia, or seizures (*see* chapters by Vornov, by Dietrich, and by Shin and Lee). These excitatory amino acids and especially glutamate bind in excess to NMDA receptors following injury or ischemia and induce an NMDA receptor-mediated excessive influx of Ca^{2+} within neurons *(198)*, resulting in cell death. Destruction of neurons in the primary site of insult (the ischemic core) leads to further release of glutamate and its congeners and an expansion of the injury into surrounding areas (the ischemic penumbra) *(199)*. NGF and FGF-2 have been shown to reduce quinolinic acid-induced death of neostriatal and hippocampal neurons in adult rats *(200)*; NGF can also prevent the decrease in choline acetyltransferase (ChAT) activity in the excitotoxically injured striatum during development *(201)*. FGF-2 reduces excitotoxic damage to cortical and hippocampal neurons evoked by seizures *(202)*. Non-NGF neurotrophins may also prevent excitotoxic injury. For example, BDNF appears to exert broader protective effects than NGF on a variety of primary neuronal cell cultures undergoing excitotoxic injury *(203)*.

These findings indicate that FGF-2 and the neurotrophins have potential as therapeutic agents for anoxic/ischemic brain injury, especially considering the side effects of glutamate receptor antagonists *(204)*. A specific expectation of the trophic intervention would be the protection of secondarily injured neurons within the ischemic penumbra and, therefore, the restriction of the infarct area. By reducing the infarct size, trophic factors may improve the acute outcome of stroke and reduce the long-term disability associated with this condition.

Spinal Cord Injury

Several lines of experimentation suggest that trophic factors may promote the survival and assist in reestablishment of connections of damaged neurons following spinal cord injury. One set of observations comes from the retrograde death of neurons after disruption of the long tracts interconnecting brain and spinal cord. For example, following spinal cord hemisections in newborn rats, the retrograde degeneration of rubrospinal neurons and neurons within Clarke's nucleus is prevented by BDNF and NT-3, respectively *(205)*. In a similar spinal section model that focuses on the corticospinal tract, NT-3 has been found to enhance the sprouting of transected corticospinal axons, an effect augmented with the local application of antibodies against myelin-derived growth inhibitors *(206)*.

Another line of research has focused on the trophic factor-assisted restoration of damaged spinal cord connections following complete or near-complete transections; such goals are usually accomplished with grafting procedures to insure apposition of the disconnected parts of the nervous system and to provide a conduit for regeneration. In one of the most successful cases of pathway regeneration after complete spinal cord section *(207)*, investigators applied multiple small peripheral nerve bridges between disconnected gray and white matter and optimized the biochemical environment in the wound with molecules that promote axonal regeneration, including trophic factors (FGF-1) and antibodies to growth inhibitors. FGF-1 was *sine qua non* in the mediation of these effects. FGF-2 was also found to help with functional recovery in a compression model of spinal cord injury via a mechanism that did not involve proliferation of capillary vessels *(208)*.

Two other components of spinal cord injury that can be treated with trophic approaches are vascular damage and massive demyelination of axons. Because CNS trauma results in vascular damage, hemorrhage, reduced blood flow, and ischemia *(209)*, revascularization is essential for the continued survival of tissue spared by the injury. Revascularization may be also crucial for the reestablishment of the compromised BBB. FGF-2 is a reasonable candidate for the previous task, based on its effects on endothelial cell proliferation and capillary differentiation *(210,211)*. Axonal demyelination, another major complication of CNS injury *(212)*, is another potential target for FGF-2, a molecule known to stimulate the proliferation of 0–2A glial precursor cells that can subsequently differentiate into oligodendrocytes or astrocytes; other trophic factors, including cytokines, may also assist with remyelination. The ingenious use of grafting within an optimal biochemical environment that promotes the regeneration of damaged nervous and nonnervous components, including the use of one or more trophic peptides, may improve the prognosis for individuals with devastating spinal cord injuries in the foreseeable future.

NOVEL CONCEPTS IN THE NEUROBIOLOGY OF TROPHIC FACTORS

The present chapter focuses on the therapeutic applications of trophic peptides, i.e., settings in which exogenous factors are applied liberally in pharmacological doses. However, an understanding of the physiological role of these peptides, involving conditions in which trophic factors act in minimal amounts and under strict regulation, is required if we are to exploit natural mechanisms for therapeutic purposes and perhaps overcome the limitations imposed by intrinsic regulatory processes. As briefly discussed in the beginning of this chapter, trophic factors have been traditionally viewed as molecular signals through which peripheral targets retrogradely regulate the amount of innervation they receive from distinct populations of neurons during development. The principal developmental operation of these factors is to insure optimal numbers of differentiated neurons that efficiently innervate targets in the periphery and, by extension, in the CNS. It is also presumed that trophic peptides continue to play this role to various degrees into adult life via the promotion of viability and the maintenance of differentiation of specific populations of neurons.

Taken together, the above principles amount to a developmental theory that predicts a complex, but precise (perhaps even static) map for the formation and maintenance of the nervous system as a functional network *(19)*. Significant ingredients of this theory include: the expression in low abundance of trophic factors by neuronal targets *(213,214)*; the developmental upregulation of trophic factor expression to fit the excessive needs of neurons during physiological cell death *(215)*; the availability of trophic factors to responding populations of neurons principally via retrograde axonal transport, but not via local expression and secretion; and the selectivity of trophic effects for specific populations of neurons (for a more topical discussion of these matters, *see* chapter by Burek and Oppenheim, this volume). These patterns overlap significantly with the generally accepted criteria for putative responsivity of a neuronal population to a trophic peptide (reviewed above in the Models section) and were broadly held as general truisms, based primarily on observations on the biological disposition of NGF in the PNS.

In the present section, we will briefly present emerging principles in the biological disposition of trophic molecules (neurotrophins) that may invite some modifications of the classical views with significant implications for the clinical use of these molecules.

Complications in Neurotrophin Signal Transduction

The traditional view of neurotrophin transduction at the receptor level (*see* the Mechanisms section) could be summarized as the necessary result of binding of a neurotrophin to a high-affinity receptor (a trk molecule) probably via the homodimerization of the receptor and/or the facilitation of the binding by the high-capacity, low-affinity p75^{NGFR} system. The discovery of noncatalytic trk receptors that lack a tyrosine kinase domain or with kinase-blocking inserts in the catalytic domain invites a reformulation of the previous notion. Based on these discoveries, the result of the interaction of a neurotrophin molecule with specified neurons may depend not only on the dose of neurotrophins and the expression of trk and/or p75^{NGFR} receptors, but also on the existence (and level of expression) of noncatalytic trk. This qualifier is especially necessary in view of the fact that truncated trkB receptors may act as inhibitory modulators of neurotrophin responsiveness *(216)*.

Selectivity vs Redundancy in the Effects of Neurotrophins

According to the classical view of the generation of highly specific and reproducible maps of neural connections during development, trophic signals must be delivered in small amounts and only to specific populations of neurons (i.e., nerve cells endowed with the appropriate receptor mechanisms). A simple and secure molecular plan to serve this selective developmental task would appear to be the correspondence of each neurotrophin with a specific trk receptor and the expression, by a particular neuron, of only one trk receptor. Although the above plan appears to hold in the case of NGF (a neurotrophin transduced selectively and specifically via trkA that is expressed in restricted populations of neurons), non-NGF neurotrophins may be transduced by more than one trk receptor (e.g., NT-3 binds to trkA and, possibly, to trkB besides trkC) whereas more than one neurotrophin may bind to a single trk receptor (e.g., both BDNF and NT-4/5 bind to trkB). In addition, many neurons in the PNS and CNS express more than one trk receptor and appear to respond to more than one neurotrophin, although there is often preference for one neurotrophin. For example, basal forebrain cholinergic neurons express both trkA and trkB and show some responsiveness to BDNF, but NGF is by far the most potent neurotrophin for these neurons *(217)*.

The principle of selectivity holds stronger in the PNS, especially with respect to NGF signaling. First, the classical view on the specific and selective effects of NGF on developing peripheral sympathetic neurons *(6)* has been confirmed recently with evidence for a loss of sympathetic neurons in mice with targeted deletions of the NGF or trkA genes *(108,109)*. Second, recent research on developing and mature DRG has shown a remarkable degree of specificity in the trophic responses of at least some classes of sensory neurons. Developing and especially maturing DRG neurons show increasing commitment to the expression of specific trk receptors. This is especially true for neurons expressing trkA (NGF responders) and trkC (preferential NT-3 responders), whereas low-level trkB expression overlaps significantly with trkC or trkA *(218)*. Consistent with the previous trk expression patterns, *in utero* applications of NGF antibodies *(219)* as well as ablation of NGF or trkA by gene targeting *(108,109)* result in selective elimination of small-diameter (nociceptive) but not large sensory neurons. In addition, NT-3 and trkC nulls show a selective loss of muscle spindles and Merkel cells, together with their specific afferents *(114,115,220)*. The effects of BDNF or trkB ablation on the development of peripheral sensory neurons are less dramatic *(111–113,218)*. These findings, in conjunction with complementary data from other lines of investigation, strongly implicate NGF as a selective physiological signal for the development and function of sensory systems that respond to inflammation and other noxious stimuli, and NT-3 as a physiological factor for the development and proper function of sensory systems that serve proprioception and fine tactile discrimination.

As we commented in the beginning of the section, the trophic biology of CNS neurons is not as well delineated or as selective as in the PNS. For example, the universal persistence of central neurons in neurotrophin and trk null mice suggests at least some type of developmental redundancy in the trophic dependencies of these neurons, which is likely to continue into adulthood, although there is some evidence for an increasing trophic preference of some classes of central neurons *(see above)*. Recent evidence from adult models of cortical plasticity suggests distinct (even contrasting) influences of various neurotrophins on the same population of cortical neurons *(221)*.

Local (Autocrine-Paracrine) vs Long-Range Effects

In many cases, the expression and biological disposition of trophic factors are consistent with the classical view of target expression and availability by retrograde transport. As with other issues discussed in the present section, this plan is especially applicable to the trophic biology of PNS neurons and to the neurobiology of NGF. For example, most neurotrophins are avidly transported in the PNS *(169,222,223)*. The pattern of expression of most neurotrophins in the PNS is also consistent with target derivation and delivery to projection neurons via retrograde transport. In the PNS, neurotrophins (e.g., NGF, BDNF and NT-3) may still play the role of a target signal determining the density of innervation to the neurotrophin-secreting structure *(220,224)*.

For NGF, all of the above conditions (expression of the neurotrophin by target structures, expression of NGF receptors by projection neurons, and avid retrograde transport) are also present in the CNS. However, BDNF is not as intensely retrogradely transported in the brain *(222)* (Koliatsos and Winslow, personal observations). In addition, the broad and high expression of BDNF and its high-affinity receptor trkB in the CNS is not compatible with a specific trophic interrelationship between projection neurons and their targets. The truncated trkB receptors, which are expressed several-fold over catalytic trkB receptors in adult brains *(169)*, are ideally suited for fixation of secreted BDNF *in situ*, so that the peptide can then be used for autocrine or paracrine processes related to neural plasticity.

An interesting permutation of the possibly unique disposition of BDNF in the CNS is recent evidence indicating that BDNF is transported avidly in the anterograde direction within axons, especially in the context of exaggerated neuronal activity *(25,225)*. This evidence is presently limited to immunocytochemical data (Fig. 3) on a preferential axonal/terminal localization of BDNF, and there is no indication that BDNF delivered to terminals is secreted in the synaptic site or that it exerts effects on postsynaptic neurons.

Neurotrophins and Neural Development

According to the trophic theory of neural development, the release of NGF in limited amounts is the principal developmental mechanism by which targets in the skin and viscera retrogradely regulate the innervation they receive from sympathetic *(213)* and sensory *(226)* ganglia in the late phases of connection formation *(227,228)*. As shown in the case of the sensory innervation of the facial skin, there is a precise temporal coincidence of increase in NGF expression in targets (whisker pads) with the arrival of sensory axons from the trigeminal ganglion *(215)*. In the CNS (i.e., septohippocampal system), the previous temporal relationships are not as clear *(229)*. NGF gradients are not required to guide NGF-responding axons but establish the normal density and pattern of innervation by the NGF-responding system *(230)*.

BDNF may have similar effects on developing neurons with those of NGF, but the retrograde (as compared to paracrine) nature of these effects remains unclear. For example, BDNF delivered into the optic tectum on *Xenopus* tadpoles increases the branching and complexity of optic axon terminals; BDNF antibodies exert the opposite effect *(231)*. These effects of BDNF can be induced very rapidly (i.e., within 2 h), a delay consistent with a local, rather than retrograde, effect.

Although earlier theories postulated an association of specific neurotrophins with distinct neurogenetic primordia (i.e., BDNF targets placode-derived, whereas NGF targets neural crest-derived, nervous structures [*232*]), other findings support a more widespread role for these molecules in early stages of neurogenesis *(233)*. In general, the specific dispositions of different trophic factors in early neurogenesis and the importance of the total impact of these molecules for early developmental processes is still poorly understood. Some large-scale studies have indicated a preferential role of NT-3 in early stages of neurogenesis where proliferation, migration, and differentiation takes place, whereas BDNF may have a greater role in mature, differentiated neural structures *(234,235)*.

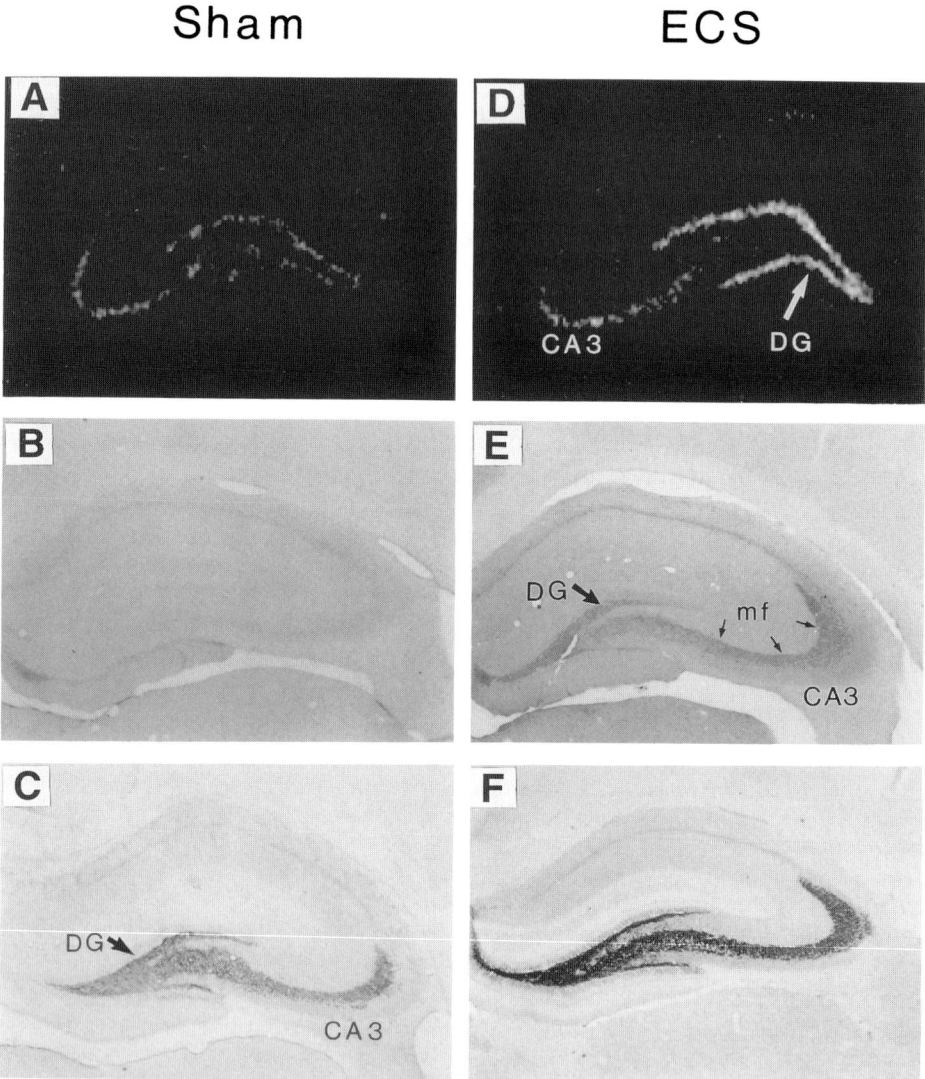

Fig. 3. Recent evidence of the activity-regulated expression and release of some trophic factors is changing the basic tenets of trophic factor neurobiology and guiding a field that traditionally focused on survival and death of neurons into the more complex territories of synaptic plasticity and behavioral neuroscience. These panels illustrate events that relate to BDNF expression and disposition under conditions of exaggerated physiological activity in the hippocampus (i.e., electroconvulsive seizures). (A–C) are taken from stimulated sham rats, whereas (D–F) are taken from rats that received chronic electroconvulsive seizures (ECS). (A,B) illustrate BDNF mRNA expression in the hippocampus by in situ hybridization. (B,E) Illustrate the localization of BDNF protein in the hippocampus by immunocytochemistry. (C,F) are stained with heavy metal stain to visualize the mossy fiber system. **(A,D)** BDNF mRNA increases in the granule cell layer of the dentate gyrus (arrow) but not in the CA3 region of Ammon's horn in an animal receiving ECS. **(B–E)** Chronic ECS causes a large increase in BDNF protein, which is localized in the mossy fiber termination zone (indicated by small arrows in panel E). **(C–F)** Note the increased intensity of Timm's staining in the animal receiving ECS (F). This stain corresponds faithfully with the pattern of BDNF immunoreactivity in (E), a similarity highly suggestive of the presence of BDNF in the mossy fiber termination zone.

ADULT PLASTICITY

Although it is generally accepted that the roles of trophic factors in adulthood are similar to those during development (i.e., the maintenance of differentiated phenotype and support of the survival of mature neurons), the only evidence to this concept originates in lesion studies (primarily axotomies or target ablations), where trophic factors are supplied exogenously in pharmacological doses (*see* the Models section). The previous models are very informative regarding the pharmacological effects of trophic factors for injured neurons but contribute little to understanding the role of trophic factors under physiological conditions.

The recent advent of soluble fusion proteins of the trk extracellular domains with IgG heavy chains that block the biological activity of their cognate neurotrophins in vitro *(236)* offers, for the first time, tools that can be used in vivo to scavenge endogenous neurotrophins and thereby allow observations of biological changes in neurotrophin-responsive neurons. The second-generation transgenics (*see* the Models section) may also allow temporary suppression of the expression of neurotrophin and neurotrophin receptor genes during the adult life of transgenic animals and the study of possible changes in dependent neurons after deprivation of endogenous trophic factors. In the mean time, the most exciting developments in the area of trophic factor-mediated plasticity of adult neurons will continue to occur, not in models that address phenotypic integrity and the survival of neurons, but in models that deal with changes occurring in the synaptic domain and use paradigms of brain activation *(see below)*.

ACTIVITY-RELATED CHANGES IN TROPHIC FACTOR EXPRESSION AND UTILIZATION

Gall and Isackson *(237)* were the first to demonstrate changes in the expression of NGF in the brain under conditions of unusual physiological activity, i.e., limbic seizures. Subsequent work on the regulation of neurotrophin and FGF expression in the CNS under a variety of conditions has confirmed and amplified these observations. For example, kindling causes reversible marked increases in NGF and BDNF expression in hippocampus, neocortex, and amygdala, with BDNF upregulated in a broader region, and NT-3 expression shows restricted or no change *(237–239)*. Lesion-induced limbic seizures cause similar increases in NGF and BDNF expression, with BDNF showing a broader upregulation in the forebrain and a faster response than NGF *(240)*. Ischemic insults cause similar increases in NGF and BDNF expression in hippocampus, whereas NT-3 expression is decreased *(241)*. FGF2 mRNA also increases after seizures in areas much broader than those showing elevations of NGF and BDNF mRNA *(242–244)*.

Activity-related alterations in neurotrophin expression are not the only aspect of neurotrophin biology that changes following seizures or lesions. Other important events observed in conjunction with the seizure-evoked, transient upregulation of BDNF include a prolonged increase in BDNF protein *(245)* and a massive anterograde transport of BDNF along intrahippocampal pathways *(25)*. TrkB expression in hippocampus also increases with kindling and following various types of lesions (mechanical injury, ischemia, and hypoglycemia); this event coincides with the upregulation of BDNF and shows the same rapid appearance 30 min after the stimulation or the injury *(239,246,247)*. TrkC shows inconsistent responses, but trkA expression does not change *(239,246)*.

Other lines of evidence indicate that activity in general, rather than seizures or lesions, can regulate the expression and secretion of some neurotrophins. Potassium-induced depolarization of hippocampal neurons in vitro results in increases of BDNF and NGF expression. Kainic acid is also very effective in upregulating NGF and BDNF in vitro and in vivo, an event that can be blocked by non-NMDA antagonists *(248)*. Very brief, noninjurious seizures evoked by focal administration of bicuculline in the amygdalopiriform region (area tempesta) also increase NGF and FGF2 mRNA *(242)*.

The upregulation, increased secretion, and possible utilization of trophic factors (especially BDNF) associated with enhanced neuronal activity may be physiologically important events leading to the increased survival of neurons and long-lasting adaptations of brain function to environmental demands. For example, the increased in vitro survival of cortical neurons mediated via activation of voltage-sensitive calcium channels appears to be blocked by specific antibodies to BDNF *(249)*; this observation is consistent with a role of BDNF as an autocrine–paracrine factor mediating the survival effects of increased neuronal activity in the CNS. In addition, exogenous BDNF and NGF have significant and opposite effects on the functional representation of a stimulated whisker in the rat barrel cortex (BDNF decreases, but NGF increases the size of a whisker representation) *(221)*. Consistent with evidence from other models discussed above, both neurotrophins show a rapid effect (within 30 min), but BDNF effects last longer. In a very interesting permutation of the activity-regulated synthesis and utilization of neurotrophins in the CNS, repeated stressful stimuli lead to reductions in BDNF and parallel increases in NT-3 mRNA levels in the dentate gyrus and Ammon's horn; BDNF, but not NT-3, mRNA changes occur also after acute stress *(250)*. Stress-induced changes in NT-3 expression are almost entirely dependent on adrenal cortical mechanisms, whereas BDNF regulation is more complex.

CHALLENGES IN THE CLINICAL PHARMACOLOGY OF TROPHIC FACTORS

Definition of the Problem

Trophic factors are large hydrophilic molecules that do not cross the BBB. Therefore, trophic peptides need to be delivered within the brain or spinal cord if their target neurons do not project outside the CNS. This condition presents formidable delivery problems and significantly complicates the management and treatment of patients; moreover, even with intrathecal administration, some trophic factors show very limited diffusion into the brain parenchyma *(251)*. These dilemmas were discussed in an advisory panel convened by the FDA in June, 1994, where enthusiasm for the use of trophic factors in degenerative diseases of the brain was tempered by skepticism concerning the need for chronic instrumentation.

It appears that delivery problems are not as severe for diseases of neurons located or projecting outside the CNS (i.e., autonomic disorders, motor neuron disease and peripheral neuropathy). For these populations of neurons, the related barriers (i.e., nerve-blood or ganglion-blood barriers) do not interfere with the optimal bioavailability of large molecules as much as the BBB *(252)*. In addition, because trophic factors are transported axonally, intramuscular or subcutaneous injection of a given factor can make it available to the spinal cord or ganglia via retrograde transport. These features of peripheral neurological diseases render them more amenable to treatment with systemically administered trophic molecules. However, the efficient delivery of trophic molecules via terminal uptake and axonal transport is uncertain in disorders where axonal and terminal pathology abounds. Until more evidence is presented that, in models recapitulating disease of the PNS, trophic factor uptake and transport is sufficient to allow for signal transduction, the results of clinical trials for these diseases must be interpreted with caution.

With recent advances in our knowledge of neurotrophic signal transduction, the decision to use a particular factor can be construed, in principle, as a selection of a particular neurotrophic cascade (e.g., selection of NGF cascade to treat injured cholinergic neurons in the forebrain). This principle implies that we can mimic the effects of a particular trophic factor in several ways, including the use of drugs that affect its synthesis, release, and uptake and the substitution of the factor with selective agonists for membrane-associated receptors or with substances that address second messenger systems. These alternatives might allow the utilization of small compounds with a potential for systemic delivery.

In this section, we review several approaches to influence trophic cascades by using small molecules. General problems anticipated with the small molecule strategies include the lack of specificity of trophic actions and an increased potential for side effects—especially with strategies that bypass the receptor site and address second messenger systems. There is also a general concern regarding the efficiency of strategies that bypass the ligand-receptor binding step in view of recent evidence that, at least in the case of NGF, the factor itself or the trkA receptor may be a crucial retrograde signal *(107)*. All of these concerns, considered along with our limited experience with small molecule strategies, make it imperative that we continue to pursue minimally invasive methods of delivering trophic peptides beyond the BBB. Such potential methods are also reviewed below.

"Precursor" Strategies: Small Agents Regulating Trophic Factor Transcription

The concept of using small molecules to influence trophic factor levels in the brain originates from evidence that expression of neurotrophic factors in certain parts of the nervous system is regulated by increased neuronal activity *(237)*. The expression of trophic factor genes in specific populations of nerve cells raises the possibility that changes in activity in specific areas of the nervous system by physiological or pharmacological methods may modulate trophic factor synthesis and release. This potential is supported by experimental evidence using steroids and β-adrenergic receptor agonists.

Corticosteroids appear to modulate the CNS expression of several trophic factors. The systemic administration of glucocorticoids upregulates the expression of NGF *(253,254)*, BDNF *(255)*, and NT-3 *(256)* in neocortex and hippocampus and the expression of FGF-2 throughout the brain *(257)*. In addition, the dexamethasone-induced upregulation of NGF may be sufficient to activate trk signaling and thereby exert trophic effects in basal forebrain cholinergic neurons *(254)*. Adrenal steroids also appear to modulate seizure-induced changes in NGF expression in the brain. For example, electrically induced convulsions *(243)* or dentate gyrus lesions *(256)* have an exacerbated effect on levels of NGF in the hippocampus following adrenalectomy. Moreover, kainic acid-evoked seizures cannot increase hippocampal BDNF and NGF mRNA in 1-wk old rats, in which the adrenal cortical response to stress is still immature *(258)*.

Although the data reviewed above indicate that adrenal steroids may play an important role in activity-regulated trophic factor expression, these data must be considered in the context of the general effects of corticosteroids on neural plasticity. These effects are complex and dose dependent; both high and low levels of glucocorticoids can damage hippocampus, whereas the mechanisms thought to operate in the execution of the previous effects are numerous and include the regulation of calcium cascades, influences on neurotrophin expression (reviewed above), and impairment of host cell defenses, including the heat shock and antioxidant systems *(259)*. In addition, glucocorticoids can exacerbate damage to certain CNS regions (e.g., hippocampus) resulting from ischemia and other insults. Given the above deleterious effects of glucocorticoids and evidence that, in high doses, these compounds fail to increase the expression of NGF and FGF-2 above levels obtained with a physiological range of plasma concentrations *(257)*, it appears that a reasonable initial approach for the use of glucocorticoids as neurotrophin modulators should involve low doses of these molecules for a limited period of time.

Beta-adrenergic receptor agonists are known to stimulate NGF expression in vitro *(260)* but penetrate the BBB poorly. The lipophilic β-agonist clenbuterol penetrates the brain exceptionally well and, when delivered systemically in rats, causes upregulation, within a few hours, of NGF expression in rat neocortex and of FGF-2 expression throughout the brain *(261,262)*. Stimulation of NGF expression may be accompanied by increased secretion of the peptide, as demonstrated by enhanced tyrosine phosphorylation of trkA (Mocchetti and Kaplan, personal observations).

"Downstream" Strategies: Small Molecules Mediating Trophic Factor Transduction or Mimicking Trophic Effects

The signal transduction of many trophic factors, especially of the non-NGF neurotrophins, is characterized by an enormous molecular complexity, evolved to suit the physiological roles of these peptides. For example, the several-fold higher expression of truncated trkB receptors in the brain may serve the purpose of fixating locally secreted BDNF to generate high local concentrations in the service of an autocrine or paracrine role *(see above)*. This transduction profile of BDNF in the brain may be different from that in the periphery, where BDNF may still play the role of a traditional target-derived, long-range signal for neuronal survival *(223)*. The existence of such physiologically advantageous molecular complexities may pose additional problems in the use of neurotrophins as drugs (e.g., the poor penetration of exogenous BDNF into brain parenchyma *[251]* may be caused by its copious binding to truncated receptors). In addition, the intricate regulation of neurotrophic effects under physiological conditions imply a need for extra caution in pharmacological approaches that bypass one or more steps in neurotrophic signal transduction.

The bacterial-derived alkaloids K252a and K252b were the first nonpolypeptide molecules found to interact specifically and probably directly with trk receptors. K252a prevents the NGF-induced differentiation of PC12 cells *(263)* as well as NGF-induced protein phosphorylation *(264)*. Both compounds inhibit *(265)* or potentiate *(266)* neurotrophin-induced effects on PC12 cells or on primary neuronal cultures, although K252b appears to be effective even at low nM concentrations and also to be less toxic than K252a. More recently, in a fibroblast cell line, it has been demonstrated that K252b specifically potentiates trkA-mediated responses induced by NT-3 not by NGF; no potentiation of BDNF-induced or trkB and trkC-mediated response was seen *(267)*. Further studies are necessary to establish the full potential and therapeutic efficiency of these compounds in vivo, especially in view of their low therapeutic index.

Great emphasis is currently placed on the engineering of neurotrophin agonists that combine features of two or more neurotrophins *(63)* or agonists of small molecular weight. The former present the advantage of combined efficacy on more populations of neurons than single native neurotrophins, an advantage confirmed in vitro for several "multifunctional" neurotrophins generated through homolog-scanning or site-directed mutagenesis *(268–270)*. Mutation analyses, however, have also confirmed earlier observations based on proteolytic cleavage of neurotrophin molecules showing that sequences far away from the receptor-binding site (e.g., the first 10 N-terminal amino acids for the neurotrophins NGF and BDNF) are crucial for receptor interactions and that the higher-order protein structure is at least as important as sequence specificity for the biological effect profile of individual neurotrophins *(63)*. The previous observations justify some skepticism over our future ability to generate small trophic agonists combining good CNS penetration with biological efficacy.

Whereas the molecules discussed above relate directly to trophic factors or to specific steps in trophic signal transduction, at least two groups of compounds appear to exert effects similar to those of trophic factors without any known relationship to trophic factors themselves or to second messengers in trophic signaling pathways. These compounds include the gangliosides and immunosuppressive drugs that bind to immunophilins. Gangliosides are a heterogeneous family of sialic-acid containing glycosphingolipids and are relatively abundant in the nervous system *(271)*. The monosialoganglioside GM1 prevents PC12 cell death after serum deprivation *(272)* and induces the tyrosine phosphorylation of trkA *(273)*. Consistent with these NGF-like effects of gangliosides, these compounds influence neuronal survival in various in vitro and in vivo models. For example, GM1 prevents loss of cholinergic neurons subsequent to hippocampal ablation *(274,275)* or to cortical lesions *(276)* and ameliorates age-associated memory impairments *(277)* in rats. GM1 also ameliorates the neurobiological and behavioral deficits associated with MPTP-induced injury to dopaminergic neurons of the substantia nigra in rats *(278)* and primates *(279)*. GM1 and other gangliosides have also been shown to protect neurons from excitotoxic injury in vitro *(280)* and after stroke or brain ischemia *(281,282)*.

Immunosuppressive drugs (such as FK506, cyclosporin A, and rapamycin) appear to have trophic effects on neurite-extension assays involving PC12 cells and chick sensory neurons *(283)*. Although the immunosuppressive properties of these drugs are dependent on calcineurin inhibition (an early calcium-dependent step in the immune response), their trophic effects are mediated through binding to specific "immunophilin" receptors such as cyclophilin (for cyclosporin A) and FK506-binding protein (FKBP) and inhibiting their rotamase activity *(284)*. It has been recently shown that nonimmunosuppressive analogs of these drugs, which lack calcineurin inhibitory activity but can still bind to immunophilins, are equally potent to their immunosuppressive relatives in neurite-extension assays and appear to have some effects on accelerating peripheral nerve regeneration in vivo and central dopamine fiber sprouting following 6-OHDA lesions *(285)*.

Delivery of Trophic Peptides to the CNS

Based on the features of small molecule approaches as delineated above, the delivery of trophic factors beyond the BBB emerges as the most straightforward method in terms of efficacy and side effects. This type of trophic factor administration can be achieved in a variety of ways, including infusions with constant flow or programmable pumps, the implantation of polymers with microencapsulated factors, the implantation of mammalian cells engineered to express and release a particular factor, the use of prodrugs that contain trophic factors bound to carrier molecules, and the transfer of trophic factor genes into neurons. Innovations in drug-delivery technologies are expected to add new methods in the near future. However, there is very limited experience with most of the previous approaches, and many of them, such as genetically engineered cells, may carry significant additional risks. Approaches using the infusion of purified or recombinant peptides via pumps and polymer microencapsulation have been discussed elsewhere *(286)*. Below, we briefly discuss some of the more recent methods in which trophic factors are delivered in a convenient fashion that minimizes chronic instrumentation.

The CNS grafting of immortalized cells (usually fibroblasts) genetically modified to overexpress trophic factors has successfully replicated the effects of chronically delivered trophic peptides in various in vivo models of neuronal injury. These cells can be delivered either as free cell suspensions *(200,287)* or as cells encapsulated in semipermeable polymer matrices *(288)*. The latter approach appears to exclude the risk of uncontrollable growth or even malignant transformation, although the issue of long-term viability of these cells has not been yet resolved. More recently, xenografts (e.g., fibroblasts obtained from the skin of the experimental animal) have been replaced by neuronal progenitors. These cells can be immortalized to be temperature sensitive (i.e., not to proliferate, but differentiate into glia and mature neurons at body temperature). An example of a successful attempt using neuronal progenitors is the transplantation of hippocampally derived, immortalized neuronal progenitors engineered to express and secrete NGF into the nucleus basalis of cognitively impaired aged rats *(289)*. This intervention restored the size of basal forebrain cholinergic neurons and improved the performance of animals on tasks of spatial memory. Although, with the developments illustrated above, grafting of trophic factor-engineered cells continues to be a powerful investigative tool for the neurobiologist, much more experience is required before these approaches can be considered for clinical testing. It is mandatory that such experience includes a better understanding of the long-term biological behavior of these grafts to preclude malignant transformation, dysfunction, or death of grafted cells.

Information on the physiology of the BBB can be used for the construction of complexes of trophic factors with molecules to facilitate diffusion across the BBB. This approach has already been tried with conjugates of NGF with antibodies against the transferrin receptor, which can cross the BBB *(290)*. The amount of NGF that penetrates into brain parenchyma via this type of delivery may be sufficient to exert biological effects on basal forebrain cholinergic neurons and to influence behaviors dependent on these nerve cells *(291)*.

As shown by recent successful efforts to treat animal models of Parkinson's disease *(166,292)*, genes of interest (e.g., trophic factor genes) can be delivered into specific CNS sites via gene transfer. The introduction of a gene into postmitotic cells such as neurons in a selective way (i.e., to specific populations of cells) is not a trivial task. Although neurotropic viruses are, in principle, the only available means by which desired pieces of DNA (i.e., trophic factor genes incorporated in the viral genome) can be introduced in mature neurons, the use of viruses as trophic gene vectors has been limited by the significant cytopathic effects of the viral genome. More recently, replication-defective adenoviruses, which do not cause any known neuropathology, have been used effectively as vectors to target GDNF overexpression in rat substantia nigra and to protect dopaminergic neurons from 6-OHDA-induced degeneration *(166)*.

Assessing and Managing Side Effects

Infusions of trophic factors in different nervous compartments, including the brain, are not without potential side effects, including known biological effects of trophic molecules and unknown or not generally appreciated complications. Although these side effects are particularly likely with pluripotential factors such as cytokines, molecules assumed to be relatively selective in their biological actions (e.g., neurotrophins) are not exempt from these considerations. It is pertinent to recall that the cachectic effects of CNTF became generally accepted after this cytokine entered clinical trials. In general, four types of problems can be expected when considering the side effects of trophic therapies.

First, there is a potential for trophic factors to induce nonphysiological, long-lasting changes in target systems. A showcase for this probability is the detrimental effect of NGF on working memory in young animals *(160)*, an action likely to occur via cholinergic hyperinnervation of neocortical and hippocampal target fields (Koliatsos, unpublished observations). It is encouraging that we have not seen such evidence in older animals, in which the plasticity of central cholinergic systems is reduced.

Second, there may be a concomitant influence on neural systems other than those that are targeted in a particular clinical trial. In recent years, there is a growing appreciation of the broader actions of trophic peptides, including neurotrophins, beyond their originally described effects on one or two model systems. For example, systems afferent to the cerebellum and inferior colliculus that participate in ocular positioning and oculocephalic reflexes may respond to NGF *(293)*. It is unknown whether a concomitant influence of these novel targets by NGF aimed at basal forebrain cholinergic neurons is associated with clinical complications. On the other hand, this finding suggests new therapeutic possibilities for NGF, including its use in Wernicke's encephalopathy and severe lithium-induced neurotoxicity. Another general observation from rodent and primate experiments of intraventricular infusions of NGF and other neurotrophins (i.e., BDNF) is that these peptides have a strong anorexic effect in these animals, which tends to subside with time, but may be a reason of concern in the first week(s) of the infusion *(294)*. Although hypothalamic mechanisms have been postulated, the exact nature of the anorexic effect of centrally delivered neurotrophins remains unclear.

Third, intracerebroventricular injections are, in all respects, equivalent to intravenous injections, and there is concern that trophic stimulation of peripheral targets may cause unacceptable side effects. For example, FGF promotes angiogenesis, which increases the risk for vascular tumors and hemorrhage. Another general concern associated with the mitogenic effects of trophic factors (i.e., neurotrophins) on nonneuronal cells in vitro (reviewed in ref. *295*) is that these molecules can behave like growth factors in cells that are still undergoing mitotic divisions. A particular concern with NGF is its recent configuration as a peptide with a special role in nociception *(296)*. Small-size (nociceptive) sensory neurons require NGF for survival during development and for normal phenotypic maturation in the early postnatal period, whereas targeted overexpression or elimination of NGF expression in the skin has profound (and contrasting) effects on the response of transgenic mice to noxious stimuli *(297)*. Pharmacological doses of NGF produce hyperalgesia in experimental animals *(298,299)*. Peripheral sympathetic neurons may cooperate with sensory neurons in the mediation of these effects. Transgenic mice that overexpress NGF show sympathetic hyperinnervation of sensory ganglia *(297)*, whereas sympathectomy reduces hyperalgesia caused by exogenous NGF *(299)*. NGF-induced hyperalgesia is clinically relevant; normal human subjects who received NGF as part of a phase-I trial for sensory neuropathy experienced diffuse muscle pain after iv injections and injection site hypersensitivity following sc injections; these effects were related to the NGF dose *(300)*.

A particularly interesting aspect of the nociceptive role of NGF is the presumed mediation, by this peptide, of inflammation-induced hypersensitivity. At least three types of peripheral cells that respond to NGF have been implicated in this function—mast cells, small-size sensory neurons, and sympathetic neurons, in that order of importance *(301)*. Inflammatory signals delivered to teeth or skin result in local upregulation in NGF expression, whereas the systemic administration of NGF antibodies prevents local tenderness and some responses of sensory neurons to the inflammatory signal *(302,303)*. High doses of both steroidal and nonsteroidal antiinflammatory compounds prevent inflammation-related increases in NGF expression/release, whereas IL-1β, a mediator increased at the site of inflammation, causes an increase in NGF levels. IL-1β-induced thermal hyperalgesia as well as complete Freund adjuvant-induced mechanical and thermal hyperalgesia are prevented by neutralizing NGF antibodies *(304)*. The importance of mast cells for NGF-induced hyperalgesia and the effects of the antiinflammatory compounds in modifying this effect of NGF suggest that simple, widely used clinical interventions may be employed to manage pain as a side effect of NGF treatment.

Finally, potential complications are associated with the delivery method itself. Any method of delivery is associated with specific side effects, but the risk of infection or malfunction in the delivery system is potentially higher when a drug is infused directly into the CNS with a pump situated under the skin for a lifetime. In the scientific community, opinions vary on the morbidity associated with chronically shunted individuals with hydrocephalus and with a previous intraventricular bethanechol trial for AD. Although some positive accounts have been published *(305)*, many clinicians and scientists remain skeptical concerning the long-term local effects of these interventions. It is our opinion that a future Advisory Panel will need to convene after the initial convention at FDA in June of 1994 to fully address the issues of intraventricular delivery and to establish appropriate guidelines.

CONCLUSIONS

The "fin-de-siècle" Neurology emerges as a brave and ambitious field that accepts no limits in the types of illnesses that it targets for therapeutic interventions. Among the most complex and difficult to understand and treat are the neurodegenerative diseases, which, with the progressive aging of the population, represent the majority of morbid conditions that neurologists will be called upon to diagnose, manage, and treat. Trophic factors, naturally occurring compounds that have been traditionally implicated as regulatory molecules in ubiquitous processes such as developmental cell death and neuronal differentiation, have been viewed by many as the most powerful modes of treating neurodegenerative diseases (i.e., illnesses featured by both dedifferentiation and degeneration/death of vulnerable populations of neurons). Although a variety of animal models based primarily on selective lesions in different parts of the CNS and PNS have amply documented the regenerative potential of trophic peptides, our limited experience with clinical trials has not confirmed, thus far, our hopes emerging from basic neurobiology and the preclinical testing.

Despite the negative results of the limited clinical trials with trophic factors, the clinical frontier of these factors is far from being closed. It is now clear that there have been specific problems in the way that these molecules were advanced into clinic in the two completed and in some of the ongoing trials. These results should teach us lessons on how to proceed in the difficult interface between laboratory and the bedside. These lessons include: our need to test, prior to clinical trial considerations, trophic molecules on animal models that are more appropriate for the disease for which a particular trophic molecule is considered; close attention to adverse effects observed in the preclinical stage and thorough investigation and characterization of other possible side effects prior to clinical testing; and a good understanding of issues of bioavailability of trophic peptides to target nerve cells, especially with the current exclusion of more invasive applications of these molecules beyond the BBB.

Besides their traditional role as molecules promoting regeneration, trophic factors (especially certain members of the neurotrophin family) have come to be increasingly recognized as potent mediators/modifiers of synaptic plasticity in the adult mammalian brain (cortex and hippocampus). These novel effects of trophic factors broaden their role as powerful, perhaps universal, mediators of both acute and chronic cellular changes that underlie adaptive functions of the forebrain, such as learning and memory. The potential role of trophic factors in these complex functions of the adult brain may open up new therapeutic possibilities for these compounds in the near future.

ACKNOWLEDGMENTS

This work was supported by Grants from the US Public Health Service, NS 20471, NS 01675, NS 29664, AG10480, and AG 01546.

Dr. Koliatsos is the recipient of a Javits Neuroscience Investigator Award (NIH NS 10580) and a Leadership and Excellence in Alzheimer's Disease (LEAD) Award (NIH AG 07914).

REFERENCES

1. Ramon y Cajal S. *Degeneration and Regeneration of the Nervous System.* Oxford University Press, London, 1928.
2. Seil FJ. *Neural Regeneration and Transplantation.* Alan R. Liss, New York, 1988.
3. Reynolds BA, Weiss S. Generation of neurons and astrocytes from isolated cells of the adult mammalian central nervous system. *Science* 1992, **255**: 1707–1710.
4. Loeb GE. Neural prosthetic interfaces with the nervous system. *Trends Neurosci* 1989, **12**: 195–201.
5. Baldessarini RJ. *Chemotherapy in Psychiatry. Principles and Practice.* Harvard University Press, Cambridge, MA, 1985.
6. Levi-Montalcini R, Booker B. Destruction of the sympathetic ganglia in mammals by an antiserum to a nerve-growth protein. *Proc Natl Acad Sci USA* 1960, **46**: 384–391.

7. Levi-Montalcini R, Hamburger V. Selective growth stimulating effects of mouse sarcoma on the sensory and sympathetic nervous system of the chick embryo. *J Exp Zool* 1951, **116**: 321–361.
8. Gibbs CJ Jr, Gajdusek DC, Epstein LG, Asher DM, Goudsmit J. Animal models of human disease. Induction of persistent human T lymphotropic retrovirus infections in nonhuman primates and equines inoculated with tissues from AIDS patients or purified virus grown in vitro, in *Animal Models of Retrovirus Infection and Their Relationship to AIDS* (Salzman LA, ed.), Academic Press, Orlando, 1986, pp. 457–462.
9. Goedert M, Fine A, Dawbarn D, Wilcock GK, Chao MV. Nerve growth factor receptor mRNA distribution in human brain: normal levels in basal forebrain in Alzheimer's disease. *Mol Brain Res* 1989, **5**: 1–7.
10. Higgins GA, Mufson EJ. NGF receptor gene expression is decreased in the nucleus basalis in Alzheimer's disease. *Exp Neurol* 1989, **106**: 222–236.
11. Ernfors P, Lindefors N, Chan-Palay V, Persson H. Cholinergic neurons of the nucleus basalis express elevated levels of nerve growth factor receptor mRNA in senile dementia of the Alzheimer type. *Dementia* 1990, **1**: 138–145.
12. Phillips HS, Hains JM, Armanini M, Laramee GR, Johnson SA, Winslow JW. BDNF mRNA is decreased in the hippocampus of individuals with Alzheimer's disease. *Neuron* 1991, **7**: 695–702.
13. Koliatsos VE, Price DL, Clatterbuck RE, Markowska AL, Olton DS, Wilcox BJ. Neurotrophic strategies for treating Alzheimer's disease: lessons from basic neurobiology and animal models. *Ann NY Acad Sci* 1993, **695**: 292–299.
14. Faradji V, Sotelo J. Low serum levels of nerve growth factor in diabetic neuropathy. *Acta Neurol Scand* 1990, **81**: 402–406.
15. Tooyama I, Kawamata Y, Walker D, Yamada T, Hanai K, Kimura H, Iwane M, Igarashi K, McGeer EG, McGeer PL. Loss of basic fibroblast growth factor in substantia nigra neurons in Parkinson's disease. *Neurology* 1993, **43**: 372–376.
16. Anand P, Terenghi G, Warner G, Kopelman P, Williams-Chestnut RE, Sinicropi DV. The role of endogenous nerve growth factor in human diabetic neuropathy. *Nature Med* 1996, **2**: 703–707.
17. Higgins GA, Koh S, Chen KS, Gage FH. NGF induction of NGF receptor gene expression and cholinergic neuronal hypertrophy within the basal forebrain of the adult rat. *Neuron* 1989, **3**: 247–256.
18. Holtzman DM, Li Y, Chen K, Gage FH, Epstein CJ, Mobley WC. Nerve growth factor reverses neuronal atrophy in a Down syndrome model of age-related neurodegeneration. *Neurology* 1993, **43**: 2668–2673.
19. Purves D. *Body and Brain*. Harvard University Press, Cambridge, 1988.
20. Lauder JM. Neurotransmitters as growth regulatory signals: role of receptors and second messengers. *Trends Neurosci* 1993, **16**: 233–240.
21. Ruegg MA, Tsim KWK, Horton SE. The agrin gene codes for a family of basal lamina proteins that differ in function and distribution. *Neuron* 1992, **8**: 691–699.
22. Reist NE, Werle MJ, McMahan UJ. Agrin released by motor neurons induces the aggregation of acetylcholine receptors at neuromuscular junctions. *Neuron* 1992, **8**: 865–868.
23. von Bartheld CS, Byers MR, Williams R, Bothwell M. Anterograde transport of neurotrophins and axodendritic transfer in the developing visual system. *Nature* 1996, **379**: 830–833.
24. Ferguson IA, Schweitzer JB, Johnson EM Jr. Basic fibroblast growth factor: receptor-mediated internalization, metabolism, and anterograde axonal transport in retinal ganglion cells. *J Neurosci* 1990, **10**: 2176–2189.
25. Smith MA, Zhang L-X, Lyons WE, Mamounas LA. Anterograde transport of endogenous brain-derived neurotrophic factor in hippocampal mossy fibers. *Neuroreport* 1997, **8**: 1829–1834.

26. Chan RS, Huey ED, Maecker HL, Cortopassi KM, Howard SA, Iyer AM, McIntosh LJ, Ajilore OA, Brooke SM, Sapolsky RM. Endocrine modulators of necrotic neuron death. *Brain Pathol* 1996, **6**: 481–491.
27. Koliatsos VE. Biological therapies for Alzheimer's disease: focus on trophic factors. *Crit Rev Neurobiol* 1996, **10**: 205–238.
28. Growdon JH. Treatment of Alzheimer's disease? *N Engl J Med* 1992, **18**: 1306–1308.
29. Growdon JH. Biological therapies for Alzheimer's disease, in *Dementia* (Whitehouse PJ, ed.), F.A. Davis, Philadelphia, 1993, pp. 375–399.
30. Marin DB, Davis KL. Experimental Therapeutics, in *Psychopharmacology: The Fourth Generation of Progress* (Bloom FE, Kupfer DJ, eds.), Raven Press, Ltd. New York, 1995, pp. 1417–1426.
31. Atack JR. Cholinesterases in human degenerative diseases, in *Cholinergic Basis for Alzheimer Therapy* (Becker R, Giacobini E, eds.), Birkhäuser, Boston, 1991, pp. 31–37.
32. Giacobini E, Linville D, Messamore E, Ogane N. Toward a third generation of cholinesterase inhibitors, in *Cholinergic Basis for Alzheimer Therapy* (Becker R, Giacobini E, eds.), Birkhäuser, Boston, 1991, pp. 477–490.
33. Nordberg A, Lilja A, Lundqvist H, Hartvig P, Amberla K, Viitanen M, Warpman U, Johansson M, Hellström-Lindahl E, Bjurling P, Fasth K-J, Långström B, Winblad B. Tacrine restores cholinergic nicotinic receptors and glucose metabolism in Alzheimer patients as visualized by positron emission tomography. *Neurobiol Aging* 1992, **13**: 747–758.
34. Davis KL, Thal LJ, Gamzu ER, Davis CS, Woolson RF, Gracon SI, Drachman DA, Schneider LS, Whitehouse PJ, Hoover TM, Morris JC, Kawas CH, Knopman DS, Earl NL, Kumar V, Doody RS, The Tacrine Collaborative Study Group. A double-blind, placebo-controlled multicenter study of tacrine for Alzheimer's disease. *N Engl J Med* 1992, **327**: 1253–1259.
35. Knapp MJ, Pharma D, Knopman DS, Solomon PR, Pendlebury WW, Davis CS, Gracon SI. A 30-week randomized controlled trial of high-dose tacrine in patients with Alzheimer's disease. *JAMA* 1994, **271**: 985–991.
36. Kang J, Lemaire H-G, Unterbeck A, Salbaum JM, Masters CL, Grzeschik K-H, Multhaup G, Beyreuther K, Müller-Hill B. The precursor of Alzheimer's disease amyloid A4 protein resembles a cell-surface receptor. *Nature* 1987, **325**: 733–736.
37. Hilbich C, Kisters-Woike B, Reed J, Masters CL, Beyreuther K. Aggregation and secondary structure of synthetic amyloid A4 peptides of Alzheimer's disease. *J Mol Biol* 1991, **218**: 149–163.
38. Sisodia SS, Koo EH, Beyreuther K, Unterbeck A, Price DL. Evidence that β-amyloid protein in Alzheimer's disease is not derived by normal processing. *Science* 1990, **248**: 492–495.
39. Chartier-Harlin M-C, Crawford F, Houlden H, Warren A, Hughes D, Fidani L, Goate A, Rossor M, Roques P, Hardy J, Mullan M. Early-onset Alzheimer's disease caused by mutations at codon 717 of the β-amyloid precursor protein gene. *Nature* 1991, **353**: 844–846.
40. Goate A, Chartier-Harlin M-C, Mullan M, Brown J, Crawford F, Fidani L, Giuffra L, Haynes A, Irving N, James L, Mant R, Newton P, Rooke K, Roques P, Talbot C, Pericak-Vance M, Roses A, Williamson R, Rossor M, Owen M, Hardy J. Segregation of a missense mutation in the amyloid precursor protein gene with familial Alzheimer's disease. *Nature* 1991, **349**: 704–706.
41. Murrell J, Farlow M, Ghetti B, Benson MD. A mutation in the amyloid precursor protein associated with hereditary Alzheimer's disease. *Science* 1991, **254**: 97–99.
42. Karlinsky H, Vaula G, Haines JL, Ridgley J, Bergeron C, Mortilla M, Tupler RG, Percy ME, Robitaille Y, Noldy NE, Yip TCK, Tanzi RE, Gusella JF, Becker R, Berg JM, Crapper McLachlan DR, St George-Hyslop PH. Molecular and prospective phe-

notypic characterization of a pedigree with familial Alzheimer's disease and a missense mutations in codon 717 of the β-amyloid precursor protein gene. *Neurology* 1992, **42**: 1445–1453.

43. Whyte S, Beyreuther K, Masters CL. Rational therapeutic strategies for Alzheimer's disease, in *Neurodegenerative Diseases* (Calne DB, ed.), W.B. Saunders, Philadelphia, 1994, pp. 647–664.
44. Haass C, Hung AY, Selkoe DJ. Processing of β-amyloid precursor protein in microglia and astrocytes favors an internal localization over constitutive secretion. *J Neurosci* 1991, **11**: 3783–3793.
45. Cole GM, Huynh TV, Saitoh T. Evidence for lysosomal processing of amyloid β-protein precursor in cultured cells. *Neurochem Res* 1989, **14**: 933–939.
46. McGeer PL, Akiyama H, Itagaki S, McGeer EG. Activation of the classical complement pathway in brain tissue of Alzheimer patients. *Neurosci Lett* 1989, **107**: 341–346.
47. Troncoso JC, Sukhov RR, Kawas CH, Koliatsos VE. In situ labeling of dying cortical neurons in normal aging and in Alzheimer's disease: correlations with senile plaques and disease progression. *J Neuropathol Exp Neurol* 1996, **55**: 1134–1142.
48. Levi-Montalcini R, Hamburger V. A diffusible agent of mouse sarcoma, producing hyperplasia of sympathetic ganglia and hyperneurotization of viscera in the chick embryo. *J Exp Zool* 1953, **123**: 233–287.
49. Cohen S. Purification and metabolic effects of a nerve growth-promoting protein from snake venom. *J Biol Chem* 1959, **234**: 1129–1137.
50. Bocchini V, Angeletti PU. The nerve growth factor: purification as a 30,000-molecular-weight protein. *Proc Natl Acad Sci USA* 1969, **64**: 787–794.
51. Angeletti RH, Bradshaw RA. Nerve growth factor from mouse submaxillary gland: amino acid sequence. *Proc Natl Acad Sci USA* 1971, **68**: 2417–2420.
52. McDonald NQ, Lapatto R, Murray-Rust J, Gunning J, Wlodawer A, Blundell TL. New protein fold revealed by a 2,3-Å resolution crystal structure of nerve growth factor. *Nature* 1991, **354**: 411–414.
53. Barde Y-A, Edgar D, Thoenen H. Purification of a new neurotrophic factor from mammalian brain. *EMBO J* 1982, **1**: 549–553.
54. Leibrock J, Lottspeich F, Hohn A, Hofer M, Hengerer B, Masiakowski P, Thoenen H, Barde Y-A. Molecular cloning and expression of brain-derived neurotrophic factor. *Nature* 1989, **341**: 149–152.
55. Ernfors P, Ibez CF, Ebendal T, Olson L, Persson H. Molecular cloning and neurotrophic activities of a protein with structural similarities to nerve growth factor: developmental and topographical expression in the brain. *Proc Natl Acad Sci USA* 1990, **87**: 5454–5458.
56. Hohn A, Liebrock J, Bailey K, Barde Y-A. Identification and characterization of a novel member of the nerve growth factor/brain-derived neurotrophic factor family. *Nature* 1990, **344**: 339–341.
57. Kaisho Y, Yoshimura K, Nakahama K. Cloning and expression of a cDNA encoding a novel human neurotrophic factor. *FEBS* 1990, **266**: 187–191.
58. Maisonpierre PC, Belluscio L, Squinto S, Ip NY, Furth ME, Lindsay RM, Yancopoulos GD. Neurotrophin-3: a neurotrophic factor related to NGF and BDNF. *Science* 1990, **247**: 1446–1451.
59. Rosenthal A, Goeddel DV, Nguyen T, Lewis M, Shih A, Laramee GR, Nikolics K, Winslow JW. Primary structure and biological activity of a novel human neurotrophic factor. *Neuron* 1990, **4**: 767–773.
60. Berkemeier LR, Winslow JW, Kaplan DR, Nikolics K, Goeddel DV, Rosenthal A. Neurotrophin-5: a novel neurotrophic factor that activates trk and trkB. *Neuron* 1991, **7**: 857–866.

61. Hallböök F, Ibáñez CF, Persson H. Evolutionary studies of the nerve growth factor family reveal a novel member abundantly expressed in Xenopus ovary. *Neuron* 1991, **6**: 845–858.
62. Ip NY, Ibáñez CF, Nye SH, McClain J, Jones PF, Gies DR, Belluscio L, Le Beau MM, Espinosa R III, Squinto SP, Persson H, Yancopoulos GD. Mammalian neurotrophin-4: structure, chromosomal localization, tissue distribution, and receptor specificity. *Proc Natl Acad Sci USA* 1992, **89**: 3060–3064.
63. Ibanez CF. Neurotrophic factors: from structure-function studies to designing effective therapeutics. *Trends BioTechnol* 1995, **13**: 217–227.
64. Hilton DJ. LIF: lots of interesting functions. *Trends Biochem Sci* 1992, **17**: 72–76.
65. Lin L-FH, Mismer D, Lile JD, Armes LG, Butler ET III, Vannice JL, Collins F. Purification, cloning, and expression of ciliary neurotrophic factor (CNTF). *Science* 1989, **246**: 1023–1025.
66. Adler R, Landa KB, Manthorpe M, Varon S. Cholinergic neuronotrophic factors: intraocular distribution of trophic activity for ciliary neurons. *Science* 1979, **204**: 1434–1436.
67. Stöckli KA, Lottspeich F, Sendtner M, Masiakowski P, Carroll P, Götz R, Lindholm D, Thoenen, H. Molecular cloning, expression and regional distribution of rat ciliary neurotrophic factor. *Nature* 1989, **342**: 920–923.
68. Masu Y, Wolf E, Holtmann B, Sendtner M, Brem G, Thoenen H. Disruption of the CNTF gene results in motor neuron degeneration. *Nature* 1993, **365**: 27–32.
69. Bazan JF. Neuropoietic cytokines in the hematopoietic fold. *Neuron* 1991, **7**: 197–208.
70. Hall AK, Rao MS. Cytokines and neurokines: related ligands and related receptors. *Trends Neurosci* 1992, **15**: 35–37.
71. Ip NY, Nye SH, Boulton TG, Davis S, Taga T, Li Y, Birren SJ, Yasukawa K, Kishimoto T, Anderson DJ, Stahl N, Yancopoulos GD. CNTF and LIF act on neuronal cells via shared signaling pathways that involve the IL-6 signal transducing receptor component gp130. *Cell* 1992, **69**: 1121–1132.
72. Henderson JT, Seniuk NA, Richardson PM, Gauldie J, Roder JC. Systemic administration of ciliary neurotrophic factor induces cachexia in rodents. *J Clin Invest* 1994, **93**: 2632–2638.
73. Kessler JA, Ludlam WH, Freidin MM, Hall DH, Michaelson MD, Spray DC, Dougherty M, Batter DK. Cytokine-induced programmed death of cultured sympathetic neurons. *Neuron* 1993, **11**: 1123–1132.
74. Pennica D, King KL, Shaw KJ, Luis E, Rullamas J, Luoh H-M, Darbonne WC, Knutzon DS, Yen R, Chien KR, Baker JB, Wood WI. Expression cloning of cardiotrophin 1, a cytokine that induces cardiac myocyte hypertrophy. *Proc Natl Acad Sci USA* 1995, **92**: 1142–1146.
75. Pennica D, Arce V, Swanson TA, Vejsada R, Pollock RA, Armanini M, Dudley K, Phillips HS, Rosenthal A, Kato AC, Henderson CE. Cardiotrophin-1, a cytokine present in embryonic muscle, supports long-term survival of spinal motoneurons. *Neuron* 1996, **17**: 63–74.
76. Abraham JA, Whang JL, Tumolo A, Mergia A, Friedman J, Gospodarowicz D, Fiddes JC. Human basic fibroblast growth factor: nucleotide sequence and genomic organization. *EMBO J* 1986, **5**: 2523–2528.
77. Baird A, Böhlen P. Fibroblast growth factors, in *Peptide Growth Factors and Their Receptors I* (Sporn MB, Roberts AB, eds.), Springer-Verlag, New York, 1991.
78. Risau W. Developing brain produces an angiogenesis factor. *Proc Natl Acad Sci USA* 1986, **83**: 3855–3859.
79. Rechler MM, Nissley SP. Insulin-like growth factors, in *Peptide Growth Factors and Their Receptors I* (Sporn MB, Roberts AB, eds.), Springer-Verlag, New York, 1991, pp. 263–367.
80. Guler H-P, Zapf J, Froesch R. Short-term metabolic effects of recombinant human insulin-like growth factor I in healthy adults. *N Engl J Med* 1987, **317**: 137–140.

81. Caroni P, Grandes P. Nerve sprouting in innervated adult skeletal muscle induced by exposure to elevated levels of insulin-like growth factors. *J Cell Biol* 1990, **110**: 1307–1317.
82. Neff NT, Prevette D, Houenou LJ, Lewis ME, Glicksman MA, Yin QW, Oppenheim RW. Insulin-like growth factors: putative muscle-derived trophic agents that promote motoneuron survival. *J Neurobiol* 1993, **24**: 1578–1588.
83. Hughes RA, Sendtner M, Thoenen H. Members of several gene families influence survival of rat motoneurons in vitro and in vivo. *J Neurosci Res* 1993, **36**: 663–671.
84. Schaar DG, Sieber B, A, Dreyfus CF, Black IB. Regional and cell-specific expression of GDNF in rat brain. *Exp Neurol* 1993, **124**: 368–371.
85. Cameron VA, Nishimura E, Mathews LS, Lewis KA, Sawchenko PE, Vale WW. Hybridization histochemical localization of activin receptor subtyped in rat brain, pituitary, ovary, and testis. *Endocrinology* 1994, **134**: 799–808.
86. Lin L-FH, Doherty DH, Lile JD, Bektesh S, Collins F. GDNF: A glial cell line-derived neurotrophic factor for midbrain dopaminergic neurons. *Science* 1993, **260**: 1130–1132.
87. Henderson CE, Phillips HS, Pollock RA, Davies AM, Lemeulle C, Armanini M, Simpson LC, Moffet B, Vandlen RA, Koliatsos VE, Rosenthal A. GDNF: a potent survival factor for motoneurons present in peripheral nerve and muscle. *Science* 1994, **266**: 1062–1064.
88. Oppenheim RW, Houenou LJ, Johnson JE, Lin L-FH, Li L, Lo AC, Newsome AL, Prevette DM, Wang S. Developing motor neurons rescued from programmed and axotomy-induced cell death by GDNF. *Nature* 1995, **373**: 344–346.
89. Yan Q, Matheson C, Lopez OT. In vivo neurotrophic effects of GDNF on neonatal and adult facial motor neurons. *Nature* 1995, **373**: 341–344.
90. Beck KD, Valverde J, Alexi T, Poulsen K, Moffat B, Vandlen RA, Rosenthal A, Hefti F. Mesencephalic dopaminergic neurons protected by GDNF from axotomy-induced degeneration in the adult brain. *Nature* 1995, **373**: 339–341.
91. Tomac A, Lindqvist E, Lin L-FH, Ögren SO, Young D, Hoffer BJ, Olson L. Protection and repair of the nigrostriatal dopaminergic system by GDNF in vivo. *Nature* 1995, **373**: 335–339.
92. Krieglstein K, Suter-Crazzolara C, Fischer WH, Unsicker K. TGF-β superfamily members promote survival of midbrain dopaminergic neurons and protect them against MPP$^+$ toxicity. *EMBO J* 1995, **14**: 736–742.
93. Kotzbauer PT, Lampe PA, Heuckeroth RO, Golden JP, Creedon DJ, Johnson EM Jr, Milbrandt J. Neurturin, a relative of glial-cell-line-derived neurotrophic factor. *Nature* 1996, **384**: 467–470.
94. Radeke MJ, Misko TP, Hsu C, Herzenberg LA, Shooter EM. Gene transfer and molecular cloning of the rat nerve growth factor receptor. *Nature* 1987, **325**: 593–597.
95. Rodriguez-Tébar A, Dechant G, Barde Y-A. Binding of brain-derived neurotrophic factor to the nerve growth factor receptor. *Neuron* 1990, **4**: 487–492.
96. Meakin SO, Shooter EM. The nerve growth factor family of receptors. *Trends Neurosci* 1992, **15**: 323–331.
97. Dobrowsky RT, Werner MH, Castellino AM, Chao MV, Hannun YA. Activation of the sphingomyelin cycle through the low-affinity neurotrophin receptor. *Science* 1994, **265**: 1596–1599.
98. Rabizadeh S, Oh J, Zhong L-T, Yang J, Bitler CM, Butcher LL, Bredesen DE. Induction of apoptosis by the low-affinity NGF receptor. *Science* 1993, **261**: 345–348.
99. Carter BD, Kaltschmidt C, Kaltschmidt B, Offenhäuser N, Böhm-Matthaei R, Baeuerle PA, Barde Y-A. Selective activation of NF-kappaB by nerve growth factor through the neurotrophin receptor p75. *Science* 1996, **272**: 542–545.
100. Frade JM, Rodríguez-Tébar A, Barde Y-A. Induction of cell death by endogenous nerve growth factor through its p75 receptor. *Nature* 1996, **383**: 166–168.
101. Lee K-F, Li E, Huber LJ, Landis SC, Sharpe AH, Chao MV, Jaenisch R. Targeted mutation of the gene encoding the low affinity NGF receptor p75 leads to deficits in the peripheral sensory nervous system. *Cell* 1992, **69**: 737–749.

102. Van der Zee CEEM, Ross GM, Riopelle RJ, Hagg T. Survival of cholinergic forebrain neurons in developing p75^{NGFR}-deficient mice. *Science* 1996, **274**: 1729–1732.
103. Martin-Zanca D, Oskam R, Mitra G, Copeland T, Barbacid M. Molecular and biochemical characterization of the human trk proto-oncogene. *Mol Cell Biol* 1989, **9**: 24–33.
104. Kaplan DR, Hempstead BL, Martin-Zanca D, Chao MV, Parada LF. The *trk* proto-oncogene product: a signal transducing receptor for nerve growth factor. *Science* 1991, **252**: 554–558.
105. Chao MV. Neurotrophin receptors: a window into neuronal differentiation. *Neuron* 1992, **9**: 583–593.
106. Davies AM, Lee K-F, Jaenisch R. p75-deficient trigeminal sensory neurons have an altered response to NGF but not to other neurotrophins. *Neuron* 1993, **11**: 565–574.
107. Ehlers MD, Kaplan DR, Price DL, Koliatsos VE. NGF-stimulated retrograde transport of trkA in the mammalian nervous system. *J Cell Biol* 1995, **130**: 149–156.
108. Smeyne RJ, Klein R, Schnapp A, Long LK, Bryant S, Lewin A, Lira SA, Barbacid M. Severe sensory and sympathetic neuropathies in mice carrying a disrupted Trk/NGF receptor gene. *Nature* 1994, **368**: 246–249.
109. Crowley C, Spencer SD, Nishimura MC, Chen KS, Pitts-Meek S, Armanini MP, Ling LH, McMahon SB, Shelton DL, Levinson AD, Phillips HS. Mice lacking nerve growth factor display perinatal loss of sensory and sympathetic neurons yet develop basal forebrain cholinergic neurons. *Cell* 1994, **76**: 1001–1011.
110. Middlemas DS, Lindberg RA, Hunter T. *trkB*, a neural receptor protein-tyrosine kinase: evidence for a full-length and two truncated receptors. *Mol Cell Biol* 1991, **11**: 143–153.
111. Klein R, Smeyne RJ, Wurst W, Long LK, Auerbach BA, Joyner AL, Barbacid M. Targeted disruption of the trkB neurotrophin receptor gene results in nervous system lesions and neonatal death. *Cell* 1993, **75**: 113–122.
112. Jones KR, Fariñas I, Backus C, Reichardt LF. Targeted disruption of the BDNF gene perturbs brain and sensory neuron development but not motor neuron development. *Cell* 1994, **76**: 989–999.
113. Conover JC, Erickson JT, Katz DM, Bianchi LM, Poueymirou WT, McClain J, Pan L, Helgren M, Ip NY, Boland P, Friedman B, Wiegand S, Vejsada R, Kato AC, DeChiara TM, Yancopoulos GD. Neuronal deficits, not involving motor neurons, in mice lacking BDNF and/or NT4. *Nature* 1995, **375**: 235–238.
114. Ernfors P, Lee KF, Jaenisch R. Mice lacking brain-derived neurotrophic factor develop with sensory deficits. *Nature* 1994, **368**: 147–150.
115. Ernfors P, Lee K-F, Kucera J, Jaenisch. Lack of neurotrophin-3 leads to deficiencies in the peripheral nervous system and loss of limb proprioceptive afferents. *Cell* 1994, **77**: 503–512.
116. Davis S, Aldrich TH, Stahl N, Pan L, Taga T, Kishimoto T, Ip NY, Yancopoulos GD. LIFRβ and gp130 as heterodimerizing signal transducers of the tripartite CNTF receptor. *Science* 1993, **260**: 1805–1808.
117. Davis S, Aldrich TH, Valenzuela DM, Wong V, Furth ME, Squinto SP, Yancopoulos GD. The receptor for ciliary neurotrophic factor. *Science* 1991, **253**: 59–63.
118. Stahl N, Boulton TG, Farruggella T, Ip NY, Davis S, Witthuhn BA, Quelle FW, Silvennoinen O, Barbieri G, Pellegrini S, Ihle JN, Yancopoulos GD. Association and activation of Jak-Tyk kinases by CNTF-LIF-OSM-IL-6 β receptor components. *Science* 1994, **263**: 92–95.
119. DeChiara TM, Vejsada R, Poueymirou WT, Acheson A, Suri C, Conover JC, Friedman B, McClain J, Pan L, Stahl N, Ip NY, Kato A, Yancopoulos GD. Mice lacking the CNTF receptor, unlike mice lacking CNTF, exhibit profound motor neuruon deficits at birth. *Cell* 1995, **83**: 313–322.
120. Jing S, Wen D, Yu Y, Holst PL, Luo Y, Fang M, Tamir R, Antonio L, Hu Z, Cupples R, Louis J-C, Hu S, Altrock BW, Fox GM. GDNF-induced activation of the ret protein tyrosine kinase is mediated by GDNFR-α, a novel receptor for GDNF. *Cell* 1996, **85**: 1113–1124.

121. Treanor JJS, Goodman L, de Sauvage F, Stone DM, Poulsen KT, Beck CD, Gray C, Armanini MP, Pollock RA, Hefti F, Phillips HS, Goddard A, Moore MW, Buj-Bello A, Davies AM, Asai N, Takahashi M, Vandlen R, Henderson CE, Rosenthal A. Characterization of a multicomponent receptor for GDNF. *Nature* 1996, **382**: 80–83.
122. Durbec P, Marcos-Gutierrez CV, Kilkenny C, Grigoriou M, Wartiowaara K, Suvanto P, Smith D, Ponder B, Costantini F, Saarma M, Sariola H, Pachnis V. GDNF signalling through the Ret receptor tyrosine kinase. *Nature* 1996, **381**: 789–792.
123. Trupp M, Arenas E, Fainzilber M, Nilsson A, Sieber B, Grigoriou M, Kilkenny C, Salazar-Grueso E, Pachnis V, Arumäe U, Sariola H, Saarma M, Ibáñez CF. Functional receptor for GDNF encoded by the *c-ret* proto-oncogene. *Nature* 1996, **381**: 785–789.
124. Edery P, Lyonnet S, Mulligan LM, Pelet A, Dow E, Abel L, Holder S, Nihoul-Fékété, C, Ponder BAJ, Munnich A. Mutations of the *RET* proto-oncogene in Hirschsprung's disease. *Nature* 1994, **367**: 378–380.
125. Romeo G, Ronchetto P, Luo Y, Barone V, Seri M, Ceccherini I, Pasini B, Bocciardi R, Lerone M, Kääriäinen H, Martucciello G. Point mutations affecting the tyrosine kinase domain of the *RET* proto-oncogene in Hirschsprung's disease. *Nature* 1994, **367**: 377–378.
126. Hofstra RMW, Landsvater RM, Ceccherini I, Stulp RP, Stelwagen T, Luo Y, Pasini B, Höppener JWM, Ploos van Amstel HK, Romeo G, Lips CJM, Buys CHCM. A mutation in the *RET* proto-oncogene associated with multiple endocrine neoplasia type 2B and sporadic medullary thyroid carcinoma. *Nature* 1994, **367**: 375–376.
127. Schuchardt A, D'Agati V, Larsson-Blomberg L, Costantini F, Pachnis V. Defects in the kidney and enteric nervous system of mice lacking the tyrosine kinase receptor Ret. *Nature* 1994, **367**: 380–383.
128. Moore MW, Klein RD, Fariñas I, Sauer H, Armanini M, Phillips H, Reichardt LF, Ryan AM, Carver-Moore K, Rosenthal A. Renal and neuronal abnormalities in mice lacking GDNF. *Nature* 1996, **382**: 76–79.
129. Popovic M, Jovanova-Nesic K, Popovic N, Bokonjic D, Dobric S, Rosic N, Rakic L. Behavioral and adaptive status in an experimental model of Alzheimer's disease in rats. *Int. J Neurosci* 1996, **86**: 281–299.
130. Sánchez M, Silos-Santiago I, Frisén J, He B, Lira SA, Barbacid M. Renal agenesis and the absence of enteric neurons in mice lacking GDNF. *Nature* 1996, **382**: 70–73.
131. Koliatsos VE, Price DL. Axotomy as an experimental model of neuronal injury and cell death. *Brain Pathol* 1996, **6**: 447–465.
132. Lieberman AR. The axon reaction: a review of the principal features of perikaryal responses to axon injury, in *International Review of Neurobiology* (Pfeiffer CC, Smythies JR, eds.), Academic Press, New York, 1971, pp. 49–124.
133. Snider WD, Elliott JL, Yan Q. Axotomy-induced neuronal death during development. *J Neurobiol* 1992, **23**: 1231–1246.
134. Fritschy JM, Grzanna R. Restoration of ascending noradrenergic projections by residual locus coeruleus neurons: compensatory response to neurotoxic-induced cell death in the adult rat brain. *J Comp Neurol* 1992, **321**: 421–441.
135. Coyle JT, Schwarcz R. Lesion of striatal neurons with kainic acid provides a model for Huntington's chorea. *Nature* 1976, **263**: 244–246.
136. Palmiter RD, Brinster RL. Germ-line transformation of mice. *Annu Rev Genet* 1986, **20**: 465–499.
137. Capecchi MR. Altering the genome by homologous recombination. *Science* 1989, **244**: 1288–1299.
138. Takahashi JS, Pinto LH, Hotz Vitaterna M. Forward and reverse genetic approaches to behavior in the mouse. *Science* 1994, **264**: 1724–1733.
139. Scott MR, Köhler R, Foster D, Prusiner SB. Chimeric prion protein expression in cultured cells and transgenic mice. *Protein Sci* 1992, **1**: 986–997.
140. Schedl A, Montoliu L, Kelsey G, Schütz G. A yeast artificial chromosome covering the tyrosinase gene confers copy number-dependent expression in transgenic mice. *Nature* 1993, **362**: 258–261.

141. Price DL, Becher MW, Wong PC, Borchelt DR, Lee MK, Sisodia SS. Inherited neurodegenerative diseases and transgenic models. *Brain Pathol* 1996, **6**: 467–480.
142. Games D, Adams D, Alessandrini R, Barbour R, Berthelette P, Blackwell C, Carr T, Clemens J, Donaldson T, Gillespie F, Guido T, Hagopian S, Johnson-Wood K, Khan K, Lee M, Leibowitz P, Lieberburg I, Little S, Masliah E, McConlogue L, Montoya-Zavala M, Mucke L, Paganini L, Penniman E, Power M, Schenk D, Seubert P, Snyder B, Soriano F, Tan H, Vitale J, Wadsworth S, Wolozin B, Zhao J. Alzheimer-type neuropathology in transgenic mice overexpressing V717F β-amyloid precursor protein. *Nature* 1995, **373**: 523–527.
143. Hsiao K, Chapman P, Nilsen S, Eckman C, Harigaya Y, Younkin S, Yang F, Cole G. Correlative memory defcits, Aβ elevation and amyloid plaques in transgenic mice. *Science* 1996, **274**: 99–102.
144. Gurney ME, Pu H, Chiu AY, Dal Canto MC, Polchow CY, Alexander DD, Caliendo J, Hentati A, Kwon YW, Deng H-X, Chen W, Zhai P, Sufit RL, Siddique T. Motor neuron degeneration in mice that express a human Cu,Zn superoxide dismutase mutation. *Science* 1994, **264**: 1772–1775.
145. Burright EN, Clark HB, Servadio A, Matilla T, Feddersen RM, Yunis WS, Duvick LA, Zoghbi HY, Orr HT. *SCA1* transgenic mice: a model for neurodegeneration caused by an expanded CAG trinucleotide repeat. *Cell* 1995, **82**: 937–948.
146. Neurath MF, Pettersson S, Meyer zum Büschenfelde K-H, Strober W. Local administration of antisense phosphorothioate oligonucleotides to the p65 subunit of NF-kappaB abrogates established experimental colitis in mice. *Nature Med* 1996, **2**: 998–1004.
147. Wilson MA, Tonegawa S. Synaptic plasticity, place cells and spatial memory: study with second generation knockouts. *Trends Neurosci* 1997, **20**: 102–106.
148. Duchen LW, Strich SJ. An hereditary motor neurone disease with progressive denervation of muscle in the mouse: the mutant "wobbler." *J Neurol Neurosurg Psychiatry* 1968, **31**: 535–542.
149. Mitsumoto H, Bradley WG. Murine motor neuron disease (the wobbler mouse). Degeneration and regeneration of the lower motor neuron. *Brain* 1982, **105**: 811–834.
150. Schmalbruch H, Jensen H-JS, Bjærg M, Kamieniecka Z, Kurland L. A new mouse mutant with progressive motor neuronopathy. *J Neuropathol Exp Neurol* 1991, **50**: 192–204.
151. Price DL, Martin LJ, Sisodia SS, Walker LC, Voytko ML, Wagster MV, Cork LC, Koliatsos VE. The aged nonhuman primate. A model for the behavioral and brain abnormalities occurring in aged humans, in *Alzheimer Disease* (Terry RD, Katzman R, Bick KL, eds.), Raven Press, New York, 1994, pp. 231–245.
152. Koliatsos VE, Price DL. Retrograde axonal transport. Applications in trophic factor research, in *Animal Models of Neurological Disorders* (Boulton AA, Baker GB, Hefti F, eds.), Humana Press, Clifton, New Jersey, 1993, pp. 247–290.
153. Hefti F. Nerve growth factor promotes survival of septal cholinergic neurons after fimbrial transections. *J Neurosci* 1986, **6**: 2155–2162.
154. Walton M, Sirimanne E, Williams C, Gluckman P, Dragunow M. The role of the cyclic AMP-responsive element binding protein (CREB) in hypoxic-ischemic brain damage and repair. *Mol Brain Res* 1996, **43**: 21–29.
155. Kromer LF. Nerve growth factor treatment after brain injury prevents neuronal death. *Science* 1987, **235**: 214–216.
156. Koliatsos VE, Nauta HJW, Clatterbuck RE, Holtzman DM, Mobley WC, Price DL. Mouse nerve growth factor prevents degeneration of axotomized basal forebrain cholinergic neurons in the monkey. *J Neurosci* 1990, **10**: 3801–3813.
157. Tuszynski MH, U HS, Amaral DG, Gage FH. Nerve growth factor infusion in the primate brain reduces lesion-induced cholinergic neuronal degeneration. *J Neurosci* 1990, **10**: 3604–3614.

158. Wilcox BJ, Applegate MD, Portera-Cailliau C, Koliatsos VE. Nerve growth factor prevents apoptotic cell death in injured central cholinergic neurons. *J Comp Neurol* 1995, **359**: 573–585.
159. Fischer W, Wictorin K, Björklund A, Williams LR, Varon S, Gage FH. Amelioration of cholinergic neuron atrophy and spatial memory impairment in aged rats by nerve growth factor. *Nature* 1987, **329**: 65–68.
160. Markowska AL, Koliatsos VE, Breckler SJ, Price DL, Olton DS. Human nerve growth factor improves spatial memory in aged but not in young rats. *J Neurosci* 1994, **14**: 4815–4824.
161. Gage FH, Batchelor P, Chen KS, Chin D, Higgins GA, Koh S, Deputy S, Rosenberg MB, Fischer W, Bjorklund A. NGF receptor reexpression and NGF-mediated cholinergic neuronal hypertrophy in the damaged adult neostriatum. *Neuron* 1989, **2**: 1177–1184.
162. Koliatsos VE, Applegate MD, Knüsel B, Junard EO, Burton LE, Mobley WC, Hefti FF, Price DL. Recombinant human nerve growth factor prevents retrograde degeneration of axotomized basal forebrain cholinergic neurons in the rat. *Exp Neurol* 1991, **112**: 161–173.
163. Clatterbuck RE, Price DL, Koliatsos VE. Ciliary neurotrophic factor prevents retrograde neuronal death in the adult central nervous system. *Proc Natl Acad Sci USA* 1993, **90**: 2222–2226.
164. Riddle DR, Lo DC, Katz LC. NT-4 mediated rescue of lateral geniculate neurons from effects of monocular deprivation. *Nature* 1995, **378**: 189–191.
165. Gash DM, Zhang Z, Ovadia A, Cass WA, Yi A, Simmerman L, Russell D, Martin D, Lapchak PA, Collins F, Hoffer BJ, Gerhardt GA. Functional recovery in parkinsonian monkeys treated with GDNF. *Nature* 1996, **380**: 252–255.
166. Choi-Lundberg DL, Lin Q, Chang Y-N, Chiang YL, Hay CM, Mohajeri H, Davidson BL, Bohn MC. Dopaminergic neurons protected from degeneration by GDNF gene therapy. *Science* 1997, **275**: 838–841.
167. Hagg T, Varon S. Ciliary neurotrophic factor prevents degeneration of adult rat substantia nigra dopaminergic neurons *in vivo*. *Proc Natl Acad Sci USA* 1993, **90**: 6315–6319.
168. Arenas E, Persson H. Neurotrophin-3 prevents the death of adult central noradrenergic neurons *in vivo*. *Nature* 1994, **367**: 368–371.
169. Koliatsos VE, Cayouette MH, Berkemeier LR, Clatterbuck RE, Price DL, Rosenthal A. Neurotrophin 4/5 is a trophic factor for mammalian facial motor neurons. *Proc Natl Acad Sci USA* 1994, **91**: 3304–3308.
170. Yan Q, Matheson C, Lopez OT, Miller JA. The biological responses of axotomized adult motoneurons to brain-derived neurotrophic factor. *J Neurosci* 1994, **14**: 5281–5291.
171. Friedman B, Kleinfeld D, Ip NY, Verge VMK, Moulton R, Boland P, Zlotchenko E, Lindsay RM, Liu L. BDNF and NT-4/5 exert neurotrophic influences on injured adult spinal motor neurons. *J Neurosci* 1995, **15**: 1044–1056.
172. Mitsumoto H, Ikeda K, Holmlund T, Greene T, Cedarbaum JM, Wong V, Lindsay RM. The effects of ciliary neurotrophic factor on motor dysfunction in wobbler mouse motor neuron disease. *Ann Neurol* 1994, **36**: 142–148.
173. Ikeda K, Wong V, Holmlund TH, Greene T, Cedarbaum JM, Lindsay RM, Mitsumoto H. Histometric effects of ciliary neurotrophic factor in wobbler mouse motor neuron disease. *Ann Neurol* 1995, **37**: 47–54.
174. Sagot Y, Tan SA, Hammang JP, Aebischer P, Kato AC. GDNF slows loss of motoneurons but not axonal degeneration or premature death of *pmn/pmn* mice. *J Neurosci* 1996, **16**: 2335–2341.
175. Wong J, Oblinger MM. NGF rescues substance P expression but not neurofilament or tubulin gene expression in axotomized sensory neurons. *J Neurosci* 1991, **11**: 543–552.
176. Yip HK, Johnson EM Jr. Developing dorsal root ganglion neurons require trophic support from their central processes: evidence for a role of retrogradely transported nerve growth factor from the central nervous system to the periphery. *Proc Natl Acad Sci USA* 1984, **81**: 6245–6249.

177. Eriksson NP, Lindsay RM, Aldskogius H. BDNF and NT-3 rescue sensory but not motoneurones following axotomy in the neonate. *Neuroreport* 1994, **5:** 1445–1448.
178. Rich KM, Disch SP, Eichler ME. The influence of regeneration and nerve growth factor on the neuronal cell body reaction to injury. *J Neurocytol* 1989, **18:** 569–576.
179. Hendry IA. The response of adrenergic neurones to axotomy and nerve growth factor. *Brain Res* 1975, **94:** 87–97.
180. Manning PT, Russell JH, Simmons B, Johnson EM Jr. Protection from guanethidine-induced neuronal destruction by nerve growth factor: effect of NGF on immune function. *Brain Res* 1985, **340:** 61–69.
181. Aloe L, Mugnaini E, Levi-Montalcini R. Light and electron microscopic studies on the excessive growth of sympathetic ganglia in rats injected daily from birth with 6-OHDA and NGF. *Archives Italiennes De Biologia* **113:** 326–353.
182. Garofalo L, Ribeiro-da-Silva A, Cuello AC. Nerve growth factor-induced synaptogenesis and hypertrophy of cortical cholinergic terminals. *Proc Natl Acad Sci USA* 1992, **89:** 2639–2643.
183. Diamond J, Holmes M, Coughlin M. Endogenous NGF and nerve impulses regulate the collateral sprouting of sensory axons in the skin of the adult rat. *J Neurosci* 1992, **12:** 1454–1466.
184. Penny GR, Afsharpour S, Kitai ST. The glutamate decarboxylase-, leucine enkephalin-, methionine enkephalin- and substance P-immunoreactive neurons in the neostriatum of the rat and cat: evidence for partial population overlap. *Neuroscience* 1986, **17:** 1011–1045.
185. Lindsay RM. Nerve growth factors (NGF, BDNF) enhance axonal regeneration but are not required for survival of adult sensory neurons. *J Neurosci* 1988, **8:** 2394–2405.
186. Helgren ME, Cliffer KD, Torrento K, Cavnor C, Curtis R, DiStefano PS, Wiegand SJ, Lindsay RM. Neurotrophin-3 administration attenuates deficits of pyridoxine-induced large-fiber sensory neuropathy. *J Neurosci* 1997, **17:** 372–382.
187. Mamounas LA, Blue ME, Siuciak JA, Altar CA. Brain-derived neurotrophic factor promotes the survival and sprouting of serotonergic axons in rat brain. *J Neurosci* 1995, **15:** 7929–7939.
188. Shigenaga MR, Park J-W, Cundy KC, Gimeno CJ, Ames BN. In vivo oxidative DNA damage: measurement of 8-hydroxy-2-deoxyguanosine in DNA and urine by high-performance liquid chormatography with electrochemical detection. *Methods Enzymol* 1990, **186:** 521–530.
189. Apfel SC, Arezzo JC, Brownlee M, Federoff H, Kessler JA. Nerve growth factor administration protects against experimental diabetic sensory neuropathy. *Brain Res* 1994, **634:** 7–12.
190. Apfel SC, Lipton RB, Arezzo JC, Kessler JA. Nerve growth factor prevents toxic neuropathy in mice. *Ann Neurol* 1991, **29:** 87–90.
191. Apfel SC, Arezzo JC, Lipson LA, Kessler JA. Nerve growth factor prevents experimental cisplatin neuropathy. *Ann Neurol* 1992, **31:** 76–80.
192. Appel SH. A unifying hypothesis for the cause of amyotrophic lateral sclerosis, parkinsonism, and Alzheimer disease. *Ann Neurol* 1981, **10:** 499–505.
193. Hefti F, Weiner WJ. Nerve growth factor and Alzheimer's disease. *Ann Neurol* 1986, **20:** 275–281.
194. Sendtner M, Kreutzberg GW, Thoenen H. Ciliary neurotrophic factor prevents the degeneration of motor neurons after axotomy. *Nature* 1990, **345:** 440–441.
195. Clatterbuck RE, Price DL, Koliatsos VE. Further characterization of the effects of brain-derived neurotrophic factor and ciliary neurotrophic factor on axotomized neonatal and adult mammalian motor neurons. *J Comp Neurol* 1994, **342:** 45–56.
196. Shapiro L, Zhang X-X, Rupp RG, Wolff SM, Dinarello CA. Ciliary neurotrophic factor is an endogenous pyrogen. *Proc Natl Acad Sci USA* 1993, **90:** 8614–8618.
197. Yuen EC, Mobley WC. Therapeutic potential of neurotrophic factors for neurological disorders. *Ann Neurol* 1996, **40:** 346–354.
198. MacDermott AB, Mayer ML, Westbrook GL, Smith SJ, Barker JL. NMDA-receptor activation increases cytoplasmic calcium concentration in cultured spinal cord neurones. *Nature* 1986, **321:** 519–522.

199. Choi DW. Excitotoxic cell death. *J Neurobiol* 1992, **23**: 1261–1276.
200. Frim DM, Uhler TA, Short MP, Ezzedine ZD, Klagsbrun M, Breakefield XO, Isacson O. Effects of biologically delivered NGF, BDNF and bFGF on striatal excitotoxic lesions. *Neuroreport* 1993, **4**: 367–370.
201. Perez-Navarro E, Alberch J. Protective role of nerve growth factor against excitatory amino acid injury during neostriatal cholinergic neurons postnatal development. *Exp Neurol* 1995, **135**: 146–152.
202. Liu Z, D'amore P, Mikati M, Gatt A, Holmes GL. Neuroprotective effect of chronic infusion of basic fibroblast growth factor on seizure-associated hippocampal damage. *Brain Res* 1993, **626**: 335–338.
203. Lindholm D, Dechant G, Heisenberg CP, Thoenen H. Brain derived neurotrophic factor is a survival factor for cultured rat cerebellar granule neurons and protects against glutamate-induced neurotoxicity. *Eur. J Neurosci* 1993, **5**: 1455–1464.
204. Albers GW, Goldberg MP, Choi DW. N-methyl-D-aspartate antagonists: ready for clinical trial in brain ischemia? *Ann Neurol* 1989, **25**: 398–403.
205. Diener PS, Bregman BS. Neurotrophic factors prevent the death of CNS neurons after spinal cord lesions in newborn rats. *Neuroreport* 1994, **5**: 1913–1917.
206. Schnell L, Schneider R, Kolbeck R, Barde Y-A, Schwab ME. Neurotrophin-3 enhances sprouting of corticospinal tract during development and after adult spinal cord lesion. *Nature* 1994, **367**: 170–173.
207. Cheng H, Cao Y, Olson L. Spinal cord repair in adult paraplegic rats: partial restoration of hind limb function. *Science* 1996, **273**: 510–513.
208. Baffour R, Achanta K, Kaufman J, Berman J, Garb JL, Rhee S, Friedmann P. Synergistic effect of basic fibroblast growth factor and methylprednisolone on neurological function after experimental spinal cord injury. *J Neurosurg* 1995, **83**: 105–110.
209. Faden AI. Experimental neurobiology of central nervous system trauma. *Crit Rev Neurobiol* 1993, **7**: 175–186.
210. Lyons WE, Fritschy JM, Grzanna R. The noradrenergic neurotoxin DSP-4 eliminates the coeruleospinal projection but spares projections of the A5 and A7 groups to the ventral horn of the rat spinal cord. *J Neurosci* 1989, **9**: 1481–1489.
211. Lyons MK, Anderson RE, Meyer FB. Basic fibroblast growth factor promotes in vivo cerebral angiogenesis in chronic forebrain ischemia. *Brain Res* 1991, **558**: 315–320.
212. Blight AR. Remyelination, revascularization, and recovery of function in experimental spinal cord injury. *Adv Neurol* 1993, **59**: 91–104.
213. Levi-Montalcini R, Angeletti PU. Nerve growth factor. *Physiol Rev* 1968, **48**: 534–569.
214. Hollyday M, Hamburger V. Reduction of the naturally occurring motor neuron loss by enlargement of the periphery. *J Comp Neurol* 1976, **170**: 311–320.
215. Davies AM, Bandtlow C, Heumann R, Korsching S, Rohrer H, Thoenen H. Timing and site of nerve growth factor synthesis in developing skin in relation to innervation and expression of the receptor. *Nature* 1987, **326**: 353–358.
216. Eide FF, Vining ER, Eide BL, Zang K, Wang X-Y, Reichardt LF. Naturally occurring truncated trkB receptors have dominant inhibitory effects on brain-derived neurotrophic factor signaling. *J Neurol Sci* 1996, **16**: 3123–3129.
217. Koliatsos VE, Price DL, Gouras GK, Cayouette MH, Burton LE, Winslow JW. Highly selective effects of nerve growth factor, brain-derived neurotrophic factor, and neurotrophin-3 on intact and injured basal forebrain magnocellular neurons. *J Comp Neurol* 1994, **343**: 247–262.
218. Snider WD, Wright DE. Neurotrophins cause a new sensation. *Neuron* 1996, **16**: 229–232.
219. Ruit KG, Elliott JL, Osborne PA, Yan Q, Snider WD. Selective dependence of mammalian dorsal root ganglion neurons on nerve growth factor during embryonic development. *Neuron* 1992, **8**: 573–587.

220. Airaksinen MS, Koltzenburg M, Lewin GR, Masu Y, Helbig C, Wolf E, Brem G, Toyka KV, Thoenen H, Meyer M. Specific subtypes of cutaneous mechanoreceptors require neurotrophin-3 following peripheral target innervation. *Neuron* 1996, **16**: 287–295.
221. Prakash N, Cohen-Cory S, Frostig RD. Rapid and opposite effects of BDNF and NGF on the functional organization of the adult cortex in vivo. *Nature* 1996, **381**: 702–706.
222. DiStefano PS, Friedman B, Radziejewski C, Alexander C, Boland P, Schick CM, Lindsay RM, Wiegand SJ. The neurotrophins BDNF, NT-3, and NGF display distinct patterns of retrograde axonal transport in peripheral and central neurons. *Neuron* 1992, **8**: 983–993.
223. Koliatsos VE, Clatterbuck RE, Winslow JW, Cayouette MH, Price DL. Evidence that brain-derived neurotrophic factor is a trophic factor for motor neurons in vivo. *Neuron* 1993, **10**: 359–367.
224. Causing CG, Gloster A, Aloyz R, Bamji SX, Chang E, Fawcett J, Kuchel G, Miller FD. Synaptic innervation density is regulated by neuron-derived BDNF. *Neuron* 1997, **18**: 257–267.
225. Conner JM, Lauterborn JC, Yan Q, Gall CM, Varon S. Distribution of brain-derived neurotrophic factor (BDNF) protein and mRNA in the normal adult rat CNS: evidence for anterograde axonal transport. *J Neurosci* 1997, **17**: 2295–2313.
226. Hamburger V, Brunso-Bechtold JK, Yip JW. Neuronal death in the spinal ganglia of the chick embryo and its reduction by nerve growth factor. *J Neurosci* 1981, **1**: 60–71.
227. Lumsden AGS, Davies AM. Earliest sensory nerve fibres are guided to peripheral targets ty attractants other than nerve growth factor. *Nature* 1983, **306**: 786–788.
228. Oppenheim RW. The neurotrophic theory and naturally occurring motoneuron death. *Trends Neurosci* 1989, **12**: 252–255.
229. Auburger G, Heumann R, Hellweg R, Korsching S, Thoenen H. Developmental changes of nerve growth factor and its mRNA in the rat hippocampus: comparison with choline acetyltransferase. *Dev Biol* 1987, **120**: 322–328.
230. Hoyle GW, Mercer EH, Palmiter RD, Brinster RL. Expression of NGF in sympathetic neurons leads to excessive axon outgrowth from ganglia but decreased terminal innervation within tissues. *Neuron* 1993, **10**: 1019–1034.
231. Cohen-Cory S, Fraser SE. Effects of brain-derived neurotrophic factor on optic axon branching and remodelling *in vivo*. *Nature* 1995, **378**: 192–196.
232. Hofer MM, Barde Y-A. Brain-derived neurotrophic factor prevents neuronal death *in vivo*. *Nature* 1988, **331**: 261–262.
233. Sieber-Blum M. Role of the neurotrophic factors BDNF and NGF in the commitment of pluripotent neural crest cells. *Neuron* 1991, **6**: 949–955.
234. Maisonpierre PC, Belluscio L, Friedman B, Alderson RF, Wiegand SJ, Furth ME, Lindsay RM, Yancopoulos GD. NT-3, BDNF, and NGF in the developing rat nervous system: parallel as well as reciprocal patterns of expression. *Neuron* 1990, **5**: 501–509.
235. ElShamy WM, Ernfors P. A local action of neurotrophin-3 prevents the death of proliferating sensory neuron precursor cells. *Neuron* 1996, **16**: 963–972.
236. Shelton DL, Sutherland J, Gripp J, Camerato T, Armanini MP, Phillips HS, Carroll K, Spencer SD, Levinson AD. Human trks: molecular cloning, tissue distribution, and expression of extracellular domain immunoadhesins. *J Neurosci* 1995, **15**: 477–491.
237. Gall CM, Isackson PJ. Limbic seizures increase neuronal production of messenger RNA for nerve growth factor. *Science* 1989, **245**: 758–761.
238. Ernfors P, Bengzon J, Kokaia Z, Persson H, Lindvall O. Increased levels of messenger RNAs for neurotrophic factors in the brain during kindling epileptogenesis. *Neuron* 1991, **7**: 165–176.
239. Bengzon J, Kokaia Z, Ernfors P, Kokaia M, Leanza G, Nilsson OG, Persson H, Lindvall O. Regulation of neurotrophin and trkA, trkB and trkC tyrosine kinase receptor messenger RNA expression in kindling. *Neuroscience* 1993, **53**: 433–446.

240. Isackson PJ, Huntsman MM, Murray KD, Gall CM. BDNF mRNA expression is increased in adult rat forebrain after limbic seizures: temporal patterns of induction distinct from NGF. *Neuron* 1991, **6**: 937–948.
241. Takeda A, Onodera H, Sugimoto A, Kogure K, Obinata M, Shibahara S. Coordinated expression of messenger RNAs for nerve growth factor, brain-derived neurotrophic factor and neurotrophin-3 in the rat hippocampus following transient forebrain ischemia. *Neuroscience* 1993, **55**: 23–31.
242. Riva M, Gale K, Mocchetti I. Basic fibroblast growth factor mRNA increases in specific brain regions following convulsive seizures. *Mol Brain Res* 1992, **15**: 311–318.
243. Follesa P, Gale K, Mocchetti I. Regional and temporal pattern of expression of nerve growth factor and basic fibroblast growth factor mRNA in rat brain following electroconvulsive shock. *Exp Neurol* 1994, **127**: 37–44.
244. Gall CM, Berschauer R, Isackson PJ. Seizures increase basic fibroblast growth factor mRNA in adult rat forebrain neurons and glia. Brain Res. *Mol Brain Res* 1994, **21**: 190–205.
245. Nawa H, Carnahan J, Gail C. BDNF protein measured by a novel enzyme immunoassay in normal brain and after seizure: partial disagreement with mRNA levels. Eur. *J Neurosci* 1995, **7**: 1527–1535.
246. Merlio J-P, Ernfors P, Kokaia Z, Middlemas DS, Bengzon J, Kokaia M, Smith M-L, Siejö, BK, Hunter T, Lindvall O, Persson H. Increased production of the TrkB protein tyrosine kinase receptor after brain insults. *Neuron* 1993, **10**: 151–164.
247. Mudó, G, Persson H, Timmusk T, Funakoshi H, Bindoni M, Belluardo N. Increased expression of trkB and trkC messenger RNAs in the rat forebrain after focal mechanical injury. *Neuroscience* 1993, **57**: 901–912.
248. Zafra F, Hengerer B, Leibrock J, Thoenen H, Lindholm D. Activity dependent regulation of BDNF and NGF mRNAs in the rat hippocampus is mediated by non-NMDA glutamate receptors. *EMBO J* 1990, **9**: 3545–3550.
249. Ghosh A, Carnahan J, Greenberg ME. Requirement for BDNF in activity-dependent survival of cortical neurons. *Science* 1994, **263**: 1618–1623.
250. Sarter M, Bruno JP. Trans-synaptic stimulation of cortical acetylcholine and enhancement of attentional functions: a rational approach for the development of cognition enhancers. *Behav Brain Res* 1997, **83**: 7–14.
251. Yan Q, Matheson C, Sun J, Radeke MJ, Feinstein SC, Miller JA. Distribution of intracerebral ventricularly administered neurotrophins in rat brain and its correlation with trk receptor expression. *Exp Neurol* 1994, **127**: 23–36.
252. Olsson Y. Vascular permeability in the peripheral nervous system, in *Peripheral Neuropathy* (Dyck PJ, Thomas PK, Lambert EH, Bunge R, eds.), W.B. Saunders, Philadelphia, 1984, pp. 579–597.
253. Fabrazzo M, Costa E, Mocchetti I. Stimulation of nerve growth factor biosynthesis in developing rat brain by reserpine: steroids as potential mediators. *Mol Pharmacol* 1991, **39**: 144–149.
254. Saporito MS, Brown ER, Hartpence KC, Wilcox HM, Robbins E, Vaught JL, Carswell S. Systemic dexamethasone administration increases septal Trk autophosphorylation in adult rats via an induction of nerve growth factor. *Mol Pharmacol* 1994, **45**: 395–401.
255. Barbany G, Persson H. Regulation of neurotrophin mRNA expression in the rat brain by glucocorticoids. *Eur J Neurosci* 1992, **4**: 396–403.
256. Lauterborn J, Berschauer R, Gall C. Cell-specific modulation of basal and seizure-induced neurotrophin expression by adrenalectomy. *Neuroscience* 1995, **68**: 363–378.
257. Mocchetti I, Spiga G, Hayes VY, Isackson PJ, Colangelo A. Glucocorticoids differentially increase nerve growth factor and basic fibroblast growth factor expression in the rat brain. *J Neurosci* 1996, **16**: 2141–2148.

258. Dugich-Djordjevic MM, Tocco G, Willoughby DA, Najm I, Pasinetti G, Thompson RF, Baudry M, Lapchak PA, Hefti F. BDNF mRNA expression in the developing rat brain following kainic acid-induced seizure activity. *Neuron* 1992, **8:** 1127–1138.
259. Chan KM, Lam DTN, Pong K, Widmer HR, Hefti F. Neurotrophin-4/5 treatment reduces infarct size in rats with middle cerebral artery occlusion. *Neurochem Res* 1996, **21:** 763–767.
260. Schwartz JP, Costa E. Regulation of nerve growth factor content in C6 glioma cells by β-adrenergic receptor stimulation. *Naunyn Schmiedebergs Arch Pharmacol* 1977, **300:** 123–129.
261. Follesa P, Mocchetti I. Regulation of basic fibroblast growth factor and nerve growth factor mRNA by β-adrenergic receptor activation and adrenal steroids in rat central nervous system. *Mol Pharmacol* 1993, **43:** 132–138.
262. Hayes VY, Isackson PJ, Fabrazzo M, Folleso P, Mocchetti I. Induction of nerve growth factor and basic fibroblast growth factor mRNA following clenbuterol: contrasting anatomical and cellular localization. *Exp Neurol* 1995, **132:** 33–41.
263. Koizumi S, Contreras ML, Matsuda Y, Hama T, Lazarovici P, Guroff G. K-252a: a specific inhibitor of the action of nerve growth factor on PC12 cells. *J Neurol Sci* 1988, **8:** 715–721.
264. Matsuda Y, Fukuda J. Inhibition by K-252a, a new inhibitor of protein kinase, of nerve growth factor-induced neurite outgrowth of chick embryo dorsal root ganglion cells. *Neurosci Lett* 1988, **87:** 11–17.
265. Knüsel B, Hefti F. K-252 compounds: modulators of neurotrophin signal transduction. *J Neurochem* 1992, **59:** 1987–1996.
266. Knüsel B, Rabin S, Widmer HR, Hefti F, Kaplan DR. Neurotrophin induced trk receptor phosphorylation and cholinergic neuron response in primary cultures of embryonic rat brain neurons. *Neuroreport* 1992, **3:** 885–888.
267. Maroney AC, Sanders C, Neff NT, Dionne CA. K-252b potentiation of neurotrophin-3 is trkA specific in cells lacking p75[NTR]. *J Neurochem* 1997, **68:** 88–94.
268. Suter U, Angst C, Tien C-L, Drinkwater CC, Lindsay RM, Shooter EM. NGF/BDNF chimeric proteins: analysis of neurotrophin specificity by homolog-scanning mutagenesis. *J Neurosci* 1992, **12:** 306–318.
269. Urfer R, Tsoulfas P, Soppet D, Escandon E, Parada LF, Presta LG. The binding epitopes of neurotrophin-3 to its receptors trkC and gp75 and the design of a multifunctional human neurotrophin. *EMBO J* 1994, **13:** 5896–5909.
270. Ilag LL, Curtis R, Glass D, Funakoshi H, Tobkes NJ, Ryan TE, Acheson A, Lindsay RM, Persson H, Yancopoulos GD, DiStefano PS, Ibáñez CF. Pan-neurotrophin 1: a genetically engineered neurotrophic factor displaying multiple specificities in peripheral neurons *in vitro* and *in vivo*. *Proc Natl Acad Sci USA* 1995, **92:** 607–611.
271. Derry DM, Wolfe LS. Gangliosides in isolated neurons and glial cells. *Science* 1967, **158:** 1450–1452.
272. Ferrari G, Batistatou A, Greene LA. Gangliosides rescue neuronal cells from death after trophic factor deprivation. *J Neurosci* 1993, **13:** 1879–1887.
273. Rabin SJ, Mocchetti I. GM1 ganglioside activates the high-affinity nerve growth factor receptor trkA. *J Neurochem* 1995, **65:** 347–354.
274. Cuello AC. Towards trophic factor pharmacology? *Neurobiol Aging* 1989, **10:** 539–540.
275. Di Patre PL, Oh JD, Simmons JM, Butcher LL. Intrafimbrial colchicine produces transient impairment of radial-arm maze performance correlated with morphologic abnormalities of septohippocampal neurons expressing cholinergic markers and nerve growth factor receptor. *Brain Res* 1990, **523:** 316–320.
276. Maysinger D, Filipovic-Greie J, Cuello CA. Effects of coencapsulated NGF and GM1 in rats with cortical lesions. *J Neurochem* 1993, **4:** 971–974.
277. Silva RH, Felicio LF, Nasello AG, Vital MA, Frussa-Filho R. Effect of ganglioside (GM1) on memory in senescent rats. *Neurobiol Aging* 1996, **17:** 583–586.

278. Hadjiconstantinou M, Weihmuller F, Neff NH. Treatment with GM1 ganglioside reverses dopamine D-2 receptor supersensitivity induced by the neurotoxin MPTP. *Eur J Pharmacol* 1989, **168**: 261–264.
279. Schneider JS, Pope A, Simpson K, Taggart J, Smith MG, DiStefano L. Recovery from experimental parkinsonism in primates with G_{M1} ganglioside treatment. *Science* 1992, **256**: 843–846.
280. Favaron M, Manev H, Alho H, Bertolino M, Ferret B, Guidotti A, Costa E. Gangliosides prevent glutamate neurotoxicity in primary cultures of neonatal rat cerebellar and cortex. *Proc Natl Acad Sci USA* 1988, **85**: 7351–7355.
281. Karpiac SE, Mahadick SP, Wakade CG. Ganglioside reduction of ischemic injury. *CRC Crit Rev Clin Neurobiol* 1990, **5**: 221–237.
282. Seren MS, Rubini R, Lazzaro A, Zanoni R, Fiori MG, Leon A. Protective effects of a monosialoganglioside derivative following transitory forebrain ischemic in rats. *Stroke* 1990, **21**: 1607–1612.
283. Lyons WE, George EB, Dawson TM, Steiner JP, Snyder SH. Immunosuppressan FK506 promotes neurite outgrowth in cultures of PC12 cells and sensory ganglia. *Proc Natl Acad Sci USA* 1994, **91**: 3191–3195.
284. Snyder SH, Sabatini DM. Immunophilins and the nervous system. *Nature Med* 1995, **1**: 32–37.
285. Steiner JP, Hamilton GS, Ross DT, Valentine HL, Guo H, Connolly MA, Liang S, Ramsey C, Li JJ, Huang W, Howorth P, Soni R, Fuller M, Sauer H, Nowotnik AC, Suzdak PD. Neurotrophic immunophilin ligands stimulate structural and functional recovery in neurodegenerative animal models. *Proc Natl Acad Sci USA* 1997, **94**: 2019–2024.
286. Koliatsos VE, Price DL. Nonhuman primate models in trophic factor research, in *Animal Models of Neurological Disorders* (Boulton AA, Baker GB, Hefti F, eds.), Humana Press, Clifton, New Jersey, 1993, pp. 331–370.
287. Rosenberg MB, Friedmann T, Robertson RC, Tuszynski M, Wolff JA, Breakefield XO, Gage FH. Grafting genetically modified cells to the damaged brain: restorative effects of NGF expression. *Science* 1988, **242**: 1575–1578.
288. Tseng JL, Baetge EE, Zurn AD, Aebischer P. GDNF reduces drug-induced rotational behavior after medial forebrain bundle transection by a mechanism not involving striatal dopamine. *J Neurosci* 1997, **17**: 325–333.
289. Martinez-Serrano A, Fischer W, Bjorklund A. Reversal of age-dependent cognitive impairments and cholinergic neuron atrophy by NGF-secreting neural progenitors grafted to the basal forebrain. *Neuron* 1995, **15**: 473–484.
290. Friden PM, Walus LR, Watson P, Doctrow SR, Kozarich JW, Bächman C, Bergman H, Hoffer B, Bloom F, Granholm A-C. Blood-brain barrier penetration and in vivo activity of an NGF conjugate. *Science* 1993, **259**: 373–377.
291. Bäckman C, Rose GM, Hoffer BJ, Henry MA, Bartus RT, Friden P, Granholm AC. Systemic administration of a nerve growth factor conjugate reverses age-related cognitive dysfunction and prevents cholinergic neuron atrophy. *J Neurosci* 1996, **16**: 5437–5442.
292. During MJ, Naegele JR, O'Malley KL, Geller AI. Long-term behavioral recovery in parkinsonian rats by an HSV vector expressing tyrosine hydroxylase. *Science* 1994, **266**: 1399–1403.
293. Sukhov RR, Cayouette MH, Radeke MJ, Feinstein SC, Blumberg D, Rosenthal A, Price DL, Koliatsos VE. Evidence that perihypoglossal neurons involved in vestibular-auditory and gaze control functions respond to nerve growth factor. *J Comp Neurol* 1997, **383**: 123–134.
294. Williams LR. Hypophagia is induced by intracerebroventricular administration of nerve growth factor. *Exp Neurol* 1991, **113**: 31–37.
295. Glass DJ, Yancopoulos GD. The neurotrophins and their receptors. *Trends Cell Biol* 1993, **3**: 262–268.

296. Lewin GR, Mendell LM. Nerve growth factor and nociception. *Trends Neurosci* 1993, **16**: 353–359.
297. Davis BM, Lewin GR, Mendell LM, Jones ME, Albers KM. Altered expression of nerve growth factor in the skin of transgenic mice leads to changes in response to mechanical stimuli. *Neuroscience* 1993, **56**: 789–792.
298. Della Seta D, de Acetis L, Aloe L, Alleva E. NGF effects on hot plate behaviors in mice. *Pharmacol Biochem Behav* 1994, **49**: 701–705.
299. Andreev NY, Dimitrieva N, Koltzenburg M, McMahon SB. Peripheral administration of nerve growth factor in the adult rat produces a thermal hyperalgesia that requires the presence of sympathetic post-ganglionic neurones. *Pain* 1995, **63**: 109–115.
300. Petty BG, Cornblath DR, Adornato BT, Chaudhry V, Flexner C, Wachsman M, Sinicropi D, Burton LE, Peroutka SJ. The effect of systematically administered recombinant human nerve growth factor in healthy human subjects. *Ann Neurol* 1994, **36**: 244–246.
301. Woolf CJ, Ma QP, Allchorne A, Poole S. Peripheral cell types contributing to the hyperalgesic action of nerve growth factor in inflammation. *J Neurosci* 1996, **16**: 2716–2723.
302. Byers MR. Dynamic plasticity of dental sensory nerve structure and cytochemistry. *Arch Oral Biol* 1994, **39**: 13–21.
303. Woolf CJ, Safieh-Garabedian B, Ma QP, Crilly P, Winter J. Nerve growth factor contributes to the generation of inflammatory sensory hypersensitivity. *Neuroscience* 1994, **62**: 327–331.
304. Safieh-Garabedian B, Poole S, Allchorne A, Winter J, Woolf CJ. Contribution of interleukin-1 beta to the inflammation-induced increase in nerve growth factor levels and inflammatory hyperalgesia. *Br J Pharmacol* 1995, **115**: 1265–1275.
305. Harbaugh RE. Intracerebroventricular bethanechol chloride administration in Alzheimer's disease. *Ann NY Acad Sci* 1988, **531**: 174–179.

29
Early Experiences with Trophic Factors as Drugs for Neurological Disease:
Brain-Derived Neurotrophic Factor and Ciliary Neurotrophic Factor for ALS

Dana C. Hilt, James A. Miller, and Errol Malta

INTRODUCTION

Neurotrophic factors (NTFs) have been proposed as potential therapeutic agents for a number of neurodegenerative diseases such as Alzheimer's disease, Parkinson's disease, and amyotrophic lateral sclerosis (ALS) (see Chapter 28). This chapter will review the rationale and clinical trial results for brain derived neurotrophic factor (BDNF) treatment of ALS as well as briefly review the clinical experience with ciliary neurotrophic factor (CNTF).

As first shown in the classical studies of Hamburger and Levi-Montalcini (and reviewed in Chapter 28), trophic factors play an important role in determining neuronal survival in the vertebrate nervous system during development (1–6). These studies were exclusively related to the role of nerve growth factor (NGF), the prototypical trophic factor that exerts specific effects on sensory and sympathetic neurons.

An activity that supported survival of dorsal root ganglion (DRG) neurons in enriched cultures, but was biologically and biochemically distinct from NGF, was identified in conditioned media of a glioma cell line (7–9). Similar activity also was found in other tissues including chicken and rat brain extracts, primary cultures of adult rat brain astrocytes, homogenates of chicken spinal cord, and embryonic chicken heart, liver, and kidney (10,11). This activity was attributed to a small basic protein first purified from porcine brain (8) and later designated as brain-derived neurotrophic factor, or BDNF (12). In DRG survival bioassays, BDNF purified from porcine brain was shown to have activity comparable to NGF. Although the physicochemical and biological properties of BDNF indicated that this protein was related to NGF, the neuronal specificities of BDNF and NGF were clearly distinct (13). Neural placode-derived sensory ganglia respond to BDNF but not NGF; NGF, but not BDNF, supports survival of sympathetic neurons (10–14). Motor neurons respond to BDNF, while NGF has no trophic effects on these cells (17).

BDNF is one member of a broader group of neurotrophic factors known as the neurotrophins. The members of this family of neurotrophic factors include NGF, BDNF, neurotrophin-3 (NT-3), neurotrophin-4/5 (NT-4/5), and neurotrophin-6 (NT-6) *(15,16)*. Each neurotrophin has selective trophic effects on specific neurons, dependent upon selective expression of specific trk receptors by target neurons; trkB is the cognate trk receptor for BDNF *(15,18)* (*see* Chapter 28). Binding of BDNF to *trk*B induces receptor dimerization and receptor autophosphorylation followed by activation of an intracellular signal transducing cascade leading to a trophic response in the target neuron. Responding neurons also retrogradely transport the neurotrophic factor to the neuron soma. Although the functional significance of retrograde neurotrophic factor transport is unclear, it is used as a marker of responsive neurons *(18)*.

BDNF: EFFECTS ON DEVELOPING MOTOR NEURONS

The first indication that motor neurons have a trophic response to BDNF was the observation that BDNF increased the expression of choline acetyltransferase (ChAT) in motor neuron cultures from E14 rat ventral spinal cord *(17)*. ChAT expression is an indicator of motor neuron survival and function. In enriched cultures of E15 spinal cord motor neurons the EC_{50} of BDNF for motor neuron survival was in the picomolar range *(19)*. It also was demonstrated that BDNF mRNA is expressed in the developing spinal cord and limb bud. Following dennervation, expression of BDNF in the limb bud increased. Thus, BDNF could be serving as a target-derived trophic factor for developing motor neurons *(19,20)*.

BDNF also has effects on the developing neuromuscular junction. In vitro, BDNF (50 μg mL^{-1}) potentiates synaptogenesis in xenopus nerve-muscle cultures as measured by an increase on spontaneous synaptic currents by approximately 7 fold *(21)* The increase was blocked by an inhibitor of trkB signal transduction, thus demonstrating that it was a specific receptor-mediated effect. In other experiments, BDNF transiently stabilized synapses in the developing neuromuscular junction *(22)*.

The role of BDNF in the development of the nervous system has been studied by targeted disruption of the BDNF and *trk*B genes in transgenic mice *(23–25)*. The BDNF gene deletion mutation is lethal with gross neurologic abnormalities in the vestibular and sensory systems. *TrkB* gene-deletion mutants have an even more profound phenotype with decreased response to noxious stimuli and death within 24–48 h. Interestingly, in neither the BDNF nor *trk*B deletion was a significant loss of motoneurons documented. This observation suggests that there may be neurotrophic factor redundancy for developing motor neurons.

During development, naturally occurring cell death leads to approx 50% motor neuron cell death in the chick embryo spinal cord *(26)*. BDNF substantially reduces the degree of cell death. BDNF (5 μg *in ovo*) rescued approximately 33% of chick motor neurons that normally die between E6 and E10. BDNF applied either locally or systemically substantially reduces facial motor neuron cell death and maintains ChAT expression after facial axotomy in the neonatal rat *(20,28)*. These findings have been extended to neonatal spinal cord motor neurons in similar experiments in which control animals had a 40% motor neuron cell loss after sciatic nerve axotomy compared to an 8% cell loss in BDNF treated animals *(27)*. Thus, BDNF has been shown to have significant trophic effects on developing motor neurons even after axotomy injury.

After axotomy *trkB* expression in facial motor neurons increases many-fold as does BDNF mRNA expression in the lesioned facial nucleus. These findings suggest that BDNF may have an autocrine/paracrine local effect in the injured motor neuron *(29)*. BDNF synthesized by the motor neuron may be secreted, bind to trkB receptors and thus have an autocrine trophic effect on the motor neuron.

BDNF: EFFECTS ON MATURE MOTOR NEURONS

Adult motor neurons express *trk*B and retrogradely transport BDNF *(30)*. These findings suggest that adult motor neurons also can respond to BDNF. In adult animals, BDNF continues to be expressed by the muscle and, therefore, could function as a target-derived maintenance factor in the mature nervous system. Adult Schwann cells also express BDNF and increased expression after axotomy has been reported *(34)*. This increase in Schwann cell-derived BDNF may play a role in motor neuron and axon regeneration. In this instance, BDNF may have a paracrine effect.

Adult motor neurons do not die after experimental peripheral axotomy. However, ChAT expression decreases and low affinity NGF receptor ($p75^{NGFR}$) expression increases. After axotomy, both expression of *trk*B by adult motor neurons and retrograde transport of BDNF are increased *(29,30)*. When the motor axon reinnervates muscle, these changes revert; therefore, these parameters provide a measure of the trophic effect on motor neurons. Local, subcutaneous, and direct CSF administration of BDNF have demonstrated positive effects in adult motor neuron injury *(29,30)*. Local application of BDNF to the cut motor nerve at the time of axotomy prevents decrease in ChAT immunoreactivity and enzyme activity and potentiates the increase in $p75^{NGFR}$ expression. Systemic administration also was efficacious in this model, as both subcutaneously and intravenously administered BDNF have positive effects. Direct infusion of BDNF into the CSF was the most potent method of administration, being approximately two orders of magnitude more potent than systemic (sc or iv) BDNF *(30)*. Doses as low as 2–4 µg/d icv significantly increased ChAT immunoreactivity in axotomized facial motor neurons.

Proximal lumbar ventral root avulsion is a more severe and noxious injury than simple axotomy and leads to substantial motor neuron cell death. Ventral root avulsion also produces an increase in nitric oxide synthase (NOS) which may play a role in nitric oxide-mediated motor neuron death *(31)*. The loss of motor neurons was markedly decreased by BDNF (10 µg/d) administration into the CSF. Vehicle-treated animals had approx 40% of normal motor neurons after lumbar ventral root avulsion, while BDNF treated animals had approx 90% of normal levels. BDNF treatment also completely blocked NOS expression in lumbar motor neurons after ventral root avulsion. These experiments show the ability of BDNF to protect motor neurons after a severe insult and suggest a possible mechanism for this effect.

The effects of BDNF in spontaneous ALS animal models have also been examined. The Wobbler mouse is an autosomal recessive disease characterized by motor neuron degeneration and muscle wasting. This leads to a progressive paralysis, paw position abnormalities, gait abnormalities, and death. BDNF (5 mg/kg/d, three times a wk) slowed the progression of degeneration in the Wobbler mouse *(32)*. BDNF-treated animals showed a decreased rate of progression to paw position abnormalities, walking pattern abnormalities, and grip strength degeneration. Biceps muscle weight and muscle twitch tension were greater in the BDNF treated mice compared to controls. Histological analysis showed that biceps muscle fiber diameter was approximately one-third greater in BDNF treated animals and the number and size of myelinated nerve fibers in the C5 and C6 ventral roots were greater. These results demonstrated that BDNF ameliorates the progressive loss of function and muscle and nerve atrophy in the Wobbler mouse.

A minority (less than 10%) of ALS patients have a familial autosomal dominant variant. Of these familial cases, approximately 20% have a defect in the gene encoding Cu^{++}/Zn^{++} superoxide dismutase (SOD). Transgenic (tg) mice expressing mutant forms of SOD develop a progressive neurologic degeneration characterized by muscle weakness, gait abnormalities, and paralysis leading to death (*see* Chapter 20). There is a decrease in motor neuron number and muscle atrophy evident on histologic examination. BDNF treatment of at least one SOD tg mouse line (G93A mutation) did not attenuate the degeneration of function nor extend the lifespan of these animals *(33)*. The lack of BDNF efficacy in the SOD tg mouse may indicate that the SOD mutation produces a defect in a cellular pathway resistant to BDNF.

BDNF, therefore, has clear trophic effects on motor neurons both in vitro and in vivo. In a number of motor neuron injury paradigms in neonatal and mature animals BDNF has shown activity in blocking or ameliorating motor neuron cell death and dysfunction. (For a more reserved position on the role of BDNF for adult motor neurons, see Chapter 28). Some of these observations supported the hypothesis that BDNF would have a therapeutic effect in patients with ALS.

BDNF PHASE I/II STUDY IN ALS

ALS is the most common degenerative motor neuron disease, with a prevalence of 2–7 per 100,000 population *(33,34)*. It is a progressive and invariably lethal disease with the majority of patients succumbing to the disease in 3–5 yr usually as the result of respiratory failure. Disease progression can be monitored by a number of quantitative and semiquantitative measures including muscle strength (isometric as well as manual testing), forced vital capacity (FVC) and testing of specific functional abilities such as walking speed, speech fluency, and peg board manipulations. Global functional rating scales, combining various measures of functional ability and daily activities, such as the ALS functional rating scale (ALSFRS) *(37)* and the Appel scale *(38)* also have been utilized.

ALS however, presents significant difficulties in designing and completing a study to test the effects of a drug. The first of these is a variable and sometimes slow progression of the disease. The lack of any known surrogate markers for the disease requires that trials monitor the actual progression of symptoms and signs. Demonstrating a benefit requires that either the drug have a large effect on disease progression or that the observation period and/or number of subjects be large in order to observe anything less than a dramatic effect. Another difficulty in performing trials in ALS is the interpatient variability in the course of the disease. The rate of deterioration and survival time for a given patient is highly variable, though parameters such as age and site of onset do have limited predictive value. Attempts have been made to find a marker of disease progression which is at least approximately linear with time, as this would make assessment of drug effects over a shorter period of time feasible. Isometric muscle dynamometry is one popular assessment method which, in large natural history groups, may provide such a measure *(39)*. However, in any given patient it may not be apparent that the deterioration in isometric strength is linear nor does the decline in isometric strength correlate with survival. The method also is quite time consuming and tiring for the patients, making its use in large studies problematic.

Most patients with ALS eventually succumb to respiratory failure, hence assessments of breathing are the best correlate with survival. Forced vital capacity (FVC) is a safe, simple, and rapid test which is performed in most clinics on a routine basis. It appears to correlate with survival, though its decline through the course of the disease, particularly in the late stages of the process, is not necessarily linear *(40)*.

BDNF PHASE I AND II TRIAL DESIGN

The studies described below assessed the safety and biological effects of subcutaneously administered BDNF in patients with ALS. They were performed at the 21 centers comprising the BDNF Study Group. Patients with diagnoses of definite or probable ALS, defined by the criteria established by the World Federation of Neurology *(41)*, were accepted into the study. Inclusion criteria also specified an FVC greater than 60% of the predicted value, age between 21 and 75 years, and an ALSFRS score greater than 18. Additional criteria excluded patients with existing medical conditions which could complicate assessment of safety.

The initial clinical program was divided into two stages: a Phase I study and a larger Phase II study. The Phase I study was a placebo-controlled, double blinded, and randomized study primarily to ascertain the safety and tolerability of daily subcutaneously administered BDNF in ALS patients. The study used six cohorts with sequentially escalating doses of 10, 25, 50, 100, 150, and 300 µg/kg/d administered as a daily subcutaneous injection. There was no intrapatient dose escalation. Intensive safety assessments were performed over the first 28 d, and patients were allowed to continue on the blinded study for up to six mo for assessment of efficacy, during which period sc drug was administered daily. There were 10 patients per dose level, two of whom were given placebo. The 300 µg/kg/d cohort was repeated, resulting in a total of 69 patients enrolled into Phase I. At the conclusion of the blinded phase, all patients were given the option of receiving open label BDNF at the dose of their assigned cohort. At the time the original draft of this chapter was prepared, many of these patients continued on drug and some had been on the drug for over two years after the conclusion of the blinded study. The open label treatment period provides long-term safety data.

BDNF was safe and well tolerated at the doses used in the Phase I study. Hence, a Phase II study was opened using doses of 10, 25, 50, 100, and 150 µg/kg, by daily sc injection. Even though the 300 µg/kg/day dose was well tolerated in Phase I, it was judged to be impractical, as it required two injections per dose in most patients. Phase II also was a 6-mo randomized, placebo controlled, double-blinded parallel dose study which enrolled 214 patients in a 4:1 drug: placebo ratio. Upon completion of the blinded phase of the study, these patients also were given the option of receiving open label BDNF treatment at the dose which their assigned cohort had been receiving.

In both studies effects of the drug on disease progression were assessed by the following:

1. Isometric muscle dynamometry, measuring three upper extremity and three lower extremity muscle groups;
2. Respiratory function, measured by %FVC (% predicted for age, sex, and height) and peak inspiratory flow rate;
3. Timed functional activities, including syllable repetition, a 15 foot (4.6m) timed walk, and peg placement in a Purdue Pegboard;
4. Activities of daily living assessed using the ALS Functional Rating Scale (ALSFRS); and
5. The Ashworth Spasticity Scale and, in Phase I, the Schwab and England Scale.

Throughout the study, safety was assessed by monitoring adverse events, clinical and blood chemistry parameters and by physical examinations.

RESULTS OF THE STUDIES

All safety and efficacy analyses presented here are on data from the combined studies, though when each phase is analyzed as an individual study the results are consistent.

There were 283 patients enrolled in the combined Phase I and Phase II studies. Of these patients, 59 were on placebo and 224 received BDNF. Of the BDNF-treated patients, 24 received 10 µg/kg/d, 45 received 25 µg/kg/d, the 50, 100, and 150 µg/kg/d cohorts each contained 46 patients and 17 patients received a dose of 300 µg/kg/d. Seventy-eight percent of placebo and 80% of BDNF patients completed the 6-mo of study treatment. Of those who failed to complete, approx 10% from each of placebo and drug groups withdrew for various reasons. In all, 11.8% of placebo patients died on the 6-mo study, as opposed to 8.5% in the treated groups. 227 patients completed the study and of these 44 placebo patients and 178 treated patients chose to continue into the open label study

Efficacy analyses included data on all patients with measurements at baseline and at least one additional timepoint, except for survival which included data from all patients who were randomized. Missing data points were estimated using the last observation carried forward method, where necessary.

Statistical analyses performed after the study revealed that the treatment groups and placebo group were balanced in terms of age, body mass, time from diagnosis, time from symptom onset and anatomical sites involved at enrollment. FVC % also was balanced in all groups. The male:female ratio was approx 2:1, consistent with the general ALS population, except for the 300 µg/kg/d cohort, which had a majority of female patients.

SAFETY AND ADVERSE EVENTS

BDNF was safe and well tolerated at the doses tested. The most commonly reported adverse events were injection site reactions, the frequency and severity of which increased in a dose dependent manner. Fifty-seven percent of all BDNF treated patients and 10% of the placebo patients experienced these reactions. The reactions were generally rated as mild and occasionally moderate in severity and typically presented as erythema, with pruritus and swelling. In most patients the injection site reactions were transitory, often disappearing after approximately two months despite continued BDNF treatment.

Changes in bowel activity characterized as increased bowel urgency and/or diarrhea were reported in 35% of the BDNF treated patients and 24% of the placebo treated patients. Conversely, reports of constipation were less frequent in the BDNF group than in the placebo group (16% and 33%, respectively). Clinical chemistry findings were unremarkable, with a notable absence of any elevation in liver enzyme levels.

During the blinded study and open label period, five patients reported disparate symptoms which included flushing, diaphoresis, headache, nausea, chills, throat tightness, and tingling sensations. In no instance was symptomatic hypotension or bronchospasm observed. These symptoms typically had a rapid onset after drug injection and rapidly resolved without medical intervention. Each of these five patients experienced a mixture of these symptoms and occurred after at least one month of dosing. Four of the patients subsequently discontinued drug therapy and the fifth continued dosing without further incident.

EFFECTS OF BDNF ON MEASURES OF DISEASE PROGRESSION

BDNF treatment had an effect on the %FVC decline over time. Fig. 1 shows the plot of the change (from baseline) of %FVC versus time over the 6-mo blinded therapy period. After 3 mo, many of the treated cohorts began to diverge from the placebo group. By 6 mo, the 25, 50, 100, and 150 µg/kg/d doses had substantially better %FVC than the placebo group, but the 10 and 300 µg/kg/d groups exhibited no such treatment effect. The placebo, 10 and 300 µg/kg/d groups declined approximately 20% in %FVC, while those in the 25 through 150 µg/kg/d declined from 11.5–15.5%. When analyzed by dose cohort, the benefit in %FVC in the 25 µg/kg/d dose group was significantly different from placebo ($p < 0.05$; Dunnett's t-test).

As a group, the BDNF treated patients also showed a benefit relative to placebo (Fig. 1B). The combined BDNF groups diverged from placebo, beginning at mo 3 and continuing to do so, achieving statistical significance ($p < 0.05$) at mo 4–6. This comparison, which included the apparently ineffective 10 and 300 µg/kg/d doses in the treated group, found that after six months of treatment the BDNF cohorts had lost 13% in %FVC, whereas the placebo group declined 20%.

Substantially more BDNF treated patients (10 through 300 µg/kg/d cohorts) had %FVC values after 6 mo of therapy that were better than or equal to their baseline %FVC values. At 6 mo, 18.1% of BDNF patients had stable or improved %FVC as opposed to 5.2% of the placebo patients ($p < 0.015$, chi-square).

SURVIVAL

Since respiratory failure is the most common cause of death in ALS patients, a treatment effect which slowed the deterioration in respiratory function would be expected to result in longer patient survival times. In a short-duration (6-mo) ALS study as reported here, it is difficult to demonstrate a survival effect, as few patients meeting the entry criteria (%FVC $\geq$ 60%) would be expected to succumb to the disease during the study period. However, the BDNF treated patients have a trend toward better survival at the 6-mo point of the study (Fig. 2). As the survival curves were diverging after 6 mo, we extended the survival analysis to include data from the first 3 mo of open label treatment, in addition to the 6 mo of blinded dosing. Our data indicate that it takes approx 3 mo for the BDNF to affect %FVC; thus, patients who crossed over onto active drug from the placebo arm were unlikely to display immediate survival benefit from the therapy. Accordingly, the curve from the former placebo group should provide a reasonable and conservative estimate of the curve of an untreated group.

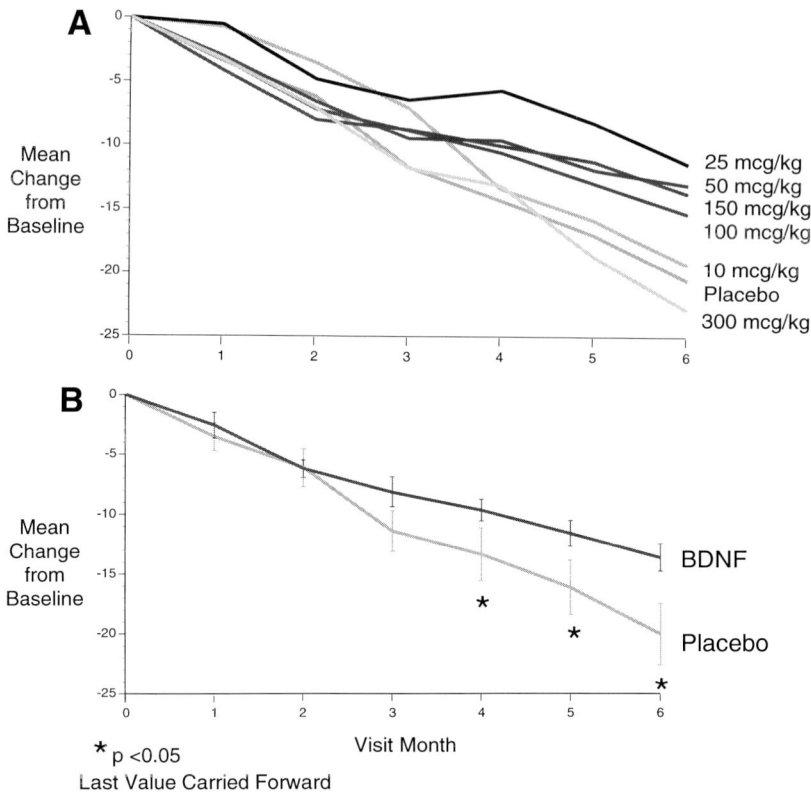

Fig. 1. (A) Mean change from baseline FVC% by dose. **(B)** Changes from baseline %FVC, all doses combined.

With this extended analysis, there was a strong trend ($p = 0.0834$) toward improved survival in the combined BDNF group using the Cox proportional hazards model adjusted for the baseline FVC% (Fig. 2). The probability of survival at 9 mo was 0.73 in the placebo group and 0.82 in the BDNF group. As in the case of the %FVC, the 10 and 300 μg/kg/d dose groups performed no better than placebo, while the 25, 50, 100, and 150 μg/kg/d doses all displayed approximately the same trend toward longer survival (data not shown). No individual dose group was statistically significantly different from placebo.

ADJUSTMENTS FOR PROGNOSTIC VARIABLES

The possibility existed that the effects on %FVC and survival could, at least in part, be explained by mismatches at the beginning of the study in patient characteristics known to affect the progression of the disease. We examined the effects of some of these factors in the analyses. Our observed treatment effects of BDNF could not be explained by differences in any of the following covariates: %FVC, weight, or age at baseline, enrollment in Phase I versus Phase II, time from onset of initial symptoms, bulbar versus limb symptoms at disease onset, and bulbar and limb versus limb-only involvement at enrollment.

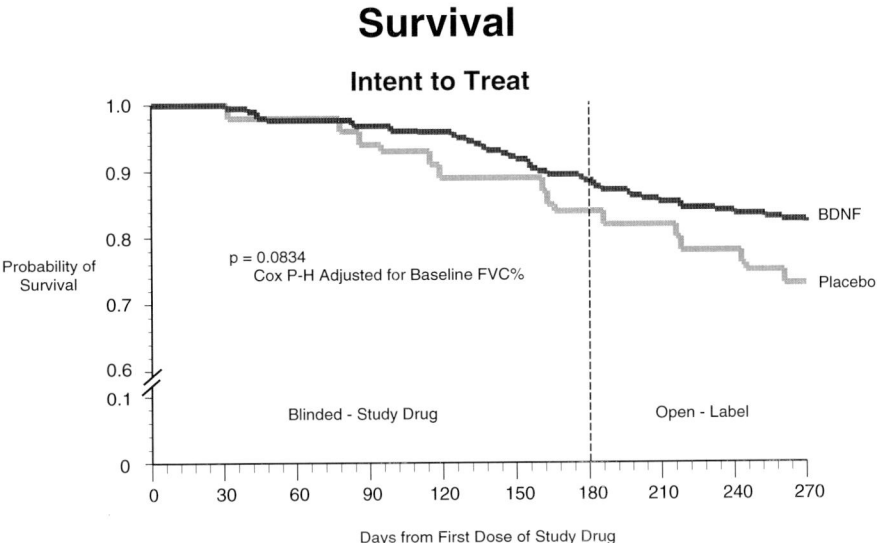

Fig. 2. Kaplan–Meier survival analysis of all BDNF dose groups combined versus placebo. Difference of all BDNF dose groups versus placebo at 270 d (9 mo) using Cox P-H analysis adjusted for baseline %FVC ($p = 0.0834$).

OTHER MEASURES OF DISEASE PROGRESSION

We measured the time to walk 15 ft (4.6 m) by the patients at monthly intervals. While walking speed declined progressively in both the BDNF and the placebo groups, there was a slower decline in the BDNF groups. When adjusted for center effects, the change from baseline to mo 6 was statistically significant for the combined BDNF group compared to the placebo (-0.96 ± 0.09 f/s vs -1.34 ± 0.16 f/s; $p < 0.05$). While the individual dose cohorts generally lost less walking speed than placebo treated patients, none of the individual doses achieved statistical significance.

There were no statistically significant treatment effects in isometric muscle strength, body mass index, peak inspiratory flow, syllable repetition, Purdue pegboard, Ashworth spasticity scale, ALSFRS, and Schwab and England score measurements. Some of these, however, displayed trends toward a favorable effect of BDNF relative to placebo but none reached statistical significance. None of these measures suggested a deleterious effect of BDNF treatment.

INTERPRETATION OF THE BDNF STUDIES

Our studies demonstrated that BDNF is a safe and well-tolerated drug when administered sc within the range of doses tested. Injection site reactions were the most common adverse event, becoming more frequent at the highest doses. These visible reactions raise the concern that they may have unblinded the study, however, we do not believe the observed effects on %FVC and survival can be attributed to a hypothetical unblinding. Ninety percent of placebo patients had no injection site reactions and so did 60% of the 25μg/kg/d patients, yet this latter group displayed a significant %FVC effect. The cohort where unblinding was most likely was the 300 μg/kg/d group, where injection-site reactions were experienced by all treated patients, yet this group failed to show any treatment effect. In addition, the reactions were transitory and in most patients they had ceased by month two, yet the %FVC effects we observed became apparent only at mo 3 and beyond. All dose groups showed a similar decline in FVC during the first 2–3 mo, a time when injection-site reactions peaked. Hence, it is difficult to ascribe the observed positive effects to bias due to unblinding.

The other common side effect was diarrhea and/or bowel urgency. These symptoms were generally mild to moderate and diminished with time. BDNF is not known to act upon autonomic neurons, but recent studies do point to a role in neurons innervating the lower digestive tract *(42,43)*.

BDNF treatment significantly decreased the decline in %FVC in the 6-mo observation period. It also produced a trend toward longer survival. In both these measures, the 25 µg/kg/d dose was the most effective, though it was not substantially better than the 50–150 µg/kg/d doses. The 10 and 300 µg/kg/d doses were apparently ineffective. While the lack of an effect at the lowest dose is not surprising, the loss of effectiveness at the 300 µg/kg/d dose is unexpected. There were only 17 patients in this dose level, hence, the variance is relatively large and could be obscuring an effect. Alternatively, high doses could result in downregulation of the *trkB* receptor for BDNF, as observed in the laboratory *(44)*. It is also possible that the complexity of the final response measured obscured a linear effect of the drug on the motor neurons themselves. Another complication is that the action of the BDNF on motor neuron cell bodies probably requires retrograde transport of the molecule to the neuronal soma once it binds at a receptor on the nerve terminal, which may produce a threshold response instead of a linear one.

We observed a significant effect of BDNF on loss of walking ability, but failed to detect one on a quantitative measure of muscle strength. This is not necessarily a paradox, as isotonic strength, as measured by walking, need not be correlated with isometric strength and it is possible that BDNF could affect one differentially. It also is possible that the entry criteria, which specified a relatively high value for %FVC but no *a priori* specification for isometric muscle strength, allowed patients to enter the study with isometric strength too low to be affected or improved by BDNF.

These studies demonstrated that sc administered BDNF is safe and well tolerated. The positive effects of BDNF we observed on respiratory function, ambulation and survival were promising. In order to confirm these findings, at the time the first draft of this chapter was prepared, a larger Phase III study was underway to prospectively examine these endpoints.

ALS TRIALS WITH CILIARY NEUROTROPHIC FACTOR (CNTF)

At least two other recombinant proteins, IGF-1 and ciliary neurotrophic factor (CNTF), have been tested in ALS trials. Results of the IGF-1 trials have been summarized in Chapter 28. Here, we briefly review the results of CNTF studies.

Two major clinical trials have been conducted in the US with sc CNTF administered to patients with ALS *(45,46)*. Although these two trials differed with respect to the dosing frequency employed, both used the highest tolerated doses determined from preliminary safety studies. In one study (ALS CNTF Treatment Study Group, 1996) the patients were dosed three times weekly with 15 µg/kg or 30 µg/kg, the latter being the maximum tolerated dose in preliminary studies. The other study *(47)* employed daily dosing with doses up to 5 µg/kg.

The ALS CNTF Treatment Study Group enrolled 730 patients into three arms (placebo, 15 and 30 µg/kg given three times weekly) in a randomized, double-blind trial with a 9-mo dosing period. The primary end-point was the slope of the loss of muscle strength determined from a combined arm and leg megascore measured by isometric muscle dynamometry. In addition, a number of secondary end-points also were defined (see ALS CNTF Treatment Study Group, 1996 for details). At the end of the study, there was no significant difference in the slopes of muscle strength loss or in the decline in pulmonary function among the three treatment arms. Patients in the CNTF cohorts lost more strength than placebo patients in the first 2 mo but thereafter showed a slower rate of decline than placebo patients. There was no difference in survival among the three arms. Adverse effects noted were cough, asthenia, anorexia, weight loss, aphthous stomatitis, injection-site reactions and fever—all most prominent in the first few months. The majority of CNTF treated patients developed antibodies which could account for the diminution of adverse effects over time.

A second CNTF study enrolled a total of 570 ALS patients into four arms (placebo, 0.5, 2, and 5 µg/kg/d) in a randomized, double blind trial of 6 mo duration *(46)*. The primary end-point was the change from baseline in a combined score derived from isometric muscle dynamometry of the arm and leg together with FVC. A number of secondary variables also were defined for the study. There was no significant difference in the change from baseline between the four treatment arms. Adverse effects in the high dose group (5 µg/kg/d) were generally similar to those reported in the ALS CNTF Treatment Study Group (1996) but doses of 2 µg/kg/d and lower were well tolerated. At the end of 6 mo treatment there was no statistically significant difference in survival between the groups but at follow-up 30 d after completing the study there was a significant difference with higher mortality in the high dose CNTF group. This increased mortality may reflect a cumulative toxic effect which was manifest during drug withdrawal or a withdrawal effect *per se.*

Thus, both studies clearly demonstrated that sc CNTF failed to show any evidence of efficacy at doses up to the highest tolerated doses in patients with ALS. Although CNTF clearly has neurotrophic effects on motor neurons and activity in preclinical ALS animal models, doses sufficient to exert these effects in patients may not be achievable due to systemic toxicity. Local, instead of systemic, delivery could possibly alleviate this problem.

Aebischer and recently reported preliminary studies with an ex vivo gene therapy approach for the use of CNTF in ALS *(47)*. In this study, six patients were implanted with polymer capsules containing baby hamster kidney cells synthesizing approximately 0.5 µg CNTF per day in the lumbar intrathecal space. While this study demonstrated the feasability of this delivery system, data from IT studies in animals defining efficacious concentrations of CNTF in CSF have not been reported. This CNTF trial in ALS patient was ongoing at the time the initial draft of this chapter was prepared.

NOTE ADDED IN PROOF

The BDNF Phase I and II results were followed by a large multicenter, placebo-controlled Phase III trial (greater than 1100 patients) of subcutaneous BDNF (cohorts of 25 and 150 µg/kg/d) versus placebo. The inclusion/exclusion criteria were similar to previous BDNF trials. The primary efficacy endpoints for the trial were survival (at 9 mo) and percent predicted FVC decline (at 6 mo).

The safety profile of subcutaneous BDNF in the Phase III trial was similar to that previously observed. However, the Phase III BDNF trial failed to document any slowing of disease progression. Percent predicted FVC and survival were not significantly different in the placebo versus the BDNF-treated groups even when corrected for baseline FVC.

With the benefit of retrospect how can the failed BDNF Phase III trial be explained? Possible reasons include the likely fact that the earlier trial results incorrectly indicated a therapeutic effect of BDNF given by the subcutaneous route. In the Phase I and II trials there was no clear BDNF dose response relationship, the highest BDNF dose (300 µg/kg/d) was not effective while lower doses suggested efficacy, and BDNF doses from 25 to 150 µg/kg/d were roughly equivalent in their effects. There were no positive effects on muscle strength measurements in the BDNF Phase I and II trials, indicating either that BDNF had selective effects on respiration and survival or these beneficial effects were due to chance. These observations suggest that the positive benefits seen on FVC and survival in the Phase I/II trials were chance events.

In addition, in the Phase III BDNF trial the placebo survival at 9 mo was higher than the placebo rate observed in previous trials thus the trial was either not adequately powered to demonstrate a survival benefit at 9 mo (should such a benefit actually exist) or would have had to be carried out for a longer period of time in order to detect a possible benefit. The different placebo cohort survival rates in the BDNF Phase I/II versus Phase III trials also strongly suggest that ALS trials must be performed with concurrent placebo control and that historical control is inadequate.

Does the failed subcutaneous BDNF Phase III trial mean that BDNF can no longer be considered a possible therapeutic candidate for ALS? The proposed mechanism by which BDNF gains access to the motor neuron cell soma is by retrograde transport after binding to trkB receptors at the neuromuscular junction. This mechanism is both inefficient and likely defective in patients with ALS. Is it possible to administer BDNF in a manner to attain higher BDNF concentrations at the level of the motor neuron? A recently completed Phase I trial of intrathecal BDNF administered by constant intrathecal infusion *(48)* has shown that significant levels of BDNF (e.g., 100–400 ng/mL) can be achieved in CSF by this method. By comparison, BDNF has in vitro trophic effects on motor neurons in the dose range of 1 to 10 ng/mL. In addition, patients reported sensory symptoms (a sense of warmth) and, at higher doses, agitation and insomnia. These results suggest that intrathecal BDNF administration can both achieve potentially physiologic levels of the growth factor in the CNS/CSF and that BDNF is inducing symptoms referable to populations of central nervous system neurons. Additional trials of BDNF given by the intrathecal route are being planned. A conclusion on the possible utility of BDNF as a therapeutic agent in ALS must be deferred until these trials have been completed.

REFERENCES

1. Hamburger V, Oppenheim RW. Naturally-occurring neuronal death in vertebrates. *Neurosci Comment* 1982, **1:** 38-55.
2. Oppenheim RW. Cell death during development of the nervous system. *Ann Rev Neurosci* 1991, **14:** 453–501.
3. Hamburger V, Levi-Montalcini R. Proliferation, differentiation and degeneration in the spinal ganglia of the chick embryo under normal and experimental conditions. *J Exp Zool* 1949, **111:** 457–501.

4. Hamburger V. Regression versus peripheral control of differentiation in motor hypoplasia. *Am J Anat* 1958, **102**: 365–409.
5. Hollyday M, Hamburger V. Reduction of the naturally occurring motoneuron loss by enlargement of the periphery. *J Comp Neurol* 1976, **170**: 311–320.
6. Levi-Montalcini R, Angeletti PU. Nerve growth factor. *Physiol Rev* 1968, **48**: 534–569.
7. Monard D, Solomon F, Rentsch M, Gysin R. Glia-induced morphological differentiation in neuroblastoma cells. *Proc Natl Acad Sci USA* 1973, **70**: 1894–1897.
8. Monard D, Stockel K, Goodman R, Thoenen H. Distinction between nerve growth factor and glial factor. *Nature* 1975, **258**: 444–445.
9. Barde YA, Lindsay RM, Monard D, Thoenen H. New factor released by cultured glioma cells supporting survival and growth of sensory neurones. *Nature* 1978, **274**: 818.
10. Lindsay RM. Adult rat brain astrocytes support survival of both NGF–dependent and NGF-insensitive neurones. *Nature* 1979, **282**: 80–82.
11. Lindsay RM, Tarbit J. Developmentally regulated induction of neurite outgrowth from immature chick sensory neurons (DRG) by homogenates of avian or mammalian heart, liver and brain. *Neurosci Lett* 1979, 12: 195–200.
12. Barde YA, Edgar D, Thoenen H. Purification of a new neurotrophic factor from mammalian brain. *EMBO J* 1982, **1**: 549–553.
13. Leibrock J, Lottspeich F, Hohn A, et al. Molecular cloning and expression of brain-derived neurotrophic factor. *Nature* 1989, **341**: 149–152.
14. Lindsay RM, Thoenen H, Barde YA. Placode and neural crest-derived sensory neurons are responsive at early developmental stages to brain-derived neurotrophic factor. *Dev Biol* 1985, **112**: 319–328.
15. Snider WD. Functions of the neurotrophins during nervous system development: What the knockouts are teaching us. *Cell* 1994, **77**: 627–638.
16. Goetz R, Koester R, Winkler C, Raulf F, Lottspeich F, Schartl M, Thoenen H. Neurotrophin-6 is a new member of the nerve growth factor family. *Nature* 1994, **372**: 266–269.
17. Wong V, Arriaga R, Ip NY, Lindsay RM. The neurotrophins BDNF, NT-3, NT-4/5 but not NGF, up-regulate the cholinergic phenotype of developing motor neurons. *Eur J Neurosci* 1993, **5**: 466–474.
18. DiStefano PS, Friedman B, Radziejewski C, et al. The neurotrophins BDNF, NT–3, and NGF display distinct patterns of retrograde axonal transport in peripheral and central neurons. *Neuron* 1992, **8**: 983–993.
19. Henderson CE, Camu W, Mettling C, Gouin A, Poulsen K, Karihaloo M, Rullamas J, Evans T, McMahon SB, Armanini P, Berkemeler L, Phillips HS, Rosenthal A. Neurotrophins promote motor neuron survival and are present in embryonic limb bud. *Nature* 1993, **363**: 266–270.
20. Koliatsos VE, Clatterbuck RE, Winslow JW, Cayouette MH, Price DL. Evidence that brain–derived neurotrophic factor is a trophic factor for motor neurons in vivo. *Neuron* 1993, **10**: 359–367.
21. Lohof AM, Ip NY, Poo M. Potentiation of developing neuromuscular synapses by the neurotrophins NT–3 and BDNF. *Nature* 1993, **363**: 350–353.
22. Kwon YW, Gurney ME. Brain-derived neurotrophic factor transiently stabilizes silent synapses on developing neuromuscular-junctions. *J Neurobiol* 1996, **29**: 503–516.
23. Jones KR, Farinas I, Backus C, Reichardt LF. Targeted disruption of the BDNF gene perturbs brain and sensory neuron development but not motor neuron development. *Cell.* 1994, **76**: 989–999.
24. Ernfors P, Lee K-F, Jaenisch R. Mice lacking brain-derived neurotrophic factor develop with sensory deficits. *Nature* 1994, **368**: 147–150.
25. Klein R, Smeyne RJ, Wurst W, et al. Targeted disruption of the trkB neurotrophin receptor gene results in nervous system lesions and neonatal death. *Cell* 1993, **75**: 113–122.

26. Oppenheim RW, Qin-Wei Y, Prevette D, Yan Q. Brain-derived neurotrophic factor rescues developing avian motor neurons from cell death. *Nature* 1992, **360**: 755–757.
27. Yan Q, Elliott JL, Matheson C, et al. Influences of neurotrophins on mammalian motoneurons in vivo. *J Neurobiol* 1993, **24**(12): 1555–1577.
28. Sendtner M, Holtmann B, Kolbeck R, Thoenen H, Barde YA. Brain-derived neurotrophic factor prevents the death of motoneurons in newborn rats after nerve section. *Nature* 1992, **360**: 757.
29. Kobayashi NR, Bedard AM, Hincke MT, Tetzlaff W. Increased expression of BDNF and *trk*B messenger-RNA in rat facial motoneurons after axotomy. *Euro J Neurosci* 1996, **8**: 1018–1029.
30. Yan Q, Matheson C, Lopez OT, Miller JA. The biological responses of axotomized adult motoneurons to brain-derived neurotrophic factor. *J Neurosci* 1994, **14**: 5281–5291.
31. Novidov L, Novikova L, Kellerth JO. Brain-derived neurotrophic factor promotes survival and blocks nitric oxide synthase expression in adult rat spinal motoneurons after ventral root avulsion. *Neuroscience Lett* 1995, **200**: 45–48.
32. Ikeda K, Klinkosz B, Greene T, et al. Effects of brain-derived neurotrophic factor (BDNF) on motor dysfunction in wobbler mouse motor neuron disease. *Ann Neurol* 1995, **37**: 505–511.
33. Yan, Q. Unpublished observations.
34. Munsat TL, Bradley WG. Amyotrophic lateral sclerosis. *Curr Neurol* 1979, **2**: 79–103.
35. Kurtzke JF. Epidemiology of amyotrophic lateral sclerosis human, in *Human Motor Neuron Diseases* (Rowland LP, ed), Raven Press, NY, 1982, pp. 281–302.
36. Norris F, Shepherd R, Denys E, et al. Onset, natural history and outcome in idiopathic adult motor neuron disease. *J Neurol Sci* 1993, **118**: 48–55.
37. ACTS Phase I-II Study Group. The amyotrophic lateral sclerosis functional rating scale assessment of activities of daily living in patients with amyotrophic lateral sclerosis. *Arch Neurol* 1996, **53**: 141–147.
38. Appel V, Stewart SS, Smith G, Appel SH. A rating scale for amyotrophic lateral sclerosis: description and preliminary experience. *Ann Neurol* 1987, **22**: 328–333.
39. Andres PL, Finison LJ, Conlon T, Thibodeau LM, Munsat TL. Use of composite scores (megascores) to measure deficit in amyotrophic lateral sclerosis. *Neurology* 1988, **38**: 405–408.
40. Fallat RJ, Jewitt B, Bass M, et al. Spirometry in amyotrophic lateral sclerosis. *Arch Neurol* 1979, **36**: 74–8017.
41. Subcommittee on Motor Neuron Diseases/Amyotrophic Lateral Sclerosis of the World Federation of Neurology Research Group on Neuromuscular Diseases and the El Escorial "Clinical Limits of Amyotrophic Lateral Sclerosis" Workshop Contributors. El Escorial World Federation of Neurology criteria for the diagnosis of amyotrophic lateral sclerosis. *J Neurol Sci* 1994, **124**(Suppl): 96–107.
42. Erickson J, Conover J, Borday V, Champagnat J, Katz D. Mice lacking brain-derived neurotrophic factor exhibit visceral sensory neruon losses distinct from mice lacking NT-4 and display a severe developmental deficit in control of breathing. *J Neurosci* 1996, **16**: 5361–5371.
43. Hoehner J, Wester T, Pahlman S, Olsen L. Localization of neurotrophins and their high affinity receptors during human enteric nervous system development. *Gastroenterol* 1996, **110**: 756–757.
44. Frank L, Ventimiglia R, Anderson K, Lindsay R, Rudge R. BDNF down-regulates neurotrophin responsiveness, takes protein and trkB mRNA levels in cultured rat hippocampal neurons. *Eur J Neurosci* 1996, **8**: 1220–1230.
45. ALS CNTF Treatment Study Group. A double-blind placebo-controlled clinical trial of subcutaneous recombinant human ciliary neurotrophic factor (rHCNTF) in amyotrophic lateral sclerosis. *Neurology* 1996, **46**: 1244–1249.

46. Miller RG, Petajan JH, Bryan WW, Armon C, Barohn RJ, Goodpasture JC, Hoagland RJ, Parry GJ, Ross MA, Stromatt SC, rhCNTF ALS Study Group. A Placebo-controlled Trial of Recombinant Human Ciliary Neurotrophic (rhCNTF) Factor in Amyotrophic Lateral Sclerosis. *Ann Neurol* 1996, **39**: 256–260.
47. Aebischer P, Schluep M, Deglon N, Joseph JM, Hirt L, Heyd B, Goddard M, Hammang JP, Zurn AD, Kato AC, Regli F, Baetge EE. Intrathecal delivery of CNTF using encapsulated genetically modified xenogeneic cells in amyotrophic lateral sclerosis patients. *Nature Medicine* 1996, **2**: 696–699.
48. Ochs G, Penn R, Giess R, Schrank B, Sendtner M. Toyka K. A trial of recombinant BDNF given by the intrathecal route to patients with ALS. *Neurology* 1998, **50**: A246.

30
Approaches in Treating Nerve Cell Death with Calcium Chelators

Michael Tymianski

INTRODUCTION

Calcium ions (Ca^{2+}) are ubiquitous intracellular second messengers which regulate numerous cellular functions. For this reason, neurons possess a complex homeostatic machinery which tightly controls the temporal and spatial distribution of Ca^{2+} within the cell. It is widely believed that a disruption of Ca^{2+} homeostasis is a major contributing factor to neuronal cell death occurring in many of the nervous system disorders including cerebral hypoxia/ischemia, brain trauma, inflammatory, and degenerative diseases (also *see* chapters by Leist and Nicotera, Dykens, Olney and Ishimaru, Dietrich, Wood, Vornov, Shin and Lee, and Bar-Peled and Rothstein). Therefore, many therapeutic strategies for these disorders aim at preventing either the processes leading to Ca^{2+} homeostatic failure, or the consequences of Ca^{2+} excess. Recent insights into mechanisms triggering and perpetuating disturbances of cellular calcium regulation have led to new approaches to the study of cellular calcium homeostasis, and to treatment of neuronal injury. One such approach, which illustrates the complexities associated with Ca^{2+} regulation, has been the use of exogenous Ca^{2+}-chelating agents. These compounds have been employed in recent years to study normal and pathological cellular Ca^{2+} signaling, to evaluate the role of Ca^{2+} buffering in neuronal vulnerability to injury, and to treat neuronal injuries thought to be associated with Ca^{2+} homeostatic failure. The purpose of this chapter, therefore, is to examine current knowledge on the role of Ca^{2+} buffers as probes of mechanisms leading to nerve cell death, and as potential therapeutic agents for Ca^{2+}-dependent neurotoxicity.

CALCIUM LOADING AS A MECHANISM OF NERVE CELL INJURY

The notion that Ca^{2+} excess is deleterious to cells has been popular for at least two decades. Although recent years have seen significant refinements in theories relating calcium loading with neuronal cell death, a number of early fundamental observations leading to the "calcium hypothesis" must be reviewed (also *see* chapters by Leist and Nicotera).

The observation that disturbances in cellular Ca^{2+} homeostasis may lead to subsequent destruction of the cell was initially made in nonneuronal tissues, i.e., when pathologists noted that calcium was deposited in areas of tissue necrosis. McLean et al.

(1) observed that livers that had been damaged by toxins accumulated calcium, and suggested that calcium entry may be responsible for tissue damage. Subsequently, Schanne et al. *(2)* revealed that adult hepatocytes in primary cultures were killed when exposed to various toxins (believed to affect plasma membrane integrity) in the presence, but not the absence, of extracellular Ca^{2+}. These authors inferred that Ca^{2+} influx into cells was an absolute requirement for the expression of toxicity, and termed this process the "final common pathway of cell death."

Although the notion that cell death is always Ca^{2+} dependent may be challenged (for review, see *[3]*), considerable evidence indicates that many forms of neuronal injury such as hypoxia/ischemia or trauma involve Ca^{2+} overload. For example, numerous studies in vitro have shown that neurodegeneration following traumatic, anoxic, or excitotoxic insults to cultured neurons is attenuated in the absence of Ca^{2+} *(4–13)*. Animal studies also support the association between Ca^{2+} influx and damage to neural tissues: Experimental spinal cord injury produces significant Ca^{2+} accumulation in white matter axons *(14–16)*. Similarly, cerebral ischemia and epileptic seizures precipitate intracellular Ca^{2+} accumulation *(17–22)*. These studies, and many others, provided the foundation for what has been termed "the calcium hypothesis" which states that neuronal Ca^{2+} overload leads to subsequent neurodegeneration.

Despite overwhelming evidence implicating Ca^{2+} ions in the neurotoxicity process, actual measurements of intracellular calcium ions $[Ca^{2+}]_i$, made possible with fluorescent calcium indicators such as fura-2, have been poor predictors of neuronal survival outcome. For example, Michaels and Rothman *(23)*, examined the impact of a number of EAA receptor agonists on $[Ca^{2+}]_i$ and the survival of hippocampal neurons. They found a poor correlation between the observed $[Ca^{2+}]_i$ rise and cell death, because certain insults could raise $[Ca^{2+}]_i$, but did not kill cells. Similar findings have been reported by others *(4,24)*.

$[Ca^{2+}]_i$ measurements with fluorescent indicators reflect only free Ca^{2+} concentrations. However, in most cells, over 95% of Ca^{2+} ions are buffered by endogenous buffering mechanisms. Thus, measuring $[Ca^{2+}]_i$ may not reflect total cellular Ca^{2+} loading. Therefore, some investigators have measured total Ca^{2+} fluxes using radiolabeled $^{45}Ca^{2+}$ in order to establish quantitative relationships between Ca^{2+} load and excitotoxic/anoxic neurotoxicity. Several reports showed that, in neurons exposed to excitotoxins or anoxia, $^{45}Ca^{2+}$ measurements correlated strongly with the extent of cell death *(25–31)*. However, these total Ca^{2+} flux measurements do not distinguish between the Ca^{2+} influx which triggers neurodegeneration, and the influx which occurs as a consequence of cell membrane disruption and a resultant passive movement of Ca^{2+} down an electrochemical gradient. Thus, direct measurements of either $[Ca^{2+}]_i$ or total Ca^{2+} have yet to quantitatively reveal the Ca^{2+} requirement for neurotoxicity.

The experiments described above have primarily addressed the issue of Ca^{2+} neurotoxicity as a function of bulk calcium loading, attempting to correlate cell death with global changes in cellular calcium content. However, it is now appreciated that calcium fluxes in cells are thoroughly compartmentalized. For example, voltage-gated Ca^{2+} channels are strategically distributed throughout the plasma membrane so as to maximize Ca^{2+} influx in subcellular regions serving specific cellular functions *(32–36)*. Similarly, N-methyl-D-aspartate (NMDA) and non-NMDA receptors are clustered on neuronal synapses, dendrites, and somata *(37–42)*, thereby optimizing Ca^{2+} fluxes in relevant areas. Furthermore, subcellular Ca^{2+} transients can have different temporal profiles in different compartments. For example, in hippocampal dendritic spines, intense afferent stimulation can evoke long-lasting $[Ca^{2+}]_i$ gradients even in the absence of a physical diffusion barrier *(40,43,44)*. Also, in some neurons, different intracellular Ca^{2+} stores may coexist, producing spatially and temporally distinct Ca^{2+} release profiles *(45)*.

Compartmentalization of Ca^{2+} fluxes allows large and highly localized rises in $[Ca^{2+}]_i$ to occur in neurons so that Ca^{2+} ions can separately activate multiple processes within the same cell *(44,46–49)*. For example, NMDA and non-NMDA receptors can trigger completely distinct Ca^{2+} signaling pathways which regulate different forms of gene expression *(50,51)*. Presumably, this occurs by activating spatially distinct sites of Ca^{2+} entry, resulting in the activation of different enzymes located at corresponding sites in the cell.

The compartmentalization of Ca^{2+} fluxes and calcium regulatory mechanisms makes it highly likely that Ca^{2+} neurotoxicity cannot be predicted by bulk measurements of either free or total cellular Ca^{2+} ion content. Some evidence now suggests that Ca^{2+} neurotoxicity may be triggered in certain specific subcellular compartments where $[Ca^{2+}]_i$ exceeds that of other areas within the cell. An example is the case of NMDA receptor channels, which are a major source of Ca^{2+} influx during certain forms of hypoxic/ischemic neuronal damage *(4,52–56)*. NMDA receptors, more than other Ca^{2+}-permeable ionic channels, trigger rapid neurodegeneration in some experimental paradigms. Some quantitative studies suggest that this is, at least in part, a consequence of some attribute specifically related to NMDA-mediated Ca^{2+} influx. It is possible that Ca^{2+} neurotoxicity is not only a function of neuronal Ca^{2+} loading, but also a function of the interactions between Ca^{2+} and specific subcellular compartments *(4)*. Under this hypothesis, Ca^{2+} influx through NMDA receptor channels is more toxic because the rate-limiting processes which trigger some forms of Ca^{2+}-dependent neurotoxicity may be preferentially associated with NMDA receptors. More specifically, it is possible that the molecules involved in triggering secondary, Ca^{2+}-dependent neurotoxic phenomena are physically located in regions close to NMDA channels, and are preferentially activated by Ca^{2+} influx through this source. Under this hypothesis, Ca^{2+} influx through alternate sources such as voltage-gated Ca^{2+} channels would be insufficient to trigger cell death because these influx pathways are not linked to neurotoxic processes. By the time Ca^{2+} diffuses from these alternate sources, the resultant $[Ca^{2+}]_i$ in the vicinity of NMDA receptors and their associated trigger sites for Ca^{2+} neurotoxicity would not suffice to cause neurotoxicity. This hypothesis may explain why certain situations in which Ca^{2+} loading occurs are not accompanied by cell death *(23,24)*.

The above augment underscores the potential importance of the spatial and temporal distribution of Ca^{2+} in the process of Ca^{2+} neurotoxicity. However, the physical and temporal dimensions of subcellular Ca^{2+} ion distributions are difficult to study directly. For this reason, investigators have used strategies in which well-characterized Ca^{2+}-dependent physiological phenomena are perturbed by manipulations which modify the concentrations, distributions, and diffusion characteristics of Ca^{2+} ions. One such approach is to evaluate the effect of various Ca^{2+} buffering compounds on phenomena such as neurotransmitter release, exocytosis, Ca^{2+}-dependent membrane currents, agonist evoked $[Ca^{2+}]_i$ changes, and neurotoxicity. The effects of endogenous and exogenous Ca^{2+} buffers on such phenomena are discussed in the sections below, with emphasis on phenomena thought to be related to nerve cell death.

PHYSIOLOGICAL ROLES OF ENDOGENOUS CALCIUM BUFFERS

Neuronal [Ca^{2+}] dynamics are dictated by a delicate interplay between calcium influx, calcium efflux, and calcium buffering. The latter process is a potential mechanism by which the intracellular spread of Ca^{2+} is regulated. Two classes of Ca^{2+} buffers exist: mobile buffers, comprised primarily of Ca^{2+}-binding proteins, and immobile buffers comprised of larger cellular components such as mitochondria. Each class of buffer likely serves different roles in Ca^{2+} regulation (*see* also chapters by Leist and Nicotera).

Mobile Calcium Buffers

As Ca^{2+} ions diffuse into the cell, they are rapidly buffered by a number of cytoplasmic proteins such as calmodulin, calbindin and parvalbumin *(57)*. There is heterogeneity in the location and quantity of Ca^{2+}-binding proteins in different types of neurons, presumably reflecting variations in the neurons' functional roles. About 95–99% of Ca^{2+} ions entering the cell under physiological conditions are buffered *(58,59)*. Although the precise role of Ca^{2+} buffering substances is controversial, recent evidence indicates that they may act to keep [Ca^{2+}]$_i$ at high levels in localized areas within cells, to limit these high [Ca^{2+}]$_i$ levels to those specific areas, and to rapidly dissipate these Ca^{2+} gradients and thus limit the time course of activation of Ca^{2+}-dependent processes *(60,63)*. For example, there is some evidence that the protein calbindin D28K acts as a fast, high-affinity Ca^{2+} buffer which helps to localize subcellular Ca^{2+} gradients in certain types of neurons *(63,64)*. This function, therefore, is to modulate rapid Ca^{2+} signaling, such as that which occurs during synaptic transmitter release. It is known from direct experiments, and from mathematical modeling, that Ca^{2+} buffers are able to increase the apparent diffusion of Ca^{2+} ions within the cell. Thus they may act as Ca^{2+} shuttles, carrying Ca^{2+} ions from their site of influx to and from their site of action [63,65–68]. These effects, which occur on a miniscule scale, likely depend on the distribution, type, and concentration of the Ca^{2+} buffers within the cell.

Physiological experiments suggest that endogenous Ca^{2+} buffers in certain neurons modulate Ca^{2+} influx *(58)*, and become maximally effective when neurons are challenged with high Ca^{2+} loads *(69)*. They may also allow cells to attenuate [Ca^{2+}]$_i$ increases following Ca^{2+} entry through Ca^{2+} channels *(60,70)*. This pattern has been taken by some as an indicator that endogenous Ca^{2+} buffers might act to protect neurons against the toxicity of Ca^{2+} excess. However, this hypothesis is unproven (*see* below). It is equally likely that endogenous Ca^{2+} buffers evolved to subserve physiological roles in modulating Ca^{2+}-dependent signaling (e.g., [63,64]) rather than as a primary neuroprotective mechanism.

Immobile Calcium Buffers

The Ca^{2+}-buffering capacity of cytosolic Ca^{2+}-binding proteins is limited. Therefore, neurons also possess mechanisms for sequestering Ca^{2+} ions into organelles in situations where Ca^{2+} loads exceed the ability of Ca^{2+} buffers to maintain [Ca^{2+}]$_i$ at tolerable levels. These include the smooth endoplasmic reticulum, mitochondria, and even synaptic vesicles *(71–74)*. These organelles can sequester large quantities of Ca^{2+} under a variety of conditions, using active and passive Ca^{2+} transport mechanisms similar to those found in the plasma membrane (e.g., a Ca^{2+} ATPase, *see* [75]). Although Ca^{2+} storage in organelles is an efficient homeostatic mechanism, this system operates at a much slower rate than cytoplasmic Ca^{2+}-binding proteins. Therefore, immobile calcium buffers are less capable of modulating rapidly changing or highly localized changes in $[Ca^{2+}]_i$. Nevertheless, there is evidence that mitochondria participate in Ca^{2+} uptake under a variety of conditions, including physiological and pathological Ca^{2+} loading *(72,73)*. Disorders of Ca^{2+} sequestration have been implicated in a variety of neurological disorders ranging from cerebral ischemia to mitochondrial diseases *(72,76)*. However, little is known about the potential for experimentally manipulating nonmobile Ca^{2+}-buffering mechanisms in order to protect neurons from injury. The remaining focus of this chapter, therefore, will be on the role of the mobile Ca^{2+}-buffering component.

ENDOGENOUS CALCIUM BUFFERING AND NEURONAL VULNERABILITY TO INJURY

The hypothesis that endogenous Ca^{2+}-buffering proteins impart neurons with increased resilience against various injuries is attractive, and has been examined repeatedly. However, the quantity of endogenous Ca^{2+} buffers in neurons is difficult to manipulate experimentally. For this reason, most studies of the relationship between Ca^{2+} buffer content and cell death have been correlative. Typically, in such studies, a defined insult to cultured neurons or brain tissues is followed by immunohistochemical staining for the presence or absence of Ca^{2+}-binding proteins. A relationship between the numbers of cells surviving and Ca^{2+}-binding protein content is then sought. The current literature provides equally divided, conflicting opinions on whether or not the presence of endogenous Ca^{2+} buffers protects neurons against a variety of insults. For example, experiments in cultured neurons or brain slices have suggested that neurons expressing calbindin D28K, calretinin, or parvalbumin are more resistant to excitatory amino acid toxicity *(77–79)*, as well as the toxicity by other neurotoxins such as the HIV-1 envelope (gp120) and β-amyloid proteins *(80,81)*. However, in many other reports, it is suggested that neurons containing Ca^{2+}-binding proteins are more vulnerable to excitotoxicity (e.g., *[82,83]*). Interestingly, in lymphocytes, Ca^{2+}-binding proteins have also been suggested to be protective against apoptosis by some *(84,85)*, and to promote apoptosis by others *(86)*.

Results from animal experiments have also been controversial. Some studies report a sparing of neurons containing Ca^{2+}-binding proteins following cerebral ischemia *(87–89)*, or in neurodegenerative diseases *(90–94)*. However, other studies have failed to demonstrate such correlations *(95,96)*.

Recently, using recombinant DNA techniques, it has been possible to manipulate intracellular Ca^{2+}-binding proteins. Lledo et al. *(70)* showed that transfecting GH3 cells with calbindin-D28K reduced the magnitude of Ca^{2+} currents and attenuated depolarization-evoked $[Ca^{2+}]_i$ transients. Chard et al. *(60)* directly injected the proteins calbindin and parvalbumin into neurons via patch pipets, and showed unequivocally that these proteins attenuated $[Ca^{2+}]_i$ increases within cells. However, experiments to date have not addressed, using such direct molecular approaches, the possibility that calcium binding proteins subserve a neuroprotective function. Thus, mechanisms by which Ca^{2+}-binding proteins may protect neurons are unknown.

Proponents of the idea that Ca^{2+} binding proteins enhance neuronal resilience to injury propose that these cellular components, by buffering Ca^{2+} loads, reduce the potential deleterious effects of Ca^{2+} excess. However, opponents of the neuroprotection hypothesis propose that Ca^{2+} buffers, by reducing the quantity of free $[Ca^{2+}]_i$, may actually enhance Ca^{2+} loading by reducing Ca^{2+}-dependent inactivation of Ca^{2+} influx and/or the extrusion of Ca^{2+} ions by mechanisms such as the Na^+/Ca^{2+} exchanger which is driven by ionic gradients. There is currently little or no experimental evidence to support either point of view.

NONFLUORESCENT EXOGENOUS CALCIUM BUFFERS AS PROBES OF CELLULAR CALCIUM HOMEOSTASIS

Fluorescent Ca^{2+} buffers such as fura-2 and fluo-3 have, in the past decade, contributed significantly to the understanding of cellular Ca^{2+} dynamics. Their use as indicators of cytoplasmic $[Ca^{2+}]_i$ concentration has revolutionized our understanding of Ca^{2+} homeostasis, and is covered in extensive reviews elsewhere (97–99). However, fluorescent and nonfluorescent Ca^{2+} buffers also have additional physical properties which have been exploited to probe various aspects of $[Ca^{2+}]_i$ homeostasis. Those properties which are believed to contribute to the neuroprotective effects of exogenous Ca^{2+} chelators are reviewed next.

As discussed above, experiments investigating the relationship between endogenous Ca^{2+} buffering and neuronal vulnerability have yielded controversial results. An alternative approach to further investigate this relationship has been to use artificial Ca^{2+} buffers, which can be more easily manipulated experimentally. These compounds can be introduced into cells either directly (by microinjections), or as cell-membrane permeant forms. The most commonly used exogenous Ca^{2+} buffers (Ca^{2+} chelators) are derivatives of ethylene-glycol tetraacetic acid (EGTA) or 1,2 bis-(2-aminophenoxy ethane-N,N,N',N'-tetraacetic acid (BAPTA). Slight structural modifications of these molecules can alter their physical properties as chelators, making this class of compounds ideal for dissecting out mechanisms by which intracellular Ca^{2+} buffers (endogenous or artificial) exert physiological and (potentially) neuroprotective effects.

The structures of BAPTA, and its cell-membrane permeant form BAPTA-AM, are shown in Fig. 1. Cell-membrane permeability is a useful feature of these compounds, which is achieved by esterifying the carboxylic acid moieties of the Ca^{2+} chelating site to acetoxymethyl (AM) esters which render the compound lipid soluble (100). Once inside the cell, the AM esters are cleaved by intracellular esterases, leaving the chelator salt trapped within the cell. This approach allows the loading of chelator molecules into many cells simultaneously, rather than having to microinject the compound into one cell at a time.

BAPTA and its analogs have a high selectivity for Ca^{2+} ions over other monovalent and divalent ions such as magnesium, Mg^{2+} (101). This makes these chelators useful to test the Ca^{2+} dependence of a particular physiological process. For example, the postsynaptic injection of BAPTA or its analog quin-2 has been shown to prevent the induction of long-term potentiation (LTP) in CA3 pyramidal neurons (102), revealing that postsynaptic Ca^{2+} increases are required for this process.

A number of additional properties of Ca^{2+} chelators have been identified as determinants of their physiological actions. Among the most important, clearly, is their intracellular concentration. Too little buffer, regardless of its other properties, will have no

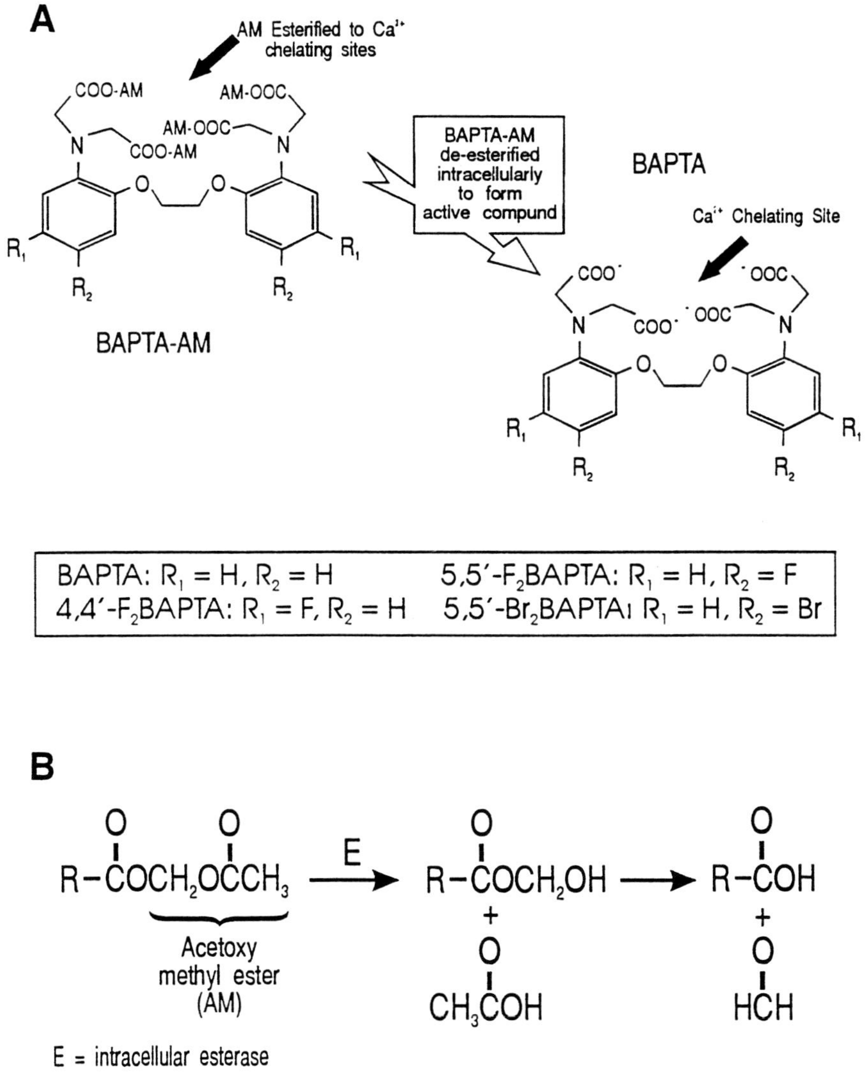

Fig. 1. (A) Structures of BAPTA-AM (1,2 bis-(2-aminophenoxy)ethane-N,N,N',N'-tetraacetic acid acetoxymethyl ester), BAPTA, and location of Ca^{2+} chelating site. The AM esters must be hydrolyzed from the molecule before it becomes an active Ca^{2+} chelator. The chelator's Ca^{2+} affinity can be altered by substituting various groups at the sites shown (R1 and R2). (B) Process of de-esterification of the AM ester from the BAPTA analog (R) by intracellular esterases.

effect. Experiments in which known amounts of calcium chelators were delivered into cells through micropipettes have shown that even small amounts of certain chelators (in the order of 10–100 μM) can dramatically alter Ca^{2+}-dependent processes such as synaptic transmitter release, Ca^{2+}-dependent membrane currents, or stimulus evoked changes in $[Ca^{2+}]_i$ *(58,59,103–105)*. The indirect delivery of chelators into cells as the cell-permeant forms results in less predictable intracellular concentrations. Nevertheless, experiments using this approach have also shown concentration-dependent physiological effects of these agents *(106)*. It is notable that many of the chelators' physiological effects, such as the modulation of synaptic transmitter release, occur at concentrations thought to be insufficient to significantly alter average cytoplasmic calcium concentrations or to effect bulk changes in cellular Ca^{2+} fluxes. Thus, alternative mechanisms of actions must be responsible for the chelators' physiological actions.

Another important property of Ca^{2+} chelators is their affinity for Ca^{2+} (K_d for Ca^{2+}), because this determines the extent to which a given quantity of the buffer will bind Ca^{2+} ions. Compounds having too low a Ca^{2+} affinity (high K_d), even when present at high concentrations, will interact minimally with Ca^{2+} ions and will therefore have lesser effects than chelators having a high Ca^{2+} affinity. Such high K_d compounds serve as useful controls in experiments in which it is desirable to demonstrate that the effect of a given chelator is truly due to its Ca^{2+}-binding ability rather than a nonspecific pharmacological effect referable to the parent BAPTA structure. For example, BAPTA, a chelator with high Ca^{2+} affinity, attenuates synaptic transmission in hippocampal CA1 neurons, whereas dinitro-BAPTA, a chelator with a similar structure but a much lower Ca^{2+} affinity, does not *(106)*. Attenuation of synaptic transmission by the buffers occurs presumably as a consequence of their ability to blunt the very high submembrane Ca^{2+} increases needed to trigger transmitter release *(107,108)*. By utilizing a range of structurally similar compounds having a range of Ca^{2+} affinities, investigators have been able to infer the actual magnitude of Ca^{2+} elevations necessary to initiate synaptic transmission *(105,109)*, to activate Ca^{2+}-dependent membrane currents *(104)*, and to trigger neurodegeneration following excitotoxicity *(110)*. These studies are discussed in greater detail in subsequent sections of this chapter.

A property of Ca^{2+} buffers which has been exploited to infer the molecular dimensions within which a Ca^{2+}-dependent cellular process occurs is the rate of Ca^{2+} binding. Typical examples are experiments in which the effects of EGTA are compared with that of BAPTA. EGTA is a compound which has a similar affinity for Ca^{2+} ions as BAPTA, but has a much slower Ca^{2+} association rate (100–400 times slower). Therefore, because of the lengthy period required for Ca^{2+} binding, EGTA is less likely to affect nonequilibrium processes which are critically dependent on the diffusion of Ca^{2+} from a given source, such as an ion channel pore, to an intrinsic Ca^{2+} binding site located a very short distance away *(66)*. For example, BAPTA is more effective than EGTA in attenuating neurotransmitter release in a number of preparations *(105,106,109)*. It is assumed that, because Ca^{2+} in synaptic terminals must diffuse very short distances from their points of entry (Ca^{2+} channels) to the site of active zones, they can reach their targets before they can be captured by EGTA molecules. BAPTA molecules, however, bind Ca^{2+} sufficiently rapidly to interfere with this diffusion process. Similar approaches have been used to infer the dimensions of distribution of Ca^{2+}-dependent potassium channels in synapses *(34)*, and the physical scale of the localized diffusion of Ca^{2+} ions involved in regenerative feedback mechanisms governing the propagation of intracellular Ca^{2+} waves *(111)*. Interestingly, BAPTA is also more effective at protecting neurons against excitotoxic insults than EGTA (*see* below), a pattern suggesting that Ca^{2+} diffusion-dependent mechanisms may be involved in the process of Ca^{2+} neurotoxicity *(110)*.

Calcium Chelators

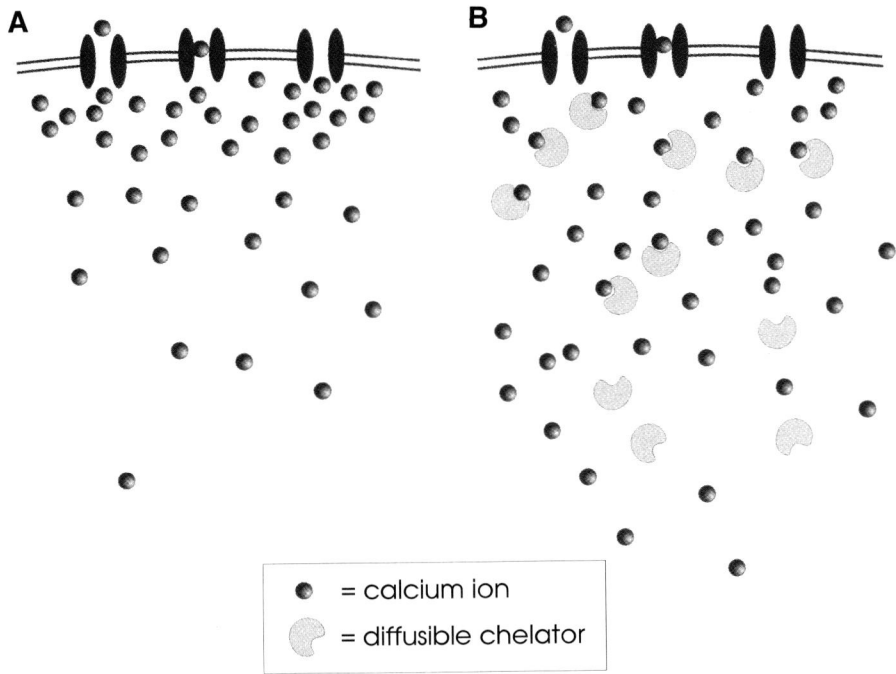

Fig. 2. Facilitation of Ca^{2+} diffusion in cytoplasm and attenuation of sub-membrane Ca^{2+} gradient by a diffusible Ca^{2+} chelator. **(A)** Opening of Ca^{2+} permeable channels results in Ca^{2+} influx, and subsequent diffusion away from the ion channel pore. During this time, there exists a gradient of higher $[Ca^{2+}]_i$ near the influx site, and lower $[Ca^{2+}]_i$ away from influx sites. **(B)** The presence of a diffusible chelator results in one. Binding of Ca^{2+} which is maximal where $[Ca^{2+}]_i$ is highest, and diffusion of the bound chelator-Ca^{2+} complex away from sites of high $[Ca^{2+}]_i$. This effectively reduces the concentration of free Ca^{2+} at the peak of the gradient, and increases the effective diffusion distances of Ca^{2+} which are released by the buffer at a distance from the sites of influx.

Finally, another important property of Ca^{2+} chelators is their cytoplasmic mobility. Briefly, immobile Ca^{2+} buffers would be quickly saturated at sites of high $[Ca^{2+}]_i$ and be rendered ineffective. Therefore, for buffers to have an effect, they must have sufficient mobility to be able to diffuse in- and out of sites of high $[Ca^{2+}]_i$ elevations (Fig. 2). This property allows Ca^{2+} chelators to facilitate the diffusion of bound Ca^{2+} ions away from local sites where $[Ca^{2+}]_i$ is high. By some estimates, BAPTA and its analogs diffuse in cytoplasm almost as readily as Ca^{2+} ions themselves.

EXOGENOUS INTRACELLULAR CALCIUM BUFFERS AS PROBES OF CA^{2+}-DEPENDENT CELL DEATH

The selectivity of BAPTA and its analogs for Ca^{2+} ions over other monovalent and divalent ions has been used to examine the dependence of a number of potentially cytotoxic processes on Ca^{2+}. For example., BAPTA-AM, quin-2 AM, or EGTA-AM have been used extensively in non-neuronal cells to investigate the dependence of free radical production and free radical-mediated cytotoxicity on Ca^{2+}. Results suggest that loading cells with these chelators attenuates the production of arachidonic acid metabolites such as prostaglandin E2 *(112)*, and can prevent free radical-mediated toxicity *(113,114)*, possibly by preventing endonuclease activation and subsequent DNA fragmentation *(115,116)*. The production of phospholipases may also be blocked by BAPTA and its analogs, although this may be a direct pharmacological effect of these compounds *(117)*. Since glutamate receptor activation is known to trigger the formation of arachidonic acids and free radicals *(53,55,118–122)*, attenuating such reactions in neurons may be a mechanism by which Ca^{2+} buffers protect neurons from hypoxic/ischemic or excitotoxic injuries.

The permeant chelators BAPTA-AM and quin-2-AM have also been used extensively to investigate the Ca^{2+} dependence of apoptotic mechanisms. They have been shown in a number of lymphocytic cell types to prevent or attenuate apoptotic death *(123–126)*, although some authors have reported the converse: that Ca^{2+} chelators promote, rather than attenuate, apoptosis *(127,128)*. Presently, the actual mechanisms by which apoptotic mechanisms are modulated by Ca^{2+} chelators are unknown.

In recent years, experiments have also been performed to examine the Ca^{2+} dependence of cell injury, particularly injury due to excitotoxic or anoxic/ischemic mechanisms. Reports have suggested that loading neurons with BAPTA analogs may protect neurons from excessive afferent stimulation *(77,129)*, glutamate toxicity *(110,130)*, or anoxia (Abdel-Hamid and Tymianski, personal observations). An example is the experiment shown in Fig. 3, performed in cultured spinal neurons: When challenged with glutamate, $[Ca^{2+}]_i$ in neurons increases transiently to a peak, followed by a return to a plateau levels. Neurons which subsequently succumb to the excitotoxic insult, however, undergo secondary, irreversible increases in $[Ca^{2+}]_i$, which herald an impending damage to the cell membrane integrity (Fig. 3A; *[4,131,132]*). Pretreatment with cell-permeant chelators such as BAPTA-AM or difluoro-BAPTA-AM can significantly attenuate the initial rise in $[Ca^{2+}]_i$. Also, fewer neurons undergo secondary $[Ca^{2+}]_i$ increases and cell death (Fig. 3B).

Studies relating Ca^{2+} chelation to cell survival have primarily exploited the selectivity of BAPTA-like compounds for Ca^{2+} compared to other ions. However, some studies of neurotoxicity have also exploited the additional physical properties of Ca^{2+} chelators described in the previous section. As discussed above, neuronal injury may occur as a consequence of excessive localized $[Ca^{+2}]i$ increases in subcellular compartments, rather than excess bulk Ca^{2+} loading. This hypothesis can be tested with the use of several different chelators having a range of Ca^{2+} affinities and Ca^{2+}-binding rates. As shown in the experiments described in Fig. 3, pretreating neurons with BAPTA-AM dramatically attenuates glutamate-induced, peak $[Ca^{2+}]_i$ levels (Fig. 3B and C), whereas pretreatment with dibromo-BAPTA-AM, a chelator having a much lower Ca^{2+} affinity (100 nM vs. 4 µM for BAPTA and Br_2BAPTA, respectively), has less impact on glutamate-evoked $[Ca^{2+}]_i$ increases. Difluoro-BAPTA, a chelator with intermediate Ca^{2+} affinity, has intermediate effects on $[Ca^{2+}]_i$. Despite this, all three chelators prevented cell death to an equal extent (Fig. 3D; *[110]*). Furthermore, in similar experiments, BAPTA-AM pretreatment was much more neuroprotective than pretreatment with EGTA-AM, a slower Ca^{2+} buffer. These results are contrary to the notion that neurotoxicity is simply a function of cellular Ca^{2+} loading because, then, a buffer's protective capacity should parallel its ability to decrease $[Ca^{2+}]_i$. Since low-affinity

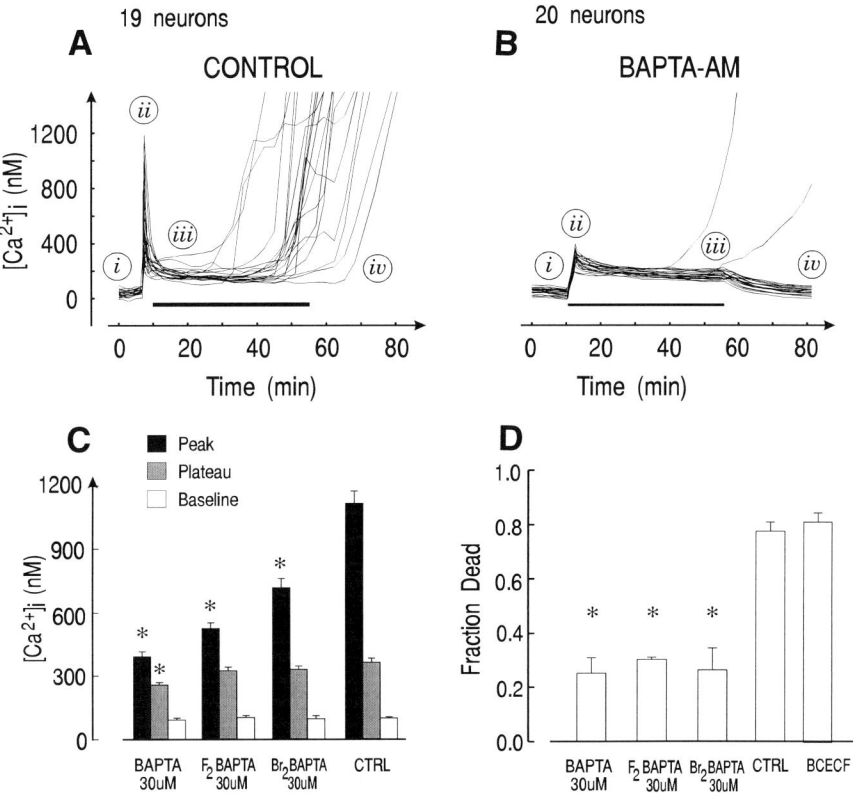

Fig. 3. Effects of intracellular Ca^{2+} chelators on $[Ca^{2+}]_i$ changes and neuronal survival in cultured spinal neurons challenged with 250 μM L-glutamate for 50 min (black bar). **(A,B)** Superimposed tracings of the time-course of $[Ca^{2+}]_i$ from individual neurons in the experimental field. A: In control neurons, applying glutamate causes $[Ca^{2+}]_i$ to increase from baseline *i* to a peak level *ii*. Ca^{2+} homeostatic mechanisms then cause $[Ca^{2+}]_i$ to decrease again to a lower plateau *(iii)*. However, neurons which are damaged by the glutamate insult sustain a secondary, irreversible rise in $[Ca^{2+}]_i$ which indicates cell death *iv*. **(B)** Pretreating neurons with BAPTA-AM (100 μM) has no effect on baseline $[Ca^{2+}]_i$ but significantly attenuates the peak $[Ca^{2+}]_i$ rise evoked by glutamate. Also, fewer neurons undergo secondary Ca^{2+} increases. **(C,D)** Different permeant BAPTA analogs attenuate glutamate-evoked $[Ca^{2+}]_i$ increases to a degree which parallels their Ca^{2+} affinity, but have equivalent neuroprotective potency. **(C)** Effects of the different permeant analogs on peak, plateau, and baseline $[Ca^{2+}]_i$ in neurons challenged with 250 μm L-glutamate for 50 min. Asterisks indicate significant differences from controls at $p < 0.0001$. The peak $[Ca^{2+}]_i$ values were also statistically different from each other ($p < 0.05$). Note that equal concentrations of dibromo-BAPTA, a chelator with a low affinity for Ca^{2+}, is least effective at attenuating peak $[Ca^{2+}]_i$. **(D)** Protective effects of different BAPTA-AM analogs on the same neurons as in (C), challenged with 250 μM glutamate. The number of surviving neurons was markedly increased by pretreatment with Ca^{2+} chelators. Note that protection was not a function of the -AM moiety as neurons loaded with BCECF-AM prior to the glutamate challenged died as frequently as the controls. Groups with asterisks are statistically equivalent, and differ from controls at $p < 0.0001$. (Panels A and B, adapted from ref. 130, panels C,D adapted from ref. 110.)

Ca^{2+} buffers are most efficient at buffering Ca^{2+} when $[Ca^{2+}]_i$ rises near their K_d, these results suggest that glutamate-triggered Ca^{2+} neurotoxicity occurs at $[Ca^{2+}]_i$ levels which exceed several micromolar. This event, combined with the inferior protective capacity of EGTA the (slow buffer) suggests that Ca^{2+} neurotoxicity may depend, in part, on Ca^{2+} diffusion-coupled processes in subcellular domains too small to be resolved by conventional imaging techniques and that fast Ca^{2+} buffers may protect neurons, at least in part, by dissipating Ca^{2+} gradients within these microdomains.

Microdomains of high $[Ca^{2+}]_i$ occur in presynaptic terminals during neurotransmitter release *(109)*, and in postsynaptic sites such as dendritic spines *(40,133)*. Exogenous chelators affect Ca^{2+} ions in these domains, and attenuate synaptic transmission. Given the excitotoxicity hypothesis, the chelators' effects on synaptic transmission may underlie their mechanism of protective action. To examine this possibility studies have been performed to determine whether there exist parallels between the protective capacity of chelators and their effects on synaptic transmission. We have used a well-characterized physiological model system, i.e., the hippocampal brain slice (Fig. 4A). Studies were performed at the level of multiple cells (field recordings, Fig. 4B and C, from refs. *106* and *134*) and at the level of single cells (intracellular recordings, Fig. 4D and E, and from ref. *105*). Briefly, field potential recordings confirmed that fast Ca^{2+} chelators such as BAPTA attenuated excitatory neurotransmission in this model (Figs. 4B and C). This effect was dependent on the intracellular action of the agent, because it could not be reproduced with extracellular applications of BAPTA salt, which is cell impermeant. The slower buffer EGTA-AM had no effect, despite its similar Ca^{2+} affinity to BAPTA, a result suggesting that subcellular Ca^{2+} diffusion is important in mediating the effect. Furthermore, dinitro-BAPTA, a chelator having a K_d for Ca^{2+} which far exceeds physiological $[Ca^{2+}]_i$ changes, also had no effect, ruling out pharmacological actions of BAPTA-related factors other than Ca^{2+} chelation or actions of the AM moiety. Intracellular recordings of excitatory postsynaptic potentials (Fig. 4D and E) confirmed the ability of BAPTA, but not EGTA, to attenuate excitatory transmission. Control experiments in these studies showed that the observed effects were entirely presynaptic *(105,106,135)*. Interestingly, APTRA, a fast chelator having Kd in the 25 μM range, was as effective as BAPTA, whereas chelators with an even lesser Ca^{2+} affinity ceased to be effective. This result confirms that $[Ca^{2+}]_i$ must rise to micromolar levels in order for transmitter release to occur, since APTRA can only significantly affect $[Ca^{2+}]_i$ when it reaches levels in the order of magnitude of its K_d for Ca^{2+}. Also, this result predicts that APTRA should be neuroprotective, a hypothesis recently confirmed in our laboratory (Abdel-Hamid and Tymianski, personal observation). Overall, these data parallel the neuroprotective features of the chelators illustrated in Fig. 3, and strengthen the association between excessive excitatory neurotransmission and neurotoxicity in several models of neuronal injury.

Fig. 4. Fast exogenous Ca^{2+} chelators having a range of Ca^{2+} affinities attenuate synaptic transmission in the hippocampal slice. **(A)** Method of stimulation and recording: the Schaeffer collaterals were stimulated with a stimulating microelectrode (S), and extracellular recordings were made of dendritic field excitatory post synaptic potentials (fEPSPs, electrode R_1, for data in panels **B,C**). Sharp electrode intracellular recordings were made of excitatory post-synaptic potentials from CA1 neurons (EPSPs, electrode R_2, for data in panels **D,E**). Unless indicated, all of the chelators were applied onto the brain slice as the cell-permeant acetoxymethyl esters. (B) Only a cell permeant, fast Ca^{2+} chelator attenuates fEPSPs in the hippocampal slice: application

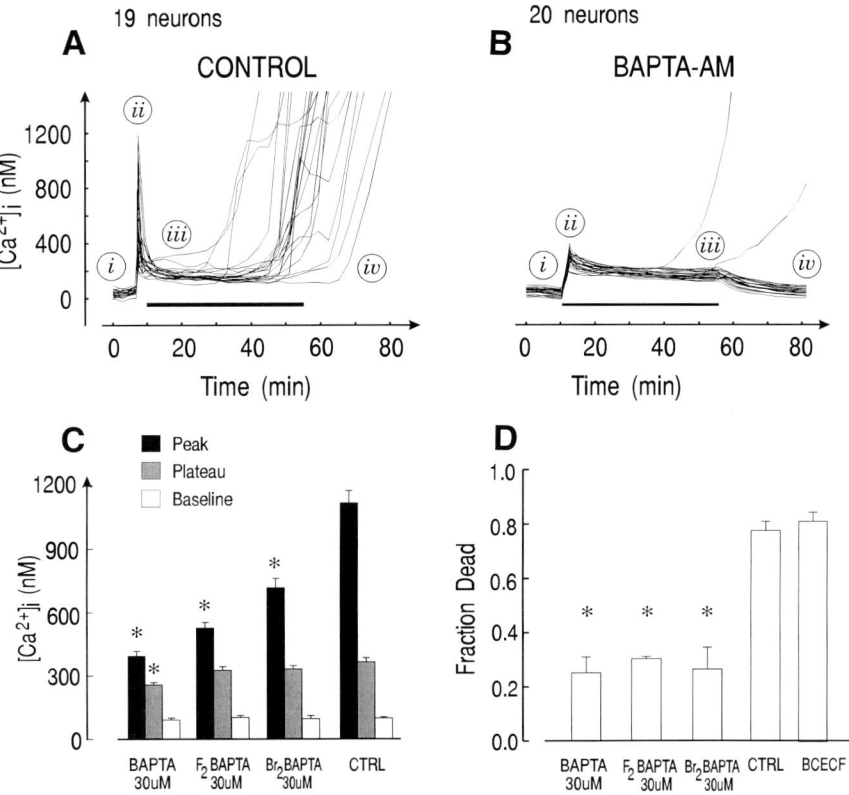

Fig. 3. Effects of intracellular Ca^{2+} chelators on $[Ca^{2+}]_i$ changes and neuronal survival in cultured spinal neurons challenged with 250 μM L-glutamate for 50 min (black bar). **(A,B)** Superimposed tracings of the time-course of $[Ca^{2+}]_i$ from individual neurons in the experimental field. A: In control neurons, applying glutamate causes $[Ca^{2+}]_i$ to increase from baseline *i* to a peak level *ii*. Ca^{2+} homeostatic mechanisms then cause $[Ca^{2+}]_i$ to decrease again to a lower plateau *(iii)*. However, neurons which are damaged by the glutamate insult sustain a secondary, irreversible rise in $[Ca^{2+}]_i$ which indicates cell death *iv*. **(B)** Pretreating neurons with BAPTA-AM (100 μM) has no effect on baseline $[Ca^{2+}]_i$ but significantly attenuates the peak $[Ca^{2+}]_i$ rise evoked by glutamate. Also, fewer neurons undergo secondary Ca^{2+} increases. **(C,D)** Different permeant BAPTA analogs attenuate glutamate-evoked $[Ca^{2+}]_i$ increases to a degree which parallels their Ca^{2+} affinity, but have equivalent neuroprotective potency. **(C)** Effects of the different permeant analogs on peak, plateau, and baseline $[Ca^{2+}]_i$ in neurons challenged with 250 μm L-glutamate for 50 min. Asterisks indicate significant differences from controls at $p < 0.0001$. The peak $[Ca^{2+}]_i$ values were also statistically different from each other ($p < 0.05$). Note that equal concentrations of dibromo-BAPTA, a chelator with a low affinity for Ca^{2+}, is least effective at attenuating peak $[Ca^{2+}]_i$. **(D)** Protective effects of different BAPTA-AM analogs on the same neurons as in (C), challenged with 250 μM glutamate. The number of surviving neurons was markedly increased by pretreatment with Ca^{2+} chelators. Note that protection was not a function of the -AM moiety as neurons loaded with BCECF-AM prior to the glutamate challenged died as frequently as the controls. Groups with asterisks are statistically equivalent, and differ from controls at $p < 0.0001$. (Panels A and B, adapted from ref. 130, panels C,D adapted from ref. 110.)

Ca^{2+} buffers are most efficient at buffering Ca^{2+} when $[Ca^{2+}]_i$ rises near their K_d, these results suggest that glutamate-triggered Ca^{2+} neurotoxicity occurs at $[Ca^{2+}]_i$ levels which exceed several micromolar. This event, combined with the inferior protective capacity of EGTA the (slow buffer) suggests that Ca^{2+} neurotoxicity may depend, in part, on Ca^{2+} diffusion-coupled processes in subcellular domains too small to be resolved by conventional imaging techniques and that fast Ca^{2+} buffers may protect neurons, at least in part, by dissipating Ca^{2+} gradients within these microdomains.

Microdomains of high $[Ca^{2+}]_i$ occur in presynaptic terminals during neurotransmitter release *(109)*, and in postsynaptic sites such as dendritic spines *(40,133)*. Exogenous chelators affect Ca^{2+} ions in these domains, and attenuate synaptic transmission. Given the excitotoxicity hypothesis, the chelators' effects on synaptic transmission may underlie their mechanism of protective action. To examine this possibility studies have been performed to determine whether there exist parallels between the protective capacity of chelators and their effects on synaptic transmission. We have used a well-characterized physiological model system, i.e., the hippocampal brain slice (Fig. 4A). Studies were performed at the level of multiple cells (field recordings, Fig. 4B and C, from refs. *106* and *134*) and at the level of single cells (intracellular recordings, Fig. 4D and E, and from ref. *105*). Briefly, field potential recordings confirmed that fast Ca^{2+} chelators such as BAPTA attenuated excitatory neurotransmission in this model (Figs. 4B and C). This effect was dependent on the intracellular action of the agent, because it could not be reproduced with extracellular applications of BAPTA salt, which is cell impermeant. The slower buffer EGTA-AM had no effect, despite its similar Ca^{2+} affinity to BAPTA, a result suggesting that subcellular Ca^{2+} diffusion is important in mediating the effect. Furthermore, dinitro-BAPTA, a chelator having a K_d for Ca^{2+} which far exceeds physiological $[Ca^{2+}]_i$ changes, also had no effect, ruling out pharmacological actions of BAPTA-related factors other than Ca^{2+} chelation or actions of the AM moiety. Intracellular recordings of excitatory postsynaptic potentials (Fig. 4D and E) confirmed the ability of BAPTA, but not EGTA, to attenuate excitatory transmission. Control experiments in these studies showed that the observed effects were entirely presynaptic *(105,106,135)*. Interestingly, APTRA, a fast chelator having Kd in the 25 μ*M* range, was as effective as BAPTA, whereas chelators with an even lesser Ca^{2+} affinity ceased to be effective. This result confirms that $[Ca^{2+}]_i$ must rise to micromolar levels in order for transmitter release to occur, since APTRA can only significantly affect $[Ca^{2+}]_i$ when it reaches levels in the order of magnitude of its K_d for Ca^{2+}. Also, this result predicts that APTRA should be neuroprotective, a hypothesis recently confirmed in our laboratory (Abdel-Hamid and Tymianski, personal observation). Overall, these data parallel the neuroprotective features of the chelators illustrated in Fig. 3, and strengthen the association between excessive excitatory neurotransmission and neurotoxicity in several models of neuronal injury.

Fig. 4. Fast exogenous Ca^{2+} chelators having a range of Ca^{2+} affinities attenuate synaptic transmission in the hippocampal slice. **(A)** Method of stimulation and recording: the Schaeffer collaterals were stimulated with a stimulating microelectrode (S), and extracellular recordings were made of dendritic field excitatory post synaptic potentials (fEPSPs, electrode R_1, for data in panels **B,C**). Sharp electrode intracellular recordings were made of excitatory post-synaptic potentials from CA1 neurons (EPSPs, electrode R_2, for data in panels **D,E**). Unless indicated, all of the chelators were applied onto the brain slice as the cell-permeant acetoxymethyl esters. (B) Only a cell permeant, fast Ca^{2+} chelator attenuates fEPSPs in the hippocampal slice: application

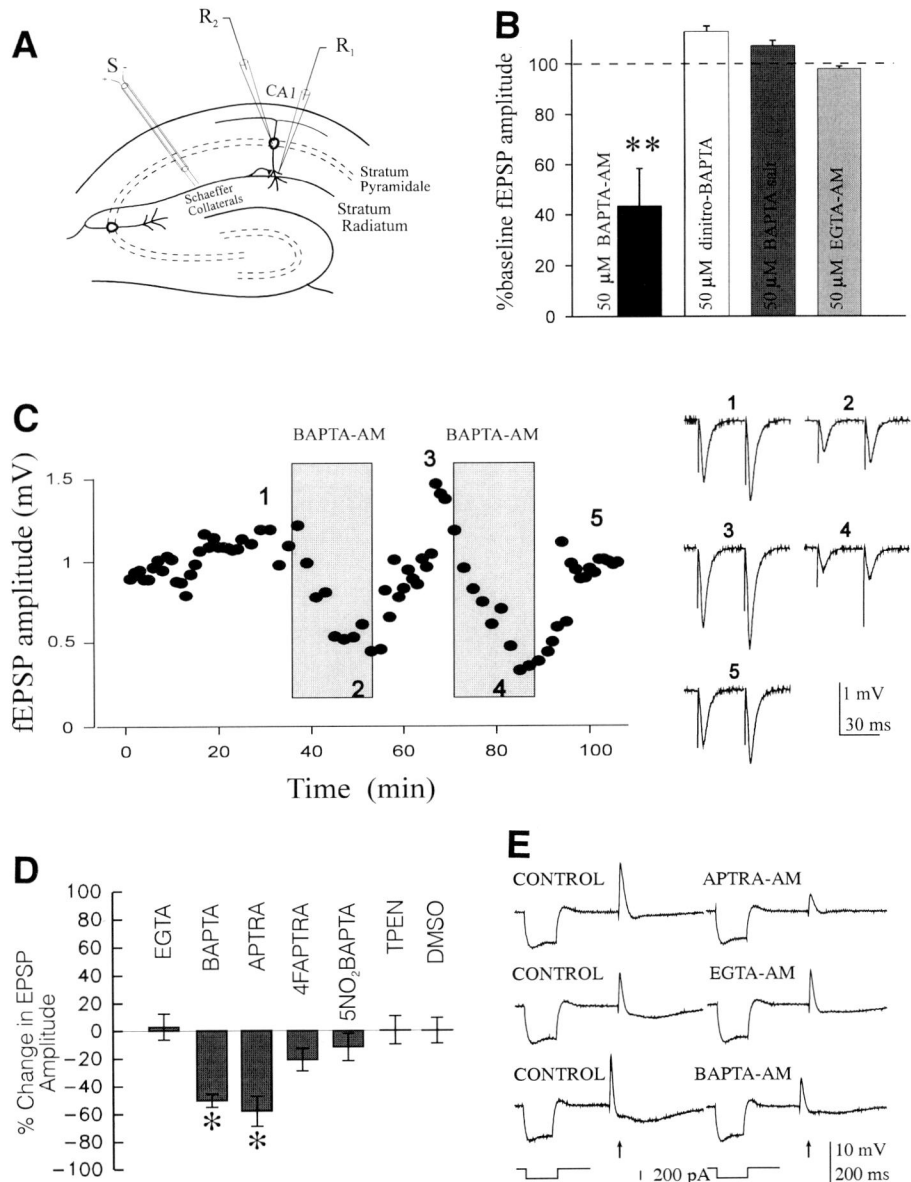

of BAPTA-AM but not BAPTA salt (cell impermeant) or EGTA-AM (slow Ca^{2+} buffer) reduces the size of the fEPSP. Dinitro-BAPTA-AM, a chelator with an extremely low Ca^{2+} affinity (K_d in the millimolar range) is ineffective. (C) Left panel: Sample experiment illustrating the reversible effects of BAPTA-AM application on the fEPSP amplitude. Shaded bars indicate the time of chelator application. Right panel: Sample tracings of fEPSPs taken at the times indicated (numbers 1–5). (D) Results of intracellular recordings of the effects of a range of exogenous Ca^{2+} chelators on EPSP amplitude. EPSPs were unaffected by TPEN, a Zn^{2+} chelator, EGTA, a slow Ca^{2+} chelator, nor by 4-fluoro APTRA ($K_d \geq 150$ μM) or 5-mononitro- BAPTA ($K_d \geq 90$ μM) whose Ca^{2+} affinity is very low. However, APTRA still effectively attenuates EPSPs even though its Ca^{2+} affinity is much lower than BAPTA (K_d about 25 μM). Asterisks indicate significant differences from zero. (E) Sample intracellular recordings illustrating the effects of applying three different chelators on EPSPs. (Panels A–C adapted from ref. *106*, D–E adapted from ref. *105*.)

The above discussion illustrates a classical use of Ca^{2+} buffers to define the boundaries of the $[Ca^{2+}]_i$ changes required for synaptic transmitter release in the mammalian brain. In addition to their presynaptic effects, however, some chelators also have marked effects on postsynaptic Ca^{2+} signaling. For example, intracellular waves of Ca^{2+} have been detected in a number of eukaryotic cells, including neuronal cell lines. These may constitute an important mode of signaling between cellular compartments. In mouse neuroblastoma cells, these Ca^{2+} waves are attenuated by BAPTA, but not EGTA, suggesting that the Ca^{2+} wave phenomenon includes a step involving rapid diffusion of Ca^{2+} on a minuscule scale (111). The $[Ca^{2+}]_i$ requirements for these types of intra- and intercellular Ca^{2+} signaling are amenable to probing with various Ca^{2+} chelators in a manner analogous to that employed in the studies of transmitter release as described above. Preliminary experiments on Ca^{2+} wave propagation in astrocytes have shown that the buffers have considerable impact on this phenomenon (Tymianski and Nedergaard, personal observations). However, partly owing to the complexity of Ca^{2+} dependent postsynaptic mechanisms, it is currently unknown which of the chelators' many postsynaptic effects contribute to their neuroprotective efficacy.

EXOGENOUS CALCIUM BUFFERS AS POTENTIAL THERAPEUTIC AGENTS FOR NEURONAL INJURY

The discussions above have been focused on the use of exogenous Ca^{2+} chelators as probes of intracellular Ca^{2+} homeostasis and of mechanisms having a neurotoxic potential. A desirable by-product of the above research has been the realization that exogenous Ca^{2+} buffers may constitute a novel approach to protecting neurons against excitotoxic, anoxic, or ischemic insults. Several findings from mechanistic experiments may have applicability in the neuroprotective use of these compounds. For example, since chelators having a relatively low affinity for Ca^{2+} attenuate the high $[Ca^{2+}]_i$ which may be involved in excitotoxicity (see above), it might be possible to tailor a Ca^{2+} buffer to affect only neurotoxic $[Ca^{2+}]_i$ elevations, and to allow normal Ca^{2+} signaling in the cell to continue unimpeded. A number of studies have examined the possibility of using cell-permeant Ca^{2+} buffers as neuroprotectants in vitro or in vivo.

Experiments in Vitro

Several experiments in cultured cells, including neurons, have been summarized above (e.g., Fig. 3). Briefly, these experiments suggested that loading cells with the permeant chelators prevented neurotoxicity triggered by excitotoxins such as glutamate. A range of Ca^{2+} chelators have this property, although the effects of some of the compounds cannot be explained based on conventional measurements of $[Ca^{2+}]_i$ using fluorescent dyes such as fura-2 (see above). It should be noted that these protective effects noted thus far in vitro have exploited experimental paradigms in which Ca^{2+}-dependent cell death occurred rapidly (within several minutes). Thus, cell death in these reports was gauged within 1–2 h of the onset of Ca^{2+} loading. However, in a number of studies in which Ca^{2+} chelators such as BAPTA-AM were applied to cells, protective effects were not noted. For example, Dubinsky (136) performed experiments in which hippocampal cultures were pretreated with BAPTA-AM, and cell death was gauged at 24 h following a 5 min insult with 500 μM glutamate. Low BAPTA-AM concentrations (4 μM)

failed to protect the cells, whereas high concentrations (100 μm) were found to be toxic at the 24 h time point. Similar observations have been reported by other investigators *(137)*.

A weakness of all the reports of the neuroprotective efficacy of Ca^{2+} chelators thus far has been the fact that the intracellular concentrations and subcellular location of the buffer were unknown. Reported results to date may suffer from significant variability owing to the fact that the physiological actions of Ca^{2+} chelators depend on their intracellular concentration (*see* above). However, when the cell-permeant forms are used, their penetration into the cell consists of a nonequilibrium process (Fig. 1A) in which the final cytoplasmic concentration of the exogenous buffer is difficult to measure and to predict. Buffer loading into the cell depends on numerous variables, including the extracellular concentration, the loading time, the temperature at which loading is performed, the cell type, and intracellular esterase activity. Also, the cell permeant buffers are prone to undergo spontaneous hydrolysis of the AM moieties, raising the possibility that the method of reconstituting and storing the chelators may also introduce variability in their final intracellular concentrations. It is likely that, like all drugs, exogenous Ca^{2+} chelators must be delivered to their target sites of action at optimal concentrations, and that excessive quantities are toxic, whereas low quantities may be insufficient to protect neurons from a given insult. Presently, no data exist for any Ca^{2+} chelator to serve as a guide to the relationships between the intracellular chelator concentration and its neuroprotective or neurotoxic potential.

Experiments in Vivo

Few reports exist on the utility of cell-permeant Ca^{2+} chelators following neuronal injury in animals. Here, the delivery of the compounds into central neurons is even more problematic and uncertain than in tissue cultures. However, recently it has been shown that small quantities of cell-permeant Ca^{2+} chelating compounds can reach brain neurons following their intravenous infusion in dimethyl sulfoxide (DMSO) vehicle. When they accumulate in the brain at sufficient concentrations, they produce physiological effects compatible with intracellular Ca^{2+} chelation, and protect neurons following ischemic brain injury *(130)*.

Since low concentrations of a cell-permeant chelator (in the micromolar range) are sufficient to profoundly affect intracellular $[Ca^{2+}]_i$ dynamics (*[59]; see* above), the drug can be administered systemically at concentrations which would have no effect on extracellular Ca^{2+} (which is in the millimolar range), but which suffices to affect intracellular Ca^{2+} ion concentration (normally about 100 nM). Recent experiments in rats using radiolabeled ^{14}C-BAPTA-AM have shown that, after an intravenous infusion which produces peak serum concentrations of 10–15 μm of the compound, sufficient BAPTA eventually accumulates in the brain to produce protective effects (Shiino, Tymianski, and Wallace, personal observation). No toxicity attributable to BAPTA-AM administration in vivo has been reported.

In a model of focal cerebral ischemia in rats, administration of BAPTA-AM resulted in a reduction of the infarcted brain volume. This effect was also reproduced using the compounds 5,5'-difluoro BAPTA (K_d similar to BAPTA) and its structural isomer 4,4'-difluoro-BAPTA-AM (K_d about 5 μM; see ref. *130*). This latter finding parallels results from in vitro experiments (Fig. 3D; *see* ref. 110) in which chelators having a relatively low Ca^{2+} affinity sufficed to protect neurons from excitotoxicity. In a further study, the ability of BAPTA-AM to reduce focal infarction size was shown to persist for at least 24 h *(134)*. Recently, intravenous infusion of BAPTA-AM in rats was also shown to attenuate neuronal damage in a model of global cerebral ischemia (Shiino, Tymianski, and Wallace, personal observation). These few studies have only begun to reveal that intracellular Ca^{2+} buffering using exogenous buffers may be a useful neuroprotective approach. However, considerable future research is required to ascertain the utility of these compounds as neuroprotectants, and to optimize their use.

Pharmacokinetic data on the behavior of BAPTA and BAPTA-like compounds in vivo are lacking. No dose-response studies exist to suggest what might be an optimal regimen for administering Ca^{2+} buffers in stroke. Studies of neuroprotection in vivo have been restricted to paradigms in which the animal is pretreated with the permeant chelator prior to induction of cerebral ischemia, because experience has shown that current methods of administration require several hours for the drug to accumulate within central neurons. Studies in which the ischemic insult is produced before drug administration (post-treatment paradigm) have not appeared. Experimental work suggests that any freely diffusible Ca^{2+} chelating compound having a Ca^{2+} affinity in the range of 0.1–25 μm and a rapid Ca^{2+} binding rate may be neuroprotective. However, a systemic exploration of alternatives to BAPTA-like compounds is only now underway. Other possible improvements might include drug vehicles which make the systemic administration of acetoxymethyl ester compounds simpler, and their delivery into the brain more predictable.

CONCLUSIONS

Cell-permeant Ca^{2+} chelators have revolutionized our ability to study Ca^{2+} homeostasis in the nervous system. They possess a number of physical properties which can be exploited to probe aspects of Ca^{2+} signaling which cannot be resolved by other means, and to set temporal and spatial constraints on the subcellular compartmentalization of Ca^{2+}-dependent processes. The phenomenon of Ca^{2+} loading following nerve cell injury is a recognized mechanism which may lead to neurotoxicity once Ca^{2+} regulatory mechanisms fail in critical subcellular compartments. Studies of Ca^{2+} loading and neurotoxicity using cell-permeant Ca^{2+} chelators confirm the notion that $[Ca^{2+}]_i$ must rise to previously unexpected levels in critical subcellular compartments in order for cell death to occur. Additional studies suggest that these exogenous Ca^{2+} buffers may protect neurons from excitotoxic, anoxic or ischemic injuries in vitro and in vivo, and thus, may have future utility as a distinct class of neuroprotective compounds.

ACKNOWLEDGMENTS

M.T. is a clinician scientist of the Medical Research Council of Canada. Some of the work described in this chapter was funded by grants from the Ontario Heart and Stroke Foundation, and from an Ontario Technology Fund grant in collaboration with Allelix Biopharmaceuticals.

REFERENCES

1. McLean AEM, McLean E, Judah JD. Cellular necrosis in the liver induced and modified by drugs. *Int Rev Exp Pathol* 1965, **4**: 127–157.

2. Schanne FAX, Kane AB, Young EA, Farber JL. Calcium dependence of toxic cell death: A final common pathway. *Science* 1979, **206:** 700–702.
3. Cheung JY, Bonventre JV, Malis CD, Leaf A. Calcium and ischemic injury. *N Eng J Med* 1986, **314:** 1670–1676.
4. Tymianski M, Charlton MP, Carlen PL, Tator CH. Source specificity of early calcium neurotoxicity in cultured embryonic spinal neurons. *J Neurosci* 1993, **13:** 2085–2104.
5. Waxman SG, Black JA, Ransom BR, Stys PK. Protection of the axonal cytoskeleton in anoxic optic nerve by decreased extracellular calcium. *Brain Res* 1993, **614:** 137–145.
6. Stys PK, Ransom BR, Waxman SG, Davis PK. Role of extracellular calcium in anoxic injury of mammalian central white matter. *Proc Natl Acad Sci USA* 1990, **87:** 4212–4216.
7. Goldberg MP, Giffard RG, Kiurth MC, Choi DW. Role of extracellular calcium and magnesium in ischemic neuronal injury in vitro. *Neurology* 1989, **39(suppl):** 217 (Abstract).
8. Choi DW. Calcium-mediated neurotoxicity: Relationship to specific channel types and role in ischemic damage. *TINS* 1988, **11:** 465–467.
9. Choi DW. Ionic dependence of glutamate neurotoxicity. *J Neurosci* 1987, **7:** 369–379.
10. Garthwaite G, Hajos F, Garthwaite J. Ionic requirements for neurotoxic effects of excitatory amino acid analogues in rat cerebellar slices. *Neuroscience* 1986, **18:** 437–447.
11. Kass IS, Lipton P. Calcium and long-term transmission damage following anoxia in dentate gyrus and CA1 regions of rat hippocampal slice. *J Physiol* 1986, **378:** 313–334.
12. Choi DW. Glutamate neurotoxicity in cortical cell culture is calcium dependent *Neurosci Lett* 1985, **58:** 293–297.
13. Schlaepfer WW, Bunge RP. Effects of calcium ion concentration on the degeneration of amputated axons in tissue culture *J Cell Biol* 1973, **59:** 456–470.
14. Balentine JD, Paris DU, Dean DL. Calcium-induced spongiform and necrotizing myelopathy. *Lab Invest* 1982, **47:** 286–295.
15. Balentine JD, Paris DU, Greene WB. Ultrastructural pathology of nerve fibers in calcium-induced myelopathy. *J Neuropathol Exp Neurol* 1984, **43:** 500–510.
16. Balentine JD. Spinal cord trauma: In search of the meaning of granular axoplasm and vesicular myelin. *J Neuropathol Exp Neurol* 1988, **47:** 77–92.
17. Hansen AJ, Zeuthen T. Extracellular ion concentrations during spreading depression and ischemia in the rat brain cortex. *Acta Physiol Scand* 1981, **113:** 437–445.
18. Simon RP, Griffiths T, Evans MC, Swan JH, Meldrum BS. Calcium overload in selectively vulnerable neurons of the hippocampus during and after ischemia: an electron microscopy study in the rat. *J Cereb Blood Flow Metab* 1984, **4:** 350–361.
19. Chen ST, Hsu CY, Hogan EL, Juan HY, Banik NL, Balentine JD. Brain calcium content in ischemic infarction. *Neurology* 1987, **37:** 1227–1229.
20. Uematsu D, Greenberg JEI, Mickey WF, Reivich M. Nimodipine attenuates both increase in cytosolic free calcium and histologic damage following focal cerebral ischemia and reperfusion in cats. *Stroke* 1989, **20:** 1531–1537.
21. Benveniste H, Diemer NH. Early postischemic ^{45}Ca accumulation in rat dentate hilus. *J Cereb Blood Flow Metab* 1988, **8:** 713–719.
22. Silver IA, Erecinska M. Intracellular and exctracellular changes of [Ca^{2+}] in hypoxia and ischemia in rat brain in vivo. *J Gen Physiol* 1990, **95:** 837–866.
23. Michaels RL, Rothman SM. Glutamate neurotoxicity in vitro: antagonist pharmacology and intracellular calcium concentrations. *J Neurosci* 1990, **10:** 283–292.
24. Dubinsky JM, Rothman SM. Intracellular calcium concentration during "chemical hypoxia" end excitotoxic neuronal injury. *J Neurosci* 1991, **11:** 2545–2551.
25. Kurth MC, Weiss JH, Choi DW. Relationship between glutamate-induced 45-Calcium influx and resultant neuronal injury in cultured cortical neurons. *Neurology* 1989, **39(suppl):** 217 (Abstract).

26. Hartley DM, Kurth MC, Bjerkness L, Weiss JH, Choi DW. Glutamate receptor- induced $^{45}Ca^{2+}$ accumulation in cortical cell culture correlates with subsequent neuronal degeneration. *J Neurosci* 1993, **13**: 1993–2000.
27. Schramm M, Eimerl S. The quantity of Ca that enters a neuron to cause its death in glutamate toxicity. *Soc Neurosci Abstr* 1993, **19**: 1501 (Abstract).
28. Eimrel S, Schramm M. The quantity of calcium that appears to induce neuronal death *J Neurochem* 1994, **62**: 1223–1226.
29. Goldberg MP, Kurth MC, Giffard RG, Choi DW. ^{45}Calcium accumulation and intracellular calcium during in vitro "ischemia". *Soc Neurosci Abstr* 1989, **15**: 803(Abstract).
30. Lobner D, Lipton P. Intracellular calcium levels and calcium fluxes in the CA1 region of the rat hippocampal slice during in vitro ischemia: Relationship to electrophysiological cell damage *J Neurosci* 1993, **13**, 4861–4871.
31. Lu YM, Yin HZ, Weiss JH. Ca^{2+} permeable AMPA/kainate channels permit rapid injurious Ca^{2+} entry *Neuroreport* 1995, **6**: 1089–1092.
32. Lipscombe D, Madison D, Poenie M, Reuter H, Tsien RY, Tsien RW. Spatial distribution of calcium channels and cytosolic transients in growth cones and cell bodies of sympathetic neurons. *Proc Natl Acad Sci USA* 1988, **85**: 2398–2402.
33. Robitaille R, Adler EM, Charlton MP. Strategic location of calcium channels at transmitter release sites of frog neuromuscular synapses. *Neuron* 1990, **5**: 773–779.
34. Robitaille R, Garcia ML, Kaczorowski GJ, Charlton MP Functional colocalization of calcium and calcium-gated potassium channels in control of transmitter release *Neuron* 1993, 11 645-655.
35. Westenbroek RE, Ahlijanian MK, Catterall WA. Clustering of L-type Ca^{2+} channels at the base of major dendrites in hippocampal neurons. *Nature* 1990, **347**: 281–284.
36. Pollock JA, Assaf A, Peretz A, Nichols CD, Mojet MH, Hardie RC, Minke B. TRP, a protein essential for inositide-mediated Ca^{2+} influx is localized adjacent to the calcium stores in Drosophila photoreceptors. *J Neurosci* 1995, **15**: 3747–3760.
37. Connor JA, Wadman WJ, Hockberger PE, Wong RKS. Sustained dendritic gradients of calcium induced by excitatory amino acids in CA1 hippocampal neurons. *Science* 1988, **240**: 649–653.
38. Jones OT, McGurk JF, Bennett MVL, Zuckin RS, Collinridge G, Benke T, Angelides KJ. Distribution of NMDA receptors on hippocampal neurons. *Soc Neurosci Abstr* 1990, **2**: No 3963 (Abstract).
39. Jones KA, Baughman RW. Both NMDA and non-NMDA subtypes of glutamate receptors are concentrated at synapses on cerebral cortical neurons in culture. *Neuron* 1991, **7**: 593–603.
40. Muller W, Connor JA. Dendritic spines as individual neuronal compartments for synaptic Ca^{2+} responses. *Nature* 1991, **354**: 73–80.
41. Craig AM, Blackstone CD, Hijganir RL, Banker G. The distribution of glutamate receptors in cultured rat hippocampal neurons: Postsynaptic clustering of AMPA- selective subunits. *Neuron* 1993, **10**: 1055–1068.
42. Benke TA, Jones OT, Collingridge GL, Angelides KJ. N-Methyl-D-aspartate receptors are clustered and immobilized on dendrites of living cortical neurons. *Proc Natl Acad Sci USA* 1993, **30**: 7819–7823.
43. Guthrie PB, Segal M, Kater SB. Independent regulation of calcium revealed by imaging dendritic spines. *Nature* 1991, **354**: 76–80.
44. Yuste R, Gutnick MJ, Saar D, Delaney KR, Tank DW. Ca^{2+} accumulations in dendrites of neocortiical pyramidal neurons: An apical band and evidence for two functional compartments. *Neuron* 1995, **13**: 23–43.
45. Seymour-Laurent KJ, Barish ME. Inositol 1,4,5-triphosphate and ryanodine receptor distribution and patterns of acetylcholine- and caffeine-induced calcium release in cultured mouse hippocampal neurons. *J Neurosci* 1995, **15**: 2592–2608.

46. Hernandez-Cruz A, Sala F, Adams PR. Subcellular calcium transients visualized by confocal microscopy in a voltage-clamped vertebrate neuron. *Science* 1990, **247**: 858–862.
47. Delisle S, Welsh MJ. Inositol triphosphate is required for the propagation of calcium waves in xenopus oocytes. *J Biol Chem* 1992, **267**: 7963–7966.
48. Lechleiter JD, Clapham DE. Molecular mechanisms of intracellular calcium excitability in X. laevis oocytes *Cell* 1992, **69**: 283–294.
49. Connor JA. Intracellular calcium mobilization by inositol 1,4,5-trisphosphate: intracellular movements and compartmentalization. *Cell Calcium* 1993, **14**: 185–200.
50. Lerea LS, McNamara JO. Ionotropic glutamate receptor subtypes activate c-fos transcription by distinct calcium-requiring intracellular signalling pathways. *Neuron* 1993, **10**: 31–41.
51. Bading H, Ginty DD, Greeneberg ME. Regulation of gene expression in hippocampal neurons by distinct calcium signalling pathways. *Science* 1993, **260**: 181–186.
52. Hajimohammadreza I, Probert AW, Coughenour LL, Borosky SA, Marcoux FW, Boxer PA, Wang KKW. A specific inhibitor of calcium/calmodulin-dependent protein kinase-II provides neuroprotection against NMDA- and hypoxia/hypoglycemia-induced cell death. *J Neurosci* 1995, **15**: 4093–4101.
53. Reynolds IJ, Hastings TG. Glutamate induces the production of reactive oxygen species in cultured forebrain neurons following NMDA receptor activation. *J Neurosci* 1995, **15**: 3318–3327.
54. Wahlestedt C, Golanov E, Yamamoto S, Yee F, Ericson H, Yoo H, Inturrisi CE, Reis DJ. Antisense oligonucleotides to NMDA-R1 receptor channel protect cortical neurons from excitotoxicity and reduce focal ischaemic infarctions. *Nature* 1993, **363**: 260–263.
55. Lafon-Cazal M, Pietri S, Culcasi M, Bockaert J. NMDA-dependent superoxide production and neurotoxicity. *Nature* 1993, **364**: 535–537.
56. Roberts-Lewis JM, Marcy VR, Zhao Y, Vaught JL, Siman R, Lewis ME. Aurintricarboxylic acid protects hippocampal neurons from NMDA-and ischemia-induced toxicity in vivo. *J Neurochem* 1993, **61**: 378–381.
57. Baimbridge KG, Celio MR, Rogers JH. Calcium-binding proteins in the nervous system. *TINS* 1992, **15**: 303–308.
58. Neher E, Augustine GJ. Calcium gradients and buffers on bovine chromaffin cells. *J Physiol* 1992, **450**: 273–301.
59. Zhou Z, Neher E. Mobile and immobile calcium buffers in bovine adrenal chromaffin cells. *J Physiol* 1993, **469**: 245–273.
60. Chard PS, Bleakman D, Christakos S, Fullmer CS, Miller RJ. Calcium buffering properties of calbindin D28k and parvalbumin in rat sensory neurones. *J Physiol* 1993, **472**: 341–357.
61. Nowycky MC, Pinter MJ. Tiime courses of calcium and calcium-bound buffers following calcium influx in a model cell. *Biophys J* 1993, **64**: 77–91.
62. Kasai H, Peterson OH. Spati.al dynamics of second messengers: IP3 and cAMP as long-range and associative messengers. *TINS* 1994, **17**: 95–101.
63. Roberts WM. Localization of carlcium signals by a mobile calcium buffer in frog saccular hair cells. *J Neurosci* 1994, **15**: 3246–3262.
64. Roberts WM. Spatial calcium bruffering in saccular hair cells. *Nature* 1993, **363**: 74–76.
65. Speksnijder JE, Miller AL, Weisenseel MH, Chen TH, Jaffe LF. Calcium buffer injections block fucoid egg development by facilitating calcium diffusion. *Proc Natl Acad Sci USA* 1989, **86**: 6607–6611.
66. Stern MD. Buffering of calcium in the vicinity of a channel pore. *Cell Calcium* 1992, **13**: 183–192.
67. Sala F, Hernandez-Cruz A. Calcium diffusion modelling in a spherical neuron: relevance of buffering properties. *Biophys J 1990,* **57**: 313–324.

68. Neher E. Concentration profiles of intracellular calcium in the presence of a diffusible chelator, in *Calcium Electrogenesis and Neuronal Functioning,* 1st ed. (Heinemann U, Klee M, Neher E, eds.), Springer-Verlag, Berlin-Heidelberg: Exp Brain Res, Series 14, 1986, pp. 80–96.
69. Thayer SA, Miller RJ. Regulation of the intracellular free calcium concentration in single rat dorsal root ganglion neurones in vitro. *J Physiol* 1990, **425**: 85–115.
70. Lledo PM, Somasundram B, Morton AJ, Emson PC, Mason WT. Stable transfection of calbindin-D28k into the GH3 cell line alters calcium currents and intracellular calcium homeostasis. *Neuron* 1992, **9**: 943–954.
71. Pozzan T, Rizzuto R, Volpe, P, Meldolesi J. Molecular and cellular physiology of intracellular calcium stores. *Physiolgical Rev* 1994, **74**: 595–635.
72. Werth JL, Thayer SA. Mitochondria buffer physiological calcium loads in cultured rat dorsal root ganglion neurons. *J Neurosci* 1994, **14**: 348–356.
73. White RJ, Reynolds IJ. Mitochondria and Na+/Ca^{2+} exchange buffer glutamate-induced calciumloads in cultured cortical neurons. *J Neurosci* 1995, **15**: 1318–1328.
74. Blaustein MP. Calcium transport and buffering in neurons. *TINS* 1988, **11**: 438–443.
75. Gunter TE, Pfeiffer DR. Mechanisms by which mitochondria transport calcium. [Review]. *Am J Physiol* 1990, **258**: C755–C786.
76. Moudy AM, Handran SD, Goldberg MP, Ruffin N, Karl I, Kranzeble P, DeVivo DC, Rothman SM. Abnormal calcium homeostasis and mitochondrial polarization in a human encephalomyopathy. *Proc Natl Acad Sci USA* 1995, **92**: 729–733.
77. Scharfman HE, Schwartzkroin PA. Protection of dentate hilar cells from prolonged stimulation by intracellular calcium chelation. *Science* 1989, **246**: 257–260.
78. Lukas W, Jones KA. Cortical neurons containing calretinin are selectively resistant to calcium overload and excitotoxicity in vitro. *Neuroscience* 1994, **61**: 307–316.
79. Mattson MP, Rychlik B, Chu JC, Christakos S. Evidence for calcium-reducing and excito-protective roles for the calcium-binding protein calbindin-D28k in cultured hippocampal neurons. *Neuron* 1991, **6**: 41–51.
80. Diop AG, Lesort M, Esclaire F, Dumas M, Hugon J. Calbindin D28K-containing neurons, and not HSP70-expressing neurons, are more resistant to HIV-1 envelope (gp120) toxicity in cortical cell cultures. *J Neurosci Res* 1995, **42**: 252–258.
81. Pike CJ, Cotman CW. Calretinin-immunoreactive neurons are resistant to beta-amyloid toxicity in vitro. *Brain Res* 1995, **671**: 293–298.
82. Mockel V, Fischer G. Vulnerability to excitotoxic stimuli of cultured rat hippocampal neurons containing the calcium-binding proteins calretinin and calbindin D28K. *Brain Res* 1994, **648**: 109–120.
83. Weiss JH, Koh JY, Baimbridge KG, Choi DW. Cortical neurons containing somatostatin- or parvalbumin-like immunoreactivity are atypically vulnerable to excitotoxic injury in vitro. *Neurology* 1990, **40**: 1288–1292.
84. Dowd DR, MacDonald PN, Komm BS, Haussler MR, Miesfeld RL. Stable expression of the calbindin-D2XK complementary DNA interferes with the apoptotic pathway in Iymphocytes. *Mol Endocrinol* 1992, **6**:1843–1848.
85. Vito P, Lacana E, D'Adamio L. Interfering with apoptosis: $Ca^{(2+)}$-binding protein ALG-2 and Alzheimer's disease gene ALG-3. *Science* 1996, **271**: 521–525.
86. Yui S, Mikami M, Yamazaki M. Induction of apoptotic cell death in mouse Iymphoma and human leukemia cell lines by a calcium-binding protein complex, calprotectin, derived from inflammatory peritoneal exudate cells. *J Leukoc Biol* 1995, **58**: 650–658.
87. Goodman JH, Wasterlain CG, Massarweh WF, Dean E, Sollas AL, Sloviter RS. Calbindin-D28K immunoreactivity and selective vulnerability to ischemia in the dentate gyrus of the developing rat. *Brain Res* 1993, **606**: 309–314.
88. Nitsch C, Scotti A, Sommacal A, Kalt G. GABAergic hippocampal neurons resistant to ischemia-induced neuronal death contain the Ca2(+)-binding protein parvalbumin. *Neurosci Lett* 1989, **105**: 263–268.

89. Rami A, Rabie A, Thomasset M, Krieglstein J. Calbindin-D28K and ischemic damage of pyramidal cells in rat hippocampus. *J Neurosci Res* 1992, **31:** 89–95.
90. German DC, Manaye KF, Sonsalla PK, Brooks BA. Midbrain dopaminergic cell loss in Parkinson's disease and MPTP-induced parkinsonism: sparing of calbindin-D28k-containing cells. *Ann NY Acad Sci* 1992, **648:** 42–62.
91. Mouatt-Prigent A, Agid Y, Hirsch EC. Does the calcium binding protein calretinin protect dopaminergic neurons against degeneration in Parkinson's disease? *Brain Res* 1994, **668:** 62–70.
92. Yamada T, McGeer PL, Baimbridge KG, McGeer KG. Relative sparing in Parkinson's disease of substantia nigra dopamine neurons containing calbindin-D28K. *Brain Res* 1990, **526:** 303–307.
93. Reiner A, Medina L, Figueredo-Cardenas G, Anfinson S. Brainstem motoneuron pools that are selectively resistant in amyotrophic lateral sclerosis are preferentially enriched in parvalbumin: Evidence from monkey brainstem for a calcium-mediated mechanism in sporadic ALS. *Exp Neurol* 1995, **131:** 239–250.
94. Harkany T, DeJong GI, Soos K, Penke B, Luiten PG, Gulya K. Beta-amyloid (1-42) affects cholinergic but not parvalbumin-containing neurons in the septal complex of the rat. *Brain Res* 1995, **698:** 270–274.
95. Freund TF, Buzsaki G, Leon A, Baimbridge KG, Somogyi P. Relationship of neuronal vulnerability and calcium binding protein immunoreactivity in ischemia. *Exp Brain Res* 1990, **83:** 55–66.
96. Tortosa A, Ferrer I. Poor correlation between delayed neuronal death induced by transient forebrain ischemia, and immunoreactivity for parvalbumin and calbindin D-28k in developing gerbil hippocampus. *Acta Neuropathol (Berl)* 1994, **88:** 67–74.
97. Tsien RY. Fluorescence measurement and photochemical manipulation of cytosolic free calcium. *TINS* 1988, **11:** 419–424.
98. Roe MW, Lemasters JJ, Herman B. Assessment of fura-2 for measurements of cytosolic free calcium. *Cell Calcium* 1990, **11:** 63–73.
99. Moore EDW, Becker PL, Fogarty KE, Williams DA, Fay FS. Ca^{2+} imaging in single living cells: Theoretical and practical issues. *Cell Calcium* 1990, **11:** 157–179.
100. Tsien RY. A non-disruptive technique for loading calcium buffers and indicators into cells. *Nature* 1981, **290:** 577–528.
101. Tsien RY. New calcium indicators and buffers with high selectivity against magnesium and protons: Design, synthesis, and properties of prototype structures. *Biochemistry* 1980, **19:** 2396.
102. Williams S, Johnston D. Long-term potentiation of hippocampal mossy fiber synapses is blocked by postsynaptic injection of calcium chelators. *Neuron* 1989, **3:** 583–588.
103. Winslow JL, Duffy SN, Charlton MP. Homosynaptic facilitation of transmitter release in crayfish is not affected by mobile calcium chelators: Implications for the residual ionized calcium hypothesis from electrophysiological and computational analyses. *J Neurophysiol* 1994, **72:** 1769–1793.
104. Zhang L, Pennefather P, Velumian A, Tymianski M, Charlton M, Carlen PL. Potentiation of a slow Ca^{2+}-dependent K+ current by intracellular Ca^{2+} chelators in hippocampal CA1 neurons of rat brain slices *J Neurophysiol* 1995, **74:** 2225–2241 (Abstract).
105. Spigelman I, Tymianski M, Wallace MC, Carlen PL, Velumian AA. Modulation of hippocampal synaptic transmission by low concentrations of cell-permeant calcium chelators: Effects of calcium affinity, structure, and binding kinetics. *Neuroscience* 1996, (in press).
106. Ouanounou A, Zhang L, Tymianski M, Charlton MP, Wallace MC, Carlen PL. Accumulation and extrusion of permeant Ca^{2+} chelators in attenuation of synaptic transmission at hippocampal CA1 neurons. *Neuroscience* 1996, (in press).
107. Smith SJ, Augustine GJ. Calcium ions, active zones, and synaptic transmitter release. *TINS* 1988, **11:** 458–464.

108. Augustine GJ, Adler EM, Charlton MP. The calcium signal for transmitter secretion from presynaptic nerve terminals. *Ann NY Acad Sci* 1991, **635**: 365–381.
109. Adler EM, Augustine GJ, Duffy SN, Charlton MP. Alien intracellular calcium chelators attenuate neurotransmitter release at the squid giant synapse. *J Neurosci* 1991, **11**: 1496–1507.
110. Tymianski M, Charlton MP, Carlen PL, Tator CH. Properties of neuroprotective cell-permeant Ca^{2+} chelators: Effects on $[Ca^{2+}]i$ and glutamate neurotoxicity in vitro *J Neurophysiol* 1994, **267**: 1973–1992.
111. Wang SS, Thompson SH. Local positive feedback by calcium in the propagation of intracellular calcium warves. *Biophys J* 1995, **69**: 1683–1697.
112. Penning LC, Keirse MJ, Vansteveninck J, Dubbelman TM. $Ca(^{2+})$mediated prostaglandin E2 induction reduces haematoporphyrin-derivative-induced cytotoxicity of T24 human bladder transitional carcinoma cells in vitro. *Biochem J* 1993, **292**: 237–240.
113. Ueda N, Shah SV. Role of intracellular calcium in hydrogen peroxideinduced renal tubular cell injury. *Am J Physiol* 1992, **263**: Pt 2):F214–21.
114. Babich H, Palace MR, Stern A. Oxidative stress in fish cells: In vitro studies. *Arch Environ Contam Toxicol* 1993, **24**: 173–178.
115. Fernandez A, Kiefer J, Fosdick L, McConkey DJ. Oxygen radical production and thiol depletion are required for $Ca(^{2+})$-mediated endogenous endonuclease activation in apoptotic thymocytes. *J Immunol* 1995, **155**: 5133–5139.
116. Cantoni O, Sestili P, Cattalbeni F, Bellomo G, Pou S, Cohen M, Cerutti P. Calcium chelator Quin 22 prevents hydrogen-peroxide-induced DNA breakage and cytotoxicity. *Eur J Biochem* 1989, **182**: 209–212.
117. Coorssen JR, Haslam RJ. GTPgammaS and phorbol ester act synergistically to stimulate both Ca^{2+}-independent secretion and phospholipase D activity in permeabilized human platelets. *FEBS Lett* 1993, **316**: 170–174.
118. Pellegrini-Giampietro DE, Cherici G, Alesiani M, Carla V, Moroni F. Excitatory amino acid release and free radical formation may cooperate in the genesis of ischemia-induced neuronal damage. *J Neurosci* 1990, **10**: 1035–1041.
119. Dugan LL, Sensi SL, Canzoniero LMT, Handran SD, Rothman SM, Lin TS, Goldberg MP, Choi DW. Mitochondrial production of reactive oxygen species in cortical neurons following exposure to N-methyl-D-aspartate. *J Neurosci* 1995, **15**: 6377–6388.
120. Sanfeliu C, Hunt A, Patel AJ. Exposure to N-methyl-D-aspartate increases release of arachidonic acid in primary cultures of rat hippocampal neurons and not in astrocytes. *Brain Res* 1990, **526**: 241–248.
121. Lazarewicz JW, Wroblewski JT, Costa E. N-Methyl-D-Aspartate sensitive glutamate receptors induce calcium-mediated arachidonic acid release in primary cultures of cerebellar granule cells. *J Neurochem* 1990, **55**: 1875–1881.
122. Dumuis A, Sebben M, Haynes L, Pin JP, Bockaert J: NMDA receptors activate the arachidonic acid cascade system instriatal neurons. *Nature* 1988, **336**: 68–70.
123. Yoshida A, Ueda T, Takauji R, Liu YP, Fukushima T, Inuzuka M, Nakamura T. Role of calcium ion in induction of apoptosis by etoposide in human leukemia HL-60 cells. *Biochem Biophys Res Commun* 1993, **196**: 927–934.
124. Zhivotovsky B, Cedervall, B, Jiang S, Nicotera P, Orrenius S. Involvement of Ca^{2+} in the formation of high molecular weight DNA fragments in thymocyte apoptosis. *Biochem Biophys Res Commun* 1994, **202**: 120–127.
125. Robertson LE, Chubb S, Meyn RE, Story M, Ford R, Hittelman WN, Plunkett W. Induction of alpoptotic cell death in chronic lymphocytic leukemia by 2-chloro-2'-deoxyedenosine and 9-beta-D-arabinosyl-2-fluoroadenine. *Blood* 1993, **81**: 143–150.
126. Jiang S, Chow SC, Nicotera P, Orrenius S. Intracellular Ca^{2+} signals activate apoptosis in thymocytes: studies using the $Ca(^{2+})$-ATPase inhibitor thapsigargin. *Exp Cell Res* 1994, **212**: 84–92.

127. Penning LC, Rasch MH, Ben-Hur E, Dubbelman TM, Havelaar AC, Van Der Zee J, Van Steveninck J. A role for the transient increase of cytoplasmic free calcium in cell rescue after photodynamic treatment. *Biochim Biophys Acta* 1992, **1107**: 255–260.
128. Whyte MK, Hardwick SJ, Meagher LC, Savill JS, Haslett C. Transient elevations of cytosolic free calcium retard subsequent apoptosis in neutrophils in vitro. *J Clin Invest* 1993, **92**: 446–455.
129. Kudo Y, Takeda K, Yamazaki K. Quin-2 protects neurons against cell death due to Ca^{2+} overload. *Brain Res* 1990, **528**: 48–54.
130. Tymianski M, Wallace MC, Spigelman I, Uno M, Carlen PL, Tator CH, Charlton MP. Cell permeant Ca^{2+} chelators reduce early excitotoxic and ischemic neuronal injury in vitro and in vivo. *Neuron* 1993, **11**: 221–235.
131. Randall RD, Thayer SA. Glutamate-induced calcium transient triggers delayed calcium overload and neurotoxicity in rat hippocampal neurons. *J Neurosci* 1992, **12**: 1882–1895.
132. Tymiaski M, Charlton MP, Carlen PL, Tator CH. Secondary Ca^{2+} overload indicates early neuronal injury which precedes staining with viability indicators. *Brain Res* 1993, **607**: 319–323.
133. Petrozzino JJ, Pozzo Miller LD, Connor JA. Micromolar Ca^{2+} transients in dendritic spines of hippocampal pyramidal neurons in brain slice. *Neuron* 1995, **14**: 1223–1231.
134. Tymianski M, Spigelman I, Zhang L, Carlen PL, Tator CH, Charlton MP, Wallace MC. Mechanism of action and persistence of neuroprotection by cell permeant Ca^{2+} chelators. *J Cereb Blood Flow Metab* 1994, **14**: 911–923.
135. Niesen C, Charlton MP, Carlen PL. Postsynaptic and presynaptic effects of the calcium chelator BAPTA on synaptic transmission in rat hippocampal dentate granule neurons. *Brain Res* 1991, **555**: 319–325.
136. Dubinsky JM. Effects of calcium chelators on intracellular calcium and excitotoxicity. *Neurosci Lett* 1993, **150**: 129–132.
137. Baimbridge KG, Abdel-Hamid KM. Intra-neuronal Ca^{2+}-buffering with BAPTA enhances glutamate excitotoxicity *in vitro* and ischemic damage *in vivo*. *Soc Neurosci Abstr* 1992, **18**.

31
Antiglutamate Therapies for Neurodegenerative Disease
The Case of Amyotrophic Lateral Sclerosis

Osnat Bar-Peled and Jeffrey D. Rothstein

INTRODUCTION

Neurodegenerative diseases are a heterogenous group of disorders that are typically adult onset, progressively debilitating, and/or fatal. They are often restricted to the central nervous system (CNS), e.g., Parkinson's disease, but in some cases can affect multiple organ systems, e.g., ataxia telangectasia. They are typified by selective neural cell degeneration that may affect cells that often share biological or biochemical functions. For example, upper and lower motor neuron degeneration in amyotrophic lateral sclerosis (ALS), dopaminergic neurons in Parkinson's disease, cerebellar Purkinje cells and olivary neurons in certain olivopontocerebellar atrophy disorders. In this chapter, the authors will review how the excitatory neurotransmitter glutamate may participate in the cellular degenerative process and discuss therapeutic interventions designed to slow the process. Importantly, glutamate-mediated neurotoxicity may act as either the primary or a secondary toxic agent in these diseases. As will be discussed below, the involvement of glutamate as a secondary event in neurodegeneration may provide a window of therapeutic opportunity to slow neuronal death in these disorders.

GLUTAMATE NEUROTRANSMISSION

Glutamate Receptors

Glutamate is a major excitatory neurotransmitter in the mammalian nervous system; it plays an important role in neurotransmission in most brain regions. Post synaptic glutamate receptors have been classified into two main types: ionotropic receptors, which are ligand-gated ion channels, and metabotropic receptors, which are linked through G proteins to secondary messenger systems *(1)*. The ionotropic receptors have been subdivided according to the pharmacological profile of their preferred agonists into two subtypes: *N*-methyl-D-aspartate (NMDA) receptors and alpha-amino-3-hydroxy-5methyl-4 isoxazole propionic acid/kainate (AMPA/kainate) receptors. Each ionotropic glutamate receptor is composed of four or five subunits arranged as hetro- or homomultimers. The variation of subunit assembly to form receptor complexes contribute to the diversity of glutamate receptors.

The NMDA receptor ion channel is highly permeable to calcium ions. Two principle subunits of the NMDA receptor have been cloned: NMDAR1 and NMDAR2. NMDAR1 consists of seven isoforms generated by alternative mRNA splicing, whereas NMDAR2 consists of four subtypes encoded by different genes

The AMPA/kainate receptors exist physiologically as heteromeric combinations of multiple subunits (GluR1-GluR7 and KA1-KA2). Depending on the particular subunit composition, the AMPA/kainate receptor may or may not be permeable to calcium; specifically the presence of GluR2 subunits inhibits calcium permeability (2). The functional properties of AMPA/kainate receptors can be further modified by post-transcriptional RNA editing of receptor subunits. For example, a change of a single amino acid in the GluR2 subunit determines whether or not the AMPA receptor channel is permeable to Ca^{+2}. In addition, alternative splicing of AMPA/kainate receptor genes generates multiple subunit variants. The flip and flop variants of the AMPA receptor subunits are the result of alternative splicing of adjacent exons (3,4). The variation in glutamate ionotropic receptors is very large, resulting from combination of known gene products and their splice variants, as well as RNA editing products. It is very possible that specific population of neurons within the CNS, such as motor neurons, will be characterized by a unique profile of glutamate receptors (5).

The metatrophic receptor consist of at least two types with at least eight subtypes mGluR1 - mGluR8. The first induces a G-protein-mediated activation of phospholipase C and inositol triphosphate metabolism and the second produces a G-protein-mediated inhibition of adenylate cyclase.

Glutamate Transporters

As a neurotransmitter, glutamate is stored in vesicles within the glutamatergic nerve terminals. After its release from nerve terminals, glutamate is actively removed from the synaptic cleft by uptake into the nerve terminals or the surrounding astrocytes. This process is handled by several types of high affinity glutamate transporters. Three rat and five human glutamate transporters have recently been cloned: GLT-1 (human excitatory amino acid transporter 2, EAAT2) which is an astroglial-specific glutamate transporter, GLAST (human EAAT1) which is another astroglial transporter, EAAC 1 (human EAAT3) which has a neuronal localization and is also expressed outside the nervous system, EAAT4 which expressed mostly in the cerebellum, and EAAT5 which is localized to the retina (6–9). Although all the transporters are sodium dependent, EAAT4 and EAAT5 also have a significant chloride conductance properties (10,11). Recent studies indicate that GLT-1 (EAAT2) may be the principal protein responsible for the bulk of glutamate transport in the brain, accounting for greater than 50% (12), and in some regions, up to 95% of all glutamate transport.

Glutamate Metabolism

Following its uptake by synaptic astrocytes, glutamate is converted to glutamine via the activation of the glial enzyme, glutamine synthetase. Glutamine is transported back to the nerve terminals where it can be recycled back to glutamate via the neuronal enzyme, glutaminase (13–16). In addition, glutamate taken up by astrocytes can be metabolized to alpha-ketoglutarate, a reaction catalyzed by glutamate dehydrogenase (GDH). The glutamate-recycling process by glutaminase is an important mechanism for replenishing glutamate levels within neuronal terminals, since new synthesis of glutamate from glucose in the brain is limited *(17–18)*. Therefore, modulation of glutamate metabolism by GDH may provide a mean for influencing glutamate recycling, as glutamate metabolites are usually not used for new glutamate synthesis in the brain.

GLUTAMATE TOXICITY AS A MEDIATOR OF NEURODEGENERATION

Amyotrophic lateral sclerosis (ALS) is largely a sporadic disease, with only 5% of all cases having dominant inheritance. Great progress in understanding the pathogenesis of ALS was made when some forms of familial ALS (FALS) were found to be due to mutations of the antioxidant enzyme superoxide dismutase 1 (SOD1). The SOD1 mutations appear to account for about 15–20% of FALS, but have not been found in the sporadic population. Thus, many other causes of ALS must exist, associated with diverse etiopathogenic mechanisms. Some of the mechanisms that have been proposed to cause or contribute to motor neuron degeneration in ALS include excitotoxicity, oxidative stress, cytoskeletal abnormalities, and autoimmunity. Chapters elsewhere in this volume have addressed the role of mutant SOD1 and neurofilaments in ALS. Below, we will outline some of the data suggesting that excitotoxicity contributes to motor neuron degeneration in ALS as well. Reviews on the role of excitotoxicity in ALS are also available *(19–22)*.

Defects in Glutamate Regulation in ALS

Studies from multiple laboratories have documented a series of abnormalities of glutamate metabolism all which point to the possible role of excess glutamate as a *cause* or, more likely, a *contributor* to motor neuron degeneration. There is no data, yet, which suggest that the loss of glutamate transport or even excess extracellular glutamate are the primary defect in the disease, but as will be discussed below, there are a series of studies that suggest that glutamate could contribute to, and propagate, the process of selective motor neuron degeneration.

Evidence that glutamate could participate in the pathogenesis of ALS came from prospective studies in which a threefold increase in mean CSF levels of glutamate and aspartate were found in ALS patients compared to appropriate controls *(23–25)*. These changes appear to be restricted to a subset of ALS patients. One possible mechanism for a chronic elevation of extracellular/CSF glutamate and aspartate could be the altered function of the glutamate transporter, which transports both amino acids with a low K_m and high capacity.

Subsequently, glutamate transport was measured in ALS postmortem tissue from a variety of brain and spinal cord regions in refs. *26* and *27*. A large loss of glutamate transport was observed only in the motor cortex and spinal cord, with preservation of transport activity in areas not typically affected in the disease, such as caudate and cerebellum. Because these are the most affected brain regions, several important controls were included in those studies to account for the possibility that changes were secondary to neuronal degeneration or prolonged agonal states associated with advanced stages of ALS. The same brain regions were also examined in Huntington's disease and Alzheimer's disease—no changes in glutamate transport were observed in these disease controls. Furthermore, the loss of transport was specific for the sodium-dependent glutamate carrier, because the transport of both phenylalanine and γ-amino butyric acid were normal in ALS motor cortex. Notably, a similar loss of glutamate transport was not seen in spinal muscular atrophy, another motor neuron disease distinct from ALS (27).

What accounted for the dramatic loss of this critically important transporter function? Perhaps it merely reflected the degeneration of cortical glutamatergic motor neurons. However, it was known that glutamate transport was present in both astroglia and neurons. To determine the transporter subtypes responsible for this dramatic loss of functional glutamate transport in ALS motor cortex and spinal cord, the transporter subtypes were individually studied in ALS and control CNS tissues. As described above, both GLAST and GLT-1 are astroglia-specific, whereas EAAC 1 is specific for neurons, including cortical motor neurons. Immunoblot analysis of brain regions in ALS reveals that the loss of glutamate transport appears to be due to a relatively selective loss of the astroglial, GLT-1 (EAAT2) subtype (9). There is a dramatic loss of the transporter protein in ALS motor cortex and spinal cord, while there are no substantial changes in either the other astroglial transporter, GLAST (EAAT1), or the neuronal transporter (*see* Fig. 1 for representative data). Approximately 25–30% of postmortem brains with sporadic ALS have greater than 75% loss of GLT-1 protein in motor cortex, with another approx 30% of specimens having 75–30% loss of the transporter protein, by semiquantitative immunoblot analysis. However, in some cases, no loss of protein has been seen. As yet, there is no clinical correlation between the loss of transport and any clinical parameter, such as rate of disease progression or age of onset.

Thus, it appears that the loss of functional glutamate transport in ALS motor cortex and spinal cord may be due to the selective loss of the astroglial GLT-1 glutamate transporter. This loss is not due to a loss of astroglia, as the glial marker GFAP and the other astroglial glutamate transporter, GLAST, are not decreased in ALS motor cortex. In fact, previous neuropathological studies had suggested that astrogliosis, and not a loss of astrocytes, may be present in ALS motor cortex (28). These results clearly suggest that a major abnormality of glutamatergic neurotransmission may exist in ALS, and that the loss of glutamate transport is not merely do to the degeneration of upper or lower motor neurons.

What could account for a selective loss of EAAT2 (GLT-1) in ALS? The decrease in GLT- 1 protein could still be secondary to motor neuron degeneration. As cortical glutamatergic motor neurons degenerate, alterations in glutamatergic neurotransmission could lead to a down regulation of the glutamate transporters. That hypothesis has been tested in an animal model of cortical neuron loss. Rat cortical neurons project, in part, to the underlying striatum. Following cortical ablation, glutamate transport does decrease in the rat striatum, but the decrease is only transient, and by one month after lesion, functional transport returns to normal (29,30), probably due to axonal sprouting. Recent studies indicate that neuron-glial interaction appear to regulate transporter expression (31). Thus, glial transporters demonstrate interesting injury responses, but these changes do not explain the persistent loss of EAAT2 (GLT-1) in ALS. Assuming one can extrapolate this experimental paradigm to human brain, the loss of cortical neurons may not be a sufficient explanation for the persistent loss of EAAT2 (GLT-1).

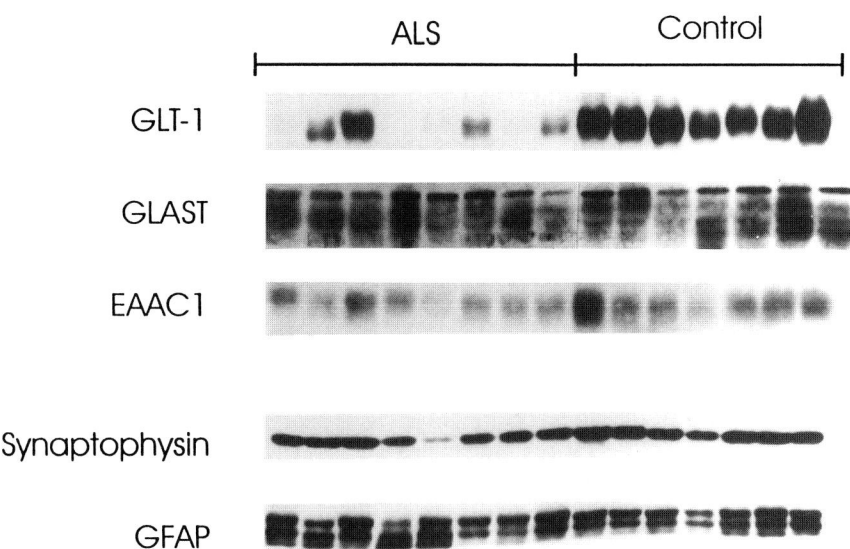

Fig. 1. Loss of GLT-1 (EAAT2) protein in ALS motor cortex. Each lane represents individual patients studied for three glutamate transporter subtypes (GLT-1, GLAST, EAAC1). In addition, the analysis was controlled for a general neuronal protein (synaptophysin) and an astroglial protein (GFAP). There was a dramatic decrease in GLT-1 protein in motor cortex and spinal cord (not shown), but not in other pathologically unaffected brain regions such as the cerebellum (not shown). Adapted from ref. 9.

Perhaps the loss of EAAT2 (GLT- 1) reflects decreased gene expression, for example, due to a defect in transcriptional processes. To address that possibility, motor cortex mRNA from over a dozen ALS (including several non-SOD1 familial ALS patients) and control patients was analyzed by Northern blot. There were no significant changes in any of the three (EAAT1, EAAT2, EAAT3) transporters studied (32). It is, therefore, unlikely that a loss of mRNA in ALS motor cortex can explain the selective loss of EAAT2 (GLT-1). More recent studies examining the gene for EAAT2 (chromosome 11) have failed to show linkage in non-SOD1 mutant familial ALS (Robert Brown, personal communication). Furthermore, extensive exon analysis of genomic DNA using SSCP analysis have not revealed germline mutations in the EAAT2 gene (Robert Brown, personal communication).

Our laboratory has recently described evidence for multiple aberrant EAAT2 RNA species in ALS brain that could provide an explanation for the loss of EAAT2 protein and function in ALS patients (33). Two abnormal EAAT2 mRNA species, one retaining an unspliced intron and the other an exon-skipping form have been found only in neuropathologically affected areas of ALS neural tissue. These aberrant species were found in 20 out of 30 ALS neural tissue specimens, but were not found in either non-neurologic disease or neurodegenerative disease controls. They were also detected in 60% of cerebrospinal fluid specimens from patients with ALS. The presence and relative quantity of these abnormal species correlated with the degree of EAAT2 protein loss in ALS tissue. When these abnormal RNA species were expressed in COS7 cells both proteins were unstable and failed to form functional transporters. Furthermore, the retained intron RNA dominantly downregulated normal EAAT2 protein.

These results suggest that the loss of EAAT2 in ALS is due, at least in part, to aberrant mRNA, and these mutants could result from a common defective mechanism, such as RNA processing errors, or acquired mutations. These results suggest that aberrant RNA processing and/or DNA damage could be important processes in the pathophysiology of neurodegenerative disease and in excitotoxicity. The presence of these mRNA species in ALS cerebrospinal fluid may also have future diagnostic utility.

Although these mutant species appear to account for much of the loss of glutamate transport in ALS, other possible explanations for a loss of EAAT2 (GLT-1) could involve translational mechanisms. The gene may be transcribed to mRNA, but defects in translation could ultimately interfere with the synthesis of the protein. There are also posttranslational mechanisms that could result in decreased tissue protein levels, for example oxidative insults from peroxynitrite, superoxide anions, hydroxyl radicals, all of which are known to modify proteins. These modifications, e.g., nitration of tyrosine residues following peroxynitrite, or protein carbonyl formation, could lead to increased degradation and dysfunction of the protein *(34,35)*. A series of studies have already demonstrated that glutamate transporters can be either reversibly (arachidonic acid) or irreversibly (peroxynitrite, hydrogen peroxide) "damaged" by these oxidative insults *(36–39)*. Nitration of glutamate transporter has been observed in transgenic mice overexpressing mutant SOD1 *(40)* in the tissue sample in which loss of glutamate transport had been found. Furthermore, several groups have documented evidence for oxidative damage in sporadic ALS, by the detection of increased protein carbonyls in motor cortex and spinal cord. Whether such processes actually lead to selective damage and loss of GLT-1 in ALS is not yet known.

SELECTIVE VULNERABILITY OF MOTOR NEURONS TO EXCITOTOXICITY

Motor neurons differ from other neurons in the CNS by their large size, their high ratio of axonal length to cell body diameter, their high metabolic rate, and their high content of neurofilament proteins and free radical scavenging enzymes *(19,20,41–44)*. At least two specific features of motor neurons identified in human spinal cord may render these cells susceptible to excitotoxic cell death. The first feature is the low expression of calcium binding proteins parvalbumin and calbindin D28K in motor neurons *(44)*. These calcium-buffering proteins play an important role in the protection of neurons from glutamate-mediated calcium load. The second feature is the low expression of GluR2 AMPA subunit which results in atypical, calcium-permeable AMPA receptors (Bar-Paled and Rothstein, personal observations) *(45–47)*.

Various in vivo experimental studies have provided evidence that glutamate receptor agonists may contribute to selective motor neuron degeneration. Intrathecal injection of kainate in mice selectively damages anterior horn cells and induces a cytoskeletal abnormality associated with motor neuron disease *(48)*. Also, excitotoxic agonists are capable of inducing motor neuron degeneration in vivo.

Studies of organotypic spinal cord cultures and other models have also revealed that motor neurons are preferentially susceptible to non-NMDA (AMPA and kainate)-mediated toxicity *(49–52)*. As described above, certain subtypes of non-NMDA receptors are permeable to calcium, and calcium influx through these receptors could target a cell for excitotoxicity. Living cells that are highly permeable to calcium via non-NMDA receptors can be identified in organotypic cultures through a recently developed histochemical method (53). Using this cobalt based stain, the most sensitive population of neurons permeable to calcium via non-NMDA glutamate receptors were motor neurons in the ventral horn (46). Thus, localization of calcium-permeable non-NMDA glutamate receptors are among the many factors that may make motor neurons selectively vulnerable in ALS.

GLUTAMATE RECEPTOR DISTRIBUTION IN MOTOR NEURONS

Animal and human motor neurons appear to have high density of glutamate receptors by *in situ* and immunocytochemical analyses (5,45,54–58). The distribution and density of ionotropic glutamate receptor subtypes has been studied in human motor system using quantitative autoradiography, immunohistochemistry, and *in situ* hybridization. Autoradiographic analysis of glutamate receptors using specific ligands revealed the following patterns: NMDA receptors labeled with ^{3}H-MK-801 were colocalized with motor neuron cell bodies in the spinal-cord. Motor neurons that tend to be spared in ALS patients express lower density of NMDA receptor binding sites and higher density of AMPA binding sites compared with motor neurons that are vulnerable to the disease (5,47,56,59). Immunohistochemical studies with subunit-specific antibodies have shown that motor neurons have a relatively distinct pattern of AMPA receptor subunits, with a low level of the GluR1 subunit and a high level of the GluR2/3 subunits. A recent *in situ* hybridization study has shown that human spinal motor neurons express GluR1, GluR3, and GluR4 subunits but have no detectable expression of GluR2 mRNA (47). Complementary studies in cultured motor neurons have confirmed the paucity of GluR2 protein in living motor neurons in vitro by using GluR2 monospecific antibodies (Bar-Peled, personal observations). The GluR2 subunit has a very important role in determining calcium permeability of AMPA receptors (2); AMPA receptors that include the edited form of GluR2 are impermeable to calcium. Thus, the lack of GluR2 expression by motor neurons and the resulting increased permeability of their AMPA receptors to calcium could render motor neurons more vulnerable to excitotoxic injury. Further work needs to be done to determine the profile of NMDA and AMPA/kainate receptors in motor neurons. However, these data strongly suggest that glutamate plays a role in motor neuron degeneration, which is very relevant for therapeutic considerations.

GLUTAMATE TRANSPORT AND NEURON/MOTOR NEURON TOXICITY

As described above, ALS is associated with a large loss of EAAT2 (GLT-1) protein. Can just the loss of EAAT2 lead to increased extracellular glutamate and to degeneration of motor neurons? Recently, molecular knockout studies of each glutamate transporter subtype revealed the importance of EAAT2 (GLT-1) for motor systems and its predominant role in excitotoxicity and maintenance of extracellular glutamate (60,61). When antisense oligonucleotides to individual glutamate transporters were administered intraventricularly to rats, the loss of either astroglial glutamate transporter produced a progressive *paresis* beginning first in hind limbs, whereas the knockout of the neuronal subtype EAAC1 produced epilepsy *(60)*. Studies employing lumbar intrathecal infusion of antisense oligonucleotides to the astroglial glutamate transporters produced paralysis along with evidence of muscle denervation and motor neuron degeneration (JDR, unpublished observations).

These studies demonstrated that astroglial glutamate transport accounted for approx 80% of all glutamate transport, with greater than 50% attributable to GLT-1. Extracellular glutamate, measured by microdialysis, was markedly increased following the loss of the astroglial glutamate transporters GLAST or GLT-1, but not following the loss of the neuronal transporter subtype. Thus, the increases in CSF glutamate in ALS patients and loss of tissue glutamate transport are likely due to the selective loss of GLT-1.

Because glutamate transporter must maintain low extracellular glutamate concentrations, it has been theorized that they may be important contributors to glutamate mediated toxicity. Antisense and GLT-1 mutant mice studies both document a predominant role for GLT-1 in prevention of excitotoxicity. Taken together, both antisense and GLT-1 mutant mice studies provide compelling evidence suggesting a central role of glial transporters in the maintenance of low extracellular glutamate and in excitotoxicity. Furthermore, the motor syndrome that develops following their loss in adult animals adds to the growing body of evidence suggesting a link between transporter dysfunction and motor neuron degeneration.

THERAPEUTIC INTERVENTIONS

In the last 10 yr, there have been more than 29 randomized trials, involving over 6100 ALS patients, of drugs designed to either increase survival and/or alter the natural decline in motor function associated with the disease. As discussed above, the synaptic action of glutamate appears to be abnormally potentiated in ALS. Therefore, attenuation of glutamatergic transmission may be relevant as therapeutic intervention in this disease. In fact, to date, only one drug, the antiglutamate agent riluzole, has been found to reliably alter the disease progression. Several strategies have been proposed to alter glutamate transmission using both presynaptic or postsynaptic modification of the glutamatergic system. Depending on where the initial or secondary damage is occurring in ALS, as shown in Fig. 2, there are multiple steps where one could intervene to halt or retard motor neuron degeneration. Louvel at al. have provided an excellent overview of recent therapies for ALS *(62)*.

PRESYNAPTIC MODIFICATION OF THE GLUTAMATERGIC SYSTEM

To reduce the level of glutamate in the synaptic cleft, one can modulate the release of glutamate from nerve terminals, enhance glutamate re-uptake into astrocytes, or reduce glutamate recycling in astrocytes.

Agents Affecting Glutamate Release

Decreasing the level of glutamate released from nerve terminals should reduce the activation of postsynaptic neurons. A number of drugs can affect glutamate release. These include some anticonvulsant drugs, adenosine receptor agonists, and autoreceptors specific for kainate.

Riluzole, originally developed for treatment of epilepsy, was found to effectively inhibit glutamate release *(63)*. It also can block GABA uptake by striatal synaptosomes,

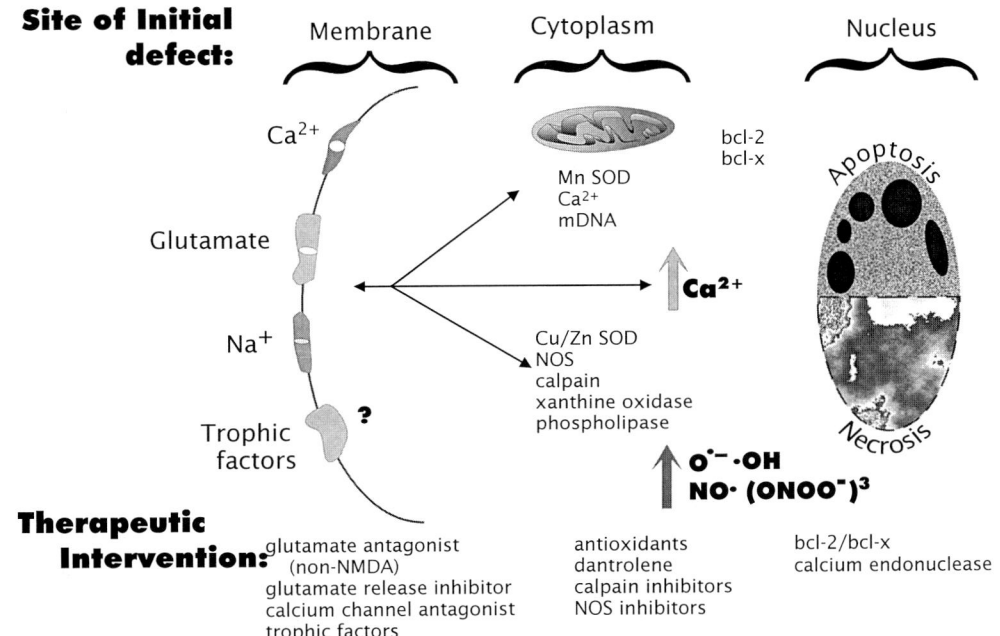

Fig. 2. Pathways of motor neuron necrotic and apoptotic degeneration and possible therapeutic interventions.

block voltage-dependent sodium channels *(63)*, and antagonize NMDA-mediated calcium entry in granule cells in culture *(63)*. Two randomized large clinical trials have both demonstrated that riluzole is efficacious at increasing survival in ALS patients *(64,65)*. Although the effect on survival was modest, this wars the first drug shown to significantly alter the natural history of the disease.

Lamotrigine, an anticonvulsant that also blocks presynaptic release of glutamate has been studied in ALS. In a small randomized trial (18 patients) lamotrigine did not alter ALS survival *(66)* after 18 mo on the drug. However, because the drug was used at a dose that is 10 times lower than that used in epilepsy, it still remains possible that the drug could have a beneficial effect in ALS.

Glutamate Receptor Antagonists

To date, few potent glutamate receptor antagonists have been available for widespread human trial in chronic diseases. Dextromethorphan, a weak NMDA antagonist has been studied in four ALS trials, in doses of 120–150 mg po and was not effective in alering survival or muscle strength *(62,67)*. Laboratory studies would suggest that at least for ALS, non-NMDA antagonists may be more beneficial at protecting against chronic excitotoxic motor neuron degeneration. Future trials with non-NMDA receptor antagonists are planned (e.g., derivatives of GYKI-42466)

Agents Enhancing Glutamate Re-Uptake

Re-uptake of glutamate into astrocytes is the major mechanism for the inactivation of glutamate transmission. Therefore, enhancing this process could be beneficial in diseases where reduction in glutamate transmission is desirable. The astrocytic glutamate transporter EAAT2 (GLT-1) which is decreased in spinal cord and motor cortex of ALS patients, is regulated by protein kinase C *(68,69)*, and by additional factors probably released from neurons *(29–31)*. The mechanisms regulate glutamate transport activity and protein expression are only now being investigated. Understanding these processes has the potential to provide important information on modulating excitotoxicity and thereby catalyze the development of therapeutic interventions.

Agents Modulating Glutamate Metabolism

Normally, re-uptake of glutamate in synaptic astrocytes is followed by its conversion to glutamine (reaction catalyzed by glutamine synthetase), which is then transported back to neurons. Studies in vitro indicated that, when extracellular glutamate levels are high, a large fraction of glutamate is being metabolized in the tricarboxytic acid cycle (TCA) cycle to alpha-ketoglutarate. This reaction is catalyzed by glutamate dehydrogenase (GDH), an enzyme present in very high levels in synaptic astrocytes. Oxidative deamination of glutamate is advantageous for two main reasons: first, it may reduce glutamate recycling, since glutamate metabolites are usually not used for new glutamate synthesis in the brain *(70–72)*. This may reduce synaptic glutamate that could be present in high levels in ALS patients. Second, the reaction product, alpha-ketoglutarate, can be used for ATP generation in the TCA cycle.

Earlier studies suggested that branched-chain amino acids such L-leucine stimulate GDH activity. For this reason, several trials of branched-chain amino acids administered in large oral doses were performed in ALS. Although one trial indicated benefit in altering the rate of muscle decline *(73)*, subsequent studies failed to replicate the original encouraging result *(74,75)*.

CONCLUSIONS

ALS as a Composite of Initiation and Propagation

It would be very simplistic to believe that glutamate toxicity is the cause of ALS. It is more likely that multiple different insults, which could overlap, may lead to the common phenotype of motor neuron death. For example, mutant SOD1 may produce excess free radical species. These radical species could damage proteins and DNA and RNA processing elements. Damage to these cytosolic and nuclear elements could generate aberrant glutamate transporter mRNA and/or directly damage glutamate transport proteins. Ultimately, loss of glutamate transport will produce chronic elevations of extracellular glutamate leading to excitotoxicity. Excitotoxicity will increase the generation of free radicals. Thus, a self-propagating cyclical cascade may follow. The slowly developing and spreading pattern of motor neuron degeneration in ALS may reflect this multi-step process of an initiating event, possibly toxic, followed by self-propagating cascades as outlined above.

If the hypothesis of initiator and propagators is correct, then drugs may be designed for ALS that could interfere with either step. As diagnosis of ALS is often made well into the disease process, drugs that block the initiating steps may not be as effective at halting the disease propagation as those directed toward altering the cyclical, self-propagating steps. Is the evidence for such processes in ALS? Recently, experiments with FALS transgenic mice suggest this hypothesis may have merit. When transgenic

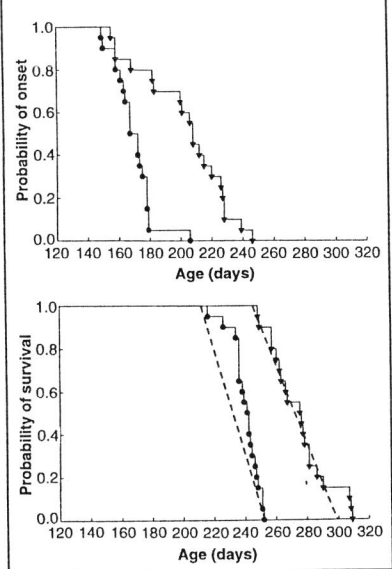 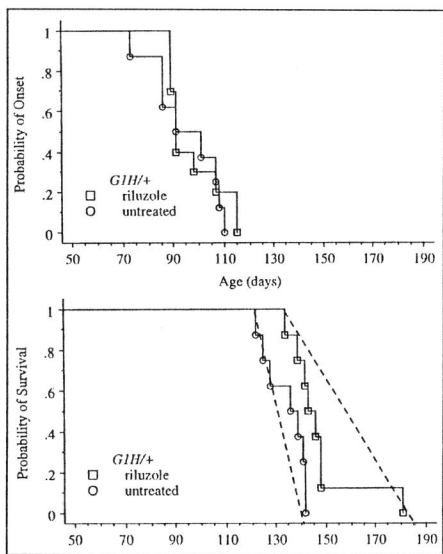

Fig. 3. Experimental differential between propagating and initiating events in motor neuron degeneration in experimental models of ALS. Transgenic mice overexpressing the G93A mutation were either genetically provided with excess bcl-2 (via crossing with bcl-2 overexpressing mice) or treated with riluzole. The bcl-2 protein appears to delay disease onset, but not the rate of disease progression (*see* angle of dotting line); Riluzole (and the antiglutamate agent gabapentin, not shown) had no effect on disease onset, but appeared to increase survival, thus changing the slope of disease progression (dotted line). Modified from refs. *76* and *77.*

mice with human SOD 1 mutation G93A were crossed with mice over expressing the anti-apoptotic protein bcl-2, the onset of the disease was significantly delayed, perhaps reflecting an effect on disease initiation *(76)* (*see* Fig. 3). However, once the disease began, the rate of progression (reflecting disease propagation) was the same in both groups. Similarly, when the SOD1 mutant mice (G93A) were treated with the antioxidant vitamin E, disease onset was also delayed. Treatment of the SOD1 mutant mice with riluzole or the gabapentin had no effect on onset, but significantly prolonged the survival (propagation) of disease in the mice *(77).* Thus, these mice may provide evidence for a multi-step process in motor neuron degeneration (initiation and propagation), and for the role of glutamate toxicity in the disease process.

ACKNOWLEDGMENTS

Some of the work described in this review were funded by grants from the NIH-NINDS (NS33958; AG12992; NS34100), the Muscular Dystrophy Association, ALS Association, and the Cal Ripken–Lou Gehrig Fund for Neuromuscular Research.

REFERENCES

1. Hollmann M, Heinemann S. Cloned glutamate receptors. *Ann Rev Neurosci* 1994, **17**: 31–108.
2. Jonas P, Racca C, Sakmann B, Seeburg PH, Monyer H. Differences in Ca2+ permeability of AMPA-type glutamate receptor channels in neocortical neurons caused by differential GluR-B subunit expression. *Neuron* 1994, **12**:

3. Sommer B, Keinanen K, Verdoorn T, et al. Flip and flop: a cell specific functional switch in glutamate operated channels in the CNS. *Science* 1990; **249**: 1580–1585.
4. Herb A, Higuchi M, Sprengel R, Seeburg PH. Q/R site editing in kainate receptor GluR5 and GluRG pre-mRNAs requires distant intronic sequences. *Proc Natl Acad Sci USA* 1996, **93**: 1875–1880.
5. Tolle TR, Berthele A, Zieglglinsberger W, Seeburg PH, Wisden W. The differential expression of 16 NMDA and non-NMDA receptor subunits in the rat spinal cord and in periaqueductal gray. *J Neurosci* 1993, **13**: 5009–5028.
6. Arriza JL, Fairman WA, Wadiche JI, Murdoch GH, Kavanaugh MP, Amara SG. Functional comparisons of three glutamate transporter subtypes cloned from human motor cortex. *J Neurosci* 1994, **14**: 5559–5569.
7. Fairman WA, Vandenberg RJ, Arriza JL, Kavanaugh MP, Amara SG. An excitatory amino-acid transporter with properties of a ligand-gated chloride channel. *Nature* 1995, **375**: 599–603.
8. Rothstein JD, Martin L, Levey AI, et al. Localization of neuronal and glial glutamate transporters. *Neuron* 1994, **13**: 713–725.
9. Rothstein JD, Van Kammen M, Levey AI, Martin LJ, Kuncl RW. Selective loss of glial glutamate transporter GLT-1 in amyotrophic lateral sclerosis. *Ann Neurol* 1995, **38**: 73–84.
10. Arriza JL, Eliasof S, Kavanaugh MP, Amara SG. Excitatory amino acid transporter 5, a retinal glutamate transporter coupled to a chloride conductance. *Proc Natl Acad Sci USA* 1997, **94**: 4155–4160.
11. Arriza JL, Kavanaugh MP, Fairman WA, et al. Cloning and expression of a human neutral amino acid transporter with structural similarity to the glutamate transporter gene family. *J Biol Chem* 1993; **268**: 15,329–15,332.
12. Rothstein JD, Dykes-Hoberg M, Pardo CA, et al. Knockout of glutamate transporters reveals a major role for astroglial transport in excitotoxicity and clearance of glutamate. *Neuron* 1996, **16**: 675–686.
13. Rothstein JD, Tabakoff B. Alteration of striatal glutamate release after glutamine synthetase inhibition. *J Neurochem* 1984; **43**:1438–1446.
14. Shank RP, Campbell GL. Amino acid uptake, content, and metabolism by neuronal and glial enriched cellular fractions from mouse cerebellum. *J Neurosci* 1984, **4**: 8–69.
15. Shank RP, Campbell GL. Glutamine and alpha-ketoglutarate uptake and metabolismby nerve terminal enriched material from mouse cerebellum. *Neurochem Res* 1982, **7**: 601–616.
16. Shank RP, Aprison MH. Present status and significance of the glutamine cycle in neural tissues. [Review] *Life Sci* 1981, **28**: 837–842.
17. Schousboe A, Westergaard N, Hertz L. Neuronal-astrocytic interactlions in glutamate metabolism. *Biochem Soc Trans* 1993, **21**: 49–53.
18. Hertz L, Drejer J, Schousboe A. Energy metabolism in glutamatergic neurons, GABAergic neurons and astrocytes in primary cultures. *Neurochem Res* 1988, **13**: 605–610.
19. Shaw PJ. Excitotoxicity and motor neuron disease: a review of the evidence. *J Neurol Sci* 1994, **124**, 6–13.
20. al-Chalabi A, Powell JF, Leigh PN. Neurofilaments, free radicals, excitotoxins, and amyotrophic lateral sclerosis. *Muscle Nerve* 1995, **18**: 540–545.
21. Rothstein JD. Therapeutic horizons for amyotrophic lateral sclerosis. *Curr Opin Neurobiol* 1996, **6**: 679–687.
22. Rothstein JD. Excitotoxicity hypothesis. *Neurology* 1996, **47**: Sl9–S26.
23. Rothstein JD, Kuncl R, Chaudhry V, et al. Excitatory amino acids in amyotrophic lateral sclerosis: an update. *Ann Neurol* 1991 **230**: 224–225.
24. Rothstein JD, Tsai G, Kuncl RW, et al. Abnormal excitatory amino acid metabolism in amyotrophic lateral sclerosis. *Ann Neurol* 1990, **28**: 18–25.
25. Shaw PJ, Forrest V, Ince PG, Richardson JP, Wastell HJ. CSF and plasma amino acid levels in motor neuron disease—elevation of CSF glutamate in a subset of patients. *Neurodegeneration* 1995, **4**: 209–216.

26. Rothstein JD, Martin LJ, Kuncl RW. Decreased glutamate transport by the brain and spinal cord in amyotrophic lateral sclerosis. *N Engl J Med* 1992, **326**: 1464–1468.
27. Shaw PJ, Chinnery RM, Ince PG. [3H]D-aspartate binding sites in the normal human spinal cord and changes in motor neuron disease: a quantitative autoradiographic study. *Brain Res* 1994, **655**: 195–201.
28. Nagy D, Kato T, Kushner PD. Reactive astrocytes are widespread in the cortical gray matter of amyotrophic lateral sclerosis. *J Neurosci Res* 1994, **38**: 336–347.
29. Ginsberg SD, Martin LJ, Rothstein JD. Regional deafferentation down-regulates subtypes of glutamate transporter proteins. *J Neurochem* 1995, **65**: 2800–2803.
30. Ginsberg SD, Rothstein JD, Price DL, Martin LJ. Fimbria-fornix transections selectively down-regulate subtypes of glutamate transporter and glutamate receptorproteins in septum and hippocampus. *J Neurochem* 1996, **67**: 1208–1216.
31. Swanson RA, Miller JW, Rothstein JD, Farrell K, Stein BA, Longuemare MC. Neuronal regulation of glutamate transporter subtype expression in astrocytes. *J Neurosci* 1997; **7**: 932–940.
32. Bristol LA, Rothstein JD. Glutamate transporter gene expression in amyotrophic lateral sclerosis motor cortex. *Ann Neurol* 1996 **339**: 676–679.
33. Lin G, Bristol LA, Rothstein JD. An abnormal mRNA leads to downregulation of glutamate transporter EAAT2 (GLT-1) expression in amyotrophic lateral sclerosis. *Ann Neurol* 1996, **740**: 540–541.
34. Beckman JS, Carson M, Smith CD, Koppenol WH. ALS, SOD and peroxynitrite. *Nature* 1993, **364**: 584.
35. Ischiropoulos H, Zhu L, Chen J, et al. Peroxynitrite-mediated tyrosine nitration catalyzed by superoxide dismutase. *Arch Biochem Biophys* 1992, **298**: 431–437.
36. Trotti D, Rossi D, Gjesdal O, et al. Peroxynitrite inhibits glutamate transporter subtypes. *J Biol Chem* 1996, **271**: 976–5979.
37. Volterra A. Inhibition of high-affinity glutamate transport in neuronal and glial cells by arachidonic acid and oxygen-free radicals. Molecular mechanisms and neuropathological relevance. *Renal Physiol Biochem* 1994; **17**: 165–167.
38. Volterra A, Trotti D, Tromba C, Floridi S, Racagni G. Glutamate uptake inhibition by oxygen free radicals in rat cortical astrocytes. *J Neurosci* 1994, **74**: 924–2932.
39. Trotti D, Volterra A, Lehre KP, et al. Arachidonic acid inhibits a purlfied and reconstituted glutamate transporter directly from the water phase and not via the phospholipid membrane. *J Biol Chem* 1995, **270**: 9890–9895.
40. Nagano I, Wong PC, Rothstein JD. Nitration of glutamate transporters in transgenic mice with a familial amyotrophic lateral sclerosis-linked SOD1 mutation. *Ann Neurol* 1996, **40**: 542.
41. Price DL, Martin LJ, Clatterbuck RE, et al. Neuronal clegeneration in human diseases and animal models. *J Neurobiol* 1992, **23**: 1277–1294.
42. Lee MK, Borchelt DR, Wong PC, Sisodia SS, Price DL. Transgenic models of neurodegenerative diseases. *Curr Opin Neurobiol* 1996; **6**: 651–660.
43. Bruijn LI, Cleveland DW. Mechanisms of selective motor neuron death in ALS: Insights from transgenic mouse models of motor neuron disease. *Neuropathol Appl Neurobiol* 1996; **22**: 373–387.
44. Ince P, Stout N, Shaw P, et al. Parvalbumin and calbiT din D-28k in the human motor system and in motor neuron disease. *Neuropathol Appl Neurobiol* 1993; **19**: 291–299.
45. Shaw PJ, Chinnery RM, Ince PG. Non-NMDA receptors in motor neuron disease (MND): a quantitative autoradiographic study in spinal cord and motor cortex using [3H]CNQX and [3H]kainate. *Brain Res* 1994; **655**: 186–194.
46. Carriedo SG, Yin H-Z, Lamberta R, Weiss JH. In vitro kainate injury to large, SMI32(+) spinal neurons is CA2+ dependent. *Neuroreport* 1995, **45**, 548.
47. Williams TL, Day NC, Ince PG, Kamboj RK, Shaw PJ. Calcium-permeable alpha-amino-3-hydroxy-5-methyl-4-isoxazole proprionic acid receptors: a molecular determinant of selective vulnerability in amyotrophic lateral sclerosis. *Ann Neurol* 1997 **42**: 200–207.

48. Hugon J, Vallet JM. Abnormal distribution of phosphorylated neurofilaments in neuronal dgeration induced by kainic acid. *Neurosci Lett* 1990 **19**: 548.
49. Ikonomidou C, Qin Qin Y, Labruyere J, Olney JW. Motor neuron degeneration induced by excitotoxin agonists has features in common with those seen in the SOD1 transgenic mouse model of amyotrophic lateral sclerosis. *J Neuropathol Exp Neurol* 1996, **55**: 211–224.
50. Rothstein JD, Kuncl RW. Neuroprotective strategies in a model of chronic glutamate-mediated motor neuron toxicity. *J Neurochem* 1995, **65**: 643–651.
51. Rothstein JD, Jin L, Dykes-Hoberg M, Kuncl RW. Chronic inhibition of glutamate uptake produces a model of slow neurotoxicity. *Proc Natl Acad Sci USA* 1993 **90**: 6591–6595.
52. Rothstein JD, Bristol LA, Hosler B, Brown RH, Jr., Kuncl RW. Chronic inhibition of superoxide dismutase produces apoptotic death of spinal neurons. *Proc Natl Acad Sci USA* 1994 **91**: 4155–4159.
53. Pruss RM, Akeson RL, Racke MM, Wilburn JL. Agonist-activated cobalt uptake identifies divalent cation-permeable kainate receptors on neurons and glial cells. *Neuron* 1991, **7**: 509–518.
54. Shaw PJ, Ince PG, Matthews JN, Johnson M, Candy JM. N-methyl-D-aspartate (NMDA) receptors in the spinal cord and motor cortex in motor neuron disease: a quantitative autoradiographic study using [3H]MK-801. *Brain Res* 1994, **637**: 297–302.
55. Chinnery RM, Shaw PJ, Ince PG, Johnson M. Autoradiographic distribution of binding sites for the non-NMDA receptor antagonist [3H]CNQX in human motor cortex, brainstem and spinal cord. *Brain Res* 1993, **630**: 75–81.
56. Shaw PJ, Ince PG, Johnson M, Perry EK, Candy JM. The quantitative autoradiographic distribution of [3H]MK-801 binding sites in the normal human brainstem in relation to motor neuron disease. *Brain Res* 1992, **572**: 276–280.
57. Tolle TR, Berthele A, Zieglgansberger W, Seeburg PH, Wisden W. Flip and Flop variants of AMPA receptors in the rat lumbar spinal cord. *Eur J Neurosci* 1995, **7**: 1414–1419.
58. Tolle TR, Berthele A, Laurie DJ, Seeburg PH, Zieglgansberger W. Cellular and subcellular distribution of NMDAR1 splice variant mRNA in the rat lumbar spinal cord. *Eur J Neurosci* 1995, **7**: 1235–1244.
59. Shaw PJ, Ince PG. A quantitative autoradiographic study of [3H]kainate binding sites in the normal human spinal cord, brainstem and motor cortex. *Brain Res* 1994, **641**: 39–45.
60. Rothstein JD, Dykes-Hoberg M, Pardo CA, et al. Antisense knockout of glutamate transporters reveals a predominant role for astroglial glutamate transport in excitotoxicity and clearance of extracellular glutamate. *Neuron* 1996; **16**: 675–686.
61. Tanaka K, Watase K, Manabe T, et al. Epilepsy and exacerbation of brain injury in mice lacking the glutamate transporter GLT-1. *Science* 1997, **276**: 1699–1702.
62. Louvel E, Hugon J, Doble A. Therapeutic advances in amyotrophic lateral sclerosis. *Trends Pharm Sci* 1997, **18**: 196–203.
63. Bryson HM, Fulton B, Benfield P. Riluzole—A review of its pharmacodynamic and pharmacokinetic properties and therapeutic potential in amyotrophic lateral sclerosis. *Drugs* 1996, **52**: 549–563.
64. Bensimon G, Lacomblez L, Meininger V, Group TARS. A controlled trial of riluzole in amyotrophic lateral sclerosis. N Engl J Med 1994; **330**: 585–591.
65. Lacomblez L, Bensimon G, Leigh PN, Guillet P, Meininger V. Dose-ranging study of riluzole in amyotrophic lateral sclerosis. *Lancet* 1996, **347**: 1425–1431.
66. Eisen A, Stewart H, Schulzer M, Cameron D. Anti-glutamate therapy in amyotrophic lateral sclerosis: a trial using lamotrigine. *Can J Neurol Sci* 1993, **20**: 297–301.
67. Hollander D, Pradas J, Kaplan R, McLeod HL, Evans WE, Munsat TL. High-dose dextromethorphan in amyotrophic lateral sclerosis: phase I safety and pharmacokinetic studies. *Ann Neurol* 1994; **36**: 920–924.
68. Casado M, Zafra F, Aragon C, Gimenez C. Activation of high-affinity uptake of glutamate by phorbol esters in primary glial cell cultures. *J Neurochem* 1991, **57**: 1185–1190.

69. Casado M, Bendahan A, Zafra F, et al. Phosphorylation and modulation of brain glutamate transporters by protein kinase C. *J Biol Chem* 1993, **268**: 27,313–27,317.
70. Erecinska M, Nelson D. Activation of glutamate debydrogenase by leucine and its nonmetabolizable analogue in rat brain synaptosomes. *J Neurochem* 1990, **54**: 1335–1343.
71. Yudkoff M, Nissim I, Nelson D, Lin ZP, Erecinska M. Glutamate dehydrogenase reaction as a source of glutamic acid in synaptosomes. *J Neurochem* 1991, **57**: 153–160.
72. Erecinska M, Silver IA. Metabolism and role of glutamate in mammalian brain. *Prog Neurobiol* 1990, **35**: 245–296.
73. Plaitakis A, Smith J, Mandeli J, Yahr MD. Pilot trial of branched-chain aminoacids in amyotrophic lateral sclerosis. *Lancet* 1988, **1**: 1015–1018.
74. Testa D, Caraceni T, Fetoni V. Branched-chain amino acids in the treatment of amyotrophic lateral sclerosis. *J Neurol* 1989, **236**: 445–447.
75. Tandan R, Bromberg MB, Forshew D, et al. A controlled trial of amino acid therapy in amyotrophic lateral sclerosis: I. Clinical, functional, and maximum isometric torque data. *Neurology* 1996, **47**: 1220–1226.
76. Kostic V, Jackson-Lewis V, Bilbao F, Dubois-Dauphin M, Przedborski S. Bcl-2: prolonging life in a transgenic mouse model of familial amyotrophic lateral sclerosis. *Science* 1997, **277**: 559–562.
77. Gurney ME, Cuttings FB, Zhai P, et al. Benefit of vitamin E, riluzole, and gabapentin in a transgenic model of familial amyotrophic lateral sclerosis. *Ann Neurol* 1996, **39**: 147–157.

32
Antioxidants and the Treatment of Neurological Disease

Rajiv R. Ratan

INTRODUCTION

Nothing in excess—Solon quoted
by Diogenes Laertius in Lives of the Philosophers

A major tenet of modern industrial society is that more is better, that one cannot "get enough of a good thing." As members of such a society, clinicians and scientists have not been immune to its enculturations and often adopt the "more is better" approach in the pharmacological treatment of human disease. However, anyone who has prescribed drugs to patients in the clinic or applied drugs to cells or animals in the laboratory appreciates that therapy represents a balancing act between the salutary and the toxic effects of the agent utilized. In some cases, the window between these two effects is small and minor increases in dosage can lead to unwanted consequences. Narrow therapeutic windows are particularly apparent under circumstances where the molecular targets of drugs simultaneously play a role in disease pathology and normal cell function. One class of therapeutic agents where this dilemma is becoming more evident is antioxidants. Indeed, recent evidence has begun to define roles for the primary cellular targets of antioxidants, free radicals, not only as mediators of neuronal injury, but also as intracellular second messengers *(1–3)* that are important in growth factor signaling and vasoregulation as well as human defense against invading micro-organisms *(4)*. Furthermore, some antioxidants become prooxidants under appropriate conditions *(5)*. The question arises: *What are the factors one must consider in identifying and developing antioxidants appropriate for use in the clinic?* The first part of this chapter will outline one strategy for defining whether an antioxidant is viable as a neuroprotective agent. It will then look toward the future and discuss novel approaches to minimizing free radical toxicity in the nervous system.

OXIDANTS AND ANTIOXIDANTS: SOME DEFINITIONS

The putative role of oxidants in mediating cell death and neurological disease has been cogently outlined in numerous other chapters of this book and will not be

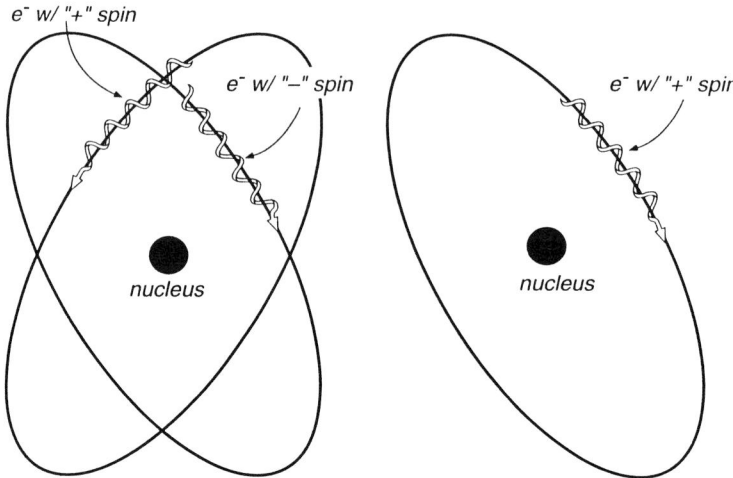

Fig. 1. (left) A nonradical contains two electrons of opposite spin in each orbital. (Right) A radical contains an unpaired electron in one or more orbitals. A free radical refers to a radical that is stable enough that it is capable of independent existence.

discussed in detail here (*see* Chapters 3 and 19, this volume). Briefly, most, but not all, *oxidants generated* in vivo *are radicals which, by definition, contain one or more orbitals with an unpaired electron* (Fig. 1). Because a molecule is most stable when all of its orbitals are filled with two electrons of opposite spin, radicals in cells react with other biomolecules such as DNA, lipid, and protein, to extract an electron from these substrates, leaving them irreversibly altered, but stabilizing the radical species. Although the precise substrate(s) that must be oxidized (i.e., which must lose an electron) to cause neuronal injury or death have yet to be defined, current data suggests that oxidative damage of DNA, lipids and proteins is likely to be involved *(6–11)*. Of course, the extent to which each of these broad classes of cell constituents is damaged varies according to the oxidant generated and its subcellular localization, the capacities of cellular defenses to neutralize radicals once they are formed, and the ability of the cell to repair oxidative damage once it has occurred. Nevertheless, it follows from these definitions of oxidants and their cellular targets, that *antioxidants are substances that prevent or repair oxidative damage of cell constituents*.

EVIDENCE FAVORING THE USE OF ANTIOXIDANTS IN THE TREATMENT OF NEUROLOGIC DISEASES IN HUMANS

Anyone who ventures into a local drugstore or nutrition center is likely to leave convinced that antioxidants, including nutrients such as vitamin E or vitamin C, are good for you and your brain. However, it is important to point out that the healthful effects of supplemental antioxidants in human neurological disease are far from axiomatic, and that further study is required to define a causal link between radicals and central nervous system (CNS) injury. Nevertheless, based on the following observations, antioxidants represent a promising means to reduce the risks of or prevent neurologic illness.

Increased Production of Oxidants is Associated with Neurologic Dysfunction

It is well established that exposure of humans to concentrations of oxygen above that found in air can lead to central nervous system *(12)*, pulmonary and retinal toxicity. In vivo and in vitro studies suggest that exposure of neurons to hyperoxia induces cell damage and death as a result of increased free radical formation *(13,14)*. Specifically, increased oxygen concentrations in cells lead to increased production of superoxide in organelles like mitochondria through the interaction of oxygen with electron rich intermediates in the electron transport chain. Increased superoxide production can ultimately lead to DNA damage, energy decline, and loss of clonal survival.

Mutations in Antioxidant Defense Genes Have Been Associated with Neurologic Dysfunction.

Genetic studies of patients with familial ataxia and familial motor neuron disease have identified a subset in each group with mutations in antioxidant defense genes. For example, inherited abetalipoproteinemia and familial isolated vitamin E deficiency, have been associated with vitamin E responsive syndromes phenotypically similar to Friedreich's ataxia (FA) *(15)*. The clinical phenotype of the vitamin E responsive syndromes include cerebellar ataxia, dysarthria, pes cavus, scoliosis, absent deep tendon reflexes, and Babinski signs.

In addition to the vitamin E-responsive inherited ataxias, missense mutations in the gene encoding cytosolic copper-zinc superoxide dismutase (SOD), were associated in a subset of patients with familial Amyotrophic Lateral Sclerosis (fALS), a motor neuron disease *(see* Chapter 20, this volume*)*. The primary function of Cu/Zn SOD is to catalyze the dismutation of superoxide into peroxide. Peroxide, in turn, can be detoxified by other cellular antioxidant enzymes. It was initially assumed that motor neuron loss in patients with inherited SOD mutations resulted from a loss of superoxide scavenging capacity. Subsequent studies have suggested that SOD mutations lead to oxidative stress and cell death through a novel gain of function *(17,18)*. Further understanding of the nature of the novel activity of mutant SOD is likely to enlighten therapeutic approaches to a subset of fALS patients and to provide insight into the selective vulnerability of motor neuron populations in sporadic ALS *(19,20)*.

Oxidative Damage Has Been Causally Linked to Aging, and May Therefore Underlie Neurodegenerative Diseases Associated with Aging.

Under steady-state conditions, cells exhibit signs of oxidative damage to DNA and proteins. These observations have led to the notion that antioxidant enzymes do not adequately neutralize basal oxidant production, that cells are constantly under mild oxidative stress, and that such stress may account for age-associated degeneration in cell function and viability. Support for this notion comes from studies demonstrating that overexpression of the antioxidant enzymes catalase and superoxide dismutase decreases age-associated oxidant damage and increases life span in Drosophila *(21)*. Because neurons are long-lived and postmitotic, this cell type appears to suffer a greater accumulation of oxidative damage than dividing cells. Accumulation of oxidative damage in neurons may thus account for the increased incidence of neurodegenerative diseases such as Alzheimer's disease *(see* Chapter 24, this volume*)*,

Table 1
Questions to be Addressed in Considering "Antioxidant" therapy in the CNS

"Antioxidants" are substances which prevent or repair oxidative damage of cell constituents

a. What biomolecule(s) is the antioxidant supposed
   to protect- lipid, DNA, and/or protein and in which subcellular local?
b. Does the "antioxidant" cross the blood–brain barrier?
c. Are concentrations of "antioxidant" required
   to achieve protection of oxidizable biomolecules attainable in the CNS in vivo?
d. If the "antioxidant" acts by scavenging oxidants,
   do the resulting antioxidant radicals themselves cause damage?
   Alternatively, does the disease process potentially convert
   an antioxidant into a pro-oxidant?
e. What is the effect of the antioxidant on normal neuronal function
   or on organs outside of the nervous system?

Parkinson's disease (see Chapter 25, this volume), Amyotrophic lateral sclerosis (see Chapters 20 and 31, this volume), and stroke in aged populations (11).

Numerous studies have also documented the salutary effects of antioxidants in attenuating neuronal loss and prolonging survival in animal models (6,9,22) and in response to excitotoxicity in vitro (23). Accordingly, antioxidants are promising agents for the treatment of neurologic disease in humans. In the next section, a strategy will be presented for identifying appropriate antioxidant(s) for treating diverse neurologic conditions.

PART I: ANTIOXIDANTS IN THE TREATMENT OF CNS DISORDERS: IDENTIFYING THE APPROPRIATE AGENTS

Several excellent reviews have been published by Halliwell and Aruoma which provide guidelines for antioxidant characterization in vivo (24,25,25a). Based on these guidelines and some unique properties of the central nervous system, the following proposed sequence of questions to be addressed in considering antioxidant therapy in the CNS (Table 1).

WHAT BIOMOLECULE(S) IS THE "ANTIOXIDANT" SUPPOSED TO PROTECT-LIPID, DNA, AND/OR PROTEIN AND IN WHICH SUBCELLULAR LOCAL?

As sites mentioned, in cells, radicals are capable of oxidizing DNA (mitochondrial or nuclear), RNA, protein, or lipid. Because each of these cell constituents and the radicals that damage them (Fig. 2) are represented in environments of distinct lipophilicity and/or subcellular local, an antioxidant which is capable of inhibiting lipid peroxidation at the plasma membrane is not necessarily going to prevent protein damage in the cytosol. Similarly, an antioxidant which inhibits protein oxidation in the cytosol, may not reach adequate levels in the nucleus or mitochondria to inhibit DNA damage (26). Thus, it is not adequate to treat antioxidants as an undifferentiated whole that are uniformly capable of abrogating all types of cellular oxidative stress.

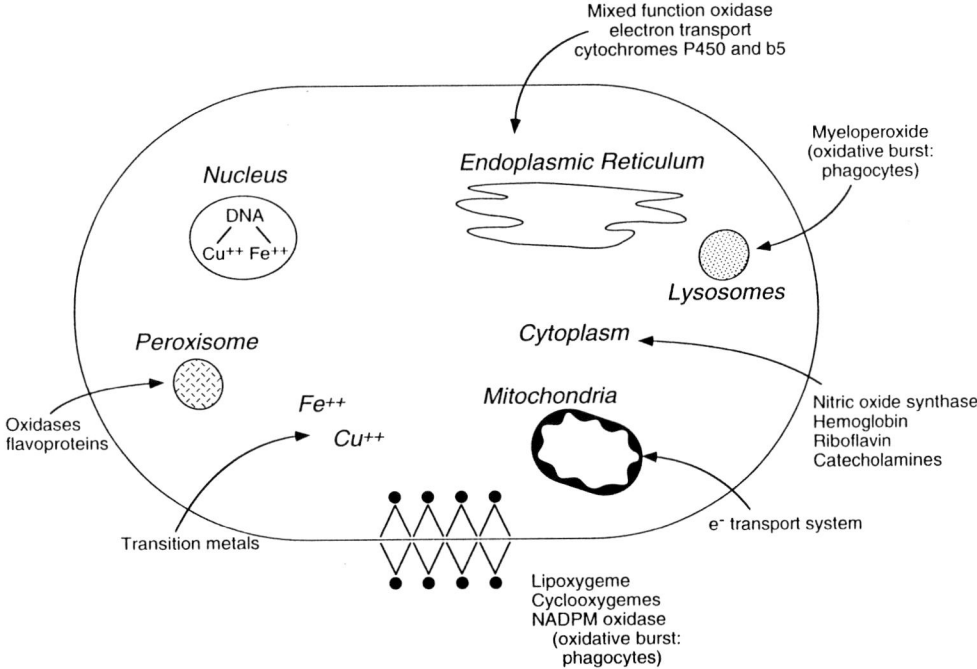

Fig. 2. Some sources of damaging free radicals in the central nervous system.

This prinicple is supported by the distinct clinical phenotypes that are seen in patients with familial isolated vitamin E deficiency (i.e., ataxia, sensory neuronal loss) and patients with missense mutations in superoxide dismutase (i.e., motor neuron disease), two diseases involving defects in antioxidant defenses and which have been linked to oxidative stress. The necessity of designing antioxidant therapeutic approaches which inhibit oxidative damage of multiple cell components may also explain why regimens involving vitamin E, a lipid peroxidation inhibitor, alone, have been ineffective or incompletely effective in the treatment of Parkinson's Disease in humans *(27,27a)* and of ALS modeled in transgenic animals *(22)*, respectively.

In order to assess the burden of oxidation born by cell constituents, methods have been developed for monitoring oxidative damage to DNA, protein and lipid in biopsy, and autopsy tissue *(28,29)*. However, these tissues often become available only at the very late stages of disease progression and thus it may be difficult to distinguish primary from secondary oxidative damage. Currently, intensive efforts are being devoted to identifying markers of oxidative damage that can be measured in the CSF, plasma, urine, or even, in the breath *(30–33)*. However, the validity of such approaches for studying CNS disease is currently unclear.

Given the limitations of monitoring oxidative damage in humans, it is preferable to define the nature of oxidative damage, if present, in animal models in addition to autopsy tissue. Clearly, the use of transgenic animals as discussed in other chapters of this book (*see* Chapters 9 and 10) will permit monitoring of oxidative damage at the earliest possible timepoints and will be valuable in defining conditions where antioxidants may be effective *(34)*. The untested assumption is that currently available markers of oxidative damage are sensitive and reasonably specific.

Once it has been shown, in an animal model, that oxidative damage precedes or coincides with the onset or progression of clinical manifestations, the results can be used to choose an agent or cocktail of antioxidant agents that could be used for treatment. It is reasonable to select agents based on their ability to inhibit oxidative damage in vitro models. Several models of oxidative stress in neurons have been characterized and several of these models are currently being utilized to define which antioxidants are effective in preventing damage to lipid, to protein and/or to DNA in neurons as well the subcellular sites of action of antioxidants *(12, 35–38)*. In the meantime, one can use the abundance of data available on the sites and targets of action of antioxidants to develop regimens of antioxidants appropriate for the disease of interest *(5)*.

Does the "Antioxidant" Cross the Blood–Brain Barrier?

The ability of drugs taken orally or intravenously to penetrate into the CNS is determined in part by the blood–brain barrier (BBB). The cellular basis of the BBB is the cerebrovascular endothelial cell *(39)*. Cerebral endothelial cells differ in several important respects from endothelial cells found in peripheral organs. Whereas fenestra between endothelial cells in peripheral microvessels are large enough to permit the free exchange of water and solutes, cerebral endothelial cells are joined by tight junctions which prevent paracellular exchange. Additionally, cerebral endothelial cells have little capacity to undergo fluid phase endocytosis and they appear to have the latent ability to actively transport drugs and other agents back into the intravascular space.

To counteract the effectiveness of the BBB in limiting "antioxidant" penetration into the CNS, several strategies have been proposed. Increased brain delivery can be achieved by using drugs of moderate lipophilicity, such as vitamin E or by adding a chemical moiety such as an ester to a parent drug to make it more lipophilic. The problem with these approaches, however, is that lipophilic drugs penetrate all organs of the body, rendering these organs susceptible to side effects and slowing the rate at which therapeutic concentrations can be achieved in brain. Additionally, many therapeutic molecules are hydrophilic and cannot be modified to make them lipid soluble.

To address these problems, delivery systems have been developed which take advantage of carrier or transport systems unique to the BBB. An example of such an approach is the delivery of L-DOPA into the CNS by the large, neutral amino acid carrier at the BBB to treat dopamine deficiency in Parkinson's disease *(40)*. Theoretically, blood–brain barrier carriers capable of transporting antioxidant precursors such as cysteine may be useful for neutralizing CNS "oxidant stress." In addition to carriers, receptor mediated-transport mechanisms have also been utilized. In fact, it has been found that antibodies that bind to the transferrin receptor selectively target BBB endothelial cells, undergo transcytosis and are effective agents for the transfer of compounds into the CNS. Proteins such as nerve growth factor (NGF) can be linked to the transferrin antibody and can be delivered past the BBB in quantities sufficient to maintain survival of transplanted tissue *(41)*. In addition to the strategies discussed above, other novel approaches that may be useful for delivering antioxidants past the BBB will be discussed in more detail later in this chapter.

Are the Concentrations of "Antioxidant" Required to Achieve Protection of Oxidizable Biomolecules Attainable in the CNS in Vivo?

Because it is often impossible to obtain brain biopsy tissue from patients on a therapeutic agent, our ability to define CNS tissue concentrations of drugs is limited. For hydrophilic agents, one may be able to sample the CSF to attempt to gain an estimate of CNS extracellular fluid levels. For hydrophobic compounds, the measurements of antioxidant levels in membranes of CSF born cells may be an option. It is not clear, however, that monitoring tissue concentrations directly or indirectly, is going to be a valuable use of resources. Rather, it seems that efforts should be focused on identifying markers of oxidative CNS damage in CSF, urine, blood, and in the breath. Once these have been defined, then "therapeutic" antioxidant levels can be determined by using dosages that inhibit oxidative damage without causing untoward side-effects. In this way, antioxidant treatment, much like current anticonvulsant treatment, can be titrated for each individual patient.

Such a strategy relies heavily on the existence of sensitive and specific markers of oxidative damage in the CNS. As mentioned before, the use of breath ethane or pentane has been proposed as a marker of lipid peroxidation, but its ability to monitor CNS lipid peroxidation remains unclear. Another marker of oxidative protein damage that has shown some promise in clinical case studies and animal models is CSF dityrosine (Cortes et al., 1996). Definition of additional sensitive and specific footprints of oxidative damage in the nervous system that can be measured readily in humans is a high priority.

If The "Antioxidant" Acts by Scavenging Oxidants, do the Resulting Antioxidant Radicals Themselves Cause Damage? Alternatively, Does the Disease Process Potentially Convert an Antioxidant Into a Pro-oxidant?

Some antioxidants neutralize free radicals by donating an electron to them. The result of this reaction is that another radical, an *antioxidant radical* is formed. Often, as is the case with α-tocopherol, the resultant radical is poorly reactive. However, in some circumstances, the resultant radicals can be highly reactive and can lead to damaging consequences and worsening of the clinical condition *(43)*.

Antioxidants can also become pro-oxidants because of cell injury associated with disease states. For example, patients with strokes or traumatic brain injury sustain rapid changes in neuronal viability during the presenting stages of their illness (*see* Chapter 16). Neuronal necrosis during this period leads to release of cellular contents including metals such as iron into the extracelluar fluid and CSF. In blood, the iron binding capacity is only 30% saturated, thus any increase in free iron is rapidly neutralized by chelation. In contrast, the iron binding capacity of the CSF and presumably, the CNS extracellular fluid is nearly saturated. Thus, release of cellular iron will lead to an increase in free iron (Fe^{3+}). A number of antioxidants, including ascorbate, glutathione, vitamin E, and plant phenolics can reduce iron from the Fe^{3+} form to the Fe^{2+}. Reduced iron (Fe^{2+}) can then interact with other oxidants such as peroxide or superoxide to form highly reactive hydroxyl radicals, which are capably of interacting with biomolecules at diffusion limited rates and could potentiate injury *(44)*. This scheme illustrates how an antioxidant could be converted into a pro-oxidant in the context of a disease state. A possible way to prevent such a scenario might be to include an iron chelator along with vitamin E or ascorbate. However, in some cases, vitamin E and ascorbate reduce metals bound to low molecular weight chelator drugs and thus continue to be capable of stimulating hydroxyl radical generation. Methods for measuring CSF iron *(45)* and copper *(46)* have been developed and may help avoid the use of antioxidants which can reduce metals under pathological conditions where increased free metal concentrations are found.

What is the Effect of the Antioxidant on Normal Neuronal Function or on Organs Outside the Nervous System?

Free radicals are produced deliberately in the nervous system by macrophages and brain microglia in the defense of the organism against invading bacterial and viral organisms. Theoretically, chronic use of antioxidants which could diminish the ability of these cells to protect the brain against invading organisms, may prevent radical injury, but may also increase susceptibility of the host to infection. Indeed, administration of vitamin E to premature infants has been reported to decrease the severity of retrolental fibroplasia and prevent to some degree hemolytic syndrome of prematurity and intraventricular hemorrhage, but it has also been shown to increase the infection rate *(47)*.

In addition to fending off invading micro-organisms, there is also growing appreciation for the role that free radicals such as superoxide, peroxide or nitric oxide, may play in normal neuronal function as second messengers. Superoxide and peroxide have been implicated in growth factor signaling *(3)* and transcription factor activation *(1)* whereas nitric oxide is thought to regulate vascular tone as well as to mediate, in part, electrophysiologic changes that subserve complex behaviors such as learning *(47,2)*. In this context, it is proposed that radicals interact with proteins such as kinases or phosphatases, modify their redox state, and thereby alter their function *(49)*.

How can this second messenger role be reconciled with radicals known toxicity to cells? The dual role of intracellular radicals as messengers and as toxins may be analogous to intracellular calcium's role as a messenger in the high nanomolar range and as a toxin at levels above several micromolar (*see* Chapters 3 and 30, this volume). Specifically, controlled production of radicals in a defined range may be good for the cell, but unopposed production of high levels of radicals is clearly deleterious.

The emerging second messenger roles of radicals must be kept in mind as we design antioxidant regimens. Such a role for radicals underscores the need to use antioxidants at levels which return the cell to its premorbid oxidant homeostasis or to develop antioxidant strategies that focus on augmenting repair mechanisms rather than decreasing oxidant production or scavenging oxidants directly.

PART II: NOVEL APPROACHES TO MINIMIZING FREE RADICAL TOXICITY IN THE NERVOUS SYSTEM

Antioxidant Vitamins, Diet, and Neurological Disease

The past several decades have witnessed a dramatic decline in the incidence of neurologic conditions such as stroke. These remarkable advances have been attributed to modification of risk factors for stroke, including the advent of better therapies for hypertension. However, recent descriptive and analytical epidemiological studies have indicated that a diet high in fruits and vegetables may also account for decreases in stroke morbidity and mortality over the past few decades *(50)*. Subsequent studies have suggested that it is the antioxidant vitamins in fruits and vegetables that may account for the neuroprotective properties of such a diet *(51)*. In particular, high plasma levels of antioxidants such as flavonoids and Beta-carotene have been associated with a decreased stroke morbidity and mortality risk. By contrast, these studies have failed to correlate high plasma levels of the antioxidants, vitamin E and vitamin C, with decreased stroke mortality or morbidity, suggesting that oral supplementation with some, but not all antioxidants may confer protection from stroke *(51a)*. The final word on the role of oral supplementation of antioxidant vitamins in stroke prevention rests on large scale randomized trials of sufficient sample size, dose and duration of treatment, and follow-up *(52)*.

One neurologic condition for which a randomized, controlled trial of orally supplemented antioxidant vitamins has been performed is Parkinson's disease. The deprenyl and tocopherol antioxidative therapy of Parkinsonism (DATATOP) trial involved α tocopherol (vitamin E) and was undertaken because of the data suggesting that oxidative stress plays a role in the pathogenesis of PD *(53)*. Unfortunately, oral α tocopherol given to early PD patients failed to result in symptomatic improvement or a delay in clinical progression *(54)*. A recent study, however, showed that oral α tocopherol failed to alter levels of α tocopherol in ventricular CSF of PD patients with Ommaya reservoirs *(55)*. To the extent that vCSF reflects extracellular brain fluid, these studies suggest that orally administered α tocopherol may have limited access across the BBB or that it may be too rapidly metabolized to increase brain α tocopherol to levels required to observe clinically beneficial effects. These important observations raise questions about the validity of prior negative effects of antioxidant vitamins in the progression of PD and highlight the need not only for measuring CSF levels of antioxidants vitamins prior to initiation of double blinded, randomized trials, but also the need for CSF, plasma and urine markers of oxidative damage that can be utilized to monitor the efficacy of orally administered antioxidants.

It is clear that major questions still remain regarding antioxidants in the diet and their role in neurologic disease. However, two promising, novel dietary strategies warrant our attention as we consider options for the future-caloric restriction and phytochemicals.

Caloric Restriction

It has been known for sometime that moderate restriction of caloric intake (CR) in phylogenetically diverse populations, extends the maximum life span of these organisms. The effects of CR appeared to be attributable to a general caloric reduction rather than decreased intake of a one or a few nutrients, however until recently, little has been known about the possible mechanism by which caloric restriction increases lifespan *(56)*. Several lines of investigation now indicate that caloric restriction may prolong lifespan, in part, by lowering the steady state level of oxidative stress and retarding age associated oxidative damage. Of note, for those interested in treating neurodegenerative disorders, the effects of CR in attenuating oxidative damage are highest in tissues such as brain and the salutary effects of CR appear to be rapidly inducible. It thus would be worthwhile to consider caloric intake not only as a primary strategy but also as an important variable in the treatment of free radical-mediated neurologic diseases.

Phytochemicals

Recognition of the anticarcinogenic effects of diets enriched in vegetables and certain spices has fueled investigation of the precise biochemical and molecular mechanisms underlying these protective actions. Candidate substances also known has "phytochemicals" have been purified from some of these vegetables, and their effects on cells in vitro and in vivo are a source of active study *(57)*. Among the substances identified, the isothiocyanates (ITCs) have figured prominently as they are widely distributed in plants consumed by man. The anticarcinogenic effects of ITCs and their main cellular metabolites, dithiocarbamates (DTCs), appear to be mediated, in part, through the induction of cellular antioxidant defenses; including the versatile antioxidant glutathione, and enzymes such as quinone reductase, glutathione transferase, and heme oxygenase. Induction of these defenses has been correlated with protection from carcinogens in animals and from oxidative stress induced by glutathione depletion *(58)* or dopamine *(59)* in a neuronal cell line. The ability of phytochemicals such as isothiocyanates to augment antioxidant defenses raises the possibility that these agents may have a role in preventing neurological diseases associated with oxidative damage including PD and familial ALS. The ability to augment CNS antioxidant enzyme activities using small molecules such as isothiocyanates which are known to penetrate the BBB, would have clear advantages over strategies which seek to deliver large antioxidant enzymes directly past the BBB into the CNS.

Inhibitors of Apoptosis as Therapeutic Agents for Oxidative Injury

An exciting development in our understanding of oxidative neuronal death is the observation, by several laboratories, that oxidative stress can initiate the controlled events of apoptosis—cytoplasmic shrinkage, chromatin condensation and DNA fragmentation—in vitro *(12, 35–38)* and in vivo *(60,61)*. These observations suggested the possibility that pluripotent inhibitors of apoptosis, including the proto-oncogene *bcl-2* and cysteine aspartase (caspases) inhibitors (*see* ref. *16* and Chapter 5, this volume) might abrogate neuronal cell death due to oxidative stress. Subsequent studies verified the ability of *bcl-2* overexpression to abrogate neuronal death induced by multiple pro-oxidants *(62,63)*. Additionally, peptide inhibitors of the caspases have been shown to block death induced by depletion of the antioxidant enzyme, superoxide dismutase (SOD) in a PC12 neuronal cell line as well as in primary sympathetic neurons *(63)*. These observations argue that caspase inhibitors will be useful therapeutic agents in diseases believed to be mediated by oxidative stress. Indeed, studies by Moskowitz, Yuan, and colleagues indicate that caspase inhibitors decrease stroke volume in an animal model of ischemia. Because the caspases have been implicated in apoptosis induced by multiple physiological and pathological stimuli, an immediate concern of using general caspase inhibitors in vivo, is that apoptosis would be inhibited in multiple organ systems in the periphery as well as the nervous system. Thus, while the goal of inhibiting neuronal oxidative death may be achieved, suppression of apoptosis in other organ systems could lead to oncogenesis and autoimmunity. These problems would limit the use of caspases to acute neurodegenerative conditions. However, exciting observations by Troy and coworkers suggested that these initial concerns may be unfounded. By comparing growth factor deprivation-induced apoptosis and oxidative stress induced apoptosis in PC12 cells, they discovered that distinct caspases are activated by distinct apoptotic stimuli *(64)*. These findings raise hope that inhbitors that specifically target oxidative stress-induced caspases can be identified and that these agents will be applicable to a host of chronic neurodegenerative conditions.

Another problem posed by the used of peptides inhibitors of caspases in in treating human neurodegenerative disease is delivery of these macromolecules past the blood-brain barrier into neurons. One promising strategy for surmounting the problem of macromolecule delivery into the CNS is the use of a peptide fragment from the Anntenapedia homeodomain protein. Prochiantz et al. serendipitously discovered that the homoedomain of Antennapedia, a 60-amino acid DNA binding sequence known to regulate multiple morphological processes, is capable of traversing neuronal membranes in an energy-independent manner *(65)*. Subsequent mutagenesis studies revealed that the translocating capacity could be ascribed to a 16 amino acid-long peptide which corresponded to the third helix of homeodomain deleted of its amino terminal glutamate. Furthermore, this 16-amino acid peptide was devoid of the biological activity of the parent homeodomain, and when linked by a disulfide bond to antisense oligo-nucleotides or small peptides, it could facilitate delivery of these macromolecules into neurons as well. The precise size limit of the macromolecule that can be delivered using this method remains unclear and in vivo studies using the Antennapedia peptide to deliver macromolecules into the brain are currently underway. However, this approach promises to enhance our ability to deliver therapeutic agents into cells in the clinic and at the bench. Indeed, it has been demonstrated that the Anntenapedia peptide (vector peptide) can be used to efficiently deliver caspases into neuronal cell lines and primary neurons to prevent oxidative death *(64)*.

Growth Factors and Oxidative Stress-Induced Injury

Another class of anti-apoptotic agents that have received significant attention for their ability to counter oxidative death is endogenous peptide factors, known as growth factors, which can activate differentiation as well as survival pathways in neurons. These factors, which include neurotrophins such as brain-derived neurotrophic factor (BDNF) and nerve growth factor (NGF) as well as mitogens such as platelet derived growth factor (PDGF) and basic fibroblast growth factor (bFGF) interact specfiically with cognate cell surface receptors on neurons *(66,67)*.

NGF, PDGF, BDNF, bFGF, and IGF-1 have been documented to abrogate oxidative stress-mediated injury induced by a variety of stimuli. For example, Perez-Polo and colleagues demonstrated that NGF diminishes cell death induced by peroxide in PC12 cells *(68,68a)*. Similarly, Mattson's group has shown that PDGF, BDNF, and IGF-1 prevent cell death in cultured neurons due to the pro-oxidant, iron sulfate *(69)*. In vivo, neurotrophic factors have been shown attenuate striatal MPP+ toxicity in neonatal rat pups *(70)*. In this paradigm, MPP+ was shown to increase hydroxyl radical formation as measured by the the conversion of salicylate to 2,3 dihydrobenzoic acid and this increase was suppressed by systemic administration of protective growth factors. These findings support the notion that growth factors can enhance resistance to oxidative stress in vivo and in vitro.

By contrast to these tantalizing observations, a parallel line of investigation from several laboratories has led to the opposite conclusion—*growth factors potentiate neuronal cell death*. In primary cortical cultures, Choi and colleagues demonstrated

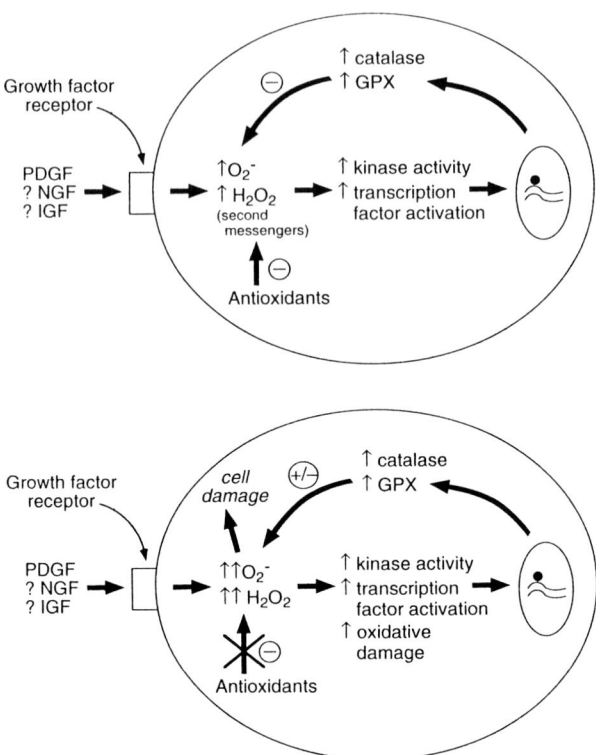

Fig. 3. A model for the cytoprotective and cytotoxic actions of peptide growth factors. (Top) Normal cells. Able to neutralize low level of oxidant messenger. (Bottom) Cells experiencing oxidative stress. 2° to ↑ radical production ↓'d antioxidant defenses cannot adequately neutralize ↑ free radical load from growth factor signaling and therefore sustain cell damage.

that pretreatment with the neurotrophins BDNF, NT-3, or NT4/5 potentiates necrosis induced by excitotoxins or oxygen-glucose deprivation *(71)*. Whereas the precise mechanism by which growth factors potentiate cell death remain unclear, studies from our laboratory and others indicate that growth factors can potentiate cell death induced by glutathione depletion and oxidative stress in cortical neurons *(72)*. These results suggest that growth factors actually induce an oxidative stress in neurons. Consistent with this notion, addition of an antioxidant together with a growth factor has been shown to result in superior levels of survival as compared to either agent alone *(72,73)*.

How can growth factors simultaneously induce and prevent oxidative stress-induced toxicity? A plausible scheme (Fig. 3) emerging from recent studies on growth factor signaling is as follows: Growth factors, such as PDGF, may induce the generation of peroxide for use as a second messenger (to inhibit tyrosine phosphatases) and to activate signaling cascades involved in cell division, differentiation and survival *(3)*. Under conditions where antioxidant defenses are depleted (i.e., ischemia or glutathione depletion), the growth factor-induced increase in peroxide (or superoxide) would lead to oxidative stress and induce or potentiate cell death. However, if the growth factors are given to cells well in advance of the stimulus (e.g., 48–72 h), at a time when antioxidant capacity is replete, then the growth factor induced changes in peroxide can be used as a second messenger to engage a panoply of protective enzymes including catalase, glutathione transport enzymes and glutathione peroxidase and thus to protect the neuron from injury. Because many of the clinical trials using growth factors as therapeutic agents involve diseases where antioxidant defenses may be defective or in limited supply, it may be prudent in the future to couple growth factor treatment of acute and chronic neurodegenerative diseases with one or several antioxidant agents. One agent that is generating significant interest will be discussed in the next section of this chapter.

Orally Active Iron Chelators: Prototype Small Molecule Antioxidants

One class of agents that hold significant promise for attenuating oxidant injury in the nervous system in humans are compounds which bind up free iron, known as iron chelators. Current evidence suggests that under conditions of oxidative stress, free radicals cause iron to be released from iron storage proteins and iron-containing enzymes. Because of its ability to readily accept and donate electrons, free iron can serve as a catalyst for the formation of lipid radicals as well as hydroxyl radicals. Of these products, hydroxyl radicals are particularly notorious as they react avidly with and damage biological constituents at diffusion limited rates. Indirect support for a central role for iron-catalyzed radicals in neuronal death comes from studies which have demonstrated that iron chelators prevent neuronal loss due to oxidative stress, mitochondrial transport inhibition, and serum deprivation in vitro and in stroke *(74)*, multiple sclerosis *(75)*, and PD *(76)* models in vivo. Collectively, these studies have focused attention on the development of iron chelators for use in the clinic.

The prototype iron chelator is deferoxamine (DFO). It is a siderophore (iron carrier) derived from the bacterium, *Streptomyces pilosus*, with a molecular weight of 657. In addition to binding iron, DFO can also act as an antioxidant by directly scavenging reactive oxygen and nitrogen species because of its hydroxamic acid moiety *(77)*. It has been used extensively in humans to treat chronic iron overload and under these conditions, it appears to be relatively safe. The non toxic nature of DFO probably derives from its ability to primarily bind the Iron (III) oxidation state. Iron (III) favoring ligands bind other not physiologically important trivalent positive cations such as aluminum (III), gadolinium (III), and indium (III). Despite many attributes, DFO also has some disadvantages. Side effects of prolonged treatment with DFO include growth failure, ocular and kidney disturbances, as well as anorexia and nausea *(78)*. Fortunately, all of these side effects can be minimized by decreasing the dosage. Because of its large molecular weight and hydrophilic nature, it is poorly absorbed through the gut and therefore must be administered by intramuscular injection or subcutaneous infusion pump. Once absorbed into the circulation, DFO has a short half life. Finally, again because of its hydrophilicity, unless the blood brain barrier is disrupted, it penetrates poorly into the CNS. These problems have been addressed, in part, by conjugating DFO to hydroxyethylstarch (HES) *(79)*. In dogs, HES-DFO has been shown to have a longer half life, to reach higher peak plasma concentrations, and to have fewer side effects. The decreased side effect profile may result from the inability of HES-DFO to penetrate into tissues. Unfortunately, DFO localized to the intravascular compartment by conjugation to HES is less effective than free DFO in preventing acidosis-mediated mechanisms of ischemic brain injury in dogs *(80)*. These observations, in conjunction with the inability to deliver DFO orally has stimulated interest in the development of novel, orally active, nontoxic iron chelators capable of penetrating the BBB that would be more useful for chronic as well as acute neurodegenerative diseases.

Significant advances in the development of orally active iron chelators have come from the laboratory of Hider and colleagues *(81)*. They have focused on a class of compounds known as 3-hydroxypyridin-4-ones (HPO), which like DFO, have high affinity for the Iron (III) oxidation state. These agents have a lower molecular weight than DFO (HPO [150–200] vs DFO [657]), and therefore enjoy significant nonfacilitated uptake in the gut through transcellular pathways. Additionally, uptake of HPO from the gut is quite efficient, thus obviating microbial flora problems customarily associated with residual chelator in the gut lumen. It should be pointed out however, that efficient absorption from the gut does not automatically translate into CNS bioavailability. In addition to having a low molecular weight (100–300), some HPO derivatives have distribution coefficients (octanol/H_2O) which optimize CNS bioavailability. Too high a distribution coefficient (octanol/H_2O ratio >10) leads to almost complete absorption of the drug into hepatocytes from the portal bloodstream, whereas lower distribution coefficients (0.05–1) lead to good penetration of the blood brain barrier. Distribution coefficients less than 0.05 predict also predict poor penetration into the CNS.

The advantages of oral CNS delivery of some HPO compounds must be balanced with higher propensity for HPO chelators to induce toxicity as compared with DFO. The HPOs are kinetically labile and thus iron redistribution is possible. Furthermore, HPO ligands, unlike DFO, form iron complexes which have been shown to possess some toxicity by forming unstable iron-chelate complexes which can participate in free radical generation. Despite these caveats, HPO chelators are ripe for testing in animal models of stroke, Alzheimer's disease and Amyotrophic lateral sclerosis and illustrate a number of the obstacles that must be surmounted in developing orally active, nontoxic antioxidants for acute and chronic use in humans. In addition to iron chelators, other small molecule compounds such as N-tert-butyl-α-phenylnitrone *(9)*, 21-aminosteroids *(6)*, and recently described carboxy-Buckminsterfullerenes *(82)* are generating significant interest as antioxidants for attenuating central nervous system disease.

CONCLUSIONS

Over the past three decades the role that free radicals may play in normal and pathological neuronal processes has come to be fully appreciated. The charge for clinicians and scientists alike is to identify agents which nullify the adverse effects of radicals while preserving their participation in normal cell physiology. For CNS disease, these agents must be able to traverse the blood brain barrier. Further improvements in methods for sensitively and specifically monitoring oxidative damage in the CNS in vivo are also required to monitor drug efficacy. Despite current limitations in all of these arenas, viable CNS antioxidants are rapidly moving from the lab shelf to the bedside and promise to lessen functional debility and loss of life associated with a spectrum of neurological diseases.

REFERENCES

1. Schreck R, Baeuerle PA. A role for radicals as second messengers. *Trends in Cell Biol* **1**: 39–42.
2. Dawson VL, Dawson TM. Physiological and toxicological actions of nitric oxide in the central nervous system. *Adv Pharmacol* 1995, **34**: 323–342.

3. Sundaresan M, Yu ZX, Ferrans VJ, Irani K, Finkel T. Requirement for generation of H_2O_2 for platelet-derived growth factor signal transduction *Science* 1995, **270**: 296–299.
4. Meloni-Bruneri LH, Campa A, Abdalla DS, Calich VL, Lenzi HL, Burger E. Neutrophil oxidative metabolism and killing of P. Brasilensis after air pouch infection of susceptible and resistant mice. *J Leukoc Biol* 1996, **59**: 526–533.
5. Thomas SR, Neuzil J, Stocker R. Cosupp lementation with coenzyme Q prevents the prooxidant effect of alpha-tocopherol and increases the resistance of LDL to transition metal-dependent oxidation initiation. *Arterioscler Thromb Vasc Biol* 1996, **16**: 687–696.
6. Hall ED, McCall JM, Means ED. Therapeutic potential of the lazaroids (21 aminosteroids) in acute central nervous system trauma, ischemia and subarachnoid hemorrhage. *Adv Pharmacol* 1994, **28**: 221–268.
7. Levine RL, Williams JA, Stadtman ER, Shacter E. Carbonyl assays for determination of oxidatively modified proteins. *Methods Enzymol* 1994, **233**: 346–357.
8. Carney JM, Carney AM. Role of protein oxidation in aging and in age-associated neurodegenerative diseases. *Life Sci* 1994, **55**:2 5–26.
9. Floyd RA, Carney JM. Free radical damage to protein and DNA: Mechanisms involved and relevant observations on brain undergoing oxidative stress. *Ann Neurol* 1992, **32**: S22–S27.
10. Halliwell B. Oxygen radicals as key mediators in neurological disease: Fact or fiction? *Ann Neurol* 1992, **32**(Suppl.): S10–S15.
11. Smith MA, Perry G, Richey PL, Sayre LM, Anderson VE, Beal MF, Kowall N. Oxidative damage in Alzheimer's [letter] *Nature* 1996, **382**: 120–121.
12. Enokido Y, Hatanaka H. Apoptotic cell death occurs in hippocampal neurons cultured in a high oxygen atmosphere. *Neuroscience* 1993, **57**(4): 965–972.
13. Freeman BA, Crapo JD. Hyperoxia increases oxygen radical production in rat lungs and lung mitochondria. *J Biol Chem* 1981, **256**: 10986–10992.
14. Crapo JD, McCord JM. Oxygen induced changes in pulmonary superoxide dismutase assayed by antibody titrations. *Am J Physiol* 1976, **231**: 1196–1203.
15. Kayden H.J. The neurologic syndrome of vitamin E deficiency: A significant cause of ataxia *Neurology* 1993, **43**: 2167–2169.
16. Rosen DR, Siddique T, Patterson D. et al. Mutations in Cu/Zn superoxide dismutase gene are associated with familial amyotrophic lateral sclerosis. *Nature,* 1993, **362**: 59–62.
17. Borchelt DR, Lee MK, Slunt HS, Guarnieri M, Xu Z-S, Wong PC, Brown RH Jr., Price DL, Sisodia SS, Cleveland DW. Superoxide dismutase 1 with mutations linked to familial amyotrophic lateral sclerosis possess significant activity. *Proc Natl Acad Sci USA* 1994, **91**: 8292.
18. Gurney ME, Pu H, Chiu AY, Dal Canto MC, Polchow CY, Alexander DD, Caliendo J, Hentati A, Kwon KY, Deng H-X, Chen W, Zhai P, Suffit RL, Siddique T. Motor neuron degeneration in mice expressing a human Cu,Zn superoxide dismutase mutation. *Science* 1994, **264**: 1772–1775.
19. Wiedau-Pazos M, Goto JJ, Rabizadeh S, Gralla EB, Roe JA, Lee MK, Valentine JS, Bredesen DE. Altered reactivity of superoxide dismutase in familial amyotrophic lateral sclerosis. *Science* 1996, **271**: 515–518.
20. Yim MB, Kang JH, Yim HS, Kwak HS, Chock PB, Stadtman ER. A gain-of-function of an amyotrophic lateral sclerosis-associated Cu,Zn superoxide dismutase mutant: An enhancement of free radical formation due to a decrease in Km for hydrogen peroxide. *Proc Natl Acad Sci USA* 1996, **93**: 5709–5714.
21. Orr WC, Sohal RS. Extension of life-span by overexpression of superoxide dismutase and catalase in Drosophila Melanogaster. *Science* 1994, **263**: 1128–1130.
22. Gurney ME, Cutting FB, Zhai PZ, Doble A, Taylor CP, Andrus PK, Hall ED. Benefit of vitamin E, riluzole, and gabapentin in a transgenic model of familial amyotrophic lateral sclerosis. *Ann of Neurol* 1996, **39**:147–157.

23. Dykens, JA, Stern A, Trenkner E. Mechanism of kainate toxicity to cerebellar neurons in vitro is analogous to reperfusion injury. *J Neurochem* 1987, **49:** 1222–1228.
24. Halliwell B. Free radicals and antioxidants: A personal view. *Nutr Rev* 1991, **52:** 253–265.
25. Aruoma, OI. Characterization of drugs as antioxidant prophylactics. *Free Radical Biol Med* 1996, **20:** 675–705.
25a. Halliwell, B. Antioxidant characterization: Methodology and mechanism. *Biochem Pharmacol* 1995, **49:** 1341–1348.
26. Muse KE, Oberley TD, Sempf JM, Oberley LW. Immunolocalization of antioxidant enzymes in adult hamster kidney. *Histochem J* 1994, **26:** 734–753.
27. LeWitt PA. Neuroprotection by anti-oxidant strategies in Parkinson's Disease. *Eur Neurol* 1993, **33:** 24–30.
27a. Parkinson's Study Group. Effects of tocopherol and deprenyl on the progression of disability in early Parkinson's disease. *N Engl J Med* 1989, **321:** 1364–1371.
28. Mecocci P, MacGarvey U, Beal MF. Oxidative damage to mitochondrial DNA is increased in Alzheimer's disease. *Ann Neurol* 1994, **36:** 747–751.
29. Beal, MF. Energy, oxidative damage, and Alzheimer's disease: Clues to the underlying puzzle. *Neurobiol Aging* 1994, **15,** Suppl **2:** S171–S174.
30. Giulivi C. Davies KJ. Dityrosine: A marker for oxidatively modified proteins and selective proteolysis. Methods Enzymol 1994, **233:** 363–371.
31. Shigenaga MK, Aboujaoude EN, Chen Q, Ames BN. Assays of oxidative DNA damage biomarkers 8-oxo-2'-deoxyguanosine and 8-oxoguanine in nuclear DNA and biological fluids by high performance liquid chromatography with electrochemical detection. *Methods Enzymol* 1994, **234:** 16–33.
32. Mendis S, Sobotka PA, Leja FL, Euler DE. Breath pentane and plasma lipid peroxides in ischemic heart disease. *Free Radical Biol Med* 1992, **5:** 679–684.
33. Arterbery VE, Pryor WA, Jiang L, Sehnert SS, Foster WM, Abrams RA, Williams JR, Wharam MD Jr., Risby TH. Breath ethane generation during clinical total body irradiation as a marker of oxygen-free-radical mediated lipid peroxidation: A case study. *Free Radical Biol Med* 1994, **17:** 569–576.
34. Price DL, Koliatsos VE, Wong PC, Pardo CA, Borchelt DR, Lee MK, Cleveland DW, Griffin, JW, Hoffman PN, Cork LC, Sisodia SS. Motor neuron disease and model systems: Aetiologies, mechanisms, and therapies. *Ciba Found Symp* 1996, **196:** 3–13.
35. Ratan RR, Murphy TH, Baraban JM. Oxidative stress-induces apoptosis in embryonic cortical neurons. *J Neurochem* 1994, **62:** 376–379.
36. Ratan RR, Murphy TH, Baraban JM. Macromolecular synthesis inhibitors prevent oxidative stress induced apoptosis in embryonic cortical neurons by shunting cysteine from protein synthesis to glutathione. *J Neurosci* 1994, **14:** 4385–4392.
37. Troy CM, Shelanski ML. Down regulation of copper/zinc superoxide dismutase causes apoptotic death in neuronal PC12 cells. *Proc Natl Acad Sci USA* 1994, **89:** 9005–9009.
38. Manev H, Cagnoli CM, Atabay C, Kharlamov E, Ikonomovic MD, Grayson DR. Neuronal apoptosis in an in vitro model of photochemically induced oxidative stress. *Exp Neurol* 1995, **133:** 198–206.
39. De Boer AG, Breimer DD. The blood brain barrier: Clinical implications for drug delivery to the brain. *J Royal College Physicians of London* 1994, **28:** 502–506.
40. Pardridge WM. *Peptide Drug Delivery to the Brain,* Raven Press, NY, 1991.
41. Friden PM. Receptor-mediated transport of therapeutics across the blood brain barrier. 1994, *Neurosurg* **35:** 294–298.
42. Cortes CM, Shea PA, Ahlers ST, Auker CA, Verma A, Elayan I, Schrot J. Measurement of dityrosine after exposure to hyperbaric oxygen and chronic administration of corticotrophin releasing factor. *Soc Neurosci Abst* 1996, **22:** 1921.

43. Laughton MJ, Halliwell B, Evans PJ, Hoult JRS. Antioxidant and pro-oxidant actions of the plant phenolics quercitin, gossypol and myricetin: Effects on lipid peroxidation, hydroxyl radical generation and bleomycin-dependent damage to DNA. *Biochem Pharmacol* 1989, **38**: 2859–2865.
44. Halliwell B. Drug antioxidants effects: A basis for drug selection? *Drugs* 1994, **42**: 569–605.
45. River Y, Honigman S, Gomori JM, Reches A. Superficial hemosiderosis of the central nervous system. *Mov Disord* 1994, **9**: 559–562.
46. Hartard C, Weisner B, Dieu C, Kunze K. Wilson's disease with cerebral manifestations: Monitoring therapy by CSF concentration. *J Neurol* 1993, **241**: 101–107.
47. Mino M. Clinical uses and abuses of vitamin E in children. *Proc Soc Exp Biol Med* 1992, **200**: 266–270.
48. Bredt DS, Snyder SH. Nitric oxide: A physiologic messenger molecule. *Annu Rev Biochem* 1994, **63**:175–195.
49. Cabiscol E., Levine RL. The phosphatase activity of carbonic anhydrase III is reversibly regulated by glutathiolation. *Proc Natl Acad Sci USA* 1996, **93**(9): 4170–4174.
50. Acheson RM, Williams DRR. Does consumption of fruit and vegetables protect against stroke? *Lancet* 1993, 1191–1193.
51. Keli SO, Hertog MGL, Feskens EJM, Kromhout D. Dietary flavonoids, antioxidant vitamins and incidence of stroke—The Zutphen Study. *Arch Int Med* 1996, 637–642.
51a. DeKeyser J, De Kippel N, Merkx H, Vervaeck M, Herroelen L. Serum concentrations of vitamins A and E and early outcome after ischemic stroke. *Lancet* 1992, **339**: 1562–1565.
52. Hennekens, C. Antioxidant vitamins and cancer. *AJM* 1994, **97**(Suppl 3A.): 2S–4S.
53. Fahn S, Cohen G. The oxidative stress hypothesis in Parkinson's Disease: evidence supporting it. *Ann Neurol* 1992, **32**: 804–812.
54. Parkinson's Study Group. Effects of tocopherol and deprenyl on the progression of disability in early Parkinson's disease. *N Engl J Med* 1989, **321**: 1364–1371.
55. Pappert EJ, Tangney CC, Goetz CG, Ling ZD, Lipton JW, Stebbins GT, Carvey PM. Alpha-tocopherol in the ventricular cerebrospinal fluid of Parkinson's disease patients: Dose response study and correlations with plasma levels. *Neurology* 1996, **47**: 1037–1042.
56. Sohal RS, Weindruch R. Oxidative stress, caloric restriction and aging. *Science* 1996, **273**: 59–63.
57. Prostera T, Zhang Y, Spencer SR, Wilczak CA, Talalay P. The electrophile counterattack response:protection against neoplasia and toxicity. *Advan Enzym Regul* 1993, **33**: 281–296.
58. Murphy TH, De Long MJ, Coyle JT. Enhanced NAD(P)H:quinone reductase activity prevents glutamate toxicity produced by oxidative stress. *J Neurochem* 1991, **56**: 990–995.
59. Duffy S, Murphy TH. Dopamine toxicity and the protective efficacy of recombinant antioxidant enzymes. *Soc Neurosci Abst* 1996, **22**:1481.
60. Wood KA, Youle RJ. Apoptosis and free radicals. *Ann NY Acad Sci* 1994, **738**: 400–407.
61. Wood KA, Youle RJ. The role of free radicals and p53 in neuron apoptosis in vivo. *J Neurosci* 1995, **15**: 5851–5857.
62. Zhong LT, Sarafian T, Kane DJ, Charles C, Mah SP, Edwards RH, Bredesen DE. Bcl-2 inhibits death of central neural cells induced by multiple agents. *Proc Natl Acad Sci USA* 1993, **90**: 4533–4544
63. Wiedau-Pazos M. Trudell JR, Altenbach C, Kane DJ, Hubbell WL, Bredesen DE. Expression of bcl-2 inhibits cellular radical generation. *Free Radic Res* 1996, **24**: 205–212.
64. Troy CM, Stefanis L, Prochiantz A, Greene LA, Shelanski ML. The contrasting role of ICE family proteases and interleukin-1 beta in apoptosis induced by trophic factor withdrawal and by copper/zinc superoxide dismutase down-regulation. *Proc Natl Acad Sci USA* 1996, **93**: 5365–5340.
65. Derossi D, Joiliot AH, Chassaing G, and Prochiantz A. The third helix of the antennapedia homeodomain translocates through biological membranes. *J of Biol Chem.* 1994, **269**: 10444–10450.

66. Mattson MP, Cheng B, Smith-Swintowsky VL. (1993) Growth factor mediated protection from excitotoxicity and disturbances in calcium and free radical metabolism. *Semin Neurosci* 1993, **5:** 295–307.
67. Sampath D, Jackson GR, Werrbach-Perez K, Perez-Polo JR. Effects of nerve growth factor on glutathione peroxidase and catalase in PC12 cells *J Neurochem* 1994, **62:** 2476–2479.
68. Jackson GR, Werrbach-Perez K, Perez-Polo JR. Role of nerve growth factor in oxidant-antioxidant balance and neuronal injury. I. Stimulation of hydrogen peroxide resistance. *J Neurosci Res* 1990, **25:** 360–368.
68a. Tiffany-Castiglioni E, Perez-Polo J. Stimulation of resistance to 6-hydroxydopamine in a human neuroblastoma cell line by nerve growth factor. *Neurosci Lett* 1981, **26:** 157–161.
69. Cheng B, Mattson MP. PDGFs protect hippocampal neurons against energy deprivation and oxidative injury: Evidence for induction of antioxidant pathways. *J Neurosci* 1995, **15:** 7095–7104.
70. Kirschner PB, Jenkins BG, Schulz JB, Finkelstein SP, Mathews RT, Rosen BR, Beal MF. NGF, BDNF, and NT-5, but not NT-3 protect against MPP+ toxicity and oxidative stress in neonatal animals. *Brain Res* 1996, **713:** 1–2,178–185.
71. Koh J-Y, Gwag BJ, Lobner D, Choi DW. Potentiated necrosis of cultured cortical cells by neurotrophins. *Science* 1996, **268:** 573–575.
72. Ratan RR, Lee PJ, Baraban JM. Serum deprivation inhibits glutathione depletion-induced death in embryonic cortical neurons: Evidence against oxidative stress as a common final mediator of neuronal apoptosis. *Neurochem Int* 1996, **29:** 153–157.
73. Mayer M, Noble M. N-acetylcysteine is a pluripotent protector against cell death and enhancer of trophic factor-mediated survival in vitro. *Proc Natl Acad Sci USA* 1994, **91:** 7596–7500.
74. Davis S, Helfaer MA, Traystman RJ, Hurn PJ. Parallel antioxidant and antiexcitotoxic therapy improves outcome after incomplete global cerebral ischemia in dogs. *Stroke* 1997, **28:** 198–205.
75. Willenborg, D.O., Bowern, N.A., Danta, G. and Doherty, P.C. Inhibition of allergic encephalomyelitis by the iron chelating agent desferrioxamine:differential effect depending on the type of sensitizing encephalitogen. *J Neuroimmun* 1988, **17:** 127–135.
76. Ben-Shachar, D., Eshel, G., Finberg, J.P.M., and Youdim, M.B.H. The iron chelator desferrioxamine (Desferal) retards 6-hydroxydopamine-induced degeneration of nigrostriatal dopamine neurons. *J Neurochem* 1991, **54:** 1441–144.
77. Keberle H. The biochemistry of desferrioxamine and its relation to iron metabolism. *Ann NY Acad Sci* 1964, **119:** 758–768.
78. Oliveri NF, Buncic JR, Chew E, Gallant T, Harrison RV, Keenan N, Logan W, Mitchell D, Ricci G, Skarf B, Taylor M, Freedman MH. Visual and auditory neurotoxicity in patients receiving subcutaneous deferoxamine infusions. *N Engl J Med* 1986, **314:** 869–873.
79. Maruyama M, Pieper GM, Kalyanaraman B, Hallaway PE, Hedlund BE, Gross GJ. Effects on hydroxyethyl starch conjugated deferoxamine on myocardial functional recovery following coronary occlusion and reperfusion in dogs. *J Cardiovas Pharm* 1991, **17:** 166–175.
80. Hurn PD, Koehler RC, Blizzard KK, Traystman RJ. Deferoxamine reduces early metabolic failure associated with severe cerebral ischemic acidosis in dogs. *Stroke* 1995, **26:** 688–694.
81. Hider RC, Choudhury R, Rai BL, Dehkordi LS, Singh S. Design of orally active iron chelators. *Acta Haematol* 1996, **95:** 6–12.
82. Dugan LL, Gabrielson JK, Yu SP, Lin TS, Choi DW. Buckminsterfullerenol free radical scavengers reduce excitotoxic and apoptotic death of cultured cortical neurons. *Neurobiol Dis* 1996, **3(2):** 129–135.

Index

A

Acute central nervous system injury,
excitotoxic neuron death,
endogenous excitotoxins, 205
exogenous excitotoxins, 204, 205
Adenovirus, E1A promotion of apoptosis, 111
AIF, *see* Apoptosis inducing factor
Alcohol, cerebellar damage, 455
Allopurinol, protection against excitotoxic neuron death, 49, 50
ALS, *see* Amyotrophic lateral sclerosis
Alzheimer's disease,
    amyloid β peptide,
        deposition and aging, 498, 499, 548
        neurotoxicity and calcium regulation, 80
    cortical anatomy, 497, 501, 502
    cortical myelination and lesional pattern, 505
    excitotoxic neuron death, 206, 207
    free radical inactivation of enzymes, 48
    mitochondrial respiration, 58, 59
    neurofibrillary tangles, 499–502, 505, 506
    neuronal loss, significance, 501
    pathology of lesions, 502, 504–506
    transgenic mouse model, 554
    treatment, 547, 548
    trophic factor receptors in brain, 546
γ-Aminobutyric acid (GABA) receptor, schizophrenia role, 536
α-Amino-3-hydroxy-5-methylisoxazole-4-propioninc acid (AMPA) receptor,
    antagonist proection against neuron death, 198, 199
    distribution in motor neurons, 639
    ibogaine-induced Purkinje cell degeneration role, 239
    subunits, 199, 633, 634
AMPA receptor, *see* α-Amino-3-hydroxy-5-methylisoxazole-4-propioninc acid receptor
Amyloid β peptide,
    deposition and aging, 498, 499, 548
    neurotoxicity and calcium regulation, 80
Amyloid precursor protein (APP),
    diffuse axonal injury marker, 381
    processing, 548
Amyotrophic lateral sclerosis (ALS),
    axotomy model of neuron death, 181, 184, 191, 192
    brain-derived neurotrophic factor therapy,
        administration routes, 604
        clinical trial design, 597, 598
        efficacy, 558, 598, 599, 602, 604
        prognostic variable adjustments, 600
        rationale, 593
        safety and adverse events, 598, 599, 601, 602, 604
        survival effect, 599, 600
        tolerance, 597, 598
    ciliary neurotrophic factor therapy,
        clinical trial design, 602, 603
        efficacy, 603
        side effects and safety, 558, 559, 603
    excitotoxic neuron death,
        glutamate dehydrogenase activators in therapy, 642
        glutamate receptor antagonist therapy, 641
        glutamate receptor distribution in motor neurons, 639
        glutamate reuptake agents in therapy, 641, 642

glutamate transporter GLT-1 defects, 435, 636–640
lamotrigine therapy, 641
riluzole therapy, 206, 435, 640, 641
susceptibility of neurons, 206, 638, 639
initiation and propagation in motor neuron degeneration, 642, 643
motor neuron degeneration mouse, 439, 440
neurofilament expression, 434, 435
neuron growth factor therapy, 184
neuropathology, 429, 430
neurotoxin induction, 436
prevalence, 596
progression monitoring, 596–599, 601
sporadic disease, 429
superoxide dismutase,
    mechanisms in pathogenesis, 433, 434
    mutations in familial disease, 191, 206, 430, 433, 434, 635, 651
    transgenic mouse models, 431, 433, 434, 554, 596
trophic factor expression, 435, 436
Antioxidants,
blood-brain barrier permeability, 654
definition, 650
iron,
    chelator therapy, 661, 662
    effects, 655
macromolecule protection in subcellular locales, 652–654
markers of efficacy, 655
modulation of growth factor activity, 659–661
phytochemicals in therapy, 658
pro-oxidant activity, 649, 655, 656
rationale for neurologic disease therapy,
    aging linked to free radicals and disease, 651, 652
    antioxidant defense gene mutations and neurologic dysfunction, 651
    oxidative stress and neurologic dysfunction, 651
therapeutic window, 649
vitamin therapy,

Parkinson's disease, 657
stroke, 656, 657
Apoptosis,
asynchronous death, see Asynchronous death
calcium role, 80, 81
cerebral ischemia recovery, 354, 355
clusters of cells, 15
commitment, 35–37
definition, 3
DNA fragmentation, 18–20, 277, 295
epilepsy, 369–371
excitotoxic neuron death, 12
execution, 36
induction compared to necrosis, 13
initiation, 34, 35
morphological phases, 31–33, 103, 468
neurons versus other cells, 4
Parkinson's disease,
    evidence, 464, 465, 468, 469
    induction in animal models, 466, 467
    overview of hypothesis, 464, 468, 469
programming evidence, 17
propagation, 35
similarity with necrosis, 20, 21
spilling of cell contents, 15, 16
Apoptosis inducing factor (AIF), apoptosis signaling, 278, 286
APP, see Amyloid precursor protein
Arbovirus,
clinical features of viral encephalitis,
    age dependence, 297
    alphaviruses, 299
    bunyaviruses, 299
    flaviviruses, 297–299
transmission, 296
Asynchronous death,
causes, 29
individual cell techniques for study, 41
measurement, 29, 30, 37–42
mechanisms, 38, 39
morphological phases, 31–33
synchronization of cells for apoptosis analysis, 39–42
time course, 30, 31

Index

AT, *see* Ataxia telangiectasia
Ataxia telangiectasia (AT),
   ATM gene, 270, 456, 457
   diagnosis, 269, 270
   incidence, 269
   neurologic degeneration, 270
Ataxin 1, *see* Spinocerebellar ataxia type 1
ATP, measurement in apoptosis commitment, 38
Atrophin-1, *see* Dentato-Rubro-Pallido-Luysian atrophy
Axotomy, induction of neuron death,
   adult neurons, 9, 10, 189–192
   genetic regulation, 187–189
   growth factor prevention, 182–184
   mechanism, 9, 10
   neonatal neuron apoptosis, 181, 182, 184–186
   trophic factor therapy model, 553, 556, 557

B

Bak, Bcl-2 dimerization, 306, 307
BAPTA, *see* 1,2-Bis-(2-aminophenoxy)ethane-*N,N,N',N'*-tetraacetic acid
Base excision repair, *see* DNA repair,
Basic fibroblast growth factor (bFGF),
   activity-related changes in expression and utilization, 565, 566
   anterograde transport, 546
   corticosteroid upregulation of expression, 567, 568
   functions, 549, 550
   oligodendrocytes, regulation of survival, 411
   side effects of therapy, 571
   traumatic brain injury treatment, 391, 392
Bax,
   apoptosis after neonatal axotomy, role, 187, 188, 306
   Bcl-2 dimerization, 306, 307
   brain expression in human immunodeficiency virus dementia, 515, 516
BBB, *see* Blood–brain barrier
Bcl-2
   apoptosis after neonatal axotomy, role, 187, 188
   brain expression in human immunodeficiency virus dementia, 515, 516
   mitochondrial-induced apoptosis prevention, 278, 279, 286, 287
   mitochondrial membrane potential preservation, 308
   nuclear gatekeeping, 308
   prevention of oxidant-induced cell death, 658
   programmed cell death role, 161, 169–171
   protection against apoptosis, 36, 127, 128, 277, 278, 306–308
   protein interactions,
      Bak, 306, 307
      Bax, 306, 307
      Ras, 307
   Sindbis virus infection modulation, 308, 309
   synchronization of cells for apoptosis analysis, 40
   transgenic mouse studies, 161, 169, 171, 306
   translocation in cancer, 305
BDNF, *see* Brain-derived neurotrophic factor
bFGF, *see* Basic fibroblast growth factor
Biological therapy,
   advantages, 544, 545
   comparison to organic pharmacology, 543, 544
   neuronal regeneration as goal, 546
1,2-Bis-(2-aminophenoxy)ethane-*N,N,N',N'*-tetraacetic acid (BAPTA),
   acetoxymethyl ester and cell permeability, 614
   calcium binding,
      affinity, 614, 616
      rate, 616
   cytoplasmic mobility, 717
   neuronal injury experiments,
      in vitro, 622, 623
      in vivo, 623, 624
   physiological effects,
      calcium homeostasis, 614–616
      cell death probing, 617–620, 622
   structure, 614, 615

Blood–brain barrier (BBB),
    antioxidant permeability, 654
    dysfunction and brain swelling, 381, 382
    trophic factor permeability in drug delivery, 566, 570, 573
Bovine spongiform encephalopathy (BSE),
    transgenic mouse studies, 330
    transmission, 325, 330
Brain-derived neurotrophic factor (BDNF),
    activity-related changes in expression and utilization, 565, 566
    amyotrophic lateral sclerosis treatment,
        administration routes, 604
        clinical trial design, 597, 598
        efficacy, 558, 598, 599, 602, 604
        prognostic variable adjustments, 600
        rationale, 593
        safety and adverse events, 598, 599, 601, 602, 604
        survival effect, 599, 600
        tolerance, 597, 598
    anterograde transport, 546
    apoptosis prevention after neonatal axotomy, 182–184, 190, 191
    corticosteroid upregulation of expression, 567
    motor neuron effects,
        developing neurons, 594, 595
        mature neurons, 595
    programmed cell death protection, 148, 150, 158, 167, 170
    purification, 548, 549
    signal transduction, 562, 563, 568
    trkB receptor, 551, 552, 562, 568, 594
    Wobbler mouse effects, 595, 596
Brain injury, *see* Traumatic brain injury
BSE, *see* Bovine spongiform encephalopathy

C

Caenorhabditis elegans,
    applicability to cell death in higher organisms, 138, 139
    features, 123, 124
    genetics and molecular biology, 124
    life cycle, 123
    programmed cell death,
        cell-specific regulation, 125, 126
        corpse removal,
            DNA degradation by NUC-1, 133
            engulfment genes, 132
        developmental changes, 124, 125
        functions,
            CED-3, 36, 91, 128–132
            CED-4, 128–132
            CED-9, 127, 128, 130, 131
            CES-1, 125, 126
            CES-2, 125, 126
            DAD-1, 128
            mutant analysis, 130–132
        genetic pathway, 125, 130, 131
        morphology, 124, 125
        mutant pathologies,
            deg-1 induced neurodegeneration, 134–137
            egl-1 and egg expulsion neuron death, 133
            ion channel mutations, 136, 137
            lin-24 and *lin-33* engulfment abnormalities, 133–134
            mec-4 induced neurodegeneration, 134–137
Calcium,
    assays of flux,
        fluorescence probes, 70, 610, 614, 622
        radiolabel assays, 70, 610
    cell injury during ischemia recovery, 353, 354
    channels for influx, overview, 71
    compartmentalization of fluxes, 611
    concentrations, intracellular and extracellular, 69
    decision point in neuronal apoptosis versus necrosis, 80, 81
    effector systems of neuronal death,
        cytoskeleton alterations, 77
        free radical generation in mitochondria, effects, 53–56, 58, 59, 74, 76
        hydrolytic enzymes, 73
        nitric oxide synthase, 73
        phosphorylation of proteins, 76, 77

transglutaminase, 77
xanthine dehydrogenase, 74
endogenous buffers,
binding protein effects on neuronal injury, 613
immobile organelle buffers, 612, 613
mobile buffers, 512
excitotoxic neuron death role, 49, 50, 78–80, 200, 201, 609–611, 624
exogenous buffers, see 1,2-Bis-(2-aminophenoxy)ethane-N,N,N',N'-tetraacetic acid; Ethylene-glycol tetraacetic acid
history of cytotoxicity studies, 69, 70
intracellular changes in apoptosis, 41, 46
intracellular release and translocation, 72
mitochondrial homeostasis and apoptosis, 284, 285, 613
neuronal set-point hypothesis, 72, 167, 168
neurotoxin modulation,
amyloid β peptide, 80
heavy metal toxicity, 77, 78
human immunodeficiency virus gp120, 78
sequestration and export, 71, 72
Caloric restriction, effect on neurologic disease, 657
Calpain, calcium stimulation and neuron death, 73
Cardiotrophin-1 (CT-1), apoptosis prevention after neonatal axotomy, 183, 184
Caspase-1
apoposis induction, 36, 91
apoptosis after neonatal axotomy, role, 189
autoprocessing, 93
inhibitors, 93, 658, 659
interleukin 1β processing, 92
structure, 92
substrate specificity, 92, 93, 95
Caspase-3
functions in nervous system, 91
substrates, 95, 96
Caspase-8, receptor signaling of apoptosis, 94

Cdc25, cyclin-dependent kinase activation, 110, 114
CDK, see Cyclin-dependent kinase
CED-3
caspase-1 homolog, 91, 128, 189, 277
programmed cell death role, 36, 91, 128–132
substrates, 129
CED-4, programmed cell death regulation, 128–132
CED-9
mammalian homolog, see Bcl-2
programmed cell death regulation, 127, 128, 130, 131, 171
Cell cycle,
overview, 104, 105
phases, 104
Central European encephalitis, clinical features, 298, 299
Cerebellum,
cell death,
apoptosis in development, 447–449, 457
cell culture models, 449, 450
mouse models for developmental disorders,
lurcher, 450, 451
Purkinje cell degeneration mouse, 451
reeler, 450
staggerer, 450
weaver, 450
cortical layers, 445
development, 446–449
human disease, see also Ibogaine-induced Purkinje cell degeneration,
degenerative diseases,
ataxia telangiectasia, 456, 457
spinocerebellar ataxia type 1, 456
inducers,
alcohol, 455
cytosine arabinoside, 454
heavy metal poisoning, 455
methylazoxymethanol, 453, 454
phenytoin, 453
radiation therapy, 454, 455
overview, 451, 452

paraneoplastic disease, 452, 453
neuron types, 446
Cerebral cortex, anatomy, 497, 501, 502
Cerebral ischemia,
   antioxidant therapy, 656, 657
   diagnosis, 355
   energy failure,
      acidosis, 351
      glutamate release, 343, 351
      hippocampal vulnerability, 345, 346
      ion flux as cause, 349, 350
      membrane depolarization, 351
      tissue culture models, 350
   excitotoxic neuron death, 352, 559
   focal ischemia, 346–348
   global ischemia, 343–345
   reperfusion and recovery, 351
   reperfusion injury,
      apoptosis, 354, 355
      calcium-mediated injury, 353, 354
      oxidative stress, 354
      tissue culture models, 352, 353
   sequence of ischemic failure, 348, 349
   therapeutic windows, duration, 344
   traumatic brain injury, 388
   trophic factor therapy, 559
Cerebral palsy (CP),
   economic impact, 401
   incidence, 401
   oligodendrocyte death, *see* Oligodendrocyte
   preventricular leukomalacia, 413
CES-1, programmed cell death regulation, 125, 126
CES-2, programmed cell death regulation, 125, 126
c-fos,
   apoptosis after neonatal axotomy, role, 185, 186
   apoptosis propagation, 35
   cell death signaling in epilepsy, 368
Ciliary neurotrophic factor (CNTF),
   amyotrophic lateral sclerosis treatment,
      clinical trial design, 602, 603
      efficacy, 603
      side effects and safety, 558, 559, 603
   apoptosis prevention after neonatal axotomy, 183, 184
   functions, 549
   oligodendrocytes, regulation of survival, 412
   programmed cell death protection, 159
   receptor, 552
   side effects of therapy, 571
CJD, *see* Creutzfeldt-Jakob disease
c-jun,
   apoptosis after neonatal axotomy, role, 185, 186
   cell death signaling in epilepsy, 368
c-Jun N-terminal protein kinase (JNK), apoptosis signaling, 35
CK, *see* Creatine kinase
c-Myc,
   apoptosis regulation, 103, 104, 110
   cyclin-dependent kinase activation, 110
   dimerization with max, 110
CNTF, *see* Ciliary neurotrophic factor
Cockayne's syndrome (CS),
   diagnosis, 266, 267
   genes, 267, 268
   incidence, 266
   xeroderma pigmentosum in combined disease, 266
Complex I,
   assay, 281
   deficiency in Parkinson's disease, 463
   free radical generation, 52
CP, *see* Cerebral palsy
CPP32, *see* Caspase-3
Creatine kinase (CK), free radical inactivation, 47
Creutzfeldt-Jakob disease (CJD),
   pathogenesis, 337
   transgenic mouse studies, 330, 331
   transmission, 325
   types, 325, 331
CS, *see* Cockayne's syndrome
CT-1, *see* Cardiotrophin-1
Cyclin D1, role in neurotrophic factor deprivation apoptosis, 108, 109
Cyclin-dependent kinase (CDK),

Index

apoptosis regulation, 104, 108–111, 114, 115
cell cycle control, 104–106
inhibitors and apoptosis prevention, 106, 111
regulation, 105, 106
substrates,
  lamins, 107, 108
  retinoblastoma protein, 106, 107
Cycloheximide, mechanism of apoptosis prevention, 17, 18
Cysteinyl aspartate-specific proteases, see Caspases
Cystine, deprivation and oligodendrocyte death, 413, 414
Cytochrome C, apoptosis signaling, 278, 286
Cytosine arabinoside, cerebellar damage, 454
Cytoskeleton, alterations induced by calcium, 77

D

DAD-1, programmed cell death regulation, 128
Deferoxamine (DFO),
  hydroxyethylstarch conjugation, 661, 662
  iron chelation, 661
  side effects of therapy, 661
deg-1, neurodegeneration induction,
  mechanism, 135
  morphology and ultrastructure, 134, 135
  sodium channel expression, 136, 137
  timing, 134
Dentate gyrus,
  anatomy, 363
  cell types, 363
  epilepsy effects, 371, 372
  neuron damage in traumatic brain injury, 385, 386
Dentato-Rubro-Pallido-Luysian atrophy (DRPLA),
  adult disease, 480
  atrophin-1
    defects, 485, 486
    protein–protein interactions, 488, 489
  CAG repeats, 477

excitotoxicity, 482, 483
juvenile disease, 480–482
selective vulnerability of neurons, 477, 479, 480
Development, *see also* Programmed cell death,
  apoptosis versus necrosis roles, 13, 14
  cerebellum, 446—449
  neuron death,
    type 1, 5, 6
    type 2, 5, 6, 8, 9
    type 3A, 5, 9
    type 3B, 5, 6, 9
DFO, see Deferoxamine
Diet, *see* Antioxidants; Caloric restriction
DNA ligase, DNA repair in brain, 259
DNA polymerase, DNA repair in brain, 259
DNA repair,
  aging effects, 249
  base excision repair, 251, 252
  brain,
    anatomic barrier to DNA damage, 271
    autoradiography studies, 256, 257
    gene-specific repair, 260, 261
    mismatch repair, 260
    repair enzymes,
      DNA ligase, 259
      DNA polymerase, 259
      ERCC1, 259, 260
      methyl transferase, 257, 258
      nuclease, 258, 259
      photolyase, 257
      uracyl glycosylase, 258
      XPB, 260
      XPD, 259, 260
      XRCC1, 260
  double-strand break repair, 255, 256
  human diseases,
    ataxia telangiectasia, 269, 270
    Cockayne's syndrome, 266–268
    trichothiodystrophy, 268, 269
    xeroderma pigmentosum, 261–264, 266
  mismatch repair, 251
  nucleotide excision repair,
    cellular hierarchy, 253, 254
    gene-specific repair, 253

steps, 252, 253
strand bias, 254, 255
transcription role, 252–255
reversal of damage, 251
sources of damage, 250
Dopamine, degradation and free radical production, 461
Double-strand break repair, see DNA repair
DRPLA, see Dentato-Rubro-Pallido-Luysian atrophy

E

EAE, see Experimental autoimmune encephalitis
Eastern equine encephalitis virus, clinical features of encephalitis, 299
EGTA, see Ethylene-glycol tetraacetic acid
Encephalitis, see Viral encephalitis
Epilepsy,
    cell death,
        apoptosis versus necrosis, 369–371
        hyperexcitability association, 371–373
        mechanism, 367, 368
        signaling, 368, 369
    classification of seizures, 361, 362
    dentate gyrus effects, 371, 372
    diagnosis, 361
    hippocampal effects, 363, 364, 370
    models of status epilepticus,
        chemoconvulsant induction, 366
        electrical stimulation, 366, 367
    neurotransmitter receptor effects, 365, 366
    pathology of temporal lobe epilepsy, 365, 366
    prevalence, 361
    treatment, 373, 374
ERCC1, DNA repair in brain, 259, 260
Ethylene-glycol tetraacetic acid (EGTA),
    calcium binding, 616,
    physiological effects,
        calcium homeostasis, 614–616
        cell death probing, 617–620
Excitotoxic amino acid receptors, see Amino-3-hydroxy-5-methylisoxazole-4-propioninc acid receptor; Kainic acid receptor; N-Methyl-D aspartate receptor
Excitotoxic neuron death,
    allopurinol protection, 49, 50
    amyotrophic lateral sclerosis,
        glutamate dehydrogenase activators in therapy, 642
        glutamate receptor antagonist therapy, 641
        glutamate receptor distribution in motor neurons, 639
        glutamate reuptake agents in therapy, 641, 642
        glutamate transporter GLT-1 defects, 435, 636–640
        lamotrigine therapy, 641
        riluzole therapy, 206, 435, 640, 641
        susceptibility of neurons, 206, 638, 639
    animal models, see Ibogaine-induced Purkinje cell degeneration
    autophagic death, 13
    calcium role, 49, 50, 78–80, 200, 201, 609–611, 624
    cerebral ischemia, 352
    classical excitotoxicity, 200–202
    clinical importance, 11–13
    Dentato-Rubro-Pallido-Luysian atrophy, 482, 483
    excitotoxic amino acid receptors, 198–200
    free radical role, 49, 50
    Huntington's disease, 482, 483
    metabotropic receptor hypoactivity, neurodevelopmental failure, 203, 204
    neoclassical excitotoxicity, 202, 203
    Parkinson's disease, 463, 464
    role in human disease,
        acute central nervous system injury, 204, 205
        Alzheimer's disease, 206, 207
        amyotrophic lateral sclerosis, 206
        Huntington's disease, 206
        neurolathyrism, 205, 206
        schizophrenia, 207, 536, 537
    time to death, 48
    trophic factor therapy, 559

Index   675

types of death,
   apoptosis, 12, 207–210
   excitotoxic cell death-1, 210, 212–214
   excitotoxic cell death-2, 212, 214
   excitotoxic cell death-3, 212–214
   necrosis, 12, 208, 215
ultrastructural analysis, 198
Experimental autoimmune encephalitis (EAE), multiple sclerosis model, 408
Extracellular signal regulated (ERK) kinase, apoptosis signaling, 35

F

Familial fatal insomnia (FFI), gene mutations, 332
Far Eastern encephalitis, clinical features, 298, 299
FFI, *see* Familial fatal insomnia
Fibroblast growth factor-2, *see* Basic fibroblast growth factor
Free radicals, *see also* Oxidative stress,
   antioxidant defense of cells, 46, 47
   definition, 650
   enzyme inactivation, 47, 48, 56
   excitotoxic neuron death role, 49, 50
   iron in formation, 414–416
   metal catalysis, 48
   methamphetamine neurotoxicity role, 521, 522
   mitochondria,
      effects of free radicals,
         enzyme inactivation, 56
         lipid peroxidation, 57
         respiration, 57, 58
      permeability transition and mitochondrial failure, 55, 56
      production of free radicals,
         aging effects, 60
         calcium effects, 53–56, 58, 59
         complex I, 52
         ubiquinone, 50–52
   Parkinson's disease role, 461–463
   reactive nitrogen species, 47
   signal transduction, 656

   xanthine oxidase generation, 49

G

GABA receptor, *see* γ-Aminobutyric acid
GDNF, *see* Glial-derived neurotrophic factor
Gerstmann-Straussler-Scheinker disease (GSS),
   gene mutations, 331, 332
   transgenic mouse studies, 330
GFAP, *see* Glial fibrillary acidic protein
Glial cell, programmed cell death, 163–165
Glial-derived neurotrophic factor (GDNF),
   apoptosis prevention after axotomy, 183, 184, 190, 191
   gene therapy, 571
   receptor, 552, 553
Glial fibrillary acidic protein (GFAP), neural injury marker in schizophrenia, 533, 534
Glutamate,
   metabolism, 634, 642
   receptors, *see* α-Amino-3-hydroxy-5-methylisoxazole-4-propioninc acid receptor; Metatrophic glutamate receptor; *N*-Methyl-D aspartate receptor
   reuptake agents in therapy, 641, 642
   transporters,
      GLT-1 defects in amyotrophic lateral sclerosis, 435, 636–640
      types, 634
Granzyme B, caspase activation, 94
GSS, *see* Gerstmann-Straussler-Scheinker disease

H

Harmaline, tremor induction, 222, 223
Heavy metal poisoning, cerebellar damage, 455
Herpes simplex virus (HSV),
   clinical features of encephalitis, 300
   neuron infection, 295, 300
   regulation of apoptosis, 312, 313
   types, 300
Hippocampus,
   cell types, 362, 363

cerebral ischemia vulnerability, 345, 346
epilepsy effects, 363, 364, 370
neuron damage in traumatic brain injury, 385, 386
subfields, 362
HIV, see Human immunodeficiency virus
HPO, see 3-Hydroxypyridin-4-one
HPV, see Human papillomavirus
HSV, see Herpes simplex virus
Human immunodeficiency virus (HIV),
   apoptosis induction,
      Bax expression, 515, 516
      Bcl-2 expression, 515, 516
      central nervous system, 513, 515
      lymphocytes, 513
   dementia,
      clinical features, 511
      incidence, 511
   gp120 neurotoxicity and calcium regulation, 78
   neuron infection, 511, 512
Human papillomavirus (HPV), E7 promotion of apoptosis, 111
Huntingtin, see Huntington's disease
Huntington's disease,
   adult disease, 480
   CAG repeats, 477, 484
   excitotoxic neuron death, 206, 482, 483
   huntingtin,
      cleavage by caspases, 95
      function, 484, 485
      structure, 484
      subcellular and tissue localization, 484, 485
      protein–protein interactions,
         E2 ubiquitin-conjugating enzyme, 488
         glutamine repeats, role, 486, 487
         huntingtin interactor protein 1, 487
         huntingtin-associated protein-1, 487, 488
      pathogenesis role, 489
   juvenile disease, 480–482

mitochondrial respiration, 58, 59
3-nitropropionic acid model, 483, 484
selective vulnerability of neurons, 477, 479, 480
3-Hydroxypyridin-4-one (HPO), iron chelation therapy, 662
Hyperthermia, traumatic brain injury, 389
Hypotension, traumatic brain injury, 388, 389
Hypothermia therapy, traumatic brain injury, 391, 392

I

Ibogaine-induced Purkinje cell degeneration,
   drug addiction treatment, 222, 454
   excitotoxicity model, 237, 238
   microglial cell activation,
      assays, 229, 231
      time course, 232, 233
   nitric oxide role, 233, 235
   olivocerebellar projection mediation, 221, 235, 237–240
   staining of Purkinje cells, 225, 227–229, 240, 241
   tremor induction, 222, 223
ICE, see Interleukin 1b-converting enzyme
IGF, see Insulin-like growth factor
Inflammation, induction by necrosis versus apoptosis, 15, 16
Injury, see Acute central nervous system injury; Spinal cord injury; Traumatic brain injury
Insulinlike growth factor (IGF),
   functions, 550
   oligodendrocytes, regulation of survival, 411, 412
Interferon-γ, oligodendrocyte toxicity, 410
Interleukin 1β-converting enzyme (ICE), see Caspase-1
Iron,
   antioxidant effects on, 655
   chelator therapy, 661, 662

Index

free radical formation and oligodendrocyte death, 414–416
metabolism in Parkinson's disease, 462, 463
Ischemia, *see* Cerebral ischemia

J

Japanese encephalitis, clinical features, 297, 298
JNK, *see* c-Jun N-terminal protein kinase

K

KA receptor, *see* Kainic acid receptor
Kainic acid (KA) receptor,
    antagonist proection against neuron death, 198, 199
    calcium channel, 71
    domoate and acute central nervous system injury, 204
    subunits, 199, 633, 634
Kearns-Sayre syndrome (KSS),
    ATP production defects, 282
    genetic mutations, 277
    morphological changes, 279

L

LaCrosse virus, clinical features of encephalitis, 299
Lactate dehydrogenase (LDH), marker of excitotoxic neuron death, 201, 202
Lamotrigine, amyotrophic lateral sclerosis therapy, 641
LDH, *see* Lactate dehydrogenase
Leber's hereditary optic neuropathy (LHON),
    electron transport dysfunction, 281
    genetic mutations, 276, 286
Leukemia inhibitory factor (LIF),
    apoptosis prevention after neonatal axotomy, 183
    functions, 549
    receptor, 552
LHON, *see* Leber's hereditary optic neuropathy
LIF, *see* Leukemia inhibitory factor
Lou Gehrig's disease, *see* Amyotrophic lateral sclerosis
Lurcher mouse, 450, 451

M

Max, dimerization with myc, 110
MDMA, *see* Methyldioxymethamphetamine
mec-4, neurodegeneration induction,
    mechanism, 135
    morphology and ultrastructure, 134, 135
    sodium channel
        expression, 136, 137
    timing, 134
MELAS, *see* Mitochondrial encephalomyopathy, lactic acidosis, and stroke-like episodes
MERFF, *see* Myoclonic epilepsy and ragged-red fibers
Mesial temporal sclerosis (MTS), epilepsy association, 365, 366
Metatrophic glutamate receptor, subunits, 634
Methamphetamine,
    dopaminergic neurotoxicity, 521
    mediation of neurotoxicity,
        mechanism of cell death, 524, 525
        nitric oxide, 522
        poly(ADP-ribose) polymerase activation, 522, 524
        superoxide radicals, 521, 522
    trophic factor therapy model, 554
Methyl-4-phenyl-1,2,3,6-tetrahydropyridine (MPTP),
    Parkinson's disease model, 460–463, 467, 470
    trophic factor therapy model, 554, 557
Methyl transferase, DNA repair in brain, 257, 258
Methylazoxymethanol, cerebellar damage, 453, 454
Methyldioxymethamphetamine (MDMA),
    dopaminergic neurotoxicity, 521
    mediation of neurotoxicity,
        mechanism of cell death, 524, 525
        nitric oxide, 522
        poly(ADP-ribose) polymerase activation, 522, 524
        superoxide radicals, 521, 522
    trophic factor therapy model, 554

Mismatch repair, *see* DNA repair
Mitochondria, *see also specific diseases*,
   apoptosis signaling,
      apoptosis inducing factor, 278
      Bcl-2 prevention, 278, 279, 286, 287
      cytochrome C, 278
      free radicals, 278
   calcium,
      extrusion, 74, 76
      homeostasis and apoptosis, 284, 285
      sequestration, 74, 613
   electron transport in disease, 280–286
   free radical effects,
      enzyme inactivation, 56
      lipid peroxidation, 57
      respiration, 57, 58
      tumor necrosis factor-mediated cytotoxicity assay, 283
   free radical production,
      aging effects, 60
      calcium effects, 53–56, 58, 59, 74, 76
      complex I, 52
      disease effects, 282, 283
      mitochondrial DNA mutation effects, assays, 283, 284
      ubiquinone, 50–52
   genetic defects of electron transfer, 60, 61
   genome,
      depletion in ρ^0 cells, 276, 283
      mutations and disease, 275–277
      structure, 275
   permeability transition and mitochondrial failure, 55, 56
   respiration activity in neurodegenerative disease, 58, 59
   transmembrane potential loss a apoptosis marker, 284, 285, 308
Mitochondrial encephalomyopathy, lactic acidosis, and stroke-like episodes (MELAS),
   ATP production defects, 282
   calcium, mitochondrial homeostasis and apoptosis, 285
   genetic mutations, 276, 280
   morphological changes, 279
   protein translation deficiencies, 282
Mitotic catastrophe, 107, 108
Motor neuron disease, *see* Amyotrophic lateral sclerosis; Spinal muscular atrophy
MPTP, *see* Methyl-4-phenyl-1,2,3,6-tetrahydropyridine
MTS, *see* Mesial temporal sclerosis
Multiple sclerosis,
   autoimmunity, 409
   experimental autoimmune encephalitis model, 408
   oligodendrocyte death, *see* Oligodendrocyte
Myoclonic epilepsy and ragged-red fibers (MERFF),
   genetic mutations, 276
   morphological changes, 279
   protein translation deficiencies, 282
N-Methyl-D aspartate (NMDA) receptor,
   antagonists,
      neurotoxicity, 202, 203
      proection against neuron death, 78, 79, 191, 198, 199
      traumatic brain injury treatment, 390
   brain distribution, 363
   calcium channel, 71, 77
   distribution in motor neurons, 639
   gp120 neurotoxicity and calcium regulation, 78
   magnesium block, 205
   schizophrenia role, 536, 537
   subunits, 199, 633, 634

N

NAIP, *see* Neuronal apoptosis inhibitor protein
Necrosis,
   antagonists, apoptosis prevention after neonatal axotomy, 184
   calcium role, 80, 81
   definition, 3
   DNA fragmentation, 18–20
   epilepsy, 369–371
   excitotoxic neuron death, 12
   genetic control, 16–18

induction compared to apoptosis, 13
inflammation induction, 15, 16
similarity with apoptosis, 20, 21
Nerve growth factor (NGF),
   activity-related changes in expression and utilization, 565, 566
   anterograde transport, 546
   corticosteroid upregulation of expression, 567, 568
   grafting of immortalized cells for expression, 570
   p75 receptor, 550, 551
   prevention of oxidant-induced cell death, 659
   programmed cell death protection, 148, 167, 169
   purification, 548
   removal and apoptosis initiation, 34, 35, 37, 40, 108, 109
   side effects of therapy, 571, 572
   signal transduction, 561–563
   trkA receptor, 551
Neurolathyrism, excitotoxic neuron death, 205, 206
Neuronal apoptosis inhibitor protein (NAIP), mutation in spinal muscular atrophy, 170, 171, 436–438
Neuronal set-point hypothesis, 72, 167, 168
Neurotrophin-3 (NT-3),
   activity-related changes in expression and utilization, 566
   anterograde transport, 546
   corticosteroid upregulation of expression, 567
   oligodendrocytes, regulation of survival, 411, 412
   programmed cell death protection, 148, 155, 158
   signal transduction, 562, 563
   trkC receptor, 551, 552, 562
Neurotrophin-4 (NT-4), apoptosis prevention after axotomy, 182, 183, 190
NGF, see Nerve growth factor
Nitric oxide (NO),
   ibogaine-induced Purkinje cell degeneration role, 233, 235
   methamphetamine neurotoxicity role, 522
   reactive nitrogen species, 47
Nitric oxide synthase (NOS),
   apoptosis after axotomy, role, 191
   calcium stimulation and neuron death, 73
3-Nitropropionic acid, Huntington's disease model, 483, 484
NM, see Nucleus magnocellularis
NMDA receptor, see N-Methyl-D aspartate receptor
NO, see Nitric oxide
NOS, see Nitric oxide synthase
NT-3, see Neurotrophin-3
NT-4, see Neurotrophin-4
NUC-1, DNA degradation in *Caenorhabditis* corpse, 133
Nucleotide excision repair, see DNA repair
Nucleus magnocellularis (NM), afferent input and cell death, 155–157, 169

O

OL, see Oligodendrocyte
Olfactory receptor neuron (ORN), turnover, 165, 166
Oligodendrocyte (OL),
   culture, 405, 406
   death assessment methods,
      immunocytochemistry, 404
      viability assays, 404, 405
   death mechanisms,
      apoptosis versus necrosis, 417–419
      cytotoxic cytokines, 409, 410
      glutamate receptors and toxicity, 412, 413
      growth factor withdrawal, 410–412
      oxidative stress, 413–419
   developmental lineage, 402, 403
   model systems,
      experimental autoimmune encephalitis, 408
      mutant mouse models, 407, 408
      optic nerve, 406, 407
   white matter disorders, 401, 402, 419

ORN, *see* Olfactory receptor neuron
Oxidative stress, *see also* Free radicals,
   cerebral ischemia recovery, 354
   markers, 655
   oligodendrocyte death role, 413, 414
   spectrum of cellular responses, 48, 49

P

$p16^{INK4A}$,
   apoptosis prevention, 111
   cyclin-dependent kinase inhibition, 106
$p21^{CIP1}$
   apoptosis prevention, 111
   cyclin-dependent kinase inhibition, 106
p38 kinase, apoptosis signaling, 35
p53, apoptosis regulation, 103, 104, 113
Paraneoplastic cerebellar degeneration (PNCD),
   associated cancers, 453
   diagnosis, 452
   etiology, 452, 453
Parkinson's disease (PD),
   6-hydroxydopamine model, 466, 467
   apoptosis,
      evidence, 464, 465, 468, 469
      induction in animal models, 466, 467
      overview of hypothesis, 464, 468, 469
   diagnosis, 459
   dopamine neuron loss sites, 459, 460
   methyl-4-phenyl-1,2,3,6-tetrahydropyridine model, 460–463, 467, 470
   mitochondrial respiration, 58, 59
   pathogenesis,
      aging role, 460
      environmental factors, 460
      excitotoxicity, 463, 464
      free radical injury, 461–463
      genetic factors, 460, 461
   prevalence, 459
   treatments, 459, 460, 469
   Weaver mouse model, 468
PARP, *see* Poly(ADP-ribose) polymerase
PCD, *see* Programmed cell death
PCP, *see* Phencyclidine
PCR, *see* Polymerase chain reaction
PD, *see* Parkinson's disease
PDGF, *see* Platelet-derived growth factor
Phencyclidine (PCP), schizophrenia induction, 536
Phenytoin, cerebellar damage, 453
Phospholipase A_2 (PLA_2), calcium stimulation and neuron death, 73
Phosphorylation, *see* Protein phosphorylation
Photolyase, DNA repair in brain, 257
PKR,
   mechanism of Sindbis virus-induced apoptosis, 304, 305
   nuclear factor-κB regulation, 305
PLA_2, *see* Phospholipase A_2
Plasticity, trophic factor mediation in adult neurons, 565
Platelet-derived growth factor (PDGF), oligodendrocytes, regulation of survival, 411, 412
PNCD, *see* Paraneoplastic cerebellar degeneration
Poly(ADP-ribose) polymerase (PARP),
   cleavage by caspases, 95, 96
   methamphetamine neurotoxicity role, 522, 524
Polymerase chain reaction (PCR), asynchronous death assay, 38, 39
pRb, *see* Retinoblastoma protein
Prion disease, *see also* Bovine spongiform encephalopathy; Creutzfeldt-Jakob disease; Familial fatal insomnia; Gerstmann-Straussler-Scheinker disease; Prion protein,
   pathogenesis, 335, 337
   transgenic mouse models, 332–335
   yeast model, 328, 329, 333, 335
Prion protein (PrP),
   accumulation in neurons, 335, 337
   cell injury mechanisms, 333–335
   expression in scrapie, 326
   gene, 326
   processing, 326, 328
   protein interactions in disease, 329
   replication, 327–330
   structure, 328, 338

Index 681

transgenic mouse studies, 326, 327, 329–331
Programmed cell death (PCD), *see also Caenorhabditis elegans*,
    adult nervous system,
        functions, 165, 166
        survival signs of neurons, 168–170
    afferent input and neuron survival, 152–157
    countdown to cell death, 166–168
    definition, 146
    dysregulation and pathology, 170, 171
    functions in developing nervous system, 159–162
    glial cells, 163–165
    neuron-target interaction in survival control, 146–148, 150–152
    neuronal populations susceptible in vertebrate development, 149
    neurotrophic factors in protection, 146–148, 150–152, 157–159
    novel forms in developing nervous system, 162, 163
    species distribution, 146
Proteasome, caspase activation, 94
Protein phosphorylation, calcium-dependent neurotoxicity role, 76, 77
PrP, *see* Prion protein
Purkinje cell, *see* Cerebellum; Ibogaine-induced Purkinje cell degeneration
Purkinje cell degeneration mouse, 451

R

Radiation therapy, cerebellar damage, 454, 455
Ras, Bcl-2 interactions, 307
reaper, triggering of apoptosis, 18
Reeler mouse, 450
Reperfusion injury
    cerebral ischemia,
        apoptosis, 354, 355
        calcium-mediated injury, 353, 354
        oxidative stress, 354
        tissue culture models, 352, 353

    traumatic brain injury, 389
Retinoblastoma protein (pRb),
    apoptosis prevention, 111–114
    binding by viral oncoproteins, 111, 113
    phosphorylation, 106, 107, 110, 111
    tumor suppression, 112
Riluzole, amyotrophic lateral sclerosis therapy, 206, 435, 640, 641

S

St. Louis encephalitis, clinical features, 298
SCA1, *see* Spinocerebellar ataxia type 1
Schizophrenia,
    degeneration model, 529, 530, 532–534
    developmental model, 528, 529, 534, 535
    excitotoxic neuron death, 207, 536, 537
    history of research, 527, 528
    phencyclidine induction, 536
    postmortem findings,
        diagnostic neuropathology, 530, 531
        macroscopic findings, 531
        neural injury markers, 533, 534
        neurodegeneration, 532–534
        neurodevelopmental abnormalities, 534, 535
        neuron quantitation, 531, 532
    prevalence, 527
Simian virus 40 (SV40), T-antigen promotion of apoptosis, 113
Sindbis virus,
    apoptosis induction,
        age dependence, 303, 304
        mechanism, 304, 305
        neurovirulence correlation, 303, 304
    Bcl-2, modulation of infection, 308, 309
    immune-mediated regulation of apoptosis during encephalitis,
        antibody downregulation of infection, 309–311
        cell-mediated immune reponses, 311
        innate immunity protection, 311, 312
    mouse model of human encephalitis, 301
    neuron infection, 295
    persistence versus apoptosis, 309
    replication, 301–303

SMA, *see* Spinal muscular atrophy
SMN, *see* Survival motor neuron
SOD, *see* Superoxide dismutase
Spinal cord injury, trophic factor therapy, 559, 560
Spinal muscular atrophy (SMA),
    dysregulation of programmed cell death, 170
    genes,
        neuronal apoptosis inhibitor protein, 170, 171, 436–438
        RNA processing, 438, 439
        splicing, 437
        survival motor neuron, 436–439
    models,
        hereditary canine model, 440
        progressive motor neuropathy mouse, 440
        wobbler mouse, 440
    pathology, 436
    types, 436
Spinocerebellar ataxia type 1 (SCA1),
    ataxin 1
        gene mutations, 456
        protein-protein interactions, 487
        transgenic mouse model, 554
    diagnosis, 456
Staggerer mouse, 450
Steroid hormones, programmed cell death protection, 158, 159
Stroke, *see* Cerebral ischemia
Superoxide dismutase (SOD),
    amyotrophic lateral sclerosis and SOD1
        mechanisms in pathogenesis, 433, 434
        mutations in familial disease, 191, 206, 430, 433, 434, 635, 651
        transgenic mouse models, 431, 433, 434
    antioxidant defense of cells, 46, 47
    expression and oligodendrocyte death, 416, 417
    protection against apoptosis after axotomy, 191
    traumatic brain injury treatment, 390
Survival motor neuron (SMN),
    function, 439
    homology between species, 437
    mutation in spinal muscular atrophy, 436–438
    RNA processing, 438, 439
    splicing, 437
SV40, *see* Simian virus 40

T

Tau,
    neurofibrillary deposition, 499
    phosphorylation state,
        Alzheimer's disease, 498
        apoptosis, 41, 42
TBI, *see* Traumatic brain injury
Tetrodotoxin (TTX), neuron death induction in chicks, 155, 156
TGFβ, *see* Transforming growth factor β
TNF, *see* Tumor necrosis factor
Transforming growth factor β (TGFβ), functions, 550
Transglutaminase, calcium stimulation and neuron death, 77
Traumatic brain injury (TBI),
    classification, 379, 380
    diffuse brain damage,
        brain swelling and secondary complications, 381, 382
        diffuse axonal injury, 381, 392
        evolution of axonal lesions, 381
    experimental models,
        acceleration concussion, 383
        cell culture models, 383, 384
        percussion concussion, 382, 383
    focal brain injury,
        cerebral contusion, 380
        intracranial hematoma, 380
    imaging, 380
    incidence, 379
    lesion classification, 382
    mechanisms,
        cerebral hypoxia/ischemia, 388
        hyperthermia, 389
        hypotension, 388, 389
        reperfusion injury, 389
        shearing, 387
    neuropathological sequelae, 384–386

selective neuronal vulnerability, 386
treatment,
  glutamate antagonists, 390
  hypothermia, 391, 392
  neurotrophic factors, 391
  radical scavengers, 390
ventriculr dilation, 386, 387
Trichothiodystrophy (TTD),
  diagnosis, 268
  genes, 269
Trk receptors,
  ligands in therapy, 568, 569
  neurotrophin affinities and expression, 551, 552, 561, 562
TTD, see Trichothiodystrophy
TTX, see Tetrodotoxin
Tumor necrosis factor (TNF),
  cytotoxicity assay of free radical effects, 283
  oligodendrocyte toxicity, 410
TUNEL assay, DNA fragmentation analysis of cell death, 208, 209, 212–214, 303, 417, 513, 515, 516

U

Ubiquinone, free radical generation, 50–52
Uracyl glycosylase, DNA repair in brain, 258

V

Viral encephalitis, *see also specific viruses*,
  animal models, 295, 301
  apoptosis in Sindbis virus infection, 301–312
  clinical features,
    alphaviruses, 299
    arboviruses, 296–299
    bunyaviruses, 299
    flaviviruses, 297–299
    herpes simplex virus, 300
  herpes simplex virus regulation of apoptosis, 312–313

Sindbis virus and apoptosis induction,
  age dependence, 303, 304
  Bcl-2, modulation of infection, 308, 309
  immune-mediated regulation of apoptosis,
    antibody downregulation of infection, 309–311
    cell-mediated immune reponses, 311
    innate immunity protection, 311, 312
  mechanism, 304, 305
  neurovirulence correlation, 303, 304
  persistence versus apoptosis, 309

W

Weaver mouse, 450, 468
Western equine encephalitis virus, clinical features of encephalitis, 299
Wobbler mouse, 440, 556, 595

X

Xanthine dehydrogenase, calcium stimulation and neuron death, 74
Xanthine oxidase, free radical generation, 49
Xeroderma pigmentosum (XP),
  Cockayne's syndrome in combined disease, 266
  diagnosis, 262
  genes and complementation groups,
    XPA, 262, 264
    XPB, 263
    XPC, 263
    XPD, 263
    XPE, 263
    XPF, 263
    XPG, 263, 264
  incidence, 261
  neurodegeneration, 261, 264, 266
XP, see Xeroderma pigmentosum
XPB, DNA repair in brain, 260
XPD, DNA repair in brain, 259, 260
XRCC1, DNA repair in brain, 260